Perspectives in Mathematical Logic

Ω-Group:

R. O. Gandy H. Hermes A. Levy G. H. Müller
G. E. Sacks D. S. Scott

Ω-Bibliography of Mathematical Logic

Edited by Gert H. Müller

In Collaboration with Wolfgang Lenski

Volume VI

Proof Theory
Constructive Mathematics

Jane E. Kister;
Dirk van Dalen & Anne S. Troelstra (Editors)

Springer-Verlag Berlin Heidelberg New York
London Paris Tokyo

Gert H. Müller
Wolfgang Lenski
Mathematisches Institut, Universität Heidelberg
Im Neuenheimer Feld 288, D-6900 Heidelberg

Jane E. Kister
Mathematical Reviews
P.O. Box 8604, Ann Arbor, MI 48107
U.S.A.

Anne S. Troelstra
Department of Mathematics, University of Amsterdam
Roeterstraat 15, NL-1018 WB Amsterdam
The Netherlands

Dirk van Dalen
Mathematisch Instituut, Rijksuniversiteit Utrecht
Budapestlaan 6, Postbus 80010, 3508 TA Utrecht
The Netherlands

The series *Perspectives in Mathematical Logic* is edited by the Ω-group of the Heidelberger Akademie der Wissenschaften. The Group initially received a generous grant (1970–1973) from the Stiftung Volkswagenwerk and since 1974 its work has been incorporated into the general scientific program of the Heidelberger Akademie der Wissenschaften (Math. Naturwiss. Klasse).

ISBN 3-540-15524-4 Springer-Verlag Berlin Heidelberg New York
ISBN 0-387-15524-4 Springer-Verlag New York Berlin Heidelberg

Library of Congress Cataloging in Publication Data
[Omega]-bibliography of mathematical logic.
(Perspectives in mathematical logic)
Includes indexes.
Contents: v. 1. Classical logic / Wolfgang Rautenberg, ed. - v. 2. Non-classical logics / Wolfgang Rautenberg, ed. - v. 3. Model theory / Heinz-Dieter Ebbinghaus, ed. [etc.]
1. Logic, Symbolic and mathematical-Bibliography. I. Müller, G. H. (Gert Heinz), 1923 - II. Lenski, Wolfgang, 1952 -. III. Title: Bibliography of mathematical logic. IV. Series.
Z6654.M26047 1987 [QA9] 016.5113 86-31426

Media conversion, printing and bookbinding: Appl, Wemding
2141/3140-543210

Dedicated
to
Alonzo Church

whose bibliographic work for the Journal of Symbolic Logic was a milestone in the development of modern logic.

Table of Contents

Preface IX

Introduction XV

User's Guide XXI

Ω-Classification Scheme XXIX

Subject Index 1

History and philosophy 3
1. Historical, critical and philosophical publications
(A05 ∪ A10) ∩ (F ∪ B40) 3

The lambda-calculus and related systems 14
2. Combinatory logic and lambda-calculus B40 14

Proof theory and systems of arithmetic 26
3. Cut elimination and normal form theorems F05 26
4. Structure of proofs F07 34
5. Functionals in proof theory F10 39
6. Recursive ordinals and ordinal notations F15 42
7. Complexity of proofs F20 49
8. Relative consistency and interpretations F25 52
9. First-order arithmetic and fragments F30 56
10. Second- and higher-order arithmetic and fragments . F35 76
11. Gödel numberings in proof theory F40 88

Constructive mathematics 89
12. Metamathematics of constructive systems F50 89
13. Constructive and intuitionistic mathematics F55 113
14. Constructive recursive analysis F60 121
15. Other constructive mathematics F65 129

General 133
16. Proceedings F97 133
17. Textbooks, surveys F98 135
18. Miscellaneous . . . . F99 \ (A05 ∪ A10) 139

Author Index 141

Source Index 329

Journals 331
Series 348
Proceedings 351
Collection volumes 367
Publishers 374

Miscellaneous Indexes 381

External classifications 383
Alphabetization and alternative spellings of author names 401
International vehicle codes 403
Transliteration scheme for Cyrillic 405

Preface

Gert H. Müller

The growth of the number of publications in almost all scientific areas, as in the area of (mathematical) logic, is taken as a sign of our scientifically minded culture, but it also has a terrifying aspect. In addition, given the rapidly growing sophistication, specialization and hence subdivision of logic, researchers, students and teachers may have a hard time getting an overview of the existing literature, particularly if they do not have an extensive library available in their neighbourhood: they simply do not even know what to ask for! More specifically, if someone vaguely knows that something vaguely connected with his interests exists somewhere in the literature, he may not be able to find it even by searching through the publications scattered in the review journals. Answering this challenge was and is the central motivation for compiling this Bibliography.

The Bibliography comprises (presently) the following six volumes (listed with the corresponding Editors):

I.	Classical Logic	W. Rautenberg
II.	Non-classical Logics	W. Rautenberg
III.	Model Theory	H.-D. Ebbinghaus
IV.	Recursion Theory	P. G. Hinman
V.	Set Theory	A. R. Blass
VI.	Proof Theory; Constructive Mathematics	J. E. Kister; D. van Dalen & A. S. Troelstra.

Each volume is divided into four main parts:

1) The *Subject Index* is arranged in sections by topics, usually corresponding to sections in the classification scheme; each section is ordered chronologically by year, and within a given year the items are listed alphabetically by author with the titles of the publications and their full classifications added.

2) The *Author Index* is ordered alphabetically by author, and contains the full bibliographical data of each publication together with its review numbers in Mathematical Reviews (MR), Zentralblatt für Mathematik und ihre Grenzgebiete (Zbl), Journal of Symbolic Logic (JSL), and Jahrbuch über die Fortschritte der Mathematik (FdM). We much regret that we were not able to include reviews from Referativnyj Zhurnal Matematika in this edition.

3) The *Source Index* gives the full bibliographical data of each source (journals and books) for which only abbreviated forms are used in the Author Index.

4) The *Miscellaneous Indexes* contain various further indexes and tables to aid the reader in using the Bibliography.

For a more detailed technical description of the Bibliography see the *Table of Contents* and the *User's Guide.*

The uniform classification of all entries is a central feature of the Bibliography. The basic framework is the 03 section of the (1985 version of the) 1980 classification scheme of Mathematical Reviews and Zentralblatt für Mathematik und ihre Grenzgebiete. However, this has been modified in a number of ways. Indeed, the 1980 scheme was designed for the classification of works written after 1980, whereas the majority of entries in the Bibliography come before this date. In some areas this has made the classification of older works difficult, and we have tried to cope with this by adding a few new sections and altering slightly the interpretation of others. We have not designated the classifications assigned to a work as primary and secondary, because of the difficulty in doing so in many cases. Each volume contains the full annotated classification scheme together with a description of its general features. In their *introductions* the Editors discuss specifically their interpretations of the classification sections falling in their respective volumes.

The Subject Index is another central feature of the Bibliography. Reading through this Index gives a *historical perspective* for each classification section and provides a rather quick overview of the literature in it. By browsing through the entries of a specific area the reader may be rewarded by finding things (literature, subjects, questions) he was not aware of or had forgotten.

An obvious question now is the extent to which one can rely on the *completeness* and *correctness* of the Bibliography and on the *accuracy of the classifications.* We comment on each of these aspects separately.

In an effort to be as complete as possible, we consulted all sources available to us and decided in favour of inclusion in doubtful cases (so that certainly some papers with little bearing on mathematical logic are listed here and there). As the historical starting point for the Bibliography we chose the appearance of Frege's *Begriffsschrift* (1879). A certain restriction on scope stems from our decision to concentrate on mathematical logic and in particular on those areas defined by the titles of the six volumes. A major source of material was provided by the review journals mentioned above; we used them both to identify publications in the less known journals and to find review numbers and other bibliographical data of items found in other sources. We also made use of various lists of literature contained in books, survey articles, mimeographed notes, etc. Some especially valuable newer sources were:

W. Hodges: A Thousand Papers in Model Theory and Algebra

M. A. McRobbie, A. Barcan and P. B. Thistlewaite: Interpolation Theorems: A Bibliography

D. S. Scott and J. M. B. Moss †: A Bibliography of Books on Symbolic Logic, Foundations of Mathematics and Related Subjects

C. A. B. Peacocke and D. S. Scott: A Selective Bibliography of Philosophical Logic.

Various strategies and crosschecks were used to ensure the completeness of the bibliographical data and in particular of the reviews mentioned above. For each item listed in the Bibliography we tried to include any translations, reprintings in alternative sources and errata, and to give cross references for a work appearing in several parts.

On the whole this Bibliography was compiled and organized for use by the practising mathematician; there is no claim that the most rigorous standards of librarianship are met.

It is hard to say how successful our striving for completeness was. This is especially true for the most recent literature. No 1986 items were included. We checked all the main journals in logic, the reviews in MR, Zbl and JSL and Current Mathematical Publications for literature published up to the end of 1985, but undoubtedly some gaps remain.

As for correctness, in any ordinary book we can tolerate a number of printing errors because of our knowledge of the language and the context, but, when one organizes data connected by (abstract) pointers in a computer program, almost every typing error has far-reaching consequences. Various consistency tests were used to check the program and the input data. There are, however, many other sources for mistakes and errors.

For some items our references contained incomplete or ambiguous information. Although we tried to complete the bibliographical data, this was often difficult, particularly in cases where, for example, the source was obscure or the publisher was given only by location. Another source of errors lies in the identification of author names. An author may publish using abbreviations of his first, his second or both of his given names. This is generally not a problem for authors with uncommon surnames, but if the surname is, e.g., Smith or Brown the possibility of misidentification arises. We may have identified two different authors or failed to identify two or more different forms of an author name.

It is unavoidable in a project of this scope that there will be errors, particularly in the classification, so perhaps it is worthwhile to explain briefly the process by which the classification was done. Items entered before 1981 were originally classified according to a scheme unrelated to the current one. To begin the conversion to the 1980 scheme we used the computer to change old categories to their new versions wherever there was a well-defined correspondence. Then every entry was checked and if necessary reclassified by hand. From 1981 each new entry was classified shortly after being entered in the database. For the most part this was done on the basis of titles, reviews, and other information, but without consulting the works themselves. This was necessary to preserve the finiteness of the enterprise, but it has inevitably led to errors, certainly in some cases egregious ones. These were constantly being corrected during the final editing process, but many will remain.

Although the Editors have to some extent used different strategies in classifying the entries falling into their respective volumes, finally a reasonable degree of uniformity has arisen. The user is referred to the Editors' introductions for further details on the classifying procedure.

A special apology goes to the native speakers of languages with diacritical marks. Our central difficulty was to get the right spelling of names used in different forms in such a variety of sources. In addition, entering diacritical marks in a computer introduces yet another source of errors; so they have almost all been ignored (the User's Guide and the Miscellaneous Indexes contain details of those that have been transliterated). We appreciate that, although the absence of, for example, accents in the text of a French title may not create undue problems, the lack of diacritical marks in author names is particularly unfortunate. We hope that this omission will not be too misleading.

The future

By its nature a bibliography has lasting value to the extent it succeeds in "completing the past". But it should also serve for some years as an aid to current research. We have various plans to extend the scope of the present Bibliography by including new areas such as universal algebra, sheaves, philosophical logic (subdividing the present volumes I and II appropriately), and philosophy of mathematics. The present six volumes cover only approximately 80% of the data on our computer files.

The possibility of extending the classification scheme by developing a so-called thesaurus system was discussed on several occasions. Certainly this would be desirable; to some extent *Alonzo Church* tried to create such a system in connection with his bibliography in the Journal of Symbolic Logic. However there are difficult scientific problems connected with the creation of such a system and their solution requires much time and expertise.

Another way to extend the Bibliography which would perhaps better serve the purpose of providing an overview of certain special areas would be to commission a series of survey papers to appear from time to time as, say, an additional issue of the Journal of Symbolic Logic; each paper would include an annotated listing of the literature taken from the Bibliography.

There are plans to establish a bibliographical centre for Mathematical Logic and adjacent areas. A central function of such a centre would be to collect information on all new publications (including mimeographed notes, theses, etc.) as well as to correct errors and omissions in the current data. It is hoped that all logicians would provide information concerning their own publications as they appear. A continuation of the Bibliography together with supplements (to appear periodically) would be prepared at the centre. We also hope to make available an on-line system. From these activities and the flow of information from the individual logician to the centre and vice versa a "living Bibliography" would emerge. This would provide a way to determine the main trends in the progress (or decline) of specified directions of work. So a centre would exist at which it would be possible to gain some oversight of the rapidly developing field of mathematical logic.

Acknowledgements

Work on the Bibliography started at the same time as the Ω-Group came into being, early in 1969. To begin with, index cards were used for storing bibliographical data; it was *Horst Zeitler* and *Diana Schmidt* who convinced me that we are living in the 20th century and that the data should be computerized. They, together with *Ann Singleterry-Ferebee,* first brought the Bibliography to a workable computerized form at the end of the seventies. In this period I also had the help of *Ulrich Felgner* and *Klaus Gloede,* in particular in classifying the literature. At about this time, others contributed in many useful ways. In particular, important problems of principle were highlighted by a long list of intriguing questions from *Dana Scott:* "How do you classify this or that item . . . ?" *Robert Harrison* worked faithfully collecting data for the Source Index.

The second period, beginning in the early eighties, was characterized by the programming necessary to manage the data. This was carried out by *Ulrich Burkhardt* (†) and *Werner Wolf* and finally by the *outstanding work* of *Rolf Bogus.* In this period we also changed the classification system and for this I had the continuous and intensive help of *Andreas Blass* and *Peter Hinman.* In addition, both of them, together with *Heinz-Dieter Ebbinghaus,* gave me much advice about organization and technical arrangements. Over the last four years the work of large groups of students has been essential for collecting reviews, entering corrections and new items into the computer, etc. Again and again I have been overwhelmed by their idealism and energy. Among them I wish to mention particularly the continuous help of *Elisabeth Wette* and *Ulrike Wieland.*

The Bibliography would not have reached publishable form without the work of my collaborator *Wolfgang Lenski* (in the second period). It would have been unthinkable for me to interfere anywhere in the process of the growth of the Bibliography without discussing the matter with him beforehand. He has accumulated a detailed knowledge of every aspect of the project and has devoted his talents for many years to the common enterprise.

My secretary, *Elfriede Ihrig,* has willingly assisted in the work of the Ω-Group and the Bibliography from the beginning, over many years, filled with ups and downs and with all kinds of tasks. She has always maintained her warmhearted balance.

To all I express my personal warm thanks!

The *Journal of Symbolic Logic* sent information concerning papers for which reviews were never published. We also acknowledge permission to use computer tapes with lists of literature covering certain periods of time from *Mathematical Reviews* and from *Zentralblatt für Mathematik und ihre Grenzgebiete.*

Yuzuru Kakuda and *Tosiyuki Tugue* collected and prepared the Japanese literature for us. *Petr Hajek* and *Gerd Wechsung* helped us with updating the bibliographical references of so many sources not available to us. *Mo Shaokui* corrected data on the Chinese literature and added items of which we were not aware.

The Editors filled many gaps, corrected all mistakes which came to their attention and undertook the burden of checking – and changing if necessary – the classification of the entries in their special areas. Here again I would like to mention *Rolf Bogus* and *Wolfgang Lenski* who organized the enormous exchange service for the transfer of literature among the Editors and the inputting of the many changes and corrections. *Andreas Blass* and *Peter Hinman* were also instrumental in this exchange; their preliminary classification of each item added to the Bibliography during the final editing process meant that the Editors had mainly to look at items inside their own areas. *Jane Kister* read through the whole Source Index correcting mistakes and suggesting valuable changes in it.

In collecting and organizing the data for the Bibliography we have received much help from various sources, and especially in letters from colleagues all over the world, containing information and suggestions. I apologize for being unable to answer them all individually, but all were read carefully.

We thank all those concerned.

As everybody can guess, the whole enterprise was indeed expensive. *Financial support was provided by the Heidelberger Akademie der Wissenschaften in the framework of the Ω-Group project.*

Special thanks go to the firm APPL, who transformed our computer tapes to the present printed form, and to the editorial and production staff of SPRINGER-Verlag for their continuous help, notably in the traditionally fine realization of the six volumes.

Finally, through working so many years on this project I have come to understand and appreciate more and more the immense work of Alonzo Church in building his Bibliography of Logic and its adjacent areas, together with a detailed classification, that is contained in so many volumes of the Journal of Symbolic Logic. Understanding comes from doing.

Introduction

Jane E. Kister (Proof Theory)
Dirk van Dalen & Anne S. Troelstra (Constructive Mathematics)

Preliminary Remarks

The present volume covers the literature on proof theory, constructive mathematics, lambda-calculus and combinatory logic. "Constructive" is here to be understood in the broad sense, and includes intuitionism, finitism, recursive mathematics and predicative mathematics.

The contents thus correspond roughly to section 03F plus 03B40 of the AMS (1980)-classification. The subsections within 03F are interpreted along the lines of the Ω-classification scheme and are discussed in greater detail below. B40 has been incorporated in this volume because it fitted in rather well with section F; there are manifold connections between B40 and section F.

We have divided the items into 18 chapters, most of which correspond to a single subsection of F each; one chapter is devoted to combinatory logic and the lambda-calculus (B40). The chapters are grouped under five headings.

Any attempt to classify a large number of items, produced over almost 100 years, is full of pitfalls. The classification scheme itself is in some sense a compromise. The continuing development of a subject requires a constant revision of the classification; papers belonging to the early history of the subject are often difficult to fit into a classification based on work of the last decade, say; on the other hand future developments may call for a refinement of the existing classification scheme. We have tried to annotate the F-section so as to accommodate the main concerns of proof theory and constructive mathematics throughout this century. However, as the scheme was originally conceived for items appearing in the 1980s, it was often difficult to find precisely the right niche for papers written in the 1920s and 1930s.

Decisions on the classifications to be assigned to a particular paper will always retain a subjective element. For one thing, one's views as to what exactly belongs under a particular heading of the classification develop and change during the process of classifying a large number of items - one is often induced to look back a second time at items classified early on, in order to check consistency and to modify, possibly, the original classifications. The reader should not expect one hundred percent consistency in our decisions.

One is also faced with a choice between a broad view and a narrow view of the classification. On the broad view, one lists an item under every subject to which it is possibly relevant; on the narrow view, an item is listed only under the topics which take up a substantial part of the item concerned. Both extremes may result in loss of information. We have tried to steer a middle course; thus, for example, a passing remark on the intuitionistic case in a paper dealing with classical formalisms does not, as a rule, lead to inclusion under F50. On the other hand, we have not tried to isolate the single most important aspect of a given item.

Since it was impossible for us to inspect each individual item, we had to rely in most cases on the available reviews to suggest appropriate classifications. If the reviews seemed to give too little information for a decision, we tried to look up the item itself. However, this was not always possible, partly because of the inaccessibility of some of the older or more obscure journals, but also because of time. In certain doubtful cases, when we had little to go on other than the title, we assigned the classification F99. Such items all appear in the 'Miscellaneous' section of the Subject Index.

Comments on the Classification Scheme

The following remarks are intended to supplement and to some extent explain the B40 and F-sections of the fully annotated classification scheme.

1. Historical, critical and philosophical publications (A05 ∪ A10) ∩ (F ∪ B40)

All items with an F classification or in B40 which have a substantial historical and/or philosophical component have been collected here. Papers which do not contain substantial technical results but rather general discussions pertinent to the present volume, for example on Hilbert's programme, are classified A05 ∩ F99.

2. Combinatory logic and lambda-calculus (B40)

Items dealing with, or containing a non-trivial application of the lambda-calculus and/or combinatory logic are listed here. Typed lambda-calculus and combinatory logic are also included but items using lambda abstraction for comprehension terms only are not.

3–4. Cut elimination and normal form theorems, structure of proofs (F05 and F07)

"Structure of proofs" (F07) is interpreted to include the study of those formal systems (generally not of the axiomatic type) in which the logical rules generate proofs with interesting structure: the prime examples of such systems are Gentzen-style natural deduction systems and sequent calculi. This section thus includes analyses of introduction and elimination rules for systems of propositional and predicate logic. Studies of various forms of the induction rule or ω-rule in systems of arithmetic are generally not in the F07 section: they are classified as F30.

F05 is also concerned with the structure of proofs but is reserved for those items treating "normal form" theorems. We have included under this heading cut-elimination theorems for sequent calculi, normalisation theorems for systems of natural deduction and Herbrand's theorem. Applications of normal form theorems also appear under F05. Normal form theorems for systems of both classical and intuitionistic logic are in this section (those for intuitionistic systems are also under F50) but those for which the underlying system is a modal, tense or relevance logic have not generally been included; they are classified as B45 or B46. In many cases the methodology behind normalisation theorems for terms in typed systems is very similar to that for natural deduction systems (W. A. Howard's "formulae-as-types" isomorphism 1980, ID 74281); however, items dealing with terms rather than proofs generally do not appear in F05, but in B40 or F10.

For practical reasons, the few items dealing with Hilbert's ε-symbol are also in F05.

5. Functionals in proof theory (F10)

The use of functionals in proof theory stems from G.Kreisel's no-counterexample interpretation and Gödel's "Dialectica interpretation" (Kreisel 1951, ID 28465; Gödel 1958, ID 20826). This section includes work on primitive recursive functionals of finite type and on systems of typed terms including a recursor, explicitly or implicitly relating to functional interpretations of arithmetic. Items concerning terms in the pure typed lambda-calculus are in B40 and not in F10.

6. Recursive ordinals and ordinal notations (F15)

This section includes those items in proof theory in which ordinals play a significant role.

Ordinal notations, both as the prime object of study and in connection with the proof-theoretic study of strengthenings of arithmetic and subsystems of analysis, are to be found here. Some systems of notations are based on a classical set-theoretical study of functions on the ordinals. An item in which the classical theory is considered is in both F15 and E10 but more constructively based studies of notations have no E classification. For example, items concerning the theory of Kleene's notation system O (1938, ID 07166) for the constructive ordinals based on recursion-theoretic ideas are in F15 and often in D as well but not in E. Proof-theoretic studies of first-order arithmetic in which ordinals $\alpha \leqslant \varepsilon_0$ are mentioned have generally not been included here unless the notation system or accessibility (well-ordering) proof for α constitutes a substantial part of the paper.

Hierarchies of number-theoretic functions indexed by (notations for) ordinals and, more generally, functions and functionals defined by transfinite recursion (using a specific well-ordering) are classified in F15. Orderings defined on classes of number-theoretic functions are another topic that appears in this section. A final topic included in F15 (as well as other F-sections) is that of transfinite progressions of theories (e.g., Turing's ordinal logics (1939, ID 13726), Feferman's systems of ramified analysis (1964, ID 03695)), where again the well-ordering used in the definition plays a critical role.

7. Complexity of proofs (F20)

This section contains those items dealing with the quantitative analysis of proof structures. Thus, work on the length or complexity of proofs appears here but papers in which the computational complexity (e.g. time and space complexity) of other processes such as testing for validity or performing reductions on terms have been excluded; they are classified under D15.

8. Relative consistency and interpretations (F25)

This section consists of two major (related) parts, on relative consistency and on interpretations. We have been rather restrictive in assigning this classification to items in which relative consistency is discussed. This is more for practical than methodological reasons. So much of the work that has been done in proof theory has been directly or indirectly concerned with questions of consistency that a liberal interpretation of this classification would include a very large number of items. Items concerning consistency which could properly be assigned alternative classifications do not generally appear here. Thus the portion of F25 dealing with relative consistency is reserved for papers concerning the relative consistency of two *arbitrary* theories. Those concerning the (relative) consistency of set theory (with additional assumptions) are in E35 and not here. Similarly papers on consistency proofs for arithmetic or (subsystems of) analysis occur under F30 or F35 and possibly elsewhere in F depending on the logical basis and the techniques used in the proofs, but usually not in F25. Philosophically oriented work on the consistency problem related to Hilbert's programme and Gödel's second incompleteness theorem is usually in A05 ∩ F99.

The Tarski notion of interpretation of one theory into another (Tarski 1953, ID 28815) has generated a number of papers on interpretations; these have been classified in F25 even if the approach is not strictly proof-theoretic, and again this is for a practical reason: there is no other obvious classification for these items.

9. First-order arithmetic and fragments (F30)

This section deals with formal systems for classical and intuitionistic arithmetic and fragments thereof. Of the many aspects of the metamathematics of arithmetic in this section, one of the largest under the F-classification, we mention some of the prominent themes.

Gödel's incompleteness theorems form the basis of one of the main topics in this section. Thus items concerning undecidable sentences appear here, as do those in which the consistency of arithmetic (in the spirit of Hilbert's programme) is discussed. Gentzen-type results on the consistency of arithmetic are also in F30 (and often in another F-section depending on the method of proof).

The more recent topic of provability logic of arithmetic, an offshoot of the incompleteness theorems, is also in this section (and usually also in B45, indicating the use of modal operators).

Papers in which (nonstandard) models of some specific formal system for arithmetic are studied are classified both under F30 and under C62 (and/or H15). However, those concerning models of "true arithmetic" or full Peano arithmetic (where the formal system is not the main focus of study) do not appear in this section; they are in C62.

One final word of caution: many of the standard introductory textbooks on logic include chapters on formal systems of arithmetic and Gödel's incompleteness theorems. Although we have tried to add F30 to the classification of such items, we suspect that there are many which will only be found elsewhere (e.g. under B98).

10. Second- and higher-order arithmetic and fragments (F35)

This section comprises metamathematical work on systems of second- and higher-order arithmetic and generalized inductive definitions. It includes both classical and intuitionistic analysis; those items dealing with the metamathematics of intuitionistic analysis appear here and under F50. Even though intuitionistic analysis with variables for sequences, rather than for sets of natural numbers, is in many ways more akin to arithmetic than to a second-order system, we have included it since many users of the bibliography will expect to find it here. Papers on the ramified hierarchy and the metamathematics of predicativism often appear in this section as well as under F65.

Systems in the language of set theory, considered as a first-order language, are not included. However, the proof theory of higher-order logic and type theory (even when the arithmetic properties of the system are not explicitly considered) appears in this section (see also B15).

Items on the model theory of analysis (β-models, etc.) are treated as are those on the model theory of arithmetic: when the underlying formal system plays a role, the two-fold classification, F35 and C62, is used; otherwise, only C62 is used.

11. Gödel numberings in proof theory (F40)

"Gödel numberings in proof theory" is intended for those papers in which the form of the Gödel numbering of syntax plays an important and specific role. It does not include the many papers in which standard properties of the usual Gödel numberings are implicitly assumed in the course of establishing incompleteness theorems and related results.

12. Metamathematics of constructive systems (F50)

This section deals with the metamathematics of constructive and intuitionistic mathematics (F55 and F60). The metamathematics of predicative mathematics is not included; such items are found under F65.

Thus in this section one finds all items dealing with formal systems for intuitionistic propositional and predicate logic, arithmetic, analysis, etc. Intuitionistic arithmetic also appears under F30, intuitionistic analysis and type theory also under F35. Subjects such as realizability, Kripke and Beth semantics for intuitionistic logic appear in this section.

13. Constructive and intuitionistic mathematics (F55)

This section contains items dealing with intuitionistic mathematics ("IM") in the spirit of L.E.J. Brouwer and A. Heyting, as well as constructive mathematics in the spirit of E. Bishop ("BCM"). These traditions have much in common, and have therefore been brought together in a single section. All BCM is acceptable as IM; conversely, many papers on IM written before Bishop's book (1967, ID 01525) can also be counted as contributions to BCM, in particular when no use has been made of special properties of choice sequences.

Constructive mathematics in the sense of A.A. Markov (based on Church's thesis) is not included; this is found under F60. Finitism, although intuitionistically acceptable, is listed under F65.

14. Constructive recursive analysis (F60)

This section includes constructive recursive mathematics ("CRM") in the sense of A.A. Markov and N.A.Shanin, as well as classical recursive analysis. Although classical recursive analysis has classical logic as its basis, many of its results are directly relevant to, and can be lifted to CRM, and vice versa.

15. Other constructive mathematics (F65)

This section contains papers based on specific constructive foundational views that could not be subsumed under F55 or F60, for example finitism, predicativism (also when the logical basis is classical) and ultra-intuitionism (ultra-finitism). The papers in this section may be either mathematical or metamathematical in nature. The reason for including papers on the metamathematics of these forms of constructivism is a pragmatical one: section F50 is already quite large as it is. Papers on primitive recursive arithmetic, the prime example of a finitist formalism, listed under F30 are also listed here when the foundational aspects are considered; papers on predicative analysis appear here and under F35. For papers of an overall historical or philosophical nature one should consult (A05 ∪ A10) ∩ (F ∪ B40).

16-17. Proceedings, textbooks, surveys (F97 and F98)

Only those volumes with a substantial percentage (more than forty percent, say) of papers falling under the F-section or B40 appear here.

18. Miscellaneous (F99 \ (A05 ∪ A10))

This section includes only those items which could not be classified more accurately, e.g. for lack of detailed information, but which nevertheless were thought to belong to the F-section. Papers with a substantial historical and/or philosophical component appear under (A05 ∪ A10) ∩ (F ∪ B40).

User's Guide

Wolfgang Lenski & Gert H. Müller

§1. Introduction

After some opening remarks, the organization of this Guide follows the main division of the volume: *Subject Index, Author Index, Source Index, Miscellaneous Indexes.* For each part we give first a general explanation followed by a more detailed description of typical entries in the index in question. The reader will probably find the User's Guide most helpful when he comes across an unclear entry in the Bibliography: he can then turn directly to the corresponding section in this Guide for an explanation of the abbreviations and conventions used.

§2. General remarks

The main languages of the Bibliography are English, French, German, Italian, and Spanish. For other languages translations (of titles, names of sources, etc.) are used - with some few exceptions in cases for which we had no translation. These translations were taken from various available sources or made by the Editors.

For practical reasons, all entries are in the Roman alphabet and diacritical marks have not been used. Thus, for languages other than English certain conventions have been adopted.

The *transliteration of Cyrillic* names and titles, the treatment of *diacritical marks* and the *alphabetization and alternative spelling* of author names are explained in detail in the Miscellaneous Indexes.

The *abbreviations* of *sources* were either taken from one of the various reviewing journals or invented by us. Although we had to abbreviate long titles, we hope that in most cases the abbreviation will suggest the full title in a sufficiently understandable way. How successful we were is left to the user to decide.

The *review numbers* given with the entries in the Author Index are from *Mathematical Reviews* (MR), *Zentralblatt für Mathematik und ihre Grenzgebiete* (Zbl), *Journal of Symbolic Logic* (JSL), and *Jahrbuch über die Fortschritte der Mathematik* (FdM). We made a serious attempt to include all reviews of any given item but we have doubts concerning our success. We also tried to avoid listing two reviews for a given item in those cases in which the second "review" simply points to the original review and does not give any additional information.

In case of *multipart publications* pointers are given to the other parts, as far as they are known, in the *Remarks* to the publication in question. It is not always the case that the different parts of a publication all have the same classifications. Thus it may happen that, for example, part I has a classification in this volume and part II does not. In this case the Remarks for part I indicate the author(s) and year of publication of part II. The user will need to consult the other volumes for further bibliographic information on part II.

The general way to search through the Bibliography is to use certain *pointers:* From the Subject Index to the Author Index the pointer is [Author, Year, Title]; from the Author Index to the Source Index the pointers are 5-digit codes; e.g. (**J** 1234) is a code which sends the user to the J-section of the Source Index.

A *word* of *caution*: In order to use the Bibliography for quotations in future publications it is necessary to use both the Author Index and the Source Index; it is not generally sufficient to quote just from the Author Index. For example, for the paper "AANDERAA, S.O. [1974] *On k-tape versus (k−1)-tape real time computation*" a quotation of the source of this paper as given in the Author Index listing of this item, "Complexity of Computation; 1973 New York 75–96", would with high probability be misleading: one might try to find a book of this title published in New York in 1973 whereas in fact "1973 New York" denotes the date and place of the conference in the proceedings of which Aanderaa's paper appears. The volume was actually published in 1974. The source code (**P** 0761) for "Complexity of Computation; 1973 New York" should be used to find the full details of the source in the Source Index. The abbreviations of the sources may themselves be misleading without the corresponding additional details (e.g. country codes) given in the Source Index. For example, many abbreviations for conference proceedings do not include an abbreviation for "Proceedings of ...". Thus "Proceedings of the Third Brazilian Conference on Mathematical Logic" is abbreviated by "Brazil Conf Math Logic (3); 1979 Recife"; a reader without the Bibliography at hand might search in vain for the volume under "Brazil" in his library whereas in fact it might be found alphabetically under "Proceedings".

The Source Index includes, as far as they are known to us, *International Standard Serial Numbers* (ISSN) or *Book Numbers* (ISBN) and *Library of Congress* (LC) numbers. They may help in finding the source in question in libraries or bookstores.

To facilitate searches for works spanning two or more of the major subfields of logic, the first of the Miscellaneous Indexes lists the entries in the present volume that also occur in other volumes of the Bibliography.

Accidental occurrences of features not explained in the User's Guide are left as exercises to the user. HINT: Write to us (in any case), please.

§3. Subject Index

This is a listing of publication items ordered

first by the (special) *classification sections,*
then by the *year of appearance,*
and finally *alphabetically* by *author,*

showing the author, title and the codes of all classification sections which apply to the given publication.

- The *titles* are given in the main languages of the Bibliography; if the original title is in another language, this is indicated in parentheses, e.g. (Russian), but only a translation of the title is given. Information on summaries in languages other than the original is included.
- If a publication is by *multiple authors,* it occurs only *once,* under the alphabetically first name. (But see also the Author Index.)
- In order to get the full bibliographical data of a publication, use the author, year and title to find the item in the Author Index.
- The classification sections listed in each volume have been selected by the individual editors. Sections B96–F96 have been systematically omitted; for the collected works of an author refer to the Author Index.

§4. Author Index

This is a listing of publication items ordered

> alphabetically by *author,* and for a given author
> chronologically by the *year of appearance,* and therein
> alphabetically by *title* of the item.

• The titles are given in the main languages as in the Subject Index.
• The *names* of the *authors* are written in the Roman alphabet using the Transliteration Table (see Miscellaneous Indexes) if necessary. There may be many versions of the name in use for a given author; (e.g. different combinations of the given name(s) or initials; different names used before and after marriage; different transliterations). The Miscellaneous Indexes include a table of different versions (known to us) and the corresponding form used in this Bibliography.
• Here publications with *multiple authors* are listed under *each* author but in the alphabetically later cases only the year is given and there is a pointer to the full entry given under the first author.
• The last entries for an author may contain a *reference to other name(s)* under which he/she also has publications in the Bibliography or *to other volumes* of the Bibliography where he/she has publications not mentioned in the present volume. A complete list of the author's papers contained in the six volumes is obtained by consulting the other volumes.
• In the following we explain the *individual entries* in more detail by giving an *idealized* example using fictitious names and sources showing all features that might occur; in a given case some features may not appear either because they do not apply or because our information is incomplete. The typefaces of the example and the order of its fields are as in the Author Index but, for *expository reasons only, here* we list all features on separate lines numbered by (1), (2), ...; we list explicitly those fields that begin a new line in the Author Index itself. (The foregoing description applies not only to the explanation of the Author Index treated here but also to the explanation of the Source Index later on.)

Example

(1) AUTHOR, K.J. & COMPANION, CECIL X. [1972]
(2) *On coding and decoding (Russian) (English and French summaries)*
(3) (**J** 9999) or (**S** 9998) or (**P** 9997) or (**C** 9996) or (**X** 9995)
(4) J Math 1*1-10
or Math Logic Series 1
or Logic Conf; 1999 London 3-10
or Math Publ xxv+200pp
(5) • ERR/ADD ibid 2*3-4 or (**J** 8888) Arch of Logic 2*3-4
(A new line begins here.)
(6) • LAST ED [1983] (**X** 9900) Logic Publ xx+100pp
(7) • REPR [1981] (**J** 9901) Math Logic J 2*3-8
(8) • TRANSL [1979] (**J** 9902) Math Transl 1*4-8
(A new line begins here.)
(9) ◇ B05 B20 C12 ◇
(10) • REV MR 99a:03001 Zbl 999 # 03001 JSL 99.321 FdM 99.123
(11) • REM This is an illustrative example
(12) • ID 12345

Explanations

(1) lists the authors followed by the year (in brackets) of publication of the item. Exceptionally a full given name (e.g. CECIL) is used to distinguish several authors with the same surname and initials.

(2) gives the title of the item followed, if the original is not in an official language of the bibliography, by the original language in parentheses and an indication of summaries in languages other than the original.

(3) is a pointer (or "source code") to the *Source Index;* there are five types: *Journal* (**J**), *Series* (**S**), *Proceedings Volume* (**P**), *Collection Volume* (**C**), and *Publisher* (**X**); one such code appears in (3). In order to find the full bibliographical data of the source use the pointer to locate the source in the Source Index; e.g. (**J** 9999) is given in the J-section of the Source Index.

Note: For a small number of items the source code is 0000, 1111, 2222 or 3333 (*not* preceded by J,S,P,C, or X). The code 0000, respectively 1111, indicates that the item is a thesis, respectively technical report. The code 2222, respectively 3333, is used for those cases in which the source, respectively publisher, is unknown. In each such case any further source information available is given in the Remarks (see line (11)).

(4) contains the abbreviation of the source indicated by the code in (3) followed by the paging as appropriate. Certain uniform features of the form of abbreviation used for proceedings and collection volumes should help the reader to recognise the volume. Abbreviations for proceedings (**P**) volumes end with an indication of the year and place of the corresponding conference, e.g. 1973 New York. Likewise, a name in parentheses, e.g. (Goedel), in an abbreviation of a collection (**C**) volume indicates the honorand to whom the volume is dedicated. A name followed by a colon, e.g. "Wang:", at the beginning of a collection volume abbreviation, indicates the author of all papers in the collection. The paging takes one of the following forms:

1*1-10 : Volume 1, pages 1-10 (for journals or series)
1/2*1-10 : Volume 1, Issue 2, pages 1-10 (for journals)
3-10 : pages 3-10 (for proceedings or collection volumes)
xx+200pp: initial paging + paging of a book (following a publisher or series)

(5) The • here and later is intended to make the entries easier to read. It is used to separate different types of information. After the • is the bibliographical information for published errata or addenda to the item. The two ways ERR/ADD can be given correspond to the cases in which its source is the same as in (4) (indicated by "ibid") and that in which it is in a different source; in the latter case the entry is of the same form as in (3) and (4).

The remaining information is not strictly part of the bibliographical data but contains useful additions.

(6), (7), (8) list the most recent edition, reprintings and translations, respectively, given by source as in (3) and (4); note that (7) and/or (8) may contain several entries for one publication.

(9) The classification codes enclosed in ◇ always begin a new line. Note that the codes are given in alphabetical/numerical order; no distinction of primary and secondary classification is made. (The classifications often differ from those assigned to the item in MR or Zbl.)

(10) lists the reviews. Sometimes two reviews are given from one reviewing journal. This may happen, e.g., when an item and its erratum/addendum are reviewed separately or when two different editions of a book have independent reviews.

(11) contains additional information not appropriate for coding in one of the standard fields.

(12) Each entry ends with its *identification number*. It is not used elsewhere in the main body of this volume except occasionally in the Introduction and the Remarks of another item where it may be used to pinpoint an item not uniquely identified by author(s) and year. The identification number is used (together with author(s) and year) as a pointer in the External Classification Code Index. We ask that the identification number be used in any correspondence with the Editors concerning this publication, as the bibliographical data base is indexed by these numbers.

§5. Source Index

This index contains the bibliographical data of the sources of the publications listed in this volume. It is subdivided into the following parts.

J (*Journals*), **S** (*Series*), **C** (*Collection volumes*), **P** (*Proceedings*), **X** (*Publishers*).

• Each part is ordered by the 4-digit source code numbers. (There is no significance to the particular 4-digit number assigned to a given source other than as a way to find the entry in the source index. Numbers were assigned as the sources were entered into the data base and so the numbering does not correspond to alphabetical order or order of publication.) Each 4-digit number is used *only once* as a source code so that, e.g., 0007 is a source code for a journal and the number 0007 is not used as a code for a series, proceedings, collection volume or publisher.
• Titles are given in the original language, using the transliteration system (see Miscellaneous Indexes) where necessary, followed, if necessary, by a translation into one of the main languages in parentheses. Sometimes if the original title is unknown to us, we give only a translated title in parentheses. Sometimes a source, e.g., a journal, has more than one title (English, French, German); in this case all titles are given, separated by *. These measures were taken to ease the search in libraries. In order to explain the entries in the Source Index we again use idealized examples and apply the conventions described in §4 above.

Journals

Example of a journal entry:

(1) **J** 8888 Math Div • F
(A new line begins here)
(2) *Mathematica Diversa * Mathematiques Diverses*
(3) [1900ff] or [1905-1935] ISSN 0007-0882
(A new line begins here.)
(4) • CONT OF (**J** 8885) J Math Ser A
(5) • CONT AS (**J** 8887) J Math Ser C
(6) • TRANSL IN (**J** 9904) Math Transl
(7) • TRANSL OF (**J** 9905) Matemat
((4) - (7) may contain more than one entry)
(A new line begins here.)
(8) • REL PUBL (**J** 9903) Mathematica (Subseria)
(9) • REM This journal is a fiction

Explanations

(1) Source code and abbreviation of the journal as used in the Author Index followed by the *international vehicle code* of the country in which the journal is published. A list of these codes is included in the Miscellaneous Indexes.
(2) The form of title(s) (and translations) are explained above.

(3) [1900ff] indicates that this journal has appeared continuously since 1900; [1905-1935] indicates that the journal appeared from 1905 to 1935. The International Standard Serial Number (ISSN) is given whenever possible.

(4), (5) give the predecessors (continuation of) and successors (continued as) of the journal in (2). In some cases in (4) or (5) the source code may be missing; this means that there are no entries in the Author Index which refer to the continued source. (It is mentioned, however, for the convenience of the user.)

(6) lists the translation journals of (2) and (7) gives the journal of which (2) is a translation; the source code is shown only if the translation in question is used as a source in this Bibliography. (6) and (7) do not both occur in a single journal entry.

(8) lists further entries in the Bibliography related to this journal, e.g. a subseries of the journal.

(9) is intended for additional information of various kinds.

Series

It is often hard to determine what should and what should not be characterised as a series. Some serials that we have chosen to treat as series may elsewhere be considered to be journals. In other cases, in particular certain publication series of university mathematics departments, the series includes all publications of its publisher and so might reasonably be identified with the publisher. Despite these considerations, we have chosen to list series separately to accord with the form of quotation often used in the modern literature.

Example of a series entry:

(1) **S** 8999 Notae Log • NL
(A new line begins here)
(2) *Notae Logicae * Notas Logicas*
(3) [1900ff] or [1905-1935]
(4) • ED: EDITOR, A.A. & COEDITOR, B.B.
(5) • SER (**S** 8998) Notes in Phil
(6) • PUBL (**X** 9950) Logic Publ Co: Heidelberg
(7) • ALT PUBL (**X** 9951) Math Publ Inc: London
(A new line begins here.)
(8) • CONT OF (**S** 9975) Notes in Logic A
(9) • CONT AS (**S** 9901) Notes in Logic B
(10) • TRANSL IN (**S** 9902) Notes de Logique
(11) • TRANSL OF (**S** 9903) Logical Notes
(A new line begins here.)
(12) • ISSN 0011-11122 (or ISBN 0011-11123) LC-No 73-10000
(13) • REL PUBL (**S** 9900) Notae Logicae (Subseria)
(14) • REM The origins of this series are somewhat obscure

Explanations

Entries (1), (2), (3), (8)-(11), (13), and (14) correspond to (1), (2), (3), (4)-(7), (8), (9), respectively, of the *journal entry* described above.

(4) lists the editors of the series (given in the same form as in line (1) of the Author Index example).

(5) Occasionally a series is itself a subseries of another series or journal. This is indicated in (5) (with an **S** or **J** as appropriate).

(6) gives the publisher of (2). For those publishers not listed in the publisher section of the Source Index, an abbreviation is sometimes used if either the abbreviation is readily understandable or the full name is not known.

(7) Some sources are published by two or more publishers; ALT PUBL lists the alternative publisher(s).

(12) lists the ISSN (or ISBN) and the Library of Congress number.

Proceedings and Collection Volumes

Example of a proceedings or collection volume:

(1) **P** 9920 Atti Congr Mat; 1971 London, ON • CDN
or
C 9921 Atti Congr Mat • D

(A new line begins here.)

(2) [1972]
(3) *Atti del Congresso di Matematica * Actes du Congres de Mathematique*
(4) • ED: EDITOR, A.A. & COEDITOR, B.B.
(5) • SER (**S** 8999) Notes in Logic
(6) • PUBL (**X** 9950) Logic Publ Co: Heidelberg
(7) • ALT PUBL (**X** 9951) Math Publ Co: London

(A new line begins here.)

(8) • DAT&PL 1971 Aug;London, ON, CDN
(9) • ISBN 0-012-34567-X, LC-No 84-98765
(10) • REL PUBL (**P** 9947) Atti Congr Mat Vol Spez

(A new line begins here.)

(11) • TRANSL IN [1973] Conf de Logique Math (3); London, ON, CDN
• PUBL (**X** 9949) Livres: Paris
(12) • TRANSL OF [1971] Konf Math Logik (3); London, ON, CDN
• PUBL (**X** 9948) Buchverlag: Stuttgart

(A new line begins here.)

(13) • REM Not all the articles appear in the translation

Explanations

(1), (3), (4) - (7), (9), (11), (12), (13) correspond to (1), (2), (4)-(7), (12), (10), (11) and (14), respectively, of a *series entry*. In (11), (12) PUBL denotes the publisher of the translation or original, respectively.

(2) denotes the year of publication of the volume (and not, in the case of a proceedings, the year of the conference).

(8) is used for proceedings volumes to indicate the date (year and month) and place of the conference, given by the city, the state (for the USA and elsewhere) and the country using its code as defined above. Note in case of *Proceedings* (**P**) volumes in (1) the country code of the place of the conference is repeated for conformity reasons, whereas for *Collection* (**C**) volumes the country code in (1) refers to the location of the publisher as in the case of *Journals* and *Series*.

(10) lists further entries in the Bibliography related to this volume, e.g. another proceedings volume of the same conference or a journal of which the volume is a special issue.

Publisher

Example of a publisher entry:

(1) **X** 9950 *Logic Publishing Company* (Heidelberg, D & London, GB) ISBN 0-01
(2) • REL PUBL (**X** 9930) Editions Logiques: Paris, F
(3) • REM In London called Logic Publishing Corporation

Explanations

(1) lists the source code and full name of the publisher followed, in parenthesis, by the cities from which the publisher publishes and the ISBN. As in (8) of a **P** or **C** entry, codes are used for countries (see Miscellaneous Indexes).

(2) lists those publishers who have connections with the publisher listed in (1).

§6. Miscellaneous Indexes

This part contains the following indexes:

1. External classifications
2. Alphabetization and alternative spellings of author names
3. International vehicle codes
4. Transliteration scheme for Cyrillic

In each case a description of the contents and use are given in the corresponding introductory texts.

Ω-Classification Scheme

Andreas R. Blass
Peter G. Hinman

The classification scheme used for the Ω-Bibliography is a modified version of the section "03: Mathematical Logic and Foundations" of the 1985 Mathematics Subject Classification of *Mathematical Reviews* and *Zentralblatt für Mathematik und ihre Grenzgebiete*. For the sake of uniformity we have labeled all sections with a letter followed by a two-digit number; the prefix 03 is superfluous and therefore omitted. This decision has led to the creation of new sections to replace 03-01 through 03-06 (cf. X96-X98 and A10) and several sections with prefix other than 03 which have substantial logical content. Examples of the latter sort are B70 (to replace 94C10) and B75 (to replace the "logical part" of 68B10) (68Q55 and 68Q60 since 1985).

An important category of differences between the two schemes arises from the fact that whereas the MR/Zbl system is intended to classify works written after 1980, the majority of entries in the Ω-Bibliography were written before 1980. The subject matter of Mathematical Logic has, of course, changed immensely over the years, and today's categories are not always sufficient to distinguish properly important lines of earlier research. To deal with this problem we have added a few new sections (e.g. B22, B28, B65, C07, E07, and E47), renamed others (e.g. B35, C35, and E10), and altered slightly the interpretation of others (e.g. B25 and D65). To aid the reader in learning our conventions we have added descriptors to the section names. Topics preceded by a + (−) sign are specifically included (excluded) from a section. When this is in conflict with current MR/Zbl practice, this fact is also noted.

A

A05 Philosophical and critical

A10 History, Biography, Bibliography
MR uses 03-03 and 01A for history and biography
MR puts bibliography under specific fields.

B GENERAL LOGIC

B03 Syntax of logical languages

B05 Classical propositional logic and boolean functions

+ Axiomatizations of classical propositional logic
+ Boolean functions (machine manipulation is also in B35); MR puts these in G05 and in 06E30 and 94C10.
− Fragments of propositional logic: see B20
− Switching circuits: see B70; MR also uses 94C10

B10 Classical first-order logic

+ Many-sorted logic
+ Syntax and semantics up to the Completeness Theorem
− Model theory: see Cn, particularly C07
− Proof theory: see Fn

B15 Higher-order logic and type theory
+ Higher-order algebraic and other theories
– Higher-order model theory: see C85
– Set theory with classes: see E30 and E70
– Intuitionistic theory of types: see F35

B20 Fragments of classical logic
+ Fragments of propositional and of first-order logic
+ Fragments used in model theory, set theory, etc.
+ Syllogistic
– Classical propositional logic: see B05
– Weak axiomatizations without restrictions on formulas: see B55, B60, F50 ("Fragment" refers to reduced expressive power, not reduced deductive power; MR heading "Subsystems of classical logic" includes both)

B22 Abstract deductive systems
+ Consequence relations
MR uses B99

B25 Decidability of theories and sets of sentences
+ Decidability of satisfiability
+ Decidable Diophantine problems
– Decidable word problems: see D40
– Other decidability results: see subject of problem, e.g. D05, or D80; MR includes these results here.
– Undecidability results: see D35, D40, D80, etc.

B28 Classical foundations of number systems
+ Natural numbers, real numbers, ordinal numbers
+ Axiomatic foundations and set-theoretic foundations
MR uses B30

B30 Logical foundations of other classical theories; axiomatics
+ Axiomatic method
+ Geometry, probability, physics, etc.
+ Models for non-mathematical theories
– Foundations of parts of logic: see that part.
MR heading: "Foundations and axiomatics of classical theories" includes also B28

B35 Mechanization of proofs and logical operations
+ Theorem proving, proof checking by machine
+ Minimization algorithms for Boolean functions
+ Optimization of logical operations
MR sometimes uses 03-04 or 68G15 (68T15 since 1985)

B40 Combinatory logic and lambda-calculus
+ Models of lambda-calculus

B45 Modal and tense logic
+ Intensional logic; see also A05
+ Normative and deontic logic
+ Other non-truth-functional systems

B46 Relevance and entailment
+ Fragments
– Primarily modal logic
MR uses B45

B48 Probability and inductive logic
See also A05 and C90
+ Confirmation theory
– Foundations of probability: see B30; MR uses B48

B50 Many-valued logic
+ Matrix interpretations of propositional connectives unless used only as a tool for investigating classical propositional logic.
– Boolean valued set theory: see E40
– Probability logic: see B48 or C90

B51 Quantum logic
– Algebraic study of Quantum logic: see G12
MR uses only G12

B52 Fuzzy logic
+ Vagueness logic
– Papers demonstrating the fuzziness of the author's thought processes

B53 Paraconsistent logic
+ Discussive and dialectical logic
MR uses B60

B55 Intermediate and related logics
+ (Fragments of) propositional and predicate logics between intuitionistic or minimal and classical

B60 Other logics
– Intuitionistic logic: see F50 (MR uses B20)

B65 Logic of natural languages
– Computer languages: see B75
– Formal grammars unless applied to natural languages: see D05
– Natural language as a tool for the study of thought, reality, etc.: see A05
MR uses B65, B99, and 68Fn (68Sn since 1985)

B70 Logic in computer design; switching circuits
+ Hardware related to logic
MR uses 94Cn

B75 Logic of algorithmic and programming languages
+ Algorithmic and dynamic logic; MR uses B70 (formerly B45)
+ Logical analysis of programs
+ Logical aspects of database query languages and information retrieval
+ Semantics of programming languages related to logic
+ Software related to logic
– Specific algorithms: see subject of algorithm
MR uses B60, B70, 68Bn, 68Fn, and 68H05 (68Pn, 68Qn, and 68Tn since 1985)

B80 Other applications of logic
MR uses B99

B96 Collected works
+ Selected works
– Collections (almost) entirely in one subfield: see that subfield
MR uses 01A75, 03-03, and 03-06

B97 Proceedings
+ Collections of papers by various authors, even if they do not derive any actual conference
– Proceedings (almost) entirely in one subfield: see that subfield
– Proceedings not concentrated in this field: see Source Index

MR uses 03-06

B98 Textbooks, surveys

MR uses 03-01 and 03-02

B99 None of the above or uncertain, but in this section

C MODEL THEORY

C05 Equational classes, universal algebra
+ Quasi-varieties, if the emphasis is algebraic
– Word problems: see D40

C07 Basic properties of first-order languages and structures
+ Completeness, compactness, Löwenheim-Skolem, and omitting types theorems for ordinary first-order logic; MR uses C50 for omitting types
+ General properties of first-order theories
+ Homomorphisms, automorphisms, and isomorphisms of first-order structures
– Analogues of these for stronger languages: see C55, C70, C75, etc.

C10 Quantifier elimination and related topics

C13 Finite structures
+ The spectrum problem
+ Probabilities of sentences being true in finite structures

C15 Denumerable structures

C20 Ultraproducts and related constructions
+ Applications of ultraproducts
+ Reduced products, limit ultrapowers, etc.
– General products: see C30

C25 Model-theoretic forcing
+ Existentially closed structures, model companions, etc.
– Model complete theories: see C35
– Set-theoretic forcing: see E35, E40

C30 Other model constructions
+ Contructions involving indiscernibles
+ Products, diagrams

C35 Categoricity and completeness of theories
+ Model completeness
– Gödel's completeness theorem: see C07
– Completeness of axiomatizations of other logics: see those logics, e.g., B45

C40 Interpolation, preservation, definability
+ Definability in classes of structures
– Definability in recursion theory: see appropriate Dn.
– Definability in set theory: see E15, E45, and E47

C45 Stability and related concepts
+ Rank, total transcendence (even before stability was defined)

C50 Models with special properties
+ Saturated, rigid, etc.

C52 Properties of classes of models

C55 Set-theoretic model theory
+ Cardinality and ordering of models
+ Generalized Löwenheim-Skolem results
– Applications of set theory to some part of model theory: see that part
– Models of set theory: see C62
– Original Löwenheim-Skolem theorem: see C07

C57 Recursion-theoretic model theory
+ Model theory of recursive, arithmetical, etc. structures, types, etc.
– Recursion theory without substantial model-theoretic content: see D45
MR uses D45

C60 Model-theoretic algebra
+ Applications of model theory to specific algebraic theories
– Applications of set theory to algebra: see E75
– Decidability questions for algebraic theories: see B25, D35, and D40
– Model theory of orderings: see C65
– Universal algebra: see C05

C62 Models of arithmetic and set theory
+ Admissible sets as models: see also C70 and D60
+ Nonstandard models of arithmetic, when model theory is emphasized
+ Omega-models of higher-order arithmetic
– Models introduced only for consistency results: see F25 and E35
– Nonstandard models of arithmetic, when non-standardness is emphasized: see H15 or H20
MR uses C62, C65, F30, or H15

C65 Models of other mathematical theories
+ Other applications of model theory outside logic
+ Theories of orderings
– Uses of models for purely foundational studies: see B30

C70 Logic on admissible sets
+ All sorts of "effective" infinitary logic

C75 Other infinitary logic
+ Infinitary logic even if not model theory, e.g., infinite terms in proof theory and infinitary definability in set theory

C80 Logic with extra quantifiers and operators
– Hilbert epsilon-theorems: see B10
– Modal or many-valued operators: see B45 or B50

C85 Second- and higher-order model theory
+ Weak second-order theories (quantification over finite sets)

C90 Nonclassical models
+ Boolean-valued models
+ Sheaf models
+ Kripke models (also in B45 or F50)
+ Probability models (often also in B48)
+ Topological models (unless the topological structure is condensed into a quantifier: see C80); MR uses C85
– Models of lambda calculus: see B40

C95 Abstract model theory
+ Lindström's theorem, delta-logics, etc.

C96 Collected works
+ Selected works
– Collections (almost) entirely in one subfield: see that subfield

MR uses 01A75, 03-03, and 03-06

C97 Proceedings
+ Collections of papers by various authors, even if they do not derive from any actual conference
– Proceedings (almost) entirely in one subfield: see that subfield
– Proceedings not concentrated in this field: see Source Index

MR uses 03-06

C98 Textbooks, surveys

MR uses 03-01 and 03-02

C99 None of the above or uncertain, but in this section

D RECURSION THEORY

D03 Thue and Post systems, etc.
+ Markov's normal algorithms

D05 Automata and formal grammars in connection with logical questions
+ Cellular automata
+ Finite automata
+ Generalized automata
+ Regular events
– Grammar of natural languages: see B65

MR uses 68 for most of these topics

D10 Turing machines and related notions
+ Potentially infinite automata
+ Probabilistic Turing machines

D15 Complexity of computation
+ Chaitin-Kolmogorov-Solomonoff complexity
+ Finer classification of decidable problems
+ Generalized complexity
+ Resource-bounded computability and reducibility
+ Speed-up theorems
– Complexity of derivations and proofs: see F20
– Complexity of specific non-logical problems (excluded from the Ω-Bibliography)
– Syntactic complexity, complexity of Boolean functions, etc.

MR uses also 68Q15

D20 Recursive functions and relations, subrecursive hierarchies
+ Computable functions of real numbers; MR uses D65 and F60
+ General theory of algorithms
+ Partial recursive functions
+ Primitive recursion

D25 Recursively enumerable sets and degrees
+ Finer classification of undecidable r.e. problems
+ Many-one, truth table, etc., degrees of r.e. sets
+ Sets whose theory is closely related to that of r.e. sets, e.g., productive sets: see also D50
– Generalizations of recursive enumerability: see D60 and D65
– Partial functions with r.e. graphs: see D20

D30 Other degrees; reducibilities
+ Degrees in generalized recursion and constructibility: see also D55, D60, D65, and E45
+ Jump operators
– Subrecursive reducibilities: see D15 and D20

D35 Undecidability and degrees of sets of sentences
+ Hilbert's tenth problem and extensions
+ Reduction classes of the predicate calculus (also in B20)
– Decidability results: see B25
– Halting problems, word problems, etc.: see D03, D05, D10, D30, D40, or D80

D40 Word problems, etc.
+ Conjugacy, isomorphism, and other algorithmic problems in algebra
+ Decidability and undecidability
+ Other algorithmic questions in classical algebra
– Problems concerning production systems or formal grammar: see D03 and D05
– Recursive functions on words: see D20

D45 Theory of numerations, effectively presented structures
+ Numberings of (partial) recursive functions
+ Numerations in the sense of Ershov
+ Recursive algebra, except when it is about recursive equivalence types: see D50
+ Recursive order types
– Classical recursive analysis: see F60
– Model theory of recursive structures: see C57
– Recursive arithmetic: see F30

D50 Recursive equivalence types of sets and structures, isols
+ Concepts traditionally associated with isols, e.g., regressiveness and immuneness

D55 Hierarchies
+ Arithmetical, Borel, analytical, projective, etc. hierarchies
– Descriptive Set Theory in which hierarchical questions are not central: see E15
– Hierarchies of definability in set theory: see E47
– Incidental use of hierarchies outside recursion theory
– Subrecursive hierarchies: see D15 and D20

D60 Recursion theory on ordinals, admissible sets, etc.
+ Beta-recursion on inadmissible ordinals
– Classification of ordinary recursive functions using ordinals: see D20
– Ordinal notations: see D45 and F15
– Other aspects of admissibility: see C62, C70, or E45

D65 Higher-type and set recursion
+ Primitive recursive set functions
– Functionals in Proof Theory: see F10
– Recursion on the hereditarily finite sets: see D20
– Recursion with all arguments and parameters of type $\leqslant 1$: see D20; MR includes this in D65 as long as there are type 1 arguments

D70 Inductive definability
+ Constructions equivalent to inductive definitions, e.g. set derivatives, game sentences, etc.
+ Recursion theory of inductive definitions and their duals
– Inductive definitions in proof theory: see F35 and F50
– Mechanics of inductive definitions: see B28, E20, or E30

D75 Abstract and axiomatic recursion theory
+ Algebras of (partial) recursive functions; MR uses D20
+ Recursion over general structures

D80 Applications
+ Decidability or undecidability results in areas outside logic and algebra
+ Effective versions of problems outside logic and algebra

D96 Collected works
+ Selected works
– Collections (almost) entirely in one subfield: see that subfield
MR uses 01A75, 03-03, and 03-06

D97 Proceedings
+ Collections of papers by various authors, even if they do not derive from any actual conference
– Proceedings (almost) entirely in one subfield: see that subfield
– Proceedings not concentrated in this field: see Source Index
MR uses 03-06

D98 Textbooks, surveys
MR uses 03-01 and 03-02

D99 None of the above or uncertain, but in this section

E SET THEORY

E05 Combinatorial set theory
+ Partition relations, ideals, ultrafilters, trees named after people; MR uses also 04A20
– Finite combinatorics (excluded from the Ω-Bibliography); MR uses 05Xn

E07 Relations and orderings
+ Relation algebras: see also G15; MR uses G15
– Theories about ordering: see C65
MR uses E20, 04A05, 04A20, or 06An

E10 Ordinal and cardinal numbers
+ Cardinal algebras, ordinal algebras
+ Dedekind finite cardinals
– Cardinal exponentiation and the (generalized) continuum hypothesis: see E50; MR sometimes uses 04A10
– Combinatorial aspects of cardinals and ordinals: see E05
– Large cardinals: see E55

E15 Descriptive set theory
+ Definability properties of sets (in the real line or similar spaces)
+ Effective descriptive set theory
– General topology, measure theory, etc.: see E75
MR sometimes uses 04A15
See also D55

E20 Other classical set theory
+ Set algebra

E25 Axiom of choice and related propositions
+ Weak axioms of choice and their negations
MR sometimes uses 04A25

E30 Axiomatics of classical set theory and its fragments
+ Zermelo-Fraenkel set theory and minor variants
+ Gödel-Bernays set theory (also in E70)
– Morse-Kelley set theory (a second order theory: see E70)
– New Foundations, etc.: see E70

E35 Consistency and independence results
+ Forcing used to prove consistency

E40 Other aspects of forcing and Boolean-valued models
+ Forcing in generalized recursion theory: see also D60 and D65
– Model theoretic forcing: see C25

E45 Constructibility, ordinal definability and related notions
+ Other inner models, e.g. the core model

E47 Other notions of set-theoretic definability
+ Lévy hierarchy, indescribability
– Formalization of branches of mathematics within set theory

E50 Continuum hypothesis and Martin's axiom
+ Cardinal exponentiation
+ Variants of Martin's axiom
MR sometimes uses 04A30

E55 Large cardinals
+ Effective (denumerable) analogues of large cardinals
+ Weakly inaccessible and larger cardinals
– Axioms of infinity provable in ZFC
– Large proof-theoretic ordinals: see F15

E60 Determinacy and related principles which contradict the axiom of choice
+ Infinite exponent partition relations
+ Projective determinacy, definable determinacy
+ Other uses of infinite games in set theory and logic
– Applications of games outside set theory and logic
– Weak axioms that merely contradict choice

E65 Other hypotheses and axioms
+ Reflection principles
+ Combinatorial principles

E70 Nonclassical and second-order set theories
+ Leśniewski's Ontology and Mereology; MR uses B60
+ Nonstandard theories, e.g. New Foundations, Ackermann
+ Set theories formulated in non-classical logic
+ Theory of real classes (Morse-Kelley, and Gödel-Bernays set theory); MR uses E30

E72 Fuzzy sets

E75 Applications
+ Independence from set theory of mathematical propositions (also in E35)
+ Results in other branches of mathematics obtained by set theoretic methods
– Set-theoretical foundations of mathematics: see B28 and B30

E96 Collected works
+ Selected works
– Collections (almost) entirely in one subfield: see that subfield

MR uses 01A75, 03-03, and 03-06

E97 Proceedings
+ Collections of papers by various authors, even if they do not derive from any actual conference
– Proceedings (almost) entirely in one subfield: see that subfield
– Proceedings not concentrated in this field: see Source Index

MR uses 03-06

E98 Textbooks, surveys

MR uses 03-01 and 03-02

E99 None of the above or uncertain, but in this section

F PROOF THEORY AND CONSTRUCTIVE MATHEMATICS

F05 Cut elimination and normal form theorems
+ Hilbert's epsilon symbol
– Cut elimination and normal form theorems for modal systems: see B45

F07 Structure of proofs
– Proof schemas used rather than studied: see B10, C07, etc.

F10 Functionals in proof theory
– Typed lambda-calculus: see B40

F15 Recursive ordinals and ordinal notations
+ Ordinal notations even if not proof theory
+ Transfinite progressions of theories (Turing, Feferman; also in F30)

F20 Complexity of proofs
– Complexity of non-proof-theoretic procedures: see D15
– Purely qualitative (rather than quantitative) properties of proofs: see F07

F25 Relative consistency and interpretations
- – Consistency of systems of arithmetic: see F30 and F35
- – Set theoretic consistency results: see E35

F30 First-order arithmetic and fragments
- + Gödel incompleteness theorems
- + Metamathematics of intuitionistic arithmetic
- + Provability logic; MR uses also B45 and F40
- + Provably recursive functions; MR uses also D20
- + Recursive arithmetic
- – Model theory of arithmetic: see C62 and H15

F35 Second- and higher-order arithmetic and fragments
- + Metamathematics of intuitionistic analysis
- + Proof theory of systems of type theory
- + Proof theory of generalized inductive definitions
- – Model theory : see C62

F40 Gödel numberings in proof theory
- + Any use of Gödel numbering of syntax
- – Gödel numberings in recursion theory: see D20 and D45

F50 Metamathematics of constructive systems
- + Intuitionistic logic and subsystems; MR uses also B20
- + Model theoretic methods applied to constructive systems
- + Realizability
- – Metamathematics of predicative systems: see F65

F55 Constructive and intuitionistic mathematics
- + Bishop school of constructivism
- – Metamathematics: see F50

F60 Constructive recursive analysis
- + Classical recursive analysis
- + Soviet school of constructivism
- – Metamathematics: see F50

F65 Other constructive mathematics
- + Constructive trends not covered by F55 or F60
- + Predicative mathematics
- + Metamathematics of predicative systems
- – Other metamathematics: see F50

F96 Collected works
- + Selected works
- – Collections (almost) entirely in one subfield: see that subfield

MR uses 01A75, 03-03, and 03-06

F97 Proceedings
- + Collections of papers by various authors, even if they do not derive from any actual conference
- – Proceedings (almost) entirely in one subfield: see that subfield
- – Proceedings not concentrated in this field: see Source Index

MR uses 03-06

F98 Textbooks, surveys

MR uses 03-01 and 03-02

F99 None of the above or uncertain, but in this section

G ALGEBRAIC LOGIC

G05 Boolean algebras
+ Boolean rings, etc.
– Boolean functions : see B05; MR puts Boolean functions in G05, 06E30, and sometimes 94C10
– Pseudo-Boolean algebras : see G10

G10 Lattices and related structures
+ Heyting algebras; MR uses also 06D20
+ Semilattices, continuous lattices; MR uses 06B35
– Studies of "The lattice of..." where the lattice structure is not the main point

G12 Quantum logic
See also B51

G15 Cylindric and polyadic algebras, relation algebras

G20 Łukasiewicz and Post algebras
+ Lattices (or weaker structures) corresponding to many-valued logic

G25 Other algebras related to logic
+ Boolean algebras with provability and other operators
+ Implicative algebras, BCK algebras, etc.

G30 Categorical logic, topoi
+ Almost any connection between categories and logic, e.g. categories of models, logical foundations of category theory
– Pure category theory (Excluded from the Ω-Bibliography); MR uses 18Xn

G96 Collected works
+ Selected works
– Collections (almost) entirely in one subfield: see that subfield
MR uses 01A75, 03-03, and 03-06

G97 Proceedings
+ Collections of papers by various authors, even if they do not derive from any actual conference
– Proceedings (almost) entirely in one subfield: see that subfield
– Proceedings not concentrated in this field: see Source Index
MR uses 03-06

G98 Textbooks, surveys
MR uses 03-01 and 03-02

G99 None of the above or uncertain, but in this section

H NONSTANDARD MODELS

H05 Infinitesimal analysis in pure mathematics

H10 Other applications of infinitesimal analysis
+ Economics, physics, etc.

H15 Nonstandard models of arithmetic
+ Work emphasizing nonstandard methods
– Work emphasizing model theory : see C62

H20 Other nonstandard models

H96 Collected works
+ Selected works
– Collections (almost) entirely in one subfield: see that subfield
MR uses 01A75, 03-03, and 03-06

H97 Proceedings
+ Collections of papers by various authors, even if they do not derive from any actual conference
– Proceedings (almost) entirely in one subfield: see that subfield
– Proceedings not concentrated in this field: see Source Index
MR uses 03-06

H98 Textbooks, surveys
MR uses 03-01 and 03-02

H99 None of the above or uncertain, but in this section

Subject Index

(A05 ∪ A10) ∩ (F ∪ B40) Historical, critical and philosophical publications

1899

HILBERT, D. *Grundlagen der Geometrie* ◇ A05 B30 F99 ◇

1900

HILBERT, D. *Mathematische Probleme* ◇ A05 D35 E30 E50 F25 ◇

1905

BAIRE, R. & BOREL, E. & HADAMARD, J. & LEBESGUE, H. *Cinq lettres sur la theorie des ensembles* ◇ A05 D55 E15 E25 F65 ◇

1906

HOBSON, E.W. *On the arithmetic continuum* ◇ A05 E47 F35 F55 ◇

1907

BROUWER, L.E.J. *On the foundations of mathematics (Dutch)* ◇ A05 E30 F55 ◇

1908

BROUWER, L.E.J. *On the foundations of mathematics (Dutch)* ◇ A05 E70 F55 ◇

BROUWER, L.E.J. *The unreliability of the logical principles (Dutch)* ◇ A05 F55 ◇

1910

WEYL, H. *Ueber die Definition der mathematischen Grundbegriffe* ◇ A05 B30 F99 ◇

1912

BROUWER, L.E.J. *Intuitionism and formalism (Dutch)* ◇ A05 F55 ◇

CHWISTEK, L.B. *The principle of contradiction in the light of recent investigations of Bertrand Russell (Russian)* ◇ A05 B15 F35 ◇

1914

BROUWER, L.E.J. *Review of Schoenflies's "Die Entwicklung der Mengenlehre und ihre Anwendungen, erste Haelfte"* ◇ A05 A10 E70 E98 F55 ◇

1917

BROUWER, L.E.J. *Addenda and corrigenda to: On the foundations of mathematics (Dutch)* ◇ A05 F55 ◇

1918

WEYL, H. *Das Kontinuum* ◇ A05 B28 E30 F65 ◇

1919

BERNSTEIN, F. *Die Mengenlehre Georg Cantors und der Finitismus* ◇ A05 E30 F99 ◇

BOREL, E. *Sur les ensembles effectivement enumerables et sur les definitions effectives* ◇ A05 E47 F65 ◇

BROUWER, L.E.J. *Mathematics, truth, reality (Dutch)* ◇ A05 F55 ◇

WEYL, H. *Der "Circulus vitiosus" in der heutigen Begruendung der Analysis* ◇ A05 B28 F65 ◇

1921

WEYL, H. *Ueber die neue Grundlagenkrise der Mathematik* ◇ A05 E70 F55 ◇

1923

SKOLEM, T.A. *Begruendung der elementaren Arithmetik durch die rekurrierende Denkweise ohne Anwendung scheinbarer Veraenderlichen mit unendlichem Ausdehnungsbereich* ◇ A05 B28 D20 F30 ◇

1924

BALDUS, R. *Formalismus und Intuitionismus in der Mathematik* ◇ A05 F55 F98 ◇

CHWISTEK, L.B. *The theory of constructive types (principles of logic and mathematics)* ◇ A05 B15 F65 ◇

DRESDEN, A. *Brouwer's contribution to the foundations of mathematics* ◇ A05 F55 ◇

NICOD, J. *Geometry in the sensible world (French)* ◇ A05 B80 F99 ◇

SCHOENFINKEL, M. *Ueber die Bausteine der mathematischen Logik* ◇ A05 B10 B40 ◇

1925

ACKERMANN, W. *Begruendung des "tertium non datur" mittels der Hilbertschen Theorie der Widerspruchsfreiheit* ◇ A05 B10 F30 ◇

LONDON, F. *Ueber die Irreversibilitaet deduktiver Schlussweisen* ◇ A05 B10 F07 ◇

WEYL, H. *Die heutige Erkenntnislage in der Mathematik* ◇ A05 F99 ◇

ZARISKI, O. *Gli sviluppi piu recenti della teoria degli insiemi e il principio di Zermelo* ◇ A05 B10 E25 F99 ◇

1926

FINSLER, P. *Formale Beweise und die Entscheidbarkeit* ◇ A05 D35 F25 ◇

GONSETH, F. *Les fondements des mathematiques. De la geometrie d'Euclide a la relativite generale et a l'intuitionnisme* ◇ A05 A10 F99 ◇

KHINCHIN, A.YA. *The ideas of intuitionism and the dispute on the object of contemporary mathematics (Russian)* ◇ A05 F99 ◇

LEVY, P. *Sur le principe du tiers exclu et sur les theoremes non susceptibles de demonstration* ◇ A05 F50 ◇

SKOLEM, T.A. *The most important recent discussions on the foundations of mathematics (Norwegian)* ◇ A05 F99 ◇

WAVRE, R. *Sur le principe du tiers exclu* ◇ A05 F99 ◇

1927

LEVY, P. *Logique classique, logique Brouwerienne et logique mixte* ⋄ A05 B55 F50 ⋄

NEUMANN VON, J. *Zur Hilbertschen Beweistheorie* ⋄ A05 F99 ⋄

1928

BROUWER, L.E.J. *Intuitionistische Betrachtungen ueber den Formalismus* ⋄ A05 F50 ⋄

CHURCH, A. *On the law of excluded middle* ⋄ A05 B05 F50 ⋄

HERBRAND, J. *Sur la theorie de la demonstration* ⋄ A05 B10 F07 ⋄

LEVY, P. *Logique classique, logique Brouwerienne et logique mixte* ⋄ A05 F50 ⋄

MENGER, K. *Bemerkungen zu Grundlagenfragen I* ⋄ A05 B30 E30 F99 ⋄

REYMOND, A. *L'axiomatique logique et le principe du tiers exclu* ⋄ A05 F50 ⋄

WEYL, H. *Diskussionsbemerkungen zu dem zweiten Hilbertschen Vortrag ueber die Grundlagen der Mathematik* ⋄ A05 F50 ⋄

1929

BARZIN, M. & ERRERA, A. *Sur le principe du tiers exclu* ⋄ A05 F50 ⋄

BROUWER, L.E.J. *Mathematik, Wissenschaft und Sprache* ⋄ A05 F55 ⋄

CHWISTEK, L.B. *Neue Grundlagen der Logik und Mathematik* ⋄ A05 B30 F99 ⋄

HERBRAND, J. *Sur le probleme fondamental des mathematiques* ⋄ A05 B10 F99 ⋄

KREER, L.I. *On proofs (Russian)* ⋄ A05 F99 ⋄

LUZIN, N.N. *Sur les voies de la theorie des ensembles* ⋄ A05 D55 E15 E50 F55 ⋄

WEYL, H. *Consistency in mathematics* ⋄ A05 F99 ⋄

1930

BERNAYS, P. *Die Philosophie der Mathematik und die Hilbertsche Beweistheorie* ⋄ A05 F99 ⋄

BROUWER, L.E.J. *Die Struktur des Kontinuums* ⋄ A05 F55 ⋄

BROUWER, L.E.J. *Review of A.Fraenkel "Zehn Vorlesungen ueber die Grundlagen der Mengenlehre"* ⋄ A05 F55 ⋄

MENGER, K. *Der Intuitionismus* ⋄ A05 F99 ⋄

NICOD, J. *Geometry and induction, containing "Geometry in the sensible world", and "The logical problem of induction"* ⋄ A05 B30 F98 ⋄

1931

GOEDEL, K. *Eine Interpretation des intuitionistischen Aussagenkalkuels* ⋄ A05 B45 F50 ⋄

HERBRAND, J. *Sur le probleme fondamental de la logique mathematique* ⋄ A05 B25 B30 F30 F99 ⋄

HEYTING, A. *Die intuitionistische Grundlegung der Mathematik* ⋄ A05 F50 F55 ⋄

HEYTING, A. *Die intuitionistische Mathematik* ⋄ A05 F98 ⋄

HILBERT, D. *Beweis des "tertium non datur"* ⋄ A05 B28 F50 ⋄

MENGER, K. *Ueber den Konstruktivitaetsbegriff. Zweite Mitteilung* ⋄ A05 F99 ⋄

NEUMANN VON, J. *Bemerkungen zu den Ausfuehrungen von Herrn S. Lesniewski ueber meine Arbeit "Zur Hilbertschen Beweistheorie"* ⋄ A05 F99 ⋄

WEYL, H. *Die Stufen des Unendlichen* ⋄ A05 E10 F99 ⋄

1932

BARZIN, M. & ERRERA, A. *Note sur la logique de M.Heyting* ⋄ A05 F50 ⋄

BARZIN, M. & ERRERA, A. *Sur la logique de M.Heyting* ⋄ A05 F50 ⋄

BARZIN, M. & ERRERA, A. *Sur la logique intuitionniste. Reponse a M.A.Heyting* ⋄ A05 F50 ⋄

BERNAYS, P. *Methoden des Nachweises von Widerspruchsfreiheit und ihre Grenzen* ⋄ A05 B30 F99 ⋄

HEYTING, A. *A propos d'un article de MM. Barzin et Errera* ⋄ A05 F99 ⋄

HEYTING, A. *Reponse a MM. Barzin et Errera* ⋄ A05 F99 ⋄

KOLMOGOROV, A.N. *Zur Deutung der intuitionistischen Logik* ⋄ A05 F50 ⋄

TARSKI, A. *Der Wahrheitsbegriff in den Sprachen der deduktiven Disziplinen* ⋄ A05 B30 C07 C40 F30 ⋄

1933

BARZIN, M. & ERRERA, A. *La logique de M.Brouwer. Etat de la question* ⋄ A05 F50 ⋄

BROUWER, L.E.J. *Volition, knowledge, speech (Dutch)* ⋄ A05 F55 ⋄

TARSKI, A. *On the notion of truth in reference to formalized deductive sciences (Polish)* ⋄ A05 B15 B30 C07 C40 F30 ⋄

WAJSBERG, M. *Beitrag zur Metamathematik* ⋄ A05 B10 B30 B96 C35 F99 ⋄

1934

HEYTING, A. *Mathematische Grundlagenforschung. Intuitionismus. Beweistheorie* ⋄ A05 F50 F98 ⋄

HILBERT, D. & BERNAYS, P. *Grundlagen der Mathematik I* ⋄ A05 B98 F05 F30 F98 ⋄

1935

AMBROSE, A. *Finitism in mathematics I,II* ⋄ A05 F99 ⋄

BARZIN, M. *Sur les demonstrations de non-contradiction des axiomes* ⋄ A05 B30 F99 ⋄

BERNAYS, P. *Quelques points essentiels de la metamathematique* ⋄ A05 F55 ⋄

BERNAYS, P. *Sur le platonisme dans les mathematiques* ⋄ A05 A10 E30 E70 F55 ⋄

HEYTING, A. *The debate concerning the foundations of mathematics (Dutch)* ⋄ A05 F99 ⋄

1936

AMBROSE, A. *The nature of the question "are there three consecutive 7's in the expansion of π ?"* ⋄ A05 F99 ⋄

BARZIN, M. *Note sur la demonstration de M.E.W.Beth* ⋄ A05 F25 F50 ⋄

DASSEN, C.C. *Sobre una objecion a la Logica Brouweriana* ⋄ A05 F50 ⋄

MANIA, B. *L'infini mathematique et l'evolution de la logique* ◇ A05 F55 F65 ◇
REYMOND, A. *La negation et le principe du tiers exclu* ◇ A05 F50 ◇

1937

FREUDENTHAL, H. *Zur intuitionistischen Deutung logischer Formeln* ◇ A05 F50 ◇
GENTZEN, G. *Der Unendlichkeitsbegriff in der Mathematik* ◇ A05 F50 F99 ◇
GENTZEN, G. *Unendlichkeitsbegriff und Widerspruchsfreiheit der Mathematik* ◇ A05 F99 ◇
HELMER, O. *Perelman versus Goedel* ◇ A05 F30 ◇
HEYTING, A. *Bemerkungen zu dem Aufsatz von Herrn Freudenthal "Zur intuitionistischen Deutung logischer Formeln"* ◇ A05 F50 ◇
HEYTING, A. *The development of intuitionistic mathematics (Dutch)* ◇ A05 F55 ◇
NINK, C. *Der Satz vom ausgeschlossenen Dritten. Ein philosophischer Beitrag zur Grundlagenkrise in der Mathematik* ◇ A05 F99 ◇
TURING, A.M. *Computability and λ-definability* ◇ A05 B40 D05 D10 D20 F99 ◇

1938

DIENES, P. *Logic of algebra* ◇ A05 B28 F55 G05 ◇
GENTZEN, G. *Die gegenwaertige Lage in der mathematischen Grundlagenforschung* ◇ A05 A10 F55 F65 ◇
MOLODSHIJ, V.N. *Effectivism in mathematics (Russian)* ◇ A05 F99 ◇

1939

HEYTING, A. *Les fondements des mathematiques du point de vue intuitionniste* ◇ A05 F55 ◇
HILBERT, D. & BERNAYS, P. *Grundlagen der Mathematik II* ◇ A05 B98 F05 F15 F30 F40 F98 ◇
ROSSER, J.B. *An informal exposition of proofs of Goedel's theorems and Church's theorem* ◇ A05 B20 D35 F30 ◇

1940

BARZIN, M. *Logique symbolique. Sur la portee du theoreme de M.Goedel* ◇ A05 F30 ◇
LARGUIER, E.H. *Brouwerian philosophy of mathematics* ◇ A05 F55 ◇

1941

BERNAYS, P. *Sur les questions methodologiques actuelles de la theorie hilbertienne de la demonstration* ◇ A05 F99 ◇
CURRY, H.B. *The paradox of Kleene and Rosser* ◇ A05 B40 ◇
KALMAR, L. *Aims, methods and results of Hilbert proof theory (Hungarian) (German summary)* ◇ A05 F99 ◇
TOMS, E. *The law of excluded middle* ◇ A05 F99 ◇

1942

FINDLAY, J. *Goedelian sentences: a non-numerical approach* ◇ A05 F30 ◇

1944

FINSLER, P. *Gibt es unentscheidbare Saetze?* ◇ A05 F30 ◇
WEYL, H. *David Hilbert and his mathematical work* ◇ A10 F98 ◇

1945

DESTOUCHES-FEVRIER, P. *Rapport entre le calcul des problemes et le calcul des propositions* ◇ A05 B05 F50 ◇

1946

LOEWENHEIM, L. *On making indirect proofs direct* ◇ A05 F99 ◇
ZAWIRSKI, Z. *Origin and development of intuitionistic logic (Polish)* ◇ A05 A10 F50 ◇

1947

CARRUCCIO, E. *Alcune conseguenze di un risultato del Goedel e la razionalita del reale* ◇ A05 F30 ◇
GOODMAN, NELSON & QUINE, W.V.O. *Steps toward a constructive nominalism* ◇ A05 F99 ◇
POPPER, K.R. *Functional logic without axioms or primitive rules of inference* ◇ A05 B20 F50 ◇
VALPOLA, V. *The position of negation in a language which expresses knowledge (Finnish)* ◇ A05 B28 F30 F50 ◇

1948

BOREL, E. *A propos de l'axiome du choix* ◇ A05 E25 F99 ◇
GRISS, G.F.C. *On negation (Dutch)* ◇ A05 F99 ◇
GRISS, G.F.C. *Sur la negation (dans les mathematiques et la logique)* ◇ A05 F50 ◇
KURODA, S. *On the logic of Aristotle and the logic of Brouwer (Japanese)* ◇ A05 A10 F07 F50 ◇

1949

BOCKSTAELE, P. *Intuitionism among the French mathematicians (Dutch)* ◇ A10 F65 ◇
BROUWER, L.E.J. *Consciousness, philosophy, and mathematics* ◇ A05 F55 ◇
DANTZIG VAN, D. *Comments on Brouwer's theorem on essentially-negative predicates* ◇ A05 F55 ◇
DESTOUCHES-FEVRIER, P. *Connexion entre les calculs des constructions, des problemes, des propositions* ◇ A05 F50 ◇
IONGH DE, J.J. *Restricted forms of intuitionistic mathematics* ◇ A05 F50 ◇
LADRIERE, J. *Le role du theoreme de Goedel dans le developpement de la theorie de la demonstration* ◇ A05 A10 F99 ◇
VREDENDUIN, P.G.J. *The constructive method* ◇ A05 F99 ◇

1950

BERNAYS, P. *Mathematische Existenz und Widerspruchsfreiheit* ◇ A05 B30 F99 ◇
HENKIN, L. *Completeness in the theory of types* ◇ A05 B15 C85 F35 ◇
HEYTING, A. *Tensions in mathematics (Dutch)* ◇ A05 F55 ◇
LORENZEN, P. *Konstruktive Begruendung der Mathematik* ◇ A05 F65 ◇

MATSUMOTO, K. *Sur la structure concernant la logique moderne* ◇ A05 B45 F50 G05 ◇
SCHMIDT, H.A. *Wie duerfen wir mit dem Unendlichen umgehen? (die Grundlage des mathematischen Intuitionismus)* ◇ A05 F55 ◇

1951

CURRY, H.B. *La logique combinatoire et les antinomies* ◇ A05 B40 ◇
DANTZIG VAN, D. *Mathematique stable et mathematique affirmative* ◇ A05 F55 ◇
FRAENKEL, A.A. *On the crisis of the principle of the excluded middle* ◇ A05 B20 F99 ◇
GOODSTEIN, R.L. *Constructive formalism: Essays on the foundations of mathematics* ◇ A05 B98 F30 F60 F98 ◇
GOODSTEIN, R.L. *The foundations of mathematics* ◇ A05 A10 F98 ◇
LADRIERE, J. *Le theoreme fondamental de Gentzen* ◇ A05 A10 F05 ◇
LORENZEN, P. *Algebraische und logistische Untersuchungen ueber freie Verbaende* ◇ A05 F35 G10 ◇
LORENZEN, P. *Ueber das Prinzip "ex falso quodlibet"* ◇ A05 F99 ◇
NEVES-REAL, L. *Kurt Goedel and the problems of the foundations of mathematics and the theory of sets (Portuguese)* ◇ A10 E25 E35 E45 E50 F30 ◇
SUETUNA, Z. *Ueber die Grundlagen der Mathematik* ◇ A05 F65 ◇

1952

BROUWER, L.E.J. *Historical background, principles and methods of intuitionism* ◇ A05 F55 ◇
CURRY, H.B. *The inferential theory of negation* ◇ A05 B20 F50 ◇
KLEENE, S.C. *Introduction to metamathematics* ◇ A05 B98 D98 F30 F50 F98 ◇
KREISEL, G. *On the concepts of completeness and interpretation of formal systems* ◇ A05 F25 F30 ◇

1953

BERNAYS, P. *Existence et non-contradiction en mathematiques. Avec une note de G.Bouligand* ◇ A05 B30 F99 ◇
FREUDENTHAL, H. *Zur Geschichte der vollstaendigen Induktion* ◇ A05 A10 F30 ◇
LORENZEN, P. *Die Allgemeingueltigkeit der logischen Regeln* ◇ A05 F50 F65 ◇

1954

BERNAYS, P. *Zur Beurteilung der Situation in der beweistheoretischen Forschung* ◇ A05 F99 ◇
FRAENKEL, A.A. *The intuitionistic revolution in mathematics and logic* ◇ A05 F55 ◇
HADAMARD, J. *Sur l'impossibilite de demontrer la compatibilite des axiomes de l'arithmetique* ◇ A05 B28 F99 ◇
HEYTING, A. *G.F.C. Griss and his negationless intuitionistic mathematics* ◇ A05 F50 ◇
HEYTING, A. *Logique et intuitionnisme* ◇ A05 F50 ◇
SKOLEM, T.A. *Results in investigations in the foundations (Norwegian)* ◇ A05 B98 D03 F55 ◇
WANG, HAO *The formalization of mathematics* ◇ A05 B30 E70 F35 F65 ◇

1955

BETH, E.W. *Remarks on natural deduction* ◇ A05 B10 F07 ◇
HEYTING, A. *On the significance of the mathematical work of G.F.C. Griss (Dutch)* ◇ A05 F99 ◇
LORENZEN, P. *Einfuehrung in die operative Logik und Mathematik* ◇ A05 B98 F50 F65 F98 ◇
SHEN, YOUDING *Two semantical paradoxes* ◇ A05 F99 ◇
WANG, HAO *Undecidable sentences generated by semantic paradoxes* ◇ A05 B28 F30 ◇

1956

BETH, E.W. *L'existence en mathematiques* ◇ A05 B98 F55 ◇
FEVRIER, P. *Tendances constructives en logique moderne* ◇ A05 F50 ◇
HEYTING, A. *La conception intuitionniste de la logigue* ◇ A05 F50 ◇
HINTIKKA, K.J.J. *Identity, variables and impredicative definitions* ◇ A05 B20 F65 ◇
KREISEL, G. *Some uses of metamathematics* ◇ A05 C98 F99 ◇
LADRIERE, J. *La notion de constructivite en metamathematique* ◇ A05 F99 ◇

1957

BLANCHE, R. *Introduction a la logique contemporaine* ◇ A05 B98 F50 ◇
KRASNER, M. *Theorie de la definition I* ◇ A05 E70 F99 ◇
LADRIERE, J. *Les limitations internes des formalismes. Etude sur la signification du theoreme de Goedel et des theoremes apparentes dans la theorie des fondements des mathematiques* ◇ A05 B98 F30 F98 ◇
ROOTSELAAR VAN, B. *Intuitionismus und Arithmetik (Dutch)* ◇ A05 F55 ◇
SMULLYAN, R.M. *Languages in which self reference is possible* ◇ A05 B10 B28 F30 F40 ◇
WANG, HAO *The axiomatization of arithmetic* ◇ A05 B28 F30 ◇

1958

BAR-HILLEL, Y. & DALEN VAN, D. & FRAENKEL, A.A. & LEVY, A. *Foundations of set theory* ◇ A05 B98 E30 E70 E98 F98 ◇
BETH, E.W. *Construction semantique de la logique intuitionniste* ◇ A05 F50 ◇
GODDARD, L. *"True" and "provable"* ◇ A05 F30 ◇
GOEDEL, K. *Ueber eine bisher noch nicht benuetzte Erweiterung des finiten Standpunktes* ◇ A05 F10 F30 F50 ◇
HEYTING, A. *Blick von der intuitionistischen Warte* ◇ A05 F50 ◇
HEYTING, A. *Intuitionism in mathematics* ◇ A05 B98 F50 F55 F98 ◇
HEYTING, A. *Intuitionism and school mathematics (Dutch)* ◇ A05 F55 ◇
KRASNER, M. *Theorie de la definition II* ◇ A05 E70 F99 ◇

KREISEL, G. *Hilbert's programme* ◇ A05 F99 ◇
MOSTOWSKI, ANDRZEJ *Quelques observations sur l'usage des methodes non finitistes dans la meta-mathematique* ◇ A05 C30 C62 C80 E30 E35 F99 ◇
SHANIN, N.A. *On the constructive interpretation of mathematical judgements (Russian)* ◇ A05 F50 ◇
SHANIN, N.A. *Ueber einen Algorithmus zur konstruktiven Dechiffrierung mathematischer Urteile (Russian) (German summary)* ◇ A05 F50 F60 ◇

1959

HEYTING, A. *Some remarks on intuitionism* ◇ A05 F50 ◇
KALMAR, L. *An argument against the plausibility of Church's thesis* ◇ A05 D20 F99 ◇
LERDA, F. *Analisi critica dei fondamenti dell'intuizionismo* ◇ A05 F50 F55 ◇
LOEB, M.H. *Constructive truth* ◇ A05 F99 ◇
LORENZEN, P. *Ueber die Begriffe "Beweis" und "Definition"* ◇ A05 F50 ◇
MEIGNE, M. *La consistance des theories formelles et le fondement des mathematiques* ◇ A05 F25 F98 ◇
PETER, R. *Rekursivitaet und Konstruktivitaet* ◇ A05 D20 F60 ◇
STEGMUELLER, W. *Unvollstaendigkeit und Unentscheidbarkeit. Die metamathematischen Resultate von Goedel, Church, Kleene, Rosser und ihre erkenntnistheoretische Bedeutung* ◇ A05 D35 F30 F98 ◇
WANG, HAO *Ordinal numbers and predicative set theory* ◇ A05 E70 F15 F65 ◇

1960

BEHMANN, H. *Das Russell'sche Paradoxon und die formale Logik* ◇ A10 B40 E70 ◇
HEYTING, A. *Remarques sur le constructivisme* ◇ A05 F55 ◇
KOHL, H.R. & PARSONS, C. *Self-reference, truth, and provability* ◇ A05 F25 F30 ◇
KREISEL, G. *La predicativite* ◇ A05 D55 E45 F35 F65 ◇
KREISEL, G. *Wittgenstein's theory and practice of philosophy* ◇ A05 F99 ◇
LORENZEN, P. *Constructive and axiomatic mathematics* ◇ A05 F65 ◇

1961

ESENIN-VOL'PIN, A.S. *Le programme ultra-intuitionniste des fondements des mathematiques* ◇ A05 E70 F65 ◇
HEYTING, A. *Infinitistic methods from a finitist point of view* ◇ A05 F50 ◇
LORENZEN, P. *Ein dialogisches Konstruktivitaetskriterium* ◇ A05 F50 F65 ◇
RIEGER, L. *On a critique of Church's thesis concerning general recursive functions in arithmetic (Czech)* ◇ A05 D20 F99 ◇
RIEGER, L. *Sur le probleme des nombres naturels* ◇ A05 A10 F30 ◇
SMART, J.J.C. *Goedel's theorem, Church's theorem and mechanism* ◇ A05 B10 D35 F30 ◇

1962

HEYTING, A. *After thirty years* ◇ A05 A10 F99 ◇
HEYTING, A. *Methodes et problemes de l'intuitionnisme* ◇ A05 F50 F55 ◇
HOERING, W. *Absolut unentscheidbare Saetze der Mathematik* ◇ A05 F30 ◇
LORENZEN, P. *Metamathematik* ◇ A05 B98 C60 D20 D35 D98 F98 ◇
MARKOV, A.A. *On constructive mathematics (Russian)* ◇ A05 F60 ◇
SCHUETTE, K. *Logische Abgrenzung des Transfiniten* ◇ A05 E10 F15 ◇

1963

LAKATOS, I. *Proofs and refutations I,II,III,IV* ◇ A05 F99 ◇
MENDELSON, E. *On some recent criticism of Church's thesis* ◇ A05 D20 F99 ◇
PRAWITZ, D. *Concerning constructive logic and the concept of implication (Swedish)* ◇ A05 F07 F50 ◇
TUCKER, J.V. *Constructivity and grammar* ◇ A05 B65 F99 ◇

1964

FUENTES MIRAS, J.R. *On the foundations of mathematics (Spanish)* ◇ A05 A10 F98 ◇
HEYTING, A. *Disputation* ◇ A05 F55 ◇
RIVETTI BARBO, F. *Il teorema e il corollario di Goedel* ◇ A05 F30 F98 ◇
STEGMUELLER, W. *Remarks on the completeness of logical systems relative to the validity concepts of P.Lorenzen and K.Lorenz* ◇ A05 F99 ◇

1965

DAVIS, MARTIN D. *Introduction to Goedel's paper: On formally undecidable propositions of the Principia Mathematica and related systems I* ◇ A10 D20 F30 ◇
MOSTOWSKI, ANDRZEJ *Thirty years of foundational studies. Lectures on the development of mathematical logic and the studies of the foundations of mathematics in 1930-1964* ◇ A10 B98 C98 D98 E98 F98 ◇
ROOTSELAAR VAN, B. *Intuitives ueber den Intuitionismus* ◇ A05 F55 ◇
SCHUETTE, K. *Probleme und Methoden der Beweistheorie* ◇ A05 F98 ◇

1966

GHOSE, A. *Konstruktive Grundlagen der Logik* ◇ A05 F65 ◇
KUTSCHERA VON, F. *Zur semantischen Begruendung der klassischen und der intuitionistischen Logik* ◇ A05 F50 ◇
LEBLANC, H. & THOMASON, R.H. *The demarcation line between intuitionistic logic and classical logic* ◇ A05 F50 ◇
ROOTSELAAR VAN, B. *Intuition und Konstruktion* ◇ A05 F98 ◇

1967

AGAZZI, E. *L'aritmetica come strumento di ricerca logico-matematica (English summary)* ◇ A05 F30 ◇

CURRY, H.B. *Combinatory logic* ◇ A05 B40 B98 ◇
FENSTAD, J.E. *Subsets of natural numbers (Norwegian) (English summary)* ◇ A05 E35 E50 F30 F35 F65 ◇
GRZEGORCZYK, A. *Non-classical propositional calculi in relation to methodological patterns of scientific investigation (Polish) (Russian and English summaries)* ◇ A05 B45 F50 ◇
HEIJENOORT VAN, J. (ED.) *From Frege to Goedel: a source book in mathematical logic, 1879-1931* ◇ A10 B97 F97 ◇
HEIJENOORT VAN, J. *Goedel's theorem* ◇ A05 A10 F30 ◇
HEYTING, A. *Informal rigour and intuitionism* ◇ A05 F99 ◇
HEYTING, A. *L.E.J. Brouwer (February 27,1881 - December 2,1966) Innovator at the pinnacles and of the ultimate foundations of mathematics (Dutch)* ◇ A05 A10 F99 ◇
KREISEL, G. *Informal rigour and completeness proofs* ◇ A05 B30 C07 E30 E50 F50 ◇
WANG, HAO *Introductory note* ◇ A10 F50 ◇

1968

GUPTA, H.N. *On the rule of existential specification in systems of natural deduction* ◇ A05 B10 F07 ◇
HERBRAND, J. *Ecrits logiques* ◇ A05 A10 B96 F96 ◇
HEYTING, A. *Intuitionism in mathematics* ◇ A05 F50 F98 ◇
KREISEL, G. *Foundations of mathematics: 1900-1950* ◇ A05 A10 F98 ◇
LORENZEN, P. *Constructive mathematics as a philosophical problem* ◇ A05 F65 ◇
LORENZEN, P. *Operative Logik. Eine Uebersicht 1956-1966* ◇ A05 F50 ◇
TROELSTRA, A.S. *The scientific work of A.Heyting* ◇ A10 F99 ◇

1969

BARZIN, M. *Sur le mecanisme des paradoxes et le theoreme de M.K.Goedel* ◇ A05 F30 ◇
BULLOFF, J.J. & HAHN, S.W. & HOLYOKE, T.C. *Bibliography of Kurt Goedel* ◇ A10 D99 E99 F99 ◇
ESENIN-VOL'PIN, A.S. *On Hilbert's second problem (Russian)* ◇ A05 F30 F65 ◇
HEIDEMA, J. *History, principles and methods of intuitionistic mathematics (Africaans) (English summary)* ◇ A05 F55 ◇
KREISEL, G. *Appendix II. Category theory and the foundations of mathematics* ◇ A05 F99 G30 ◇
KREISEL, G. & NEWMAN, M.H.A. *Luitzen Egbertus Jan Brouwer (1881 - 1966)* ◇ A05 A10 F99 ◇
SZABO, M.E. *Introduction* ◇ A10 F05 F30 F98 ◇

1970

BERNAYS, P. *The original Gentzen consistency proof for number theory* ◇ A10 B28 F30 F50 ◇
BISHOP, E.A. *Mathematics as a numerical language* ◇ A05 F50 F55 ◇
CIAMPA, S. *I numeri, l'aritmetica, gli insiemi* ◇ A05 F30 ◇
ESENIN-VOL'PIN, A.S. *The ultra-intuitionistic criticism and the anti-traditional program for foundations of mathematics* ◇ A05 F65 ◇
FITTING, M. *Intuitionistic model theory and the Cohen independence proofs* ◇ A05 C90 E35 E40 F50 ◇
GERMANO, G. *Metamathematische Begriffe in Standardtheorien* ◇ A05 B30 F30 ◇
KREISEL, G. *Hilbert's programme and the search for automatic proof procedures* ◇ A05 B35 F99 ◇
KREISEL, G. *Principles of proof and ordinals implicit in given concepts* ◇ A05 F15 ◇
KREISEL, G. *The formalist-positivist doctrine of mathematical precision in the light of experience* ◇ A05 F99 ◇
KUZICHEVA, Z.A. *On the history of incompleteness theorems (Russian)* ◇ A10 F30 ◇
LACEY, H.M. *The consequence of Goedel's theorem* ◇ A05 F30 ◇
MAYOH, B.H. *The relation between an object and its name: notation systems and their fixed point theorems* ◇ A05 B03 D20 D35 D45 F30 F60 ◇
PERZANOWSKI, J. *On the origin of Hilbert's programme (Polish) (English summary)* ◇ A10 F99 ◇
STOLZENBERG, G. *Review: Foundations of constructive analysis by Errett Bishop* ◇ A05 F55 ◇
THIEL, C. *Was will die operative Mathematik ?* ◇ A05 F65 ◇
WETTE, E. *Vom Unendlichen zum Endlichen* ◇ A05 B28 E30 F30 ◇

1971

BERNAYS, P. *Zum Symposium ueber die Grundlagen der Mathematik* ◇ A05 F99 ◇
CORCORAN, J. *Discourse grammars and the structure of mathematical reasoning III: Two theories of proof* ◇ A05 A10 F07 ◇
DREBEN, B. & GOLDFARB, W.D. *Note N* ◇ A10 B10 F30 F35 ◇
DUMITRIU, A. *Fermat's method of infinite descent compared with the method of mathematical induction (Romanian) (French summary)* ◇ A10 F30 ◇
FRIEDMAN, H.M. *A more explicit set theory* ◇ A05 C62 E30 E35 E45 F30 F35 ◇
FRIEDMAN, H.M. *Algorithmic procedures, generalized Turing algorithms, and elementary recursion theory* ◇ A05 D10 D20 D75 F99 ◇
HANSON, W.H. *Mechanism and Goedel's theorems* ◇ A05 F30 ◇
HENKIN, L. *Mathematical foundations for mathematics* ◇ A10 C98 E30 E55 F55 G05 G15 ◇
KREISEL, G. *The collected works of Gerhard Gentzen* ◇ A10 F99 ◇
PARIKH, R. *Existence and feasibility in arithmetic* ◇ A05 D15 F30 F65 H15 ◇
POZSGAY, L.J. *Liberal intuitionism as a basis for set theory* ◇ A05 E30 F65 ◇
TROELSTRA, A.S. *Intuitionisme (Dutch)* ◇ A05 F99 ◇

1972

FREGUGLIA, P. *Platonismo ed intuizionismo e loro superamento* ◇ A05 F98 ◇

KREISEL, G. *Which number theoretic problems can be solved in recursive progressions on Π^1_1-paths through* $\mathcal{O}$? ◇ A05 D20 D55 F15 F30 F99 ◇
MAINZER, K. *Mathematischer Konstruktivismus im Lichte kantischer Philosophie* ◇ A05 F99 ◇
POZSGAY, L.J. *Semi-intuitionistic set theory* ◇ A05 E30 E70 F65 ◇
PRAWITZ, D. *The philosophical position of proof theory* ◇ A05 F98 ◇
QUINE, W.V.O. *Algebraic logic and predicate functors* ◇ A05 B40 G15 ◇
TAKAHASHI, A. *Mathematical concepts I,II,III,IV,V (Spanish)* ◇ A10 E98 F98 F99 ◇

1973

BOWIE, G.L. *An argument against Church's thesis* ◇ A05 D20 F99 ◇
CHRISTIAN, C.C. *Inhaltliche und formale Wahrheit* ◇ A05 E30 E35 E45 F30 ◇
EDGAR, W.J. *Is intuitionism the epistemically serious foundation for mathematics?* ◇ A05 F99 ◇
KREISEL, G. *Perspectives in the philosophy of pure mathematics* ◇ A05 F99 ◇
LARGEAULT, J. *La theorie de la demonstration de Hilbert et les theoremes d'impossibilite de Goedel* ◇ A05 A10 F99 ◇
LORENZ, K. *Rules versus theorems. A new approach for mediation between intuitionistic and two-valued logic* ◇ A05 B55 F50 ◇
PRAWITZ, D. *Towards a foundation of a general proof theory* ◇ A05 F05 F07 ◇
SCHROETER, K. *Interpretation der intuitionistischen Mathematik mit Hilfe des Beweisbarkeitsbegriffs* ◇ A05 F50 ◇
THOMAS, WILLIAM J. *Doubts about some standard arguments for Church's Thesis* ◇ A05 D10 D20 F99 ◇
ZARNECKA-BIALY, E. *Negation in Ch.S.Peirce's propositional calculus* ◇ A10 B20 F50 ◇

1974

CELLUCCI, C. *On the role of reducibility principles* ◇ A05 E30 F50 ◇
COCCHIARELLA, N.B. *A new formulation of predicative second order logic* ◇ A05 B15 F65 ◇
HEYTING, A. *Intuitionistic views on the nature of mathematics* ◇ A05 F55 ◇
METEAU, G. *The foundations of mathematics - a new analysis showing Goedel's theorem based on fallacy* ◇ A05 F30 ◇
MINTS, G.E. *Proof theory (Russian)* ◇ A05 F99 ◇
POSY, C.J. *Brouwer's constructivism* ◇ A05 F99 ◇
PRAWITZ, D. *On the idea of a general proof theory* ◇ A05 F05 F07 ◇
RESNIK, M.D. *On the philosophical significance of consistency proofs* ◇ A05 F25 F30 ◇
SMIRNOVA, E.D. *Consistency and eliminability in proof theory (Russian)* ◇ A05 B10 F99 ◇

1975

BERG, J. & CHIHARA, C.S. *Church's thesis misconstrued* ◇ A05 D20 F99 ◇
BISHOP, E.A. *The crisis in contemporary mathematics* ◇ A05 F99 ◇
BROUWER, L.E.J. *Collected works I* ◇ A05 A10 F55 F96 ◇
DUMMETT, M. *The justification of deduction* ◇ A05 F50 ◇
DUMMETT, M. *The philosophical basis of intuitionistic logic* ◇ A05 F99 ◇
EMDE BOAS VAN, P. *Ten years of speedup* ◇ A10 D15 D20 F20 G05 ◇
KREISEL, G. *Observations on a recent generalization of completeness theorems due to Schuette* ◇ A05 C07 C57 F05 F20 F35 F50 ◇
KREISEL, G. & MINTS, G.E. & SIMPSON, S.G. *The use of abstract language in elementary metamathematics: Some pedagogic examples* ◇ A05 C07 C57 C75 F05 F07 F20 F50 ◇
LUCKHARDT, H. *Ueber Hilberts reale und ideale Elemente* ◇ A05 F99 ◇
MAC LANE, S. *Sets, topoi, and internal logic in categories* ◇ A05 E70 F50 G30 ◇
MAGARI, R. *Significato e verita nell'aritmetica peaniana (English summary)* ◇ A05 F30 ◇
MAINZER, K. *Zur Begruendung von Freges analytischen Begriffen 2.Stufe in konstruktiver Absicht* ◇ A05 F99 ◇
RADU, E. *Le "sujet creatif" dans les mathematiques intuitionnistes* ◇ A05 F50 ◇
SCOTT, D.S. *Some philosophical issues concerning theories of combinators* ◇ A05 B40 ◇
TAKEUTI, G. *Consistency proofs and ordinals* ◇ A05 F15 ◇
WANDSCHNEIDER, D. *Zur Eliminierung des Goedelschen Unvollstaendigkeitsproblems im Zusammenhang mit dem Antinomien-Problem* ◇ A05 E30 F30 ◇
XENAKIS, J. *Natural deduction "puzzle"* ◇ A05 F07 ◇

1976

APERY, R. *Mathematique constructive* ◇ A05 F98 ◇
CORCORAN, J. *The nature of a correct theory of proof and its value* ◇ A05 F07 ◇
CORCORAN, J. *Two theories of proof* ◇ A05 F07 ◇
HERMES, H. *Dialog games* ◇ A05 B60 F50 ◇
KREISEL, G. *What have we learnt from Hilbert's second problem?* ◇ A05 F98 ◇
WANDSCHNEIDER, D. *Reflexive Unbeweisbarkeitsaussagen. Anmerkungen zur Grundsatzdiskussion um Goedels Unvollstaendigkeitsproblem* ◇ A05 E30 F30 ◇

1977

BIELTZ, P. *On the controversies of the excluded middle (Romanian)* ◇ A05 A10 F55 ◇
BISHOP, E.A. *Review of H. Jerome Keisler's "Elementary calculus"* ◇ A05 F55 ◇
GUREVICH, Y. *Intuitionistic logic with strong negation* ◇ A05 F50 ◇
HALMOS, P.R. *Logic from A to G* ◇ A10 F30 ◇

KREISEL, G. *From foundations to science: Justifying and unwinding proofs* ◇ A05 B35 F07 ◇
KREISEL, G. *On the kind of data needed for a theory of proofs* ◇ A05 A10 F07 ◇
KREISEL, G. *Review of "L.E.J.Brouwer collected works, Volume I, Philosophy and foundations of mathematics"* ◇ A05 A10 E30 E70 F99 ◇
LORENZ, K. *On the relation between the partition of a whole into parts and the attribution of properties to an object* ◇ A05 F99 ◇
MAINZER, K. *Is the intuitionistic bar induction a constructive principle ?* ◇ A05 F50 ◇
PRAWITZ, D. *Meaning and proofs: On the conflict between classical and intuitionistic logic* ◇ A05 F99 ◇
SEPER, K. *Constructive mathematics – an essay, I, II* ◇ A05 F55 F65 ◇
SEPER, K. *Contribution to the discussion of Markov principle* ◇ A05 F65 ◇
SHAPIRO, S. *On Church's thesis* ◇ A05 D20 F99 ◇
THIEL, C. *Que significa "constructivismo" ?* ◇ A05 F65 ◇
TROELSTRA, A.S. *Axioms for intuitionistic mathematics incompatible with classical logic* ◇ A05 F35 F50 ◇
TROELSTRA, A.S. *Choice sequences, a chapter of intuitionistic mathematics* ◇ A05 A10 F35 F50 F55 F98 ◇

1978

DALEN VAN, D. *Brouwer: the genesis of his intuitionism* ◇ A05 A10 F99 ◇
DALEN VAN, D. *Philosophical foundations of mathematics (Dutch)* ◇ A05 F98 ◇
GHOSE, A. *Different notions of constructivity – their ontology* ◇ A05 F99 ◇
KEARNS, J.T. *Intuitionist logic, a logic of justification* ◇ A05 C90 F50 ◇
KREISEL, G. *The motto of 'Philosophical Investigations' and the philosophy of proofs and rules* ◇ A05 F99 ◇
KREISEL, G. *Wittgenstein's lectures on the foundations of mathematics, Cambridge 1939* ◇ A05 F99 ◇
LAUENER, H. *Semantique et methode constructiviste* ◇ A05 F99 ◇
LORENZ, K. & LORENZEN, P. *Dialogische Logik* ◇ A05 B60 F99 ◇
SCHUETTE, K. *Die Entwicklung der Beweistheorie* ◇ A10 F98 ◇
TRAGESSER, R.S. & ZUCKER, J.I. *The adequacy problem for inferential logic* ◇ A05 B55 F50 ◇
TROELSTRA, A.S. *A. Heyting on the formalization of intuitionistic mathematics* ◇ A10 F50 ◇

1979

CHAUVIN, A. *Theory of objects and set theory: Introduction and semantics* ◇ A05 B40 E30 E70 ◇
DALEN VAN, D. *Interpreting intuitionistic logic* ◇ A05 F50 F98 ◇
DAWSON JR., J.W. *The Goedel incompleteness theorem from a length-of-proof perspective* ◇ A10 F20 F30 H15 ◇
DRAGALIN, A.G. *Intuitionism (Russian)* ◇ A05 F50 F98 ◇
FLEISCHHACKER, L. *Is Russell's vicious circle principle false or meaningless?* ◇ A05 B40 ◇
GRATTAN-GUINNESS, I. *In memoriam Kurt Goedel: His 1931 correspondence with Zermelo on his incompletability theorem* ◇ A10 F30 ◇
HUMPHRIES, J. *Goedel's proof and the liar paradox* ◇ A05 F30 ◇
KREISEL, G. *Formal rules and questions of justifying mathematical practice* ◇ A05 F99 ◇
KREISEL, G. *Some facts from the theory of proofs and some fictions from general proof theory* ◇ A05 F05 F15 ◇
MANIN, YU.I. *Provable and unprovable (Russian)* ◇ A05 B98 E35 E50 F30 F98 G12 ◇
MENGER, K. *Selected papers in logic and foundations, didactics, economics* ◇ A05 A10 F96 ◇
POTTS, T.C. *"The grossest confusion possible?" – Frege and the lambda-calculus* ◇ A05 B40 ◇
PRAWITZ, D. *Proofs and the meaning and completeness of the logical constants* ◇ A05 F07 F50 ◇
SAMOKHVALOV, K.F. *Hilbert's program and Goedel's theorems (Russian)* ◇ A05 F99 ◇
SIKIC, Z. *Towards the distinction of pure and applied mathematics in light of the problem of mathematical existence (Croatian)* ◇ A05 F99 ◇
STIGT VAN, W.P. *L.E.J.Brouwer: intuitionism and topology* ◇ A10 F99 ◇
STIGT VAN, W.P. *The rejected parts of Brouwer's dissertation on the foundations of mathematics* ◇ A10 F99 ◇
TSITKIN, A.I. *On the question of an error in a famous work due to Wajsberg (Russian)* ◇ A10 B20 F50 ◇
VAEAENAENEN, J. *A new incompleteness in arithmetic (Finnish) (English summary)* ◇ A10 C62 D25 E05 F30 ◇

1980

BEESON, M.J. *Some problems in constructive mathematics* ◇ A10 F50 ◇
CORCORAN, J. *On definitional equivalence and related topics* ◇ A10 C40 F25 ◇
CURRY, H.B. *Some philosophical aspects of combinatory logic* ◇ A05 B40 ◇
DETLEFSEN, M. *On a theorem of Feferman* ◇ A05 F25 F30 ◇
GANDY, R.O. *Church's thesis and principles for mechanisms* ◇ A05 D05 D10 D20 F99 ◇
GARBAYO, E. *Ideological ignorance of constructivism (Spanish)* ◇ A05 F99 ◇
HINDLEY, J.R. & SELDIN, J.P. (EDS.) *To H. B. Curry: Essays on combinatory logic, lambda-calculus and formalism* ◇ A05 B40 ◇
KREISEL, G. *Finitismus* ◇ A05 F65 ◇
KREISEL, G. *Widerspruchsfreiheit* ◇ A05 F99 ◇
MOSCHOVAKIS, J.R. *Kleene's realizability and "divides" notions for formalized intuitionistic mathematics* ◇ A10 F50 ◇
POSY, C.J. *On Brouwer's definition of unextendable order* ◇ A10 F55 ◇
SCOTT, D.S. *Lambda calculus: some models, some philosophy* ◇ A05 B40 C90 ◇

SELDIN, J.P. *Curry's program* ◇ A05 B40 ◇
VESLEY, R.E. *Intuitionistic analysis: the search for axiomatization and understanding* ◇ A10 F35 F50 ◇
WEBB, J.C. *Mechanism, mentalism, and metamathematics. An essay on finitism* ◇ A05 D20 F99 ◇

1981

ABRUSCI, V.M. *Due note sul principio di induzione matematica* ◇ A10 F30 ◇
BATLE, N. & PLA I CARRERA, J. & SALES VALLES, F.A. & VERDU I SOLANS, V. *To Kurt Goedel, in memoriam (Catalan)* ◇ A10 C99 E35 E50 F99 ◇
BEESON, M.J. *Formalizing constructive mathematics: why and how?* ◇ A05 F50 ◇
BROUWER, L.E.J. *Brouwer's Cambridge lectures on intuitionism* ◇ A05 A10 F50 F55 F98 ◇
BROUWER, L.E.J. *On the foundation of mathematics (Dutch)* ◇ A05 A10 F55 ◇
CELLUCCI, C. *Proof theory and theory of meaning* ◇ A05 F07 F30 ◇
CHRISTIAN, C.C. *Das rekursive Inaccessibilitaetstheorem und der Goedelsche Unvollstaendigkeitssatz in ihrer Bedeutung fuer die Informatik* ◇ A05 D35 D80 F30 ◇
DALEN VAN, D. *Brouwer and the solipsistic conception of the world (Dutch)* ◇ A05 A10 F99 ◇
ESENIN-VOL'PIN, A.S. *About infinity, finiteness and finitization (in connection with the foundations of mathematics)* ◇ A05 E30 F65 ◇
FREUDENTHAL, H. *L.E.J.Brouwer -topologist, intuitionist, philosopher (Dutch)* ◇ A10 F99 ◇
GALVAN, S. *Nota sull'aritmetica formalizzata di Peano (Dimostrazione del piccolo teorema di Fermat mediante una procedura ristretta ai numerali)* ◇ A10 F30 ◇
GIARETTA, P. & MARTINO, E. *Brouwer, Dummett and the bar theorem* ◇ A10 F50 ◇
GOODMAN, NICOLAS D. *Reflections on Bishop's philosophy of mathematics* ◇ A05 F55 ◇
GREENLEAF, N. *Liberal constructive set theory* ◇ A05 E70 F55 F65 ◇
HELLMAN, GEOFFREY *How to Goedel a Frege-Russell: Goedel's incompleteness theorems and logicism* ◇ A05 F30 ◇
HEYTING, A. *Continuum and choice sequence in Brouwer (Dutch)* ◇ A10 F55 ◇
ISLES, D. *Remarks on the notion of standard non-isomorphic natural number series* ◇ A05 F15 F65 ◇
KOSHELEVA, O.M. & KREJNOVICH, V.YA. *What can physics contribute to constructive mathematics? (Russian)* ◇ A05 F99 ◇
KRAJEWSKI, S. *Kurt Goedel and his work (Polish)* ◇ A10 C98 D98 E98 F99 ◇
KREISEL, G. *Constructivist approaches to modern logic* ◇ A05 F99 ◇
KREISEL, G. *Kurt Goedel* ◇ A10 E99 F99 ◇
KREISEL, G. *Zur Bewertung mathematischer Definitionen* ◇ A05 F99 ◇
MANIN, YU.I. *Expanding constructive universes* ◇ A05 F99 ◇
NEGRI, M. *Constructive sequent reduction in Gentzen's first consistency proof for arithmetic* ◇ A10 F05 F30 ◇
NIEKUS, N.H. & RIEMSDIJK VAN, H. & TROELSTRA, A.S. *Bibliography of A. Heyting* ◇ A10 F99 ◇
PAEPPINGHAUS, P. *Was ist konstruktive Mathematik?* ◇ A05 F99 ◇
PEACOCKE, C. *Hacking on logic: two comments* ◇ A05 F50 ◇
POSY, C.J. & SWART DE, H.C.M. *Validity and quantification in intuitionism* ◇ A05 F50 ◇
SHAPIRO, S. *Understanding Church's thesis* ◇ A05 D20 F99 ◇
SMORYNSKI, C.A. *Fifty years of self-reference in arithmetic* ◇ A10 F30 F40 ◇
STAHL, G. *The diagonal method in set theory and metamathematics (Spanish)* ◇ A05 E10 F99 ◇
TAIT, W.W. *Finitism* ◇ A05 F65 ◇
TROELSTRA, A.S. *Arend Heyting and his contribution to intuitionism* ◇ A10 F55 ◇
TROELSTRA, A.S. *The interplay between logic and mathematics: Intuitionism* ◇ A05 F50 F98 ◇
VOGEL, HELMUT *Der finite Standpunkt in Hilberts Programm* ◇ A05 F65 F99 ◇
WANG, HAO *Popular lectures on mathematical logic (Chinese)* ◇ A05 B98 C98 E98 F30 F98 ◇
XU, LIZHI & YUAN, XIANGWAN & ZHENG, YUXIN & ZHU, WUJIA *On Goedel's incompleteness theorem (Chinese) (English summary)* ◇ A05 F30 ◇

1982

APERY, R. *Mathematique constructive* ◇ A05 F99 ◇
CHAITIN, G.J. *Goedel's theorem and information* ◇ A05 F30 ◇
DALEN VAN, D. *Braucht die konstruktive Mathematik Grundlagen?* ◇ A05 F98 ◇
DANCY, J. *Intuitionism in meta-epistemology* ◇ A05 F99 ◇
DELZELL, C.N. *Case distinctions are necessary for representing polynomials as sums of squares* ◇ A10 C60 C65 F55 ◇
FRIEDMANN, J. *Bemerkungen zur Logikbegruendung im deutschen Konstruktivismus* ◇ A05 F65 ◇
GANDY, R.O. *Limitations to mathematical knowledge* ◇ A05 F65 ◇
GANDY, R.O. *Symposium on constructivity in mathematics: Introductory remarks* ◇ A05 F99 ◇
KOLATA, G. *Does Goedel's theorem matter to mathematics?* ◇ A05 F99 ◇
KREISEL, G. *Brouwer's Cambridge lectures on intuitionism* ◇ A05 A10 F55 ◇
KREISEL, G. & MACINTYRE, A. *Constructive logic versus algebraization I* ◇ A05 F05 F50 ◇
LUCI, F. *Una teoria dell'infinito potenziale. II (English summary)* ◇ A05 F99 ◇
NEPEJVODA, N.N. *On some possibilities of applying constructive analysis to classical analysis (Russian)* ◇ A05 F99 ◇

PRAWITZ, D. *Beweise und die Bedeutung und Vollstaendigkeit der logischen Konstanten* ◇ A05 B20 F07 ◇
REZUS, A. *A bibliography of lambda-calculi, combinatory logics and related topics* ◇ A10 B40 ◇
RICHMAN, F. *Meaning and information in constructive mathematics* ◇ A05 F55 ◇
SLEZAK, P. *Goedel's theorem and the mind* ◇ A05 F99 ◇
STIGT VAN, W.P. *L.E.J.Brouwer, the signific interlude* ◇ A05 A10 F99 ◇
TROELSTRA, A.S. *On the origin and development of Brouwer's concept of choice sequence* ◇ A10 F55 ◇
TROELSTRA, A.S. & DALEN VAN, D. (EDS.) *The L.E.J. Brouwer centenary symposium* ◇ A05 A10 B97 F50 F55 F97 ◇
WRIGHT, C. *Strict finitism* ◇ A05 F65 ◇

1983

BOZZI, S. *Principi ideali e logica intuizionista* ◇ A05 F50 ◇
DRAGO, A. *Dimensional theory and constructive mathematics* ◇ A05 F99 ◇
GIL, D. *Intuitionism, transformational generative grammar and mental acts* ◇ A05 F99 ◇
KREISEL, G. *Einige Erlaeuterungen zu Wittgensteins Kummer mit Hilbert und Goedel* ◇ A05 F99 ◇
KREISEL, G. *Wittgenstein on rules and private language* ◇ A05 F99 ◇
MCCARTY, C. *Intuitionism: an introduction to a seminar* ◇ A05 F50 F98 ◇
MORICONI, E. *Hilbert and the "omega rule" (Italian)* ◇ A05 A10 F99 ◇
PRIEST, G. *The logical paradoxes and the law of excluded middle* ◇ A05 F99 ◇
SHAPIRO, S. *Remarks on the development of computability* ◇ A10 D98 F99 ◇
SUNDHOLM, G. *Constructions, proofs and the meaning of logical constants* ◇ A05 F50 ◇
TAIT, W.W. *Against intuitionism: constructive mathematics is part of classical mathematics* ◇ A05 F99 ◇
TROELSTRA, A.S. *Analysing choice sequences* ◇ A05 F35 F50 ◇
TROELSTRA, A.S. *Logic in the writings of Brouwer and Heyting* ◇ A10 F99 ◇
WEBB, J.C. *Goedel's theorems and Church's thesis: A prologue to mechanism* ◇ A05 D20 F99 ◇
WEINSTEIN, S. *The intended interpretation of intuitionistic logic* ◇ A05 F50 ◇

1984

BARRICELLI, N.A. *A few trivialities about Goedel's proof and the rules of definition* ◇ A05 F30 ◇
BRIDGES, D.S. & MINES, R. *What is constructive mathematics?* ◇ A05 F55 F98 ◇
BURGESS, J.P. *Dummett's case for intuitionism* ◇ A05 F50 ◇
CHIHARA, C.S. *Priest, the Liar, and Goedel* ◇ A05 F30 ◇
DALEN VAN, D. *Four letters from Edmund Husserl to Hermann Weyl* ◇ A10 F99 ◇
DILLER, J. & TROELSTRA, A.S. *Realizability and intuitionistic logic* ◇ A05 F30 F35 F50 ◇
FEFERMAN, S. *Between constructive and classical mathematics* ◇ A05 F35 F50 F55 ◇
FEFERMAN, S. *Kurt Goedel: conviction and caution* ◇ A05 A10 F99 ◇
FOURMAN, M.P. *Continuous truth. I. Nonconstructive objects* ◇ A05 E70 F35 F50 G30 ◇
IYANAGA, S. *Problems in foundations of mathematics. 4. Proof theory (Japanese)* ◇ A05 F99 ◇
KREISEL, G. *Frege's foundations and intuitionistic logic* ◇ A05 F50 ◇
NORMANN, D. *The infinite - a mathematical necessity* ◇ A05 D20 F30 ◇
PANOV, M.I. *Methodological problems of intuitionistic mathematics (Russian)* ◇ A05 F98 ◇
ROSSER, J.B. *Highlights of the history of the lambda-calculus* ◇ A10 B40 ◇
SCHNEIDER, B. *Klassische und intuitionistische Mathematik bei L.E.J. Brouwer- dargestellt am Cantorschen Haupttheorem* ◇ A10 F55 ◇
SCHREIBER, P. *Constructivity in arbitrary mathematical theories* ◇ A05 F99 ◇
SIEG, W. *Foundations of analysis and proof theory* ◇ A10 F30 F35 F98 ◇
SLEZAK, P. *Minds, machines and self-reference (French and German summaries)* ◇ A05 D99 F30 ◇
SMORYNSKI, C.A. *Lectures on nonstandard models of arithmetic* ◇ A10 C62 C98 F30 F98 H15 ◇
TABAKOV, M. *Goedel's theorem in retrospect* ◇ A05 F99 ◇
WESSEL, H. *Kritische Bemerkungen zur inuitionistischen Logikkonzeption* ◇ A05 F99 ◇
ZHU, SHUILIN *An essay on Goedel's incompleteness theorem (Chinese)* ◇ A05 F30 ◇

1985

AUERBACH, D.D. *Intensionality and the Goedel theorems* ◇ A05 F99 ◇
BISHOP, E.A. *Schizophrenia in contemporary mathematics* ◇ A05 F99 ◇
FEFERMAN, S. *High-level programs and typed vs. untyped constructive foundations* ◇ A05 F35 F50 ◇
FINE, K. *Natural deduction and arbitrary objects* ◇ A05 F07 ◇
KRABBE, E.C.W. *Noncumulative dialectical models and formal dialectics* ◇ A05 B46 F99 ◇
KREISEL, G. *Proof theory and the synthesis of programs: Potential and limitations* ◇ A05 F99 ◇
KUTSCHERA VON, F. *Der Satz vom ausgeschlossenen Dritten. Untersuchungen ueber die Grundlagen der Logik* ◇ A05 F99 ◇
MARTINO, E. *On the Brouwerian concept of negative continuity* ◇ A10 F50 F55 ◇
MARTINO, E. *On Brower's concept of negative continuity (Italian)* ◇ A05 A10 F50 F55 ◇
PRAWITZ, D. *Remarks on some approaches to the concept of logical consequence* ◇ A05 B22 F07 ◇
RABINOWICZ, W. *Intuitionistic truth* ◇ A05 F50 ◇
ROYDEN, H.L. *Aspects of constructive analysis* ◇ A05 F98 ◇

SHAPIRO, S. *Introduction - intensional mathematics and constructive mathematics* ◇ A05 F50 ◇
TROELSTRA, A.S. *Choice sequences and informal rigour* ◇ A05 F35 F50 ◇
WEYL, H. *Axiomatic versus constructive procedures in mathematics* ◇ A05 B30 F99 ◇
WU, WENJUN *Renaisance of constructive mathematics (Chinese)* ◇ A05 F55 ◇

B40 Combinatory logic and lambda-calculus

1924

SCHOENFINKEL, M. *Ueber die Bausteine der mathematischen Logik* ◇ A05 B10 B40 ◇

1929

CURRY, H.B. *An analysis of logical substitution* ◇ B40 ◇

1930

CURRY, H.B. *Grundlagen der kombinatorischen Logik* ◇ B40 ◇

1931

CURRY, H.B. *The universal quantifier in combinatory logic* ◇ B40 ◇

1932

CHURCH, A. *A set of postulates for the foundation of logic* ◇ B10 B40 D20 ◇

CURRY, H.B. *Some additions to the theory of combinators* ◇ B40 ◇

1933

CHURCH, A. *A set of postulates for the foundation of logic (second paper)* ◇ B10 B40 D20 ◇

CURRY, H.B. *Apparent variables from the standpoint of combinatory logic* ◇ B40 ◇

1934

CURRY, H.B. *Functionality in combinatory logic* ◇ B40 ◇

CURRY, H.B. *Some properties of equality and implication in combinatory logic* ◇ B40 ◇

KLEENE, S.C. *Proof by cases in formal logic* ◇ B40 ◇

1935

CHURCH, A. *A proof of freedom from contradiction* ◇ B40 D20 ◇

CHURCH, A. *An unsolvable problem of elementary number theory* ◇ B40 D20 D35 ◇

KLEENE, S.C. *A theory of positive integers in formal logic I,II* ◇ B28 B40 ◇

KLEENE, S.C. & ROSSER, J.B. *The inconsistency of certain formal logics* ◇ B40 ◇

ROSSER, J.B. *A mathematical logic without variables. I* ◇ B40 ◇

1936

CHURCH, A. *Mathematical logic (mimeographed notes)* ◇ B40 B98 D20 ◇

CHURCH, A. & ROSSER, J.B. *Some properties of conversion* ◇ B40 D20 ◇

CURRY, H.B. *First properties of functionality in combinatory logic* ◇ B40 ◇

FITCH, F.B. *A system of formal logic without an analogue to the Curry W operator* ◇ B40 ◇

KLEENE, S.C. *λ-definability and recursiveness* ◇ B40 ◇

QUINE, W.V.O. *A reinterpretation of Schoenfinkel's logical operators* ◇ B40 ◇

1937

TURING, A.M. *Computability and λ-definability* ◇ A05 B40 D05 D10 D20 F99 ◇

TURING, A.M. *The 𝔭-function in λ-K-conversion* ◇ B40 ◇

1939

TURING, A.M. *Systems of logic based on ordinals* ◇ B40 D25 D30 F15 F30 ◇

1941

CHURCH, A. *The calculi of λ conversion* ◇ B40 D20 ◇

CURRY, H.B. *A revision of the fundamental rules of combinatory logic* ◇ B40 ◇

CURRY, H.B. *Consistency and completeness of the theory of combinators* ◇ B40 ◇

CURRY, H.B. *The paradox of Kleene and Rosser* ◇ A05 B40 ◇

1942

CURRY, H.B. *Some advances in the combinatory theory of quantification* ◇ B40 ◇

CURRY, H.B. *The combinatory foundations of mathematical logic* ◇ B40 ◇

CURRY, H.B. *The inconsistency of certain formal logics* ◇ B40 ◇

FITCH, F.B. *A basic logic* ◇ B40 ◇

NEWMAN, M.H.A. *On theories with a combinatorial definition of "equivalence"* ◇ B10 B40 ◇

ROSSER, J.B. *New sets of postulates for combinatory logics* ◇ B40 ◇

1944

FITCH, F.B. *A minimum calculus for logic* ◇ B20 B40 ◇

FITCH, F.B. *Representation of calculi* ◇ B20 B40 ◇

1946

FEYS, R. *La technique de la logique combinatoire* ◇ B40 ◇

1948

FITCH, F.B. *An extension of basic logic* ◇ B20 B40 F35 ◇

1949

CURRY, H.B. *A simplification of the theory of combinators* ◇ B40 ◇

FITCH, F.B. *A further consistent extension of basic logic* ◇ B20 B40 ◇

FITCH, F.B. *The Heine-Borel theorem in extended basic logic* ◇ B28 B40 E70 F35 ◇

1950

Fitch, F.B. *A demonstrably consistent mathematics. Part I* ◇ B30 B40 E70 ◇

1951

Curry, H.B. *La logique combinatoire et les antinomies* ◇ A05 B40 ◇
Curry, H.B. *La theorie des combinateurs* ◇ B40 ◇
Fitch, F.B. *A demonstrably consistent mathematics. Part II* ◇ B30 B40 E70 ◇

1952

Curry, H.B. *A new proof of the Church-Rosser theorem* ◇ B40 ◇
Fitch, F.B. *Symbolic logic. An introduction* ◇ B40 B98 ◇
Myhill, J.R. *A finitary metalanguage for extended basic logic* ◇ B40 ◇

1953

Fitch, F.B. *A simplification of basic logic* ◇ B40 ◇
Fitch, F.B. *Self-referential relations* ◇ B40 ◇
L'Abbe, M. *Systems of transfinite types involving λ-conversion* ◇ B15 B40 ◇
Schuette, K. *Zur Widerspruchsfreiheit einer typenfreien Logik* ◇ B15 B40 F05 F65 ◇

1954

Fitch, F.B. *A definition of negation in extended basic logic* ◇ B40 ◇

1955

Cogan, E.J. *A formalization of the theory of sets from the point of view of combinatory logic* ◇ B40 E70 ◇
Hermes, H. *Vorlesung ueber Entscheidungsprobleme in Mathematik und Logik* ◇ B25 B40 B65 D10 D20 D35 D98 ◇
Rosser, J.B. *Deux esquisses de logique* ◇ B10 B40 D20 E30 ◇
Rosser, J.B. *Logique combinatoire et λ-conversion* ◇ B40 D20 ◇

1956

Fitch, F.B. *Recursive functions in basic logic* ◇ B40 ◇

1957

Curry, H.B. *Combinatory logic* ◇ B40 B98 ◇
Fitch, F.B. *A definition of existence in terms of abstraction and disjunction* ◇ B40 ◇

1958

Craig, W. *Definitional independence of combinators* ◇ B40 ◇
Craig, W. *Modifications and generalizations* ◇ B40 ◇
Curry, H.B. & Feys, R. *Combinatory logic. Volume I* ◇ B40 B98 ◇
Fitch, F.B. *An extensional variety of extended basic logic* ◇ B40 ◇
Fitch, F.B. *Representation of sequential circuits in combinatory logic* ◇ B40 B70 ◇
Stahl, G. *Ideas generales de la logica combinatoria* ◇ B40 ◇

1959

Fitch, F.B. *Quasi-constructive foundations for mathematics* ◇ B40 ◇

1960

Behmann, H. *Das Russell'sche Paradoxon und die formale Logik* ◇ A10 B40 E70 ◇
Curry, H.B. *The deduction theorem in the combinatory theory of restricted generality* ◇ B40 ◇
Dopp, J. *Essai d'une presentation de la logique combinatoire* ◇ B40 ◇

1961

Curry, H.B. *Basic verifiability in the combinatory theory of restricted generality* ◇ B40 ◇
Ladriere, J. *Expression de la recursion primitive dans le calcul-λK* ◇ B40 D20 ◇
Titgemeyer, R. *Ueber einen Widerspruch in Cogans Darstellung der Mengenlehre* ◇ B40 E70 ◇

1962

Kleene, S.C. *λ-definable functionals of finite types* ◇ B40 D30 D55 D65 ◇

1963

Fitch, F.B. *The system $C\Delta$ of combinatory logic* ◇ B40 ◇
Skordev, D.G. *Rekursiv vollstaendige arithmetische Operationen (Russian summary)* ◇ B40 ◇

1964

Curry, H.B. *Combinatory recursive objects of all finite types* ◇ B40 D65 ◇
Curry, H.B. *The elimination of variables by regular combinators* ◇ B40 ◇
Sanchis, L.E. *Types in combinatory logic* ◇ B40 ◇

1965

Andrews, P.B. *A transfinite type theory with type variables* ◇ B15 B40 ◇
Sanchis, L.E. *A property of combinations in normal form* ◇ B40 ◇

1966

Boehm, C. & Gross, W.F. *Introduction to the CUCH* ◇ B40 B75 ◇
Boehm, C. *The CUCH as a formal and descriptive language* ◇ B40 B70 ◇

1967

Curry, H.B. *Combinatory logic* ◇ A05 B40 B98 ◇
Fitch, F.B. *A theory of logical essences* ◇ B40 ◇
Hindley, J.R. *Axioms for strong reduction in combinatory logic* ◇ B40 ◇
Lercher, B. *Strong reduction and normal form in combinatory logic* ◇ B40 ◇
Lercher, B. *The decidability of Hindley's axiom for strong reduction* ◇ B40 ◇

1968

Curry, H.B. *Recent advances in combinatory logic* ◇ B40 ◇
Fitch, F.B. *A note on recursive relations* ◇ B40 D20 ◇
Loewen, K. *A standardization theorem for strong reduction* ◇ B40 ◇

LOEWEN, K. *Modified strong reduction in combinatory logic* ◇ B40 ◇

LOEWEN, K. *The Church-Rosser theorem for strong reduction in combinatory logic* ◇ B40 ◇

SELDIN, J.P. *Note on definitional reductions* ◇ B40 ◇

TAIT, W.W. *Constructive reasoning* ◇ B40 F10 F15 F30 F35 F50 F65 ◇

1969

CURRY, H.B. *Modified basic functionality in combinatory logic* ◇ B40 ◇

CURRY, H.B. *The undecidability of λK-conversion* ◇ B40 ◇

FITCH, F.B. & ORGASS, R.J. *A theory of computing machines* ◇ B40 D10 ◇

HINDLEY, J.R. *An abstract form of the Church-Rosser theorem I* ◇ B40 ◇

HINDLEY, J.R. *The principal type-scheme of an object in combinatory logic* ◇ B40 ◇

KEARNS, J.T. *Combinatory logic with discriminators* ◇ B40 ◇

ROBINSON, JOHN ALAN *Mechanizing higher-order logic* ◇ B35 B40 F35 ◇

WAGNER, E.G. *Uniformly reflexive structures: on the nature of Goedelization and relative computability* ◇ B40 D45 D75 F40 ◇

1970

BUNDER, M.W. *A paradox in illative combinatory logic* ◇ B40 ◇

GILMORE, P.C. *The lambda-gamma calculus: a language adequate for defining recursive functions* ◇ B40 ◇

HINDLEY, J.R. & LERCHER, B. *A short proof of Curry's normal form theorem* ◇ B40 ◇

KUZICHEV, A.S. & NIKOGOSOV, S.L. *Implicative systems of combinatory logic (Russian) (English summary)* ◇ B40 ◇

STENLUND, S. *Combinators as effectively calculable functions* ◇ B40 D20 ◇

1971

BAKKER DE, J.W. *Recursive procedures* ◇ B40 B75 D20 ◇

GRIZE, J.-B. *Quelques problemes logico-linguistiques* ◇ B40 ◇

KUZICHEV, A.S. *F^n-systems of combinatory logic. Generalized arithmetic operators (Russian)* ◇ B40 ◇

KUZICHEV, A.S. *Certain properties of Schoenfinkel-Curry combinators (Russian)* ◇ B40 ◇

KUZICHEV, A.S. & NIKOGOSOV, S.L. *Implicative systems of combinatory logic with a universal quantifier (Russian)* ◇ B40 ◇

NEBESKY, L. *A combinatory approach to grammar* ◇ B40 ◇

NOLIN, L. *Logique combinatoire et algorithmes* ◇ B40 ◇

SHABUNIN, L.V. *On the consistency of certain systems of combinatory logic (Russian) (English summary)* ◇ B40 ◇

STENLUND, S. *Introduction to combinatory logic* ◇ B40 ◇

VUCKOVIC, V. *Combinatorial operators and their quasi-inverses* ◇ B40 ◇

1972

BOEHM, C. & DEZANI, M. *A CUCH-machine: The automatic treatment of bound variables* ◇ B35 B40 ◇

BRUIJN DE, N.G. *Lambda calculus notation with nameless dummies, a tool for automatic formula manipulation, with application to the Church-Rosser theorem* ◇ B40 ◇

BUNDER, M.W. *An introduction to and trends in combinatory logic* ◇ B40 ◇

CASTANET, R. *Sur la semantique des operateurs d'insertion dans une arborescence* ◇ B40 ◇

CASTANET, R. *Une formalisation de la semantique des operateurs d'extraction dans une arborescence* ◇ B40 ◇

CURRY, H.B. & HINDLEY, J.R. & SELDIN, J.P. *Combinatory logic. Vol.II* ◇ B40 ◇

GOODMAN, NICOLAS D. *A simplification of combinatory logic* ◇ B40 ◇

HINDLEY, J.R. & LERCHER, B. & SELDIN, J.P. *Introduction to combinatory logic* ◇ B40 B98 F98 ◇

KUZICHEV, A.S. *The functionality operator in implicative systems of combinatory logic (Russian)* ◇ B40 ◇

LAMBEK, J. *Deductive systems and categories III. Cartesian closed categories, intuitionist propositional calculus, and combinatory logic* ◇ B40 F50 G30 ◇

LEDGARD, H.F. *Embedding Markov normal algorithms within the λ-calculus* ◇ B40 D03 D15 ◇

MITSCHKE, G. *λ-definierbare Funktionen auf Peanoalgebren* ◇ B40 D75 ◇

MITSCHKE, G. *Ein algebraischer Beweis fuer das Church-Rosser-Theorem* ◇ B40 ◇

QUINE, W.V.O. *Algebraic logic and predicate functors* ◇ A05 B40 G15 ◇

SCOTT, D.S. *Continuous lattices* ◇ B40 G10 ◇

SCOTT, D.S. *Lattice theory, data types and semantics* ◇ B40 B75 G10 ◇

SHABUNIN, L.V. *A variant of combinatory logic with discriminators (Russian)* ◇ B40 ◇

STENLUND, S. *Combinators, λ-terms and proof theory* ◇ B40 F05 F10 F50 F98 ◇

1973

BARENDREGT, H.P. *A characterization of terms of the lambda-calculus having a normal form* ◇ B40 ◇

BARENDREGT, H.P. *Combinatory logic and the axiom of choice* ◇ B40 E25 F50 ◇

BOEHM, C. & DEZANI, M. *Notes on "A CUCH-machine: The automatic treatment of bound variables"* ◇ B40 B75 ◇

BUNDER, M.W. *A deduction theorem for restricted generality* ◇ B40 ◇

BUNDER, M.W. *A generalized Kleene-Rosser paradox for a system containing the combinator K* ◇ B40 ◇

CURRY, H.B. *The consistency of a system of combinatory restricted generality* ◇ B40 ◇

GRIZE, J.-B. *Logique moderne. Fasc. III: Implications-modalites, logiques polyvalentes, logique combinatoire, ontologie et mereologie de Lesniewski* ◇ B40 ◇

KEARNS, J.T. *The completeness of combinatory logic with discriminators* ◇ B40 ◇

KUZICHEV, A.S. *A system of λ-conversion with a deductive formal implication operator (Russian)* ◇ B40 ◇
KUZICHEV, A.S. *Consistent extensions of pure combinatory logic (Russian) (English summary)* ◇ B40 ◇
KUZICHEV, A.S. *Deductive-combinatorial construction of the theory of functionality (Russian)* ◇ B40 ◇
KUZICHEV, A.S. *The subject and the methods of combinatory logic (Russian)* ◇ B40 ◇
LEWIS, H.A. *Combinators and deep structure, syntactic and semantic functions* ◇ B40 B65 ◇
MICHEL, J.Y. *APLAMBDA: A lambda-calculus machine for APL-operators* ◇ B40 ◇
MORAVEK, J. *Computational optimality of a dynamic programming method* ◇ B40 D15 ◇
NOLIN, L. *Systemes algorithmiques, systemes fonctionnels* ◇ B40 B75 ◇
OWINGS JR., J.C. *Diagonalization and the recursion theorem* ◇ B40 D20 F30 ◇
ROBINET, B. *Un modele semantique pour un langage simple de programmation* ◇ B40 B75 ◇
ROSEN, B.K. *Tree-manipulating systems and Church-Rosser-theorems* ◇ B40 ◇
RUGGIU, G. *Les types et les appels de procedures* ◇ B40 ◇
RUGGIU, G. & VASSEUR, J. *Structure et proprietes d'une famille de langages de programmation, EXEL* ◇ B40 ◇
SCOTT, D.S. *Models for various type-free calculi* ◇ B40 C90 ◇
SELDIN, J.P. *Equality in* $\mathfrak{F}_{21}$ ◇ B40 F05 ◇
SHABUNIN, L.V. *Simple combinatory calculi (Russian) (English summary)* ◇ B40 ◇
SHABUNIN, L.V. *The undecidability of certain formal systems of combinatory logic (Russian)* ◇ B40 D35 ◇
STAPLES, J. *Combinator realizability of constructive finite type analysis* ◇ B40 F35 F50 ◇

1974

AUSIELLO, G. *Relations between semantics and complexity of recursive programs* ◇ B40 B75 D15 D20 ◇
BARENDREGT, H.P. *Combinatory logic and the ω-rule* ◇ B40 ◇
BARENDREGT, H.P. *Pairing without conventional restraints* ◇ B40 F35 ◇
BISHOP, P. & GREIF, I. & HALE, R.L.V. & HEWITT, C. & MATSON, T. & SMITH, BRIAN & STEIGER, R. *Behavioral semantics of nonrecursive control structures* ◇ B35 B40 ◇
BOEHM, C. & DEZANI-CIANCAGLINI, M. *Combinatorial problems, combinator equations and normal forms* ◇ B40 ◇
BUNDER, M.W. *Propositional and predicate calculuses based on combinatory logic* ◇ B40 ◇
BUNDER, M.W. *Some inconsistencies in illative combinatory logic* ◇ B40 ◇
BUNDER, M.W. *Various systems of set theory based on combinatory logic* ◇ B40 E30 E70 ◇
CASTANET, R. *Semantique formelle des operateurs d'un langage de listes (English summary)* ◇ B40 B75 ◇
FITCH, F.B. *Elements of combinatory logic* ◇ B40 B98 ◇
HINDLEY, J.R. *An abstract Church-Rosser theorem II. Applications* ◇ B40 ◇
HOSONO, C. *On the cardinality of some lattices* ◇ B40 ◇
IGARASHI, S. *Admissibility of fixed-point induction in first-order logic of typed theories* ◇ B40 ◇
ITO, T. *A formal approximation theory of semantic data types* ◇ B40 ◇
JACOPINI, G. *Principio di estensionalita nel calcolo dei combinatori (English summary)* ◇ B40 ◇
JACOPINI, G. *Una condizione per l'identificabilita di due elementi di un generico modello della logica combinatoria* ◇ B40 ◇
KUZICHEV, A.S. *Das Prinzip der kombinatorischen Vollstaendigkeit in der mathematischen Logik (Russian)* ◇ B40 ◇
KUZICHEV, A.S. *Deductive operators of combinatory logic (Russian) (English summary)* ◇ B40 ◇
KUZICHEV, A.S. *On the expressive potentialities of deductive systems of λ-conversion and combinatory logic (Russian) (English summary)* ◇ B40 ◇
LAMBEK, J. *Functional completeness of cartesian categories* ◇ B40 G30 ◇
MEREDITH, D. *Combinatory and propositional logic* ◇ B40 ◇
MILNER, R. *A calculus for the mathematical theory of computation* ◇ B40 ◇
PLOTKIN, G.D. *The λ-calculus is ω-incomplete* ◇ B40 ◇
REYNOLDS, J.C. *Towards a theory of type structure* ◇ B40 ◇
ROBINET, B. *Permutations et logique combinatoire* ◇ B40 B70 ◇
ROEVER DE, W.P. *Recursion and parameter mechanisms: An axiomatic approach* ◇ B40 B75 D75 ◇
SETHI, R. *Testing for the Church-Rosser property* ◇ B40 E07 ◇
SHABUNIN, L.V. *Some algorithmic problems of calculi of combinatory logic (Russian) (English summary)* ◇ B40 D35 ◇
SIMONE, C. *Un modello di λ-calcolo fortemente equivalente agli schemi ricorsivi (English summary)* ◇ B40 ◇
STAPLES, J. *Combinator realizability of constructive Morse theory* ◇ B40 F35 F50 ◇
VENTURINI ZILLI, M. *On different kinds of indefinite* ◇ B40 D75 ◇

1975

ABDALI, S.K. *A lambda-calculus model of programming languages I: Simple constructs. II: Jumps and procedures* ◇ B40 ◇
AIELLO, L. & AIELLO, M. *Programming language semantics in a typed lambda-calculus* ◇ B40 B75 ◇
AMER, M.A. *Parentheses in combinatory logic* ◇ B40 ◇
ANDO, S. & ITO, T. *On applying Scott's logic to termination problems* ◇ B40 D05 ◇
AUSIELLO, G. *On the description of time varying systems in λ-calculus* ◇ B40 ◇
BAKKER DE, J.W. *Least fixed points revisited* ◇ B40 B75 ◇
BARALT-TORRIJOS, J. & CHIARAVIGLIO, L. & GROSKY, W. *The programmatic semantics of binary predicator calculi* ◇ B40 D05 ◇

Subject Index

BARENDREGT, H.P. *Normed uniformly reflexive structures* ◇ B40 D75 ◇
BARENDREGT, H.P. *Open problems* ◇ B40 ◇
BATINI, C. & PETTOROSSI, A. *On subrecursiveness in weak combinatory logic* ◇ B40 ◇
BOEHM, C. & DEZANI-CIANCAGLINI, M. *λ-terms as total or partial functions on normal forms* ◇ B40 ◇
BOEHM, C. (ED.) *Lambda-calculus and computer science theory* ◇ B40 ◇
CURRY, H.B. *A study of generalized standardization in combinatory logic* ◇ B40 ◇
CURRY, H.B. *Representation of Markov algorithms by combinators* ◇ B40 D03 ◇
EGLI, H. *Typed meaning in Scott's λ-calculus models* ◇ B40 ◇
FEHR, E. *Eine universelle λ-Kalkuel-Programmiersprache und ihr Interpreter* ◇ B40 B75 ◇
FEHR, E. *Eine universelle Lambda-Kalkuel Programmiersprache und ihr Interpreter* ◇ B40 B75 ◇
FRIEDMAN, H.M. *Equality between functionals* ◇ B40 F10 ◇
HUET, G. *A unification algorithm for typed $\overline{\lambda}$-calculus* ◇ B40 ◇
HUET, G. *Unification in typed lambda calculus* ◇ B40 ◇
HYLAND, J.M.E. *A survey of some useful partial order relations on terms of the lambda calculus* ◇ B40 ◇
KLOP, J.W. *On solvability by λI-terms* ◇ B40 ◇
LEVY, J.-J. *An algebraic interpretation of the λβk-calculus and a labelled λ-calculus* ◇ B40 ◇
MAASS, W. *Church-Rosser-Theorem fuer λ-Kalkuele mit unendlich langen Termen* ◇ B40 ◇
MARTIN-LOEF, P. *About models for intuitionistic type theories and the notion of definitional equality* ◇ B40 F35 F50 ◇
MEREDITH, D. *Combinator operations* ◇ B40 ◇
MILNER, R. *Processes: a mathematical model of computing agents* ◇ B40 ◇
MYHILL, J.R. *Levels of implication* ◇ B40 ◇
NAKAJIMA, R. *Infinite normal forms for the λ-calculus* ◇ B40 ◇
NOLIN, L. *Algorithmes seriels, algorithmes paralleles* ◇ B40 B75 D20 ◇
NOLIN, L. *Les modeles informatiques des λ-calculs* ◇ B40 B75 ◇
NOLIN, L. *Theorie des algorithmes et semantique des langages de programmation* ◇ B40 B75 D20 ◇
ORGASS, R.J. *Extended basic logic and ordinal numbers* ◇ B40 F15 ◇
PLOTKIN, G.D. *Call-by-name, call-by-value and the λ-calculus* ◇ B40 ◇
POEL VAN DER, W.L. *Combinators and lambda forms (Dutch)* ◇ B40 ◇
RAULEFS, P. *The D-calculus: a system to describe the semantics of programs involving complex data types* ◇ B40 B75 ◇
SAZONOV, V.YU. *Sequentially and parallelly computable functionals* ◇ B40 B75 ◇
SCOTT, D.S. *Combinators and classes* ◇ B40 B50 E70 ◇
SCOTT, D.S. *Data types as lattices* ◇ B40 B70 G10 ◇
SCOTT, D.S. *Lambda calculus and recursion theory* ◇ B40 B75 D25 D75 ◇
SCOTT, D.S. *Some philosophical issues concerning theories of combinators* ◇ A05 B40 ◇
SELDIN, J.P. *Equality in $\mathfrak{F}_{22}$* ◇ B40 F05 ◇
SHABUNIN, L.V. *Combinatory calculi I,II (Russian)* ◇ B40 D35 ◇
SHABUNIN, L.V. *On an interpretation of α-words that is connected with simple combinatory calculi I (Russian) (English summary)* ◇ B40 ◇
SKORDEV, D.G. *Some topological examples of iterative combinatory spaces (Russian)* ◇ B40 D75 ◇
STAPLES, J. *Church-Rosser theorems for replacement systems* ◇ B40 ◇
VANDERVEKEN, D.R. *An extension of Lesniewski-Curry's formal theory of syntactical categories adequate for the categorially open functors* ◇ B40 B60 ◇
VRIJER DE, R. *Big trees in a λ-calculus with λ-expressions as types* ◇ B40 ◇
WELCH, P. *Continuous semantics and inside-out reductions* ◇ B40 ◇

1976

ABDALI, S.K. *An abstraction algorithm for combinatory logic* ◇ B40 ◇
AMER, M.A. *Typed boolean structures I (Arabic summary)* ◇ B15 B40 C90 E40 G05 ◇
ASHCROFT, E.A. & HENNESSY, M. *The semantics of nondeterminism* ◇ B40 B75 ◇
BARENDREGT, H.P. & BERGSTRA, J.A. & KLOP, J.W. & VOLKEN, H. *Representability in lambda-algebras* ◇ B40 ◇
BIELA, A. & HALLALA, P. *Studies on Church's calculus* ◇ B40 ◇
BUNDER, M.W. *Some notes on "A deduction theorem for restricted generality"* ◇ B40 ◇
BUNDER, M.W. *The inconsistency of $\mathscr{F}^*_{21}$* ◇ B40 ◇
COURVOISIER, M. & LAGASSE, J. & RICHARD, J. *Logique combinatoire. 3e ed* ◇ B40 B98 ◇
CRABBE, M. *Prelogic of logoi* ◇ B40 F07 G30 ◇
DEZANI-CIANCAGLINI, M. *Characterization of normal forms possessing inverse in the λ-β-η-calculus* ◇ B40 ◇
HYLAND, J.M.E. *A syntactic characterization of the equality in some models for the lambda-calculus* ◇ B40 ◇
KUZICHEV, A.S. *Combinatorially complete systems with operators Ξ, F, Q, Π, ∃, P, ¬, &, ∨, ≡ (Russian) (English summary)* ◇ B40 F05 ◇
KUZICHEV, A.S. *Substitution operation in systems with unrestricted principle of combinatorial completeness (Russian) (English summary)* ◇ B40 ◇
LERCHER, B. *λ-calculus terms that reduce to themselves* ◇ B40 ◇
LEVY, J.-J. *An algebraic interpretation of the λβK-calculus and an application of the labelled λ-calculus* ◇ B40 ◇
LONGO, G. *On the problem of deciding equality in partial combinatory algebras and in a formal system* ◇ B40 D75 ◇

MILNER, R. *Models of LCF* ◇ B40 D20 ◇
POTTINGER, G. *A new way of normalizing intuitionist propositional logic* ◇ B40 F05 F50 ◇
ROBINET, B. *About the logical foundations of data types* ◇ B40 B75 ◇
SAZONOV, V.YU. *Degrees of parallelism in computations* ◇ B40 B75 D30 D65 ◇
SAZONOV, V.YU. *Expressibility of functions in D.Scott's LFC language (Russian)* ◇ B40 D20 ◇
SAZONOV, V.YU. *Functionals computable in series and in parallel (Russian)* ◇ B40 D20 ◇
SCHWICHTENBERG, H. *Definierbare Funktionen im λ-Kalkuel mit Typen* ◇ B40 ◇
SELDIN, J.P. *Recent advances in Curry's program* ◇ B40 F05 ◇
SHABUNIN, L.V. *An interpretation of α-words connected with simple combinatoral calculi. II (Russian) (English summary)* ◇ B40 ◇
SKORDEV, D.G. *Certain combinatory spaces that are connected with the complexity of data processing (Russian)* ◇ B40 B75 D15 D75 ◇
SKORDEV, D.G. *Recursion theory on iterative combinatory spaces* ◇ B40 D75 ◇
SKORDEV, D.G. *Some models of combinatory logic (Russian)* ◇ B40 ◇
SKORDEV, D.G. *The concept of search computability from the point of view of the theory of combinatory spaces (Russian)* ◇ B40 D75 ◇
SKORDEV, D.G. *The partial order on the set* $\mathfrak{C}$ *in combinatory spaces (Russian)* ◇ B40 ◇
WADSWORTH, C.P. *The relation between computational and denotational properties for Scott's* D_∞*-models of the lambda-calculus* ◇ B40 ◇

1977

AMER, M.A. *Typed boolean structures II (Arabic summary)* ◇ B15 B40 C90 E40 G05 ◇
BARENDREGT, H.P. *A global representation of the recursive functions in the λ calculus* ◇ B40 ◇
BARENDREGT, H.P. *Solvability in lambda calculi* ◇ B40 ◇
BARENDREGT, H.P. *The type free lambda calculus* ◇ B40 ◇
BEL, M. *An intuitionistic combinatory theory not closed under the rule of choice* ◇ B40 F50 ◇
BOEHM, C. & COPPO, M. & DEZANI-CIANCAGLINI, M. *Termination tests inside λ-calculus* ◇ B40 ◇
BUNDER, M.W. *Consistency notions in illative combinatory logic* ◇ B40 ◇
COPPO, M. & DEZANI-CIANCAGLINI, M. *Trees and λ-terms* ◇ B40 ◇
ENGELER, E. *A new type of models of computation* ◇ B40 ◇
HELMAN, GLEN *Completeness of the normal typed fragment of the λ-system U* ◇ B40 ◇
HINDLEY, J.R. *Combinatory reductions and λ reductions compared* ◇ B40 ◇
HINDLEY, J.R. & MITSCHKE, G. *Some remarks about the connections between combinatory logic and axiomatic recursion theory* ◇ B40 D75 ◇
HINDLEY, J.R. *The equivalence of complete reductions* ◇ B40 ◇
HOSONO, C. & SATO, M. *The retracts in Pω do not form a continuous lattice – a solution to Scott's problem* ◇ B40 B75 G10 ◇
HUET, G. *Confluent reductions: abstract properties and applications to term rewriting systems* ◇ B40 ◇
KEARNS, J.T. *The logic of calculation* ◇ B40 ◇
KUZICHEV, A.S. *A system of λ-conversion with logical operators and an equality operator (Russian)* ◇ B10 B40 F05 ◇
KUZICHEV, A.S. *Cut theorem for R-theories in combinatorially complete systems (Russian) (English summary)* ◇ B40 F05 ◇
KUZICHEV, A.S. *Formal arithmetic in the* $\mathfrak{A}$*-system of λ-conversion (Russian)* ◇ B40 F05 F30 ◇
LEDGARD, H.F. *Production systems: A notation for defining syntax and translation* ◇ B40 ◇
LIPTON, R.J. & MILLER, R.E. & SNYDER, LAWRENCE *Synchronization and computing capabilities of linear asynchronous structures* ◇ B40 B70 D05 ◇
MILNER, R. *Fully abstract models of typed λ-calculi* ◇ B40 B75 ◇
MIRKOWSKA, G. *Algorithmic logic and its application in the theory of programs I,II* ◇ B40 B75 ◇
OBTULOWICZ, A. *Functorial semantics of the type free λ-β-η calculus* ◇ B40 ◇
O'DONNELL, M.J. *Computing in systems described by equations* ◇ B40 B75 ◇
O'DONNELL, M.J. *Subtree replacement systems: a unifying theory for recursive equations, LISP, lucid and combinatory logic* ◇ B40 B75 D05 D20 ◇
PETTOROSSI, A. *Combinators as tree transducers* ◇ B40 ◇
ROBINET, B. *Un modele fonctionnel des structures de controle* ◇ B40 B75 ◇
SANCHIS, L.E. *Data types as lattices: Retractions, closures and projections* ◇ B40 B75 D25 G10 ◇
SELDIN, J.P. *A sequent calculus for type assignment* ◇ B40 ◇
SELDIN, J.P. *The Q-consistency of* F_{22} ◇ B40 F05 ◇
SKORDEV, D.G. *Simplification of some definitions in the theory of combinatory spaces* ◇ B40 D75 ◇
STAPLES, J. *A class of replacement systems with simple optimality theory* ◇ B40 ◇
STAPLES, J. *Combinatory logics and lambda calculi as model programming languages* ◇ B40 B75 ◇
STAPLES, J. *Optimal reductions in replacement systems* ◇ B40 ◇
STATMAN, R. *The typed λ-calculus is not elementary recursive* ◇ B40 D15 D20 ◇

1978

BARENDREGT, H.P. & BERGSTRA, J.A. & KLOP, J.W. & VOLKEN, H. *Degrees of sensible lambda theories* ◇ B40 D30 ◇
BAXTER, L.D. *The undecidability of the third order dyadic unification problem* ◇ B15 B40 D80 ◇
BERRY, GERARD *Sequentialite de l'evaluation formelle des λ-expressions* ◇ B40 ◇

BERRY, GERARD *Stable models of typed λ-calculi* ◇ B40 ◇

BRUIJN DE, N.G. *A namefree lambda calculus with facilities for internal definition of expressions and segments* ◇ B40 ◇

BRUIJN DE, N.G. *Lambda-calculus with namefree formulas involving symbols that represent reference transforming mappings* ◇ B35 B40 ◇

BUNDER, M.W. *Equality in $\mathscr{I}_{21}^*$ with restricted subjects* ◇ B40 ◇

BUNDER, M.W. & MEYER, R.K. *On the inconsistency of systems similar to $\mathscr{F}_{21}^*$* ◇ B40 ◇

BUNDER, M.W. & SELDIN, J.P. *Some anomalies in Fitch's system QD* ◇ B40 ◇

CANAL, R. *Complexite de la reduction en logique combinatoire* ◇ B40 ◇

COPPO, M. & DEZANI-CIANCAGLINI, M. & RONCHI DELLA ROCCA, S. *(Semi)-separability of finite sets of terms in Scott's D_∞-models of the λ-calculus* ◇ B40 ◇

COPPO, M. & DEZANI-CIANCAGLINI, M. *Syntactic and semantics trees versus type trees inside lambda-calculus* ◇ B40 ◇

DAMM, W. & FEHR, E. & INDERMARK, K. *Higher type recursion and self-application as control structures* ◇ B40 B75 D20 ◇

DAMM, W. & FEHR, E. *On the power of self-application and higher type recursion* ◇ B40 ◇

FRIEDMAN, D.P. & WAND, M. *Compiling lambda-expressions using continuations and factorizations* ◇ B40 ◇

HINDLEY, J.R. *Reductions of residuals are finite* ◇ B40 ◇

HINDLEY, J.R. *Standard and normal reductions* ◇ B40 ◇

IVANOV, L.L. *Natural combinatory spaces* ◇ B40 D75 ◇

JACOPINI, G. & VENTURINI ZILLI, M. *Equating for recurrent terms of lambda-calculus and combinatory logic* ◇ B40 ◇

KUZICHEV, A.S. *A theorem on the consistency of formal arithmetic (Russian)* ◇ B40 F05 F30 ◇

KUZICHEV, A.S. *Formal arithmetic in combinatorially complete systems I,II (Russian) (English summary)* ◇ B40 F05 F30 ◇

KUZICHEV, A.S. *On the consistency of formal arithmetic (Russian)* ◇ B40 F30 ◇

KUZICHEV, A.S. *The theorem on the midsequent in the $\mathfrak{A}$-system of λ-conversion (Russian)* ◇ B40 F05 ◇

KUZICHEV, A.S. *The Curry paradox in combinatorially complete systems (Russian) (English summary)* ◇ B40 ◇

LEVY, J.-J. *Le probleme du partage dans l'evaluation des λ-expressions* ◇ B40 ◇

MEREDITH, D. *Positive logic and λ-constants* ◇ B20 B25 B40 ◇

OBTULOWICZ, A. & WIWEGER, A. *Church algebraic theories* ◇ B40 ◇

PLOTKIN, G.D. *$\mathbb{T}^\omega$ as a universal domain* ◇ B40 B75 ◇

POTTINGER, G. *Proofs of the normalization and Church-Rosser theorems for the typed λ-calculus* ◇ B40 ◇

ROBINET, B. *Logique combinatoire et programmation: Une autre approche de la theorie de la programmation* ◇ B40 B75 ◇

SALLE, P. *Une extension de la theorie des types en λ-calcul* ◇ B15 B40 ◇

SCIORE, E. & TANG, A. *Admissible coherent c.p.o.'s* ◇ B40 B75 E07 ◇

SELDIN, J.P. *A sequent calculus formulation of type assignment with equality rules for the $\lambda\beta$-calculus* ◇ B40 F05 ◇

SHUMIKHIN, V.P. *About the inconsistency of the system $\mathscr{F}_{21}^{*A}$ of combinatory logic (Russian)* ◇ B40 ◇

SILVERSTEIN, A. *A generalization of combinatorial operators* ◇ B40 D50 ◇

SKORDEV, D.G. *A normal form theorem for recursive operators in iterative combinatory spaces* ◇ B40 D75 ◇

VENTURINI ZILLI, M. *Head recurrent terms in combinatory logic: A generalization of the notion of head normal form* ◇ B40 ◇

WADSWORTH, C.P. *Approximate reduction and λ calculus models* ◇ B40 ◇

1979

ASTESIANO, E. & COSTA, G. *Sharing in nondeterminism* ◇ B40 B75 ◇

BAETEN, J. & BOERBOOM, B. *Ω can be anything it should not be* ◇ B40 ◇

BERGSTRA, J.A. & KLOP, J.W. *Church-Rosser strategies in the lambda calculus* ◇ B40 ◇

BERRY, GERARD & LEVY, J.-J. *A survey of some syntactic results in the λ-calculus* ◇ B40 ◇

BERRY, GERARD & LEVY, J.-J. *Minimal and optimal computations of recursive programs* ◇ B40 B75 ◇

BOEHM, C. & DEZANI-CIANCAGLINI, M. & PERETTI, P. & RONCHI DELLA ROCCA, S. *A discrimination algorithm inside λ-β-calculus* ◇ B40 ◇

BUNDER, M.W. *Λ-elimination in illative combinatory logic* ◇ B40 B50 ◇

BUNDER, M.W. *Generalised restricted generality* ◇ B40 ◇

BUNDER, M.W. *On the equivalence of systems of rules and systems of axioms in illative combinatory logic* ◇ B40 ◇

BUNDER, M.W. *Scott's models and illative combinatory logic* ◇ B40 ◇

BUNDER, M.W. *Variable binding term operators in λ-calculus* ◇ B40 ◇

CHAUVIN, A. *Theory of objects and set theory: Introduction and semantics* ◇ A05 B40 E30 E70 ◇

COMYN, G. & WERNER, G. *Computable data types* ◇ B40 ◇

COPPO, M. & DEZANI-CIANCAGLINI, M. *A new type assignment for λ-terms* ◇ B40 ◇

COPPO, M. & DEZANI-CIANCAGLINI, M. & SALLE, P. *Functional characterization of some semantic equalities inside λ-calculus* ◇ B40 ◇

CURRY, H.B. *On a polynomial representation of $\lambda\beta$-normal forms* ◇ B40 ◇

DEZANI-CIANCAGLINI, M. & RONCHI DELLA ROCCA, S. & SAITTA, L. *Complexity of λ-term reductions* ◇ B40 D15 ◇
DONAHUE, J. *On the semantics of "data type"* ◇ B40 ◇
FLEISCHHACKER, L. *Is Russell's vicious circle principle false or meaningless?* ◇ A05 B40 ◇
GEORGIEVA, N.V. *Ueber gewisse McCarthysche Aequivalenzen (Russisch)* ◇ B40 ◇
HINDLEY, J.R. *The discrimination theorem holds for combinatory weak reduction* ◇ B40 ◇
KUZICHEV, A.S. *Classes of objects having normal forms in the system of λ-conversion with logical operators (Russian)* ◇ B40 ◇
KUZICHEV, A.S. *Formale Arithmetik in einem System der λ-Konversion mit logischen Operatoren (Russisch)* ◇ B40 F05 F30 ◇
KUZNETSOV, P.A. *A method of analysis for consistency in the description of a combinative scheme (Russian)* ◇ B40 ◇
LONGO, G. *Ricorsivita nei tipi superiori: un'introduzione alle caratterizzazioni di Ershov ed Hyland* ◇ B40 D65 G30 ◇
MILNER, R. *ICF: A way of doing proofs with a machine* ◇ B35 B40 ◇
MITSCHKE, G. *The standardization theorem for λ-calculus* ◇ B40 ◇
MOULOUD, N. *Quelques hypotheses sur la porte des notions operatoires dans les langues naturelles et techniques* ◇ B40 B65 ◇
O'DONNELL, M.J. *A practical programming theorem which is independent of Peano arithmetic* ◇ B40 B75 D20 F05 F30 ◇
PETROV, V.P. & SKORDEV, D.G. *Combinatory structures* ◇ B40 D75 ◇
PETTOROSSI, A. *A hierarchy of infinite computations in type free subtree replacement systems* ◇ B40 ◇
PETTOROSSI, A. *On the definition of hierarchies of infinite sequential computations* ◇ B40 D20 ◇
POTTS, T.C. *"The grossest confusion possible?" - Frege and the lambda-calculus* ◇ A05 B40 ◇
RICCI, G. *P-algebras and combinatory notation* ◇ B40 C05 G30 ◇
SANCHIS, L.E. *Reducibilities in two models for combinatory logic* ◇ B40 D30 ◇
SELDIN, J.P. *Progress report on generalized functionality* ◇ B40 ◇
SKORDEV, D.G. *The first recursion theorem for iterative combinatory spaces* ◇ B40 D75 ◇
STAPLES, J. *A graph-like lambda-calculus for which leftmost-outermost reduction is optimal* ◇ B40 ◇
STAPLES, J. *A lambda calculus with naive substitution* ◇ B40 ◇
STATMAN, R. *Intuitionistic propositional logic is polynomial-space complete* ◇ B40 D15 F20 F50 ◇
STATMAN, R. *The typed λ-calculus is not elementary recursive* ◇ B40 D15 D20 ◇
TANG, A. *Chain properties in Pω* ◇ B40 B75 E20 ◇
TURNER, D.A. *Another algorithm for bracket abstraction* ◇ B40 ◇

1980

ASHCROFT, E.A. & HENNESSY, M. *A mathematical semantics for a nondeterministic typed lambda-calculus* ◇ B40 B75 ◇
ASTESIANO, E. & COSTA, G. *Delayed evaluation trees and their application to semantics* ◇ B40 B75 ◇
ASTESIANO, E. & COSTA, G. *Languages with reducing reflexive types* ◇ B40 B75 ◇
ASTESIANO, E. & COSTA, G. *Nondeterminism and fully abstract models* ◇ B40 B75 ◇
BARENDREGT, H.P. & KOYMANS, K. *Comparing some classes of lambda calculus models* ◇ B40 ◇
BARENDREGT, H.P. & LONGO, G. *Equality of lambda terms in the model* T^{ω} ◇ B40 ◇
BERGSTRA, J.A. & KLOP, J.W. *Invertible terms in the lambda-calculus* ◇ B40 ◇
BOEHM, C. *An abstract approach to (hereditary) finite sequences of combinators* ◇ B40 ◇
BOEHM, C. & MICALI, S. *Minimal forms in λ-calculus computations* ◇ B40 ◇
BRACHO, F. *Continuously generated fixed points in Pω* ◇ B40 D70 ◇
BRUIJN DE, N.G. *A survey of the project AUTOMATH* ◇ B35 B40 ◇
BUNDER, M.W. *A note on quantified significance logics* ◇ B40 B45 ◇
BUNDER, M.W. *Significance and illative combinatory logics* ◇ B40 B45 ◇
BUNDER, M.W. *The consistency of a higher order predicate calculus and set theory based on combinatory logic* ◇ B15 B40 F05 ◇
BUNDER, M.W. *The naturalness of illative combinatory logic as a basis for mathematics* ◇ B40 ◇
COPPO, M. *An extended polymorphic type system for applicative languages* ◇ B40 B75 ◇
COPPO, M. & DEZANI-CIANCAGLINI, M. *An extension of basic functionality theory for λ-calculus* ◇ B40 ◇
COPPO, M. & DEZANI-CIANCAGLINI, M. & VENNERI, B.M. *Principal type schemes and lambda-calculus semantics* ◇ B40 ◇
CURRY, H.B. *Some philosophical aspects of combinatory logic* ◇ A05 B40 ◇
DAALEN VAN, D.T. *The language theory of Automath (Dissertation)* ◇ B40 ◇
FEHR, E. *Lambda-calculus as control structures of programming languages* ◇ B40 B75 ◇
FITCH, F.B. *A consistent combinatory logic with an inverse to equality* ◇ B40 ◇
FITCH, F.B. *An extension of a system of combinatory logic* ◇ B40 ◇
FRIEDMAN, JOYCE & WARREN, D.S. *λ-normal forms in an intensional logic for English* ◇ B40 ◇
GANDY, R.O. *An early proof of normalization by A. M. Turing* ◇ B40 F05 ◇
GANDY, R.O. *Proofs of strong normalization* ◇ B40 F05 F50 ◇
HALSTEAD, R.H.J. & WARD, S.A. *A syntactic theory of message passing* ◇ B40 ◇
HENNESSY, M. & PLOTKIN, G.D. *A term model for CCS* ◇ B40 ◇

HINDLEY, J.R. & LONGO, G. *Lambda-calculus models and extensionality* ◇ B40 ◇

HINDLEY, J.R. & SELDIN, J.P. (EDS.) *To H. B. Curry: Essays on combinatory logic, lambda-calculus and formalism* ◇ A05 B40 ◇

HOWARD, W.A. *The formulae-as-types notion of construction* ◇ B40 F05 F10 F50 ◇

KEARNS, J.T. *Fully explicit deductive systems* ◇ B40 ◇

KLOP, J.W. *Combinatory reduction systems* ◇ B40 ◇

KLOP, J.W. *Reduction cycles in combinatory logic* ◇ B40 ◇

KUZICHEV, A.A. & KUZICHEV, A.S. *On the embedding of formal arithmetic in combinatorially complete systems (Russian)* ◇ B40 F05 F30 ◇

KUZICHEV, A.S. *On consistency proofs for first-order theories (Russian) (English summary)* ◇ B40 ◇

KUZICHEV, A.S. *Permissibility of a cut in combinatorially complete systems with equality (Russian) (English summary)* ◇ B40 F05 ◇

KUZICHEV, A.S. *Sequential systems of λ-conversion and of combinatory logic* ◇ B40 F30 ◇

LAMBEK, J. *From lambda-calculus to Cartesian closed categories* ◇ B40 F07 G30 ◇

LAWSON, J.D. *Algebraic conditions leading to continuous lattices* ◇ B40 ◇

LEVY, J.-J. *Optimal reductions in the lambda-calculus* ◇ B40 ◇

LONGO, G. & VENTURINI ZILLI, M. *A λδ-calculus with an algorithmic δ** ◇ B40 ◇

MEREDITH, D. *A positive logic proof procedure* ◇ B20 B35 B40 F50 ◇

MEY VAN DER, G. & POEL VAN DER, W.L. & SCHAAP, C.E. *New arithmetical operators in the theory of combinators. I,II,III* ◇ B40 ◇

MITSCHKE, G. *Infinite terms and infinite reductions* ◇ B40 C75 ◇

NEDERPELT, R.P. *An approach to theorem proving on the basis of a typed lambda-calculus* ◇ B35 B40 F60 ◇

PETTOROSSI, A. *Synthesis of subtree rewriting systems behaviour by solving equations* ◇ B40 ◇

PLOTKIN, G.D. *λ-definability in the full type hierarchy* ◇ B15 B40 ◇

POTTINGER, G. *A type assignment for the strongly normalizable lambda-terms* ◇ B40 F07 ◇

RIMSCHA VON, M. *Mengentheoretische Modelle des λK-Kalkuels* ◇ B40 C65 E65 E70 ◇

ROBINET, B. *Les langages de Backus sont des systemes de manipulation d'arbres* ◇ B40 ◇

SALLE, P. *Une generalisation de la theorie des types en λ-calcul I,II* ◇ B15 B40 ◇

SANCHIS, L.E. *Reflexive domains* ◇ B40 ◇

SCOTT, D.S. *Lambda calculus: some models, some philosophy* ◇ A05 B40 C90 ◇

SCOTT, D.S. *Relating theories of the λ-calculus* ◇ B40 C90 F50 G30 ◇

SELDIN, J.P. *Curry's program* ◇ A05 B40 ◇

SKORDEV, D.G. *Combinatory spaces and recursiveness in them (Russian) (English summary)* ◇ B40 D75 ◇

SKORDEV, D.G. *Semi-combinatory spaces (Russian)* ◇ B40 D75 ◇

STAPLES, J. *Computation on graph-like expressions* ◇ B40 ◇

STAPLES, J. *Optimal evaluations of graph-like expressions* ◇ B40 ◇

STAPLES, J. *Speeding up subtree replacement systems* ◇ B40 ◇

STATMAN, R. *On the existence of closed terms in the typed λ-calculus. I* ◇ B40 ◇

VISSER, A. *Numerations, λ-calculus & arithmetic* ◇ B40 D45 F30 ◇

WADSWORTH, C.P. *Some unusual lambda-calculus numeral systems* ◇ B40 ◇

1981

AERTS, J.P.H. *Implementing SASL without garbage collection* ◇ B40 ◇

AIBA, A. & NAKANISHI, M. *General calculator based on lambda conversion* ◇ B40 ◇

AIELLO, L. & PRINI, G. *An efficient interpreter for the lambda-calculus* ◇ B40 ◇

BACKUS, J. *The algebra of functional programs: function level reasoning, linear equations, and extended definitions* ◇ B40 B75 ◇

BARENDREGT, H.P. *The lambda calculus, its syntax and semantics* ◇ B40 B98 ◇

BEN-YELLES, C.B. *G-stratification is equivalent to F-stratification* ◇ B40 ◇

BERRE LE, F. & NOLIN, L. *L'existence d'espaces informatiques (English summary)* ◇ B40 ◇

BERRY, GERARD *On the definition of lambda-calculus models* ◇ B40 ◇

BOEHM, C. *Logic and computers* ◇ B40 ◇

BUNDER, M.W. *Predicate calculus and naive set theory in pure combinatory logic* ◇ B10 B40 E70 ◇

BUNDER, M.W. *Simpler axioms for BCK algebras and the connection between the axioms and the combinators B, C and K* ◇ B40 G25 ◇

COPPO, M. & DEZANI-CIANCAGLINI, M. & VENNERI, B.M. *Functional characters of solvable terms* ◇ B40 ◇

COPPO, M. & DEZANI-CIANCAGLINI, M. & VENNERI, B.M. *Un'estensione della teoria della funzionalita di Curry* ◇ B40 ◇

DOMRACHEV, V.N. *An example of proof of the correctness of a program, using its representation by a λ-term (Russian)* ◇ B40 B75 ◇

DOMRACHEV, V.N. & KAPITONOVA, YU.V. & SAMOJLENKO, L.G. *Definition of the semantics of programming language constructs in terms of λ-calculus I (Russian)* ◇ B40 B75 ◇

ENGELER, E. *Algebras and combinators* ◇ B40 ◇

FITCH, F.B. *The consistency of system Q* ◇ B40 ◇

IVANOV, L.L. *P-recursiveness in iterative combinatorial spaces (Russian)* ◇ B40 D75 ◇

KOZEN, D. *Semantics of probabilistic programs* ◇ B40 B75 ◇

KUZICHEV, A.S. *Arithmetic theories constructed on the basis of λ- conversion (Russian)* ◇ B40 F30 ◇

MANCA, V. *Computational formalism: Abstract combinatory view-point and related first order logical framework* ◇ B40 D75 ◇

OBTULOWICZ, A. & WIWEGER, A. *Functional interpretation of λ-terms* ◇ B40 ◇
PETTOROSSI, A. *A property which guarantees termination in weak combinatory logic and subtree replacement systems* ◇ B40 ◇
PETTOROSSI, A. *Comparing and putting together recursive path ordering, simplification orderings and non-ascending property for termination proofs of term rewriting systems* ◇ B40 ◇
POTTINGER, G. *The Church-Rosser theorem for the typed λ-calculus with surjective pairing* ◇ B40 ◇
QUINE, W.V.O. *Predicate functors revisited* ◇ B10 B40 ◇
REVESZ, G. *Axioms for the theory of lambda-conversion* ◇ B40 ◇
RONCHI DELLA ROCCA, S. *Discriminability of infinite sets of terms in the D_∞-models of the λ-calculus* ◇ B40 ◇
STAPLES, J. *Efficient combinatory reduction* ◇ B40 ◇
STATMAN, R. *On the existence of closed terms in the typed λ calculus. II: Transformations of unification problems* ◇ B40 ◇
STEIN, M. *A general theorem on existence theorems* ◇ B40 F10 F35 F50 ◇
STOY, J.E. *Denotational semantics: the Scott-Strachey approach to programming language theory* ◇ B40 B75 ◇

1982

BERGSTRA, J.A. & KLOP, J.W. *Strong normalization and perpetual reduction in the lambda calculus (German and Russian summaries)* ◇ B40 ◇
BERKLING, K.J. & FEHR, E. *A consistent extension of the lambda-calculus as a base for functional programming languages* ◇ B40 B60 B75 ◇
BERKLING, K.J. & FEHR, E. *A modification of the λ-calculus as a base for functional programming languages* ◇ B40 B75 ◇
BERRY, GERARD & CURIEN, P.-L. *Sequential algorithms on concrete data structures* ◇ B40 B75 ◇
BUNDER, M.W. *Illative combinatory logic without equality as a primitive predicate* ◇ B40 ◇
BURTON, F.W. *A linear space translation of functional programs to Turner combinators* ◇ B40 ◇
BYERLY, R.E. *Recursion theory and the lambda-calculus* ◇ B40 D20 ◇
DEZANI-CIANCAGLINI, M. & ERMINE, F. *Maximal monoids of normal form* ◇ B40 C05 ◇
DOMRACHEV, V.N. & KAPITONOVA, YU.V. & SAMOJLENKO, L.G. *On a method for the definition of semantics of constructions of programming languages in terms of typed λ-calculus (Russian)* ◇ B40 B75 ◇
HINDLEY, J.R. *The simple semantics for Coppo-Dezani-Salle types* ◇ B40 ◇
KOYMANS, C.P.J. *Models of the lambda-calculus* ◇ B40 ◇
KUZICHEV, A.S. *Arithmetically consistent λ-theories (Russian)* ◇ B40 F05 F30 ◇
KUZICHEV, A.S. *Axiomatic theories in combinatorially complete systems (Russian)* ◇ B40 ◇
KUZICHEV, A.S. *On the representation of first-order theories in type-free combinatorially complete systems (Russian)* ◇ B40 ◇
MEYER, A.R. *What is a model of the lambda calculus?* ◇ B40 ◇
OBTULOWICZ, A. & WIWEGER, A. *Categorical, functorial and algebraic aspects of the type-free lambda calculus* ◇ B40 G30 ◇
OKITA, K. *A valuation for the primitive logic* ◇ B40 ◇
PETTOROSSI, A. *Tree rewriting systems, combinatory weak reduction systems and type free languages* ◇ B40 ◇
REZUS, A. *A bibliography of lambda-calculi, combinatory logics and related topics* ◇ A10 B40 ◇
ROBINET, B. *Sur des sequences iteratives de combinateurs (English summary)* ◇ B40 ◇
RONCHI DELLA ROCCA, S. *Characterization theorems for a filter lambda model* ◇ B40 ◇
SCHWICHTENBERG, H. *Complexity of normalization in the pure typed lambda-calculus* ◇ B40 ◇
SCOTT, D.S. *Domains for denotational semantics* ◇ B22 B40 ◇
SCOTT, D.S. *Lectures on a mathematical theory of computation* ◇ B40 B98 D75 ◇
SKORDEV, D.G. *An algebraic treatment of flow diagrams and its application to generalized recursion theory* ◇ B40 B75 D75 ◇
STATMAN, R. *Completeness, invariance and λ-definability* ◇ B40 D15 D20 ◇

1983

ASTESIANO, E. & COSTA, G. *The intensitivity theorem for nonreducing reflexive types* ◇ B40 B75 ◇
BARENDREGT, H.P. & COPPO, M. & DEZANI-CIANCAGLINI, M. *A filter lambda model and the completeness of type assignment* ◇ B40 ◇
BARENDREGT, H.P. & REZUS, A. *Semantics for classical AUTOMATH and related systems* ◇ B40 B75 ◇
BUNDER, M.W. *A one axiom set theory based on higher order predicate calculus* ◇ B15 B40 E70 ◇
BUNDER, M.W. *A weak absolute consistency proof for some systems of illative combinatory logic* ◇ B40 ◇
COPPO, M. & GIOVANNETTI, E. *Completeness results for a polymorphic type system* ◇ B40 ◇
COPPO, M. *On the semantics of polymorphism* ◇ B40 ◇
DOMRACHEV, V.N. & KAPITONOVA, YU.V. & SAMOJLENKO, L.G. *One method of defining the semantics of programming languages constructs in terms of lambda calculus. II (Russian) (English summary)* ◇ B40 B75 ◇
GIOVANNETTI, E. *Une caracterisation des termes types dans un langage applicatif (English summary)* ◇ B10 B40 ◇
HAGIYA, M. *A proof description language and its reduction system* ◇ B40 F05 ◇
HINDLEY, J.R. *Curry's type-rules are complete with respect to the F-semantics too* ◇ B40 ◇
HINDLEY, J.R. *The completeness theorem for typing λ-terms* ◇ B40 ◇
IVANOV, L.L. *Iterative operatory spaces and the system of Scott and de Bakker* ◇ B40 ◇
KUZICHEV, A.S. *An arithmetically consistent λ-theory (Russian)* ◇ B40 F30 ◇
KUZICHEV, A.S. *Arithmetically consistent λ-theories of type-free logic (Russian)* ◇ B40 F05 F30 ◇

KUZICHEV, A.S. *Set theory in type-free combinatorially complete systems (Russian) (English summary)* ◇ B40 E30 E70 ◇
KUZICHEV, A.S. *The consistency of Quine's system NF (Russian)* ◇ B40 E35 E70 ◇
LONGO, G. *Set-theoretical models of λ-calculus: theories, expansions, isomorphisms* ◇ B40 ◇
MARGARIA, I. & ZACCHI, M. *Right and left invertibility in λ−β-calculus* ◇ B40 ◇
MIRASYEDIOGLU, S. *λ-functional EXCLUSIVE-OR and LOGICAL EQUIVALENCE* ◇ B40 ◇
SHABUNIN, L.V. *On the interpretaion of combinators with weak reduction* ◇ B40 ◇
STAPLES, J. *Two-level expression representation for faster evaluation* ◇ B35 B40 D03 ◇
STATMAN, R. *λ-definable functionals and βη conversion* ◇ B40 ◇
VENTURINI ZILLI, M. *Cofinality in reduction graphs* ◇ B40 ◇

1984

ASTESIANO, E. & COSTA, G. *Distributive semantics for nondeterministic typed λ-calculi* ◇ B40 B75 ◇
BAKER-FINCH, C.A. *Algebraic, operational and denotational semantics of the lambda calculus* ◇ B40 ◇
BARENDREGT, H.P. *Introduction to lambda calculus* ◇ B40 ◇
BARENDREGT, H.P. *Lambda calculus and its models* ◇ B40 ◇
BRUCE, KIM B. & LONGO, G. *On combinatory algebras and their expansions* ◇ B40 ◇
BRUCE, KIM B. & MEYER, A.R. *The semantics of second order polymorphic lambda calculus* ◇ B40 ◇
BUNDER, M.W. *Category theory based on combinatory logic* ◇ B40 G30 ◇
BUNDER, M.W. *Conjunction without conditions in illative combinatory logic* ◇ B40 ◇
BUNDER, M.W. *Some problems and inadequacies in the van der Poel-Schaap-van der Mey theory of combinators* ◇ B40 ◇
BUSZKOWSKI, W. *A note on Lambek - van Benthem calculus* ◇ B40 ◇
COPPO, M. *Completeness of type assignment in continuous lambda models* ◇ B40 F07 ◇
COPPO, M. & DEZANI-CIANCAGLINI, M. & HONSELL, F. & LONGO, G. *Extended type structures and filter lambda models* ◇ B40 ◇
DEZANI-CIANCAGLINI, M. & MARGARIA, I. *F-semantics for intersection type discipline* ◇ B40 B75 ◇
GEORGEFF, M. *Transformations and reduction strategies for typed lambda expressions* ◇ B40 ◇
GIANNINI, P. & LONGO, G. *Effectively given domains and lambda-calculus models* ◇ B40 ◇
HINDLEY, J.R. *Coppo-Dezani types do not correspond to propositional logic* ◇ B40 ◇
KANDA, A. *Numeration models of λ-calculus* ◇ B40 D45 ◇
KOYMANS, C.P.J. *Models of the lambda-calculus* ◇ B40 ◇
KUZICHEV, A.A. *Addition of noncombinatory relations to the calculus of λ-conversion (Russian)* ◇ B40 ◇
KUZICHEV, A.A. *On the representability of algebras in λ-conversion calculus (Russian)* ◇ B40 ◇
KUZICHEV, A.A. *The representation of arithmetic in combinatorially complete systems (Russian)* ◇ B40 ◇
LONGO, G. & MOGGI, E. *Goedel numberings, principal morphisms, combinatory algebras* ◇ B40 F40 G30 ◇
MEZGHICHE, M. *Une nouvelle Cβ-reduction dans la logique combinatoire (English summary)* ◇ B40 ◇
PLESNEVICH, G.S. *The semantics of schemes of structure-free notions (Russian)* ◇ B40 B75 ◇
POIGNE, A. *Higher order data structures - Cartesian closure versus λ-calculus* ◇ B40 B75 G30 ◇
REVESZ, G. *An extension of lambda-calculus for functional programming* ◇ B40 B75 ◇
ROSSER, J.B. *Highlights of the history of the lambda-calculus* ◇ A10 B40 ◇
SKORDEV, D.G. *First reduction theorem for iterative semicombinatory spaces (Russian)* ◇ B40 ◇
VENTURINI ZILLI, M. *Reduction graphs in the lambda calculus* ◇ B40 ◇
WIWEGER, A. *Pre-adjunctions and lambda-algebraic theories* ◇ B40 G30 ◇
ZASHEV, J.A. *∇-combinatory algebras and abstract Plotkin's models* ◇ B40 ◇
ZASHEV, J.A. *Basic recursion theory in partially ordered models of some fragments of the combinatory logic* ◇ B40 D20 ◇

1985

BELLOT, P. *A new proof for Craig's theorem* ◇ B10 B40 ◇
BERCOVICI, I. *Unsolvable terms in typed lambda calculus with fix-point operators* ◇ B40 ◇
BREAZU-TANNEN, V. & MEYER, A.R. *Lambda calculus with constrained types* ◇ B40 ◇
BUNDER, M.W. *An extension of Klop's counterexample to the Church-Rosser property to λ-calculus with other ordered pair combinators* ◇ B40 ◇
COCCHIARELLA, N.B. *Two λ-extensions of the theory of homogeneous simple types as a second-order logic* ◇ B15 B40 ◇
CURIEN, P.-L. *Categorical combinatory logic* ◇ B40 G30 ◇
CURIEN, P.-L. *Typed categorical combinatory logic* ◇ B40 G30 ◇
JACOPINI, G. & VENTURINI ZILLI, M. *Easy terms in the lambda calculus* ◇ B40 ◇
KANDA, A. *Numeration models of λ-calculus* ◇ B40 B60 C57 D45 ◇
KARR, M. *"Delayability" in proofs of strong normalizability in the typed lambda calculus* ◇ B40 ◇
KUZICHEV, A.A. *On a criterion of λ-representability (Russian)* ◇ B40 ◇
MEYER, A.R. & WAND, M. *Continuation semantics in typed lambda-calculi* ◇ B40 ◇
PAEPPINGHAUS, P. *A typed λ-calculus and Girard's model of ptykes* ◇ B40 F10 F15 F35 ◇

REVESZ, G. *Axioms for the theory of lambda-conversion* ◇ B40 ◇
STATMAN, R. *Logical relations and the typed λ-calculus* ◇ B40 ◇
VRIJER DE, R. *A direct proof of the finite developments theorem* ◇ B40 ◇
ZAIONC, M. *Nondeterministic programs definable in typed λ-calculus* ◇ B40 ◇
ZAIONC, M. *The set of unifers in typed λ-calculus as regular expression* ◇ B40 ◇
ZLATUSKA, J. *Normal forms in the typed λ-calculus with tuple type* ◇ B40 ◇

F05 Cut elimination and normal form theorems

1930

HERBRAND, J. *Recherches sur la theorie de la demonstration* ◇ B10 B25 F05 F07 F25 F30 ◇

1931

HERBRAND, J. *Sur la non-contradiction de l'arithmetique* ◇ F05 F30 ◇

1934

HILBERT, D. & BERNAYS, P. *Grundlagen der Mathematik I* ◇ A05 B98 F05 F30 F98 ◇

1935

GENTZEN, G. *Untersuchungen ueber das logische Schliessen I,II* ◇ B10 B25 F05 F07 F30 F50 ◇

1936

GENTZEN, G. *Die Widerspruchsfreiheit der Stufenlogik* ◇ B15 F05 F35 ◇

GENTZEN, G. *Die Widerspruchsfreiheit der reinen Zahlentheorie* ◇ B28 F05 F07 F30 ◇

1938

GENTZEN, G. *Neue Fassung des Widerspruchsfreiheitsbeweises fuer die reine Zahlentheorie* ◇ F05 F15 F30 ◇

ONO, K. *Logische Untersuchungen ueber die Grundlagen der Mathematik* ◇ F05 F30 ◇

1939

HILBERT, D. & BERNAYS, P. *Grundlagen der Mathematik II* ◇ A05 B98 F05 F15 F30 F40 F98 ◇

1950

SCHUETTE, K. *Beweistheoretische Erfassung der unendlichen Induktion in der Zahlentheorie* ◇ F05 F30 ◇

SCHUETTE, K. *Schlussweisen-Kalkuele der Praedikatenlogik* ◇ B10 B25 F05 F07 F50 ◇

1951

LADRIERE, J. *Le theoreme fondamental de Gentzen* ◇ A05 A10 F05 ◇

1952

KLEENE, S.C. *Finite axiomatizability of theories in the predicate calculus using additional predicate symbols* ◇ F05 ◇

KLEENE, S.C. *Permutability of inferences in Gentzen's calculi LK and LJ* ◇ F05 F07 ◇

SCHUETTE, K. *Beweistheoretische Untersuchungen der verzweigten Analysis* ◇ F05 F35 F65 ◇

1953

SCHUETTE, K. *Zur Widerspruchsfreiheit einer typenfreien Logik* ◇ B15 B40 F05 F65 ◇

1954

BERNAYS, P. *Ueber den Zusammenhang des Herbrand'schen Satzes mit den neueren Ergebnissen von Schuette und Stenius* ◇ F05 ◇

MAEHARA, S. *Gentzen's theorem on an extended predicate calculus* ◇ B15 F05 ◇

1955

TAKEUTI, G. *On the fundamental conjecture of GLC I,II* ◇ B15 F05 F35 ◇

1956

ISEKI, K. *On the cut operation in Gentzen calculi I* ◇ F05 F07 ◇

KLEENE, S.C. *Decision procedure* ◇ F05 ◇

LOS, J. & MOSTOWSKI, ANDRZEJ & RASIOWA, H. *A proof of Herbrand's theorem* ◇ B10 C07 F05 F50 ◇

SIMAUTI, T. *Proof of a special case of the fundamental conjecture of Takeuti's GLC* ◇ B15 F05 F35 ◇

TAKEUTI, G. *A metamathematical theorem on functions* ◇ B15 F05 F10 F35 ◇

TAKEUTI, G. *Construction of ramified real numbers* ◇ B15 F05 F35 F65 ◇

TAKEUTI, G. *On the fundamental conjecture of GLC III,IV* ◇ B15 F05 F35 ◇

1957

CRAIG, W. *Analysis of first-order implications* ◇ B10 F05 ◇

CRAIG, W. *Linear reasoning. A new form of the Herbrand-Gentzen theorem* ◇ B10 C07 C40 F05 ◇

CRAIG, W. *Three uses of the Herbrand-Gentzen theorem in relating model theory and proof theory* ◇ B10 C07 C40 F05 ◇

ESENIN-VOL'PIN, A.S. *Consistency proof of classical arithmetics with the aid of induction to ε_0 (following Schuette) (Russian)* ◇ F05 F30 ◇

ISEKI, K. *On the cut operation in Gentzen calculi II* ◇ F05 F07 ◇

KANGER, S. *Provability in logic* ◇ B10 F05 F07 F98 ◇

TAKEUTI, G. *On the theory of ordinal numbers* ◇ B28 E10 E30 E35 E45 F05 ◇

1958

KREISEL, G. *Mathematical significance of consistency proofs* ◇ C57 C60 F05 F25 F50 ◇

SCHROETER, K. *Theorie des logischen Schliessens II* ◇ B10 C07 F05 F07 ◇

SCHUETTE, K. *Aussagenlogische Grundeigenschaften formaler Systeme* ◇ B05 F05 F07 ◇

SIKORSKI, R. *On Herbrand's theorem* ◇ B10 F05 G05 ◇
TAKEUTI, G. *On the fundamental conjecture of GLC V* ◇ B15 F05 F35 ◇
TAKEUTI, G. *Remark on the fundamental conjecture of GLC* ◇ B15 F05 F35 ◇

1959

KHLODOVSKIJ, I.N. *A new proof of the consistency of arithmetic (Russian)* ◇ F05 F30 ◇
KURODA, S. *An investigation on the logical structure of mathematics. II. Transformation of the proof* ◇ B28 B30 F05 ◇

1960

CRAIG, W. *Bases for first-order theories and subtheories* ◇ B10 D25 F05 ◇
MAEHARA, S. & NISHIMURA, T. & SEKI, S. *Non-constructive proofs of a metamathematical theorem concerning the consistency of analysis and its extension* ◇ B15 F05 F35 ◇
MAEHARA, S. *On the interpolation theorem of Craig (Japanese)* ◇ B10 C40 F05 ◇
RASIOWA, H. & SIKORSKI, R. *On the Gentzen theorem* ◇ B10 C07 F05 ◇
SCHUETTE, K. *Beweistheorie* ◇ B98 F05 F15 F30 F35 F98 ◇
SCHUETTE, K. *Syntactical and semantical properties of simple type theory* ◇ B15 F05 F35 ◇
TAKEUTI, G. *An example on the fundamental conjecture of GLC* ◇ B15 F05 F35 ◇

1961

MAEHARA, S. & TAKEUTI, G. *A formal system of first-order predicate calculus with infinitely long expressions* ◇ C40 C75 F05 ◇
TAKEUTI, G. *On the fundamental conjecture of GLC VI* ◇ B15 F05 F35 ◇

1962

DOPP, J. *Logiques construites par une methode de deduction naturelle* ◇ B98 F05 F07 F50 F98 ◇
MAEHARA, S. *Cut-elimination theorem concerning a formal system for ramified theory of types which admits quantification on types* ◇ B15 F05 F30 F35 F65 ◇
MINTS, G.E. *An analogue of Herbrand's theorem for a constructive predicate calculus (Russian)* ◇ F05 F50 ◇
PARSONS, C. *The ω-consistency of ramified analysis* ◇ F05 F10 F35 F65 ◇

1963

AANDERAA, S.O. & ANDREWS, P.B. & DREBEN, B. *False lemmas in Herbrand* ◇ F05 F07 ◇
SMULLYAN, R.M. *A unifying principle in quantification theory* ◇ B10 F05 F07 ◇
TAIT, W.W. *Appendix B: An interpretation of functionals by convertible terms* ◇ F05 F35 F50 ◇

1964

AANDERAA, S.O. & DREBEN, B. *Herbrand analyzing function* ◇ F05 F07 F20 ◇

1965

BERNAYS, P. *Betrachtungen zum Sequenzen-Kalkuel* ◇ B20 F05 F07 ◇
NAMBA, K. *On a comprehension axiom without negation* ◇ E70 F05 ◇
PRAWITZ, D. *Natural deduction. A proof-theoretical study* ◇ B15 B98 F05 F07 F50 F98 ◇
RAGGIO, A.R. *Gentzen's Hauptsatz for the systems NI and NK* ◇ F05 F50 ◇
SANCHIS, L.E. *A predicative extension of elementary logic I* ◇ B15 F05 F65 ◇
SMULLYAN, R.M. *A unifying principle in quantification theory* ◇ B10 F05 F07 ◇
SMULLYAN, R.M. *Analytic natural deduction* ◇ B10 F05 F07 ◇
TITANI, S. *An algebraic formulation of cut-elimination theorem* ◇ F05 G05 G25 ◇

1966

BARTH, E.M. *On the transformation of closed semantic tableaus into natural and axiomatic deductions* ◇ F05 F07 ◇
DENTON JR., J.S. & DREBEN, B. *A supplement to Herbrand* ◇ F05 F07 ◇
MINTS, G.E. *Herbrand's theorem for the predicate calculus with equality and functional symbols (Russian)* ◇ B10 F05 ◇
TAIT, W.W. *A nonconstructive proof of Gentzen's Hauptsatz for second order predicate logic* ◇ B15 F05 F35 ◇
YASUHARA, M. *The cut elimination theorem in the unary second order language* ◇ F05 ◇

1967

CRAIG, W. *Modus ponens and derivations from Horn formulas* ◇ B20 F05 F50 ◇
LIFSCHITZ, V. *A normal form for conclusions in the predicate calculus with equality and functional symbols (Russian)* ◇ F05 ◇
LUCKHARDT, H. *Aussagenlogisch fundierte Theorien* ◇ B25 F05 ◇
MINTS, G.E. *Analog of Herbrand's theorem for non-prenex formulas of the constructive predicate calculus (Russian)* ◇ F05 F50 ◇
MINTS, G.E. *Herbrand's theorem (Russian)* ◇ B10 F05 F07 ◇
PRAWITZ, D. *Completeness and Hauptsatz for second order logic* ◇ F05 F35 ◇
SAKAI, H. *Ueber den Relativisationssatz* ◇ F05 ◇
TAKAHASHI, MOTO-O *A proof of cut-elimination in simple type theory* ◇ F05 F35 ◇
TAKEUTI, G. *Consistency proofs of subsystems of classical analysis* ◇ F05 F15 F35 ◇

1968

FEFERMAN, S. *Lectures on proof theory* ◇ C40 C70 C75 F05 F15 F98 ◇
HANATANI, Y. *Demonstration de l'ω-non-contradiction de l'arithmetique* ◇ F05 F15 F30 F50 ◇
KINO, A. *On provably recursive functions and ordinal recursive functions* ◇ D20 F05 F30 F35 ◇

MALMNAES, P.E. & PRAWITZ, D. *A survey of some connections between classical, intuitionistic and minimal logic* ◇ B55 F05 F25 F50 ◇
PRAWITZ, D. *Hauptsatz for higher order logic* ◇ B15 F05 F35 ◇
SCHUETTE, K. *Neuere Ergebnisse der Beweistheorie* ◇ F05 F10 F15 F30 F35 F98 ◇
SMULLYAN, R.M. *Analytic cut* ◇ B10 F05 F07 ◇
TAIT, W.W. *Normal derivability in classical logic* ◇ C75 F05 F15 F30 F35 F60 ◇
TAKAHASHI, MOTO-O *On simple type theory (Japanese)* ◇ B15 F05 F35 ◇
TAKAHASHI, MOTO-O *Simple type theory of Gentzen style with the inference of extensionality* ◇ F05 F35 ◇
TAKEUTI, G. *The Π^1_1-comprehension schema and ω-rules* ◇ F05 F15 F35 ◇

1969

BIBEL, W. *Schnittelimination in einem Teilsystem der einfachen Typenlogik* ◇ B15 F05 F15 F35 ◇
GENTZEN, G. *The collected papers of Gerhard Gentzen* ◇ B96 F05 F07 F30 F50 F96 ◇
HAILPERIN, T. *A form of Herbrand's theorem* ◇ B10 F05 F07 ◇
LEISENRING, A.C. *Mathematical logic and Hilbert's ε-symbol* ◇ B10 E25 F05 F98 ◇
OREVKOV, V.P. *On nonlengthening applications of equality rules (Russian)* ◇ B25 B35 F05 F07 ◇
PLIUSKEVICIENE, A. *Elimination of cut type rules in axiomatic theories with equality (Russian)* ◇ F05 ◇
SCARPELLINI, B. *Some applications of Gentzen's second consistency proof* ◇ F05 F30 F35 F50 ◇
SZABO, M.E. *Introduction* ◇ A10 F05 F30 F98 ◇
UESU, T. *Gentzen's Hauptsatz for simple type theory with the infinite induction* ◇ F05 F35 ◇
ZAHN, P. *Eine Verallgemeinerung des Hauptsatzes von Gentzen* ◇ F05 ◇

1970

DENTON JR., J.S. & DREBEN, B. *Herbrand-style consistency proofs* ◇ F05 F30 F35 ◇
OHYA, T. *On recursive restriction of proofs in a system with constructive infinitely long expressions* ◇ F05 F35 ◇
PRAWITZ, D. *Some results for intuitionistic logic with second order quantification rules* ◇ F05 F35 F50 ◇
SCARPELLINI, B. *On cut elimination in intuitionistic systems of analysis* ◇ F05 F35 F50 ◇
TAIT, W.W. *Applications of the cut elimination theorem to some subsystems of classical analysis* ◇ F05 F35 ◇
TAKAHASHI, MOTO-O *A system of simple type theory of Gentzen style with inference on extensionality, and the cut-elimination in it* ◇ F05 F35 ◇
YASUGI, M. *Cut elimination theorem for second order arithmetic with the Π^1_1 comprehension axiom and the ω-rule* ◇ F05 F35 ◇

1971

ANDREWS, P.B. *Resolution in type theory* ◇ B15 F05 F35 ◇
CARSTENGERDES, W. *Mehrsortige logische Systeme mit unendlich langen Formeln I, II* ◇ C40 C75 F05 ◇
CAVINESS, B.F. & POLLACK, P.L. & RUBALD, C.M. *An existence lemma for canonical forms in symbolic mathematics* ◇ D25 F05 ◇
DREBEN, B. & GOLDFARB, W.D. *Note J* ◇ B10 F05 F07 ◇
DREBEN, B. *Notes E-I* ◇ F05 F07 ◇
GIRARD, J.-Y. *Une extension de l'interpretation de Goedel a l'analyse et son application a l'elimination des coupures dans l'analyse et la theorie des types* ◇ F05 F10 F35 F50 ◇
JERVELL, H.R. *A normalform in first order arithmetic* ◇ F05 F30 F50 ◇
MARTIN-LOEF, P. *Hauptsatz for the intuitionistic theory of iterated inductive definitions* ◇ F05 F35 F50 ◇
MARTIN-LOEF, P. *Hauptsatz for the theory of species* ◇ F05 F35 F50 ◇
MINTS, G.E. *Exact estimation of the provability of transfinite induction in the initial segments of arithmetic (Russian) (English summary)* ◇ F05 F15 F30 ◇
PLIUSKEVICIENE, A. *Elimination of cut-type rules from the Robinson and Presburger axiomatic systems (Russian)* ◇ F05 F30 ◇
PRAWITZ, D. *Ideas and results in proof theory* ◇ F05 F07 F10 F30 F35 F50 F98 ◇
SANCHIS, L.E. *A generalization of the Gentzen Hauptsatz* ◇ B10 F05 ◇
SANCHIS, L.E. *Cut elimination, consistency and completeness in classical logic* ◇ B10 F05 ◇
SATINDER, P.K. *Completeness and cut-elimination in constructive ω-rule arithmetics* ◇ F05 F30 ◇
SCARPELLINI, B. *Proof theory and intuitionistic systems* ◇ F05 F15 F30 F35 F50 F98 ◇
TAIT, W.W. *Normal form theorem for bar recursive functions of finite type* ◇ F05 F10 F35 F50 ◇
TAKAHASHI, MOTO-O *Cut-elimination theorem and Brouwerian-valued models for intuitionistic type theory* ◇ F05 F35 F50 ◇
UESU, T. *Simple type theory with constructive infinitely long expressions* ◇ B15 C75 F05 F35 ◇
UESU, T. *Two formal systems of simple type theory with type variables* ◇ F05 F35 ◇

1972

ARZARELLO, F. *Il teorema di normalizzazione per il sistema di deduzione naturale classico (English summary)* ◇ F05 ◇
BOWEN, K.A. *A note on cut elimination and completeness in first order theories* ◇ B10 C07 F05 ◇
JERVELL, H.R. *Herbrand and Skolem theorems in infinitary languages* ◇ C75 F05 ◇
JERVELL, H.R. *On Skolem and Herbrand theorems for intuitionistic logic* ◇ F05 F50 ◇
KELLY, G.M. *A cut-elimination theorem* ◇ F05 G30 ◇
MARTIN-LOEF, P. *Infinite terms and a system of natural deduction* ◇ C75 F05 F50 ◇
MOTOHASHI, N. *Interpolation theorem and characterization theorem* ◇ C40 C75 F05 ◇
MOTOHASHI, N. *Object logic and morphism logic* ◇ F05 ◇
NEBRES, B.F. *Herbrand uniformity theorems for infinitary languages* ◇ C40 C70 C75 F05 ◇

OREVKOV, V.P. *A specialization of the form of deductions in Gentzen calculi and its applications (Russian)* ◇ F05 F07 F50 ◇

OSSWALD, H. *Vollstaendigkeit und Schnittelimination in der intuitionistischen Typenlogik (English summary)* ◇ F05 F35 F50 ◇

PLIUSKEVICIENE, A. *A sequential variant of R. M. Robinson's arithmetic system not containing cut rules (Russian)* ◇ F05 F30 ◇

SMIRNOV, V.A. *Formal derivation and logical calculi (Russian)* ◇ B98 C07 C40 C98 F05 F07 F98 ◇

STENLUND, S. *Combinators, λ-terms and proof theory* ◇ B40 F05 F10 F50 F98 ◇

1973

BOWEN, K.A. *Cut elimination in transfinite type theory* ◇ B15 F05 F35 ◇

JERVELL, H.R. *Skolem and Herbrand theorems in first order logic* ◇ B10 C07 F05 ◇

MARTIN-LOEF, P. *Hauptsatz for intuitionistic simple type theory* ◇ B15 F05 F35 F50 ◇

OSSWALD, H. *Ein syntaktischer Beweis fuer die Zuverlaessigkeit der Schnittregel im Kalkuel von Schuette fuer die intuitionistische Typenlogik* ◇ F05 F35 F50 ◇

POHLERS, W. *Ein starker Normalisationssatz fuer die intuitionistische Typentheorie* ◇ F05 F35 F50 ◇

PRAWITZ, D. *Towards a foundation of a general proof theory* ◇ A05 F05 F07 ◇

SCANLON, T.M. *The consistency of number theory via Herbrand's theorem* ◇ B10 F05 F30 ◇

SELDIN, J.P. *Equality in $\mathfrak{F}_{21}$* ◇ B40 F05 ◇

SMIRNOV, V.A. *An absolute first order predicate calulus* ◇ B10 F05 ◇

TAKEUTI, G. & YASUGI, M. *The ordinals of the systems of second order arithmetic with the provably Δ^1_2-comprehension axiom and with the Δ^1_2-comprehension axiom respectively* ◇ F05 F15 F35 ◇

TITANI, S. *A proof of the cut-elimination theorem in simple type theory* ◇ B15 F05 F35 ◇

TROELSTRA, A.S. (ED.) *Metamathematical investigation of intuitionistic arithmetic and analysis* ◇ F05 F10 F30 F35 F50 F55 F98 ◇

TROELSTRA, A.S. *Models and computability* ◇ F05 F10 F30 F35 F50 ◇

TROELSTRA, A.S. *Normalization theorems for systems of natural deduction* ◇ F05 F30 F35 F50 ◇

VESSEL', KH.A. *Ueber eine logische explizite Beschreibung der Entwicklungsterme (Russian)* ◇ F05 ◇

1974

ANDREWS, P.B. *Resolution and the consistency of analysis* ◇ B15 F05 F25 F35 ◇

BOWEN, K.A. *Systems of transfinite type theory based on intuitionistic and modal logics* ◇ B15 F05 F35 F50 ◇

CELLUCCI, C. *Una dimostrazione del teorema di uniformita* ◇ F05 ◇

GENTZEN, G. *Der erste Widerspruchsfreiheitsbeweis fuer die klassische Zahlentheorie* ◇ F05 F15 F25 F30 ◇

GOLDFARB, W.D. & SCANLON, T.M. *The ω-consistency of number theory via Herbrand's theorem* ◇ F05 F15 F30 ◇

KREISEL, G. & TAKEUTI, G. *Formally self-referential propositions for cut free classical analysis and related systems* ◇ F05 F35 ◇

MINTS, G.E. *E theorems (Russian) (English summary)* ◇ F05 F30 F50 ◇

MINTS, G.E. *Functional form. Herbrand's theorem for non-prenex formulas (Russian)* ◇ B10 F05 ◇

MINTS, G.E. *Gentzen's formal systems (Russian)* ◇ B10 F05 F07 ◇

MINTS, G.E. *Heyting predicate calculus with ε symbol (Russian) (English summary)* ◇ F05 F50 ◇

MINTS, G.E. *Normalization of proofs (Russian)* ◇ F05 ◇

NISHIMURA, T. *Gentzen-style formulation of systems of set-calculus* ◇ C75 E30 F05 ◇

NISHIMURA, T. *Gentzen-style formal system representing properties of functions* ◇ C75 E30 F05 ◇

PRAWITZ, D. *On the idea of a general proof theory* ◇ A05 F05 F07 ◇

RAGGIO, A.R. *A simple proof of Herbrand's theorem* ◇ B10 F05 F07 ◇

SZABO, M.E. *A categorical equivalence of proofs* ◇ F05 F07 F50 G05 G30 ◇

UMEZAWA, T. *Disjunction property in higher order number theory with intuitionistic rules of inference I* ◇ F05 F35 F50 ◇

YASUHARA, M. *Completeness of cut-free type theories* ◇ F05 ◇

ZUCKER, J.I. *The correspondence between cut-elimination and normalization I,II* ◇ F05 F30 F50 ◇

1975

ANDERSON, A.R. *Fitch on consistency* ◇ B15 F05 F25 ◇

ARZARELLO, F. *Un sistema di deduzione naturale per linguaggi infinitari (English summary)* ◇ C75 F05 ◇

BUCHHOLZ, W. *Ein ausgezeichnetes Modell fuer die intuitionistische Typenlogik* ◇ B15 F05 F35 F50 ◇

CELLUCCI, C. *Teoremi di normalizzazione per alcuni sistemi funzionali* ◇ F05 F10 F15 F30 ◇

CELLUCCI, C. *Teoremi di normalizzazione per alcuni sistemi funzionali II* ◇ F05 F10 F15 F30 ◇

FELSCHER, W. *Kombinatorische Konstruktionen mit Beweisen und Schnittelimination* ◇ F05 F20 ◇

FLANNAGAN, T.B. *On an extension of Hilbert's second ε-theorem* ◇ F05 ◇

KREISEL, G. *Observations on a recent generalization of completeness theorems due to Schuette* ◇ A05 C07 C57 F05 F20 F35 F50 ◇

KREISEL, G. & MINTS, G.E. & SIMPSON, S.G. *The use of abstract language in elementary metamathematics: Some pedagogic examples* ◇ A05 C07 C57 C75 F05 F07 F20 F50 ◇

LEIVANT, D. *Strong normalization for arithmetic (Variations on a theme of Prawitz.)* ◇ F05 F30 F50 ◇

MANN, C.R. *The connection between equivalence of proofs and Cartesian closed categories* ◇ F05 F07 G30 ◇
MARTIN-LOEF, P. *An intuitionistic theory of types: predicative part* ◇ F05 F35 F50 ◇
MEZHLUMBEKOVA, V.F. *Cut-elimination in a system of negationless arithmetic (Russian) (English summary)* ◇ B20 F05 F30 F50 ◇
MINTS, G.E. *Finite investigations of transfinite derivations (Russian) (English summary)* ◇ F05 F07 F10 F20 F30 F35 F50 ◇
MINTS, G.E. *Proof theory (arithmetic and analysis) (Russian)* ◇ F05 F10 F30 F35 F50 F98 ◇
NISHIMURA, T. *Gentzen-type formal system representing properties of functions* ◇ E30 F05 ◇
PRAWITZ, D. *Comments on Gentzen-type procedures and the classical notion of truth* ◇ B10 F05 F07 F30 ◇
ROGAVA, M.G. *Cut elimination in SCI* ◇ B60 F05 ◇
SELDIN, J.P. *Equality in* $\mathfrak{F}_{22}$ ◇ B40 F05 ◇
SHIRAI, K. *Intuitionistic version of the Los-Tarski-Robinson theorem* ◇ C40 F05 F50 ◇
TAIT, W.W. *A realizability interpretation of the theory of species* ◇ D65 F05 F35 F50 ◇
TAKEUTI, G. *Proof theory* ◇ F05 F07 F30 F35 F98 ◇
TOLEDO, S. *Tableau systems for first order number theory and certain higher order theories* ◇ F05 F07 F30 F35 F98 ◇
UMEZAWA, T. *Disjunction property in higher order number theory with intuitionistic rules of inference II* ◇ F05 F35 F50 ◇

1976

ARZARELLO, F. *Alcune questioni riguardanti l'aritmetica e i linguaggi infinitari (English summary)* ◇ C75 F05 ◇
BOWEN, K.A. *An Herbrand theorem for prenex formulas of LJ* ◇ F05 F50 ◇
DRAGALIN, A.G. *Cut-elimination in the theory of definable sets of natural numbers (Russian)* ◇ F05 F35 F65 ◇
GIRARD, J.-Y. *Three-valued logic and cut-elimination: The actual meaning of Takeuti's conjecture* ◇ B50 C85 C90 F05 F35 F50 ◇
KREISEL, G. *Wie die Beweistheorie zu ihren Ordinalzahlen kam und kommt* ◇ F05 F15 F30 F35 ◇
KUZICHEV, A.S. *Combinatorially complete systems with operators* $\Xi, F, Q, \Pi, \exists, P, \neg, \&, \vee, \equiv$ *(Russian) (English summary)* ◇ B40 F05 ◇
LOPEZ-ESCOBAR, E.G.K. *On an extremely restricted* ω*-rule* ◇ F05 F15 F30 F50 ◇
MEYER, R.K. *Ackermann, Takeuti, und Schnitt:* γ *for higher-order relevant logic* ◇ B46 F05 F35 ◇
MINTS, G.E. *What can be done with PRA (Russian) (English summary)* ◇ F05 F30 F35 ◇
NISHIMURA, T. *On cut-elimination in simple type theory (on the work of Moto-o Takahashi, winner of the third Mathematical Society Prize) (Japanese)* ◇ F05 F35 ◇
POTTINGER, G. *A new way of normalizing intuitionist propositional logic* ◇ B40 F05 F50 ◇
SELDIN, J.P. *Recent advances in Curry's program* ◇ B40 F05 ◇
SLONNEGER, K. *A complete infinitary logic* ◇ C75 F05 ◇
SZABO, M.E. *Quantifier complete categories* ◇ F05 F50 G10 G30 ◇

1977

CAGNONI, D. *A note on the elimination rules* ◇ F05 F07 ◇
CELLUCCI, C. *Proprieta di uniformita e 1-coerenza dell'aritmetica del primo ordine* ◇ F05 F30 ◇
DRAGALIN, A.G. *Cut-elimination in the theory of definable sets of natural numbers (Russian)* ◇ F05 F35 F65 ◇
HAYASHI, S. *On derived rules of intuitionistic second order arithmetic* ◇ F05 F35 F50 ◇
HAYASHI, S. *Some derived rules of intuitionistic second order arithmetic* ◇ F05 F35 F50 ◇
HINATA, S. *A normalization theorem in formal theories of natural numbers* ◇ F05 F30 F50 ◇
JUKNA, S. *On cut-elimination in Hoare's system (Russian)* ◇ B75 F05 F50 ◇
KREISEL, G. *Some uses of proof theory for finding computer programs* ◇ B75 F05 F20 ◇
KUZICHEV, A.S. *A system of* λ*-conversion with logical operators and an equality operator (Russian)* ◇ B10 B40 F05 ◇
KUZICHEV, A.S. *Cut theorem for R-theories in combinatorially complete systems (Russian) (English summary)* ◇ B40 F05 ◇
KUZICHEV, A.S. *Formal arithmetic in the* $\mathfrak{A}$*-system of* λ*-conversion (Russian)* ◇ B40 F05 F30 ◇
MASLOV, S.YU. & NORGELA, S.A. *Herbrand strategies and the "greater deducibility" relation (Russian) (English summary)* ◇ B25 B35 F05 ◇
MINTS, G.E. *Closed categories and the theory of proofs (Russian)* ◇ F05 F07 G30 ◇
MINTS, G.E. *Gentzen's formal system (Russian)* ◇ F05 F07 ◇
POTTINGER, G. *Normalization as a homomorphic image of cut-elimination* ◇ F05 F07 F50 ◇
SCHWICHTENBERG, H. *Proof theory: Some applications of cut-elimination* ◇ F05 F10 F15 F30 F35 F98 ◇
SELDIN, J.P. *The Q-consistency of* F_{22} ◇ B40 F05 ◇
STATMAN, R. *Herbrand's theorem and Gentzen's notion of a direct proof* ◇ D15 F05 F07 F20 ◇
WESSELS, L. *Cut elimination in a Gentzen-style* ε*-calculus without identity* ◇ F05 ◇

1978

CELLUCCI, C. *Teoria della dimostrazione. Normalizzazioni e assegnazioni di numeri ordinali* ◇ F05 F07 F15 F30 F98 ◇
HABERTHUER, R. *Choice sequences and reduction processes* ◇ F05 F35 F50 ◇
HAYASHI, S. *Existence property by means of a normalization method* ◇ F05 F50 ◇
JERVELL, H.R. *Constructive universes. I* ◇ F05 F15 F35 F50 ◇
KUZICHEV, A.S. *A theorem on the consistency of formal arithmetic (Russian)* ◇ B40 F05 F30 ◇
KUZICHEV, A.S. *Formal arithmetic in combinatorially complete systems I,II (Russian) (English summary)* ◇ B40 F05 F30 ◇

KUZICHEV, A.S. *The theorem on the midsequent in the 𝔄-system of λ-conversion (Russian)* ◇ B40 F05 ◇
POHLERS, W. *Ordinals connected with formal theories for transfinitely iterated inductive definitions* ◇ F05 F15 F35 ◇
RICHTER, M.M. *Logikkalkuele* ◇ B35 B98 F05 G10 ◇
SELDIN, J.P. *A sequent calculus formulation of type assignment with equality rules for the λβ-calculus* ◇ B40 F05 ◇
STATMAN, R. *Bounds for proof-search and speed-up in the predicate calculus* ◇ B35 F05 F20 ◇
SZABO, M.E. *Algebra of proofs* ◇ F05 F07 F50 F98 G30 ◇
TAKEUTI, G. *Two applications of logic to mathematics* ◇ B98 C90 E40 E75 F05 F30 F35 G12 ◇
ZASLAVSKIJ, I.D. *Symmetric constructive logic (Russian)* ◇ B55 F05 F30 F50 ◇

1979

BABAEV, A.A. & SOLOV'EV, S.V. *A coherence theorem for canonical maps in Cartesian closed categories (Russian) (English summary)* ◇ F05 G30 ◇
BIBEL, W. *Tautology testing with a generalized matrix reduction method* ◇ B35 D15 F05 ◇
CAPORASO, S. *Consistency proof without transfinite induction for a formal system for Turing machines* ◇ D10 F05 ◇
DALEN VAN, D. & STATMAN, R. *Equality in the presence of apartness* ◇ F05 F50 ◇
DRAGALIN, A.G. *Mathematical intuitionism. Introduction to proof theory (Russian)* ◇ F05 F30 F35 F50 F98 ◇
DRAGALIN, A.G. *Strong normalization theorem for derivations in Gentzen's sequent calculus (Russian)* ◇ F05 F50 ◇
GRISHIN, V.N. *Herbrand's theorem for the logic without contraction rule (Russian)* ◇ F05 ◇
HANAZAWA, M. *An interpretation of Skolem's paradox in the predicate calculus with ε-symbol* ◇ B10 C07 F05 F25 ◇
KREISEL, G. *Some facts from the theory of proofs and some fictions from general proof theory* ◇ A05 F05 F15 ◇
KUR'EROV, YU.N. *Normal form of mutual-absorption tactics (Russian)* ◇ B35 F05 ◇
KUZICHEV, A.S. *Formale Arithmetik in einem System der λ-Konversion mit logischen Operatoren (Russisch)* ◇ B40 F05 F30 ◇
LEIVANT, D. *Absoluteness of intuitionistic logic* ◇ F05 F30 F50 ◇
LEIVANT, D. *Assumption classes in natural deduction* ◇ F05 F07 ◇
MINTS, G.E. *A new reduction sequence for arithmetic (Russian) (English summary)* ◇ F05 F15 F30 ◇
MINTS, G.E. *A primitive recursive bound of strong normalization for predicate calculus (Russian) (English summary)* ◇ F05 ◇
MINTS, G.E. *Normalisation of natural deduction and effectiveness of classical existence (Russian)* ◇ F05 F10 F30 F35 ◇
O'DONNELL, M.J. *A practical programming theorem which is independent of Peano arithmetic* ◇ B40 B75 D20 F05 F30 ◇
OREVKOV, V.P. *Lower bounds for lengthening of proofs after cut-elimination (Russian) (English summary)* ◇ F05 F07 F20 ◇
ROGAVA, M.G. *A new decision procedure for SCI (Russian)* ◇ B25 B60 F05 ◇
SCHWICHTENBERG, H. *Logic and the axiom of choice* ◇ E25 F05 F50 ◇
SEREBRYANNIKOV, O.F. *Normal forms of logical proofs (Russian)* ◇ F05 ◇
SMULLYAN, R.M. *Trees and ball games* ◇ E05 F05 ◇
STATMAN, R. *Lower bounds on Herbrand's theorem* ◇ F05 F07 F20 ◇
UMEZAWA, T. *A method for cut elimination in intuitionistic predicate logic and classical predicate logic* ◇ B10 F05 F50 ◇
VENNERI, B.M. *Arbres de deduction naturelle et programmation en LCF: possibilites de "normalisation"* ◇ F05 ◇
WHITE, R.B. *The consistency of the axiom of comprehension in the infinite-valued predicate logic of Lukasiewicz* ◇ B50 E35 E70 F05 ◇

1980

BUNDER, M.W. *The consistency of a higher order predicate calculus and set theory based on combinatory logic* ◇ B15 B40 F05 ◇
DILLER, J. *Modified realization and the formulae-as-types notion* ◇ F05 F10 F30 F35 F50 ◇
DRAGALIN, A.G. *Higher-order predicate logic in the form of calculus realization (Russian)* ◇ B15 F05 F35 F50 ◇
GANDY, R.O. *An early proof of normalization by A. M. Turing* ◇ B40 F05 ◇
GANDY, R.O. *Proofs of strong normalization* ◇ B40 F05 F50 ◇
HOWARD, W.A. *The formulae-as-types notion of construction* ◇ B40 F05 F10 F50 ◇
JAEGER, G. *Beweistheorie von KPN* ◇ E30 F05 F15 ◇
KODERA, H. *Remark on classical logic and intuitionistic logic* ◇ B20 F05 F50 ◇
KUZICHEV, A.A. & KUZICHEV, A.S. *On the embedding of formal arithmetic in combinatorially complete systems (Russian)* ◇ B40 F05 F30 ◇
KUZICHEV, A.S. *Permissibility of a cut in combinatorially complete systems with equality (Russian) (English summary)* ◇ B40 F05 ◇
MINTS, G.E. *Cartesian closed categories and proof theory (Russian)* ◇ F05 F07 G30 ◇
MINTS, G.E. *Category theory and proof theory (Russian)* ◇ F05 F07 G30 ◇
SAMBIN, G. & VALENTINI, S. *A modal sequent calculus for a fragment of arithmetic* ◇ B45 F05 F30 ◇
TENNANT, N. *A proof-theoretic approach to entailment* ◇ B46 F05 ◇
TOSI, P. *Normal derivability and first-order arithmetic* ◇ F05 F30 F50 ◇
UMEZAWA, T. *Cut elimination in intuitionistic and some intermediate predicate logics* ◇ B55 F05 F50 ◇

Subject Index

YASUGI, M. *A progression of consistency proofs* ◇ F05 F15 F35 ◇

YASUGI, M. *Gentzen reduction revisited* ◇ F05 F15 F30 F35 ◇

1981

BUCHHOLZ, W. & FEFERMAN, S. & POHLERS, W. & SIEG, W. *Iterated inductive definitions and subsystems of analysis: recent proof-theoretical studies* ◇ F05 F15 F35 F50 F98 ◇

BUCHHOLZ, W. *The $\Omega_{\mu+1}$-rule* ◇ F05 F15 F35 F50 ◇

GIRARD, J.-Y. & PAEPPINGHAUS, P. *A result on implications of Σ_1-sentences and its application to normal form theorems* ◇ F05 F30 F35 ◇

KOGAN-BERNSHTEIJN, L.M. *Simplification of Gentzen's reductions in classical arithmetic (Russian)* ◇ F05 F30 ◇

LEIVANT, D. *On the proof theory of the modal logic for arithmetic provability* ◇ B45 F05 F30 ◇

LYALETSKIJ, A.V. *A variant of Herbrand's theorem for formulas in prenex form (Russian)* ◇ B35 F05 ◇

NEGRI, M. *Constructive sequent reduction in Gentzen's first consistency proof for arithmetic* ◇ A10 F05 F30 ◇

POHLERS, W. *Cut-elimination for impredicative infinitary systems I. Ordinal-analysis for ID_1* ◇ C75 F05 F15 F35 ◇

POHLERS, W. *Proof-theoretical analysis of ID_ν by the method of local predicativity* ◇ F05 F15 F35 F50 ◇

PRAWITZ, D. *Validity and normalizability of proofs in 1st and 2nd order classical and intuitionistic logic* ◇ F05 F07 F35 F50 ◇

SCHULTE-MOENTING, J. *Cut elimination and word problems for varieties of lattices* ◇ F05 G10 ◇

SIEG, W. *Inductive definitions, constructive ordinals, and normal derivations* ◇ F05 F15 F35 F50 ◇

SUNDHOLM, G. *Hacking's logic* ◇ F05 F55 ◇

TANG, TONGGAO *A note on Herbrand's theorem (Chinese) (English summary)* ◇ B10 F05 ◇

TOSI, P. *Forme normali* ◇ F05 F30 ◇

TOSI, P. *Intuitive proofs and first-order derivations: Some notes on the metamathematics of first-order number theory* ◇ F05 F30 ◇

1982

GIRARD, J.-Y. *Herbrand's theorem and proof-theory* ◇ F05 F07 F10 F35 ◇

GIRARD, J.-Y. *Proof-theoretic investigations of inductive definitions I* ◇ F05 F35 F50 ◇

GLUBRECHT, J.-M. *Ein Vollstaendigkeitsbeweis fuer schnittfreie Kalkuele mit der Maximalisierungsmethode von Henkin* ◇ B10 C07 F05 ◇

KREISEL, G. & MACINTYRE, A. *Constructive logic versus algebraization I* ◇ A05 F05 F50 ◇

KREISEL, G. *Finiteness theorems in arithmetic: an application of Herbrand's theorem for Σ_2-formulas* ◇ F05 F30 ◇

KUZICHEV, A.S. *Arithmetically consistent λ-theories (Russian)* ◇ B40 F05 F30 ◇

LOPEZ-ESCOBAR, E.G.K. *Further applications of ultra-conservative ω-rules* ◇ F05 F30 ◇

MAC LANE, S. *Why commutative diagrams coincide with equivalent proofs* ◇ F05 G30 ◇

MINTS, G.E. *A simplified consistency proof for arithmetic (Russian) (English and Estonian summaries)* ◇ F05 F30 ◇

MOTOHASHI, N. *ε-theorems and elimination theorems of uniqueness conditions* ◇ B10 C40 F05 ◇

POHLERS, W. *Cut elimination for impredicative infinitary systems II. Ordinal analysis for iterated inductive definitions* ◇ C75 F05 F15 F35 ◇

SAMBIN, G. & VALENTINI, S. *The modal logic of provability. The sequential approach* ◇ B45 F05 F30 ◇

YASUHARA, M. *Cut elimination in ε-calculi* ◇ F05 ◇

1983

BORGA, M. *On some proof theoretical properties of the modal logic GL* ◇ B45 F05 F30 ◇

HAGIYA, M. *A proof description language and its reduction system* ◇ B40 F05 ◇

KUZICHEV, A.S. *Arithmetically consistent λ-theories of type-free logic (Russian)* ◇ B40 F05 F30 ◇

LOPEZ-ESCOBAR, E.G.K. *A second paper on the interpolation theorem for the logic of constant domains* ◇ B55 C40 C90 F05 F50 ◇

PAEPPINGHAUS, P. *Completeness properties of classical theories of finite type and the normal form theorem* ◇ B15 F05 F35 ◇

SZABO, M.E. *A sequent calculus for Kroeger logic* ◇ B75 F05 F07 ◇

VALENTINI, S. *The modal logic of provability: cut-elimination* ◇ B45 F05 F30 ◇

1984

AVRON, A. *On modal systems having arithmetical interpretations* ◇ B45 F05 F30 ◇

BOOLOS, G. *Don't eliminate cut* ◇ F05 F07 ◇

FERBUS, M.-C. *Functorial bounds for cut elimination in $L_{\beta\omega}$ I* ◇ C75 F05 F15 G30 ◇

GIRARD, J.-Y. & VAUZEILLES, J. *Les premiers recursivement inaccessible et Mahlo et la theorie des dilatateurs* ◇ D60 E47 E55 F05 F15 ◇

GOODMAN, NICOLAS D. *Epistemic arithmetic is a conservative extension of intuitionistic arithmetic* ◇ B45 F05 F30 F50 ◇

OREVKOV, V.P. *Upper bounds for lengthening of proofs after cut elimination (Russian) (English summary)* ◇ F05 F07 F20 ◇

PEARCE, J. *A constructive consistency proof of a fragment of set theory* ◇ E30 E35 F05 F15 F35 ◇

PETKOV, P.P. *The uniqueness of syntactical analysis for some calculi that are similar to Post's calculus. A generalization of the cut elimination theorem for classical propositional calculus (Russian)* ◇ B50 F05 ◇

POPOV, S.V. *Diagrams of deductions in sequential calculi (Russian)* ◇ F05 F07 ◇

UESU, T. *An axiomatization of the apartness fragment of the theory DLO^+ of dense linear order* ◇ F05 F50 ◇

WOJTYLAK, P. *A proof of Herbrand's theorem* ◇ C10 F05 ◇

YASHIN, A.D. *Completeness of the intuitionistic predicate calculus with the concept of "bar" (Russian)* ◇ C90 F05 F50 ◇

1985

ARAI, T. *A subsystem of classical analysis proper to Takeuti's reduction method for* Π*-analysis* ◇ F05 F35 ◇

CELLUCCI, C. *Proof theory and complexity* ◇ D15 F05 F20 F98 ◇

FELSCHER, W. *Dialogues, strategies, and intuitionistic provability* ◇ F05 F07 F50 ◇

FERBUS, M.-C. *Functorial bounds for cut elimination in* $L_{\beta\omega}$ *II* ◇ F05 F35 ◇

HAIMAN, M. *Linear lattice proof theory: an overview* ◇ F05 F07 ◇

HAIMAN, M. *Proof theory for linear lattices* ◇ F05 F07 ◇

PRAWITZ, D. *Normalizations of proofs in set theory* ◇ E30 F05 ◇

SCHWICHTENBERG, H. *A normal form for natural deductions in a type theory with realizing terms* ◇ F05 F35 F50 ◇

TAKANO, M. *Completeness of a cut-free calculus with equality and function constants* ◇ F05 ◇

F07 Structure of proofs

1925

LONDON, F. *Ueber die Irreversibilitaet deduktiver Schlussweisen* ◇ A05 B10 F07 ◇

1928

HERBRAND, J. *Sur la theorie de la demonstration* ◇ A05 B10 F07 ◇

1930

HERBRAND, J. *Recherches sur la theorie de la demonstration* ◇ B10 B25 F05 F07 F25 F30 ◇

1935

CARNAP, R. *Ein Gueltigkeitskriterium fuer die Saetze der klassischen Mathematik* ◇ B30 F07 ◇

GENTZEN, G. *Untersuchungen ueber das logische Schliessen I,II* ◇ B10 B25 F05 F07 F30 F50 ◇

1936

GENTZEN, G. *Die Widerspruchsfreiheit der reinen Zahlentheorie* ◇ B28 F05 F07 F30 ◇

1948

KURODA, S. *On the logic of Aristotle and the logic of Brouwer (Japanese)* ◇ A05 A10 F07 F50 ◇

1950

ACKERMANN, W. *Widerspruchsfreier Aufbau der Logik I. Typenfreies System ohne tertium non datur* ◇ B15 E70 F07 F25 F65 ◇

SCHUETTE, K. *Schlussweisen-Kalkuele der Praedikatenlogik* ◇ B10 B25 F05 F07 F50 ◇

1952

KLEENE, S.C. *Permutability of inferences in Gentzen's calculi LK and LJ* ◇ F05 F07 ◇

1955

BETH, E.W. *Remarks on natural deduction* ◇ A05 B10 F07 ◇

SCHROETER, K. *Theorie des logischen Schliessens I* ◇ B10 F07 F50 ◇

1956

ISEKI, K. *On the cut operation in Gentzen calculi I* ◇ F05 F07 ◇

SCHUETTE, K. *Ein System des verknuepfenden Schliessens* ◇ B10 F07 F30 ◇

1957

ISEKI, K. *On the cut operation in Gentzen calculi II* ◇ F05 F07 ◇

KALISH, D. & MONTAGUE, R. *Remarks on descriptions and natural deduction I,II* ◇ B10 F07 ◇

KANGER, S. *Provability in logic* ◇ B10 F05 F07 F98 ◇

KREISEL, G. *Sums of squares* ◇ C57 C60 D20 F07 F99 ◇

TAKEUTI, G. *Ordinal diagrams* ◇ F07 F15 ◇

1958

GUILLAUME, M. *Les tableaux semantiques du calcul des predicats restreint* ◇ B20 F07 F98 ◇

KURODA, S. *An investigation on the logical structure of mathematics. I. A logical system* ◇ B15 B28 E70 F07 ◇

KURODA, S. *An investigation on the logical structure of mathematics. III. Fundamental deductions. IV. Compendium for deductions* ◇ B15 B28 E70 F07 ◇

LAMBEK, J. *The mathematics of sentence structures* ◇ B30 B65 F07 ◇

SCHROETER, K. *Theorie des logischen Schliessens II* ◇ B10 C07 F05 F07 ◇

SCHUETTE, K. *Aussagenlogische Grundeigenschaften formaler Systeme* ◇ B05 F05 F07 ◇

1960

FEFERMAN, S. *Arithmetization of metamathematics in a general setting* ◇ D55 F07 F25 F30 F40 ◇

KREISEL, G. *Ordinal logics and the characterization of informal concepts of proof* ◇ F07 F15 F30 F65 ◇

1961

ENGELER, E. *Zur Beweistheorie von Sprachen mit unendlich langen Formeln* ◇ C75 F07 ◇

LAMBEK, J. *On the calculus of syntactic types* ◇ F07 ◇

OGLESBY, F.C. *Report: an examination of a decision procedure* ◇ B25 F07 ◇

PRICE, ROBERT *The stroke function in natural deduction* ◇ B05 F07 ◇

1962

ANDERSON, J.M. & JOHNSTONE JR., H.W. *Natural deduction. The logical basis of axiom systems* ◇ B98 F07 F98 ◇

BETH, E.W. *Umformung einer abgeschlossenen deduktiven oder semantischen Tafel in eine natuerliche Ableitung auf Grund der derivativen bzw. klassischen Implikationslogik* ◇ B10 F07 ◇

DOPP, J. *Logiques construites par une methode de deduction naturelle* ◇ B98 F05 F07 F50 F98 ◇

LEBLANC, H. *Etudes sur les regles d'inference dites regles de Gentzen I* ◇ F07 ◇

LEBLANC, H. *Structural rules of inference* ◇ F07 ◇

ONO, K. *On a practical way of describing formal deductions* ◇ F07 ◇

1963

AANDERAA, S.O. & ANDREWS, P.B. & DREBEN, B. *False lemmas in Herbrand* ◇ F05 F07 ◇

BELNAP JR., N.D. & THOMASON, R.H. *A rule-completeness theorem* ◇ B22 F07 F50 ◇

ENGELER, E. *A reduction-principle for infinite formulas* ◇ C75 F07 ◇

KANGER, S. *A simplified proof method for elementary logic* ◇ B10 F07 ◇

LEBLANC, H. *Etudes sur les regles d'inference dites regles de Gentzen II* ◇ F07 ◇

MATULIS, V.A. *Variants of the classical predicate calculus with a single deduction tree (Russian)* ◇ B10 F07 ◇

PRAWITZ, D. *Concerning constructive logic and the concept of implication (Swedish)* ◇ A05 F07 F50 ◇

SMULLYAN, R.M. *A unifying principle in quantification theory* ◇ B10 F05 F07 ◇

SMULLYAN, R.M. *First order logic* ◇ B98 F07 F98 ◇

1964

AANDERAA, S.O. & DREBEN, B. *Herbrand analyzing function* ◇ F05 F07 F20 ◇

RAGGIO, A.R. *Direct consistency proof of Gentzen's system of natural deduction* ◇ B10 F07 ◇

1965

BERNAYS, P. *Betrachtungen zum Sequenzen-Kalkuel* ◇ B20 F05 F07 ◇

LEBLANC, H. *Marginalia on Gentzen's Sequenzen-Kalkule* ◇ B10 F07 F50 ◇

PARRY, W.T. *Comments on a variant form of natural deduction* ◇ B10 F07 ◇

PRAWITZ, D. *Natural deduction. A proof-theoretical study* ◇ B15 B98 F05 F07 F50 F98 ◇

SMULLYAN, R.M. *A unifying principle in quantification theory* ◇ B10 F05 F07 ◇

SMULLYAN, R.M. *Analytic natural deduction* ◇ B10 F05 F07 ◇

1966

BARTH, E.M. *On the transformation of closed semantic tableaus into natural and axiomatic deductions* ◇ F05 F07 ◇

DENTON JR., J.S. & DREBEN, B. *A supplement to Herbrand* ◇ F05 F07 ◇

FITCH, F.B. *Natural deduction rules for obligation* ◇ B45 F07 ◇

LEBLANC, H. *Two separation theorems for natural deduction* ◇ B10 F07 F50 ◇

1967

GUILLAUME, M. *Quelques remarques sur les "tableaux de Beth"* ◇ F07 ◇

MINTS, G.E. *Herbrand's theorem (Russian)* ◇ B10 F05 F07 ◇

MINTS, G.E. *Variation in the deduction search tactics in sequential calculi (Russian)* ◇ B35 F07 F50 ◇

NIELAND, J.J.F. *Beth's tableau-method* ◇ B05 B25 B50 F07 ◇

REICHBACH, J. *On generalizations of the satisfiability definition and Gentzen-Jaskowski's sequent proof rules* ◇ B10 F07 ◇

REICHBACH, J. *Some methods of formal proofs. II: Generalization of the satisfiability definition* ◇ F07 ◇

1968

BING, K. *Natural deduction with few restrictions on variables* ◇ B10 F07 ◇

GUPTA, H.N. *On the rule of existential specification in systems of natural deduction* ◇ A05 B10 F07 ◇

LAMBEK, J. *Deductive systems and categories I: Syntactic calculi and residuated categories* ◇ F07 G30 ◇

LEBLANC, H. *Subformula theorems for N-sequents* ◇ F07 F50 ◇

LIFSCHITZ, V. *A specialization of the form of deduction in the predicate calculus with equality and functional symbols I (Russian)* ◇ F07 ◇

MINTS, G.E. *Admissible and deductive rules (Russian)* ◇ B10 F07 F50 ◇

MINTS, G.E. *Implicative complexity of axiomatic systems (Russian)* ◇ F07 F50 ◇

MINTS, G.E. *Independence of the postulates of natural calculi (Russian)* ◇ F07 F50 ◇

MINTS, G.E. *The construction of conservative logical inferences (Russian)* ◇ F07 F50 ◇

SMULLYAN, R.M. *Analytic cut* ◇ B10 F05 F07 ◇

1969

CORCORAN, J. & WEAVER, G.E. *Logical consequences in modal logic: natural deduction in S5* ◇ B45 F07 ◇

GENTZEN, G. *The collected papers of Gerhard Gentzen* ◇ B96 F05 F07 F30 F50 F96 ◇

HAILPERIN, T. *A form of Herbrand's theorem* ◇ B10 F05 F07 ◇

LAMBEK, J. *Deductive systems and categories II. Standard constructions and closed categories* ◇ F07 G30 ◇

ONO, K. *On a method of describing formal deductions convenient for theoretical purposes* ◇ F07 ◇

OREVKOV, V.P. *On nonlengthening applications of equality rules (Russian)* ◇ B25 B35 F05 F07 ◇

1970

GREGG, J.R. *Axiomatic quasi-natural deduction* ◇ B10 F07 ◇

KLEINBERG, E.M. *Recursion theory and formal deducibility* ◇ D25 F07 F40 ◇

SMULLYAN, R.M. *Abstract quantification theory* ◇ B10 C07 F07 ◇

1971

BENEYTO, R. *Analytic labyrinths (Spanish)* ◇ B10 F07 ◇

CORCORAN, J. *Discourse grammars and the structure of mathematical reasoning III: Two theories of proof* ◇ A05 A10 F07 ◇

DREBEN, B. & GOLDFARB, W.D. *Note J* ◇ B10 F05 F07 ◇

DREBEN, B. *Notes E-I* ◇ F05 F07 ◇

PARKS, R.Z. & RESCHER, N. *Restricted inference* ◇ B10 F07 ◇

PRAWITZ, D. *Ideas and results in proof theory* ◇ F05 F07 F10 F30 F35 F50 F98 ◇

ROBITASHVILI, N.G. *A combination of the inverse method and the method of resolution (Russian) (Georgian and English summaries)* ◇ B35 F07 ◇

1972

AFRICK, H. *A proof theoretic proof of Scott's general interpolation theorem* ◇ C40 F07 ◇

CHUBARYAN, A.A. & TSEJTIN, G.S. *Certain estimates of the length of logical deductions in classical propositional calculus (Russian) (Armenian summary)* ◇ B05 F07 F20 ◇

LEBLANC, H. & SNYDER, D.P. *Duals of Smullyan trees* ◇ B10 F07 ◇

LEBLANC, H. & MEYER, R.K. *Matters of separation* ◇ B10 F07 F50 ◇

MASLOV, S.YU. *Deduction search in calculi of general type (Russian) (English summary)* ◇ B35 D03 F07 ◇

OREVKOV, V.P. *A specialization of the form of deductions in Gentzen calculi and its applications (Russian)* ◇ F05 F07 F50 ◇

ROGAVA, M.G. *Sequential variants of applied predicate calculi without structural deductive rules (Russian)* ◇ B10 F07 ◇

SLAGLE, J.R. *An approach for finding C-linear complete inference systems* ◇ F07 ◇

SMIRNOV, V.A. *Formal derivation and logical calculi (Russian)* ◇ B98 C07 C40 C98 F05 F07 F98 ◇

1973

APT, K.R. *Infinitistic rules of proof and their semantics (Russian summary)* ◇ C62 F07 F35 ◇

BENNETT, D.W. *An elementary completeness proof for a system of natural deduction* ◇ B10 C07 F07 ◇

HINTIKKA, K.J.J. & NIINILUOTO, I. *On the surface semantics of quantificational proof procedures* ◇ B35 C07 F07 ◇

JENSEN, D.C. *On local proof restrictions for strong theories* ◇ F07 ◇

PRAWITZ, D. *Towards a foundation of a general proof theory* ◇ A05 F05 F07 ◇

TALYSHLY, S.M. *The problem of inverse logical operations (Russian)* ◇ F07 ◇

1974

BENTHEM VAN, J.F.A.K. *Semantic tableaus* ◇ B10 C07 F07 ◇

GLEASON, G.G. *Normal and skew systems* ◇ B05 F07 ◇

GRISHIN, V.N. *A nonstandard logic, and its application to set theory (Russian)* ◇ B60 E70 F07 ◇

MASLOV, S.YU. & NORGELA, S.A. *Cut-type rules for calculi of general type (Russian) (English summary)* ◇ F07 ◇

MAYOH, B.H. *Extracting information from logical proofs* ◇ B35 F07 ◇

MINTS, G.E. *Gentzen's formal systems (Russian)* ◇ B10 F05 F07 ◇

POPOV, S.V. *An equivalence relation and a complete system of schemes of equivalent transformations of deductions in propositional calculus (Russian)* ◇ B05 F07 ◇

PRAWITZ, D. *On the idea of a general proof theory* ◇ A05 F05 F07 ◇

RAGGIO, A.R. *A simple proof of Herbrand's theorem* ◇ B10 F05 F07 ◇

SZABO, M.E. *A categorical equivalence of proofs* ◇ F05 F07 F50 G05 G30 ◇

URQUHART, A.I.F. *Proofs, snakes and ladders* ◇ B22 F07 ◇

1975

BIBEL, W. & SCHREIBER, J. *Proof search in a Gentzen-like system of first-order logic* ◇ B35 F07 ◇

HENDRY, H.E. *Another system of natural deduction* ◇ B10 F07 ◇

KREISEL, G. & MINTS, G.E. & SIMPSON, S.G. *The use of abstract language in elementary metamathematics: Some pedagogic examples* ◇ A05 C07 C57 C75 F05 F07 F20 F50 ◇

LYALETSKIJ, A.V. & MALASHONOK, A.I. *A calculus of k-disjuncts with the rule of a latent clash-resolution (Russian)* ◇ B35 F07 ◇

LYALETSKIJ, A.V. *A k-disjunct calculus (Russian)* ◇ B35 F07 ◇

MANN, C.R. *The connection between equivalence of proofs and Cartesian closed categories* ◇ F05 F07 G30 ◇

MINTS, G.E. *Finite investigations of transfinite derivations (Russian) (English summary)* ◇ F05 F07 F10 F20 F30 F35 F50 ◇

PRAWITZ, D. *Comments on Gentzen-type procedures and the classical notion of truth* ◇ B10 F05 F07 F30 ◇

SCARPELLINI, B. *Bemerkungen zu Regel und Schema* ◇ F07 F30 F50 ◇

SZABO, M.E. *Polycategories* ◇ F07 G30 ◇

TAKEUTI, G. *Proof theory* ◇ F05 F07 F30 F35 F98 ◇

TOLEDO, S. *Tableau systems for first order number theory and certain higher order theories* ◇ F05 F07 F30 F35 F98 ◇

XENAKIS, J. *Natural deduction "puzzle"* ◇ A05 F07 ◇

1976

APT, K.R. *Semantics of the infinitistic rules of proof* ◇ C62 D55 E45 F07 F35 ◇

CORCORAN, J. *The nature of a correct theory of proof and its value* ◇ A05 F07 ◇

CORCORAN, J. *Two theories of proof* ◇ A05 F07 ◇

CRABBE, M. *Prelogic of logoi* ◇ B40 F07 G30 ◇

DARDZHANIYA, G.K. *On a variant of classical sequent calculus (Russian) (English summary)* ◇ F07 ◇

GILBERT, M.A. *A heuristic procedure for natural deduction derivations using reductio ad absurdum* ◇ B35 F07 ◇

WASILEWSKA, A. *A sequence formalization for SCI* ◇ B25 F07 ◇

WASILEWSKA, A. *On the decidability theorems* ◇ B25 F07 ◇

1977

BENNETT, D.W. *A note on the completeness proof for natural deduction* ◇ B10 C07 F07 ◇

CAGNONI, D. *A note on the elimination rules* ◇ F05 F07 ◇

CZERMAK, J. *A remark on Gentzen's calculus of sequents* ◇ B20 F07 ◇

KREISEL, G. *From foundations to science: Justifying and unwinding proofs* ◇ A05 B35 F07 ◇

KREISEL, G. *On the kind of data needed for a theory of proofs* ◇ A05 A10 F07 ◇

LAMBERT JR., W.M. *Un tipo de solidez para un sistema de deduccion natural* ◇ B10 F07 ◇
MINTS, G.E. *Closed categories and the theory of proofs (Russian)* ◇ F05 F07 G30 ◇
MINTS, G.E. *Gentzen's formal system (Russian)* ◇ F05 F07 ◇
MISAWA, T. & YASUDA, Y. *Suslin logics of Gentzen style* ◇ C75 F07 ◇
NEDERPELT, R.P. *Presentation of natural deduction* ◇ B10 F07 ◇
POTTINGER, G. *Normalization as a homomorphic image of cut-elimination* ◇ F05 F07 F50 ◇
RAGGIO, A.R. *Semi-formal Beth tableaux* ◇ B10 F07 ◇
STATMAN, R. *Herbrand's theorem and Gentzen's notion of a direct proof* ◇ D15 F05 F07 F20 ◇

1978

CELLUCCI, C. *Teoria della dimostrazione. Normalizzazioni e assegnazioni di numeri ordinali* ◇ F05 F07 F15 F30 F98 ◇
GABBAY, D.M. *What is a classical connective?* ◇ B22 F07 ◇
NEPEJVODA, N.N. *A relation between the natural deduction rules and operators of higher level algorithmic languages (Russian)* ◇ B75 F07 F50 ◇
NORGELA, S.A. *Herbrand strategies of deduction-search in predicate calculus I (Russian) (English and Lithuanian summaries)* ◇ B25 B35 D35 F07 ◇
SHOESMITH, D.J. & SMILEY, T.J. *Multiple-conclusion logic* ◇ B22 F07 F98 ◇
SZABO, M.E. *Algebra of proofs* ◇ F05 F07 F50 F98 G30 ◇
THIEL, C. *Duality lost? Transforming Gentzen derivations into winning strategies for dialogue games* ◇ B10 F07 ◇

1979

FUNAHASHI, S. & NAGATA, S. & OSHIBA, T. *A method for obtaining proof figures of valid formulas in the first order predicate calculus* ◇ B10 F07 ◇
LEIVANT, D. *Assumption classes in natural deduction* ◇ F05 F07 ◇
MASLOV, S.YU. *Deductive systems with thinnings (Russian)* ◇ F07 ◇
MELONI, G.C. & RADAELLI, G. *Calcolo naturale per categorie con residui* ◇ B60 F07 G30 ◇
OREVKOV, V.P. *Lower bounds for lengthening of proofs after cut-elimination (Russian) (English summary)* ◇ F05 F07 F20 ◇
PRAWITZ, D. *Proofs and the meaning and completeness of the logical constants* ◇ A05 F07 F50 ◇
SEELY, R.A.G. *Weak adjointness in proof theory* ◇ F07 G30 ◇
SOLOV'EV, S.V. *Preservation of the equivalence of proofs under reduction of the formula depth (Russian) (English summary)* ◇ F07 F20 ◇
SOLOV'EV, S.V. *The increase in length of an L-derivation transformed into a natural deduction (Russian) (English summary)* ◇ F07 F20 F50 ◇
STATMAN, R. *Lower bounds on Herbrand's theorem* ◇ F05 F07 F20 ◇
SUVOROV, P.YU. *Representation of proofs by colored graphs and the Hadwiger conjecture (Russian)* ◇ B10 F07 ◇

1980

GOAD, C.A. *Proofs as descriptions of computation* ◇ B35 F07 F50 ◇
KROEGER, F. *Infinite proof rules for loops* ◇ B75 F07 ◇
LAMBEK, J. *From lambda-calculus to Cartesian closed categories* ◇ B40 F07 G30 ◇
LIFSCHITZ, V. *Semantical completeness theorems in logic and algebra* ◇ B10 F07 ◇
MINTS, G.E. *Cartesian closed categories and proof theory (Russian)* ◇ F05 F07 G30 ◇
MINTS, G.E. *Category theory and proof theory (Russian)* ◇ F05 F07 G30 ◇
POTTINGER, G. *A type assignment for the strongly normalizable lambda-terms* ◇ B40 F07 ◇
WASILEWSKA, A. *On the Gentzen-type formalizations* ◇ B25 D05 F07 ◇

1981

BABAEV, A.A. *Equality of morphisms in closed categories. I,II (Russian) (Azerbaijani and English summaries)* ◇ F07 G30 ◇
CELLUCCI, C. *Proof theory and theory of meaning* ◇ A05 F07 F30 ◇
GRISHIN, V.N. *Predicate and set-theoretic calculi based on a logic without contractions (Russian)* ◇ B53 B60 E30 E70 F07 ◇
MIGLIOLI, P.A. & ORNAGHI, M. *A logically justified model of computation. I,II* ◇ B75 D80 F07 F50 ◇
PLIUSKEVICIUS, R. *On the Gentzen type proof theory for program analysis* ◇ B75 F07 F35 F50 ◇
PRAWITZ, D. *Validity and normalizability of proofs in 1st and 2nd order classical and intuitionistic logic* ◇ F05 F07 F35 F50 ◇
RESL, M. & SOCHOR, A. *Provability in the alternative set theory* ◇ E70 F07 ◇
SOLOV'EV, S.V. *The category of finite sets and Cartesian closed categories (Russian) (English summary)* ◇ F07 F25 G30 ◇

1982

BIBEL, W. *A comparative study of several proof procedures* ◇ B35 F07 ◇
BIBEL, W. *Deduktionsverfahren* ◇ B35 F07 ◇
GIRARD, J.-Y. *Herbrand's theorem and proof-theory* ◇ F05 F07 F10 F35 ◇
PRAWITZ, D. *Beweise und die Bedeutung und Vollstaendigkeit der logischen Konstanten* ◇ A05 B20 F07 ◇
SCHROEDER-HEISTER, P. *Logische Konstanten und Regeln. Zur Deutung von Aussagenoperatoren* ◇ F07 F50 ◇

1983

CASANOVA, M.A. *The theory of functional and subset dependencies over relational expressions* ◇ F07 ◇
FITTING, M. *Proof methods for modal and intuitionistic logics* ◇ B45 B55 F07 F50 F98 ◇
GRUENBERG, T. *A tableau system of proof for predicate-functor logic with identity* ◇ B10 F07 ◇

MADDUX, R. *A sequent calculus for relation algebras* ◇ F07 G15 ◇
PLIUSKEVICIUS, R. *Some disjunction properties for propositional classical and constructive dynamic logic (Russian) (English and Lithuanian summaries)* ◇ B75 F07 ◇
SCHOENFELD, W. *Proof search for unprovable formulas* ◇ B35 F07 F20 ◇
SCHROEDER-HEISTER, P. *The completeness of intuitionistic logic with respect to a validity concept based on an inversion principle* ◇ F07 F50 ◇
SEELY, R.A.G. *Hyperdoctrines, natural deduction and the Beck condition* ◇ F07 F50 G30 ◇
SZABO, M.E. *A sequent calculus for Kroeger logic* ◇ B75 F05 F07 ◇

1984

BOOLOS, G. *Don't eliminate cut* ◇ F05 F07 ◇
BOOLOS, G. *Trees and finite satisfiability: proof of a conjecture of Burgess* ◇ B10 C13 F07 ◇
CALL, R.L. *Constructing sequent rules for generalized propositional logics* ◇ B05 F07 ◇
COPPO, M. *Completeness of type assignment in continuous lambda models* ◇ B40 F07 ◇
KAPUR, D. & KRISHNAMURTHY, B. *A natural proof sytem based on rewriting techniques* ◇ B35 F07 ◇
MILLER, DALE A. *Expansion tree proofs and their conversion to natural deduction proofs* ◇ B35 F07 ◇
MOTOHASHI, N. *Approximation theory of uniqueness conditions by existence conditions* ◇ C40 C75 C80 F07 F50 ◇
NIKOLENKO, A.B. *Method of invariant transformations and logical deduction (Russian)* ◇ F07 ◇
OREVKOV, V.P. *Upper bounds for lengthening of proofs after cut elimination (Russian) (English summary)* ◇ F05 F07 F20 ◇
POPOV, S.V. *Diagrams of deductions in sequential calculi (Russian)* ◇ F05 F07 ◇
RICHTER, M.M. *Some reordering properties for inequality proof trees* ◇ F07 ◇
SCHROEDER-HEISTER, P. *A natural extension of natural deduction* ◇ B22 F07 F50 ◇
SCHROEDER-HEISTER, P. *Generalized rules for quantifiers and the completeness of the intuitionistic operators* &, ∨, ⊃, ⊥, ∀, ∃ ◇ F07 F50 ◇
SIKIC, Z. *Multiple forms of Gentzen's rules and some intermediate forms* ◇ B55 F07 ◇

1985

BORICIC, B.R. *On sequent-conclusion natural deduction systems* ◇ F07 ◇
FELSCHER, W. *Dialogues, strategies, and intuitionistic provability* ◇ F05 F07 F50 ◇
FINE, K. *Natural deduction and arbitrary objects* ◇ A05 F07 ◇
GIAMBRONE, S. *On purported Gentzen formulations of two positive relevant logics* ◇ B46 F07 ◇
HAIMAN, M. *Linear lattice proof theory: an overview* ◇ F05 F07 ◇
HAIMAN, M. *Proof theory for linear lattices* ◇ F05 F07 ◇
LOPEZ-ESCOBAR, E.G.K. *Proof functional connectives* ◇ B55 F07 F30 ◇
PRAWITZ, D. *Remarks on some approaches to the concept of logical consequence* ◇ A05 B22 F07 ◇
SARKARIA, K.S. *Formal theories are acyclic* ◇ F07 ◇

F10 Functionals in proof theory

1951

KREISEL, G. *On the interpretation of non-finitist proofs I* ◇ F10 F25 F30 F50 ◇

1952

KREISEL, G. *On the interpretation of non-finitist proofs II. Interpretation of number theory. Applications* ◇ F10 F25 F30 F50 ◇

KREISEL, G. *Some concepts concerning formal systems of number theory* ◇ F10 F25 F30 ◇

1955

KREISEL, G. *Models, translations and interpretations* ◇ F10 F25 ◇

1956

TAKEUTI, G. *A metamathematical theorem on functions* ◇ B15 F05 F10 F35 ◇

1957

KREISEL, G. *Goedel's interpretation of Heyting's arithmetic* ◇ F10 F30 F50 ◇

KREISEL, G. *Relations between classes of constructive functionals* ◇ F10 F50 ◇

1958

GOEDEL, K. *Ueber eine bisher noch nicht benuetzte Erweiterung des finiten Standpunktes* ◇ A05 F10 F30 F50 ◇

1959

KREISEL, G. *Interpretation of analysis by means of constructive functionals of finite types* ◇ D65 F10 F30 F35 F50 ◇

1962

PARSONS, C. *The ω-consistency of ramified analysis* ◇ F05 F10 F35 F65 ◇

SPECTOR, C. *Provably recursive functionals of analysis: a consistency proof of analysis by an extension of principles formulated in current intuitionistic mathematics* ◇ F10 F35 F50 ◇

1963

HOWARD, W.A. *Appendices to section 2, vol. 1* ◇ F10 F35 F50 ◇

HOWARD, W.A. *Transfinite induction and transfinite recursion* ◇ F10 F35 F50 ◇

TAIT, W.W. *A second order theory of functionals of higher type* ◇ F10 F35 F50 ◇

TAIT, W.W. *Appendix A: Intensional functionals* ◇ F10 F35 F50 ◇

YASUGI, M. *Intuitionistic analysis and Goedel's interpretation* ◇ F10 F35 F50 ◇

1964

GRZEGORCZYK, A. *Recursive objects in all finite types* ◇ D65 F10 ◇

1965

KREISEL, G. *Mathematical logic* ◇ B98 F10 F35 F50 F98 ◇

TAIT, W.W. *Functionals defined by transfinite recursion* ◇ E30 F10 F15 F35 ◇

TAIT, W.W. *Infinitely long terms of transfinite type* ◇ C75 F10 F15 F50 ◇

TAIT, W.W. *The substitution method* ◇ F10 F15 F30 F35 ◇

1966

HANATANI, Y. *Calculabilite des fonctionelles recursives primitives de type fini sur les nombres naturels* ◇ D65 F10 F35 F50 ◇

HOWARD, W.A. & KREISEL, G. *Transfinite induction and bar induction of types zero and one, and the role of continuity in intuitionistic analysis* ◇ F10 F35 F50 ◇

1967

MINTS, G.E. *Translator's supplements to K.Goedel's paper (Russian)* ◇ F10 ◇

SANCHIS, L.E. *Functionals defined by recursion* ◇ D20 F10 ◇

TAIT, W.W. *Intensional interpretation of functionals of finite type I* ◇ F10 F30 F35 F50 ◇

1968

DILLER, J. *Zur Berechenbarkeit primitiv-rekursiver Funktionale endlicher Typen* ◇ F10 F15 F50 ◇

DRAGALIN, A.G. *The computability of primitive recursive terms of finite type, and primitive recursive realization (Russian)* ◇ F10 F35 F50 ◇

HINATA, S. *Calculability of primitive recursive functionals of finite type* ◇ F10 F50 ◇

HOWARD, W.A. *Functional interpretation of bar induction by bar recursion* ◇ F10 F35 F50 ◇

KREISEL, G. *Functions, ordinals, species* ◇ F10 F35 F50 ◇

SCHUETTE, K. *Neuere Ergebnisse der Beweistheorie* ◇ F05 F10 F15 F30 F35 F98 ◇

TAIT, W.W. *Constructive reasoning* ◇ B40 F10 F15 F30 F35 F50 F65 ◇

1969

WETTE, E. *Definition eines (relativ vollstaendigen) formalen Systems konstruktiver Arithmetik* ◇ F10 F15 F30 F35 F65 ◇

1970

FEFERMAN, S. *Hereditarily replete functionals over the ordinals* ◇ F10 F15 ◇

HOWARD, W.A. *Assignment of ordinals to terms for primitive recursive functionals of finite type* ◇ F10 F15 F50 ◇

PARSONS, C. *On a number-theoretic choice schema and its relation to induction* ◇ F10 F30 ◇

1971

DILLER, J. & SCHUETTE, K. *Simultane Rekursionen in der Theorie der Funktionale endlicher Typen* ◇ F10 F35 F50 ◇

FEFERMAN, S. *Ordinals and functionals in proof theory* ◇ F10 F15 F35 F50 ◇

GIRARD, J.-Y. *Une extension de l'interpretation de Goedel a l'analyse et son application a l'elimination des coupures dans l'analyse et la theorie des types* ◇ F05 F10 F35 F50 ◇

PRAWITZ, D. *Ideas and results in proof theory* ◇ F05 F07 F10 F30 F35 F50 F98 ◇

TAIT, W.W. *Normal form theorem for bar recursive functions of finite type* ◇ F05 F10 F35 F50 ◇

1972

ACZEL, P. *Describing ordinals using functionals of transfinite type* ◇ F10 F15 ◇

ERSHOV, YU.L. *Everywhere-defined continuous functionals (Russian)* ◇ D65 F10 ◇

HOWARD, W.A. *A system of abstract constructive ordinals* ◇ F10 F15 F35 F50 ◇

PARSONS, C. *On n-quantifier induction* ◇ F10 F30 ◇

STENLUND, S. *Combinators, λ-terms and proof theory* ◇ B40 F05 F10 F50 F98 ◇

1973

GIRARD, J.-Y. *Quelques resultats sur les interpretations fonctionelles* ◇ F10 F35 F50 ◇

HOWARD, W.A. *Appendix: Hereditarily majorizable functionals of finite type* ◇ E25 F10 F35 F50 ◇

LEVITZ, H. *A characterisation of the Veblen-Schuette functions by means of functionals* ◇ F10 F15 ◇

LUCKHARDT, H. *Extensional Goedel functional interpretation. A consistency proof of classical analysis* ◇ F10 F35 F50 F98 ◇

TROELSTRA, A.S. (ED.) *Metamathematical investigation of intuitionistic arithmetic and analysis* ◇ F05 F10 F30 F35 F50 F55 F98 ◇

TROELSTRA, A.S. *Models and computability* ◇ F05 F10 F30 F35 F50 ◇

TROELSTRA, A.S. *Realizability and functional interpretations* ◇ F10 F30 F35 F50 ◇

ZUCKER, J.I. *Iterated inductive definitions, trees, and ordinals* ◇ F10 F15 F35 ◇

1974

DILLER, J. & NAHM, W. *Eine Variante zur Dialectica-Interpretation der Heyting-Arithmetik endlicher Typen* ◇ F10 F35 F50 ◇

ERSHOV, YU.L. *The model G of the theory BR (Russian)* ◇ F10 F50 ◇

1975

CELLUCCI, C. *Teoremi di normalizzazione per alcuni sistemi funzionali* ◇ F05 F10 F15 F30 ◇

CELLUCCI, C. *Teoremi di normalizzazione per alcuni sistemi funzionali II* ◇ F05 F10 F15 F30 ◇

DILLER, J. & VOGEL, HELMUT *Intensionale Funktionalinterpretation der Analysis* ◇ F10 F35 F50 ◇

FRIEDMAN, H.M. *Equality between functionals* ◇ B40 F10 ◇

HANATANI, Y. *Calculability of the primitive recursive functionals of finite type over the natural numbers* ◇ D65 F10 F35 F50 ◇

LUCKHARDT, H. *The real elements in a consistency proof for simple type theory I* ◇ F10 F35 F50 ◇

MINTS, G.E. *Finite investigations of transfinite derivations (Russian) (English summary)* ◇ F05 F07 F10 F20 F30 F35 F50 ◇

MINTS, G.E. *Proof theory (arithmetic and analysis) (Russian)* ◇ F05 F10 F30 F35 F50 F98 ◇

SCHWICHTENBERG, H. *Elimination of higher type levels in definitions of primitive recursive functionals by means of transfinite recursion* ◇ C75 F10 F15 F35 F50 ◇

SCHWICHTENBERG, H. & WAINER, S.S. *Infinite terms and recursion in higher types* ◇ C75 D65 F10 ◇

1976

GOODMAN, NICOLAS D. *The theory of the Goedel functionals* ◇ F10 F35 F50 ◇

MAASS, W. *Eine Funktionalinterpretation der praedikativen Analysis* ◇ D20 F10 F35 F65 ◇

VOGEL, HELMUT *Ein starker Normalisationssatz fuer die bar-rekursiven Funktionale* ◇ F10 F35 ◇

1977

ERSHOV, YU.L. *Model* $\mathbb{C}$ *of partial continuous functionals* ◇ D45 D65 F10 F50 ◇

FEFERMAN, S. *Theories of finite type related to mathematical practice* ◇ B15 D55 D65 F10 F15 F35 F98 ◇

GIRARD, J.-Y. *Functional interpretation and Kripke models* ◇ C90 F10 F30 F50 ◇

GIRARD, J.-Y. *Functionals and ordinoids* ◇ D65 F10 F15 ◇

SCHWICHTENBERG, H. *Proof theory: Some applications of cut-elimination* ◇ F05 F10 F15 F30 F35 F98 ◇

TAKANO, M. *On Goedel's primitive recursive functionals (Japanese)* ◇ F10 ◇

1978

SCOTT, P.J. *The "Dialectica" interpretation and categories* ◇ F10 F35 F50 G30 ◇

VOGEL, HELMUT *Eine beweistheoretische Anwendung partieller stetiger Funktionale* ◇ F10 F35 F50 ◇

1979

DILLER, J. *Functional interpretations of Heyting's arithmetic in all finite types* ◇ F10 F35 F50 ◇

GOTO, S. *Program synthesis through Goedel's interpretation* ◇ B75 F10 F50 ◇

MINTS, G.E. *Normalisation of natural deduction and effectiveness of classical existence (Russian)* ◇ F05 F10 F30 F35 ◇

SCHWICHTENBERG, H. *On bar recursion of types 0 and 1* ◇ F10 F35 F50 ◇
STEIN, M. *Interpretationen der Heyting-Arithmetik endlicher Typen* ◇ F10 F35 F50 ◇
VOGEL, HELMUT *Ueber die mit dem Bar-Rekursor vom Typ 0 definierbaren Ordinalzahlen* ◇ F10 F15 ◇
VOGEL, HELMUT *Ueber ein mit der Bar-Induktion verwandtes Schema* ◇ F10 F35 F50 ◇

1980

DILLER, J. *Modified realization and the formulae-as-types notion* ◇ F05 F10 F30 F35 F50 ◇
HOWARD, W.A. *Ordinal analysis of terms of finite type* ◇ F10 F15 F35 F50 ◇
HOWARD, W.A. *Ordinal analysis of bar recursion of type zero* ◇ F10 F15 F35 F50 ◇
HOWARD, W.A. *The formulae-as-types notion of construction* ◇ B40 F05 F10 F50 ◇
STEIN, M. *Interpretations of Heyting's arithmetic - An analysis by means of a language with set symbols* ◇ F10 F30 F50 ◇

1981

HOWARD, W.A. *Computability of ordinal recursion of type level two* ◇ D20 F10 F15 F35 F50 ◇
HOWARD, W.A. *Ordinal analysis of simple cases of bar recursion* ◇ F10 F15 F35 F50 ◇
NEGRI, M. *L'aritmetizzazione della computabilita dei funzionali di tipo finito* ◇ F10 F35 F50 ◇
STEIN, M. *A general theorem on existence theorems* ◇ B40 F10 F35 F50 ◇

1982

GIRARD, J.-Y. *Herbrand's theorem and proof-theory* ◇ F05 F07 F10 F35 ◇
KOLETSOS, G. *A Goedel-functional interpretation of the ω-rule* ◇ F10 F35 ◇
YASUGI, M. *Construction principle and transfinite induction up to ε_0* ◇ F10 F15 ◇

1983

CANTINI, A. *A note on the theory of admissible sets with ε-induction restricted to formulas with one quantifier and related systems (Italian summary)* ◇ C62 E30 E45 E70 F10 F15 F35 ◇

1984

FRIEDRICH, W. *Spielquantorinterpretation unstetiger Funktionale der hoeheren Analysis* ◇ F10 F35 ◇

1985

BEZEM, M. *Stongly majorizable functionals of finite type: A model for barrecursion containing discontinuous functionals* ◇ F10 F35 F50 ◇
FRIEDRICH, W. *Goedelsche Funktionalinterpretation fuer eine Erweiterung der klassischen Analysis* ◇ F10 F35 F50 ◇
KOLETSOS, G. *Functional interpretation of the β-rule* ◇ F10 F30 F35 ◇
MAYBERRY, J. *Global quantification in Zermelo-Fraenkel set theory* ◇ E30 E70 F10 F50 ◇
PAEPPINGHAUS, P. *A typed λ-calculus and Girard's model of ptykes* ◇ B40 F10 F15 F35 ◇
TERLOUW, J. *Reduction of higher type levels by means of an ordinal analysis of finite terms* ◇ F10 F15 ◇

F15 Recursive ordinals and ordinal notations

1904

HARDY, G.H. *A theorem concerning the infinite cardinal numbers* ◇ E10 F15 ◇

1908

VEBLEN, O. *Continuous increasing functions of finite and transfinite ordinals* ◇ E05 E10 F15 ◇

1928

ACKERMANN, W. *Zum Hilbertschen Aufbau der reellen Zahlen* ◇ B28 D20 F15 ◇

1937

CHURCH, A. & KLEENE, S.C. *Formal definitions in the theory of ordinal numbers* ◇ F15 ◇

1938

CHURCH, A. *The constructive second number class* ◇ F15 ◇

GENTZEN, G. *Neue Fassung des Widerspruchsfreiheitsbeweises fuer die reine Zahlentheorie* ◇ F05 F15 F30 ◇

KLEENE, S.C. *On notation for ordinal numbers* ◇ F15 ◇

1939

HILBERT, D. & BERNAYS, P. *Grundlagen der Mathematik II* ◇ A05 B98 F05 F15 F30 F40 F98 ◇

TURING, A.M. *Systems of logic based on ordinals* ◇ B40 D25 D30 F15 F30 ◇

1943

GENTZEN, G. *Beweisbarkeit und Unbeweisbarkeit von Anfangsfaellen der transfiniten Induktion in der reinen Zahlentheorie* ◇ F15 F30 ◇

1944

GOODSTEIN, R.L. *On the restricted ordinal theorem* ◇ F15 F30 ◇

KLEENE, S.C. *On the forms of predicates in the theory of constructive ordinals* ◇ D55 D70 F15 ◇

1947

GOODSTEIN, R.L. *Transfinite ordinals in recursive number theory* ◇ E10 F15 F30 ◇

1950

BACHMANN, H. *Die Normalfunktionen und das Problem der ausgezeichneten Folgen von Ordnungszahlen* ◇ E10 F15 ◇

1951

ACKERMANN, W. *Konstruktiver Aufbau eines Abschnitts der zweiten Cantorschen Zahlenklasse* ◇ F15 ◇

FINSLER, P. *Eine transfinite Folge arithmetischer Operationen* ◇ E10 F15 ◇

1952

BACHMANN, H. *Vergleich und Kombination zweier Methoden von Veblen und Finsler zur Loesung des Problems der ausgezeichneten Folgen von Ordnungszahlen* ◇ E10 F15 ◇

TAKEUTI, G. *A metamathematical theorem on the theory of ordinal numbers* ◇ E10 F15 ◇

1953

NEUMER, W. *Zur Konstruktion von Ordnungszahlen I,II* ◇ E10 F15 ◇

ROUTLEDGE, N.A. *Ordinal recursion* ◇ D20 F15 ◇

1954

BACHMANN, H. *Normalfunktionen und Hauptfolgen* ◇ E10 E25 F15 ◇

MARKWALD, W. *Zur Theorie der konstruktiven Wohlordnungen* ◇ C80 D55 F15 ◇

NEUMER, W. *Zur Konstruktion von Ordnungszahlen III,IV* ◇ E10 F15 ◇

SCHUETTE, K. *Kennzeichnung von Ordnungszahlen durch rekursiv erklaerte Funktionen* ◇ D20 F15 ◇

1955

BACHMANN, H. *Transfinite Zahlen* ◇ E10 E25 E98 F15 ◇

KLEENE, S.C. *On the forms of the predicates in the theory of constructive ordinals II* ◇ D55 D70 F15 ◇

SPECTOR, C. *Recursive well-orderings* ◇ D30 D45 D55 F15 ◇

1956

NEUMER, W. *Zur Konstruktion von Ordnungszahlen V* ◇ E10 F15 ◇

SKOLEM, T.A. *An ordered set of arithmetic functions representing the least ε-number* ◇ F15 F30 ◇

1957

NEUMER, W. *Algorithmen fuer Ordnungszahlen und Normalfunktionen I* ◇ E10 F15 ◇

SPECTOR, C. *Recursive ordinals and predicative set theory* ◇ D20 D55 E70 F15 F65 ◇

TAKEUTI, G. *Ordinal diagrams* ◇ F07 F15 ◇

WANG, HAO *Remarks on constructive ordinals and set theory* ◇ D55 E45 F15 ◇

1958

KINO, A. *A consistency-proof of a formal theory of Ackermann's ordinal numbers* ◇ F15 F25 ◇

TAKEUTI, G. *On the formal theory of the ordinal diagrams* ◇ F15 F35 ◇

WANG, HAO *Alternative proof of a theorem of Kleene* ◇ F15 ◇

1959

KREISEL, G. & SHOENFIELD, J.R. & WANG, HAO *Number theoretic concepts and recursive well-orderings* ◇ F15 F30 ◇

WANG, HAO *Ordinal numbers and predicative set theory* ◇ A05 E70 F15 F65 ◇

1960

GANDY, R.O. *Proof of Mostowski's conjecture* ◇ F15 ◇

KREISEL, G. *Non-uniqueness results for transfinite progressions* ◇ F15 ◇

KREISEL, G. *Ordinal logics and the characterization of informal concepts of proof* ◇ F07 F15 F30 F65 ◇

NEUMER, W. *Algorithmen fuer Ordnungszahlen und Normalfunktionen II* ◇ E10 F15 ◇

SCHUETTE, K. *Beweistheorie* ◇ B98 F05 F15 F30 F35 F98 ◇

TAKEUTI, G. *Ordinal diagrams II* ◇ F15 ◇

1961

KINO, A. *On ordinal diagrams* ◇ F15 ◇

KREIDER, D.L. & ROGERS JR., H. *Constructive versions of ordinal number classes* ◇ D55 D70 F15 ◇

PUTNAM, H. *Uniqueness ordinals in higher constructive number classes* ◇ D55 F15 ◇

TAKEUTI, G. *On the inductive definition with quantifiers of second order* ◇ B15 F15 F35 ◇

WERMUS, H. *Eine konstruktiv-figuerliche Begruendung eines Abschnittes der zweiten Zahlklasse* ◇ F15 ◇

1962

CHAUVIN, A. *Sur les ensembles arithmetiques constructibles* ◇ D55 E45 F15 ◇

FEFERMAN, S. & SPECTOR, C. *Incompleteness along paths in progressions of theories* ◇ D55 F15 F30 ◇

FEFERMAN, S. *Transfinite recursive progressions of axiomatic theories* ◇ D55 F15 F30 ◇

KINO, A. & TAKEUTI, G. *On hierarchies of predicates of ordinal numbers* ◇ D55 D60 F15 ◇

SAARNIO, U. *Von den Rechenoperationen hoeherer Ordnung bei der Darstellung der transfiniten Ordnungszahlen* ◇ E10 F15 ◇

SCHUETTE, K. *Lecture notes in mathematical logic. Vol.I* ◇ B98 F15 ◇

SCHUETTE, K. *Logische Abgrenzung des Transfiniten* ◇ A05 E10 F15 ◇

1963

KINO, A. *A note on constructive ordinals* ◇ F15 ◇

KINO, A. & TAKEUTI, G. *On predicates with constructive infinitely long expressions* ◇ C70 F15 ◇

SCHUETTE, K. *Lecture notes in mathematical logic. Vol.II* ◇ B98 F15 ◇

TAKEUTI, G. *A remark on Gentzen's paper "Beweisbarkeit und Unbeweisbarkeit von Anfangsfaellen der transfiniten Induktion in der reinen Zahlentheorie" I-II* ◇ F15 F35 ◇

1964

PUTNAM, H. *On hierarchies and systems of notations* ◇ D55 D70 F15 ◇

SIMAUTI, T. *A note on the construction of ordinal numbers* ◇ E10 F15 ◇

1965

HENSEL, G. & PUTNAM, H. *On the notational independence of various hierarchies of degrees of unsolvability* ◇ D30 D55 F15 ◇

KURATA, R. *Recursive progression of intuitionistic number theories* ◇ F15 F30 F50 ◇

LUCKHAM, D.C. & PUTNAM, H. *On minimal and almost-minimal systems of notations* ◇ F15 ◇

RICHTER, W.H. *Extensions of the constructive ordinals* ◇ D70 F15 ◇

SCHUETTE, K. *Eine Grenze fuer die Beweisbarkeit der transfiniten Induktion in der verzweigten Typenlogik* ◇ B15 F15 F35 F65 ◇

SCHUETTE, K. *Predicative well-orderings* ◇ F15 F35 F65 ◇

TAIT, W.W. *Functionals defined by transfinite recursion* ◇ E30 F10 F15 F35 ◇

TAIT, W.W. *Infinitely long terms of transfinite type* ◇ C75 F10 F15 F50 ◇

TAIT, W.W. *The substitution method* ◇ F10 F15 F30 F35 ◇

1966

ACZEL, P. & CROSSLEY, J.N. *Constructive order types III* ◇ D45 F15 ◇

CROSSLEY, J.N. *Constructive order types II* ◇ D20 D45 F15 ◇

CROSSLEY, J.N. & SCHUETTE, K. *Non-uniqueness at ω^2 in Kleene's O* ◇ D45 F15 ◇

LEVITZ, H. *Ueber die Finslerschen hoeheren arithmetischen Operationen* ◇ E10 F15 ◇

PAOLA DI, R.A. *Pseudo-complements and ordinal logics based on consistency statements* ◇ D25 F15 F30 ◇

RITTER, W.E. *Notation systems and an effective fixed point property* ◇ D20 D30 F15 ◇

1967

ACZEL, P. *Paths in Kleene's O* ◇ F15 ◇

GERBER, H. *An extension of Schuette's Klammer-symbols* ◇ F15 ◇

PAOLA DI, R.A. *Some theorems on extensions of arithmetic* ◇ D25 D35 F15 F30 ◇

PARIKH, R. *On nonuniqueness in transfinite progressions* ◇ F15 ◇

RICHTER, W.H. *Constructive transfinite number classes* ◇ D65 D70 F15 ◇

ROGERS JR., H. *Theory of recursive functions and effective computability* ◇ D25 D30 D55 D98 F15 ◇

SAARNIO, U. *Ueber die Positionsdarstellung der Ordnungszahlen der Cantorschen zweiten Zahlenklasse* ◇ E10 F15 ◇

TAKEUTI, G. *Consistency proofs of subsystems of classical analysis* ◇ F05 F15 F35 ◇

1968

DILLER, J. *Zur Berechenbarkeit primitiv-rekursiver Funktionale endlicher Typen* ◇ F10 F15 F50 ◇

FEFERMAN, S. *Autonomous transfinite progressions and the extent of predicative mathematics* ◇ F15 F30 F35 F65 ◇

FEFERMAN, S. *Lectures on proof theory* ◇ C40 C70 C75 F05 F15 F98 ◇

FEFERMAN, S. *Systems of predicative analysis II : representation of ordinals* ◇ F15 F35 F65 ◇
FENSTAD, J.E. *On the completeness of some transfinite recursive progressions of axiomatic theories* ◇ F15 F30 ◇
HANATANI, Y. *Demonstration de l' ω -non-contradiction de l'arithmetique* ◇ F05 F15 F30 F50 ◇
RICHTER, W.H. *Constructively accessible ordinal numbers* ◇ D65 D70 F15 ◇
SAARNIO, U. *Die kritischen Zahlen hoeherer Ordnung innerhalb der zweiten Cantorschen Zahlenklasse* ◇ E10 F15 ◇
SAARNIO, U. *Ein Zeichensystem fuer die Darstellung der transfiniten Ordnungszahlen der zweiten Zahlenklasse* ◇ E10 F15 ◇
SCHUETTE, K. *Ein konstruktives System von Ordinalzahlen I* ◇ F15 ◇
SCHUETTE, K. *Neuere Ergebnisse der Beweistheorie* ◇ F05 F10 F15 F30 F35 F98 ◇
TAIT, W.W. *Constructive reasoning* ◇ B40 F10 F15 F30 F35 F50 F65 ◇
TAIT, W.W. *Normal derivability in classical logic* ◇ C75 F05 F15 F30 F35 F60 ◇
TAKEUTI, G. & YASUGI, M. *Reflection principles of subsystems of analysis* ◇ F15 F35 ◇
TAKEUTI, G. *The Π^1_1-comprehension schema and ω -rules* ◇ F05 F15 F35 ◇

1969

BELYAKIN, N.V. *A variant of Richter's constructive ordinals (Russian)* ◇ D70 F15 ◇
BIBEL, W. *Schnittelimination in einem Teilsystem der einfachen Typenlogik* ◇ B15 F05 F15 F35 ◇
LEVITZ, H. *A simplification of Takeuti's ordinal diagrams of finite order* ◇ F15 ◇
LEVITZ, H. *On the Finsler and Doner-Tarski arithmetical hierarchies* ◇ E10 F15 ◇
PFEIFFER, H. *Ein Bezeichnungssystem fuer Ordinalzahlen* ◇ F15 ◇
SCHUETTE, K. *Ein konstruktives System von Ordinalzahlen II* ◇ F15 ◇
WETTE, E. *Definition eines (relativ vollstaendigen) formalen Systems konstruktiver Arithmetik* ◇ F10 F15 F30 F35 F65 ◇

1970

ERSHOV, YU.L. *On a hierarchy of sets III (Russian)* ◇ D30 D55 F15 ◇
FEFERMAN, S. *Hereditarily replete functionals over the ordinals* ◇ F10 F15 ◇
GERBER, H. *Brouwer's bar theorem and a system of ordinal notations* ◇ F15 F35 F50 ◇
HOWARD, W.A. *Assignment of ordinals to terms for primitive recursive functionals of finite type* ◇ F10 F15 F50 ◇
ISLES, D. *Regular ordinals and normal forms* ◇ E10 F15 ◇
KINO, A. *Formalization of the theory of ordinal diagrams of infinite order* ◇ F15 F35 ◇
KREISEL, G. *Principles of proof and ordinals implicit in given concepts* ◇ A05 F15 ◇
LEVITZ, H. *On the relationship between Takeuti's ordinal diagrams $O(n)$ and Schuette's system of ordinal notations $\Sigma(n)$* ◇ F15 ◇
NEUMER, W. *Algorithmen fuer Ordnungszahlen und Normalfunktionen III* ◇ E10 F15 ◇
PFEIFFER, H. *Ein Bezeichnungssystem fuer Ordinalzahlen* ◇ F15 ◇

1971

FEFERMAN, S. *Ordinals and functionals in proof theory* ◇ F10 F15 F35 F50 ◇
FUCHS, HARTWIG *Algorithmic representation of Wermus' constructions of ordinal numbers* ◇ F15 ◇
GASS, F.S. *Generalized ordinal notation* ◇ D55 F15 ◇
ISLES, D. *Natural well-orderings* ◇ F15 ◇
LEVITZ, H. & SCHUETTE, K. *A characterisation of Takeuti's ordinal diagrams of finite order* ◇ F15 ◇
MINTS, G.E. *Exact estimation of the provability of transfinite induction in the initial segments of arithmetic (Russian) (English summary)* ◇ F05 F15 F30 ◇
RICHTER, W.H. *Recursively Mahlo ordinals and inductive definitions* ◇ D60 D65 D70 E55 F15 ◇
SCARPELLINI, B. *Proof theory and intuitionistic systems* ◇ F05 F15 F30 F35 F50 F98 ◇
SWARD, G.L. *Transfinite sequences of axiom systems for set theory* ◇ E30 E65 F15 ◇

1972

ACZEL, P. *Describing ordinals using functionals of transfinite type* ◇ F10 F15 ◇
FEFERMAN, S. *Infinitary properties, local functors, and systems of ordinal functions* ◇ C30 C40 C75 E10 F15 G30 ◇
GASS, F.S. *A note on Π^1_1 ordinals* ◇ D55 F15 ◇
HOWARD, W.A. *A system of abstract constructive ordinals* ◇ F10 F15 F35 F50 ◇
KREISEL, G. *Which number theoretic problems can be solved in recursive progressions on Π^1_1-paths through $\mathcal{O}$?* ◇ A05 D20 D55 F15 F30 F99 ◇
LUCIAN, M. *Systems of notations and the constructible hierarchy* ◇ D55 E45 F15 ◇
PFEIFFER, H. *Vergleich zweier Bezeichnungssysteme fuer Ordinalzahlen* ◇ F15 ◇
SCHWICHTENBERG, H. *Beweistheoretische Charakterisierung einer Erweiterung der Grzegorczyk-Hierarchie* ◇ D20 F15 F30 ◇

1973

LEVITZ, H. *A characterisation of the Veblen-Schuette functions by means of functionals* ◇ F10 F15 ◇
MEDVEDEV, YU.T. *An interpretation of intuitionistic number theory* ◇ F15 F30 F50 ◇
NEPEJVODA, N.N. *On a generalization of the Kleene-Mostowski hierarchy (Russian)* ◇ D55 F15 F35 ◇
PARIKH, R. *A note on paths through O* ◇ F15 ◇
PFEIFFER, H. *Bezeichnungssysteme fuer Ordinalzahlen* ◇ E10 F15 ◇
TAKEUTI, G. & YASUGI, M. *The ordinals of the systems of second order arithmetic with the provably*

Δ_2^1*-comprehension axiom and with the* Δ_2^1*-comprehension axiom respectively* ◇ F05 F15 F35 ◇
ZUCKER, J.I. *Iterated inductive definitions, trees, and ordinals* ◇ F10 F15 F35 ◇

1974

BOWEN, K.A. *Primitive recursive notations for infinitary formulas* ◇ C75 F15 ◇
DEREVYANKINA, E.A. *Some problems of completeness of arithmetic (Russian)* ◇ F15 F30 ◇
GENTZEN, G. *Der erste Widerspruchsfreiheitsbeweis fuer die klassische Zahlentheorie* ◇ F05 F15 F25 F30 ◇
GOLDFARB, W.D. *Ordinal bounds for k-consistency* ◇ F15 F30 ◇
GOLDFARB, W.D. & SCANLON, T.M. *The* ω*-consistency of number theory via Herbrand's theorem* ◇ F05 F15 F30 ◇
HICKMAN, J.L. *Concerning the number of sums obtainable from a countable series of ordinals by permutations that preserve the order-type* ◇ E10 F15 ◇
LOEB, M.H. & WAINER, S.S. *Hierarchies of number-theoretic functions (Russian)* ◇ D20 F15 ◇
LUKAS, J.D. & PUTNAM, H. *Systems of notations and the ramified analytical hierarchy* ◇ D30 D55 E45 F15 F35 ◇
PFEIFFER, H. *Ueber zwei Bezeichnungssysteme fuer Ordinalzahlen* ◇ F15 ◇
YASUGI, M. *On ordinal diagrams (Japanese)* ◇ F15 ◇

1975

BOOLOS, G. *On Kalmar's consistency proof and a generalization of the notion of* ω*-consistency* ◇ F15 F30 ◇
BRIDGE, J. *A simplification of the Bachmann method for generating large countable ordinals* ◇ E10 F15 ◇
BUCHHOLZ, W. *Normalfunktionen und konstruktive Systeme von Ordinalzahlen* ◇ E10 F15 ◇
CELLUCCI, C. *Teoremi di normalizzazione per alcuni sistemi funzionali* ◇ F05 F10 F15 F30 ◇
CELLUCCI, C. *Teoremi di normalizzazione per alcuni sistemi funzionali II* ◇ F05 F10 F15 F30 ◇
ISLES, D. *The accessibility of* ε_0 ◇ F15 ◇
JOCKUSCH JR., C.G. *Recursiveness of initial segments of Kleene's O* ◇ D20 F15 ◇
KINO, A. & MYHILL, J.R. *A hierarchy of languages with infinitely long expressions* ◇ C40 C75 E10 F15 ◇
LEVITZ, H. *An ordered set of arithmetic functions representing the least* ε*-number* ◇ C62 F15 F30 ◇
ORGASS, R.J. *Extended basic logic and ordinal numbers* ◇ B40 F15 ◇
PFEIFFER, H. *Eine Variante des Bezeichnungssystems* $W(X)$ *fuer Ordinalzahlen* ◇ F15 ◇
PINUS, A.G. *Effective linear orders (Russian)* ◇ C57 D45 E07 F15 ◇
POHLERS, W. *An upper bound for the provability of transfinite induction in systems with N-times iterated inductive definitions* ◇ F15 F35 F50 ◇
SCHMIDT, DIANA *Bounds for the closure ordinals of replete monotonic increasing functions* ◇ E10 F15 ◇
SCHWICHTENBERG, H. *Elimination of higher type levels in definitions of primitive recursive functionals by means of transfinite recursion* ◇ C75 F10 F15 F35 F50 ◇
TAKEUTI, G. *Consistency proofs and ordinals* ◇ A05 F15 ◇

1976

BUCHHOLZ, W. & SCHUETTE, K. *Die Beziehungen zwischen den Ordinalzahlsystemen* Σ *und* $\overline{\Theta}(\omega)$ ◇ F15 ◇
BUCHHOLZ, W. *Ueber Teilsysteme von* $\Theta(\{g\})$ ◇ F15 ◇
FRIEDMAN, H.M. *Recursiveness in* Π_1^1 *paths through* $\mathcal{O}$ ◇ D30 D55 F15 ◇
KREISEL, G. *Wie die Beweistheorie zu ihren Ordinalzahlen kam und kommt* ◇ F05 F15 F30 F35 ◇
LOPEZ-ESCOBAR, E.G.K. *On an extremely restricted* ω*-rule* ◇ F05 F15 F30 F50 ◇
MCBETH, R. *Fundamental sequences for initial ordinals smaller than a certain* Θ_0 ◇ E10 F15 ◇
MILLER, L.W. *Normal functions and constructive ordinal notations* ◇ E10 F15 ◇
MINTS, G.E. *The universality of the canonical tree (Russian)* ◇ F15 F30 ◇
SCHMIDT, DIANA *Built-up systems of fundamental sequences and hierarchies of number theoretic functions* ◇ D20 E10 F15 ◇
SCHUETTE, K. *Einfuehrung der Normalfunktionen* Θ_α *ohne Auswahlaxiom und ohne Regularitaetsbedingung* ◇ E10 E25 F15 ◇
SCHUETTE, K. *Primitiv-rekursive Ordinalzahlfunktionen* ◇ D60 F15 ◇
TAKEUTI, G. & YASUGI, M. *Fundamental sequences of ordinal diagrams* ◇ F15 ◇

1977

FEFERMAN, S. *Theories of finite type related to mathematical practice* ◇ B15 D55 D65 F10 F15 F35 F98 ◇
GIRARD, J.-Y. *Functionals and ordinoids* ◇ D65 F10 F15 ◇
GORDEEV, L.N. *A majorizing semantics for hyperarithmetic sentences (Russian) (English summary)* ◇ D55 F15 F50 ◇
HAYASHI, S. *A note on provable well-orderings in first order systems with infinitary inference rules* ◇ F15 ◇
SCHWICHTENBERG, H. *Proof theory: Some applications of cut-elimination* ◇ F05 F10 F15 F30 F35 F98 ◇
SHIRAI, K. *A relation between transfinite induction and mathematical induction in elementary number theory* ◇ F15 F30 ◇
SORNIKOV, YA.A. *On quantifier-free* ε_0*-recursive arithmetic (Russian) (English summary)* ◇ F15 F30 ◇
VOGEL, HELMUT *Ausgezeichnete Folgen fuer praedikative Ordinalzahlen und praedikativ-rekursive Funktionen* ◇ D20 F15 ◇
ZEMKE, F. *P.R.-regulated systems of notation and the subrecursive hierarchy equivalence property* ◇ D20 F15 ◇

Subject Index

1978

BUCHHOLZ, W. & POHLERS, W. *Provable well orderings of formal theories for transfinitely iterated inductive definitions* ◇ F15 F30 F35 F50 ◇

CELLUCCI, C. *Teoria della dimostrazione. Normalizzazioni e assegnazioni di numeri ordinali* ◇ F05 F07 F15 F30 F98 ◇

CHEN, KEHSUN *Recursive well-founded orderings* ◇ D45 F15 ◇

HICKMAN, J.L. *On R. McBeth's paper: "Fundamental sequences for initial ordinals smaller than a certain Θ_0"* ◇ E10 F15 ◇

JERVELL, H.R. *Constructive universes. I* ◇ F05 F15 F35 F50 ◇

LEVITZ, H. *An ordinal bound for the set of polynomial functions with exponentiation* ◇ D15 D20 E07 E10 F15 F30 ◇

PARLAMENTO, F. *Sui sistemi metaformali e le progressioni ricorsive di teorie* ◇ F15 F30 ◇

POHLERS, W. *Ordinals connected with formal theories for transfinitely iterated inductive definitions* ◇ F05 F15 F35 ◇

SCHMIDT, DIANA *Associative ordinal functions, well partial orderings and a problem of Skolem* ◇ E10 F15 ◇

YASUGI, M. *A formalization of $Od(\Omega)$* ◇ F15 F35 ◇

YASUGI, M. *Self-iterating schemes of ordinal diagrams* ◇ F15 ◇

YASUGI, M. *Some properties of ordinal diagrams* ◇ F15 ◇

1979

FELSCHER, W. *Eine weitreichende Wohlordnung der natuerlichen Zahlen* ◇ E10 F15 ◇

JAEGER, G. & SCHUETTE, K. *Eine syntaktische Abgrenzung der $(\Delta^1_1 - CA)$-Analysis* ◇ F15 F35 ◇

KREISEL, G. *Some facts from the theory of proofs and some fictions from general proof theory* ◇ A05 F05 F15 ◇

MCBETH, R. *A note on Hardy's persistent numbers* ◇ E10 F15 ◇

MINTS, G.E. *A new reduction sequence for arithmetic (Russian) (English summary)* ◇ F05 F15 F30 ◇

SCHMIDT, DIANA *A partition theorem for ordinals* ◇ E05 E10 F15 ◇

VOGEL, HELMUT *Ueber die mit dem Bar-Rekursor vom Typ 0 definierbaren Ordinalzahlen* ◇ F10 F15 ◇

1980

HOWARD, W.A. *Ordinal analysis of terms of finite type* ◇ F10 F15 F35 F50 ◇

HOWARD, W.A. *Ordinal analysis of bar recursion of type zero* ◇ F10 F15 F35 F50 ◇

JAEGER, G. *Beweistheorie von KPN* ◇ E30 F05 F15 ◇

MCALOON, K. *Progressions transfinies de theories axiomatiques, formes combinatoires du theoreme d'incompletude et fonctions recursives a croissance rapide* ◇ C62 D20 F15 F30 ◇

MCBETH, R. *Fundamental sequences for exponential polynomials* ◇ E07 E10 F15 ◇

SCHUETTE, K. *Beweistheoretische Abgrenzung von Teilsystemen der Analysis* ◇ F15 F35 ◇

SIKIC, Z. *A note on sum-eliminator* ◇ F15 ◇

YASUGI, M. *A progression of consistency proofs* ◇ F05 F15 F35 ◇

YASUGI, M. *Gentzen reduction revisited* ◇ F05 F15 F30 F35 ◇

1981

ACZEL, P. *Two notes on the Paris independence result I: A generalization of Ramsey's theorem. II: The ordinal height of a density* ◇ C62 E05 F15 F30 ◇

BUCHHOLZ, W. & FEFERMAN, S. & POHLERS, W. & SIEG, W. *Iterated inductive definitions and subsystems of analysis: recent proof-theoretical studies* ◇ F05 F15 F35 F50 F98 ◇

BUCHHOLZ, W. *Ordinal analysis of ID_ν* ◇ F15 F35 F50 ◇

BUCHHOLZ, W. & SCHUETTE, K. *Syntaktische Abgrenzungen von formalen Systemen der Π^1_1-Analysis und Δ^1_2-Analysis* ◇ F15 F35 ◇

BUCHHOLZ, W. *The $\Omega_{\mu+1}$-rule* ◇ F05 F15 F35 F50 ◇

GIRARD, J.-Y. *Π^1_2-logic I. Dilators* ◇ D55 E10 F15 G30 ◇

HOWARD, W.A. *Computability of ordinal recursion of type level two* ◇ D20 F10 F15 F35 F50 ◇

HOWARD, W.A. *Ordinal analysis of simple cases of bar recursion* ◇ F10 F15 F35 F50 ◇

ISLES, D. *Remarks on the notion of standard non-isomorphic natural number series* ◇ A05 F15 F65 ◇

MAEHARA, S. *Semi-formal finitist proof of the transfinite induction in an initial segment of Cantor's second number class* ◇ F15 ◇

MCBETH, R. *A note on exponential polynomials and prime factors* ◇ D20 F15 ◇

POHLERS, W. *Cut-elimination for impredicative infinitary systems I. Ordinal-analysis for ID_1* ◇ C75 F05 F15 F35 ◇

POHLERS, W. *Proof-theoretical analysis of ID_ν by the method of local predicativity* ◇ F05 F15 F35 F50 ◇

SIEG, W. *Inductive definitions, constructive ordinals, and normal derivations* ◇ F05 F15 F35 F50 ◇

TAKEUTI, G. & YASUGI, M. *An accessibility proof of ordinal diagrams* ◇ F15 ◇

VISSER, A. *An incompleteness result for paths through or within O* ◇ F15 ◇

1982

FEFERMAN, S. *Iterated inductive fixed-point theories: application to Hancock's conjecture* ◇ F15 F35 F65 ◇

FRIEDMAN, H.M. & MCALOON, K. & SIMPSON, S.G. *A finite combinatorial principle which is equivalent to the 1-consistency of predicative analysis* ◇ F15 F35 F65 ◇

GIRARD, J.-Y. *A survey of Π^1_2-logic* ◇ F15 F35 ◇

JAEGER, G. *Zur Beweistheorie der Kripke-Platek-Mengenlehre ueber den natuerlichen Zahlen* ◇ C70 E30 F15 F35 ◇

JERVELL, H.R. *Introducing homogeneous trees* ◇ F15 ◇
POHLERS, W. *Admissibility in proof theory; a survey* ◇ C70 F15 F35 F98 ◇
POHLERS, W. *Cut elimination for impredicative infinitary systems II. Ordinal analysis for iterated inductive definitions* ◇ C75 F05 F15 F35 ◇
RESSAYRE, J.-P. *Bounding generalized recursive functions of ordinals by effective functors: a complement to the Girard theorem* ◇ D60 F15 ◇
SCHMERL, U.R. *A proof theoretical fine structure in systems of ramified analysis* ◇ F15 F35 F65 ◇
SCHMERL, U.R. *Number theory and the Bachmann/Howard ordinal* ◇ F15 F30 ◇
VAUZEILLES, J. *Functors and ordinal notations. III: dilators and gardens* ◇ E10 F15 G30 ◇
VOGEL, HELMUT *Eine Variante des Ordinalzahlbezeichungssystems* Σ ◇ F15 ◇
WIELE VAN DE, J. *Recursive dilators and generalized recursions* ◇ D60 D65 F15 F35 ◇
YASUGI, M. *Construction principle and transfinite induction up to* ε_0 ◇ F10 F15 ◇

1983

ABRUSCI, V.M. *Sistemi di notazione ordinale e dilatori* ◇ F15 ◇
BUCHHOLZ, W. & SCHUETTE, K. *Ein Ordinalzahlensystem fuer die beweistheoretische Abgrenzung der* Π^1_2*-Separation und Bar-Induktion* ◇ F15 F35 ◇
CANTINI, A. *A note on the theory of admissible sets with* ε*-induction restricted to formulas with one quantifier and related systems (Italian summary)* ◇ C62 E30 E45 E70 F10 F15 F35 ◇
CICHON, E.A. & WAINER, S.S. *The slow-growing and the Grzegorczyk hierarchies* ◇ D20 F15 ◇
FEFERMAN, S. & JAEGER, G. *Choice principles, the bar rule and autonomously iterated comprehension schemes in analysis* ◇ F15 F35 ◇
JAEGER, G. *A well ordering proof for Feferman's theory* T_0 ◇ F15 F35 ◇
JAEGER, G. & POHLERS, W. *Eine beweistheoretische Untersuchung von* $(\Delta^1_2\text{-}CA)+(BI)$ *und verwandter Systeme* ◇ F15 F35 ◇
MASSERON, M. *Rungs and trees* ◇ D55 E05 F15 ◇

1984

ABRUSCI, V.M. *Recenti risultati e metodi nella teoria della dimostrazione: dilatatori,* β*-logica* ◇ F15 F35 G30 ◇
ARAI, T. *An accessibility proof of ordinal diagrams in intuitionistic theories for iterated inductive definitions* ◇ F15 F35 F50 ◇
DENNIS-JONES, E.C. & WAINER, S.S. *Subrecursive hierarchies via direct limits* ◇ D20 F15 G30 ◇
DRIES VAN DEN, L. & LEVITZ, H. *On Skolem's exponential functions below* 2^{2^X} ◇ E07 E75 F15 F30 ◇
FERBUS, M.-C. *Functorial bounds for cut elimination in* $L_{\beta\omega}I$ ◇ C75 F05 F15 G30 ◇
GASS, F.S. *Constructive ordinal notation systems* ◇ F15 ◇
GIRARD, J.-Y. & VAUZEILLES, J. *Functors and ordinal notations I: A functorial construction of the Veblen hierarchy. II: A functorial construction of the Bachmann hierarchy* ◇ E10 F15 G30 ◇
GIRARD, J.-Y. & VAUZEILLES, J. *Les premiers recursivement inaccessible et Mahlo et la theorie des dilatateurs* ◇ D60 E47 E55 F05 F15 ◇
JAEGER, G. ρ*-inaccessible ordinals, collapsing functions and a recursive notation system* ◇ E10 F15 ◇
JAEGER, G. *The strength of admissibility without foundation* ◇ E30 F15 F35 ◇
MCBETH, R. *A second normal form for functions of the system EP* ◇ F15 ◇
PEARCE, J. *A constructive consistency proof of a fragment of set theory* ◇ E30 E35 F05 F15 F35 ◇
SLESSENGER, P.H. *On subsets of the Skolem class of exponential polynomials* ◇ F15 F30 ◇
YASUGI, M. *Projection and elevation of ordinal diagrams* ◇ F15 ◇

1985

ABRUSCI, V.M. *Bachmann collections. Gardens. Recent applications (Italian)* ◇ F15 F35 ◇
ABRUSCI, V.M. *Uses of dilators: combinatorial problems and combinatorial results not provable in PA or in* ID_1 *(Italian)* ◇ F15 F30 F35 ◇
CANTINI, A. *A note on a predicatively reducible theory of iterated elementary induction (Italian summary)* ◇ F15 F35 ◇
CANTINI, A. *Majorizing provably recursive functions in fragments of PA* ◇ D20 F15 F30 ◇
CANTINI, A. *The Veblen hierarchy and the stability theorem for L* ◇ F15 ◇
GIRARD, J.-Y. & RESSAYRE, J.-P. *Elements de logique* Π^1_n ◇ D60 F15 F35 ◇
GIRARD, J.-Y. *Introduction to* Π^1_2*-logic* ◇ E10 F15 F35 ◇
GIRARD, J.-Y. & NORMANN, D. *Set recursion and* Π^1_2*-logic* ◇ D65 F15 ◇
JERVELL, H.R. *Large finite sets* ◇ E20 F15 F30 ◇
JERVELL, H.R. *Recursion on homogeneous trees* ◇ D75 F15 ◇
MCALOON, K. *Paris-Harrington incompleteness and progressions of theories* ◇ F15 F30 ◇
PAEPPINGHAUS, P. *A typed* λ*-calculus and Girard's model of ptykes* ◇ B40 F10 F15 F35 ◇
SLESSENGER, P.H. *A height restricted generation of a set of arithmetic functions of order-type* ε_0 ◇ F15 F30 ◇
TAKEUTI, G. *Proof theory and set theory* ◇ E55 F15 F98 ◇
TERLOUW, J. *Reduction of higher type levels by means of an ordinal analysis of finite terms* ◇ F10 F15 ◇
VAUZEILLES, J. *Functors and ordinal notations. IV: the Howard ordinal and the functor* Λ ◇ F15 G30 ◇
WAINER, S.S. *Subrecursive ordinals* ◇ D20 F15 ◇
WAINER, S.S. *The "slow-growing"* Π^1_2 *approach to hierarchies* ◇ D20 F15 ◇

Subject Index

YASUGI, M. *Groundedness property and accessibility of ordinal diagrams* ◇ F15 ◇

YASUGI, M. *Hyperprinciple and the functional structure of ordinal diagrams. I* ◇ F15 ◇

F20 Complexity of proofs

1934

GOEDEL, K. *Ueber die Laenge von Beweisen* ◇ B10 F20 F30 F35 ◇

1964

AANDERAA, S.O. & DREBEN, B. *Herbrand analyzing function* ◇ F05 F07 F20 ◇

1968

GHOSE, A. *The length of proofs* ◇ F20 ◇

TSEJTIN, G.S. *The complexity of a deduction in the propositional calculus (Russian)* ◇ F20 ◇

1971

EHRENFEUCHT, A. & MYCIELSKI, J. *Abbreviating proofs by adding new axioms* ◇ D35 F20 ◇

SLISENKO, A.O. *A property of enumerable sets containing "complexly deducible" formulas (Russian) (English summary)* ◇ D25 F20 ◇

1972

ANIKEEV, A.S. *Classification of derivable propositional formulas (Russian)* ◇ B15 D05 F20 F50 ◇

CHUBARYAN, A.A. & TSEJTIN, G.S. *Certain estimates of the length of logical deductions in classical propositional calculus (Russian) (Armenian summary)* ◇ B05 F07 F20 ◇

1973

CHERNIAVSKY, J.C. *The complexity of some non-classical logics* ◇ B45 D15 F20 F50 ◇

PARIKH, R. *Some results on the length of proofs* ◇ B25 F20 F30 ◇

1974

CHAITIN, G.J. *Information-theoretic limitations of formal systems* ◇ D15 D80 F20 ◇

CHUBARYAN, A.A. *The complexity of deductions in extensions of formal arithmetic (Russian) (Armenian summary)* ◇ F20 F30 ◇

CHUBARYAN, A.A. *The length of derivations of formulae in extensions of formal arithmetic (Russian) (Armenian and English summaries)* ◇ F20 F30 ◇

COOK, S.A. & RECKHOW, R.A. *On the length of proofs in the propositional calculus: preliminary version* ◇ B05 D15 F20 ◇

FISCHER, MICHAEL J. & RABIN, M.O. *Super-exponential complexity of Presburger arithmetic* ◇ B25 D15 F20 F30 ◇

LONGO, G. *I problemi di decisione e la loro complessita* ◇ B25 B35 D15 F20 ◇

RICHARDSON, D.B. *Sets of theorems with short proofs* ◇ F20 F30 ◇

ZAMOV, N.K. *Beschraenkung der Kompliziertheit der Terme in Ableitungen und aufloesbare Fragmente der Praedikatenkalkuele (Russisch)* ◇ B35 F20 ◇

1975

CHUBARYAN, A.A. & NGUEN VAN TIN' *Some estimates of the complexity characteristics of deductions in classical propositional calculus (Russian)* ◇ B05 F20 ◇

COOK, S.A. *Feasibly constructive proofs and the propositional calculus* ◇ B05 F20 ◇

EMDE BOAS VAN, P. *Ten years of speedup* ◇ A10 D15 D20 F20 G05 ◇

FELSCHER, W. *Kombinatorische Konstruktionen mit Beweisen und Schnittelimination* ◇ F05 F20 ◇

GOLDFARB, W.D. *On the effective ω-rule* ◇ F20 F30 ◇

KREISEL, G. *Observations on a recent generalization of completeness theorems due to Schuette* ◇ A05 C07 C57 F05 F20 F35 F50 ◇

KREISEL, G. & MINTS, G.E. & SIMPSON, S.G. *The use of abstract language in elementary metamathematics: Some pedagogic examples* ◇ A05 C07 C57 C75 F05 F07 F20 F50 ◇

MINTS, G.E. *Finite investigations of transfinite derivations (Russian) (English summary)* ◇ F05 F07 F10 F20 F30 F35 F50 ◇

PAOLA DI, R.A. *A theorem on shortening the length of proof in formal systems of arithmetic* ◇ F20 F30 ◇

POPOV, S.V. *The complexity of deductions in certain propositional calculi (Russian)* ◇ B05 F20 ◇

1976

HARTMANIS, J. *On effective speed-up and long proofs of trivial theorems in formal theories (French summary)* ◇ B25 B35 D15 D20 F20 ◇

POPOV, S.V. *On the complexity of derivations in classical propositional calculus (Russian)* ◇ B05 F20 ◇

1977

CHUBARYAN, A.A. *The complexity of deductions in formal arithmetic and predicate calculus (Russian)* ◇ B10 F20 F30 ◇

KOZEN, D. *Lower bounds for natural proof systems* ◇ D15 F20 ◇

KREISEL, G. *Some uses of proof theory for finding computer programs* ◇ B75 F05 F20 ◇

NORGELA, S.A. *Minus-normal sequential calculi (Russian) (English summary)* ◇ F20 ◇

POPOV, S.V. *Some complexity characteristics of derivations in propositional calculus (Russian) (English summary)* ◇ B05 B70 F20 ◇

POPOV, S.V. *The complexity of derivations in intuitionistic propositional calculus (Russian)* ◇ F20 F50 ◇

STATMAN, R. *Complexity and derivations from quantifier-free Horn formulae, mechanical introduction of explicit definitions, and refinement of completeness theorems* ◇ B35 F20 ◇

STATMAN, R. *Herbrand's theorem and Gentzen's notion of a direct proof* ◇ D15 F05 F07 F20 ◇

YUKAMI, T. *A theorem on the formalized arithmetic with function symbols ' and +* ◇ F20 F30 ◇

1978

JONES, JAMES P. *Three universal representations of recursively enumerable sets* ◇ D25 F20 ◇

KANOVICH, M.I. *An estimate of the complexity of arithmetic incompleteness (Russian)* ◇ D15 F20 F30 ◇

KHACHATRYAN, M.A. *On the derivation complexity of some formulas in sequential calculi (Russian)* ◇ F20 F50 ◇

LOVELAND, D.W. & REDDY, C.R. *Presburger arithmetic with bounded quantifier alternation* ◇ B25 C10 D15 F20 F30 ◇

PARLAMENTO, F. *Sulla completezza della ω-regola (English summary)* ◇ F20 F30 ◇

SOLOMON, M.K. *Some results on measure independent Goedel speed-ups* ◇ D15 F20 ◇

STATMAN, R. *Bounds for proof-search and speed-up in the predicate calculus* ◇ B35 F05 F20 ◇

YUKAMI, T. *A note on a formalized arithmetic with function symbols ' and +* ◇ F20 F30 ◇

1979

COOK, S.A. & RECKHOW, R.A. *The relative efficiency of propositional proof systems* ◇ B05 D15 F20 ◇

DARDZHANIYA, G.K. *Polynomial complexities of deduction of some logical calculi (Russian) (English summary)* ◇ B35 F20 ◇

DAWSON JR., J.W. *The Goedel incompleteness theorem from a length-of-proof perspective* ◇ A10 F20 F30 H15 ◇

DEKHTYAR', M.I. *Complexity spectra of recursive sets of approximability of initial segments of complete problems* ◇ D10 D15 D30 F20 F30 ◇

HAFNER, I. *On proof length in the equivalential calculus* ◇ B20 F20 ◇

KANOVICH, M.I. *A complexity version of Goedel's incompleteness theorem* ◇ D15 F20 F30 ◇

OREVKOV, V.P. *Lower bounds for lengthening of proofs after cut-elimination (Russian) (English summary)* ◇ F05 F07 F20 ◇

SOLOV'EV, S.V. *Preservation of the equivalence of proofs under reduction of the formula depth (Russian) (English summary)* ◇ F07 F20 ◇

SOLOV'EV, S.V. *The increase in length of an L-derivation transformed into a natural deduction (Russian) (English summary)* ◇ F07 F20 F50 ◇

STATMAN, R. *Intuitionistic propositional logic is polynomial-space complete* ◇ B40 D15 F20 F50 ◇

STATMAN, R. *Lower bounds on Herbrand's theorem* ◇ F05 F07 F20 ◇

WOEHL, K. *Zur Komplexitaet der Presburger Arithmetik und des Aequivalenz-Problems einfacher Programme* ◇ B25 B75 D15 F20 F30 ◇

1980

BEN-ARI, M. *A simplified proof that regular resolution is exponential* ◇ B35 D15 F20 ◇

KRAMOSIL, I. *Statistical testing procedure for lengths of formalized proofs* ◇ B35 F20 ◇

MIYATAKE, T. *On the length of proofs in formal systems* ◇ F20 F30 ◇

PLAISTED, D.A. *The application of multivariate polynomials to inference rules and partial tests for unsatisfiability* ◇ B35 D15 F20 ◇

1981

DANTSIN, E.YA. *Two systems for proving tautologies based on the splitting method (Russian) (English summary)* ◇ B35 D15 F20 ◇

GOLDFARB, W.D. *The undecidability of the second-order unification problem* ◇ B15 D35 F20 ◇

HATCHER, W.S. & HODGSON, B.R. *Complexity bounds on proofs* ◇ D25 F20 ◇

JUSTEN, K. *A note on regular resolution* ◇ B35 F20 ◇

MIYATAKE, T. *On the length of proofs in a formal system of recursive arithmetic* ◇ F20 F30 ◇

SAZONOV, V.YU. *On existence of complete predicate calculus in metamathematics without exponentiation* ◇ F20 F30 ◇

STATMAN, R. *Speed-up by theories with infinite models* ◇ F20 ◇

1982

CHUBARYAN, A.A. *Complexity characteristics of inferences in systems of predicate calculus and formal arithmetic (Russian) (Armenian summary)* ◇ B10 F20 F30 ◇

EVANGELIST, M. *Non-standard propositional logics and their application to complexity theory* ◇ B80 D15 F20 ◇

NORWOOD, F.H. *Long proofs* ◇ B35 F20 ◇

PLOTKIN, J.M. & ROSENTHAL, J.W. *The expected complexity of analytic tableaux analysis in propositional calculus* ◇ B05 B35 D15 F20 ◇

SCHOENFELD, W. *Upper bounds for proof-search in a sequent calculus for relational equations* ◇ B35 F20 ◇

1983

CELLUCCI, C. *The complexity of proofs in first-order predicate logic (Italian)* ◇ F20 ◇

HARTMANIS, J. *On Goedel speed-up and succinctness of language representation* ◇ D15 F20 ◇

SCHOENFELD, W. *Proof search for unprovable formulas* ◇ B35 F07 F20 ◇

SPENCER, J.H. *Short theorems with long proofs* ◇ F20 ◇

YUKAMI, T. *A theorem on lengths of proof of Presburger formulas* ◇ B35 F20 ◇

1984

GAVRILENKO, YU.V. *Monotonic theories of accessible numbers (Russian)* ◇ F20 F30 F65 ◇

HONG, JIAWEI *The complexity of formal proving (Chinese)* ◇ F20 ◇

OREVKOV, V.P. *Upper bounds for lengthening of proofs after cut elimination (Russian) (English summary)* ◇ F05 F07 F20 ◇

SCARPELLINI, B. *Complexity of subcases of Presburger arithmetic* ◇ B25 D15 F20 F30 ◇

YUKAMI, T. *Some results on speed-up* ◇ D15 F20 F30 ◇

1985

CELLUCCI, C. *Complexity of reduction trees in predicate calculus* ◇ F20 ◇

CELLUCCI, C. *Proof theory and complexity* ◇ D15 F05 F20 F98 ◇

SCARPELLINI, B. *Lower bound results on lengths of second-order formulas* ◇ B15 C13 D10 F20 F35 ◇

YOUNG, P. *Goedel theorems, exponential difficulty and undecidability of arithmetic theories: an exposition* ◇ D15 D35 F20 F30 ◇

F25 Relative consistency and interpretations

1900

HILBERT, D. *Mathematische Probleme* ◇ A05 D35 E30 E50 F25 ◇

1926

FINSLER, P. *Formale Beweise und die Entscheidbarkeit* ◇ A05 D35 F25 ◇

1930

HERBRAND, J. *Recherches sur la theorie de la demonstration* ◇ B10 B25 F05 F07 F25 F30 ◇

1931

GOEDEL, K. *Ueber formal unentscheidbare Saetze der "Principia Mathematica" und verwandter Systeme I* ◇ B28 B30 D20 D35 F25 F30 F35 ◇

1934

ITO, MAKOTO *Einige Anwendungen der Theorie des Entscheidungsproblems zur Axiomatik I* ◇ B30 C75 F25 ◇

ROESSLER, K. *Beweis der Widerspruchsfreiheit des Funktionenkalkuels der mathematischen Logik (French summary)* ◇ B10 F25 ◇

1935

BETH, E.W. *La metamathematique et ses applications au probleme de la non-contradiction de la logique et de l'arithmetique* ◇ F25 F30 ◇

BETH, E.W. *Sur un theoreme concernant le principe du tiers exclu et ses applications dans la theorie de la non-contradiction* ◇ F25 F50 ◇

ITO, MAKOTO *Einige Anwendungen der Theorie des Entscheidungsproblems zur Axiomatik. II* ◇ B30 F25 ◇

1936

BARZIN, M. *Note sur la demonstration de M.E.W.Beth* ◇ A05 F25 F50 ◇

1937

ERRERA, A. *Sur les demonstrations de non-contradiction* ◇ F25 ◇

1938

FITCH, F.B. *The consistency of the ramified Principia* ◇ B15 F25 ◇

1950

ACKERMANN, W. *Widerspruchsfreier Aufbau der Logik I. Typenfreies System ohne tertium non datur* ◇ B15 E70 F07 F25 F65 ◇

MYHILL, J.R. *A system which can define its own truth* ◇ B30 D20 F25 F30 ◇

ROSSER, J.B. & WANG, HAO *Non-standard models for formal logics* ◇ B15 C62 E30 E70 F25 H15 H20 ◇

WANG, HAO *A formal system of logic* ◇ E70 F25 ◇

WANG, HAO *Remarks on the comparison of axiom systems* ◇ B15 F25 F30 ◇

1951

KREISEL, G. *On the interpretation of non-finitist proofs I* ◇ F10 F25 F30 F50 ◇

WANG, HAO *Arithmetic models for formal systems* ◇ F25 F30 ◇

WANG, HAO *Arithmetic translation of axiom systems* ◇ F25 F30 ◇

1952

ACKERMANN, W. *Widerspruchsfreier Aufbau einer typenfreien Logik I (erweitertes System)* ◇ B10 B15 E70 F25 F65 ◇

KREISEL, G. *On the concepts of completeness and interpretation of formal systems* ◇ A05 F25 F30 ◇

KREISEL, G. *On the interpretation of non-finitist proofs II. Interpretation of number theory. Applications* ◇ F10 F25 F30 F50 ◇

KREISEL, G. *Some concepts concerning formal systems of number theory* ◇ F10 F25 F30 ◇

MOSTOWSKI, ANDRZEJ *On models of axiomatic systems* ◇ C62 F25 F30 F35 ◇

SZMIELEW, W. & TARSKI, A. *Mutual interpretability of some essentially undecidable theories* ◇ D35 F25 ◇

WANG, HAO *Truth definitions and consistency proofs* ◇ F25 F30 F65 ◇

1953

MCNAUGHTON, R. *Some formal relative consistency proofs* ◇ F25 F30 F35 ◇

MOSTOWSKI, ANDRZEJ & ROBINSON, R.M. & TARSKI, A. *Undecidable theories* ◇ D35 F25 F30 ◇

TARSKI, A. *A general method in proofs of undecidability* ◇ D35 F25 ◇

1954

KREISEL, G. *Remark on complete interpretations by models* ◇ B15 F25 F30 ◇

QUINE, W.V.O. *Interpretations of sets of conditions* ◇ F25 ◇

QUINE, W.V.O. *Reduction to a dyadic predicate* ◇ F25 ◇

SHOENFIELD, J.R. *A relative consistency proof* ◇ B30 E30 E35 E70 F25 ◇

1955

KREISEL, G. *Models, translations and interpretations* ◇ F10 F25 ◇

KREISEL, G. & WANG, HAO *Some applications of formalized consistency proofs* ◇ F25 F30 ◇

1956

QUINE, W.V.O. *On formulas with valid cases* ◇ F25 ◇

1957

FEFERMAN, S. *Formal consistency proofs and interpretability of theories* ◇ F25 ◇

MONTAGUE, R. *Non-finite axiomatizability* ◇ B30 F25 F30 ◇

MONTAGUE, R. *Two theorems on relative interpretability* ◇ F25 ◇

TAKEUTI, G. *On Skolem's theorem* ◇ C07 F25 ◇

TAKEUTI, G. *Remark on my paper: On Skolem's theorem* ◇ C07 F25 F30 ◇

1958

KINO, A. *A consistency-proof of a formal theory of Ackermann's ordinal numbers* ◇ F15 F25 ◇

KREISEL, G. *Mathematical significance of consistency proofs* ◇ C57 C60 F05 F25 F50 ◇

MAEHARA, S. *Another proof of Takeuti's theorems on Skolem's paradox* ◇ C07 F25 F30 ◇

1959

MEIGNE, M. *La consistance des theories formelles et le fondement des mathematiques* ◇ A05 F25 F98 ◇

1960

FEFERMAN, S. *Arithmetization of metamathematics in a general setting* ◇ D55 F07 F25 F30 F40 ◇

KOHL, H.R. & PARSONS, C. *Self-reference, truth, and provability* ◇ A05 F25 F30 ◇

1961

MONTAGUE, R. *Semantical closure and non-finite axiomatizability I* ◇ B15 E30 F25 F30 ◇

OREY, S. *Relative interpretations* ◇ C62 F25 ◇

TAKEUTI, G. *Remarks on the truth definition* ◇ F25 F30 ◇

1962

FEFERMAN, S. & KREISEL, G. & OREY, S. *1-consistency and faithful interpretations* ◇ F25 F30 F35 ◇

MONTAGUE, R. *Theories incomparable with respect to relative interpretability* ◇ C07 F25 F30 ◇

1963

HATCHER, W.S. *La notion d'equivalence entre systemes formels et une generalisation inductive du systeme dit "New Foundations" de Quine* ◇ E70 F25 ◇

KUBINSKI, T. *A proof of consistency of Borkowski's logical system containing Peano's arithmetic (Polish and Russian summaries)* ◇ B15 F25 F30 ◇

WANG, HAO *Ackermann's consistency proof* ◇ F25 F30 ◇

WANG, HAO *Relative strength and reducibility* ◇ B10 E30 E70 F25 F30 ◇

WANG, HAO *The arithmetization of metamathematics* ◇ F25 F30 ◇

1964

HATCHER, W.S. *La notion d'equivalence entre systemes formels* ◇ B22 F25 ◇

1965

GUREVICH, Y. *Existential interpretation (Russian)* ◇ B20 B25 C60 D35 F25 ◇

HAJEK, P. *Syntactic models of axiomatic theories* ◇ E30 E70 F25 ◇

HANF, W.P. *Model-theoretic methods in the study of elementary logic* ◇ B25 C07 D20 D25 D35 F25 ◇

MONTAGUE, R. *Interpretability in terms of models* ◇ C07 F25 ◇

RABIN, M.O. *A simple method for undecidability proofs and some applications* ◇ D35 F25 ◇

1966

HAJEK, P. *Generalized interpretability in terms of models. Note to a paper of R.Montague (Czech and Russian summaries)* ◇ C07 F25 ◇

1967

LAMBERT, K. & SCHARLE, T.W. *A translation theorem for two systems of free logic* ◇ B20 B60 F25 ◇

MINTS, G.E. & OREVKOV, V.P. *On imbedding operators (Russian)* ◇ F25 F30 F50 ◇

PREVIALE, F. *Teorie equipollenti e K-equipollenti* ◇ F25 ◇

1968

CLAY, R.E. *The consistency of Lesniewski's mereology relative to the real number system* ◇ B28 E35 E70 F25 ◇

DALLA CHIARA SCABIA, M.L. *Modelli sintattici e semantici delle teorie elementari* ◇ C07 C98 E35 E45 F25 F98 ◇

KREISEL, G. & LEVY, A. *Reflection principles and their use for establishing the complexity of axiomatic systems* ◇ E30 E47 F25 F30 F35 F50 ◇

MALMNAES, P.E. & PRAWITZ, D. *A survey of some connections between classical, intuitionistic and minimal logic* ◇ B55 F05 F25 F50 ◇

1970

COLLINS, G.E. & HALPERN, J.D. *On the interpretability of arithmetic in set theory* ◇ B28 D35 E30 F25 F30 ◇

HAJEK, P. *Logische Kategorien* ◇ B10 C40 E40 F25 G30 ◇

HAUSCHILD, K. & RAUTENBERG, W. *Universelle Interpretierbarkeit in Verbaenden (Russian, English and French summaries)* ◇ D35 F25 G10 ◇

LATREILLE, F. & PAILLET, J.L. *Rapports entre l'interpretabilite d'une relation par une autre et l'interpretabilite locale* ◇ F25 ◇

1971

HAJEK, P. *On interpretability in set theories I* ◇ F25 ◇

HAUSCHILD, K. & RAUTENBERG, W. *Interpretierbarkeit in der Gruppentheorie* ◇ C60 D35 F25 ◇

HAUSCHILD, K. & RAUTENBERG, W. *Interpretierbarkeit und Entscheidbarkeit in der Graphentheorie I* ◇ B25 C65 D35 F25 ◇

1972

HAJEK, P. *Contributions to the theory of semisets I* ◇ E30 E35 E70 F25 ◇

HAJEK, P. & HAJKOVA, M. *On interpretability in theories containing arithmetic* ◇ F25 F30 ◇

HAJEK, P. *On interpretability in set theories II* ◇ F25 ◇

HAUSCHILD, K. & HERRE, H. & RAUTENBERG, W. *Interpretierbarkeit und Entscheidbarkeit in der Graphentheorie II* ◇ B25 C65 D35 F25 ◇

HAUSCHILD, K. *Universalitaet in der Ringtheorie* ◇ C60 D35 F25 ◇

VOPENKA, P. & HAJEK, P. *The theory of semisets* ◇ E35 E40 E70 E98 F25 ◇

1973

BARWISE, J. *A preservation theorem for interpretations* ◇ C40 F25 ◇

1974

ANDREWS, P.B. *Resolution and the consistency of analysis* ◇ B15 F05 F25 F35 ◇

APT, K.R. & MAREK, W. *Second order arithmetic and related topics* ◇ C62 D55 E45 F25 F35 ◇

GENTZEN, G. *Der erste Widerspruchsfreiheitsbeweis fuer die klassische Zahlentheorie* ◇ F05 F15 F25 F30 ◇

KOREC, I. & PERETYAT'KIN, M.G. & RAUTENBERG, W. *Definability in structures of finite valency* ◇ C60 C65 F25 ◇

KRAJEWSKI, S. *Predicative expansions of axiomatic theories* ◇ C30 C62 E30 E70 F25 ◇

PERRIN, M.J. & ZALC, A. *Sous-systemes bicommutables d'analyse et de theorie des ensembles* ◇ E25 E30 F25 F35 ◇

RESNIK, M.D. *On the philosophical significance of consistency proofs* ◇ A05 F25 F30 ◇

1975

ANDERSON, A.R. *Fitch on consistency* ◇ B15 F05 F25 ◇

MCALOON, K. *Formules de Rosser pour ZF (English summary)* ◇ C62 C70 E30 E35 E45 F25 ◇

1976

FLANNAGAN, T.B. *A new finitary proof of a theorem of Mostowski* ◇ E30 E70 F25 ◇

KOREC, I. & RAUTENBERG, W. *Model-interpretability into trees and applications* ◇ B25 C45 C60 C65 F25 ◇

MAREK, W. & SREBRNY, M. *Urelements and extendability* ◇ C62 E30 E70 F25 F30 F35 ◇

PRAZMOWSKI, K. & SZCZERBA, L.W. *Interpretability and categoricity (Russian summary)* ◇ C35 F25 ◇

SHELAH, S. *Interpreting set theory in the endomorphism semi-group of a free algebra or in a category* ◇ C55 E05 F25 G30 ◇

WEESE, M. *The universality of boolean algebras with the Haertig quantifier* ◇ C55 C80 D35 F25 G05 ◇

1977

ARTIGUE, M. & ISAMBERT, E. *Quelques resultats de bicommutabilite en arithmetique du troisieme ordre (English summary)* ◇ E30 F25 F35 ◇

MOTOHASHI, N. *Partially ordered interpretations* ◇ C90 E40 F25 F50 ◇

MYCIELSKI, J. *A lattice of interpretability types of theories* ◇ F25 G10 ◇

SMORYNSKI, C.A. *The incompleteness theorems* ◇ C62 F25 F30 F98 ◇

SZCZERBA, L.W. *Interpretability of elementary theories* ◇ B30 F25 ◇

1978

ARTIGUE, M. & ISAMBERT, E. & PERRIN, M.J. & ZALC, A. *Some remarks on bicommutability* ◇ C62 E30 F25 F35 ◇

CRABBE, M. *Ambiguity and stratification* ◇ B15 E35 E70 F25 ◇

DELON, F. *Definition de l'arithmetique dans la theorie des anneaux de series formelles (English summary)* ◇ C60 D35 F25 F35 ◇

MAREK, W. *Some comments on the paper by Artigue, Isambert, Perrin and Zalc: "Some remarks on bicommutability"* ◇ C62 E30 E45 F25 F35 ◇

PINTER, C. *Properties preserved under definitional equivalence and interpretations* ◇ C25 C35 C50 F25 ◇

PRAZMOWSKI, K. *Conciseness and hierarchy of theories (Russian summary)* ◇ C07 F25 ◇

SETTE, A.-M. & SZCZERBA, L.W. *Algebraic characterization of interpretability* ◇ F25 G30 ◇

SVEJDAR, V. *Degrees of interpretability* ◇ F25 F30 G10 ◇

1979

EPSTEIN, R.L. & SZCZERBA, L.W. *Relatedness and interpretability* ◇ F25 ◇

GUASPARI, D. *Partially conservative extensions of arithmetic* ◇ C62 F25 F30 ◇

HANAZAWA, M. *An interpretation of Skolem's paradox in the predicate calculus with ε-symbol* ◇ B10 C07 F05 F25 ◇

HANAZAWA, M. *On a problem about Skolem's paradox of Takeuti's version* ◇ E30 E35 E65 E70 F25 ◇

LINDSTROEM, P. *Some results on interpretability* ◇ F25 F30 ◇

SZCZERBA, L.W. *Interpretability and axiomatizability* ◇ B30 F25 ◇

1980

CORCORAN, J. *On definitional equivalence and related topics* ◇ A10 C40 F25 ◇

DETLEFSEN, M. *On a theorem of Feferman* ◇ A05 F25 F30 ◇

GUREVICH, Y. *Two notes on formalized topology* ◇ C65 C85 F25 ◇

HODGES, W. *Interpreting number theory in nilpotent groups* ◇ C25 C60 C62 F25 F35 ◇

JAMBU-GIRAUDET, M. *Theorie des modeles de groupes d'automorphismes d'ensembles totalement ordonnes 2-homogenes (English summary)* ◇ C07 C60 C62 C65 D35 E07 F25 F35 ◇

KHAKHANYAN, V.KH. *The consistency of intuitionistic set theory with Church's principle and the uniformization principle (Russian)* ◇ D20 E35 E70 F25 F50 ◇

KHAKHANYAN, V.KH. *The consistency of intuitionistic set theory with formal mathematical analysis (Russian)* ◇ D20 E35 E70 F25 F35 F50 ◇

PABION, J.F. *TT_3I est equivalent a l'arithmetique du second ordre (English summary)* ◇ B15 F25 F35 ◇
SETTE, A.-M. *Functorial approach to interpretability* ◇ C90 F25 G30 ◇
SZCZERBA, L.W. *Interpretations with parameters* ◇ B25 C07 C35 C45 F25 ◇

1981

HAJEK, P. *On interpretability in theories containing arithmetics II* ◇ B28 B45 F25 F30 ◇
JAMBU-GIRAUDET, M. *Interpretations d'arithmetiques dans des groupes et des treillis* ◇ C07 C62 C65 D35 E07 F25 F30 G10 ◇
KANOVEJ, V.G. *The theory of Zermelo without the power set axiom and the theory of Zermelo-Fraenkel without the power set axiom are equiconsistent (Russian)* ◇ E30 E35 F25 ◇
SOLOV'EV, S.V. *The category of finite sets and Cartesian closed categories (Russian) (English summary)* ◇ F07 F25 G30 ◇
SVEJDAR, V. *A sentence that is difficult to interpret* ◇ E30 F25 ◇
WOJTYLAK, P. *Mutual interpretability of sentential logics. I,II* ◇ B22 F25 ◇

1982

FURMANOWSKI, T. *Adjoint interpretations of sentential calculi* ◇ B22 F25 G30 ◇
GUREVICH, Y. *Existential interpretation II* ◇ B20 D35 F25 ◇
KOWALCZYK, W. *A sufficient condition for the consistency of P=NP with Peano arithmetic* ◇ D15 F25 F30 ◇
PINUS, A.G. *Complete embeddings of categories of algebraic systems and definability of a model for its semigroup of endomorphisms (Russian)* ◇ C07 F25 G30 ◇

1983

DANKO, W. *Interpretability of algorithmic theories* ◇ B25 B75 C35 F25 ◇
GUREVICH, Y. & SHELAH, S. *Interpreting second-order logic in the monadic theory of order* ◇ B15 C65 C85 D35 E07 E50 F25 ◇
PUDLAK, P. *Some prime elements in the lattice of interpretability types* ◇ E30 F25 F30 G10 ◇
SVEJDAR, V. *Modal analysis of generalized Rosser sentences* ◇ B45 F25 F30 ◇
SZCZERBA, L.W. *Interpretations* ◇ B30 C07 F25 ◇

1984

BENTHEM VAN, J.F.A.K. & PEARCE, D.A. *A mathematical characterization of interpretation between theories* ◇ C20 F25 ◇
GUREVICH, Y. & SHELAH, S. *The monadic theory and the "next world"* ◇ C62 C65 C85 E40 F25 ◇
LINDSTROEM, P. *On certain lattices of degrees of interpretability* ◇ F25 F30 ◇
LINDSTROEM, P. *On faithful interpretability* ◇ F25 F30 ◇
LINDSTROEM, P. *On partially conservative sentences and interpretability* ◇ F25 F30 ◇
SHELAH, S. *On logical sentences in PA* ◇ C62 C80 E05 F25 F30 ◇
SZCZERBA, L.W. *Some remarks on interpretations with parameters* ◇ F25 ◇

1985

HOOK, J.L. *A note on interpretations of many-sorted theories* ◇ B10 C07 F25 ◇
KRAJICEK, J. *Some theorems on the lattice of local interpretability types* ◇ F25 ◇
KRYNICKI, M. *Interpretations in nonelementary languages* ◇ C80 C95 F25 ◇
LINDSTROEM, P. *Provability and interpretability in theories containing arithmetic* ◇ F25 F30 ◇
PUDLAK, P. *Cuts, consistency statements and interpretations* ◇ C62 E30 F25 ◇
SETTE, A.-M. & SZCZERBA, L.W. *Characterizations of elementary interpretations in category theory* ◇ C07 F25 G30 ◇

F30 First-order arithmetic and fragments

1889

PEANO, G. *The principles of arithmetic, presented by a new method (Latin)* ◇ B03 B28 E75 F30 ◇

1923

SKOLEM, T.A. *Begruendung der elementaren Arithmetik durch die rekurrierende Denkweise ohne Anwendung scheinbarer Veraenderlichen mit unendlichem Ausdehnungsbereich* ◇ A05 B28 D20 F30 ◇

1925

ACKERMANN, W. *Begruendung des "tertium non datur" mittels der Hilbertschen Theorie der Widerspruchsfreiheit* ◇ A05 B10 F30 ◇

1928

KALMAR, L. *Eine Bemerkung zur Entscheidungstheorie* ◇ B20 B25 F30 ◇

1929

HERBRAND, J. *Non-contradiction des axiomes arithmetiques* ◇ F30 ◇

HERBRAND, J. *Sur quelques proprietes des propositions vraies et leurs applications* ◇ B10 F30 ◇

1930

GOEDEL, K. *Einige metamathematische Resultate ueber Entscheidungsdefinitheit und Widerspruchsfreiheit* ◇ F30 ◇

HERBRAND, J. *Recherches sur la theorie de la demonstration* ◇ B10 B25 F05 F07 F25 F30 ◇

HEYTING, A. *Die formalen Regeln der intuitionistischen Mathematik* ◇ F30 F35 F50 ◇

PRESBURGER, M. *Ueber die Vollstaendigkeit eines gewissen Systems der Arithmetik ganzer Zahlen, in welchem die Addition als einzige Operation hervortritt* ◇ B25 B28 C10 C35 F30 ◇

1931

GOEDEL, K. *Ueber formal unentscheidbare Saetze der "Principia Mathematica" und verwandter Systeme I* ◇ B28 B30 D20 D35 F25 F30 F35 ◇

GOEDEL, K. *Ueber Vollstaendigkeit und Widerspruchsfreiheit* ◇ F30 F35 ◇

HERBRAND, J. *Sur la non-contradiction de l'arithmetique* ◇ F05 F30 ◇

HERBRAND, J. *Sur le probleme fondamental de la logique mathematique* ◇ A05 B25 B30 F30 F99 ◇

SKOLEM, T.A. *Ueber einige Satzfunktionen in der Arithmetik* ◇ B25 B28 C10 F30 ◇

1932

GOEDEL, K. *Zur intuitionistischen Arithmetik und Zahlentheorie* ◇ F30 F50 ◇

LINDENBAUM, A. *Sur les constructions non-effectives dans l'arithmetique elementaire* ◇ F30 ◇

TARSKI, A. *Der Wahrheitsbegriff in den Sprachen der deduktiven Disziplinen* ◇ A05 B30 C07 C40 F30 ◇

1933

SKOLEM, T.A. *Ueber die Unmoeglichkeit einer Charakterisierung der Zahlenreihe mittels eines endlichen Axiomensystems* ◇ B28 C20 C62 F30 H15 ◇

TARSKI, A. *Einige Betrachtungen ueber die Begriffe der ω-Widerspruchsfreiheit und der ω-Vollstaendigkeit* ◇ B28 C62 F30 ◇

TARSKI, A. *On the notion of truth in reference to formalized deductive sciences (Polish)* ◇ A05 B15 B30 C07 C40 F30 ◇

1934

GOEDEL, K. *Ueber die Laenge von Beweisen* ◇ B10 F20 F30 F35 ◇

HETPER, W. *Semantische Arithmetik* ◇ B28 F30 ◇

HILBERT, D. & BERNAYS, P. *Grundlagen der Mathematik I* ◇ A05 B98 F05 F30 F98 ◇

SKOLEM, T.A. *Ueber die Nichtcharakterisierbarkeit der Zahlenreihe mittels endlich oder abzaehlbar unendlich vieler Aussagen mit ausschliesslich Zahlenvariablen* ◇ B28 C20 C62 F30 H15 ◇

1935

BETH, E.W. *La metamathematique et ses applications au probleme de la non-contradiction de la logique et de l'arithmetique* ◇ F25 F30 ◇

GENTZEN, G. *Untersuchungen ueber das logische Schliessen I,II* ◇ B10 B25 F05 F07 F30 F50 ◇

1936

GENTZEN, G. *Die Widerspruchsfreiheit der reinen Zahlentheorie* ◇ B28 F05 F07 F30 ◇

ROSSER, J.B. *Extensions of some theorems of Goedel and Church* ◇ D25 D35 F30 ◇

SKOLEM, T.A. *Ueber die Zurueckfuehrbarkeit einiger durch Rekursionen definierter Relationen auf "arithmetische"* ◇ B28 F30 ◇

1937

HELMER, O. *Perelman versus Goedel* ◇ A05 F30 ◇

KLEIN, H. *Einfuehrung von Ordnung und Rechenoperationen bei natuerlichen Zahlen auf Grund der Peano-Axiome* ◇ F30 ◇

ROSSER, J.B. *Goedel theorems for non-constructive logics* ◇ D70 F30 ◇

SECKENDORFF VON, V. *Beweis des Induktionsschlusses der natuerlichen Zahlen aus der Dedekindschen Definition endlicher Mengen* ◇ B28 E10 F30 ◇

1938

GENTZEN, G. *Neue Fassung des Widerspruchsfreiheitsbeweises fuer die reine Zahlentheorie* ◇ F05 F15 F30 ◇

ONO, K. *Logische Untersuchungen ueber die Grundlagen der Mathematik* ◇ F05 F30 ◇

1939

CHWISTEK, L.B. *A formal proof of Goedel's theorem* ◇ F30 ◇

HILBERT, D. & BERNAYS, P. *Grundlagen der Mathematik II* ◇ A05 B98 F05 F15 F30 F40 F98 ◇

MOSTOWSKI, ANDRZEJ *Bemerkungen zum Begriff der inhaltlichen Widerspruchsfreiheit* ◇ F30 ◇

ROSSER, J.B. *An informal exposition of proofs of Goedel's theorems and Church's theorem* ◇ A05 B20 D35 F30 ◇

SKOLEM, T.A. *Eine Bemerkung ueber die Induktionsschemata in der rekursiven Zahlentheorie* ◇ F30 ◇

TARSKI, A. *On undecidable statements in enlarged systems of logic and the concept of truth* ◇ B10 F30 ◇

TURING, A.M. *Systems of logic based on ordinals* ◇ B40 D25 D30 F15 F30 ◇

1940

ACKERMANN, W. *Zur Widerspruchsfreiheit der Zahlentheorie* ◇ F30 ◇

BARZIN, M. *Logique symbolique. Sur la portee du theoreme de M.Goedel* ◇ A05 F30 ◇

PETER, R. *Contribution to recursive number theory* ◇ D20 F30 ◇

SKOLEM, T.A. *Einfacher Beweis der Unmoeglichkeit eines allgemeinen Loesungsverfahren fuer arithmetische Probleme* ◇ D35 F30 ◇

1941

CURRY, H.B. *A formalization of recursive arithmetic* ◇ B28 D20 F30 ◇

1942

FINDLAY, J. *Goedelian sentences: a non-numerical approach* ◇ A05 F30 ◇

1943

GENTZEN, G. *Beweisbarkeit und Unbeweisbarkeit von Anfangsfaellen der transfiniten Induktion in der reinen Zahlentheorie* ◇ F15 F30 ◇

KALMAR, L. *Ein einfaches Beispiel fuer ein unentscheidbares arithmetisches Problem (Hungarian) (German summary)* ◇ D20 F30 ◇

1944

FINSLER, P. *Gibt es unentscheidbare Saetze?* ◇ A05 F30 ◇

GOODSTEIN, R.L. *On the restricted ordinal theorem* ◇ F15 F30 ◇

SKOLEM, T.A. *A note on recursive arithmetic* ◇ D20 F30 ◇

SKOLEM, T.A. *Some remarks on recursive arithmetic* ◇ D20 F30 ◇

1945

GOODSTEIN, R.L. *Function theory in an axiom-free equation calculus* ◇ B28 F30 F60 ◇

1946

MOSTOWSKI, ANDRZEJ *On undecidable sentences in formalized mathematical systems (Polish)* ◇ F30 F98 ◇

SKOLEM, T.A. *Recursive arithmetic (Norwegian)* ◇ F30 ◇

1947

CARRUCCIO, E. *Alcune conseguenze di un risultato del Goedel e la razionalita del reale* ◇ A05 F30 ◇

GOODSTEIN, R.L. *Transfinite ordinals in recursive number theory* ◇ E10 F15 F30 ◇

MOSTOWSKI, ANDRZEJ *On definable sets of positive integers* ◇ D55 F30 ◇

NELSON, D. *Recursive functions and intuitionistic number theory* ◇ F30 F50 ◇

SKOLEM, T.A. *The development of recursive arithmetic* ◇ F30 ◇

VALPOLA, V. *The position of negation in a language which expresses knowledge (Finnish)* ◇ A05 B28 F30 F50 ◇

1948

SCHROETER, K. *Die Arithmetik der natuerlichen Zahlen im Rahmen der Theorie der Verbaende* ◇ F30 G10 ◇

1949

CHAUVIN, A. *Generalisation du theoreme de Goedel* ◇ B15 F30 ◇

CHAUVIN, A. *Structures logiques* ◇ F30 ◇

KALMAR, L. *Quelques formes generales du theoreme de Goedel* ◇ F30 ◇

KALMAR, L. *Une forme du theoreme de Goedel sous des hypotheses minimales* ◇ F30 ◇

MOSTOWSKI, ANDRZEJ *A classification of logical systems* ◇ D55 F30 ◇

MOSTOWSKI, ANDRZEJ *An undecidable arithmetical statement* ◇ F30 ◇

MYHILL, J.R. *Note on an idea of Fitch* ◇ D55 F30 ◇

NOVIKOV, P.S. *On the axiom of complete induction (Russian)* ◇ B25 B28 D35 F30 ◇

QUINE, W.V.O. *On decidability and completeness* ◇ B25 D35 F30 ◇

ROBINSON, JULIA *Definability and decision problems in arithmetic* ◇ D35 F30 ◇

1950

FOGELIS, E. *On finite proofs of arithmetical theorems* ◇ B28 F30 ◇

GOODSTEIN, R.L. *The formal structure of a denumerable system* ◇ F30 F60 ◇

KALMAR, L. *A simple construction of undecidable propositions in formal systems (German and English)* ◇ F30 ◇

KALMAR, L. *Another proof of the Goedel-Rosser incompletability theorem* ◇ F30 ◇

KREISEL, G. *Note on arithmetic models for consistent formulae of the predicate calculus I* ◇ B10 C57 D45 D55 F30 ◇

LORENZEN, P. *Einfuehrung in die Logik* ◇ F30 ◇
MYHILL, J.R. *A system which can define its own truth* ◇ B30 D20 F25 F30 ◇
SCHUETTE, K. *Beweistheoretische Erfassung der unendlichen Induktion in der Zahlentheorie* ◇ F05 F30 ◇
SKOLEM, T.A. *A remark on the induction scheme* ◇ F30 ◇
WANG, HAO *Remarks on the comparison of axiom systems* ◇ B15 F25 F30 ◇

1951

BERNAYS, P. *Ueber das Induktionsschema in der rekursiven Zahlentheorie* ◇ F30 ◇
GOODSTEIN, R.L. *Constructive formalism: Essays on the foundations of mathematics* ◇ A05 B98 F30 F60 F98 ◇
GRENIEWSKI, H. *Arithmetics of natural numbers as part of the bi-valued propositional calculus* ◇ B05 F30 ◇
KREISEL, G. *On the interpretation of non-finitist proofs I* ◇ F10 F25 F30 F50 ◇
NEVES-REAL, L. *Kurt Goedel and the problems of the foundations of mathematics and the theory of sets (Portuguese)* ◇ A10 E25 E35 E45 E50 F30 ◇
ROBINSON, R.M. *Arithmetical definitions in the ring of integers* ◇ C65 F30 ◇
WANG, HAO *Arithmetic models for formal systems* ◇ F25 F30 ◇
WANG, HAO *Arithmetic translation of axiom systems* ◇ F25 F30 ◇

1952

HASENJAEGER, G. *Ueber ω-Unvollstaendigkeit in der Peano-Arithmetik* ◇ F30 ◇
IZUMI, Y. *Ueber den Begriff der ω-Vollstaendigkeit* ◇ F30 ◇
KLEENE, S.C. *Introduction to metamathematics* ◇ A05 B98 D98 F30 F50 F98 ◇
KREISEL, G. *On the concepts of completeness and interpretation of formal systems* ◇ A05 F25 F30 ◇
KREISEL, G. *On the interpretation of non-finitist proofs II. Interpretation of number theory. Applications* ◇ F10 F25 F30 F50 ◇
KREISEL, G. *Some concepts concerning formal systems of number theory* ◇ F10 F25 F30 ◇
MOLODSHIJ, V.N. *On interrelations of certain assertions of generality with the induction axiom in Peano's system of axioms (Russian)* ◇ C62 F30 ◇
MOSTOWSKI, ANDRZEJ *On models of axiomatic systems* ◇ C62 F25 F30 F35 ◇
MOSTOWSKI, ANDRZEJ *Sentences undecidable in formalized arithmetic. An exposition of the theory of Kurt Goedel* ◇ F30 F98 ◇
ROBINSON, R.M. *An essentially undecidable axiom system* ◇ D35 F30 ◇
SKOLEM, T.A. *Consideraciones sobre los fundamentos de la matematica I* ◇ B98 F30 F50 ◇
STENIUS, E. *Das Interpretationsproblem der formalisierten Zahlentheorie und ihre formale Widerspruchsfreiheit* ◇ F30 ◇
WANG, HAO *Truth definitions and consistency proofs* ◇ F25 F30 F65 ◇

1953

ACKERMANN, W. *Widerspruchsfreier Aufbau einer typenfreien Logik II* ◇ B10 B15 E35 E70 F30 F35 F65 ◇
FREUDENTHAL, H. *Zur Geschichte der vollstaendigen Induktion* ◇ A05 A10 F30 ◇
GOODSTEIN, R.L. *Permutation in recursive arithmetic* ◇ F30 ◇
KREISEL, G. *Note on arithmetic models for consistent formulae of the predicate calculus II* ◇ B10 C57 D45 D55 F30 ◇
KREISEL, G. *On a problem of Henkin's* ◇ F30 ◇
KREISEL, G. *The diagonal method in formalized arithmetic* ◇ F30 ◇
MCNAUGHTON, R. *Some formal relative consistency proofs* ◇ F25 F30 F35 ◇
MOSTOWSKI, ANDRZEJ & ROBINSON, R.M. & TARSKI, A. *Undecidability and essential undecidability in arithmetic* ◇ D35 F30 ◇
MOSTOWSKI, ANDRZEJ & ROBINSON, R.M. & TARSKI, A. *Undecidable theories* ◇ D35 F25 F30 ◇
MYHILL, J.R. *Arithmetic with creative definitions by induction* ◇ F30 ◇
QUINE, W.V.O. *On ω-inconsistency and a so-called axiom of infinity* ◇ E70 F30 ◇
SHANIN, N.A. *On some operations on logico-arithmetical formulae (Russian)* ◇ F30 F50 ◇
SKOLEM, T.A. *Consideraciones sobre los fundamentos de la matematica II* ◇ B98 F30 F50 ◇
SKOLEM, T.A. *Some considerations concerning recursive arithmetic* ◇ F30 ◇
USPENSKIJ, V.A. *The Goedel theorem and the theory of algorithms (Russian)* ◇ D20 D35 F30 ◇
USPENSKIJ, V.A. *Theorem of Goedel and theory of algorithms (Russian)* ◇ D20 D25 F30 ◇
WANG, HAO *Between number theory and set theory* ◇ B28 E30 F30 ◇
WANG, HAO *Certain predicates defined by induction schemata* ◇ D70 F30 F35 ◇
WANG, HAO *Quelques notions d'axiomatique* ◇ B25 B30 C07 C35 E30 F30 ◇

1954

GENTZEN, G. *Zusammenfassung von mehreren vollstaendigen Induktionen zu einer einzigen* ◇ B28 F30 ◇
GOODSTEIN, R.L. *Logic-free formalisations of recursive arithmetic* ◇ F30 F50 ◇
GOODSTEIN, R.L. *The recursive irrationality of π* ◇ F30 F60 ◇
GOODSTEIN, R.L. *The relatively exponential, logarithmic and circular functions in recursive function theory* ◇ D20 F30 F60 ◇
HENKIN, L. *A generalization of the notion of ω-consistency* ◇ B10 C07 F30 ◇
KREISEL, G. *Discussion sur divers themes* ◇ F30 F50 ◇
KREISEL, G. *Remark on complete interpretations by models* ◇ B15 F25 F30 ◇
LOEB, M.H. *Solution of a problem by Leon Henkin* ◇ F30 ◇

SHANIN, N.A. *On imbeddings of the classical logico-arithmetical calculus into the constructive logico-arithmetical calculus (Russian)* ◇ F30 F50 ◇

1955

GOODSTEIN, R.L. *On non-constructive theorems of analysis and the decision problem* ◇ B25 B28 F30 F60 ◇

KREISEL, G. & WANG, HAO *Some applications of formalized consistency proofs* ◇ F25 F30 ◇

LOEB, M.H. *Solution of a problem of Leon Henkin* ◇ F30 ◇

ROSE, A. *A Goedel theorem for an infinite-valued erweiterter Aussagenkalkuel* ◇ B50 F30 ◇

SHANIN, N.A. *On some logical problems of arithmetic (Russian)* ◇ D20 F30 F50 ◇

SKOLEM, T.A. *Peano's axioms and models of arithmetic* ◇ B28 C20 C62 F30 H15 ◇

WANG, HAO *Undecidable sentences generated by semantic paradoxes* ◇ A05 B28 F30 ◇

1956

GRZEGORCZYK, A. *Some proofs of undecidability of arithmetic* ◇ D35 F30 ◇

HARROP, R. *On disjunctions and existential statements in intuitionistic systems of logic* ◇ F30 F50 ◇

OREY, S. *On ω-consistency and related properties* ◇ C07 F30 ◇

ROBINSON, R.M. *Arithmetical representation of recursively enumerable sets* ◇ D25 F30 ◇

SCHUETTE, K. *Ein System des verknuepfenden Schliessens* ◇ B10 F07 F30 ◇

SKOLEM, T.A. *A version of the proof of equivalence between complete induction and the uniqueness of primitive recursion* ◇ B28 F30 ◇

SKOLEM, T.A. *An ordered set of arithmetic functions representing the least ε-number* ◇ F15 F30 ◇

1957

CHURCH, A. *Binary recursive arithmetic* ◇ D20 F30 ◇

COBHAM, A. *Effectively decidable theories* ◇ B25 F30 ◇

ESENIN-VOL'PIN, A.S. *Consistency proof of classical arithmetics with the aid of induction to ε_0 (following Schuette) (Russian)* ◇ F05 F30 ◇

FRIEDMAN, JOYCE *Some results in Church's restricted recursive arithmetic* ◇ F30 ◇

GOODSTEIN, R.L. *Recursive number theory. A development of recursive arithmetic in a logic-free equation calculus* ◇ B28 F30 F50 F60 F98 ◇

GRZEGORCZYK, A. *Decision problems* ◇ B25 C10 D35 D98 F30 ◇

HENKIN, L. *A generalization of the concept of ω-completeness* ◇ B10 C07 F30 ◇

KREISEL, G. *Goedel's interpretation of Heyting's arithmetic* ◇ F10 F30 F50 ◇

KUZNETSOV, A.V. *Completeness of the axiomatic system of arithmetics with the rules of constructive-infinite induction (Russian)* ◇ F30 ◇

LADRIERE, J. *Les limitations internes des formalismes. Etude sur la signification du theoreme de Goedel et des theoremes apparentes dans la theorie des fondements des mathematiques* ◇ A05 B98 F30 F98 ◇

MAEHARA, S. *General recursive functions in the number-theoretic formal system* ◇ D20 F30 ◇

MAEHARA, S. *Remark on Skolem's theorem concerning the impossibility of characterization of the natural number sequence* ◇ B28 C62 F30 ◇

MONTAGUE, R. *Non-finite axiomatizability* ◇ B30 F25 F30 ◇

OHASHI, K. *On undecidable theorems (Japanese)* ◇ F30 ◇

ROBINSON, R.M. *Restricted set-theoretical definitions in arithmetic* ◇ B28 F30 F35 ◇

SMULLYAN, R.M. *Languages in which self reference is possible* ◇ A05 B10 B28 F30 F40 ◇

SPECKER, E. *Eine Verschaerfung des Unvollstaendigkeitssatzes der Zahlentheorie (Russian summary)* ◇ F30 ◇

TAKEUTI, G. *Remark on my paper: On Skolem's theorem* ◇ C07 F25 F30 ◇

WANG, HAO *Symbolic representations of calculating machines* ◇ B70 D05 D10 F30 ◇

WANG, HAO *The axiomatization of arithmetic* ◇ A05 B28 F30 ◇

1958

GODDARD, L. *"True" and "provable"* ◇ A05 F30 ◇

GOEDEL, K. *Ueber eine bisher noch nicht benuetzte Erweiterung des finiten Standpunktes* ◇ A05 F10 F30 F50 ◇

GOODSTEIN, R.L. *Models of propositional calculi in recursive arithmetic* ◇ B05 B50 F30 F50 ◇

GRZEGORCZYK, A. & MOSTOWSKI, ANDRZEJ & RYLL-NARDZEWSKI, C. *The classical and the ω-complete arithmetic* ◇ C62 D55 D70 F30 ◇

KEMENY, J.G. *Undecidable problems of elementary number theory* ◇ F30 H15 ◇

KIREEVSKIJ, N.N. *Ueber die Allgemeingueltigkeit gewisser Zaehlausdruecke (Russian)* ◇ B25 F30 ◇

LEVY, A. *Comparison of subtheories* ◇ C07 C40 E30 F30 ◇

MAEHARA, S. *Another proof of Takeuti's theorems on Skolem's paradox* ◇ C07 F25 F30 ◇

NAGEL, E. & NEWMAN, J.R. *Goedel's proof* ◇ B28 D35 F30 F98 ◇

ROBINSON, R.M. *Restricted set-theoretical definitions in arithmetic* ◇ B28 F30 F35 ◇

SCHACH, A. *Two forms of mathematical induction* ◇ F30 ◇

SCHMIDT, J. *Einige Prinzipien der doppelten Induktion* ◇ F30 ◇

SHOENFIELD, J.R. *Open sentences and the induction axiom* ◇ F30 ◇

1959

BOUVERE DE, K.L. *A method in proofs of undefinability, with applications to functions in the arithmetic of natural numbers* ◇ C40 F30 ◇

EHRENFEUCHT, A. & FEFERMAN, S. *Representability of recursively enumerable sets in formal theories* ◇ D25 F30 ◇

ESENIN-VOL'PIN, A.S. *On Goedel's second theorem (Russian)* ◇ F30 ◇

FALEVICH, B.YA. *Incompleteness theorems in systems with infinite induction (Russian)* ◇ F30 ◇
KHLODOVSKIJ, I.N. *A new proof of the consistency of arithmetic (Russian)* ◇ F05 F30 ◇
KREISEL, G. *Interpretation of analysis by means of constructive functionals of finite types* ◇ D65 F10 F30 F35 F50 ◇
KREISEL, G. & SHOENFIELD, J.R. & WANG, HAO *Number theoretic concepts and recursive well-orderings* ◇ F15 F30 ◇
KURODA, S. *An investigation on the logical structure of mathematics. VI. Consistent V-systems T(V). VII. Set-theoretical contradictions* ◇ B28 B30 E70 F30 ◇
KURODA, S. *An investigation on the logical structure of mathematics. VIII. Consistency of the natural-number theory $T_1(N)$* ◇ B28 B30 F30 ◇
KURODA, S. *An investigation on the logical struture of mathematics. IX. Deductions in the natural-number theory $T_1(N)$. X. Concepts and sets* ◇ B15 B28 B30 F30 ◇
NISHIMURA, T. *On Goedel's theorem (Japanese)* ◇ D35 F30 ◇
SHOENFIELD, J.R. *On a restricted ω-rule (Russian summary)* ◇ F30 ◇
STEGMUELLER, W. *Unvollstaendigkeit und Unentscheidbarkeit. Die metamathematischen Resultate von Goedel, Church, Kleene, Rosser und ihre erkenntnistheoretische Bedeutung* ◇ A05 D35 F30 F98 ◇
TRAKHTENBROT, B.A. *Descriptive classifications in recursive arithmetics (Russian)* ◇ D20 F30 ◇
VUCKOVIC, V. *Partially ordered recursive arithmetics* ◇ D20 F30 ◇

1960

FEFERMAN, S. *Arithmetization of metamathematics in a general setting* ◇ D55 F07 F25 F30 F40 ◇
HARROP, R. *Concerning formulas of the types $A \rightarrow B \vee C$, $A \rightarrow (Ex)B(x)$ in intuitionistic formal systems* ◇ F30 F50 ◇
KOHL, H.R. & PARSONS, C. *Self-reference, truth, and provability* ◇ A05 F25 F30 ◇
KREISEL, G. *Ordinal logics and the characterization of informal concepts of proof* ◇ F07 F15 F30 F65 ◇
NISHIMURA, T. *On a certain system with infinite induction* ◇ F30 ◇
ROSE, H.E. *Independence of induction schemas in recursive arithmetic* ◇ F30 ◇
SCHUETTE, K. *Beweistheorie* ◇ B98 F05 F15 F30 F35 F98 ◇
SOUDIEUX, C. *De l'infini arithmetique* ◇ B28 F30 F98 ◇
VUCKOVIC, V. *Rekursive Modelle einiger nichtklassischer Aussagenkalkuele (Serbo-Croatian)* ◇ B22 F30 ◇

1961

ELGOT, C.C. *Decision problems of finite automata design and related arithmetics* ◇ B25 D03 D05 D35 F30 ◇
GOODSTEIN, R.L. *Recursive analysis* ◇ F30 F60 F98 ◇
HANSON, N.R. *The Goedel theorem* ◇ D20 D35 F30 ◇
MONTAGUE, R. *Semantical closure and non-finite axiomatizability I* ◇ B15 E30 F25 F30 ◇
MOSTOWSKI, ANDRZEJ *A generalization of the incompleteness theorem* ◇ F30 ◇
MUELLER, GERT H. *Nicht-Standardmodelle der Zahlentheorie* ◇ C62 F30 ◇
MUELLER, GERT H. *Ueber die unendliche Induktion* ◇ F30 ◇
NAGEL, E. & NEWMAN, J.R. *Discussion: Putnam's review of Goedel's proof* ◇ B28 F30 ◇
RABIN, M.O. *Non-standard models and independence of the induction axiom* ◇ B28 C62 F30 H15 ◇
RIEGER, L. *Sur le probleme des nombres naturels* ◇ A05 A10 F30 ◇
ROSE, H.E. *On the consistency and undecidability of recursive arithmetic* ◇ D35 F30 ◇
SHEPHERDSON, J.C. *Representability of recursively enumerable sets in formal theories* ◇ D25 F30 ◇
SMART, J.J.C. *Goedel's theorem, Church's theorem and mechanism* ◇ A05 B10 D35 F30 ◇
TAKEUTI, G. *Remarks on the truth definition* ◇ F25 F30 ◇

1962

BETH, E.W. *Formal methods. An introduction to symbolic logic and to the study of effective operations in arithmetic and logic* ◇ B28 B98 F30 F50 F98 ◇
CANNONITO, F.B. *The Goedel incompleteness theorem and intelligent machines* ◇ B35 D10 D25 F30 ◇
FEFERMAN, S. & SPECTOR, C. *Incompleteness along paths in progressions of theories* ◇ D55 F15 F30 ◇
FEFERMAN, S. *Transfinite recursive progressions of axiomatic theories* ◇ D55 F15 F30 ◇
FEFERMAN, S. & KREISEL, G. & OREY, S. *1-consistency and faithful interpretations* ◇ F25 F30 F35 ◇
HOERING, W. *Absolut unentscheidbare Saetze der Mathematik* ◇ A05 F30 ◇
KRIPKE, S.A. *"Flexible" predicates of formal number theory* ◇ C90 F30 ◇
MAEHARA, S. *Cut-elimination theorem concerning a formal system for ramified theory of types which admits quantification on types* ◇ B15 F05 F30 F35 F65 ◇
MONTAGUE, R. *Theories incomparable with respect to relative interpretability* ◇ C07 F25 F30 ◇
MOSTOWSKI, ANDRZEJ *Representability of sets in formal systems* ◇ D55 E15 F30 F35 ◇
POGORZELSKI, H.A. *Recursive arithmetic of Skolem. I,II* ◇ F30 ◇
ROSE, H.E. *Ternary recursive arithmetic* ◇ F30 ◇
SCOTT, D.S. *Algebras of sets binumerable in complete extensions of arithmetic* ◇ C40 C62 F30 ◇
VUCKOVIC, V. *Einfuehrung von $\Sigma_f(x)$ und $\Pi_f(x)$ in der rekursiven Gitterpunktarithmetik* ◇ D20 F30 ◇

1963

BORGERS, A. *Quelques propositions arithmetiques equivalentes au principe de l'induction complete* ◇ F30 ◇

GOODSTEIN, R.L. *A decidable fragment of recursive arithmetic* ◇ B20 B25 F30 ◇
HU, SHIHUA & HUANG, ZULIANG *Addition and multiplication (Chinese)* ◇ D20 F30 ◇
KUBINSKI, T. *A proof of consistency of Borkowski's logical system containing Peano's arithmetic (Polish and Russian summaries)* ◇ B15 F25 F30 ◇
MONTAGUE, R. *Syntactical treatments of modality, with corollaries on reflexion principles and finite axiomatizability* ◇ B45 B65 C62 F30 ◇
WANG, HAO *Ackermann's consistency proof* ◇ F25 F30 ◇
WANG, HAO *Partial systems of number theory* ◇ C62 F30 ◇
WANG, HAO *Relative strength and reducibility* ◇ B10 E30 E70 F25 F30 ◇
WANG, HAO *The arithmetization of metamathematics* ◇ F25 F30 ◇

1964

ARBIB, M.A. *Brains, machines, and mathematics* ◇ D05 D10 D80 F30 ◇
BURGER, E. *Bemerkungen zu einigen Fassungen des Goedelschen Unvollstaendigkeitssatzes* ◇ D80 F30 ◇
ELGOT, C.C. & RUTLEDGE, J.D. *RS-machines with almost blank tapes* ◇ D05 D10 F30 ◇
GAO, HENGSHAN *Remarks on Ryll-Nardzewski's proof of the impossibility of finite axiomatization of Peano arithmetic (Chinese)* ◇ F30 ◇
HERMES, H. *Unentscheidbarkeit der Arithmetik* ◇ D35 F30 ◇
NAGORNYJ, N.M. *On realizable and completable logico-arithmetical formulas (Russian)* ◇ F30 F50 ◇
PUTNAM, H. *On families of sets represented in theories* ◇ C40 D25 F30 ◇
RIVETTI BARBO, F. *Il teorema e il corollario di Goedel* ◇ A05 F30 F98 ◇
SHEPHERDSON, J.C. *A non-standard model for a free variable fragment of number theory* ◇ C62 F30 ◇

1965

ACKERMANN, W. *Der Aufbau einer hoeheren Logik* ◇ B15 E70 F30 F35 ◇
CHURCH, A. *An independence question in recursive arithmetic* ◇ F30 ◇
DAVIS, MARTIN D. *Introduction to Goedel's paper: On formally undecidable propositions of the Principia Mathematica and related systems I* ◇ A10 D20 F30 ◇
DAVIS, MARTIN D. (ED.) *The undecidable. Basic papers on undecidable propositions, unsolvable problems and computable functions* ◇ D20 D25 D35 D97 F30 ◇
FINSLER, P. *Zur Goldbachschen Vermutung* ◇ F30 ◇
FISCHER, P.C. *Theory of provably recursive functions* ◇ D20 D30 F30 ◇
GOEDEL, K. *On undecidable propositions of formal mathematical systems* ◇ D20 D35 F30 ◇
GOODSTEIN, R.L. *Multiple successor arithmetics* ◇ C62 F30 ◇
KURATA, R. *Recursive progression of intuitionistic number theories* ◇ F15 F30 F50 ◇
LEE, R.D. *The substitution schema in recursive arithmetic* ◇ F30 ◇
MAKKAI, M. *Ueber die transfinite Induktion in zahlentheoretischen Formalismen* ◇ F30 ◇
MO, SHAOKUI & SHEN, BAIYING *New systems of primitive recursive arithmetic (Chinese)* ◇ F30 ◇
MO, SHAOKUI & SHEN, BAIYING *Systems of primitive recursive arithmetic (Chinese)* ◇ D20 F30 ◇
MOLLER, S. *Induction models (Danish)* ◇ C62 F30 ◇
NISHIMURA, T. & TANAKA, H. *On two systems for arithmetic* ◇ C75 F30 F35 ◇
ROBINSON, T.T. *Interpretations of Kleene's metamathematical predicate* $\Gamma | A$ *in intuitionistic arithmetic* ◇ F30 F50 ◇
ROSE, H.E. *A note on reducible induction schemata* ◇ F30 ◇
SHEPHERDSON, J.C. *Non-standard models for fragments of number theory* ◇ C62 F30 ◇
STRALBERG, A.H. *On the reduction of classical logic. An extension of some theorems of Glivenko and Goedel* ◇ B55 F30 F50 ◇
TAIT, W.W. *The substitution method* ◇ F10 F15 F30 F35 ◇

1966

APELT, H. *Axiomatische Untersuchungen ueber einige mit der Presburgerschen Arithmetik verwandte Systeme* ◇ B28 C35 C80 F30 ◇
ARBIB, M.A. *Speed-up theorems and incompleteness theorems* ◇ D15 F30 ◇
ELGOT, C.C. & RABIN, M.O. *Decidability and undecidability of extensions of second (first) order theory of (generalized) successor* ◇ B15 B25 C85 D05 D35 F30 F35 ◇
GINSBURG, S. & SPANIER, E.H. *Semigroups, Presburger formulas, and languages* ◇ D05 F30 ◇
GOODSTEIN, R.L. & LEE, R.D. *A decidable class of equations in recursive arithmetic* ◇ B20 B25 F30 ◇
GRANDY, R.E. *A note on the recursive unsolvability of primitive recursive arithmetic* ◇ D35 F30 ◇
MO, SHAOKUI & SHEN, BAIYING *A new system for primitive recursive arithmetic I,II (Chinese)* ◇ F30 ◇
PAOLA DI, R.A. *On sets represented by the same formula in distinct consistent axiomatizable Rosser theories* ◇ D25 F30 ◇
PAOLA DI, R.A. *Pseudo-complements and ordinal logics based on consistency statements* ◇ D25 F15 F30 ◇
PAOLA DI, R.A. *Some properties of pseudo-complements of recursively enumerable sets* ◇ D25 D30 F30 ◇
ROUSSEAU, G. *A decidable class of number theoretic equations* ◇ B20 B25 F30 ◇
VOPENKA, P. *A new proof of the Goedel's result on non-provability of consistency (Russian summary)* ◇ E30 E70 F30 ◇

1967

AGAZZI, E. *L'aritmetica come strumento di ricerca logico-matematica (English summary)* ◇ A05 F30 ◇
BELYAKIN, N.V. *A complete system of classical arithmetic (Russian) (English summary)* ◇ F30 ◇
BELYAKIN, N.V. *On the completeness of arithmetic (Russian)* ◇ F30 ◇

BOTH, N. *On the bases of complete induction. Simultaneous induction (Romanian)* ◇ B28 F30 ◇
CLEAVE, J.P. & ROSE, H.E. *$\mathscr{E}_n$-arithmetic* ◇ D20 F30 ◇
FENSTAD, J.E. *Subsets of natural numbers (Norwegian) (English summary)* ◇ A05 E35 E50 F30 F35 F65 ◇
HEATH, I.J. *Omitting the replacement schema in recursive arithmetic* ◇ F30 ◇
HEIJENOORT VAN, J. *Goedel's theorem* ◇ A05 A10 F30 ◇
JENSEN, R.B. *Unabhaengigkeitsbeweise in Teilsystemen der elementaren Zahlentheorie* ◇ F30 F35 ◇
KLEENE, S.C. *Mathematical logic* ◇ B98 D20 D25 D55 D98 F30 F98 ◇
KOSOVSKIJ, N.K. *Sufficient conditions of incompleteness for the formalization of parts of arithmetic (Russian)* ◇ F30 ◇
KRIPKE, S.A. & POUR-EL, M.B. *Deduction-preserving "recursive isomorphisms" between theories* ◇ B30 D25 D35 F30 ◇
MINTS, G.E. & OREVKOV, V.P. *On imbedding operators (Russian)* ◇ F25 F30 F50 ◇
MULLIN, A.A. *On new theorems for elementary number theory* ◇ F30 ◇
NAGEL, E. & NEWMAN, J.R. *Goedel's proof* ◇ D35 F30 ◇
PAOLA DI, R.A. *Some theorems on extensions of arithmetic* ◇ D25 D35 F15 F30 ◇
RITTER, W.E. *Representability of partial recursive functions in formal theories* ◇ D20 F30 ◇
ROSE, H.E. *Some metamathematical results in recursive arithmetic* ◇ F30 ◇
SHEPHERDSON, J.C. *The rule of induction in the free variable arithmetic based on + and ∗* ◇ C62 F30 ◇
SINGH, S. *The natural number arithmetic in Goedel's axiomatic set theory* ◇ B28 F30 ◇
TAIT, W.W. *Intensional interpretation of functionals of finite type I* ◇ F10 F30 F35 F50 ◇
TSINMAN, L.L. *On the complete induction axiom (Russian)* ◇ B28 F30 ◇
VUCKOVIC, V. *A recursive model for the extended system A of B. Sobocinski* ◇ B45 F30 ◇
VUCKOVIC, V. *Mathematics of incompleteness and undecidability* ◇ D35 F30 ◇
VUCKOVIC, V. *Recursive models for three-valued propositional calculi with classical implication* ◇ B50 C57 C90 F30 ◇

1968

ACZEL, P. *Saturated intuitionistic theories* ◇ F30 F50 ◇
ENDERTON, H.B. *On provable recursive functions* ◇ D20 F30 ◇
FEFERMAN, S. *Autonomous transfinite progressions and the extent of predicative mathematics* ◇ F15 F30 F35 F65 ◇
FENSTAD, J.E. *On the completeness of some transfinite recursive progressions of axiomatic theories* ◇ F15 F30 ◇
FOL'K, N.F. & SHESTOPAL, G.A. *The solvability of elementary theories of integral and natural numbers with addition (Russian)* ◇ B25 C10 F30 ◇
HANATANI, Y. *Demonstration de l'ω-non-contradiction de l'arithmetique* ◇ F05 F15 F30 F50 ◇
KINO, A. *On provably recursive functions and ordinal recursive functions* ◇ D20 F05 F30 F35 ◇
KIPNIS, M.M. *The constructive classification of arithmetic predicates and the semantic bases of arithmetic (Russian)* ◇ D55 F30 F50 ◇
KREISEL, G. & LEVY, A. *Reflection principles and their use for establishing the complexity of axiomatic systems* ◇ E30 E47 F25 F30 F35 F50 ◇
LOPEZ-ESCOBAR, E.G.K. *A decision method for the intuitionistic theory of successor* ◇ B25 C10 F30 F50 ◇
NJASTAD, O. *Note on the Peano axioms* ◇ F30 ◇
POZSGAY, L.J. *Goedel's second theorem for elementary arithmetic* ◇ F30 ◇
RITCHIE, R.W. & YOUNG, P. *Strong representability of partial functions in arithmetic theories* ◇ D20 F30 ◇
ROEDDING, W. *Eine Art von Gleichgewicht zahlentheoretischer und mengentheoretischer Axiomensysteme* ◇ B30 E30 F30 ◇
ROUSSEAU, G. *Note on the decidability of a certain class of number theoretic equations* ◇ B20 B25 F30 ◇
SCHUETTE, K. *Neuere Ergebnisse der Beweistheorie* ◇ F05 F10 F15 F30 F35 F98 ◇
TAIT, W.W. *Constructive reasoning* ◇ B40 F10 F15 F30 F35 F50 F65 ◇
TAIT, W.W. *Normal derivability in classical logic* ◇ C75 F05 F15 F30 F35 F60 ◇
TOMPKINS, R.R. *On Kleene's recursive realizability as an interpretation for intuitionistic elementary number theory* ◇ F30 F50 ◇
TSEJTIN, G.S. *The disjunctive rank of the formulas of constructive arithmetic (Russian)* ◇ F30 F50 ◇
TSINMAN, L.L. *The role of the principle of induction in a formal arithmetical system (Russian)* ◇ B28 F30 ◇

1969

BARZIN, M. *Sur le mecanisme des paradoxes et le theoreme de M.K.Goedel* ◇ A05 F30 ◇
DOU, A. *El teorema de incomplitud de Goedel* ◇ F30 ◇
ESENIN-VOL'PIN, A.S. *On Hilbert's second problem (Russian)* ◇ A05 F30 F65 ◇
GENTZEN, G. *The collected papers of Gerhard Gentzen* ◇ B96 F05 F07 F30 F50 F96 ◇
GOODSTEIN, R.L. *Decision methods in recursive arithmetic* ◇ B25 F30 ◇
JONES, JAMES P. *Independent recursive axiomatizability in arithmetic* ◇ D25 F30 ◇
POGORZELSKI, H.A. *Goldbach sentences in abstract arithmetics $\mathscr{A}^k(A)$ I* ◇ D20 F30 ◇
SCARPELLINI, B. *Some applications of Gentzen's second consistency proof* ◇ F05 F30 F35 F50 ◇
SHEPHERDSON, J.C. *Weak and strong induction* ◇ C62 F30 ◇
SZABO, M.E. *Introduction* ◇ A10 F05 F30 F98 ◇
TSINMAN, L.L. *A theorem on normal proofs in the theory of regular formulae (Russian)* ◇ F30 ◇
TSINMAN, L.L. *Certain algorithms in a formal arithmetic system (Russian)* ◇ B25 F30 ◇

WETTE, E. *Definition eines (relativ vollstaendigen) formalen Systems konstruktiver Arithmetik* ◇ F10 F15 F30 F35 F65 ◇

1970

BERNAYS, P. *The original Gentzen consistency proof for number theory* ◇ A10 B28 F30 F50 ◇
CIAMPA, S. *I numeri, l'aritmetica, gli insiemi* ◇ A05 F30 ◇
COLLINS, G.E. & HALPERN, J.D. *On the interpretability of arithmetic in set theory* ◇ B28 D35 E30 F25 F30 ◇
DENTON JR., J.S. & DREBEN, B. *Herbrand-style consistency proofs* ◇ F05 F30 F35 ◇
EBBINGHAUS, H.-D. *Aufzaehlbarkeit* ◇ D03 D10 D20 D25 F30 ◇
GERMANO, G. *Metamathematische Begriffe in Standardtheorien* ◇ A05 B30 F30 ◇
GOODMAN, NICOLAS D. *A theory of constructions equivalent to arithmetic* ◇ F30 F50 ◇
KABAKOV, F.A. *On modelling of pseudo-boolean algebras by realizability (Russian)* ◇ F30 F50 G10 ◇
KUZICHEVA, Z.A. *On the history of incompleteness theorems (Russian)* ◇ A10 F30 ◇
LACEY, H.M. *The consequence of Goedel's theorem* ◇ A05 F30 ◇
MAYOH, B.H. *The relation between an object and its name: notation systems and their fixed point theorems* ◇ A05 B03 D20 D35 D45 F30 F60 ◇
PARSONS, C. *On a number-theoretic choice schema and its relation to induction* ◇ F10 F30 ◇
POGORZELSKI, H.A. *Goldbach sentences in abstract arithmetics* $\mathscr{A}^k(A)$ *II* ◇ F30 ◇
SHMAIN, I.KH. *Imbeddings of classical systems in intuitionistic and minimal ones (Russian)* ◇ F30 F50 ◇
UESU, T. *On the recursively restricted rules* ◇ C75 F30 ◇
WETTE, E. *Vom Unendlichen zum Endlichen* ◇ A05 B28 E30 F30 ◇

1971

BECVAR, J. *Hilbert's second problem (the question of the consistency of arithmetic) (Czech)* ◇ F30 ◇
CANNONITO, F.B. *A note on inverses of elementary permutations* ◇ D20 D45 F30 ◇
DALEN VAN, D. & GORDON, C.E. *Independence problems in subsystems of intuitionistic arithmetic* ◇ F30 F50 ◇
DREBEN, B. & GOLDFARB, W.D. *Note N* ◇ A10 B10 F30 F35 ◇
DUMITRIU, A. *Fermat's method of infinite descent compared with the method of mathematical induction (Romanian) (French summary)* ◇ A10 F30 ◇
ELLENTUCK, E. *Incompleteness via simple sets* ◇ D25 F30 ◇
FRIEDMAN, H.M. *A more explicit set theory* ◇ A05 C62 E30 E35 E45 F30 F35 ◇
GERMANO, G. *Incompleteness and truth definitions* ◇ C40 D35 F30 ◇
GRZEGORCZYK, A. *Outline of theoretical arithmetic (Polish)* ◇ B28 C62 F30 F35 F98 ◇
HAJKOVA, M. *The lattice of bi-numerations of arithmetic. I,II* ◇ D45 F30 ◇
HANSON, W.H. *Mechanism and Goedel's theorems* ◇ A05 F30 ◇
HAUSCHILD, K. *Nichtaxiomatisierbarkeit von Satzmengen durch Ausdruecke spezieller Gestalt* ◇ B28 C20 C40 C52 F30 ◇
JEROSLOW, R. *Consistency statements in formal theories* ◇ F30 ◇
JERVELL, H.R. *A normalform in first order arithmetic* ◇ F05 F30 F50 ◇
LOLLI, G. *Formule universali dell'aritmetica di Peano* ◇ F30 ◇
MANGIONE, C. *Su alcune questioni connesse con i principi di riflessione* ◇ F30 ◇
MCKENZIE, R. *Negative solution of the decision problem for sentences true in every subalgebra of* $\langle N,+\rangle$ ◇ D35 F30 ◇
MINTS, G.E. *Exact estimation of the provability of transfinite induction in the initial segments of arithmetic (Russian) (English summary)* ◇ F05 F15 F30 ◇
MINTS, G.E. *Quantifier-free and one-quantifier systems (Russian) (English summary)* ◇ B20 D20 F30 ◇
NELSON, GEORGE C. *A further restricted* ω*-rule* ◇ F30 ◇
PARIKH, R. *Existence and feasibility in arithmetic* ◇ A05 D15 F30 F65 H15 ◇
PLIUSKEVICIENE, A. *Elimination of cut-type rules from the Robinson and Presburger axiomatic systems (Russian)* ◇ F05 F30 ◇
PRAWITZ, D. *Ideas and results in proof theory* ◇ F05 F07 F10 F30 F35 F50 F98 ◇
RICHARDSON, D.B. *The simple exponential constant problem* ◇ D80 F30 ◇
SATINDER, P.K. *Completeness and cut-elimination in constructive* ω*-rule arithmetics* ◇ F05 F30 ◇
SCARPELLINI, B. *Proof theory and intuitionistic systems* ◇ F05 F15 F30 F35 F50 F98 ◇
TROELSTRA, A.S. *Notions of realizability for intuitionistic arithmetic and intuitionistic arithmetic in all finite types* ◇ F30 F35 F50 ◇

1972

CHRISTIAN, C.C. *Konsistenzkriterien fuer formale Theorien und ihre Anwendung auf Zahlen- und Mengentheorie* ◇ B28 E30 F30 ◇
COOPER, D.C. *Theorem proving in arithmetic without multiplication* ◇ B25 B35 F30 ◇
DRAGALIN, A.G. *On the use of classical calculi for establishing constructive truth (Russian) (English summary)* ◇ B10 F30 F50 ◇
ENDERTON, H.B. *A mathematical introduction to logic* ◇ B98 C98 F30 ◇
HAJEK, P. & HAJKOVA, M. *On interpretability in theories containing arithmetic* ◇ F25 F30 ◇
HAUCK, J. & HERRE, H. & POSEGGA, M. *Zur Metatheorie formaler Systeme* ◇ B10 C07 F30 ◇
JOCKUSCH JR., C.G. & SOARE, R.I. *Degrees of members of* Π^0_1 *classes* ◇ D25 D30 F30 ◇

KREISEL, G. *Which number theoretic problems can be solved in recursive progressions on Π^1_1-paths through $\mathcal{O}$?* ◇ A05 D20 D55 F15 F30 F99 ◇
LOPEZ-ESCOBAR, E.G.K. *Refutability and elementary number theory* ◇ F30 F50 ◇
MEDVEDEV, YU.T. *Locally finitary algorithmic problems (Russian)* ◇ D20 F30 F50 ◇
MYHILL, J.R. *An absolutely independent set of Σ^0_1 sentences* ◇ F30 ◇
PARSONS, C. *On n-quantifier induction* ◇ F10 F30 ◇
PLIUSKEVICIENE, A. *A sequential variant of R. M. Robinson's arithmetic system not containing cut rules (Russian)* ◇ F05 F30 ◇
ROSE, H.E. *$\mathcal{E}^\alpha$-arithmetic and transfinite induction* ◇ D20 F30 ◇
SCHWICHTENBERG, H. *Beweistheoretische Charakterisierung einer Erweiterung der Grzegorczyk-Hierarchie* ◇ D20 F15 F30 ◇
STEEN, S.W.P. *Mathematical logic, with special reference to the natural numbers* ◇ B98 F30 F98 ◇

1973

BOLLMAN, D.A. & LAPLAZA, M.L. *A set-theoretic model for nonassociative number theory* ◇ B28 E75 F30 ◇
BUNDY, A. *A note on omitting the replacement schema* ◇ D20 F30 ◇
CHRISTIAN, C.C. *Inhaltliche und formale Wahrheit* ◇ A05 E30 E35 E45 F30 ◇
EHRENFEUCHT, A. *Polynomial functions with exponentiation are well ordered* ◇ E07 F30 ◇
FRIDMAN, EH.I. & SLOBODSKOJ, A.M. *The theory of the additive group of the integers with an arbitrary number of predicates that define maximal subgroups (Russian)* ◇ C10 F30 ◇
GARRO, I. *Independence proofs in arithmetic theories with very weak induction* ◇ F30 ◇
GERMANO, G. *Incompleteness theorem via weak definability of truth: a short proof* ◇ C40 F30 ◇
GOODMAN, NICOLAS D. *The arithmetic theory of constructions* ◇ F30 F50 ◇
GOODMAN, NICOLAS D. *The faithfulness of the interpretation of arithmetic in the theory of constructions* ◇ F30 F50 ◇
JACOBS, K. (ED.) *Selecta mathematica II* ◇ D97 F30 ◇
JEROSLOW, R. *Redundancies in the Hilbert-Bernays derivability conditions for Goedel's second incompleteness theorem* ◇ F30 ◇
KENT, C.F. *The relation of A to Prov⌜A⌝ in the Lindenbaum sentence algebra* ◇ F30 G05 ◇
KHOMICH, V.I. *The complexity of recognition of the realizability of logico-arithmetic formulas (Russian)* ◇ F30 F50 ◇
KOLODZIEJ, R. *On a certain subclass of formulas of recursive arithmetics (Polish) (Russian and English summaries)* ◇ F30 ◇
LIU, SHICHAO *Ordering formulas and quasi-unsecuredness* ◇ F30 ◇
MACINTYRE, A. & SIMMONS, H. *Goedel's diagonalization technique and related properties of theories* ◇ F30 G05 ◇
MAGARI, R. *Meaning and truth in the Peano arithmetic (Italian summary)* ◇ F30 ◇
MEDVEDEV, YU.T. *An interpretation of intuitionistic number theory* ◇ F15 F30 F50 ◇
METROPOLIS, N. & ROTA, G.-C. & TANNY, S. *Significance arithmetic: the carrying algorithm* ◇ B75 F30 ◇
OPPEN, D.C. *Elementary bounds for Presburger arithmetic* ◇ B25 C10 D15 F30 ◇
OWINGS JR., J.C. *Diagonalization and the recursion theorem* ◇ B40 D20 F30 ◇
PALUCH, S. *The lattices of numerations of theories containing Peano's arithmetic* ◇ F30 G10 ◇
PARIKH, R. *Some results on the length of proofs* ◇ B25 F20 F30 ◇
PENZIN, YU.G. *Decidability of certain theories of integers (Russian)* ◇ B25 F30 ◇
PENZIN, YU.G. *Decidability of the theory of integers with addition, order and multiplication by an arbitrary number (Russian)* ◇ B25 F30 ◇
PENZIN, YU.G. *Decidability of a theory of the integers with addition, order and predicates that distinguish a chain of subgroups (Russian)* ◇ B25 F30 ◇
ROBINSON, JULIA *Axioms for number theoretic functions (Russian)* ◇ C62 D20 D75 F30 ◇
SCANLON, T.M. *The consistency of number theory via Herbrand's theorem* ◇ B10 F05 F30 ◇
SHEPARD, P.T. *A finite arithmetic* ◇ F30 F65 ◇
SMORYNSKI, C.A. *Applications of Kripke models* ◇ F30 F50 ◇
TROELSTRA, A.S. *Intuitionistic formal systems* ◇ F30 F35 F50 ◇
TROELSTRA, A.S. (ED.) *Metamathematical investigation of intuitionistic arithmetic and analysis* ◇ F05 F10 F30 F35 F50 F55 F98 ◇
TROELSTRA, A.S. *Models and computability* ◇ F05 F10 F30 F35 F50 ◇
TROELSTRA, A.S. *Normalization theorems for systems of natural deduction* ◇ F05 F30 F35 F50 ◇
TROELSTRA, A.S. *Realizability and functional interpretations* ◇ F10 F30 F35 F50 ◇
WOLTER, H. *Eine Erweiterung der elementaren Praedikatenlogik: Anwendungen in der Arithmetik und anderen mathematischen Theorien* ◇ B28 C10 C80 F30 ◇

1974

BARNES, D.W. & MONRO, G.P. *A simple model for a weak system of arithmetic* ◇ C62 F30 ◇
CHUBARYAN, A.A. *The complexity of deductions in extensions of formal arithmetic (Russian) (Armenian summary)* ◇ F20 F30 ◇
CHUBARYAN, A.A. *The length of derivations of formulae in extensions of formal arithmetic (Russian) (Armenian and English summaries)* ◇ F20 F30 ◇
DEREVYANKINA, E.A. *Some problems of completeness of arithmetic (Russian)* ◇ F15 F30 ◇
DRAGALIN, A.G. *The completeness of an arithmetic with a constructive rule of infinite induction (Russian)* ◇ F30 F50 ◇
FISCHER, MICHAEL J. & RABIN, M.O. *Super-exponential complexity of Presburger arithmetic* ◇ B25 D15 F20 F30 ◇

GENTZEN, G. *Der erste Widerspruchsfreiheitsbeweis fuer die klassische Zahlentheorie* ◇ F05 F15 F25 F30 ◇
GENTZEN, G. *Ueber das Verhaeltnis zwischen intuitionistischer und klassischer Arithmetik* ◇ B30 F30 F50 ◇
GOLDFARB, W.D. *Ordinal bounds for k-consistency* ◇ F15 F30 ◇
GOLDFARB, W.D. & SCANLON, T.M. *The ω-consistency of number theory via Herbrand's theorem* ◇ F05 F15 F30 ◇
KENT, C.F. *"Disorder" in lattices of binumerations* ◇ F30 G05 ◇
KRAJEWSKI, S. *Mutually inconsistent satisfaction classes* ◇ C62 F30 ◇
LOLLI, G. *Una dimostrazione di non contraddittorieta per l'aritmetica (English summary)* ◇ F30 ◇
LOPEZ-ESCOBAR, E.G.K. *Elementary interpretations of negationless arithmetic* ◇ F30 F50 ◇
MAGARI, R. *Sur certe teorie non enumerabili. Sulle limitazioni dei sistemi formali I* ◇ B53 B60 F30 ◇
MARONGIU, G. *Sequenze di predicati aritmetici di tipo "teor" (English summary)* ◇ F30 ◇
MCMINN, T.J. *A formal number-termed number system based on recursion* ◇ B28 F30 ◇
METEAU, G. *The foundations of mathematics - a new analysis showing Goedel's theorem based on fallacy* ◇ A05 F30 ◇
MINTS, G.E. *E theorems (Russian) (English summary)* ◇ F05 F30 F50 ◇
PABION, J.F. *L'axiomatisation de la syntaxe et le second theoreme de Goedel* ◇ B30 F30 ◇
POGORZELSKI, H.A. *On the Goldbach conjecture and the consistency of general recursive arithmetic* ◇ F30 ◇
RESNIK, M.D. *On the philosophical significance of consistency proofs* ◇ A05 F25 F30 ◇
RICHARDSON, D.B. *Sets of theorems with short proofs* ◇ F20 F30 ◇
SAKAI, H. *On necessary but not-sufficient conditions* ◇ B10 F30 F35 G05 ◇
SAKAI, H. *On numerations of a formal system* ◇ D45 F30 ◇
SAMBIN, G. *Un'estensione del teorema di Loeb (English summary)* ◇ F30 ◇
TROELSTRA, A.S. *Note on the fan theorem* ◇ F30 F35 F50 ◇
USPENSKIJ, V.A. *An elementary exposition of Goedel's incompleteness theorem (Russian)* ◇ F30 ◇
WETTE, E. *Contradiction within pure number theory because of a system-internal "consistency"-deduction* ◇ B10 F30 ◇
WETTE, E. *The refutation of number theory. I* ◇ D10 F30 ◇
ZINOV'EV, A.A. *Certain systems of formal arithmetic (Russian)* ◇ B45 F30 ◇
ZUCKER, J.I. *The correspondence between cut-elimination and normalization I,II* ◇ F05 F30 F50 ◇

1975

ACZEL, P. *Recursive density types and Nerode extensions of arithmetic* ◇ D45 D50 F30 ◇
BERNARDI, C. *On the equational class of diagonalizable algebras (the algebraization of the theories which express Theor. VI)* ◇ B25 F30 G05 G25 ◇
BERNARDI, C. *The fixed-point theorem for diagonalizable algebras (the algebraization of the theories which express Theor. III)* ◇ B45 F30 G05 G25 ◇
BOOLOS, G. *On Kalmar's consistency proof and a generalization of the notion of ω-consistency* ◇ F15 F30 ◇
CELLUCCI, C. *Teoremi di normalizzazione per alcuni sistemi funzionali* ◇ F05 F10 F15 F30 ◇
CELLUCCI, C. *Teoremi di normalizzazione per alcuni sistemi funzionali II* ◇ F05 F10 F15 F30 ◇
CROSSLEY, J.N. *What is mathematical logic ?* ◇ B98 F30 ◇
DAWES, A.M. & FLORENCE, J.B. *Independent Goedel sentences and independent sets* ◇ D25 F30 ◇
DEUTSCH, M. *Zur Benutzung der Verkettung als Basis fuer die Arithmetik* ◇ B28 F30 ◇
DEUTSCH, M. *Zur Darstellung koaufzaehlbarer Praedikate bei Verwendung eines einzigen unbeschraenkten Quantors* ◇ D25 F30 ◇
FRIEDMAN, H.M. *Provable equality in primitive recursive arithmetic with and without induction* ◇ F30 ◇
GOLDFARB, W.D. *On the effective ω-rule* ◇ F20 F30 ◇
LEIVANT, D. *Strong normalization for arithmetic (Variations on a theme of Prawitz.)* ◇ F05 F30 F50 ◇
LEVITZ, H. *An ordered set of arithmetic functions representing the least ε-number* ◇ C62 F15 F30 ◇
MAGARI, R. *Metodi algebrici in teoria della dimostrazione* ◇ F30 G05 ◇
MAGARI, R. *Representation and duality theory for diagonalizable algebras (The algebraization of theories which express Theor; IV.)* ◇ F30 G25 ◇
MAGARI, R. *Significato e verita nell'aritmetica peaniana (English summary)* ◇ A05 F30 ◇
MAGARI, R. *The diagonalizable algebras (The algebraization of the theories which express Theor. II)* ◇ F30 G25 ◇
MCALOON, K. *Applications alternees de theoremes d'incompletude et des theoremes de completude (English summary)* ◇ C62 E35 F30 ◇
MEZHLUMBEKOVA, V.F. *Cut-elimination in a system of negationless arithmetic (Russian) (English summary)* ◇ B20 F05 F30 F50 ◇
MEZHLUMBEKOVA, V.F. *Deductive capabilities of negationless intuitionistic arithmetic (Russian) (English summary)* ◇ F30 F50 ◇
MINTS, G.E. *Finite investigations of transfinite derivations (Russian) (English summary)* ◇ F05 F07 F10 F20 F30 F35 F50 ◇
MINTS, G.E. *Proof theory (arithmetic and analysis) (Russian)* ◇ F05 F10 F30 F35 F50 F98 ◇
MINTS, G.E. *Transfinite expansions of arithmetic formulas (Russian) (English summary)* ◇ F30 F50 ◇
MONTAGNA, F. *For every n, the n-freely generated algebra is not functionally free in the equational class of diagonalizable algebras (the algebraization of theories which express Theor. V)* ◇ F30 G05 ◇

PAOLA DI, R.A. *A theorem on shortening the length of proof in formal systems of arithmetic* ◇ F20 F30 ◇
PODNIEKS, K.M. *The double-incompleteness theorem (Russian) (English summary)* ◇ D35 F30 ◇
PRAWITZ, D. *Comments on Gentzen-type procedures and the classical notion of truth* ◇ B10 F05 F07 F30 ◇
SCARPELLINI, B. *Bemerkungen zu Regel und Schema* ◇ F07 F30 F50 ◇
SIMMONS, H. *Topological aspects of suitable theories* ◇ B45 F30 G05 ◇
TAKEUTI, G. *Proof theory* ◇ F05 F07 F30 F35 F98 ◇
TOLEDO, S. *Tableau systems for first order number theory and certain higher order theories* ◇ F05 F07 F30 F35 F98 ◇
TOMAS, F. *A test for consistency and its application to recursive arithmetic (Spanish)* ◇ B28 F30 ◇
WANDSCHNEIDER, D. *Zur Eliminierung des Goedelschen Unvollstaendigkeitsproblems im Zusammenhang mit dem Antinomien-Problem* ◇ A05 E30 F30 ◇
WAUW-DE KINDER VAN DE, G. *Arithmetique de premier ordre dans les topos (English summary)* ◇ F30 F50 G30 H15 ◇
WOLTER, H. *Entscheidbarkeit der Arithmetik mit Addition und Ordnung in Logiken mit verallgemeinerten Quantoren* ◇ B25 C10 C55 C80 F30 ◇

1976

BARENDREGT, H.P. *The incompleteness theorems* ◇ D35 F30 ◇
BEL'TYUKOV, A.P. *Decidability of the universal theory of natural numbers with addition and divisibility (Russian)* ◇ B25 B28 F30 ◇
BERNARDI, C. *The uniqueness of the fixed-point in every diagonalizable algebra (The algebraization of the theories which express Theor. VIII.)* ◇ B45 F30 G05 G25 ◇
BEZBORUAH, A. & SHEPHERDSON, J.C. *Goedel's second incompleteness theorem for Q* ◇ C62 F30 ◇
BOOLOS, G. *On deciding the truth of certain statements involving the notion of consistency* ◇ B25 B45 F30 ◇
BORGA, M. *On a preliminary reduction step in Gentzen's second consistency proof* ◇ F30 ◇
CHERNIAVSKY, J.C. *Simple programs realize exactly Presburger formulas* ◇ B75 D15 D20 F30 ◇
CHERNYAKHOVSKIJ, N.P. *The expressibility of realizability, in the language of formal arithmetic (Russian)* ◇ D80 F30 F50 ◇
CHRISTIAN, C.C. *Peano-Systeme* ◇ B28 F30 F35 ◇
EHRENFEUCHT, A. & JENSEN, D.C. *Some problems in elementary arithmetics* ◇ C62 F30 H15 ◇
FRIEDMAN, H.M. *Uniformly defined descending sequences of degrees* ◇ D30 D55 F30 F35 ◇
FRIEDRICHSDORF, U. *Einige Bemerkungen zur Peano-Arithmetik* ◇ C25 C52 C62 F30 ◇
GERMANO, G. *An arithmetical reconstruction of the liar's antinomy using addition and multiplication* ◇ F30 ◇
HISCHER, HORST & LUCHT, L. *Zum Verstaendnis des Induktionsaxioms* ◇ B28 F30 ◇
KREISEL, G. *Wie die Beweistheorie zu ihren Ordinalzahlen kam und kommt* ◇ F05 F15 F30 F35 ◇
LIVCHAK, A.B. *Definable sets of integers (Russian)* ◇ C60 C62 F30 ◇
LOPEZ-ESCOBAR, E.G.K. *On an extremely restricted ω-rule* ◇ F05 F15 F30 F50 ◇
MAREK, W. & SREBRNY, M. *Urelements and extendability* ◇ C62 E30 E70 F25 F30 F35 ◇
MEYER, R.K. *Relevant arithmetic* ◇ B46 F30 ◇
MINTS, G.E. *The universality of the canonical tree (Russian)* ◇ F15 F30 ◇
MINTS, G.E. *Unification of the two Peano axioms (Russian)* ◇ F30 ◇
MINTS, G.E. *What can be done with PRA (Russian) (English summary)* ◇ F05 F30 F35 ◇
MLCEK, J. *Twin prime problem in an arithmetic without induction* ◇ F30 ◇
NAGORNYJ, N.M. *A variant of the definition of realization of a logical arithmetical formula (Russian)* ◇ F30 F50 ◇
PENZIN, YU.G. *Undecidability of a theory of the integers with addition and predicate mutually disjoint (Russian)* ◇ D35 F30 ◇
POGORZELSKI, H.A. *Dirichlet theorems and prime number hypotheses of a conditional Goldbach theorem* ◇ F30 ◇
RYAN, W.J. *The equivalence of equational and sentential general recursive arithmetics* ◇ F30 ◇
SAMBIN, G. *An effective fixed-point theorem in intuitionistic diagonalizable algebras (The algebraisation of the theories which express Theor. IX)* ◇ F30 F50 G10 G25 ◇
SOLOVAY, R.M. *Provability interpretations of modal logic* ◇ B45 E30 F30 ◇
STANFORD, P.H. *A formalisation of the integers in a multi-successor arithmetic* ◇ F30 ◇
WANDSCHNEIDER, D. *Reflexive Unbeweisbarkeitsaussagen. Anmerkungen zur Grundsatzdiskussion um Goedels Unvollstaendigkeitsproblem* ◇ A05 E30 F30 ◇

1977

BERMAN, L. *Precise bounds for Presburger arithmetic and the reals with addition: preliminary report* ◇ B25 D15 F30 ◇
BOOLOS, G. *On deciding the provability of certain fixed point statements* ◇ B25 B45 F30 ◇
BUTRICK, R. *The numeral axioms* ◇ F30 ◇
CARSTENS, H.G. *The theorem of Matijasevic is provable in Peano's arithmetic by finitely many axioms* ◇ C52 D25 D35 F30 H15 ◇
CELLUCCI, C. *Proprieta di uniformita e 1-coerenza dell'aritmetica del primo ordine* ◇ F05 F30 ◇
CHUBARYAN, A.A. *The complexity of deductions in formal arithmetic and predicate calculus (Russian)* ◇ B10 F20 F30 ◇
FRANZ, W. *Ueber mathematische Aussagen, die samt ihrer Negation nachweislich unbeweisbar sind. Der Unvollstaendigkeitssatz von Goedel* ◇ F30 ◇
GIRARD, J.-Y. *Functional interpretation and Kripke models* ◇ C90 F10 F30 F50 ◇
HAFNER, I. *On some subtheory of formal arithmetic* ◇ F30 ◇

HAJEK, P. *Experimental logics and Π_3^0 theories* ◇ B60 C62 D55 F30 ◇
HALMOS, P.R. *Logic from A to G* ◇ A10 F30 ◇
HARRINGTON, L.A. & PARIS, J.B. *A mathematical incompleteness in Peano arithmetic* ◇ C30 C62 E05 F30 ◇
HINATA, S. *A normalization theorem in formal theories of natural numbers* ◇ F05 F30 F50 ◇
KIRBY, L.A.S. & PARIS, J.B. *Initial segments of models of Peano's axioms* ◇ C62 F30 ◇
KUZICHEV, A.S. *Formal arithmetic in the $\mathfrak{A}$-system of λ-conversion (Russian)* ◇ B40 F05 F30 ◇
LIVCHAK, A.B. *Expressible subsets of the ordered group of integers (Russian)* ◇ C60 C62 F30 ◇
LOPEZ-ESCOBAR, E.G.K. *Infinite rules in finite systems* ◇ F30 F35 ◇
MANIN, YU.I. *A course in mathematical logic* ◇ B98 C07 D98 E35 E50 F30 G12 ◇
MART'YANOV, V.I. *Extended universal theories of the integers (Russian)* ◇ B25 D35 F30 ◇
MCALOON, K. *Consistency statements and number theories* ◇ C62 F30 ◇
MINTS, G.E. *Formal arithmetic (Russian)* ◇ F30 ◇
MINTS, G.E. *Goedel's incompleteness theorem (Russian)* ◇ F30 ◇
NAKAMURA, A. & ONO, H. *Two-dimensional finite automata and their application to the decision problem of monadic first-order arithmetic $A[P, F(x), G(x)]$* ◇ D05 D35 F30 F35 ◇
NEPEJVODA, N.N. *Level Beth models and realizability (Russian)* ◇ F30 F50 ◇
POGORZELSKI, H.A. *Semisemiological structure of the prime numbers and conditional Goldbach theorems* ◇ F30 ◇
SCHWICHTENBERG, H. *Proof theory: Some applications of cut-elimination* ◇ F05 F10 F15 F30 F35 F98 ◇
SEMENOV, A.L. *Presburgerness of predicates regular in two number systems (Russian)* ◇ B10 D20 F30 ◇
SHIRAI, K. *A relation between transfinite induction and mathematical induction in elementary number theory* ◇ F15 F30 ◇
SHOSTAK, R.E. *On the SUP-INF method for proving Presburger formulas* ◇ B25 B35 F30 ◇
SMORYNSKI, C.A. *ω-consistency and reflection* ◇ F30 ◇
SMORYNSKI, C.A. *The incompleteness theorems* ◇ C62 F25 F30 F98 ◇
SORNIKOV, YA.A. *On quantifier-free ε_0-recursive arithmetic (Russian) (English summary)* ◇ F15 F30 ◇
TAKAHASHI, MOTO-O *A foundation of finite mathematics* ◇ F30 ◇
TAKAHASHI, MOTO-O *An abstract form of Goedel's theorem on consistency and Loeb's* ◇ F30 ◇
TOMAS, F. *A free formalism for the general recursive arithmetic (Spanish)* ◇ F30 ◇
TROELSTRA, A.S. *Aspects of constructive mathematics* ◇ F30 F35 F50 F55 F60 F98 ◇
URSINI, A. *A sequence of theories for arithmetic whose union is complete* ◇ C62 F30 ◇
YANOV, YU.I. *Computations in a class of programs (Russian)* ◇ B75 D20 F30 ◇
YUKAMI, T. *A theorem on the formalized arithmetic with function symbols ' and +* ◇ F20 F30 ◇

1978

BUCHHOLZ, W. & POHLERS, W. *Provable well orderings of formal theories for transfinitely iterated inductive definitions* ◇ F15 F30 F35 F50 ◇
CELLUCCI, C. *Teoria della dimostrazione. Normalizzazioni e assegnazioni di numeri ordinali* ◇ F05 F07 F15 F30 F98 ◇
FRIEDMAN, H.M. *Classically and intuitionistically provably recursive functions* ◇ B15 D20 E70 F30 F50 ◇
GOODMAN, NICOLAS D. *The nonconstructive content of sentences of arithmetic* ◇ D20 D30 D55 F30 F50 ◇
KANOVICH, M.I. *An estimate of the complexity of arithmetic incompleteness (Russian)* ◇ D15 F20 F30 ◇
KIRBY, L.A.S. & PARIS, J.B. *Σ_n-collection schemas in arithmetic* ◇ C62 F30 ◇
KUZICHEV, A.S. *A theorem on the consistency of formal arithmetic (Russian)* ◇ B40 F05 F30 ◇
KUZICHEV, A.S. *Formal arithmetic in combinatorially complete systems I,II (Russian) (English summary)* ◇ B40 F05 F30 ◇
KUZICHEV, A.S. *On the consistency of formal arithmetic (Russian)* ◇ B40 F30 ◇
LEVITZ, H. *An ordinal bound for the set of polynomial functions with exponentiation* ◇ D15 D20 E07 E10 F15 F30 ◇
LIPTON, R.J. *Model theoretic aspects of computational complexity* ◇ C62 D15 F30 F65 H15 ◇
LOVELAND, D.W. & REDDY, C.R. *Presburger arithmetic with bounded quantifier alternation* ◇ B25 C10 D15 F20 F30 ◇
MCALOON, K. *Completeness theorems, incompleteness theorems and models of arithmetic* ◇ C25 C62 F30 ◇
MCALOON, K. *Diagonal methods and strong cuts in models of arithmetic* ◇ C57 C62 D80 F30 ◇
MLCEK, J. *End-extensions of countable structures and the induction schema* ◇ C15 C62 F30 ◇
MONTAGNA, F. *On the algebraization of a Feferman's predicate (the algebraization of theories, which express Theor. X)* ◇ C05 F30 G05 G25 ◇
OPPEN, D.C. *A $2^{2^{2^{pn}}}$ upper bound on the complexity of Presburger arithmetic* ◇ B25 C10 D15 F30 ◇
PARIS, J.B. *Note on an induction axiom* ◇ B28 C62 F30 ◇
PARIS, J.B. *Some independence results for Peano arithmetic* ◇ C62 F30 ◇
PARLAMENTO, F. *Sui sistemi metaformali e le progressioni ricorsive di teorie* ◇ F15 F30 ◇
PARLAMENTO, F. *Sulla completezza della ω-regola (English summary)* ◇ F20 F30 ◇
PRIDA, J.F. *Una nueva prueba del teorema de incompletitud de la aritmetica* ◇ F30 ◇

RYAN, W.J. *Goedel's second incompleteness theorem for general recursive arithmetic* ◇ D20 F30 ◇
SAMBIN, G. *Fixed points through the finite model property (The algebraization of the theories which express Theor. XI)* ◇ B45 F30 G25 ◇
SMORYNSKI, C.A. *Avoiding self-referential statements* ◇ D25 F30 ◇
SMORYNSKI, C.A. *Beth's theorem and self-referential sentences* ◇ B45 C40 F30 ◇
SVEJDAR, V. *Degrees of interpretability* ◇ F25 F30 G10 ◇
TAKEUTI, G. *Two applications of logic to mathematics* ◇ B98 C90 E40 E75 F05 F30 F35 G12 ◇
URSINI, A. *On the set of "meaningful" sentences of arithmetic* ◇ F30 ◇
WILKIE, A.J. *Some results and problems on weak systems of arithmetic* ◇ C62 F30 H15 ◇
YUKAMI, T. *A note on a formalized arithmetic with function symbols ' and +* ◇ F20 F30 ◇
ZASLAVSKIJ, I.D. *Symmetric constructive logic (Russian)* ◇ B55 F05 F30 F50 ◇

1979

BAKER, T.P. *On "provable" analogs of $\mathscr{P}$ and $\mathscr{NP}$* ◇ D15 F30 ◇
BOOLOS, G. *Reflection principles and iterated consistency assertions* ◇ B45 F30 ◇
BOOLOS, G. *The unprovability of consistency. An essay in modal logic* ◇ B25 B28 B45 B98 F30 F98 ◇
BORGA, M. *Alcuni risultati di teoria della dimostrazione* ◇ F30 ◇
CSIRMAZ, L. *On definability in Peano arithmetic* ◇ C40 C62 F30 ◇
DAWSON JR., J.W. *The Goedel incompleteness theorem from a length-of-proof perspective* ◇ A10 F20 F30 H15 ◇
DEKHTYAR', M.I. *Complexity spectra of recursive sets of approximability of initial segments of complete problems* ◇ D10 D15 D30 F20 F30 ◇
DEMILLO, R.A. & LIPTON, R.J. *Some connections between mathematical logic and complexity theory* ◇ C57 D15 F30 H15 ◇
DRAGALIN, A.G. *Mathematical intuitionism. Introduction to proof theory (Russian)* ◇ F05 F30 F35 F50 F98 ◇
GLUSHKOV, V.M. *Incompleteness theorem of formal theories from programmer's viewpoint (Russian)* ◇ B75 F30 ◇
GORDON, D. *Complexity classes of provable recursive functions* ◇ D15 D20 F30 ◇
GRATTAN-GUINNESS, I. *In memoriam Kurt Goedel: His 1931 correspondence with Zermelo on his incompletability theorem* ◇ A10 F30 ◇
GUASPARI, D. *Partially conservative extensions of arithmetic* ◇ C62 F25 F30 ◇
GUASPARI, D. & SOLOVAY, R.M. *Rosser sentences* ◇ B45 F30 ◇
HAJEK, P. *On partially conservative extensions of arithmetic* ◇ F30 ◇
HUMPHRIES, J. *Goedel's proof and the liar paradox* ◇ A05 F30 ◇
KANOVICH, M.I. *A complexity version of Goedel's incompleteness theorem* ◇ D15 F20 F30 ◇
KREISEL, G. *Comment on Zinoviev's paper* ◇ F30 ◇
KUZICHEV, A.S. *Formale Arithmetik in einem System der λ-Konversion mit logischen Operatoren (Russisch)* ◇ B40 F05 F30 ◇
LAVROV, I.A. *Computability of partial functions and enumerability of sets in Peano's arithmitic (Russian)* ◇ D20 D25 F30 ◇
LEIVANT, D. *Absoluteness of intuitionistic logic* ◇ F05 F30 F50 ◇
LIFSCHITZ, V. CT_0 *is stronger than* $CT_0!$ ◇ F30 F50 ◇
LIFSCHITZ, V. *An intuitionistic definition of classical natural numbers* ◇ F30 F50 ◇
LINDSTROEM, P. *Some results on interpretability* ◇ F25 F30 ◇
LIPTON, R.J. *On the consistency of $P = NP$ and fragments of arithmetic* ◇ D15 F30 ◇
LOURENCO, M. (ED.) *Goedel's theorem and the continuum hypothesis (Portuguese)* ◇ B97 E50 F30 F97 ◇
MANIN, YU.I. *Provable and unprovable (Russian)* ◇ A05 B98 E35 E50 F30 F98 G12 ◇
MINTS, G.E. *A new reduction sequence for arithmetic (Russian) (English summary)* ◇ F05 F15 F30 ◇
MINTS, G.E. *Normalisation of natural deduction and effectiveness of classical existence (Russian)* ◇ F05 F10 F30 F35 ◇
MONTAGNA, F. *On the formulas of Peano arithmetic which are provably closed under modus ponens* ◇ F30 ◇
NAKAMURA, A. & ONO, H. *Undecidability of the first-order arithmetic $A[P(x),2x,x+1]$* ◇ D05 D35 F30 F35 ◇
O'DONNELL, M.J. *A practical programming theorem which is independent of Peano arithmetic* ◇ B40 B75 D20 F05 F30 ◇
O'DONNELL, M.J. *A programming language theorem which is independent of Peano arithmetic* ◇ B75 D15 F30 ◇
OHAMA, S. *On the results of Paris and Harrington (Japanese)* ◇ F30 ◇
PENZIN, YU.G. *Twins problem in formal arithmetic (Russian)* ◇ F30 H15 ◇
RYAN, W.J. *Proof of the quadratic reciprocity law in primitive recursive arithmetic* ◇ F30 ◇
SCHMERL, U.R. *A fine structure generated by reflection formulas over primitive recursive arithmetic* ◇ F30 ◇
SEMENOV, A.L. *On certain extensions of the arithmetic of addition of natural numbers (Russian)* ◇ B25 F30 ◇
SHOSTAK, R.E. *A practical decision procedure for arithmetic with function symbols* ◇ B25 B35 F30 ◇
SMORYNSKI, C.A. *Calculating self-referential statements. I: Explicit calculations* ◇ B45 F30 G25 ◇
SMORYNSKI, C.A. *Some rapidly growing functions* ◇ D20 D80 F30 ◇
SVENONIUS, L. *Two kinds of extensions of primitive recursive arithmetic* ◇ D20 F30 F35 ◇
VAEAENAENEN, J. *A new incompleteness in arithmetic (Finnish) (English summary)* ◇ A10 C62 D25 E05 F30 ◇

WOEHL, K. *Zur Komplexitaet der Presburger Arithmetik und des Aequivalenz-Problems einfacher Programme* ◇ B25 B75 D15 F20 F30 ◇
ZINOV'EV, A.A. *Complete (rigorous) induction & Fermat's great theorem* ◇ F30 ◇

1980

ABERTH, O. *Computable analysis* ◇ F30 F60 F98 ◇
BENDA, M. *On strong axioms of induction in set theory and arithmetic* ◇ E35 F30 ◇
BENDA, M. *On Harrington's partition relation* ◇ F30 ◇
BERMAN, L. *The complexity of logical theories* ◇ B25 D10 D15 F30 ◇
BOOLOS, G. *ω-consistency and the diamond* ◇ B28 F30 ◇
BOOLOS, G. *On systems of modal logic with provability interpretations* ◇ B45 F30 ◇
BORGA, M. *Un raffronto fra il teorema di incompletezza di Goedel e il teorema di Tarski sulla nozione di verita* ◇ F30 ◇
CANTINI, A. *A note on three-valued logic and Tarski theorem on truth definitions* ◇ B50 C40 F30 ◇
CEGIELSKI, P. *La theorie elementaire de la multiplication (English summary)* ◇ B25 C10 C80 F30 ◇
CELLUCCI, C. *A note on Goedel's second incompleteness theorem* ◇ F30 ◇
CHAPMAN, T. *An apparent contradiction in Goedel's proof* ◇ F30 ◇
CLOTE, P. *Weak partition relations, finite games, and independence results in Peano arithmetic* ◇ C62 E05 F30 ◇
DETLEFSEN, M. *On a theorem of Feferman* ◇ A05 F25 F30 ◇
DILLER, J. *Modified realization and the formulae-as-types notion* ◇ F05 F10 F30 F35 F50 ◇
DRAGALIN, A.G. *New forms of realizability and Markov's rule (Russian)* ◇ F30 F50 ◇
DRIES VAN DEN, L. *Some model theory and number theory for models of weak systems of arithmetic* ◇ B25 C62 F30 H15 ◇
FRIEDMAN, H.M. *A strong conservative extension of Peano arithmetic* ◇ B28 F30 F35 ◇
HIRSCHFELD, J. *A lower bound for Ramsey's theorem* ◇ F30 ◇
KIRBY, L.A.S. *La methode des indicatrices et le theoreme d'incompletude* ◇ C62 F30 ◇
KUZICHEV, A.A. & KUZICHEV, A.S. *On the embedding of formal arithmetic in combinatorially complete systems (Russian)* ◇ B40 F05 F30 ◇
KUZICHEV, A.S. *Sequential systems of λ-conversion and of combinatory logic* ◇ B40 F30 ◇
LASCAR, D. *Une indicatrice de type "Ramsey" pour l'arithmetique de Peano et la formule de Paris-Harrington* ◇ C62 F30 ◇
LEIVANT, D. *Innocuous substitutions* ◇ F30 F50 ◇
MACINTYRE, A. *Ramsey quantifiers in arithmetic* ◇ C62 C80 F30 F35 ◇
MANEVITZ, L.M. & STAVI, J. *Δ^2_0 operators and alternating sentences in arithmetic* ◇ C62 F30 H15 ◇
MCALOON, K. *Les rapports entre la methode des indicatrices et la methode de Goedel pour obtenir des resultats d'independance* ◇ C62 E35 F30 ◇
MCALOON, K. (ED.) *Modeles de l'arithmetique* ◇ C62 C97 F30 ◇
MCALOON, K. *Progressions transfinies de theories axiomatiques, formes combinatoires du theoreme d'incompletude et fonctions recursives a croissance rapide* ◇ C62 D20 F15 F30 ◇
MILLS, G. *A tree analysis of unprovable combinatorial statements* ◇ F30 ◇
MISERCQUE, D. *Sur le treillis distributif des $\forall_1$-formules fermees de l'arithmetique de Peano (English summary)* ◇ B20 F30 G10 ◇
MIYATAKE, T. *On the length of proofs in formal systems* ◇ F20 F30 ◇
MONTAGNA, F. *Interpretations of the first-order theory of diagonalizable algebras in Peano arithmetic* ◇ F30 G25 ◇
PAEPPINGHAUS, P. *A version of the Σ_1-reflection principle for CFA provable in PRA* ◇ F30 F35 ◇
PARIS, J.B. *A hierarchy of cuts in models of arithmetic* ◇ C15 C62 F30 ◇
SAMBIN, G. & VALENTINI, S. *A modal sequent calculus for a fragment of arithmetic* ◇ B45 F05 F30 ◇
SIMMONS, H. *The lattice of universal sentences modulo Peano arithmetic* ◇ F30 G10 ◇
SMORYNSKI, C.A. *Calculating self-referential statements* ◇ B45 F30 G25 ◇
STEIN, M. *Interpretations of Heyting's arithmetic - An analysis by means of a language with set symbols* ◇ F10 F30 F50 ◇
TOSI, P. *Normal derivability and first-order arithmetic* ◇ F05 F30 F50 ◇
TVERSKOJ, A.A. *A sequence of combinatorial judgements which are independent of Peano arithmetic (Russian) (English summary)* ◇ C62 E05 F30 ◇
VISSER, A. *Numerations, λ-calculus & arithmetic* ◇ B40 D45 F30 ◇
WETTE, E. *Inconsistenza dell'induzione matematica* ◇ F30 ◇
WILKIE, A.J. *Applications of complexity theory to Σ_0-definability problems in arithmetic* ◇ C62 D15 F30 ◇
YASUGI, M. *Gentzen reduction revisited* ◇ F05 F15 F30 F35 ◇
ZBIERSKI, P. *Indicators and incompleteness of Peano arithmetic* ◇ C62 F30 ◇

1981

ABRUSCI, V.M. *Due note sul principio di induzione matematica* ◇ A10 F30 ◇
ACZEL, P. *Two notes on the Paris independence result I: A generalization of Ramsey's theorem. II: The ordinal height of a density* ◇ C62 E05 F15 F30 ◇
BELLISSIMA, F. & MIROLLI, M. *Metodi algebrici nella teoria della dimostrazione* ◇ F30 ◇
BERLINE, C. & MCALOON, K. & RESSAYRE, J.-P. (EDS.) *Model theory and arithmetic. Comptes rendus d'une action thematique programmee du C.N.R.S. sur la*

theorie des modeles et l'arithmetique
◇ C62 C97 F30 ◇

BERNARDI, C. *On the relation provable equivalence and on partitions in effectively inseparable sets*
◇ D25 D35 F30 ◇

CEGIELSKI, P. *Theorie elementaire de la multiplication des entiers naturels* ◇ B25 C10 C80 F30 ◇

CELLUCCI, C. *Proof theory and theory of meaning*
◇ A05 F07 F30 ◇

CHARLESWORTH, A. *A proof of Goedel's theorem in terms of computer programs* ◇ D10 F30 ◇

CHRISTIAN, C.C. *Das rekursive Inaccessibilitaetstheorem und der Goedelsche Unvollstaendigkeitssatz in ihrer Bedeutung fuer die Informatik*
◇ A05 D35 D80 F30 ◇

CLOTE, P. *Anti-basis theorems and their relation to independence results in Peano arithmetic*
◇ C62 C98 E05 F30 ◇

ERDOES, P. & MILLS, G. *Some bounds for the Ramsey-Paris-Harrington numbers* ◇ F30 ◇

ESAKIA, L.L. *Diagonal constructions, the Loeb formula and rarefied Cantor spaces* ◇ F30 ◇

GALVAN, S. *Nota sull'aritmetica formalizzata di Peano (Dimostrazione del piccolo teorema di Fermat mediante una procedura ristretta ai numerali)* ◇ A10 F30 ◇

GIRARD, J.-Y. & PAEPPINGHAUS, P. *A result on implications of Σ_1-sentences and its application to normal form theorems* ◇ F05 F30 F35 ◇

GURARI, E.M. & IBARRA, O.H. *The complexity of the equivalence problem for two characterizations of Presburger sets* ◇ B25 D05 D15 F30 ◇

HAJEK, P. *Completion closed algebras and models of Peano arithmetic* ◇ C62 F30 ◇

HAJEK, P. *On interpretability in theories containing arithmetics II* ◇ B28 B45 F25 F30 ◇

HELLMAN, GEOFFREY *How to Goedel a Frege-Russell: Goedel's incompleteness theorems and logicism*
◇ A05 F30 ◇

IBARRA, O.H. & LEININGER, B.S. *Characterizations of Presburger functions* ◇ D20 F30 ◇

JAMBU-GIRAUDET, M. *Interpretations d'arithmetiques dans des groupes et des treillis*
◇ C07 C62 C65 D35 E07 F25 F30 G10 ◇

JOSEPH, D. & YOUNG, P. *A survey of some recent results on computational complexity in weak theories of arithmetic*
◇ D15 F30 ◇

KETONEN, J. & SOLOVAY, R.M. *Rapidly growing Ramsey functions* ◇ F30 ◇

KOGAN-BERNSHTEIJN, L.M. *Simplification of Gentzen's reductions in classical arithmetic (Russian)*
◇ F05 F30 ◇

KOSOVSKIJ, N.K. *Elements of mathematical logic and its applications to the theory of subrecursive algorithms (Russian)* ◇ B98 D20 D98 F30 F60 F98 ◇

KOSSAK, R. & PARIS, J.B. *Subsets of models of arithmetic*
◇ C62 F30 ◇

KURATA, R. *The reflection principle,transfinite induction, and the Paris-Harrington principle (Japanese)*
◇ C62 F30 ◇

KUZICHEV, A.S. *Arithmetic theories constructed on the basis of λ-conversion (Russian)* ◇ B40 F30 ◇

LEIVANT, D. *Implicational complexity in intuitionistic arithmetic* ◇ F30 F50 ◇

LEIVANT, D. *On the proof theory of the modal logic for arithmetic provability* ◇ B45 F05 F30 ◇

MCALOON, K. & RESSAYRE, J.-P. *Les methodes de Kirby-Paris et la theorie des ensembles*
◇ C62 E05 E30 E55 F30 ◇

MISERCQUE, D. *Answer to a problem of D. Guaspari*
◇ F30 ◇

MISERCQUE, D. *Solutions de deux problemes poses par H. Simmons* ◇ C62 F30 H15 ◇

MIYATAKE, T. *On the length of proofs in a formal system of recursive arithmetic* ◇ F20 F30 ◇

NADEL, M.E. *The completeness of Peano multiplication*
◇ F30 ◇

NEGRI, M. *Constructive sequent reduction in Gentzen's first consistency proof for arithmetic*
◇ A10 F05 F30 ◇

PABION, J.F. & RICHARD, D. *Synonymy and re-interpretation for some sublanguages of Peano arithmetic* ◇ B28 C62 F30 ◇

PARIS, J.B. & WILKIE, A.J. *Δ_0 sets and induction*
◇ C62 F30 ◇

PARIS, J.B. *Some conservation results for fragments of arithmetic* ◇ C62 F30 ◇

PARLAMENTO, F. *Binumerability in a sequence of theories*
◇ F30 ◇

PODNIEKS, K.M. *Around Goedel's theorem (Russian)*
◇ F30 ◇

SAZONOV, V.YU. *On existence of complete predicate calculus in metamathematics without exponentiation*
◇ F20 F30 ◇

SHANIN, N.A. *Role of a notion algorithm in the arithmetic language semantics* ◇ B75 F30 F50 ◇

SHORE, R.A. *The theory of the degrees below O'*
◇ D25 D30 D35 F30 ◇

SMORYNSKI, C.A. *Calculating self-referential statements: Guaspari sentences of the first kind* ◇ F30 G25 ◇

SMORYNSKI, C.A. *Fifty years of self-reference in arithmetic*
◇ A10 F30 F40 ◇

STEPANOV, V.I. *Self-referential propositions for standard and Rosser provability predicates (Russian)* ◇ F30 ◇

TOSI, P. *Forme normali* ◇ F05 F30 ◇

TOSI, P. *Intuitive proofs and first-order derivations: Some notes on the metamathematics of first-order number theory* ◇ F05 F30 ◇

TWER VON DER, T. *Some remarks on the mathematical incompleteness of Peano's arithmetic found by Paris and Harrington* ◇ C62 E05 F30 ◇

WANG, HAO *Popular lectures on mathematical logic (Chinese)* ◇ A05 B98 C98 E98 F30 F98 ◇

XU, LIZHI & YUAN, XIANGWAN & ZHENG, YUXIN & ZHU, WUJIA *On Goedel's incompleteness theorem (Chinese) (English summary)* ◇ A05 F30 ◇

1982

ARTEMOV, S.N. *Applications of modal logic in proof theory (Russian)* ◇ B45 F30 ◇

BENDOVA, K. & HAJEK, P. *A logical analysis of the truth-reaction paradox* ◇ B45 F30 ◇

BERNARDI, C. & PAGLI, P. *Logical and algebraic versions of diagonalization lemma (Italian summary)* ◇ F30 G25 ◇

BOOLOS, G. *Extremely undecidable sentences* ◇ F30 ◇

BOOLOS, G. *On the nonexistence of certain normal forms in the logic of provability* ◇ B45 F30 ◇

BUNDER, M.W. *A formal metatheory in which Goedel's first but not his second incompleteness theorem can be proved* ◇ F30 ◇

CANTINI, A. *Non-extensional theories of predicative classes over PA* ◇ F30 F35 ◇

CEGIELSKI, P. & MCALOON, K. & WILMERS, G.M. *Modeles recursivement satures de l'addition et de la multiplication des entiers naturels (English summary)* ◇ C50 C57 C62 F30 H15 ◇

CHAITIN, G.J. *Goedel's theorem and information* ◇ A05 F30 ◇

CHUBARYAN, A.A. *Complexity characteristics of inferences in systems of predicate calculus and formal arithmetic (Russian) (Armenian summary)* ◇ B10 F20 F30 ◇

DALEN VAN, D. *The creative subject and Heyting's arithmetic* ◇ F30 F55 ◇

DAPUETO, C. *Sull' aritmetizzazione della sintassi del primo ordine (French summary)* ◇ B28 F30 F40 ◇

DIMITRACOPOULOS, C. & GAIFMAN, H. *Fragments of Peano's arithmetic and the MRDP theorem* ◇ F30 ◇

FAHMY, M.H. *A formal proof of Ackermann's theorem (Arabic summary)* ◇ D20 F30 ◇

FEFERMAN, S. *Inductively presented systems and the formalization of meta-mathematics* ◇ F30 F35 F40 ◇

FISHER, A. *Formal number theory and computability* ◇ B98 D20 F30 F98 ◇

HOTOMSKI, P. *An induction law in proofs by contradiction with an application to automatic theorem proving (Russian)* ◇ B35 F30 ◇

HUBER-DYSON, V. & JONES, JAMES P. & SHEPHERDSON, J.C. *Some diophantine forms of Goedel's theorem* ◇ D35 F30 ◇

JONGH DE, D.H.J. *Formulas of one propositional variable in intuitionistic arithmetic* ◇ F30 F50 ◇

KIRBY, L.A.S. & PARIS, J.B. *Accessible independence results for Peano arithmetic* ◇ C62 F30 ◇

KIRBY, L.A.S. *Flipping properties in arithmetic* ◇ C62 E05 F30 ◇

KOWALCZYK, W. *A sufficient condition for the consistency of P = NP with Peano arithmetic* ◇ D15 F25 F30 ◇

KREISEL, G. *Finiteness theorems in arithmetic: an application of Herbrand's theorem for Σ_2-formulas* ◇ F05 F30 ◇

KUZICHEV, A.S. *Arithmetically consistent λ-theories (Russian)* ◇ B40 F05 F30 ◇

LEIVANT, D. *Unprovability of theorems of complexity theory in weak number theories* ◇ D15 F30 ◇

LOPEZ-ESCOBAR, E.G.K. *Further applications of ultra-conservative ω-rules* ◇ F05 F30 ◇

MAGARI, R. *Primi risultati sulla varieta di Boolos (English summary)* ◇ B45 F30 G25 ◇

MALIAUKIENE, L. *Equivalence of some arithmetical systems (Russian) (English and Lithuanian summaries)* ◇ F30 ◇

MALIAUKIENE, L. *Free variable additive arithmetic with restricted difference (Russian) (English and Lithuanian summaries)* ◇ F30 ◇

MINTS, G.E. *A simplified consistency proof for arithmetic (Russian) (English and Estonian summaries)* ◇ F05 F30 ◇

MISERCQUE, D. *The nonhomogeneity of the E-tree – answer to a problem raised by D. Jensen and A. Ehrenfeucht* ◇ C62 F30 ◇

MO, SHAOKUI & SHEN, BAIYING *Inverse functions of number-theoretic functions I,II (Chinese) (English summaries)* ◇ F30 ◇

MONTAGNA, F. *Relatively precomplete numerations and arithmetic* ◇ D45 F30 ◇

MORGENSTERN, C.F. *On generalized quantifiers in arithmetic* ◇ C62 C80 F30 ◇

PAGLI, P. *Alcune estensioni del lemma di diagonalizzazione (English summary)* ◇ F30 G25 ◇

PFENDER, M. & REITER, R. & SARTORIUS, M. *Constructive arithmetics* ◇ F30 G30 ◇

PLA I CARRERA, J. *On the R-representability of primitive recursive functions (Catalan)* ◇ D20 F30 ◇

RATAJCZYK, Z. *Satisfaction classes and combinatorial sentences independent from PA* ◇ C62 E05 F30 ◇

RICHARD, D. *La theorie sans egalite du successeur et de la coprimarite des entiers naturels est indecidable. Le predicat de primarite est definissable dans le langage de cette theorie (English summary)* ◇ D35 F30 ◇

ROSENBERG, R. *Recursively enumerable images of arithmetic sets* ◇ D25 D55 F30 ◇

ROSINGER, E.E. *The algebraic uniqueness of the addition of natural numbers* ◇ F30 ◇

SAMBIN, G. & VALENTINI, S. *The modal logic of provability. The sequential approach* ◇ B45 F05 F30 ◇

SAMOKHVALOV, K.F. *Refinement of the usual interpretation of Goedel's theorems on incompleteness and concepts of recursive enumerability (Russian)* ◇ F30 ◇

SCHMERL, U.R. *Iterated reflection principles and the ω-rule* ◇ F30 ◇

SCHMERL, U.R. *Number theory and the Bachmann/Howard ordinal* ◇ F15 F30 ◇

SEMENOV, A.L. *On the definability of arithmetic in its fragments (Russian)* ◇ F30 ◇

SMORYNSKI, C.A. *Commutativity and self-reference* ◇ B45 F30 ◇

SMORYNSKI, C.A. *Fixed point algebras* ◇ E30 F30 G25 ◇

SMORYNSKI, C.A. *Nonstandard models and constructivity* ◇ C62 F30 F50 ◇

SMORYNSKI, C.A. *The finite inseparability of the first-order theory of diagonalisable algebras* ◇ B45 C13 D35 F30 ◇

SMORYNSKI, C.A. *The varieties of arboreal experience* ◇ D20 F30 ◇

SORBI, A. *Σ_0^n-equivalence relations* ◇ D55 F30 ◇

SORBI, A. *Numerazioni positive, r.e. classi e formule (English summary)* ◇ D20 F30 ◇
STEPANOV, V.I. *Second-order arithmetic and the consistency of first-order theories (Russian)* ◇ F30 F35 ◇
TOMAS, F. *A recursive, finitarily consistent formalization of a fragment of nonconstructive arithmetic, and its relation to recursive analysis (Spanish)* ◇ F30 ◇
TOMAS, F. *About a recursive formalism for arithmetic and its relation to a recursive construction of analysis (Catalan)* ◇ F30 ◇
TSUBOI, A. *On reflection principles* ◇ F30 ◇
USPENSKIJ, V.A. *Goedel's incompleteness theorem (Russian)* ◇ F30 F98 ◇
VISSER, A. *On the completeness principle: a study of provability in Heyting's arithmetic and extensions* ◇ F30 F50 ◇
WILKIE, A.J. *On core structures for Peano arithmetic* ◇ C50 C62 F30 H15 ◇

1983

ANSHEL, M. & MCALOON, K. *Reducibilities among decision problems for HNN groups, vector addition systems and subsystems of Peano arithmetic* ◇ D20 D40 D80 F30 ◇
BARRICELLI, N.A. *Challenge to Goedel's proof* ◇ F30 ◇
BERGSTRA, J.A. & TUCKER, J.V. *Hoare's logic and Peano's arithmetic* ◇ B75 F30 ◇
BERNARDI, C. & SORBI, A. *Classifying positive equivalence relations* ◇ F30 ◇
BERNARDI, C. *Relazioni di equivalenza positive rappresentate da formule* ◇ F30 ◇
BORGA, M. *On some proof theoretical properties of the modal logic GL* ◇ B45 F05 F30 ◇
CALUDE, C. & PAUN, G. *Independent instances for some undecidable problems (French summary)* ◇ D20 F30 ◇
CICHON, E.A. *A short proof of two recently discovered independence results using recursion theoretic methods* ◇ F30 ◇
CLOTE, P. & MCALOON, K. *Two further combinatorial theorems equivalent to the 1-consistency of Peano arithmetic* ◇ B28 C62 F30 ◇
CUDA, K. *Nonstandard models of arithmetic as an alternative basis for continuum considerations* ◇ E70 F30 H15 ◇
DIMITRACOPOULOS, C. & PARIS, J.B. *A note on the undefinability of cuts* ◇ C40 C62 F30 H15 ◇
FLANNAGAN, T.B. *A note on a proof of Shepherdson* ◇ F30 ◇
GUASPARI, D. *Sentences implying their own provability* ◇ F30 ◇
JANICKA-ZUK, I. *Strong amalgamation property of diagonalizable algebras* ◇ F30 G25 ◇
JONES, JAMES P. & SHEPHERDSON, J.C. *Variants of Robinson's essentially undecidable theory R* ◇ F30 ◇
JOSEPH, D. *Polynomial time computations in models of ET* ◇ D15 F30 ◇
KHAKHANYAN, V.KH. *Set theory and Church's thesis (Russian)* ◇ D20 E35 E70 F30 F35 F50 ◇
KROL', M.D. *Various forms of the continuity principle (Russian)* ◇ F30 F50 ◇
KUZICHEV, A.S. *An arithmetically consistent λ-theory (Russian)* ◇ B40 F30 ◇
KUZICHEV, A.S. *Arithmetically consistent λ-theories of type-free logic (Russian)* ◇ B40 F05 F30 ◇
KUZICHEV, A.S. *Arithmetic completeness of type-free logic (Russian)* ◇ F30 ◇
LEIVANT, D. *The optimality of induction as an axiomatization of arithmetic* ◇ F30 ◇
MAGARI, R. *Verita non Tarskiana nelle algebre diagonalizzabili (English summary)* ◇ F30 G25 ◇
MALIAUKIENE, L. *A constructive proof of the replaceability on the induction scheme for quantifier-free formulas in additive arithmetic (Russian) (English and Lithuanian summaries)* ◇ F30 ◇
MALIAUKIENE, L. *A constructive proof of the replaceability of the induction axiom in the quantifier-free multiplicative arithmetic (Russian) (English and Lithuanian summaries)* ◇ F30 ◇
MALIAUKIENE, L. *On the relationship of certain weak inductions (Russian)(English and Lithuanian summaries)* ◇ F30 ◇
MIKOLAJEWICZ, B. *Some classes of models for Peano arithmetic and for some related theories (Polish) (English summary)* ◇ B28 F30 ◇
MONTAGNA, F. *Il teorema di Solovay e l'aritmetica di Heyting* ◇ B45 F30 ◇
MONTAGNA, F. *ZFC-models as Kripke-models* ◇ B45 C62 E30 F30 ◇
MUNDICI, D. *Natural limitations of decision procedures for arithmetic with bounded quantifiers* ◇ B25 D10 D15 F30 ◇
MURAVITSKIJ, A.YU. *Extensions of the provability logic (Russian)* ◇ B45 F30 ◇
ODIFREDDI, P. *Forcing and reducibilities* ◇ D30 D55 E40 F30 ◇
PUDLAK, P. *A definition of exponentiation by a bounded arithmetical formula* ◇ B28 C62 F30 ◇
PUDLAK, P. *Some prime elements in the lattice of interpretability types* ◇ E30 F25 F30 G10 ◇
SHEN, BAIYING *Studies on higher rules (Chinese) (English summary)* ◇ F30 F99 ◇
SOLITRO, U. & VALENTINI, S. *The modal logic of consistency assertions of Peano arithmetic* ◇ B45 F30 ◇
SPENCER, J.H. *Large numbers and unprovable theorems* ◇ D20 F30 ◇
SVEJDAR, V. *Modal analysis of generalized Rosser sentences* ◇ B45 F25 F30 ◇
UESU, T. *On a problem of Smorynski (Japanese)* ◇ F30 ◇
VALENTINI, S. *A "canonical" model for GL (Italian summary)* ◇ B45 F30 ◇
VALENTINI, S. *The modal logic of provability: cut-elimination* ◇ B45 F05 F30 ◇

1984

ADAMOWICZ, Z. *Axiomatization of the forcing relation with an application to Peano arithmetic* ◇ C62 E40 F30 ◇

ARTEMOV, S.N. *On modal representations of extensions of Peano arithmetic* ◇ B45 F30 ◇
AVRON, A. *On modal systems having arithmetical interpretations* ◇ B45 F05 F30 ◇
BARRICELLI, N.A. *A few trivialities about Goedel's proof and the rules of definition* ◇ A05 F30 ◇
BENCIVENGA, E. *Finitary consistency of a free arithmetic* ◇ F30 ◇
BERNARDI, C. *A shorter proof of a recent result by R. Di Paola* ◇ B45 F30 ◇
BERNARDI, C. & MONTAGNA, F. *Equivalence relations induced by extensional formulae: classification by means of a new fixed point property* ◇ F30 ◇
BOFFA, M. *Arithmetic and the theory of types* ◇ B15 F30 F35 ◇
BOOLOS, G. *The logic of provability* ◇ B45 F30 ◇
CARTWRIGHT, ROBERT *Nonstandard fixed points in first order logic* ◇ B75 F30 ◇
CEGIELSKI, P. *La theorie elementaire de la divisibilite est finiment axiomatisable (English summary)* ◇ B25 B28 F30 ◇
CHIHARA, C.S. *Priest, the Liar, and Goedel* ◇ A05 F30 ◇
COHEN, D.E. *Modular machines, undecidability and incompleteness* ◇ D20 D25 D35 F30 ◇
DILLER, J. & TROELSTRA, A.S. *Realizability and intuitionistic logic* ◇ A05 F30 F35 F50 ◇
DING, DECHENG & MO, SHAOKUI *Simplification of the axioms of recursive arithmetic (Chinese)* ◇ B28 F30 ◇
DRIES VAN DEN, L. & LEVITZ, H. *On Skolem's exponential functions below 2^{2^X}* ◇ E07 E75 F15 F30 ◇
DYWAN, Z. *An interpretation of a certain fragment of arithmetic in some propositional calculus* ◇ B20 F30 ◇
GARGOV, G.K. *A note on the provability logics of certain extensions of Heyting's arithmetic* ◇ B45 F30 ◇
GAVRILENKO, YU.V. *Monotonic theories of accessible numbers (Russian)* ◇ F20 F30 F65 ◇
GOODMAN, NICOLAS D. *Epistemic arithmetic is a conservative extension of intuitionistic arithmetic* ◇ B45 F05 F30 F50 ◇
HAJEK, P. *On a new notion of partial conservativity* ◇ C62 F30 ◇
HENSON, C.W. & RUBEL, L.A. *Some applications of Nevanlinna theory to mathematical logic:identities of exponential functions* ◇ B20 C65 F30 ◇
HENSON, C.W. & KAUFMANN, M. & KEISLER, H.J. *The strength of nonstandard methods in arithmetic* ◇ C62 E30 F30 F35 H05 H15 ◇
IVANOV, A.A. *Decidability of extended theories of addition of the natural numbers and the integers (Russian)* ◇ B25 C65 F30 ◇
JAEGER, G. *A version of Kripke-Platek set theory which is conservative over Peano arithmetic* ◇ E30 F30 F35 ◇
KIROV, K.A. *An intuitionistic analogue of the modal logic of provability in Peano's arithmetic* ◇ B45 F30 ◇
KRIVTSOV, V.N. *A formal system of negationless arithmetic is conservative with respect to Heyting arithmetic (Russian)* ◇ F30 F50 ◇
KUBINSKI, T. & MIKOLAJEWICZ, B. *Finite arithmetics, their categoricity and trees (Polish)* ◇ F30 ◇
KURATA, R. *Paris-Harrington theory and reflection principles* ◇ C62 F30 ◇
LINDSTROEM, P. *On certain lattices of degrees of interpretability* ◇ F25 F30 ◇
LINDSTROEM, P. *On faithful interpretability* ◇ F25 F30 ◇
LINDSTROEM, P. *On partially conservative sentences and interpretability* ◇ F25 F30 ◇
MAGARI, R. *Algebraic logic and diagonal phenomena* ◇ F30 G25 ◇
MELLIS, W. *PA-beweisbare ∀∃-Formeln* ◇ F30 ◇
MILLS, G. & PARIS, J.B. *Regularity in models of arithmetic* ◇ C62 C80 F30 ◇
MONTAGNA, F. *A completeness result for fixed-point algebras* ◇ F30 G10 G25 ◇
MONTAGNA, F. *The predicate modal logic of provability* ◇ B45 C90 F30 ◇
MURAWSKI, R. *Mathematical incompleteness of arithmetic (Polish)* ◇ D35 F30 ◇
NEGRI, M. *An application of recursive saturation* ◇ C50 C57 C62 F30 ◇
NORMANN, D. *The infinite - a mathematical necessity* ◇ A05 D20 F30 ◇
PAOLA DI, R.A. *A uniformly, extremely nonextensional formula of arithmetic with many undecidable fixed points in many theories* ◇ F30 G25 ◇
PARIS, J.B. *On the structure of models with restricted E_1-induction (Czech) (Russian and English summaries)* ◇ C62 F30 ◇
PARIS, J.B. & WILKIE, A.J. *Some results on bounded induction* ◇ F30 ◇
PARLAMENTO, F. *PRA provability of Schmerl's fine structure theorem* ◇ F30 ◇
RICHARD, D. *Les relations arithmetiques sur les entiers primaires sont definissables au premier ordre par successeur et coprimarite (English summary)* ◇ B28 F30 ◇
RICHARD, D. *The arithmetics as theories of two orders (English and French summaries)* ◇ B28 C62 D35 F30 ◇
SCARPELLINI, B. *Complexity of subcases of Presburger arithmetic* ◇ B25 D15 F20 F30 ◇
SCHMERL, U.R. *Diophantine equations in a fragment of number theory* ◇ C62 F30 ◇
SHELAH, S. *On logical sentences in PA* ◇ C62 C80 E05 F25 F30 ◇
SHEN, BAIYING *Primitive recursive arithmetic in the second class $A^0(D)$ (Chinese)* ◇ D20 F30 ◇
SHEN, BAIYING *The primitive recursive arithmetic $B^1 1$ of the first class (Chinese)* ◇ F30 ◇
SIEG, W. *Foundations of analysis and proof theory* ◇ A10 F30 F35 F98 ◇
SLESSENGER, P.H. *On subsets of the Skolem class of exponential polynomials* ◇ F15 F30 ◇
SLEZAK, P. *Minds, machines and self-reference (French and German summaries)* ◇ A05 D99 F30 ◇

SMORYNSKI, C.A. *Lectures on nonstandard models of arithmetic* ◇ A10 C62 C98 F30 F98 H15 ◇
SMORYNSKI, C.A. *Modal logic and self-reference* ◇ B45 F30 ◇
SMULLYAN, R.M. *Fixed points and self-reference* ◇ F30 ◇
TOMAS, F. *Formally recursive arithmetic and analysis (Catalan)* ◇ F30 F35 ◇
VISSER, A. *The provability logic of recursively enumerable theories extending Peano arithmetic and arbitrary theories extending Peano arithmetic* ◇ B45 C62 F30 ◇
YUKAMI, T. *Some results on speed-up* ◇ D15 F20 F30 ◇
ZHU, SHUILIN *An essay on Goedel's incompleteness theorem (Chinese)* ◇ A05 F30 ◇

1985

ABRUSCI, V.M. *Uses of dilators: combinatorial problems and combinatorial results not provable in PA or in ID_1 (Italian)* ◇ F15 F30 F35 ◇
ADAMOWICZ, Z. & MORALES-LUNA, G. *A recursive model for arithmetic with weak induction* ◇ C57 F30 H15 ◇
ADAMOWICZ, Z. *Algebraic approach to $\exists_1$ induction* ◇ C60 C62 F30 ◇
ARTEMOV, S.N. *Modal logics axiomatizing provability (Russian)* ◇ B45 F30 ◇
ARTEMOV, S.N. *Nonarithmeticity of truth predicate logics of provability (Russian)* ◇ B45 D35 D55 F30 ◇
BOOLOS, G. *1-consistency and the diamond* ◇ B45 F30 ◇
BOSCH, J.E. *Sur les nombres natureles et les fondements de l'arithmetique et la geometrie* ◇ B28 F30 ◇
BOZHICH, E.S. *Local consistency of an arithmetic with an "attainability" predicate (Russian)* ◇ F30 ◇
CANTINI, A. *Majorizing provably recursive functions in fragments of PA* ◇ D20 F15 F30 ◇
CLOTE, P. *Applications of the low-basis theorem in arithmetic* ◇ C62 D20 D30 D55 F30 ◇
CLOTE, P. *Partition relations in arithmetic* ◇ C62 E05 F30 ◇
DIMITRACOPOULOS, C. *A generalization of a theorem of H.Friedman* ◇ C62 D15 F30 ◇
FLAGG, R.C. *Church's thesis is consistent with epistemic arithmetic* ◇ F30 F50 ◇
FRIEDMAN, H.M. & SCEDROV, A. *Arithmetic transfinite induction and recursive well-orderings* ◇ D45 F30 F50 ◇
JERVELL, H.R. *Gentzen games* ◇ F30 ◇
JERVELL, H.R. *Large finite sets* ◇ E20 F15 F30 ◇
KOLETSOS, G. *Functional interpretation of the β-rule* ◇ F10 F30 F35 ◇
KOSSAK, R. *A note on satisfaction classes* ◇ C50 C57 C62 F30 ◇
KUCERA, A. *Measure, Π^0_1-classes and complete extensions of PA* ◇ D30 E15 E75 F30 ◇
LEIVANT, D. *Syntactic translations and provably recursive functions* ◇ D20 F30 F50 ◇
LIFSCHITZ, V. *Calculable natural numbers* ◇ B28 F30 F50 ◇
LINDSTROEM, P. *Provability and interpretability in theories containing arithmetic* ◇ F25 F30 ◇
LOEBL, M. *Hercules and Hydra, a game on rooted finite trees* ◇ F30 ◇
LOPEZ-ESCOBAR, E.G.K. *Koenig's lemma, the ω-rule and primitive recursive arithmetic* ◇ F30 ◇
LOPEZ-ESCOBAR, E.G.K. *Proof functional connectives* ◇ B55 F07 F30 ◇
MALIAUKIENE, L. *Elimination of the induction axiom in the multiplicative arithmetic with restricted difference (Russian) (English and Lithuanian summaries)* ◇ B28 F30 ◇
MCALOON, K. *Paris-Harrington incompleteness and progressions of theories* ◇ F15 F30 ◇
MEYER, R.K. & MORTENSEN, C. *Relevant quantum arithmetic* ◇ B46 F30 ◇
MISERCQUE, D. *Branches of the E-trees which are not isomorphic* ◇ F30 ◇
NEGRI, M. *On the nonfinite axiomatizability of P (Italian)* ◇ F30 ◇
PARIS, J.B. & WILKIE, A.J. *Counting problems in bounded arithmetic* ◇ B28 C62 F30 ◇
PARLAMENTO, F. *Rules and principles of restricted induction relative to PRA (Italian)* ◇ F30 ◇
RAUZY, G. *Mots infinis en arithmetique* ◇ C05 F30 ◇
REINHARDT, W.N. *Absolute versions of incompleteness theorems* ◇ F30 ◇
RICHARD, D. *All arithmetical sets of powers of primes are first-order definable in terms of the successor function and the coprimeness predicate* ◇ B28 F30 ◇
RICHARD, D. *Answer to a problem raised by J.Robinson: the arithmetic of positive or negative integers is definable from successor and divisibility* ◇ F30 ◇
RICHARD, D. *Definissabilite de l'arithmetique par successeur, coprimarite et puissance* ◇ B28 F30 ◇
SCHUETTE, K. & SIMPSON, S.G. *Ein in der reinen Zahlentheorie unbeweisbarer Satz ueber endliche Folgen von natuerlichen Zahlen* ◇ F30 ◇
SHAPIRO, S. *Epistemic and intuitionistic arithmetic* ◇ B28 F30 F50 ◇
SHEN, BAIYING *Primitive recursive arithmetic in the first class A^0. I (Chinese)* ◇ D20 F30 ◇
SIEG, W. *Fragments of arithmetic* ◇ F30 F35 ◇
SIMPSON, S.G. *Nichtbeweisbarkeit von gewissen kombinatorischen Eigenschaften endlicher Baeume (English summary)* ◇ F30 ◇
SLESSENGER, P.H. *A height restricted generation of a set of arithmetic functions of order-type ε_0* ◇ F15 F30 ◇
SMORYNSKI, C.A. *Nonstandard models and related developments* ◇ C62 C98 F30 F98 ◇
SMORYNSKI, C.A. *Self-reference and modal logic* ◇ B45 F30 F98 ◇
SMULLYAN, R.M. *Modality and self-reference* ◇ B45 F30 ◇
SMULLYAN, R.M. *Some principles related to Loeb's theorem* ◇ B45 F30 ◇
SOLOVAY, R.M. *Explicit Henkin sentences* ◇ F30 F40 ◇

STEPANOV, V.A. *Propositional logic of reflexive sentences. II (Russian)* ◇ B05 F30 ◇

STEPANOV, V.I. *The incompleteness theorems and related results for nonconstructive theories* ◇ F30 ◇

WILMERS, G.M. *Bounded existential induction* ◇ C62 F30 ◇

YOUNG, P. *Goedel theorems, exponential difficulty and undecidability of arithmetic theories: an exposition* ◇ D15 D35 F20 F30 ◇

F35 Second- and higher-order arithmetic and fragments

1906

HOBSON, E.W. *On the arithmetic continuum* ◇ A05 E47 F35 F55 ◇

1912

CHWISTEK, L.B. *The principle of contradiction in the light of recent investigations of Bertrand Russell (Russian)* ◇ A05 B15 F35 ◇

1928

BERNAYS, P. *Zusatz zu Hilberts Vortrag ueber "Die Grundlagen der Mathematik"* ◇ F35 ◇

1930

HEYTING, A. *Die formalen Regeln der intuitionistischen Mathematik* ◇ F30 F35 F50 ◇

1931

GOEDEL, K. *Ueber formal unentscheidbare Saetze der "Principia Mathematica" und verwandter Systeme I* ◇ B28 B30 D20 D35 F25 F30 F35 ◇

GOEDEL, K. *Ueber Vollstaendigkeit und Widerspruchsfreiheit* ◇ F30 F35 ◇

TARSKI, A. *Sur les ensembles definissables de nombres reels I* ◇ B25 B28 C10 C40 C60 C65 D55 E15 E47 F35 ◇

1934

GOEDEL, K. *Ueber die Laenge von Beweisen* ◇ B10 F20 F30 F35 ◇

1936

GENTZEN, G. *Die Widerspruchsfreiheit der Stufenlogik* ◇ B15 F05 F35 ◇

1937

BETH, E.W. *Une demonstration de la non-contradiction de la logique des types au point de vue fini* ◇ B15 F35 ◇

1938

CHWISTEK, L.B. & HETPER, W. *New foundation of formal metamathematics* ◇ B30 F35 ◇

1948

FITCH, F.B. *An extension of basic logic* ◇ B20 B40 F35 ◇

1949

FITCH, F.B. *The Heine-Borel theorem in extended basic logic* ◇ B28 B40 E70 F35 ◇

WANG, HAO *A theory of constructive types* ◇ B15 F35 F65 ◇

1950

HENKIN, L. *Completeness in the theory of types* ◇ A05 B15 C85 F35 ◇

1951

LORENZEN, P. *Algebraische und logistische Untersuchungen ueber freie Verbaende* ◇ A05 F35 G10 ◇

LORENZEN, P. *Die Widerspruchsfreiheit der klassischen Analysis* ◇ F35 F65 ◇

MYHILL, J.R. *Report on some investigations concerning the consistency of the axiom of reducibility* ◇ B15 F35 ◇

SCHUETTE, K. *Die Eliminierbarkeit des bestimmten Artikels in Kodifikaten der Analysis* ◇ F35 ◇

1952

GOETLIND, E. *A note on Chwistek and Hetper's foundation of formal metamathematics* ◇ B30 F35 ◇

MOSTOWSKI, ANDRZEJ *On models of axiomatic systems* ◇ C62 F25 F30 F35 ◇

SCHUETTE, K. *Beweistheoretische Untersuchungen der verzweigten Analysis* ◇ F05 F35 F65 ◇

1953

ACKERMANN, W. *Widerspruchsfreier Aufbau einer typenfreien Logik II* ◇ B10 B15 E35 E70 F30 F35 F65 ◇

MCNAUGHTON, R. *Some formal relative consistency proofs* ◇ F25 F30 F35 ◇

TAKEUTI, G. *On a generalized logic calculus* ◇ B15 F35 ◇

WANG, HAO *Certain predicates defined by induction schemata* ◇ D70 F30 F35 ◇

1954

SCHUETTE, K. *Ein widerspruchsloses System der Analysis auf typenfreier Grundlage* ◇ E70 F35 F65 ◇

SEKI, S. *A metatheorem on SLK* ◇ F35 ◇

WANG, HAO *The formalization of mathematics* ◇ A05 B30 E70 F35 F65 ◇

1955

TAKEUTI, G. *On the fundamental conjecture of GLC I,II* ◇ B15 F05 F35 ◇

1956

SIMAUTI, T. *Proof of a special case of the fundamental conjecture of Takeuti's GLC* ◇ B15 F05 F35 ◇

TAKEUTI, G. *A metamathematical theorem on functions* ◇ B15 F05 F10 F35 ◇

TAKEUTI, G. *Construction of ramified real numbers* ◇ B15 F05 F35 F65 ◇

TAKEUTI, G. *On the fundamental conjecture of GLC III,IV* ◇ B15 F05 F35 ◇

1957

ROBINSON, R.M. *Restricted set-theoretical definitions in arithmetic* ◇ B28 F30 F35 ◇

1958

BORKOWSKI, L. *Reduction of arithmetic to logic based on the theory of types without the axiom of infinity and the typical ambiguity of arithmetical constant (Polish and Russian summaries)* ◇ B15 F35 ◇

FALEVICH, B.YA. *A new method of proving incompleteness theorems for systems with Carnap's rule, and its application to the problem of interrelation between classical and constructive analysis (Russian)* ◇ F35 F60 ◇

ROBINSON, R.M. *Restricted set-theoretical definitions in arithmetic* ◇ B28 F30 F35 ◇

TAKEUTI, G. *On the formal theory of the ordinal diagrams* ◇ F15 F35 ◇

TAKEUTI, G. *On the fundamental conjecture of GLC V* ◇ B15 F05 F35 ◇

TAKEUTI, G. *Remark on the fundamental conjecture of GLC* ◇ B15 F05 F35 ◇

1959

KREISEL, G. *Interpretation of analysis by means of constructive functionals of finite types* ◇ D65 F10 F30 F35 F50 ◇

1960

BUECHI, J.R. *Weak second-order arithmetic and finite automata* ◇ B15 B25 C85 D05 F35 ◇

GANDY, R.O. & KREISEL, G. & TAIT, W.W. *Set existence* ◇ C62 D55 F35 ◇

KREISEL, G. *La predicativite* ◇ A05 D55 E45 F35 F65 ◇

MAEHARA, S. & NISHIMURA, T. & SEKI, S. *Non-constructive proofs of a metamathematical theorem concerning the consistency of analysis and its extension* ◇ B15 F05 F35 ◇

SCHUETTE, K. *Beweistheorie* ◇ B98 F05 F15 F30 F35 F98 ◇

SCHUETTE, K. *Syntactical and semantical properties of simple type theory* ◇ B15 F05 F35 ◇

TAKEUTI, G. *An example on the fundamental conjecture of GLC* ◇ B15 F05 F35 ◇

1961

GANDY, R.O. & KREISEL, G. & TAIT, W.W. *Set existence II* ◇ C62 D55 F35 ◇

MOSTOWSKI, ANDRZEJ *Formal systems of analysis based on an infinitistic rule of proof* ◇ C62 E45 F35 ◇

OSHIBA, T. *On elimination of function-types of GLC* ◇ F35 ◇

TAKEUTI, G. *On the fundamental conjecture of GLC VI* ◇ B15 F05 F35 ◇

TAKEUTI, G. *On the inductive definition with quantifiers of second order* ◇ B15 F15 F35 ◇

TRAKHTENBROT, B.A. *Certain constructions in the logic of one-place predicates (Russian)* ◇ B20 F35 ◇

TRAKHTENBROT, B.A. *Finite automata and the logic of one-place predicates (Russian)* ◇ B20 D05 F35 ◇

1962

BUECHI, J.R. *On a decision method in restricted second order arithmetic* ◇ B25 C85 D05 F35 ◇

FEFERMAN, S. & KREISEL, G. & OREY, S. *1-consistency and faithful interpretations* ◇ F25 F30 F35 ◇

KREISEL, G. *The axiom of choice and the class of hyperarithmetic functions* ◇ D55 F35 ◇

MAEHARA, S. *Cut-elimination theorem concerning a formal system for ramified theory of types which admits quantification on types* ◇ B15 F05 F30 F35 F65 ◇

MOSTOWSKI, ANDRZEJ *Representability of sets in formal systems* ◇ D55 E15 F30 F35 ◇

PARSONS, C. *The ω-consistency of ramified analysis* ◇ F05 F10 F35 F65 ◇

SPECTOR, C. *Provably recursive functionals of analysis: a consistency proof of analysis by an extension of principles formulated in current intuitionistic mathematics* ◇ F10 F35 F50 ◇

TRAKHTENBROT, B.A. *Finite automata and the logic of one-place predicates (Russian)* ◇ B20 D05 F35 ◇

1963

HOWARD, W.A. *Appendices to section 2, vol. 1* ◇ F10 F35 F50 ◇

HOWARD, W.A. *The axiom of choice ($\Sigma^1_1 - AC_{01}$), bar induction and bar recursion* ◇ F35 F50 ◇

HOWARD, W.A. *Transfinite induction and transfinite recursion* ◇ F10 F35 F50 ◇

KREISEL, G. *Axiomatic results of second-order arithmetic* ◇ C62 F35 ◇

KREISEL, G. *Generalized inductive definitions* ◇ F35 F50 ◇

KREISEL, G. *Introduction (to volume 1)* ◇ F35 F50 ◇

KREISEL, G. *Introduction (to volume 2)* ◇ F35 ◇

KREISEL, G. (ED.) *Reports of the seminar on foundations of analysis* ◇ F35 ◇

KREISEL, G. *Theory of free choice sequences of natural numbers* ◇ F35 F50 ◇

PARIKH, R. *Some generalisations of the notion of well-ordering* ◇ F35 F50 ◇

TAIT, W.W. *A second order theory of functionals of higher type* ◇ F10 F35 F50 ◇

TAIT, W.W. *Appendix A: Intensional functionals* ◇ F10 F35 F50 ◇

TAIT, W.W. *Appendix B: An interpretation of functionals by convertible terms* ◇ F05 F35 F50 ◇

TAKEUTI, G. *A remark on Gentzen's paper "Beweisbarkeit und Unbeweisbarkeit von Anfangsfaellen der transfiniten Induktion in der reinen Zahlentheorie" I-II* ◇ F15 F35 ◇

YASUGI, M. *Intuitionistic analysis and Goedel's interpretation* ◇ F10 F35 F50 ◇

1964

FEFERMAN, S. *Systems of predicative analysis* ◇ F35 F65 ◇

1965

ACKERMANN, W. *Der Aufbau einer hoeheren Logik* ◇ B15 E70 F30 F35 ◇

FEFERMAN, S. *Some applications of the notions of forcing and generic sets* ◇ C62 D55 E25 E35 E40 E45 F35 ◇

KLEENE, S.C. *Classical extensions of intuitionistic mathematics* ◇ F35 F50 ◇

KLEENE, S.C. & VESLEY, R.E. *The foundations of intuitionistic mathematics, especially in relation to recursive functions* ◇ F35 F50 F98 ◇

KREISEL, G. *Mathematical logic* ◇ B98 F10 F35 F50 F98 ◇

NISHIMURA, T. & TANAKA, H. *On two systems for arithmetic* ◇ C75 F30 F35 ◇

SCHUETTE, K. *Eine Grenze fuer die Beweisbarkeit der transfiniten Induktion in der verzweigten Typenlogik* ◇ B15 F15 F35 F65 ◇

SCHUETTE, K. *Predicative well-orderings* ◇ F15 F35 F65 ◇

TAIT, W.W. *Functionals defined by transfinite recursion* ◇ E30 F10 F15 F35 ◇

TAIT, W.W. *The substitution method* ◇ F10 F15 F30 F35 ◇

1966

ELGOT, C.C. & RABIN, M.O. *Decidability and undecidability of extensions of second (first) order theory of (generalized) successor* ◇ B15 B25 C85 D05 D35 F30 F35 ◇

HANATANI, Y. *Calculabilite des fonctionelles recursives primitives de type fini sur les nombres naturels* ◇ D65 F10 F35 F50 ◇

HOWARD, W.A. & KREISEL, G. *Transfinite induction and bar induction of types zero and one, and the role of continuity in intuitionistic analysis* ◇ F10 F35 F50 ◇

MYHILL, J.R. *Notes towards an axiomatization of intuitionistic analysis* ◇ F35 F50 ◇

PARIKH, R. *Some generalisations of the notion of well ordering* ◇ D20 D45 E07 F35 ◇

STEINER, H.-G. *Aequivalente Fassungen des Vollstaendigkeitsaxioms fuer die Theorie der reellen Zahlen* ◇ F35 ◇

TAIT, W.W. *A nonconstructive proof of Gentzen's Hauptsatz for second order predicate logic* ◇ B15 F05 F35 ◇

VENNE, M. *Langues et theories d'ordres superieurs* ◇ F35 ◇

1967

ENDERTON, H.B. *An infinitistic rule of proof* ◇ C62 F35 ◇

FENSTAD, J.E. *Subsets of natural numbers (Norwegian) (English summary)* ◇ A05 E35 E50 F30 F35 F65 ◇

JENSEN, R.B. *Unabhaengigkeitsbeweise in Teilsystemen der elementaren Zahlentheorie* ◇ F30 F35 ◇

LOPEZ-ESCOBAR, E.G.K. *A complete, infinitary axiomatization of weak second-order logic* ◇ B15 C75 C85 F35 ◇

MOSCHOVAKIS, J.R. *Disjunction and existence in formalized intuitionistic analysis* ◇ F35 F50 ◇

PRAWITZ, D. *Completeness and Hauptsatz for second order logic* ◇ F05 F35 ◇

SCOTT, D.S. *A proof of the independence of the continuum hypothesis* ◇ E35 E50 F35 ◇

TAIT, W.W. *Intensional interpretation of functionals of finite type I* ◇ F10 F30 F35 F50 ◇

TAKAHASHI, MOTO-O *A proof of cut-elimination in simple type theory* ◇ F05 F35 ◇

TAKEUTI, G. *Consistency proofs of subsystems of classical analysis* ◇ F05 F15 F35 ◇

UESU, T. *Zermelo's set theory and G^*LC* ◇ E30 E35 F35 ◇

ZAHN, P. *Eine Einfuehrung der reellen Zahlen in der operativen Mathematik ohne die Unterscheidung von Sprachschichten* ◇ B28 F35 F65 ◇

1968

DRAGALIN, A.G. *The computability of primitive recursive terms of finite type, and primitive recursive realization (Russian)* ◇ F10 F35 F50 ◇

FEFERMAN, S. *Autonomous transfinite progressions and the extent of predicative mathematics* ◇ F15 F30 F35 F65 ◇

FEFERMAN, S. *Persistent and invariant formulas for outer extensions* ◇ C30 C40 C70 C75 F35 ◇

FEFERMAN, S. *Systems of predicative analysis II : representation of ordinals* ◇ F15 F35 F65 ◇

HOWARD, W.A. *Functional interpretation of bar induction by bar recursion* ◇ F10 F35 F50 ◇

KINO, A. *Foundations of mathematics: consistency* ◇ F35 F98 ◇

KINO, A. *On provably recursive functions and ordinal recursive functions* ◇ D20 F05 F30 F35 ◇

KOGALOVSKIJ, S.R. *Some remarks on higher order logic (Russian)* ◇ B15 C40 C85 F35 ◇

KREISEL, G. *Functions, ordinals, species* ◇ F10 F35 F50 ◇

KREISEL, G. & LEVY, A. *Reflection principles and their use for establishing the complexity of axiomatic systems* ◇ E30 E47 F25 F30 F35 F50 ◇

LOEB, M.H. *Die Vollstaendigkeit der verzweigten Typenlogik mit unendlicher Terminduktion* ◇ B15 F35 ◇

MYHILL, J.R. *Formal systems of intuitionistic analysis I* ◇ F35 F50 ◇

PRAWITZ, D. *Hauptsatz for higher order logic* ◇ B15 F05 F35 ◇

SCHUETTE, K. *Neuere Ergebnisse der Beweistheorie* ◇ F05 F10 F15 F30 F35 F98 ◇

SCOTT, D.S. *Extending the topological interpretation to intuitionistic analysis I* ◇ F35 F50 ◇

TAIT, W.W. *Constructive reasoning* ◇ B40 F10 F15 F30 F35 F50 F65 ◇

TAIT, W.W. *Normal derivability in classical logic* ◇ C75 F05 F15 F30 F35 F60 ◇

TAKAHASHI, MOTO-O *On simple type theory (Japanese)* ◇ B15 F05 F35 ◇

TAKAHASHI, MOTO-O *Simple type theory of Gentzen style with the inference of extensionality* ◇ F05 F35 ◇

TAKEUTI, G. & YASUGI, M. *Reflection principles of subsystems of analysis* ◇ F15 F35 ◇

TAKEUTI, G. *The Π^1_1-comprehension schema and ω-rules* ◇ F05 F15 F35 ◇

TROELSTRA, A.S. *The theory of choice sequences* ◇ F35 F50 ◇

1969

BIBEL, W. *Schnittelimination in einem Teilsystem der einfachen Typenlogik* ◇ B15 F05 F15 F35 ◇
BUECHI, J.R. & LANDWEBER, L.H. *Definability in the monadic second-order theory of successor* ◇ B15 B25 C40 C85 D05 F35 ◇
FRIEDMAN, H.M. *Bar induction and Π^1_1-CA* ◇ F35 ◇
HULL, R.G. *Counterexamples in intuitionistic analysis using Kripke's schema* ◇ F35 F50 ◇
KLEENE, S.C. *Formalized recursive functionals and formalized realizability* ◇ D65 F35 F50 ◇
KREISEL, G. *Axiomatisations of nonstandard analysis that are conservative extensions of formal systems for classical standard analysis* ◇ B30 F35 H05 ◇
ROBINSON, JOHN ALAN *Mechanizing higher-order logic* ◇ B35 B40 F35 ◇
ROYSE, JAMES R. *Mathematical induction in ramified type theory* ◇ B15 F35 F65 ◇
SCARPELLINI, B. *Some applications of Gentzen's second consistency proof* ◇ F05 F30 F35 F50 ◇
TROELSTRA, A.S. *Informal theory of choice sequences (Polish and Russian summaries)* ◇ F35 F50 F55 ◇
TROELSTRA, A.S. *Notes on the intuitionistic theory of sequences I* ◇ F35 F50 ◇
TROELSTRA, A.S. *Principles of intuitionism* ◇ F35 F50 F55 F98 ◇
UESU, T. *Gentzen's Hauptsatz for simple type theory with the infinite induction* ◇ F05 F35 ◇
WETTE, E. *Definition eines (relativ vollstaendigen) formalen Systems konstruktiver Arithmetik* ◇ F10 F15 F30 F35 F65 ◇

1970

ACZEL, P. *Representability in some systems of second order arithmetic* ◇ D65 D70 F35 ◇
BELYAKIN, N.V. *Generalized computations and second order arithmetic (Russian)* ◇ C62 D10 D45 D65 D70 F35 ◇
DALEN VAN, D. & TROELSTRA, A.S. *Projections of lawless sequences* ◇ F35 F50 ◇
DENTON JR., J.S. & DREBEN, B. *Herbrand-style consistency proofs* ◇ F05 F30 F35 ◇
FEFERMAN, S. *Formal theories for transfinite iteration of generalized inductive definitions and some subsystems of analysis* ◇ F35 F50 ◇
FRIEDMAN, H.M. *Higher set theory and mathematical practice* ◇ D55 E15 E35 E45 E60 F35 ◇
FRIEDMAN, H.M. *Iterated inductive definitions and Σ^1_2-AC* ◇ C62 D55 E25 F35 ◇
GASTEV, YU.A. *The expressible and deductive possibilities of logico-arithmetic calculi on the basis of the theory of types (Russian)* ◇ B15 F35 ◇
GERBER, H. *Brouwer's bar theorem and a system of ordinal notations* ◇ F15 F35 F50 ◇
GOGUADZE, D.F. *The finite axiomatizability of a finite fragment of type theory (Russian) (Georgian and English summaries)* ◇ B15 F35 ◇
KINO, A. *Formalization of the theory of ordinal diagrams of infinite order* ◇ F15 F35 ◇
KOGALOVSKIJ, S.R. *Some reduction theorems for higher order logic (Russian)* ◇ B15 F35 ◇
KREISEL, G. & TROELSTRA, A.S. *Formal systems for some branches of intuitionistic analysis* ◇ F35 F50 ◇
MUELLER, GUENTER *Groessen ueber arithmetisierbaren Gesamtheiten* ◇ F35 ◇
MYHILL, J.R. *Formal systems of intuitionistic analysis II. The theory of species* ◇ F35 F50 ◇
OHYA, T. *On recursive restriction of proofs in a system with constructive infinitely long expressions* ◇ F05 F35 ◇
PRAWITZ, D. *On the proof theory of mathematical analysis* ◇ F35 ◇
PRAWITZ, D. *Some results for intuitionistic logic with second order quantification rules* ◇ F05 F35 F50 ◇
SCARPELLINI, B. *A model for intuitionistic analysis* ◇ F35 F50 ◇
SCARPELLINI, B. *On cut elimination in intuitionistic systems of analysis* ◇ F05 F35 F50 ◇
SCOTT, D.S. *Extending the topological interpretation to intuitionistic analysis, II* ◇ F35 F50 ◇
SIEFKES, D. *Decidable theories I. Buechi's monadic second order successor arithmetic* ◇ B15 B25 C85 D05 F35 F98 ◇
SIEFKES, D. *Decidable extensions of monadic second order successor arithmetic* ◇ B15 B25 C85 D05 F35 ◇
TAIT, W.W. *Applications of the cut elimination theorem to some subsystems of classical analysis* ◇ F05 F35 ◇
TAKAHASHI, MOTO-O *A system of simple type theory of Gentzen style with inference on extensionality, and the cut-elimination in it* ◇ F05 F35 ◇
TAKAHASHI, MOTO-O *A theorem on the second order arithmetic with the ω-rule* ◇ F35 ◇
TROELSTRA, A.S. *Notes on the intuitionistic theory of sequences II,III* ◇ F35 F50 ◇
VESLEY, R.E. *A palatable substitute for Kripke's schema* ◇ F35 F50 ◇
YASUGI, M. *Cut elimination theorem for second order arithmetic with the Π^1_1 comprehension axiom and the ω-rule* ◇ F05 F35 ◇

1971

ANDREWS, P.B. *Resolution in type theory* ◇ B15 F05 F35 ◇
DILLER, J. & SCHUETTE, K. *Simultane Rekursionen in der Theorie der Funktionale endlicher Typen* ◇ F10 F35 F50 ◇
DREBEN, B. & GOLDFARB, W.D. *Note N* ◇ A10 B10 F30 F35 ◇
ENDERTON, H.B. & FRIEDMAN, H.M. *Approximating the standard model of analysis* ◇ C62 E45 F35 ◇
FEFERMAN, S. *Ordinals and functionals in proof theory* ◇ F10 F15 F35 F50 ◇
FRIEDMAN, H.M. *A more explicit set theory* ◇ A05 C62 E30 E35 E45 F30 F35 ◇
GIRARD, J.-Y. *Une extension de l'interpretation de Goedel a l'analyse et son application a l'elimination des coupures dans l'analyse et la theorie des types* ◇ F05 F10 F35 F50 ◇

GRZEGORCZYK, A. *Outline of theoretical arithmetic (Polish)* ⋄ B28 C62 F30 F35 F98 ⋄
MAEHARA, S. & TAKEUTI, G. *Two interpolation theorems for a Π^1_1 predicate calculus* ⋄ B15 C40 F35 ⋄
MARTIN-LOEF, P. *Hauptsatz for the intuitionistic theory of iterated inductive definitions* ⋄ F05 F35 F50 ⋄
MARTIN-LOEF, P. *Hauptsatz for the theory of species* ⋄ F05 F35 F50 ⋄
MOSCHOVAKIS, J.R. *Can there be no nonrecursive functions?* ⋄ F35 F50 ⋄
MYHILL, J.R. *Embedding classical type theory in "intuitionistic" type theory* ⋄ B15 F35 F50 ⋄
PRAWITZ, D. *Ideas and results in proof theory* ⋄ F05 F07 F10 F30 F35 F50 F98 ⋄
SCARPELLINI, B. *A model for barrecursion of higher types* ⋄ F35 F50 ⋄
SCARPELLINI, B. *Proof theory and intuitionistic systems* ⋄ F05 F15 F30 F35 F50 F98 ⋄
SIEFKES, D. *Undecidable extensions of monadic second order successor arithmetic* ⋄ B15 B25 C85 D35 F35 ⋄
TAIT, W.W. *Normal form theorem for bar recursive functions of finite type* ⋄ F05 F10 F35 F50 ⋄
TAKAHASHI, MOTO-O *Cut-elimination theorem and Brouwerian-valued models for intuitionistic type theory* ⋄ F05 F35 F50 ⋄
TROELSTRA, A.S. *Computability of terms and notions of realizability for intuitionistic analysis* ⋄ F35 F50 ⋄
TROELSTRA, A.S. *Notions of realizability for intuitionistic arithmetic and intuitionistic arithmetic in all finite types* ⋄ F30 F35 F50 ⋄
UESU, T. *Simple type theory with constructive infinitely long expressions* ⋄ B15 C75 F05 F35 ⋄
UESU, T. *Two formal systems of simple type theory with type variables* ⋄ F05 F35 ⋄
ZBIERSKI, P. *Models for higher order arithmetics (Russian summary)* ⋄ C62 E30 F35 ⋄

1972

ANDREWS, P.B. *General models, descriptions, and choice in type theory* ⋄ B15 C85 E25 E35 F35 ⋄
ANDREWS, P.B. *General models and extensionality* ⋄ B15 C85 E35 F35 ⋄
APT, K.R. *Non-finite axiomatizability of the second order arithmetic* ⋄ F35 ⋄
FRIEDRICH, U. *Entscheidbarkeit der monadischen Nachfolgerarithmetik mit endlichen Automaten ohne Analysetheorem (Russian, English and French summaries)* ⋄ B25 C85 D05 F35 ⋄
GEL'FOND, M.G. *Relationship between the classical and constructive developments of mathematical analysis (Russian) (English summary)* ⋄ B28 F35 F60 ⋄
GOODMAN, NICOLAS D. & MYHILL, J.R. *The formalization of Bishop's constructive mathematics* ⋄ F35 F50 F55 ⋄
HOWARD, W.A. *A system of abstract constructive ordinals* ⋄ F10 F15 F35 F50 ⋄
LUCCHESI, C.L. *The undecidability of the unification problem for 3rd order languages* ⋄ F35 ⋄
OSSWALD, H. *Vollstaendigkeit und Schnittelimination in der intuitionistischen Typenlogik (English summary)* ⋄ F05 F35 F50 ⋄
SCARPELLINI, B. *Formally constructive model for barrecursion of higher types* ⋄ F35 F50 ⋄
SCARPELLINI, B. *Induction and transfinite induction in intuitionistic systems* ⋄ F35 F50 ⋄
VESLEY, R.E. *Choice sequences and Markov's principle* ⋄ F35 F50 ⋄

1973

APT, K.R. *Infinitistic rules of proof and their semantics (Russian summary)* ⋄ C62 F07 F35 ⋄
BELYAKIN, N.V. *Generalized computations (Russian)* ⋄ F35 ⋄
BOWEN, K.A. *Cut elimination in transfinite type theory* ⋄ B15 F05 F35 ⋄
FRIEDMAN, H.M. *Countable models of set theories* ⋄ C15 C62 E45 F35 H20 ⋄
FRIEDMAN, H.M. *Some applications of Kleene's methods for intuitionistic systems* ⋄ F35 F50 ⋄
GIRARD, J.-Y. *Quelques resultats sur les interpretations fonctionelles* ⋄ F10 F35 F50 ⋄
HOWARD, W.A. *Appendix: Hereditarily majorizable functionals of finite type* ⋄ E25 F10 F35 F50 ⋄
HUET, G. *The undecidability of unification on third order logic* ⋄ B15 F35 ⋄
LEVIN, A.M. *Some applications of truth definitions (Russian) (English summary)* ⋄ E35 E70 F35 ⋄
LUCKHARDT, H. *Extensional Goedel functional interpretation. A consistency proof of classical analysis* ⋄ F10 F35 F50 F98 ⋄
MARTIN-LOEF, P. *Hauptsatz for intuitionistic simple type theory* ⋄ B15 F05 F35 F50 ⋄
MOSCHOVAKIS, J.R. *A topological interpretation of second-order intuitionistic arithmetic* ⋄ F35 F50 ⋄
NEPEJVODA, N.N. *A new concept of predicative truth and definability (Russian)* ⋄ F35 F65 ⋄
NEPEJVODA, N.N. *On a generalization of the Kleene-Mostowski hierarchy (Russian)* ⋄ D55 F15 F35 ⋄
OSSWALD, H. *Ein syntaktischer Beweis fuer die Zuverlaessigkeit der Schnittregel im Kalkuel von Schuette fuer die intuitionistische Typenlogik* ⋄ F05 F35 F50 ⋄
PIETRZYKOWSKI, T. *A complete mechanization of second-order type theory* ⋄ B15 B35 F35 ⋄
POHLERS, W. *Ein starker Normalisationssatz fuer die intuitionistische Typentheorie* ⋄ F05 F35 F50 ⋄
SCARPELLINI, B. *On barinduction of higher types for decidable predicates* ⋄ F35 F50 ⋄
STAPLES, J. *Combinator realizability of constructive finite type analysis* ⋄ B40 F35 F50 ⋄
TAKEUTI, G. & YASUGI, M. *The ordinals of the systems of second order arithmetic with the provably Δ^1_2-comprehension axiom and with the Δ^1_2-comprehension axiom respectively* ⋄ F05 F15 F35 ⋄
TITANI, S. *A proof of the cut-elimination theorem in simple type theory* ⋄ B15 F05 F35 ⋄

TROELSTRA, A.S. *Intuitionistic formal systems* ◇ F30 F35 F50 ◇
TROELSTRA, A.S. (ED.) *Metamathematical investigation of intuitionistic arithmetic and analysis* ◇ F05 F10 F30 F35 F50 F55 F98 ◇
TROELSTRA, A.S. *Models and computability* ◇ F05 F10 F30 F35 F50 ◇
TROELSTRA, A.S. *Normalization theorems for systems of natural deduction* ◇ F05 F30 F35 F50 ◇
TROELSTRA, A.S. *Notes on intuitionistic second order arithmetic* ◇ F35 F50 ◇
TROELSTRA, A.S. *Realizability and functional interpretations* ◇ F10 F30 F35 F50 ◇
ZUCKER, J.I. *Iterated inductive definitions, trees, and ordinals* ◇ F10 F15 F35 ◇

1974

ANDREWS, P.B. *Provability in elementary type theory* ◇ B15 B25 D35 F35 ◇
ANDREWS, P.B. *Resolution and the consistency of analysis* ◇ B15 F05 F25 F35 ◇
APT, K.R. & MAREK, W. *Second order arithmetic and related topics* ◇ C62 D55 E45 F25 F35 ◇
BARENDREGT, H.P. *Pairing without conventional restraints* ◇ B40 F35 ◇
BELYAKIN, N.V. *Generalized computations, and third order arithmetic (Russian)* ◇ D65 F35 ◇
BILY, J. & BUKOVSKY, L. *On expansion of β-models* ◇ C62 E25 E40 E70 F35 ◇
BOWEN, K.A. *Systems of transfinite type theory based on intuitionistic and modal logics* ◇ B15 F05 F35 F50 ◇
DALEN VAN, D. *A model for HAS. A topological interpretation of the theory of species of natural numbers* ◇ F35 F50 ◇
DILLER, J. & NAHM, W. *Eine Variante zur Dialectica-Interpretation der Heyting-Arithmetik endlicher Typen* ◇ F10 F35 F50 ◇
DRAGALIN, A.G. *Constructive model of intuitionistic analysis (Russian)* ◇ F35 F50 ◇
GABBAY, D.M. *On 2nd order intuitionistic propositional calculus with full comprehension* ◇ F35 F50 ◇
GUZICKI, W. *Elementary extensions of Levy's model of A_2^-* ◇ C55 C62 C80 E25 E35 E40 F35 ◇
KOLODZIEJ, R. *Reducibility of formulae of weak second order arithmetic to pseudo-canonical forms I,II,III* ◇ F35 ◇
KREISEL, G. & TAKEUTI, G. *Formally self-referential propositions for cut free classical analysis and related systems* ◇ F05 F35 ◇
KREJNOVICH, V.YA. *From what does the law of the excluded middle follow? (Russian) (English summary)* ◇ F35 F50 ◇
LUKAS, J.D. & PUTNAM, H. *Systems of notations and the ramified analytical hierarchy* ◇ D30 D55 E45 F15 F35 ◇
MYHILL, J.R. *Embedding classical type theory in "intuitionistic" type theory, a correction* ◇ B15 F35 F50 ◇
PERRIN, M.J. & ZALC, A. *Sous-systemes bicommutables d'analyse et de theorie des ensembles* ◇ E25 E30 F25 F35 ◇
ROBERTSON, E.L. *Structure of complexity in the weak monadic second-order theories of the natural numbers* ◇ D15 D55 F35 ◇
SAKAI, H. *On necessary but not-sufficient conditions* ◇ B10 F30 F35 G05 ◇
STAPLES, J. *Combinator realizability of constructive Morse theory* ◇ B40 F35 F50 ◇
TANAKA, H. *Some analytical rules of inference in the second-order arithmetic* ◇ C62 D55 F35 ◇
TROELSTRA, A.S. *Note on the fan theorem* ◇ F30 F35 F50 ◇
UMEZAWA, T. *Disjunction property in higher order number theory with intuitionistic rules of inference I* ◇ F05 F35 F50 ◇

1975

BUCHHOLZ, W. *Ein ausgezeichnetes Modell fuer die intuitionistische Typenlogik* ◇ B15 F05 F35 F50 ◇
DILLER, J. & VOGEL, HELMUT *Intensionale Funktionalinterpretation der Analysis* ◇ F10 F35 F50 ◇
FEFERMAN, S. *A language and axioms for explicit mathematics* ◇ D55 F35 F50 F65 ◇
FERRANTE, J. & RACKOFF, C.W. *A decision procedure for the first order theory of real addition with order* ◇ B25 C10 C60 D15 F35 ◇
FRIEDMAN, H.M. *Some systems of second order arithmetic and their use* ◇ E30 F35 ◇
HANATANI, Y. *Calculability of the primitive recursive functionals of finite type over the natural numbers* ◇ D65 F10 F35 F50 ◇
INOUE, K. & NAKAMURA, A. *On the expressive power of logical metalanguages I^n and I_+* ◇ B15 F35 ◇
KANOVEJ, V.G. *On the independence of some propositions of descriptive set theory and second-order arithmetic (Russian)* ◇ D55 E15 E35 F35 ◇
KREISEL, G. *Observations on a recent generalization of completeness theorems due to Schuette* ◇ A05 C07 C57 F05 F20 F35 F50 ◇
LEVIN, A.M. *A comparison of various forms of the axiom of choice in classical analysis (Russian)* ◇ E25 E35 F35 ◇
LEVIN, A.M. *The axiom of choice in classical analysis (Russian) (English summary)* ◇ B30 E15 E25 E35 E75 F35 ◇
LUCKHARDT, H. *The real elements in a consistency proof for simple type theory I* ◇ F10 F35 F50 ◇
MAREK, W. & SREBRNY, M. *No minimal transitive model of Z^-* ◇ C62 E30 E70 F35 ◇
MARTIN-LOEF, P. *About models for intuitionistic type theories and the notion of definitional equality* ◇ B40 F35 F50 ◇
MARTIN-LOEF, P. *An intuitionistic theory of types: predicative part* ◇ F05 F35 F50 ◇
MINTS, G.E. *Finite investigations of transfinite derivations (Russian) (English summary)* ◇ F05 F07 F10 F20 F30 F35 F50 ◇

MINTS, G.E. *Proof theory (arithmetic and analysis) (Russian)* ◇ F05 F10 F30 F35 F50 F98 ◇
NAGATA, M. & NAKANISHI, M. & NISHIMURA, T. *TKP 1.2 – the extension of TKP 1 by adding some facilities* ◇ B35 F35 ◇
POHLERS, W. *An upper bound for the provability of transfinite induction in systems with N-times iterated inductive definitions* ◇ F15 F35 F50 ◇
SCHWICHTENBERG, H. *Elimination of higher type levels in definitions of primitive recursive functionals by means of transfinite recursion* ◇ C75 F10 F15 F35 F50 ◇
SIEFKES, D. *The recursive sets in certain monadic second order fragments of arithmetic* ◇ D20 F35 ◇
TAIT, W.W. *A realizability interpretation of the theory of species* ◇ D65 F05 F35 F50 ◇
TAKEUTI, G. *Proof theory* ◇ F05 F07 F30 F35 F98 ◇
THOMAS, WOLFGANG *A note on undecidable extensions of monadic second order successor arithmetic* ◇ D35 F35 ◇
TOLEDO, S. *Tableau systems for first order number theory and certain higher order theories* ◇ F05 F07 F30 F35 F98 ◇
TROELSTRA, A.S. *Markov's principle and Markov's rule for theories of choice sequences* ◇ F35 F50 ◇
UMEZAWA, T. *Disjunction property in higher order number theory with intuitionistic rules of inference II* ◇ F05 F35 F50 ◇

1976

APT, K.R. *Semantics of the infinitistic rules of proof* ◇ C62 D55 E45 F07 F35 ◇
BERNINI, S. *A very strong intuitionistic theory* ◇ F35 F50 ◇
CHRISTIAN, C.C. *Peano-Systeme* ◇ B28 F30 F35 ◇
DRAGALIN, A.G. *Cut-elimination in the theory of definable sets of natural numbers (Russian)* ◇ F05 F35 F65 ◇
FRIEDMAN, H.M. *Uniformly defined descending sequences of degrees* ◇ D30 D55 F30 F35 ◇
GIRARD, J.-Y. *Three-valued logic and cut-elimination: The actual meaning of Takeuti's conjecture* ◇ B50 C85 C90 F05 F35 F50 ◇
GOODMAN, NICOLAS D. *The theory of the Goedel functionals* ◇ F10 F35 F50 ◇
GUZICKI, W. *On weaker forms of choice in second order arithmetic* ◇ E25 E35 F35 ◇
JONGH DE, D.H.J. & SMORYNSKI, C.A. *Kripke models and the intuitionistic theory of species* ◇ C90 F35 F50 ◇
KANOVEJ, V.G. *Definability with the help of degrees of constructibility (Russian)* ◇ E25 E35 E45 F35 ◇
KREISEL, G. *Wie die Beweistheorie zu ihren Ordinalzahlen kam und kommt* ◇ F05 F15 F30 F35 ◇
KROL', M.D. *The topological models of intuitionistic analysis. One counter-example (Russian)* ◇ C90 F35 F50 ◇
MAASS, W. *Eine Funktionalinterpretation der praedikativen Analysis* ◇ D20 F10 F35 F65 ◇
MAREK, W. & SREBRNY, M. *Urelements and extendability* ◇ C62 E30 E70 F25 F30 F35 ◇
MEYER, R.K. *Ackermann, Takeuti, und Schnitt: γ for higher-order relevant logic* ◇ B46 F05 F35 ◇
MINTS, G.E. *What can be done with PRA (Russian) (English summary)* ◇ F05 F30 F35 ◇
NISHIMURA, T. *On cut-elimination in simple type theory (on the work of Moto-o Takahashi, winner of the third Mathematical Society Prize) (Japanese)* ◇ F05 F35 ◇
POSY, C.J. *Varieties of indeterminacy in the theory of general choice sequences* ◇ F35 F50 ◇
VOGEL, HELMUT *Ein starker Normalisationssatz fuer die bar-rekursiven Funktionale* ◇ F10 F35 ◇
ZARACH, A. *Generic extension of admissible sets* ◇ C62 E25 E35 F35 ◇

1977

ACZEL, P. *The strength of Martin-Loef's intuitionistic type theory with one universe* ◇ F35 F50 ◇
ARTIGUE, M. & ISAMBERT, E. *Quelques resultats de bicommutabilite en arithmetique du troisieme ordre (English summary)* ◇ E30 F25 F35 ◇
BEESON, M.J. *Principles of continuous choice and continuity of functions in formal systems for constructive mathematics* ◇ F35 F50 ◇
BERNINI, S. *Interpretazione intuizionista di teorie a logica classica; una particolare applicazione del metodo della traduzione negativa via concetti empirici e anomici (lawless)* ◇ F35 F50 ◇
DALEN VAN, D. *The use of Kripke's schema as a reduction principle* ◇ F35 F50 ◇
DRAGALIN, A.G. *Cut-elimination in the theory of definable sets of natural numbers (Russian)* ◇ F05 F35 F65 ◇
FEFERMAN, S. *Theories of finite type related to mathematical practice* ◇ B15 D55 D65 F10 F15 F35 F98 ◇
HAYASHI, S. *On derived rules of intuitionistic second order arithmetic* ◇ F05 F35 F50 ◇
HAYASHI, S. *Some derived rules of intuitionistic second order arithmetic* ◇ F05 F35 F50 ◇
KROL', M.D. *Disjunctive and existential properties of intuitionistic analysis with Kripke's scheme (Russian)* ◇ C90 F35 F50 ◇
LEVIN, A.M. *A conservative extension of formal mathematical analysis with a scheme of dependent choice (Russian)* ◇ E25 F35 ◇
LOPEZ-ESCOBAR, E.G.K. *Infinite rules in finite systems* ◇ F30 F35 ◇
NABEBIN, A.A. *Expressibility in restricted second-order arithmetic (Russian)* ◇ B15 C40 D05 F35 ◇
NAKAMURA, A. & ONO, H. *Two-dimensional finite automata and their application to the decision problem of monadic first-order arithmetic* $A[P, F(x), G(x)]$ ◇ D05 D35 F30 F35 ◇
SCARPELLINI, B. *A new realizability notion for intuitionistic analysis* ◇ F35 F50 ◇
SCHWICHTENBERG, H. *Proof theory: Some applications of cut-elimination* ◇ F05 F10 F15 F30 F35 F98 ◇
SIMPSON, S.G. *Degrees of unsolvability: A survey of results* ◇ D25 D30 D35 F35 ◇
SIMPSON, S.G. *First order theory of the degrees of recursive unsolvability* ◇ D30 D35 F35 ◇
TROELSTRA, A.S. *A note on non-extensional operations in connection with continuity and recursiveness* ◇ F35 F50 ◇

Troelstra, A.S. *Aspects of constructive mathematics* ◇ F30 F35 F50 F55 F60 F98 ◇

Troelstra, A.S. *Axioms for intuitionistic mathematics incompatible with classical logic* ◇ A05 F35 F50 ◇

Troelstra, A.S. *Choice sequences, a chapter of intuitionistic mathematics* ◇ A05 A10 F35 F50 F55 F98 ◇

Troelstra, A.S. *Some models for intuitionistic finite type arithmetic with fan functional* ◇ F35 F50 ◇

Troelstra, A.S. *Special instances of generalized continuity which are conservative over intuitionistic arithmetic* ◇ F35 F50 ◇

1978

Aczel, P. *The type theoretic interpretation of constructive set theory* ◇ E70 F35 F50 ◇

Apt, K.R. *Inductive definitions, models of comprehension and invariant definability* ◇ C40 C62 D55 D70 F35 ◇

Artigue, M. & Isambert, E. & Perrin, M.J. & Zalc, A. *Some remarks on bicommutability* ◇ C62 E30 F25 F35 ◇

Bernini, S. *A note on my paper: "A very strong intuitionistic theory"* ◇ F35 F50 ◇

Buchholz, W. & Pohlers, W. *Provable well orderings of formal theories for transfinitely iterated inductive definitions* ◇ F15 F30 F35 F50 ◇

Dalen van, D. *An interpretation of intuitionistic analysis* ◇ C90 F35 F50 ◇

Delon, F. *Definition de l'arithmetique dans la theorie des anneaux de series formelles (English summary)* ◇ C60 D35 F25 F35 ◇

Friedman, H.M. *A proof of foundation from the axioms of cumulation* ◇ E30 F35 ◇

Goodman, Nicolas D. *Relativized realizability in intuitionistic arithmetic of all finite types* ◇ F35 F50 ◇

Haberthuer, R. *Choice sequences and reduction processes* ◇ F05 F35 F50 ◇

Jervell, H.R. *Constructive universes. I* ◇ F05 F15 F35 F50 ◇

Kanovej, V.G. *The significance of the parameters and of the complexity of the basic formula in the comprehension axiom schema for second order arithmetic (Russian)* ◇ F35 ◇

Krol', M.D. *A topological model for intuitionistic analysis with Kripke's scheme* ◇ C90 F35 F50 ◇

Krol', M.D. *Distinct variants of Kripke's scheme in intuitionistic analysis (Russian)* ◇ C90 F35 F50 ◇

Levin, A.M. *A fragment of classical analysis (Russian) (English summary)* ◇ E25 E35 F35 ◇

Marek, W. *ω-models of second order arithmetic and admissible sets* ◇ C62 E30 E45 F35 ◇

Marek, W. *Some comments on the paper by Artigue, Isambert, Perrin and Zalc: "Some remarks on bicommutability"* ◇ C62 E30 E45 F25 F35 ◇

Pohlers, W. *Ordinals connected with formal theories for transfinitely iterated inductive definitions* ◇ F05 F15 F35 ◇

Scott, P.J. *The "Dialectica" interpretation and categories* ◇ F10 F35 F50 G30 ◇

Steel, J.R. *Forcing with tagged trees* ◇ C62 D30 D55 E40 F35 ◇

Takeuti, G. *Two applications of logic to mathematics* ◇ B98 C90 E40 E75 F05 F30 F35 G12 ◇

Troelstra, A.S. *Appendix to D. van Dalen's "An interpretation of intuitionistic analysis"* ◇ F35 F50 ◇

Vogel, Helmut *Eine beweistheoretische Anwendung partieller stetiger Funktionale* ◇ F10 F35 F50 ◇

Wendel, N. *The inconsistency of Bernini's very strong intuitionistic theory* ◇ B55 F35 F50 ◇

Yasugi, M. *A formalization of $Od(\Omega)$* ◇ F15 F35 ◇

Zbierski, P. *Axiomatizability of second order arithmetic with ω-rule* ◇ B15 C62 E45 E70 F35 ◇

1979

Beeson, M.J. *Goodman's theorem and beyond* ◇ F35 F50 ◇

Diller, J. *Functional interpretations of Heyting's arithmetic in all finite types* ◇ F10 F35 F50 ◇

Dragalin, A.G. *Mathematical intuitionism. Introduction to proof theory (Russian)* ◇ F05 F30 F35 F50 F98 ◇

Feferman, S. *A more perspicuous system for predicativity* ◇ F35 F65 ◇

Fourman, M.P. & Hyland, J.M.E. *Sheaf models for analysis* ◇ C65 C90 F35 F50 G10 G30 ◇

Fourman, M.P. & Scott, D.S. *Sheaves and logic* ◇ C90 C98 E70 F35 F50 G30 ◇

Goldblatt, R.I. *Topoi. The categorial analysis of logic* ◇ B98 C98 E98 F35 F50 F98 G30 ◇

Hoeven van der, G.F. & Troelstra, A.S. *Projections of lawless sequences II* ◇ F35 F50 ◇

Jaeger, G. & Schuette, K. *Eine syntaktische Abgrenzung der $(\Delta^1_1 - CA)$-Analysis* ◇ F15 F35 ◇

Kanovej, V.G. *On descriptive forms of the countable axiom of choice (Russian)* ◇ D55 E15 E25 E35 F35 ◇

Kanovej, V.G. *The definability of forcing in analysis (Russian) (English summary)* ◇ D55 E40 F35 ◇

Levin, A.M. *On an interesting axiomatic theory (Russian)* ◇ B28 F35 ◇

Mints, G.E. *Normalisation of natural deduction and effectiveness of classical existence (Russian)* ◇ F05 F10 F30 F35 ◇

Nakamura, A. & Ono, H. *Undecidability of the first-order arithmetic $A[P(x),2x,x+1]$* ◇ D05 D35 F30 F35 ◇

Nepejvoda, N.N. *Stable truth and computability (Russian)* ◇ C90 F35 F50 ◇

Schwabhaeuser, W. *Non-finitizability of a weak second-order theory* ◇ B15 C62 C65 C85 F35 ◇

Schwichtenberg, H. *On bar recursion of types 0 and 1* ◇ F10 F35 F50 ◇

Scott, D.S. *Identity and existence in intuitionistic logic* ◇ F35 F50 ◇

Shvarts, G.F. *Some extensions of intuitionistic type theory (Russian) (English summary)* ◇ B15 F35 F50 ◇

Stein, M. *Interpretationen der Heyting-Arithmetik endlicher Typen* ◇ F10 F35 F50 ◇

SVENONIUS, L. *Two kinds of extensions of primitive recursive arithmetic* ◇ D20 F30 F35 ◇
TROELSTRA, A.S. *On Ashvinikumar's principle of microscopic completeness* ◇ F35 F50 ◇
VOGEL, HELMUT *Ueber ein mit der Bar-Induktion verwandtes Schema* ◇ F10 F35 F50 ◇
WEINSTEIN, S. *Some applications of Kripke models to formal systems of intuitionistic analysis* ◇ C90 F35 F50 ◇
ZASHEV, J.A. *On a stepwise semantic system with deductive understanding of the second order universal quantifier (Russian)* ◇ F35 ◇

1980

COSTE-ROY, M.-F. & COSTE, M. & MAHE, L. *Contribution to the study of the natural number object in elementary topoi* ◇ D20 F35 F50 G30 ◇
DILLER, J. *Modified realization and the formulae-as-types notion* ◇ F05 F10 F30 F35 F50 ◇
DRAGALIN, A.G. *Higher-order predicate logic in the form of calculus realization (Russian)* ◇ B15 F05 F35 F50 ◇
FRIEDMAN, H.M. *A strong conservative extension of Peano arithmetic* ◇ B28 F30 F35 ◇
FRIEDRICH, W. & LUCKHARDT, H. *Intuitionistic uniformity principles for propositions and some applications* ◇ F35 F50 ◇
HAYASHI, S. *Derived rules related to a constructive theory of metric spaces in intuitionistic higher order arithmetic without countable choice* ◇ F35 F50 G30 ◇
HODGES, W. *Interpreting number theory in nilpotent groups* ◇ C25 C60 C62 F25 F35 ◇
HOWARD, W.A. *Ordinal analysis of terms of finite type* ◇ F10 F15 F35 F50 ◇
HOWARD, W.A. *Ordinal analysis of bar recursion of type zero* ◇ F10 F15 F35 F50 ◇
JAMBU-GIRAUDET, M. *Theorie des modeles de groupes d'automorphismes d'ensembles totalement ordonnes 2-homogenes (English summary)* ◇ C07 C60 C62 C65 D35 E07 F25 F35 ◇
KHAKHANYAN, V.KH. *The consistency of intuitionistic set theory with formal mathematical analysis (Russian)* ◇ D20 E35 E70 F25 F35 F50 ◇
LAMBEK, J. *From types to sets* ◇ F35 F50 G30 ◇
LAMBEK, J. & SCOTT, P.J. *Intuitionist type theory and the free topos* ◇ F35 F50 G30 ◇
MACINTYRE, A. *Ramsey quantifiers in arithmetic* ◇ C62 C80 F30 F35 ◇
NERODE, A. & SHORE, R.A. *Second order logic and first order theories of reducibility orderings* ◇ B10 B15 D30 D35 F35 F40 ◇
PABION, J.F. *TT_3I est equivalent a l'arithmetique du second ordre (English summary)* ◇ B15 F25 F35 ◇
PAEPPINGHAUS, P. *A version of the Σ_1-reflection principle for CFA provable in PRA* ◇ F30 F35 ◇
PRANK, R.K. *Semantics of realizability for a language with variables for recursively enumerable sets (Russian)* ◇ B60 D25 F35 F50 ◇
SCHUETTE, K. *Beweistheoretische Abgrenzung von Teilsystemen der Analysis* ◇ F15 F35 ◇
SCHULTZ, KONRAD *A topological model for Troelstra's system CS of intuitionistic analysis* ◇ C90 E70 F35 F50 ◇
SHVARTS, G.F. *The existence property with parameters for some extensions of the intuitionistic theory of types (Russian)* ◇ F35 F50 ◇
TROELSTRA, A.S. *Extended bar induction of type zero* ◇ F35 F50 ◇
TROELSTRA, A.S. *Intuitionistic extensions of the reals* ◇ F35 F50 F55 ◇
VESLEY, R.E. *Intuitionistic analysis: the search for axiomatization and understanding* ◇ A10 F35 F50 ◇
YASUGI, M. *A progression of consistency proofs* ◇ F05 F15 F35 ◇
YASUGI, M. *Gentzen reduction revisited* ◇ F05 F15 F30 F35 ◇

1981

BUCHHOLZ, W. & FEFERMAN, S. & POHLERS, W. & SIEG, W. *Iterated inductive definitions and subsystems of analysis: recent proof-theoretical studies* ◇ F05 F15 F35 F50 F98 ◇
BUCHHOLZ, W. *Ordinal analysis of ID_ν* ◇ F15 F35 F50 ◇
BUCHHOLZ, W. & SCHUETTE, K. *Syntaktische Abgrenzungen von formalen Systemen der Π^1_1-Analysis und Δ^1_2-Analysis* ◇ F15 F35 ◇
BUCHHOLZ, W. *The $\Omega_{\mu+1}$-rule* ◇ F05 F15 F35 F50 ◇
FEFERMAN, S. & SIEG, W. *Inductive definitions and subsystems of analysis* ◇ F35 F50 ◇
FEFERMAN, S. & SIEG, W. *Proof-theoretic equivalence between classical and constructive theories for analysis* ◇ F35 F50 ◇
GAVRILENKO, YU.V. *Recursive realizability from the intuitionistic point of view (Russian)* ◇ F35 F50 ◇
GIRARD, J.-Y. & PAEPPINGHAUS, P. *A result on implications of Σ_1-sentences and its application to normal form theorems* ◇ F05 F30 F35 ◇
GRAYSON, R.J. *Concepts of general topology in constructive mathematics and in sheaves* ◇ C90 E35 E70 E75 F35 F50 F55 G30 ◇
GUZICKI, W. *The equivalence of definable quantifiers in second order arithmetic* ◇ C30 C55 C62 C80 E40 E45 F35 ◇
HOWARD, W.A. *Computability of ordinal recursion of type level two* ◇ D20 F10 F15 F35 F50 ◇
HOWARD, W.A. *Ordinal analysis of simple cases of bar recursion* ◇ F10 F15 F35 F50 ◇
LAMBEK, J. & SCOTT, P.J. *Independence of premisses and the free topos* ◇ F35 F50 G30 ◇
LAMBEK, J. & SCOTT, P.J. *Intuitionist type theory and foundations* ◇ F35 F50 ◇
MOSCHOVAKIS, J.R. *A disjunctive decomposition theorem for classical theories* ◇ F35 F50 ◇
MOSTOWSKI, A.WLODZIMIERZ *The complexity of automata and subtheories of monadic second order arithmetics* ◇ B15 B25 D05 D15 F35 ◇

NEGRI, M. *L'aritmetizzazione della computabilita dei funzionali di tipo finito* ◇ F10 F35 F50 ◇
PLIUSKEVICIUS, R. *On the Gentzen type proof theory for program analysis* ◇ B75 F07 F35 F50 ◇
POHLERS, W. *Cut-elimination for impredicative infinitary systems I. Ordinal-analysis for ID_1* ◇ C75 F05 F15 F35 ◇
POHLERS, W. *Proof-theoretical analysis of ID_ν by the method of local predicativity* ◇ F05 F15 F35 F50 ◇
PRAWITZ, D. *Validity and normalizability of proofs in 1st and 2nd order classical and intuitionistic logic* ◇ F05 F07 F35 F50 ◇
REYES, G.E. *Logic and category theory* ◇ F35 F50 G30 ◇
SHORE, R.A. *The degrees of unsolvability: global results* ◇ C40 D30 D35 F35 ◇
SIEG, W. *Inductive definitions, constructive ordinals, and normal derivations* ◇ F05 F15 F35 F50 ◇
STEIN, M. *A general theorem on existence theorems* ◇ B40 F10 F35 F50 ◇
TROELSTRA, A.S. *Lawless sequences and their uses* ◇ F35 F50 ◇
TROELSTRA, A.S. *On a second order propositional operator in intuitionistic logic* ◇ B55 F35 F50 ◇
YAKUBOVICH, A.M. *On the consistency of the theory of types with the axiom of choice relative to type theory (Russian)* ◇ B15 E25 E35 F35 ◇
YASUGI, M. *The Hahn-Banach theorem and a restricted inductive definition* ◇ F35 F65 ◇
ZBIERSKI, P. *Nonstandard interpretations of higher order theories* ◇ C62 F35 H15 ◇
ZHANG, JINWEN *Higher order and transfinite incompleteness (Chinese)* ◇ F35 ◇

1982

ACZEL, P. *The type theoretic interpretation of constructive set theory: choice principles* ◇ E25 E70 F35 F50 ◇
BEESON, M.J. *Recursive models for constructive set theories* ◇ E70 F35 F50 ◇
CANTINI, A. *Non-extensional theories of predicative classes over PA* ◇ F30 F35 ◇
DALEN VAN, D. & LODDER, J.S. *Lawlessness and independence* ◇ F35 F50 ◇
FEFERMAN, S. *Inductively presented systems and the formalization of meta-mathematics* ◇ F30 F35 F40 ◇
FEFERMAN, S. *Iterated inductive fixed-point theories: application to Hancock's conjecture* ◇ F15 F35 F65 ◇
FEFERMAN, S. *Monotone inductive definitions* ◇ F35 F50 ◇
FOURMAN, M.P. *Notions of choice sequence* ◇ F35 F50 G30 ◇
FRIEDMAN, H.M. & MCALOON, K. & SIMPSON, S.G. *A finite combinatorial principle which is equivalent to the l-consistency of predicative analysis* ◇ F15 F35 F65 ◇
GIRARD, J.-Y. *A survey of Π^1_2-logic* ◇ F15 F35 ◇
GIRARD, J.-Y. *Herbrand's theorem and proof-theory* ◇ F05 F07 F10 F35 ◇
GIRARD, J.-Y. *Proof-theoretic investigations of inductive definitions I* ◇ F05 F35 F50 ◇
HAYASHI, S. *A note on the bar induction rule* ◇ F35 F50 ◇
HOEVEN VAN DER, G.F. *An application of projections of lawless sequences* ◇ F35 F50 ◇
HOEVEN VAN DER, G.F. *Models for the theory of lawless sequences constructed from a single generator* ◇ F35 F50 ◇
HOEVEN VAN DER, G.F. *Projections of lawless sequences* ◇ F35 F50 ◇
HYLAND, J.M.E. *Applications of constructivity* ◇ F35 F50 ◇
HYLAND, J.M.E. *The effective topos* ◇ D75 D80 F35 F50 G30 ◇
JAEGER, G. *Iterating admissibility in proof theory* ◇ F35 F50 ◇
JAEGER, G. *Zur Beweistheorie der Kripke-Platek-Mengenlehre ueber den natuerlichen Zahlen* ◇ C70 E30 F15 F35 ◇
JANSOHN, H.-S. & LANDWEHR, R. & WRIGHTSON, G. *An interactive proof system for higher order logic* ◇ B15 B35 F35 ◇
KOLETSOS, G. *A Goedel-functional interpretation of the ω-rule* ◇ F10 F35 ◇
MARTIN-LOEF, P. *Constructive mathematics and computer programming* ◇ B75 F35 F50 ◇
MOERDIJK, I. *Glueing topoi and higher-order disjunction and existence* ◇ F35 F50 G30 ◇
POHLERS, W. *Admissibility in proof theory; a survey* ◇ C70 F15 F35 F98 ◇
POHLERS, W. *Cut elimination for impredicative infinitary systems II. Ordinal analysis for iterated inductive definitions* ◇ C75 F05 F15 F35 ◇
SCEDROV, A. & SCOTT, P.J. *A note on the Friedman slash and Freyd covers* ◇ F35 F50 G30 ◇
SCEDROV, A. *Independence of the fan theorem in the presence of continuity principles* ◇ F35 F50 G30 ◇
SCHMERL, U.R. *A proof theoretical fine structure in systems of ramified analysis* ◇ F15 F35 F65 ◇
SEELY, R.A.G. *Locally Cartesian closed categories and type theory 1* ◇ B15 F35 F50 G30 ◇
SIMPSON, S.G. *Σ^1_1 and Π^1_1 transfinite induction* ◇ F35 ◇
SIMPSON, S.G. *Set theoretic aspects of ATR_0* ◇ C62 E30 F35 ◇
STEPANOV, V.I. *Second-order arithmetic and the consistency of first-order theories (Russian)* ◇ F30 F35 ◇
TROELSTRA, A.S. *Intuitionistic extensions of the reals II* ◇ F35 F50 F55 ◇
WIELE VAN DE, J. *Recursive dilators and generalized recursions* ◇ D60 D65 F15 F35 ◇

1983

ARZARELLO, F. *Un' interpretazione categoriale degli oggetti anomici* ◇ F35 F50 G30 ◇
BELYAKIN, N.V. *A means of modeling a classical second-order arithmetic (Russian)* ◇ B15 C62 D55 F35 ◇

BOFFA, M. *Arithmetic and theory of types*
◇ B15 C62 E70 F35 ◇
BUCHHOLZ, W. & SCHUETTE, K. *Ein Ordinalzahlensystem fuer die beweistheoretische Abgrenzung der Π^1_2-Separation und Bar-Induktion* ◇ F15 F35 ◇
CANTINI, A. *A note on the theory of admissible sets with ε-induction restricted to formulas with one quantifier and related systems (Italian summary)*
◇ C62 E30 E45 E70 F10 F15 F35 ◇
FEFERMAN, S. & JAEGER, G. *Choice principles, the bar rule and autonomously iterated comprehension schemes in analysis* ◇ F15 F35 ◇
FRIEDMAN, H.M. & SIMPSON, S.G. & SMITH, RICK L. *Countable algebra and set existence axioms* ◇ E30 F35 ◇
FRIEDMAN, H.M. & SCEDROV, A. *Set existence property for intuitionistic theories with dependent choice*
◇ E70 F35 F50 ◇
FRIEDMAN, H.M. *Unary Borel functions and second-order arithmetic* ◇ E30 E40 F35 ◇
GRAYSON, R.J. *Forcing in intuitionistic systems without power-set* ◇ E40 E70 F35 F50 ◇
GRAYSON, R.J. & MOERDIJK, I. *Some remarks on extending bar induction* ◇ F35 F50 ◇
GUZICKI, W. *Definable quantifiers in second order arithmetic and elementary extensions of ω-models*
◇ C30 C55 C62 C80 E40 E50 F35 ◇
HAYASHI, S. *Extracting Lisp programs from constructive proofs: a formal theory of constructive mathematics based on Lisp* ◇ B35 F35 F50 ◇
HOEVEN VAN DER, G.F. & MOERDIJK, I. *On an independence result in the theory of lawless sequences*
◇ F35 F50 ◇
JAEGER, G. *A well ordering proof for Feferman's theory T_0*
◇ F15 F35 ◇
JAEGER, G. & POHLERS, W. *Eine beweistheoretische Untersuchung von $(\Delta^1_2\text{-}CA)+(BI)$ und verwandter Systeme* ◇ F15 F35 ◇
KHAKHANYAN, V.KH. *Set theory and Church's thesis (Russian)* ◇ D20 E35 E70 F30 F35 F50 ◇
KOVTUN, M.R. *A system of intuitionistic analysis that is equivalent to classical analysis (Russian) (English summary)* ◇ F35 F50 ◇
LAMBEK, J. & SCOTT, P.J. *New proofs of some intuitionistic principles* ◇ F35 F50 G30 ◇
MOERDIJK, I. *On the Freyd cover of a topos*
◇ F35 F50 G30 ◇
NORMANN, D. *General type-structures of continuous and countable functionals* ◇ D55 D65 D75 F35 ◇
PAEPPINGHAUS, P. *Completeness properties of classical theories of finite type and the normal form theorem*
◇ B15 F05 F35 ◇
SCEDROV, A. & VESLEY, R.E. *On a weakening of Markov's principle* ◇ F35 F50 ◇
SCHWICHTENBERG, H. *On Martin-Loef's theory of types*
◇ B15 F35 F50 ◇
SHVARTS, G.F. *Properties of effectiveness of logical connectives in the intuitionistic theory of types (Russian)*
◇ F35 F50 ◇
SMITH, JAN *The identification of propositions and types in Martin-Loef's type theory: a programming example*
◇ B75 F35 F50 ◇
TROELSTRA, A.S. *Analysing choice sequences*
◇ A05 F35 F50 ◇
TROELSTRA, A.S. *Definability of finite sum types in Martin-Loef's type theories* ◇ F35 F50 ◇
WENDEL, N. *Kontextuelle Definition neuer Wahlfolgen*
◇ F35 F50 ◇
YASUGI, M. *Definability in L^p-spaces* ◇ F35 ◇

1984

ABRUSCI, V.M. *Recenti risultati e metodi nella teoria della dimostrazione: dilatatori, β-logica*
◇ F15 F35 G30 ◇
ARAI, T. *A subsystem of classical analysis proper to Takeuti's reduction method for Π^1_1-analysis* ◇ F35 ◇
ARAI, T. *An accessibility proof of ordinal diagrams in intuitionistic theories for iterated inductive definitions*
◇ F15 F35 F50 ◇
BOFFA, M. *Arithmetic and the theory of types*
◇ B15 F30 F35 ◇
DALEN VAN, D. *How to glue analysis models*
◇ C65 C90 F35 F50 ◇
DILLER, J. & TROELSTRA, A.S. *Realizability and intuitionistic logic* ◇ A05 F30 F35 F50 ◇
FARRINGTON, PADDY *The first-order theory of the c-degrees* ◇ D30 D35 E45 E55 F35 ◇
FEFERMAN, S. *Between constructive and classical mathematics* ◇ A05 F35 F50 F55 ◇
FOURMAN, M.P. *Continuous truth. I. Nonconstructive objects* ◇ A05 E70 F35 F50 G30 ◇
FRIEDRICH, W. *Spielquantorinterpretation unstetiger Funktionale der hoeheren Analysis* ◇ F10 F35 ◇
GIRARD, J.-Y. *The Ω-rule* ◇ B15 C62 C75 F35 ◇
GRAYSON, R.J. *Heyting-valued semantics*
◇ B15 C90 F35 F50 G30 ◇
HENSON, C.W. & KAUFMANN, M. & KEISLER, H.J. *The strength of nonstandard methods in arithmetic*
◇ C62 E30 F30 F35 H05 H15 ◇
HOEVEN VAN DER, G.F. & MOERDIJK, I. *Constructing choice sequences from lawless sequences of neighbourhood functions* ◇ C90 F35 F50 ◇
HOEVEN VAN DER, G.F. & MOERDIJK, I. *On choice sequences determined by spreads*
◇ D30 D55 F35 F50 G30 ◇
HOEVEN VAN DER, G.F. & MOERDIJK, I. *Sheaf models for choice sequences* ◇ C90 F35 F50 G30 ◇
JAEGER, G. *A version of Kripke-Platek set theory which is conservative over Peano arithmetic*
◇ E30 F30 F35 ◇
JAEGER, G. *The strength of admissibility without foundation* ◇ E30 F15 F35 ◇
JENSEN, C.U. *Theorie des modeles pour des anneaux de fonctions entieres et des corps de fonctions meromorphes*
◇ C60 C65 D35 F35 ◇
KASHAPOVA, F.R. *Constructive set theory with types, and consistency with Church's thesis (Russian)*
◇ D20 E70 F35 F50 ◇

KASHAPOVA, F.R. *Determination of classes of constructively derivable theorems in a many-sorted intuitionistic set theory equivalent to second-order arithmetic (Russian)* ◇ F35 F50 ◇

KRIVTSOV, V.N. *Deductive possiblitities of intuitionistic analysis without negation (Russian)* ◇ F35 F50 ◇

KRIVTSOV, V.N. *Imbedding of the intuitionistic theory of types into the negationless intuitionistic theory of types (Russian)* ◇ F35 F50 ◇

LAMBEK, J. & SCOTT, P.J. *Aspects of higher order categorical logic* ◇ B15 F35 F50 G30 ◇

MARTIN-LOEF, P. *Intuitionistic type theory. Notes by Giovanni Sambin of a series of lectures given in Padua, June 1980* ◇ F35 F50 F98 ◇

MOERDIJK, I. *Heine-Borel does not imply the Fan theorem* ◇ C90 E35 F35 F50 G30 ◇

MOERDIJK, I. & REYES, G.E. *Smooth spaces versus continuous spaces in models for synthetic differential geometry* ◇ F35 F50 G30 ◇

NORDSTROEM, B. & SMITH, JAN *Propositions and specifications of programs in Martin-Loef's type theory* ◇ B75 F35 F50 ◇

PEARCE, J. *A constructive consistency proof of a fragment of set theory* ◇ E30 E35 F05 F15 F35 ◇

SCOWCROFT, P. *The real algebraic structure of Scott's model of intuitionistic analysis* ◇ B25 C90 F35 F50 ◇

SEELY, R.A.G. *Locally Cartesian closed categories and type theory* ◇ F35 F50 ◇

SIEG, W. *Foundations of analysis and proof theory* ◇ A10 F30 F35 F98 ◇

SIMPSON, S.G. *Which set existence axioms are needed to prove the Cauchy/Peano theorem for ordinary differential equations?* ◇ E30 F35 ◇

TOMAS, F. *Formally recursive arithmetic and analysis (Catalan)* ◇ F30 F35 ◇

VETULANI, Z. *Ramified analysis and the minimal β-models of higher order arithmetics* ◇ C62 D55 E45 F35 F65 ◇

1985

ABRUSCI, V.M. *Bachmann collections. Gardens. Recent applications (Italian)* ◇ F15 F35 ◇

ABRUSCI, V.M. *Uses of dilators: combinatorial problems and combinatorial results not provable in PA or in ID_1 (Italian)* ◇ F15 F30 F35 ◇

ARAI, T. *A subsystem of classical analysis proper to Takeuti's reduction method for Π-analysis* ◇ F05 F35 ◇

BARR, M. & WELLS, C. *Toposes, triples and theories* ◇ C98 E98 F35 F50 G30 ◇

BAUVAL, A. *Polynomial rings and weak second-order logic* ◇ C60 C85 F35 ◇

BEZEM, M. *Stongly majorizable functionals of finite type: A model for barrecursion containing discontinuous functionals* ◇ F10 F35 F50 ◇

CANTINI, A. *A note on a predicatively reducible theory of iterated elementary induction (Italian summary)* ◇ F15 F35 ◇

CONSTABLE, R.L. & MENDLER, N.P. *Recursive definitions in type theory* ◇ B15 F35 ◇

FEFERMAN, S. *High-level programs and typed vs. untyped constructive foundations* ◇ A05 F35 F50 ◇

FERBUS, M.-C. *Functorial bounds for cut elimination in $L_{\beta\omega}$ II* ◇ F05 F35 ◇

FRIEDRICH, W. *Goedelsche Funktionalinterpretation fuer eine Erweiterung der klassischen Analysis* ◇ F10 F35 F50 ◇

GIRARD, J.-Y. & RESSAYRE, J.-P. *Elements de logique Π^1_n* ◇ D60 F15 F35 ◇

GIRARD, J.-Y. *Introduction to Π^1_2-logic* ◇ E10 F15 F35 ◇

HARNIK, V. *Stability theory and set existence axioms* ◇ C45 F35 ◇

JANKOWSKI, A.W. & ZAWADOWSKI, M. *Sheaves over Heyting lattices* ◇ C90 F35 F50 G10 G30 ◇

KOLETSOS, G. *Functional interpretation of the β-rule* ◇ F10 F30 F35 ◇

PAEPPINGHAUS, P. *A typed λ-calculus and Girard's model of ptykes* ◇ B40 F10 F15 F35 ◇

ROSENZWEIG, D. *Inductive definitions in ML_0* ◇ F35 ◇

SCARPELLINI, B. *Lower bound results on lengths of second-order formulas* ◇ B15 C13 D10 F20 F35 ◇

SCEDROV, A. *Extending Goedel's modal interpretation to type theory and set theory* ◇ B15 B45 E70 F35 F50 ◇

SCHWICHTENBERG, H. *A normal form for natural deductions in a type theory with realizing terms* ◇ F05 F35 F50 ◇

SIEG, W. *Fragments of arithmetic* ◇ F30 F35 ◇

SIEG, W. *Reductions of theories for classical analysis* ◇ F35 ◇

SIMPSON, S.G. *Friedman's research on subsystems of second order arithmetic* ◇ C62 C98 F35 F98 ◇

SIMPSON, S.G. *Reverse mathematics* ◇ E30 F35 ◇

STRAUSS, P. *Number-theoretic set theories* ◇ E70 F35 ◇

TROELSTRA, A.S. *Choice sequences and informal rigour* ◇ A05 F35 F50 ◇

F40 Goedel numberings in proof theory

1939

HILBERT, D. & BERNAYS, P. *Grundlagen der Mathematik II* ◇ A05 B98 F05 F15 F30 F40 F98 ◇

1957

SMULLYAN, R.M. *Languages in which self reference is possible* ◇ A05 B10 B28 F30 F40 ◇

1958

ROGERS JR., H. *Goedel numberings of partial recursive functions* ◇ D20 D45 F40 ◇

1960

FEFERMAN, S. *Arithmetization of metamathematics in a general setting* ◇ D55 F07 F25 F30 F40 ◇

1969

WAGNER, E.G. *Uniformly reflexive structures: on the nature of Goedelization and relative computability* ◇ B40 D45 D75 F40 ◇

1970

KLEINBERG, E.M. *Recursion theory and formal deducibility* ◇ D25 F07 F40 ◇

1973

HELM, J.P. & MEYER, A.R. & YOUNG, P. *On orders of translations and enumerations* ◇ D15 D20 D25 D45 F40 ◇

1977

CALUDE, C. *Une construction grammaticale des codages de Goedel (English summary)* ◇ F40 ◇

1978

CALUDE, C. *On the category of recursive languages* ◇ B03 D45 F40 G30 ◇

MACHTEY, M. & WINKLMANN, K. & YOUNG, P. *Simple Goedel numberings, isomorphisms, and programming properties* ◇ D05 D10 D15 D20 D45 F40 ◇

1980

NERODE, A. & SHORE, R.A. *Second order logic and first order theories of reducibility orderings* ◇ B10 B15 D30 D35 F35 F40 ◇

1981

SMORYNSKI, C.A. *Fifty years of self-reference in arithmetic* ◇ A10 F30 F40 ◇

1982

DAPUETO, C. *Sull' aritmetizzazione della sintassi del primo ordine (French summary)* ◇ B28 F30 F40 ◇

FEFERMAN, S. *Inductively presented systems and the formalization of meta-mathematics* ◇ F30 F35 F40 ◇

1983

STREINU, I. *Grammar directed Goedel numberings* ◇ F40 ◇

1984

LONGO, G. & MOGGI, E. *Goedel numberings, principal morphisms, combinatory algebras* ◇ B40 F40 G30 ◇

1985

SOLOVAY, R.M. *Explicit Henkin sentences* ◇ F30 F40 ◇

F50 Metamathematics of constructive systems

1925

KOLMOGOROV, A.N. *On the principle of excluded middle (Russian)* ◇ F50 ◇

1926

LEVY, P. *Sur le principe du tiers exclu et sur les theoremes non susceptibles de demonstration* ◇ A05 F50 ◇

1927

BARZIN, M. & ERRERA, A. *Sur la logique de M.Brouwer* ◇ F50 ◇

LEVY, P. *Logique classique, logique Brouwerienne et logique mixte* ◇ A05 B55 F50 ◇

1928

BROUWER, L.E.J. *Intuitionistische Betrachtungen ueber den Formalismus* ◇ A05 F50 ◇

CHURCH, A. *On the law of excluded middle* ◇ A05 B05 F50 ◇

GLIVENKO, V.I. *Sur la logique de M. Brouwer* ◇ F50 ◇

LEVY, P. *Logique classique, logique Brouwerienne et logique mixte* ◇ A05 F50 ◇

REYMOND, A. *L'axiomatique logique et le principe du tiers exclu* ◇ A05 F50 ◇

WEYL, H. *Diskussionsbemerkungen zu dem zweiten Hilbertschen Vortrag ueber die Grundlagen der Mathematik* ◇ A05 F50 ◇

1929

BARZIN, M. & ERRERA, A. *Sur le principe du tiers exclu* ◇ A05 F50 ◇

GLIVENKO, V.I. *Sur quelques points de la logique de M.Brouwer* ◇ F50 ◇

1930

HEYTING, A. *Die formalen Regeln der intuitionistischen Logik* ◇ F50 ◇

HEYTING, A. *Die formalen Regeln der intuitionistischen Mathematik* ◇ F30 F35 F50 ◇

HEYTING, A. *Sur la logique intuitionniste* ◇ F50 ◇

1931

GOEDEL, K. *Eine Interpretation des intuitionistischen Aussagenkalkuels* ◇ A05 B45 F50 ◇

GOEDEL, K. *Zum intuitionistischen Aussagenkalkuel* ◇ F50 F55 ◇

HEYTING, A. *Die intuitionistische Grundlegung der Mathematik* ◇ A05 F50 F55 ◇

HILBERT, D. *Beweis des "tertium non datur"* ◇ A05 B28 F50 ◇

1932

BARZIN, M. & ERRERA, A. *Note sur la logique de M.Heyting* ◇ A05 F50 ◇

BARZIN, M. & ERRERA, A. *Sur la logique de M.Heyting* ◇ A05 F50 ◇

BARZIN, M. & ERRERA, A. *Sur la logique intuitionniste. Reponse a M.A.Heyting* ◇ A05 F50 ◇

GOEDEL, K. *Zur intuitionistischen Arithmetik und Zahlentheorie* ◇ F30 F50 ◇

HEYTING, A. *Anwendung der intuitionistischen Logik auf die Definition der Vollstaendigkeit eines Kalkuels* ◇ F50 ◇

KOLMOGOROV, A.N. *Zur Deutung der intuitionistischen Logik* ◇ A05 F50 ◇

1933

BARZIN, M. & ERRERA, A. *La logique de M.Brouwer. Etat de la question* ◇ A05 F50 ◇

1934

HEYTING, A. *Mathematische Grundlagenforschung. Intuitionismus. Beweistheorie* ◇ A05 F50 F98 ◇

ZHEGALKIN, I.I. *On the decision problem in Brouwer's propositional logic (Russian)* ◇ F50 ◇

1935

BETH, E.W. *Sur un theoreme concernant le principe du tiers exclu et ses applications dans la theorie de la non-contradiction* ◇ F25 F50 ◇

GENTZEN, G. *Untersuchungen ueber das logische Schliessen I,II* ◇ B10 B25 F05 F07 F30 F50 ◇

HUNTINGTON, E.V. *Effective equality and effective implication in formal logic* ◇ F50 ◇

1936

BARZIN, M. *Note sur la demonstration de M.E.W.Beth* ◇ A05 F25 F50 ◇

BETH, E.W. *Demonstration d'un theoreme concernant le principe du tiers exclu* ◇ F50 ◇

DASSEN, C.C. *Sobre una objecion a la Logica Brouweriana* ◇ A05 F50 ◇

JASKOWSKI, S. *Recherches sur le systeme de la logique intuitionniste* ◇ B55 F50 ◇

REYMOND, A. *La negation et le principe du tiers exclu* ◇ A05 F50 ◇

1937

ERRERA, A. *Sur la notion de compatibilite et les rapports entre l'intuitionnisme et le formalisme* ◇ F50 ◇

FREUDENTHAL, H. *Zur intuitionistischen Deutung logischer Formeln* ◇ A05 F50 ◇

GENTZEN, G. *Der Unendlichkeitsbegriff in der Mathematik* ◇ A05 F50 F99 ◇

HEYTING, A. *Bemerkungen zu dem Aufsatz von Herrn Freudenthal "Zur intuitionistischen Deutung logischer Formeln"* ◇ A05 F50 ◇

JOHANSSON, I. *Der Minimalkalkuel, ein reduzierter intuitionistischer Formalismus* ◊ F50 ◊
STONE, M.H. *Topological representations of distributive lattices and Brouwerian logics* ◊ F50 G10 ◊

1938

FRINK, O. *New algebras of logic* ◊ B50 F50 G05 ◊
MOISIL, G.C. *Sur le syllogisme hypothetique dans la logique intuitionniste* ◊ F50 ◊
TARSKI, A. *Der Aussagenkalkuel und die Topologie* ◊ B05 F50 G05 G10 ◊
WAJSBERG, M. *Untersuchungen ueber den Aussagenkalkuel von A.Heyting* ◊ B25 F50 ◊

1939

CURRY, H.B. *A note on the reduction of Gentzen's calculus LJ* ◊ F50 ◊
MCKINSEY, J.C.C. *Proof of the independence of the primitive symbols of Heyting's calculus of propositions* ◊ F50 ◊
OGASAWARA, T. *Relation between intuitionistic logic and lattice* ◊ F50 G10 ◊

1943

VAIDYANATHASWAMY, R. *On disjunction in intuitionistic logic* ◊ F50 ◊

1944

GRISS, G.F.C. *Negationless intuitionistic mathematics* ◊ F50 F55 ◊

1945

DESTOUCHES-FEVRIER, P. *Rapport entre le calcul des problemes et le calcul des propositions* ◊ A05 B05 F50 ◊
KLEENE, S.C. *On the interpretation of intuitionistic number theory* ◊ B28 F50 ◊

1946

GRISS, G.F.C. *Negationless intuitionistic mathematics I* ◊ F50 F55 ◊
HEYTING, A. *On weakened quantification* ◊ F50 ◊
HIZ, H. *Remarques sur le degre de completude* ◊ B22 F50 ◊
ZAWIRSKI, Z. *Origin and development of intuitionistic logic (Polish)* ◊ A05 A10 F50 ◊

1947

BETH, E.W. *Semantical considerations on intuitionistic mathematics* ◊ F50 ◊
NELSON, D. *Recursive functions and intuitionistic number theory* ◊ F30 F50 ◊
POPPER, K.R. *Functional logic without axioms or primitive rules of inference* ◊ A05 B20 F50 ◊
VALPOLA, V. *The position of negation in a language which expresses knowledge (Finnish)* ◊ A05 B28 F30 F50 ◊

1948

BENGY-PUYVALLEE DE, R. *Sur les regles de composabilite dans la logique de la mathematique intuitionniste sans negation* ◊ F50 ◊
DESTOUCHES-FEVRIER, P. *Le calcul des constructions* ◊ F50 ◊
DESTOUCHES-FEVRIER, P. *Logique de l'intuitionisme sans negation et logique de l'intuitionisme positif* ◊ B55 F50 ◊
FITCH, F.B. *Intuitionistic modal logic with quantifiers* ◊ B45 F50 ◊
GRISS, G.F.C. *Logique des mathematiques intuitionnistes sans negation* ◊ F50 ◊
GRISS, G.F.C. *Sur la negation (dans les mathematiques et la logique)* ◊ A05 F50 ◊
KURODA, S. *On the logic of Aristotle and the logic of Brouwer (Japanese)* ◊ A05 A10 F07 F50 ◊
MOSTOWSKI, ANDRZEJ *Proofs of non-deducibility in intuitionistic functional calculus* ◊ F50 ◊

1949

DESTOUCHES-FEVRIER, P. *Connexion entre les calculs des constructions, des problemes, des propositions* ◊ A05 F50 ◊
IONGH DE, J.J. *Restricted forms of intuitionistic mathematics* ◊ A05 F50 ◊
KLEENE, S.C. *On the intuitionistic logic* ◊ F50 ◊
NELSON, D. *Constructible falsity* ◊ F50 ◊
RIEGER, L. *On the lattice theory of brouwerian propositional logic* ◊ F50 G10 ◊

1950

CURRY, H.B. *A theory of formal deducibility* ◊ B10 F50 F98 ◊
GRISS, G.F.C. *Negationless intuitionistic mathematics II* ◊ F50 F55 ◊
MARKOV, A.A. *A constructive logic (Russian)* ◊ F50 ◊
MATSUMOTO, K. *On a lattice relating to the intuitionistic logic* ◊ F50 G10 ◊
MATSUMOTO, K. *Sur la structure concernant la logique moderne* ◊ A05 B45 F50 G05 ◊
PIL'CHAK, B.YU. *On the decision problem for the calculus of problems (Russian)* ◊ B25 F50 ◊
RIDDER, J. *Formalistische Betrachtungen ueber intuitionistische und verwandte logische Systeme I, II, III, IV* ◊ F50 ◊
SCHUETTE, K. *Schlussweisen-Kalkuele der Praedikatenlogik* ◊ B10 B25 F05 F07 F50 ◊

1951

DESTOUCHES-FEVRIER, P. *Sur l'intuitionisme et la conception strictement constructive* ◊ F50 ◊
ERRERA, A. *Observations sur la communication du Professeur Heyting* ◊ F50 ◊
GRISS, G.F.C. *The logic of negationless intuitionistic mathematics* ◊ F50 ◊
HEYTING, A. *L'axiomatique intuitionniste* ◊ F50 F55 ◊
KREISEL, G. *On the interpretation of non-finitist proofs I* ◊ F10 F25 F30 F50 ◊
KURODA, S. *Intuitionistische Untersuchungen der formalistischen Logik* ◊ F50 ◊
RASIOWA, H. *Algebraic treatment of the functional calculi of Heyting and Lewis* ◊ B45 F50 G10 G25 ◊
RIDDER, J. *Formalistische Betrachtungen ueber intuitionistische und verwandte logische Systeme V, VI, VII* ◊ F50 ◊

1952

CURRY, H.B. *The inferential theory of negation* ◇ A05 B20 F50 ◇

CURRY, H.B. *The system LD* ◇ B10 F50 ◇

DIENES, P. *On H-matrices* ◇ F50 ◇

KLEENE, S.C. *Introduction to metamathematics* ◇ A05 B98 D98 F30 F50 F98 ◇

KLEENE, S.C. *Recursive functions and intuitionistic mathematics* ◇ F50 ◇

KREISEL, G. *On the interpretation of non-finitist proofs II. Interpretation of number theory. Applications* ◇ F10 F25 F30 F50 ◇

LUKASIEWICZ, J. *On the intuitionistic theory of deduction* ◇ F50 ◇

PIL'CHAK, B.YU. *On the calculus of problems (Russian)* ◇ F50 ◇

PIL'CHAK, B.YU. *On the role of the law of the excluded middle in mathematics (Russian)* ◇ F50 ◇

SKOLEM, T.A. *Consideraciones sobre los fundamentos de la matematica I* ◇ B98 F30 F50 ◇

VOROB'EV, N.N. *A constructive propositional calculus with strong negation (Russian)* ◇ F50 ◇

VOROB'EV, N.N. *The problem of deducibility in the constructive propositional calculus with strong negation (Russian)* ◇ F50 ◇

1953

COHEN, K.J. *A remark on Lukasiewicz's "On the intuitionistic theory of deduction"* ◇ F50 ◇

GILMORE, P.C. *Griss' criticism of the intuitionistic logic and the theory of order* ◇ F50 ◇

GILMORE, P.C. *The effect of Griss' criticism of the intuitionistic logic on deductive theories formalized within the intuitionistic logic* ◇ F50 ◇

JOHANSSON, I. *Sur le concept de "le" (ou de "ce qui") dans le calcul affirmatif et dans les calculs intuitionnistes* ◇ B20 F50 ◇

LORENZEN, P. *Die Allgemeingueltigkeit der logischen Regeln* ◇ A05 F50 F65 ◇

LUKASIEWICZ, J. *Comment on K.J.Cohen's remark* ◇ B05 F50 ◇

MEREDITH, C.A. *A single axiom of positive logic* ◇ B20 F50 ◇

MOSTOWSKI, ANDRZEJ & RASIOWA, H. *A geometric interpretation of logical formulae (Polish) (English and Russian summaries)* ◇ B10 F50 G05 G10 G15 ◇

OHNISHI, M. *On intuitionistic functional calculus* ◇ F50 ◇

RASIOWA, H. & SIKORSKI, R. *On satisfiability and decidability in non-classical functional calculi* ◇ B10 B25 B45 C90 F50 ◇

ROSE, G.F. *Propositional calculus and realizibility* ◇ F50 ◇

SHANIN, N.A. *On some operations on logico-arithmetical formulae (Russian)* ◇ F30 F50 ◇

SKOLEM, T.A. *Consideraciones sobre los fundamentos de la matematica II* ◇ B98 F30 F50 ◇

1954

GOODSTEIN, R.L. *Logic-free formalisations of recursive arithmetic* ◇ F30 F50 ◇

HEYTING, A. *G.F.C. Griss and his negationless intuitionistic mathematics* ◇ A05 F50 ◇

HEYTING, A. *Logique et intuitionnisme* ◇ A05 F50 ◇

KREISEL, G. *Discussion sur divers themes* ◇ F30 F50 ◇

MAEHARA, S. *Eine Darstellung der intuitionistischen Logik in der klassischen* ◇ F50 ◇

RASIOWA, H. *Constructive theories* ◇ F50 G05 G10 G25 ◇

SHANIN, N.A. *On imbeddings of the classical logico-arithmetical calculus into the constructive logico-arithmetical calculus (Russian)* ◇ F30 F50 ◇

VREDENDUIN, P.G.J. *The logic of negationless mathematics* ◇ F50 ◇

1955

BROUWER, L.E.J. *The effect of intuitionism on classical algebra of logic* ◇ B05 F50 ◇

GRISS, G.F.C. *La mathematique intuitionniste sans negation* ◇ F50 ◇

LORENZEN, P. *Einfuehrung in die operative Logik und Mathematik* ◇ A05 B98 F50 F65 F98 ◇

MONTEIRO, A. *Axiomes independants pour les algebres de Brouwer* ◇ F50 G10 ◇

RASIOWA, H. *Algebraic models of axiomatic theories* ◇ B30 C90 F50 G05 G10 G25 ◇

RASIOWA, H. & SIKORSKI, R. *On existential theorems in non-classical functional calculi* ◇ B45 C90 F50 G05 ◇

SCHROETER, K. *Theorie des logischen Schliessens I* ◇ B10 F07 F50 ◇

SHANIN, N.A. *On some logical problems of arithmetic (Russian)* ◇ D20 F30 F50 ◇

TEENSMA, E. *The intuitionistic interpretation of analysis* ◇ F50 ◇

VALPOLA, V. *Ein System der negationslosen Logik mit ausschliesslich realisierbaren Praedikaten* ◇ B20 F50 ◇

VALPOLA, V. *Eine Eigenschaft gewoehnlicher negationsloser Kalkuele der Propositionen- und Praedikatenlogik* ◇ B20 F50 ◇

1956

BETH, E.W. *Semantic construction of intuitionistic logic* ◇ F50 ◇

BETH, E.W. *Semantische Tafeln fuer die intuitionistische Praedikatenlogik erster Ordnung* ◇ F50 ◇

FEVRIER, P. *Tendances constructives en logique moderne* ◇ A05 F50 ◇

HARROP, R. *On disjunctions and existential statements in intuitionistic systems of logic* ◇ F30 F50 ◇

HENKIN, L. *La structure algebrique des theories mathematiques* ◇ C07 F50 G05 G10 G15 ◇

HEYTING, A. *La conception intuitionniste de la logigue* ◇ A05 F50 ◇

LOS, J. & MOSTOWSKI, ANDRZEJ & RASIOWA, H. *A proof of Herbrand's theorem* ◇ B10 C07 F05 F50 ◇

SCHROETER, K. *Ueber den Zusammenhang der in den Implikationsaxiomen vollstaendigen Axiomensysteme des zweiwertigen mit denen des intuitionistischen Aussagenkalkuel* ◇ B20 F50 ◇

SIKORSKI, R. *A theorem on non-classical functional calculi* ◇ B45 F50 G10 ◇

1957

BETH, E.W. *Intuitionistic predicate logic* ◇ F50 ◇
BLANCHE, R. *Introduction a la logique contemporaine* ◇ A05 B98 F50 ◇
COSTA DA, N.C.A. *Considerations on the Heyting calculus (Portuguese) (English summary)* ◇ F50 ◇
GOODSTEIN, R.L. *Recursive number theory. A development of recursive arithmetic in a logic-free equation calculus* ◇ B28 F30 F50 F60 F98 ◇
KLEENE, S.C. *Realizability* ◇ D80 F50 ◇
KREISEL, G. & PUTNAM, H. *Eine Unableitbarkeitsbeweismethode fuer den intuitionistischen Aussagenkalkuel* ◇ B55 F50 ◇
KREISEL, G. *Goedel's interpretation of Heyting's arithmetic* ◇ F10 F30 F50 ◇
KREISEL, G. *Relations between classes of constructive functionals* ◇ F10 F50 ◇
MO, SHAOKUI *Simplified introduction to intuitionistic logic (Chinese)* ◇ F50 ◇
SCHROETER, K. *Eine Umformung des heytingschen Axiomensystems fuer den intuitionistischen Aussagenkalkuel* ◇ F50 ◇
SCOTT, D.S. *Completeness proofs for the intuitionistic sentential calculus* ◇ F50 ◇

1958

BETH, E.W. *Construction semantique de la logique intuitionniste* ◇ A05 F50 ◇
BIALYNICKI-BIRULA, A. & RASIOWA, H. *On constructible falsity in the constructive logic with strong negation* ◇ F50 ◇
COSTA DA, N.C.A. *Note on the logic of Brouwer-Heyting (Portuguese) (English summary)* ◇ F50 ◇
GAL, I.L. & ROSSER, J.B. & SCOTT, D.S. *Generalization of a lemma of G.F.Rose* ◇ F50 ◇
GOEDEL, K. *Ueber eine bisher noch nicht benuetzte Erweiterung des finiten Standpunktes* ◇ A05 F10 F30 F50 ◇
GOODSTEIN, R.L. *Models of propositional calculi in recursive arithmetic* ◇ B05 B50 F30 F50 ◇
HEYTING, A. *Blick von der intuitionistischen Warte* ◇ A05 F50 ◇
HEYTING, A. *Intuitionism in mathematics* ◇ A05 B98 F50 F55 F98 ◇
KREISEL, G. *A remark on free choice sequences and the topological completeness proofs* ◇ F50 ◇
KREISEL, G. *Constructive mathematics* ◇ F50 F55 F65 ◇
KREISEL, G. *Elementary completeness properties of intuitionistic logic with a note on negations of prenex formulae* ◇ F50 ◇
KREISEL, G. *Mathematical significance of consistency proofs* ◇ C57 C60 F05 F25 F50 ◇
LORENZEN, P. *Formale Logik* ◇ F50 F98 ◇
PORTE, J. *Une propriete du calcul propositionnel intuitionniste* ◇ F50 ◇
RASIOWA, H. *$\mathcal{N}$-lattices and constructive logic with strong negation* ◇ F50 G10 ◇
SCHMIDT, H.A. *Un procede maniable de decision pour la logique propositionnelle intuitionniste* ◇ B25 F50 ◇
SHANIN, N.A. *On the constructive interpretation of mathematical judgements (Russian)* ◇ A05 F50 ◇
SHANIN, N.A. *Ueber einen Algorithmus zur konstruktiven Dechiffrierung mathematischer Urteile (Russian) (German summary)* ◇ A05 F50 F60 ◇
SIKORSKI, R. *Some applications of interior mappings* ◇ B45 E75 F50 G05 G10 G25 ◇
SKOLEM, T.A. *Remarks on the connection between intuitionistic logic and a certain class of lattices* ◇ F50 G10 ◇
VOROB'EV, N.N. *A new algorithm of deducibility in the constructive propositional calculus (Russian)* ◇ F50 ◇

1959

BETH, E.W. *Remarks on intuitionistic logic* ◇ F50 ◇
HERMES, H. *Zum Inversionsprinzip der operativen Logik* ◇ F50 ◇
HEYTING, A. *Some remarks on intuitionism* ◇ A05 F50 ◇
KREISEL, G. *Interpretation of analysis by means of constructive functionals of finite types* ◇ D65 F10 F30 F35 F50 ◇
LERDA, F. *Analisi critica dei fondamenti dell'intuizionismo* ◇ A05 F50 F55 ◇
LORENZEN, P. *Ueber die Begriffe "Beweis" und "Definition"* ◇ A05 F50 ◇
MARKOV, A.A. *On a principle of constructive mathematical logic (Russian)* ◇ F50 ◇
MOSTOWSKI, ANDRZEJ *On various degrees of constructivism* ◇ D55 F50 F60 F98 ◇
NELSON, D. *Negation and separation of concepts in constructive systems* ◇ F50 ◇
RASIOWA, H. *Algebraische Charakterisierung der intuitionistischen Logik mit starker Negation* ◇ F50 G10 ◇
RASIOWA, H. & SIKORSKI, R. *Formalisierte intuitionistische elementare Theorien* ◇ F50 ◇
SHANIN, N.A. *On the constructive concept of mathematical judgements (Russian)* ◇ F50 ◇
SIKORSKI, R. *Der Heyting'sche Praedikatenkalkuel und metrische Raeume* ◇ F50 ◇
SMETANICH, YA.S. *Ueber eine Eigenschaft der Formeln des konstruktiven Aussagenkalkuels, die von einer Variablen abhaengen (Russisch)* ◇ F50 ◇

1960

BETH, E.W. & LEBLANC, H. *A note on the intuitionist and the classical propositional calculus* ◇ B05 F50 ◇
HARROP, R. *Concerning formulas of the types $A \to B \vee C$, $A \to (Ex)B(x)$ in intuitionistic formal systems* ◇ F30 F50 ◇
KLEENE, S.C. *Realizability and Shanin's algorithm for the constructive deciphering of mathematical sentences* ◇ D80 F50 ◇
NISHIMURA, I. *On formulas of one variable in intuitionistic propositional calculus* ◇ F50 ◇
POGORZELSKI, W.A. & SLUPECKI, J. *Basic properties of deductive systems based on nonclassical logics I,II (Polish) (Russian and English summaries)* ◇ B22 F50 ◇
SCHMIDT, H.A. *Mathematische Gesetze der Logik. I. Vorlesungen ueber Aussagenlogik* ◇ B05 B98 F50 F98 ◇

WETTE, E. *Intuitionistisch-rekursiver Konsistenzbeweis fuer die axiomatische Mengenlehre* ◇ E30 E35 F50 ◇

1961

HEYTING, A. *Infinitistic methods from a finitist point of view* ◇ A05 F50 ◇

HUBER-DYSON, V. & KREISEL, G. *Analysis of Beth's semantic construction of intuitionistic logic* ◇ F50 F55 ◇

LORENZEN, P. *Ein dialogisches Konstruktivitaetskriterium* ◇ A05 F50 F65 ◇

MO, SHAOKUI *N-generalizable, intuitionistic, co-denial, pseudo-modal and co-Δ systems (Chinese)* ◇ F50 ◇

TEIXEIRA, M.T. *Der Tarskische Konsequenz-Operator und mit dem intuitionistischen Aussagenkalkuel verknuepfte algebraische Strukturen (Portugiesisch)* ◇ F50 G10 ◇

1962

BELNAP JR., N.D. & LEBLANC, H. *Intuitionism reconsidered* ◇ F50 ◇

BERTOLINI, F. *Giustificazione semantica secondo Lorenzen del calcolo proposizionale intuizionistico* ◇ F50 ◇

BETH, E.W. *Formal methods. An introduction to symbolic logic and to the study of effective operations in arithmetic and logic* ◇ B28 B98 F30 F50 F98 ◇

DIEGO, A. *On implicative algebras (Spanish)* ◇ C05 F50 G10 G25 ◇

DOPP, J. *Logiques construites par une methode de deduction naturelle* ◇ B98 F05 F07 F50 F98 ◇

HEYTING, A. *Methodes et problemes de l'intuitionnisme* ◇ A05 F50 F55 ◇

KLEENE, S.C. *Disjunction and existence under implication in elementary intuitionistic formalisms* ◇ F50 ◇

KREISEL, G. *Foundations of intuitionistic logic* ◇ F50 ◇

KREISEL, G. *On weak completeness of intuitionistic predicate logic* ◇ F50 ◇

MCCALL, S. *A simple decision procedure for one-variable implication/negation formulae in intuitionistic logic* ◇ B25 F50 ◇

MEDVEDEV, YU.T. *Finite problems (Russian)* ◇ D80 F50 ◇

MINTS, G.E. *An analogue of Herbrand's theorem for a constructive predicate calculus (Russian)* ◇ F05 F50 ◇

SCHUETTE, K. *Der Interpolationssatz der intuitionistischen Praedikatenlogik* ◇ C40 F50 ◇

SIKORSKI, R. *Applications of topology to foundations of mathematics* ◇ B30 E75 F50 ◇

SPECTOR, C. *Provably recursive functionals of analysis: a consistency proof of analysis by an extension of principles formulated in current intuitionistic mathematics* ◇ F10 F35 F50 ◇

1963

BELNAP JR., N.D. & THOMASON, R.H. *A rule-completeness theorem* ◇ B22 F07 F50 ◇

BELNAP JR., N.D. & LEBLANC, H. & THOMASON, R.H. *On not strengthening intuitionistic logic* ◇ F50 ◇

HARRISON, J. *Appendix C: Equivalence of the effective operations and the hereditarily continuous functionals* ◇ D65 F50 ◇

HOWARD, W.A. *Appendices to section 2, vol. 1* ◇ F10 F35 F50 ◇

HOWARD, W.A. *The axiom of choice ($\Sigma_1^1 - AC_{01}$), bar induction and bar recursion* ◇ F35 F50 ◇

HOWARD, W.A. *Transfinite induction and transfinite recursion* ◇ F10 F35 F50 ◇

KABAKOV, F.A. *On the derivability of some realizable formulas of the sentential calculus (Russian) (German summary)* ◇ F50 ◇

KREISEL, G. *Generalized inductive definitions* ◇ F35 F50 ◇

KREISEL, G. *Introduction (to volume 1)* ◇ F35 F50 ◇

KREISEL, G. *Theory of free choice sequences of natural numbers* ◇ F35 F50 ◇

MEDVEDEV, YU.T. *Interpretation of logical formulae by means of finite problems and its relation to the realizability theory (Russian)* ◇ D80 F50 ◇

MINTS, G.E. & OREVKOV, V.P. *An extension of the theorems of Glivenko and Kreisel to a certain class of formulae of predicate calculus (Russian)* ◇ F50 ◇

MYHILL, J.R. *The invalidity of Markov's schema* ◇ F50 ◇

PARIKH, R. *Some generalisations of the notion of well-ordering* ◇ F35 F50 ◇

PRAWITZ, D. *Concerning constructive logic and the concept of implication (Swedish)* ◇ A05 F07 F50 ◇

RASIOWA, H. & SIKORSKI, R. *The mathematics of metamathematics* ◇ B98 C98 F50 F98 G05 G10 G98 ◇

REZNIKOFF, I. *Chaines de formules* ◇ B05 C40 C90 F50 ◇

TAIT, W.W. *A second order theory of functionals of higher type* ◇ F10 F35 F50 ◇

TAIT, W.W. *Appendix A: Intensional functionals* ◇ F10 F35 F50 ◇

TAIT, W.W. *Appendix B: An interpretation of functionals by convertible terms* ◇ F05 F35 F50 ◇

VESLEY, R.E. *On strengthening intuitionistic logic* ◇ F50 ◇

YANKOV, V.A. *Realizable formulas of propositional logic (Russian)* ◇ B55 F50 ◇

YANKOV, V.A. *The relationship between deducibility in the intuitionistic propositional calculus and finite implicational structures (Russian)* ◇ B55 F50 G25 ◇

YASUGI, M. *Intuitionistic analysis and Goedel's interpretation* ◇ F10 F35 F50 ◇

ZASLAVSKIJ, I.D. *Ueber gewisse Unterschiede zwischen den Basis- und den untergeordneten Variablen in logisch-mathematischen Sprachen (Russisch)* ◇ F50 ◇

1964

GRZEGORCZYK, A. *A philosophically plausible formal interpretation of intuitionistic logic* ◇ F50 ◇

HAO, K. *The location in the Kleene hierarchy of some fundamental predicates of constructive analysis (Chinese)* ◇ F50 ◇

IDEL'SON, A.V. *Calculi of constructive logic with subordinate variables (Russian)* ◇ F50 ◇

MINTS, G.E. *On predicate and operator variants of the formation of theories of constructive mathematics (Russian)* ◇ F50 ◇

NAGORNYJ, N.M. *On realizable and completable logico-arithmetical formulas (Russian)* ◇ F30 F50 ◇

SHANIN, N.A. *Concerning the constructive interpretation of auxiliary formulas I (Russian)* ◇ F50 ◇

VOROB'EV, N.N. *A constructive calculus of statements with strong negation (Russian)* ◇ B55 F50 ◇

1965

HE, KEGANG *On the types of some basic predicates of constructive real number theory in the S.C.Kleene classifications (Chinese)* ◇ F50 F60 ◇

KLEENE, S.C. *Classical extensions of intuitionistic mathematics* ◇ F35 F50 ◇

KLEENE, S.C. *Logical calculus and realizability* ◇ F50 ◇

KLEENE, S.C. & VESLEY, R.E. *The foundations of intuitionistic mathematics, especially in relation to recursive functions* ◇ F35 F50 F98 ◇

KREISEL, G. *Mathematical logic* ◇ B98 F10 F35 F50 F98 ◇

KRIPKE, S.A. *Semantical analysis of intuitionistic logic I* ◇ B55 C90 F50 ◇

KURATA, R. *Recursive progression of intuitionistic number theories* ◇ F15 F30 F50 ◇

KUZNETSOV, A.V. *Analogs of the "Sheffer stroke" in constructive logic (Russian)* ◇ F50 ◇

LEBLANC, H. *Marginalia on Gentzen's Sequenzen-Kalkule* ◇ B10 F07 F50 ◇

MASLOV, S.YU. & MINTS, G.E. & OREVKOV, V.P. *Unsolvability in the constructive predicate calculus of certain classes of formulas containing only monadic predicate variables (Russian)* ◇ B20 D35 F50 ◇

OREVKOV, V.P. *Certain reduction classes and solvable classes of sequents for the constructive predicate calculus (Russian)* ◇ F50 ◇

OREVKOV, V.P. *Unsolvability of the class of formulas of the type* $\neg\neg\forall\exists$ *in the constructive predicate calculus (Russian)* ◇ F50 ◇

PLIUSKEVICIUS, R. *A version of the constructive calculus of predicates without structural deduction rules (Russian)* ◇ F50 ◇

PRAWITZ, D. *Natural deduction. A proof-theoretical study* ◇ B15 B98 F05 F07 F50 F98 ◇

RAGGIO, A.R. *Gentzen's Hauptsatz for the systems NI and NK* ◇ F05 F50 ◇

ROBINSON, T.T. *Interpretations of Kleene's metamathematical predicate* $\Gamma | A$ *in intuitionistic arithmetic* ◇ F30 F50 ◇

STEGMUELLER, W. *Die Aequivalenz des klassischen und intuitionistischen Ableitungsbegriffs im Gentzen-Quine-Kalkuel und in Kleenes Kalkuel H* ◇ F50 ◇

STRALBERG, A.H. *On the reduction of classical logic. An extension of some theorems of Glivenko and Goedel* ◇ B55 F30 F50 ◇

TAIT, W.W. *Infinitely long terms of transfinite type* ◇ C75 F10 F15 F50 ◇

VARPAKHOVSKIJ, F.L. *The nonrealizability of a disjunction of nonrealizable formulas of propositional logic (Russian)* ◇ B55 F50 ◇

1966

HANATANI, Y. *Calculabilite des fonctionelles recursives primitives de type fini sur les nombres naturels* ◇ D65 F10 F35 F50 ◇

HANAZAWA, M. *A characterization of axiom schema playing the role of tertium non dactur in intuitionistic logic* ◇ F50 ◇

HOSOI, T. *The separation theorem on the classical system* ◇ B10 F50 ◇

HOWARD, W.A. & KREISEL, G. *Transfinite induction and bar induction of types zero and one, and the role of continuity in intuitionistic analysis* ◇ F10 F35 F50 ◇

IMAI, Y. & ISEKI, K. *On Griss algebra. I* ◇ F50 G10 ◇

JONGH DE, D.H.J. & TROELSTRA, A.S. *On the connection of partially ordered sets with some pseudo-boolean algebras* ◇ F50 G10 ◇

KOTAS, J. & PIECZKOWSKI, A. *On a generalized cylindrical algebra and intuitionistic logic (Polish and Russian summaries)* ◇ F50 G15 ◇

KUTSCHERA VON, F. *Zur semantischen Begruendung der klassischen und der intuitionistischen Logik* ◇ A05 F50 ◇

LEBLANC, H. & THOMASON, R.H. *The demarcation line between intuitionistic logic and classical logic* ◇ A05 F50 ◇

LEBLANC, H. *Two separation theorems for natural deduction* ◇ B10 F07 F50 ◇

MEDVEDEV, YU.T. *Interpretation of logical formulae by means of finite problems (Russian)* ◇ D80 F50 ◇

MINTS, G.E. *Skolem's method of elimination of positive quantifiers in sequential calculi (Russian)* ◇ B25 C10 F50 ◇

MYHILL, J.R. *Notes towards an axiomatization of intuitionistic analysis* ◇ F35 F50 ◇

NAGASHIMA, T. *An extension of the Craig-Schuette interpolation theorem* ◇ C40 F50 ◇

NELSON, D. *Non-null implication* ◇ B20 B28 F50 ◇

ONO, K. *Taboo versus axiom* ◇ F50 ◇

RESCHER, N. *On modal renderings of intuitionistic propositional logic* ◇ B45 F50 ◇

REZNIKOFF, I. *Sur les ensembles denombrables de formules en logique intuitionniste* ◇ C40 C90 F50 ◇

SCHULTZ, KONRAD *Zur Darstellung der intuitionistischen Mathematik im Rahmen eines Beweisbarkeitskalkuels* ◇ F50 ◇

1967

CRAIG, W. *Modus ponens and derivations from Horn formulas* ◇ B20 F05 F50 ◇

DRABBE, J. *Quelques notions de la logique algebrique* ◇ F50 G15 ◇

DRAGALIN, A.G. *Constructive transfinite systems and the construction of an algorithm by transfinite recursion (Russian)* ◇ F50 ◇

DRAGALIN, A.G. *Justification of A.A.Markov's constructive selection principle (Russian)* ◇ F50 ◇

GRZEGORCZYK, A. *Non-classical propositional calculi in relation to methodological patterns of scientific investigation (Polish) (Russian and English summaries)* ◇ A05 B45 F50 ◇

IDEL'SON, A.V. *Remarks on the calculi of constructive logic with subordinate variables and the Axiom of complete induction (Russian)* ◇ F50 ◇
KAWADA, K. & MUTI, N. *A generalization of Curry's theorem* ◇ F50 ◇
KREISEL, G. *Informal rigour and completeness proofs* ◇ A05 B30 C07 E30 E50 F50 ◇
LIFSCHITZ, V. *Constructive mathematical theories consistent with classical logic (Russian)* ◇ F50 ◇
LIFSCHITZ, V. *The decision problem for some constructive theories of equality (Russian)* ◇ B25 D35 F50 ◇
LOPEZ-ESCOBAR, E.G.K. *Remarks on an infinitary language with constructive formulas* ◇ C40 C70 C75 F50 ◇
LORENTS, A.A. *Certain problems in the theory of finite probabilistic automata (Russian)* ◇ D05 F50 ◇
MASLOV, S.YU. *An invertible sequential version of the constructive predicate calculus (Russian)* ◇ F50 ◇
MCKAY, C.G. *A note on the Jaskowski sequence* ◇ B55 F50 ◇
MCKAY, C.G. *Implicationless WFFS in IC* ◇ F50 ◇
MINTS, G.E. *Analog of Herbrand's theorem for non-prenex formulas of the constructive predicate calculus (Russian)* ◇ F05 F50 ◇
MINTS, G.E. *Choice of terms in quantifier rules of the constructive predicate calculus (Russian)* ◇ F50 ◇
MINTS, G.E. & OREVKOV, V.P. *On imbedding operators (Russian)* ◇ F25 F30 F50 ◇
MINTS, G.E. *Variation in the deduction search tactics in sequential calculi (Russian)* ◇ B35 F07 F50 ◇
MOSCHOVAKIS, J.R. *Disjunction and existence in formalized intuitionistic analysis* ◇ F35 F50 ◇
PHAN DINH DIEU *A language of constructive mathematics involving systems of sets (Russian)* ◇ F50 ◇
PLIUSKEVICIUS, R. *A sequential version of the calculus of constructive logic for normal formulas without structural rules of interference (Russian)* ◇ F50 ◇
REZNIKOFF, I. *On independent recursive axiomatisation in intuitionistic logic (Russian)* ◇ F50 ◇
TAIT, W.W. *Intensional interpretation of functionals of finite type I* ◇ F10 F30 F35 F50 ◇
WANG, HAO *Introductory note* ◇ A10 F50 ◇
YANKOV, V.A. *Finite validity of formulas of a special form (Russian)* ◇ B55 C13 F50 ◇

1968

ACZEL, P. *Saturated intuitionistic theories* ◇ F30 F50 ◇
DILLER, J. *Zur Berechenbarkeit primitiv-rekursiver Funktionale endlicher Typen* ◇ F10 F15 F50 ◇
DRAGALIN, A.G. *The computability of primitive recursive terms of finite type, and primitive recursive realization (Russian)* ◇ F10 F35 F50 ◇
GHOSE, A. *Free choice sequences I,II* ◇ F50 ◇
HANATANI, Y. *Demonstration de l'ω-non-contradiction de l'arithmetique* ◇ F05 F15 F30 F50 ◇
HEYTING, A. *Intuitionism in mathematics* ◇ A05 F50 F98 ◇
HINATA, S. *Calculability of primitive recursive functionals of finite type* ◇ F10 F50 ◇
HOWARD, W.A. *Functional interpretation of bar induction by bar recursion* ◇ F10 F35 F50 ◇
JONGH DE, D.H.J. *Essay on i-valuations* ◇ F50 ◇
KIPNIS, M.M. *The constructive classification of arithmetic predicates and the semantic bases of arithmetic (Russian)* ◇ D55 F30 F50 ◇
KLEENE, S.C. *Constructive functions in "The foundations of intuitionistic mathematics"* ◇ F50 ◇
KREISEL, G. *Functions, ordinals, species* ◇ F10 F35 F50 ◇
KREISEL, G. *Lawless sequences of natural numbers* ◇ F50 ◇
KREISEL, G. & LEVY, A. *Reflection principles and their use for establishing the complexity of axiomatic systems* ◇ E30 E47 F25 F30 F35 F50 ◇
KUTSCHERA VON, F. *Die Vollstaendigkeit des Operatorensystems* $\{\neg, \wedge, \vee, \rightarrow\}$ *fuer die intuitionistische Aussagenlogik im Rahmen der Gentzensemantik* ◇ F50 ◇
LEBLANC, H. *Subformula theorems for N-sequents* ◇ F07 F50 ◇
LOPEZ-ESCOBAR, E.G.K. *A decision method for the intuitionistic theory of successor* ◇ B25 C10 F30 F50 ◇
LORENZ, K. *Dialogspiele als semantische Grundlage von Logikkalkuelen I,II* ◇ B60 F50 ◇
LORENZEN, P. *Operative Logik. Eine Uebersicht 1956-1966* ◇ A05 F50 ◇
MALMNAES, P.E. & PRAWITZ, D. *A survey of some connections between classical, intuitionistic and minimal logic* ◇ B55 F05 F25 F50 ◇
MARKOV, A.A. *An approach to constructive mathematical logic* ◇ F50 ◇
MEREDITH, C.A. & PRIOR, A.N. *Equational logic* ◇ B20 F50 G05 ◇
MINTS, G.E. *Admissible and deductive rules (Russian)* ◇ B10 F07 F50 ◇
MINTS, G.E. *Disjunctive interpretation of the LJ calculus (Russian)* ◇ F50 ◇
MINTS, G.E. *Implicative complexity of axiomatic systems (Russian)* ◇ F07 F50 ◇
MINTS, G.E. *Independence of the postulates of natural calculi (Russian)* ◇ F07 F50 ◇
MINTS, G.E. *Solvability of the problem of deducibility in LJ for a class of formulas which do not contain negative occurences of quantors (Russian)* ◇ B25 F50 ◇
MINTS, G.E. *The construction of conservative logical inferences (Russian)* ◇ F07 F50 ◇
MYHILL, J.R. *Formal systems of intuitionistic analysis I* ◇ F35 F50 ◇
MYHILL, J.R. *The formalization of intuitionism* ◇ F50 ◇
OREVKOV, V.P. *Glivenko classes of sequents (Russian)* ◇ B10 F50 ◇
OREVKOV, V.P. *Glivenko's sequence classes (Russian)* ◇ B10 B20 F50 ◇
PERLATOV, G.N. *The law of the excluded middle, and intuitionistic propositional logic (Russian)* ◇ F50 ◇
PLIUSKEVICIUS, R. *A sequential variant of constructive logic calculi for normal formulas not containing structural rules (Russian)* ◇ F50 ◇
REZNIKOFF, I. *Axiomatisation independante des ensembles denombrables de formules en logique intuitionniste* ◇ F50 ◇

ROUSSEAU, G. *Note on a problem of Porte* ◇ F50 ◇
ROUSSEAU, G. *Sheffer functions in intuitionistic logic* ◇ F50 ◇
SCHUETTE, K. *Vollstaendige Systeme modaler und intuitionistischer Logik* ◇ B45 C90 F50 F98 ◇
SCHUETTE, K. *Zur Semantik der intuitionistischen Aussagenlogik* ◇ F50 ◇
SCOTT, D.S. *Extending the topological interpretation to intuitionistic analysis I* ◇ F35 F50 ◇
SEGERBERG, K. *Propositional logics related to Heyting and Johansson* ◇ B55 F50 ◇
TAIT, W.W. *Constructive reasoning* ◇ B40 F10 F15 F30 F35 F50 F65 ◇
THOMASON, R.H. *On the strong semantical completeness of the intuitionistic predicate calculus* ◇ F50 ◇
TOMPKINS, R.R. *On Kleene's recursive realizability as an interpretation for intuitionistic elementary number theory* ◇ F30 F50 ◇
TROELSTRA, A.S. *The theory of choice sequences* ◇ F35 F50 ◇
TSEJTIN, G.S. *The disjunctive rank of the formulas of constructive arithmetic (Russian)* ◇ F30 F50 ◇
YANKOV, V.A. *On the extension of the intuitionist propositional calculus to the classical calculus, and the minimal calculus to the intuitionist calculus* ◇ B55 F50 ◇
ZEMAN, J.J. *Some calculi with strong negation primitive* ◇ B45 F50 ◇

1969

CELLUCCI, C. *Un'osservazione sul teorema di Minc-Orevkov (English summary)* ◇ F50 ◇
DRAGALIN, A.G. *Transfinite completions of constructive arithmetical calculus (Russian)* ◇ F50 ◇
FITTING, M. *Intuitionistic logic, model theory, and forcing* ◇ B98 C90 C98 E25 E35 E45 E50 F50 F98 ◇
GENTZEN, G. *The collected papers of Gerhard Gentzen* ◇ B96 F05 F07 F30 F50 F96 ◇
GOLOTA, YA.YA. *Nets of marks and deducibility in the intuitionistic propositional calculus (Russian)* ◇ B55 B75 F50 ◇
GOLOTA, YA.YA. *Some techniques for simplifying the construction of nets of marks (Russian)* ◇ B25 B75 F50 ◇
HULL, R.G. *Counterexamples in intuitionistic analysis using Kripke's schema* ◇ F35 F50 ◇
KLEENE, S.C. *Formalized recursive functionals and formalized realizability* ◇ D65 F35 F50 ◇
LEVIN, L.A. *Some syntactic theorems on the calculus of finite problems of Ju.T.Medvedev (Russian)* ◇ B55 F50 ◇
LUCKHARDT, H. *Kodifikation und Aussagenlogik* ◇ B05 B25 F50 ◇
MEDVEDEV, YU.T. *A method for proving the unsolvability of algorithmic problems (Russian)* ◇ D35 F50 ◇
MINICHIELLO, J.K. *An extension of negationless logic* ◇ F50 ◇
SCARPELLINI, B. *Some applications of Gentzen's second consistency proof* ◇ F05 F30 F35 F50 ◇
TROELSTRA, A.S. *Informal theory of choice sequences (Polish and Russian summaries)* ◇ F35 F50 F55 ◇
TROELSTRA, A.S. *Notes on the intuitionistic theory of sequences I* ◇ F35 F50 ◇
TROELSTRA, A.S. *Principles of intuitionism* ◇ F35 F50 F55 F98 ◇
YANKOV, V.A. *Conjunctively indecomposable formulas in propositional calculi (Russian)* ◇ B55 F50 G10 ◇
ZASLAVSKIJ, I.D. *The axiomatic determination of constructive objects and operations (Russian) (Armenian and English summaries)* ◇ D20 F50 ◇

1970

BERNAYS, P. *The original Gentzen consistency proof for number theory* ◇ A10 B28 F30 F50 ◇
BISHOP, E.A. *Mathematics as a numerical language* ◇ A05 F50 F55 ◇
DALEN VAN, D. & TROELSTRA, A.S. *Projections of lawless sequences* ◇ F35 F50 ◇
FEFERMAN, S. *Formal theories for transfinite iteration of generalized inductive definitions and some subsystems of analysis* ◇ F35 F50 ◇
FITTING, M. *Intuitionistic model theory and the Cohen independence proofs* ◇ A05 C90 E35 E40 F50 ◇
GABBAY, D.M. *The decidability of the Kreisel-Putnam system* ◇ B25 B55 F50 ◇
GERBER, H. *Brouwer's bar theorem and a system of ordinal notations* ◇ F15 F35 F50 ◇
GOODMAN, NICOLAS D. *A theory of constructions equivalent to arithmetic* ◇ F30 F50 ◇
HOMAGK, F. *Zur klassischen Deutung der intuitionistischen Speziestheorie und Arithmetik (Russian) (English and French summaries)* ◇ F50 ◇
HOWARD, W.A. *Assignment of ordinals to terms for primitive recursive functionals of finite type* ◇ F10 F15 F50 ◇
JONGH DE, D.H.J. *A characterization of the intuitionistic propositional calculus* ◇ F50 ◇
KABAKOV, F.A. *Intuitionistic deducibility of some realizable formulae of propositional logic (Russian)* ◇ F50 ◇
KABAKOV, F.A. *On modelling of pseudo-boolean algebras by realizability (Russian)* ◇ F30 F50 G10 ◇
KHOMICH, V.I. *On the complexity of realization of propositional formulae (Russian)* ◇ B35 F50 ◇
KREISEL, G. *Church's thesis: a kind of reducibility axiom for constructive mathematics* ◇ F50 ◇
KREISEL, G. & TROELSTRA, A.S. *Formal systems for some branches of intuitionistic analysis* ◇ F35 F50 ◇
LAEUCHLI, H. *An abstract notion of realizability for which intuitionistic predicate calculus is complete* ◇ F50 ◇
LUCKHARDT, H. *Ein Henkin-Vollstaendigkeitsbeweis fuer die intuitionistische Praedikatenlogik bezueglich der Kripke-Semantik* ◇ C90 F50 ◇
LUCKHARDT, H. *Kripke-Semantik der derivativen Praedikatenlogik* ◇ C90 F50 ◇
MAEHARA, S. *A general theory of completeness proofs* ◇ C07 C90 F50 ◇
MARKOV, A.A. *On the logic of constructive mathematics (Russian)* ◇ F50 ◇

MARTIN-LOEF, P. *Notes on constructive mathematics* ◇ F50 F60 F98 ◇

MCCALL, S. *A non-classical theory of truth, with an application to intuitionism* ◇ F50 ◇

MUTI, N. *A characterization of the intuitionistic propositional logic* ◇ F50 ◇

MYHILL, J.R. *Formal systems of intuitionistic analysis II. The theory of species* ◇ F35 F50 ◇

OSSWALD, H. *Modelltheoretische Untersuchungen in der Kripke-Semantik* ◇ C20 C90 F50 ◇

PRAWITZ, D. *Constructive semantics* ◇ F50 ◇

PRAWITZ, D. *Some results for intuitionistic logic with second order quantification rules* ◇ F05 F35 F50 ◇

ROOTSELAAR VAN, B. *On subjective mathematical assertions* ◇ F50 ◇

ROUSSEAU, G. *The separation theorem for fragments of the intuitionistic propositional calculus* ◇ B55 F50 ◇

SCARPELLINI, B. *A model for intuitionistic analysis* ◇ F35 F50 ◇

SCARPELLINI, B. *On cut elimination in intuitionistic systems of analysis* ◇ F05 F35 F50 ◇

SCOTT, D.S. *Constructive validity* ◇ F50 ◇

SCOTT, D.S. *Extending the topological interpretation to intuitionistic analysis, II* ◇ F35 F50 ◇

SEMENENKO, M.I. *Properties of some subsystems of classical and intuitionistic propositional calculi (Russian)* ◇ B55 F50 ◇

SHMAIN, I.KH. *Imbeddings of classical systems in intuitionistic and minimal ones (Russian)* ◇ F30 F50 ◇

SMIRNOV, V.A. *A syllogistics without the law of the excluded middle and its inclusion into the predicate calculus (Russian)* ◇ F50 ◇

TROELSTRA, A.S. *Notes on the intuitionistic theory of sequences II,III* ◇ F35 F50 ◇

VESLEY, R.E. *A palatable substitute for Kripke's schema* ◇ F35 F50 ◇

WOJCICKI, R. *On reconstructability of the classical propositional logic in intuitionistic logic (Russian summary)* ◇ B55 F50 ◇

1971

ASHVINIKUMAR *Another proof of the intuitionistic inconsistency of the axiom of choice (Dutch)* ◇ E25 F50 F55 ◇

BERTOLINI, F. *Kripke models and manyvalued logics* ◇ B50 F50 ◇

BOWEN, K.A. *An extension of the intuitionistic propositional calculus* ◇ B55 F50 ◇

CELLUCCI, C. *Operazioni di Brouwer e realizzabilita formalizzata (English summary)* ◇ F50 ◇

DALEN VAN, D. & GORDON, C.E. *Independence problems in subsystems of intuitionistic arithmetic* ◇ F30 F50 ◇

DILLER, J. & SCHUETTE, K. *Simultane Rekursionen in der Theorie der Funktionale endlicher Typen* ◇ F10 F35 F50 ◇

FEFERMAN, S. *Ordinals and functionals in proof theory* ◇ F10 F15 F35 F50 ◇

GABBAY, D.M. *Semantic proof of the Craig interpolation theorem for intuitionistic logic and extensions I,II* ◇ B55 C40 C90 F50 ◇

GIRARD, J.-Y. *Une extension de l'interpretation de Goedel a l'analyse et son application a l'elimination des coupures dans l'analyse et la theorie des types* ◇ F05 F10 F35 F50 ◇

GOERNEMANN, S. *A logic stronger than intuitionism* ◇ B55 C90 F50 ◇

GRZEGORCZYK, A. *Classical relativistic and constructivistic ways of assertions of theorems (Polish) (Russian and English summaries)* ◇ F50 ◇

JERVELL, H.R. *A normalform in first order arithmetic* ◇ F05 F30 F50 ◇

KIPNIS, M.M. *On the realizations of predicate formulas (Russian) (English summary)* ◇ F50 ◇

KLEMKE, D. *Ein Henkin-Beweis fuer die Vollstaendigkeit eines Kalkuels relativ zur Grzegorczyk-Semantik* ◇ B55 C90 F50 ◇

KREISEL, G. *A survey of proof theory II* ◇ F50 F98 ◇

LEIVANT, D. *A note on translations of C into I* ◇ F50 ◇

MARKOV, A.A. *Essai de construction d'une logique de la mathematique constructive* ◇ F50 ◇

MARTIN, N.M. *Systems of logic* ◇ F50 ◇

MARTIN-LOEF, P. *Hauptsatz for the intuitionistic theory of iterated inductive definitions* ◇ F05 F35 F50 ◇

MARTIN-LOEF, P. *Hauptsatz for the theory of species* ◇ F05 F35 F50 ◇

MCCULLOUGH, D.P. *Logical connectives for intuitionistic propositional logic* ◇ F50 ◇

MOSCHOVAKIS, J.R. *Can there be no nonrecursive functions?* ◇ F35 F50 ◇

MYHILL, J.R. *Embedding classical type theory in "intuitionistic" type theory* ◇ B15 F35 F50 ◇

NADIU, G.S. *Sur la logique de Heyting* ◇ C90 F50 G10 G30 ◇

PETKOV, P.P. *On the question of the possibility of introducing disjunction on the lower levels of A.A. Markov's hierarchy of mathematical logic (Russian)* ◇ B60 F50 ◇

PRAWITZ, D. *Ideas and results in proof theory* ◇ F05 F07 F10 F30 F35 F50 F98 ◇

RATSA, M.F. *A criterion for functional completeness in the intuitionistic propositional logic (Russian)* ◇ F50 ◇

REZNIKOFF, I. *Logique mathematique. II Theorie de la demonstration et intuitionnisme* ◇ F50 F98 ◇

SCARPELLINI, B. *A model for barrecursion of higher types* ◇ F35 F50 ◇

SCARPELLINI, B. *Proof theory and intuitionistic systems* ◇ F05 F15 F30 F35 F50 F98 ◇

SHIRAI, K. *Intuitionistic predicate calculus with ε-symbol* ◇ F50 ◇

SMIRNOV, V.A. *Elimination des termes ε dans la logique intuitionniste* ◇ F50 ◇

SURMA, S.J. *Jaskowski's matrix criterion for the intuitionistic propositional calculus (Polish summary)* ◇ F50 ◇

TAIT, W.W. *Normal form theorem for bar recursive functions of finite type* ◇ F05 F10 F35 F50 ◇

TAKAHASHI, MOTO-O *Cut-elimination theorem and Brouwerian-valued models for intuitionistic type theory* ◇ F05 F35 F50 ◇

THARP, L.H. *A quasi-intuitionistic set theory* ◇ E35 E70 F50 ◇

TOKARZ, M. & WOJCICKI, R. *The problem of reconstructability of propositional calculi (Polish and Russian summaries)* ◇ B22 F50 ◇

TOLSTOVA, YU.N. *A weakening of intuitionistic logic (Russian) (English summary)* ◇ B60 F50 ◇

TROELSTRA, A.S. *Computability of terms and notions of realizability for intuitionistic analysis* ◇ F35 F50 ◇

TROELSTRA, A.S. *Notions of realizability for intuitionistic arithmetic and intuitionistic arithmetic in all finite types* ◇ F30 F35 F50 ◇

VARPAKHOVSKIJ, F.L. *A class of realizable propositional formulae (Russian) (English summary)* ◇ F50 ◇

VARPAKHOVSKIJ, F.L. *The nonrealizability of the disjunction of nonrealizable formulas of propositional logic (Russian)* ◇ F50 ◇

WRONSKI, A. *Axiomatization of the implicational Goedel's matrices by Kalmar's method (Polish summary)* ◇ B55 F50 ◇

1972

ANIKEEV, A.S. *Classification of derivable propositional formulas (Russian)* ◇ B15 D05 F20 F50 ◇

DRAGALIN, A.G. *On the use of classical calculi for establishing constructive truth (Russian) (English summary)* ◇ B10 F30 F50 ◇

GABBAY, D.M. *Decidability of some intuitionistic predicate theories* ◇ B25 C90 F50 ◇

GABBAY, D.M. *Model theory for intuitionistic logic* ◇ B55 C20 C90 F50 ◇

GABBAY, D.M. *Sufficient conditions for the undecidability of intuitionistic theories with applications* ◇ D35 F50 ◇

GOODMAN, NICOLAS D. & MYHILL, J.R. *The formalization of Bishop's constructive mathematics* ◇ F35 F50 F55 ◇

GRZEGORCZYK, A. *An approach to logical calculus (Polish and Russian summaries)* ◇ B22 F50 ◇

HOMAGK, F. *Zur Axiomatisierung der Nichtidentitaeten des intuitionistischen Aussagenkalkuels (Russian, English and French summaries)* ◇ F50 ◇

HOWARD, W.A. *A system of abstract constructive ordinals* ◇ F10 F15 F35 F50 ◇

JERVELL, H.R. *On Skolem and Herbrand theorems for intuitionistic logic* ◇ F05 F50 ◇

KURKA, P. *The Heyting doctrines* ◇ F50 G30 ◇

LAMBEK, J. *Deductive systems and categories III. Cartesian closed categories, intuitionist propositional calculus, and combinatory logic* ◇ B40 F50 G30 ◇

LAWVERE, F.W. (ED.) *Toposes, algebraic geometry and logic* ◇ F50 G30 G97 ◇

LEBLANC, H. & MEYER, R.K. *Matters of separation* ◇ B10 F07 F50 ◇

LIFSCHITZ, V. *Metamathematical interpretation of the fan theorem (Russian) (English summary)* ◇ F50 ◇

LOPEZ-ESCOBAR, E.G.K. *Constructions and negationless logic (Polish and Russian summaries)* ◇ B20 E70 F50 ◇

LOPEZ-ESCOBAR, E.G.K. *Refutability and elementary number theory* ◇ F30 F50 ◇

LORENZEN, P. *Dialogkalkuele* ◇ B60 F50 ◇

LORENZEN, P. *Zur konstruktiven Deutung der semantischen Vollstaendigkeit klassischer Quantoren- und Modalkalkuele* ◇ B45 F50 ◇

MARTIN-LOEF, P. *About models for intuitionistic type theories and the notion of definitional equality* ◇ C90 F50 ◇

MARTIN-LOEF, P. *Infinite terms and a system of natural deduction* ◇ C75 F05 F50 ◇

MEDVEDEV, YU.T. *Locally finitary algorithmic problems (Russian)* ◇ D20 F30 F50 ◇

MINTS, G.E. *Derivability of addmissible rules (Russian) (English summary)* ◇ B55 F50 ◇

MINTS, G.E. *The Skolem method in intuitionistic calculi (Russian)* ◇ F50 ◇

MOTOHASHI, N. *A note on Schuette's interpolation theorem* ◇ F50 ◇

OKEE, J. *Untersuchungen ueber den einstelligen intuitionistischen Praedikatenkalkuel der ersten Stufe* ◇ F50 ◇

OREVKOV, V.P. *A specialization of the form of deductions in Gentzen calculi and its applications (Russian)* ◇ F05 F07 F50 ◇

OREVKOV, V.P. *Undecidable classes of formulas for the constructive predicate calculus I (Russian)* ◇ D35 F50 ◇

OSSWALD, H. *Homomorphie-invariante Formeln in der intuitionistischen Logik* ◇ C40 C90 F50 ◇

OSSWALD, H. *Unterstruktur-invariante Formeln in der intuitionistischen Logik* ◇ F50 ◇

OSSWALD, H. *Vollstaendigkeit und Schnittelimination in der intuitionistischen Typenlogik (English summary)* ◇ F05 F35 F50 ◇

SCARPELLINI, B. *Formally constructive model for barrecursion of higher types* ◇ F35 F50 ◇

SCARPELLINI, B. *Induction and transfinite induction in intuitionistic systems* ◇ F35 F50 ◇

SHIRAI, K. *On the intuitionistic predicate calculus with the ε-symbol* ◇ F50 ◇

STENLUND, S. *Combinators, λ-terms and proof theory* ◇ B40 F05 F10 F50 F98 ◇

TAUTS, A. *A connection between generalized Beth models and topological pseudo-Boolean algebras (Russian) (Estonian and German summaries)* ◇ C90 F50 G10 ◇

TAUTS, A. *A semantic interpretation of formulas in generalized Beth models and in pseudo-boolean algebras (Russian) (Estonian and German summaries)* ◇ C90 F50 G10 ◇

VESLEY, R.E. *Choice sequences and Markov's principle* ◇ F35 F50 ◇

VIER, L.C. *Church's thesis in northern Dutch constructivism* ◇ D20 F50 ◇

ZARNECKA-BIALY, E. *The deduction theorems for propositional calculi when implication and falsum is present (Polish summary)* ◇ B20 F50 ◇

1973

BARENDREGT, H.P. *Combinatory logic and the axiom of choice* ◇ B40 E25 F50 ◇

CHERNIAVSKY, J.C. *The complexity of some non-classical logics* ◇ B45 D15 F20 F50 ◇

DALEN VAN, D. *Lectures on intuitionism* ◇ F50 F55 F98 ◇
DRAGALIN, A.G. *Constructive mathematics and models of intuitionistic theories* ◇ F50 ◇
DRAGALIN, A.G. *Intuitionistic model theory (Russian)* ◇ C90 F50 ◇
ERSHOV, YU.L. *Skolem functions and constructive models (Russian)* ◇ C57 C65 F50 ◇
FITTING, M. *Model existence theorems for modal and intuitionistic logics* ◇ B45 F50 ◇
FRIEDMAN, H.M. *Some applications of Kleene's methods for intuitionistic systems* ◇ F35 F50 ◇
FRIEDMAN, H.M. *The consistency of classical set theory relative to a set theory with intuitionistic logic* ◇ B15 E30 E35 F50 ◇
GABBAY, D.M. *A survey of decidability results for modal, tense and intermediate logics* ◇ B25 B45 B55 C98 F50 ◇
GABBAY, D.M. *The undecidability of intuitionistic theories of algebraically closed fields and real closed fields* ◇ B25 C60 C90 D35 F50 ◇
GEORGIEVA, N.V. *On the logical signs in Schuette's interpolation formulae (Bulgarian) (Russian and English summaries)* ◇ C40 F50 ◇
GIMON, V.V. *The dependence of the second level rules of inference in Markov's stratified semantic system (Russian)* ◇ F50 ◇
GIRARD, J.-Y. *Quelques resultats sur les interpretations fonctionelles* ◇ F10 F35 F50 ◇
GOODMAN, NICOLAS D. *The arithmetic theory of constructions* ◇ F30 F50 ◇
GOODMAN, NICOLAS D. *The faithfulness of the interpretation of arithmetic in the theory of constructions* ◇ F30 F50 ◇
HOWARD, W.A. *Appendix: Hereditarily majorizable functionals of finite type* ◇ E25 F10 F35 F50 ◇
IBRAGIMOV, S.G. *Consequence logic of Paul Lorenzen (Russian)* ◇ B60 F50 ◇
KABZINSKI, J.K. *The Wajsberg's results connected with separability of the intuitionistic propositional logic* ◇ F50 ◇
KANOVICH, M.I. *Irreducibility of languages of the stepped semantic system (Russian)* ◇ D15 D30 F50 ◇
KHOMICH, V.I. *The complexity of recognition of the realizability of logico-arithmetic formulas (Russian)* ◇ F30 F50 ◇
KLEENE, S.C. *Realizability: a retrospective survey* ◇ F50 ◇
LORENZ, K. *Rules versus theorems. A new approach for mediation between intuitionistic and two-valued logic* ◇ A05 B55 F50 ◇
LUCKHARDT, H. *Extensional Goedel functional interpretation. A consistency proof of classical analysis* ◇ F10 F35 F50 F98 ◇
MARTIN-LOEF, P. *Hauptsatz for intuitionistic simple type theory* ◇ B15 F05 F35 F50 ◇
MEDVEDEV, YU.T. *An interpretation of intuitionistic number theory* ◇ F15 F30 F50 ◇
MEYER, R.K. *Intuitionism, entailment, negation* ◇ B46 F50 ◇
MOSCHOVAKIS, J.R. *A topological interpretation of second-order intuitionistic arithmetic* ◇ F35 F50 ◇
MOTOHASHI, N. *A faithful interpretation of intuitionistic predicate logic in classical predicate logic* ◇ F50 ◇
MYHILL, J.R. *A note on indicator functions* ◇ F50 ◇
MYHILL, J.R. *Embedding classical logic in intuitionistic logic* ◇ F50 ◇
MYHILL, J.R. *Some properties of intuitionistic Zermelo-Fraenkel set theory* ◇ E35 E70 F50 ◇
NELSON, D. *A complete negationless system (Polish and Russian summaries)* ◇ F50 ◇
NEPEJVODA, N.N. *The relationship between predicative significance and the intuition of universality (Russian)* ◇ F50 F65 ◇
OSIUS, G. *The internal and external aspect of logic and set theory in elementary topoi* ◇ C90 E70 E75 F50 G30 ◇
OSSWALD, H. *Ein syntaktischer Beweis fuer die Zuverlaessigkeit der Schnittregel im Kalkuel von Schuette fuer die intuitionistische Typenlogik* ◇ F05 F35 F50 ◇
PETKOV, P.P. *The possibility of introducing disjunction at the level ω in A.A. Markov's graduated semantic system* ◇ F50 ◇
PLISKO, V.E. *On realizable predicate formulae (Russian)* ◇ F50 ◇
POHLERS, W. *Ein starker Normalisationssatz fuer die intuitionistische Typentheorie* ◇ F05 F35 F50 ◇
RADU, E. *On Goedel's argument for the completeness of intuitionistic logic (Romanian) (French summary)* ◇ F50 ◇
REICHBACH, J. *Generalized models for intuitionistic and classical predicate calculi with ultraproducts* ◇ C20 C25 C80 C90 F50 ◇
SANCHIS, L.E. *Formally defined operations in Kripke models* ◇ C90 F50 ◇
SCARPELLINI, B. *On barinduction of higher types for decidable predicates* ◇ F35 F50 ◇
SCHROETER, K. *Interpretation der intuitionistischen Mathematik mit Hilfe des Beweisbarkeitsbegriffs* ◇ A05 F50 ◇
SHANIN, N.A. *On a hierarchy of methods of interpreting propositions in constructive mathematics (Russian)* ◇ F50 ◇
SMORYNSKI, C.A. *Applications of Kripke models* ◇ F30 F50 ◇
SMORYNSKI, C.A. *Elementary intuitionistic theories* ◇ B25 C90 D35 F50 ◇
SMORYNSKI, C.A. *Investigations of intuitionistic formal systems by means of Kripke models* ◇ B55 C90 F50 ◇
SMULLYAN, R.M. *A generalization of intuitionistic and modal logics* ◇ B45 F50 ◇
STAPLES, J. *Combinator realizability of constructive finite type analysis* ◇ B40 F35 F50 ◇
SURMA, S.J. *Jaskowski's matrix criterion for the intuitionistic propositional calculus* ◇ B55 F50 ◇
TROELSTRA, A.S. *Intuitionistic formal systems* ◇ F30 F35 F50 ◇
TROELSTRA, A.S. (ED.) *Metamathematical investigation of intuitionistic arithmetic and analysis* ◇ F05 F10 F30 F35 F50 F55 F98 ◇

TROELSTRA, A.S. *Models and computability* ◇ F05 F10 F30 F35 F50 ◇
TROELSTRA, A.S. *Normalization theorems for systems of natural deduction* ◇ F05 F30 F35 F50 ◇
TROELSTRA, A.S. *Notes on intuitionistic second order arithmetic* ◇ F35 F50 ◇
TROELSTRA, A.S. *Realizability and functional interpretations* ◇ F10 F30 F35 F50 ◇
VERKHOZINA, M.I. *The undecidability of the separation problem for positive fragments of logical calculi (Russian)* ◇ D35 F50 ◇
WASILEWSKA, A. *The diagrams of formulas of the intuitionistic propositional calculus (Polish and Russian summaries)* ◇ F50 ◇
ZARNECKA-BIALY, E. *Modal functors and their definability in propositional calculus systems (Polish) (English summary)* ◇ B45 F50 ◇
ZARNECKA-BIALY, E. *Negation in Ch.S.Peirce's propositional calculus* ◇ A10 B20 F50 ◇
ZASLAVSKIJ, I.D. *On predicate and arithmetic calculi of symmetric constructive logic (Russian)* ◇ F50 ◇

1974

BARTHELEMY, J.-P. *Sur la refutabilite* ◇ B20 F50 G30 ◇
BOWEN, K.A. *Systems of transfinite type theory based on intuitionistic and modal logics* ◇ B15 F05 F35 F50 ◇
CELLUCCI, C. *On the role of reducibility principles* ◇ A05 E30 F50 ◇
CELLUCCI, C. *Un connettivo per la logica intuizionista (English summary)* ◇ F50 ◇
DALEN VAN, D. *A model for HAS. A topological interpretation of the theory of species of natural numbers* ◇ F35 F50 ◇
DILLER, J. & NAHM, W. *Eine Variante zur Dialectica-Interpretation der Heyting-Arithmetik endlicher Typen* ◇ F10 F35 F50 ◇
DRAGALIN, A.G. *Constructive models of the theories of intuitionistic sequences of sampling (Russian)* ◇ F50 ◇
DRAGALIN, A.G. *Constructive model of intuitionistic analysis (Russian)* ◇ F35 F50 ◇
DRAGALIN, A.G. *Die intuitionistische Logik und das ε-Symbol Hilberts (Russian)* ◇ F50 ◇
DRAGALIN, A.G. *The completeness of an arithmetic with a constructive rule of infinite induction (Russian)* ◇ F30 F50 ◇
EPSTEIN, G. & HORN, A. *Propositional calculi based on subresiduation* ◇ B50 F50 G10 ◇
ERSHOV, YU.L. *The model G of the theory BR (Russian)* ◇ F10 F50 ◇
ERSHOV, YU.L. *Theory of numerations, III: Constructive models (Russian)* ◇ B25 C57 C98 D45 F50 ◇
FRIEDMAN, H.M. *On existence proofs of Hanf numbers* ◇ C55 C75 E30 E70 F50 ◇
GABBAY, D.M. *On 2nd order intuitionistic propositional calculus with full comprehension* ◇ F35 F50 ◇
GEISER, J.R. *A formalization of Essenin-Volpin's proof theoretical studies by means of nonstandard analysis* ◇ C75 F50 H10 H15 ◇
GENTZEN, G. *Ueber das Verhaeltnis zwischen intuitionistischer und klassischer Arithmetik* ◇ B30 F30 F50 ◇
GIMON, V.V. *Version of the semantical stratified system for formulas without quantifiers (Russian)* ◇ F50 ◇
GOBLE, L.F. *Gentzen systems for modal logic* ◇ B45 F50 ◇
GOLOTA, YA.YA. *A matrix notation for nets of marks (Russian) (English summary)* ◇ B25 F50 ◇
GORGY, F.W. *The dependence and independence of the rules of inference of the step-by-step semantic system of constructive mathematical logic (Russian)* ◇ F50 ◇
KANOVICH, M.I. *A certain extension of the step-by-step semantic system of A. A. Markov (Russian)* ◇ D15 F50 ◇
KREJNOVICH, V.YA. *From what does the law of the excluded middle follow? (Russian) (English summary)* ◇ F35 F50 ◇
LOPEZ-ESCOBAR, E.G.K. *Elementary interpretations of negationless arithmetic* ◇ F30 F50 ◇
MARKOV, A.A. *On the language* Я_0 *(Russian)* ◇ B25 F50 F65 ◇
MARKOV, A.A. *On the language* Я_1 *(Russian)* ◇ D03 D35 F50 ◇
MARKOV, A.A. *On the language* Я_2 *(Russian)* ◇ F50 ◇
MARKOV, A.A. *On the language* Я_3 *(Russian)* ◇ F50 ◇
MARKOV, A.A. *On the languages* $\text{Я}_4, \text{Я}_5, \ldots$ *(Russian)* ◇ F50 ◇
MARKOV, A.A. *On the language* Я_ω *(Russian)* ◇ F50 ◇
MARKOV, A.A. *On the language* $\text{Я}_{\omega|}$ *(Russian)* ◇ F50 ◇
MARKOV, A.A. *The completeness of the classical predicate calculus in constructive logic (Russian)* ◇ C07 F50 ◇
MINTS, G.E. *E theorems (Russian) (English summary)* ◇ F05 F30 F50 ◇
MINTS, G.E. *Heyting predicate calculus with ε symbol (Russian) (English summary)* ◇ F05 F50 ◇
MULVEY, C.J. *Intuitionistic algebra and representations of rings* ◇ C60 C90 F50 F55 G30 ◇
MYHILL, J.R. *Embedding classical type theory in "intuitionistic" type theory, a correction* ◇ B15 F35 F50 ◇
PLISKO, V.E. *A certain formal system that is connected with realizability (Russian)* ◇ F50 ◇
PLISKO, V.E. *On interpretations of predicate formulae that are connected with constructive logic (Russian)* ◇ F50 ◇
PLISKO, V.E. *Recursive realizability and constructive predicate logic (Russian)* ◇ F50 ◇
POREBSKA, M. & WRONSKI, A. *A characterization of fragments of the intuitionistic propositional logic* ◇ F50 ◇
RASIOWA, H. *An algebraic approach to non-classical logics* ◇ B98 F50 G05 G10 G20 G25 G98 ◇
RAUSZER, C. *A formalization of the propositional calculus of H-B logic* ◇ B55 F50 G10 ◇
RAUSZER, C. *Semi-boolean algebras and their applications to intuitionistic logic with dual operations* ◇ F50 G10 ◇

REYES, G.E. *From sheaves to logic* ◇ C90 F50 G30 ◇
ROUTLEY, R. *Semantical analyses of propositional systems of Fitch and Nelson* ◇ B25 B46 C90 F50 ◇
SCARPELLINI, B. *Disjunctive properties of intuitionistic systems* ◇ F50 ◇
SEGERBERG, K. *Proof of a conjecture of McKay* ◇ B55 F50 ◇
SHANIN, N.A. *The hierarchy of constructive Brouwer functionals (Russian) (English summary)* ◇ F50 ◇
SHIMADA, K. *A theorem prover for intuitionistic propositional logic* ◇ B35 F50 ◇
SHMAIN, I.KH. *Extended calculus of recursive functions I (Russian)* ◇ D20 D75 F50 ◇
SMYTH, M.B. *Involution as a basis for propositional calculi* ◇ B05 F50 ◇
STAPLES, J. *Combinator realizability of constructive Morse theory* ◇ B40 F35 F50 ◇
SURMA, S.J. & WRONSKI, A. & ZACHOROWSKI, S. *On Jaskowski-type semantics for the intuitionistic propositional logic* ◇ B55 F50 ◇
SUSZKO, R. *A note on intuitionistic calculus (ISC)* ◇ F50 G05 ◇
SZABO, M.E. *A categorical equivalence of proofs* ◇ F05 F07 F50 G05 G30 ◇
TAKAMATSU, T. *Elements of linear algebra (Japanese)* ◇ F50 ◇
TAUTS, A. *A formal deduction of tautological formulas in pseudo-boolean algebras (Russian) (Estonian and German summaries)* ◇ B60 C90 F50 G05 G10 ◇
TAUTS, A. *A game for the construction of a propositional semantics in generalized Beth models (Russian) (Estonian and German summaries)* ◇ C75 C90 E60 F50 ◇
TROELSTRA, A.S. *Non-extensional equality* ◇ F50 ◇
TROELSTRA, A.S. *Note on the fan theorem* ◇ F30 F35 F50 ◇
UMEZAWA, T. *Disjunction property in higher order number theory with intuitionistic rules of inference I* ◇ F05 F35 F50 ◇
URQUHART, A.I.F. *Implicational formulas in intuitionistic logic* ◇ F50 G05 G25 ◇
VAKARELOV, D. *Representation theorems for semi-boolean algebras and semantics for Heyting-Brouwer predicate logic* ◇ F50 G05 G10 ◇
WRONSKI, A. *On cardinalities of matrices strongly adequate for the intuitionistic propositional logic* ◇ B55 F50 ◇
WRONSKI, A. & ZYGMUNT, J. *Remarks on the free pseudo-boolean algebra with one-element free-generating set* ◇ F50 G10 ◇
WRONSKI, A. *The degree of completeness of some fragments of the intuitionistic propositional logic* ◇ B20 F50 ◇
ZUCKER, J.I. *The correspondence between cut-elimination and normalization I,II* ◇ F05 F30 F50 ◇

1975

BEESON, M.J. *The nonderivability in intuitionistic formal systems of theorems on the continuity of effective operations* ◇ D20 F50 ◇
BUCHHOLZ, W. *Ein ausgezeichnetes Modell fuer die intuitionistische Typenlogik* ◇ B15 F05 F35 F50 ◇
DIACONESCU, R. *Axiom of choice and complementation* ◇ E25 F50 G30 ◇
DILLER, J. & VOGEL, HELMUT *Intensionale Funktionalinterpretation der Analysis* ◇ F10 F35 F50 ◇
DUMMETT, M. *The justification of deduction* ◇ A05 F50 ◇
DZIK, W. *On structural completeness of some nonclassical predicate calculi* ◇ B22 F50 ◇
FEFERMAN, S. *A language and axioms for explicit mathematics* ◇ D55 F35 F50 F65 ◇
FRIEDMAN, H.M. *The disjunction property implies the numerical existence property* ◇ F50 ◇
GABBAY, D.M. *Decidability results in non-classical logics I* ◇ B25 B45 C90 F50 ◇
GABBAY, D.M. *The decision problem for some finite extensions of the intuitionistic theory of abelian groups* ◇ B25 F50 ◇
GARGOV, G.K. *Intuitionistic analysis, and intermediate logics (Russian)* ◇ B55 F50 ◇
HANATANI, Y. *Calculability of the primitive recursive functionals of finite type over the natural numbers* ◇ D65 F10 F35 F50 ◇
JOYAL, A. *Les theoremes de Chevalley-Tarski et remarque sur l'algebre constructive* ◇ C60 F50 G30 ◇
KABZINSKI, J.K. & WRONSKI, A. *On equivalential algebras* ◇ F50 G25 ◇
KABZINSKI, J.K. & POREBSKA, M. *Proof of the separability of the intuitionistic propositional logic by the Wajsberg method* ◇ F50 ◇
KANOVICH, M.I. *A hierachical semantic system with set variables (Russian)* ◇ B20 B65 C40 D05 F50 ◇
KHOMICH, V.I. *Weakly and strongly nonrealizable propositional formulae (Russian)* ◇ B20 D15 D20 F50 ◇
KOCK, A. & LECOUTURIER, P. & MIKKELSEN, C.J. *Some topos theoretic concepts of finiteness* ◇ F50 G30 ◇
KREISEL, G. *Observations on a recent generalization of completeness theorems due to Schuette* ◇ A05 C07 C57 F05 F20 F35 F50 ◇
KREISEL, G. & MINTS, G.E. & SIMPSON, S.G. *The use of abstract language in elementary metamathematics: Some pedagogic examples* ◇ A05 C07 C57 C75 F05 F07 F20 F50 ◇
LEIVANT, D. *Strong normalization for arithmetic (Variations on a theme of Prawitz.)* ◇ F05 F30 F50 ◇
LOPEZ-ESCOBAR, E.G.K. & VELDMAN, W. *Intuitionistic completeness of a restricted second-order logic* ◇ B15 C90 F50 ◇
LUCKHARDT, H. *The real elements in a consistency proof for simple type theory I* ◇ F10 F35 F50 ◇
MAC LANE, S. *Sets, topoi, and internal logic in categories* ◇ A05 E70 F50 G30 ◇
MARINI, D. & MIGLIOLI, P.A. & ORNAGHI, M. *First order logic as a tool to solve and classify problems* ◇ B10 B25 F50 ◇
MARTIN-LOEF, P. *About models for intuitionistic type theories and the notion of definitional equality* ◇ B40 F35 F50 ◇

MARTIN-LOEF, P. *An intuitionistic theory of types: predicative part* ◇ F05 F35 F50 ◇
MEZHLUMBEKOVA, V.F. *Cut-elimination in a system of negationless arithmetic (Russian) (English summary)* ◇ B20 F05 F30 F50 ◇
MEZHLUMBEKOVA, V.F. *Deductive capabilities of negationless intuitionistic arithmetic (Russian) (English summary)* ◇ F30 F50 ◇
MEZHLUMBEKOVA, V.F. *On systems of the negationless calculus of predicates (Russian) (English summary)* ◇ B20 F50 ◇
MINTS, G.E. *Finite investigations of transfinite derivations (Russian) (English summary)* ◇ F05 F07 F10 F20 F30 F35 F50 ◇
MINTS, G.E. *Proof theory (arithmetic and analysis) (Russian)* ◇ F05 F10 F30 F35 F50 F98 ◇
MINTS, G.E. *Transfinite expansions of arithmetic formulas (Russian) (English summary)* ◇ F30 F50 ◇
MOISIL, G.C. *Sur une logique intuitionniste* ◇ F50 ◇
MYHILL, J.R. *Constructive set theory* ◇ E70 F50 ◇
NEPEJVODA, N.N. *The language Δ with intuitionistic connectives (Russian)* ◇ E70 F50 ◇
OKEE, J. *A semantical proof of the undecidability of the monadic intuitionistic predicate calculus of first order* ◇ D35 F50 ◇
OSIUS, G. *A note on Kripke-Joyal semantics for the internal language of topoi* ◇ C90 F50 G30 ◇
OSIUS, G. *Logical and set theoretical tools in elementary topoi* ◇ C90 E70 F50 G30 ◇
OSSWALD, H. *Ueber Skolemerweiterungen in der intuitionistischen Logik mit Gleichheit* ◇ F50 ◇
POHLERS, W. *An upper bound for the provability of transfinite induction in systems with N-times iterated inductive definitions* ◇ F15 F35 F50 ◇
POREBSKA, M. & WRONSKI, A. *A characterization of fragments of the intuitionistic propositional logic* ◇ B55 F50 ◇
POWELL, W.C. *Extending Goedel's negative interpretation to ZF* ◇ E35 E70 F50 ◇
PREVIALE, F. *Tavole semantiche per sistemi astratti di logica estensionale* ◇ B50 F50 ◇
RADU, E. *Le "sujet creatif" dans les mathematiques intuitionnistes* ◇ A05 F50 ◇
REICHBACH, J. *Generalized models for classical and intuitionistic predicate calculi* ◇ B10 B25 C90 F50 ◇
REYES, G.E. *Faisceaux et concepts* ◇ C90 F50 G30 ◇
SCARPELLINI, B. *Bemerkungen zu Regel und Schema* ◇ F07 F30 F50 ◇
SCHWICHTENBERG, H. *Elimination of higher type levels in definitions of primitive recursive functionals by means of transfinite recursion* ◇ C75 F10 F15 F35 F50 ◇
SHIRAI, K. *Intuitionistic version of the Los-Tarski-Robinson theorem* ◇ C40 F05 F50 ◇
SONOBE, O. *Bi-relational frameworks for minimal and intuitionistic logics* ◇ B45 C90 F50 ◇
STENLUND, S. *Description in intuitionistic logic* ◇ F50 ◇
STOUT, L.N. *Topological space objects in a topos. II: $\mathfrak{E}$-completeness and $\mathfrak{E}$-cocompleteness* ◇ E75 F50 G30 ◇
SURMA, S.J. & WRONSKI, A. & ZACHOROWSKI, S. *On Jaskowski-type semantics for the intuitionistic propositional logic* ◇ B55 F50 ◇
TAIT, W.W. *A realizability interpretation of the theory of species* ◇ D65 F05 F35 F50 ◇
TAUTS, A. *A semantic model for infinite formulas (Russian) (Estonian and German summaries)* ◇ C75 C90 F50 G10 ◇
TAUTS, A. *Search for deduction by means of a semantic model (Russian) (Estonian and German summaries)* ◇ B55 C75 C90 F50 ◇
TAUTS, A. *Strong tautology and construction of a countermodel for nonderivable propositions (Russian) (Estonian and German summaries)* ◇ C75 C90 F50 ◇
TROELSTRA, A.S. *Markov's principle and Markov's rule for theories of choice sequences* ◇ F35 F50 ◇
UMEZAWA, T. *Disjunction property in higher order number theory with intuitionistic rules of inference II* ◇ F05 F35 F50 ◇
WAUW-DE KINDER VAN DE, G. *Arithmetique de premier ordre dans les topos (English summary)* ◇ F30 F50 G30 H15 ◇
YEVICK, M.L. *Holographic or Fourier logic* ◇ F50 ◇
ZACHOROWSKI, S. *A proof of a conjecture of Roman Suszko* ◇ B45 B50 F50 ◇
ZASLAVSKIJ, I.D. *On constructive truth of judgements and some untraditional systems of constructive logic (Russian)* ◇ F50 ◇

1976

BEESON, M.J. *Derived rules of inference related to the continuity of effective operations* ◇ D20 F50 ◇
BEESON, M.J. *The unprovability in intuitionistic formal systems of the continuity of effective operations on the reals* ◇ F50 ◇
BERNINI, S. *A very strong intuitionistic theory* ◇ F35 F50 ◇
BOWEN, K.A. *An Herbrand theorem for prenex formulas of LJ* ◇ F05 F50 ◇
CHERNYAKHOVSKIJ, N.P. *The expressibility of realizability, in the language of formal arithmetic (Russian)* ◇ D80 F30 F50 ◇
EPSTEIN, G. & HORN, A. *Logics which are characterized by subresiduated lattices* ◇ B50 F50 G10 ◇
FELSCHER, W. *On interpolation when function symbols are present* ◇ C40 C75 F50 ◇
FISCHER SERVI, G. *Un'algebrizzazione del calcolo intuizionista monadico (English summary)* ◇ B55 F50 G10 ◇
GABBAY, D.M. *Completeness properties of Heyting's predicate calculus with respect to RE models* ◇ B55 F50 ◇
GABBAY, D.M. *On Kreisel's notion of validity in Post systems* ◇ B50 D03 F50 ◇
GIRARD, J.-Y. *Three-valued logic and cut-elimination: The actual meaning of Takeuti's conjecture* ◇ B50 C85 C90 F05 F35 F50 ◇
GOODMAN, NICOLAS D. *The theory of the Goedel functionals* ◇ F10 F35 F50 ◇

HERMES, H. *Dialog games* ◇ A05 B60 F50 ◇
JONGH DE, D.H.J. & SMORYNSKI, C.A. *Kripke models and the intuitionistic theory of species* ◇ C90 F35 F50 ◇
KOCK, A. *Universal projective geometry via topos theory* ◇ E75 F50 G30 ◇
KROL', M.D. *The topological models of intuitionistic analysis. One counter-example (Russian)* ◇ C90 F35 F50 ◇
LEIVANT, D. *Failure of completeness properties of intuitionistic predicate logic for constructive models* ◇ F50 ◇
LEVIN, L.A. *On the principle of conservation of information in intuitionistic mathematics (Russian)* ◇ D15 D80 F50 F55 ◇
LOEB, M.H. *Embedding first order predicate logic in fragments of intuitionistic logic* ◇ B10 D35 F50 ◇
LOPEZ-ESCOBAR, E.G.K. *On an extremely restricted ω-rule* ◇ F05 F15 F30 F50 ◇
LUCKHARDT, H. *Ueber das Markov-Prinzip. I,II* ◇ F50 ◇
MARKOV, A.A. *An attempt to construct a logic of constructive mathematics (Russian)* ◇ F50 ◇
MEYER, R.K. *Metacompleteness* ◇ B45 F50 ◇
MORGAN, C.G. *Many-valued propositional intuitionism* ◇ B50 F50 ◇
MORGAN, C.G. *Methods for automated theorem proving in nonclassical logics* ◇ B35 F50 ◇
NAGORNYJ, N.M. *A variant of the definition of realization of a logical arithmetical formula (Russian)* ◇ F30 F50 ◇
OKEE, J. *A species algebraic interpretation of the intuitionistic propositional calculus* ◇ F50 ◇
OKEE, J. *Completeness of the algebra of species* ◇ F50 G15 ◇
OKEE, J. *On the independence of the fundamental operations of the algebra of species* ◇ F50 ◇
OREVKOV, V.P. *Solvable classes of pseudoprenex formulas (Russian) (English summary)* ◇ B25 F50 ◇
PLISKO, V.E. *Some variants of the notion of realizability for predicate formulas (Russian)* ◇ F50 ◇
POSY, C.J. *Varieties of indeterminacy in the theory of general choice sequences* ◇ F35 F50 ◇
POTTINGER, G. *A new way of normalizing intuitionist propositional logic* ◇ B40 F05 F50 ◇
RAUSZER, C. *On the strong semantical completeness of any extension of the intuitionistic predicate calculus* ◇ B55 F50 ◇
SAMBIN, G. *An effective fixed-point theorem in intuitionistic diagonalizable algebras (The algebraisation of the theories which express Theor. IX)* ◇ F30 F50 G10 G25 ◇
SHANIN, N.A. *On the quantifier of limiting realizability (Russian)* ◇ C57 C80 F50 ◇
SKVORTSOV, D.P. *The occurrence of an implication in finitely valid, intuitively unprovable formulas of propositional logic (Russian)* ◇ F50 ◇
STOUT, L.N. *Topological properties of the real numbers object in a topos* ◇ E70 E75 F50 G30 ◇
SWART DE, H.C.M. *Another intuitionistic completeness proof* ◇ F50 ◇
SZABO, M.E. *Quantifier complete categories* ◇ F05 F50 G10 G30 ◇
TOKARZ, M. *Definitions by context in propositional calculi* ◇ F50 ◇
ULRICH, D. *On a property of matrices for subsystems of IC^+* ◇ B20 F50 ◇
VELDMAN, W. *An intuitionistic completeness theorem for intuitionistic predicate logic* ◇ F50 ◇

1977

ACZEL, P. *The strength of Martin-Loef's intuitionistic type theory with one universe* ◇ F35 F50 ◇
BEESON, M.J. *Continuity and comprehension in intuitionistic formal systems* ◇ F50 ◇
BEESON, M.J. *Principles of continuous choice and continuity of functions in formal systems for constructive mathematics* ◇ F35 F50 ◇
BEL, M. *An intuitionistic combinatory theory not closed under the rule of choice* ◇ B40 F50 ◇
BERNINI, S. *Interpretazione intuizionista di teorie a logica classica; una particolare applicazione del metodo della traduzione negativa via concetti empirici e anomici (lawless)* ◇ F35 F50 ◇
BESSONOV, A.V. *New operations in intuitionistic calculus (Russian)* ◇ B55 F50 ◇
BLOOM, S.L. *A note on Ψ-consequences* ◇ B22 F50 ◇
BOZZI, S. & MELONI, G.C. *Ideal properties and intuitionistic algebra* ◇ C60 C90 F50 F55 ◇
CONSTABLE, R.L. *A constructive programming logic* ◇ F50 ◇
DALEN VAN, D. *The use of Kripke's schema as a reduction principle* ◇ F35 F50 ◇
DARDZHANIYA, G.K. *Intuitionistic system without contraction* ◇ B55 F50 ◇
DUMMETT, M. *Elements of intuitionism* ◇ F50 F55 F98 ◇
ERSHOV, YU.L. *Model $\mathbb{C}$ of partial continuous functionals* ◇ D45 D65 F10 F50 ◇
FISCHER SERVI, G. *On modal logic with an intuitionistic base* ◇ B45 F50 ◇
FOURMAN, M.P. *The logic of topoi* ◇ F50 G30 ◇
FRIEDMAN, H.M. *On the derivability of instantiation properties* ◇ F50 ◇
FRIEDMAN, H.M. *Set theoretic foundations for constructive analysis* ◇ E70 F50 ◇
GABBAY, D.M. *A new version of Beth semantics for intuitionistic logic* ◇ C90 F50 ◇
GABBAY, D.M. *Craig interpolation theorem for intuitionistic logic and extensions part III* ◇ B55 C40 C90 F50 ◇
GABBAY, D.M. *On some new intuitionistic propositional connectives I* ◇ B45 B55 F50 ◇
GABBAY, D.M. *Undecidability of intuitionistic theories formulated with the apartness relation* ◇ D35 F50 ◇
GAUTHIER, Y. *Intuitionistic logic and local mathematical theories* ◇ F50 ◇
GIRARD, J.-Y. *Functional interpretation and Kripke models* ◇ C90 F10 F30 F50 ◇
GORDEEV, L.N. *A majorizing semantics for hyperarithmetic sentences (Russian) (English summary)* ◇ D55 F15 F50 ◇

GORDEEV, L.N. *An approach to the constructivization of Cantor's set theory (Russian) (English summary)* ◇ E70 F50 ◇
GUREVICH, Y. *Intuitionistic logic with strong negation* ◇ A05 F50 ◇
HAYASHI, S. *On derived rules of intuitionistic second order arithmetic* ◇ F05 F35 F50 ◇
HAYASHI, S. *Some derived rules of intuitionistic second order arithmetic* ◇ F05 F35 F50 ◇
HINATA, S. *A normalization theorem in formal theories of natural numbers* ◇ F05 F30 F50 ◇
HYLAND, J.M.E. *Aspects of constructivity in mathematics* ◇ F50 ◇
JOHNSTONE, P.T. *Rings, fields, and spectra* ◇ C90 F50 G30 ◇
JUKNA, S. *On cut-elimination in Hoare's system (Russian)* ◇ B75 F05 F50 ◇
KHACHATRYAN, M.A. *On certain theorems of analysis in the formal system Kleene-Vesley (Russian)* ◇ F50 ◇
KOCK, A. & REYES, G.E. *Doctrines in categorical logic* ◇ F50 G30 ◇
KROL', M.D. *Disjunctive and existential properties of intuitionistic analysis with Kripke's scheme (Russian)* ◇ C90 F35 F50 ◇
MACNAB, D.S. *Some applications of double-negation sheafification* ◇ C30 C90 F50 G30 ◇
MAINZER, K. *Is the intuitionistic bar induction a constructive principle ?* ◇ A05 F50 ◇
MAKKAI, M. & REYES, G.E. *First order categorical logic. Model-theoretical methods in the theory of topoi and related categories* ◇ C75 C90 F50 G30 ◇
MARKOV, A.A. *On a semantical language hierarchy in a constructive mathematical logic* ◇ F50 ◇
MARKOVIC, Z. *Reduced products of saturated intuitionistic theories* ◇ C20 C90 F50 ◇
MEYER, R.K. *First degree formulas in Curry's LD* ◇ B46 F50 ◇
MINTS, G.E. *Heytings's formal system (Russian)* ◇ F50 ◇
MOTOHASHI, N. *Partially ordered interpretations* ◇ C90 E40 F25 F50 ◇
NEPEJVODA, N.N. *Level Beth models and realizability (Russian)* ◇ F30 F50 ◇
ORLOWSKA, E. *On a problem posed by Horn* ◇ F50 ◇
PLISKO, V.E. *The nonarithmeticity of the class of realizable predicate formulas (Russian)* ◇ D35 D55 F50 ◇
POPOV, S.V. *The complexity of derivations in intuitionistic propositional calculus (Russian)* ◇ F20 F50 ◇
POTTINGER, G. *Normalization as a homomorphic image of cut-elimination* ◇ F05 F07 F50 ◇
RAUSZER, C. *An algebraic approach to the Heyting-Brouwer predicate calculus* ◇ F50 G10 ◇
RAUSZER, C. *Applications of Kripke models to Heyting-Brouwer logic* ◇ B55 C90 F50 ◇
RAUSZER, C. *Model theory for an extension of intuitionistic logic* ◇ B55 C90 F50 ◇
RAUSZER, C. *The Craig interpolation theorem for an extension of intuitionistic logic* ◇ F50 ◇
REYES, G.E. *Sheaves and concepts: a model-theoretic interpretation of Grothendieck topoi* ◇ C75 C90 F50 G30 ◇
ROMANSKI, J. *On connections between algebraic and Kripke semantics* ◇ C90 F50 G10 G25 ◇
SCARPELLINI, B. *A new realizability notion for intuitionistic analysis* ◇ F35 F50 ◇
SCHLOMIUK, D.I. *Logique des topos* ◇ E70 F50 G30 ◇
SHANIN, N.A. *On the quantifier of limiting realizability* ◇ B55 C57 C80 F50 ◇
SMORYNSKI, C.A. *On axiomatizing fragments* ◇ C90 F50 ◇
SOBOLEV, S.K. *On finite-dimensional superintuitionistic logics (Russian)* ◇ B55 D05 F50 G10 G25 ◇
SOBOLEV, S.K. *The intuitionistic propositional calculus with quantifiers (Russian)* ◇ B55 C80 C90 D35 F50 ◇
SWART DE, H.C.M. *An intuitionistically plausible interpretation of intuitionistic logic* ◇ C90 F50 ◇
SZABO, M.E. *The logic of closed categories* ◇ F50 G30 ◇
TROELSTRA, A.S. *A note on non-extensional operations in connection with continuity and recursiveness* ◇ F35 F50 ◇
TROELSTRA, A.S. *Aspects of constructive mathematics* ◇ F30 F35 F50 F55 F60 F98 ◇
TROELSTRA, A.S. *Axioms for intuitionistic mathematics incompatible with classical logic* ◇ A05 F35 F50 ◇
TROELSTRA, A.S. *Choice sequences, a chapter of intuitionistic mathematics* ◇ A05 A10 F35 F50 F55 F98 ◇
TROELSTRA, A.S. *Completeness and validity for intuitionistic predicate logic* ◇ C90 F50 ◇
TROELSTRA, A.S. *Some models for intuitionistic finite type arithmetic with fan functional* ◇ F35 F50 ◇
TROELSTRA, A.S. *Special instances of generalized continuity which are conservative over intuitionistic arithmetic* ◇ F35 F50 ◇
TSITKIN, A.I. *On admissible rules of intuitionistic propositional logic (Russian)* ◇ B55 F50 ◇
VAKARELOV, D. *Notes on $\mathcal{N}$-lattices and constructive logic with strong negation* ◇ B55 F50 G10 ◇

1978

ACZEL, P. *The type theoretic interpretation of constructive set theory* ◇ E70 F35 F50 ◇
ARRUDA, A.I. *Some remarks on Griss' logic of negationless intuitionistic mathematics* ◇ B55 F50 F55 G25 ◇
BEESON, M.J. *A type-free Goedel interpretation* ◇ F50 ◇
BEESON, M.J. *Some relations between classical and constructive mathematics* ◇ F50 F55 ◇
BERNINI, S. *A note on my paper: "A very strong intuitionistic theory"* ◇ F35 F50 ◇
BRIDGES, D.S. *A note on Morse's lambda-notation in set theory* ◇ F50 ◇
BUCHHOLZ, W. & POHLERS, W. *Provable well orderings of formal theories for transfinitely iterated inductive definitions* ◇ F15 F30 F35 F50 ◇
CRISCUOLO, G. & TORTORA, R. *Duals of intuitionistic tableaus* ◇ F50 ◇
DALEN VAN, D. *An interpretation of intuitionistic analysis* ◇ C90 F35 F50 ◇
DRUGUSH, YA.M. *A class of logics without disjunction property (Russian)* ◇ B55 F50 ◇

FIDEL, M.M. *An algebraic study of a propositional system of Nelson* ◇ B55 F50 G10 G25 ◇

FISCHER SERVI, G. *The finite model property for MIPQ and some consequences* ◇ B55 F50 G10 ◇

FRIEDMAN, H.M. *Classically and intuitionistically provably recursive functions* ◇ B15 D20 E70 F30 F50 ◇

GOAD, C.A. *Monadic infinitary propositional logic: a special operator* ◇ B25 B60 C75 F50 ◇

GOLDBLATT, R.I. *Arithmetical necessity, provability and intuitionistic logic* ◇ B45 F50 ◇

GOODMAN, NICOLAS D. & MYHILL, J.R. *Choice implies excluded middle* ◇ E25 F50 ◇

GOODMAN, NICOLAS D. *Relativized realizability in intuitionistic arithmetic of all finite types* ◇ F35 F50 ◇

GOODMAN, NICOLAS D. *The nonconstructive content of sentences of arithmetic* ◇ D20 D30 D55 F30 F50 ◇

GREENLEAF, N. *Linear order in lattices: A constructive study* ◇ F50 F55 ◇

HABERTHUER, R. *Choice sequences and reduction processes* ◇ F05 F35 F50 ◇

HART, A.M. & HENDRY, H.E. *Some observations on a method of McKinsey* ◇ B50 F50 ◇

HAYASHI, S. *Existence property by means of a normalization method* ◇ F05 F50 ◇

JERVELL, H.R. *Constructive universes. I* ◇ F05 F15 F35 F50 ◇

KEARNS, J.T. *Intuitionist logic, a logic of justification* ◇ A05 C90 F50 ◇

KHACHATRYAN, M.A. *On the derivation complexity of some formulas in sequential calculi (Russian)* ◇ F20 F50 ◇

KHOSROVSHAHI, G.B. *A few words on intuitionism (Persian)* ◇ F50 ◇

KROL', M.D. *A topological model for intuitionistic analysis with Kripke's scheme* ◇ C90 F35 F50 ◇

KROL', M.D. *Distinct variants of Kripke's scheme in intuitionistic analysis (Russian)* ◇ C90 F35 F50 ◇

LUCKHARDT, H. *A fundamental effect in computations on real numbers* ◇ F50 ◇

NADEL, M.E. *Infinitary intuitionistic logic from a classical point of view* ◇ C75 C90 F50 ◇

NEPEJVODA, N.N. *A relation between the natural deduction rules and operators of higher level algorithmic languages (Russian)* ◇ B75 F07 F50 ◇

NEPEJVODA, N.N. *The construction of correct programs (Russian)* ◇ F50 ◇

PLISKO, V.E. *Some variants of the notion of realizability for predicate formulas (Russian)* ◇ B10 F50 ◇

REYES, G.E. *Theorie des modeles et faisceaux* ◇ C90 F50 G30 ◇

ROUSSEAU, C. *Topos theory and complex analysis* ◇ F50 G30 H05 ◇

ROUTLEY, R. *An inadequacy in Kripke-semantics for intuitionistic quantificational logic* ◇ B55 C90 F50 ◇

SCOTT, P.J. *The "Dialectica" interpretation and categories* ◇ F10 F35 F50 G30 ◇

SMORYNSKI, C.A. *The axiomatization problem for fragments* ◇ B55 F50 ◇

STAPLES, J. *Truth in constructive metamathematics* ◇ F50 ◇

STOUT, L.N. *Quels sont les espaces topologiques dans les topos?* ◇ F50 G30 ◇

SWART DE, H.C.M. *First steps in intuitionistic model theory* ◇ C90 F50 ◇

SZABO, M.E. *Algebra of proofs* ◇ F05 F07 F50 F98 G30 ◇

TRAGESSER, R.S. & ZUCKER, J.I. *The adequacy problem for inferential logic* ◇ A05 B55 F50 ◇

TROELSTRA, A.S. *A. Heyting on the formalization of intuitionistic mathematics* ◇ A10 F50 ◇

TROELSTRA, A.S. *Appendix to D. van Dalen's "An interpretation of intuitionistic analysis"* ◇ F35 F50 ◇

TROELSTRA, A.S. *Some remarks on the complexity of Henkin-Kripke models* ◇ C90 D55 F50 ◇

VOGEL, HELMUT *Eine beweistheoretische Anwendung partieller stetiger Funktionale* ◇ F10 F35 F50 ◇

WENDEL, N. *The inconsistency of Bernini's very strong intuitionistic theory* ◇ B55 F35 F50 ◇

ZASLAVSKIJ, I.D. *Symmetric constructive logic (Russian)* ◇ B55 F05 F30 F50 ◇

1979

AKIMOV, A.P. *On imbeddings of classical modal calculi in constructive calculi (Russian)* ◇ B45 F50 ◇

BEESON, M.J. *Continuity in intuitionistic set theories* ◇ E35 E70 F50 ◇

BEESON, M.J. *Goodman's theorem and beyond* ◇ F35 F50 ◇

DALEN VAN, D. & STATMAN, R. *Equality in the presence of apartness* ◇ F05 F50 ◇

DALEN VAN, D. *Interpreting intuitionistic logic* ◇ A05 F50 F98 ◇

DARDZHANIYA, G.K. *On the complexity of countermodels for intuitionistic propositional calculus (Russian) (English summary)* ◇ B55 F50 ◇

DILLER, J. *Functional interpretations of Heyting's arithmetic in all finite types* ◇ F10 F35 F50 ◇

DOMBROVSKIJ-KABANCHENKO, M.N. *A transfinite extension of the stepped semantic system (Russian)* ◇ D55 F50 ◇

DRAGALIN, A.G. *Algebraic models of intuitionistic theories (Russian)* ◇ F50 G10 ◇

DRAGALIN, A.G. *An algebraic approach to intuitionistic models of the realizability type (Russian)* ◇ F50 G25 ◇

DRAGALIN, A.G. *Calculus of constructions, equivalent to the intuitionistic predicate calculus (Russian)* ◇ F50 ◇

DRAGALIN, A.G. *Intuitionism (Russian)* ◇ A05 F50 F98 ◇

DRAGALIN, A.G. *Mathematical intuitionism. Introduction to proof theory (Russian)* ◇ F05 F30 F35 F50 F98 ◇

DRAGALIN, A.G. *Strong normalization theorem for derivations in Gentzen's sequent calculus (Russian)* ◇ F05 F50 ◇

FEFERMAN, S. *Constructive theories of functions and classes* ◇ F50 F55 F60 F98 ◇

FOURMAN, M.P. & HYLAND, J.M.E. *Sheaf models for analysis* ◇ C65 C90 F35 F50 G10 G30 ◇

FOURMAN, M.P. & SCOTT, D.S. *Sheaves and logic* ◇ C90 C98 E70 F35 F50 G30 ◇
GIMON, V.V. *Quantifier-free formulas in A. A. Markov's ramified semantics (Russian)* ◇ B55 F50 ◇
GOLDBLATT, R.I. *Topoi. The categorial analysis of logic* ◇ B98 C98 E98 F35 F50 F98 G30 ◇
GORGY, F.W. *The independence of the rule of syllogism in* S_2 ◇ B45 F50 ◇
GOTO, S. *Program synthesis through Goedel's interpretation* ◇ B75 F10 F50 ◇
GRAYSON, R.J. *Heyting-valued models for intuitionistic set theory* ◇ E35 E70 F50 G30 ◇
GUPTA, H.N. *On a minimal system of intuitionistically acceptable propositional calculus* ◇ F50 ◇
HEIJENOORT VAN, J. *Introduction a la semantique des logiques non-classiques* ◇ B45 C90 F50 ◇
HOEVEN VAN DER, G.F. & TROELSTRA, A.S. *Projections of lawless sequences II* ◇ F35 F50 ◇
HYLAND, J.M.E. *Continuity in spatial toposes* ◇ F50 G30 ◇
JOHNSTONE, P.T. *Another condition equivalent to De Morgan's law* ◇ C60 C90 F50 G05 G30 ◇
JOHNSTONE, P.T. *Conditions related to de Morgan's law* ◇ B55 C65 C90 F50 G30 ◇
KIRK, R.E. *Some classes of Kripke frames characteristic for the intuitionistic logic* ◇ B55 C90 F50 ◇
KOCK, A. & REYES, G.E. *Connections in formal differential geometry* ◇ F50 G30 ◇
LEIVANT, D. *Absoluteness of intuitionistic logic* ◇ F05 F30 F50 ◇
LIFSCHITZ, V. CT_0 *is stronger than* CT_0*!* ◇ F30 F50 ◇
LIFSCHITZ, V. *An intuitionistic definition of classical natural numbers* ◇ F30 F50 ◇
MARKOVIC, Z. *An intuitionistic omitting types theorem* ◇ C07 C90 F50 ◇
MEREDITH, D. *Axiomatics for implication* ◇ B20 F50 ◇
MINTS, G.E. *Constructive logic (Russian)* ◇ F50 ◇
MINTS, G.E. *Constructive semantics (Russian)* ◇ F50 ◇
NEPEJVODA, N.N. *Application of proof theory to the problem of construction of correct programs (Russian)* ◇ B75 F50 ◇
NEPEJVODA, N.N. *Stable truth and computability (Russian)* ◇ C90 F35 F50 ◇
OREVKOV, V.P. *Three ways of recognizing unessential formulas in sequences (Russian) (English summary)* ◇ F50 ◇
PRANK, R.K. *Expressibility in the elementary theory of recursive sets with realizability logic (Russian) (English summary)* ◇ B60 D20 F50 ◇
PRAWITZ, D. *Proofs and the meaning and completeness of the logical constants* ◇ A05 F07 F50 ◇
ROUSSEAU, C. *Topos theory and complex analysis* ◇ F50 F55 G30 H05 ◇
SCHWICHTENBERG, H. *Logic and the axiom of choice* ◇ E25 F05 F50 ◇
SCHWICHTENBERG, H. *On bar recursion of types 0 and 1* ◇ F10 F35 F50 ◇
SCOTT, D.S. *Identity and existence in intuitionistic logic* ◇ F35 F50 ◇
SHVARTS, G.F. *Some extensions of intuitionistic type theory (Russian) (English summary)* ◇ B15 F35 F50 ◇
SKVORTSOV, D.P. *Logic of infinite problems and Kripke models on atomic semilattices of sets (Russian)* ◇ B55 C90 F50 G20 ◇
SKVORTSOV, D.P. *Two generalizations of the concept of a finite problem (Russian)* ◇ F50 ◇
SMIRNOV, V.A. *Theory of quantification and* $\mathfrak{E}$*-calculi* ◇ B10 F50 ◇
SOLOV'EV, S.V. *The increase in length of an L-derivation transformed into a natural deduction (Russian) (English summary)* ◇ F07 F20 F50 ◇
STATMAN, R. *Intuitionistic propositional logic is polynomial-space complete* ◇ B40 D15 F20 F50 ◇
STEIN, M. *Interpretationen der Heyting-Arithmetik endlicher Typen* ◇ F10 F35 F50 ◇
TAKEUTI, G. & TITANI, S. *Heyting valued set theory (Japanese)* ◇ E40 E70 F50 G30 ◇
TROELSTRA, A.S. *On Ashvinikumar's principle of microscopic completeness* ◇ F35 F50 ◇
TSITKIN, A.I. *On the question of an error in a famous work due to Wajsberg (Russian)* ◇ A10 B20 F50 ◇
UMEZAWA, T. *A method for cut elimination in intuitionistic predicate logic and classical predicate logic* ◇ B10 F05 F50 ◇
VOGEL, HELMUT *Ueber ein mit der Bar-Induktion verwandtes Schema* ◇ F10 F35 F50 ◇
WEINSTEIN, S. *Some applications of Kripke models to formal systems of intuitionistic analysis* ◇ C90 F35 F50 ◇

1980

ARTEMOV, S.N. *Arithmetically complete modal theories (Russian)* ◇ B45 C90 F50 ◇
BEESON, M.J. *Extensionality and choice in constructive mathematics* ◇ E25 E70 F50 ◇
BEESON, M.J. *Some problems in constructive mathematics* ◇ A10 F50 ◇
BOZZI, S. & MELONI, G.C. *Representation of Heyting algebras with covering and propositional intuitionistic logic with local operator* ◇ B25 B55 C90 F50 G10 ◇
COSTE-ROY, M.-F. & COSTE, M. & MAHE, L. *Contribution to the study of the natural number object in elementary topoi* ◇ D20 F35 F50 G30 ◇
DILLER, J. *Modified realization and the formulae-as-types notion* ◇ F05 F10 F30 F35 F50 ◇
DRAGALIN, A.G. *Higher-order predicate logic in the form of calculus realization (Russian)* ◇ B15 F05 F35 F50 ◇
DRAGALIN, A.G. *New forms of realizability and Markov's rule (Russian)* ◇ F30 F50 ◇
FOURMAN, M.P. *Sheaf models for set theory* ◇ C62 C90 E25 E35 E70 F50 G30 ◇
FRIEDRICH, W. & LUCKHARDT, H. *Intuitionistic uniformity principles for propositions and some applications* ◇ F35 F50 ◇
GANDY, R.O. *Proofs of strong normalization* ◇ B40 F05 F50 ◇

GOAD, C.A. *Proofs as descriptions of computation* ◇ B35 F07 F50 ◇
HAYASHI, S. *Derived rules related to a constructive theory of metric spaces in intuitionistic higher order arithmetic without countable choice* ◇ F35 F50 G30 ◇
HOWARD, W.A. *Ordinal analysis of terms of finite type* ◇ F10 F15 F35 F50 ◇
HOWARD, W.A. *Ordinal analysis of bar recursion of type zero* ◇ F10 F15 F35 F50 ◇
HOWARD, W.A. *The formulae-as-types notion of construction* ◇ B40 F05 F10 F50 ◇
HYLAND, J.M.E. & JOHNSTONE, P.T. & PITTS, A.M. *Tripos theory* ◇ E70 F50 G30 ◇
JONGH DE, D.H.J. *A class of intuitionistic connectives* ◇ B55 C75 C90 F50 ◇
KALICKI, C. *Infinitary propositional intuitionistic logic* ◇ B55 C75 F50 ◇
KHAKHANYAN, V.KH. *Comparative strength of variants of Church's thesis at the level of set theory (Russian)* ◇ D20 E70 F50 ◇
KHAKHANYAN, V.KH. *The consistency of intuitionistic set theory with Church's principle and the uniformization principle (Russian)* ◇ D20 E35 E70 F25 F50 ◇
KHAKHANYAN, V.KH. *The consistency of intuitionistic set theory with formal mathematical analysis (Russian)* ◇ D20 E35 E70 F25 F35 F50 ◇
KIRK, R.E. *A characterization of the classes of finite tree frames which are adequate for the intuitionistic logic* ◇ B55 C90 F50 ◇
KODERA, H. *Remark on classical logic and intuitionistic logic* ◇ B20 F05 F50 ◇
LAMBEK, J. *From types to sets* ◇ F35 F50 G30 ◇
LAMBEK, J. & SCOTT, P.J. *Intuitionist type theory and the free topos* ◇ F35 F50 G30 ◇
LEIVANT, D. *Innocuous substitutions* ◇ F30 F50 ◇
LOPEZ-ESCOBAR, E.G.K. *Semantical models for intuitionistic logics* ◇ C90 C95 F50 ◇
LUCKHARDT, H. *On constructive functions ranging over propositions* ◇ F50 ◇
MATSUMOTO, K. & SHIRAI, K. *A note on elimination of function symbols in the intuitionistic predicate calculus* ◇ F50 ◇
MEREDITH, D. *A positive logic proof procedure* ◇ B20 B35 B40 F50 ◇
MOSCHOVAKIS, J.R. *Kleene's realizability and "divides" notions for formalized intuitionistic mathematics* ◇ A10 F50 ◇
POPOV, S.V. & ZAKHAR'YASHCHEV, M.V. *On the power of countermodels in intuitionistic calculus (Russian)* ◇ C90 F50 ◇
PRANK, R.K. *Semantics of realizability for a language with variables for recursively enumerable sets (Russian)* ◇ B60 D25 F35 F50 ◇
RAUSZER, C. *An algebraic and Kripke-style approach to a certain extension of intuitionistic logic* ◇ B55 C20 C90 F50 G25 ◇
SCHULTZ, KONRAD *A topological model for Troelstra's system CS of intuitionistic analysis* ◇ C90 E70 F35 F50 ◇
SCOTT, D.S. *Relating theories of the λ-calculus* ◇ B40 C90 F50 G30 ◇
SHVARTS, G.F. *The existence property with parameters for some extensions of the intuitionistic theory of types (Russian)* ◇ F35 F50 ◇
SKVORTSOV, D.P. *On the connection of finitary general validity of certain propositional formulas with derivability in the Kreisel-Putnam logic (Russian)* ◇ B55 F50 ◇
SOYKA, D. *Metamathematische Methoden in der konstruktiven Mathematik (Russian) (English and French summaries)* ◇ F50 ◇
STEIN, M. *Interpretations of Heyting's arithmetic - An analysis by means of a language with set symbols* ◇ F10 F30 F50 ◇
TOSI, P. *Normal derivability and first-order arithmetic* ◇ F05 F30 F50 ◇
TROELSTRA, A.S. *Extended bar induction of type zero* ◇ F35 F50 ◇
TROELSTRA, A.S. *Intuitionistic extensions of the reals* ◇ F35 F50 F55 ◇
UMEZAWA, T. *Cut elimination in intuitionistic and some intermediate predicate logics* ◇ B55 F05 F50 ◇
VESLEY, R.E. *Intuitionistic analysis: the search for axiomatization and understanding* ◇ A10 F35 F50 ◇
WRAITH, G.C. *Intuitionistic algebra: some recent developments in topos theory* ◇ F50 G30 ◇

1981

ASANIDZE, G.Z. *A dialogue justification of logic (Russian)* ◇ B60 F50 F65 ◇
BEESON, M.J. *Formalizing constructive mathematics: why and how?* ◇ A05 F50 ◇
BOILEAU, A. & JOYAL, A. *La logique des topos* ◇ F50 G30 ◇
BORICIC, B.R. *Equational reformulations of intuitionistic propositional calculus and classical first-order predicate calculus* ◇ B55 F50 ◇
BROUWER, L.E.J. *Brouwer's Cambridge lectures on intuitionism* ◇ A05 A10 F50 F55 F98 ◇
BUCHHOLZ, W. & FEFERMAN, S. & POHLERS, W. & SIEG, W. *Iterated inductive definitions and subsystems of analysis: recent proof-theoretical studies* ◇ F05 F15 F35 F50 F98 ◇
BUCHHOLZ, W. *Ordinal analysis of ID_ν* ◇ F15 F35 F50 ◇
BUCHHOLZ, W. *The $\Omega_{\mu+1}$-rule* ◇ F05 F15 F35 F50 ◇
BURGESS, J.P. *The completeness of intuitionistic propositional calculus for its intended interpretation* ◇ B55 F50 ◇
CELLUCCI, C. *A calculus of constructions as a representation of intuitionistic logical proofs* ◇ F50 ◇
DAHN, B.I. *Partial isomorphisms and intuitionistic logic* ◇ C75 C90 F50 ◇
DOSEN, K. *A reduction of classical propositional logic to the conjunction-negation fragment of an intuitionistic relevant logic* ◇ B46 B55 F50 ◇
DOSEN, K. *Minimal modal systems in which Heyting and classical logic can be embedded* ◇ B45 F50 ◇

FEFERMAN, S. & SIEG, W. *Inductive definitions and subsystems of analysis* ◇ F35 F50 ◇
FEFERMAN, S. & SIEG, W. *Proof-theoretic equivalence between classical and constructive theories for analysis* ◇ F35 F50 ◇
FISCHER SERVI, G. *Semantics for a class of intuitionistic modal calculi* ◇ B45 B55 F50 ◇
GABBAY, D.M. *Semantical investigations in Heyting's intuitionistic logic* ◇ B55 F50 F98 ◇
GAVRILENKO, YU.V. *Recursive realizability from the intuitionistic point of view (Russian)* ◇ F35 F50 ◇
GEL'FOND, M.G. *A class of theorems with valid constructive counterparts* ◇ F50 ◇
GIARETTA, P. & MARTINO, E. *Brouwer, Dummett and the bar theorem* ◇ A10 F50 ◇
GIELEN, W. & SWART DE, H.C.M. & VELDMAN, W. *The continuum hypothesis in intuitionism* ◇ E15 E50 F50 ◇
GOLDBLATT, R.I. *Grothendieck topology as geometric modality* ◇ B25 B45 C90 F50 G30 ◇
GORGY, F.W. *L'independance de quelques regles de deduction du systeme $S_{\omega+n}$ de A. A. Markov* ◇ F50 ◇
GORGY, F.W. & SAHYOUN, A.H. *Mutual transformability of the formulas of the basic languages of constructive mathematical logic* ◇ F50 ◇
GRAYSON, R.J. *Concepts of general topology in constructive mathematics and in sheaves* ◇ C90 E35 E70 E75 F35 F50 F55 G30 ◇
HAYASHI, S. *On set theory in toposes* ◇ E70 F50 G30 ◇
HOWARD, W.A. *Computability of ordinal recursion of type level two* ◇ D20 F10 F15 F35 F50 ◇
HOWARD, W.A. *Ordinal analysis of simple cases of bar recursion* ◇ F10 F15 F35 F50 ◇
HYLAND, J.M.E. *Function spaces in the category of locales* ◇ F50 G30 ◇
KHAKHANYAN, V.KH. *The consistency of some intuitionistic and constructive principles with a set theory* ◇ D20 E35 E70 F50 ◇
KOCK, A. *Synthetic differential geometry* ◇ F50 G30 ◇
KREISEL, G. *Monadic operators defined by means of propositional quantification in intuitionistic logic* ◇ F50 ◇
LAMBEK, J. & SCOTT, P.J. *Independence of premisses and the free topos* ◇ F35 F50 G30 ◇
LAMBEK, J. & SCOTT, P.J. *Intuitionist type theory and foundations* ◇ F35 F50 ◇
LEIVANT, D. *Implicational complexity in intuitionistic arithmetic* ◇ F30 F50 ◇
LOPEZ-ESCOBAR, E.G.K. *Equivalence between semantics for intuitionism I* ◇ F50 ◇
LOPEZ-ESCOBAR, E.G.K. *Integrating intuitionistic and classical theories* ◇ F50 ◇
LOPEZ-ESCOBAR, E.G.K. *On the interpolation theorem for the logic of constant domains* ◇ B55 C40 C90 F50 ◇
LOPEZ-ESCOBAR, E.G.K. *Variations on a system of Gentzen* ◇ B55 F50 ◇
MIGLIOLI, P.A. & ORNAGHI, M. *A logically justified model of computation. I,II* ◇ B75 D80 F07 F50 ◇
MIGLIOLI, P.A. & MOSCATO, U. & ORNAGHI, M. *Trees in Kripke models and in an intuitionistic refutation system* ◇ C90 F50 ◇
MORICONI, E. *Sulla completezza del calcolo dei predicati intuizionista* ◇ F50 ◇
MOSCHOVAKIS, J.R. *A disjunctive decomposition theorem for classical theories* ◇ F35 F50 ◇
MURAVITSKIJ, A.YU. *Strong equivalence on an intuitionistic Kripke model and assertorically equivolumetric logics (Russian)* ◇ B55 C90 F50 ◇
NEGRI, M. *L'aritmetizzazione della computabilita dei funzionali di tipo finito* ◇ F10 F35 F50 ◇
PEACOCKE, C. *Hacking on logic: two comments* ◇ A05 F50 ◇
PENON, J. *Infinitesimaux et intuitionnisme* ◇ E75 F50 G30 ◇
PLIUSKEVICIUS, R. *On the Gentzen type proof theory for program analysis* ◇ B75 F07 F35 F50 ◇
POHLERS, W. *Proof-theoretical analysis of ID_ν by the method of local predicativity* ◇ F05 F15 F35 F50 ◇
POSY, C.J. & SWART DE, H.C.M. *Validity and quantification in intuitionism* ◇ A05 F50 ◇
PRANK, R.K. *Expressibility in the elementary theory of recursively enumerable sets with realizability logic (Russian)* ◇ B60 D25 F50 ◇
PRAWITZ, D. *Validity and normalizability of proofs in 1st and 2nd order classical and intuitionistic logic* ◇ F05 F07 F35 F50 ◇
RENARDEL DE LAVALETTE, G.R. *The interpolation theorem in fragments of logics* ◇ B20 B55 C40 F50 ◇
REYES, G.E. *Logic and category theory* ◇ F35 F50 G30 ◇
SCEDROV, A. *Consistency and independence results in intuitionistic set theory* ◇ E35 E70 F50 G30 ◇
SHANIN, N.A. *Role of a notion algorithm in the arithmetic language semantics* ◇ B75 F30 F50 ◇
SHIMODA, M. *Categorical aspects of Heyting-valued models for intuitionistic set theory* ◇ C90 E40 E70 F50 G30 ◇
SIEG, W. *Inductive definitions, constructive ordinals, and normal derivations* ◇ F05 F15 F35 F50 ◇
SPRING, D. *The Riemann integral in constructive mathematics* ◇ F50 ◇
STEIN, M. *A general theorem on existence theorems* ◇ B40 F10 F35 F50 ◇
TAKEUTI, G. *Logic and set theory* ◇ B51 B98 E40 E70 F50 ◇
TAUTS, A. *The connection between semantic models and pseudo-boolean algebras (Russian) (Estonian and German summaries)* ◇ C75 C90 F50 G10 ◇
TROELSTRA, A.S. *Lawless sequences and their uses* ◇ F35 F50 ◇
TROELSTRA, A.S. *On a second order propositional operator in intuitionistic logic* ◇ B55 F35 F50 ◇
TROELSTRA, A.S. *The interplay between logic and mathematics: Intuitionism* ◇ A05 F50 F98 ◇
UESU, T. *Intuitionistic theories and toposes* ◇ E70 F50 G30 ◇

UESU, T. *Intuitionistic theories and topoi (Japanese)* ◇ E70 F50 G30 ◇

1982

ACZEL, P. *The type theoretic interpretation of constructive set theory: choice principles* ◇ E25 E70 F35 F50 ◇

BEESON, M.J. *Problematic principles in constructive mathematics* ◇ F50 ◇

BEESON, M.J. *Recursive models for constructive set theories* ◇ E70 F35 F50 ◇

BOZZI, S. *Formule attuali ed ereditarie nella logica intuizionista* ◇ C90 F50 ◇

DALEN VAN, D. & LODDER, J.S. *Lawlessness and independence* ◇ F35 F50 ◇

FEFERMAN, S. *Monotone inductive definitions* ◇ F35 F50 ◇

FOURMAN, M.P. & GRAYSON, R.J. *Formal spaces* ◇ F50 G30 ◇

FOURMAN, M.P. *Notions of choice sequence* ◇ F35 F50 G30 ◇

FOURMAN, M.P. & SCEDROV, A. *The "world's simplest axiom of choice" fails* ◇ E25 E35 F50 G30 ◇

GABBAY, D.M. *Intuitionistic basis for non-monotonic logic* ◇ B55 C90 F50 ◇

GEORGACARAKOS, G.N. *The semantics of minimal intuitionism* ◇ B55 F50 ◇

GIRARD, J.-Y. *Proof-theoretic investigations of inductive definitions I* ◇ F05 F35 F50 ◇

GORDEEV, L.N. *Constructive models for set theory with extensionality* ◇ C62 E70 F50 ◇

GRAYSON, R.J. *Concepts of general topology in constructive mathematics and in sheaves II* ◇ E75 F50 F55 G30 ◇

GRAYSON, R.J. *Constructive well-orderings* ◇ E10 E70 F50 ◇

HAYASHI, S. *A note on the bar induction rule* ◇ F35 F50 ◇

HOEVEN VAN DER, G.F. *An application of projections of lawless sequences* ◇ F35 F50 ◇

HOEVEN VAN DER, G.F. *Models for the theory of lawless sequences constructed from a single generator* ◇ F35 F50 ◇

HOEVEN VAN DER, G.F. *Projections of lawless sequences* ◇ F35 F50 ◇

HYLAND, J.M.E. *Applications of constructivity* ◇ F35 F50 ◇

HYLAND, J.M.E. *The effective topos* ◇ D75 D80 F35 F50 G30 ◇

JAEGER, G. *Iterating admissibility in proof theory* ◇ F35 F50 ◇

JOHNSTONE, P.T. *Stone spaces* ◇ F50 G30 ◇

JONGH DE, D.H.J. *Formulas of one propositional variable in intuitionistic arithmetic* ◇ F30 F50 ◇

KREISEL, G. & MACINTYRE, A. *Constructive logic versus algebraization I* ◇ A05 F05 F50 ◇

LIFSCHITZ, V. *Constructive assertions in an extension of classical mathematics* ◇ F50 ◇

MARCHINI, C. *Realizations and witnesses for Kripke models* ◇ C90 F50 G30 ◇

MARTIN-LOEF, P. *Constructive mathematics and computer programming* ◇ B75 F35 F50 ◇

MARTINO, E. *Creative subject and bar theorem* ◇ F50 ◇

MINTS, G.E. & TYUGU, E.KH. *The completeness of structural synthesis rules (Russian)* ◇ B20 B75 D20 F50 ◇

MIZUTANI, C. *Definability theorem for the intuitionistic predicate logic with equality* ◇ F50 ◇

MOERDIJK, I. *Glueing topoi and higher-order disjunction and existence* ◇ F35 F50 G30 ◇

MOERDIJK, I. *Some topological spaces which are universal for intuitionistic predicate logic* ◇ F50 ◇

MOTOHASHI, N. *An elimination theorem of uniqueness conditions in the intuitionistic predicate calculus* ◇ F50 ◇

NEPEJVODA, N.N. *Constructive logics (Russian)* ◇ B60 C90 C95 F50 ◇

RATSA, M.F. *Functional completeness in intuitionistic propositional logic (Russian)* ◇ F50 ◇

SCEDROV, A. & SCOTT, P.J. *A note on the Friedman slash and Freyd covers* ◇ F35 F50 G30 ◇

SCEDROV, A. *Independence of the fan theorem in the presence of continuity principles* ◇ F35 F50 G30 ◇

SCHROEDER-HEISTER, P. *Logische Konstanten und Regeln. Zur Deutung von Aussagenoperatoren* ◇ F07 F50 ◇

SEELY, R.A.G. *Locally Cartesian closed categories and type theory 1* ◇ B15 F35 F50 G30 ◇

SMORYNSKI, C.A. *Nonstandard models and constructivity* ◇ C62 F30 F50 ◇

TROELSTRA, A.S. *Intuitionistic extensions of the reals II* ◇ F35 F50 F55 ◇

TROELSTRA, A.S. & DALEN VAN, D. (EDS.) *The L.E.J. Brouwer centenary symposium* ◇ A05 A10 B97 F50 F55 F97 ◇

UMEZAWA, T. *Definability and strict definability of logical symbols in intuitionistic predicate logics of first and higher order* ◇ F50 ◇

VELDMAN, W. *On the constructive contrapositions of two axioms of countable choice* ◇ E25 E70 F50 F55 ◇

VISSER, A. *On the completeness principle: a study of provability in Heyting's arithmetic and extensions* ◇ F30 F50 ◇

1983

ARZARELLO, F. *Un' interpretazione categoriale degli oggetti anomici* ◇ F35 F50 G30 ◇

BLASS, A.R. *Words, free algebras, and coequalizers* ◇ C05 C75 E25 E35 F50 F55 G30 ◇

BORICIC, B.R. *Equational reformulation of the Heyting first-order predicate calculus* ◇ F50 ◇

BOZIC, M. & DOSEN, K. *Axiomatizations of intuitionistic double negation* ◇ B55 F50 ◇

BOZZI, S. *Principi ideali e logica intuizionista* ◇ A05 F50 ◇

CONSTABLE, R.L. *Constructive mathematics as a programming logic I: some principles of theory* ◇ B75 F50 ◇

FABIANO, A. *Completezza della logica intuizionista rispetto a modelli con coperture (English summary)* ◇ F50 ◇

FITTING, M. *Proof methods for modal and intuitionistic logics* ◇ B45 B55 F07 F50 F98 ◇

FRIEDMAN, H.M. & SCEDROV, A. *Set existence property for intuitionistic theories with dependent choice* ◇ E70 F35 F50 ◇

GIARETTA, P. *On the construction of sequences in the theory of the creative subject (Italian)* ◇ F50 ◇

GORGY, F.W. & SAHYOUN, A.H. *Mutual transformability of the formulas of the languages of Markov $L_{\omega N}$ and A_{N1}* ◇ F50 ◇

GORGY, F.W. & SAHYOUN, A.H. *Transformability of the formulas of the languages of Markov L_N, L_ω into formulas of the language A_2* ◇ F50 ◇

GRAYSON, R.J. *Forcing in intuitionistic systems without power-set* ◇ E40 E70 F35 F50 ◇

GRAYSON, R.J. *On closed subsets of the intuitionistic reals* ◇ F50 ◇

GRAYSON, R.J. & MOERDIJK, I. *Some remarks on extending bar induction* ◇ F35 F50 ◇

HAYASHI, S. *Constructive mathematics and program synthesis (Japanese)* ◇ B75 F50 ◇

HAYASHI, S. *Extracting Lisp programs from constructive proofs: a formal theory of constructive mathematics based on Lisp* ◇ B35 F35 F50 ◇

HOEVEN VAN DER, G.F. & MOERDIJK, I. *On an independence result in the theory of lawless sequences* ◇ F35 F50 ◇

KHAKHANYAN, V.KH. *Set theory and Church's thesis (Russian)* ◇ D20 E35 E70 F30 F35 F50 ◇

KOVTUN, M.R. *A system of intuitionistic analysis that is equivalent to classical analysis (Russian) (English summary)* ◇ F35 F50 ◇

KROL', M.D. *Various forms of the continuity principle (Russian)* ◇ F30 F50 ◇

LAMBEK, J. & SCOTT, P.J. *New proofs of some intuitionistic principles* ◇ F35 F50 G30 ◇

LATOCHA, P. *The problem of structural completeness of the intuitionistic propositional logic and its fragments* ◇ F50 ◇

LAVENDHOMME, R. & LUCAS, T. *A note on intuitionistic models of ZF* ◇ E70 F50 G30 ◇

LEBLANC, H. & MORGAN, C.G. *Probabilistic semantics for intuitionistic logic* ◇ B48 F50 ◇

LEBLANC, H. & MORGAN, C.G. *Probability theory, intuitionism, semantics, and the Dutch book argument* ◇ B48 C90 F50 ◇

LOPEZ-ESCOBAR, E.G.K. *A second paper on the interpolation theorem for the logic of constant domains* ◇ B55 C40 C90 F05 F50 ◇

MARKOVIC, Z. *Some preservation results for classical and intuitionistic satisfiability in Kripke models* ◇ B20 C25 C90 F50 ◇

MARTINO, E. *Connection between the "principle of inductive evidence" and the bar theorem* ◇ F50 ◇

MARTINO, E. *Semantica intuizionista naturale e semantica di Beth generallizata* ◇ F50 ◇

MCCARTY, C. *Intuitionism: an introduction to a seminar* ◇ A05 F50 F98 ◇

MOERDIJK, I. *On the Freyd cover of a topos* ◇ F35 F50 G30 ◇

MUCHNIK, A.A. *Supplement of the translator to the paper "On alternation. I, II" (Russian)* ◇ B45 D05 D10 D15 F50 ◇

NEPEJVODA, N.N. *Notes concerning constructive implicative logics (Russian)* ◇ B55 F50 ◇

PLISKO, V.E. *Absolute realizability of predicate formulas (Russian)* ◇ F50 ◇

ROSOLINI, G. *Un modello per la teoria intuizionista degli insiemi* ◇ E70 F50 G30 ◇

SCEDROV, A. & VESLEY, R.E. *On a weakening of Markov's principle* ◇ F35 F50 ◇

SCHROEDER-HEISTER, P. *The completeness of intuitionistic logic with respect to a validity concept based on an inversion principle* ◇ F07 F50 ◇

SCHWICHTENBERG, H. *On Martin-Loef's theory of types* ◇ B15 F35 F50 ◇

SEELY, R.A.G. *Hyperdoctrines, natural deduction and the Beck condition* ◇ F07 F50 G30 ◇

SHVARTS, G.F. *Properties of effectiveness of logical connectives in the intuitionistic theory of types (Russian)* ◇ F35 F50 ◇

SKVORTSOV, D.P. *The intuitionistic propositional calculus with an additional logical connective (Russian)* ◇ F50 ◇

SMITH, JAN *The identification of propositions and types in Martin-Loef's type theory: a programming example* ◇ B75 F35 F50 ◇

STEPANOV, V.I. *On model theory for intuitionistic logic (Russian)* ◇ C90 F50 ◇

SUNDHOLM, G. *Constructions, proofs and the meaning of logical constants* ◇ A05 F50 ◇

TROELSTRA, A.S. *Analysing choice sequences* ◇ A05 F35 F50 ◇

TROELSTRA, A.S. *Definability of finite sum types in Martin-Loef's type theories* ◇ F35 F50 ◇

WEINSTEIN, S. *The intended interpretation of intuitionistic logic* ◇ A05 F50 ◇

WENDEL, N. *Kontextuelle Definition neuer Wahlfolgen* ◇ F35 F50 ◇

WOJTYLAK, P. *Collapse of a class of infinite disjunctions in intuitionistic propositional logic* ◇ F50 ◇

1984

ALMUKDAD, A. & NELSON, D. *Constructible falsity and inexact predicates* ◇ B52 F50 ◇

ARAI, T. *An accessibility proof of ordinal diagrams in intuitionistic theories for iterated inductive definitions* ◇ F15 F35 F50 ◇

BEESON, M.J. & SCEDROV, A. *Church's thesis, continuity, and set theory* ◇ D20 E70 F50 ◇

BERTONI, A. & MAURI, G. & MIGLIOLI, P.A. & ORNAGHI, M. *Abstract data types and their extensions within a constructive logic* ◇ B75 F50 ◇

BURGESS, J.P. *Dummett's case for intuitionism* ◇ A05 F50 ◇

DALEN VAN, D. *How to glue analysis models* ◇ C65 C90 F35 F50 ◇

DILLER, J. & TROELSTRA, A.S. *Realizability and intuitionistic logic* ◇ A05 F30 F35 F50 ◇

DOSEN, K. *Intuitionistic double negation as a necessity operator* ◇ B45 F50 ◇

DOSEN, K. *Negative modal operators in intuitionistic logic* ◇ B55 F50 ◇

FEFERMAN, S. *Between constructive and classical mathematics* ◇ A05 F35 F50 F55 ◇

FOURMAN, M.P. *Continuous truth. I. Nonconstructive objects* ◇ A05 E70 F35 F50 G30 ◇

FRIEDMAN, H.M. & SCEDROV, A. *Large sets in intuitionistic set theory* ◇ E55 E70 F50 ◇

GOODMAN, NICOLAS D. *Epistemic arithmetic is a conservative extension of intuitionistic arithmetic* ◇ B45 F05 F30 F50 ◇

GRAYSON, R.J. *Heyting-valued semantics* ◇ B15 C90 F35 F50 G30 ◇

HAYASHI, S. *Constructive mathematics and program synthesis* ◇ B75 F50 ◇

HOEVEN VAN DER, G.F. & MOERDIJK, I. *Constructing choice sequences from lawless sequences of neighbourhood functions* ◇ C90 F35 F50 ◇

HOEVEN VAN DER, G.F. & MOERDIJK, I. *On choice sequences determined by spreads* ◇ D30 D55 F35 F50 G30 ◇

HOEVEN VAN DER, G.F. & MOERDIJK, I. *Sheaf models for choice sequences* ◇ C90 F35 F50 G30 ◇

JOZSA, R. *Sheaf models and massless fields* ◇ B30 C90 F50 G30 ◇

KASHAPOVA, F.R. *Constructive set theory with types, and consistency with Church's thesis (Russian)* ◇ D20 E70 F35 F50 ◇

KASHAPOVA, F.R. *Determination of classes of constructively derivable theorems in a many-sorted intuitionistic set theory equivalent to second-order arithmetic (Russian)* ◇ F35 F50 ◇

KONDO, M. *The completeness theorems for some intuitionistic epistemic logics in terms of interval semantics* ◇ B45 F50 ◇

KREISEL, G. *Frege's foundations and intuitionistic logic* ◇ A05 F50 ◇

KRIVTSOV, V.N. *A formal system of negationless arithmetic is conservative with respect to Heyting arithmetic (Russian)* ◇ F30 F50 ◇

KRIVTSOV, V.N. *Deductive possiblitities of intuitionistic analysis without negation (Russian)* ◇ F35 F50 ◇

KRIVTSOV, V.N. *Imbedding of the intuitionistic theory of types into the negationless intuitionistic theory of types (Russian)* ◇ F35 F50 ◇

LAMBEK, J. & SCOTT, P.J. *Aspects of higher order categorical logic* ◇ B15 F35 F50 G30 ◇

LAVENDHOMME, R. & LUCAS, T. *Une interpretation modale de la logique intuitionniste (English summary)* ◇ B45 F50 ◇

LIU, SHICHAO *A proof-theoretic approach to nonstandard analysis (continued)* ◇ E70 F50 H05 ◇

MARGENSTERN, M. *Sur une extension simple du calcul intuitionniste des predicats du premier ordre applique a l'analyse* ◇ F50 ◇

MARKOVIC, Z. *Kripke models for intuitionistic theories with decidable atomic formulas* ◇ C90 F50 ◇

MARTIN, J.N. *Epistemic semantics for classical and intuitionistic logic* ◇ B50 F50 ◇

MARTIN-LOEF, P. *Intuitonistic type theory. Notes by Giovanni Sambin of a series of lectures given in Padua, June 1980* ◇ F35 F50 F98 ◇

MCCARTY, C. *Information systems, continuity and realizability* ◇ B75 F50 ◇

MOERDIJK, I. & REYES, G.E. *De Rham's theorem in a smooth topos* ◇ F50 G30 ◇

MOERDIJK, I. *Heine-Borel does not imply the Fan theorem* ◇ C90 E35 F35 F50 G30 ◇

MOERDIJK, I. & REYES, G.E. *Smooth spaces versus continuous spaces in models for synthetic differential geometry* ◇ F35 F50 G30 ◇

MOTOHASHI, N. *Approximation theory of uniqueness conditions by existence conditions* ◇ C40 C75 C80 F07 F50 ◇

NORDSTROEM, B. & SMITH, JAN *Propositions and specifications of programs in Martin-Loef's type theory* ◇ B75 F35 F50 ◇

RENARDEL DE LAVALETTE, G.R. *Descriptions in mathematical logic* ◇ B10 F50 ◇

RUITENBURG, W. *On the period of sequences $(A^n(p))$ in intuitionistic propositional calculus* ◇ F50 ◇

SCEDROV, A. *Forcing and classifying topoi* ◇ C90 E35 E40 F50 G30 ◇

SCEDROV, A. *On some nonclassical extensions of second-order intuitionistic propositional calculus* ◇ B55 F50 ◇

SCHROEDER-HEISTER, P. *A natural extension of natural deduction* ◇ B22 F07 F50 ◇

SCHROEDER-HEISTER, P. *Generalized rules for quantifiers and the completeness of the intuitionistic operators* $\&, \vee, \supset, \perp, \forall, \exists$ ◇ F07 F50 ◇

SCOWCROFT, P. *The real algebraic structure of Scott's model of intuitionistic analysis* ◇ B25 C90 F35 F50 ◇

SEELY, R.A.G. *Locally Cartesian closed categories and type theory* ◇ F35 F50 ◇

UESU, T. *A method of axiomatizing fragments of intuitionistic theories* ◇ F50 ◇

UESU, T. *An axiomatization of the apartness fragment of the theory DLO^+ of dense linear order* ◇ F05 F50 ◇

WOJTYLAK, P. *A recursive theory for the* $\{\neg, \wedge, \vee, \rightarrow, o\}$ *fragment of intuitionistic logic* ◇ F50 ◇

YASHIN, A.D. *Completeness of the intuitionistic predicate calculus with the concept of "bar" (Russian)* ◇ C90 F05 F50 ◇

YASHIN, A.D. *Intuitionistic logical connectives on linear structures (Russian)* ◇ F50 ◇

YASHIN, A.D. *Nishimura's formulas as one-place logical connectives in the elementary theory of Kripke models (Russian)* ◇ C90 F50 ◇

1985

ARZARELLO, F. *Methods of realizability and forcing in the metamathematics of constructive formal systems (Italian)* ◇ F50 ◇

BARR, M. & WELLS, C. *Toposes, triples and theories* ◇ C98 E98 F35 F50 G30 ◇

BEESON, M.J. *Foundations of constructive mathematics* ◇ F50 F55 F60 F98 ◇

BEZEM, M. *Stongly majorizable functionals of finite type: A model for barrecursion containing discontinuous functionals* ◇ F10 F35 F50 ◇

DOSEN, K. *An intuitionistic Sheffer function* ◇ B55 F50 ◇
DYMENT, E.Z. *On the interpretation of the intuitionistic predicate calculus by means of the problem calculus with parameters (Russian)* ◇ F50 ◇
FEFERMAN, S. *High-level programs and typed vs. untyped constructive foundations* ◇ A05 F35 F50 ◇
FELSCHER, W. *Dialogues, strategies, and intuitionistic provability* ◇ F05 F07 F50 ◇
FLAGG, R.C. *Church's thesis is consistent with epistemic arithmetic* ◇ F30 F50 ◇
FLAGG, R.C. *Epistemic set theory is a conservative extension of intuitionistic set theory* ◇ B45 E35 E70 F50 ◇
FRIEDMAN, H.M. & SCEDROV, A. *Arithmetic transfinite induction and recursive well-orderings* ◇ D45 F30 F50 ◇
FRIEDMAN, H.M. & SCEDROV, A. *The lack of definable witnesses and provably recursive functions in intuitionistic set theories* ◇ E70 F50 ◇
FRIEDRICH, W. *Goedelsche Funktionalinterpretation fuer eine Erweiterung der klassischen Analysis* ◇ F10 F35 F50 ◇
GOODMAN, NICOLAS D. *Replacement and collection in intuitionistic set theory* ◇ E35 E70 F50 ◇
JANKOWSKI, A.W. & ZAWADOWSKI, M. *Sheaves over Heyting lattices* ◇ C90 F35 F50 G10 G30 ◇
KUZNETSOV, A.V. *Proof-intuitionistic propositional calculus (Russian)* ◇ F50 ◇
LEIVANT, D. *Syntactic translations and provably recursive functions* ◇ D20 F30 F50 ◇
LIFSCHITZ, V. *Calculable natural numbers* ◇ B28 F30 F50 ◇
MARTINO, E. *On the Brouwerian concept of negative continuity* ◇ A10 F50 F55 ◇
MARTINO, E. *On Brower's concept of negative continuity (Italian)* ◇ A05 A10 F50 F55 ◇
MAYBERRY, J. *Global quantification in Zermelo-Fraenkel set theory* ◇ E30 E70 F10 F50 ◇
MURAVITSKIJ, A.YU. *Correspondence of proof-intuitionistic logic extensions to proof-logic extensions (Russian)* ◇ F50 ◇
ONO, H. *Semantical analysis of predicate logics without the contraction rule* ◇ B10 C90 F50 ◇
RABINOWICZ, W. *Intuitionistic truth* ◇ A05 F50 ◇
RENARDEL DE LAVALETTE, G.R. *A type-free system for constructive metamathematics* ◇ F50 ◇
SCEDROV, A. *Extending Goedel's modal interpretation to type theory and set theory* ◇ B15 B45 E70 F35 F50 ◇
SCHROEDER-HEISTER, P. *Proof-theoretic validity and the completeness of intuitionistic logic* ◇ F50 ◇
SCHWICHTENBERG, H. *A normal form for natural deductions in a type theory with realizing terms* ◇ F05 F35 F50 ◇
SHAPIRO, S. *Epistemic and intuitionistic arithmetic* ◇ B28 F30 F50 ◇
SHAPIRO, S. *Introduction – intensional mathematics and constructive mathematics* ◇ A05 F50 ◇
TROELSTRA, A.S. *Choice sequences and informal rigour* ◇ A05 F35 F50 ◇

F55 Constructive and intuitionistic mathematics

1885

MOLK, J. *Sur une notion qui comprend celle de la divisibilite et sur la theorie generale de l'elimination* ◇ F55 F65 ◇

1887

KRONECKER, L. *Ueber den Zahlbegriff* ◇ B28 F55 F65 ◇

1906

HOBSON, E.W. *On the arithmetic continuum* ◇ A05 E47 F35 F55 ◇

1907

BROUWER, L.E.J. *On the foundations of mathematics (Dutch)* ◇ A05 E30 F55 ◇

1908

BROUWER, L.E.J. *On the foundations of mathematics (Dutch)* ◇ A05 E70 F55 ◇

BROUWER, L.E.J. *The unreliability of the logical principles (Dutch)* ◇ A05 F55 ◇

1909

BROUWER, L.E.J. *Die moeglichen Maechtigkeiten* ◇ E10 F55 ◇

1911

BOCKWINKEL, H.B.A. *A constructive proof of the theorem of Borel (Dutch)* ◇ E75 F55 ◇

1912

BROUWER, L.E.J. *Intuitionism and formalism (Dutch)* ◇ A05 F55 ◇

1914

BROUWER, L.E.J. *Review of Schoenflies's "Die Entwicklung der Mengenlehre und ihre Anwendungen, erste Haelfte"* ◇ A05 A10 E70 E98 F55 ◇

1917

BROUWER, L.E.J. *Addenda and corrigenda to: On the foundations of mathematics (Dutch)* ◇ A05 F55 ◇

1918

BROUWER, L.E.J. *Begruendung der Mengenlehre unabhaengig vom logischen Satz vom ausgeschlossenen Dritten. I: Allgemeine Mengenlehre* ◇ E70 F55 ◇

1919

BROUWER, L.E.J. *Begruendung der Mengenlehre unabhaengig vom logischen Satz vom ausgeschlossenen Dritten. II: Theorie der Punktmengen* ◇ E70 F55 ◇

BROUWER, L.E.J. *Intuitionistische Mengenlehre* ◇ E70 F55 ◇

BROUWER, L.E.J. *Mathematics, truth, reality (Dutch)* ◇ A05 F55 ◇

1921

BROUWER, L.E.J. *Besitzt jede reelle Zahl eine Dezimalbruchentwicklung?* ◇ F55 ◇

WEYL, H. *Ueber die neue Grundlagenkrise der Mathematik* ◇ A05 E70 F55 ◇

1923

BROUWER, L.E.J. *Begruendung der Funktionenlehre unabhaengig vom logischen Satz vom ausgeschlossenen Dritten. Erster Teil: Stetigkeit, Messbarkeit, Derivierbarkeit* ◇ F55 ◇

BROUWER, L.E.J. *On the role of the principle of the excluded middle in mathematics, especially in function theory (Dutch)* ◇ F55 ◇

1924

BALDUS, R. *Formalismus und Intuitionismus in der Mathematik* ◇ A05 F55 F98 ◇

BROUWER, L.E.J. *Bemerkungen zum Beweise der gleichmaessigen Stetigkeit voller Funktionen* ◇ F55 ◇

BROUWER, L.E.J. & LOOR DE, B. *Intuitionistischer Beweis des Fundamentalsatzes der Algebra* ◇ F55 ◇

BROUWER, L.E.J. *Intuitionistische Ergaenzung des Fundamentalsatzes der Algebra* ◇ F55 ◇

BROUWER, L.E.J. *Perfect sets of primes with positively irrational distances* ◇ F55 ◇

DRESDEN, A. *Brouwer's contribution to the foundations of mathematics* ◇ A05 F55 ◇

WEYL, H. *Randbemerkungen zu Hauptproblemen der Mathematik* ◇ B30 F55 ◇

1925

BROUWER, L.E.J. *Intuitionistische Zerlegung mathematischer Grundbegriffe* ◇ F55 ◇

BROUWER, L.E.J. *Intuitionistischer Beweis des Jordanschen Kurvensatzes* ◇ F55 ◇

BROUWER, L.E.J. *Zur Begruendung der intuitionistischen Mathematik I* ◇ F55 ◇

HEYTING, A. *Intuitionistic axiomatics of projective geometry (Dutch)* ◇ F55 ◇

1926

BROUWER, L.E.J. *Die intuitionistische Form des Heine-Borelschen Theorems* ◇ F55 ◇

BROUWER, L.E.J. *Intuitionistische Einfuehrung des Dimensionsbegriffs* ◇ F55 ◇

BROUWER, L.E.J. *Zur Begruendung der intuitionistischen Mathematik II* ◇ F55 ◇

BROUWER, L.E.J. *Zur Begruendung der intuitionistischen Mathematik III* ◇ F55 ◇

HERMANN, G. *Die Frage der endlich vielen Schritte in der Theorie der Polynomideale* ◇ C57 C60 D45 F55 ◇

1927

BROUWER, L.E.J. *Ueber Definitionsbereiche von Funktionen* ◇ F55 ◇

BROUWER, L.E.J. *Virtuelle Ordnung und unerweiterbare Ordnung* ◇ F55 ◇

BROUWER, L.E.J. *Zur intuitionistischen Zerlegung mathematischer Grundbegriffe* ◇ F55 ◇

1928

BROUWER, L.E.J. *Beweis dass jede Menge in einer individualisierten Menge enthalten ist* ◇ E70 F55 ◇

HEYTING, A. *Die Theorie der linearen Gleichungen in einer Zahlenspezies mit nicht-kommutativer Multiplikation* ◇ F55 ◇

HEYTING, A. *Zur intuitionistischen Axiomatic der projektiven Geometrie* ◇ F55 ◇

1929

BELINFANTE, M.J. *Ueber einen Grenzwertsatz aus der Theorie der unendlichen Folgen* ◇ F55 ◇

BELINFANTE, M.J. *Zur intuitionistischen Theorie der unendlichen Reihen* ◇ F55 ◇

BROUWER, L.E.J. *Mathematik, Wissenschaft und Sprache* ◇ A05 F55 ◇

EUWE, M. *Mengentheoretische Betrachtungen ueber das Schachspiel* ◇ E60 F55 ◇

HEYTING, A. *The countability predicates of Professor Brouwer (Dutch)* ◇ F55 F99 ◇

LUZIN, N.N. *Sur les voies de la theorie des ensembles* ◇ A05 D55 E15 E50 F55 ◇

1930

BELINFANTE, M.J. *Absolute Konvergenz in der intuitionistischen Mathematik* ◇ F55 ◇

BELINFANTE, M.J. *Ueber eine besondere Klasse von non-oszillierenden Reihen* ◇ F55 ◇

BROUWER, L.E.J. *Die Struktur des Kontinuums* ◇ A05 F55 ◇

BROUWER, L.E.J. *Review of A.Fraenkel "Zehn Vorlesungen ueber die Grundlagen der Mengenlehre"* ◇ A05 F55 ◇

WAERDEN VAN DER, B.L. *Eine Bemerkung ueber die Unzerlegbarkeit von Polynomen* ◇ B25 C60 D45 F55 ◇

1931

BELINFANTE, M.J. *Die Hardy-Littlewoodsche Umkehrung des Abelschen Stetigkeitssatzes in der intuitionistischen Mathematik* ◇ F55 ◇

BELINFANTE, M.J. *Ueber die Elemente der Funktionentheorie und die Picardschen Saetze in der intuitionistischen Mathematik* ◇ F55 ◇

GOEDEL, K. *Zum intuitionistischen Aussagenkalkuel* ◇ F50 F55 ◇

HEYTING, A. *Die intuitionistische Grundlegung der Mathematik* ◇ A05 F50 F55 ◇

1932

BELINFANTE, M.J. *Ueber den intuitionistischen Beweis der Picardschen Saetze* ◇ F55 ◇

1933

BROUWER, L.E.J. *Volition, knowledge, speech (Dutch)* ◇ A05 F55 ◇

1934

VANDIVER, H.S. *On the foundations of a constructive theory of discrete commutative algebra* ◇ F55 ◇

1935

BERNAYS, P. *Quelques points essentiels de la metamathematique* ◇ A05 F55 ◇

BERNAYS, P. *Sur le platonisme dans les mathematiques* ◇ A05 A10 E30 E70 F55 ◇

VANDIVER, H.S. *On the foundations of a constructive theory of discrete commutative algebra II* ◇ F55 ◇

1936

HEYTING, A. *Intuitionistic mathematics I,II,III (Dutch)* ◇ F55 ◇

MANIA, B. *L'infini mathematique et l'evolution de la logique* ◇ A05 F55 F65 ◇

VANDIVER, H.S. *Constructive derivation of the decomposition-field of a polynomial* ◇ F55 ◇

VANDIVER, H.S. *On the ordering of real algebraic numbers by constructive methods* ◇ F55 ◇

1937

FREUDENTHAL, H. *Zum intuitionistischen Raumbegriff* ◇ F55 ◇

HEYTING, A. *Intuitionistic mathematics IV,V (Dutch)* ◇ F55 ◇

HEYTING, A. *The development of intuitionistic mathematics (Dutch)* ◇ A05 F55 ◇

1938

DIENES, P. *Logic of algebra* ◇ A05 B28 F55 G05 ◇

GENTZEN, G. *Die gegenwaertige Lage in der mathematischen Grundlagenforschung* ◇ A05 A10 F55 F65 ◇

1939

BELINFANTE, M.J. *Das Riemannsche Umordnungsprinzip in der intuitionistischen Theorie der unendlichen Reihen* ◇ F55 ◇

BELINFANTE, M.J. *Der Levysche Umordnungssatz und seine intuitionistische Uebertragung* ◇ F55 ◇

HEYTING, A. *Intuitionistic mathematics VI (Dutch)* ◇ F55 ◇

HEYTING, A. *Les fondements des mathematiques du point de vue intuitionniste* ◇ A05 F55 ◇

1940

KNESER, H. *Der Fundamentalsatz der Algebra und der Intuitionismus* ◇ F55 ◇

LARGUIER, E.H. *Brouwerian philosophy of mathematics* ◇ A05 F55 ◇

1941

BELINFANTE, M.J. *Elemente der intuitionistischen Funktionentheorie. Erste Mitteilung: Die Cauchyschen Integralsaetze und die Taylorsche Reihe. Zweite, dritte Mitteilung: Der Satz vom Integral der logarithmischen Ableitung I,II* ◇ F55 ◇

BELINFANTE, M.J. *Elemente der intuitionistischen Funktionentheorie. Vierte Mitteilung: Der*

Weierstrass'sche Unbestimmtheitssatz. Fuenfte Mitteilung: Die intuitionistische Uebertragung des Picardschen Satzes ◇ F55 ◇

CHANDRASEKHARAN, K. *The logic of intuitionistic mathematics* ◇ E70 F55 ◇

HEYTING, A. *Untersuchungen ueber intuitionistische Algebra* ◇ F55 ◇

1942

BROUWER, L.E.J. *Beweis dass der Begriff der Menge hoeherer Ordnung nicht als Grundbegriff der intuitionistischen Mathematik in Betracht kommt* ◇ F55 ◇

BROUWER, L.E.J. *Die repraesentierende Menge der stetigen Funktionen des Einheitskontinuums* ◇ F55 ◇

BROUWER, L.E.J. *Zum freien Werden von Mengen und Funktionen* ◇ E70 F55 ◇

CHANDRASEKHARAN, K. *Intuitionistic theory of linear order* ◇ E07 E70 F55 ◇

DANTZIG VAN, D. *A remark and a problem concerning the intuitionistic form of Cantor's intersection theorem* ◇ E20 F55 ◇

DANTZIG VAN, D. *On the affirmative content of Peano's theorem on differential equations* ◇ F55 ◇

1944

GRISS, G.F.C. *Negationless intuitionistic mathematics* ◇ F50 F55 ◇

1945

CHANDRASEKHARAN, K. *A further note on intuitionistic set theory* ◇ E70 F55 ◇

ROSENBLOOM, P.C. *An elementary constructive proof of the fundamental theorem of algebra* ◇ F55 ◇

1946

DIJKMAN, J.G. *Einige Saetze ueber mehrfach negativ-konvergente Reihen in der intuitionistischen Mathematik* ◇ F55 ◇

GRISS, G.F.C. *Negationless intuitionistic mathematics I* ◇ F50 F55 ◇

1947

BILLING, J. *A failure of the Bolzano-Weierstrass lemma* ◇ F55 ◇

BROUWER, L.E.J. *Directives of intuitionistic mathematics (Dutch)* ◇ F55 ◇

DANTZIG VAN, D. *On the principles of intuitionistic and affirmative mathematics* ◇ F55 ◇

DESTOUCHES-FEVRIER, P. *Esquisse d'une mathematique intuitionniste positive* ◇ F55 ◇

1948

BROUWER, L.E.J. *Essentially-negative properties (Dutch)* ◇ F55 ◇

BROUWER, L.E.J. *Remarks on the law of the excluded third and on negative propositions (Dutch)* ◇ F55 ◇

DIJKMAN, J.G. *Recherche de la convergence negative dans les mathematiques intuitionnistes* ◇ F55 ◇

1949

BILLING, J. *The principle of the excluded third and the Bolzano-Weierstrass lemma* ◇ F55 ◇

BROUWER, L.E.J. *Consciousness, philosophy, and mathematics* ◇ A05 F55 ◇

BROUWER, L.E.J. *Contradictority of elementary geometry (Dutch)* ◇ F55 ◇

BROUWER, L.E.J. *The non-equivalence of the constructive and the negative order relation on the continuum (Dutch)* ◇ F55 ◇

DANTZIG VAN, D. *Comments on Brouwer's theorem on essentially-negative predicates* ◇ A05 F55 ◇

DEQUOY, N. *La geometrie projective plane en mathematique intuitionniste sans negation* ◇ F55 ◇

1950

BROUWER, L.E.J. *Remarques sur la notion d'ordre* ◇ F55 ◇

BROUWER, L.E.J. *Sur la possibilite d'ordonner le continu* ◇ F55 ◇

DEQUOY, N. *Expose d'un type de raisonnement en mathematique intuitionniste sans negation et resultats obtenus pour la geometrie projective plane* ◇ F55 ◇

GRISS, G.F.C. *Negationless intuitionistic mathematics II* ◇ F50 F55 ◇

HEYTING, A. *Tensions in mathematics (Dutch)* ◇ A05 F55 ◇

SCHMIDT, H.A. *Wie duerfen wir mit dem Unendlichen umgehen? (die Grundlage des mathematischen Intuitionismus)* ◇ A05 F55 ◇

1951

BROUWER, L.E.J. *On order in the continuum, and the relation of truth to non-contradictority* ◇ F55 ◇

DANTZIG VAN, D. *Mathematique stable et mathematique affirmative* ◇ A05 F55 ◇

DESTOUCHES, J.-L. *Sur la mecanique classique et l'intuitionnisme* ◇ F55 ◇

GRISS, G.F.C. *Negationless intuitionistic mathematics III,IV* ◇ F55 ◇

HEYTING, A. *L'axiomatique intuitionniste* ◇ F50 F55 ◇

HEYTING, A. *Note on the Riesz-Fischer theorem* ◇ F55 ◇

1952

BROUWER, L.E.J. *An intuitionistic correction of the fixed-point theorem on the sphere* ◇ F55 ◇

BROUWER, L.E.J. *Fixed cores which cannot be found, though they are claimed to exist by classical theorems (Dutch)* ◇ F55 ◇

BROUWER, L.E.J. *Historical background, principles and methods of intuitionism* ◇ A05 F55 ◇

BROUWER, L.E.J. *On accumulation cores of infinite core species (Dutch)* ◇ F55 ◇

KREISEL, G. *Some elementary inequalities* ◇ F55 ◇

ROOTSELAAR VAN, B. *Un probleme de M. Dijkman* ◇ F55 ◇

1953

HEYTING, A. *Espace de Hilbert et intuitionnisme* ◇ F55 ◇

1954

BROUWER, L.E.J. *Addenda and corrigenda on the role of the principium tertii exclusi in mathematics (Dutch)* ◇ F55 ◇

BROUWER, L.E.J. *An example of contradictority in classical theory of functions* ◇ F55 ◇
BROUWER, L.E.J. *Further addenda and corrigenda concerning the role of the principle tertii exclusi in mathematics (Dutch)* ◇ F55 ◇
BROUWER, L.E.J. *Intuitionistic Differentiability (Dutch)* ◇ F55 ◇
BROUWER, L.E.J. *Ordnungswechsel in Bezug auf eine coupierbare geschlossene stetige Kurve* ◇ F55 ◇
BROUWER, L.E.J. *Points and spaces* ◇ F55 ◇
FRAENKEL, A.A. *The intuitionistic revolution in mathematics and logic* ◇ A05 F55 ◇
GOODSTEIN, R.L. *A free variable function theory* ◇ F55 F60 ◇
HEYTING, A. *Sur la theorie intuitionnististe de la mesure* ◇ F55 ◇
SKOLEM, T.A. *Results in investigations in the foundations (Norwegian)* ◇ A05 B98 D03 F55 ◇

1955

DEQUOY, N. *Axiomatique intuitionniste sans negation de la geometrie projective* ◇ F55 ◇
ROOTSELAAR VAN, B. *Generating schemes for full mappings* ◇ F55 ◇
ROOTSELAAR VAN, B. *On the mapping of spreads* ◇ F55 ◇
ROSENBLOOM, P.C. *Konstruktive Aequivalente fuer Saetze aus der klassischen Analysis* ◇ F55 ◇

1956

BETH, E.W. *L'existence en mathematiques* ◇ A05 B98 F55 ◇
HEYTING, A. *Intuitionism. An introduction* ◇ F55 F98 ◇
ROOTSELAAR VAN, B. *A remark on Brouwer measurable functions* ◇ F55 ◇

1957

HEYTING, A. *Axioms for intuitionistic plane affine geometry. The axiomatic method. With special reference to geometry and physics* ◇ F55 ◇
KURODA, S. *On the intuitionistic and formalistic theory of real numbers* ◇ F55 F65 ◇
ROOTSELAAR VAN, B. *Intuitionismus und Arithmetik (Dutch)* ◇ A05 F55 ◇

1958

HEYTING, A. *Intuitionism in mathematics* ◇ A05 B98 F50 F55 F98 ◇
HEYTING, A. *Intuitionism and school mathematics (Dutch)* ◇ A05 F55 ◇
HEYTING, A. *La theorie elementaire de l'integration en mathematiques intuitionnistes* ◇ F55 ◇
KREISEL, G. *Constructive mathematics* ◇ F50 F55 F65 ◇

1959

HEYTING, A. *Axioms for intuitionistic plane affine geometry* ◇ B30 F55 ◇
LERDA, F. *Analisi critica dei fondamenti dell'intuizionismo* ◇ A05 F50 F55 ◇

1960

HEYTING, A. *Remarques sur le constructivisme* ◇ A05 F55 ◇
ROOTSELAAR VAN, B. *On intuitionistic difference relations* ◇ F55 ◇

1961

DIJKMAN, J.G. *On Markov chains and intuitionism* ◇ F55 ◇
DIJKMAN, J.G. *Some intuitionistic remarks about transformations of sequences* ◇ F55 ◇
HEYTING, A. *Axiomatic method and intuitionism* ◇ B30 F55 ◇
HUBER-DYSON, V. & KREISEL, G. *Analysis of Beth's semantic construction of intuitionistic logic* ◇ F50 F55 ◇

1962

DIJKMAN, J.G. *A note on intuitionistic divergence theory* ◇ F55 ◇
HEYTING, A. *Methodes et problemes de l'intuitionnisme* ◇ A05 F50 F55 ◇

1963

DALEN VAN, D. *Extension problems in intuitionistic plane projective geometry I,II* ◇ F55 ◇
DIJKMAN, J.G. *On Markov chains and intuitionism. II: Discrete state-space and continous parameter* ◇ F55 ◇

1964

DIJKMAN, J.G. *Markov chains and intuitionism III: Note on continuous functions with an application to Markov chains* ◇ F55 ◇
HEYTING, A. *Disputation* ◇ A05 F55 ◇
ROOTSELAAR VAN, B. *On intuitionistic notions of divergence and convergence* ◇ F55 ◇

1965

ROOTSELAAR VAN, B. *Intuitives ueber den Intuitionismus* ◇ A05 F55 ◇

1966

DIJKMAN, J.G. *Probability theory and intuitionism. Discrete state space* ◇ F55 ◇
HEYTING, A. *Remarques sur la theorie intuitionniste des espaces lineaires* ◇ F55 ◇

1967

BISHOP, E.A. *Foundations of constructive analysis* ◇ F55 F98 ◇
SCHULTZ, KONRAD *Speziestheorie und intuitionistische Topologie* ◇ F55 ◇
SCHULTZ, KONRAD *Spreadtheorie und intuitionistische Topologie* ◇ F55 ◇
TROELSTRA, A.S. *Finite and infinite in intuitionistic mathematics* ◇ E10 F55 ◇
TROELSTRA, A.S. *Intuitionistic continuity* ◇ F55 ◇
TROELSTRA, A.S. *Intuitionistic connectedness* ◇ F55 ◇

1968

BISHOP, E.A. *The constructivization of abstract mathematical analysis* ◇ F55 ◇
DALEN VAN, D. *A note on spread-cardinals* ◇ E10 F55 ◇

DALEN VAN, D. *Fans generated by nondeterministic automata* ◇ D05 F55 ◇
DALEN VAN, D. *Reducibilities in intuitionistic topology* ◇ F55 ◇
STOLZENBERG, G. *Constructive normalization of an algebraic variety* ◇ F55 ◇
TROELSTRA, A.S. *New sets of postulates for intuitionistic topology* ◇ F55 ◇
TROELSTRA, A.S. *One-point compactifications of intuitionistic locally compact spaces* ◇ F55 ◇
TROELSTRA, A.S. *The use of "Brouwer's principle" in intuitionistic topology* ◇ F55 ◇

1969

ASHVINIKUMAR *The intuitionist contradictority of certain classical set theoretic results* ◇ F55 ◇
ASHVINIKUMAR *Ueber katalogisierte Raeume* ◇ F55 ◇
GIBSON, C.G. *The intuitionist measure* ◇ F55 ◇
HEIDEMA, J. *History, principles and methods of intuitionistic mathematics (Africaans) (English summary)* ◇ A05 F55 ◇
HEYTING, A. *Remark on the paper by Ashvinikumar "The intuitionist contradictority of certain classical set-theoretic results"* ◇ F55 ◇
HOMAGK, F. *Ein intuitionistischer Beweis fuer den Graphensatz von D.Koenig* ◇ F55 ◇
TROELSTRA, A.S. *Informal theory of choice sequences (Polish and Russian summaries)* ◇ F35 F50 F55 ◇
TROELSTRA, A.S. *Principles of intuitionism* ◇ F35 F50 F55 F98 ◇

1970

ASHVINIKUMAR *On the intuitionist theory of Stieltjes integration and its applications* ◇ F55 ◇
ASHVINIKUMAR *On Brouwer-Stieltjes integration* ◇ F55 ◇
BELDING, W.R. *A note on the intuitionistic fan theorem* ◇ F55 ◇
BISHOP, E.A. *Mathematics as a numerical language* ◇ A05 F50 F55 ◇
HEYTING, A. *Recent progress in intuitionistic analysis* ◇ F55 ◇
SEIDENBERG, A. *Construction of the integral closure of a finite integral domain* ◇ C57 C60 F55 ◇
STOLZENBERG, G. *Review: Foundations of constructive analysis by Errett Bishop* ◇ A05 F55 ◇

1971

ASHVINIKUMAR *Another proof of the intuitionistic inconsistency of the axiom of choice (Dutch)* ◇ E25 F50 F55 ◇
ASHVINIKUMAR & SHUKLA, S.L. *Intuitionist determination of dual spaces of certain catalogued linear spaces I, II* ◇ F55 ◇
BELDING, W.R. *Intuitionistic negation* ◇ B55 F55 ◇
CHAN, Y.K. *A constructive proof of Sard's theorem* ◇ F55 ◇
GIBSON, C.G. *On the definition of an infinite species* ◇ F55 ◇
HENKIN, L. *Mathematical foundations for mathematics* ◇ A10 C98 E30 E55 F55 G05 G15 ◇
HOMAGK, F. *Intuitionistische Kennzeichnung der endlichen Spezies (Polish and Russian summaries)* ◇ F55 ◇
SEIDENBERG, A. *On the length of a Hilbert ascending chain* ◇ C60 F55 ◇
STAPLES, J. *On constructive fields* ◇ F55 ◇

1972

BISHOP, E.A. & CHENG, HENRY *Constructive measure theory* ◇ F55 ◇
CHAN, Y.K. *A constructive study of measure theory* ◇ F55 ◇
CHAN, Y.K. *A constructive approach to the theory of stochastic processes* ◇ F55 ◇
CONSTABLE, R.L. *Constructive mathematics and automatic program writers* ◇ B75 D15 F55 ◇
GIBSON, C.G. *On the almost periodicity of trigonometric polynomials in constructive mathematics* ◇ F55 ◇
GOODMAN, NICOLAS D. & MYHILL, J.R. *The formalization of Bishop's constructive mathematics* ◇ F35 F50 F55 ◇
LEE, PENGYEE *A constructive proof of a Differential inequality* ◇ F55 ◇
MYHILL, J.R. *What is a real number ?* ◇ F55 F60 ◇
NUBER, J.A. *A constructive ergodic theorem* ◇ F55 ◇
SEIDENBERG, A. *Constructive proof of Hilbert's theorem on ascending chains* ◇ F55 ◇
SHUKLA, S.L. *Intuitionist treatment of some spaces of sequences* ◇ F55 ◇
SHUKLA, S.L. *On intuitionist analogues of classically inseparable spaces* ◇ F55 ◇
SHUKLA, S.L. *On some linear spaces which coincide classically but are different intuitionistically* ◇ F55 ◇

1973

CHENG, HENRY *A constructive Riemann mapping theorem* ◇ F55 ◇
CHENG, HENRY *A constructive intermediate value theorem* ◇ F55 ◇
DALEN VAN, D. *Lectures on intuitionism* ◇ F50 F55 F98 ◇
RICHMAN, F. *The constructive theory of countable abelian p-groups* ◇ F55 ◇
SEIDENBERG, A. *On the impossibility of some constructions in polynomial rings* ◇ C60 F55 ◇
SHUKLA, S.L. *On a new method in intuitionist linear analysis* ◇ F55 ◇
STAPLES, J. *Axioms for constructive fields* ◇ F55 ◇
TROELSTRA, A.S. (ED.) *Metamathematical investigation of intuitionistic arithmetic and analysis* ◇ F05 F10 F30 F35 F50 F55 F98 ◇

1974

ASHVINIKUMAR *A new version of the proof of the fan theorem* ◇ F55 ◇
CHAN, Y.K. *Notes on constructive probability theory* ◇ F55 ◇
CHAN, Y.K. *On constructive convergence of measures on the real line* ◇ F55 ◇
HEYTING, A. *Intuitionistic views on the nature of mathematics* ◇ A05 F55 ◇

MULVEY, C.J. *Intuitionistic algebra and representations of rings* ◇ C60 C90 F50 F55 G30 ◇
RICHMAN, F. *Constructive aspects of noetherian rings* ◇ F55 ◇
SEIDENBERG, A. *Constructions in algebra* ◇ F55 ◇
SEIDENBERG, A. *What is noetherian ? (Italian summary)* ◇ F55 ◇
WIEDMER, E. *Ein neuer negationsloser Beweis eines Satzes von G.F.C.Griss* ◇ B20 F55 ◇

1975

BERG, G. & JULIAN, W. & MINES, R. & RICHMAN, F. *The constructive Jordan curve theorem* ◇ F55 ◇
BROUWER, L.E.J. *Collected works I* ◇ A05 A10 F55 F96 ◇
CHAN, Y.K. *A short proof of an existence theorem in constructive measure theory* ◇ F55 ◇
JAMBON, M. *Sur la notion de la variabilite locale constructive (English summary)* ◇ F55 ◇
LUCKHARDT, H. *A short proof of a well-known theorem of intuitionistic analysis* ◇ F55 ◇
RICHMAN, F. *The constructive theory of KT-modules* ◇ F55 ◇
SEIDENBERG, A. *Construction of the integral closure of a finite integral domain. II* ◇ C57 C60 F55 ◇

1976

BERG, G. & CHENG, HENRY & MINES, R. & RICHMAN, F. *Constructive dimension theory* ◇ F55 ◇
BRIDGES, D.S. *Some notes on continuity in constructive analysis* ◇ F55 ◇
LEVIN, L.A. *On the principle of conservation of information in intuitionistic mathematics (Russian)* ◇ D15 D80 F50 F55 ◇
MANDELKERN, M. *Connectivity of an interval* ◇ F55 ◇
RICHMAN, F. *A constructive modification of Vietoris homology* ◇ F55 ◇
SWART DE, H.C.M. *Elements of intuitionistic analysis I: Rolle's theorem and complete totally bounded, metric spaces. II: The Stone-Weierstrass theorem and Ascoli's theorem* ◇ F55 ◇

1977

BERG, G. & JULIAN, W. & MINES, R. & RICHMAN, F. *The constructive equivalence of covering and inductive dimensions* ◇ F55 ◇
BIELTZ, P. *On the controversies of the excluded middle (Romanian)* ◇ A05 A10 F55 ◇
BISHOP, E.A. *Review of H. Jerome Keisler's "Elementary calculus"* ◇ A05 F55 ◇
BOZZI, S. & MELONI, G.C. *Ideal properties and intuitionistic algebra* ◇ C60 C90 F50 F55 ◇
BRIDGES, D.S. *A constructive look at orthonormal bases in Hilbert space* ◇ F55 ◇
BRIDGES, D.S. *The constructive Radon-Nikodym theorem* ◇ F55 ◇
BROM, J. *The theory of almost periodic functions in constructive mathematics* ◇ F55 ◇
CHAN, Y.K. *Constructive foundations of potential theory* ◇ F55 ◇
DUMMETT, M. *Elements of intuitionism* ◇ F50 F55 F98 ◇
RICHMAN, F. *Computing heights in Tor* ◇ F55 ◇
SEPER, K. *Constructive mathematics - an essay, I, II* ◇ A05 F55 F65 ◇
SURESHKUMAR *A note on Ashvinikumar's paper:"The intuitionist contradictority of certain classical set-theoretic results"* ◇ E70 E75 F55 ◇
TROELSTRA, A.S. *Aspects of constructive mathematics* ◇ F30 F35 F50 F55 F60 F98 ◇
TROELSTRA, A.S. *Choice sequences, a chapter of intuitionistic mathematics* ◇ A05 A10 F35 F50 F55 F98 ◇

1978

ARRUDA, A.I. *Some remarks on Griss' logic of negationless intuitionistic mathematics* ◇ B55 F50 F55 G25 ◇
BEESON, M.J. *Some relations between classical and constructive mathematics* ◇ F50 F55 ◇
BRIDGES, D.S. *More on the connectivity of convex sets* ◇ F55 ◇
BRIDGES, D.S. *On continuous mappings between locally compact metric spaces* ◇ F55 ◇
BRIDGES, D.S. *On the connectivity of convex sets* ◇ F55 ◇
BRIDGES, D.S. *On weak operator compactness of the unit ball of L(H)* ◇ F55 ◇
GREENLEAF, N. *Linear order in lattices: A constructive study* ◇ F50 F55 ◇
JULIAN, W. & MINES, R. & RICHMAN, F. *Algebraic numbers, a constructive development* ◇ F55 ◇
SEIDENBERG, A. *Constructions in a polynomial ring over the ring of integers* ◇ D45 F55 ◇

1979

ASHVINIKUMAR *Microscopic completeness of full indications and intuitionist contradictory proofs* ◇ F55 ◇
BRIDGES, D.S. *A criterion for compactness in metric spaces?* ◇ F55 ◇
BRIDGES, D.S. *Connectivity properties of metric spaces* ◇ F55 ◇
BRIDGES, D.S. *Constructive functional analysis* ◇ F55 F98 ◇
BRIDGES, D.S. *Geometric intuition and elementary constructive analysis* ◇ F55 ◇
BRIDGES, D.S. *On the constructive convergence of series of independent functions* ◇ F55 ◇
CARTIER, P. *Logique, categories et faisceaux (d'apres F. Lawvere et M. Tierney)* ◇ E35 E50 F55 G10 G30 ◇
FEFERMAN, S. *Constructive theories of functions and classes* ◇ F50 F55 F60 F98 ◇
GIRSTMAIR, K. *Ueber konstruktive Methoden der Galoistheorie* ◇ C57 C60 F55 ◇
HOMAGK, F. *Zur intuitionistischen Kennzeichnung reeller Funktionen auf substantiellen Intervallen* ◇ F55 ◇
MANDELKERN, M. *Suprema of located sets* ◇ F55 ◇
ROUSSEAU, C. *Topos theory and complex analysis* ◇ F50 F55 G30 H05 ◇

1980

BRIDGES, D.S. *A constructive development of Chebyshev approximation theory* ◇ F55 ◇

BRIDGES, D.S. *On the foundations of best-approximation theory* ◇ F55 ◇

BURGESS, J.P. *Brouwer and Souslin on transfinite cardinals* ◇ E15 F55 ◇

MANDELKERN, M. *Resolutions on the line* ◇ F55 ◇

POSY, C.J. *On Brouwer's definition of unextendable order* ◇ A10 F55 ◇

ROGERS, L. *Basic subgroups from a constructive viewpoint* ◇ F55 ◇

TROELSTRA, A.S. *Intuitionistic extensions of the reals* ◇ F35 F50 F55 ◇

1981

BRIDGES, D.S. *A constructive look at positive linear functionals on $\mathscr{L}(H)$* ◇ F55 ◇

BRIDGES, D.S. *A constructive analysis of the Remes algorithm* ◇ F55 ◇

BRIDGES, D.S. *A constructive proximinality property of finite-dimensional linear subspaces* ◇ F55 ◇

BRIDGES, D.S. & CALDER, A. & JULIAN, W. & MINES, R. & RICHMAN, F. *Bounded linear mappings of finite rank* ◇ F55 ◇

BRIDGES, D.S. & CALDER, A. & JULIAN, W. & MINES, R. & RICHMAN, F. *Compactly generated Banach spaces* ◇ F55 ◇

BRIDGES, D.S. & CALDER, A. & JULIAN, W. & MINES, R. & RICHMAN, F. *Locating metric complements in R^n* ◇ F55 ◇

BRIDGES, D.S. *On the isolation of zeroes of an analytic function* ◇ F55 ◇

BRIDGES, D.S. *On Montel's proof of the Great Picard Theorem* ◇ F55 ◇

BROUWER, L.E.J. *Brouwer's Cambridge lectures on intuitionism* ◇ A05 A10 F50 F55 F98 ◇

BROUWER, L.E.J. *On the foundation of mathematics (Dutch)* ◇ A05 A10 F55 ◇

CHAN, Y.K. *On some open problems in constructive probability theory* ◇ F55 ◇

GIBSON, C.G. & JOHNS, D.L. *A constructive approach to the duality theorem for certain Orlicz spaces* ◇ F55 ◇

GOODMAN, NICOLAS D. *Reflections on Bishop's philosophy of mathematics* ◇ A05 F55 ◇

GRAYSON, R.J. *Concepts of general topology in constructive mathematics and in sheaves* ◇ C90 E35 E70 E75 F35 F50 F55 G30 ◇

GREENLEAF, N. *Liberal constructive set theory* ◇ A05 E70 F55 F65 ◇

HEYTING, A. *Continuum and choice sequence in Brouwer (Dutch)* ◇ A10 F55 ◇

JOHNSTONE, P.T. *Tychonoff's theorem without the axiom of choice* ◇ E25 E75 F55 G30 ◇

KNESER, M. *Ergaenzung zu einer Arbeit von Hellmuth Kneser ueber den Fundamentalsatz der Algebra* ◇ F55 ◇

MANDELKERN, M. *Located sets on the line* ◇ F55 ◇

RICHMAN, F. *Seidenberg's condition P* ◇ F55 ◇

RUITENBURG, W. *Field extensions* ◇ F55 ◇

SMITH, RICK L. *Effective valuation theory* ◇ C57 C60 D45 F55 ◇

SUNDHOLM, G. *Hacking's logic* ◇ F05 F55 ◇

TROELSTRA, A.S. *Arend Heyting and his contribution to intuitionism* ◇ A10 F55 ◇

1982

BRIDGES, D.S. *Lipschitz constants and moduli of continuity for the Chebyshev projection* ◇ F55 ◇

BRIDGES, D.S. & CALDER, A. & JULIAN, W. & MINES, R. & RICHMAN, F. *Picard's theorem* ◇ F55 ◇

BRIDGES, D.S. *Preference and utility: a constructive development* ◇ F55 ◇

BRIDGES, D.S. *Recent progress in constructive approximation theory* ◇ F55 ◇

DALEN VAN, D. *Singleton reals* ◇ F55 ◇

DALEN VAN, D. *The creative subject and Heyting's arithmetic* ◇ F30 F55 ◇

DELZELL, C.N. *Case distinctions are necessary for representing polynomials as sums of squares* ◇ A10 C60 C65 F55 ◇

DELZELL, C.N. *Continuous sums of squares of forms* ◇ C60 C65 F55 ◇

DRAGO, A. *Caratheodory's thermodynamics and constructive mathematics* ◇ F55 ◇

GRAYSON, R.J. *Concepts of general topology in constructive mathematics and in sheaves II* ◇ E75 F50 F55 G30 ◇

KREISEL, G. *Brouwer's Cambridge lectures on intuitionism* ◇ A05 A10 F55 ◇

MANDELKERN, M. *Components of an open set* ◇ F55 ◇

MANDELKERN, M. *Continuity of monotone functions* ◇ F55 ◇

MINES, R. *Algebraic number theory, a survey* ◇ F55 ◇

MINES, R. & RICHMAN, F. *Separability and factoring polynomials* ◇ F55 ◇

RICHMAN, F. *Finite dimensional algebras over discrete fields* ◇ F55 ◇

RICHMAN, F. *Meaning and information in constructive mathematics* ◇ A05 F55 ◇

RUITENBURG, W. *Primality and invertibility of polynomials* ◇ F55 ◇

TROELSTRA, A.S. *Intuitionistic extensions of the reals II* ◇ F35 F50 F55 ◇

TROELSTRA, A.S. *On the origin and development of Brouwer's concept of choice sequence* ◇ A10 F55 ◇

TROELSTRA, A.S. & DALEN VAN, D. (EDS.) *The L.E.J. Brouwer centenary symposium* ◇ A05 A10 B97 F50 F55 F97 ◇

VELDMAN, W. *On the constructive contrapositions of two axioms of countable choice* ◇ E25 E70 F50 F55 ◇

1983

BLASS, A.R. *Words, free algebras, and coequalizers* ◇ C05 C75 E25 E35 F50 F55 G30 ◇

ESPANOL, L. *Le spectre d'un anneau dans l'algebre constructive et applications a la dimension* ◇ F55 G30 ◇

JULIAN, W. & MINES, R. & RICHMAN, F. *Alexander duality* ◇ F55 ◇

MANDELKERN, M. *Constructive continuity* ◇ F55 ◇

MITANI, S. *Dimension theory constructed on the basis of LJ* ◇ B35 F55 ◇

RICHMAN, F. *Church's thesis without tears* ◇ D20 D75 F55 F60 ◇

1984

BRIDGES, D.S. & MINES, R. *What is constructive mathematics?* ◇ A05 F55 F98 ◇

DELZELL, C.N. *A continuous, constructive solution to Hilbert's 17th problem* ◇ C60 F55 ◇

DRAGO, A. *Constructive analysis and thermodynamics formulations* ◇ F55 ◇

FEFERMAN, S. *Between constructive and classical mathematics* ◇ A05 F35 F50 F55 ◇

JULIAN, W. *ε-continuity and monotone operators* ◇ F55 ◇

JULIAN, W. & RICHMAN, F. *A uniformly continuous function on* $[0,1]$ *that is everywhere different from its infimum* ◇ F55 ◇

MINES, R. & RICHMAN, F. *Valuation theory: A constructive view* ◇ F55 ◇

SCHNEIDER, B. *Klassische und intuitionistische Mathematik bei L.E.J. Brouwer- dargestellt am Cantorschen Hauptheorem* ◇ A10 F55 ◇

SEIDENBERG, A. *On the Lasker-Noether decomposition theorem* ◇ F55 ◇

1985

BEESON, M.J. *Foundations of constructive mathematics* ◇ F50 F55 F60 F98 ◇

BISHOP, E.A. & BRIDGES, D.S. *Constructive analysis* ◇ D80 F55 F98 ◇

BRIDGES, D.S. *Operator ranges, integrable sets, and the functional calculus* ◇ F55 ◇

DALEN VAN, D. *A comment on the essay "The fundamental theorem of algebra and intuitionism"* ◇ F55 ◇

JULIAN, W. & PHILLIPS, K. *Constructive bounded sequences and Lipschitz functions* ◇ F55 ◇

MANDELKERN, M. *Constructive mathematics* ◇ F55 ◇

MARTINO, E. *On the Brouwerian concept of negative continuity* ◇ A10 F50 F55 ◇

MARTINO, E. *On Brower's concept of negative continuity (Italian)* ◇ A05 A10 F50 F55 ◇

WU, WENJUN *Renaisance of constructive mathematics (Chinese)* ◇ A05 F55 ◇

F60 Constructive recursive analysis

1936

TURING, A.M. *On computable numbers, with an application to the "Entscheidungsproblem"* ◇ D05 D10 D20 D35 F60 ◇

1937

HERMES, H. *Definite Begriffe und berechenbare Zahlen* ◇ D05 D10 D20 F60 ◇

1945

GOODSTEIN, R.L. *Function theory in an axiom-free equation calculus* ◇ B28 F30 F60 ◇

1949

SPECKER, E. *Nicht konstruktiv beweisbare Saetze der Analysis* ◇ F60 ◇

1950

GOODSTEIN, R.L. *Mean value theorems in recursive function theory. part I, Differential mean value theorems* ◇ F60 ◇

GOODSTEIN, R.L. *The formal structure of a denumerable system* ◇ F30 F60 ◇

GOODSTEIN, R.L. *The Gauss test for relative convergence* ◇ F60 ◇

PETER, R. *Zum Begriff der rekursiven reellen Zahl* ◇ F60 ◇

1951

GOODSTEIN, R.L. *Constructive formalism: Essays on the foundations of mathematics* ◇ A05 B98 F30 F60 F98 ◇

1953

GOODSTEIN, R.L. *A problem in recursive function theory* ◇ F60 ◇

MYHILL, J.R. *Criteria of constructibility for real numbers* ◇ F60 ◇

1954

GOODSTEIN, R.L. *A free variable function theory* ◇ F55 F60 ◇

GOODSTEIN, R.L. *The recursive irrationality of* π ◇ F30 F60 ◇

GOODSTEIN, R.L. *The relatively exponential, logarithmic and circular functions in recursive function theory* ◇ D20 F30 F60 ◇

MARKOV, A.A. *On the continuity of constructive functions (Russian)* ◇ F60 ◇

RICE, H.G. *Recursive real numbers* ◇ F60 ◇

1955

GOODSTEIN, R.L. *On non-constructive theorems of analysis and the decision problem* ◇ B25 B28 F30 F60 ◇

GRZEGORCZYK, A. *Elementarily definable analysis* ◇ F60 ◇

GRZEGORCZYK, A. *On the definition of computable functionals* ◇ D20 F60 ◇

LACOMBE, D. *Extension de la notion de fonction recursive aux fonctions d'une ou plusieurs variables reelles. I* ◇ F60 ◇

LACOMBE, D. *Extension de la notion de fonction recursive aux fonctions d'une ou plusieurs variables reelles II,III* ◇ F60 ◇

LACOMBE, D. *Remarque sur les operateurs recursifs et sur les fonctions recursives d'une variable reelle* ◇ F60 ◇

TSEJTIN, G.S. *On Cauchy's theorem in constructive analysis (Russian)* ◇ F60 ◇

ZASLAVSKIJ, I.D. *Disproof of some theorems of classical analysis in constructive analysis (Russian)* ◇ F60 ◇

1956

KLAUA, D. *Berechenbare Analysis* ◇ F60 ◇

KONDO, M. *Sur la nommabilite d'ensembles* ◇ D55 E15 F60 F65 ◇

MESCHKOWSKI, H. *Rekursive reelle Zahlen* ◇ F60 ◇

MESCHKOWSKI, H. *Zur rekursiven Funktionentheorie* ◇ F60 ◇

SHANIN, N.A. *Some problems of mathematical analysis in the light of constructive logic (Russian) (German summary)* ◇ F60 ◇

1957

GOODSTEIN, R.L. *Recursive number theory. A development of recursive arithmetic in a logic-free equation calculus* ◇ B28 F30 F50 F60 F98 ◇

GRZEGORCZYK, A. *On the definitions of computable real continuous functions* ◇ D20 F60 ◇

KREISEL, G. & LACOMBE, D. & SHOENFIELD, J.R. *Effective operations and partial recursive functionals* ◇ D20 F60 ◇

KREISEL, G. & LACOMBE, D. *Ensembles recursivement mesurables et ensembles recursivement ouverts ou fermes* ◇ D80 F60 ◇

KREISEL, G. & LACOMBE, D. & SHOENFIELD, J.R. *Fonctionnelles recursivement definissables et fonctionnelles recursives* ◇ D20 F60 ◇

LACOMBE, D. *Les ensembles recursivement ouverts ou fermes, et leurs applications a l'analyse recursive* ◇ D55 F60 ◇

LACOMBE, D. *Quelques proprietes d'analyse recursive* ◇ F60 ◇

MARKOV, A.A. *Mathematical logic and numerical analysis (Russian)* ◇ F60 ◇

MOSTOWSKI, ANDRZEJ *On computable sequences* ◇ D20 F60 ◇

USPENSKIJ, V.A. *On the theorem of uniform continuity (Russian)* ◇ F60 ◇

1958

FALEVICH, B.YA. *A new method of proving incompleteness theorems for systems with Carnap's rule, and its application to the problem of interrelation between classical and constructive analysis (Russian)* ◇ F35 F60 ◇

LACOMBE, D. *Sur les possibilites d'extension de la notion de fonction recursive aux fonctions d'une ou plusieurs variables reelles* ◇ F60 ◇

MARKOV, A.A. *On constructive functions (Russian)* ◇ F60 ◇

SHANIN, N.A. *Ueber einen Algorithmus zur konstruktiven Dechiffrierung mathematischer Urteile (Russian) (German summary)* ◇ A05 F50 F60 ◇

1959

GOODSTEIN, R.L. *Recursive analysis* ◇ F60 ◇

GRZEGORCZYK, A. *Some approaches to constructive analysis* ◇ F60 ◇

KLAUA, D. *Die Praezisierung des Berechenbarkeitsbegriffs in der Analysis mit Hilfe rationaler Funktionale* ◇ F60 ◇

KREISEL, G. & LACOMBE, D. & SHOENFIELD, J.R. *Partial recursive functionals and effective operations* ◇ D20 F60 ◇

MOSTOWSKI, ANDRZEJ *On various degrees of constructivism* ◇ D55 F50 F60 F98 ◇

PETER, R. *Rekursivitaet und Konstruktivitaet* ◇ A05 D20 F60 ◇

SHANIN, N.A. *On constructive mathematical analysis (Russian)* ◇ F60 ◇

SHANIN, N.A. *Ueber konstruktive lineare Funktionale in einem konstruktiven Hilbertschen Raum (Russian) (German summary)* ◇ F60 ◇

SPECKER, E. *Der Satz vom Maximum in der rekursiven Analysis* ◇ F60 ◇

TSEJTIN, G.S. *Algorithmic operators in constructive complete separable metric spaces (Russian)* ◇ F60 ◇

TSEJTIN, G.S. & ZASLAVSKIJ, I.D. *On relations between fundamental properties of constructive functions (Russian)* ◇ F60 ◇

TSEJTIN, G.S. *The theorem concerning nested segments, Cauchy's theorem, and Rolle's theorem in constructive analysis (Russian)* ◇ F60 ◇

TSEJTIN, G.S. *Uniform recursiveness of algorithmic operators on general recursive functions and a canonical representation for constructive functions of a real argument (Russian)* ◇ D20 F60 ◇

USPENSKIJ, V.A. *Computable operations, computable operators and constructively continuous functions (Russian)* ◇ D20 F60 ◇

ZASLAVSKIJ, I.D. *On constructive Dedekind cuts (Russian)* ◇ F60 ◇

ZASLAVSKIJ, I.D. *Some peculiarities of constructive functions of a real variable in comparison with classical ones (Russian)* ◇ F60 ◇

1960

GOODSTEIN, R.L. & HOOLEY, J. *On recursive transcendence* ◇ F60 ◇

KLAUA, D. *Berechenbare Reihen* ◇ F60 ◇

LEHMAN, R.S. *On primitive recursive real numbers* ◇ F60 ◇

MIHALJINEC, M. *On the continuity of constructive transformations. I,II (Serbo-Croatian summaries)* ◇ F60 ◇

USPENSKIJ, V.A. *On the relation between various systems of constructive real numbers (Russian)* ◇ F60 ◇

1961

GOODSTEIN, R.L. *Recursive analysis* ◇ F30 F60 F98 ◇

HERMES, H. *Aufzaehlbarkeit, Entscheidbarkeit, Berechenbarkeit: Einfuehrung in die Theorie der rekursiven Funktionen* ◇ D98 F60 ◇

KLAUA, D. *Konstruktive Analysis* ◇ F60 F98 ◇

MAL'TSEV, A.I. *Constructive algebra I (Russian)* ◇ C57 C98 F60 F98 ◇

1962

MARKOV, A.A. *On constructive mathematics (Russian)* ◇ A05 F60 ◇

MINTS, G.E. *The differentiability predicate and the differentiation operator in constructive mathematical analysis (Russian)* ◇ F60 ◇

SHANIN, N.A. *Constructive real numbers and constructive function spaces (Russian)* ◇ F60 ◇

TSEJTIN, G.S. *Algorithmic operators in constructive metric spaces (Russian)* ◇ F60 ◇

TSEJTIN, G.S. *Mean value theorems in constructive analysis (Russian)* ◇ F60 ◇

TSEJTIN, G.S. & ZASLAVSKIJ, I.D. *On singular coverings and properties of constructive functions connected with them (Russian)* ◇ F60 ◇

ZASLAVSKIJ, I.D. *Some properties of constructive real numbers and constructive functions (Russian)* ◇ F60 ◇

1963

LACHLAN, A.H. *Recursive real numbers* ◇ F60 ◇

MAZUR, S. *Computable analysis* ◇ F60 F98 ◇

OREVKOV, V.P. *A constructive mapping of the square onto itself displacing every constructive point (Russian)* ◇ F60 ◇

OREVKOV, V.P. *Constructive mappings of polyhedra (Russian)* ◇ F60 ◇

PARTIS, M.T. *Commutative partially ordered recursive arithmetics* ◇ D20 F60 ◇

SLISENKO, A.O. *Certain properties of arithmetic operations on duplexes (Russian)* ◇ F60 ◇

1964

KUSHNER, B.A. *Riemann integration in constructive analysis (Russian)* ◇ F60 ◇

MANUKYAN, S.N. & ZASLAVSKIJ, I.D. *Jordan theorem in constructive analysis* ◇ F60 ◇

MOSCHOVAKIS, Y.N. *Recursive metric spaces* ◇ C57 C65 D45 F60 ◇

OREVKOV, V.P. *Certain questions of the theory of polynomials with constructive real coefficients (Russian)* ◇ F60 ◇

OREVKOV, V.P. *On constructive mappings of a disk into itself (Russian)* ◇ F60 ◇

SLISENKO, A.O. *Example of a non-discontinuous but not continuous constructive operator in a metric space (Russian)* ◇ F60 ◇

SLISENKO, A.O. *On certain algorithmic problems connected with arithmetic operations on duplexes (Russian)* ◇ F60 ◇

SLISENKO, A.O. *On constructive non-separable spaces (Russian)* ◇ F60 ◇

TSEJTIN, G.S. & ZASLAVSKIJ, I.D. *Concerning a generalized principle of constructive selection (Russian)* ◇ F60 ◇

TSEJTIN, G.S. *Three theorems on constructive functions (Russian)* ◇ F60 ◇

ZASLAVSKIJ, I.D. *Differentation and integration of constructive functions (Russian)* ◇ F60 ◇

1965

DEMUTH, O. *On Lebesgue integration in constructive analysis (Russian)* ◇ F60 ◇

HE, KEGANG *On the types of some basic predicates of constructive real number theory in the S.C.Kleene classifications (Chinese)* ◇ F50 F60 ◇

ILSE, D. *Zur Stetigkeit berechenbarer reeller Funktionen* ◇ F60 ◇

KUSHNER, B.A. *Constructive theory of the Riemann integral (Russian)* ◇ F60 ◇

KUSHNER, B.A. *On the existence of unbounded analytic constructive functions (Russian)* ◇ F60 ◇

MAYOH, B.H. *Unsolvable problems in the theory of computable numbers* ◇ D20 D35 F60 ◇

MIHALJINEC, M. *Inverse upper bound theorems for constructive real functions (Serbo-Croatian summary)* ◇ F60 ◇

PHAN DINH DIEU *Constructive locally convex linear topological spaces (Russian)* ◇ F60 ◇

PHAN DINH DIEU *The metrizability, normability and multinormability of constructive locally convex spaces (Russian)* ◇ F60 ◇

1966

KUSHNER, B.A. *Certain properties of quasinumbers and of operators from quasinumbers into quasinumbers (Russian)* ◇ F60 ◇

MOSCHOVAKIS, Y.N. *Notation systems and recursive ordered fields* ◇ C57 C60 D45 F60 ◇

PHAN DINH DIEU *On spaces adjoint to constructive locally convex spaces (Russian)* ◇ F60 ◇

SHURYGIN, V.A. *Nontrivial constructive mappings of certain sets (Russian)* ◇ D20 D45 F60 ◇

1967

DEMUTH, O. *Lebesgue integration in constructive analysis (Russian)* ◇ F60 ◇

DEMUTH, O. *Necessary and sufficient conditions for Riemann integrability of constructive functions (Russian)* ◇ F60 H05 ◇

KUSHNER, B.A. *Certain relations among the properties of constructive functions and operators from quasinumbers into quasinumbers (Russian)* ◇ F60 ◇

KUSHNER, B.A. *On constructive antiderivatives (Russian)* ◇ F60 ◇

LIFSCHITZ, V. *Constructive groups (Russian)* ◇ D45 F60 ◇

OREVKOV, V.P. *Constructive mappings of finite polyhedra (Russian)* ◇ F60 ◇

OREVKOV, V.P. *On certain types of continuity of constructive operators (Russian)* ◇ F60 ◇

OREVKOV, V.P. & SHANIN, N.A. (EDS.) *Problems in the constructive trend in mathematics IV (Russian)* ◇ F60 F97 ◇

PARTIS, M.T. *Limited universal and existential quantifiers in commutative partially ordered recursive arithmetics* ◇ D20 F60 ◇

PHAN DINH DIEU *Certain properties of constructive generalized functions (Russian)* ◇ F60 ◇

PHAN DINH DIEU *Constructive generalized functions (Russian)* ◇ F60 ◇

PHAN DINH DIEU *On closed and open sets in constructive topological spaces (Russian)* ◇ F60 ◇

SHURYGIN, V.A. *Constructive sets with equality and their mappings (Russian)* ◇ D03 F60 ◇

SLISENKO, A.O. *Arithmetical operations on certain sets of duplexes (Russian)* ◇ F60 ◇

SLISENKO, A.O. *On constructing of maximal continuity regulators of constructive functions (Russian)* ◇ F60 ◇

SLISENKO, A.O. *On maximal continuity regulators of constructive functions (Russian)* ◇ F60 ◇

ZASLAVSKIJ, I.D. *Rectifiability of constructive plane curves (Russian) (English and Armenian summaries)* ◇ F60 ◇

1968

ABERTH, O. *Analysis in the computable number field* ◇ F60 ◇

DEMUTH, O. *Fubini's theorem for the Riemann integral in constructive mathematics (Russian)* ◇ F60 ◇

DEMUTH, O. *The connection between Riemann and Lebesgue integrability of constructive functions (Russian)* ◇ F60 H05 ◇

DEMUTH, O. *The Lebesgue integral and the concept of measureability of functions in constructive analysis (Russian)* ◇ F60 H05 ◇

KHACHATRYAN, M.A. *Constructive series of numbers (Russian) (Armenian summary)* ◇ F60 ◇

KUSHNER, B.A. *A remark on the domains of definition of constructive functions (Russian)* ◇ F60 ◇

KUSHNER, B.A. *Some examples of quasidense but not dense sets of duplexes (Russian)* ◇ F60 ◇

KUSHNER, B.A. & TSEJTIN, G.S. *Some properties of F-numbers (Russian)* ◇ F60 ◇

LIFSCHITZ, V. *Constructive analytic functions of one real variable (Russian)* ◇ F60 ◇

MANUKYAN, S.N. & ZASLAVSKIJ, I.D. *Partitionings of the plane by constructive curves (Russian)* ◇ F60 ◇

MAYOH, B.H. *Semi-effective numberings and definitions of the computable numbers* ◇ D25 D30 F60 ◇

PHAN DINH DIEU *On spaces of constructive infinitely differentiable functions and on functionals in them (Russian)* ◇ F60 ◇

SHANIN, N.A. & TSEJTIN, G.S. & ZASLAVSKIJ, I.D. *Peculiarities in constructive mathematical analysis (Russian)* ◇ F60 ◇

SHURYGIN, V.A. *Complete constructive sets with equality, and some of their properties (Russian)* ◇ D03 F60 ◇

TAIT, W.W. *Normal derivability in classical logic* ◇ C75 F05 F15 F30 F35 F60 ◇

1969

ABERTH, O. *A chain of inclusion relations in computable analysis* ◇ F60 ◇

CLEAVE, J.P. *The primitive recursive analysis of ordinary differential equations and the complexity of their solutions* ◇ D20 D80 F60 ◇

DEMUTH, O. *Linear functionals in the constructive spaces L_r (Russian)* ◇ F60 ◇

DEMUTH, O. *Note on the paper "Fubini's theorem for the Riemann integral in constructive mathematics" (Russian)* ◇ F60 ◇

DEMUTH, O. *The differentiability of constructive functions (Russian)* ◇ F60 ◇

DEMUTH, O. *The spaces L_r and S in constructive mathematics (Russian)* ◇ F60 ◇

DEMUTH, O. *The Lebesgue measurability of sets in constructive mathematics (Russian)* ◇ F60 ◇

GEL'FOND, M.G. *On constructive pseudofunctions (Russian)* ◇ F60 ◇

GOODSTEIN, R.L. *A constructive form of the second Gauss proof of the fundamental theorem of algebra* ◇ F60 ◇

KANOVICH, M.I. & KUSHNER, B.A. *Estimating the complexity of certain algorithmic problems of analysis (Russian)* ◇ D15 D20 D80 F60 ◇

KHACHATRYAN, M.A. *An example of a constructive nondifferentiable monotone function (Russian) (Armenian and English summaries)* ◇ F60 ◇

KHACHATRYAN, M.A. *The differentiation of constructive monotone functions (Russian)* ◇ F60 ◇

KOSOVSKIJ, N.K. *Integrable FR-constructs over a probability space (Russian)* ◇ F60 ◇

KOSOVSKIJ, N.K. *Laws of large numbers in constructive probability theory (Russian)* ◇ F60 ◇

KOSOVSKIJ, N.K. *Necessary and sufficient conditions for a probability space to have Specker properties (Russian)* ◇ F60 ◇

LIFSCHITZ, V. *The set of zeros of a constructive power series in the real domain (Russian)* ◇ F60 ◇

LORENTS, A.A. *Elements of the constructive probability theory (Russian) (German summary)* ◇ F60 ◇

OREVKOV, V.P. *Some properties of homeomorphisms of constructive metric spaces (Russian)* ◇ F60 ◇

SHAPIRO, N.Z. *Real numbers and functions in the Kleene hierarchy and limits of recursive, rational functions* ◇ D55 F60 ◇

SHAPIRO, N.Z. *Recursively countable subsets of recursive metric spaces (Russian summary)* ◇ F60 ◇

SOARE, R.I. *Recursion theory and Dedekind cuts* ◇ D25 D30 F60 ◇

SPECKER, E. *The fundamental theorem of algebra in recursive analysis* ◇ F60 ◇

ZASLAVSKIJ, I.D. *On Shannon pseudofunctions (Russian)* ◇ D05 D15 F60 ◇

1970

ABERTH, O. *Computable analysis and differential equations* ◇ F60 ◇

DEMUTH, O. *Mittelwertsaetze fuer konstruktive Lebesgueintegrale (Russisch)* ◇ F60 ◇

DEMUTH, O. *Necessary and sufficient conditions for the absolute continuity of constructive functions (Russian)* ◇ F60 ◇

DEMUTH, O. *The integrability of derivatives of constructive functions (Russian)* ◇ F60 ◇

DEMUTH, O. *The representability of constructive functions of weakly bounded variation (Russian)* ◇ F60 ◇

GOODSTEIN, R.L. *Polynomials with computable coefficients* ◇ F60 ◇

KOSOVSKIJ, N.K. *Some questions in the constructive theory of normed boolean algebras (Russian)* ◇ C57 F60 G05 ◇

KUCERA, A. *Weak convergence in constructive mathematics (Russian)* ◇ F60 ◇

KUSHNER, B.A. *Some mass problems connected with the integration of constructive functions (Russian)* ◇ D30 F60 ◇

MADISON, E.W. *A note on computable real fields* ◇ C57 C60 D45 F60 ◇

MARTIN-LOEF, P. *Notes on constructive mathematics* ◇ F50 F60 F98 ◇

MAYOH, B.H. *The relation between an object and its name: notation systems and their fixed point theorems* ◇ A05 B03 D20 D35 D45 F30 F60 ◇

MINTS, G.E. (ED.) *Recursive mathematical analysis (Russian)* ◇ F60 ◇

OREVKOV, V.P. & SHANIN, N.A. (EDS.) *Problems in the constructive trend in mathematics V (Russian)* ◇ F60 F97 ◇

PHAN DINH DIEU *Some questions in constructive functional analysis (Russian)* ◇ F60 ◇

SHURYGIN, V.A. *Constructive sets with equality and their mappings (Russian)* ◇ D03 F60 ◇

SLISENKO, A.O. *Some questions of the approximation of the maximal regulators of continuity (Russian)* ◇ F60 ◇

TSEJTIN, G.S. & ZASLAVSKIJ, I.D. *A criterion of the rectifiability of constructive plane curves (Russian) (Armenian and English summaries)* ◇ F60 ◇

TSEJTIN, G.S. *On upper bounds of recursively enumerable sets of constructive real numbers (Russian)* ◇ D25 F60 ◇

1971

ABERTH, O. *The failure in computable analysis of a classical existence theorem for differential equations* ◇ F60 ◇

DEMUTH, O. *A certain condition for the differentiability of constructive functions of bounded variation (Russian)* ◇ F60 ◇

DEMUTH, O. *A necessary and sufficient condition for the representability of constructive functions in the form of the sum of a singular and an absolutely continuous function (Russian)* ◇ F60 ◇

DEMUTH, O. *The superposition of absolutely continuous functions (Russian)* ◇ F60 ◇
HAUCK, J. *Ein Kriterium fuer die Annahme des Maximums in der berechenbaren Analysis* ◇ F60 ◇
HAUCK, J. *Zur Praezisierung des Begriffes berechenbare reelle Funktion* ◇ F60 ◇
KUCERA, A. *Sufficient conditions for the normability of linear operators in constructive mathematics (Russian)* ◇ F60 ◇
LIFSCHITZ, V. *Investigations of constructive functions by the method of fillings (Russian) (English summary)* ◇ F60 ◇
MANUKYAN, S.N. *On interior points of nondegenerate constructive curves (Russian)* ◇ F60 ◇
MATIYASEVICH, YU.V. *A sufficient condition for the convergence of monotone sequences (Russian) (English summary)* ◇ F60 ◇
MYHILL, J.R. *A recursive function, defined on a compact interval and having a continuous derivative that is not recursive* ◇ F60 ◇
OREVKOV, V.P. *Equivalence of two definitions of continuity (Russian) (English summary)* ◇ F60 ◇
OREVKOV, V.P. *On the continuity of constructive functionals (Russian) (English summary)* ◇ F60 ◇
SPECKER, E. *Ramsey's theorem does not hold in recursive set theory* ◇ D25 D80 E05 F60 ◇
TSEJTIN, G.S. *A pseudofundamental sequence that is not equivalent to any monotone sequence (Russian) (English summary)* ◇ F60 ◇
TSEJTIN, G.S. & ZASLAVSKIJ, I.D. *Another constructive version of the Cauchy theorems (Russian) (English summary)* ◇ F60 ◇

1972

CHERNOV, V.P. *Classification of spaces of operators of finite types (Russian)* ◇ D45 D65 F60 ◇
CHERNOV, V.P. *Constructive operators of finite types (Russian) (English summary)* ◇ D45 D65 F60 ◇
CHERNOV, V.P. & LIFSCHITZ, V. *Noncompact closed ball in a constructive compact metric space (Russian) (English summary)* ◇ F60 ◇
CHERNOV, V.P. *Topological variants of the continuity theorem for mappings and related theorems theorems (Russian) (English summary)* ◇ D20 F60 ◇
DEMUTH, O. *A necessary and sufficient condition for the representability of constructive functions in the form of a superposition of absolutely continuous functions (Russian)* ◇ F60 ◇
DEMUTH, O. *A sufficient condition for the representability of constructive functions in the form of the sum of two superpositions of absolutely continuous functions (Russian)* ◇ F60 ◇
FREJDZON, R.I. *Families of recursive predicates of measure zero (Russian) (English summary)* ◇ D20 F60 ◇
GEL'FOND, M.G. *Relationship between the classical and constructive developments of mathematical analysis (Russian) (English summary)* ◇ B28 F35 F60 ◇
HAUCK, J. *Berechenbare gleichmaessige Stetigkeit in der konstruktiven Analysis von Klaua (Russian, English and French summaries)* ◇ F60 ◇
HAUCK, J. *Zur berechenbaren gleichmaessigen Finitheit rekursiver Funktionale* ◇ D20 F60 ◇
LIFSCHITZ, V. *A locally analytic constructive function that is not analytic (Russian)* ◇ F60 ◇
LORENTS, A.A. *Elemente der konstruktiven Theorie stochastischer Automaten (Russian)* ◇ D05 D10 D98 F60 F98 ◇
MIHALJINEC, M. *Some applications of the theory of categories in partially constructive mathematics (Serbo-Croatian summary)* ◇ F60 G30 ◇
MYHILL, J.R. *What is a real number ?* ◇ F55 F60 ◇
PODNIEKS, K.M. *Constructive decomposition of stochastic matrices (Russian)* ◇ F60 ◇
SKARBEK, W. & ZEMBRZUSKI, K. *Computable real functions and their relation to the analog computer* ◇ D20 F60 ◇
STRAUCH, O. *Minimal covering of a closed interval (Slovak) (English summary)* ◇ E75 F60 ◇

1973

DEMUTH, O. *The constructive analogue of the connection between the Lebesgue measurability of sets and of functions (Russian)* ◇ F60 ◇
DEMUTH, O. & NEMECKOVA, L. *The constructive analogue of the property* (T_1) *(Russian)* ◇ F60 ◇
DEMUTH, O. & NEMECKOVA, L. *The constructive analogues of the properties (N) and (S) (Russian)* ◇ F60 ◇
DEMUTH, O. *The representability of uniformly continuous constructive functions (Russian)* ◇ F60 ◇
GANOV, V.A. *A generalized constructive continuum (Russian)* ◇ F60 ◇
HAUCK, J. *Berechenbare reelle Funktionen* ◇ F60 ◇
KOSOVSKIJ, N.K. *Constructive versions of the laws of large numbers (Russian)* ◇ F60 ◇
KOSOVSKIJ, N.K. *Some problems in the constructive probability theory* ◇ F60 ◇
KUSHNER, B.A. *A certain problem of Mostowski (Russian)* ◇ D20 F60 ◇
KUSHNER, B.A. *Computationally complex real numbers (Russian)* ◇ D20 F60 ◇
KUSHNER, B.A. *Continuity theorems for some types of computable operators (Russian)* ◇ F60 ◇
KUSHNER, B.A. *Coverings of separable sets (Russian)* ◇ F60 ◇
KUSHNER, B.A. *Lectures on constructive mathematical analysis (Russian)* ◇ F60 F98 ◇
KUSHNER, B.A. *Two theorems on Riemann-integrable constructive functions (Russian)* ◇ F60 ◇
MANUKYAN, S.N. *Constructive everywhere dense simple arcs (Russian) (Armenian and English summaries)* ◇ E75 F60 ◇
OREVKOV, V.P. & SHANIN, N.A. (EDS.) *Problems in the constructive trend in mathematics VI (Russian)* ◇ F60 F97 ◇
OSTROUKHOV, D.A. *Linearization of constructive sequences of normal algorithms (Russian)* ◇ D03 F60 ◇
SHURYGIN, V.A. *Invertibility and regularity of constructive operators (Russian)* ◇ F60 ◇

1974

CHERNOV, V.P. *Some properties of mappings of sheaf-spaces (Russian) (English summary)* ◇ E75 F60 G30 ◇

DEMUTH, O. *The connection between the representablility of a constructive function in the form of a superposition of two absolutely continuous functions and the differentiability of this function (Russian)* ◇ F60 ◇

DEMUTH, O. *The representability of constructive functions possessing the properties (S) and (T_1) in the form of superpositions (Russian)* ◇ F60 ◇

KANOVICH, M.I. *Complexity of the limit of Specker sequences (Russian)* ◇ D15 D20 F60 ◇

KHISAMIEV, N.G. *Strongly constructive models of a decidable theory (Russian)* ◇ C15 C57 F60 ◇

KREJNOVICH, V.YA. *Constructivization of the concepts of ε entropy and ε capacity (Russian) (English summary)* ◇ F60 ◇

KRYL, R. *The constructive analogue of a certain theorem of Luzin (Russian)* ◇ F60 ◇

KUCERA, A. *Locally convex topologies of finite-dimensional constructive spaces (Russian)* ◇ F60 ◇

KUSHNER, B.A. *A constructive version of Koenig's theorem; functions that are computable in the sense of Markov, Grzegorczyk and Lacombe (Russian)* ◇ F60 ◇

KUSHNER, B.A. *On a type of computable real functions (Russian)* ◇ F60 ◇

MARGENSTERN, M. *Constructive functionals in spaces of almost periodic functions (Russian) (English summary)* ◇ F60 ◇

MARGENSTERN, M. *Proprietes topologiques constructives des espaces de fonctions presque periodiques* ◇ F60 ◇

MELIKYAN, S.M. *Constructive transfinite hierarchies of pseudonumbers (Russian)* ◇ D55 F60 ◇

OREVKOV, V.P. *A new proof of the uniqueness theorem for constructive differentiable functions of a complex variable (Russian) (English summary)* ◇ F60 ◇

PETRI, N.V. *Effektive Nichtaufzaehlbarkeit von Pseudozahlen (Russian)* ◇ F60 ◇

SHURYGIN, V.A. *Einige Eigenschaften der Kompliziertheit konstruktiver reeller Zahlen (Russian)* ◇ D15 F60 ◇

ZHAROV, V.G. *On an analog of a theorem of Specker (Russian)* ◇ D20 D80 F60 ◇

1975

BLOOM, S.L. *A note on the predicatively definable sets of N.N.Nepeivoda* ◇ F60 F65 ◇

DEMUTH, O. *Constructive pseudonumbers (Russian)* ◇ D55 F60 ◇

DEMUTH, O. *The differentiability of constructive functions of weakly bounded variation on pseudo-numbers (Russian)* ◇ F60 ◇

HAUCK, J. *Iterationsverfahren in der konstruktiven Analysis* ◇ F60 ◇

HAUCK, J. *Turingmaschinen und berechenbare reelle Funktionen* ◇ D10 D20 F60 ◇

MADISON, E.W. & NELSON, GEORGE C. *Some examples of constructive and non-constructive extension of the countable atomless boolean algebra* ◇ C57 F60 G05 ◇

MANUKYAN, S.N. *Some properties of partitioning of the plane by non-degenerate constructive curves (Russian)* ◇ F60 ◇

MANUKYAN, S.N. *The stability of constructive simple arcs with respect to algorithmic transformations (Russian)* ◇ F60 ◇

VERSHININ, V.A. *On the question of superposability of constructive sets with equality (Russian)* ◇ D25 F60 ◇

1976

BREMER, H. & REYNVAAN, H.H. *Zur konstruktiven Differenzierbarkeit von monotonen berechenbaren Funktionen* ◇ F60 ◇

CHERNOV, V.P. *Example of a constructive separable space with an inseparable completion (Russian)* ◇ F60 ◇

DANTSIN, E.YA. *On an approximative version of the notion of constructive analytic function (Russian) (English summary)* ◇ F60 ◇

DEGTYAREV, A.I. *A class of algorithmic operators (Russian)* ◇ F60 ◇

DEMUTH, O. *The constructive analogue of the Denjoy-Young theorem on derived numbers (Russian)* ◇ F60 ◇

DEMUTH, O. *The domains of definition of effective operators over general recursive functions and of constructive functions of a real variable (Russian)* ◇ D20 F60 ◇

FILIPEC, P. *On convergence of the Fourier series of a constructive function of weakly bounded variation* ◇ F60 ◇

GONCHAROV, S.S. *Restricted theories of constructive boolean algebras (Russian)* ◇ B25 C57 D35 F60 G05 ◇

HAUCK, J. *Berechenbare reelle Funktionenfolgen* ◇ F60 ◇

KOSOVSKIJ, N.K. *On constructive distribution functions (Russian) (English summary)* ◇ F60 ◇

KUSHNER, B.A. *On the computation of isolated roots of constructive functions (Russian)* ◇ F60 ◇

KUSHNER, B.A. *On Grzegorczyk's theorem on the computability of an isolated extremum (Russian)* ◇ D20 D80 F60 ◇

MANUKYAN, S.N. *Some topological peculiarities of constructive simple arcs (Russian)* ◇ F60 ◇

MARGENSTERN, M. *Some constructive topological properties of function spaces* ◇ F60 ◇

NAGORNYJ, N.M. *Algorithms and real numbers* ◇ F60 ◇

PAKHOMOV, S.V. *Approximability of operators in constructive metric spaces (Russian) (English summary)* ◇ F60 ◇

PAKHOMOV, S.V. *Continuity of operators in separable constructive metric spaces (Russian) (English summary)* ◇ F60 ◇

PAKHOMOV, S.V. *Hierarchies of operators in constructive metric spaces (Russian) (English summary)* ◇ D80 F60 ◇

SHEPHERDSON, J.C. *On the definition of computable function of a real variable* ◇ D20 F60 ◇

SHURYGIN, V.A. *Pseudoregular and quasiregular constructive operators (Russian)* ◇ F60 ◇

SHURYGIN, V.A. *Traceable operators in constructive analysis (Russian)* ◇ F60 ◇

VAJNBERG, YU.R. *On the domains of definition of computable mappings of metric spaces (Russian)* ◇ F60 ◇

1977

DEMUTH, O. *A constructive analogue of the functions of the n-th Baire class* ◇ D55 E15 F60 ◇

DEMUTH, O. *On a generalization of the constructive Lebesgue integral (Russian)* ◇ F60 ◇

HAUCK, J. *Rekursive Funktionale und der Fixpunktsatz von Banach* ◇ F60 ◇

HAUCK, J. *Zum Zwischenwertsatz in der konstruktiven Analysis* ◇ F60 ◇

KUCERA, A. *The algorithmic nonapproximability of the least upper bounds of constructive pseudosections (Russian)* ◇ F60 ◇

MANUKYAN, S.N. *On a property of the notion of interior and exterior point with respect to constructive closed curve (Russian)(English and Armenian summaries)* ◇ F60 ◇

SHURYGIN, V.A. *On estimates of the complexity of algorithmic problems in constructive analysis (Russian)* ◇ D15 D20 F60 ◇

TROELSTRA, A.S. *Aspects of constructive mathematics* ◇ F30 F35 F50 F55 F60 F98 ◇

VERSHININ, V.A. *On certain operations over constructive sets with equality (Russian)* ◇ F60 ◇

1978

COLLINS, W.J. *Provably recursive real numbers* ◇ D20 F60 ◇

DEMUTH, O. & KRYL, R. & KUCERA, A. *An application of the theory of functionals partial recursive relative to number sets in constructive mathematics (Russian)* ◇ D30 D55 F60 ◇

DEMUTH, O. *Constructive analogues of generalized absolutely continuous functions of generalized bounded variation (Russian)* ◇ F60 ◇

DEMUTH, O. *Some questions of the theory of constructive functions of a real variable (Russian) (English summary)* ◇ F60 ◇

DEMUTH, O. *The pseudodifferentiability of uniformly continuous constructive functions on constructive real numbers (Russian)* ◇ F60 ◇

DOBRITSA, V.P. & KHISAMIEV, N.G. & NURTAZIN, A.T. *Constructive periodic abelian groups (Russian)* ◇ C57 C60 F60 ◇

HAUCK, J. *Konstruktive Darstellungen reeller Zahlen und Folgen* ◇ F60 ◇

KAROL', A.M. *Complexity of the set of constructive real numbers (Russian)* ◇ D15 F60 ◇

KUCERA, A. & KUSHNER, B.A. *A type of recursive isomorphism of certain concepts of constructive analysis (Russian)* ◇ D55 F60 ◇

KUSHNER, B.A. *On some systems of computable real numbers (Russian)* ◇ D45 F60 ◇

MARGENSTERN, M. *On a variant of constructivisation of the theory of almost periodic functions* ◇ F60 ◇

MIHALJINEC, M. *Endomorphisms of sheafs of constructive functions* ◇ F60 G30 ◇

POUR-EL, M.B. & RICHARDS, I. *Differentiability properties of computable functions – a summary* ◇ D20 D80 F60 ◇

1979

DEMUTH, O. *Constructive Denjoy integrals (Russian)* ◇ F60 ◇

DEMUTH, O. & KUCERA, A. *Remarks on constructive mathematical analysis* ◇ F60 ◇

DEMUTH, O. & POLIVKA, I. *The representation of linear functionals in the space of codes of constructive functions that are uniformly continuous on the segment 0Δ1 (Russian)* ◇ F60 ◇

DEMUTH, O. *The use of the Riemann-Stieltjes integral in the theory of the constructive Lebesgue integral, and its generalizations (Russian)* ◇ F60 ◇

FEFERMAN, S. *Constructive theories of functions and classes* ◇ F50 F55 F60 F98 ◇

KHISAMIEV, N.G. *On subgroups of finite index of abelian groups (Russian) (Kazakh summary)* ◇ C57 C60 D45 F60 ◇

KUSHNER, B.A. *Segment coverings and the uniform continuity of constructive functions (Russian)* ◇ F60 ◇

MURZIN, F.A. *One result on finite approximability (Russian)* ◇ F60 ◇

POUR-EL, M.B. & RICHARDS, I. *A computable ordinary differential equation which possesses no computable solution* ◇ D80 F60 ◇

SCIENZA, G. *Elementary mathematics from an information-theoretic viewpoint: the real numbers* ◇ B28 F60 ◇

SHURYGIN, V.A. *On the structure of constructive sets with equality (Russian)* ◇ F60 ◇

ZHAROV, V.G. *On the complexity of the terms of the constructive sequences of Turing machines (Russian)* ◇ D10 D45 F60 ◇

1980

ABERTH, O. *Computable analysis* ◇ F30 F60 F98 ◇

DEMUTH, O. *A constructive analogue of K.M.Garg's theorem on derivative numbers (Russian)* ◇ F60 ◇

DEMUTH, O. *On constructive Perron integral (Russian) (English summary)* ◇ F60 ◇

DEMUTH, O. & KMINEK, V. *On properties of indefinite constructive Lebesgue-Stieltjes integrals (Russian)* ◇ F60 ◇

DEMUTH, O. *Pseudodifferentiability of constructive functions on constructive real numbers (Russian)* ◇ F60 ◇

HAUCK, J. *Dezimalbrueche und Berechenbarkeit* ◇ F60 ◇

HAUCK, J. *Stetigkeitseigenschaften berechenbarer reeller Funktionen* ◇ D20 F60 ◇

NEDERPELT, R.P. *An approach to theorem proving on the basis of a typed lambda-calculus* ◇ B35 B40 F60 ◇

1981

DEMUTH, O. *A generalization of a constructive analogue of K.M. Garg's theorem (Russian)* ◇ F60 ◇

DEMUTH, O. *On the pseudodifferentiability of uniformly continuous constructive functions from functions of the same type (Russian)* ◇ F60 ◇

KOSOVSKIJ, N.K. *Elements of mathematical logic and its applications to the theory of subrecursive algorithms (Russian)* ◇ B98 D20 D98 F30 F60 F98 ◇
KUSHNER, B.A. *Behavior of the general term of a Specker series (Russian)* ◇ F60 ◇
KUSHNER, B.A. *On some topological properties of constructive plane curves (Russian)* ◇ F60 ◇
MELIKYAN, S.M. *On some properties of classes of pseudonumbers (Russian) (Armenian summary)* ◇ F60 ◇
MELIKYAN, S.M. *Two theorems on classes of pseudonumbers (Russian) (Armenian summary)* ◇ F60 ◇
PINUS, A.G. *Constructivization of Boolean algebras (Russian)* ◇ F60 G05 ◇
ROCHE LA, P. *Effective Galois theory* ◇ C57 C60 D45 F60 ◇
WU, WENJUN *A constructive theory of algebraic topology I. Notions of measure and calculability (Chinese summary)* ◇ F60 ◇

1982

DEMUTH, O. *Borel types of some classes of arithmetical real numbers (Russian)* ◇ F60 ◇
DEMUTH, O. *Some classes of arithmetical real numbers (Russian)* ◇ F60 ◇
GONCHAROV, S.S. *Limiting equivalent constructivizations (Russian)* ◇ C57 D45 F60 ◇
KOCHUBIEVSKIJ, I.P. & SATAEV, A.G. *Potential realizability: pragmatic aspect (Russian)* ◇ F60 ◇
KUSHNER, B.A. *Some extensions of Markov's constructive continuum and their applications to the theory of constructive functions* ◇ F60 ◇
MANUKYAN, S.N. *Algorithms transforming constructive curves into points (Russian) (Armenian summary)* ◇ F60 ◇
MELIKYAN, S.M. *Classes of pseudonumbers (Russian) (Armenian summary)* ◇ F60 ◇
METAKIDES, G. & NERODE, A. *The introduction of non-recursive methods into mathematics* ◇ C57 D45 D80 F60 ◇
SHURYGIN, V.A. *Construction of eigenvectors of completely continuous selfadjoint linear operators in constructive analysis for an eigenvalue of given multiplicity (Russian)* ◇ F60 ◇
SHURYGIN, V.A. *On eigenvectors of completely continuous linear operators in constructive analysis (Russian)* ◇ F60 ◇
SHURYGIN, V.A. *On the eigenvalue problem for linear operators in constructive analysis (Russian)* ◇ F60 ◇
VERSHININ, V.A. *Some limit theorems in the constructive probability theory (Russian)* ◇ F60 ◇

1983

CHERNOV, V.P. *Dialogue spaces of trivial type (Russian)* ◇ F60 ◇
COLLINS, W.J. & YOUNG, P. *Discontinuities of provably correct operators on the provably recursive real numbers* ◇ D20 F60 ◇
DEMUTH, O. *Arithmetic complexity of differentiation in constructive mathematics (Russian)* ◇ D55 F60 ◇
DEMUTH, O. *On the pseudodifferentiability of pseudo-uniformly continuous constructive functions from functions of the same type (Russian)* ◇ F60 ◇
HAUCK, J. *Konstruktive reelle Funktionale und Operatoren* ◇ F60 ◇
KHISAMIEV, N.G. *Strongly constructive abelian p-groups (Russian)* ◇ C57 C60 D45 F60 ◇
KUSHNER, B.A. *A class of Specker sequences (Russian)* ◇ F60 ◇
MOROZOV, A.S. *Groups of recursive automorphisms of constructive boolean algebras (Russian)* ◇ C07 C57 D45 F60 G05 ◇
RICHMAN, F. *Church's thesis without tears* ◇ D20 D75 F55 F60 ◇

1984

HAUCK, J. *Zur Wellengleichung mit konstruktiven Randbedingungen* ◇ F60 ◇
KREITZ, C. & WEIHRAUCH, K. *A unified approach to constructive and recursive analysis* ◇ F60 ◇
KUDAJBERGENOV, K.ZH. *Autostability and extensions of constructivizations (Russian)* ◇ C50 C57 F60 ◇
KUDAJBERGENOV, K.ZH. *Constructivizability of a prime model (Russian)* ◇ C50 C57 F60 ◇
MANUKYAN, S.N. *The Dirichlet problem in a constructive analysis (Russian) (Armenian summary)* ◇ F60 ◇
MOROZOV, A.S. *Group $Aut_r(Q, \leqslant)$ is not constructivizable (Russian)* ◇ C07 C57 F60 ◇
POUR-EL, M.B. & RICHARDS, I. *L^p-computability in recursive analysis* ◇ D80 F60 ◇
SCEDROV, A. *Differential equations in constructive analysis and in the recursive realizability topos* ◇ F60 G30 ◇

1985

ANDRIEVSKIJ, V.V. *Geometric structure of domains and direct theorems of the constructive theory of functions (Russian)* ◇ F60 ◇
BEESON, M.J. *Foundations of constructive mathematics* ◇ F50 F55 F60 F98 ◇
CHERNOV, V.P. *Nonpoint and continuous mappings (Russian)* ◇ F60 ◇
HUANG, WENGI & NERODE, A. *Application of pure recursion theory in recursive analysis (Chinese)* ◇ D20 F60 ◇
KUSHNER, B.A. *Differentiability and uniform continuity of constructive functions (Russian)* ◇ F60 ◇
LORENTS, P.P. *Constructive and enumerable sets (Russian)* ◇ F60 ◇

F65 Other constructive mathematics

1885

Molk, J. *Sur une notion qui comprend celle de la divisibilite et sur la theorie generale de l'elimination* ◇ F55 F65 ◇

1887

Kronecker, L. *Ueber den Zahlbegriff* ◇ B28 F55 F65 ◇

1905

Baire, R. & Borel, E. & Hadamard, J. & Lebesgue, H. *Cinq lettres sur la theorie des ensembles* ◇ A05 D55 E15 E25 F65 ◇

1918

Weyl, H. *Das Kontinuum* ◇ A05 B28 E30 F65 ◇

1919

Borel, E. *Sur les ensembles effectivement enumerables et sur les definitions effectives* ◇ A05 E47 F65 ◇

Weyl, H. *Der "Circulus vitiosus" in der heutigen Begruendung der Analysis* ◇ A05 B28 F65 ◇

1924

Chwistek, L.B. *The theory of constructive types (principles of logic and mathematics)* ◇ A05 B15 F65 ◇

1932

Sierpinski, W. *Sur les ensembles de la meme puissance qui ne sont pas effectivement de la meme puissance* ◇ E10 F65 ◇

1936

Mania, B. *L'infini mathematique et l'evolution de la logique* ◇ A05 F55 F65 ◇

1938

Gentzen, G. *Die gegenwaertige Lage in der mathematischen Grundlagenforschung* ◇ A05 A10 F55 F65 ◇

1949

Bockstaele, P. *Intuitionism among the French mathematicians (Dutch)* ◇ A10 F65 ◇

Wang, Hao *A theory of constructive types* ◇ B15 F35 F65 ◇

1950

Ackermann, W. *Widerspruchsfreier Aufbau der Logik I. Typenfreies System ohne tertium non datur* ◇ B15 E70 F07 F25 F65 ◇

Lorenzen, P. *Konstruktive Begruendung der Mathematik* ◇ A05 F65 ◇

1951

Lorenzen, P. *Die Widerspruchsfreiheit der klassischen Analysis* ◇ F35 F65 ◇

Lorenzen, P. *Mass und Integral in der konstruktiven Analysis* ◇ F65 ◇

Suetuna, Z. *Ueber die Grundlagen der Mathematik* ◇ A05 F65 ◇

1952

Ackermann, W. *Widerspruchsfreier Aufbau einer typenfreien Logik I (erweitertes System)* ◇ B10 B15 E70 F25 F65 ◇

Lorenzen, P. *Konstruktive Begruendung der klassischen Mathematik* ◇ F65 ◇

Schuette, K. *Beweistheoretische Untersuchungen der verzweigten Analysis* ◇ F05 F35 F65 ◇

Wang, Hao *The irreducibility of impredicative principles* ◇ E30 E70 F65 ◇

Wang, Hao *Truth definitions and consistency proofs* ◇ F25 F30 F65 ◇

1953

Ackermann, W. *Widerspruchsfreier Aufbau einer typenfreien Logik II* ◇ B10 B15 E35 E70 F30 F35 F65 ◇

Lorenzen, P. *Die Allgemeingueltigkeit der logischen Regeln* ◇ A05 F50 F65 ◇

Schuette, K. *Zur Widerspruchsfreiheit einer typenfreien Logik* ◇ B15 B40 F05 F65 ◇

1954

Schuette, K. *Ein widerspruchsloses System der Analysis auf typenfreier Grundlage* ◇ E70 F35 F65 ◇

Wang, Hao *The formalization of mathematics* ◇ A05 B30 E70 F35 F65 ◇

1955

Lorenzen, P. *Einfuehrung in die operative Logik und Mathematik* ◇ A05 B98 F50 F65 F98 ◇

Mannoury, G. *Finitistic-formalistic development of projective geometry* ◇ F65 ◇

1956

Goodstein, R.L. *A constructivist theory of plane curves* ◇ F65 ◇

Hintikka, K.J.J. *Identity, variables and impredicative definitions* ◇ A05 B20 F65 ◇

Kondo, M. *Sur la nommabilite d'ensembles* ◇ D55 E15 F60 F65 ◇

Kondo, M. *Sur la notion du transfini* ◇ E10 E47 F65 ◇

Kondo, M. *Sur les analyses relatives* ◇ D55 E15 F65 ◇

KONDO, M. *Sur les nombres reels et nommables* ◇ B28 D55 E15 F65 ◇
LOEB, M.H. *Formal systems of constructive mathematics* ◇ F65 ◇
TAKEUTI, G. *Construction of ramified real numbers* ◇ B15 F05 F35 F65 ◇

1957

KURODA, S. *On the intuitionistic and formalistic theory of real numbers* ◇ F55 F65 ◇
SPECTOR, C. *Recursive ordinals and predicative set theory* ◇ D20 D55 E70 F15 F65 ◇

1958

KREISEL, G. *Constructive mathematics* ◇ F50 F55 F65 ◇

1959

FALEVICH, B.YA. *On the non-existence of models of classical analysis in W. Ackermann's constructive analysis (Russian)* ◇ F65 ◇
WANG, HAO *Ordinal numbers and predicative set theory* ◇ A05 E70 F15 F65 ◇

1960

KONDO, M. *Le fondement constructif du calcul infinitesimal* ◇ F65 ◇
KREISEL, G. *La predicativite* ◇ A05 D55 E45 F35 F65 ◇
KREISEL, G. *Ordinal logics and the characterization of informal concepts of proof* ◇ F07 F15 F30 F65 ◇
LORENZEN, P. *Constructive and axiomatic mathematics* ◇ A05 F65 ◇

1961

ESENIN-VOL'PIN, A.S. *Le programme ultra-intuitionniste des fondements des mathematiques* ◇ A05 E70 F65 ◇
KONDO, M. *Sur la nommabilite d'etres mathematiques* ◇ E47 F65 ◇
LORENZEN, P. *Ein dialogisches Konstruktivitaetskriterium* ◇ A05 F50 F65 ◇

1962

MAEHARA, S. *Cut-elimination theorem concerning a formal system for ramified theory of types which admits quantification on types* ◇ B15 F05 F30 F35 F65 ◇
PARSONS, C. *The ω-consistency of ramified analysis* ◇ F05 F10 F35 F65 ◇

1963

WANG, HAO *Some formal details on predicative set theories* ◇ B15 E70 F65 ◇

1964

FEFERMAN, S. *Systems of predicative analysis* ◇ F35 F65 ◇

1965

LORENZEN, P. *Die klassische Analysis als eine konstruktive Theorie* ◇ B28 F65 ◇
LORENZEN, P. *Differential und Integral. Eine konstruktive Einfuehrung in die klassische Analysis (Loose errata)* ◇ F65 F98 ◇
SANCHIS, L.E. *A predicative extension of elementary logic I* ◇ B15 F05 F65 ◇
SCHUETTE, K. *Eine Grenze fuer die Beweisbarkeit der transfiniten Induktion in der verzweigten Typenlogik* ◇ B15 F15 F35 F65 ◇
SCHUETTE, K. *Predicative well-orderings* ◇ F15 F35 F65 ◇

1966

AMSTISLAVSKIJ, V.I. *Recursive sieves (Russian)* ◇ D55 E15 F65 ◇
FEFERMAN, S. *Predicative provability in set theory* ◇ E30 F65 ◇
GHOSE, A. *Konstruktive Grundlagen der Logik* ◇ A05 F65 ◇

1967

FENSTAD, J.E. *Subsets of natural numbers (Norwegian) (English summary)* ◇ A05 E35 E50 F30 F35 F65 ◇
KLAUA, D. *Bemerkungen zur finiten Mathematik* ◇ E30 F65 ◇
ZAHN, P. *Eine Einfuehrung der reellen Zahlen in der operativen Mathematik ohne die Unterscheidung von Sprachschichten* ◇ B28 F35 F65 ◇

1968

BURTON, W.K. *Constructive thermodynamics* ◇ F65 ◇
FEFERMAN, S. *Autonomous transfinite progressions and the extent of predicative mathematics* ◇ F15 F30 F35 F65 ◇
FEFERMAN, S. *Systems of predicative analysis II : representation of ordinals* ◇ F15 F35 F65 ◇
LORENTS, A.A. *Some problems in the constructive theory of finite probabilistic automata (Russian) (German summary)* ◇ D05 F65 ◇
LORENZEN, P. *Constructive mathematics as a philosophical problem* ◇ A05 F65 ◇
TAIT, W.W. *Constructive reasoning* ◇ B40 F10 F15 F30 F35 F50 F65 ◇

1969

ESENIN-VOL'PIN, A.S. *On Hilbert's second problem (Russian)* ◇ A05 F30 F65 ◇
HERMES, H. *On the notion of constructivity* ◇ D20 F65 ◇
ROYSE, JAMES R. *Mathematical induction in ramified type theory* ◇ B15 F35 F65 ◇
SCHNORR, C.-P. *Eine Bemerkung zum Begriff der zufaelligen Folge (English summary)* ◇ D20 D80 F65 ◇
WETTE, E. *Definition eines (relativ vollstaendigen) formalen Systems konstruktiver Arithmetik* ◇ F10 F15 F30 F35 F65 ◇

1970

ESENIN-VOL'PIN, A.S. *The ultra-intuitionistic criticism and the anti-traditional program for foundations of mathematics* ◇ A05 F65 ◇
THIEL, C. *Was will die operative Mathematik ?* ◇ A05 F65 ◇

1971

JELTSCH, R. *Der Riemannsche Abbildungssatz in der konstruktiven Analysis. Diss* ◇ F65 ◇
PARIKH, R. *Existence and feasibility in arithmetic* ◇ A05 D15 F30 F65 H15 ◇
POZSGAY, L.J. *Liberal intuitionism as a basis for set theory* ◇ A05 E30 F65 ◇
SCHREIBER, P. *Konstruktive Teiltheorien* ◇ F65 ◇

1972

POZSGAY, L.J. *Semi-intuitionistic set theory* ◇ A05 E30 E70 F65 ◇

1973

HAJEK, P. *Why semisets ?* ◇ E35 E70 F65 ◇
NEPEJVODA, N.N. *A new concept of predicative truth and definability (Russian)* ◇ F35 F65 ◇
NEPEJVODA, N.N. *The relationship between predicative significance and the intuition of universality (Russian)* ◇ F50 F65 ◇
POUR-EL, M.B. *Abstract computability versus analog-generability. A survey* ◇ D75 D80 F65 ◇
POUR-EL, M.B. *Analog computers, digital computers, mathematical logic, differential equations - interrelations* ◇ D20 D75 D80 F65 ◇
SHEPARD, P.T. *A finite arithmetic* ◇ F30 F65 ◇

1974

COCCHIARELLA, N.B. *A new formulation of predicative second order logic* ◇ A05 B15 F65 ◇
EHRENFEUCHT, A. *Logic without iterations* ◇ B60 F65 ◇
FEFERMAN, S. *Predicatively reducible systems of set theory* ◇ E25 E30 E70 F65 ◇
MARKOV, A.A. *On the language $Я_0$ (Russian)* ◇ B25 F50 F65 ◇
METROPOLIS, N. & ROTA, G.-C. *Significance arithmetic on the algebra of binary strings* ◇ D15 F65 G05 ◇
POUR-EL, M.B. *Abstract computability and its relation to the general purpose analog computer (some connections between logic, differential equations and analog computers)* ◇ D20 D75 D80 F65 ◇

1975

BLOOM, S.L. *A note on the predicatively definable sets of N.N.Nepeivoda* ◇ F60 F65 ◇
FEFERMAN, S. *A language and axioms for explicit mathematics* ◇ D55 F35 F50 F65 ◇
FEFERMAN, S. *Impredicativity of the existence of the largest divisible subgroup of an abelian p-group* ◇ C57 C60 F65 ◇

1976

CRABBE, M. *La predicativite dans les theories elementaires* ◇ F65 ◇
DRAGALIN, A.G. *Cut-elimination in the theory of definable sets of natural numbers (Russian)* ◇ F05 F35 F65 ◇
MAASS, W. *Eine Funktionalinterpretation der praedikativen Analysis* ◇ D20 F10 F35 F65 ◇

1977

BECK, JON M. *From the pocket calculator to de Rham's theorem* ◇ F65 ◇
CHRISTIAN, C.C. *Eine Note zum System PC* ◇ E70 F65 ◇
DRAGALIN, A.G. *Cut-elimination in the theory of definable sets of natural numbers (Russian)* ◇ F05 F35 F65 ◇
HARMS, S. & KLIX, W.-D. *Ein Algorithmus zur automatischen Loesung konstruktiver Problemstellungen* ◇ B35 F65 ◇
SEPER, K. *Constructive mathematics - an essay, I, II* ◇ A05 F55 F65 ◇
SEPER, K. *Contribution to the discussion of Markov principle* ◇ A05 F65 ◇
THIEL, C. *Que significa "constructivismo" ?* ◇ A05 F65 ◇

1978

FEFERMAN, S. *Recursion theory and set theory: a marriage of convenience* ◇ D75 E30 E70 F65 ◇
LIPTON, R.J. *Model theoretic aspects of computational complexity* ◇ C62 D15 F30 F65 H15 ◇
ZAHN, P. *Ein konstruktiver Weg zur Masstheorie und Funktionalanalysis* ◇ F65 F98 ◇

1979

BECK, JON M. *Simplicial sets and the foundations of analysis* ◇ B28 F65 ◇
DEMILLO, R.A. & LIPTON, R.J. *A constructive generalization of the Borel-Cantelli lemma with application to the complexity of infinite strings* ◇ F65 ◇
FEFERMAN, S. *A more perspicuous system for predicativity* ◇ F35 F65 ◇
REYES, G.E. *Cramer's rule in the Zariski topos* ◇ F65 G30 ◇

1980

KREISEL, G. *Finitismus* ◇ A05 F65 ◇

1981

ASANIDZE, G.Z. *A dialogue justification of logic (Russian)* ◇ B60 F50 F65 ◇
ENGELER, E. *An algorithmic model of strict finitism* ◇ D20 F65 ◇
ESENIN-VOL'PIN, A.S. *About infinity, finiteness and finitization (in connection with the foundations of mathematics)* ◇ A05 E30 F65 ◇
GEISER, J.R. *Rational constructive analysis* ◇ F65 ◇
GREENLEAF, N. *Liberal constructive set theory* ◇ A05 E70 F55 F65 ◇
ISLES, D. *Remarks on the notion of standard non-isomorphic natural number series* ◇ A05 F15 F65 ◇
MYCIELSKI, J. *Analysis without actual infinity* ◇ F65 ◇
MYCIELSKI, J. *Finitistic real analysis* ◇ B28 F65 ◇
RICHMAN, F. (ED.) *Constructive mathematics. Proceedings of the New Mexico State University Conference* ◇ F65 F97 ◇
TAIT, W.W. *Finitism* ◇ A05 F65 ◇
VOGEL, HELMUT *Der finite Standpunkt in Hilberts Programm* ◇ A05 F65 F99 ◇
YASUGI, M. *The Hahn-Banach theorem and a restricted inductive definition* ◇ F35 F65 ◇

1982

FEFERMAN, S. *Iterated inductive fixed-point theories: application to Hancock's conjecture* ◇ F15 F35 F65 ◇

FRIEDMAN, H.M. & MCALOON, K. & SIMPSON, S.G. *A finite combinatorial principle which is equivalent to the l-consistency of predicative analysis* ◇ F15 F35 F65 ◇

FRIEDMANN, J. *Bemerkungen zur Logikbegruendung im deutschen Konstruktivismus* ◇ A05 F65 ◇

GANDY, R.O. *Limitations to mathematical knowledge* ◇ A05 F65 ◇

MIGLIOLI, P.A. & MOSCATO, U. & ORNAGHI, M. & USBERTI, G. *Constructive validity and classical truth to assign meaning to programs* ◇ B75 F65 ◇

MIGLIOLI, P.A. & ORNAGHI, M. *Constructive proofs and logical computations* ◇ F65 ◇

SCHMERL, U.R. *A proof theoretical fine structure in systems of ramified analysis* ◇ F15 F35 F65 ◇

WRIGHT, C. *Strict finitism* ◇ A05 F65 ◇

1983

MIGLIOLI, P.A. & MOSCATO, U. & ORNAGHI, M. & USBERTI, G. *Alcuni calcoli intermedi costruttivi* ◇ B55 F65 ◇

YASUGI, M. *Definability problems in elementary topology* ◇ F65 ◇

YASUGI, M. *Definability in the abstract theory of integration* ◇ E30 F65 ◇

1984

GAVRILENKO, YU.V. *Monotonic theories of accessible numbers (Russian)* ◇ F20 F30 F65 ◇

HINMAN, P.G. *Finitely approximable sets* ◇ D65 F65 ◇

VETULANI, Z. *Ramified analysis and the minimal β-models of higher order arithmetics* ◇ C62 D55 E45 F35 F65 ◇

YASUGI, M. *A definable interpretation of metric spaces* ◇ F65 ◇

YASUGI, M. *Definability in L^p-spaces* ◇ F65 ◇

1985

DRAGALIN, A.G. *Correctness of inconsistent theories with notions of feasibility* ◇ F65 ◇

ZAHN, P. *Gilt das Lemma von Koenig "konstruktiv"?* ◇ F65 ◇

F97 Proceedings

1941

GONSETH, F. (ED.) *Les entretiens de Zuerich sur les fondements et la methode des sciences mathematiques: Exposes et discussions* ◇ B97 E97 F97 ◇

1959

HEYTING, A. (ED.) *Constructivity in mathematics. Proceedings of the colloquium held at Amsterdam 1957* ◇ F97 ◇

1967

HEIJENOORT VAN, J. (ED.) *From Frege to Goedel: a source book in mathematical logic, 1879-1931* ◇ A10 B97 F97 ◇

OREVKOV, V.P. & SHANIN, N.A. (EDS.) *Problems in the constructive trend in mathematics IV (Russian)* ◇ F60 F97 ◇

SLISENKO, A.O. (ED.) *Studies in constructive mathematics and mathematical logic I (Russian)* ◇ F97 ◇

1968

DALEN VAN, D. & DYKMAN, J.G. & KLEENE, S.C. & TROELSTRA, A.S. (EDS.) *Logic and foundations of mathematics. Papers dedicated to A.Heyting on the occasion of his 70th birthday* ◇ F97 ◇

LOEB, M.H. (ED.) *Proceedings of the summer school in logic, Leeds, 1967* ◇ B97 C97 F97 ◇

OREVKOV, V.P. (ED.) *The calculi of symbolic logic I (Russian)* ◇ B97 F97 ◇

SLISENKO, A.O. (ED.) *Studies in constructive mathematics and mathematical logic II (Russian)* ◇ F97 ◇

1969

DEJON, B. & HENRICI, P. (EDS.) *Constructive aspects of the fundamental theorem of algebra: proceedings of a symposium, IBM* ◇ B28 C97 F97 ◇

SLISENKO, A.O. (ED.) *Studies in constructive mathematics and mathematical logic III (Russian)* ◇ F97 ◇

1970

KINO, A. & MYHILL, J.R. & VESLEY, R.E. (EDS.) *Intuitionism and proof theory. Proceedings of the Summer Conference at Buffalo, N.Y., 1968* ◇ F97 ◇

LACOMBE, D. & LAUDET, M. & NOLIN, L. & SCHUETZENBERGER, M.-P. (EDS.) *Symposium on automatic demonstration* ◇ B35 B97 F97 ◇

OREVKOV, V.P. & SHANIN, N.A. (EDS.) *Problems in the constructive trend in mathematics V (Russian)* ◇ F60 F97 ◇

1971

MATIYASEVICH, YU.V. & SLISENKO, A.O. (EDS.) *Studies in constructive mathematics and mathematical logic IV (Russian)* ◇ F97 ◇

1972

MATIYASEVICH, YU.V. & SLISENKO, A.O. (EDS.) *Studies in constructive mathematics and mathematical logic V (Russian)* ◇ F97 ◇

1973

OREVKOV, V.P. & SHANIN, N.A. (EDS.) *Problems in the constructive trend in mathematics VI (Russian)* ◇ F60 F97 ◇

1974

MATIYASEVICH, YU.V. & SLISENKO, A.O. (EDS.) *Studies in constructive mathematics and mathematical logic VI (Russian)* ◇ F97 ◇

1975

DILLER, J. & MUELLER, GERT H. (EDS.) ⊧ *ISILC proof theory symposium* ◇ C97 F97 ◇

LAWVERE, F.W. & MAURER, C. & WRAITH, G.C. (EDS.) *Model theory and topoi* ◇ C97 F97 G97 ◇

1976

MATIYASEVICH, YU.V. & SLISENKO, A.O. (EDS.) *Studies in constructive mathematics and mathematical logic VII (Russian)* ◇ F97 ◇

1979

LOURENCO, M. (ED.) *Goedel's theorem and the continuum hypothesis (Portuguese)* ◇ B97 E50 F30 F97 ◇

MATIYASEVICH, YU.V. & SLISENKO, A.O. (EDS.) *Studies in constructive mathematics and mathematical logic VIII (Russian)* ◇ B97 F97 ◇

1980

SENDOV, B. & VACHOV, D. (EDS.) *Constructive function theory '77. Proceedings* ◇ F97 ◇

1981

RICHMAN, F. (ED.) *Constructive mathematics. Proceedings of the New Mexico State University Conference* ◇ F65 F97 ◇

1982

TROELSTRA, A.S. & DALEN VAN, D. (EDS.) *The L.E.J. Brouwer centenary symposium* ◇ A05 A10 B97 F50 F55 F97 ◇

1984

BOERGER, E. & OBERSCHELP, W. & RICHTER, M.M. & SCHINZEL, B. & THOMAS, WOLFGANG (EDS.) *Computation and proof theory* ◇ B97 D97 F97 ◇

LOLLI, G. & LONGO, G. & MARCJA, A. (EDS.) *Logic colloquium '82. Proceedings of the colloquium held in Florence, 23-28 August, 1982* ◇ B97 C97 D97 F97 ◇

1985

HARRINGTON, L.A. & MORLEY, M.D. & SCEDROV, A. & SIMPSON, S.G. *Harvey Friedman's research on the foundations of mathematics*
◇ B97 C97 D97 E97 F97 ◇

F98 Textbooks, surveys

1924

BALDUS, R. *Formalismus und Intuitionismus in der Mathematik* ◇ A05 F55 F98 ◇

1930

NICOD, J. *Geometry and induction, containing "Geometry in the sensible world", and "The logical problem of induction"* ◇ A05 B30 F98 ◇

1931

HEYTING, A. *Die intuitionistische Mathematik* ◇ A05 F98 ◇

1934

HEYTING, A. *Mathematische Grundlagenforschung. Intuitionismus. Beweistheorie* ◇ A05 F50 F98 ◇

HILBERT, D. & BERNAYS, P. *Grundlagen der Mathematik I* ◇ A05 B98 F05 F30 F98 ◇

1939

HILBERT, D. & BERNAYS, P. *Grundlagen der Mathematik II* ◇ A05 B98 F05 F15 F30 F40 F98 ◇

1944

WEYL, H. *David Hilbert and his mathematical work* ◇ A10 F98 ◇

1946

MOSTOWSKI, ANDRZEJ *On undecidable sentences in formalized mathematical systems (Polish)* ◇ F30 F98 ◇

1950

CURRY, H.B. *A theory of formal deducibility* ◇ B10 F50 F98 ◇

1951

GOODSTEIN, R.L. *Constructive formalism: Essays on the foundations of mathematics* ◇ A05 B98 F30 F60 F98 ◇

GOODSTEIN, R.L. *The foundations of mathematics* ◇ A05 A10 F98 ◇

1952

KLEENE, S.C. *Introduction to metamathematics* ◇ A05 B98 D98 F30 F50 F98 ◇

MOSTOWSKI, ANDRZEJ *Sentences undecidable in formalized arithmetic. An exposition of the theory of Kurt Goedel* ◇ F30 F98 ◇

1955

LORENZEN, P. *Einfuehrung in die operative Logik und Mathematik* ◇ A05 B98 F50 F65 F98 ◇

1956

HEYTING, A. *Intuitionism. An introduction* ◇ F55 F98 ◇

1957

GOODSTEIN, R.L. *Recursive number theory. A development of recursive arithmetic in a logic-free equation calculus* ◇ B28 F30 F50 F60 F98 ◇

KANGER, S. *Provability in logic* ◇ B10 F05 F07 F98 ◇

LADRIERE, J. *Les limitations internes des formalismes. Etude sur la signification du theoreme de Goedel et des theoremes apparentes dans la theorie des fondements des mathematiques* ◇ A05 B98 F30 F98 ◇

1958

BAR-HILLEL, Y. & DALEN VAN, D. & FRAENKEL, A.A. & LEVY, A. *Foundations of set theory* ◇ A05 B98 E30 E70 E98 F98 ◇

GUILLAUME, M. *Les tableaux semantiques du calcul des predicats restreint* ◇ B20 F07 F98 ◇

HEYTING, A. *Intuitionism in mathematics* ◇ A05 B98 F50 F55 F98 ◇

LORENZEN, P. *Formale Logik* ◇ F50 F98 ◇

NAGEL, E. & NEWMAN, J.R. *Goedel's proof* ◇ B28 D35 F30 F98 ◇

1959

MEIGNE, M. *La consistance des theories formelles et le fondement des mathematiques* ◇ A05 F25 F98 ◇

MOSTOWSKI, ANDRZEJ *On various degrees of constructivism* ◇ D55 F50 F60 F98 ◇

STEGMUELLER, W. *Unvollstaendigkeit und Unentscheidbarkeit. Die metamathematischen Resultate von Goedel, Church, Kleene, Rosser und ihre erkenntnistheoretische Bedeutung* ◇ A05 D35 F30 F98 ◇

1960

SCHMIDT, H.A. *Mathematische Gesetze der Logik. I. Vorlesungen ueber Aussagenlogik* ◇ B05 B98 F50 F98 ◇

SCHUETTE, K. *Beweistheorie* ◇ B98 F05 F15 F30 F35 F98 ◇

SOUDIEUX, C. *De l'infini arithmetique* ◇ B28 F30 F98 ◇

1961

GOODSTEIN, R.L. *Recursive analysis* ◇ F30 F60 F98 ◇

KLAUA, D. *Konstruktive Analysis* ◇ F60 F98 ◇

MAL'TSEV, A.I. *Constructive algebra I (Russian)* ◇ C57 C98 F60 F98 ◇

1962

ANDERSON, J.M. & JOHNSTONE JR., H.W. *Natural deduction. The logical basis of axiom systems* ◇ B98 F07 F98 ◇

BETH, E.W. *Formal methods. An introduction to symbolic logic and to the study of effective operations in arithmetic and logic* ◇ B28 B98 F30 F50 F98 ◇

DOPP, J. *Logiques construites par une methode de deduction naturelle* ◇ B98 F05 F07 F50 F98 ◇

KREISEL, G. *Beweistheorie* ◇ F98 ◇

LORENZEN, P. *Metamathematik* ◇ A05 B98 C60 D20 D35 D98 F98 ◇

1963

MAZUR, S. *Computable analysis* ◇ F60 F98 ◇

RASIOWA, H. & SIKORSKI, R. *The mathematics of metamathematics* ◇ B98 C98 F50 F98 G05 G10 G98 ◇

SMULLYAN, R.M. *First order logic* ◇ B98 F07 F98 ◇

1964

FUENTES MIRAS, J.R. *On the foundations of mathematics (Spanish)* ◇ A05 A10 F98 ◇

RIVETTI BARBO, F. *Il teorema e il corollario di Goedel* ◇ A05 F30 F98 ◇

1965

GEYMONAT, L. *La metamatematica dopo Hilbert* ◇ F98 ◇

KLEENE, S.C. & VESLEY, R.E. *The foundations of intuitionistic mathematics, especially in relation to recursive functions* ◇ F35 F50 F98 ◇

KREISEL, G. *Mathematical logic* ◇ B98 F10 F35 F50 F98 ◇

LORENZEN, P. *Differential und Integral. Eine konstruktive Einfuehrung in die klassische Analysis (Loose errata)* ◇ F65 F98 ◇

MOSTOWSKI, ANDRZEJ *Thirty years of foundational studies. Lectures on the development of mathematical logic and the studies of the foundations of mathematics in 1930-1964* ◇ A10 B98 C98 D98 E98 F98 ◇

PRAWITZ, D. *Natural deduction. A proof-theoretical study* ◇ B15 B98 F05 F07 F50 F98 ◇

SCHUETTE, K. *Probleme und Methoden der Beweistheorie* ◇ A05 F98 ◇

1966

ROOTSELAAR VAN, B. *Intuition und Konstruktion* ◇ A05 F98 ◇

1967

BISHOP, E.A. *Foundations of constructive analysis* ◇ F55 F98 ◇

KLEENE, S.C. *Mathematical logic* ◇ B98 D20 D25 D55 D98 F30 F98 ◇

SHOENFIELD, J.R. *Mathematical logic* ◇ B98 C98 D98 E98 F98 ◇

1968

DALLA CHIARA SCABIA, M.L. *Modelli sintattici e semantici delle teorie elementari* ◇ C07 C98 E35 E45 F25 F98 ◇

FEFERMAN, S. *Lectures on proof theory* ◇ C40 C70 C75 F05 F15 F98 ◇

HEYTING, A. *Intuitionism in mathematics* ◇ A05 F50 F98 ◇

KINO, A. *Foundations of mathematics: consistency* ◇ F35 F98 ◇

KREISEL, G. *A survey of proof theory* ◇ F98 ◇

KREISEL, G. *Foundations of mathematics: 1900-1950* ◇ A05 A10 F98 ◇

SCHUETTE, K. *Neuere Ergebnisse der Beweistheorie* ◇ F05 F10 F15 F30 F35 F98 ◇

SCHUETTE, K. *Vollstaendige Systeme modaler und intuitionistischer Logik* ◇ B45 C90 F50 F98 ◇

1969

FITTING, M. *Intuitionistic logic, model theory, and forcing* ◇ B98 C90 C98 E25 E35 E45 E50 F50 F98 ◇

LEISENRING, A.C. *Mathematical logic and Hilbert's ε-symbol* ◇ B10 E25 F05 F98 ◇

SZABO, M.E. *Introduction* ◇ A10 F05 F30 F98 ◇

TROELSTRA, A.S. *Principles of intuitionism* ◇ F35 F50 F55 F98 ◇

1970

MARTIN-LOEF, P. *Notes on constructive mathematics* ◇ F50 F60 F98 ◇

SIEFKES, D. *Decidable theories I. Buechi's monadic second order successor arithmetic* ◇ B15 B25 C85 D05 F35 F98 ◇

1971

GRZEGORCZYK, A. *Outline of theoretical arithmetic (Polish)* ◇ B28 C62 F30 F35 F98 ◇

KREISEL, G. *A survey of proof theory II* ◇ F50 F98 ◇

PRAWITZ, D. *Ideas and results in proof theory* ◇ F05 F07 F10 F30 F35 F50 F98 ◇

REZNIKOFF, I. *Logique mathematique. II Theorie de la demonstration et intuitionnisme* ◇ F50 F98 ◇

SCARPELLINI, B. *Proof theory and intuitionistic systems* ◇ F05 F15 F30 F35 F50 F98 ◇

1972

FREGUGLIA, P. *Platonismo ed intuizionismo e loro superamento* ◇ A05 F98 ◇

HINDLEY, J.R. & LERCHER, B. & SELDIN, J.P. *Introduction to combinatory logic* ◇ B40 B98 F98 ◇

LORENTS, A.A. *Elemente der konstruktiven Theorie stochastischer Automaten (Russian)* ◇ D05 D10 D98 F60 F98 ◇

PRAWITZ, D. *The philosophical position of proof theory* ◇ A05 F98 ◇

SMIRNOV, V.A. *Formal derivation and logical calculi (Russian)* ◇ B98 C07 C40 C98 F05 F07 F98 ◇

STEEN, S.W.P. *Mathematical logic, with special reference to the natural numbers* ◇ B98 F30 F98 ◇

STENLUND, S. *Combinators, λ-terms and proof theory* ◇ B40 F05 F10 F50 F98 ◇

TAKAHASHI, A. *Mathematical concepts I,II,III,IV,V (Spanish)* ◇ A10 E98 F98 F99 ◇

1973

DALEN VAN, D. *Lectures on intuitionism* ◇ F50 F55 F98 ◇

KUSHNER, B.A. *Lectures on constructive mathematical analysis (Russian)* ◇ F60 F98 ◇
LUCKHARDT, H. *Extensional Goedel functional interpretation. A consistency proof of classical analysis* ◇ F10 F35 F50 F98 ◇
TROELSTRA, A.S. (ED.) *Metamathematical investigation of intuitionistic arithmetic and analysis* ◇ F05 F10 F30 F35 F50 F55 F98 ◇

1975

FRIEDMAN, H.M. *One hundred and two problems in mathematical logic* ◇ B98 C98 D98 E98 F98 ◇
MINTS, G.E. *Proof theory (arithmetic and analysis) (Russian)* ◇ F05 F10 F30 F35 F50 F98 ◇
TAKEUTI, G. *Proof theory* ◇ F05 F07 F30 F35 F98 ◇
TOLEDO, S. *Tableau systems for first order number theory and certain higher order theories* ◇ F05 F07 F30 F35 F98 ◇

1976

APERY, R. *Mathematique constructive* ◇ A05 F98 ◇
KREISEL, G. *What have we learnt from Hilbert's second problem?* ◇ A05 F98 ◇
WRAITH, G.C. *Logic from topology: a survey of topoi* ◇ C98 F98 G30 ◇

1977

BARWISE, J. (ED.) *Handbook of mathematical logic* ◇ B98 C98 D98 E98 F98 H98 ◇
DUMMETT, M. *Elements of intuitionism* ◇ F50 F55 F98 ◇
FEFERMAN, S. *Theories of finite type related to mathematical practice* ◇ B15 D55 D65 F10 F15 F35 F98 ◇
NOVIKOV, P.S. *Constructive mathematical logic from the point of view of classical logic (Russian)* ◇ F98 ◇
SCHWICHTENBERG, H. *Proof theory: Some applications of cut-elimination* ◇ F05 F10 F15 F30 F35 F98 ◇
SMORYNSKI, C.A. *The incompleteness theorems* ◇ C62 F25 F30 F98 ◇
TROELSTRA, A.S. *Aspects of constructive mathematics* ◇ F30 F35 F50 F55 F60 F98 ◇
TROELSTRA, A.S. *Choice sequences, a chapter of intuitionistic mathematics* ◇ A05 A10 F35 F50 F55 F98 ◇

1978

CELLUCCI, C. *Teoria della dimostrazione. Normalizzazioni e assegnazioni di numeri ordinali* ◇ F05 F07 F15 F30 F98 ◇
DALEN VAN, D. *Philosophical foundations of mathematics (Dutch)* ◇ A05 F98 ◇
SCHUETTE, K. *Die Entwicklung der Beweistheorie* ◇ A10 F98 ◇
SHOESMITH, D.J. & SMILEY, T.J. *Multiple-conclusion logic* ◇ B22 F07 F98 ◇
SZABO, M.E. *Algebra of proofs* ◇ F05 F07 F50 F98 G30 ◇
ZAHN, P. *Ein konstruktiver Weg zur Masstheorie und Funktionalanalysis* ◇ F65 F98 ◇

1979

BOOLOS, G. *The unprovability of consistency. An essay in modal logic* ◇ B25 B28 B45 B98 F30 F98 ◇
BRIDGES, D.S. *Constructive functional analysis* ◇ F55 F98 ◇
DALEN VAN, D. *Interpreting intuitionistic logic* ◇ A05 F50 F98 ◇
DRAGALIN, A.G. *Intuitionism (Russian)* ◇ A05 F50 F98 ◇
DRAGALIN, A.G. *Mathematical intuitionism. Introduction to proof theory (Russian)* ◇ F05 F30 F35 F50 F98 ◇
DRAGALIN, A.G. *Proof theory (Russian)* ◇ F98 ◇
FEFERMAN, S. *Constructive theories of functions and classes* ◇ F50 F55 F60 F98 ◇
FEFERMAN, S. *What does logic have to tell us about mathematical proofs?* ◇ F98 ◇
GOLDBLATT, R.I. *Topoi. The categorial analysis of logic* ◇ B98 C98 E98 F35 F50 F98 G30 ◇
MANIN, YU.I. *Provable and unprovable (Russian)* ◇ A05 B98 E35 E50 F30 F98 G12 ◇

1980

ABERTH, O. *Computable analysis* ◇ F30 F60 F98 ◇

1981

BROUWER, L.E.J. *Brouwer's Cambridge lectures on intuitionism* ◇ A05 A10 F50 F55 F98 ◇
BUCHHOLZ, W. & FEFERMAN, S. & POHLERS, W. & SIEG, W. *Iterated inductive definitions and subsystems of analysis: recent proof-theoretical studies* ◇ F05 F15 F35 F50 F98 ◇
GABBAY, D.M. *Semantical investigations in Heyting's intuitionistic logic* ◇ B55 F50 F98 ◇
KOSOVSKIJ, N.K. *Elements of mathematical logic and its applications to the theory of subrecursive algorithms (Russian)* ◇ B98 D20 D98 F30 F60 F98 ◇
SCHUETTE, K. *Proof theory* ◇ F98 ◇
TROELSTRA, A.S. *The interplay between logic and mathematics: Intuitionism* ◇ A05 F50 F98 ◇
WANG, HAO *Popular lectures on mathematical logic (Chinese)* ◇ A05 B98 C98 E98 F30 F98 ◇

1982

DALEN VAN, D. *Braucht die konstruktive Mathematik Grundlagen?* ◇ A05 F98 ◇
FISHER, A. *Formal number theory and computability* ◇ B98 D20 F30 F98 ◇
POHLERS, W. *Admissibility in proof theory; a survey* ◇ C70 F15 F35 F98 ◇
USPENSKIJ, V.A. *Goedel's incompleteness theorem (Russian)* ◇ F30 F98 ◇

1983

FITTING, M. *Proof methods for modal and intuitionistic logics* ◇ B45 B55 F07 F50 F98 ◇
MCCARTY, C. *Intuitionism: an introduction to a seminar* ◇ A05 F50 F98 ◇

1984

BRIDGES, D.S. & MINES, R. *What is constructive mathematics?* ◇ A05 F55 F98 ◇

HAAS, G. *Konstruktive Einfuehrung in die formale Logik* ◇ B98 F98 ◇

MARTIN-LOEF, P. *Intuitonistic type theory. Notes by Giovanni Sambin of a series of lectures given in Padua, June 1980* ◇ F35 F50 F98 ◇

PANOV, M.I. *Methodological problems of intuitionistic mathematics (Russian)* ◇ A05 F98 ◇

SIEG, W. *Foundations of analysis and proof theory* ◇ A10 F30 F35 F98 ◇

SMORYNSKI, C.A. *Lectures on nonstandard models of arithmetic* ◇ A10 C62 C98 F30 F98 H15 ◇

1985

BEESON, M.J. *Foundations of constructive mathematics* ◇ F50 F55 F60 F98 ◇

BISHOP, E.A. & BRIDGES, D.S. *Constructive analysis* ◇ D80 F55 F98 ◇

CELLUCCI, C. *Proof theory and complexity* ◇ D15 F05 F20 F98 ◇

ROYDEN, H.L. *Aspects of constructive analysis* ◇ A05 F98 ◇

SIMPSON, S.G. *Friedman's research on subsystems of second order arithmetic* ◇ C62 C98 F35 F98 ◇

SMORYNSKI, C.A. *Nonstandard models and related developments* ◇ C62 C98 F30 F98 ◇

SMORYNSKI, C.A. *Self-reference and modal logic* ◇ B45 F30 F98 ◇

TAKEUTI, G. *Proof theory and set theory* ◇ E55 F15 F98 ◇

F99 \ (A05 ∪ A10) Miscellaneous

1927

GRUZINTSEV, G.A. *The concept of relation and the axiomatical definition of number (Russian) (French summary)* ◇ B28 F99 ◇

1929

HEYTING, A. *The countability predicates of Professor Brouwer (Dutch)* ◇ F55 F99 ◇

1935

SARANTOPOULOS, S. *Un theoreme se rattachant a la methode par recurrence (induction complete)* ◇ F99 ◇

1957

KREISEL, G. *Sums of squares* ◇ C57 C60 D20 F07 F99 ◇

1959

WETTE, E. *Von operativen Modellen der axiomatischen Mengenlehre* ◇ C62 E35 F99 ◇

1964

TIRNOVEANU, M. *Ueber ein logisches semikonstruktives globales System (Rumaenisch)* ◇ F99 ◇

1968

LEVY, A. *The effectivity of existential statements in axiomatic set theory* ◇ E25 E30 E47 F99 ◇

1970

TIRNOVEANU, M. *Sur quelques notions de proto-logistique semi-constructive* ◇ B60 F99 ◇

TIRNOVEANU, M. *Ueber einen Begriff der semikonstruktiven Protologik. I (Rumaenisch)* ◇ F99 ◇

1973

ANISHEV, P.A. & KASHIN, V.I. & SIDRISTYJ, B. *Sprache der konstruktiven Definitionen (Russisch)* ◇ F99 ◇

ENGELER, E. *Towards a Galois theory of algorithmic problems* ◇ B75 F99 ◇

1975

LINDNER, R. *On the theory of inference operators* ◇ B48 F99 ◇

MIJOULE, R. *Une formalisation des fondements par les logiques internes (English summary)* ◇ F99 G30 ◇

1978

SCHREIBER, P. *Zur optimalen Loesung von Konstruktionsaufgaben* ◇ F99 ◇

1979

KATERINAK, J. *The theory of finite structures of G-numbers* ◇ F99 ◇

NEPEJVODA, N.N. *Some remarks about constructive non-standard analysis (Russian)* ◇ F99 H05 ◇

1980

BACK, R.J.R. *Correctness preserving program refinements: proof theory and applications* ◇ B75 F99 ◇

FAL'KOVICH, M.A. & GALIN, A.B. *Diagnosis of memoryless digital automata using a modified technique of natural logical deduction (Russian)* ◇ B70 D05 F99 ◇

LIU, SHICHAO *A proof-theoretic approach to nonstandard analysis with emphasis on distinguishing between constructive and nonconstructive results* ◇ E70 F99 H05 H20 ◇

1981

LEIVANT, D. *Proof theoretic methodology for propositional dynamic logic* ◇ B75 F99 ◇

SELDIN, J.P. *A constructive approach to classical mathematics* ◇ F99 ◇

YASUGI, M. *Definability problems in metric spaces; a summary* ◇ E15 F99 ◇

1982

CONSTABLE, R.L. *Partial functions in constructive formal theories* ◇ F99 ◇

LUCI, F. *Una teoria dell'infinito potenziale (English summary)* ◇ E70 F99 ◇

1983

JUKNA, S. *Arithmetization of calculation (Russian) (English and Lithuanian summaries)* ◇ F99 ◇

KUZNETSOV, V.E. *K-systems (Russian)* ◇ F99 ◇

SHEN, BAIYING *Studies on higher rules (Chinese) (English summary)* ◇ F30 F99 ◇

SMULLYAN, R.M. *Dame oder Tiger? Logische Denkspiele und eine mathematische Novelle ueber Goedels grosse Entdeckung* ◇ B98 F99 ◇

1984

SHEN', A.KH. *Algorithmic variants of the notion of entropy (Russian)* ◇ F99 ◇

SOTO-ANDRADE, J. & VARELA GARCIA, F.J. *Self-reference and fixed points: a discussion and an extension of Lawvere's theorem* ◇ F99 G30 ◇

TAKEUTI, G. & TITANI, S. *Intuitionistic fuzzy logic and intuitionistic fuzzy set theory* ◇ E70 F99 ◇

1985

PERLIS, D. *Languages with self-reference. I. Foundations* ◇ F99 ◇

Author Index

AANDERAA, S.O. & ANDREWS, P.B. & DREBEN, B. [1963] *False lemmas in Herbrand* (**J** 0015) Bull Amer Math Soc 69*699-706
◇ F05 F07 ◇ REV MR 27 # 2401 Zbl 126.10 JSL 31.657 • ID 00001

AANDERAA, S.O. & DREBEN, B. [1964] *Herbrand analyzing function* (**J** 0015) Bull Amer Math Soc 70*697-698
◇ F05 F07 F20 ◇ REV MR 30 # 1924 Zbl 221 # 02011 JSL 32.521 • ID 00005

AANDERAA, S.O. see Vol. I, III, IV for further entries

ABDALI, S.K. [1975] *A lambda-calculus model of programming languages I: Simple constructs. II: Jumps and procedures* (**J** 3144) Comput Lang, Int J 1*287-301,303-320
◇ B40 ◇ REV Zbl 356 # 68041 Zbl 356 # 68042 • ID 50334

ABDALI, S.K. [1976] *An abstraction algorithm for combinatory logic* (**J** 0036) J Symb Logic 41*222-224
◇ B40 ◇ REV MR 53 # 12888 Zbl 331 # 02012 • ID 14802

ABERTH, O. [1968] *Analysis in the computable number field* (**J** 0037) ACM J 15*275-299
◇ F60 ◇ REV MR 38 # 5626 Zbl 159.12 JSL 40.84 • ID 00025

ABERTH, O. [1969] *A chain of inclusion relations in computable analysis* (**J** 0053) Proc Amer Math Soc 22*539-548
◇ F60 ◇ REV MR 39 # 3998 Zbl 283 # 02029 JSL 40.84 • ID 00026

ABERTH, O. [1970] *Computable analysis and differential equations* (**P** 0603) Intuitionism & Proof Th;1968 Buffalo 47-52
◇ F60 ◇ REV MR 43 # 1835 Zbl 283 # 02030 JSL 40.84 • ID 00088

ABERTH, O. [1971] *The failure in computable analysis of a classical existence theorem for differential equations* (**J** 0053) Proc Amer Math Soc 30*151-156
◇ F60 ◇ REV MR 46 # 2124 Zbl 219 # 02022 JSL 40.85 • ID 00027

ABERTH, O. [1980] *Computable analysis* (**X** 0822) McGraw-Hill: New York xi + 187pp
◇ F30 F60 F98 ◇ REV Zbl 461 # 03015 JSL 49.988 • ID 54492

ABERTH, O. see Vol. IV for further entries

ABRUSCI, V.M. [1981] *Due note sul principio di induzione matematica* (**P** 3092) Congr Naz Logica;1979 Montecatini Terme 55-77
◇ A10 F30 ◇ ID 48360

ABRUSCI, V.M. [1983] *Sistemi di notazione ordinale e dilatori* (**P** 3829) Atti Incontri Log Mat (1);1982 Siena 341-345
◇ F15 ◇ REV Zbl 522 # 03042 • ID 37794

ABRUSCI, V.M. [1984] *Recenti risultati e metodi nella teoria della dimostrazione: dilatatori, β-logica* (**J** 2038) Rend Sem Mat, Torino 41*39-82
◇ F15 F35 G30 ◇ ID 44760

ABRUSCI, V.M. [1985] *Bachmann collections. Gardens. Recent applications (Italian)* (**P** 4646) Atti Incontri Log Mat (2);1983/84 Siena 461-465
◇ F15 F35 ◇ ID 49670

ABRUSCI, V.M. [1985] *Uses of dilators: combinatorial problems and combinatorial results not provable in PA or in ID_1 (Italian)* (**P** 4646) Atti Incontri Log Mat (2);1983/84 Siena 309-314
◇ F15 F30 F35 ◇ ID 49667

ACKERMANN, W. [1925] *Begruendung des "tertium non datur" mittels der Hilbertschen Theorie der Widerspruchsfreiheit* (**J** 0043) Math Ann 93*1-36
◇ A05 B10 F30 ◇ ID 00104

ACKERMANN, W. [1928] *Zum Hilbertschen Aufbau der reellen Zahlen* (**J** 0043) Math Ann 99*118-133
• TRANSL [1967] (**C** 0675) From Frege to Goedel 493-507
◇ B28 D20 F15 ◇ REV FdM 54.56 • ID 00106

ACKERMANN, W. [1940] *Zur Widerspruchsfreiheit der Zahlentheorie* (**J** 0043) Math Ann 117*162-194
◇ F30 ◇ REV MR 1.322 Zbl 22.292 JSL 5.125 FdM 66.31 • ID 00118

ACKERMANN, W. [1950] *Widerspruchsfreier Aufbau der Logik I. Typenfreies System ohne tertium non datur* (**J** 0036) J Symb Logic 15*33-57
◇ B15 E70 F07 F25 F65 ◇ REV MR 12.384 Zbl 36.147 JSL 16.72 • ID 00122

ACKERMANN, W. [1951] *Konstruktiver Aufbau eines Abschnitts der zweiten Cantorschen Zahlenklasse* (**J** 0044) Math Z 53*403-413
◇ F15 ◇ REV MR 12.579 Zbl 42.50 JSL 17.152 • ID 00123

ACKERMANN, W. [1952] *Widerspruchsfreier Aufbau einer typenfreien Logik I (erweitertes System)* (**J** 0044) Math Z 55*364-384
◇ B10 B15 E70 F25 F65 ◇ REV MR 14.344 Zbl 50.245 JSL 19.295 • REM Part II 1953 • ID 00124

ACKERMANN, W. [1953] *Widerspruchsfreier Aufbau einer typenfreien Logik II* (**J** 0044) Math Z 57*155-166
◇ B10 B15 E35 E70 F30 F35 F65 ◇ REV MR 14.834 Zbl 50.245 JSL 19.295 • REM Part I 1952 • ID 00125

ACKERMANN, W. [1965] *Der Aufbau einer hoeheren Logik* (**J** 0009) Arch Math Logik Grundlagenforsch 7*5-22
◇ B15 E70 F30 F35 ◇ REV MR 34 # 26 Zbl 154.255 JSL 40.458 • ID 00135

ACKERMANN, W. see Vol. I, II, III, IV, V for further entries

Author Index

ACZEL, P. & CROSSLEY, J.N. [1966] *Constructive order types III* (J 0009) Arch Math Logik Grundlagenforsch 9*112-116
◇ D45 F15 ◇ REV MR 35 # 5319 Zbl 199.29 • REM Part II 1966 by Crossley,J.N. • ID 00137

ACZEL, P. [1967] *Paths in Kleene's O* (J 0009) Arch Math Logik Grundlagenforsch 10*8-12
◇ F15 ◇ REV MR 35 # 5311 Zbl 153.316 JSL 35.336 • ID 00139

ACZEL, P. [1968] *Saturated intuitionistic theories* (P 0608) Logic Colloq;1966 Hannover 1-11
◇ F30 F50 ◇ REV MR 38 # 5587 Zbl 198.322 • ID 21304

ACZEL, P. [1970] *Representability in some systems of second order arithmetic* (J 0029) Israel J Math 8*309-328
◇ D65 D70 F35 ◇ REV MR 42 # 4396 Zbl 216.6 • ID 00141

ACZEL, P. [1972] *Describing ordinals using functionals of transfinite type* (J 0036) J Symb Logic 37*35-47
◇ F10 F15 ◇ REV MR 48 # 72 Zbl 264 # 02025 JSL 48.876 • ID 00142

ACZEL, P. [1975] *Recursive density types and Nerode extensions of arithmetic* (J 0038) J Austral Math Soc 20*146-158
◇ D45 D50 F30 ◇ REV MR 51 # 10060 Zbl 308 # 02045 • ID 21303

ACZEL, P. [1977] *The strength of Martin-Loef's intuitionistic type theory with one universe* (P 1629) Symp Math Log;1974 Oulo;1975 Helsinki 1-32
◇ F35 F50 ◇ REV JSL 49.313 • ID 47386

ACZEL, P. [1978] *The type theoretic interpretation of constructive set theory* (P 1897) Logic Colloq;1977 Wroclaw 55-66
◇ E70 F35 F50 ◇ REV MR 80f:03068 Zbl 481 # 03035 JSL 49.313 • ID 70298

ACZEL, P. [1981] *Two notes on the Paris independence result I: A generalization of Ramsey's theorem. II: The ordinal height of a density* (P 3404) Model Th & Arithm;1979/80 Paris 21-31
◇ C62 E05 F15 F30 ◇ REV MR 84i:03071 Zbl 487 # 03025 • ID 34552

ACZEL, P. [1982] *The type theoretic interpretation of constructive set theory: choice principles* (P 3638) Brouwer Centenary Symp;1981 Noordwijkerhout 1-40
◇ E25 E70 F35 F50 ◇ REV MR 85g:03085 Zbl 529 # 03035 • ID 37664

ACZEL, P. see Vol. II, III, IV, V for further entries

ADAMOWICZ, Z. [1984] *Axiomatization of the forcing relation with an application to Peano arithmetic* (J 0027) Fund Math 120*167-186
◇ C62 E40 F30 ◇ REV MR 86h:03090 • ID 45406

ADAMOWICZ, Z. & MORALES-LUNA, G. [1985] *A recursive model for arithmetic with weak induction* (J 0036) J Symb Logic 50*49-54
◇ C57 F30 H15 ◇ REV MR 86d:03058 Zbl 576 # 03045 • ID 39780

ADAMOWICZ, Z. [1985] *Algebraic approach to $\exists_1$ induction* (P 4310) Easter Conf on Model Th (3);1985 Gross Koeris 5-15
◇ C60 C62 F30 ◇ ID 49053

ADAMOWICZ, Z. also published under the name SMOLSKA-ADAMOWICZ, Z.

ADAMOWICZ, Z. see Vol. III, IV, V for further entries

AERTS, J.P.H. [1981] *Implementing SASL without garbage collection* (1111) Preprints, Manuscr., Techn. Reports etc. 81-WSK-05*70pp
◇ B40 ◇ REV Zbl 476 # 68016 • REM Technological University, Department of Mathematics. Eindhoven • ID 55587

AFRICK, H. [1972] *A proof theoretic proof of Scott's general interpolation theorem* (J 0036) J Symb Logic 37*683-695
◇ C40 F07 ◇ REV MR 51 # 7833 Zbl 258 # 02037 • ID 00199

AFRICK, H. see Vol. III for further entries

AGAZZI, E. [1967] *L'aritmetica come strumento di ricerca logico-matematica (English summary)* (J 0059) Rend Sem Mat Fis Milano 37*191-219
◇ A05 F30 ◇ REV MR 37 # 3917 Zbl 202.3 • ID 00202

AGAZZI, E. see Vol. I, II for further entries

AIBA, A. & NAKANISHI, M. [1981] *General calculator based on lambda conversion* (J 2770) Keio Math Sem Rep (Yokohama) 6*1-9
◇ B40 ◇ REV MR 83a:68029 • ID 38840

AIELLO, L. & AIELLO, M. [1975] *Programming language semantics in a typed lambda-calculus* (P 1603) λ-Calc & Comput Sci Th;1975 Roma 240-251
◇ B40 B75 ◇ REV MR 58 # 31994 Zbl 331 # 68006 • ID 60076

AIELLO, L. & PRINI, G. [1981] *An efficient interpreter for the lambda-calculus* (J 0119) J Comp Syst Sci 23*383-424
◇ B40 ◇ REV MR 82m:68061 Zbl 472 # 03012 • ID 55278

AIELLO, L. see Vol. I for further entries

AIELLO, M. [1975] see AIELLO, L.

AKIMOV, A.P. [1979] *On imbeddings of classical modal calculi in constructive calculi (Russian)* (J 0023) Dokl Akad Nauk SSSR 247*777-778
• TRANSL [1979] (J 0062) Sov Math, Dokl 20*781-782
◇ B45 F50 ◇ REV MR 81d:03015 Zbl 426 # 03028 • ID 70347

AKIMOV, A.P. see Vol. II for further entries

ALMUKDAD, A. & NELSON, D. [1984] *Constructible falsity and inexact predicates* (J 0036) J Symb Logic 49*231-233
◇ B52 F50 ◇ REV MR 86c:03020 Zbl 575 # 03016 • ID 42414

AMBROSE, A. [1935] *Finitism in mathematics I,II* (J 0094) Mind 44*186-203,317-340
◇ A05 F99 ◇ REV Zbl 12.146 FdM 61.48 • ID 37299

AMBROSE, A. [1936] *The nature of the question "are there three consecutive 7's in the expansion of π?"* (J 3915) Pap Michigan Acad Sci 22*505-513
◇ A05 F99 ◇ ID 37300

AMBROSE, A. see Vol. I, II for further entries

AMER, M.A. [1975] *Parentheses in combinatory logic* (P 0775) Logic Colloq;1973 Bristol 429-432
◇ B40 ◇ REV MR 52 # 48 Zbl 309 # 02022 • ID 17216

AMER, M.A. [1976] *Typed boolean structures I (Arabic summary)* (J 0397) Proc Math Phys Soc Egypt 41*15-22
◇ B15 B40 C90 E40 G05 ◇ REV MR 52 # 7860 Zbl 418 # 03009 • REM Part II 1977 • ID 17946

AMER, M.A. [1977] *Typed boolean structures II (Arabic summary)* (J 0397) Proc Math Phys Soc Egypt 44*11-15 ◇ B15 B40 C90 E40 G05 ◇ REV MR 56 # 2784 Zbl 427 # 03011 • REM Part I 1976 • ID 30619

AMER, M.A. see Vol. II, III for further entries

AMSTISLAVSKIJ, V.I. [1966] *Recursive sieves (Russian)* (J 0092) Sib Mat Zh 7*233-241 • TRANSL [1966] (J 0475) Sib Math J 7*187-193 ◇ D55 E15 F65 ◇ REV MR 33 # 5477 Zbl 216.291 JSL 33.295 • ID 00303

AMSTISLAVSKIJ, V.I. see Vol. IV, V for further entries

ANDERSON, A.R. [1975] *Fitch on consistency* (C 1856) Log Enterprise 123-141 ◇ B15 F05 F25 ◇ REV Zbl 361 # 02040 • ID 50675

ANDERSON, A.R. see Vol. I, II, III for further entries

ANDERSON, J.M. & JOHNSTONE JR., H.W. [1962] *Natural deduction. The logical basis of axiom systems* (X 0821) Wadsworth Publ: Belmont xii + 418pp ◇ B98 F07 F98 ◇ REV MR 25 # 4986 JSL 29.93 • ID 00341

ANDO, S. & ITO, T. [1975] *On applying Scott's logic to termination problems* (P 3299) Progr Kiso Riron, Algor Okeru Shomei Ron;1973/74 Kyoto 1-7 ◇ B40 D05 ◇ REV MR 58 # 8427 Zbl 384 # 03007 • ID 52049

ANDO, S. see Vol. IV for further entries

ANDREWS, P.B. [1963] see AANDERAA, S.O.

ANDREWS, P.B. [1965] *A transfinite type theory with type variables* (X 0809) North Holland: Amsterdam xv + 143pp ◇ B15 B40 ◇ REV MR 37 # 51 Zbl 132.245 JSL 33.112 • ID 00402

ANDREWS, P.B. [1971] *Resolution in type theory* (J 0036) J Symb Logic 36*414-432 • REPR [1983] (C 4659) Autom of Reasoning 2*487-507 ◇ B15 F05 F35 ◇ REV MR 46 # 1551 Zbl 231 # 02038 • ID 00365

ANDREWS, P.B. [1972] *General models, descriptions, and choice in type theory* (J 0036) J Symb Logic 37*385-394 ◇ B15 C85 E25 E35 F35 ◇ REV MR 47 # 6433 Zbl 264 # 02049 • ID 00367

ANDREWS, P.B. [1972] *General models and extensionality* (J 0036) J Symb Logic 37*395-397 ◇ B15 C85 E35 F35 ◇ REV MR 47 # 6434 Zbl 264 # 02050 • ID 00366

ANDREWS, P.B. [1974] *Provability in elementary type theory* (J 0068) Z Math Logik Grundlagen Math 20*411-418 ◇ B15 B25 D35 F35 ◇ REV MR 52 # 7867 Zbl 306 # 02017 • ID 03832

ANDREWS, P.B. [1974] *Resolution and the consistency of analysis* (J 0047) Notre Dame J Formal Log 15*73-84 ◇ B15 F05 F25 F35 ◇ REV MR 52 # 7886 Zbl 226 # 02042 • ID 00368

ANDREWS, P.B. see Vol. I for further entries

ANDRIEVSKIJ, V.V. [1985] *Geometric structure of domains and direct theorems of the constructive theory of functions (Russian)* (J 0142) Mat Sb, Akad Nauk SSSR, NS 126(168)*41-58,143 • TRANSL [1986] (J 0349) Math of USSR, Sbor 54*39-56 ◇ F60 ◇ ID 45413

ANIKEEV, A.S. [1972] *Classification of derivable propositional formulas (Russian)* (J 0087) Mat Zametki (Akad Nauk SSSR) 11*165-174 • TRANSL [1972] (J 1044) Math Notes, Acad Sci USSR 11*106-110 ◇ B15 D05 F20 F50 ◇ REV MR 45 # 6568 Zbl 239 # 02003 • ID 00375

ANISHEV, P.A. & KASHIN, V.I. & SIDRISTYJ, B. [1973] *Sprache der konstruktiven Definitionen (Russisch)* (S 0507) Vychisl Sist (Akad Nauk SSSR Novosibirsk) 54*36-47 ◇ F99 ◇ REV Zbl 288 # 68007 • ID 60146

ANSHEL, M. & MCALOON, K. [1983] *Reducibilities among decision problems for HNN groups, vector addition systems and subsystems of Peano arithmetic* (J 0053) Proc Amer Math Soc 89*425-429 ◇ D20 D40 D80 F30 ◇ REV MR 85d:03096 Zbl 523 # 20025 • ID 36994

ANSHEL, M. see Vol. III, IV for further entries

APELT, H. [1966] *Axiomatische Untersuchungen ueber einige mit der Presburgerschen Arithmetik verwandte Systeme* (J 0068) Z Math Logik Grundlagen Math 12*131-168 ◇ B28 C35 C80 F30 ◇ REV MR 36 # 6279 Zbl 149.7 • ID 00403

APERY, R. [1976] *Mathematique constructive* (P 1748) Lang & Pensee Math;1976 Luxembourg 391-410 ◇ A05 F98 ◇ REV Zbl 419 # 00011 • ID 53343

APERY, R. [1982] *Mathematique constructive* (P 3753) Penser Math;1981 Paris 58-72 ◇ A05 F99 ◇ REV MR 84a:03002 • ID 34892

APT, K.R. [1972] *Non-finite axiomatizability of the second order arithmetic* (J 0014) Bull Acad Pol Sci, Ser Math Astron Phys 20*347-348 ◇ F35 ◇ REV MR 46 # 5118 Zbl 241 # 02014 • ID 00430

APT, K.R. [1973] *Infinitistic rules of proof and their semantics (Russian summary)* (J 0014) Bull Acad Pol Sci, Ser Math Astron Phys 21*879-886 ◇ C62 F07 F35 ◇ REV MR 51 # 5291 Zbl 272 # 02077 • ID 00431

APT, K.R. & MAREK, W. [1974] *Second order arithmetic and related topics* (J 0007) Ann Math Logic 6*177-229 ◇ C62 D55 E45 F25 F35 ◇ REV MR 51 # 12512 Zbl 299 # 02066 • ID 00432

APT, K.R. [1976] *Semantics of the infinitistic rules of proof* (J 0036) J Symb Logic 41*121-138 ◇ C62 D55 E45 F07 F35 ◇ REV MR 55 # 10230 Zbl 328 # 02016 • ID 14789

APT, K.R. [1978] *Inductive definitions, models of comprehension and invariant definability* (J 0029) Israel J Math 29*221-238 ◇ C40 C62 D55 D70 F35 ◇ REV MR 81e:03048 Zbl 379 # 02014 • ID 29131

APT, K.R. see Vol. I, II, III, IV for further entries

ARAI, T. [1984] *A subsystem of classical analysis proper to Takeuti's reduction method for Π_1^1-analysis* (P 3668) Log & Founds of Math;1983 Kyoto 92-110 ◇ F35 ◇ ID 42935

Author Index

ARAI, T. [1984] *An accessibility proof of ordinal diagrams in intuitionistic theories for iterated inductive definitions* (**J** 2606) Tsukuba J Math 8*209-218
◇ F15 F35 F50 ◇ REV MR 86a:03064 Zbl 564 # 03041 • ID 44971

ARAI, T. [1985] *A subsystem of classical analysis proper to Takeuti's reduction method for Π-analysis* (**J** 2606) Tsukuba J Math 9*21-29
◇ F05 F35 ◇ ID 47595

ARBIB, M.A. [1964] *Brains, machines, and mathematics* (**X** 0822) McGraw-Hill: New York xiv+152pp
• TRANSL [1968] (**X** 2027) Nauka: Moskva
◇ D05 D10 D80 F30 ◇ REV MR 31 # 5766 Zbl 174.24 JSL 35.482 • ID 00454

ARBIB, M.A. [1966] *Speed-up theorems and incompleteness theorems* (**P** 0746) Automata Th;1964 Ravello 6-24
◇ D15 F30 ◇ REV Zbl 202.316 • ID 27533

ARBIB, M.A. see Vol. II, IV, V for further entries

ARRUDA, A.I. [1978] *Some remarks on Griss' logic of negationless intuitionistic mathematics* (**P** 1800) Brazil Conf Math Log (1);1977 Campinas 9-29
◇ B55 F50 F55 G25 ◇ REV MR 80b:03095 Zbl 383 # 03014 • ID 31617

ARRUDA, A.I. see Vol. I, II, III, V for further entries

ARTEMOV, S.N. [1980] *Arithmetically complete modal theories (Russian)* (**S** 2582) Semiotika & Inf, Akad Nauk SSSR 14*115-133
◇ B45 C90 F50 ◇ REV MR 82a:03056 Zbl 463 # 03006 • ID 54546

ARTEMOV, S.N. [1982] *Applications of modal logic in proof theory (Russian)* (**S** 2874) Vopr Kibern, Akad Nauk SSSR 75*3-22
◇ B45 F30 ◇ REV Zbl 499 # 03046 • ID 38128

ARTEMOV, S.N. [1984] *On modal representations of extensions of Peano arithmetic* (**J** 2128) C R Math Acad Sci, Soc Roy Canada 6*129-132
◇ B45 F30 ◇ REV Zbl 555 # 03010 • ID 45832

ARTEMOV, S.N. [1985] *Modal logics axiomatizing provability (Russian)* (**J** 0216) Izv Akad Nauk SSSR, Ser Mat 49*1123-1154,1342
◇ B45 F30 ◇ ID 49322

ARTEMOV, S.N. [1985] *Nonarithmeticity of truth predicate logics of provability (Russian)* (**J** 0023) Dokl Akad Nauk SSSR 284*270-271
• TRANSL [1985] (**J** 0062) Sov Math, Dokl 32*403-405
◇ B45 D35 D55 F30 ◇ ID 48972

ARTIGUE, M. & ISAMBERT, E. [1977] *Quelques resultats de bicommutabilite en arithmetique du troisieme ordre (English summary)* (**J** 2313) C R Acad Sci, Paris, Ser A-B 284*A1425-A1428
◇ E30 F25 F35 ◇ REV MR 55 # 12521 Zbl 362 # 02063 • ID 27310

ARTIGUE, M. & ISAMBERT, E. & PERRIN, M.J. & ZALC, A. [1978] *Some remarks on bicommutability* (**J** 0027) Fund Math 101*207-226
◇ C62 E30 F25 F35 ◇ REV MR 80j:03019a Zbl 401 # 03024 • REM Comments have been published by Marek ibid 101*227-228 • ID 70505

ARTIGUE, M. see Vol. III, IV, V for further entries

ARZARELLO, F. [1972] *Il teorema di normalizzazione per il sistema di deduzione naturale classico (English summary)* (**J** 0220) Atti Accad Sci Torino, Fis Mat Nat 106*705-723
◇ F05 ◇ REV MR 51 # 118 Zbl 249 # 02015 • ID 17958

ARZARELLO, F. [1975] *Un sistema di deduzione naturale per linguaggi infinitari (English summary)* (**J** 0220) Atti Accad Sci Torino, Fis Mat Nat 109*633-641
◇ C75 F05 ◇ REV MR 55 # 7744 Zbl 367 # 02006 • ID 51166

ARZARELLO, F. [1976] *Alcune questioni riguardanti l'aritmetica e i linguaggi infinitari (English summary)* (**J** 2038) Rend Sem Mat, Torino 34*369-380
◇ C75 F05 ◇ REV MR 55 # 12463 Zbl 359 # 02008 • ID 50556

ARZARELLO, F. [1983] *Un' interpretazione categoriale degli oggetti anomici* (**P** 3829) Atti Incontri Log Mat (1);1982 Siena 345-349
◇ F35 F50 G30 ◇ REV MR 84k:03006 Zbl 547 # 03039 • ID 43237

ARZARELLO, F. [1985] *Methods of realizability and forcing in the metamathematics of constructive formal systems (Italian)* (**P** 4646) Atti Incontri Log Mat (2);1983/84 Siena 315-318
◇ F50 ◇ ID 49672

ASANIDZE, G.Z. [1981] *A dialogue justification of logic (Russian)* (**C** 3810) Log-Semat Issl (Tbilisi) 36-53
◇ B60 F50 F65 ◇ REV MR 84j:03061 • ID 34653

ASHCROFT, E.A. & HENNESSY, M. [1976] *The semantics of nondeterminism* (**P** 1870) Automata, Lang & Progr (3);1976 Edinburgh 478-493
◇ B40 B75 ◇ REV Zbl 379 # 02015 • ID 51848

ASHCROFT, E.A. & HENNESSY, M. [1980] *A mathematical semantics for a nondeterministic typed lambda-calculus* (**J** 1426) Theor Comput Sci 11*227-245
◇ B40 B75 ◇ REV MR 82a:68049 Zbl 462 # 68006 • ID 54532

ASHCROFT, E.A. see Vol. IV for further entries

ASHVINIKUMAR [1969] *The intuitionist contradictority of certain classical set theoretic results* (**J** 3077) Nieuw Arch Wisk, Ser 3 17*218-222
◇ F55 ◇ REV MR 41 # 1507 Zbl 188.317 • ID 00527

ASHVINIKUMAR [1969] *Ueber katalogisierte Raeume* (**J** 0020) Compos Math 21*431-456
◇ F55 ◇ REV MR 41 # 3241 Zbl 215.321 • ID 00525

ASHVINIKUMAR [1970] *On the intuitionist theory of Stieltjes integration and its applications* (**J** 0028) Indag Math 32*62-76
◇ F55 ◇ REV MR 41 # 3703 Zbl 201.12 • ID 00529

ASHVINIKUMAR [1970] *On Brouwer-Stieltjes integration* (**J** 0028) Indag Math 32*161-171
◇ F55 ◇ REV MR 41 # 7040 Zbl 216.7 • ID 00528

ASHVINIKUMAR [1971] *Another proof of the intuitionistic inconsistency of the axiom of choice (Dutch)* (**J** 3077) Nieuw Arch Wisk, Ser 3 19*193-195
◇ E25 F50 F55 ◇ REV MR 49 # 2270 Zbl 226 # 02033 • ID 03841

ASHVINIKUMAR & SHUKLA, S.L. [1971] *Intuitionist determination of dual spaces of certain catalogued linear spaces I, II* (J 0028) Indag Math 33*240-260
◇ F55 ◇ REV MR 44 # 5744a MR 44 # 5744b Zbl 215.194
• ID 00530

ASHVINIKUMAR [1974] *A new version of the proof of the fan theorem* (J 0248) Math Student 42*1-4
◇ F55 ◇ REV MR 53 # 2649 Zbl 384 # 03041 • ID 21533

ASHVINIKUMAR [1979] *Microscopic completeness of full indications and intuitionist contradictory proofs* (J 0028) Indag Math 41*95-103
◇ F55 ◇ REV MR 80i:03066a Zbl 405 # 03031 • ID 70524

ASTESIANO, E. & COSTA, G. [1979] *Sharing in nondeterminism* (P 1873) Automata, Lang & Progr (6);1979 Graz 71*1-15
◇ B40 B75 ◇ REV MR 82b:68006 Zbl 416 # 68033
• ID 53230

ASTESIANO, E. & COSTA, G. [1980] *Delayed evaluation trees and their application to semantics* (P 3057) CAAP'80 Arbres en Algeb & Progr (5);1980 Lille 110-118
◇ B40 B75 ◇ REV MR 82e:68004 Zbl 462 # 68004
• ID 54531

ASTESIANO, E. & COSTA, G. [1980] *Languages with reducing reflexive types* (P 2904) Automata, Lang & Progr (7);1980 Noordwijkerhout 38-50
◇ B40 B75 ◇ REV MR 81j:68082 Zbl 444 # 68026
• ID 80581

ASTESIANO, E. & COSTA, G. [1980] *Nondeterminism and fully abstract models* (J 3441) RAIRO Inform Theor 14*323-347
◇ B40 B75 ◇ REV MR 82j:68084 Zbl 463 # 03024
• ID 54564

ASTESIANO, E. & COSTA, G. [1983] *The intensitivity theorem for nonreducing reflexive types* (J 0119) J Comp Syst Sci 27*389-399
◇ B40 B75 ◇ REV MR 85j:03013 Zbl 537 # 68035
• ID 41324

ASTESIANO, E. & COSTA, G. [1984] *Distributive semantics for nondeterministic typed λ-calculi* (J 1426) Theor Comput Sci 32*121-156
◇ B40 B75 ◇ REV Zbl 555 # 03007 • ID 46124

AUERBACH, D.D. [1985] *Intensionality and the Goedel theorems* (J 0095) Philos Stud 48*337-351
◇ A05 F99 ◇ ID 49285

AUSIELLO, G. [1974] *Relations between semantics and complexity of recursive programs* (P 1869) Automata, Lang & Progr (2);1974 Saarbruecken 129-140
◇ B40 B75 D15 D20 ◇ REV MR 55 # 13868
Zbl 288 # 68006 • ID 80587

AUSIELLO, G. [1975] *On the description of time varying systems in λ-calculus* (P 1603) λ-Calc & Comput Sci Th;1975 Roma 177-191
◇ B40 ◇ REV MR 57 # 5495 Zbl 333 # 02022 • ID 60215

AUSIELLO, G. see Vol. I, III, IV for further entries

AVRON, A. [1984] *On modal systems having arithmetical interpretations* (J 0036) J Symb Logic 49*935-942
◇ B45 F05 F30 ◇ REV MR 86g:03028 • ID 42415

AVRON, A. see Vol. II for further entries

BABAEV, A.A. & SOLOV'EV, S.V. [1979] *A coherence theorem for canonical maps in Cartesian closed categories (Russian) (English summary)* (S 0228) Zap Nauch Sem Leningrad Otd Mat Inst Steklov 88*3-29,236
• TRANSL [1982] (J 1531) J Sov Math 20*2263-2279
◇ F05 G30 ◇ REV MR 81g:03074 Zbl 429 # 03037
• ID 53868

BABAEV, A.A. [1981] *Equality of morphisms in closed categories. I,II (Russian) (Azerbaijani and English summaries)* (J 0135) Izv Akad Nauk Azerb SSR, Ser Fiz-Tekh Mat 2/1*3-10,2/2*3-9
◇ F07 G30 ◇ REV MR 83d:18007 MR 83d:18008
Zbl 461 # 18004 Zbl 487 # 18001 • ID 40067

BACHMANN, H. [1950] *Die Normalfunktionen und das Problem der ausgezeichneten Folgen von Ordnungszahlen* (J 0123) Vierteljahrschr Naturforsch Ges Zuerich 95*115-147
◇ E10 F15 ◇ REV MR 12.165 Zbl 41.21 JSL 22.336
• ID 00608

BACHMANN, H. [1952] *Vergleich und Kombination zweier Methoden von Veblen und Finsler zur Loesung des Problems der ausgezeichneten Folgen von Ordnungszahlen* (J 2022) Comm Math Helvetici 26*55-67
◇ E10 F15 ◇ REV MR 13.728 Zbl 46.51 JSL 22.336
• ID 00609

BACHMANN, H. [1954] *Normalfunktionen und Hauptfolgen* (J 2022) Comm Math Helvetici 28*9-16
◇ E10 E25 F15 ◇ REV MR 16.20 Zbl 55.49 JSL 22.336
• ID 00610

BACHMANN, H. [1955] *Transfinite Zahlen* (X 0811) Springer: Heidelberg & New York vii+204pp
◇ E10 E25 E98 F15 ◇ REV MR 17.134 MR 36 # 2506
Zbl 65.35 JSL 24.223 • REM 2nd rev. ed. 1967 • ID 00611

BACHMANN, H. see Vol. I, II, V for further entries

BACK, R.J.R. [1980] *Correctness preserving program refinements: proof theory and applications* (S 1605) Math Centr Tracts iii+118pp
◇ B75 F99 ◇ REV MR 83k:68008 Zbl 451 # 68018
• ID 40291

BACK, R.J.R. see Vol. I, III for further entries

BACKUS, J. [1981] *The algebra of functional programs: function level reasoning, linear equations, and extended definitions* (P 2930) Formal of Progr Concepts;1981 Peniscola 1-43
◇ B40 B75 ◇ REV MR 82j:68003 Zbl 474 # 68009
• ID 80594

BAETEN, J. & BOERBOOM, B. [1979] *Ω can be anything it should not be* (J 0028) Indag Math 41*111-120
◇ B40 ◇ REV MR 80h:03018 Zbl 417 # 03006 • ID 53243

BAETEN, J. see Vol. III, IV, V for further entries

BAIRE, R. & BOREL, E. & HADAMARD, J. & LEBESGUE, H. [1905] *Cinq lettres sur la theorie des ensembles* (J 0353) Bull Soc Math Fr 33*261-273
◇ A05 D55 E15 E25 F65 ◇ REV FdM 36.99 • ID 05473

BAIRE, R. see Vol. V for further entries

BAKER, T.P. [1979] *On "provable" analogs of $\mathcal{P}$ and $\mathcal{NP}$* (J 0041) Math Syst Theory 12*213-218
◇ D15 F30 ◇ REV MR 80g:68051 Zbl 405 # 68046
• ID 80606

BAKER, T.P. see Vol. IV for further entries

BAKER-FINCH, C.A. [1984] *Algebraic, operational and denotational semantics of the lambda calculus* (**J** 3116) Austral Comp J 16*96-101
◇ B40 ◇ REV Zbl 569 # 68023 • ID 44683

BAKKER DE, J.W. [1971] *Recursive procedures* (**X** 1121) Math Centr: Amsterdam 24*viii + 108pp
◇ B40 B75 D20 ◇ REV MR 49 # 7121 Zbl 274 # 02015 JSL 40.83 • ID 28688

BAKKER DE, J.W. [1975] *Least fixed points revisited* (**P** 1603) λ-Calc & Comput Sci Th;1975 Roma 27-61
◇ B40 B75 ◇ REV MR 57 # 14568 Zbl 328 # 68009 • ID 60260

BAKKER DE, J.W. see Vol. IV for further entries

BALDUS, R. [1924] *Formalismus und Intuitionismus in der Mathematik* (**X** 1240) Braun: Karlsruhe 45pp
◇ A05 F55 F98 ◇ REV FdM 50.26 • ID 25506

BAR-HILLEL, Y. & DALEN VAN, D. & FRAENKEL, A.A. & LEVY, A. [1958] *Foundations of set theory* (**X** 0809) North Holland: Amsterdam x + 415pp
◇ A05 B98 E30 E70 E98 F98 ◇ REV MR 21 # 648 MR 49 # 10546 Zbl 248 # 02071 Zbl 82.262 JSL 29.141 • REM 2nd ed. 1973; x + 404pp; 1st ed. by Fraenkel,A.A. & Bar-Hillel,Y. • ID 00805

BAR-HILLEL, Y. see Vol. I, II, III, IV, V for further entries

BARALT-TORRIJOS, J. & CHIARAVIGLIO, L. & GROSKY, W. [1975] *The programmatic semantics of binary predicator calculi* (**J** 0047) Notre Dame J Formal Log 16*591-596
◇ B40 D05 ◇ REV MR 52 # 7861 Zbl 283 # 02038 • ID 17965

BARENDREGT, H.P. [1973] *A characterization of terms of the lambda-calculus having a normal form* (**J** 0036) J Symb Logic 38*441-445
◇ B40 ◇ REV MR 49 # 24 Zbl 279 # 02013 • ID 00796

BARENDREGT, H.P. [1973] *Combinatory logic and the axiom of choice* (**J** 0028) Indag Math 35*203-221
◇ B40 E25 F50 ◇ REV MR 58 # 163 Zbl 274 # 02010 • ID 29015

BARENDREGT, H.P. [1974] *Combinatory logic and the ω-rule* (**J** 0027) Fund Math 82*199-215
◇ B40 ◇ REV MR 51 # 72 Zbl 299 # 02028 • ID 00798

BARENDREGT, H.P. [1974] *Pairing without conventional restraints* (**J** 0068) Z Math Logik Grundlagen Math 20*289-306
◇ B40 F35 ◇ REV MR 51 # 10039 Zbl 299 # 02030 • ID 03861

BARENDREGT, H.P. [1975] *Normed uniformly reflexive structures* (**P** 1603) λ-Calc & Comput Sci Th;1975 Roma 272-286
◇ B40 D75 ◇ REV MR 57 # 12197 Zbl 333 # 02021 • ID 27590

BARENDREGT, H.P. [1975] *Open problems* (**P** 1603) λ-Calc & Comput Sci Th;1975 Roma 367-370
◇ B40 ◇ REV MR 58 # 16175 • ID 70673

BARENDREGT, H.P. & BERGSTRA, J.A. & KLOP, J.W. & VOLKEN, H. [1976] *Representability in lambda-algebras* (**J** 0028) Indag Math 38*377-387
◇ B40 ◇ REV MR 55 # 2512 Zbl 399 # 03013 • ID 26057

BARENDREGT, H.P. [1976] *The incompleteness theorems* (**X** 1051) Univ Utrecht Math Inst: Utrecht iv + 59pp
◇ D35 F30 ◇ REV MR 53 # 2611 Zbl 323 # 02057 • ID 21504

BARENDREGT, H.P. [1977] *A global representation of the recursive functions in the λ calculus* (**J** 1426) Theor Comput Sci 3*225-242
◇ B40 ◇ REV MR 58 # 5078 Zbl 375 # 02013 • ID 31337

BARENDREGT, H.P. [1977] *Solvability in lambda calculi* (**P** 1729) Colloq Int Log;1975 Clermont-Ferrand 209-219
◇ B40 ◇ REV MR 80f:03020 Zbl 445 # 03005 • ID 56469

BARENDREGT, H.P. [1977] *The type free lambda calculus* (**C** 1523) Handb of Math Logic 1091-1132
◇ B40 ◇ REV MR 58 # 10343 JSL 49.980 • ID 27333

BARENDREGT, H.P. & BERGSTRA, J.A. & KLOP, J.W. & VOLKEN, H. [1978] *Degrees of sensible lambda theories* (**J** 0036) J Symb Logic 43*45-55
◇ B40 D30 ◇ REV MR 57 # 15981 Zbl 408 # 03012 • ID 31152

BARENDREGT, H.P. & KOYMANS, K. [1980] *Comparing some classes of lambda calculus models* (**C** 3050) Essays Combin Log, Lambda Calc & Formalism (Curry) 287-301
◇ B40 ◇ REV MR 82b:03038 Zbl 469 # 03006 • ID 70683

BARENDREGT, H.P. & LONGO, G. [1980] *Equality of lambda terms in the model T^{ω}* (**C** 3050) Essays Combin Log, Lambda Calc & Formalism (Curry) 303-337
◇ B40 ◇ REV MR 82g:03017 Zbl 469 # 03006 • ID 70684

BARENDREGT, H.P. [1981] *The lambda calculus, its syntax and semantics* (**S** 3303) Stud Logic Found Math 103*xiv + 615pp
• TRANSL [1985] (**X** 0885) Mir: Moskva 606pp
◇ B40 B98 ◇ REV MR 83b:03016 MR 86a:03012 Zbl 467 # 03010 Zbl 551 # 03007 JSL 49.301 • REM Revised ed.1984, xv + 621pp • ID 55008

BARENDREGT, H.P. & COPPO, M. & DEZANI-CIANCAGLINI, M. [1983] *A filter lambda model and the completeness of type assignment* (**J** 0036) J Symb Logic 48*931-940
◇ B40 ◇ REV MR 85j:03014 Zbl 545 # 03004 • ID 39449

BARENDREGT, H.P. & REZUS, A. [1983] *Semantics for classical AUTOMATH and related systems* (**J** 0194) Inform & Control 59*127-147
◇ B40 B75 ◇ REV Zbl 564 # 68060 • ID 39456

BARENDREGT, H.P. [1984] *Introduction to lambda calculus* (**J** 3929) Nieuw Arch Wisk, Ser 4 4*337-372
◇ B40 ◇ REV Zbl 549 # 03012 • ID 39458

BARENDREGT, H.P. [1984] *Lambda calculus and its models* (**P** 3710) Logic Colloq;1982 Firenze 209-239
◇ B40 ◇ REV MR 86f:03025 Zbl 558 # 03006 • ID 39461

BARENDREGT, H.P. see Vol. IV for further entries

BARNES, D.W. & MONRO, G.P. [1974] *A simple model for a weak system of arithmetic* (**J** 0016) Bull Austral Math Soc 11*321-323
◇ C62 F30 ◇ REV MR 50 # 9571 Zbl 284 # 02030 • ID 03863

BARNES, D.W. see Vol. I, II for further entries

BARR, M. & WELLS, C. [1985] *Toposes, triples and theories* (**X** 0811) Springer: Heidelberg & New York xiii+345pp ◇ C98 E98 F35 F50 G30 ◇ REV MR 86f:18001 Zbl 567 # 18001 • ID 40111

BARR, M. see Vol. V for further entries

BARRICELLI, N.A. [1983] *Challenge to Goedel's proof* (**X** 3805) Blindern Theoretic Res Team: Oslo i+14 pp ◇ F30 ◇ REV MR 86f:03008 • ID 43987

BARRICELLI, N.A. [1984] *A few trivialities about Goedel's proof and the rules of definition* (**J** 3099) Theoretic Papers 1∗87-121 ◇ A05 F30 ◇ REV MR 86f:03009 • ID 44376

BARRICELLI, N.A. see Vol. I for further entries

BARTH, E.M. [1966] *On the transformation of closed semantic tableaus into natural and axiomatic deductions* (**J** 0079) Logique & Anal, NS 9∗147-170 ◇ F05 F07 ◇ REV MR 35 # 21 Zbl 149.245 • ID 00819

BARTH, E.M. see Vol. II, V for further entries

BARTHELEMY, J.-P. [1974] *Sur la refutabilite* (**J** 0306) Cah Topol & Geom Differ 15∗21-46 ◇ B20 F50 G30 ◇ REV MR 50 # 1852 Zbl 327 # 18008 • ID 04140

BARTHELEMY, J.-P. see Vol. V for further entries

BARWISE, J. [1973] *A preservation theorem for interpretations* (**P** 0713) Cambridge Summer School Math Log;1971 Cambridge GB 618-621 ◇ C40 F25 ◇ REV MR 51 # 5285 Zbl 276 # 02034 • ID 04141

BARWISE, J. (ED.) [1977] *Handbook of mathematical logic* (**X** 0809) North Holland: Amsterdam xi+1165pp • TRANSL [1982] (**X** 2027) Nauka: Moskva ◇ B98 C98 D98 E98 F98 H98 ◇ REV MR 56 # 15351 MR 84g:03004 MR 84j:03006 Zbl 443 # 03001 JSL 49.968 JSL 49.971 JSL 49.975 JSL 49.980 • REM 3rd ed 1982. Transl. in 4 parts. Russian suppl. by Mints,G.E. & Orevkov,V.P • ID 70117

BARWISE, J. see Vol. I, II, III, IV, V for further entries

BARZIN, M. & ERRERA, A. [1927] *Sur la logique de M.Brouwer* (**J** 0150) Acad Roy Belg Bull Cl Sci (5) 13∗56-71 ◇ F50 ◇ REV FdM 53.40 • ID 00846

BARZIN, M. & ERRERA, A. [1929] *Sur le principe du tiers exclu* (**J** 0151) Arch Soc Belg Philos 1∗26 ◇ A05 F50 ◇ REV FdM 55.633 • ID 00848

BARZIN, M. & ERRERA, A. [1932] *Note sur la logique de M.Heyting* (**J** 0152) Enseign Math 31∗122-124 ◇ A05 F50 ◇ REV FdM 59.56 • ID 00851

BARZIN, M. & ERRERA, A. [1932] *Sur la logique de M.Heyting* (**J** 0152) Enseign Math 30∗248-250 ◇ A05 F50 ◇ REV FdM 58.62 • ID 00849

BARZIN, M. & ERRERA, A. [1932] *Sur la logique intuitionniste. Reponse a M.A.Heyting* (**J** 0152) Enseign Math 31∗273-274 ◇ A05 F50 ◇ REV FdM 59.56 • ID 00853

BARZIN, M. & ERRERA, A. [1933] *La logique de M.Brouwer. Etat de la question* (**J** 0494) Bull Math Soc Sci Roumanie 35∗51-52 ◇ A05 F50 ◇ REV Zbl 8.290 FdM 59.867 • ID 00855

BARZIN, M. [1935] *Sur les demonstrations de non-contradiction des axiomes* (**P** 0639) Congr Nat des Sci (2);1935 Bruxelles 156-159 ◇ A05 B30 F99 ◇ ID 00914

BARZIN, M. [1936] *Note sur la demonstration de M.E.W.Beth* (**J** 0150) Acad Roy Belg Bull Cl Sci (5) 22∗582-583 ◇ A05 F25 F50 ◇ REV Zbl 15.194 JSL 1.118 FdM 62.30 • REM Cf an article by Beth ibid 22∗580-581 • ID 00857

BARZIN, M. [1940] *Logique symbolique. Sur la portee du theoreme de M.Goedel* (**J** 0150) Acad Roy Belg Bull Cl Sci (5) 26∗230-239 ◇ A05 F30 ◇ REV MR 7.185 Zbl 25.4 FdM 66.27 • ID 00858

BARZIN, M. [1969] *Sur le mecanisme des paradoxes et le theoreme de M.K.Goedel* (**J** 0147) An Univ Bucuresti, Acta Logica 12∗193-197 ◇ A05 F30 ◇ REV MR 46 # 1579 Zbl 247 # 02007 • ID 00860

BARZIN, M. see Vol. V for further entries

BATINI, C. & PETTOROSSI, A. [1975] *On subrecursiveness in weak combinatory logic* (**P** 1603) λ-Calc & Comput Sci Th;1975 Roma 297-311 ◇ B40 ◇ REV MR 58 # 188 Zbl 332 # 02033 • ID 60342

BATLE, N. & PLA I CARRERA, J. & SALES VALLES, F.A. & VERDU I SOLANS, V. [1981] *To Kurt Goedel, in memoriam (Catalan)* (**J** 1854) Bull Sec Mat Soc Catalana Cienc 1∗261-276 ◇ A10 C99 E35 E50 F99 ◇ REV MR 83k:01043 • ID 40048

BATLE, N. see Vol. I for further entries

BAUVAL, A. [1985] *Polynomial rings and weak second-order logic* (**J** 0036) J Symb Logic 50∗953-972 ◇ C60 C85 F35 ◇ ID 48368

BAUVAL, A. see Vol. III for further entries

BAXTER, L.D. [1978] *The undecidability of the third order dyadic unification problem* (**J** 0194) Inform & Control 38∗170-178 ◇ B15 B40 D80 ◇ REV MR 80m:03077 Zbl 387 # 03006 • ID 52222

BECK, JON M. [1977] *From the pocket calculator to de Rham's theorem* (**X** 1599) Aarhus Univ Mat Inst: Aarhus 15pp ◇ F65 ◇ REV Zbl 363 # 57011 • ID 50915

BECK, JON M. [1979] *Simplicial sets and the foundations of analysis* (**P** 2901) Appl Sheaves;1977 Durham 113-124 ◇ B28 F65 ◇ REV MR 80m:03029 Zbl 438 # 03057 • ID 55969

BECK, JON M. see Vol. I for further entries

BECVAR, J. [1971] *Hilbert's second problem (the question of the consistency of arithmetic) (Czech)* (**J** 1527) Pokroky Mat Fyz Astron (Prague) 16∗225-237 ◇ F30 ◇ REV MR 56 # 15373 Zbl 218 # 02046 • ID 26286

BECVAR, J. see Vol. I, II, IV for further entries

BEESON, M.J. [1975] *The nonderivability in intuitionistic formal systems of theorems on the continuity of effective operations* (**J** 0036) J Symb Logic 40∗321-346 ◇ D20 F50 ◇ REV MR 51 # 10050 Zbl 316 # 02038 • ID 17547

BEESON, M.J. [1976] *Derived rules of inference related to the continuity of effective operations* (**J** 0036) J Symb Logic 41*328-336
◇ D20 F50 ◇ REV MR 54#7208 Zbl 333#02027 • ID 14764

BEESON, M.J. [1976] *The unprovability in intuitionistic formal systems of the continuity of effective operations on the reals* (**J** 0036) J Symb Logic 41*18-24
◇ F50 ◇ REV MR 54#2425 Zbl 333#02026 • ID 14756

BEESON, M.J. [1977] *Continuity and comprehension in intuitionistic formal systems* (**J** 0048) Pac J Math 68*29-40
◇ F50 ◇ REV MR 58#21508 Zbl 323#02048 • ID 50098

BEESON, M.J. [1977] *Principles of continuous choice and continuity of functions in formal systems for constructive mathematics* (**J** 0007) Ann Math Logic 12*249-322
◇ F35 F50 ◇ REV MR 58#5106 Zbl 393#03012 • ID 27948

BEESON, M.J. [1978] *A type-free Goedel interpretation* (**J** 0036) J Symb Logic 43*213-227
◇ F50 ◇ REV MR 58#21507 Zbl 393#03042 • ID 29255

BEESON, M.J. [1978] *Some relations between classical and constructive mathematics* (**J** 0036) J Symb Logic 43*228-246
◇ F50 F55 ◇ REV MR 58#5094 Zbl 393#03041 • ID 29256

BEESON, M.J. [1979] *Continuity in intuitionistic set theories* (**P** 2627) Logic Colloq;1978 Mons 1-52
◇ E35 E70 F50 ◇ REV MR 82f:03053 Zbl 429#03041 • ID 53872

BEESON, M.J. [1979] *Goodman's theorem and beyond* (**J** 0048) Pac J Math 84*1-16
◇ F35 F50 ◇ REV MR 81b:03062 Zbl 433#03039 • ID 70829

BEESON, M.J. [1980] *Extensionality and choice in constructive mathematics* (**J** 0048) Pac J Math 88*1-28
◇ E25 E70 F50 ◇ REV MR 82g:03100 Zbl 452#03046 • ID 54110

BEESON, M.J. [1980] *Some problems in constructive mathematics* (**J** 2038) Rend Sem Mat, Torino 38/1*13-21
◇ A10 F50 ◇ REV MR 82k:03093 Zbl 452#03047 • ID 54111

BEESON, M.J. [1981] *Formalizing constructive mathematics: why and how?* (**P** 3146) Constr Math;1980 Las Cruces 146-190
◇ A05 F50 ◇ REV MR 83e:03097 Zbl 464#03050 • ID 35247

BEESON, M.J. [1982] *Problematic principles in constructive mathematics* (**P** 3623) Logic Colloq;1980 Prague 11-55
◇ F50 ◇ REV MR 84h:03134 Zbl 523#03048 • ID 34320

BEESON, M.J. [1982] *Recursive models for constructive set theories* (**J** 0007) Ann Math Logic 23*127-178
◇ E70 F35 F50 ◇ REV MR 85d:03111 Zbl 514#03039 • ID 37415

BEESON, M.J. & SCEDROV, A. [1984] *Church's thesis, continuity, and set theory* (**J** 0036) J Symb Logic 49*630-643
◇ D20 E70 F50 ◇ REV MR 86f:03097 • ID 42422

BEESON, M.J. [1985] *Foundations of constructive mathematics* (**X** 0811) Springer: Heidelberg & New York xxiii+466pp
◇ F50 F55 F60 F98 ◇ REV Zbl 565#03028 • ID 45120

BEHMANN, H. [1960] *Das Russell'sche Paradoxon und die formale Logik* (**P** 0560) Int Congr Philos (12);1958 Venezia 5*45-54
◇ A10 B40 E70 ◇ REV JSL 40.583 • ID 14635

BEHMANN, H. see Vol. I, II, III, IV, V for further entries

BEL, M. [1977] *An intuitionistic combinatory theory not closed under the rule of choice* (**J** 0028) Indag Math 39*69-72
◇ B40 F50 ◇ REV MR 57#75 Zbl 382#03012 • ID 26063

BELDING, W.R. [1970] *A note on the intuitionistic fan theorem* (**J** 0047) Notre Dame J Formal Log 11*484-486
◇ F55 ◇ REV MR 47#1580 Zbl 169.297 • ID 00956

BELDING, W.R. [1971] *Intuitionistic negation* (**J** 0047) Notre Dame J Formal Log 12*183-187
◇ B55 F55 ◇ REV MR 45#3160 Zbl 177.8 • ID 00957

BELDING, W.R. see Vol. V for further entries

BELINFANTE, M.J. [1929] *Ueber einen Grenzwertsatz aus der Theorie der unendlichen Folgen* (**J** 0043) Math Ann 101*312-315
◇ F55 ◇ REV FdM 55.121 • ID 43862

BELINFANTE, M.J. [1929] *Zur intuitionistischen Theorie der unendlichen Reihen* (**J** 0277) Sitzb Preuss Akad Wiss Phys Math Kl 29*639-660
◇ F55 ◇ REV FdM 55.121 • ID 37317

BELINFANTE, M.J. [1930] *Absolute Konvergenz in der intuitionistischen Mathematik* (**J** 0278) Nederl Akad Wetensch Proc 33*1180-1184
◇ F55 ◇ REV FdM 56.904 • ID 43863

BELINFANTE, M.J. [1930] *Ueber eine besondere Klasse von non-oszillierenden Reihen* (**J** 0278) Nederl Akad Wetensch Proc 33*1170-1179
◇ F55 ◇ REV FdM 56.904 • ID 37318

BELINFANTE, M.J. [1931] *Die Hardy-Littlewoodsche Umkehrung des Abelschen Stetigkeitssatzes in der intuitionistischen Mathematik* (**J** 0278) Nederl Akad Wetensch Proc 34*401-412
◇ F55 ◇ REV Zbl 1.332 FdM 57.260 • ID 43864

BELINFANTE, M.J. [1931] *Ueber die Elemente der Funktionentheorie und die Picardschen Saetze in der intuitionistischen Mathematik* (**J** 0278) Nederl Akad Wetensch Proc 34*1395-1397
◇ F55 ◇ REV Zbl 4.64 FdM 57.342 • ID 37319

BELINFANTE, M.J. [1932] *Ueber den intuitionistischen Beweis der Picardschen Saetze* (**P** 0653) Int Congr Math (II, 4);1932 Zuerich 2*345-346
◇ F55 ◇ REV FdM 58.70 • ID 42964

BELINFANTE, M.J. [1939] *Das Riemannsche Umordnungsprinzip in der intuitionistischen Theorie der unendlichen Reihen* (**J** 0020) Compos Math 6*118-123
◇ F55 ◇ REV Zbl 19.98 • ID 00967

BELINFANTE, M.J. [1939] *Der Levysche Umordnungssatz und seine intuitionistische Uebertragung* (**J** 0020) Compos Math 6*124-135
◇ F55 ◇ REV Zbl 19.98 • ID 00968

Belinfante, M.J. [1941] *Elemente der intuitionistischen Funktionentheorie. Erste Mitteilung: Die Cauchyschen Integralsaetze und die Taylorsche Reihe. Zweite, dritte Mitteilung: Der Satz vom Integral der logarithmischen Ableitung I,II* (J 0028) Indag Math 3*64-76,90-99,170-175 ◇ F55 ◇ REV MR 2.355 MR 3.76 Zbl 25.163 Zbl 25.164 • REM Vierte Mitteilung 1941 • ID 43858

Belinfante, M.J. [1941] *Elemente der intuitionistischen Funktionentheorie. Vierte Mitteilung: Der Weierstrass'sche Unbestimmtheitssatz. Fuenfte Mitteilung: Die intuitionistische Uebertragung des Picardschen Satzes* (J 0028) Indag Math 3*226-230,322-328 ◇ F55 ◇ REV MR 3.76 Zbl 25.164 Zbl 25.165 • REM Dritte Mitteilung 1941 • ID 43861

Bellissima, F. & Mirolli, M. [1981] *Metodi algebrici nella teoria della dimostrazione* (P 3092) Congr Naz Logica;1979 Montecatini Terme 47-53 ◇ F30 ◇ ID 48369

Bellissima, F. see Vol. II, III for further entries

Bellot, P. [1985] *A new proof for Craig's theorem* (J 0036) J Symb Logic 50*395-396 ◇ B10 B40 ◇ ID 42554

Belnap Jr., N.D. & Leblanc, H. [1962] *Intuitionism reconsidered* (J 0047) Notre Dame J Formal Log 3*79-82 ◇ F50 ◇ REV MR 27 # 1367 Zbl 113.5 JSL 28.256 • ID 01005

Belnap Jr., N.D. & Thomason, R.H. [1963] *A rule-completeness theorem* (J 0047) Notre Dame J Formal Log 4*39-43 ◇ B22 F07 F50 ◇ REV MR 27 # 28 Zbl 118.13 • ID 01010

Belnap Jr., N.D. & Leblanc, H. & Thomason, R.H. [1963] *On not strengthening intuitionistic logic* (J 0047) Notre Dame J Formal Log 4*313-320 ◇ F50 ◇ REV MR 29 # 4680 Zbl 131.6 • ID 01007

Belnap Jr., N.D. see Vol. I, II for further entries

Bel'tyukov, A.P. [1976] *Decidability of the universal theory of natural numbers with addition and divisibility (Russian)* (S 0228) Zap Nauch Sem Leningrad Otd Mat Inst Steklov 60*15-28,221 • TRANSL [1980] (J 1531) J Sov Math 14*1436-1444 ◇ B25 B28 F30 ◇ REV MR 58 # 27419 Zbl 345 # 02035 • ID 29789

Bel'tyukov, A.P. see Vol. IV for further entries

Belyakin, N.V. [1967] *A complete system of classical arithmetic (Russian) (English summary)* (J 0003) Algebra i Logika 6/2*5-11 ◇ F30 ◇ REV MR 36 # 1293 Zbl 165.18 • ID 00980

Belyakin, N.V. [1967] *On the completeness of arithmetic (Russian)* (J 0023) Dokl Akad Nauk SSSR 175*975-977 • TRANSL [1967] (J 0062) Sov Math, Dokl 8*926-928 ◇ F30 ◇ REV MR 37 # 30 Zbl 167.274 • ID 00981

Belyakin, N.V. [1969] *A variant of Richter's constructive ordinals (Russian)* (J 0003) Algebra i Logika 8*154-171 • TRANSL [1969] (J 0069) Algeb and Log 8*86-96 ◇ D70 F15 ◇ REV MR 41 # 8223 Zbl 212.25 JSL 40.626 • ID 00982

Belyakin, N.V. [1970] *Generalized computations and second order arithmetic (Russian)* (J 0003) Algebra i Logika 9*375-405 • TRANSL [1970] (J 0069) Algeb and Log 9*225-243 ◇ C62 D10 D45 D65 D70 F35 ◇ REV MR 46 # 32 Zbl 278 # 02036 • ID 29065

Belyakin, N.V. [1973] *Generalized computations (Russian)* (S 0066) Tr Mat Inst Steklov 133*59-64,274 • TRANSL [1973] (S 0055) Proc Steklov Inst Math 133*57-61 ◇ F35 ◇ REV MR 49 # 2318 Zbl 293 # 02030 • ID 03884

Belyakin, N.V. [1974] *Generalized computations, and third order arithmetic (Russian)* (J 0003) Algebra i Logika 13*132-144,234 • TRANSL [1974] (J 0069) Algeb and Log 13*71-78 ◇ D65 F35 ◇ REV MR 51 # 2898 Zbl 296 # 02023 • ID 17342

Belyakin, N.V. [1983] *A means of modeling a classical second-order arithmetic (Russian)* (J 0003) Algebra i Logika 22*3-25 • TRANSL [1983] (J 0069) Algeb and Log 22*1-18 ◇ B15 C62 D55 F35 ◇ REV MR 85h:03046 Zbl 538 # 03040 • ID 41479

Belyakin, N.V. see Vol. I, III, IV, V for further entries

Ben-Ari, M. [1980] *A simplified proof that regular resolution is exponential* (J 0232) Inform Process Lett 10*96-98 ◇ B35 D15 F20 ◇ REV MR 81g:68062 Zbl 438 # 03054 • ID 55966

Ben-Ari, M. see Vol. I, II for further entries

Ben-Yelles, C.B. [1981] *G-stratification is equivalent to F-stratification* (J 0068) Z Math Logik Grundlagen Math 27*141-150 ◇ B40 ◇ REV MR 82i:03022 Zbl 481 # 03009 • ID 70908

Bencivenga, E. [1984] *Finitary consistency of a free arithmetic* (J 0047) Notre Dame J Formal Log 25*224-226 ◇ F30 ◇ REV MR 85e:03139 Zbl 545 # 03034 • ID 40753

Bencivenga, E. see Vol. I, II, V for further entries

Benda, M. [1980] *On strong axioms of induction in set theory and arithmetic* (P 2958) Latin Amer Symp Math Log (4);1978 Santiago 43-56 ◇ E35 F30 ◇ REV MR 81j:03087 Zbl 429 # 03030 • ID 53861

Benda, M. [1980] *On Harrington's partition relation* (J 0164) J Comb Th, Ser A 28*338-350 ◇ F30 ◇ REV MR 82a:05065 Zbl 461 # 03011 • ID 54488

Benda, M. see Vol. III, V for further entries

Bendova, K. & Hajek, P. [1982] *A logical analysis of the truth-reaction paradox* (J 0140) Comm Math Univ Carolinae (Prague) 23*699-713 ◇ B45 F30 ◇ REV MR 84e:03068 Zbl 514 # 03037 • ID 34402

Bendova, K. see Vol. II, IV for further entries

Beneyto, R. [1971] *Analytic labyrinths (Spanish)* (J 0162) Teorema (Valencia) 1971/4*19-30 ◇ B10 F07 ◇ REV MR 47 # 4770 • ID 01031

Beneyto, R. see Vol. I for further entries

BENGY-PUYVALLEE DE, R. [1948] *Sur les regles de composabilite dans la logique de la mathematique intuitionniste sans negation* (J 0109) C R Acad Sci, Paris 226*454-456
◇ F50 ◇ REV MR 9.322 Zbl 41.352 • ID 01033

BENGY-PUYVALLEE DE, R. see Vol. II for further entries

BENNETT, D.W. [1973] *An elementary completeness proof for a system of natural deduction* (J 0047) Notre Dame J Formal Log 14*430-432
◇ B10 C07 F07 ◇ REV MR 47 # 8284 Zbl 258 # 02048
• ID 01038

BENNETT, D.W. [1977] *A note on the completeness proof for natural deduction* (J 0047) Notre Dame J Formal Log 18*145-146
◇ B10 C07 F07 ◇ REV MR 56 # 8355 Zbl 338 # 02014
• ID 21957

BENNETT, D.W. see Vol. I for further entries

BENTHEM VAN, J.F.A.K. [1974] *Semantic tableaus* (J 3077) Nieuw Arch Wisk, Ser 3 22*44-59
◇ B10 C07 F07 ◇ REV MR 49 # 2282 Zbl 285 # 02048
• ID 03889

BENTHEM VAN, J.F.A.K. & PEARCE, D.A. [1984] *A mathematical characterization of interpretation between theories* (J 0063) Studia Logica 43*295-303
◇ C20 F25 ◇ REV Zbl 573 # 03009 • ID 39892

BENTHEM VAN, J.F.A.K. see Vol. II, III, V for further entries

BERCOVICI, I. [1985] *Unsolvable terms in typed lambda calculus with fix-point operators* (P 4571) Log of Progr;1985 Brooklyn 16-22
◇ B40 ◇ ID 49184

BERG, G. & JULIAN, W. & MINES, R. & RICHMAN, F. [1975] *The constructive Jordan curve theorem* (J 0308) Rocky Mountain J Math 5*225-236
◇ F55 ◇ REV MR 53 # 14448 Zbl 298 # 54020 • ID 23220

BERG, G. & CHENG, HENRY & MINES, R. & RICHMAN, F. [1976] *Constructive dimension theory* (J 0020) Compos Math 33*161-177
◇ F55 ◇ REV MR 55 # 1329 Zbl 337 # 55008 • ID 27167

BERG, G. & JULIAN, W. & MINES, R. & RICHMAN, F. [1977] *The constructive equivalence of covering and inductive dimensions* (J 0254) Gen Topology Appl 7*99-108
◇ F55 ◇ REV MR 58 # 2751 Zbl 359 # 54025 • ID 50621

BERG, J. & CHIHARA, C.S. [1975] *Church's thesis misconstrued* (J 0095) Philos Stud 28*357-362
◇ A05 D20 F99 ◇ REV MR 58 # 27199 • ID 70955

BERG, J. see Vol. I, II for further entries

BERGSTRA, J.A. [1976] see BARENDREGT, H.P.

BERGSTRA, J.A. [1978] see BARENDREGT, H.P.

BERGSTRA, J.A. & KLOP, J.W. [1979] *Church-Rosser strategies in the lambda calculus* (J 1426) Theor Comput Sci 9*27-38
◇ B40 ◇ REV MR 80i:03024 Zbl 404 # 03015 • ID 54802

BERGSTRA, J.A. & KLOP, J.W. [1980] *Invertible terms in the lambda-calculus* (J 1426) Theor Comput Sci 11*19-37
◇ B40 ◇ REV MR 81k:03016 Zbl 446 # 03011 • ID 56540

BERGSTRA, J.A. & KLOP, J.W. [1982] *Strong normalization and perpetual reduction in the lambda calculus (German and Russian summaries)* (J 0129) Elektr Informationsverarbeitung & Kybern 18*403-417
◇ B40 ◇ REV MR 85e:03028 Zbl 523 # 03009 • ID 37020

BERGSTRA, J.A. & TUCKER, J.V. [1983] *Hoare's logic and Peano's arithmetic* (J 1426) Theor Comput Sci 22*265-284
◇ B75 F30 ◇ REV MR 84k:68021 Zbl 455 # 68019
• ID 54304

BERGSTRA, J.A. see Vol. I, II, III, IV for further entries

BERKLING, K.J. & FEHR, E. [1982] *A consistent extension of the lambda-calculus as a base for functional programming languages* (J 0194) Inform & Control 55*89-101
◇ B40 B60 B75 ◇ REV Zbl 553 # 68025 • ID 43431

BERKLING, K.J. & FEHR, E. [1982] *A modification of the λ-calculus as a base for functional programming languages* (P 3836) Automata, Lang & Progr (9);1982 Aarhus 35-47
◇ B40 B75 ◇ REV MR 83m:68013 Zbl 489 # 68023
• ID 36600

BERLINE, C. & MCALOON, K. & RESSAYRE, J.-P. (EDS.) [1981] *Model theory and arithmetic. Comptes rendus d'une action thematique programmee du C.N.R.S. sur la theorie des modeles et l'arithmetique* (S 3301) Lect Notes Math 890*vi+306pp
◇ C62 C97 F30 ◇ REV MR 82m:03004 Zbl 465 # 00004
• ID 54902

BERLINE, C. see Vol. III for further entries

BERMAN, L. [1977] *Precise bounds for Presburger arithmetic and the reals with addition: preliminary report* (P 3572) IEEE Symp Found of Comput Sci (18);1977 Providence 95-99
◇ B25 D15 F30 ◇ REV MR 58 # 5116 • ID 70972

BERMAN, L. [1980] *The complexity of logical theories* (J 1426) Theor Comput Sci 11*71-77
◇ B25 D10 D15 F30 ◇ REV MR 82c:03061b
Zbl 475 # 03017 • ID 55471

BERMAN, L. see Vol. IV for further entries

BERNARDI, C. [1975] *On the equational class of diagonalizable algebras (the algebraization of the theories which express Theor. VI)* (J 0063) Studia Logica 34*321-331
◇ B25 F30 G05 G25 ◇ REV MR 57 # 109
Zbl 322 # 02033 • REM Part V 1975 by Montagna,F. Part VII 1976 by Magari,R. • ID 31012

BERNARDI, C. [1975] *The fixed-point theorem for diagonalizable algebras (the algebraization of the theories which express Theor. III)* (J 0063) Studia Logica 34*239-251
◇ B45 F30 G05 G25 ◇ REV MR 57 # 106
Zbl 318 # 02031 • REM Part II 1975 by Magari,R. Part IV 1975 by Magari,R. • ID 31011

BERNARDI, C. [1976] *The uniqueness of the fixed-point in every diagonalizable algebra (The algebraization of the theories which express Theor. VIII.)* (J 0063) Studia Logica 35*335-343
◇ B45 F30 G05 G25 ◇ REV MR 57 # 111
Zbl 345 # 02020 • REM Part VII 1976 by Magari,R. Part IX 1976 by Sambin,G. • ID 60478

BERNARDI, C. [1981] *On the relation provable equivalence and on partitions in effectively inseparable sets* (J 0063) Studia Logica 40*29-37
◇ D25 D35 F30 ◇ REV MR 82m:03057 Zbl 468 # 03020
• ID 55085

BERNARDI, C. & PAGLI, P. [1982] *Logical and algebraic versions of diagonalization lemma (Italian summary)* (J 2100) Boll Unione Mat Ital, VI Ser, B 1*627-639
◇ F30 G25 ◇ REV MR 83j:03029 Zbl 496 # 03044 • ID 35339

BERNARDI, C. & SORBI, A. [1983] *Classifying positive equivalence relations* (J 0036) J Symb Logic 48*529-538
◇ F30 ◇ REV MR 85i:03148 Zbl 528 # 03030 • ID 37642

BERNARDI, C. [1983] *Relazioni di equivalenza positive rappresentate da formule* (P 3829) Atti Incontri Log Mat (1);1982 Siena 197-200
◇ F30 ◇ REV Zbl 514 # 03008 • ID 36982

BERNARDI, C. [1984] *A shorter proof of a recent result by R. Di Paola* (J 0047) Notre Dame J Formal Log 25*390-393
◇ B45 F30 ◇ REV MR 85m:03038 Zbl 562 # 03029 • ID 42581

BERNARDI, C. & MONTAGNA, F. [1984] *Equivalence relations induced by extensional formulae: classification by means of a new fixed point property* (J 0027) Fund Math 124*221-233
◇ F30 ◇ REV MR 86g:03074 Zbl 564 # 03043 • ID 45384

BERNARDI, C. see Vol. II, III, IV for further entries

BERNAYS, P. [1928] *Zusatz zu Hilberts Vortrag ueber "Die Grundlagen der Mathematik"* (J 0107) Abh Math Sem Univ Hamburg 6*89-92
• TRANSL [1967] (C 0675) From Frege to Goedel 486-489
◇ F35 ◇ REV FdM 54.56 • ID 14930

BERNAYS, P. [1930] *Die Philosophie der Mathematik und die Hilbertsche Beweistheorie* (J 0168) Blaetter Deutsch Philos 4*326-367
◇ A05 F99 ◇ REV FdM 56.44 • ID 01078

BERNAYS, P. [1932] *Methoden des Nachweises von Widerspruchsfreiheit und ihre Grenzen* (P 0653) Int Congr Math (II, 4);1932 Zuerich 2*342-343
◇ A05 B30 F99 ◇ REV FdM 58.70 • ID 01289

BERNAYS, P. [1934] see HILBERT, D.

BERNAYS, P. [1935] *Quelques points essentiels de la metamathematique* (J 0152) Enseign Math 34*70-95
◇ A05 F55 ◇ REV Zbl 14.97 FdM 61.48 • ID 01080

BERNAYS, P. [1935] *Sur le platonisme dans les mathematiques* (J 0152) Enseign Math 34*52-69
• TRANSL [1964] (C 1105) Phil of Math. Sel Readings 274-288
◇ A05 A10 E30 E70 F55 ◇ REV Zbl 14.1 FdM 61.47 • ID 01079

BERNAYS, P. [1939] see HILBERT, D.

BERNAYS, P. [1941] *Sur les questions methodologiques actuelles de la theorie hilbertienne de la demonstration* (P 0652) Entretiens Zuerich Fond & Method Sci Math;1938 Zuerich 144-152, discussion 153-161
◇ A05 F99 ◇ REV MR 2.339 Zbl 61.9 JSL 7.35 • ID 01287

BERNAYS, P. [1950] *Mathematische Existenz und Widerspruchsfreiheit* (C 1627) Etud Phil Sci (Gonseth) 11-25
◇ A05 B30 F99 ◇ REV JSL 22.210 • ID 37320

BERNAYS, P. [1951] *Ueber das Induktionsschema in der rekursiven Zahlentheorie* (C 0621) Kontrolliertes Denken (Britzelmayr) 10-17
◇ F30 ◇ REV Zbl 45.297 JSL 17.272 • ID 01086

BERNAYS, P. [1953] *Existence et non-contradiction en mathematiques. Avec une note de G.Bouligand* (J 0180) Rev Phil France & Etranger 143*85-87
◇ A05 B30 F99 ◇ REV Zbl 52.9 • ID 47892

BERNAYS, P. [1954] *Ueber den Zusammenhang des Herbrand'schen Satzes mit den neueren Ergebnissen von Schuette und Stenius* (P 0575) Int Congr Math (II, 7);1954 Amsterdam 2*397
◇ F05 ◇ ID 29462

BERNAYS, P. [1954] *Zur Beurteilung der Situation in der beweistheoretischen Forschung* (J 0169) Theoria (Madrid) 2*153-154
◇ A05 F99 ◇ REV JSL 21.106 • ID 01087

BERNAYS, P. [1965] *Betrachtungen zum Sequenzen-Kalkuel* (C 0749) Contrib Logic & Methodol (Bochenski) 1-44
◇ B20 F05 F07 ◇ REV MR 51 # 10043 Zbl 192.29 • ID 16287

BERNAYS, P. [1970] *The original Gentzen consistency proof for number theory* (P 0603) Intuitionism & Proof Th;1968 Buffalo 409-417
◇ A10 B28 F30 F50 ◇ REV MR 43 # 1810 Zbl 204.309 JSL 40.95 • ID 01284

BERNAYS, P. [1971] *Zum Symposium ueber die Grundlagen der Mathematik* (J 0076) Dialectica 25*171-195
◇ A05 F99 ◇ REV Zbl 246 # 02005 JSL 43.152 • ID 27408

BERNAYS, P. see Vol. I, II, III, IV, V for further entries

BERNINI, S. [1976] *A very strong intuitionistic theory* (J 0063) Studia Logica 35*377-385
◇ F35 F50 ◇ REV MR 56 # 5229 Zbl 348 # 02029 • ID 60491

BERNINI, S. [1977] *Interpretazione intuizionista di teorie a logica classica; una particolare applicazione del metodo della traduzione negativa via concetti empirici e anomici (lawless)* (J 0088) Ann Univ Ferrara, NS, Sez 7 22*21-41
◇ F35 F50 ◇ REV MR 55 # 10246 Zbl 355 # 02024 • ID 50222

BERNINI, S. [1978] *A note on my paper: "A very strong intuitionistic theory"* (J 0063) Studia Logica 37*349-350
◇ F35 F50 ◇ REV MR 83b:03069 Zbl 416 # 03052 • REM Published ibid. 35*377-385 • ID 53218

BERNINI, S. see Vol. II for further entries

BERNSTEIN, F. [1919] *Die Mengenlehre Georg Cantors und der Finitismus* (J 0157) Jbuchber Dtsch Math-Ver 28*63-78
◇ A05 E30 F99 ◇ REV FdM 47.895 • ID 01136

BERNSTEIN, F. see Vol. V for further entries

BERRE LE, F. & NOLIN, L. [1981] *L'existence d'espaces informatiques (English summary)* (J 3364) C R Acad Sci, Paris, Ser 1 292*499-502
◇ B40 ◇ REV MR 83f:68023 • ID 40209

BERRY, GERARD [1978] *Sequentialite de l'evaluation formelle des λ-expressions* (P 3247) Transform de Progr;1978 Paris 67-80
◇ B40 ◇ REV MR 80k:68017 Zbl 416 # 03017 • ID 53183

BERRY, GERARD [1978] *Stable models of typed λ-calculi* (P 1872) Automata, Lang & Progr (5);1978 Udine 72-89
◇ B40 ◇ REV MR 80d:68034 Zbl 382 # 68041 • ID 51981

BERRY, GERARD & LEVY, J.-J. [1979] *A survey of some syntactic results in the λ-calculus* (**P** 2059) Math Founds of Comput Sci (8);1979 Olomouc 552-566
◇ B40 ◇ REV MR 81f:68013 Zbl 416 # 03016 • ID 53182

BERRY, GERARD & LEVY, J.-J. [1979] *Minimal and optimal computations of recursive programs* (**J** 0037) ACM J 26*148-175
◇ B40 B75 ◇ REV MR 80b:68007 Zbl 388 # 68012 • ID 52285

BERRY, GERARD [1981] *On the definition of lambda-calculus models* (**P** 2930) Formal of Progr Concepts;1981 Peniscola 218-230
◇ B40 ◇ REV MR 82h:68044 Zbl 469 # 03007 • ID 55135

BERRY, GERARD & CURIEN, P.-L. [1982] *Sequential algorithms on concrete data structures* (**J** 1426) Theor Comput Sci 20*265-321
◇ B40 B75 ◇ REV MR 83h:68040 Zbl 497 # 68012 • ID 39192

BERRY, GERARD see Vol. IV for further entries

BERTOLINI, F. [1962] *Giustificazione semantica secondo Lorenzen del calcolo proposizionale intuizionistico* (**J** 2038) Rend Sem Mat, Torino 22*39-53
◇ F50 ◇ REV MR 30 # 20 Zbl 143.250 • ID 01540

BERTOLINI, F. [1971] *Kripke models and manyvalued logics* (**P** 0669) Conv Teor Modelli & Geom;1969/70 Roma 113-131
◇ B50 F50 ◇ REV MR 45 # 29 Zbl 212.19 • REM Part I. Part II 1970 • ID 01565

BERTOLINI, F. see Vol. II for further entries

BERTONI, A. & MAURI, G. & MIGLIOLI, P.A. & ORNAGHI, M. [1984] *Abstract data types and their extensions within a constructive logic* (**P** 3090) Semant of Data Types;1984 Sophia-Antipolis 177-195
◇ B75 F50 ◇ ID 45072

BERTONI, A. see Vol. III, IV for further entries

BESSONOV, A.V. [1977] *New operations in intuitionistic calculus (Russian)* (**J** 0087) Mat Zametki (Akad Nauk SSSR) 22*23-28
• TRANSL [1977] (**J** 1044) Math Notes, Acad Sci USSR 22*503-506
◇ B55 F50 ◇ REV MR 56 # 11762 Zbl 365 # 02014 • ID 53417

BESSONOV, A.V. see Vol. I, II for further entries

BETH, E.W. [1935] *La metamathematique et ses applications au probleme de la non-contradiction de la logique et de l'arithmetique* (**J** 1379) Christiaan Huygens Internat Math Tijdschr 14*141-168
◇ F25 F30 ◇ REV Zbl 15.49 • ID 28659

BETH, E.W. [1935] *Sur un theoreme concernant le principe du tiers exclu et ses applications dans la theorie de la non-contradiction* (**P** 0639) Congr Nat des Sci (2);1935 Bruxelles 160-164
◇ F25 F50 ◇ REV Zbl 14.385 JSL 1.118 • ID 01149

BETH, E.W. [1936] *Demonstration d'un theoreme concernant le principe du tiers exclu* (**J** 0150) Acad Roy Belg Bull Cl Sci (5) 22*580-581
◇ F50 ◇ REV Zbl 15.194 JSL 1.118 FdM 62.30 • ID 01150

BETH, E.W. [1937] *Une demonstration de la non-contradiction de la logique des types au point de vue fini* (**J** 1793) Nieuw Arch Wisk, Ser 2 19*59-62
◇ B15 F35 ◇ REV FdM 63.822 • ID 41045

BETH, E.W. [1947] *Semantical considerations on intuitionistic mathematics* (**J** 0028) Indag Math 9*572-577
◇ F50 ◇ REV MR 9.322 Zbl 35.148 JSL 13.173 • ID 01153

BETH, E.W. [1955] *Remarks on natural deduction* (**J** 0028) Indag Math 17*322-325
◇ A05 B10 F07 ◇ REV MR 17.4 Zbl 67.251 JSL 22.360 • ID 01163

BETH, E.W. [1956] *L'existence en mathematiques* (**X** 0834) Gauthier-Villars: Paris 60pp
◇ A05 B98 F55 ◇ REV MR 19.625 Zbl 75.231 • ID 01166

BETH, E.W. [1956] *Semantic construction of intuitionistic logic* (**J** 0182) Kon Nederl Akad Wetensch Afd Let Med N S 19/11*357-388
◇ F50 ◇ REV MR 19.625 Zbl 73.249 JSL 22.363 • ID 01270

BETH, E.W. [1956] *Semantische Tafeln fuer die intuitionistische Praedikatenlogik erster Ordnung* (**P** 1645) Oesterr Math Kongr (4);1956 Wien 502
◇ F50 ◇ ID 29423

BETH, E.W. [1957] *Intuitionistic predicate logic* (**P** 1675) Summer Inst Symb Log;1957 Ithaca 113-121
◇ F50 ◇ ID 29328

BETH, E.W. [1958] *Construction semantique de la logique intuitionniste* (**P** 0576) Raisonn en Math & Sci Exper;1955 Paris 77-83
◇ A05 F50 ◇ REV MR 21 # 4100 JSL 25.287 • ID 01539

BETH, E.W. [1959] *Remarks on intuitionistic logic* (**P** 0634) Constructivity in Math;1957 Amsterdam 15-21
◇ F50 ◇ REV MR 21 # 4101 Zbl 91.9 JSL 39.609 • ID 01167

BETH, E.W. & LEBLANC, H. [1960] *A note on the intuitionist and the classical propositional calculus* (**J** 0079) Logique & Anal, NS 3*174-176
◇ B05 F50 ◇ REV JSL 25.351 • ID 01171

BETH, E.W. [1962] *Formal methods. An introduction to symbolic logic and to the study of effective operations in arithmetic and logic* (**X** 0835) Reidel: Dordrecht xiv + 170pp
◇ B28 B98 F30 F50 F98 ◇ REV MR 28 # 3920 Zbl 105.245 JSL 30.235 • ID 01176

BETH, E.W. [1962] *Umformung einer abgeschlossenen deduktiven oder semantischen Tafel in eine natuerliche Ableitung auf Grund der derivativen bzw. klassischen Implikationslogik* (**C** 0712) Logik & Logikkalkuel 49-55
◇ B10 F07 ◇ REV Zbl 126.9 • ID 04250

BETH, E.W. see Vol. I, II, III, IV, V for further entries

BEZBORUAH, A. & SHEPHERDSON, J.C. [1976] *Goedel's second incompleteness theorem for Q* (**J** 0036) J Symb Logic 41*503-512
◇ C62 F30 ◇ REV MR 53 # 7756 Zbl 328 # 02017 • ID 14785

BEZEM, M. [1985] *Stongly majorizable functionals of finite type: A model for barrecursion containing discontinuous functionals* (**J** 0036) J Symb Logic 50*652-660
◇ F10 F35 F50 ◇ ID 47370

BEZEM, M. see Vol. IV for further entries

BIALYNICKI-BIRULA, A. & RASIOWA, H. [1958] *On constructible falsity in the constructive logic with strong negation* (**S** 0019) Colloq Math (Warsaw) 6*287-310
◇ F50 ◇ REV MR 21 # 650 Zbl 87.10 JSL 35.138
• ID 01195

BIBEL, W. [1969] *Schnittelimination in einem Teilsystem der einfachen Typenlogik* (**J** 0009) Arch Math Logik Grundlagenforsch 12*159-178
◇ B15 F05 F15 F35 ◇ REV MR 42 # 37 • ID 01198

BIBEL, W. & SCHREIBER, J. [1975] *Proof search in a Gentzen-like system of first-order logic* (**P** 3525) Int Comput Symp;1975 Antibes 205-212
◇ B35 F07 ◇ REV MR 58 # 32125 Zbl 324 # 68053
• ID 60499

BIBEL, W. [1979] *Tautology testing with a generalized matrix reduction method* (**J** 1426) Theor Comput Sci 8*31-44
◇ B35 D15 F05 ◇ REV MR 80i:03023 Zbl 421 # 03011
• ID 53410

BIBEL, W. [1982] *A comparative study of several proof procedures* (**J** 0503) Artif Intell 18*269-293
◇ B35 F07 ◇ REV MR 83f:68106 Zbl 505 # 68041
• ID 40212

BIBEL, W. [1982] *Deduktionsverfahren* (**P** 4072) Kuenstl Intell;1982 Teisendorf 99-140
◇ B35 F07 ◇ REV MR 85d:68004 • ID 40242

BIBEL, W. see Vol. I, II for further entries

BIELA, A. & HALLALA, P. [1976] *Studies on Church's calculus* (**J** 0387) Bull Sect Logic, Pol Acad Sci 5*57-62
◇ B40 ◇ REV MR 55 # 5396 • ID 27110

BIELA, A. see Vol. I, II for further entries

BIELTZ, P. [1977] *On the controversies of the excluded middle (Romanian)* (**S** 1613) Probl Logic (Bucharest) 7*169-182
◇ A05 A10 F55 ◇ REV MR 58 # 27201 • ID 71023

BIELTZ, P. see Vol. I for further entries

BILLING, J. [1947] *A failure of the Bolzano-Weierstrass lemma* (**J** 3970) Ark Mat, Astron & Fys 34B/11*2pp
◇ F55 ◇ REV MR 9.488 Zbl 29.113 JSL 12.94 • ID 01535

BILLING, J. [1949] *The principle of the excluded third and the Bolzano-Weierstrass lemma* (**J** 0189) Ark Mat 1*59
◇ F55 ◇ REV MR 11.412 Zbl 35.8 • ID 01531

BILY, J. & BUKOVSKY, L. [1974] *On expansion of β-models* (**J** 0027) Fund Math 82*239-244
◇ C62 E25 E40 E70 F35 ◇ REV MR 51 # 149
Zbl 311 # 02062 • ID 01205

BING, K. [1968] *Natural deduction with few restrictions on variables* (**J** 0191) Inform Sci 1*381-402
◇ B10 F07 ◇ REV MR 41 # 6664 • ID 01210

BING, K. see Vol. I, III for further entries

BISHOP, E.A. [1967] *Foundations of constructive analysis* (**X** 0822) McGraw-Hill: New York xiii+370pp
◇ F55 F98 ◇ REV MR 36 # 4930 Zbl 183.15 JSL 37.744
• ID 01525

BISHOP, E.A. [1968] *The constructivization of abstract mathematical analysis* (**P** 0657) Int Congr Math (II,10);1966 Moskva 308-313
◇ F55 ◇ REV MR 38 # 2009 Zbl 195.304 • ID 01526

BISHOP, E.A. [1970] *Mathematics as a numerical language* (**P** 0603) Intuitionism & Proof Th;1968 Buffalo 53-71
◇ A05 F50 F55 ◇ REV MR 42 # 5777 Zbl 205.12
JSL 37.744 • ID 01276

BISHOP, E.A. & CHENG, HENRY [1972] *Constructive measure theory* (**S** 0167) Mem Amer Math Soc 116*v+85pp
◇ F55 ◇ REV MR 58 # 17016 Zbl 239 # 02019 • ID 23324

BISHOP, E.A. [1975] *The crisis in contemporary mathematics* (**J** 1648) Hist Math 2*507-517
◇ A05 F99 ◇ REV MR 58 # 16192a Zbl 361 # 02001
• ID 50636

BISHOP, E.A. [1977] *Review of H. Jerome Keisler's "Elementary calculus"* (**J** 0015) Bull Amer Math Soc 83*205-208
◇ A05 F55 ◇ ID 41687

BISHOP, E.A. & BRIDGES, D.S. [1985] *Constructive analysis* (**X** 0811) Springer: Heidelberg & New York xii+477pp
◇ D80 F55 F98 ◇ ID 48575

BISHOP, E.A. [1985] *Schizophrenia in contemporary mathematics* (**P** 4271) E.Bishop-Reflection on Him & Research;1983 San Diego 1-32
◇ A05 F99 ◇ ID 45520

BISHOP, P. & GREIF, I. & HALE, R.L.V. & HEWITT, C. & MATSON, T. & SMITH, BRIAN & STEIGER, R. [1974] *Behavioral semantics of nonrecursive control structures* (**P** 3013) Progr Symp;1974 Paris 385-407
◇ B35 B40 ◇ REV Zbl 316 # 68013 • ID 62501

BLANCHE, R. [1957] *Introduction a la logique contemporaine* (**X** 0850) Colin: Paris 208pp
◇ A05 B98 F50 ◇ REV Zbl 292 # 02001 JSL 24.71
• ID 01538

BLANCHE, R. see Vol. I, II, III for further entries

BLASS, A.R. [1983] *Words, free algebras, and coequalizers* (**J** 0027) Fund Math 117*117-160
◇ C05 C75 E25 E35 F50 F55 G30 ◇ REV
MR 85d:03127 Zbl 542 # 08005 • ID 40316

BLASS, A.R. see Vol. I, III, IV, V for further entries

BLOOM, S.L. [1975] *A note on the predicatively definable sets of N.N.Nepeivoda* (**J** 0068) Z Math Logik Grundlagen Math 21*427-431
◇ F60 F65 ◇ REV MR 52 # 10386 Zbl 326 # 02020
• ID 03922

BLOOM, S.L. [1977] *A note on Ψ-consequences* (**J** 0302) Rep Math Logic, Krakow & Katowice 8*3-9
◇ B22 F50 ◇ REV MR 58 # 218 Zbl 377 # 02027
• ID 28146

BLOOM, S.L. see Vol. I, II, III, IV, V for further entries

BOCKSTAELE, P. [1949] *Intuitionism among the French mathematicians (Dutch)* (**S** 0200) Acad Roy Belg Cl Sci Mem, Ser 2 11(32)*123pp
◇ A10 F65 ◇ REV MR 11.305 Zbl 41.353 • ID 01550

BOCKWINKEL, H.B.A. [1911] *A constructive proof of the theorem of Borel (Dutch)* (**J** 3962) Handel Nederl Natuur- & Geneesk Congr 13*147-151
◇ E75 F55 ◇ REV FdM 42.91 • ID 37966

BOEHM, C. & GROSS, W.F. [1966] *Introduction to the CUCH* (**P** 0746) Automata Th;1964 Ravello 35-65
◇ B40 B75 ◇ REV Zbl 192.69 JSL 40.81 • ID 17013

Author Index

BOEHM, C. [1966] *The CUCH as a formal and descriptive language* (**P** 0747) Form Lang, Descript Langs for Comput Progr;1964 Wien 179-197
◇ B40 B70 ◇ REV JSL 40.81 • ID 17012

BOEHM, C. & DEZANI, M. [1972] *A CUCH-machine: The automatic treatment of bound variables* (**J** 0435) Int J Comput & Inf Sci 1*171-191
◇ B35 B40 ◇ REV Zbl 277 # 68026 • ID 60576

BOEHM, C. & DEZANI, M. [1973] *Notes on "A CUCH-machine: The automatic treatment of bound variables"* (**J** 0435) Int J Comput & Inf Sci 2*157-160
◇ B40 B75 ◇ REV Zbl 277 # 68027 • ID 60577

BOEHM, C. & DEZANI-CIANCAGLINI, M. [1974] *Combinatorial problems, combinator equations and normal forms* (**P** 1869) Automata, Lang & Progr (2);1974 Saarbruecken 185-199
◇ B40 ◇ REV MR 55 # 2513 Zbl 309 # 68037 • ID 60579

BOEHM, C. & DEZANI-CIANCAGLINI, M. [1975] *λ-terms as total or partial functions on normal forms* (**P** 1603) λ-Calc & Comput Sci Th;1975 Roma 96-121
◇ B40 ◇ REV MR 58 # 5143 Zbl 342 # 02017 • ID 60621

BOEHM, C. (ED.) [1975] *Lambda-calculus and computer science theory* (**X** 0811) Springer: Heidelberg & New York xxi + 370pp
◇ B40 ◇ REV MR 57 # 1957 Zbl 316 # 00005 • ID 80454

BOEHM, C. & COPPO, M. & DEZANI-CIANCAGLINI, M. [1977] *Termination tests inside λ-calculus* (**P** 1632) Automata, Lang & Progr (4);1977 Turku SF 95-110
◇ B40 ◇ REV MR 57 # 5695 Zbl 358 # 02025 • ID 50479

BOEHM, C. & DEZANI-CIANCAGLINI, M. & PERETTI, P. & RONCHI DELLA ROCCA, S. [1979] *A discrimination algorithm inside λ-β-calculus* (**J** 1426) Theor Comput Sci 8*271-291
◇ B40 ◇ REV MR 80g:03012a Zbl 405 # 03005 • ID 71182

BOEHM, C. [1980] *An abstract approach to (hereditary) finite sequences of combinators* (**C** 3050) Essays Combin Log, Lambda Calc & Formalism (Curry) 231-242
◇ B40 ◇ REV MR 82b:03039 Zbl 469 # 03006 • ID 71183

BOEHM, C. & MICALI, S. [1980] *Minimal forms in λ-calculus computations* (**J** 0036) J Symb Logic 45*165-171
◇ B40 ◇ REV MR 81h:03030 Zbl 451 # 03002 • ID 54017

BOEHM, C. [1981] *Logic and computers* (**C** 2617) Modern Log Survey 297-309
◇ B40 ◇ REV MR 82f:03002 Zbl 464 # 03001 • ID 42773

BOEHM, C. see Vol. IV for further entries

BOERBOOM, B. [1979] see BAETEN, J.

BOERGER, E. & OBERSCHELP, W. & RICHTER, M.M. & SCHINZEL, B. & THOMAS, WOLFGANG (EDS.) [1984] *Computation and proof theory* (**S** 3301) Lect Notes Math 1104*viii + 475pp
◇ B97 D97 F97 ◇ REV MR 85k:03002b Zbl 547 # 00036 • REM Proc. Log. Coll., Aachen 1983, Vol.II. Vol.I 1984 by Mueller,G.H. • ID 40062

BOERGER, E. see Vol. I, III, IV for further entries

BOFFA, M. [1983] *Arithmetic and theory of types* (**P** 1601) Easter Conf on Model Th (1);1983 Diedrichshagen 10-16
◇ B15 C62 E70 F35 ◇ REV MR 84i:03008 Zbl 528 # 03020 • ID 37638

BOFFA, M. [1984] *Arithmetic and the theory of types* (**J** 0036) J Symb Logic 49*621-624
◇ B15 F30 F35 ◇ REV MR 85i:03176 Zbl 566 # 03024 • ID 42423

BOFFA, M. see Vol. I, III, IV, V for further entries

BOILEAU, A. & JOYAL, A. [1981] *La logique des topos* (**J** 0036) J Symb Logic 46*6-16
◇ F50 G30 ◇ REV MR 82a:03063 Zbl 544 # 03035 • ID 71191

BOLLMAN, D.A. & LAPLAZA, M.L. [1973] *A set-theoretic model for nonassociative number theory* (**J** 0047) Notre Dame J Formal Log 14*107-110
◇ B28 E75 F30 ◇ REV MR 47 # 6487 Zbl 226 # 02046 • ID 01422

BOLLMAN, D.A. see Vol. I, IV for further entries

BOOLOS, G. [1975] *On Kalmar's consistency proof and a generalization of the notion of ω-consistency* (**J** 0009) Arch Math Logik Grundlagenforsch 17*3-7
◇ F15 F30 ◇ REV MR 52 # 7868 Zbl 324 # 02021 • ID 01437

BOOLOS, G. [1976] *On deciding the truth of certain statements involving the notion of consistency* (**J** 0036) J Symb Logic 41*779-781
◇ B25 B45 F30 ◇ REV MR 56 # 8353 Zbl 359 # 02050 • ID 14578

BOOLOS, G. [1977] *On deciding the provability of certain fixed point statements* (**J** 0036) J Symb Logic 42*191-193
◇ B25 B45 F30 ◇ REV MR 58 # 192 Zbl 381 # 03013 • ID 26442

BOOLOS, G. [1979] *Reflection principles and iterated consistency assertions* (**J** 0036) J Symb Logic 44*33-35
◇ B45 F30 ◇ REV MR 80k:03064 Zbl 409 # 03010 • ID 56313

BOOLOS, G. [1979] *The unprovability of consistency. An essay in modal logic* (**X** 0805) Cambridge Univ Pr: Cambridge, GB viii + 184pp
◇ B25 B28 B45 B98 F30 F98 ◇ REV MR 81c:03013 Zbl 409 # 03009 JSL 46.871 • ID 56312

BOOLOS, G. [1980] *ω-consistency and the diamond* (**J** 0063) Studia Logica 39*237-243
◇ B28 F30 ◇ REV MR 81m:03068 Zbl 464 # 03049 • ID 71208

BOOLOS, G. [1980] *On systems of modal logic with provability interpretations* (**J** 0105) Theoria (Lund) 46*7-18
◇ B45 F30 ◇ REV MR 83c:03013 Zbl 473 # 03010 • ID 55341

BOOLOS, G. [1982] *Extremely undecidable sentences* (**J** 0036) J Symb Logic 47*191-196
◇ F30 ◇ REV MR 83h:03022 Zbl 485 # 03001 • ID 36042

BOOLOS, G. [1982] *On the nonexistence of certain normal forms in the logic of provability* (**J** 0036) J Symb Logic 47*638-640
◇ B45 F30 ◇ REV MR 83j:03031 Zbl 491 # 03005 • ID 35340

BOOLOS, G. [1984] *Don't eliminate cut* (**J** 0122) J Philos Logic 13*373-378
◇ F05 F07 ◇ REV MR 85i:03029 Zbl 564 # 03005 • ID 42647

BOOLOS, G. [1984] *The logic of provability* (J 0005) Amer Math Mon 91*470-480
◇ B45 F30 ◇ REV MR 85m:03010 Zbl 562#03007 • ID 43982

BOOLOS, G. [1984] *Trees and finite satisfiability: proof of a conjecture of Burgess* (J 0047) Notre Dame J Formal Log 25*193-197
◇ B10 C13 F07 ◇ REV MR 85f:03005 Zbl 561#03004 • ID 40442

BOOLOS, G. [1985] *1-consistency and the diamond* (J 0047) Notre Dame J Formal Log 26*341-347
◇ B45 F30 ◇ ID 47529

BOOLOS, G. see Vol. I, II, III, IV, V for further entries

BOREL, E. [1905] see BAIRE, R.

BOREL, E. [1919] *Sur les ensembles effectivement enumerables et sur les definitions effectives* (J 1157) Rend Accad Lincei Roma, Ser 5 28/2*163-165
◇ A05 E47 F65 ◇ REV FdM 47.174 • ID 41669

BOREL, E. [1948] *A propos de l'axiome du choix* (J 3970) Ark Mat, Astron & Fys 34B/15*2pp
◇ A05 E25 F99 ◇ REV MR 9.488 Zbl 30.115 • ID 01466

BOREL, E. see Vol. II, V for further entries

BORGA, M. [1976] *On a preliminary reduction step in Gentzen's second consistency proof* (J 3285) Boll Unione Mat Ital, V Ser, A 13*667-676
◇ F30 ◇ REV MR 55#2515 Zbl 359#02025 • ID 50573

BORGA, M. [1979] *Alcuni risultati di teoria della dimostrazione* (J 0335) Atti Accad Ligure Sci Lett (Genova) 35*370-380
◇ F30 ◇ REV MR 81a:03056 Zbl 432#03033 • ID 53990

BORGA, M. [1980] *Un raffronto fra il teorema di incompletezza di Goedel e il teorema di Tarski sulla nozione di verita* (J 3197) Rass Mat Log Mat Appl Didat 25-40
◇ F30 ◇ REV MR 80g:03001 Zbl 471#03051 • ID 55246

BORGA, M. [1983] *On some proof theoretical properties of the modal logic GL* (J 0063) Studia Logica 42*453-459
◇ B45 F05 F30 ◇ REV MR 86a:03015 Zbl 547#03016 • ID 42318

BORGA, M. see Vol. I, IV for further entries

BORGERS, A. [1963] *Quelques propositions arithmetiques equivalentes au principe de l'induction complete* (J 0082) Bull Soc Math Belg 15*155-178
◇ F30 ◇ REV MR 26#6051 Zbl 106.4 • ID 01471

BORGERS, A. see Vol. I, V for further entries

BORICIC, B.R. [1981] *Equational reformulations of intuitionistic propositional calculus and classical first-order predicate calculus* (J 0400) Publ Inst Math, NS (Belgrade) 29(43)*23-28
◇ B55 F50 ◇ REV MR 83e:03022 Zbl 498#03050 • ID 35211

BORICIC, B.R. [1983] *Equational reformulation of the Heyting first-order predicate calculus* (P 3855) Algeb Conf (3);1982 Beograd 41-44
◇ F50 ◇ REV MR 85j:03030 Zbl 536#03043 • ID 37123

BORICIC, B.R. [1985] *On sequent-conclusion natural deduction systems* (J 0122) J Philos Logic 14*359-378
◇ F07 ◇ REV Zbl 572#03033 • ID 47641

BORICIC, B.R. see Vol. II, III for further entries

BORKOWSKI, L. [1958] *Reduction of arithmetic to logic based on the theory of types without the axiom of infinity and the typical ambiguity of arithmetical constant (Polish and Russian summaries)* (J 0063) Studia Logica 8*283-297
◇ B15 F35 ◇ REV MR 21#4094 • ID 01479

BORKOWSKI, L. see Vol. I, II, III, V for further entries

BOSCH, J.E. [1985] *Sur les nombres natureles et les fondements de l'arithmetique et la geometrie* (C 4181) Math Log & Formal Syst (Costa da) 17-42
◇ B28 F30 ◇ ID 48160

BOSCH, J.E. see Vol. V for further entries

BOTH, N. [1967] *On the bases of complete induction. Simultaneous induction (Romanian)* (J 0197) Stud Cercet Mat Acad Romana 19*1251-1258
◇ B28 F30 ◇ REV MR 40#4077 • ID 01491

BOTH, N. see Vol. I, V for further entries

BOUVERE DE, K.L. [1959] *A method in proofs of undefinability, with applications to functions in the arithmetic of natural numbers* (X 0809) North Holland: Amsterdam xvi+64pp
◇ C40 F30 ◇ REV MR 21#3327 Zbl 231#02061 JSL 25.271 • ID 01501

BOUVERE DE, K.L. see Vol. I, III for further entries

BOWEN, K.A. [1971] *An extension of the intuitionistic propositional calculus* (J 0028) Indag Math 33*287-294
◇ B55 F50 ◇ REV MR 45#3161 Zbl 221#02009 • ID 01504

BOWEN, K.A. [1972] *A note on cut elimination and completeness in first order theories* (J 0068) Z Math Logik Grundlagen Math 18*173-176
◇ B10 C07 F05 ◇ REV MR 46#8809 Zbl 243#02020 • ID 01505

BOWEN, K.A. [1973] *Cut elimination in transfinite type theory* (J 0068) Z Math Logik Grundlagen Math 19*141-162
◇ B15 F05 F35 ◇ REV MR 48#1893 Zbl 302#02009 • ID 01507

BOWEN, K.A. [1974] *Primitive recursive notations for infinitary formulas* (S 0019) Colloq Math (Warsaw) 30*1-5
◇ C75 F15 ◇ REV MR 50#6785 Zbl 262#02014 • ID 03942

BOWEN, K.A. [1974] *Systems of transfinite type theory based on intuitionistic and modal logics* (J 0068) Z Math Logik Grundlagen Math 20*355-372
◇ B15 F05 F35 F50 ◇ REV MR 53#102 Zbl 299#02010 • ID 03943

BOWEN, K.A. [1976] *An Herbrand theorem for prenex formulas of LJ* (J 0047) Notre Dame J Formal Log 17*263-266
◇ F05 F50 ◇ REV MR 54#4931 Zbl 313#02016 • ID 15251

BOWEN, K.A. see Vol. II, III, V for further entries

BOWIE, G.L. [1973] *An argument against Church's thesis* (J 0301) J Phil 70*66-73
◇ A05 D20 F99 ◇ ID 27103

BOZHICH, E.S. [1985] *Local consistency of an arithmetic with an "attainability" predicate (Russian)* (J 0288) Vest Ser Mat Mekh, Univ Moskva 1985/5*37-41,96
◇ F30 ◇ ID 49509

BOZIC, M. & DOSEN, K. [1983] *Axiomatizations of intuitionistic double negation* (**J** 0387) Bull Sect Logic, Pol Acad Sci 12*99-104
◇ B55 F50 ◇ REV Zbl 542 # 03007 • ID 46894

BOZIC, M. see Vol. II, III, V for further entries

BOZZI, S. & MELONI, G.C. [1977] *Ideal properties and intuitionistic algebra* (**J** 0207) Ist Lombardo Accad Sci Rend, A (Milano) 111*145-150
◇ C60 C90 F50 F55 ◇ REV MR 58 # 16827 Zbl 393 # 03026 • ID 52446

BOZZI, S. & MELONI, G.C. [1980] *Representation of Heyting algebras with covering and propositional intuitionistic logic with local operator* (**J** 3285) Boll Unione Mat Ital, V Ser, A 17*436-442
◇ B25 B55 C90 F50 G10 ◇ REV MR 82k:03099 Zbl 454 # 03008 • ID 54221

BOZZI, S. [1982] *Formule attuali ed ereditarie nella logica intuizionista* (**J** 3746) Note Math (Lecce) 2*159-166
◇ C90 F50 ◇ REV MR 85e:03024 Zbl 547 # 03038 • ID 40429

BOZZI, S. [1983] *Principi ideali e logica intuizionista* (**J** 2100) Boll Unione Mat Ital, VI Ser, B 2*879-888
◇ A05 F50 ◇ REV MR 86b:03083 Zbl 538 # 03022 • ID 41468

BOZZI, S. see Vol. II for further entries

BRACHO, F. [1980] *Continuously generated fixed points in* $P\omega$ (**J** 2095) Fund Inform, Ann Soc Math Pol, Ser 4 3*477-489
◇ B40 D70 ◇ REV MR 82c:06015 Zbl 458 # 68004 • ID 69241

BREAZU-TANNEN, V. & MEYER, A.R. [1985] *Lambda calculus with constrained types* (**P** 4571) Log of Progr;1985 Brooklyn 23-40
◇ B40 ◇ ID 49188

BREMER, H. & REYNVAAN, H.H. [1976] *Zur konstruktiven Differenzierbarkeit von monotonen berechenbaren Funktionen* (**J** 0009) Arch Math Logik Grundlagenforsch 17*135-143
◇ F60 ◇ REV MR 54 # 7220 Zbl 337 # 02022 • ID 03944

BREMER, H. see Vol. IV for further entries

BRIDGE, J. [1975] *A simplification of the Bachmann method for generating large countable ordinals* (**J** 0036) J Symb Logic 40*171-185
◇ E10 F15 ◇ REV MR 53 # 2644 Zbl 309 # 02027 JSL 48.876 • ID 04262

BRIDGE, J. see Vol. III for further entries

BRIDGES, D.S. [1976] *Some notes on continuity in constructive analysis* (**J** 0161) Bull London Math Soc 8*179-182
◇ F55 ◇ REV MR 54 # 7221 Zbl 333 # 02028 • ID 25011

BRIDGES, D.S. [1977] *A constructive look at orthonormal bases in Hilbert space* (**J** 0005) Amer Math Mon 84*189-191
◇ F55 ◇ REV MR 55 # 12483 Zbl 356 # 46026 • ID 27275

BRIDGES, D.S. [1977] *The constructive Radon-Nikodym theorem* (**J** 0048) Pac J Math 70*51-65
◇ F55 ◇ REV MR 58 # 22489 Zbl 386 # 28003 • ID 52214

BRIDGES, D.S. [1978] *A note on Morse's lambda-notation in set theory* (**J** 0068) Z Math Logik Grundlagen Math 24*113-114
◇ F50 ◇ REV MR 80b:03073 Zbl 385 # 03041 • ID 52149

BRIDGES, D.S. [1978] *More on the connectivity of convex sets* (**J** 0053) Proc Amer Math Soc 68*214-216
◇ F55 ◇ REV MR 81f:03071 Zbl 397 # 52001 • ID 52719

BRIDGES, D.S. [1978] *On continuous mappings between locally compact metric spaces* (**J** 0161) Bull London Math Soc 10*201-208
◇ F55 ◇ REV MR 80i:03069 Zbl 386 # 46024 • ID 52215

BRIDGES, D.S. [1978] *On the connectivity of convex sets* (**J** 0161) Bull London Math Soc 10*86-90
◇ F55 ◇ REV MR 57 # 17504 Zbl 378 # 46005 • ID 51820

BRIDGES, D.S. [1978] *On weak operator compactness of the unit ball of* $L(H)$ (**J** 0068) Z Math Logik Grundlagen Math 24*493-494
◇ F55 ◇ REV MR 80d:03060 Zbl 416 # 03053 • ID 53219

BRIDGES, D.S. [1979] *A criterion for compactness in metric spaces?* (**J** 0068) Z Math Logik Grundlagen Math 25*97-98
◇ F55 ◇ REV MR 81h:03115 Zbl 413 # 03036 • ID 53030

BRIDGES, D.S. [1979] *Connectivity properties of metric spaces* (**J** 0048) Pac J Math 80*325-331
◇ F55 ◇ REV MR 80h:54024 Zbl 429 # 54007 • ID 53892

BRIDGES, D.S. [1979] *Constructive functional analysis* (**X** 1330) Pitman Publ: Belmont & London v+203pp
◇ F55 F98 ◇ REV MR 82k:03094 Zbl 401 # 03027 JSL 47.703 • ID 71322

BRIDGES, D.S. [1979] *Geometric intuition and elementary constructive analysis* (**J** 0068) Z Math Logik Grundlagen Math 25*521-523
◇ F55 ◇ REV MR 81f:03070 Zbl 427 # 03052 • ID 53739

BRIDGES, D.S. [1979] *On the constructive convergence of series of independent functions* (**J** 0068) Z Math Logik Grundlagen Math 25*93-96
◇ F55 ◇ REV MR 80g:03060 Zbl 405 # 03033 • ID 71326

BRIDGES, D.S. [1980] *A constructive development of Chebyshev approximation theory* (**J** 2747) J Approx Th 30*99-120
◇ F55 ◇ REV MR 82d:41031 Zbl 458 # 41021 • ID 54433

BRIDGES, D.S. [1980] *On the foundations of best-approximation theory* (**J** 2747) J Approx Th 28*273-292
◇ F55 ◇ REV MR 81i:41027 Zbl 441 # 41013 • ID 80791

BRIDGES, D.S. [1981] *A constructive look at positive linear functionals on* $\mathscr{L}(H)$ (**J** 0048) Pac J Math 95*11-25
◇ F55 ◇ REV MR 84f:47056 Zbl 473 # 46045 • ID 39719

BRIDGES, D.S. [1981] *A constructive analysis of the Remes algorithm* (**J** 2747) J Approx Th 32*257-270
◇ F55 ◇ REV MR 83b:41027 Zbl 472 # 41024 • ID 55324

BRIDGES, D.S. [1981] *A constructive proximinality property of finite-dimensional linear subspaces* (**J** 0308) Rocky Mountain J Math 11*491-497
◇ F55 ◇ REV MR 82m:41024 Zbl 478 # 41021 • ID 55652

BRIDGES, D.S. & CALDER, A. & JULIAN, W. & MINES, R. & RICHMAN, F. [1981] *Bounded linear mappings of finite rank* (**J** 2752) J Fct Anal 43*143-148
◇ F55 ◇ REV MR 83e:03102 • ID 35249

BRIDGES, D.S. & CALDER, A. & JULIAN, W. & MINES, R. & RICHMAN, F. [1981] *Compactly generated Banach spaces* (**J** 0008) Arch Math (Basel) 36*239-243
◇ F55 ◇ REV MR 82m:46013 Zbl 439 # 46012 • ID 82608

BRIDGES, D.S. & CALDER, A. & JULIAN, W. & MINES, R. & RICHMAN, F. [1981] *Locating metric complements in R^n* (P 3146) Constr Math;1980 Las Cruces 241-249
◇ F55 ◇ REV MR 83e:03103 Zbl 463 # 03037 • ID 54577

BRIDGES, D.S. [1981] *On the isolation of zeroes of an analytic function* (J 0048) Pac J Math 96*13-22 • ERR/ADD ibid 97*487-488
◇ F55 ◇ REV MR 82k:03095 MR 84c:03103 Zbl 467 # 30004 • ID 71321

BRIDGES, D.S. [1981] *On Montel's proof of the Great Picard Theorem* (J 0005) Amer Math Mon 88*45-47
◇ F55 ◇ REV MR 82e:30032 Zbl 476 # 30021 • ID 41696

BRIDGES, D.S. [1982] *Lipschitz constants and moduli of continuity for the Chebyshev projection* (J 0053) Proc Amer Math Soc 85*557-561
◇ F55 ◇ REV MR 83h:41027 Zbl 526 # 41021 • ID 39247

BRIDGES, D.S. & CALDER, A. & JULIAN, W. & MINES, R. & RICHMAN, F. [1982] *Picard's theorem* (J 0064) Trans Amer Math Soc 269*513-520
◇ F55 ◇ REV MR 83h:03088 Zbl 492 # 30021 • ID 36088

BRIDGES, D.S. [1982] *Preference and utility: a constructive development* (J 2757) J Math Econ 9*165-185
◇ F55 ◇ REV MR 83a:90025 Zbl 471 # 90018 • ID 38886

BRIDGES, D.S. [1982] *Recent progress in constructive approximation theory* (P 3638) Brouwer Centenary Symp;1981 Noordwijkerhout 41-50
◇ F55 ◇ REV MR 85a:03075 Zbl 521 # 41022 • ID 34747

BRIDGES, D.S. & MINES, R. [1984] *What is constructive mathematics?* (J 2789) Math Intell 6*32-38
◇ A05 F55 F98 ◇ REV Zbl 561 # 03031 • ID 43974

BRIDGES, D.S. [1985] see BISHOP, E.A.

BRIDGES, D.S. [1985] *Operator ranges, integrable sets, and the functional calculus* (J 1447) Houston J Math 11*31-44
◇ F55 ◇ REV Zbl 564 # 47007 • ID 44762

BRIDGES, D.S. see Vol. II, V for further entries

BROM, J. [1977] *The theory of almost periodic functions in constructive mathematics* (J 0048) Pac J Math 70*67-81
◇ F55 ◇ REV MR 57 # 77 Zbl 446 # 42008 • ID 71344

BROUWER, L.E.J. [1907] *On the foundations of mathematics (Dutch)* (0000) Diss., Habil. etc 183pp
• TRANSL [1975] (C 4429) Brouwer: Coll Works 11-101
◇ A05 E30 F55 ◇ REV FdM 38.81 • REM Diss., Amsterdam, Leipzig. A trade edition also exists (publisher Maas & van Suchtelen) • ID 16756

BROUWER, L.E.J. [1908] *On the foundations of mathematics (Dutch)* (J 1793) Nieuw Arch Wisk, Ser 2 8*326-328
• ERR/ADD ibid 12*439-445
• TRANSL [1975] (C 4429) Brouwer: Coll Works 105-106
◇ A05 E70 F55 ◇ REV JSL 44.271 FdM 39.18 • ID 01616

BROUWER, L.E.J. [1908] *The unreliability of the logical principles (Dutch)* (J 0292) Tijdsch Wijsbegeerte 2*152-158
• TRANSL [1975] (C 4429) Brouwer: Coll Works 107-111
◇ A05 F55 ◇ REM Reprinted also in Brouwer,L.E.J. 1919 ID 23337 • ID 03598

BROUWER, L.E.J. [1909] *Die moeglichen Maechtigkeiten* (P 0689) Int Congr Math (4);1908 Roma 3*569-571
◇ E10 F55 ◇ REV JSL 44.271 FdM 40.99 • ID 02690

BROUWER, L.E.J. [1912] *Intuitionism and formalism (Dutch)* (X 1317) Noordhoff: Groningen 32pp
• TRANSL [1914] (J 0015) Bull Amer Math Soc 20*81-96 [1964] (C 1105) Phil of Math. Sel Readings 66-77 (English, fragment) • REPR [1913] (J 1784) Wisk Tijdsch 9*180-211
◇ A05 F55 ◇ REV JSL 44.271 FdM 43.111 • REM Reprinted also in Brouwer 1919 (3rd paper) • ID 42968

BROUWER, L.E.J. [1914] *Review of Schoenflies's "Die Entwicklung der Mengenlehre und ihre Anwendungen, erste Haelfte"* (J 0157) Jbuchber Dtsch Math-Ver 23*78-83,2.Abt.
◇ A05 A10 E70 E98 F55 ◇ ID 01617

BROUWER, L.E.J. [1917] *Addenda and corrigenda to: On the foundations of mathematics (Dutch)* (J 0358) Versl Gewone Vergad Afd Natuurkd 25*1418-1423
• TRANSL [1975] (C 4429) Brouwer: Coll Works 145-149 (English) • REPR [1918] (J 1793) Nieuw Arch Wisk, Ser 2 12*439-445
◇ A05 F55 ◇ REV FdM 46.69 • ID 42969

BROUWER, L.E.J. [1918] *Begruendung der Mengenlehre unabhaengig vom logischen Satz vom ausgeschlossenen Dritten. I: Allgemeine Mengenlehre* (J 0258) Nederl Akad Wetensch Verh Tweed Afd Nat 12/5*43pp
◇ E70 F55 ◇ REV FdM 47.171 • REM Part II 1919 • ID 41665

BROUWER, L.E.J. [1919] *Begruendung der Mengenlehre unabhaengig vom logischen Satz vom ausgeschlossenen Dritten. II: Theorie der Punktmengen* (J 0258) Nederl Akad Wetensch Verh Tweed Afd Nat 12/7*33pp
◇ E70 F55 ◇ REV FdM 47.171 • REM Part I 1918 • ID 41406

BROUWER, L.E.J. [1919] *Intuitionistische Mengenlehre* (J 0157) Jbuchber Dtsch Math-Ver 28*203-208
• TRANSL [1921] (J 0358) Versl Gewone Vergad Afd Natuurkd 29*797-802 (Dutch) • REPR [1922] (J 0278) Nederl Akad Wetensch Proc 23*949-954
◇ E70 F55 ◇ REV FdM 47.171 • ID 01619

BROUWER, L.E.J. [1919] *Mathematics, truth, reality (Dutch)* (X 1317) Noordhoff: Groningen 7+23+29pp
◇ A05 F55 ◇ REV FdM 47.49 • REM Contains reprints of Brouwer,L.E.J. 1908 (2nd paper) and 1912 • ID 23337

BROUWER, L.E.J. [1921] *Besitzt jede reelle Zahl eine Dezimalbruchentwicklung?* (J 0043) Math Ann 83*201-210
• TRANSL [1921] (J 0358) Versl Gewone Vergad Afd Natuurkd 29*803-812 (Dutch) • REPR [1922] (J 0278) Nederl Akad Wetensch Proc 23*955-964
◇ F55 ◇ REV FdM 48.62 • ID 01620

BROUWER, L.E.J. [1923] *Begruendung der Funktionenlehre unabhaengig vom logischen Satz vom ausgeschlossenen Dritten. Erster Teil: Stetigkeit, Messbarkeit, Derivierbarkeit* (J 0258) Nederl Akad Wetensch Verh Tweed Afd Nat 13/2*24pp
◇ F55 ◇ ID 02434

BROUWER, L.E.J. [1923] *On the role of the principle of the excluded middle in mathematics, especially in function theory (Dutch)* (J 0291) Wis-natuur Tijdsch 2*1-7
• TRANSL [1924] (J 0127) J Reine Angew Math 154*1-7 (German) [1967] (C 0675) From Frege to Goedel 334-345 (English) [1975] (C 4429) Brouwer: Coll Works 268-274 (English)
◇ F55 ◇ ID 03600

BROUWER, L.E.J. [1924] *Bemerkungen zum Beweise der gleichmaessigen Stetigkeit voller Funktionen* (J 0278) Nederl Akad Wetensch Proc 27*644-646
• TRANSL [1924] (J 0358) Versl Gewone Vergad Afd Natuurkd 33*646-648 (Dutch)
◇ F55 ◇ ID 41425

BROUWER, L.E.J. & LOOR DE, B. [1924] *Intuitionistischer Beweis des Fundamentalsatzes der Algebra* (J 0278) Nederl Akad Wetensch Proc 27*186-188
• TRANSL [1924] (J 0358) Versl Gewone Vergad Afd Natuurkd 33*82-84 (Dutch)
◇ F55 ◇ REV FdM 50.37 • ID 37334

BROUWER, L.E.J. [1924] *Intuitionistische Ergaenzung des Fundamentalsatzes der Algebra* (J 0278) Nederl Akad Wetensch Proc 27*631-634
• TRANSL [1924] (J 0358) Versl Gewone Vergad Afd Natuurkd 33*459-462 (Dutch)
◇ F55 ◇ ID 41423

BROUWER, L.E.J. [1924] *Perfect sets of primes with positively irrational distances* (J 0278) Nederl Akad Wetensch Proc 27*248
• TRANSL [1924] (J 0358) Versl Gewone Vergad Afd Natuurkd 33*81 (Dutch)
◇ F55 ◇ REV FdM 50.132 • ID 01623

BROUWER, L.E.J. [1925] *Intuitionistische Zerlegung mathematischer Grundbegriffe* (J 0157) Jbuchber Dtsch Math-Ver 33*251-256
• TRANSL [1923] (J 0358) Versl Gewone Vergad Afd Natuurkd 32*877-880 (Dutch)
◇ F55 ◇ REV FdM 51.47 • ID 01622

BROUWER, L.E.J. [1925] *Intuitionistischer Beweis des Jordanschen Kurvensatzes* (J 0278) Nederl Akad Wetensch Proc 28*503-508
• TRANSL [1968] (S 0422) Tr Vychisl Tsentra Akad Nauk Armyan SSR & Univ Erevan 5*139-146
◇ F55 ◇ REV FdM 51.459 • ID 37328

BROUWER, L.E.J. [1925] *Zur Begruendung der intuitionistischen Mathematik I* (J 0043) Math Ann 93*244-257
◇ F55 ◇ REV FdM 51.164 • REM Part II 1926 • ID 02700

BROUWER, L.E.J. [1926] *Die intuitionistische Form des Heine-Borelschen Theorems* (J 0278) Nederl Akad Wetensch Proc 35*866-867
• TRANSL [1926] (J 0358) Versl Gewone Vergad Afd Natuurkd 35*677-678 (Dutch)
◇ F55 ◇ REV FdM 52.198 • ID 41574

BROUWER, L.E.J. [1926] *Intuitionistische Einfuehrung des Dimensionsbegriffs* (J 0278) Nederl Akad Wetensch Proc 29*855-863
• TRANSL [1926] (J 0358) Versl Gewone Vergad Afd Natuurkd 35*634-642 (Dutch)
◇ F55 ◇ REV FdM 52.589 • ID 37329

BROUWER, L.E.J. [1926] *Zur Begruendung der intuitionistischen Mathematik II* (J 0043) Math Ann 95*453-472
◇ F55 ◇ REV FdM 51.164 • REM Part I 1925. Part III 1926 • ID 02701

BROUWER, L.E.J. [1926] *Zur Begruendung der intuitionistischen Mathematik III* (J 0043) Math Ann 96*451-488
◇ F55 ◇ REV FdM 52.193 • REM Part II 1926 • ID 02702

BROUWER, L.E.J. [1927] *Ueber Definitionsbereiche von Funktionen* (J 0043) Math Ann 97*60-75
• TRANSL [1967] (C 0675) From Frege to Goedel 446-463 (English, fragment)
◇ F55 ◇ REV JSL 35.332 • ID 37330

BROUWER, L.E.J. [1927] *Virtuelle Ordnung und unerweiterbare Ordnung* (J 0127) J Reine Angew Math 157*255-257
◇ F55 ◇ REV FdM 53.40 • ID 41427

BROUWER, L.E.J. [1927] *Zur intuitionistischen Zerlegung mathematischer Grundbegriffe* (J 0157) Jbuchber Dtsch Math-Ver 36*127-129
• TRANSL [1924] (J 0358) Versl Gewone Vergad Afd Natuurkd 33*479-480 (Dutch)
◇ F55 ◇ REV FdM 50.127 • ID 41424

BROUWER, L.E.J. [1928] *Beweis dass jede Menge in einer individualisierten Menge enthalten ist* (J 0278) Nederl Akad Wetensch Proc 31*380-381
◇ E70 F55 ◇ REV FdM 54.53 • REM Dutch summary in J0358 36*1189 • ID 02697

BROUWER, L.E.J. [1928] *Intuitionistische Betrachtungen ueber den Formalismus* (J 0278) Nederl Akad Wetensch Proc 31*374-379
• TRANSL [1967] (C 0675) From Frege to Goedel 490-492 (English, fragment) • REPR [1928] (J 0277) Sitzb Preuss Akad Wiss Phys Math Kl 1928*48-52
◇ A05 F50 ◇ REV FdM 54.53 • ID 42970

BROUWER, L.E.J. [1929] *Mathematik, Wissenschaft und Sprache* (J 0124) Monatsh Math-Phys 36*153-164
◇ A05 F55 ◇ REV FdM 55.28 • ID 01624

BROUWER, L.E.J. [1930] *Die Struktur des Kontinuums* (2222) See Remarks 12pp
◇ A05 F55 ◇ REV FdM 56.83 • REM Komitee zur Veranstaltung von Gastvortraegen auslaendischer Gelehrter der Exakten Wissenschaften, Wien, A • ID 39426

BROUWER, L.E.J. [1930] *Review of A.Fraenkel "Zehn Vorlesungen ueber die Grundlagen der Mengenlehre"* (J 0157) Jbuchber Dtsch Math-Ver 39*10-11,2.Abt.
◇ A05 F55 ◇ ID 02699

BROUWER, L.E.J. [1933] *Volition, knowledge, speech (Dutch)* (J 0290) Euclides 9*177-193 • ERR/ADD ibid 11*160
• TRANSL [1975] (C 4429) Brouwer: Coll Works 443-446 (English, fragment) • REPR [1933] (C 4428) Way of Express of Sci 1-19
◇ A05 F55 ◇ REV FdM 59.55 • REM Modified and expanded version of Brouwer,L.E.J. 1929 ID ID 01624
• ID 03601

BROUWER, L.E.J. [1942] *Beweis dass der Begriff der Menge hoeherer Ordnung nicht als Grundbegriff der intuitionistischen Mathematik in Betracht kommt* (J 0028) Indag Math 6*274-276
◇ F55 ◇ REV MR 6.31 Zbl 27.300 • ID 01626

BROUWER, L.E.J. [1942] *Die repraesentierende Menge der stetigen Funktionen des Einheitskontinuums* (J 0028) Indag Math 4*154
◇ F55 ◇ REV MR 6.31 Zbl 27.49 • ID 01625

BROUWER, L.E.J. [1942] *Zum freien Werden von Mengen und Funktionen* (J 0028) Indag Math 4*107-108
◇ E70 F55 ◇ REV MR 6.31 Zbl 27.49 • ID 01627

BROUWER, L.E.J. [1947] *Directives of intuitionistic mathematics (Dutch)* (**J** 0028) Indag Math 9*197
• TRANSL [1975] (**C** 4429) Brouwer: Coll Works 477
◇ F55 ◇ REV MR 8.430 Zbl 35.148 JSL 12.136 JSL 13.174 • ID 01628

BROUWER, L.E.J. [1948] *Essentially-negative properties (Dutch)* (**J** 0028) Indag Math 10*322-323
• TRANSL [1975] (**C** 4429) Brouwer: Coll Works 478-479
◇ F55 ◇ REV MR 10.421 Zbl 41.352 JSL 14.137
• ID 01630

BROUWER, L.E.J. [1948] *Remarks on the law of the excluded third and on negative propositions (Dutch)* (**J** 0028) Indag Math 10*383-387
◇ F55 ◇ REV MR 10.421 Zbl 41.352 JSL 14.138 • REM This is essentially the mathematical part of Brouwer 1949 ID 17925 • ID 01631

BROUWER, L.E.J. [1949] *Consciousness, philosophy, and mathematics* (**P** 0682) Int Congr Philos (10);1948 Amsterdam 1235-1249
• REPR [1964] (**C** 1105) Phil of Math. Sel Readings 78-84
◇ A05 F55 ◇ REV MR 10.422 JSL 14.132 • REM Reprint is only a fragment • ID 17925

BROUWER, L.E.J. [1949] *Contradictority of elementary geometry (Dutch)* (**J** 0028) Indag Math 11*89-90
• TRANSL [1975] (**C** 4429) Brouwer: Coll Works 497-498 (English)
◇ F55 ◇ REV MR 11.2 Zbl 37.215 JSL 14.262 JSL 15.262
• ID 01633

BROUWER, L.E.J. [1949] *The non-equivalence of the constructive and the negative order relation on the continuum (Dutch)* (**J** 0028) Indag Math 11*37-39
• TRANSL [1975] (**C** 4429) Brouwer: Coll Works 495-496
◇ F55 ◇ REV MR 10.499 Zbl 41.353 JSL 14.195
• ID 01632

BROUWER, L.E.J. [1950] *Remarques sur la notion d'ordre* (**J** 0109) C R Acad Sci, Paris 230*263-265
◇ F55 ◇ REV MR 11.305 JSL 19.125 • ID 01635

BROUWER, L.E.J. [1950] *Sur la possibilite d'ordonner le continu* (**J** 0109) C R Acad Sci, Paris 230*349-350
◇ F55 ◇ REV MR 11.305 Zbl 41.375 JSL 19.125
• ID 01634

BROUWER, L.E.J. [1951] *On order in the continuum, and the relation of truth to non-contradictority* (**J** 0028) Indag Math 13*357-358 • ERR/ADD ibid 14*79
◇ F55 ◇ REV MR 13.898 JSL 19.125 • ID 01636

BROUWER, L.E.J. [1952] *An intuitionistic correction of the fixed-point theorem on the sphere* (**J** 1150) Proc Roy Soc London, Ser A 213*1-2
◇ F55 ◇ REV Zbl 46.409 • ID 37332

BROUWER, L.E.J. [1952] *Fixed cores which cannot be found, though they are claimed to exist by classical theorems (Dutch)* (**J** 0028) Indag Math 14*443-445
• TRANSL [1975] (**C** 4429) Brouwer: Coll Works 519-521
◇ F55 ◇ ID 42971

BROUWER, L.E.J. [1952] *Historical background, principles and methods of intuitionism* (**J** 0279) S African J Sci 49*139-146
• TRANSL [1967] (**C** 2141) Filos Matematica 223-231 (Italian) [1956] (**J** 0597) Afr Tydsk Wetensk Kuns 1956*186-197 (Dutch)
◇ A05 F55 ◇ REV MR 14.528 JSL 19.125 JSL 34.313
• REM Reprinted also in Brouwer,L.E.J. 1981 ID 39596
• ID 02694

BROUWER, L.E.J. [1952] *On accumulation cores of infinite core species (Dutch)* (**J** 0028) Indag Math 14*439-441
• TRANSL [1975] (**C** 4429) Brouwer: Coll Works 516-518
◇ F55 ◇ REV MR 14.441 Zbl 47.59 JSL 19.125 • ID 01637

BROUWER, L.E.J. [1954] *Addenda and corrigenda on the role of the principium tertii exclusi in mathematics (Dutch)* (**J** 0028) Indag Math 16*104-105
• TRANSL [1967] (**C** 0675) From Frege to Goedel 341-342 [1975] (**C** 4429) Brouwer: Coll Works 539-540
◇ F55 ◇ REV MR 15.670 • ID 01638

BROUWER, L.E.J. [1954] *An example of contradictority in classical theory of functions* (**J** 0028) Indag Math 16*204-205
◇ F55 ◇ REV MR 16.2 Zbl 56.13 • ID 01641

BROUWER, L.E.J. [1954] *Further addenda and corrigenda concerning the role of the principle tertii exclusi in mathematics (Dutch)* (**J** 0028) Indag Math 16*109-111
• TRANSL [1967] (**C** 0675) From Frege to Goedel 342-345 [1975] (**C** 4429) Brouwer: Coll Works 541-543
◇ F55 ◇ REV MR 15.925 Zbl 59.14 • ID 01642

BROUWER, L.E.J. [1954] *Intuitionistic Differentiability (Dutch)* (**J** 0028) Indag Math 16*201-203
• TRANSL [1975] (**C** 4429) Brouwer: Coll Works 546-548
◇ F55 ◇ REV MR 16.2 Zbl 56.13 JSL 24.189 • ID 01639

BROUWER, L.E.J. [1954] *Ordnungswechsel in Bezug auf eine coupierbare geschlossene stetige Kurve* (**J** 0028) Indag Math 16*112-113
◇ F55 ◇ REV MR 15.889 Zbl 55.168 • ID 41357

BROUWER, L.E.J. [1954] *Points and spaces* (**J** 0017) Canad J Math 6*1-17
◇ F55 ◇ REV MR 15.593 Zbl 55.46 JSL 34.519 • ID 01640

BROUWER, L.E.J. [1955] *The effect of intuitionism on classical algebra of logic* (**J** 0215) Proc Irish Acad, Sect A 57*113-116
◇ B05 F50 ◇ REV MR 17.446 Zbl 66.11 JSL 24.204
• ID 02051

BROUWER, L.E.J. [1975] *Collected works I* (**X** 0809) North Holland: Amsterdam xv+628pp
◇ A05 A10 F55 F96 ◇ REV MR 58#27140 Zbl 311#01021 • REM Edited by Heyting,A. Papers in Dutch have been translated into English. • ID 27570

BROUWER, L.E.J. [1981] *Brouwer's Cambridge lectures on intuitionism* (**X** 0805) Cambridge Univ Pr: Cambridge, GB xii+109pp
• TRANSL [1983] (**X** 0905) Boringhieri: Torino 113pp (Italian)
◇ A05 A10 F50 F55 F98 ◇ REV MR 82i:03069 Zbl 476#03056 JSL 48.214 • REM Edited by van Dalen,D.
• ID 71345

BROUWER, L.E.J. [1981] *On the foundation of mathematics (Dutch)* (**X** 1121) Math Centr: Amsterdam i+267pp
◇ A05 A10 F55 ◇ REV MR 83d:01076 Zbl 484#01009
• REM With additional material; edited by van Dalen,D.
• ID 39596

Author Index

BROUWER, L.E.J. see Vol. V for further entries

BRUCE, KIM B. & LONGO, G. [1984] *On combinatory algebras and their expansions* (**J** 1426) Theor Comput Sci 31*31-40
◇ B40 ◇ REV MR 86a:03013 Zbl 555 # 03006 • ID 45459

BRUCE, KIM B. & MEYER, A.R. [1984] *The semantics of second order polymorphic lambda calculus* (**P** 3090) Semant of Data Types;1984 Sophia-Antipolis 131-144
◇ B40 ◇ REV MR 85m:68003 Zbl 554 # 03011 • ID 44935

BRUCE, KIM B. see Vol. III, V for further entries

BRUIJN DE, N.G. [1972] *Lambda calculus notation with nameless dummies, a tool for automatic formula manipulation, with application to the Church-Rosser theorem* (**J** 0028) Indag Math 34*381-392
◇ B40 ◇ REV MR 48 # 71 Zbl 253 # 68007 JSL 40.470 • ID 01660

BRUIJN DE, N.G. [1978] *A namefree lambda calculus with facilities for internal definition of expressions and segments* (1111) Preprints, Manuscr., Techn. Reports etc. WSK-03*39pp
◇ B40 ◇ REV Zbl 423 # 68048 • REM Technological University, Department of Mathematics. Eindhoven • ID 53581

BRUIJN DE, N.G. [1978] *Lambda-calculus with namefree formulas involving symbols that represent reference transforming mappings* (**J** 0028) Indag Math 40*348-356
◇ B35 B40 ◇ REV MR 80b:03016 Zbl 393 # 03009 • ID 29191

BRUIJN DE, N.G. [1980] *A survey of the project AUTOMATH* (**C** 3050) Essays Combin Log, Lambda Calc & Formalism (Curry) 579-606
◇ B35 B40 ◇ REV MR 81m:03017 Zbl 469 # 03006 • ID 72186

BRUIJN DE, N.G. see Vol. I, II, III, V for further entries

BUCHHOLZ, W. [1975] *Ein ausgezeichnetes Modell fuer die intuitionistische Typenlogik* (**J** 0009) Arch Math Logik Grundlagenforsch 17*55-60
◇ B15 F05 F35 F50 ◇ REV MR 52 # 5368 Zbl 317 # 02026 • ID 01681

BUCHHOLZ, W. [1975] *Normalfunktionen und konstruktive Systeme von Ordinalzahlen* (**P** 1440) ⊧ ISILC Proof Th Symp (Schuette);1974 Kiel 4-25
◇ E10 F15 ◇ REV MR 54 # 72 Zbl 342 # 02021 JSL 48.876 • ID 23954

BUCHHOLZ, W. & SCHUETTE, K. [1976] *Die Beziehungen zwischen den Ordinalzahlsystemen Σ und $\overline{\Theta}(\omega)$* (**J** 0009) Arch Math Logik Grundlagenforsch 17*179-189
◇ F15 ◇ REV MR 55 # 2516 Zbl 339 # 02027 JSL 48.876 • ID 03950

BUCHHOLZ, W. [1976] *Ueber Teilsysteme von $\Theta(\{g\})$* (**J** 0009) Arch Math Logik Grundlagenforsch 18*85-98
◇ F15 ◇ REV MR 57 # 12188 Zbl 366 # 02017 JSL 48.876 • ID 23718

BUCHHOLZ, W. & POHLERS, W. [1978] *Provable well orderings of formal theories for transfinitely iterated inductive definitions* (**J** 0036) J Symb Logic 43*118-125
◇ F15 F30 F35 F50 ◇ REV MR 58 # 21498 Zbl 411 # 03046 JSL 48.878 • ID 29248

BUCHHOLZ, W. & FEFERMAN, S. & POHLERS, W. & SIEG, W. [1981] *Iterated inductive definitions and subsystems of analysis: recent proof-theoretical studies* (**S** 3301) Lect Notes Math 897*v+383pp
◇ F05 F15 F35 F50 F98 ◇ REV MR 84d:03074 Zbl 489 # 03022 • ID 34107

BUCHHOLZ, W. [1981] *Ordinal analysis of ID_ν* (**C** 3588) Iter Induct Def & Subsyst of Anal: Rec Proof-Th Stud 234-260
◇ F15 F35 F50 ◇ REV MR 84d:03074 • ID 45807

BUCHHOLZ, W. & SCHUETTE, K. [1981] *Syntaktische Abgrenzungen von formalen Systemen der Π^1_1-Analysis und Δ^1_2-Analysis* (**J** 1944) Sitzb, Akad Wiss, Bayern, Math-Nat Kl 1980*1-35
◇ F15 F35 ◇ REV MR 84g:03096 Zbl 472 # 03046 • ID 34200

BUCHHOLZ, W. [1981] *The $\Omega_{\mu+1}$-rule* (**C** 3588) Iter Induct Def & Subsyst of Anal: Rec Proof-Th Stud 188-233
◇ F05 F15 F35 F50 ◇ ID 45806

BUCHHOLZ, W. & SCHUETTE, K. [1983] *Ein Ordinalzahlensystem fuer die beweistheoretische Abgrenzung der Π^1_2-Separation und Bar-Induktion* (**J** 1944) Sitzb, Akad Wiss, Bayern, Math-Nat Kl 1983*99-132
◇ F15 F35 ◇ REV Zbl 573 # 03028 • ID 39611

BUECHI, J.R. [1960] *Weak second-order arithmetic and finite automata* (**J** 0068) Z Math Logik Grundlagen Math 6*66-92
◇ B15 B25 C85 D05 F35 ◇ REV MR 23 # A2317 Zbl 103.247 JSL 28.100 • ID 01690

BUECHI, J.R. [1962] *On a decision method in restricted second order arithmetic* (**P** 0612) Int Congr Log, Meth & Phil of Sci (1,Proc);1960 Stanford 1-11
• TRANSL [1964] (**J** 1048) Kiber Sb Perevodov 1964/8*78-90 (Russian)
◇ B25 C85 D05 F35 ◇ REV MR 32 # 1116 Zbl 147.251 JSL 28.100 • ID 01693

BUECHI, J.R. & LANDWEBER, L.H. [1969] *Definability in the monadic second-order theory of successor* (**J** 0036) J Symb Logic 34*166-170
◇ B15 B25 C40 C85 D05 F35 ◇ REV MR 42 # 4387 Zbl 209.22 • ID 01700

BUECHI, J.R. see Vol. I, III, IV, V for further entries

BUKOVSKY, L. [1974] see BILY, J.

BUKOVSKY, L. see Vol. III, IV, V for further entries

BULLOFF, J.J. & HAHN, S.W. & HOLYOKE, T.C. [1969] *Bibliography of Kurt Goedel* (**C** 0705) Found of Math (Goedel) xi-xii
◇ A10 D99 E99 F99 ◇ ID 14484

BUNDER, M.W. [1970] *A paradox in illative combinatory logic* (**J** 0047) Notre Dame J Formal Log 11*467-470
◇ B40 ◇ REV MR 44 # 2611 Zbl 185.25 • ID 01759

BUNDER, M.W. [1972] *An introduction to and trends in combinatory logic* (**P** 1483) Austral Conf Combin Math (1);1972 Newcastle 183-191
◇ B40 ◇ REV MR 51 # 73 Zbl 261 # 02016 • ID 30676

BUNDER, M.W. [1973] *A deduction theorem for restricted generality* (J 0047) Notre Dame J Formal Log 14*341-346 ◇ B40 ◇ REV MR 48 # 5828 Zbl 197.282 • REM See also Bunder,M.W. 1976 • ID 01761

BUNDER, M.W. [1973] *A generalized Kleene-Rosser paradox for a system containing the combinator K* (J 0047) Notre Dame J Formal Log 14*53-54 ◇ B40 ◇ REV MR 47 # 6463 Zbl 197.282 • ID 01762

BUNDER, M.W. [1974] *Propositional and predicate calculuses based on combinatory logic* (J 0047) Notre Dame J Formal Log 15*25-34 ◇ B40 ◇ REV MR 49 # 8818 Zbl 212.25 • ID 01763

BUNDER, M.W. [1974] *Some inconsistencies in illative combinatory logic* (J 0068) Z Math Logik Grundlagen Math 20*199-201 ◇ B40 ◇ REV MR 52 # 7862 Zbl 299 # 02031 • ID 03955

BUNDER, M.W. [1974] *Various systems of set theory based on combinatory logic* (J 0047) Notre Dame J Formal Log 15*192-206 ◇ B40 E30 E70 ◇ REV MR 49 # 8819 Zbl 212.25 • ID 01764

BUNDER, M.W. [1976] *Some notes on "A deduction theorem for restricted generality"* (J 0047) Notre Dame J Formal Log 17*153-154 ◇ B40 ◇ REV MR 52 # 10377 Zbl 292 # 02025 • REM For the article see Bunder, 1973 • ID 17989

BUNDER, M.W. [1976] *The inconsistency of* $\mathscr{F}^*_{21}$ (J 0036) J Symb Logic 41*467-468 ◇ B40 ◇ REV MR 53 # 12889 Zbl 332 # 02034 • ID 19215

BUNDER, M.W. [1977] *Consistency notions in illative combinatory logic* (J 0036) J Symb Logic 42*527-529 ◇ B40 ◇ REV MR 58 # 16176 Zbl 382 # 03013 • ID 26855

BUNDER, M.W. [1978] *Equality in* $\mathscr{F}^*_{21}$ *with restricted subjects* (J 0068) Z Math Logik Grundlagen Math 24*125-127 ◇ B40 ◇ REV MR 80e:03011 Zbl 422 # 03002 • ID 53463

BUNDER, M.W. & MEYER, R.K. [1978] *On the inconsistency of systems similar to* $\mathscr{F}^*_{21}$ (J 0036) J Symb Logic 43*1-2 ◇ B40 ◇ REV MR 80a:03002a Zbl 394 # 03022 • ID 29236

BUNDER, M.W. & SELDIN, J.P. [1978] *Some anomalies in Fitch's system QD* (J 0036) J Symb Logic 43*247-249 ◇ B40 ◇ REV MR 58 # 164 Zbl 409 # 03005 • ID 29257

BUNDER, M.W. [1979] *Λ-elimination in illative combinatory logic* (J 0047) Notre Dame J Formal Log 20*628-630 ◇ B40 B50 ◇ REV MR 80h:03020 Zbl 394 # 03024 • ID 56104

BUNDER, M.W. [1979] *Generalised restricted generality* (J 0047) Notre Dame J Formal Log 20*620-624 ◇ B40 ◇ REV MR 80i:03025 Zbl 363 # 02029 • ID 56101

BUNDER, M.W. [1979] *On the equivalence of systems of rules and systems of axioms in illative combinatory logic* (J 0047) Notre Dame J Formal Log 20*603-608 ◇ B40 ◇ REV MR 80h:03019 Zbl 349 # 02020 • ID 56103

BUNDER, M.W. [1979] *Scott's models and illative combinatory logic* (J 0047) Notre Dame J Formal Log 20*609-612 ◇ B40 ◇ REV MR 80j:03022 Zbl 349 # 02021 • ID 56102

BUNDER, M.W. [1979] *Variable binding term operators in λ-calculus* (J 0047) Notre Dame J Formal Log 20*876-878 ◇ B40 ◇ REV MR 80i:03026 Zbl 394 # 03023 • ID 56188

BUNDER, M.W. [1980] *A note on quantified significance logics* (J 0387) Bull Sect Logic, Pol Acad Sci 9*159-162 ◇ B40 B45 ◇ REV MR 82d:03041 Zbl 452 # 03009 • ID 54073

BUNDER, M.W. [1980] *Significance and illative combinatory logics* (J 0047) Notre Dame J Formal Log 21*380-384 ◇ B40 B45 ◇ REV MR 81g:03011 Zbl 394 # 03025 • ID 53770

BUNDER, M.W. [1980] *The consistency of a higher order predicate calculus and set theory based on combinatory logic* (P 2958) Latin Amer Symp Math Log (4);1978 Santiago 73-82 ◇ B15 B40 F05 ◇ REV MR 81h:03106 Zbl 427 # 03012 • ID 53699

BUNDER, M.W. [1980] *The naturalness of illative combinatory logic as a basis for mathematics* (C 3050) Essays Combin Log, Lambda Calc & Formalism (Curry) 55-64 ◇ B40 ◇ REV MR 82a:03015 Zbl 469 # 03006 • ID 31629

BUNDER, M.W. [1981] *Predicate calculus and naive set theory in pure combinatory logic* (J 0009) Arch Math Logik Grundlagenforsch 21*169-177 ◇ B10 B40 E70 ◇ REV MR 84f:03012 Zbl 472 # 03011 • ID 55277

BUNDER, M.W. [1981] *Simpler axioms for BCK algebras and the connection between the axioms and the combinators B, C and K* (J 0352) Math Jap 26*415-418 ◇ B40 G25 ◇ REV MR 83f:03069 Zbl 473 # 03063 • ID 55391

BUNDER, M.W. [1982] *A formal metatheory in which Goedel's first but not his second incompleteness theorem can be proved* (J 3783) J Non-Classical Log (Univ Campinas) 1*43-69 ◇ F30 ◇ REV MR 84j:03112 Zbl 508 # 03005 • ID 34702

BUNDER, M.W. [1982] *Illative combinatory logic without equality as a primitive predicate* (J 0047) Notre Dame J Formal Log 23*62-70 ◇ B40 ◇ REV MR 83g:03015 Zbl 452 # 03008 • ID 55075

BUNDER, M.W. [1983] *A one axiom set theory based on higher order predicate calculus* (J 0009) Arch Math Logik Grundlagenforsch 23*99-107 ◇ B15 B40 E70 ◇ REV Zbl 537 # 03010 • ID 31631

BUNDER, M.W. [1983] *A weak absolute consistency proof for some systems of illative combinatory logic* (J 0036) J Symb Logic 48*771-776 ◇ B40 ◇ REV MR 86b:03021 Zbl 527 # 03003 • ID 37504

BUNDER, M.W. [1984] *Category theory based on combinatory logic* (J 0009) Arch Math Logik Grundlagenforsch 24*1-15 ◇ B40 G30 ◇ REV MR 86b:18001 Zbl 545 # 03005 • ID 41220

BUNDER, M.W. [1984] *Conjunction without conditions in illative combinatory logic* (J 0387) Bull Sect Logic, Pol Acad Sci 13*207-214 ◇ B40 ◇ REV Zbl 557 # 03008 • ID 44652

BUNDER, M.W. [1984] *Some problems and inadequacies in the van der Poel-Schaap-van der Mey theory of combinators* (J 0028) Indag Math 46*373-377 ◇ B40 ◇ REV Zbl 556 # 03013 • ID 44351

Author Index

BUNDER, M.W. [1985] *An extension of Klop's counterexample to the Church-Rosser property to λ-calculus with other ordered pair combinators* (J 1426) Theor Comput Sci 39*337-342
◇ B40 ◇ ID 49031

BUNDER, M.W. see Vol. I, II, V for further entries

BUNDY, A. [1973] *A note on omitting the replacement schema* (J 0047) Notre Dame J Formal Log 14*118-120
◇ D20 F30 ◇ REV MR 47 # 8268 Zbl 225 # 02031
• ID 01765

BUNDY, A. see Vol. I for further entries

BURGER, E. [1964] *Bemerkungen zu einigen Fassungen des Goedelschen Unvollstaendigkeitssatzes* (J 0068) Z Math Logik Grundlagen Math 10*57-63
◇ D80 F30 ◇ REV MR 28 # 2047 Zbl 146.9 • ID 01771

BURGER, E. see Vol. V for further entries

BURGESS, J.P. [1980] *Brouwer and Souslin on transfinite cardinals* (J 0068) Z Math Logik Grundlagen Math 26*209-214
◇ E15 F55 ◇ REV MR 82i:03071 Zbl 432 # 03031
• ID 53988

BURGESS, J.P. [1981] *The completeness of intuitionistic propositional calculus for its intended interpretation* (J 0047) Notre Dame J Formal Log 22*17-28
◇ B55 F50 ◇ REV MR 82g:03015 Zbl 445 # 03030
• ID 56494

BURGESS, J.P. [1984] *Dummett's case for intuitionism* (J 2028) Hist & Phil Log 5*177-194
◇ A05 F50 ◇ REV MR 86a:03002 Zbl 549 # 03006
• ID 43101

BURGESS, J.P. see Vol. II, III, IV, V for further entries

BURTON, F.W. [1982] *A linear space translation of functional programs to Turner combinators* (J 0232) Inform Process Lett 14*201-204
◇ B40 ◇ REV MR 84b:03027 Zbl 507 # 03004 • ID 35622

BURTON, W.K. [1968] *Constructive thermodynamics* (P 0608) Logic Colloq;1966 Hannover 75-89
◇ F65 ◇ REV Zbl 181.308 • ID 01797

BUSZKOWSKI, W. [1984] *A note on Lambek - van Benthem calculus* (J 0387) Bull Sect Logic, Pol Acad Sci 13*31-37
◇ B40 ◇ REV MR 85m:03009 Zbl 557 # 03017 • ID 45747

BUSZKOWSKI, W. see Vol. III, IV, V for further entries

BUTRICK, R. [1977] *The numeral axioms* (J 0047) Notre Dame J Formal Log 18*475-476
◇ F30 ◇ REV Zbl 245 # 02048 • ID 23644

BUTRICK, R. see Vol. I, III, V for further entries

BYERLY, R.E. [1982] *Recursion theory and the lambda-calculus* (J 0036) J Symb Logic 47*67-83
◇ B40 D20 ◇ REV MR 83i:03066 Zbl 488 # 03023
• ID 35525

BYERLY, R.E. see Vol. IV for further entries

CAGNONI, D. [1977] *A note on the elimination rules* (J 0122) J Philos Logic 6*269-281
◇ F05 F07 ◇ REV MR 58 # 21499 Zbl 368 # 02035
• ID 51252

CALDER, A. [1981] see BRIDGES, D.S.

CALDER, A. [1982] see BRIDGES, D.S.

CALL, R.L. [1984] *Constructing sequent rules for generalized propositional logics* (J 0047) Notre Dame J Formal Log 25*171-178
◇ B05 F07 ◇ REV MR 85e:03020 Zbl 551 # 03036
• ID 40582

CALL, R.L. see Vol. I, II, III for further entries

CALUDE, C. [1977] *Une construction grammaticale des codages de Goedel (English summary)* (J 2313) C R Acad Sci, Paris, Ser A-B 284*A525-A526
◇ F40 ◇ REV MR 55 # 87 Zbl 347 # 02026 • ID 27296

CALUDE, C. [1978] *On the category of recursive languages* (J 0517) Mathematica (Cluj) 19(24)*29-32
◇ B03 D45 F40 G30 ◇ REV MR 80b:03056
Zbl 384 # 03029 • ID 52071

CALUDE, C. & PAUN, G. [1983] *Independent instances for some undecidable problems (French summary)* (J 3441) RAIRO Inform Theor 17*49-54
◇ D20 F30 ◇ REV MR 85d:03087 Zbl 517 # 03022
• ID 37287

CALUDE, C. see Vol. I, II, IV, V for further entries

CANAL, R. [1978] *Complexite de la reduction en logique combinatoire* (J 3441) RAIRO Inform Theor 12*339-367
◇ B40 ◇ REV MR 80a:03016 Zbl 432 # 03012 • ID 53970

CANNONITO, F.B. [1962] *The Goedel incompleteness theorem and intelligent machines* (P 1197) AFIPS Spring Jt Computer Conf (21);1962 San Francisco 21*71-77
◇ B35 D10 D25 F30 ◇ REV JSL 36.693 • ID 21294

CANNONITO, F.B. [1971] *A note on inverses of elementary permutations* (J 0191) Inform Sci 3*355-359
◇ D20 D45 F30 ◇ REV MR 46 # 33 Zbl 229 # 02034
• ID 71525

CANNONITO, F.B. see Vol. IV for further entries

CANTINI, A. [1980] *A note on three-valued logic and Tarski theorem on truth definitions* (J 0063) Studia Logica 39*405-414
◇ B50 C40 F30 ◇ REV MR 83i:03019 Zbl 457 # 03024
• ID 54349

CANTINI, A. [1982] *Non-extensional theories of predicative classes over PA* (J 2038) Rend Sem Mat, Torino 40/3*47-79
◇ F30 F35 ◇ REV MR 85m:03041 Zbl 549 # 03051
• ID 43133

CANTINI, A. [1983] *A note on the theory of admissible sets with ε-induction restricted to formulas with one quantifier and related systems (Italian summary)* (J 2100) Boll Unione Mat Ital, VI Ser, B 2*721-737
◇ C62 E30 E45 E70 F10 F15 F35 ◇ REV
Zbl 558 # 03027 • ID 41846

CANTINI, A. [1985] *A note on a predicatively reducible theory of iterated elementary induction (Italian summary)* (J 2100) Boll Unione Mat Ital, VI Ser, B 4*413-430
◇ F15 F35 ◇ ID 49073

CANTINI, A. [1985] *Majorizing provably recursive functions in fragments of PA* (J 0009) Arch Math Logik Grundlagenforsch 25*21-31
◇ D20 F15 F30 ◇ ID 49071

CANTINI, A. [1985] *The Veblen hierarchy and the stability theorem for L* (P 4646) Atti Incontri Log Mat (2);1983/84 Siena 167-170
◇ F15 ◇ ID 49599

CANTINI, A. see Vol. I, III, V for further entries

CAPORASO, S. [1979] *Consistency proof without transfinite induction for a formal system for Turing machines* (J 0009) Arch Math Logik Grundlagenforsch 19*157-164
◇ D10 F05 ◇ REV MR 80j:03082 Zbl 406 # 03070
• ID 56152

CARNAP, R. [1935] *Ein Gueltigkeitskriterium fuer die Saetze der klassischen Mathematik* (J 0124) Monatsh Math-Phys 42*163-190
◇ B30 F07 ◇ REV Zbl 12.145 FdM 61.970 • ID 03962

CARNAP, R. see Vol. I, II for further entries

CARRUCCIO, E. [1947] *Alcune conseguenze di un risultato del Goedel e la razionalita del reale* (J 0385) Atti Sem Mat Fis Univ Modena 78*88-90
◇ A05 F30 ◇ REV MR 9.559 Zbl 33.341 • ID 24917

CARRUCCIO, E. see Vol. I for further entries

CARSTENGERDES, W. [1971] *Mehrsortige logische Systeme mit unendlich langen Formeln I, II* (J 0009) Arch Math Logik Grundlagenforsch 14*38-53,108-126
◇ C40 C75 F05 ◇ REV MR 44 # 2583 MR 48 # 5818 Zbl 235 # 02023 • ID 01873

CARSTENS, H.G. [1977] *The theorem of Matijasevic is provable in Peano's arithmetic by finitely many axioms* (J 0079) Logique & Anal, NS 20*116-121
◇ C52 D25 D35 F30 H15 ◇ REV MR 58 # 27400 Zbl 382 # 03037 • ID 51961

CARSTENS, H.G. see Vol. III, IV, V for further entries

CARTIER, P. [1979] *Logique, categories et faisceaux (d'apres F. Lawvere et M. Tierney)* (S 1567) Semin Bourbaki Exp.513*123-146
◇ E35 E50 F55 G10 G30 ◇ REV MR 80m:03111 Zbl 406 # 03074 • REM Springer: Heidelberg & New York; Lecture Notes Math 710 • ID 56156

CARTIER, P. see Vol. I for further entries

CARTWRIGHT, ROBERT [1984] *Nonstandard fixed points in first order logic* (P 2989) Log of Progr; 1983 Pittsburgh 87-100
◇ B75 F30 ◇ REV Zbl 566 # 68015 • ID 44538

CARTWRIGHT, ROBERT see Vol. I, IV for further entries

CASANOVA, M.A. [1983] *The theory of functional and subset dependencies over relational expressions* (J 0232) Inform Process Lett 16*153-160
◇ F07 ◇ REV MR 84e:68081 Zbl 507 # 68073 • ID 39287

CASTANET, R. [1972] *Sur la semantique des operateurs d'insertion dans une arborescence* (J 2313) C R Acad Sci, Paris, Ser A-B 275*A209-A212
◇ B40 ◇ REV MR 46 # 1561 Zbl 239 # 02016 Zbl 247 # 02029 • ID 01892

CASTANET, R. [1972] *Une formalisation de la semantique des operateurs d'extraction dans une arborescence* (J 2313) C R Acad Sci, Paris, Ser A-B 275*A135-A137
◇ B40 ◇ REV MR 46 # 80 Zbl 239 # 02016 • ID 01891

CASTANET, R. [1974] *Semantique formelle des operateurs d'un langage de listes (English summary)* (J 0205) Rev Franc Autom, Inf & Rech Operat 8/R-3*19-36
◇ B40 B75 ◇ REV MR 51 # 2878 Zbl 325 # 68012
• ID 17329

CAVINESS, B.F. & POLLACK, P.L. & RUBALD, C.M. [1971] *An existence lemma for canonical forms in symbolic mathematics* (J 0232) Inform Process Lett 1*45-46
◇ D25 F05 ◇ REV MR 45 # 1427 Zbl 221 # 68032
• ID 02214

CAVINESS, B.F. see Vol. I, IV for further entries

CEGIELSKI, P. [1980] *La theorie elementaire de la multiplication (English summary)* (J 2313) C R Acad Sci, Paris, Ser A-B 290*A935-A938
◇ B25 C10 C80 F30 ◇ REV MR 82d:03020 Zbl 435 # 03009 • ID 55770

CEGIELSKI, P. [1981] *Theorie elementaire de la multiplication des entiers naturels* (P 3404) Model Th & Arithm;1979/80 Paris 44-89
◇ B25 C10 C80 F30 ◇ REV MR 83e:03094 Zbl 478 # 03011 • ID 55626

CEGIELSKI, P. & MCALOON, K. & WILMERS, G.M. [1982] *Modeles recursivement satures de l'addition et de la multiplication des entiers naturels (English summary)* (P 3623) Logic Colloq;1980 Prague 57-68
◇ C50 C57 C62 F30 H15 ◇ REV MR 84i:03072 Zbl 527 # 03043 • ID 34553

CEGIELSKI, P. [1984] *La theorie elementaire de la divisibilite est finiment axiomatisable (English summary)* (J 3364) C R Acad Sci, Paris, Ser 1 299*367-369
◇ B25 B28 F30 ◇ REV MR 85j:03042 • ID 46773

CEGIELSKI, P. see Vol. I for further entries

CELLUCCI, C. [1969] *Un'osservazione sul teorema di Minc-Orevkov (English summary)* (J 0012) Boll Unione Mat Ital, IV Ser 2*1-8
◇ F50 ◇ REV MR 39 # 5317 Zbl 174.9 • ID 01913

CELLUCCI, C. [1971] *Operazioni di Brouwer e realizzabilita formalizzata (English summary)* (J 0315) Ann Sc Norm Sup Pisa Fis Mat, Ser 3 25*649-682
◇ F50 ◇ REV MR 51 # 12485 Zbl 242 # 02037 • ID 17273

CELLUCCI, C. [1974] *On the role of reducibility principles* (J 0154) Synthese 27*93-110
◇ A05 E30 F50 ◇ REV MR 57 # 15997 Zbl 319 # 02010
• ID 29670

CELLUCCI, C. [1974] *Un connettivo per la logica intuizionista (English summary)* (J 0319) Matematiche (Sem Mat Catania) 29*274-290
◇ F50 ◇ REV MR 53 # 12882 Zbl 334 # 02014 • ID 23140

CELLUCCI, C. [1974] *Una dimostrazione del teorema di uniformita* (J 0319) Matematiche (Sem Mat Catania) 29*373-378
◇ F05 ◇ REV MR 55 # 5401 Zbl 324 # 02022 • ID 32265

CELLUCCI, C. [1975] *Teoremi di normalizzazione per alcuni sistemi funzionali* (J 0319) Matematiche (Sem Mat Catania) 30*1-42
◇ F05 F10 F15 F30 ◇ REV MR 56 # 2799a Zbl 367 # 02013 • REM Part I. Part II 1975 • ID 32266

CELLUCCI, C. [1975] *Teoremi di normalizzazione per alcuni sistemi funzionali II* (J 0319) Matematiche (Sem Mat Catania) 30*210-240
◇ F05 F10 F15 F30 ◇ REV MR 56 # 2799b Zbl 367 # 02013 • REM Part I 1975 • ID 71606

Author Index

Cellucci, C. [1977] *Proprieta di uniformita e 1-coerenza dell'aritmetica del primo ordine* (**J** 0319) Matematiche (Sem Mat Catania) 32*35-49
◇ F05 F30 ◇ REV MR 81k:03058 Zbl 437 # 03029 • ID 55897

Cellucci, C. [1978] *Teoria della dimostrazione. Normalizzazioni e assegnazioni di numeri ordinali* (**X** 0905) Boringhieri: Torino 315pp
◇ F05 F07 F15 F30 F98 ◇ REV MR 80f:03065 JSL 47.220 • ID 32268

Cellucci, C. [1980] *A note on Goedel's second incompleteness theorem* (**J** 3285) Boll Unione Mat Ital, V Ser, A 17*137-142
◇ F30 ◇ REV MR 81a:03057 Zbl 494 # 03044 • ID 71603

Cellucci, C. [1981] *A calculus of constructions as a representation of intuitionistic logical proofs* (**P** 3092) Congr Naz Logica;1979 Montecatini Terme 175-193
◇ F50 ◇ ID 48378

Cellucci, C. [1981] *Proof theory and theory of meaning* (**C** 3515) Ital Studies in Phil of Sci 13-29
◇ A05 F07 F30 ◇ REV MR 82f:03048 Zbl 463 # 03030 • ID 54570

Cellucci, C. [1983] *The complexity of proofs in first-order predicate logic (Italian)* (**P** 3829) Atti Incontri Log Mat (1);1982 Siena 355-358
◇ F20 ◇ REV MR 84k:03006 • ID 44825

Cellucci, C. [1985] *Complexity of reduction trees in predicate calculus* (**P** 4646) Atti Incontri Log Mat (2);1983/84 Siena 319-327
◇ F20 ◇ ID 49674

Cellucci, C. [1985] *Proof theory and complexity* (**J** 0154) Synthese 62*173-189
◇ D15 F05 F20 F98 ◇ ID 45633

Cellucci, C. see Vol. III, IV, V for further entries

Chaitin, G.J. [1974] *Information-theoretic limitations of formal systems* (**J** 0037) ACM J 21*403-424
◇ D15 D80 F20 ◇ REV MR 56 # 13775 Zbl 287 # 68027 • ID 60893

Chaitin, G.J. [1982] *Goedel's theorem and information* (**J** 2736) Int J Theor Phys 21*941-954
◇ A05 F30 ◇ REV MR 84c:03007 • ID 35674

Chaitin, G.J. see Vol. IV for further entries

Chan, Y.K. [1971] *A constructive proof of Sard's theorem* (**J** 0048) Pac J Math 36*291-301
◇ F55 ◇ REV MR 43 # 2726 Zbl 191.62 • ID 01926

Chan, Y.K. [1972] *A constructive study of measure theory* (**J** 0048) Pac J Math 41*63-79
◇ F55 ◇ REV MR 46 # 9263 Zbl 243 # 02025 • ID 01927

Chan, Y.K. [1972] *A constructive approach to the theory of stochastic processes* (**J** 0064) Trans Amer Math Soc 165*37-44
◇ F55 ◇ REV MR 51 # 1894 Zbl 243 # 60021 • ID 17496

Chan, Y.K. [1974] *Notes on constructive probability theory* (**J** 2661) Ann Probab 2*51-75
◇ F55 ◇ REV MR 50 # 8669 Zbl 278 # 60045 • ID 41685

Chan, Y.K. [1974] *On constructive convergence of measures on the real line* (**J** 2661) Ann Probab 2*131-135
◇ F55 ◇ REV MR 50 # 8649 Zbl 278 # 60004 • ID 41688

Chan, Y.K. [1975] *A short proof of an existence theorem in constructive measure theory* (**J** 0053) Proc Amer Math Soc 48*435-437
◇ F55 ◇ REV MR 51 # 844 Zbl 299 # 28006 • ID 17491

Chan, Y.K. [1977] *Constructive foundations of potential theory* (**J** 0048) Pac J Math 71*405-418
◇ F55 ◇ REV MR 58 # 22615 Zbl 362 # 31006 • ID 80872

Chan, Y.K. [1981] *On some open problems in constructive probability theory* (**P** 3146) Constr Math;1980 Las Cruces 44-53
◇ F55 ◇ REV MR 83f:60097 Zbl 457 # 03057 • ID 54382

Chandrasekharan, K. [1941] *The logic of intuitionistic mathematics* (**J** 0248) Math Student 9*143-154
◇ E70 F55 ◇ REV MR 4.126 Zbl 61.10 JSL 7.171 • ID 02408

Chandrasekharan, K. [1942] *Intuitionistic theory of linear order* (**J** 0248) Math Student 10*149-162
◇ E07 E70 F55 ◇ REV MR 5.198 Zbl 61.10 • ID 02407

Chandrasekharan, K. [1945] *A further note on intuitionistic set theory* (**J** 0248) Math Student 13*49-51
◇ E70 F55 ◇ REV MR 7.406 Zbl 61.10 JSL 11.127 • ID 02409

Chandrasekharan, K. see Vol. V for further entries

Chapman, T. [1980] *An apparent contradiction in Goedel's proof* (**J** 0286) Int Logic Rev 21*25-28
◇ F30 ◇ REV MR 83i:03020 Zbl 481 # 03014 • ID 35496

Chapman, T. see Vol. II for further entries

Charlesworth, A. [1981] *A proof of Goedel's theorem in terms of computer programs* (**J** 0497) Math Mag 54*109-121
◇ D10 F30 ◇ REV MR 83e:03001 Zbl 512 # 03006 • ID 35267

Charlesworth, A. see Vol. V for further entries

Chauvin, A. [1949] *Generalisation du theoreme de Goedel* (**J** 0109) C R Acad Sci, Paris 228*1179-1180
◇ B15 F30 ◇ REV MR 10.668 Zbl 34.293 JSL 14.193 • ID 01996

Chauvin, A. [1949] *Structures logiques* (**J** 0109) C R Acad Sci, Paris 228*1085-1087
◇ F30 ◇ REV MR 10.668 Zbl 34.293 JSL 14.193 • ID 01997

Chauvin, A. [1962] *Sur les ensembles arithmetiques constructibles* (**J** 0109) C R Acad Sci, Paris 254*3615-3617
◇ D55 E45 F15 ◇ REV MR 25 # 2948 Zbl 113.6 • ID 02061

Chauvin, A. [1979] *Theory of objects and set theory: Introduction and semantics* (**J** 0047) Notre Dame J Formal Log 20*37-54
◇ A05 B40 E30 E70 ◇ REV MR 81c:03041 Zbl 285 # 02052 • ID 31368

Chauvin, A. see Vol. II, III, IV, V for further entries

Chen, Kehsun [1978] *Recursive well-founded orderings* (**J** 0007) Ann Math Logic 13*117-147
◇ D45 F15 ◇ REV MR 80e:03047 Zbl 384 # 03027 • ID 27962

Cheng, Henry [1972] see Bishop, E.A.

CHENG, HENRY [1973] *A constructive Riemann mapping theorem* (J 0048) Pac J Math 44*435-454
◇ F55 ◇ REV MR 48 # 1895 Zbl 252 # 30012 • ID 02073

CHENG, HENRY [1973] *A constructive intermediate value theorem* (J 0345) Adv Math 10*297-299
◇ F55 ◇ REV MR 49 # 7398 Zbl 249 # 26004 • ID 17159

CHENG, HENRY [1976] see BERG, G.

CHERNIAVSKY, J.C. [1973] *The complexity of some non-classical logics* (P 3062) IEEE Symp Switch & Automata Th (14);1973 Iowa City 209-213
◇ B45 D15 F20 F50 ◇ REV MR 56 # 2801 • ID 71692

CHERNIAVSKY, J.C. [1976] *Simple programs realize exactly Presburger formulas* (J 1428) SIAM J Comp 5*666-677
◇ B75 D15 D20 F30 ◇ REV MR 58 # 13841 Zbl 353 # 68018 • ID 31025

CHERNIAVSKY, J.C. see Vol. IV for further entries

CHERNOV, V.P. [1972] *Classification of spaces of operators of finite types (Russian)* (S 0228) Zap Nauch Sem Leningrad Otd Mat Inst Steklov 32*148-152,158 • ERR/ADD ibid 40*161
• TRANSL [1976] (J 1531) J Sov Math 6*471-474
◇ D45 D65 F60 ◇ REV MR 52 # 5384a MR 52 # 5384b Zbl 347 # 02024 • ID 17990

CHERNOV, V.P. [1972] *Constructive operators of finite types (Russian) (English summary)* (S 0228) Zap Nauch Sem Leningrad Otd Mat Inst Steklov 32*140-147,158 • ERR/ADD ibid 40*161
• TRANSL [1976] (J 1531) J Sov Math 6*465-470
◇ D45 D65 F60 ◇ REV MR 52 # 5383a MR 52 # 5383b Zbl 347 # 02023 • ID 17991

CHERNOV, V.P. & LIFSCHITZ, V. [1972] *Noncompact closed ball in a constructive compact metric space (Russian) (English summary)* (S 0228) Zap Nauch Sem Leningrad Otd Mat Inst Steklov 32*53-58,155
• TRANSL [1976] (J 1531) J Sov Math 6*390-394
◇ F60 ◇ REV MR 49 # 1487 Zbl 345 # 02026 • ID 22348

CHERNOV, V.P. [1972] *Topological variants of the continuity theorem for mappings and related theorems theorems (Russian) (English summary)* (S 0228) Zap Nauch Sem Leningrad Otd Mat Inst Steklov 32*129-139,158 • ERR/ADD ibid 40*161
• TRANSL [1976] (J 1531) J Sov Math 6*456-464
◇ D20 F60 ◇ REV MR 52 # 5381a MR 52 # 5381b Zbl 347 # 02022 • ID 17992

CHERNOV, V.P. [1974] *Some properties of mappings of sheaf-spaces (Russian) (English summary)* (S 0228) Zap Nauch Sem Leningrad Otd Mat Inst Steklov 40*136-141,159
• TRANSL [1977] (J 1531) J Sov Math 8*341-344
◇ E75 F60 G30 ◇ REV MR 51 # 2886 Zbl 368 # 02037 • ID 17333

CHERNOV, V.P. [1976] *Example of a constructive separable space with an inseparable completion (Russian)* (S 0228) Zap Nauch Sem Leningrad Otd Mat Inst Steklov 60*207-208,227
• TRANSL [1980] (J 1531) J Sov Math 14*1563-1564
◇ F60 ◇ REV MR 58 # 27377 Zbl 332 # 02042 • ID 60889

CHERNOV, V.P. [1983] *Dialogue spaces of trivial type (Russian)* (C 3798) Mat Log, Mat Ling & Teor Algor 83-95
◇ F60 ◇ REV MR 85e:03147 • ID 40762

CHERNOV, V.P. [1985] *Nonpoint and continuous mappings (Russian)* (J 0031) Izv Vyssh Ucheb Zaved, Mat (Kazan) 1985/8*51-59,85
◇ F60 ◇ ID 49510

CHERNYAKHOVSKIJ, N.P. [1976] *The expressibility of realizability, in the language of formal arithmetic (Russian)* (S 0554) Issl Teor Algor & Mat Logik (Moskva) 2*51-56,157
◇ D80 F30 F50 ◇ REV MR 58 # 16193 • ID 71693

CHIARAVIGLIO, L. [1975] see BARALT-TORRIJOS, J.

CHIHARA, C.S. [1975] see BERG, J.

CHIHARA, C.S. [1984] *Priest, the Liar, and Goedel* (J 0122) J Philos Logic 13*117-124
◇ A05 F30 ◇ REV MR 86d:03005a Zbl 543 # 03003 • ID 40897

CHIHARA, C.S. see Vol. I, V for further entries

CHRISTIAN, C.C. [1972] *Konsistenzkriterien fuer formale Theorien und ihre Anwendung auf Zahlen- und Mengentheorie* (J 0455) Phil Naturalis 13*405-442
◇ B28 E30 F30 ◇ REV MR 56 # 5263 • ID 71771

CHRISTIAN, C.C. [1973] *Inhaltliche und formale Wahrheit* (J 0455) Phil Naturalis 14*173-196
◇ A05 E30 E35 E45 F30 ◇ REV MR 56 # 5195 • ID 71773

CHRISTIAN, C.C. [1976] *Peano-Systeme* (J 0045) Monatsh Math 82*81-116
◇ B28 F30 F35 ◇ REV MR 57 # 88 Zbl 344 # 02038 • ID 60952

CHRISTIAN, C.C. [1977] *Eine Note zum System PC* (J 0045) Monatsh Math 83*191-200
◇ E70 F65 ◇ REV MR 58 # 10446 Zbl 362 # 02068 • ID 50790

CHRISTIAN, C.C. [1981] *Das rekursive Inaccessibilitaetstheorem und der Goedelsche Unvollstaendigkeitssatz in ihrer Bedeutung fuer die Informatik* (J 3480) Informatik & Philos 11*152-168
◇ A05 D35 D80 F30 ◇ REV Zbl 475 # 03002 • ID 55456

CHRISTIAN, C.C. see Vol. I, II, V for further entries

CHUBARYAN, A.A. & TSEJTIN, G.S. [1972] *Certain estimates of the length of logical deductions in classical propositional calculus (Russian) (Armenian summary)* (J 0346) Dokl Akad Nauk Armyan SSR 55*10-12
◇ B05 F07 F20 ◇ REV MR 50 # 12639 Zbl 246 # 02011 • ID 17157

CHUBARYAN, A.A. [1974] *The complexity of deductions in extensions of formal arithmetic (Russian) (Armenian summary)* (J 2892) Molodoj Nauch Rabotnik, Erevan 2(20)*60-64
◇ F20 F30 ◇ REV MR 58 # 168 • ID 71778

CHUBARYAN, A.A. [1974] *The length of derivations of formulae in extensions of formal arithmetic (Russian) (Armenian and English summaries)* (J 0312) Izv Akad Nauk Armyan SSR, Ser Mat 9*409-420,429
◇ F20 F30 ◇ REV MR 52 # 49 Zbl 329 # 02012 • ID 17217

CHUBARYAN, A.A. & NGUEN VAN TIN' [1975] *Some estimates of the complexity characteristics of deductions in classical propositional calculus (Russian)* (J 2892) Molodoj Nauch Rabotnik, Erevan 2(22)*47-54
◇ B05 F20 ◇ REV MR 58 # 130 Zbl 364 # 02005 • ID 76827

CHUBARYAN, A.A. [1977] *The complexity of deductions in formal arithmetic and predicate calculus (Russian)* (J 0346) Dokl Akad Nauk Armyan SSR 64*193-196
◇ B10 F20 F30 ◇ REV MR 57 # 4613 Zbl 383 # 03039 • ID 52021

CHUBARYAN, A.A. [1982] *Complexity characteristics of inferences in systems of predicate calculus and formal arithmetic (Russian) (Armenian summary)* (S 0422) Tr Vychisl Tsentra Akad Nauk Armyan SSR & Univ Erevan 10*124-139
◇ B10 F20 F30 ◇ REV MR 85e:03138 • ID 40752

CHUBARYAN, A.A. see Vol. I for further entries

CHURCH, A. [1928] *On the law of excluded middle* (J 0015) Bull Amer Math Soc 34*75-78
◇ A05 B05 F50 ◇ REV FdM 54.53 • ID 02122

CHURCH, A. [1932] *A set of postulates for the foundation of logic* (J 0120) Ann of Math, Ser 2 33*346-366
◇ B10 B40 D20 ◇ REV Zbl 4.145 JSL 23.23 FdM 58.997 • REM Part I. Part II 1933 • ID 02123

CHURCH, A. [1933] *A set of postulates for the foundation of logic (second paper)* (J 0120) Ann of Math, Ser 2 34*839-864
◇ B10 B40 D20 ◇ REV Zbl 8.289 JSL 24.94 FdM 59.52 • REM Part I 1932 • ID 02124

CHURCH, A. [1935] *A proof of freedom from contradiction* (J 0054) Proc Nat Acad Sci USA 21*275-281
◇ B40 D20 ◇ REV Zbl 12.241 FdM 61.55 • ID 02126

CHURCH, A. [1935] *An unsolvable problem of elementary number theory* (J 0015) Bull Amer Math Soc 41*332-333
◇ B40 D20 D35 ◇ REV FdM 61.62 • ID 27542

CHURCH, A. [1936] *Mathematical logic (mimeographed notes)* (X 0857) Princeton Univ Pr: Princeton iii+113pp
◇ B40 B98 D20 ◇ REV JSL 2.39 FdM 62.1048 • ID 23466

CHURCH, A. & ROSSER, J.B. [1936] *Some properties of conversion* (J 0064) Trans Amer Math Soc 39*472-482
◇ B40 D20 ◇ REV Zbl 14.385 JSL 1.74 FdM 62.37 • ID 02130

CHURCH, A. & KLEENE, S.C. [1937] *Formal definitions in the theory of ordinal numbers* (J 0027) Fund Math 28*11-21 • ERR/ADD [1959] (J 0020) Compos Math 14*40
◇ F15 ◇ REV Zbl 16.2 JSL 2.87 FdM 63.29 • ID 02135

CHURCH, A. [1938] *The constructive second number class* (J 0015) Bull Amer Math Soc 44*224-232
◇ F15 ◇ REV Zbl 18.338 JSL 3.168 FdM 64.37 • ID 02138

CHURCH, A. [1941] *The calculi of λ conversion* (X 0857) Princeton Univ Pr: Princeton 77pp
◇ B40 D20 ◇ REV MR 3.129 Zbl 26.242 JSL 17.76 JSL 6.171 • ID 02182

CHURCH, A. [1957] *Binary recursive arithmetic* (J 3941) J Math Pures Appl, Ser 9 36*39-55
◇ D20 F30 ◇ REV MR 19.239 Zbl 77.15 JSL 23.35 • ID 02402

CHURCH, A. [1965] *An independence question in recursive arithmetic* (P 0797) Fonds des Math, Machines Math & Appl;1962 Tihany 21-26
◇ F30 ◇ REV MR 33 # 45 Zbl 192.53 JSL 36.692 • ID 02150

CHURCH, A. see Vol. I, II, III, IV, V for further entries

CHWISTEK, L.B. [1912] *The principle of contradiction in the light of recent investigations of Bertrand Russell (Russian)* (J 0300) Rozpr Akad Krakow Histor-Filoz Ser 2 30*270-334
◇ A05 B15 F35 ◇ ID 04231

CHWISTEK, L.B. [1924] *The theory of constructive types (principles of logic and mathematics)* (J 0283) Ann Soc Pol Math 2*9-48
◇ A05 B15 F65 ◇ REV FdM 49.139 • ID 02670

CHWISTEK, L.B. [1929] *Neue Grundlagen der Logik und Mathematik* (J 0044) Math Z 30*704-724
◇ A05 B30 F99 ◇ REV FdM 55.626 • ID 02160

CHWISTEK, L.B. & HETPER, W. [1938] *New foundation of formal metamathematics* (J 0036) J Symb Logic 3*1-36
◇ B30 F35 ◇ REV Zbl 18.337 JSL 3.120 FdM 64.29 • ID 02162

CHWISTEK, L.B. [1939] *A formal proof of Goedel's theorem* (J 0036) J Symb Logic 4*61-68
◇ F30 ◇ REV Zbl 21.386 JSL 5.28 FdM 65.27 • ID 02165

CHWISTEK, L.B. see Vol. I, II, V for further entries

CIAMPA, S. [1970] *I numeri, l'aritmetica, gli insiemi* (J 0012) Boll Unione Mat Ital, IV Ser 3*935-952
◇ A05 F30 ◇ REV MR 43 # 21 Zbl 205.300 • ID 02168

CICHON, E.A. [1983] *A short proof of two recently discovered independence results using recursion theoretic methods* (J 0053) Proc Amer Math Soc 87*704-706
◇ F30 ◇ REV MR 84f:03049 Zbl 512 # 03028 • ID 34468

CICHON, E.A. & WAINER, S.S. [1983] *The slow-growing and the Grzegorczyk hierarchies* (J 0036) J Symb Logic 48*399-408
◇ D20 F15 ◇ REV MR 85e:03097 Zbl 567 # 03020 • ID 40697

CLAY, R.E. [1968] *The consistency of Lesniewski's mereology relative to the real number system* (J 0036) J Symb Logic 33*251-257
◇ B28 E35 E70 F25 ◇ REV MR 39 # 1321 Zbl 231 # 02030 • ID 02255

CLAY, R.E. see Vol. II, V for further entries

CLEAVE, J.P. & ROSE, H.E. [1967] *$\mathscr{E}_n$-arithmetic* (P 0691) Sets, Models & Recursion Th;1965 Leicester 297-308 • TRANSL [1974] (C 2319) Slozh Vychisl & Algor 18-32 (Russian)
◇ D20 F30 ◇ REV MR 39 # 2605 Zbl 162.315 • ID 02682

CLEAVE, J.P. [1969] *The primitive recursive analysis of ordinary differential equations and the complexity of their solutions* (J 0119) J Comp Syst Sci 3*447-455
◇ D20 D80 F60 ◇ REV MR 40 # 4512 Zbl 263 # 34001 JSL 39.345 • ID 02266

CLEAVE, J.P. see Vol. I, II, III, IV, V for further entries

CLOTE, P. [1980] *Weak partition relations, finite games, and independence results in Peano arithmetic* (P 2625) Model Th of Algeb & Arithm;1979 Karpacz 92-107
◇ C62 E05 F30 ◇ REV MR 84h:03144 Zbl 445 # 03029 • ID 56493

CLOTE, P. [1981] *Anti-basis theorems and their relation to independence results in Peano arithmetic* (P 3404) Model Th & Arithm;1979/80 Paris 115-133
◇ C62 C98 E05 F30 ◇ REV MR 83e:03095 Zbl 474 # 03021 • ID 55425

CLOTE, P. & MCALOON, K. [1983] *Two further combinatorial theorems equivalent to the 1-consistency of Peano arithmetic* (**J** 0036) J Symb Logic 48*1090-1104
◇ B28 C62 F30 ◇ REV MR 85e:03087 Zbl 545#03033
• ID 40361

CLOTE, P. [1985] *Applications of the low-basis theorem in arithmetic* (**P** 3342) Rec Th Week;1984 Oberwolfach 65-88
◇ C62 D20 D30 D55 F30 ◇ REV Zbl 574#03053
• ID 45298

CLOTE, P. [1985] *Partition relations in arithmetic* (**P** 2160) Latin Amer Symp Math Log (6);1983 Caracas 32-68
◇ C62 E05 F30 ◇ REV Zbl 567#03029 • ID 41801

CLOTE, P. see Vol. III, IV, V for further entries

COBHAM, A. [1957] *Effectively decidable theories* (**P** 1675) Summer Inst Symb Log;1957 Ithaca 391-395
◇ B25 F30 ◇ REV JSL 31.653 • ID 29385

COBHAM, A. see Vol. I, IV for further entries

COCCHIARELLA, N.B. [1974] *A new formulation of predicative second order logic* (**J** 0079) Logique & Anal, NS 17*61-87
◇ A05 B15 F65 ◇ REV MR 50#12643 Zbl 343#02010
• ID 04010

COCCHIARELLA, N.B. [1985] *Two λ-extensions of the theory of homogeneous simple types as a second-order logic* (**J** 0047) Notre Dame J Formal Log 26*377-407
◇ B15 B40 ◇ ID 47532

COCCHIARELLA, N.B. see Vol. I, II, III, V for further entries

COGAN, E.J. [1955] *A formalization of the theory of sets from the point of view of combinatory logic* (**J** 0068) Z Math Logik Grundlagen Math 1*198-240
◇ B40 E70 ◇ REV MR 17.1171 Zbl 68.11 JSL 35.147
• ID 02295

COHEN, D.E. [1984] *Modular machines, undecidability and incompleteness* (**P** 2342) Symp Rek Kombin;1983 Muenster 237-247
◇ D20 D25 D35 F30 ◇ REV Zbl 549#03028 • ID 43120

COHEN, D.E. see Vol. IV for further entries

COHEN, K.J. [1953] *A remark on Lukasiewicz's "On the intuitionistic theory of deduction"* (**J** 0028) Indag Math 15*111-112
◇ F50 ◇ REV MR 14.1053 Zbl 51.6 JSL 19.217 • ID 02298

COLLINS, G.E. & HALPERN, J.D. [1970] *On the interpretability of arithmetic in set theory* (**J** 0047) Notre Dame J Formal Log 11*477-483
◇ B28 D35 E30 F25 F30 ◇ REV MR 45#4970
Zbl 212.17 • ID 02332

COLLINS, G.E. see Vol. III, IV, V for further entries

COLLINS, W.J. [1978] *Provably recursive real numbers* (**J** 0047) Notre Dame J Formal Log 19*513-522
◇ D20 F60 ◇ REV MR 82e:03040 Zbl 305#02054
• ID 52157

COLLINS, W.J. & YOUNG, P. [1983] *Discontinuities of provably correct operators on the provably recursive real numbers* (**J** 0036) J Symb Logic 48*913-920
◇ D20 F60 ◇ REV MR 86d:03056 Zbl 546#03034
• ID 43537

COMYN, G. & WERNER, G. [1979] *Computable data types* (**P** 2059) Math Founds of Comput Sci (8);1979 Olomouc 228-236
◇ B40 ◇ REV Zbl 409#68005 • ID 56350

COMYN, G. see Vol. III, IV for further entries

CONSTABLE, R.L. [1972] *Constructive mathematics and automatic program writers* (**P** 1455) Inform Processing (5);1971 Ljubljana 229-233
◇ B75 D15 F55 ◇ REV MR 54#1696 Zbl 255#68014
• ID 61071

CONSTABLE, R.L. [1977] *A constructive programming logic* (**P** 1694) Inform Processing (7);1977 Toronto 733-738
◇ F50 ◇ REV Zbl 363#68029 • ID 31031

CONSTABLE, R.L. [1982] *Partial functions in constructive formal theories* (**P** 3862) Theor Comput Sci (6);1983 Dortmund 1-18
◇ F99 ◇ REV MR 84b:68003 Zbl 505#03027 • ID 38137

CONSTABLE, R.L. [1983] *Constructive mathematics as a programming logic I: some principles of theory* (**P** 3864) FCT'83 Found of Comput Th;1983 Borgholm 64-77
• REPR [1985] (**P** 3083) FCT'83 Found of Comput Th (Sel Pap);1983 Borgholm 21-37
◇ B75 F50 ◇ REV MR 85f:68060 Zbl 535#68009
Zbl 561#68008 • ID 38337

CONSTABLE, R.L. & MENDLER, N.P. [1985] *Recursive definitions in type theory* (**P** 4571) Log of Progr;1985 Brooklyn 61-78
◇ B15 F35 ◇ ID 49190

CONSTABLE, R.L. see Vol. I, IV for further entries

COOK, S.A. & RECKHOW, R.A. [1974] *On the length of proofs in the propositional calculus: preliminary version* (**P** 1464) ACM Symp Th of Comput (6);1974 Seattle 135-148
◇ B05 D15 F20 ◇ REV MR 54#7215 Zbl 375#02004
• ID 25005

COOK, S.A. [1975] *Feasibly constructive proofs and the propositional calculus* (**P** 1618) ACM Symp Th of Comput (7);1975 Albuquerque 83-97
◇ B05 F20 ◇ REV MR 58#19341 Zbl 357#68061
• ID 28181

COOK, S.A. & RECKHOW, R.A. [1979] *The relative efficiency of propositional proof systems* (**J** 0036) J Symb Logic 44*36-50
◇ B05 D15 F20 ◇ REV MR 80e:03007 Zbl 408#03044
• ID 56282

COOK, S.A. see Vol. IV, V for further entries

COOPER, D.C. [1972] *Theorem proving in arithmetic without multiplication* (**J** 0508) Machine Intelligence 7*91-99
◇ B25 B35 F30 ◇ REV Zbl 258#68046 • ID 61080

COOPER, D.C. see Vol. I, IV for further entries

COPPO, M. [1977] see BOEHM, C.

COPPO, M. & DEZANI-CIANCAGLINI, M. [1977] *Trees and λ-terms* (**P** 3199) CAAP'77 Arbres en Algeb & Progr (2);1977 Lille 91-120
◇ B40 ◇ REV Zbl 364#02016 • ID 50946

COPPO, M. & DEZANI-CIANCAGLINI, M. & RONCHI DELLA ROCCA, S. [1978] *(Semi)-separability of finite sets of terms in Scott's D_∞-models of the λ-calculus* (**P** 1872) Automata, Lang & Progr (5);1978 Udine 142-164
◇ B40 ◇ REV MR 80k:03018 Zbl 394#03021 • ID 52498

Author Index

COPPO, M. & DEZANI-CIANCAGLINI, M. [1978] *Syntactic and semantics trees versus type trees inside lambda-calculus* (**P** 2599) CAAP'78 Arbres en Algeb & Progr (3);1978 Lille 51-75
◇ B40 ◇ REV MR 80k:68018 Zbl 381 # 03015 • ID 51883

COPPO, M. & DEZANI-CIANCAGLINI, M. [1979] *A new type assignment for λ-terms* (**J** 0009) Arch Math Logik Grundlagenforsch 19*139-156
◇ B40 ◇ REV MR 81d:03014 Zbl 418 # 03010 • ID 53295

COPPO, M. & DEZANI-CIANCAGLINI, M. & SALLE, P. [1979] *Functional characterization of some semantic equalities inside λ-calculus* (**P** 1873) Automata, Lang & Progr (6);1979 Graz 71*133-146
◇ B40 ◇ REV MR 81d:68043 Zbl 411 # 03013 • ID 52867

COPPO, M. [1980] *An extended polymorphic type system for applicative languages* (**P** 3210) Math Founds of Comput Sci (9);1980 Rydzyna 194-204
◇ B40 B75 ◇ REV MR 81k:68003 Zbl 447 # 68021 • ID 56592

COPPO, M. & DEZANI-CIANCAGLINI, M. [1980] *An extension of basic functionality theory for λ-calculus* (**J** 0047) Notre Dame J Formal Log 21*685-693
◇ B40 ◇ REV MR 82d:03023 Zbl 423 # 03010 • ID 55932

COPPO, M. & DEZANI-CIANCAGLINI, M. & VENNERI, B.M. [1980] *Principal type schemes and lambda-calculus semantics* (**C** 3050) Essays Combin Log, Lambda Calc & Formalism (Curry) 535-560
◇ B40 ◇ REV MR 82d:03022 Zbl 469 # 03006 • ID 71932

COPPO, M. & DEZANI-CIANCAGLINI, M. & VENNERI, B.M. [1981] *Functional characters of solvable terms* (**J** 0068) Z Math Logik Grundlagen Math 27*45-58
◇ B40 ◇ REV MR 82j:03011 Zbl 479 # 03006 • ID 55668

COPPO, M. & DEZANI-CIANCAGLINI, M. & VENNERI, B.M. [1981] *Un'estensione della teoria della funzionalita di Curry* (**P** 3092) Congr Naz Logica;1979 Montecatini Terme 145-163
◇ B40 ◇ ID 48381

COPPO, M. [1983] see BARENDREGT, H.P.

COPPO, M. & GIOVANNETTI, E. [1983] *Completeness results for a polymorphic type system* (**P** 3889) CAAP'83 Arbres en Algeb & Progr (8);1983 L'Aquila 179-190
◇ B40 ◇ REV MR 86a:68064 Zbl 523 # 68011 • ID 45175

COPPO, M. [1983] *On the semantics of polymorphism* (**J** 1431) Acta Inf 20*159-170
◇ B40 ◇ REV MR 85k:03008 Zbl 515 # 03003 • ID 37836

COPPO, M. [1984] *Completeness of type assignment in continuous lambda models* (**J** 1426) Theor Comput Sci 29*309-324
◇ B40 F07 ◇ REV MR 85e:03029 Zbl 538 # 03012 • ID 40601

COPPO, M. & DEZANI-CIANCAGLINI, M. & HONSELL, F. & LONGO, G. [1984] *Extended type structures and filter lambda models* (**P** 3710) Logic Colloq;1982 Firenze 241-262
◇ B40 ◇ REV Zbl 558 # 03007 • ID 46835

CORCORAN, J. & WEAVER, G.E. [1969] *Logical consequences in modal logic: natural deduction in S5* (**J** 0047) Notre Dame J Formal Log 10*370-384
◇ B45 F07 ◇ REV MR 40 # 2524 Zbl 187.266 • REM Part I. Part II 1974 • ID 02388

CORCORAN, J. [1971] *Discourse grammars and the structure of mathematical reasoning III: Two theories of proof* (**J** 1604) J Struct Learning 3*1-24
◇ A05 A10 F07 ◇ ID 28170

CORCORAN, J. [1976] *The nature of a correct theory of proof and its value* (**P** 1792) Struct Learning;1968 Philadelphia II*195-205
◇ A05 F07 ◇ REV MR 56 # 15264 • ID 80962

CORCORAN, J. [1976] *Two theories of proof* (**P** 1792) Struct Learning;1968 Philadelphia II*207-223
◇ A05 F07 ◇ REV MR 56 # 15265 • ID 80961

CORCORAN, J. [1980] *On definitional equivalence and related topics* (**J** 2028) Hist & Phil Log 1*231-234
◇ A10 C40 F25 ◇ REV MR 83g:03014 • ID 35994

CORCORAN, J. see Vol. I, II, III for further entries

COSTA, G. [1979] see ASTESIANO, E.

COSTA, G. [1980] see ASTESIANO, E.

COSTA, G. [1983] see ASTESIANO, E.

COSTA, G. [1984] see ASTESIANO, E.

COSTA DA, N.C.A. [1957] *Considerations on the Heyting calculus (Portuguese) (English summary)* (**J** 0294) Anuar Soc Paranaense Mat 4*42-46 • ERR/ADD ibid 3*38
◇ F50 ◇ REV MR 20 # 5135 MR 24 # A43 Zbl 82.15 • ID 04200

COSTA DA, N.C.A. [1958] *Note on the logic of Brouwer-Heyting (Portuguese) (English summary)* (**J** 1831) Anuar Soc Paranaense Mat, Ser 2 1*9-10
◇ F50 ◇ REV MR 21 # 7147 Zbl 85.245 • ID 42814

COSTA DA, N.C.A. see Vol. I, II, III, V for further entries

COSTE, M. [1980] see COSTE-ROY, M.-F.

COSTE, M. see Vol. III, V for further entries

COSTE-ROY, M.-F. & COSTE, M. & MAHE, L. [1980] *Contribution to the study of the natural number object in elementary topoi* (**J** 0326) J Pure Appl Algebra 17*35-68
◇ D20 F35 F50 G30 ◇ REV MR 81h:18005 Zbl 427 # 03056 • ID 53743

COSTE-ROY, M.-F. also published under the name COSTE, M.-F.

COURVOISIER, M. & LAGASSE, J. & RICHARD, J. [1976] *Logique combinatoire. 3e ed* (**X** 0856) Dunod: Paris vi+73pp
◇ B40 B98 ◇ REV MR 57 # 15759 Zbl 376 # 02045 • ID 51687

CRABBE, M. [1976] *La predicativite dans les theories elementaires* (**J** 0079) Logique & Anal, NS 19*255-266
◇ F65 ◇ REV MR 56 # 8363 Zbl 348 # 02047 • ID 31376

CRABBE, M. [1976] *Prelogic of logoi* (**J** 0063) Studia Logica 35*219-226
◇ B40 F07 G30 ◇ REV MR 55 # 10248 Zbl 346 # 02015 • ID 31375

CRABBE, M. [1978] *Ambiguity and stratification* (**J** 0027) Fund Math 101*11-17
◇ B15 E35 E70 F25 ◇ REV MR 80h:03073 Zbl 404 # 03041 • ID 29204

CRABBE, M. see Vol. I, V for further entries

CRAIG, W. [1957] *Analysis of first-order implications* (**P** 1675) Summer Inst Symb Log;1957 Ithaca 175-180
◇ B10 F05 ◇ REV JSL 38.519 • ID 02509

CRAIG, W. [1957] *Linear reasoning. A new form of the Herbrand-Gentzen theorem* (**J** 0036) J Symb Logic 22*250-268
◇ B10 C07 C40 F05 ◇ REV MR 21 # 3317 Zbl 81.244 JSL 24.243 • ID 02510

CRAIG, W. [1957] *Three uses of the Herbrand-Gentzen theorem in relating model theory and proof theory* (**J** 0036) J Symb Logic 22*269-285
◇ B10 C07 C40 F05 ◇ REV MR 21 # 3318 Zbl 79.245 JSL 24.243 • ID 02508

CRAIG, W. [1958] *Definitional independence of combinators* (**C** 4297) Curry & Feys: Combin Logic 179-184
◇ B40 ◇ REV Zbl 81.241 • ID 30991

CRAIG, W. [1958] *Modifications and generalizations* (**C** 4297) Curry & Feys: Combin Logic 250-255
◇ B40 ◇ REV Zbl 81.241 • ID 30992

CRAIG, W. [1960] *Bases for first-order theories and subtheories* (**J** 0036) J Symb Logic 25*97-142
◇ B10 D25 F05 ◇ REV MR 24 # A1812 Zbl 108.3 JSL 37.616 • ID 02513

CRAIG, W. [1967] *Modus ponens and derivations from Horn formulas* (**J** 0068) Z Math Logik Grundlagen Math 13*33-54
◇ B20 F05 F50 ◇ REV MR 35 # 22 Zbl 165.10 • ID 02516

CRAIG, W. see Vol. I, II, III, IV, V for further entries

CRISCUOLO, G. & TORTORA, R. [1978] *Duals of intuitionistic tableaus* (**J** 2827) Ricerche Mat 27*355-365
◇ F50 ◇ REV MR 80k:03016 Zbl 419 # 03038 • ID 53381

CRISCUOLO, G. see Vol. IV for further entries

CROSSLEY, J.N. [1966] see ACZEL, P.

CROSSLEY, J.N. [1966] *Constructive order types II* (**J** 0036) J Symb Logic 31*525-538
◇ D20 D45 F15 ◇ REV MR 35 # 5318 Zbl 156.27 • REM Part I 1965. Part III 1966 by Aczel,P.H.G. & Crossley,J.N. • ID 02555

CROSSLEY, J.N. & SCHUETTE, K. [1966] *Non-uniqueness at ω^2 in Kleene's O* (**J** 0009) Arch Math Logik Grundlagenforsch 9*95-101
◇ D45 F15 ◇ REV MR 35 # 4101 Zbl 192.55 JSL 35.336 • ID 02556

CROSSLEY, J.N. [1975] *What is mathematical logic ?* (**J** 2053) Math Medley 1*3-6
◇ B98 F30 ◇ ID 31689

CROSSLEY, J.N. see Vol. I, II, III, IV, V for further entries

CSIRMAZ, L. [1979] *On definability in Peano arithmetic* (**J** 0387) Bull Sect Logic, Pol Acad Sci 8*148-153
◇ C40 C62 F30 ◇ REV MR 81b:03001 Zbl 424 # 03008 • ID 72011

CSIRMAZ, L. see Vol. I, III, IV, V for further entries

CUDA, K. [1983] *Nonstandard models of arithmetic as an alternative basis for continuum considerations* (**J** 0140) Comm Math Univ Carolinae (Prague) 24*415-430
◇ E70 F30 H15 ◇ REV MR 85g:03095 Zbl 531 # 03045 • ID 37690

CUDA, K. see Vol. I, III, V for further entries

CURIEN, P.-L. [1982] see BERRY, GERARD

CURIEN, P.-L. [1985] *Categorical combinafory logic* (**P** 4628) Automata, Lang & Progr (12);1985 Nafplion 130-139
◇ B40 G30 ◇ ID 49489

CURIEN, P.-L. [1985] *Typed categorical combinatory logic* (**P** 4627) CAAP'85 Arbres en Algeb & Progr (10);1985 Berlin 157-172
◇ B40 G30 ◇ ID 49325

CURRY, H.B. [1929] *An analysis of logical substitution* (**J** 0100) Amer J Math 51*363-384
◇ B40 ◇ REV FdM 55.33 • ID 02599

CURRY, H.B. [1930] *Grundlagen der kombinatorischen Logik* (**J** 0100) Amer J Math 52*509-536,789-834
◇ B40 ◇ REV FdM 56.48 • ID 02600

CURRY, H.B. [1931] *The universal quantifier in combinatory logic* (**J** 0120) Ann of Math, Ser 2 32*154-180
◇ B40 ◇ REV Zbl 1.261 FdM 57.57 • ID 02601

CURRY, H.B. [1932] *Some additions to the theory of combinators* (**J** 0100) Amer J Math 54*551-558
◇ B40 ◇ REV Zbl 4.387 FdM 58.69 • ID 02602

CURRY, H.B. [1933] *Apparent variables from the standpoint of combinatory logic* (**J** 0120) Ann of Math, Ser 2 34*381-404
◇ B40 ◇ REV Zbl 7.194 FdM 59.863 • ID 02603

CURRY, H.B. [1934] *Functionality in combinatory logic* (**J** 0054) Proc Nat Acad Sci USA 20*584-590
◇ B40 ◇ REV FdM 60.850 • ID 02605

CURRY, H.B. [1934] *Some properties of equality and implication in combinatory logic* (**J** 0120) Ann of Math, Ser 2 35*849-860
◇ B40 ◇ REV FdM 60.849 • ID 02604

CURRY, H.B. [1936] *First properties of functionality in combinatory logic* (**J** 0261) Tohoku Math J 41*371-401
◇ B40 ◇ REV Zbl 15.51 JSL 1.65 FdM 62.37 • ID 02739

CURRY, H.B. [1939] *A note on the reduction of Gentzen's calculus LJ* (**J** 0015) Bull Amer Math Soc 45*288-293
◇ F50 ◇ REV Zbl 20.337 JSL 4.128 FdM 65.30 • ID 02608

CURRY, H.B. [1941] *A formalization of recursive arithmetic* (**J** 0100) Amer J Math 63*263-282
• TRANSL [1970] (**C** 3626) Rekursiv Mat Analiz 3*437-461 (Russian)
◇ B28 D20 F30 ◇ REV MR 2.340 Zbl 25.5 JSL 7.42 • ID 02612

CURRY, H.B. [1941] *A revision of the fundamental rules of combinatory logic* (**J** 0036) J Symb Logic 6*41-53
◇ B40 ◇ REV MR 2.340 Zbl 60.22 JSL 6.170 • ID 04219

CURRY, H.B. [1941] *Consistency and completeness of the theory of combinators* (**J** 0036) J Symb Logic 6*54-61
◇ B40 ◇ REV MR 2.340 Zbl 60.22 JSL 6.170 • ID 04220

CURRY, H.B. [1941] *The paradox of Kleene and Rosser* (**J** 0064) Trans Amer Math Soc 50*454-516
◇ A05 B40 ◇ REV MR 3.129 Zbl 25.386 JSL 11.136 • ID 02611

CURRY, H.B. [1942] *Some advances in the combinatory theory of quantification* (**J** 0054) Proc Nat Acad Sci USA 28*564-569
◇ B40 ◇ REV MR 4.182 Zbl 69.22 JSL 8.52 • ID 02614

CURRY, H.B. [1942] *The combinatory foundations of mathematical logic* (**J** 0036) J Symb Logic 7*49-64
◇ B40 ◇ REV MR 3.289 Zbl 60.22 JSL 8.31 • ID 02615

Curry, H.B. [1942] *The inconsistency of certain formal logics* (**J** 0036) J Symb Logic 7*115–117
◇ B40 ◇ REV MR 4.125 Zbl 60.22 JSL 7.170 • ID 02613

Curry, H.B. [1949] *A simplification of the theory of combinators* (**J** 0154) Synthese 7*391–399
◇ B40 ◇ REV MR 12.384 JSL 17.76 • ID 02616

Curry, H.B. [1950] *A theory of formal deducibility* (**X** 0845) Univ Notre Dame Pr: Notre Dame ix + 126pp
◇ B10 F50 F98 ◇ REV MR 11.487 Zbl 41.348 JSL 16.56 JSL 34.113 • REM 2nd ed. 1957; 3rd ed 1966, xi + 129pp • ID 04222

Curry, H.B. [1951] *La logique combinatoire et les antinomies* (**J** 0297) Rend Mat Appl, Ser 5 10*360–370
◇ A05 B40 ◇ REV MR 14.4 Zbl 44.251 • ID 04225

Curry, H.B. [1951] *La theorie des combinateurs* (**J** 0297) Rend Mat Appl, Ser 5 10*347–359
◇ B40 ◇ REV MR 14.4 Zbl 44.250 • ID 04224

Curry, H.B. [1952] *A new proof of the Church-Rosser theorem* (**J** 0028) Indag Math 14*16–23
◇ B40 ◇ REV MR 13.715 Zbl 49.149 JSL 21.377 • ID 02619

Curry, H.B. [1952] *The inferential theory of negation* (**P** 0593) Int Congr Math (II, 6);1950 Cambridge MA 1*722
◇ A05 B20 F50 ◇ ID 28059

Curry, H.B. [1952] *The system LD* (**J** 0036) J Symb Logic 17*35–42
◇ B10 F50 ◇ REV MR 13.811 Zbl 48.2 JSL 18.266 • ID 02620

Curry, H.B. [1957] *Combinatory logic* (**P** 1675) Summer Inst Symb Log;1957 Ithaca 90–99
• REPR [1968] (**C** 0552) Phil Contemp - Chroniques 295–307
◇ B40 B98 ◇ REV Zbl 158.247 • ID 29324

Curry, H.B. & Feys, R. [1958] *Combinatory logic. Volume I* (**X** 0809) North Holland: Amsterdam xvi + 417pp
• TRANSL [1967] (**X** 1781) Tecnos: Madrid
◇ B40 B98 ◇ REV MR 20 # 817 MR 39 # 5368 Zbl 81.241 JSL 32.267 • REM With two sections by Craig,W.. 2nd ed. 1968. For vol.II 1972 see Curry,H.B. & Hindley,J.R. & Seldin,J.P. • ID 02627

Curry, H.B. [1960] *The deduction theorem in the combinatory theory of restricted generality* (**J** 0079) Logique & Anal, NS 3*15–39
◇ B40 ◇ REV JSL 35.468 • ID 02630

Curry, H.B. [1961] *Basic verifiability in the combinatory theory of restricted generality* (**C** 0622) Essays Found of Math (Fraenkel) 165–189
◇ B40 ◇ REV MR 29 # 7 Zbl 168.257 JSL 35.469 • ID 02631

Curry, H.B. [1964] *Combinatory recursive objects of all finite types* (**J** 0015) Bull Amer Math Soc 70*814–817
◇ B40 D65 ◇ REV MR 31 # 5795 Zbl 168.257 JSL 39.343 • ID 02635

Curry, H.B. [1964] *The elimination of variables by regular combinators* (**C** 1102) Critic Approach to Sci & Philos (Popper) 127–143
◇ B40 ◇ ID 21200

Curry, H.B. [1967] *Combinatory logic* (**C** 0601) Encycl of Philos 4*504–509
◇ A05 B40 B98 ◇ REV JSL 35.299 • ID 02637

Curry, H.B. [1968] *Recent advances in combinatory logic* (**J** 0082) Bull Soc Math Belg 20*288–298
◇ B40 ◇ REV MR 39 # 4007 Zbl 197.6 • ID 02638

Curry, H.B. [1969] *Modified basic functionality in combinatory logic* (**J** 0076) Dialectica 23*83–92
◇ B40 ◇ REV Zbl 257 # 02018 • ID 28992

Curry, H.B. [1969] *The undecidability of λK-conversion* (**C** 0705) Found of Math (Goedel) 10–14
◇ B40 ◇ REV MR 40 # 4118 Zbl 179.20 JSL 40.246 • ID 04284

Curry, H.B. & Hindley, J.R. & Seldin, J.P. [1972] *Combinatory logic. Vol.II* (**X** 0809) North Holland: Amsterdam xiv + 520pp
◇ B40 ◇ REV Zbl 242 # 02029 JSL 42.109 • REM Vol.I 1958 by Curry,H.B. & Feys,R. • ID 17139

Curry, H.B. [1973] *The consistency of a system of combinatory restricted generality* (**J** 0036) J Symb Logic 38*489–492
◇ B40 ◇ REV MR 49 # 2273 Zbl 274 # 02009 • ID 02644

Curry, H.B. [1975] *A study of generalized standardization in combinatory logic* (**P** 1440) ⊧ ISILC Proof Th Symp (Schuette);1974 Kiel 44–55
◇ B40 ◇ REV MR 53 # 12890 Zbl 327 # 02025 • ID 27276

Curry, H.B. [1975] *Representation of Markov algorithms by combinators* (**C** 1856) Log Enterprise 109–119
◇ B40 D03 ◇ REV Zbl 362 # 02016 • ID 50738

Curry, H.B. [1979] *On a polynomial representation of $\lambda\beta$-normal forms* (**C** 2537) Spez Wissenschaftsth 1*94–98
◇ B40 ◇ ID 33899

Curry, H.B. [1980] *Some philosophical aspects of combinatory logic* (**P** 2058) Kleene Symp;1978 Madison 85–101
◇ A05 B40 ◇ REV MR 82j:03012 Zbl 479 # 03005 • ID 55667

Curry, H.B. see Vol. I, II, IV for further entries

Czermak, J. [1977] *A remark on Gentzen's calculus of sequents* (**J** 0047) Notre Dame J Formal Log 18*471–474
◇ B20 F07 ◇ REV MR 58 # 21500 Zbl 314 # 02026 • ID 24249

Czermak, J. see Vol. II, V for further entries

Daalen van, D.T. [1980] *The language theory of Automath (Dissertation)* (**X** 3152) Techn Hogeschool: Eindhoven vi + 309pp
◇ B40 ◇ REV MR 81h:68092 Zbl 422 # 68045 • ID 53508

Dahn, B.I. [1981] *Partial isomorphisms and intuitionistic logic* (**J** 0063) Studia Logica 40*405–413
◇ C75 C90 F50 ◇ REV MR 84m:03012 Zbl 496 # 03011 • ID 35717

Dahn, B.I. see Vol. II, III, IV, V for further entries

Dalen van, D. [1958] see Bar-Hillel, Y.

Dalen van, D. [1963] *Extension problems in intuitionistic plane projective geometry I,II* (**J** 0028) Indag Math 25*349–368,369–383
◇ F55 ◇ REV MR 27 # 3531 Zbl 127.113 Zbl 132.406 • ID 02757

DALEN VAN, D. [1968] *A note on spread-cardinals* (**J** 0020) Compos Math 20*21-28
• REPR [1968] (**C** 0727) Logic Found of Math (Heyting) 21-28
◇ E10 F55 ◇ REV MR 37 # 1230 Zbl 175.263 • ID 02759

DALEN VAN, D. [1968] *Fans generated by nondeterministic automata* (**J** 0068) Z Math Logik Grundlagen Math 14*273-278
◇ D05 F55 ◇ REV MR 38 # 1961 Zbl 165.304 • ID 02761

DALEN VAN, D. & DYKMAN, J.G. & KLEENE, S.C. & TROELSTRA, A.S. (EDS.) [1968] *Logic and foundations of mathematics. Papers dedicated to A.Heyting on the occasion of his 70th birthday* (**X** 0812) Wolters-Noordhoff : Groningen 249pp
◇ F97 ◇ REV Zbl 169.293 • REM Also published as J0028 Vol. 20 • ID 48628

DALEN VAN, D. [1968] *Reducibilities in intuitionistic topology* (**J** 0036) J Symb Logic 33*412-417
◇ F55 ◇ REV MR 38 # 2010 Zbl 164.308 • ID 02760

DALEN VAN, D. & TROELSTRA, A.S. [1970] *Projections of lawless sequences* (**P** 0603) Intuitionism & Proof Th;1968 Buffalo 163-186
◇ F35 F50 ◇ REV MR 43 # 26 Zbl 205.5 • REM Part I. Part II 1979 by Hoeven van der,G.F. & Troelstra,A.S. • ID 02763

DALEN VAN, D. & GORDON, C.E. [1971] *Independence problems in subsystems of intuitionistic arithmetic* (**J** 0028) Indag Math 33*448-456
◇ F30 F50 ◇ REV MR 46 # 3272 Zbl 229 # 02030 • ID 04169

DALEN VAN, D. [1973] *Lectures on intuitionism* (**P** 0713) Cambridge Summer School Math Log;1971 Cambridge GB 1-94
◇ F50 F55 F98 ◇ REV MR 49 # 2271 Zbl 272 # 02035 • ID 04061

DALEN VAN, D. [1974] *A model for HAS. A topological interpretation of the theory of species of natural numbers* (**J** 0027) Fund Math 82*167-174
◇ F35 F50 ◇ REV MR 50 # 12671 Zbl 299 # 02038 • ID 02764

DALEN VAN, D. [1977] *The use of Kripke's schema as a reduction principle* (**J** 0036) J Symb Logic 42*238-240
◇ F35 F50 ◇ REV MR 58 # 157 Zbl 382 # 03042 • ID 26450

DALEN VAN, D. [1978] *An interpretation of intuitionistic analysis* (**J** 0007) Ann Math Logic 13*1-43
◇ C90 F35 F50 ◇ REV MR 80i:03067 Zbl 399 # 03049 • ID 27959

DALEN VAN, D. [1978] *Brouwer: the genesis of his intuitionism* (**J** 0076) Dialectica 32*291-303
◇ A05 A10 F99 ◇ REV MR 81j:03004 • ID 79612

DALEN VAN, D. [1978] *Philosophical foundations of mathematics (Dutch)* (**X** 1994) Roy van Gorcum: Assen viii + 122pp
◇ A05 F98 ◇ ID 39633

DALEN VAN, D. & STATMAN, R. [1979] *Equality in the presence of apartness* (**P** 1705) Scand Logic Symp (4);1976 Jyvaeskylae 95-116
◇ F05 F50 ◇ REV MR 80h:03089 Zbl 414 # 03036 • ID 53082

DALEN VAN, D. [1979] *Interpreting intuitionistic logic* (**P** 1947) Bicent Congr Wisk Genootschap;1978 Amsterdam 1*133-148
◇ A05 F50 F98 ◇ REV MR 80m:03093 Zbl 418 # 03038 • ID 53323

DALEN VAN, D. [1981] *Brouwer and the solipsistic conception of the world (Dutch)* (**J** 1829) Algem Nederl Tijdschr Wijsbeg 73*1-19
◇ A05 A10 F99 ◇ ID 39616

DALEN VAN, D. [1982] *Braucht die konstruktive Mathematik Grundlagen?* (**J** 0157) Jbuchber Dtsch Math-Ver 84*57-78
◇ A05 F98 ◇ REV MR 84c:03098 Zbl 523 # 03004 • ID 34032

DALEN VAN, D. & LODDER, J.S. [1982] *Lawlessness and independence* (**P** 3638) Brouwer Centenary Symp;1981 Noordwijkerhout 297-309
◇ F35 F50 ◇ REV MR 85e:03148 Zbl 523 # 03044 • ID 37032

DALEN VAN, D. [1982] *Singleton reals* (**P** 3623) Logic Colloq;1980 Prague 83-94
◇ F55 ◇ REV MR 84h:03131 Zbl 494 # 03040 • ID 34318

DALEN VAN, D. [1982] *The creative subject and Heyting's arithmetic* (**P** 3831) Universal Algeb & Appl;1978 Warsaw 379-382
◇ F30 F55 ◇ REV MR 85f:03061 Zbl 518 # 03025 • ID 37522

DALEN VAN, D. [1982] see TROELSTRA, A.S.

DALEN VAN, D. [1984] *Four letters from Edmund Husserl to Hermann Weyl* (**C** 4464) Husserl Stud 1*1-12
◇ A10 F99 ◇ ID 39623

DALEN VAN, D. [1984] *How to glue analysis models* (**J** 0036) J Symb Logic 49*1339-1349
◇ C65 C90 F35 F50 ◇ REV MR 86e:03056 Zbl 576 # 03038 • ID 39878

DALEN VAN, D. [1985] *A comment on the essay "The fundamental theorem of algebra and intuitionism"* (**J** 0009) Arch Math Logik Grundlagenforsch 25*43-44
◇ F55 ◇ REM For the article see Kneser,H. 1940 • ID 49118

DALEN VAN, D. see Vol. I, II, III, IV, V for further entries

DALLA CHIARA SCABIA, M.L. [1968] *Modelli sintattici e semantici delle teorie elementari* (**X** 0844) Feltrinelli: Milano 240pp
◇ C07 C98 E35 E45 F25 F98 ◇ REV MR 39 # 5308 Zbl 202.8 JSL 39.959 JSL 40.236 • ID 04285

DALLA CHIARA SCABIA, M.L. see Vol. I, II, V for further entries

DAMM, W. & FEHR, E. & INDERMARK, K. [1978] *Higher type recursion and self-application as control structures* (**P** 2929) Form Descr of Progr Concepts (1);1977 St.Andrews 461-489
◇ B40 B75 D20 ◇ REV MR 80e:68014 Zbl 373 # 68021 • ID 51522

DAMM, W. & FEHR, E. [1978] *On the power of self-application and higher type recursion* (**P** 1872) Automata, Lang & Progr (5);1978 Udine 177-191
◇ B40 ◇ REV MR 80g:68106 Zbl 391 # 68041 • ID 52365

DAMM, W. see Vol. IV for further entries

DANCY, J. [1982] *Intuitionism in meta-epistemology* (**J** 0095) Philos Stud 42*395-408
◇ A05 F99 ◇ ID 45863

Author Index

DANKO, W. [1983] *Interpretability of algorithmic theories* (J 2095) Fund Inform, Ann Soc Math Pol, Ser 4 6*217-233 ◇ B25 B75 C35 F25 ◇ REV MR 84h:68022 Zbl 559 # 68048 • ID 39522

DANKO, W. see Vol. II, III, IV for further entries

DANTSIN, E.YA. [1976] *On an approximative version of the notion of constructive analytic function (Russian) (English summary)* (S 0228) Zap Nauch Sem Leningrad Otd Mat Inst Steklov 60*49-58,222
• TRANSL [1980] (J 1531) J Sov Math 14*1457-1463 ◇ F60 ◇ REV MR 58 # 27378 Zbl 345 # 02024 • ID 56732

DANTSIN, E.YA. [1981] *Two systems for proving tautologies based on the splitting method (Russian) (English summary)* (S 0228) Zap Nauch Sem Leningrad Otd Mat Inst Steklov 105*24-44,198-199
• TRANSL [1983] (J 1531) J Sov Math 22*1293-1305 ◇ B35 D15 F20 ◇ REV MR 83i:68140 Zbl 476 # 03020 Zbl 509 # 03004 • ID 55532

DANTSIN, E.YA. see Vol. I, IV for further entries

DANTZIG VAN, D. [1942] *A remark and a problem concerning the intuitionistic form of Cantor's intersection theorem* (J 0028) Indag Math 4*147-148 ◇ E20 F55 ◇ REV MR 6.31 Zbl 27.97 JSL 9.21 • ID 02772

DANTZIG VAN, D. [1942] *On the affirmative content of Peano's theorem on differential equations* (J 0028) Indag Math 6*140-146 ◇ F55 ◇ REV MR 6.30 Zbl 27.104 • ID 02771

DANTZIG VAN, D. [1947] *On the principles of intuitionistic and affirmative mathematics* (J 0028) Indag Math 9*429-440 ◇ F55 ◇ REV MR 9.221 Zbl 35.149 JSL 13.173 • ID 02774

DANTZIG VAN, D. [1949] *Comments on Brouwer's theorem on essentially-negative predicates* (J 0028) Indag Math 11*347-355 ◇ A05 F55 ◇ REV MR 11.412 Zbl 38.6 JSL 21.195 • ID 02775

DANTZIG VAN, D. [1951] *Mathematique stable et mathematique affirmative* (P 0642) Congr Int Phil Sci;1949 Paris 2*123-135 ◇ A05 F55 ◇ REV MR 13.898 JSL 21.322 • ID 02776

DANTZIG VAN, D. see Vol. I, II for further entries

DAPUETO, C. [1982] *Sull' aritmetizzazione della sintassi del primo ordine (French summary)* (J 0549) Riv Mat Univ Parma, Ser 4 8*361-371 ◇ B28 F30 F40 ◇ REV MR 85e:03146 Zbl 524 # 03009 • ID 45889

DAPUETO, C. see Vol. III, V for further entries

DARDZHANIYA, G.K. [1976] *On a variant of classical sequent calculus (Russian) (English summary)* (J 0233) Soobshch Akad Nauk Gruz SSR 84*281-284 ◇ F07 ◇ REV MR 55 # 2520 Zbl 355 # 02020 • ID 50218

DARDZHANIYA, G.K. [1977] *Intuitionistic system without contraction* (J 0387) Bull Sect Logic, Pol Acad Sci 6*2-8 ◇ B55 F50 ◇ REV MR 55 # 12476 Zbl 416 # 03030 • ID 53196

DARDZHANIYA, G.K. [1979] *On the complexity of countermodels for intuitionistic propositional calculus (Russian) (English summary)* (J 0233) Soobshch Akad Nauk Gruz SSR 95*17-20 ◇ B55 F50 ◇ REV MR 81d:03012 Zbl 417 # 03032 • ID 53268

DARDZHANIYA, G.K. [1979] *Polynomial complexities of deduction of some logical calculi (Russian) (English summary)* (J 0288) Vest Ser Mat Mekh, Univ Moskva 1979/3*10-18,84
• TRANSL [1979] (J 0510) Moscow Univ Math Bull 34/3*8-17 ◇ B35 F20 ◇ REV MR 81a:03055 Zbl 413 # 03032 • ID 53026

DARDZHANIYA, G.K. see Vol. II for further entries

DASSEN, C.C. [1936] *Sobre una objecion a la Logica Brouweriana* (J 1393) Ann Acad Nac Cienc Exact Fis Nat Buenos Aires 4*7-20 ◇ A05 F50 ◇ REV JSL 6.106 • ID 41442

DAVIS, MARTIN D. [1965] *Introduction to Goedel's paper: On formally undecidable propositions of the Principia Mathematica and related systems I* (C 0718) The Undecidable 4 ◇ A10 D20 F30 ◇ REV JSL 31.484 • REM Goedel's paper was published in J0124 38(1937)*173-198 • ID 04186

DAVIS, MARTIN D. (ED.) [1965] *The undecidable. Basic papers on undecidable propositions, unsolvable problems and computable functions* (X 0887) Raven Pr: New York 440pp ◇ D20 D25 D35 D97 F30 ◇ REV MR 32 # 7412 • ID 25888

DAVIS, MARTIN D. see Vol. I, II, IV for further entries

DAWES, A.M. & FLORENCE, J.B. [1975] *Independent Goedel sentences and independent sets* (J 0036) J Symb Logic 40*159-166 ◇ D25 F30 ◇ REV MR 52 # 10383 Zbl 324 # 02027 • ID 04287

DAWES, A.M. see Vol. III, IV, V for further entries

DAWSON JR., J.W. [1979] *The Goedel incompleteness theorem from a length-of-proof perspective* (J 0005) Amer Math Mon 86*740-748 ◇ A10 F20 F30 H15 ◇ REV MR 80i:03001 Zbl 426 # 03061 • ID 53657

DAWSON JR., J.W. see Vol. IV, V for further entries

DEGTYAREV, A.I. [1976] *A class of algorithmic operators (Russian)* (S 0554) Issl Teor Algor & Mat Logik (Moskva) 2*100-111,159 ◇ F60 ◇ REV MR 58 # 16208 • ID 72232

DEGTYAREV, A.I. see Vol. I for further entries

DEJON, B. & HENRICI, P. (EDS.) [1969] *Constructive aspects of the fundamental theorem of algebra: proceedings of a symposium, IBM* (X 0827) Wiley & Sons: New York 337pp ◇ B28 C97 F97 ◇ REV Zbl 175.1 • ID 23438

DEKHTYAR', M.I. [1979] *Complexity spectra of recursive sets of approximability of initial segments of complete problems* (J 0129) Elektr Informationsverarbeitung & Kybern 15*11-32 ◇ D10 D15 D30 F20 F30 ◇ REV MR 80k:03038 Zbl 412 # 03021 • ID 52952

DEKHTYAR', M.I. see Vol. IV for further entries

DELON, F. [1978] *Definition de l'arithmetique dans la theorie des anneaux de series formelles (English summary)* (J 2313) C R Acad Sci, Paris, Ser A-B 286*A87-A89
◇ C60 D35 F25 F35 ◇ REV MR 82a:12018 Zbl 382#13010 • ID 27313

DELON, F. see Vol. I, III, IV for further entries

DELZELL, C.N. [1982] *Case distinctions are necessary for representing polynomials as sums of squares* (P 3708) Herbrand Symp Logic Colloq;1981 Marseille 87-103
◇ A10 C60 C65 F55 ◇ REV Zbl 502#03032 • ID 36898

DELZELL, C.N. [1982] *Continuous sums of squares of forms* (P 3638) Brouwer Centenary Symp;1981 Noordwijkerhout 65-75
◇ C60 C65 F55 ◇ REV MR 85g:03086 Zbl 527#10016 • ID 38556

DELZELL, C.N. [1984] *A continuous, constructive solution to Hilbert's 17th problem* (J 0305) Invent Math 76*365-384
◇ C60 F55 ◇ REV MR 86e:12003 Zbl 547#12017 • ID 43271

DEMILLO, R.A. & LIPTON, R.J. [1979] *A constructive generalization of the Borel-Cantelli lemma with application to the complexity of infinite strings* (J 0041) Math Syst Theory 13*95-104
◇ F65 ◇ REV MR 81e:68058 Zbl 434#03042 • ID 55746

DEMILLO, R.A. & LIPTON, R.J. [1979] *Some connections between mathematical logic and complexity theory* (P 3542) ACM Symp Th of Comput (11);1979 Atlanta 153-159
◇ C57 D15 F30 H15 ◇ REV MR 81h:03084 • ID 72280

DEMILLO, R.A. see Vol. IV for further entries

DEMUTH, O. [1965] *On Lebesgue integration in constructive analysis (Russian)* (J 0023) Dokl Akad Nauk SSSR 160*1239-1241
• TRANSL [1965] (J 0062) Sov Math, Dokl 6*280-282
◇ F60 ◇ REV MR 31#54 Zbl 224#02022 • ID 42977

DEMUTH, O. [1967] *Lebesgue integration in constructive analysis (Russian)* (S 0228) Zap Nauch Sem Leningrad Otd Mat Inst Steklov 4*30-43
• TRANSL [1967] (J 0521) Semin Math, Inst Steklov 4*9-14
◇ F60 ◇ REV MR 40#7112 Zbl 224#02022 • ID 04233

DEMUTH, O. [1967] *Necessary and sufficient conditions for Riemann integrability of constructive functions (Russian)* (J 0023) Dokl Akad Nauk SSSR 176*757-758
• TRANSL [1967] (J 0062) Sov Math, Dokl 8*1176-1177
◇ F60 H05 ◇ REV MR 37#65 Zbl 224#02020 • ID 02915

DEMUTH, O. [1968] *Fubini's theorem for the Riemann integral in constructive mathematics (Russian)* (J 0140) Comm Math Univ Carolinae (Prague) 9*677-686
◇ F60 ◇ REV MR 39#5768a Zbl 177.16 • ID 02928

DEMUTH, O. [1968] *The connection between Riemann and Lebesgue integrability of constructive functions (Russian)* (S 0228) Zap Nauch Sem Leningrad Otd Mat Inst Steklov 8*29-31
◇ F60 H05 ◇ REV MR 39#377 Zbl 172.299 • ID 28592

DEMUTH, O. [1968] *The Lebesgue integral and the concept of measureability of functions in constructive analysis (Russian)* (S 0228) Zap Nauch Sem Leningrad Otd Mat Inst Steklov 8*21-28
• TRANSL [1968] (J 0521) Semin Math, Inst Steklov 8*7-10
◇ F60 H05 ◇ REV MR 39#1605 Zbl 224#02024 • ID 16949

DEMUTH, O. [1969] *Linear functionals in the constructive spaces L_r (Russian)* (J 0140) Comm Math Univ Carolinae (Prague) 10*357-390
◇ F60 ◇ REV MR 41#5957 Zbl 258#02042 • ID 02918

DEMUTH, O. [1969] *Note on the paper "Fubini's theorem for the Riemann integral in constructive mathematics" (Russian)* (J 0140) Comm Math Univ Carolinae (Prague) 10*115-120
◇ F60 ◇ REV MR 39#5768a Zbl 224#02021 • ID 02929

DEMUTH, O. [1969] *The differentiability of constructive functions (Russian)* (J 0140) Comm Math Univ Carolinae (Prague) 10*167-175
◇ F60 ◇ REV MR 41#8229 Zbl 224#02019 • ID 02917

DEMUTH, O. [1969] *The spaces L_r and S in constructive mathematics (Russian)* (J 0140) Comm Math Univ Carolinae (Prague) 10*261-284
◇ F60 ◇ REV MR 39#6737 Zbl 258#02041 • ID 02930

DEMUTH, O. [1969] *The Lebesgue measurability of sets in constructive mathematics (Russian)* (J 0140) Comm Math Univ Carolinae (Prague) 10*463-492
◇ F60 ◇ REV MR 41#3697 Zbl 237#02008 • ID 02916

DEMUTH, O. [1970] *Mittelwertsaetze fuer konstruktive Lebesgueintegrale (Russisch)* (J 0140) Comm Math Univ Carolinae (Prague) 11*249-269
◇ F60 ◇ REV MR 43#4992 Zbl 257#02024 • ID 29010

DEMUTH, O. [1970] *Necessary and sufficient conditions for the absolute continuity of constructive functions (Russian)* (J 0140) Comm Math Univ Carolinae (Prague) 11*705-726
◇ F60 ◇ REV MR 43#6092 Zbl 257#02026 • ID 14503

DEMUTH, O. [1970] *The integrability of derivatives of constructive functions (Russian)* (J 0140) Comm Math Univ Carolinae (Prague) 11*667-691
◇ F60 ◇ REV MR 44#3869 Zbl 257#02027 • ID 02920

DEMUTH, O. [1970] *The representability of constructive functions of weakly bounded variation (Russian)* (J 0140) Comm Math Univ Carolinae (Prague) 11*421-434
◇ F60 ◇ REV MR 42#4397 Zbl 257#02025 • ID 02922

DEMUTH, O. [1971] *A certain condition for the differentiability of constructive functions of bounded variation (Russian)* (J 0140) Comm Math Univ Carolinae (Prague) 12*687-711
◇ F60 ◇ REV MR 45#3648 Zbl 255#02032 • ID 28963

DEMUTH, O. [1971] *A necessary and sufficient condition for the representability of constructive functions in the form of the sum of a singular and an absolutely continuous function (Russian)* (J 0140) Comm Math Univ Carolinae (Prague) 12*587-610
◇ F60 ◇ REV MR 45#2114 Zbl 264#02030 • ID 02707

DEMUTH, O. [1971] *The superposition of absolutely continuous functions (Russian)* (J 0140) Comm Math Univ Carolinae (Prague) 12*423-451
◇ F60 ◇ REV MR 45#6986 Zbl 264#02029 • ID 26104

DEMUTH, O. [1972] *A necessary and sufficient condition for the representability of constructive functions in the form of a superposition of absolutely continuous functions (Russian)* (J 0140) Comm Math Univ Carolinae (Prague) 13*227-251
◇ F60 ◇ REV MR 47#3146 Zbl 264#02031 • ID 25926

DEMUTH, O. [1972] *A sufficient condition for the representability of constructive functions in the form of the sum of two superpositions of absolutely continuous functions (Russian)* (J 0140) Comm Math Univ Carolinae (Prague) 13*265-282
◇ F60 ◇ REV MR 58#27379 Zbl 264#02032 • ID 27505

DEMUTH, O. [1973] *The constructive analogue of the connection between the Lebesgue measurability of sets and of functions (Russian)* (J 0140) Comm Math Univ Carolinae (Prague) 14*377-396
◇ F60 ◇ REV MR 48 # 6357 Zbl 272 # 02058 • ID 29785

DEMUTH, O. & NEMECKOVA, L. [1973] *The constructive analogue of the property (T_1) (Russian)* (J 0140) Comm Math Univ Carolinae (Prague) 14*421-439
◇ F60 ◇ REV MR 53 # 12898 Zbl 278 # 02031 • ID 23149

DEMUTH, O. & NEMECKOVA, L. [1973] *The constructive analogues of the properties (N) and (S) (Russian)* (J 0140) Comm Math Univ Carolinae (Prague) 14*565-582
◇ F60 ◇ REV MR 49 # 506 Zbl 282 # 02008 • ID 04075

DEMUTH, O. [1973] *The representability of uniformly continuous constructive functions (Russian)* (J 0140) Comm Math Univ Carolinae (Prague) 14*7-25
◇ F60 ◇ REV MR 49 # 2302 Zbl 264 # 02033 • ID 04077

DEMUTH, O. [1974] *The connection between the representablility of a constructive function in the form of a superposition of two absolutely continuous functions and the differentiability of this function (Russian)* (J 0140) Comm Math Univ Carolinae (Prague) 15*195-210
◇ F60 ◇ REV MR 50 # 70 Zbl 286 # 02041 • ID 04078

DEMUTH, O. [1974] *The representability of constructive functions possessing the properties (S) and (T_1) in the form of superpositions (Russian)* (J 0140) Comm Math Univ Carolinae (Prague) 15*49-64
◇ F60 ◇ REV MR 52 # 2841 Zbl 286 # 02040 • ID 17632

DEMUTH, O. [1975] *Constructive pseudonumbers (Russian)* (J 0140) Comm Math Univ Carolinae (Prague) 16*315-331
◇ D55 F60 ◇ REV MR 52 # 2844 Zbl 319 # 02029 • ID 17635

DEMUTH, O. [1975] *The differentiability of constructive functions of weakly bounded variation on pseudo-numbers (Russian)* (J 0140) Comm Math Univ Carolinae (Prague) 16*583-599
◇ F60 ◇ REV MR 57 # 16005 Zbl 325 # 02023 • ID 30438

DEMUTH, O. [1976] *The constructive analogue of the Denjoy-Young theorem on derived numbers (Russian)* (J 0140) Comm Math Univ Carolinae (Prague) 17*111-126
◇ F60 ◇ REV MR 53 # 7736 Zbl 328 # 02018 • ID 22983

DEMUTH, O. [1976] *The domains of definition of effective operators over general recursive functions and of constructive functions of a real variable (Russian)* (J 0140) Comm Math Univ Carolinae (Prague) 17*633-646
◇ D20 F60 ◇ REV MR 55 # 7755 Zbl 345 # 02025 • ID 29783

DEMUTH, O. [1977] *A constructive analogue of the functions of the n-th Baire class* (J 0140) Comm Math Univ Carolinae (Prague) 18*231-245
◇ D55 E15 F60 ◇ REV MR 57 # 5701 Zbl 359 # 02031 • ID 72287

DEMUTH, O. [1977] *On a generalization of the constructive Lebesgue integral (Russian)* (J 0140) Comm Math Univ Carolinae (Prague) 18*499-514
◇ F60 ◇ REV MR 56 # 15377 Zbl 394 # 03052 • ID 52529

DEMUTH, O. & KRYL, R. & KUCERA, A. [1978] *An application of the theory of functionals partial recursive relative to number sets in constructive mathematics (Russian)* (J 0165) Acta Univ Carolinae Math Phys (Prague) 19/1*15-60
◇ D30 D55 F60 ◇ REV MR 80a:03071 Zbl 394 # 03054 • ID 31808

DEMUTH, O. [1978] *Constructive analogues of generalized absolutely continuous functions of generalized bounded variation (Russian)* (J 0140) Comm Math Univ Carolinae (Prague) 19*471-487
◇ F60 ◇ REV MR 80b:03100 Zbl 394 # 03053 • ID 52530

DEMUTH, O. [1978] *Some questions of the theory of constructive functions of a real variable (Russian) (English summary)* (J 0165) Acta Univ Carolinae Math Phys (Prague) 19/1*61-96
◇ F60 ◇ REV MR 81g:03068 Zbl 394 # 03055 • ID 52532

DEMUTH, O. [1978] *The pseudodifferentiability of uniformly continuous constructive functions on constructive real numbers (Russian)* (J 0140) Comm Math Univ Carolinae (Prague) 19*319-333
◇ F60 ◇ REV MR 58 # 10352 Zbl 399 # 03051 • ID 52847

DEMUTH, O. [1979] *Constructive Denjoy integrals (Russian)* (J 0140) Comm Math Univ Carolinae (Prague) 20*213-227
◇ F60 ◇ REV MR 80k:03065 Zbl 414 # 03039 • ID 53085

DEMUTH, O. & KUCERA, A. [1979] *Remarks on constructive mathematical analysis* (P 2627) Logic Colloq;1978 Mons 81-129
◇ F60 ◇ REV MR 81g:03072 Zbl 437 # 03030 • ID 55898

DEMUTH, O. & POLIVKA, I. [1979] *The representation of linear functionals in the space of codes of constructive functions that are uniformly continuous on the segment 0Δ1 (Russian)* (J 0140) Comm Math Univ Carolinae (Prague) 20*765-780
◇ F60 ◇ REV MR 81h:03113 Zbl 428 # 03054 • ID 53813

DEMUTH, O. [1979] *The use of the Riemann-Stieltjes integral in the theory of the constructive Lebesgue integral, and its generalizations (Russian)* (J 0140) Comm Math Univ Carolinae (Prague) 20*781-793
◇ F60 ◇ REV MR 81h:03114 Zbl 431 # 03038 • ID 53944

DEMUTH, O. [1980] *A constructive analogue of K.M.Garg's theorem on derivative numbers (Russian)* (J 0140) Comm Math Univ Carolinae (Prague) 21*457-472
◇ F60 ◇ REV MR 83a:03064 Zbl 485 # 03034 • ID 35074

DEMUTH, O. [1980] *On constructive Perron integral (Russian) (English summary)* (J 0165) Acta Univ Carolinae Math Phys (Prague) 21/1*3-57
◇ F60 ◇ REV MR 82h:26013 Zbl 448 # 03044 • ID 56644

DEMUTH, O. & KMINEK, V. [1980] *On properties of indefinite constructive Lebesgue-Stieltjes integrals (Russian)* (J 0140) Comm Math Univ Carolinae (Prague) 21*629-644
◇ F60 ◇ REV MR 82d:03097 Zbl 479 # 03028 • ID 55690

DEMUTH, O. [1980] *Pseudodifferentiability of constructive functions on constructive real numbers (Russian)* (J 0140) Comm Math Univ Carolinae (Prague) 21*489-505
◇ F60 ◇ REV MR 83a:03065 Zbl 499 # 03047 • ID 35075

DEMUTH, O. [1981] *A generalization of a constructive analogue of K.M. Garg's theorem (Russian)* (J 0140) Comm Math Univ Carolinae (Prague) 22*607-620
◇ F60 ◇ REV MR 83e:03104b Zbl 482 # 03028 • ID 35251

DEMUTH, O. [1981] *On the pseudodifferentiability of uniformly continuous constructive functions from functions of the same type (Russian)* (**J** 0140) Comm Math Univ Carolinae (Prague) 22*497-512
◇ F60 ◇ REV MR 83e:03104a Zbl 482 # 03027 • ID 35250

DEMUTH, O. [1982] *Borel types of some classes of arithmetical real numbers (Russian)* (**J** 0140) Comm Math Univ Carolinae (Prague) 23*593-606
◇ F60 ◇ REV MR 84c:03104 Zbl 525 # 03040 • ID 34038

DEMUTH, O. [1982] *Some classes of arithmetical real numbers (Russian)* (**J** 0140) Comm Math Univ Carolinae (Prague) 23*453-465
◇ F60 ◇ REV MR 84g:03101 Zbl 519 # 03046 • ID 34205

DEMUTH, O. [1983] *Arithmetic complexity of differentiation in constructive mathematics (Russian)* (**J** 0140) Comm Math Univ Carolinae (Prague) 24*301-316
◇ D55 F60 ◇ REV MR 85i:03178 • ID 44288

DEMUTH, O. [1983] *On the pseudodifferentiability of pseudo-uniformly continuous constructive functions from functions of the same type (Russian)* (**J** 0140) Comm Math Univ Carolinae (Prague) 24*391-406
◇ F60 ◇ REV MR 85i:03179 • ID 44290

DENNIS-JONES, E.C. & WAINER, S.S. [1984] *Subrecursive hierarchies via direct limits* (**P** 2153) Logic Colloq;1983 Aachen 2*117-128
◇ D20 F15 G30 ◇ ID 45385

DENTON JR., J.S. & DREBEN, B. [1966] *A supplement to Herbrand* (**J** 0036) J Symb Logic 31*393-398
◇ F05 F07 ◇ REV MR 34 # 4107 Zbl 221 # 02012 JSL 32.521 • ID 02951

DENTON JR., J.S. & DREBEN, B. [1970] *Herbrand-style consistency proofs* (**P** 0603) Intuitionism & Proof Th;1968 Buffalo 419-434
◇ F05 F30 F35 ◇ REV MR 43 # 3090 Zbl 224 # 02016 • ID 03133

DEQUOY, N. [1949] *La geometrie projective plane en mathematique intuitionniste sans negation* (**J** 0109) C R Acad Sci, Paris 228*1098-1100
◇ F55 ◇ REV MR 10.499 • ID 02953

DEQUOY, N. [1950] *Expose d'un type de raisonnement en mathematique intuitionniste sans negation et resultats obtenus pour la geometrie projective plane* (**J** 0109) C R Acad Sci, Paris 230*357-359
◇ F55 ◇ REV MR 11.305 Zbl 36.101 • ID 41683

DEQUOY, N. [1955] *Axiomatique intuitionniste sans negation de la geometrie projective* (**X** 0834) Gauthier-Villars: Paris 108pp
◇ F55 ◇ REV MR 16.783 Zbl 64.12 JSL 20.174 • ID 02954

DEREVYANKINA, E.A. [1974] *Some problems of completeness of arithmetic (Russian)* (**J** 0092) Sib Mat Zh 15*299-317,460 • TRANSL [1974] (**J** 0475) Sib Math J 15*210-223
◇ F15 F30 ◇ REV MR 50 # 67 Zbl 281 # 02050 • ID 04082

DESTOUCHES, J.-L. [1951] *Sur la mecanique classique et l'intuitionnisme* (**J** 0028) Indag Math 13*74-79
◇ F55 ◇ REV MR 12.792 JSL 34.307 • ID 02959

DESTOUCHES, J.-L. see Vol. I, II, IV for further entries

DESTOUCHES-FEVRIER, P. [1945] *Rapport entre le calcul des problemes et le calcul des propositions* (**J** 0109) C R Acad Sci, Paris 220*484-486
◇ A05 B05 F50 ◇ REV MR 7.185 Zbl 61.471 JSL 12.133 • ID 02962

DESTOUCHES-FEVRIER, P. [1947] *Esquisse d'une mathematique intuitionniste positive* (**J** 0109) C R Acad Sci, Paris 225*1241-1243
◇ F55 ◇ REV MR 9.261 Zbl 35.149 JSL 13.163 • ID 02963

DESTOUCHES-FEVRIER, P. [1948] *Le calcul des constructions* (**J** 0109) C R Acad Sci, Paris 227*1192-1193
◇ F50 ◇ REV MR 10.585 • ID 02964

DESTOUCHES-FEVRIER, P. [1948] *Logique de l'intuitionisme sans negation et logique de l'intuitionisme positif* (**J** 0109) C R Acad Sci, Paris 226*38-39
◇ B55 F50 ◇ REV MR 10.94 Zbl 35.149 • ID 02965

DESTOUCHES-FEVRIER, P. [1949] *Connexion entre les calculs des constructions, des problemes, des propositions* (**J** 0109) C R Acad Sci, Paris 228*31-33
◇ A05 F50 ◇ REV MR 10.585 Zbl 36.8 • ID 02966

DESTOUCHES-FEVRIER, P. [1951] *Sur l'intuitionisme et la conception strictement constructive* (**J** 0028) Indag Math 13*80-86
◇ F50 ◇ REV MR 12.792 Zbl 43.250 JSL 34.306 • ID 02968

DESTOUCHES-FEVRIER, P. also published under the name FEVRIER, P.

DESTOUCHES-FEVRIER, P. see Vol. II for further entries

DETLEFSEN, M. [1980] *On a theorem of Feferman* (**J** 0095) Philos Stud 38*129-140
◇ A05 F25 F30 ◇ REV MR 81g:03002 • ID 72314

DETLEFSEN, M. see Vol. I for further entries

DEUTSCH, M. [1975] *Zur Benutzung der Verkettung als Basis fuer die Arithmetik* (**J** 0068) Z Math Logik Grundlagen Math 21*145-158
◇ B28 F30 ◇ REV MR 51 # 12498 Zbl 357 # 02036 • ID 04086

DEUTSCH, M. [1975] *Zur Darstellung koaufzaehlbarer Praedikate bei Verwendung eines einzigen unbeschraenkten Quantors* (**J** 0068) Z Math Logik Grundlagen Math 21*443-454
◇ D25 F30 ◇ REV MR 58 # 10360 Zbl 357 # 02041 • ID 04085

DEUTSCH, M. see Vol. I, III, IV, V for further entries

DEZANI, M. [1972] see BOEHM, C.

DEZANI, M. [1973] see BOEHM, C.

DEZANI, M. also published under the name DEZANI-CIANCAGLINI, M.

DEZANI-CIANCAGLINI, M. [1974] see BOEHM, C.

DEZANI-CIANCAGLINI, M. [1975] see BOEHM, C.

DEZANI-CIANCAGLINI, M. [1976] *Characterization of normal forms possessing inverse in the λ-β-η-calculus* (**J** 1426) Theor Comput Sci 2*323-337
◇ B40 ◇ REV MR 56 # 2797 Zbl 368 # 02028 • ID 51245

DEZANI-CIANCAGLINI, M. [1977] see BOEHM, C.

DEZANI-CIANCAGLINI, M. [1977] see COPPO, M.

Author Index

DEZANI-CIANCAGLINI, M. [1978] see COPPO, M.

DEZANI-CIANCAGLINI, M. [1979] see BOEHM, C.

DEZANI-CIANCAGLINI, M. [1979] see COPPO, M.

DEZANI-CIANCAGLINI, M. & RONCHI DELLA ROCCA, S. & SAITTA, L. [1979] *Complexity of λ-term reductions* (**J** 3441) RAIRO Inform Theor 13*257-287
◇ B40 D15 ◇ REV MR 81j:68016 Zbl 424#03009 • ID 81091

DEZANI-CIANCAGLINI, M. [1980] see COPPO, M.

DEZANI-CIANCAGLINI, M. [1981] see COPPO, M.

DEZANI-CIANCAGLINI, M. & ERMINE, F. [1982] *Maximal monoids of normal form* (**J** 2095) Fund Inform, Ann Soc Math Pol, Ser 4 5*129-141
◇ B40 C05 ◇ REV MR 84b:03028 Zbl 527#03002 • ID 35623

DEZANI-CIANCAGLINI, M. [1983] see BARENDREGT, H.P.

DEZANI-CIANCAGLINI, M. [1984] see COPPO, M.

DEZANI-CIANCAGLINI, M. & MARGARIA, I. [1984] *F-semantics for intersection type discipline* (**P** 3090) Semant of Data Types;1984 Sophia-Antipolis 279-300
◇ B40 B75 ◇ REV Zbl 552#68006 • ID 43462

DEZANI-CIANCAGLINI, M. also published under the name DEZANI, M.

DIACONESCU, R. [1975] *Axiom of choice and complementation* (**J** 0053) Proc Amer Math Soc 51*176-178
◇ E25 F50 G30 ◇ REV MR 51#10093 Zbl 317#02077 • ID 17526

DIACONESCU, R. see Vol. III, V for further entries

DIEGO, A. [1962] *On implicative algebras (Spanish)* (**J** 0188) Rev Union Mat Argentina 20*310-311
◇ C05 F50 G10 G25 ◇ REV MR 28#29 • ID 03002

DIEGO, A. see Vol. II for further entries

DIENES, P. [1938] *Logic of algebra* (**X** 0859) Hermann: Paris 76pp
◇ A05 B28 F55 G05 ◇ REV Zbl 20.98 JSL 4.100 FdM 64.924 • ID 03008

DIENES, P. [1952] *On H-matrices* (**J** 0028) Indag Math 14*32-36
◇ F50 ◇ REV MR 13.897 Zbl 46.6 • ID 48142

DIENES, P. see Vol. II, V for further entries

DIJKMAN, J.G. [1946] *Einige Saetze ueber mehrfach negativ-konvergente Reihen in der intuitionistischen Mathematik* (**J** 0028) Indag Math 8*532-536
◇ F55 ◇ REV MR 8.245 • ID 03012

DIJKMAN, J.G. [1948] *Recherche de la convergence negative dans les mathematiques intuitionnistes* (**J** 0028) Indag Math 10*232-243
◇ F55 ◇ REV MR 10.94 Zbl 30.101 • ID 03013

DIJKMAN, J.G. [1961] *On Markov chains and intuitionism* (**J** 0028) Indag Math 23*314-327
◇ F55 ◇ REV MR 25#3831 Zbl 108.3 • REM Part I. Part II 1963 • ID 03016

DIJKMAN, J.G. [1961] *Some intuitionistic remarks about transformations of sequences* (**J** 0020) Compos Math 15*70-87
◇ F55 ◇ REV MR 25#4272 Zbl 100.248 • ID 03015

DIJKMAN, J.G. [1962] *A note on intuitionistic divergence theory* (**J** 3077) Nieuw Arch Wisk, Ser 3 10*17-19
◇ F55 ◇ REV MR 25#1991 Zbl 105.248 • ID 03017

DIJKMAN, J.G. [1963] *On Markov chains and intuitionism. II: Discrete state-space and continous parameter* (**J** 0028) Indag Math 25*275-281
◇ F55 ◇ REV MR 27#6304 • REM Part I 1961. Part III 1964 • ID 48888

DIJKMAN, J.G. [1964] *Markov chains and intuitionism III: Note on continuous functions with an application to Markov chains* (**J** 0028) Indag Math 26*256-261
◇ F55 ◇ REV MR 29#3373 Zbl 143.201 • ID 03018

DIJKMAN, J.G. [1966] *Probability theory and intuitionism. Discrete state space* (**J** 0020) Compos Math 17*72-95
◇ F55 ◇ REV MR 31#5224 Zbl 131.7 • ID 03019

DILLER, J. [1968] *Zur Berechenbarkeit primitiv-rekursiver Funktionale endlicher Typen* (**P** 0608) Logic Colloq;1966 Hannover 109-120
◇ F10 F15 F50 ◇ REV MR 41#54 Zbl 182.17 JSL 40.453 • ID 03021

DILLER, J. & SCHUETTE, K. [1971] *Simultane Rekursionen in der Theorie der Funktionale endlicher Typen* (**J** 0009) Arch Math Logik Grundlagenforsch 14*69-74
◇ F10 F35 F50 ◇ REV MR 43#7294 Zbl 215.323 • ID 03023

DILLER, J. & NAHM, W. [1974] *Eine Variante zur Dialectica-Interpretation der Heyting-Arithmetik endlicher Typen* (**J** 0009) Arch Math Logik Grundlagenforsch 16*49-66
◇ F10 F35 F50 ◇ REV MR 51#2873 Zbl 277#02006 • ID 03025

DILLER, J. & MUELLER, GERT H. (EDS.) [1975] *⊧ISILC proof theory symposium* (**S** 3301) Lect Notes Math 500*viii+383pp
◇ C97 F97 ◇ REV MR 52#13304 Zbl 309#00006 • REM This volume contains only the proof theory part of the conference • ID 21771

DILLER, J. & VOGEL, HELMUT [1975] *Intensionale Funktionalinterpretation der Analysis* (**P** 1440) ⊧ISILC Proof Th Symp (Schuette);1974 Kiel 56-72
◇ F10 F35 F50 ◇ REV MR 53#10557 Zbl 323#02047 • ID 23066

DILLER, J. [1979] *Functional interpretations of Heyting's arithmetic in all finite types* (**P** 1947) Bicent Congr Wisk Genootschap;1978 Amsterdam I*149-176
• REPR [1979] (**J** 3077) Nieuw Arch Wisk, Ser 3 27*70-97
◇ F10 F35 F50 ◇ REV MR 80f:03066 MR 81c:03051 Zbl 401#03026 Zbl 417#03031 • ID 53267

DILLER, J. [1980] *Modified realization and the formulae-as-types notion* (**C** 3050) Essays Combin Log, Lambda Calc & Formalism (Curry) 491-501
◇ F05 F10 F30 F35 F50 ◇ REV MR 82g:03093 Zbl 469#03006 • ID 72387

DILLER, J. & TROELSTRA, A.S. [1984] *Realizability and intuitionistic logic* (**J** 0154) Synthese 60*253-282
◇ A05 F30 F35 F50 ◇ REV MR 86f:03098 Zbl 537#03043 • ID 40082

DILLER, J. see Vol. I, II, IV for further entries

DIMITRACOPOULOS, C. & GAIFMAN, H. [1982] *Fragments of Peano's arithmetic and the MRDP theorem* (**P** 3482) Logic & Algor (Specker);1980 Zuerich 187-206
◇ F30 ◇ REV MR 83j:03095 Zbl 498 # 03045 • ID 35380

DIMITRACOPOULOS, C. & PARIS, J.B. [1983] *A note on the undefinability of cuts* (**J** 0036) J Symb Logic 48*564-569
◇ C40 C62 F30 H15 ◇ REV MR 84k:03151 Zbl 529 # 03040 • ID 35043

DIMITRACOPOULOS, C. [1985] *A generalization of a theorem of H.Friedman* (**J** 0068) Z Math Logik Grundlagen Math 31*221-225
◇ C62 D15 F30 ◇ ID 47569

DIMITRACOPOULOS, C. see Vol. III, V for further entries

DING, DECHENG & MO, SHAOKUI [1984] *Simplification of the axioms of recursive arithmetic (Chinese)* (**J** 3187) Shuxue Niankan, Xi A 5*169-176
◇ B28 F30 ◇ REV MR 86a:03065 Zbl 566 # 03029 • REM English summary in J4719 5*262 • ID 44972

DING, DECHENG see Vol. IV for further entries

DOBRITSA, V.P. & KHISAMIEV, N.G. & NURTAZIN, A.T. [1978] *Constructive periodic abelian groups (Russian)* (**J** 0092) Sib Mat Zh 19*1260-1265
• TRANSL [1978] (**J** 0475) Sib Math J 19*886-890
◇ C57 C60 F60 ◇ REV MR 80b:20067 Zbl 421 # 20021 • ID 32600

DOBRITSA, V.P. see Vol. III, IV for further entries

DOMBROVSKIJ-KABANCHENKO, M.N. [1979] *A transfinite extension of the stepped semantic system (Russian)* (**S** 0554) Issl Teor Algor & Mat Logik (Moskva) 3*18-26
◇ D55 F50 ◇ REV MR 81j:03091 Zbl 443 # 03024 • ID 56430

DOMRACHEV, V.N. [1981] *An example of proof of the correctness of a program, using its representation by a λ-term (Russian)* (**C** 4349) Spec Design & Model Tools for PROEKT-ES Syst (Kiev) 40-45,104
◇ B40 B75 ◇ ID 47031

DOMRACHEV, V.N. & KAPITONOVA, YU.V. & SAMOJLENKO, L.G. [1981] *Definition of the semantics of programming language constructs in terms of λ-calculus I (Russian)* (**J** 0040) Kibernetika, Akad Nauk Ukr SSR 1981/5*19-23
• TRANSL [1981] (**J** 0021) Cybernetics 17*590-595
◇ B40 B75 ◇ REV MR 85m:68022a Zbl 491 # 68034 • REM Part II 1983 • ID 37762

DOMRACHEV, V.N. & KAPITONOVA, YU.V. & SAMOJLENKO, L.G. [1982] *On a method for the definition of semantics of constructions of programming languages in terms of typed λ-calculus (Russian)* (**C** 3832) Avtom Issl Mat 46-54
◇ B40 B75 ◇ REV MR 84j:68055 Zbl 537 # 68020 • ID 43856

DOMRACHEV, V.N. & KAPITONOVA, YU.V. & SAMOJLENKO, L.G. [1983] *One method of defining the semantics of programming languages constructs in terms of lambda calculus. II (Russian) (English summary)* (**J** 0040) Kibernetika, Akad Nauk Ukr SSR 1983/3*27-32
• TRANSL [1983] (**J** 0021) Cybernetics 19*325-333
◇ B40 B75 ◇ REV MR 85m:68022b • REM Part I 1981 • ID 45223

DONAHUE, J. [1979] *On the semantics of "data type"* (**J** 1428) SIAM J Comp 8*546-560
◇ B40 ◇ REV MR 82f:68015 Zbl 418 # 68010 • ID 53334

DOPP, J. [1960] *Essai d'une presentation de la logique combinatoire* (**J** 0079) Logique & Anal, NS 3*183-201
◇ B40 ◇ REV JSL 27.77 • ID 03084

DOPP, J. [1962] *Logiques construites par une methode de deduction naturelle* (**X** 0834) Gauthier-Villars: Paris 191pp
◇ B98 F05 F07 F50 F98 ◇ REV MR 26 # 1243 Zbl 143.8 JSL 34.502 • ID 03085

DOPP, J. see Vol. I, II for further entries

DOSEN, K. [1981] *A reduction of classical propositional logic to the conjunction-negation fragment of an intuitionistic relevant logic* (**J** 0122) J Philos Logic 10*399-408
◇ B46 B55 F50 ◇ REV MR 83b:03017 Zbl 473 # 03008 • ID 55339

DOSEN, K. [1981] *Minimal modal systems in which Heyting and classical logic can be embedded* (**J** 0400) Publ Inst Math, NS (Belgrade) 30(44)*41-52
◇ B45 F50 ◇ REV MR 84c:03034 Zbl 494 # 03008 • ID 35691

DOSEN, K. [1983] see BOZIC, M.

DOSEN, K. [1984] *Intuitionistic double negation as a necessity operator* (**J** 0482) Publ Math Univ Belgrade 35(49)*15-20
◇ B45 F50 ◇ REV Zbl 555 # 03012 • ID 44540

DOSEN, K. [1984] *Negative modal operators in intuitionistic logic* (**J** 0482) Publ Math Univ Belgrade 35(49)*3-14
◇ B55 F50 ◇ REV Zbl 555 # 03011 • ID 44539

DOSEN, K. [1985] *An intuitionistic Sheffer function* (**J** 0047) Notre Dame J Formal Log 26*479-482
◇ B55 F50 ◇ ID 47541

DOSEN, K. see Vol. II, III, IV for further entries

DOU, A. [1969] *El teorema de incomplitud de Goedel* (**J** 1576) Publ Sem Mat Garcia Galdeano (Zaragoza) 10*95-106
◇ F30 ◇ REV MR 41 # 61 Zbl 191.293 • ID 32314

DRABBE, J. [1967] *Quelques notions de la logique algebrique* (**J** 0082) Bull Soc Math Belg 19*25-41
◇ F50 G15 ◇ REV MR 36 # 39 Zbl 189.284 • ID 03099

DRABBE, J. see Vol. I, II, III, IV, V for further entries

DRAGALIN, A.G. [1967] *Constructive transfinite systems and the construction of an algorithm by transfinite recursion (Russian)* (**J** 0023) Dokl Akad Nauk SSSR 175*993-996
• TRANSL [1967] (**J** 0062) Sov Math, Dokl 8*946-949
◇ F50 ◇ REV MR 36 # 2494 Zbl 204.16 • ID 03109

DRAGALIN, A.G. [1967] *Justification of A.A.Markov's constructive selection principle (Russian)* (**J** 0023) Dokl Akad Nauk SSSR 177*13-16
• TRANSL [1967] (**J** 0062) Sov Math, Dokl 8*1526-1530
◇ F50 ◇ REV MR 38 # 5617 • ID 03110

DRAGALIN, A.G. [1968] *The computability of primitive recursive terms of finite type, and primitive recursive realization (Russian)* (**S** 0228) Zap Nauch Sem Leningrad Otd Mat Inst Steklov 8*32-45
◇ F10 F35 F50 ◇ REV MR 43 # 45 Zbl 176.279 • ID 03111

DRAGALIN, A.G. [1969] *Transfinite completions of constructive arithmetical calculus (Russian)* (**J** 0023) Dokl Akad Nauk SSSR 189*10-12
• TRANSL [1969] (**J** 0062) Sov Math, Dokl 10*1417-1420
◇ F50 ◇ REV MR 41 # 3242 Zbl 205.310 • ID 03113

Author Index

DRAGALIN, A.G. [1972] *On the use of classical calculi for establishing constructive truth (Russian) (English summary)* (J 0288) Vest Ser Mat Mekh, Univ Moskva 27/2*25-29
• TRANSL [1972] (J 0510) Moscow Univ Math Bull 27/1-2*76-79
◇ B10 F30 F50 ◇ REV MR 47 # 1581 Zbl 237 # 02007
• ID 03610

DRAGALIN, A.G. [1973] *Constructive mathematics and models of intuitionistic theories* (P 0793) Int Congr Log, Meth & Phil of Sci (4,Proc);1971 Bucharest 111-128
◇ F50 ◇ REV MR 56 # 5237 • ID 72443

DRAGALIN, A.G. [1973] *Intuitionistic model theory (Russian)* (S 2337) Istor Metodol Estest Nauk (Moskva) 14*106-126
◇ C90 F50 ◇ REV MR 58 # 27451 Zbl 275 # 02026
• ID 29696

DRAGALIN, A.G. [1974] *Constructive models of the theories of intuitionistic sequences of sampling (Russian)* (C 2577) Issl Formaliz Yazyk & Neklass Log 214-252
◇ F50 ◇ REV MR 56 # 5230 • ID 90001

DRAGALIN, A.G. [1974] *Constructive model of intuitionistic analysis (Russian)* (C 2578) Filos & Logika 55-78
◇ F35 F50 ◇ REV MR 58 # 27373 • ID 90002

DRAGALIN, A.G. [1974] *Die intuitionistische Logik und das ε-Symbol Hilberts (Russian)* (S 2337) Istor Metodol Estest Nauk (Moskva) 16*78-84
◇ F50 ◇ REV Zbl 335 # 02011 • ID 61398

DRAGALIN, A.G. [1974] *The completeness of an arithmetic with a constructive rule of infinite induction (Russian)* (C 1450) Teor Algor & Mat Logika (Markov) 14-33,213
◇ F30 F50 ◇ REV MR 54 # 66 Zbl 305 # 02045 • ID 23948

DRAGALIN, A.G. [1976] *Cut-elimination in the theory of definable sets of natural numbers (Russian)* (P 2064) All-Union Conf Math Log (4);1976 Kishinev 45
◇ F05 F35 F65 ◇ ID 33029

DRAGALIN, A.G. [1977] *Cut-elimination in the theory of definable sets of natural numbers (Russian)* (S 2579) Teor Mnozhestv & Topol (Izhevsk) 1*27-36
◇ F05 F35 F65 ◇ REV Zbl 483 # 03038 • ID 36148

DRAGALIN, A.G. [1979] *Algebraic models of intuitionistic theories (Russian)* (P 2554) All-Union Symp Th Log Infer;1974 Moskva 206-245
◇ F50 G10 ◇ REV MR 84m:03092 • ID 90005

DRAGALIN, A.G. [1979] *An algebraic approach to intuitionistic models of the realizability type (Russian)* (C 2581) Issl Neklass Log & Teor Mnozh 183-201
◇ F50 G25 ◇ REV MR 82a:03057 Zbl 427 # 03050
• ID 53737

DRAGALIN, A.G. [1979] *Calculus of constructions, equivalent to the intuitionistic predicate calculus (Russian)* (P 2570) Intens Log & Log Analiz Estestv Yazyk;1979 Moskva 23-27
◇ F50 ◇ ID 90012

DRAGALIN, A.G. [1979] *Intuitionism (Russian)* (C 2580) Mat Entsikl 2*637-643
◇ A05 F50 F98 ◇ ID 90011

DRAGALIN, A.G. [1979] *Mathematical intuitionism. Introduction to proof theory (Russian)* (X 2027) Nauka: Moskva 256pp
◇ F05 F30 F35 F50 F98 ◇ REV Zbl 439 # 03041
• ID 56032

DRAGALIN, A.G. [1979] *Proof theory (Russian)* (C 2580) Mat Entsikl 2*367-371
◇ F98 ◇ ID 90010

DRAGALIN, A.G. [1979] *Strong normalization theorem for derivations in Gentzen's sequent calculus (Russian)* (S 0554) Issl Teor Algor & Mat Logik (Moskva) 3*26-39,132
◇ F05 F50 ◇ REV MR 81j:03086 Zbl 422 # 03031
• ID 53492

DRAGALIN, A.G. [1980] *Higher-order predicate logic in the form of calculus realization (Russian)* (C 2583) Aktual Probl Log & Metodol Nauki 236-252
◇ B15 F05 F35 F50 ◇ REV Zbl 533 # 03034 • ID 36552

DRAGALIN, A.G. [1980] *New forms of realizability and Markov's rule (Russian)* (J 0023) Dokl Akad Nauk SSSR 251*534-537
• TRANSL [1980] (J 0062) Sov Math, Dokl 21*461-464
◇ F30 F50 ◇ REV MR 81k:03063 Zbl 481 # 03034
• ID 72439

DRAGALIN, A.G. [1985] *Correctness of inconsistent theories with notions of feasibility* (P 4670) Comput Th (5);1984 Zaborow 58-79
◇ F65 ◇ ID 49665

DRAGALIN, A.G. see Vol. III, IV, V for further entries

DRAGO, A. [1982] *Caratheodory's thermodynamics and constructive mathematics* (J 2778) Lett Nuovo Cimento Ser 2 34*52-56
◇ F55 ◇ REV MR 83j:80001 • ID 39900

DRAGO, A. [1983] *Dimensional theory and constructive mathematics* (J 2778) Lett Nuovo Cimento Ser 2 37*409-412
◇ A05 F99 ◇ REV MR 84k:03011 • ID 34951

DRAGO, A. [1984] *Constructive analysis and thermodynamics formulations* (J 0009) Arch Math Logik Grundlagenforsch 24*17-22
◇ F55 ◇ REV MR 85i:03180 Zbl 548 # 03035 • ID 42399

DREBEN, B. [1963] see AANDERAA, S.O.

DREBEN, B. [1964] see AANDERAA, S.O.

DREBEN, B. [1966] see DENTON JR., J.S.

DREBEN, B. [1970] see DENTON JR., J.S.

DREBEN, B. & GOLDFARB, W.D. [1971] *Note J* (C 0745) Herbrand: Log Writings 201-202
◇ B10 F05 F07 ◇ REV JSL 40.94 • REM Footnote to English translation of Herbrand 1930 • ID 17043

DREBEN, B. & GOLDFARB, W.D. [1971] *Note N* (C 0745) Herbrand: Log Writings 265-271
◇ A10 B10 F30 F35 ◇ REV JSL 40.94 • REM Footnote to English transl. of Herbrand 1931: Sur le probleme fondamental de la logique math. • ID 17045

DREBEN, B. [1971] *Notes E-I* (C 0745) Herbrand: Log Writings 190-201
◇ F05 F07 ◇ REV JSL 40.94 • REM Footnotes to English translation of Herbrand 1930 • ID 17042

DREBEN, B. see Vol. I, II, III, IV for further entries

DRESDEN, A. [1924] *Brouwer's contribution to the foundations of mathematics* (J 0015) Bull Amer Math Soc 30*31-40
◇ A05 F55 ◇ REV FdM 50.25 • ID 41693

DRESDEN, A. see Vol. I for further entries

DRIES VAN DEN, L. [1980] *Some model theory and number theory for models of weak systems of arithmetic* (**P** 2625) Model Th of Algeb & Arithm;1979 Karpacz 346-362
◇ B25 C62 F30 H15 ◇ REV MR 82f:03029 Zbl 454 # 03034 • ID 54246

DRIES VAN DEN, L. & LEVITZ, H. [1984] *On Skolem's exponential functions below* 2^{2^X} (**J** 0064) Trans Amer Math Soc 286*339-349
◇ E07 E75 F15 F30 ◇ REV MR 86g:03068 Zbl 556 # 03036 • ID 46154

DRIES VAN DEN, L. see Vol. I, III, IV for further entries

DRUGUSH, YA.M. [1978] *A class of logics without disjunction property (Russian)* (**J** 0288) Vest Ser Mat Mekh, Univ Moskva 1978/6*9-14
• TRANSL [1978] (**J** 0510) Moscow Univ Math Bull 33/6*6-9
◇ B55 F50 ◇ REV MR 80j:03039 Zbl 396 # 03025 • ID 52635

DRUGUSH, YA.M. see Vol. II, III for further entries

DUMITRIU, A. [1971] *Fermat's method of infinite descent compared with the method of mathematical induction (Romanian) (French summary)* (**S** 1613) Probl Logic (Bucharest) 3*27-52
◇ A10 F30 ◇ REV Zbl 216.3 • ID 26234

DUMITRIU, A. see Vol. I, II for further entries

DUMMETT, M. [1975] *The justification of deduction* (**J** 4274) Proc British Acad 59*201-232
◇ A05 F50 ◇ REV JSL 47.689 • ID 47403

DUMMETT, M. [1975] *The philosophical basis of intuitionistic logic* (**P** 0775) Logic Colloq;1973 Bristol 5-40
◇ A05 F99 ◇ REV MR 55 # 12449 Zbl 325 # 02005 JSL 47.689 • ID 30431

DUMMETT, M. [1977] *Elements of intuitionism* (**X** 0815) Clarendon Pr: Oxford xii + 467pp
◇ F50 F55 F98 ◇ REV MR 58 # 16194 Zbl 358 # 02032 JSL 44.276 • ID 50486

DUMMETT, M. see Vol. I, II, IV for further entries

DYKMAN, J.G. [1968] see DALEN VAN, D.

DYMENT, E.Z. [1985] *On the interpretation of the intuitionistic predicate calculus by means of the problem calculus with parameters (Russian)* (**X** 2235) VINITI: Moskva 1716-85
◇ F50 ◇ ID 46728

DYMENT, E.Z. see Vol. III, IV for further entries

DYWAN, Z. [1984] *An interpretation of a certain fragment of arithmetic in some propositional calculus* (**J** 0387) Bull Sect Logic, Pol Acad Sci 13*99-105
◇ B20 F30 ◇ REV Zbl 565 # 03007 • ID 45756

DYWAN, Z. see Vol. I, II, III, V for further entries

DZIK, W. [1975] *On structural completeness of some nonclassical predicate calculi* (**J** 0302) Rep Math Logic, Krakow & Katowice 5*19-26
◇ B22 F50 ◇ REV MR 58 # 16147 Zbl 343 # 02038 • ID 21897

DZIK, W. see Vol. I, II, V for further entries

EBBINGHAUS, H.-D. [1970] *Aufzaehlbarkeit* (**S** 1415) Sel Math 2*64-113
• TRANSL [1972] (**C** 1534) Mash Turing & Rek Funk 86-149
◇ D03 D10 D20 D25 F30 ◇ REV MR 43 # 4665 Zbl 251 # 02037 • ID 28196

EBBINGHAUS, H.-D. see Vol. I, II, III, IV, V for further entries

EDGAR, W.J. [1973] *Is intuitionism the epistemically serious foundation for mathematics?* (**J** 1099) Phil Math 10*113-133
◇ A05 F99 ◇ REV Zbl 272 # 02003 • ID 30375

EGLI, H. [1975] *Typed meaning in Scott's λ-calculus models* (**P** 1603) λ-Calc & Comput Sci Th;1975 Roma 220-239
◇ B40 ◇ REV MR 58 # 5145 Zbl 378 # 02009 • ID 51791

EGLI, H. see Vol. IV for further entries

EHRENFEUCHT, A. & FEFERMAN, S. [1959] *Representability of recursively enumerable sets in formal theories* (**J** 0009) Arch Math Logik Grundlagenforsch 5*37-41
◇ D25 F30 ◇ REV MR 23 # A3088 Zbl 118.251 JSL 32.530 • ID 03237

EHRENFEUCHT, A. & MYCIELSKI, J. [1971] *Abbreviating proofs by adding new axioms* (**J** 0015) Bull Amer Math Soc 77*366-367
• TRANSL [1974] (**C** 2319) Slozh Vychisl & Algor 172-173 (Russian)
◇ D35 F20 ◇ REV MR 43 # 1838 Zbl 216.10 • ID 29998

EHRENFEUCHT, A. [1973] *Polynomial functions with exponentiation are well ordered* (**J** 0004) Algeb Universalis 3*261-262
◇ E07 F30 ◇ REV MR 48 # 10908 Zbl 308 # 06001 • ID 23826

EHRENFEUCHT, A. [1974] *Logic without iterations* (**P** 0610) Tarski Symp;1971 Berkeley 265-268
◇ B60 F65 ◇ REV MR 50 # 9546 Zbl 345 # 02042 • ID 03261

EHRENFEUCHT, A. & JENSEN, D.C. [1976] *Some problems in elementary arithmetics* (**J** 0027) Fund Math 92*223-245
◇ C62 F30 H15 ◇ REV MR 54 # 7244 Zbl 362 # 02049 • ID 25797

EHRENFEUCHT, A. see Vol. I, III, IV, V for further entries

ELGOT, C.C. [1961] *Decision problems of finite automata design and related arithmetics* (**J** 0064) Trans Amer Math Soc 98*21-51 • ERR/ADD ibid 103*558-559
◇ B25 D03 D05 D35 F30 ◇ REV MR 25 # 2962 Zbl 111.11 JSL 34.509 • ID 03288

ELGOT, C.C. & RUTLEDGE, J.D. [1964] *RS-machines with almost blank tapes* (**J** 0037) ACM J 11*313-337
◇ D05 D10 F30 ◇ REV MR 29 # 3376 Zbl 168.257 • ID 03289

ELGOT, C.C. & RABIN, M.O. [1966] *Decidability and undecidability of extensions of second (first) order theory of (generalized) successor* (**J** 0036) J Symb Logic 31*169-181
◇ B15 B25 C85 D05 D35 F30 F35 ◇ REV Zbl 144.245 • ID 03291

ELGOT, C.C. see Vol. III, IV for further entries

ELLENTUCK, E. [1971] *Incompleteness via simple sets* (**J** 0047) Notre Dame J Formal Log 12*255-256
◇ D25 F30 ◇ REV MR 44 # 5222 Zbl 193.308 • ID 03306

ELLENTUCK, E. see Vol. III, IV, V for further entries

EMDE BOAS VAN, P. [1975] *Ten years of speedup* (**P** 0454) Math Founds of Comput Sci (4);1975 Marianske Lazne 13-29 ◇ A10 D15 D20 F20 G05 ◇ REV MR 52 # 16112 Zbl 324 # 68025 • ID 21889

EMDE BOAS VAN, P. see Vol. II, III, IV, V for further entries

ENDERTON, H.B. [1967] *An infinitistic rule of proof* (**J** 0036) J Symb Logic 32*447-451 ◇ C62 F35 ◇ REV MR 37 # 66 Zbl 157.27 • ID 03347

ENDERTON, H.B. [1968] *On provable recursive functions* (**J** 0047) Notre Dame J Formal Log 9*86-88 ◇ D20 F30 ◇ REV MR 39 # 5355 Zbl 184.18 JSL 38.526 • ID 03348

ENDERTON, H.B. & FRIEDMAN, H.M. [1971] *Approximating the standard model of analysis* (**J** 0027) Fund Math 72*175-188 ◇ C62 E45 F35 ◇ REV MR 45 # 6609 Zbl 235 # 02053 JSL 39.600 • ID 03353

ENDERTON, H.B. [1972] *A mathematical introduction to logic* (**X** 0801) Academic Pr: New York xiii + 295pp ◇ B98 C98 F30 ◇ REV MR 49 # 2239 Zbl 298 # 02002 JSL 38.340 • ID 03355

ENDERTON, H.B. see Vol. I, III, IV, V for further entries

ENGELER, E. [1961] *Zur Beweistheorie von Sprachen mit unendlich langen Formeln* (**J** 0068) Z Math Logik Grundlagen Math 7*213-218 ◇ C75 F07 ◇ REV MR 30 # 1030 Zbl 124.248 JSL 36.685 • ID 03362

ENGELER, E. [1963] *A reduction-principle for infinite formulas* (**J** 0043) Math Ann 151*296-303 ◇ C75 F07 ◇ REV MR 28 # 2974 Zbl 114.245 JSL 33.123 • ID 03364

ENGELER, E. [1973] *Towards a Galois theory of algorithmic problems* (**P** 1448) Math Founds of Comput Sci (2);1973 Strbske Pleso 51-55 ◇ B75 F99 ◇ REV MR 53 # 12901 • ID 28202

ENGELER, E. [1977] *A new type of models of computation* (**P** 1635) Math Founds of Comput Sci (6);1977 Tatranska Lomnica 52-58 ◇ B40 ◇ REV MR 58 # 3613 Zbl 365 # 68023 • ID 51072

ENGELER, E. [1981] *Algebras and combinators* (**J** 0004) Algeb Universalis 13*389-392 ◇ B40 ◇ REV MR 83f:03012 Zbl 482 # 08005 • ID 35294

ENGELER, E. [1981] *An algorithmic model of strict finitism* (**P** 3642) Colloq Math Log in Computer Sci;1978 Salgotarjan 345-357 ◇ D20 F65 ◇ REV MR 83g:68007 Zbl 498 # 03048 JSL 49.990 • ID 36907

ENGELER, E. see Vol. I, II, III, IV, V for further entries

EPSTEIN, G. & HORN, A. [1974] *Propositional calculi based on subresiduation* (**J** 0387) Bull Sect Logic, Pol Acad Sci 3/1*41-42 ◇ B50 F50 G10 ◇ REV MR 53 # 5264 • ID 22895

EPSTEIN, G. & HORN, A. [1976] *Logics which are characterized by subresiduated lattices* (**J** 0068) Z Math Logik Grundlagen Math 22*199-210 ◇ B50 F50 G10 ◇ REV MR 56 # 8360 Zbl 347 # 02040 • ID 18456

EPSTEIN, G. see Vol. I, II for further entries

EPSTEIN, R.L. & SZCZERBA, L.W. [1979] *Relatedness and interpretability* (**J** 0095) Philos Stud 36*225-231 ◇ F25 ◇ REV MR 80k:03028b • ID 72682

EPSTEIN, R.L. see Vol. II, IV for further entries

ERDOES, P. & MILLS, G. [1981] *Some bounds for the Ramsey-Paris-Harrington numbers* (**J** 0164) J Comb Th, Ser A 30*53-70 ◇ F30 ◇ REV MR 84c:03099 Zbl 471 # 05045 • ID 34033

ERDOES, P. see Vol. III, V for further entries

ERMINE, F. [1982] see DEZANI-CIANCAGLINI, M.

ERRERA, A. [1927] see BARZIN, M.

ERRERA, A. [1929] see BARZIN, M.

ERRERA, A. [1932] see BARZIN, M.

ERRERA, A. [1933] see BARZIN, M.

ERRERA, A. [1937] *Sur la notion de compatibilite et les rapports entre l'intuitionnisme et le formalisme* (**P** 1608) Int Congr Math (II, 5);1936 Oslo 2*270-271 ◇ F50 ◇ REV FdM 63.31 • ID 27936

ERRERA, A. [1937] *Sur les demonstrations de non-contradiction* (**P** 0756) Congr Int Phil (9);1937 Paris 121-127 ◇ F25 ◇ REV JSL 3.116 FdM 63.837 • ID 17049

ERRERA, A. [1951] *Observations sur la communication du Professeur Heyting* (**P** 0642) Congr Int Phil Sci;1949 Paris 2*87-89 ◇ F50 ◇ REV MR 13.898 • ID 03511

ERRERA, A. see Vol. V for further entries

ERSHOV, YU.L. [1970] *On a hierarchy of sets III (Russian)* (**J** 0003) Algebra i Logika 9*34-51 • TRANSL [1970] (**J** 0069) Algeb and Log 9*20-31 ◇ D30 D55 F15 ◇ REV MR 45 # 8526 Zbl 233 # 02017 • REM Parts I,II 1968 • ID 03540

ERSHOV, YU.L. [1972] *Everywhere-defined continuous functionals (Russian)* (**J** 0003) Algebra i Logika 11*656-665,736 • TRANSL [1972] (**J** 0069) Algeb and Log 11*363-368 ◇ D65 F10 ◇ REV MR 50 # 12689 Zbl 285 # 02041 • ID 61604

ERSHOV, YU.L. [1973] *Skolem functions and constructive models (Russian)* (**J** 0003) Algebra i Logika 12*644-654,735 • TRANSL [1973] (**J** 0069) Algeb and Log 12*368-373 ◇ C57 C65 F50 ◇ REV MR 57 # 5727 Zbl 287 # 02024 • ID 30533

ERSHOV, YU.L. [1974] *The model G of the theory BR (Russian)* (**J** 0023) Dokl Akad Nauk SSSR 217*1004-1006 • TRANSL [1974] (**J** 0062) Sov Math, Dokl 15*1158-1161 ◇ F10 F50 ◇ REV MR 50 # 6798 Zbl 312 # 02034 • ID 04118

ERSHOV, YU.L. [1974] *Theory of numerations, III: Constructive models (Russian)* (**X** 0913) Novosibirsk Gos Univ: Novosibirsk 139pp • TRANSL [1977] (**J** 0068) Z Math Logik Grundlagen Math 23*289-371 ◇ B25 C57 C98 D45 F50 ◇ REV MR 54 # 9995 Zbl 374 # 02028 • REM Part II 1973 • ID 25764

ERSHOV, YU.L. [1977] *Model $\mathbb{C}$ of partial continuous functionals* (**P** 1075) Logic Colloq;1976 Oxford 455-467
◇ D45 D65 F10 F50 ◇ REV MR 58#21541 Zbl 427#03037 • ID 16634

ERSHOV, YU.L. see Vol. I, II, III, IV, V for further entries

ESAKIA, L.L. [1981] *Diagonal constructions, the Loeb formula and rarefied Cantor spaces* (**C** 3810) Log-Semat Issl (Tbilisi) 128-143
◇ F30 ◇ ID 45313

ESENIN-VOL'PIN, A.S. [1957] *Consistency proof of classical arithmetics with the aid of induction to ε_0 (following Schuette) (Russian)* (**X** 1656) Izdat Inostr Lit: Moskva
◇ F05 F30 ◇ REM Appendix VII in the Russian edition of Kleene's "Introduction into metamathematics" • ID 40525

ESENIN-VOL'PIN, A.S. [1959] *On Goedel's second theorem (Russian)* (**P** 0607) All-Union Math Conf (3);1956 Moskva 4*84-85
◇ F30 ◇ ID 30598

ESENIN-VOL'PIN, A.S. [1961] *Le programme ultra-intuitionniste des fondements des mathematiques* (**P** 0633) Infinitist Meth;1959 Warsaw 201-223
◇ A05 E70 F65 ◇ REV MR 26#4905 Zbl 134.9 JSL 32.517 • ID 03557

ESENIN-VOL'PIN, A.S. [1969] *On Hilbert's second problem (Russian)* (**C** 3724) Probl Gil'berta 83-91
◇ A05 F30 F65 ◇ REV MR 41#5188 • ID 42984

ESENIN-VOL'PIN, A.S. [1970] *The ultra-intuitionistic criticism and the anti-traditional program for foundations of mathematics* (**P** 0603) Intuitionism & Proof Th;1968 Buffalo 3-45
◇ A05 F65 ◇ REV MR 45#4938 JSL 40.95 • ID 16297

ESENIN-VOL'PIN, A.S. [1981] *About infinity, finiteness and finitization (in connection with the foundations of mathematics)* (**P** 3146) Constr Math;1980 Las Cruces 274-313
◇ A05 E30 F65 ◇ REV MR 83b:03068 Zbl 473#03052 • ID 35110

ESENIN-VOL'PIN, A.S. see Vol. V for further entries

ESPANOL, L. [1983] *Le spectre d'un anneau dans l'algebre constructive et applications a la dimension* (**J** 0306) Cah Topol & Geom Differ 24*133-144
◇ F55 G30 ◇ REV MR 85e:13034 Zbl 547#13004 • ID 39927

ESPANOL, L. see Vol. III for further entries

EUWE, M. [1929] *Mengentheoretische Betrachtungen ueber das Schachspiel* (**J** 0278) Nederl Akad Wetensch Proc 32*633-642
◇ E60 F55 ◇ REV FdM 55.57 • ID 39332

EVANGELIST, M. [1982] *Non-standard propositional logics and their application to complexity theory* (**J** 0047) Notre Dame J Formal Log 23*384-392
◇ B80 D15 F20 ◇ REV MR 83k:03014 Zbl 464#03028 • ID 54618

FABIANO, A. [1983] *Completezza della logica intuizionista rispetto a modelli con coperture (English summary)* (**J** 2099) Boll Unione Mat Ital, VI Ser, A 2*21-28
◇ F50 ◇ REV MR 84d:03018 Zbl 526#03013 • ID 34058

FAHMY, M.H. [1982] *A formal proof of Ackermann's theorem (Arabic summary)* (**J** 0397) Proc Math Phys Soc Egypt 51*1-6
◇ D20 F30 ◇ REV MR 84f:03039 Zbl 572#03019 • ID 34459

FAHMY, M.H. see Vol. IV for further entries

FALEVICH, B.YA. [1958] *A new method of proving incompleteness theorems for systems with Carnap's rule, and its application to the problem of interrelation between classical and constructive analysis (Russian)* (**J** 0023) Dokl Akad Nauk SSSR 120*1210-1213
◇ F35 F60 ◇ REV MR 20#6345 Zbl 90.12 JSL 32.419 • ID 03651

FALEVICH, B.YA. [1959] *Incompleteness theorems in systems with infinite induction (Russian)* (**P** 0607) All-Union Math Conf (3);1956 Moskva 4*89-90
◇ F30 ◇ ID 31283

FALEVICH, B.YA. [1959] *On the non-existence of models of classical analysis in W. Ackermann's constructive analysis (Russian)* (**P** 0607) All-Union Math Conf (3);1956 Moskva 4*90
◇ F65 ◇ ID 31284

FALEVICH, B.YA. see Vol. V for further entries

FAL'KOVICH, M.A. & GALIN, A.B. [1980] *Diagnosis of memoryless digital automata using a modified technique of natural logical deduction (Russian)* (**J** 0011) Avtom Telemekh 1980/1*131-137
• TRANSL [1980] (**J** 0010) Autom & Remote Control 41*104-109
◇ B70 D05 F99 ◇ REV Zbl 439#94032 • ID 69607

FARRINGTON, PADDY [1984] *The first-order theory of the c-degrees* (**J** 0068) Z Math Logik Grundlagen Math 30*437-446
◇ D30 D35 E45 E55 F35 ◇ REV MR 86a:03057 Zbl 565#03025 • ID 42273

FEFERMAN, S. [1957] *Formal consistency proofs and interpretability of theories* (**P** 1675) Summer Inst Symb Log;1957 Ithaca 71-77
◇ F25 ◇ REV JSL 34.150 • ID 29320

FEFERMAN, S. [1959] see EHRENFEUCHT, A.

FEFERMAN, S. [1960] *Arithmetization of metamathematics in a general setting* (**J** 0027) Fund Math 49*35-92
◇ D55 F07 F25 F30 F40 ◇ REV MR 26#4913 Zbl 95.243 JSL 31.269 • ID 03686

FEFERMAN, S. & SPECTOR, C. [1962] *Incompleteness along paths in progressions of theories* (**J** 0036) J Symb Logic 27*383-390
• TRANSL [1971] (**S** 1582) Matematika - Period Sb Perevodov Inostran Statej 15/6*159-166 (Russian)
◇ D55 F15 F30 ◇ REV MR 30#3012 Zbl 117.257 JSL 32.531 • ID 03688

FEFERMAN, S. [1962] *Transfinite recursive progressions of axiomatic theories* (**J** 0036) J Symb Logic 27*259-316
• TRANSL [1979] (**C** 3351) Teorem Goedel & Hipotese Cont 573-754 (Portuguese)
◇ D55 F15 F30 ◇ REV MR 30#3011 Zbl 117.254 JSL 32.530 • ID 03693

Author Index

FEFERMAN, S. & KREISEL, G. & OREY, S. [1962] *1-consistency and faithful interpretations* (**J** 0009) Arch Math Logik Grundlagenforsch 6*52-63
◇ F25 F30 F35 ◇ REV MR 26 # 2361 Zbl 122.245 • ID 03690

FEFERMAN, S. [1964] *Systems of predicative analysis* (**J** 0036) J Symb Logic 29*1-30
• TRANSL [1979] (**C** 3351) Teorem Goedel & Hipotese Cont 790-866 (Portuguese) • REPR [1969] (**C** 0569) Phil of Math Oxford Readings 95-127
◇ F35 F65 ◇ REV MR 33 # 1228 Zbl 134.11 JSL 31.660 • REM Part I. Part II 1968 • ID 03695

FEFERMAN, S. [1965] *Some applications of the notions of forcing and generic sets* (**J** 0027) Fund Math 56*325-345
• REPR [1965] (**P** 0614) Th Models;1963 Berkeley 89-95
◇ C62 D55 E25 E35 E40 E45 F35 ◇ REV MR 31 # 1193 MR 34 # 2439 Zbl 129.264 JSL 37.612 • REM Reprint is a summary • ID 03696

FEFERMAN, S. [1966] *Predicative provability in set theory* (**J** 0015) Bull Amer Math Soc 72*486-489
◇ E30 F65 ◇ REV MR 33 # 1230 Zbl 148.254 • ID 03699

FEFERMAN, S. [1968] *Autonomous transfinite progressions and the extent of predicative mathematics* (**P** 0627) Int Congr Log, Meth & Phil of Sci (3,Proc);1967 Amsterdam 121-135
• TRANSL [1979] (**C** 3351) Teorem Goedel & Hipotese Cont 755-790 (Portuguese)
◇ F15 F30 F35 F65 ◇ REV MR 40 # 5417 Zbl 198.323 • ID 03702

FEFERMAN, S. [1968] *Lectures on proof theory* (**P** 0692) Summer School in Logic;1967 Leeds 1-107
◇ C40 C70 C75 F05 F15 F98 ◇ REV MR 38 # 4294 Zbl 248 # 02033 • ID 03700

FEFERMAN, S. [1968] *Persistent and invariant formulas for outer extensions* (**J** 0020) Compos Math 20*29-52
• REPR [1968] (**C** 0727) Logic Found of Math (Heyting) 29-52
◇ C30 C40 C70 C75 F35 ◇ REV MR 37 # 6170 Zbl 162.16 JSL 37.764 • ID 03703

FEFERMAN, S. [1968] *Systems of predicative analysis II : representation of ordinals* (**J** 0036) J Symb Logic 33*193-220
◇ F15 F35 F65 ◇ REV MR 41 # 5213 Zbl 162.22 JSL 48.876 • REM Part I 1964 • ID 03701

FEFERMAN, S. [1970] *Formal theories for transfinite iteration of generalized inductive definitions and some subsystems of analysis* (**P** 0603) Intuitionism & Proof Th;1968 Buffalo 303-326
◇ F35 F50 ◇ REV MR 46 # 1568 Zbl 218 # 02024 • ID 03704

FEFERMAN, S. [1970] *Hereditarily replete functionals over the ordinals* (**P** 0603) Intuitionism & Proof Th;1968 Buffalo 289-301
◇ F10 F15 ◇ REV MR 44 # 3863 Zbl 218 # 02025 • ID 03705

FEFERMAN, S. [1971] *Ordinals and functionals in proof theory* (**P** 0743) Int Congr Math (II,11,Proc);1970 Nice 1*229-233
◇ F10 F15 F35 F50 ◇ REV MR 55 # 10250 Zbl 244 # 02010 JSL 40.625 • ID 14688

FEFERMAN, S. [1972] *Infinitary properties, local functors, and systems of ordinal functions* (**P** 2080) Conf Math Log;1970 London 63-97
◇ C30 C40 C75 E10 F15 G30 ◇ REV MR 50 # 12646 Zbl 302 # 02018 • ID 03706

FEFERMAN, S. [1974] *Predicatively reducible systems of set theory* (**P** 0693) Axiomatic Set Th;1967 Los Angeles 2*11-32
◇ E25 E30 E70 F65 ◇ REV MR 52 # 10418 Zbl 312 # 02051 • ID 21714

FEFERMAN, S. [1975] *A language and axioms for explicit mathematics* (**P** 0765) Algeb & Log;1974 Clayton 87-139
◇ D55 F35 F50 F65 ◇ REV MR 53 # 12899 Zbl 357 # 02029 JSL 49.308 • ID 23151

FEFERMAN, S. [1975] *Impredicativity of the existence of the largest divisible subgroup of an abelian p-group* (**C** 0782) Model Th & Algeb (A. Robinson) 117-130
◇ C57 C60 F65 ◇ REV MR 53 # 5274 Zbl 316 # 02058 • ID 22905

FEFERMAN, S. [1977] *Theories of finite type related to mathematical practice* (**C** 1523) Handb of Math Logic 913-971
◇ B15 D55 D65 F10 F15 F35 F98 ◇ REV MR 58 # 10343 JSL 49.980 • ID 27330

FEFERMAN, S. [1978] *Recursion theory and set theory: a marriage of convenience* (**P** 1628) Generalized Recursion Th (2);1977 Oslo 55-98
◇ D75 E30 E70 F65 ◇ REV MR 81i:03074 Zbl 453 # 03047 • ID 72806

FEFERMAN, S. [1979] *A more perspicuous system for predicativity* (**C** 2537) Spez Wissenschaftsth I*58-93
◇ F35 F65 ◇ ID 41697

FEFERMAN, S. [1979] *Constructive theories of functions and classes* (**P** 2627) Logic Colloq;1978 Mons 159-224
◇ F50 F55 F60 F98 ◇ REV MR 81i:03093 Zbl 441 # 03022 JSL 49.308 • ID 56075

FEFERMAN, S. [1979] *What does logic have to tell us about mathematical proofs?* (**J** 2789) Math Intell 2*20-24
◇ F98 ◇ REV Zbl 426 # 03011 • ID 53607

FEFERMAN, S. & SIEG, W. [1981] *Inductive definitions and subsystems of analysis* (**C** 3588) Iter Induct Def & Subsyst of Anal: Rec Proof-Th Stud 16-77
◇ F35 F50 ◇ REV MR 84d:03074 • ID 39775

FEFERMAN, S. [1981] see BUCHHOLZ, W.

FEFERMAN, S. & SIEG, W. [1981] *Proof-theoretic equivalence between classical and constructive theories for analysis* (**C** 3588) Iter Induct Def & Subsyst of Anal: Rec Proof-Th Stud 78-142
◇ F35 F50 ◇ REV MR 84d:03074 • ID 39782

FEFERMAN, S. [1982] *Inductively presented systems and the formalization of meta-mathematics* (**P** 3623) Logic Colloq;1980 Prague 95-128
◇ F30 F35 F40 ◇ REV MR 83m:03067 Zbl 516 # 03006 • ID 35466

FEFERMAN, S. [1982] *Iterated inductive fixed-point theories: application to Hancock's conjecture* (**P** 3634) Patras Logic Symp;1980 Patras 171-196
◇ F15 F35 F65 ◇ REV MR 84f:03050 Zbl 522 # 03045 • ID 34469

FEFERMAN, S. [1982] *Monotone inductive definitions* (**P** 3638) Brouwer Centenary Symp;1981 Noordwijkerhout 77-89 ◇ F35 F50 ◇ REV MR 85h:03061 Zbl 525 # 03037 • ID 37481

FEFERMAN, S. & JAEGER, G. [1983] *Choice principles, the bar rule and autonomously iterated comprehension schemes in analysis* (**J** 0036) J Symb Logic 48*63-70 ◇ F15 F35 ◇ REV MR 84g:03097 Zbl 525 # 03038 • ID 34201

FEFERMAN, S. [1984] *Between constructive and classical mathematics* (**P** 2153) Logic Colloq;1983 Aachen 2*143-162 ◇ A05 F35 F50 F55 ◇ REV MR 86e:03057 • ID 40395

FEFERMAN, S. [1984] *Kurt Goedel: conviction and caution* (**J** 0455) Phil Naturalis 21*546-562 ◇ A05 A10 F99 ◇ ID 47923

FEFERMAN, S. [1985] *High-level programs and typed vs. untyped constructive foundations* (**P** 4646) Atti Incontri Log Mat (2);1983/84 Siena 601-605 ◇ A05 F35 F50 ◇ ID 49780

FEFERMAN, S. see Vol. I, II, III, IV, V for further entries

FEHR, E. [1975] *Eine universelle λ-Kalkuel-Programmiersprache und ihr Interpreter* (**P** 1449) Automata Th & Formal Lang;1975 Kaiserslautern 245-251 ◇ B40 B75 ◇ REV MR 53 # 14964 Zbl 312 # 68007 • ID 23225

FEHR, E. [1975] *Eine universelle Lambda-Kalkuel Programmiersprache und ihr Interpreter* (**X** 1512) Univ Bonn Inst Informatik: Bonn 126pp ◇ B40 B75 ◇ REV MR 54 # 9133 Zbl 379 # 02008 • ID 25829

FEHR, E. [1978] see DAMM, W.

FEHR, E. [1980] *Lambda-calculus as control structures of programming languages* (**S** 1642) Schr Inf Angew Math, Ber (Aachen) 57*81pp ◇ B40 B75 ◇ REV Zbl 448 # 68003 • ID 56662

FEHR, E. [1982] see BERKLING, K.J.

FELSCHER, W. [1975] *Kombinatorische Konstruktionen mit Beweisen und Schnittelimination* (**P** 1440) ⊧ ISILC Proof Th Symp (Schuette);1974 Kiel 500*119-151 ◇ F05 F20 ◇ REV MR 54 # 9984 Zbl 363 # 02038 • ID 25751

FELSCHER, W. [1976] *On interpolation when function symbols are present* (**J** 0009) Arch Math Logik Grundlagenforsch 17*145-157 ◇ C40 C75 F50 ◇ REV MR 54 # 12476 Zbl 339 # 02011 • ID 04239

FELSCHER, W. [1979] *Eine weitreichende Wohlordnung der natuerlichen Zahlen* (**J** 0056) Publ Dep Math, Lyon 16/3*47-61 ◇ E10 F15 ◇ REV MR 82d:03078 Zbl 446 # 04001 • ID 72846

FELSCHER, W. [1985] *Dialogues, strategies, and intuitionistic provability* (**J** 0073) Ann Pure Appl Logic 28*217-254 ◇ F05 F07 F50 ◇ REV Zbl 571 # 03029 • ID 42595

FELSCHER, W. see Vol. I, III, V for further entries

FENSTAD, J.E. [1967] *Subsets of natural numbers (Norwegian) (English summary)* (**J** 0311) Nordisk Mat Tidskr 15*5-24 ◇ A05 E35 E50 F30 F35 F65 ◇ REV MR 35 # 2714 Zbl 156.247 • ID 17152

FENSTAD, J.E. [1968] *On the completeness of some transfinite recursive progressions of axiomatic theories* (**J** 0036) J Symb Logic 33*69-76 ◇ F15 F30 ◇ REV MR 39 # 59 Zbl 195.304 JSL 34.131 • ID 03743

FENSTAD, J.E. see Vol. I, II, III, IV, V for further entries

FERBUS, M.-C. [1984] *Functorial bounds for cut elimination in $L_{\beta\omega}$ I* (**J** 0009) Arch Math Logik Grundlagenforsch 24*141-158 ◇ C75 F05 F15 G30 ◇ REV MR 86c:03049 Zbl 558 # 03026 • ID 42605

FERBUS, M.-C. [1985] *Functorial bounds for cut elimination in $L_{\beta\omega}$ II* (**J** 0009) Arch Math Logik Grundlagenforsch 25*13-20 ◇ F05 F35 ◇ ID 49120

FERRANTE, J. & RACKOFF, C.W. [1975] *A decision procedure for the first order theory of real addition with order* (**J** 1428) SIAM J Comp 4*69-76 ◇ B25 C10 C60 D15 F35 ◇ REV MR 52 # 10403 Zbl 277 # 02010 • ID 21702

FERRANTE, J. see Vol. III, IV for further entries

FEVRIER, P. [1956] *Tendances constructives en logique moderne* (**J** 0154) Synthese 10*52-58 ◇ A05 F50 ◇ REV MR 20 # 5136 • ID 42819

FEVRIER, P. also published under the name DESTOUCHES-FEVRIER, P.

FEVRIER, P. see Vol. I, V for further entries

FEYS, R. [1946] *La technique de la logique combinatoire* (**J** 0252) Rev Philos Louvain 44*74-103,237-270 ◇ B40 ◇ REV JSL 12.27 • ID 28457

FEYS, R. [1958] see CURRY, H.B.

FEYS, R. see Vol. I, II, V for further entries

FIDEL, M.M. [1978] *An algebraic study of a propositional system of Nelson* (**P** 1800) Brazil Conf Math Log (1);1977 Campinas 99-117 ◇ B55 F50 G10 G25 ◇ REV MR 80a:03073 Zbl 389 # 03025 • ID 52314

FIDEL, M.M. see Vol. II, III for further entries

FILIPEC, P. [1976] *On convergence of the Fourier series of a constructive function of weakly bounded variation* (**J** 0140) Comm Math Univ Carolinae (Prague) 17*755-769 ◇ F60 ◇ REV MR 56 # 5238 Zbl 346 # 02016 • ID 61680

FINDLAY, J. [1942] *Goedelian sentences: a non-numerical approach* (**J** 0094) Mind 51*259-265 ◇ A05 F30 ◇ REV JSL 7.129 • ID 03781

FINE, K. [1985] *Natural deduction and arbitrary objects* (**J** 0122) J Philos Logic 14*57-107 ◇ A05 F07 ◇ REV Zbl 571 # 03004 • ID 42652

FINE, K. see Vol. II, III, V for further entries

Author Index

FINSLER, P. [1926] *Formale Beweise und die Entscheidbarkeit* (**J** 0044) Math Z 25*676-682
• TRANSL [1967] (**C** 0675) From Frege to Goedel 440-445
◇ A05 D35 F25 ◇ REV FdM 52.49 • ID 03793

FINSLER, P. [1944] *Gibt es unentscheidbare Saetze?* (**J** 2022) Comm Math Helvetici 16*310-320
◇ A05 F30 ◇ REV MR 6.197 JSL 11.131 • ID 03797

FINSLER, P. [1951] *Eine transfinite Folge arithmetischer Operationen* (**J** 2022) Comm Math Helvetici 25*75-90
◇ E10 F15 ◇ REV MR 13.120 Zbl 42.280 JSL 22.336
• ID 03799

FINSLER, P. [1965] *Zur Goldbachschen Vermutung* (**J** 0026) Elem Math 20*121-122
◇ F30 ◇ REV MR 32 # 7528 Zbl 215.47 • ID 28007

FINSLER, P. see Vol. I, V for further entries

FISCHER, MICHAEL J. & RABIN, M.O. [1974] *Super-exponential complexity of Presburger arithmetic* (**P** 0761) Compl of Computation;1973 New York 27-41
◇ B25 D15 F20 F30 ◇ REV MR 51 # 2893
Zbl 319 # 68024 • ID 17344

FISCHER, MICHAEL J. see Vol. I, II, IV for further entries

FISCHER, P.C. [1965] *Theory of provably recursive functions* (**J** 0064) Trans Amer Math Soc 117*494-520
◇ D20 D30 F30 ◇ REV MR 31 # 47 Zbl 129.260
JSL 32.270 • ID 04301

FISCHER, P.C. see Vol. IV for further entries

FISCHER SERVI, G. [1976] *Un'algebrizzazione del calcolo intuizionista monadico (English summary)* (**J** 0319) Matematiche (Sem Mat Catania) 31*262-276
◇ B55 F50 G10 ◇ REV MR 83b:03072 Zbl 437 # 03033
• ID 55901

FISCHER SERVI, G. [1977] *On modal logic with an intuitionistic base* (**J** 0063) Studia Logica 36*141-149
◇ B45 F50 ◇ REV Zbl 364 # 02015 • ID 50945

FISCHER SERVI, G. [1978] *The finite model property for MIPQ and some consequences* (**J** 0047) Notre Dame J Formal Log 19*687-692
◇ B55 F50 G10 ◇ REV MR 80g:03015 Zbl 368 # 02024
• ID 52127

FISCHER SERVI, G. [1981] *Semantics for a class of intuitionistic modal calculi* (**C** 3515) Ital Studies in Phil of Sci 59-72
◇ B45 B55 F50 ◇ REV MR 82f:03013 Zbl 452 # 03014
• ID 54078

FISCHER SERVI, G. see Vol. II for further entries

FISHER, A. [1982] *Formal number theory and computability* (**X** 0894) Oxford Univ Pr: Oxford xiii + 190pp
◇ B98 D20 F30 F98 ◇ REV MR 85g:03001
Zbl 504 # 03002 • ID 36967

FITCH, F.B. [1936] *A system of formal logic without an analogue to the Curry W operator* (**J** 0036) J Symb Logic 1*92-100
◇ B40 ◇ REV Zbl 15.241 JSL 2.37 FdM 62.1055
• ID 04319

FITCH, F.B. [1938] *The consistency of the ramified Principia* (**J** 0036) J Symb Logic 3*140-149
◇ B15 F25 ◇ REV Zbl 20.97 JSL 4.97 FdM 64.925
• ID 04321

FITCH, F.B. [1942] *A basic logic* (**J** 0036) J Symb Logic 7*105-114
◇ B40 ◇ REV MR 4.125 Zbl 60.23 JSL 8.30 • ID 04325

FITCH, F.B. [1944] *A minimum calculus for logic* (**J** 0036) J Symb Logic 9*89-94
◇ B20 B40 ◇ REV MR 7.45 Zbl 60.23 JSL 11.127
• ID 04326

FITCH, F.B. [1944] *Representation of calculi* (**J** 0036) J Symb Logic 9*57-62
◇ B20 B40 ◇ REV MR 6.197 Zbl 60.23 JSL 11.28
• ID 04328

FITCH, F.B. [1948] *An extension of basic logic* (**J** 0036) J Symb Logic 13*95-106
◇ B20 B40 F35 ◇ REV MR 9.559 Zbl 30.193 JSL 14.68
• ID 04329

FITCH, F.B. [1948] *Intuitionistic modal logic with quantifiers* (**J** 0050) Port Math 7*113-118
◇ B45 F50 ◇ REV MR 10.669 Zbl 34.153 JSL 14.261
• ID 04332

FITCH, F.B. [1949] *A further consistent extension of basic logic* (**J** 0036) J Symb Logic 14*209-218
◇ B20 B40 ◇ REV MR 12.2 Zbl 38.5 JSL 15.219
• ID 04334

FITCH, F.B. [1949] *The Heine-Borel theorem in extended basic logic* (**J** 0036) J Symb Logic 14*9-15
◇ B28 B40 E70 F35 ◇ REV MR 10.669 Zbl 35.8
JSL 15.137 • ID 04333

FITCH, F.B. [1950] *A demonstrably consistent mathematics. Part I* (**J** 0036) J Symb Logic 15*17-24
◇ B30 B40 E70 ◇ REV MR 12.2 Zbl 39.245 JSL 16.268
• REM Part II 1951 • ID 17113

FITCH, F.B. [1951] *A demonstrably consistent mathematics. Part II* (**J** 0036) J Symb Logic 16*121-124
◇ B30 B40 E70 ◇ REV MR 13.4 Zbl 45.151 JSL 16.268
• REM Part I 1950 • ID 17114

FITCH, F.B. [1952] *Symbolic logic. An introduction* (**X** 0879) Inst Sup Philos: Louvain x + 238pp
◇ B40 B98 ◇ REV MR 15.592 Zbl 49.5 JSL 17.266
• ID 04335

FITCH, F.B. [1953] *A simplification of basic logic* (**J** 0036) J Symb Logic 18*317-325
◇ B40 ◇ REV MR 15.924 Zbl 52.251 JSL 20.81 • ID 04337

FITCH, F.B. [1953] *Self-referential relations* (**P** 0645) Int Congr Philos (11);1953 Bruxelles 5*121-127
◇ B40 ◇ REV MR 15.91 Zbl 52.251 JSL 24.240 • ID 04336

FITCH, F.B. [1954] *A definition of negation in extended basic logic* (**J** 0036) J Symb Logic 19*29-36
◇ B40 ◇ REV MR 16.2 Zbl 59.14 JSL 20.81 • ID 04338

FITCH, F.B. [1956] *Recursive functions in basic logic* (**J** 0036) J Symb Logic 21*337-346
◇ B40 ◇ REV MR 19.237 Zbl 88.12 JSL 35.152 • ID 04339

FITCH, F.B. [1957] *A definition of existence in terms of abstraction and disjunction* (**J** 0036) J Symb Logic 22*343-344
◇ B40 ◇ REV MR 20 # 5730 Zbl 88.12 • ID 04340

FITCH, F.B. [1958] *An extensional variety of extended basic logic* (**J** 0036) J Symb Logic 23*13-21
◇ B40 ◇ REV MR 20 # 6978 Zbl 218 # 02013 • ID 04341

FITCH, F.B. [1958] *Representation of sequential circuits in combinatory logic* (**J** 0153) Phil of Sci (East Lansing) 25*263-279
◇ B40 B70 ◇ REV MR 22 # 6660 JSL 30.380 • ID 04342

FITCH, F.B. [1959] *Quasi-constructive foundations for mathematics* (**P** 0634) Constructivity in Math;1957 Amsterdam 26-36
◇ B40 ◇ REV MR 23 # A1538 Zbl 109.241 JSL 37.402 • ID 04343

FITCH, F.B. [1963] *The system CΔ of combinatory logic* (**J** 0036) J Symb Logic 28*87-97
◇ B40 ◇ REV MR 31 # 1181 Zbl 129.257 JSL 29.198 • ID 04345

FITCH, F.B. [1966] *Natural deduction rules for obligation* (**J** 0325) Amer Phil Quart 3*27-38
◇ B45 F07 ◇ REV JSL 33.136 • ID 17115

FITCH, F.B. [1967] *A theory of logical essences* (**J** 0320) Monist 51*104-109
◇ B40 ◇ REV JSL 34.125 • ID 04346

FITCH, F.B. [1968] *A note on recursive relations* (**J** 0036) J Symb Logic 33*107
◇ B40 D20 ◇ REV MR 38 # 4304 Zbl 184.19 JSL 37.758 • ID 17117

FITCH, F.B. & ORGASS, R.J. [1969] *A theory of computing machines* (**J** 0178) Stud Gen 22*83-104
◇ B40 D10 ◇ REV MR 43 # 2879 • ID 14672

FITCH, F.B. [1974] *Elements of combinatory logic* (**X** 0875) Yale Univ Pr: New Haven viii + 162pp
◇ B40 B98 ◇ REV MR 54 # 2429 JSL 41.789 • ID 24019

FITCH, F.B. [1980] *A consistent combinatory logic with an inverse to equality* (**J** 0036) J Symb Logic 45*529-543
◇ B40 ◇ REV MR 82i:03023 Zbl 445 # 03003 • ID 56467

FITCH, F.B. [1980] *An extension of a system of combinatory logic* (**C** 3050) Essays Combin Log, Lambda Calc & Formalism (Curry) 125-140
◇ B40 ◇ REV MR 82g:03018 Zbl 469 # 03006 • ID 72930

FITCH, F.B. [1981] *The consistency of system Q* (**J** 0036) J Symb Logic 46*67-76
◇ B40 ◇ REV MR 82g:03019 Zbl 461 # 03001 • ID 54478

FITCH, F.B. see Vol. I, II, V for further entries

FITTING, M. [1969] *Intuitionistic logic, model theory, and forcing* (**X** 0809) North Holland: Amsterdam 191pp
◇ B98 C90 C98 E25 E35 E45 E50 F50 F98 ◇ REV MR 41 # 6666 Zbl 188.320 JSL 36.166 • ID 04349

FITTING, M. [1970] *Intuitionistic model theory and the Cohen independence proofs* (**P** 0603) Intuitionism & Proof Th;1968 Buffalo 219-226
◇ A05 C90 E35 E40 F50 ◇ REV MR 42 # 4390 Zbl 205.306 • ID 14547

FITTING, M. [1973] *Model existence theorems for modal and intuitionistic logics* (**J** 0036) J Symb Logic 38*613-627
◇ B45 F50 ◇ REV MR 52 # 7852 Zbl 286 # 02060 • ID 04355

FITTING, M. [1983] *Proof methods for modal and intuitionistic logics* (**S** 3307) Synth Libr viii + 555pp
◇ B45 B55 F07 F50 F98 ◇ REV MR 84j:03036 Zbl 523 # 03013 JSL 50.855 • ID 34628

FITTING, M. see Vol. I, II, III, IV, V for further entries

FLAGG, R.C. [1985] *Church's thesis is consistent with epistemic arithmetic* (**C** 3659) Intens Math 121-172
◇ F30 F50 ◇ ID 44895

FLAGG, R.C. [1985] *Epistemic set theory is a conservative extension of intuitionistic set theory* (**J** 0036) J Symb Logic 50*895-902
◇ B45 E35 E70 F50 ◇ ID 49605

FLAGG, R.C. see Vol. I for further entries

FLANNAGAN, T.B. [1975] *On an extension of Hilbert's second ε-theorem* (**J** 0036) J Symb Logic 40*393-397
◇ F05 ◇ REV MR 52 # 50 Zbl 327 # 02013 • ID 14814

FLANNAGAN, T.B. [1976] *A new finitary proof of a theorem of Mostowski* (**C** 1468) Sets & Classes (Bernays) 257-275
◇ E30 E70 F25 ◇ REV MR 57 # 16075 Zbl 341 # 02051 • ID 23736

FLANNAGAN, T.B. [1983] *A note on a proof of Shepherdson* (**J** 0009) Arch Math Logik Grundlagenforsch 23*55-60
◇ F30 ◇ REV MR 85e:03140 Zbl 544 # 03030 • ID 40755

FLANNAGAN, T.B. see Vol. III, V for further entries

FLEISCHHACKER, L. [1979] *Is Russell's vicious circle principle false or meaningless?* (**J** 0076) Dialectica 33*23-30
◇ A05 B40 ◇ REV MR 83i:03014b Zbl 442 # 03007 • ID 56364

FLORENCE, J.B. [1975] see DAWES, A.M.

FLORENCE, J.B. see Vol. IV for further entries

FOGELIS, E. [1950] *On finite proofs of arithmetical theorems* (**J** 0324) Izv Akad Nauk Latv SSR 6*81-86
◇ B28 F30 ◇ REV MR 13.199 • ID 17125

FOL'K, N.F. & SHESTOPAL, G.A. [1968] *The solvability of elementary theories of integral and natural numbers with addition (Russian)* (**J** 0323) Uch Zap Ped Inst, Ryazan 67*141-155
◇ B25 C10 F30 ◇ REV MR 41 # 1522 • ID 17126

FOURMAN, M.P. [1977] *The logic of topoi* (**C** 1523) Handb of Math Logic 1053-1090
◇ F50 G30 ◇ REV MR 58 # 10343 JSL 49.980 • ID 27332

FOURMAN, M.P. & HYLAND, J.M.E. [1979] *Sheaf models for analysis* (**P** 2901) Appl Sheaves;1977 Durham 280-301
◇ C65 C90 F35 F50 G10 G30 ◇ REV MR 82g:03069 Zbl 427 # 03028 • ID 53715

FOURMAN, M.P. & SCOTT, D.S. [1979] *Sheaves and logic* (**P** 2901) Appl Sheaves;1977 Durham 302-401
◇ C90 C98 E70 F35 F50 G30 ◇ REV MR 82d:03061 Zbl 415 # 03053 JSL 48.1201 • ID 53155

FOURMAN, M.P. [1980] *Sheaf models for set theory* (**J** 0326) J Pure Appl Algebra 19*91-101
◇ C62 C90 E25 E35 E70 F50 G30 ◇ REV MR 82i:03078 Zbl 446 # 03041 • ID 56570

FOURMAN, M.P. & GRAYSON, R.J. [1982] *Formal spaces* (**P** 3638) Brouwer Centenary Symp;1981 Noordwijkerhout 107-122
◇ F50 G30 ◇ REV MR 85c:03023 Zbl 537 # 03040 • ID 33229

FOURMAN, M.P. [1982] *Notions of choice sequence* (**P** 3638) Brouwer Centenary Symp;1981 Noordwijkerhout 91-105
◇ F35 F50 G30 ◇ REV MR 85d:03115 Zbl 545 # 03036 • ID 33228

FOURMAN, M.P. & SCEDROV, A. [1982] *The "world's simplest axiom of choice" fails* (**J** 0504) Manuscr Math 38∗325-332 ◇ E25 E35 F50 G30 ◇ REV MR 83k:03074 Zbl 499 # 03048 • ID 33227

FOURMAN, M.P. [1984] *Continuous truth. I. Nonconstructive objects* (**P** 3710) Logic Colloq;1982 Firenze 161-180 ◇ A05 E70 F35 F50 G30 ◇ REV MR 86b:03084 Zbl 575 # 03041 • ID 43977

FOURMAN, M.P. see Vol. III, V for further entries

FRAENKEL, A.A. [1951] *On the crisis of the principle of the excluded middle* (**J** 0287) Scripta Math 17∗5-16 ◇ A05 B20 F99 ◇ REV Zbl 54.4 • ID 47908

FRAENKEL, A.A. [1954] *The intuitionistic revolution in mathematics and logic* (**J** 0493) Bull Res Counc Israel Sect F 3∗283-289 ◇ A05 F55 ◇ REV MR 16.663 • ID 24919

FRAENKEL, A.A. [1958] see BAR-HILLEL, Y.

FRAENKEL, A.A. see Vol. I, II, V for further entries

FRANZ, W. [1977] *Ueber mathematische Aussagen, die samt ihrer Negation nachweislich unbeweisbar sind. Der Unvollstaendigkeitssatz von Goedel* (**X** 2717) Steiner: Wiesbaden 27pp ◇ F30 ◇ REV MR 80c:03004 Zbl 352 # 02010 • ID 73027

FREGUGLIA, P. [1972] *Platonismo ed intuizionismo e loro superamento* (**J** 1515) Archimede 24∗292-297 ◇ A05 F98 ◇ REV Zbl 247 # 02010 • ID 29495

FREGUGLIA, P. see Vol. I, IV, V for further entries

FREJDZON, R.I. [1972] *Families of recursive predicates of measure zero (Russian) (English summary)* (**S** 0228) Zap Nauch Sem Leningrad Otd Mat Inst Steklov 32∗121-128,158 • TRANSL [1976] (**J** 1531) J Sov Math 6∗449-455 ◇ D20 F60 ◇ REV MR 49 # 8826 Zbl 354 # 02030 • ID 04594

FREJDZON, R.I. see Vol. IV for further entries

FREUDENTHAL, H. [1937] *Zum intuitionistischen Raumbegriff* (**J** 0020) Compos Math 4∗82-111 ◇ F55 ◇ REV Zbl 15.241 • ID 04606

FREUDENTHAL, H. [1937] *Zur intuitionistischen Deutung logischer Formeln* (**J** 0020) Compos Math 4∗112-116 ◇ A05 F50 ◇ REV Zbl 15.242 JSL 2.48 FdM 62.1045 • ID 04605

FREUDENTHAL, H. [1953] *Zur Geschichte der vollstaendigen Induktion* (**J** 2341) Arch Int Hist Sci 22∗17-37 ◇ A05 A10 F30 ◇ REV MR 14.1049 Zbl 50.242 • ID 28214

FREUDENTHAL, H. [1981] *L.E.J.Brouwer -topologist, intuitionist, philosopher (Dutch)* (**J** 3077) Nieuw Arch Wisk, Ser 3 29∗249-253 ◇ A10 F99 ◇ REV MR 83m:01050 Zbl 476 # 01006 • ID 40345

FREUDENTHAL, H. see Vol. I, II, IV for further entries

FRIDMAN, EH.I. & SLOBODSKOJ, A.M. [1973] *The theory of the additive group of the integers with an arbitrary number of predicates that define maximal subgroups (Russian)* (**C** 1443) Algebra, Vyp 2 (Irkutsk) 130-137 ◇ C10 F30 ◇ REV MR 53 # 10575 • ID 23081

FRIDMAN, EH.I. see Vol. I, III, IV for further entries

FRIEDMAN, D.P. & WAND, M. [1978] *Compiling lambda-expressions using continuations and factorizations* (**J** 3144) Comput Lang, Int J 3∗241-263 ◇ B40 ◇ REV Zbl 385 # 68023 • ID 52172

FRIEDMAN, H.M. [1969] *Bar induction and Π^1_1-CA* (**J** 0036) J Symb Logic 34∗353-362 ◇ F35 ◇ REV MR 40 # 4109 Zbl 182.9 • ID 04646

FRIEDMAN, H.M. [1970] *Higher set theory and mathematical practice* (**J** 0007) Ann Math Logic 2∗325-357 ◇ D55 E15 E35 E45 E60 F35 ◇ REV MR 44 # 1556 Zbl 215.327 • ID 03773

FRIEDMAN, H.M. [1970] *Iterated inductive definitions and Σ^1_2-AC* (**P** 0603) Intuitionism & Proof Th;1968 Buffalo 435-442 ◇ C62 D55 E25 F35 ◇ REV MR 44 # 1555 Zbl 216.6 • ID 14546

FRIEDMAN, H.M. [1971] *A more explicit set theory* (**P** 0693) Axiomatic Set Th;1967 Los Angeles 1∗49-66 ◇ A05 C62 E30 E35 E45 F30 F35 ◇ REV MR 43 # 4658 Zbl 233 # 02026 • ID 17088

FRIEDMAN, H.M. [1971] *Algorithmic procedures, generalized Turing algorithms, and elementary recursion theory* (**P** 0638) Logic Colloq;1969 Manchester 361-389 ◇ A05 D10 D20 D75 F99 ◇ REV MR 46 # 3275 Zbl 221 # 02018 • ID 04648

FRIEDMAN, H.M. [1971] see ENDERTON, H.B.

FRIEDMAN, H.M. [1973] *Countable models of set theories* (**P** 0713) Cambridge Summer School Math Log;1971 Cambridge GB 539-573 ◇ C15 C62 E45 F35 H20 ◇ REV MR 50 # 102 Zbl 271 # 02036 • ID 04654

FRIEDMAN, H.M. [1973] *Some applications of Kleene's methods for intuitionistic systems* (**P** 0713) Cambridge Summer School Math Log;1971 Cambridge GB 113-170 ◇ F35 F50 ◇ REV MR 51 # 12486 Zbl 272 # 02038 • ID 17274

FRIEDMAN, H.M. [1973] *The consistency of classical set theory relative to a set theory with intuitionistic logic* (**J** 0036) J Symb Logic 38∗315-319 ◇ B15 E30 E35 F50 ◇ REV MR 50 # 68 Zbl 278 # 02045 • ID 04650

FRIEDMAN, H.M. [1974] *On existence proofs of Hanf numbers* (**J** 0036) J Symb Logic 39∗318-324 ◇ C55 C75 E30 E70 F50 ◇ REV MR 53 # 12943 Zbl 293 # 02039 • ID 04656

FRIEDMAN, H.M. [1975] *Equality between functionals* (**C** 0758) Logic Colloq Boston 1972-73 22-37 ◇ B40 F10 ◇ REV MR 52 # 2833 Zbl 311 # 02040 • ID 17627

FRIEDMAN, H.M. [1975] *One hundred and two problems in mathematical logic* (**J** 0036) J Symb Logic 40∗113-129 ◇ B98 C98 D98 E98 F98 ◇ REV MR 51 # 5254 Zbl 318 # 02002 • ID 04296

FRIEDMAN, H.M. [1975] *Provable equality in primitive recursive arithmetic with and without induction* (**J** 0048) Pac J Math 57∗379-392 ◇ F30 ◇ REV MR 52 # 5373 Zbl 354 # 02034 • ID 18149

FRIEDMAN, H.M. [1975] *Some systems of second order arithmetic and their use* (**P** 1521) Int Congr Math (II,12);1974 Vancouver 1*235-242
◇ E30 F35 ◇ REV MR 55#2521 Zbl 344#02022
• ID 61836

FRIEDMAN, H.M. [1975] *The disjunction property implies the numerical existence property* (**J** 0054) Proc Nat Acad Sci USA 72*2877-2878
◇ F50 ◇ REV MR 52#47 Zbl 342#02012 • ID 17215

FRIEDMAN, H.M. [1976] *Recursiveness in Π^1_1 paths through $\mathcal{O}$* (**J** 0053) Proc Amer Math Soc 54*311-315
◇ D30 D55 F15 ◇ REV MR 53#2663 Zbl 349#02037
• ID 21610

FRIEDMAN, H.M. [1976] *Uniformly defined descending sequences of degrees* (**J** 0036) J Symb Logic 41*363-367
◇ D30 D55 F30 F35 ◇ REV MR 55#2540
Zbl 366#02030 • ID 14767

FRIEDMAN, H.M. [1977] *On the derivability of instantiation properties* (**J** 0036) J Symb Logic 42*506-514
◇ F50 ◇ REV MR 57#9501 Zbl 436#03051 • ID 26852

FRIEDMAN, H.M. [1977] *Set theoretic foundations for constructive analysis* (**J** 0120) Ann of Math, Ser 2 105*1-28
◇ E70 F50 ◇ REV MR 55#7748 Zbl 353#02014
JSL 46.868 • ID 31761

FRIEDMAN, H.M. [1978] *A proof of foundation from the axioms of cumulation* (**P** 1864) Higher Set Th;1977 Oberwolfach 669*15-16
◇ E30 F35 ◇ REV MR 80m:03085 Zbl 381#03040
• ID 31765

FRIEDMAN, H.M. [1978] *Classically and intuitionistically provably recursive functions* (**P** 1864) Higher Set Th;1977 Oberwolfach 21-27
◇ B15 D20 E70 F30 F50 ◇ REV MR 80b:03093
Zbl 396#03045 • ID 31762

FRIEDMAN, H.M. [1980] *A strong conservative extension of Peano arithmetic* (**P** 2058) Kleene Symp;1978 Madison 113-122
◇ B28 F30 F35 ◇ REV MR 82d:03096 Zbl 471#03045
• ID 55240

FRIEDMAN, H.M. & MCALOON, K. & SIMPSON, S.G. [1982] *A finite combinatorial principle which is equivalent to the l-consistency of predicative analysis* (**P** 3634) Patras Logic Symp;1980 Patras 197-230
◇ F15 F35 F65 ◇ REV MR 84k:03138 Zbl 522#03044
• ID 35030

FRIEDMAN, H.M. & SIMPSON, S.G. & SMITH, RICK L. [1983] *Countable algebra and set existence axioms* (**J** 0073) Ann Pure Appl Logic 25*141-181 • ERR/ADD ibid 28*319-320
◇ E30 F35 ◇ REV MR 85i:03157 Zbl 575#03038
Zbl 575#03039 • ID 44230

FRIEDMAN, H.M. & SCEDROV, A. [1983] *Set existence property for intuitionistic theories with dependent choice* (**J** 0073) Ann Pure Appl Logic 25*129-140 • ERR/ADD ibid 26*101
◇ E70 F35 F50 ◇ REV MR 85f:03062a MR 85f:03062b
Zbl 539#03039 • ID 40803

FRIEDMAN, H.M. [1983] *Unary Borel functions and second-order arithmetic* (**J** 0345) Adv Math 50*155-159
◇ E30 E40 F35 ◇ REV MR 85e:03144 Zbl 543#03034
• ID 40758

FRIEDMAN, H.M. & SCEDROV, A. [1984] *Large sets in intuitionistic set theory* (**J** 0073) Ann Pure Appl Logic 27*1-24
◇ E55 E70 F50 ◇ REV MR 85j:03103 Zbl 544#03031
• ID 41019

FRIEDMAN, H.M. & SCEDROV, A. [1985] *Arithmetic transfinite induction and recursive well-orderings* (**J** 0345) Adv Math 56*283-294
◇ D45 F30 F50 ◇ ID 47284

FRIEDMAN, H.M. & SCEDROV, A. [1985] *The lack of definable witnesses and provably recursive functions in intuitionistic set theories* (**J** 0345) Adv Math 57*1-13
◇ E70 F50 ◇ ID 48256

FRIEDMAN, H.M. see Vol. III, IV, V for further entries

FRIEDMAN, JOYCE [1957] *Some results in Church's restricted recursive arithmetic* (**J** 0036) J Symb Logic 22*337-342
◇ F30 ◇ REV MR 20#3779 Zbl 81.13 JSL 24.241
• ID 04664

FRIEDMAN, JOYCE & WARREN, D.S. [1980] *λ-normal forms in an intensional logic for English* (**J** 0063) Studia Logica 39*311-324
◇ B40 ◇ REV MR 82k:03030 Zbl 457#03010 • ID 54335

FRIEDMAN, JOYCE see Vol. I, III, IV for further entries

FRIEDMANN, J. [1982] *Bemerkungen zur Logikbegruendung im deutschen Konstruktivismus* (**J** 0989) Z Allg Wissth 13*383-402
◇ A05 F65 ◇ REV MR 84c:03011 • ID 35676

FRIEDRICH, U. [1972] *Entscheidbarkeit der monadischen Nachfolgerarithmetik mit endlichen Automaten ohne Analysetheorem (Russian, English and French summaries)* (**J** 0115) Wiss Z Humboldt-Univ Berlin, Math-Nat Reihe 21*503-504
◇ B25 C85 D05 F35 ◇ REV MR 48#3724
Zbl 248#02051 • ID 04668

FRIEDRICH, W. & LUCKHARDT, H. [1980] *Intuitionistic uniformity principles for propositions and some applications* (**J** 0063) Studia Logica 39*361-369
◇ F35 F50 ◇ REV MR 82m:03068 Zbl 463#03035
• ID 54575

FRIEDRICH, W. [1984] *Spielquantorinterpretation unstetiger Funktionale der hoeheren Analysis* (**J** 0009) Arch Math Logik Grundlagenforsch 24*73-99
◇ F10 F35 ◇ REV MR 86c:03052 Zbl 544#03021
• ID 41010

FRIEDRICH, W. [1985] *Goedelsche Funktionalinterpretation fuer eine Erweiterung der klassischen Analysis* (**J** 0068) Z Math Logik Grundlagen Math 31*3-29
◇ F10 F35 F50 ◇ REV Zbl 548#03025 Zbl 565#03021
• ID 42282

FRIEDRICHSDORF, U. [1976] *Einige Bemerkungen zur Peano-Arithmetik* (**J** 0068) Z Math Logik Grundlagen Math 22*431-436
◇ C25 C52 C62 F30 ◇ REV MR 56#119
Zbl 356#02044 • ID 24288

FRIEDRICHSDORF, U. see Vol. V for further entries

FRINK, O. [1938] *New algebras of logic* (**J** 0005) Amer Math Mon 45*210-219
◇ B50 F50 G05 ◇ REV Zbl 18.337 JSL 3.117 FdM 64.30
• ID 04671

FRINK, O. see Vol. V for further entries

FUCHS, HARTWIG [1971] *Algorithmic representation of Wermus' constructions of ordinal numbers* (J 0307) Rev Colomb Mat 5*10-16
◇ F15 ◇ REV MR 49 # 2297 Zbl 266 # 04002 • ID 17097

FUENTES MIRAS, J.R. [1964] *On the foundations of mathematics (Spanish)* (J 0234) Rev Acad Cienc Exact Fis Nat Madrid 58*11-45,73-102,197-214
◇ A05 A10 F98 ◇ REV MR 30 # 4682 Zbl 131.5
• ID 22323

FUENTES MIRAS, J.R. see Vol. I, III for further entries

FUNAHASHI, S. & NAGATA, S. & OSHIBA, T. [1979] *A method for obtaining proof figures of valid formulas in the first order predicate calculus* (J 0523) Bull Nagoya Inst Tech 31*117-126
◇ B10 F07 ◇ REV MR 81j:03093 • ID 77023

FUNAHASHI, S. see Vol. I for further entries

FURMANOWSKI, T. [1982] *Adjoint interpretations of sentential calculi* (J 0063) Studia Logica 41*359-374
◇ B22 F25 G30 ◇ REV MR 85i:03195 Zbl 561 # 03013
• ID 42298

FURMANOWSKI, T. see Vol. II, III for further entries

GABBAY, D.M. [1970] *The decidability of the Kreisel-Putnam system* (J 0036) J Symb Logic 35*431-436
◇ B25 B55 F50 ◇ REV MR 58 # 16167 Zbl 228 # 02013
• ID 04721

GABBAY, D.M. [1971] *Semantic proof of the Craig interpolation theorem for intuitionistic logic and extensions I,II* (P 0638) Logic Colloq;1969 Manchester 391-401,403-410
◇ B55 C40 C90 F50 ◇ REV MR 43 # 3094a MR 43 # 3094b Zbl 234 # 02017 • REM Part III 1977
• ID 04726

GABBAY, D.M. [1972] *Decidability of some intuitionistic predicate theories* (J 0036) J Symb Logic 37*579-587
◇ B25 C90 F50 ◇ REV MR 48 # 86 Zbl 266 # 02025
• ID 04729

GABBAY, D.M. [1972] *Model theory for intuitionistic logic* (J 0068) Z Math Logik Grundlagen Math 18*49-54
◇ B55 C20 C90 F50 ◇ REV MR 45 # 4959 Zbl 242 # 02059 • ID 04730

GABBAY, D.M. [1972] *Sufficient conditions for the undecidability of intuitionistic theories with applications* (J 0036) J Symb Logic 37*375-384
◇ D35 F50 ◇ REV MR 47 # 8278 Zbl 266 # 02026
• ID 04731

GABBAY, D.M. [1973] *A survey of decidability results for modal, tense and intermediate logics* (P 0793) Int Congr Log, Meth & Phil of Sci (4,Proc);1971 Bucharest 29-43
◇ B25 B45 B55 C98 F50 ◇ REV MR 55 # 12470
• ID 17911

GABBAY, D.M. [1973] *The undecidability of intuitionistic theories of algebraically closed fields and real closed fields* (J 0036) J Symb Logic 38*86-92
◇ B25 C60 C90 D35 F50 ◇ REV MR 47 # 6466 Zbl 266 # 02027 • ID 61880

GABBAY, D.M. [1974] *On 2nd order intuitionistic propositional calculus with full comprehension* (J 0009) Arch Math Logik Grundlagenforsch 16*177-186
◇ F35 F50 ◇ REV MR 50 # 12672 Zbl 289 # 02016
• ID 17109

GABBAY, D.M. [1975] *Decidability results in non-classical logics I* (J 0007) Ann Math Logic 8*237-295
◇ B25 B45 C90 F50 ◇ REV MR 52 # 68 Zbl 309 # 02053
• REM Parts II,IV have not yet appeared. Part III 1971
• ID 04738

GABBAY, D.M. [1975] *The decision problem for some finite extensions of the intuitionistic theory of abelian groups* (J 0063) Studia Logica 34*59-67
◇ B25 F50 ◇ REV MR 51 # 12500 Zbl 306 # 02027
• ID 17286

GABBAY, D.M. [1976] *Completeness properties of Heyting's predicate calculus with respect to RE models* (J 0036) J Symb Logic 41*81-94
◇ B55 F50 ◇ REV MR 53 # 12883 Zbl 328 # 02013
• ID 14748

GABBAY, D.M. [1976] *On Kreisel's notion of validity in Post systems* (J 0063) Studia Logica 35*285-295
◇ B50 D03 F50 ◇ REV MR 55 # 12477 Zbl 363 # 02027
• ID 50854

GABBAY, D.M. [1977] *A new version of Beth semantics for intuitionistic logic* (J 0036) J Symb Logic 42*306-308
◇ C90 F50 ◇ REV MR 58 # 16169 Zbl 379 # 02007
• ID 26463

GABBAY, D.M. [1977] *Craig interpolation theorem for intuitionistic logic and extensions part III* (J 0036) J Symb Logic 42*269-271
◇ B55 C40 C90 F50 ◇ REV MR 58 # 16168 Zbl 372 # 02016 • REM Parts I,II 1971 • ID 26456

GABBAY, D.M. [1977] *On some new intuitionistic propositional connectives I* (J 0063) Studia Logica 36*127-139
◇ B45 B55 F50 ◇ REV MR 58 # 161 Zbl 363 # 02026
• ID 50853

GABBAY, D.M. [1977] *Undecidability of intuitionistic theories formulated with the apartness relation* (J 0027) Fund Math 97*57-69
◇ D35 F50 ◇ REV MR 57 # 5690 Zbl 365 # 02012
• ID 27197

GABBAY, D.M. [1978] *What is a classical connective?* (J 0068) Z Math Logik Grundlagen Math 24*37-44
◇ B22 F07 ◇ REV MR 57 # 9500 Zbl 374 # 02016
• ID 32163

GABBAY, D.M. [1981] *Semantical investigations in Heyting's intuitionistic logic* (X 0835) Reidel: Dordrecht x + 287pp
◇ B55 F50 F98 ◇ REV MR 83b:03012 Zbl 453 # 03001
• ID 54131

GABBAY, D.M. [1982] *Intuitionistic basis for non-monotonic logic* (P 3840) Autom Deduct (6);1982 New York 260-273
◇ B55 C90 F50 ◇ REV MR 85e:03062 Zbl 481 # 68091
• ID 38464

GABBAY, D.M. see Vol. I, II, III for further entries

GAIFMAN, H. [1982] see DIMITRACOPOULOS, C.

GAIFMAN, H. see Vol. I, II, III, IV, V for further entries

GAL, I.L. & ROSSER, J.B. & SCOTT, D.S. [1958] *Generalization of a lemma of G.F.Rose* (**J** 0036) J Symb Logic 23*137-138 ◇ F50 ◇ REV MR 21 # 3321 JSL 25.179 • ID 04754

GALIN, A.B. [1980] see FAL'KOVICH, M.A.

GALVAN, S. [1981] *Nota sull'aritmetica formalizzata di Peano (Dimostrazione del piccolo teorema di Fermat mediante una procedura ristretta ai numerali)* (**P** 3092) Congr Naz Logica;1979 Montecatini Terme 701-716 ◇ A10 F30 ◇ ID 48393

GANDY, R.O. [1960] *Proof of Mostowski's conjecture* (**J** 0014) Bull Acad Pol Sci, Ser Math Astron Phys 8*571-575 ◇ F15 ◇ REV MR 23 # A3679 JSL 29.103 • ID 04775

GANDY, R.O. & KREISEL, G. & TAIT, W.W. [1960] *Set existence* (**J** 0014) Bull Acad Pol Sci, Ser Math Astron Phys 8*577-583 ◇ C62 D55 F35 ◇ REV MR 28 # 2964a JSL 27.232 • REM Part I. Part II 1961 • ID 15061

GANDY, R.O. & KREISEL, G. & TAIT, W.W. [1961] *Set existence II* (**J** 0014) Bull Acad Pol Sci, Ser Math Astron Phys 9*881-882 ◇ C62 D55 F35 ◇ REV MR 28 # 2964b JSL 27.232 • REM Part I 1960 • ID 22330

GANDY, R.O. [1980] *An early proof of normalization by A. M. Turing* (**C** 3050) Essays Combin Log, Lambda Calc & Formalism (Curry) 453-455 ◇ B40 F05 ◇ REV MR 81m:03018 Zbl 469 # 03006 • ID 73215

GANDY, R.O. [1980] *Church's thesis and principles for mechanisms* (**P** 2058) Kleene Symp;1978 Madison 123-148 ◇ A05 D05 D10 D20 F99 ◇ REV MR 82h:03036 Zbl 465 # 03022 • ID 54925

GANDY, R.O. [1980] *Proofs of strong normalization* (**C** 3050) Essays Combin Log, Lambda Calc & Formalism (Curry) 457-477 ◇ B40 F05 F50 ◇ REV MR 82i:03024 Zbl 469 # 03006 • ID 73214

GANDY, R.O. [1982] *Limitations to mathematical knowledge* (**P** 3623) Logic Colloq;1980 Prague 129-146 ◇ A05 F65 ◇ REV MR 84d:00010 Zbl 503 # 03001 • ID 36917

GANDY, R.O. [1982] *Symposium on constructivity in mathematics: Introductory remarks* (**P** 3622) Int Congr Log, Meth & Phil of Sci (6,Proc);1979 Hannover 141-143 ◇ A05 F99 ◇ REV MR 85b:03104 Zbl 494 # 03042 • ID 36874

GANDY, R.O. see Vol. I, III, IV, V for further entries

GANOV, V.A. [1973] *A generalized constructive continuum (Russian)* (**J** 0092) Sib Mat Zh 14*957-977,1156 • TRANSL [1973] (**J** 0475) Sib Math J 14*666-680 ◇ F60 ◇ REV MR 51 # 2883 Zbl 279 # 02019 • ID 17331

GANOV, V.A. see Vol. IV, V for further entries

GAO, HENGSHAN [1964] *Remarks on Ryll-Nardzewski's proof of the impossibility of finite axiomatization of Peano arithmetic (Chinese)* (**J** 0420) Shuxue Jinzhan 7*43-48 ◇ F30 ◇ REV MR 38 # 3144 • ID 06909

GAO, HENGSHAN see Vol. I, II, III, IV, V for further entries

GARBAYO, E. [1980] *Ideological ignorance of constructivism (Spanish)* (**P** 1789) Congr Soc Espan Hist de Cien (1);1978 Madrid 359-364 ◇ A05 F99 ◇ REV MR 83e:00018 • ID 40092

GARGOV, G.K. [1975] *Intuitionistic analysis, and intermediate logics (Russian)* (**J** 0023) Dokl Akad Nauk SSSR 224*1245-1247 • TRANSL [1975] (**J** 0062) Sov Math, Dokl 16*1372-1374 ◇ B55 F50 ◇ REV MR 52 # 10375 Zbl 342 # 02014 • ID 21677

GARGOV, G.K. [1984] *A note on the provability logics of certain extensions of Heyting's arithmetic* (**P** 4392) Mat Logika (Markova);1980 Sofia 20-26 ◇ B45 F30 ◇ ID 46969

GARGOV, G.K. see Vol. II for further entries

GARRO, I. [1973] *Independence proofs in arithmetic theories with very weak induction* (**X** 0908) Univ Math Inst: Bonn vii+71pp ◇ F30 ◇ REV MR 49 # 2367 Zbl 279 # 02039 • ID 17918

GASS, F.S. [1971] *Generalized ordinal notation* (**J** 0047) Notre Dame J Formal Log 12*104-114 ◇ D55 F15 ◇ REV MR 44 # 1561 Zbl 221 # 02028 • ID 04808

GASS, F.S. [1972] *A note on Π^1_1 ordinals* (**J** 0047) Notre Dame J Formal Log 13*103-104 ◇ D55 F15 ◇ REV MR 47 # 1590 Zbl 228 # 02020 • ID 04809

GASS, F.S. [1984] *Constructive ordinal notation systems* (**J** 0497) Math Mag 57*131-141 ◇ F15 ◇ REV MR 85j:03097 Zbl 558 # 03028 • ID 45020

GASTEV, YU.A. [1970] *The expressible and deductive possibilities of logico-arithmetic calculi on the basis of the theory of types (Russian)* (**C** 1530) Issl Log Sist (Yanovskaya) 78-85 ◇ B15 F35 ◇ REV MR 50 # 12644 Zbl 209.8 • ID 17919

GASTEV, YU.A. see Vol. I for further entries

GAUTHIER, Y. [1977] *Intuitionistic logic and local mathematical theories* (**J** 0068) Z Math Logik Grundlagen Math 23*411-414 ◇ F50 ◇ REV MR 58 # 10326 Zbl 439 # 03047 • ID 56038

GAUTHIER, Y. see Vol. I, II for further entries

GAVRILENKO, YU.V. [1981] *Recursive realizability from the intuitionistic point of view (Russian)* (**J** 0023) Dokl Akad Nauk SSSR 256*18-22 • TRANSL [1981] (**J** 0062) Sov Math, Dokl 23*9-14 ◇ F35 F50 ◇ REV MR 82h:03067 Zbl 467 # 03055 • ID 55053

GAVRILENKO, YU.V. [1984] *Monotonic theories of accessible numbers (Russian)* (**J** 0023) Dokl Akad Nauk SSSR 276*18-22 • TRANSL [1984] (**J** 0062) Sov Math, Dokl 29*415-419 ◇ F20 F30 F65 ◇ ID 45177

GEISER, J.R. [1974] *A formalization of Essenin-Volpin's proof theoretical studies by means of nonstandard analysis* (**J** 0036) J Symb Logic 39*81-87 ◇ C75 F50 H10 H15 ◇ REV MR 55 # 5406 Zbl 296 # 02018 • ID 04831

Author Index

GEISER, J.R. [1981] *Rational constructive analysis* (**P** 3146) Constr Math;1980 Las Cruces 321-347
◇ F65 ◇ REV MR 83d:03067 Zbl 494 # 03041 • ID 35199

GEISER, J.R. see Vol. I, III, IV for further entries

GEL'FOND, M.G. [1969] *On constructive pseudofunctions (Russian)* (**S** 0228) Zap Nauch Sem Leningrad Otd Mat Inst Steklov 16*20-27
• TRANSL [1969] (**J** 0521) Semin Math, Inst Steklov 16*7-10
◇ F60 ◇ REV MR 42 # 75 Zbl 243 # 02017 • ID 04840

GEL'FOND, M.G. [1972] *Relationship between the classical and constructive developments of mathematical analysis (Russian) (English summary)* (**S** 0228) Zap Nauch Sem Leningrad Otd Mat Inst Steklov 32*5-11,153
• TRANSL [1976] (**J** 1531) J Sov Math 6*347-352
◇ B28 F35 F60 ◇ REV MR 49 # 8827 Zbl 354 # 02028
• ID 04841

GEL'FOND, M.G. [1981] *A class of theorems with valid constructive counterparts* (**P** 3146) Constr Math;1980 Las Cruces 314-320
◇ F50 ◇ REV MR 83c:03052 Zbl 517 # 03026 • ID 35146

GENTZEN, G. [1935] *Untersuchungen ueber das logische Schliessen I,II* (**J** 0044) Math Z 39*176-210,405-431
• TRANSL [1955] (**X** 0840) Pr Univ France: Paris xi + 170pp (French) [1964] (**J** 0325) Amer Phil Quart 1*288-306, 2*204-218 (English) [1969] (**C** 1409) Gentzen: Collected Papers 69-131 (English) [1978] (**J** 2751) J Fac Lib Art Yamaguchi Univ 12*1-20, 13*21-40 (Japanese)
• REPR [1969] (**X** 0890) Wiss Buchges: Darmstadt ii + 62pp
◇ B10 B25 F05 F07 F30 F50 ◇ REV MR 17.3 MR 50 # 4228 MR 82m:01080 Zbl 10.145 Zbl 10.146 JSL 1.209 JSL 22.350 JSL 35.144 FdM 60.20 • REM 1969 transl. contains additional notes • ID 24221

GENTZEN, G. [1936] *Die Widerspruchsfreiheit der Stufenlogik* (**J** 0044) Math Z 41*357-366
• TRANSL [1969] (**C** 1409) Gentzen: Collected Papers 214-222
◇ B15 F05 F35 ◇ REV Zbl 15.193 JSL 1.119 FdM 62.43
• ID 04851

GENTZEN, G. [1936] *Die Widerspruchsfreiheit der reinen Zahlentheorie* (**J** 0043) Math Ann 112*493-565
• TRANSL [1969] (**C** 1409) Gentzen: Collected Papers 132-170 • LAST ED [1967] (**X** 0890) Wiss Buchges: Darmstadt 73pp
◇ B28 F05 F07 F30 ◇ REV MR 36 # 4973 Zbl 14.38 Zbl 169.308 JSL 1.75 FdM 62.44 • REM For additional sections IV and V see 1974 • ID 22402

GENTZEN, G. [1937] *Der Unendlichkeitsbegriff in der Mathematik* (**J** 2074) Sem-ber, Muenster 9*65-80
• TRANSL [1969] (**C** 1409) Gentzen: Collected Papers 223-233
◇ A05 F50 F99 ◇ REV JSL 2.95 FdM 63.21 • ID 41002

GENTZEN, G. [1937] *Unendlichkeitsbegriff und Widerspruchsfreiheit der Mathematik* (**P** 0756) Congr Int Phil (9);1937 Paris 201-205
◇ A05 F99 ◇ REV JSL 3.94 FdM 63.837 • ID 41408

GENTZEN, G. [1938] *Die gegenwaertige Lage in der mathematischen Grundlagenforschung* (**J** 0956) Forsch Logik Grundl exakt Wiss 4*5-18
• REPR [1938] (**J** 0426) Dt Math 3*255-268
◇ A05 A10 F55 F65 ◇ REV Zbl 19.241 Zbl 19.97 JSL 3.166 FdM 64.26 FdM 64.931 • ID 16984

GENTZEN, G. [1938] *Neue Fassung des Widerspruchsfreiheitsbeweises fuer die reine Zahlentheorie* (**J** 0956) Forsch Logik Grundl exakt Wiss 4*19-44
• TRANSL [1969] (**C** 1409) Gentzen: Collected Papers 252-286
◇ F05 F15 F30 ◇ REV Zbl 19.241 JSL 4.31 FdM 64.27
• ID 16987

GENTZEN, G. [1943] *Beweisbarkeit und Unbeweisbarkeit von Anfangsfaellen der transfiniten Induktion in der reinen Zahlentheorie* (**J** 0043) Math Ann 119*140-161
◇ F15 F30 ◇ REV MR 5.198 Zbl 28.102 JSL 9.70 • REM A remark was published by Takeuti in J0081 39(1963)*263-269
• ID 04852

GENTZEN, G. [1954] *Zusammenfassung von mehreren vollstaendigen Induktionen zu einer einzigen* (**J** 0009) Arch Math Logik Grundlagenforsch 2*1-3
◇ B28 F30 ◇ REV MR 18.272 Zbl 55.5 JSL 38.157
• ID 04853

GENTZEN, G. [1969] *The collected papers of Gerhard Gentzen* (**X** 0809) North Holland: Amsterdam xii + 338pp
◇ B96 F05 F07 F30 F50 F96 ◇ REV MR 41 # 6660 Zbl 209.300 • REM Ed. by Szabo,M.E. • ID 29483

GENTZEN, G. [1974] *Der erste Widerspruchsfreiheitsbeweis fuer die klassische Zahlentheorie* (**J** 0009) Arch Math Logik Grundlagenforsch 16*97-118
• TRANSL [1969] (**C** 1409) Gentzen: Collected Papers 170-213
◇ F05 F15 F25 F30 ◇ REV MR 52 # 51 Zbl 286 # 02034
• REM These are sections IV and V of Gentzen 1936, never printed in J0043 • ID 04854

GENTZEN, G. [1974] *Ueber das Verhaeltnis zwischen intuitionistischer und klassischer Arithmetik* (**J** 0009) Arch Math Logik Grundlagenforsch 16*119-132
• TRANSL [1969] (**C** 1409) Gentzen: Collected Papers 53-67
◇ B30 F30 F50 ◇ REV MR 52 # 52 Zbl 286 # 02035
• REM Originally to appear in J0043; was withdrawn
• ID 04855

GENTZEN, G. see Vol. I for further entries

GEORGACARAKOS, G.N. [1982] *The semantics of minimal intuitionism* (**J** 0079) Logique & Anal, NS 25*383-397
◇ B55 F50 ◇ REV MR 85g:03021 Zbl 545 # 03035
• ID 43512

GEORGACARAKOS, G.N. see Vol. II for further entries

GEORGEFF, M. [1984] *Transformations and reduction strategies for typed lambda expressions* (**J** 3120) ACM Trans Program Lang & Syst 6*603-631
◇ B40 ◇ REV Zbl 555 # 68007 • ID 46139

GEORGIEVA, N.V. [1973] *On the logical signs in Schuette's interpolation formulae (Bulgarian) (Russian and English summaries)* (**J** 1156) Izv Bulgar Akad Nauk Mat Inst 14*99-103
◇ C40 F50 ◇ REV MR 50 # 9543 Zbl 334 # 02013
• ID 20996

GEORGIEVA, N.V. [1979] *Ueber gewisse McCarthysche Aequivalenzen (Russisch)* (**J** 0137) C R Acad Bulgar Sci 32*721-724
◇ B40 ◇ REV MR 80j:03066 Zbl 423 # 03047 • ID 53559

GEORGIEVA, N.V. see Vol. I, II, IV for further entries

GERBER, H. [1967] *An extension of Schuette's Klammer-symbols* (**J** 0043) Math Ann 174∗203–216
◇ F15 ◇ REV MR 36 # 6295 Zbl 162.19 JSL 34.655
• ID 04889

GERBER, H. [1970] *Brouwer's bar theorem and a system of ordinal notations* (**P** 0603) Intuitionism & Proof Th;1968 Buffalo 327–338
◇ F15 F35 F50 ◇ REV MR 43 # 6054 Zbl 205.11
• ID 17908

GERMANO, G. [1970] *Metamathematische Begriffe in Standardtheorien* (**J** 0009) Arch Math Logik Grundlagenforsch 13∗22–38
◇ A05 B30 F30 ◇ REV MR 44 # 70 Zbl 221 # 02037
• ID 04906

GERMANO, G. [1971] *Incompleteness and truth definitions* (**J** 0105) Theoria (Lund) 37∗86–90
◇ C40 D35 F30 ◇ REV MR 50 # 4274 Zbl 243 # 02038
• ID 04907

GERMANO, G. [1973] *Incompleteness theorem via weak definability of truth: a short proof* (**J** 0047) Notre Dame J Formal Log 14∗377–380
◇ C40 F30 ◇ REV MR 48 # 90 Zbl 232 # 02035 • ID 04910

GERMANO, G. [1976] *An arithmetical reconstruction of the liar's antinomy using addition and multiplication* (**J** 0047) Notre Dame J Formal Log 17∗457–461
◇ F30 ◇ REV MR 55 # 12502 Zbl 283 # 02037 • ID 18163

GERMANO, G. see Vol. IV for further entries

GEYMONAT, L. [1965] *La metamatematica dopo Hilbert* (**P** 0794) Congr Unione Mat Ital (7);1963 Genova 288–318
◇ F98 ◇ REV MR 35 # 5278 Zbl 171.258 • ID 17910

GEYMONAT, L. see Vol. II, V for further entries

GHOSE, A. [1966] *Konstruktive Grundlagen der Logik* (**X** 1170) Sonne: New Delhi vi + 42pp
◇ A05 F65 ◇ REV MR 33 # 7239 Zbl 142.247 JSL 32.518
• ID 22163

GHOSE, A. [1968] *Free choice sequences I,II* (**C** 0684) Automation in Lang Transl & Theorem Prov 245–257, 259–261
◇ F50 ◇ REV MR 39 # 1297 Zbl 195.12 • ID 16298

GHOSE, A. [1968] *The length of proofs* (**C** 0684) Automation in Lang Transl & Theorem Prov 41–46
◇ F20 ◇ REV MR 39 # 5356 Zbl 193.336 • ID 24935

GHOSE, A. [1978] *Different notions of constructivity – their ontology* (**J** 0076) Dialectica 32∗245–253
◇ A05 F99 ◇ REV MR 81k:03004 Zbl 402 # 03015
• ID 54662

GHOSE, A. see Vol. II for further entries

GIAMBRONE, S. [1985] *On purported Gentzen formulations of two positive relevant logics* (**J** 0063) Studia Logica 44∗233–236
◇ B46 F07 ◇ ID 47499

GIAMBRONE, S. see Vol. II, III for further entries

GIANNINI, P. & LONGO, G. [1984] *Effectively given domains and lambda-calculus models* (**J** 0194) Inform & Control 62∗36–63
◇ B40 ◇ ID 44586

GIARETTA, P. & MARTINO, E. [1981] *Brouwer, Dummett and the bar theorem* (**P** 3092) Congr Naz Logica;1979 Montecatini Terme 541–558
◇ A10 F50 ◇ ID 48394

GIARETTA, P. [1983] *On the construction of sequences in the theory of the creative subject (Italian)* (**P** 3829) Atti Incontri Log Mat (1);1982 Siena 369–371
◇ F50 ◇ REV MR 84k:03006 • ID 44828

GIBSON, C.G. [1969] *The intuitionist measure* (**J** 3172) J London Math Soc, Ser 2 44∗617–624
◇ F55 ◇ REV MR 40 # 27 Zbl 181.297 • ID 04924

GIBSON, C.G. [1971] *On the definition of an infinite species* (**J** 3077) Nieuw Arch Wisk, Ser 3 19∗196–197
◇ F55 ◇ REV MR 49 # 4762 Zbl 234 # 02019 • ID 04925

GIBSON, C.G. [1972] *On the almost periodicity of trigonometric polynomials in constructive mathematics* (**J** 0028) Indag Math 34∗355–361
◇ F55 ◇ REV MR 49 # 3443 Zbl 242 # 42020 • ID 04926

GIBSON, C.G. & JOHNS, D.L. [1981] *A constructive approach to the duality theorem for certain Orlicz spaces* (**J** 0332) Math Proc Cambridge Phil Soc 89∗49–69
◇ F55 ◇ REV MR 81m:46046 Zbl 475 # 46027 • ID 55503

GIELEN, W. & SWART DE, H.C.M. & VELDMAN, W. [1981] *The continuum hypothesis in intuitionism* (**J** 0036) J Symb Logic 46∗121–136
◇ E15 E50 F50 ◇ REV MR 82i:03070 Zbl 461 # 03014
• ID 54491

GIL, D. [1983] *Intuitionism, transformational generative grammar and mental acts* (**J** 3273) Stud Hist Philos Sci 14∗231–254
◇ A05 F99 ◇ REV MR 85a:03006 • ID 34757

GILBERT, M.A. [1976] *A heuristic procedure for natural deduction derivations using reductio ad absurdum* (**J** 0047) Notre Dame J Formal Log 17∗638–639
◇ B35 F07 ◇ REV Zbl 305 # 02020 • ID 62011

GILMORE, P.C. [1953] *Griss' criticism of the intuitionistic logic and the theory of order* (**P** 0645) Int Congr Philos (11);1953 Bruxelles 5∗98–104
◇ F50 ◇ REV MR 15.278 Zbl 53.6 JSL 21.93 • ID 04939

GILMORE, P.C. [1953] *The effect of Griss' criticism of the intuitionistic logic on deductive theories formalized within the intuitionistic logic* (**J** 0028) Indag Math 15∗162–186
◇ F50 ◇ REV MR 14.1053 Zbl 53.341 JSL 21.91
• ID 28433

GILMORE, P.C. [1970] *The lambda-gamma calculus: a language adequate for defining recursive functions* (**J** 0191) Inform Sci 2∗111–126
◇ B40 ◇ REV MR 43 # 57 Zbl 205.14 • ID 04947

GILMORE, P.C. see Vol. I, II, III, V for further entries

GIMON, V.V. [1973] *The dependence of the second level rules of inference in Markov's stratified semantic system (Russian)* (**J** 0023) Dokl Akad Nauk SSSR 212∗1036–1038
• TRANSL [1973] (**J** 0062) Sov Math, Dokl 14∗1504–1507
◇ F50 ◇ REV MR 49 # 2286 Zbl 308 # 02038 • ID 04949

GIMON, V.V. [1974] *Version of the semantical stratified system for formulas without quantifiers (Russian)* (**C** 1450) Teor Algor & Mat Logika (Markov) 3–13
◇ F50 ◇ REV MR 58 # 10349 Zbl 309 # 02030 • ID 62024

GIMON, V.V. [1979] *Quantifier-free formulas in A. A. Markov's ramified semantics (Russian)* (S 0554) Issl Teor Algor & Mat Logik (Moskva) 3*5-17
◇ B55 F50 ◇ REV MR 81j:03092 Zbl 422 # 03033
• ID 53494

GINSBURG, S. & SPANIER, E.H. [1966] *Semigroups, Presburger formulas, and languages* (J 0048) Pac J Math 16*285-296
◇ D05 F30 ◇ REV MR 32 # 9172 Zbl 143.16 JSL 34.137
• ID 04980

GINSBURG, S. see Vol. IV, V for further entries

GIOVANNETTI, E. [1983] see COPPO, M.

GIOVANNETTI, E. [1983] *Une caracterisation des termes types dans un langage applicatif (English summary)* (J 3364) C R Acad Sci, Paris, Ser 1 296*97-99
◇ B10 B40 ◇ REV MR 84b:03029 Zbl 567 # 68050
• ID 35624

GIRARD, J.-Y. [1971] *Une extension de l'interpretation de Goedel a l'analyse et son application a l'elimination des coupures dans l'analyse et la theorie des types* (P 0604) Scand Logic Symp (2);1970 Oslo 63-92
◇ F05 F10 F35 F50 ◇ REV MR 53 # 12895
Zbl 221 # 02013 JSL 40.503 • ID 15075

GIRARD, J.-Y. [1973] *Quelques resultats sur les interpretations fonctionelles* (P 0713) Cambridge Summer School Math Log;1971 Cambridge GB 232-252
◇ F10 F35 F50 ◇ REV MR 54 # 4932 Zbl 287 # 02017
• ID 14975

GIRARD, J.-Y. [1976] *Three-valued logic and cut-elimination: The actual meaning of Takeuti's conjecture* (J 0202) Diss Math (Warsaw) 136*45pp
◇ B50 C85 C90 F05 F35 F50 ◇ REV MR 56 # 5235
Zbl 357 # 02027 • ID 50375

GIRARD, J.-Y. [1977] *Functional interpretation and Kripke models* (P 1704) Int Congr Log, Meth & Phil of Sci (5);1975 London ON 1*33-57
◇ C90 F10 F30 F50 ◇ REV MR 58 # 159
Zbl 417 # 03029 • ID 53265

GIRARD, J.-Y. [1977] *Functionals and ordinoids* (P 1729) Colloq Int Log;1975 Clermont-Ferrand 59-71
◇ D65 F10 F15 ◇ REV MR 80h:03067 Zbl 441 # 03012
• ID 56065

GIRARD, J.-Y. [1981] *Π^1_2-logic I. Dilators* (J 0007) Ann Math Logic 21*75-219
◇ D55 E10 F15 G30 ◇ REV MR 83i:03093
Zbl 496 # 03037 • ID 35541

GIRARD, J.-Y. & PAEPPINGHAUS, P. [1981] *A result on implications of Σ_1-sentences and its application to normal form theorems* (J 0036) J Symb Logic 46*634-642
◇ F05 F30 F35 ◇ REV MR 83f:03056 Zbl 469 # 03039
• ID 55166

GIRARD, J.-Y. [1982] *A survey of Π^1_2-logic* (P 3622) Int Congr Log, Meth & Phil of Sci (6,Proc);1979 Hannover 89-107
◇ F15 F35 ◇ REV MR 84e:03073 Zbl 496 # 03038
• ID 34405

GIRARD, J.-Y. [1982] *Herbrand's theorem and proof-theory* (P 3708) Herbrand Symp Logic Colloq;1981 Marseille 29-38
◇ F05 F07 F10 F35 ◇ REV MR 85k:03035
Zbl 496 # 03036 • ID 36865

GIRARD, J.-Y. [1982] *Proof-theoretic investigations of inductive definitions I* (P 3482) Logic & Algor (Specker);1980 Zuerich 207-236
◇ F05 F35 F50 ◇ REV MR 84b:03075 Zbl 494 # 03037
• ID 35657

GIRARD, J.-Y. & VAUZEILLES, J. [1984] *Functors and ordinal notations I: A functorial construction of the Veblen hierarchy. II: A functorial construction of the Bachmann hierarchy* (J 0036) J Symb Logic 49*713-729,1079-1114
◇ E10 F15 G30 ◇ REV MR 85j:03098 Zbl 568 # 03027
Zbl 568 # 03028 • REM Part III 1982 by Vauzeilles,J.
• ID 42447

GIRARD, J.-Y. & VAUZEILLES, J. [1984] *Les premiers recursivement inaccessible et Mahlo et la theorie des dilatateurs* (J 0009) Arch Math Logik Grundlagenforsch 24*167-191
◇ D60 E47 E55 F05 F15 ◇ REV Zbl 556 # 03043
• ID 42404

GIRARD, J.-Y. [1984] *The Ω-rule* (P 4313) Int Congr Math (II,14);1983 Warsaw 1*307-321
◇ B15 C62 C75 F35 ◇ REV Zbl 569 # 03023 • ID 48565

GIRARD, J.-Y. & RESSAYRE, J.-P. [1985] *Elements de logique Π^1_n* (P 4046) Rec Th;1982 Ithaca 389-445
◇ D60 F15 F35 ◇ REV Zbl 573 # 03029 • ID 46368

GIRARD, J.-Y. [1985] *Introduction to Π^1_2-logic* (J 0154) Synthese 62*191-216
◇ E10 F15 F35 ◇ ID 45634

GIRARD, J.-Y. & NORMANN, D. [1985] *Set recursion and Π^1_2-logic* (J 0073) Ann Pure Appl Logic 28*255-286
◇ D65 F15 ◇ REV Zbl 575 # 03034 • ID 42597

GIRSTMAIR, K. [1979] *Ueber konstruktive Methoden der Galoistheorie* (J 0504) Manuscr Math 26*423-441
◇ C57 C60 F55 ◇ REV MR 80b:12016 Zbl 408 # 12024
• ID 56289

GLEASON, G.G. [1974] *Normal and skew systems* (J 0047) Notre Dame J Formal Log 15*379-401
◇ B05 F07 ◇ REV MR 50 # 4235 Zbl 281 # 02017
• ID 23612

GLIVENKO, V.I. [1928] *Sur la logique de M. Brouwer* (J 0150) Acad Roy Belg Bull Cl Sci (5) 14*225-228
◇ F50 ◇ REV FdM 54.54 • ID 40722

GLIVENKO, V.I. [1929] *Sur quelques points de la logique de M.Brouwer* (J 0150) Acad Roy Belg Bull Cl Sci (5) 15*183-188
◇ F50 ◇ REV FdM 55.30 • ID 05039

GLIVENKO, V.I. see Vol. I, V for further entries

GLUBRECHT, J.-M. [1982] *Ein Vollstaendigkeitsbeweis fuer schnittfreie Kalkuele mit der Maximalisierungsmethode von Henkin* (J 0009) Arch Math Logik Grundlagenforsch 22*159-166
◇ B10 C07 F05 ◇ REV MR 84b:03022 Zbl 496 # 03002
• ID 35619

GLUBRECHT, J.-M. see Vol. I, II, V for further entries

GLUSHKOV, V.M. [1979] *Incompleteness theorem of formal theories from programmer's viewpoint (Russian)* (**J** 0040) Kibernetika, Akad Nauk Ukr SSR 1979/2*1-5
• TRANSL [1979] (**J** 0021) Cybernetics 15*155-159
◇ B75 F30 ◇ REV MR 81g:68076 Zbl 471 # 68023
• ID 55259

GLUSHKOV, V.M. see Vol. I, II, IV for further entries

GOAD, C.A. [1978] *Monadic infinitary propositional logic: a special operator* (**J** 0302) Rep Math Logic, Krakow & Katowice 10*43-50
◇ B25 B60 C75 F50 ◇ REV MR 81e:03024 Zbl 424 # 03029 • ID 73424

GOAD, C.A. [1980] *Proofs as descriptions of computation* (**P** 3063) Autom Deduct (5);1980 Les Arcs 39-52
◇ B35 F07 F50 ◇ REV Zbl 438 # 68052 • ID 55988

GOBLE, L.F. [1974] *Gentzen systems for modal logic* (**J** 0047) Notre Dame J Formal Log 15*455-461
◇ B45 F50 ◇ REV MR 51 # 59 Zbl 226 # 02014 • ID 05060

GOBLE, L.F. see Vol. II, III for further entries

GODDARD, L. [1958] *"True" and "provable"* (**J** 0094) Mind 67*13-31
◇ A05 F30 ◇ REV JSL 25.85 • ID 31082

GODDARD, L. see Vol. I, II, IV for further entries

GOEDEL, K. [1930] *Einige metamathematische Resultate ueber Entscheidungsdefinitheit und Widerspruchsfreiheit* (**J** 0931) Anz Oesterr Akad Wiss, Math-Nat Kl 67*214-215
• TRANSL [1967] (**C** 0675) From Frege to Goedel 595-596
◇ F30 ◇ REV JSL 37.405 FdM 56.60 • ID 16829

GOEDEL, K. [1931] *Eine Interpretation des intuitionistischen Aussagenkalkuels* (**J** 1124) Ergebn Math Kolloquium 4*39-40
• TRANSL [1969] (**C** 0569) Phil of Math Oxford Readings 128-129
◇ A05 B45 F50 ◇ REV Zbl 7.193 FdM 59.866 • ID 22120

GOEDEL, K. [1931] *Ueber formal unentscheidbare Saetze der "Principia Mathematica" und verwandter Systeme I* (**J** 0124) Monatsh Math-Phys 38*173-198
• TRANSL [1965] (**C** 0718) The Undecidable 5-38 [1967] (**C** 0675) From Frege to Goedel 596-616 [1962] (**X** 1323) Oliver & Boyd: Edinburgh vii+72pp
◇ B28 B30 D20 D35 F25 F30 F35 ◇ REV MR 27 # 1373 Zbl 124.4 Zbl 2.1 FdM 57.54 • ID 15052

GOEDEL, K. [1931] *Ueber Vollstaendigkeit und Widerspruchsfreiheit* (**J** 1124) Ergebn Math Kolloquium 3*12-13
• TRANSL [1967] (**C** 0675) From Frege to Goedel 616-617
◇ F30 F35 ◇ REV JSL 37.405 FdM 57.1318 • ID 22124

GOEDEL, K. [1931] *Zum intuitionistischen Aussagenkalkuel* (**J** 1124) Ergebn Math Kolloquium 4*40
• REPR [1932] (**J** 0931) Anz Oesterr Akad Wiss, Math-Nat Kl 69*65-66
◇ F50 F55 ◇ REV FdM 58.1001 • ID 22126

GOEDEL, K. [1932] *Zur intuitionistischen Arithmetik und Zahlentheorie* (**J** 1124) Ergebn Math Kolloquium 4*34-38
• TRANSL [1965] (**C** 0718) The Undecidable 75-81
◇ F30 F50 ◇ REV Zbl 7.193 FdM 59.865 • ID 22121

GOEDEL, K. [1934] *Ueber die Laenge von Beweisen* (**J** 1124) Ergebn Math Kolloquium 7*23-24
• TRANSL [1965] (**C** 0718) The Undecidable 82-83
◇ B10 F20 F30 F35 ◇ REV Zbl 14.241 JSL 1.116 JSL 31.484 FdM 62.43 • ID 35971

GOEDEL, K. [1958] *Ueber eine bisher noch nicht benuetzte Erweiterung des finiten Standpunktes* (**J** 0076) Dialectica 12*280-287
• TRANSL [1980] (**J** 0122) J Philos Logic 9*133-142 • REPR [1959] (**C** 1173) Logica (Bernays) 76-83
◇ A05 F10 F30 F50 ◇ REV MR 21 # 1275 MR 81i:03094 Zbl 90.10 JSL 25.351 • ID 20826

GOEDEL, K. [1965] *On undecidable propositions of formal mathematical systems* (**C** 0718) The Undecidable 41-71
◇ D20 D35 F30 ◇ REM Publ. as mimeographed lecture notes in 1934 • ID 21353

GOEDEL, K. see Vol. I, II, III, IV, V for further entries

GOERNEMANN, S. [1971] *A logic stronger than intuitionism* (**J** 0036) J Symb Logic 36*249-261
◇ B55 C90 F50 ◇ REV MR 45 # 27 Zbl 276 # 02013
• ID 39863

GOERNEMANN, S. also published under the name KOPPELBERG, S.

GOETLIND, E. [1952] *A note on Chwistek and Hetper's foundation of formal metamathematics* (**P** 0098) Skand Mat Kongr (11);1949 Trondheim 268-270
◇ B30 F35 ◇ REV MR 15.846 Zbl 48.247 JSL 19.140
• ID 21134

GOETLIND, E. see Vol. I, II for further entries

GOGUADZE, D.F. [1970] *The finite axiomatizability of a finite fragment of type theory (Russian) (Georgian and English summaries)* (**J** 0233) Soobshch Akad Nauk Gruz SSR 60*541-544
◇ B15 F35 ◇ REV MR 45 # 53 Zbl 216.10 • ID 05093

GOLDBLATT, R.I. [1978] *Arithmetical necessity, provability and intuitionistic logic* (**J** 0105) Theoria (Lund) 44*38-46
◇ B45 F50 ◇ REV MR 80h:03026 Zbl 409 # 03011
• ID 56314

GOLDBLATT, R.I. [1979] *Topoi. The categorial analysis of logic* (**S** 3303) Stud Logic Found Math 98*xv+486pp
• TRANSL [1983] (**X** 0885) Mir: Moskva 488pp
◇ B98 C98 E98 F35 F50 F98 G30 ◇ REV MR 81a:03063 Zbl 434 # 03050 JSL 47.445 • REM 2nd rev. ed. 1984; xvi+552pp • ID 55754

GOLDBLATT, R.I. [1981] *Grothendieck topology as geometric modality* (**J** 0068) Z Math Logik Grundlagen Math 27*495-529
◇ B25 B45 C90 F50 G30 ◇ REV MR 83d:03069 Zbl 474 # 03018 • ID 55422

GOLDBLATT, R.I. see Vol. I, II, III, V for further entries

GOLDFARB, W.D. [1971] see DREBEN, B.

GOLDFARB, W.D. [1974] *Ordinal bounds for k-consistency* (**J** 0036) J Symb Logic 39*693-699
◇ F15 F30 ◇ REV MR 51 # 2880 Zbl 317 # 02035
• ID 05116

Goldfarb, W.D. & Scanlon, T.M. [1974] *The ω-consistency of number theory via Herbrand's theorem* (**J** 0036) J Symb Logic 39*678-692
◇ F05 F15 F30 ◇ rev MR 51 # 2879 Zbl 317 # 02034
• id 05114

Goldfarb, W.D. [1975] *On the effective ω-rule* (**J** 0068) Z Math Logik Grundlagen Math 21*409-412
◇ F20 F30 ◇ rev MR 52 # 13338 Zbl 332 # 02038
• id 05119

Goldfarb, W.D. [1981] *The undecidability of the second-order unification problem* (**J** 1426) Theor Comput Sci 13*225-230
◇ B15 D35 F20 ◇ rev MR 82c:03067 Zbl 457 # 03006
• id 54331

Goldfarb, W.D. see Vol. I, III, IV for further entries

Golota, Ya.Ya. [1969] *Nets of marks and deducibility in the intuitionistic propositional calculus (Russian)* (**S** 0228) Zap Nauch Sem Leningrad Otd Mat Inst Steklov 16*28-43
• err/add ibid 20*292-294
• transl [1969] (**J** 0521) Semin Math, Inst Steklov 16*11-19
◇ B55 B75 F50 ◇ rev MR 41 # 6667 Zbl 237 # 02003
• id 05123

Golota, Ya.Ya. [1969] *Some techniques for simplifying the construction of nets of marks (Russian)* (**S** 0228) Zap Nauch Sem Leningrad Otd Mat Inst Steklov 16*44-53 • err/add ibid 20*292-294
• transl [1969] (**J** 0521) Semin Math, Inst Steklov 16*20-25
◇ B25 B75 F50 ◇ rev MR 41 # 6668 Zbl 237 # 02004
• id 05124

Golota, Ya.Ya. [1974] *A matrix notation for nets of marks (Russian) (English summary)* (**S** 0228) Zap Nauch Sem Leningrad Otd Mat Inst Steklov 40*4-9,155
• transl [1977] (**J** 1531) J Sov Math 8*247-251
◇ B25 F50 ◇ rev MR 51 # 82 Zbl 356 # 02021 • id 15236

Goncharov, S.S. [1976] *Restricted theories of constructive boolean algebras (Russian)* (**J** 0092) Sib Mat Zh 17*797-812
• transl [1976] (**J** 0475) Sib Math J 17*601-611
◇ B25 C57 D35 F60 G05 ◇ rev MR 55 # 93 Zbl 361 # 02066 • id 52189

Goncharov, S.S. [1982] *Limiting equivalent constructivizations (Russian)* (**C** 3953) Mat Log & Teor Algor 4-12
◇ C57 D45 F60 ◇ rev MR 85d:03122 Zbl 543 # 03017
• id 40426

Goncharov, S.S. see Vol. III, IV for further entries

Gonseth, F. [1926] *Les fondements des mathematiques. De la geometrie d'Euclide a la relativite generale et a l'intuitionnisme* (**X** 0872) Blanchard: Paris xiv+243pp
◇ A05 A10 F99 ◇ rev MR 50 # 1802 JSL 46.668 FdM 52.44 • id 05126

Gonseth, F. (ed.) [1941] *Les entretiens de Zuerich sur les fondements et la methode des sciences mathematiques: Exposes et discussions* (**X** 2220) Leemen: Zuerich 209pp
◇ B97 E97 F97 ◇ id 48625

Gonseth, F. see Vol. I, II, V for further entries

Goodman, Nelson & Quine, W.V.O. [1947] *Steps toward a constructive nominalism* (**J** 0036) J Symb Logic 12*105-122
• transl [1967] (**C** 2141) Filos Matematica 269-298
◇ A05 F99 ◇ rev MR 9.262 Zbl 41.147 JSL 34.313
• id 05152

Goodman, Nelson see Vol. I, V for further entries

Goodman, Nicolas D. [1970] *A theory of constructions equivalent to arithmetic* (**P** 0603) Intuitionism & Proof Th;1968 Buffalo 101-120
◇ F30 F50 ◇ rev MR 44 # 31 Zbl 212.313 JSL 40.500
• id 05161

Goodman, Nicolas D. [1972] *A simplification of combinatory logic* (**J** 0036) J Symb Logic 37*225-246
◇ B40 ◇ rev MR 48 # 3708 Zbl 264 # 02024 • id 05162

Goodman, Nicolas D. & Myhill, J.R. [1972] *The formalization of Bishop's constructive mathematics* (**P** 0771) Toposes, Algeb Geom & Log;1971 Halifax 83-96
◇ F35 F50 F55 ◇ rev MR 50 # 6809 Zbl 247 # 02033
• id 14699

Goodman, Nicolas D. [1973] *The arithmetic theory of constructions* (**P** 0713) Cambridge Summer School Math Log;1971 Cambridge GB 274-298
◇ F30 F50 ◇ rev MR 51 # 85 Zbl 272 # 02040 • id 73511

Goodman, Nicolas D. [1973] *The faithfulness of the interpretation of arithmetic in the theory of constructions* (**J** 0036) J Symb Logic 38*453-459
◇ F30 F50 ◇ rev MR 51 # 86 Zbl 285 # 02028 • id 17324

Goodman, Nicolas D. [1976] *The theory of the Goedel functionals* (**J** 0036) J Symb Logic 41*574-582
◇ F10 F35 F50 ◇ rev MR 58 # 5095 Zbl 366 # 02021
• id 14580

Goodman, Nicolas D. & Myhill, J.R. [1978] *Choice implies excluded middle* (**J** 0068) Z Math Logik Grundlagen Math 24*461
◇ E25 F50 ◇ rev MR 80e:03068 Zbl 387 # 03017
• id 52233

Goodman, Nicolas D. [1978] *Relativized realizability in intuitionistic arithmetic of all finite types* (**J** 0036) J Symb Logic 43*23-44
◇ F35 F50 ◇ rev MR 81b:03061 Zbl 395 # 03038
• id 29239

Goodman, Nicolas D. [1978] *The nonconstructive content of sentences of arithmetic* (**J** 0036) J Symb Logic 43*497-501
◇ D20 D30 D55 F30 F50 ◇ rev MR 81e:03045 Zbl 398 # 03046 • id 29277

Goodman, Nicolas D. [1981] *Reflections on Bishop's philosophy of mathematics* (**P** 3146) Constr Math;1980 Las Cruces 135-145
• repr [1983] (**J** 2789) Math Intell 5*61-68
◇ A05 F55 ◇ rev MR 84h:03135 MR 85k:03005 Zbl 467 # 03003 • id 55001

Goodman, Nicolas D. [1984] *Epistemic arithmetic is a conservative extension of intuitionistic arithmetic* (**J** 0036) J Symb Logic 49*192-203
◇ B45 F05 F30 F50 ◇ rev MR 85g:03084 • id 42452

Goodman, Nicolas D. [1985] *Replacement and collection in intuitionistic set theory* (**J** 0036) J Symb Logic 50*344-348
◇ E35 E70 F50 ◇ id 42539

Goodman, Nicolas D. see Vol. II, V for further entries

Goodstein, R.L. [1944] *On the restricted ordinal theorem* (**J** 0036) J Symb Logic 9*33-41
◇ F15 F30 ◇ rev MR 6.30 Zbl 60.23 JSL 10.104
• id 05165

GOODSTEIN, R.L. [1945] *Function theory in an axiom-free equation calculus* (**J** 1910) Proc London Math Soc, Ser 2 48*401-434
◇ B28 F30 F60 ◇ REV MR 8.245 Zbl 60.23 JSL 11.24
• ID 05166

GOODSTEIN, R.L. [1947] *Transfinite ordinals in recursive number theory* (**J** 0036) J Symb Logic 12*123-129
◇ E10 F15 F30 ◇ REV MR 9.221 Zbl 30.4 JSL 13.171
• ID 05167

GOODSTEIN, R.L. [1950] *Mean value theorems in recursive function theory. part I, Differential mean value theorems* (**J** 1910) Proc London Math Soc, Ser 2 52*81-106
◇ F60 ◇ REV MR 12.664 Zbl 41.347 JSL 20.88 • ID 05171

GOODSTEIN, R.L. [1950] *The formal structure of a denumerable system* (**J** 0064) Trans Amer Math Soc 68*174-182
◇ F30 F60 ◇ REV MR 12.233 Zbl 39.244 JSL 23.94
• ID 05170

GOODSTEIN, R.L. [1950] *The Gauss test for relative convergence* (**J** 0100) Amer J Math 72*217-228
◇ F60 ◇ REV MR 11.636 Zbl 39.58 JSL 20.88 • ID 05169

GOODSTEIN, R.L. [1951] *Constructive formalism: Essays on the foundations of mathematics* (**X** 0886) Leicester Univ Pr: Leicester 91pp
◇ A05 B98 F30 F60 F98 ◇ REV MR 14.123 Zbl 45.150 JSL 18.258 • ID 25669

GOODSTEIN, R.L. [1951] *The foundations of mathematics* (**X** 0886) Leicester Univ Pr: Leicester 27pp
◇ A05 A10 F98 ◇ REV MR 14.527 Zbl 45.1 JSL 18.258
• ID 25670

GOODSTEIN, R.L. [1953] *A problem in recursive function theory* (**J** 0036) J Symb Logic 18*225-232
◇ F60 ◇ REV MR 15.278 Zbl 51.246 JSL 20.89 • ID 05173

GOODSTEIN, R.L. [1953] *Permutation in recursive arithmetic* (**J** 0132) Math Scand 1*222-226
◇ F30 ◇ REV MR 15.668 Zbl 52.250 • ID 05172

GOODSTEIN, R.L. [1954] *A free variable function theory* (**P** 0575) Int Congr Math (II, 7);1954 Amsterdam 2*401-402
◇ F55 F60 ◇ ID 29466

GOODSTEIN, R.L. [1954] *Logic-free formalisations of recursive arithmetic* (**J** 0132) Math Scand 2*247-261
◇ F30 F50 ◇ REV MR 19.378 Zbl 58.249 JSL 24.245
• ID 05174

GOODSTEIN, R.L. [1954] *The recursive irrationality of π* (**J** 0036) J Symb Logic 19*267-274
◇ F30 F60 ◇ REV MR 16.662 Zbl 58.249 JSL 25.166
• ID 05176

GOODSTEIN, R.L. [1954] *The relatively exponential, logarithmic and circular functions in recursive function theory* (**J** 0118) Acta Math 92*171-190
◇ D20 F30 F60 ◇ REV MR 16.783 Zbl 58.250 • ID 05175

GOODSTEIN, R.L. [1955] *On non-constructive theorems of analysis and the decision problem* (**J** 0132) Math Scand 3*261-263
◇ B25 B28 F30 F60 ◇ REV MR 17.816 Zbl 67.248
• ID 05177

GOODSTEIN, R.L. [1956] *A constructivist theory of plane curves* (**J** 0027) Fund Math 43*23-35
◇ F65 ◇ REV MR 18.2 Zbl 70.10 JSL 35.487 • ID 05178

GOODSTEIN, R.L. [1957] *Recursive number theory. A development of recursive arithmetic in a logic-free equation calculus* (**X** 0809) North Holland: Amsterdam xii+190pp
◇ B28 F30 F50 F60 F98 ◇ REV MR 21 # 1272 Zbl 77.14 JSL 23.227 • ID 05180

GOODSTEIN, R.L. [1958] *Models of propositional calculi in recursive arithmetic* (**J** 0132) Math Scand 6*293-296
◇ B05 B50 F30 F50 ◇ REV MR 21 # 5570 Zbl 88.10 JSL 28.291 • ID 05182

GOODSTEIN, R.L. [1959] *Recursive analysis* (**P** 0634) Constructivity in Math;1957 Amsterdam 37-42
◇ F60 ◇ REV MR 21 # 4102 Zbl 88.250 JSL 40.501
• ID 05179

GOODSTEIN, R.L. & HOOLEY, J. [1960] *On recursive transcendence* (**J** 0047) Notre Dame J Formal Log 1*127-137
◇ F60 ◇ REV MR 26 # 13 JSL 35.335 • ID 05184

GOODSTEIN, R.L. [1961] *Recursive analysis* (**X** 0809) North Holland: Amsterdam viii+138pp
• TRANSL [1970] (**X** 2027) Nauka: Moskva 472pp
◇ F30 F60 F98 ◇ REV MR 21 # 4102 MR 50 # 4261 Zbl 217.302 JSL 27.244 • ID 05186

GOODSTEIN, R.L. [1963] *A decidable fragment of recursive arithmetic* (**J** 0068) Z Math Logik Grundlagen Math 9*199-201
◇ B20 B25 F30 ◇ REV MR 27 # 2403 Zbl 114.248 JSL 33.618 • ID 05189

GOODSTEIN, R.L. [1965] *Multiple successor arithmetics* (**P** 0688) Logic Colloq;1963 Oxford 265-271
◇ C62 F30 ◇ REV MR 35 # 6553 Zbl 166.5 • ID 05187

GOODSTEIN, R.L. & LEE, R.D. [1966] *A decidable class of equations in recursive arithmetic* (**J** 0068) Z Math Logik Grundlagen Math 12*235-239
◇ B20 B25 F30 ◇ REV MR 33 # 5491 Zbl 148.246 JSL 33.618 • ID 05190

GOODSTEIN, R.L. [1969] *A constructive form of the second Gauss proof of the fundamental theorem of algebra* (**P** 1060) Constr Aspects Fund Thm Algeb;1967 Zuerich 69-76
◇ F60 ◇ REV MR 41 # 3275 Zbl 181.13 • ID 20953

GOODSTEIN, R.L. [1969] *Decision methods in recursive arithmetic* (**P** 1841) Fct Recurs & Appl;1967 Tihany 99-104
◇ B25 F30 ◇ ID 32559

GOODSTEIN, R.L. [1970] *Polynomials with computable coefficients* (**J** 0047) Notre Dame J Formal Log 11*447-448
◇ F60 ◇ REV MR 44 # 2606 Zbl 177.16 • ID 05193

GOODSTEIN, R.L. see Vol. I, II, III, IV, V for further entries

GORDEEV, L.N. [1977] *A majorizing semantics for hyperarithmetic sentences (Russian) (English summary)* (**S** 0228) Zap Nauch Sem Leningrad Otd Mat Inst Steklov 68*30-37,143
• TRANSL [1981] (**J** 1531) J Sov Math 15*16-21
◇ D55 F15 F50 ◇ REV MR 58 # 27410 Zbl 359 # 02032
• ID 50580

GORDEEV, L.N. [1977] *An approach to the constructivization of Cantor's set theory (Russian) (English summary)* (**S** 0228) Zap Nauch Sem Leningrad Otd Mat Inst Steklov 68*38-50,143-144
• TRANSL [1981] (**J** 1531) J Sov Math 15*22-28
◇ E70 F50 ◇ REV MR 58 # 27382 Zbl 362 # 02023
• ID 50745

GORDEEV, L.N. [1982] *Constructive models for set theory with extensionality* (**P** 3638) Brouwer Centenary Symp;1981 Noordwijkerhout 123-147
◇ C62 E70 F50 ◇ REV MR 85k:03037 Zbl 524 # 03051 • ID 37615

GORDON, C.E. [1971] see DALEN VAN, D.

GORDON, C.E. see Vol. III, IV, V for further entries

GORDON, D. [1979] *Complexity classes of provable recursive functions* (**J** 0119) J Comp Syst Sci 18*294-303
◇ D15 D20 F30 ◇ REV MR 80e:68105 Zbl 423 # 03045 • ID 53557

GORDON, D. see Vol. IV, V for further entries

GORGY, F.W. [1974] *The dependence and independence of the rules of inference of the step-by-step semantic system of constructive mathematical logic (Russian)* (**C** 1450) Teor Algor & Mat Logika (Markov) 195-210,214
◇ F50 ◇ REV MR 53 # 10560 Zbl 298 # 02025 • ID 23069

GORGY, F.W. [1979] *The independence of the rule of syllogism in* S_2 (**J** 0068) Z Math Logik Grundlagen Math 25*481-484
◇ B45 F50 ◇ REV MR 80m:03094 Zbl 432 # 03034 • ID 53991

GORGY, F.W. [1981] *L'independance de quelques regles de deduction du systeme* $S_{\omega+n}$ *de A. A. Markov* (**J** 3364) C R Acad Sci, Paris, Ser 1 292*607-610
◇ F50 ◇ REV MR 82i:03073 Zbl 457 # 03056 • ID 54381

GORGY, F.W. & SAHYOUN, A.H. [1981] *Mutual transformability of the formulas of the basic languages of constructive mathematical logic* (**J** 0068) Z Math Logik Grundlagen Math 27*489-494
◇ F50 ◇ REV MR 83f:03062 Zbl 519 # 03048 • ID 33323

GORGY, F.W. & SAHYOUN, A.H. [1983] *Mutual transformability of the formulas of the languages of Markov* $L_{\omega N}$ *and* A_{N1} (**J** 0068) Z Math Logik Grundlagen Math 29*207-212
◇ F50 ◇ REV MR 86c:03023 Zbl 527 # 03039 • ID 37583

GORGY, F.W. & SAHYOUN, A.H. [1983] *Transformability of the formulas of the languages of Markov* L_N, L_ω *into formulas of the language* A_2 (**J** 0068) Z Math Logik Grundlagen Math 29*203-206
◇ F50 ◇ REV MR 86c:03022 Zbl 527 # 03038 • ID 48802

GOTO, S. [1979] *Program synthesis through Goedel's interpretation* (**P** 3479) Math Stud of Inform Process;1978 Kyoto 75*302-325
◇ B75 F10 F50 ◇ REV Zbl 407 # 68041 • ID 56230

GRANDY, R.E. [1966] *A note on the recursive unsolvability of primitive recursive arithmetic* (**J** 0053) Proc Amer Math Soc 17*146-147
◇ D35 F30 ◇ REV MR 32 # 7404 Zbl 148.247 • ID 05303

GRANDY, R.E. see Vol. I, II, III for further entries

GRATTAN-GUINNESS, I. [1979] *In memoriam Kurt Goedel: His 1931 correspondence with Zermelo on his incompletability theorem* (**J** 1648) Hist Math 6*294-304
◇ A10 F30 ◇ REV MR 80g:01019 Zbl 416 # 01008 • ID 53166

GRATTAN-GUINNESS, I. see Vol. II, V for further entries

GRAYSON, R.J. [1979] *Heyting-valued models for intuitionistic set theory* (**P** 2901) Appl Sheaves;1977 Durham 402-414
◇ E35 E70 F50 G30 ◇ REV MR 82g:03089 Zbl 419 # 03033 • ID 53376

GRAYSON, R.J. [1981] *Concepts of general topology in constructive mathematics and in sheaves* (**J** 0007) Ann Math Logic 20*1-41 • ERR/ADD ibid 23*99
◇ C90 E35 E70 E75 F35 F50 F55 G30 ◇ REV MR 82g:03105 MR 84d:03075 Zbl 458 # 03015 • REM Part I. Part II 1982 • ID 54421

GRAYSON, R.J. [1982] *Concepts of general topology in constructive mathematics and in sheaves II* (**J** 0007) Ann Math Logic 23*55-98
◇ E75 F50 F55 G30 ◇ REV MR 84c:03102 Zbl 495 # 03040 • REM Part I 1981 • ID 34036

GRAYSON, R.J. [1982] *Constructive well-orderings* (**J** 0068) Z Math Logik Grundlagen Math 28*495-504
◇ E10 E70 F50 ◇ REV MR 84f:03053 Zbl 501 # 03041 • ID 34472

GRAYSON, R.J. [1982] see FOURMAN, M.P.

GRAYSON, R.J. [1983] *Forcing in intuitionistic systems without power-set* (**J** 0036) J Symb Logic 48*670-682
◇ E40 E70 F35 F50 ◇ REV MR 84k:03141 • ID 35033

GRAYSON, R.J. [1983] *On closed subsets of the intuitionistic reals* (**J** 0068) Z Math Logik Grundlagen Math 29*7-9
◇ F50 ◇ REV MR 84b:03079 Zbl 521 # 03045 • ID 35661

GRAYSON, R.J. & MOERDIJK, I. [1983] *Some remarks on extending bar induction* (**J** 0028) Indag Math 45*315-318
◇ F35 F50 ◇ REV MR 84m:03093 Zbl 565 # 03029 • ID 35788

GRAYSON, R.J. [1984] *Heyting-valued semantics* (**P** 3710) Logic Colloq;1982 Firenze 181-208
◇ B15 C90 F35 F50 G30 ◇ REV MR 86f:03099 Zbl 574 # 03048 • ID 41840

GREENLEAF, N. [1978] *Linear order in lattices: A constructive study* (**C** 2295) Stud Founds & Combin 11-30
◇ F50 F55 ◇ REV MR 80c:06025 Zbl 449 # 06009 • ID 56745

GREENLEAF, N. [1981] *Liberal constructive set theory* (**P** 3146) Constr Math;1980 Las Cruces 213-240
◇ A05 E70 F55 F65 ◇ REV MR 83c:03054 Zbl 519 # 03047 • ID 35148

GREENLEAF, N. see Vol. III for further entries

GREGG, J.R. [1970] *Axiomatic quasi-natural deduction* (**J** 0047) Notre Dame J Formal Log 11*221-228
◇ B10 F07 ◇ REV MR 44 # 28 Zbl 198.317 • ID 05317

GREGG, J.R. see Vol. I for further entries

GREIF, I. [1974] see BISHOP, P.

GRENIEWSKI, H. [1951] *Arithmetics of natural numbers as part of the bi-valued propositional calculus* (**S** 0019) Colloq Math (Warsaw) 2*291-297
◇ B05 F30 ◇ REV MR 14.345 Zbl 45.295 JSL 33.305 • ID 05333

GRENIEWSKI, H. see Vol. I, II, III for further entries

GRISHIN, V.N. [1974] *A nonstandard logic, and its application to set theory (Russian)* (**C** 2577) Issl Formaliz Yazyk & Neklass Log 135-171
◇ B60 E70 F07 ◇ REV MR 56 # 8331 • ID 73648

GRISHIN, V.N. [1979] *Herbrand's theorem for the logic without contraction rule (Russian)* (**C** 2581) Issl Neklass Log & Teor Mnozh 316-329
◇ F05 ◇ REV MR 82d:03021 Zbl 422 # 03030 • ID 53491

GRISHIN, V.N. [1981] *Predicate and set-theoretic calculi based on a logic without contractions (Russian)* (**J** 0216) Izv Akad Nauk SSSR, Ser Mat 45*47-68,239
• TRANSL [1981] (**J** 0448) Math of USSR, Izv 18*41-59
◇ B53 B60 E30 E70 F07 ◇ REV MR 82g:03016 Zbl 464 # 03027 Zbl 478 # 03009 • ID 54617

GRISHIN, V.N. see Vol. I, II, V for further entries

GRISS, G.F.C. [1944] *Negationless intuitionistic mathematics* (**J** 0258) Nederl Akad Wetensch Verh Tweed Afd Nat 53*261-268
◇ F50 F55 ◇ REV MR 7.405 JSL 11.24 • ID 24911

GRISS, G.F.C. [1946] *Negationless intuitionistic mathematics I* (**J** 0028) Indag Math 8*675-681
◇ F50 F55 ◇ REV MR 8.307 Zbl 61.8 JSL 13.174 • REM Part II 1950 • ID 16830

GRISS, G.F.C. [1948] *Logique des mathematiques intuitionnistes sans negation* (**J** 0109) C R Acad Sci, Paris 227*946-948
◇ F50 ◇ REV MR 10.277 Zbl 41.353 JSL 15.235 • ID 05362

GRISS, G.F.C. [1948] *On negation (Dutch)* (**C** 4219) Feestbundel an H.J.Pos 96-106
◇ A05 F99 ◇ REV JSL 15.135 • ID 41612

GRISS, G.F.C. [1948] *Sur la negation (dans les mathematiques et la logique)* (**J** 0154) Synthese 7*71-74
◇ A05 F50 ◇ REV MR 10.277 JSL 15.234 • ID 05363

GRISS, G.F.C. [1950] *Negationless intuitionistic mathematics II* (**J** 0028) Indag Math 12*108-115
◇ F50 F55 ◇ REV MR 12.3 Zbl 39.244 • REM Part I 1946. Part III 1951 • ID 05365

GRISS, G.F.C. [1951] *Negationless intuitionistic mathematics III,IV* (**J** 0028) Indag Math 13*193-199,452-462,463-471
◇ F55 ◇ REV MR 13.310 MR 14.4 Zbl 43.250 • REM Part II 1950 • ID 48146

GRISS, G.F.C. [1951] *The logic of negationless intuitionistic mathematics* (**J** 0028) Indag Math 13*41-49
◇ F50 ◇ REV MR 13.97 Zbl 43.249 JSL 20.67 • ID 05364

GRISS, G.F.C. [1955] *La mathematique intuitionniste sans negation* (**J** 3077) Nieuw Arch Wisk, Ser 3 3*134-142
◇ F50 ◇ REV MR 17.569 Zbl 66.11 JSL 22.214 • ID 05366

GRIZE, J.-B. [1971] *Quelques problemes logico-linguistiques* (**J** 0392) Math Sci Hum 35*34-50
◇ B40 ◇ REV Zbl 263 # 68035 • ID 31110

GRIZE, J.-B. [1973] *Logique moderne. Fasc. III: Implications-modalites, logiques polyvalentes, logique combinatoire, ontologie et mereologie de Lesniewski* (**X** 0834) Gauthier-Villars: Paris xxii + 104pp
◇ B40 ◇ REV Zbl 279 # 02007 • REM Part II 1971 • ID 29538

GRIZE, J.-B. see Vol. I, II for further entries

GROSKY, W. [1975] see BARALT-TORRIJOS, J.

GROSS, W.F. [1966] see BOEHM, C.

GROSS, W.F. see Vol. III, IV, V for further entries

GRUENBERG, T. [1983] *A tableau system of proof for predicate-functor logic with identity* (**J** 0036) J Symb Logic 48*1140-1144
◇ B10 F07 ◇ REV MR 86d:03011 Zbl 539 # 03003 • ID 43736

GRUZINTSEV, G.A. [1927] *The concept of relation and the axiomatical definition of number (Russian) (French summary)* (**J** 4110) Zap Inst Nar Prosv, Dnepropetrovsk 1*25-43
◇ B28 F99 ◇ REV FdM 53.182 • ID 40886

GRUZINTSEV, G.A. see Vol. V for further entries

GRZEGORCZYK, A. [1955] *Elementarily definable analysis* (**J** 0027) Fund Math 41*311-338
◇ F60 ◇ REV MR 16.891 Zbl 64.9 JSL 23.444 • ID 05391

GRZEGORCZYK, A. [1955] *On the definition of computable functionals* (**J** 0027) Fund Math 42*232-239
◇ D20 F60 ◇ REV MR 19.238 Zbl 67.3 JSL 24.50 • ID 05392

GRZEGORCZYK, A. [1956] *Some proofs of undecidability of arithmetic* (**J** 0027) Fund Math 43*166-177
◇ D35 F30 ◇ REV MR 18.552 Zbl 72.5 JSL 23.46 • ID 05395

GRZEGORCZYK, A. [1957] *Decision problems* (**X** 1034) PWN: Warsaw 142pp
◇ B25 C10 D35 D98 F30 ◇ REV MR 22 # 7936 Zbl 89.246 • ID 24864

GRZEGORCZYK, A. [1957] *On the definitions of computable real continuous functions* (**J** 0027) Fund Math 44*61-71
◇ D20 F60 ◇ REV MR 19.723 Zbl 79.248 • ID 05396

GRZEGORCZYK, A. & MOSTOWSKI, ANDRZEJ & RYLL-NARDZEWSKI, C. [1958] *The classical and the ω-complete arithmetic* (**J** 0036) J Symb Logic 23*188-206
◇ C62 D55 D70 F30 ◇ REV MR 21 # 4908 Zbl 84.248 JSL 27.80 • ID 05398

GRZEGORCZYK, A. [1959] *Some approaches to constructive analysis* (**P** 0634) Constructivity in Math;1957 Amsterdam 43-61
◇ F60 ◇ REV MR 22 # 3685 Zbl 88.17 JSL 40.501 • ID 05397

GRZEGORCZYK, A. [1964] *A philosophically plausible formal interpretation of intuitionistic logic* (**J** 0028) Indag Math 26*596-601
◇ F50 ◇ REV MR 30 # 3015 Zbl 131.7 JSL 36.329 • ID 05409

GRZEGORCZYK, A. [1964] *Recursive objects in all finite types* (**J** 0027) Fund Math 54*73-93
◇ D65 F10 ◇ REV MR 28 # 3926 Zbl 196.14 JSL 39.343 • ID 05411

GRZEGORCZYK, A. [1967] *Non-classical propositional calculi in relation to methodological patterns of scientific investigation (Polish) (Russian and English summaries)* (**J** 0063) Studia Logica 20*117-132
◇ A05 B45 F50 ◇ REV Zbl 295 # 02005 • ID 62216

GRZEGORCZYK, A. [1971] *Classical relativistic and constructivistic ways of assertions of theorems (Polish) (Russian and English summaries)* (**J** 0063) Studia Logica 27*151-161
◇ F50 ◇ REV MR 46 # 28 Zbl 245 # 02008 • ID 14701

GRZEGORCZYK, A. [1971] *Outline of theoretical arithmetic (Polish)* (**X** 1034) PWN: Warsaw 314pp
◇ B28 C62 F30 F35 F98 ◇ REV MR 53 # 12856 MR 85e:03002 Zbl 251 # 02002 • REM 2nd ed. 1983 • ID 23119

GRZEGORCZYK, A. [1972] *An approach to logical calculus (Polish and Russian summaries)* (**J** 0063) Studia Logica 30*33-43
◇ B22 F50 ◇ REV MR 47 # 4768 Zbl 286 # 02029 • ID 05418

GRZEGORCZYK, A. see Vol. I, II, III, IV, V for further entries

GUASPARI, D. [1979] *Partially conservative extensions of arithmetic* (**J** 0064) Trans Amer Math Soc 254*47-68
◇ C62 F25 F30 ◇ REV MR 80j:03083 Zbl 417 # 03030 • ID 53266

GUASPARI, D. & SOLOVAY, R.M. [1979] *Rosser sentences* (**J** 0007) Ann Math Logic 16*81-99
◇ B45 F30 ◇ REV MR 80h:03083 Zbl 426 # 03062 • ID 53658

GUASPARI, D. [1983] *Sentences implying their own provability* (**J** 0036) J Symb Logic 48*777-789
◇ F30 ◇ REV MR 85g:03027 Zbl 547 # 03035 • ID 42406

GUASPARI, D. see Vol. III, IV, V for further entries

GUILLAUME, M. [1958] *Les tableaux semantiques du calcul des predicats restreint* (**S** 1567) Semin Bourbaki Exp.153*13pp
◇ B20 F07 F98 ◇ REV MR 21 # 4902 Zbl 84.8 • ID 42813

GUILLAUME, M. [1967] *Quelques remarques sur les "tableaux de Beth"* (**P** 0683) Log & Founds of Sci (Beth);1964 Paris 39-45
◇ F07 ◇ REV MR 39 # 46 Zbl 203.9 • ID 05441

GUILLAUME, M. see Vol. I, II, V for further entries

GUPTA, H.N. [1968] *On the rule of existential specification in systems of natural deduction* (**J** 0094) Mind 77*96-103
◇ A05 B10 F07 ◇ ID 14682

GUPTA, H.N. [1979] *On a minimal system of intuitionistically acceptable propositional calculus* (**P** 2539) Frege Konferenz (1);1979 Jena 142-148
◇ F50 ◇ REV MR 82c:03015 • ID 73706

GUPTA, H.N. see Vol. III for further entries

GURARI, E.M. & IBARRA, O.H. [1981] *The complexity of the equivalence problem for two characterizations of Presburger sets* (**J** 1426) Theor Comput Sci 13*295-314
◇ B25 D05 D15 F30 ◇ REV MR 82m:68084 Zbl 454 # 03005 • ID 54218

GURARI, E.M. see Vol. III, IV for further entries

GUREVICH, Y. [1965] *Existential interpretation (Russian)* (**J** 0003) Algebra i Logika 4/4*71-85
◇ B20 B25 C60 D35 F25 ◇ REV MR 34 # 46 Zbl 294 # 02023 • REM Part II 1982 • ID 05452

GUREVICH, Y. [1977] *Intuitionistic logic with strong negation* (**J** 0063) Studia Logica 36*49-59
◇ A05 F50 ◇ REV MR 58 # 160 Zbl 366 # 02015 • ID 31106

GUREVICH, Y. [1980] *Two notes on formalized topology* (**J** 0027) Fund Math 107*145-148
◇ C65 C85 F25 ◇ REV MR 82c:03057 Zbl 362 # 54004 • ID 53951

GUREVICH, Y. [1982] *Existential interpretation II* (**J** 0009) Arch Math Logik Grundlagenforsch 22*103-120
◇ B20 D35 F25 ◇ REV MR 84b:03059 Zbl 493 # 03020 • REM Part I 1965 • ID 33757

GUREVICH, Y. & SHELAH, S. [1983] *Interpreting second-order logic in the monadic theory of order* (**J** 0036) J Symb Logic 48*816-828
◇ B15 C65 C85 D35 E07 E50 F25 ◇ REV MR 85f:03007 Zbl 559 # 03008 • ID 33765

GUREVICH, Y. & SHELAH, S. [1984] *The monadic theory and the "next world"* (**J** 0029) Israel J Math 49*55-68
◇ C62 C65 C85 E40 F25 ◇ REV Zbl 575 # 03028 • ID 38749

GUREVICH, Y. see Vol. I, II, III, IV, V for further entries

GUZICKI, W. [1974] *Elementary extensions of Levy's model of* A_2^- (**J** 0154) Synthese 27*265-270
◇ C55 C62 C80 E25 E35 E40 F35 ◇ REV MR 58 # 16268 Zbl 344 # 02045 • ID 62249

GUZICKI, W. [1976] *On weaker forms of choice in second order arithmetic* (**J** 0027) Fund Math 93*131-144
◇ E25 E35 F35 ◇ REV MR 55 # 100 Zbl 345 # 02048 • ID 26507

GUZICKI, W. [1981] *The equivalence of definable quantifiers in second order arithmetic* (**J** 0027) Fund Math 113*59-65
◇ C30 C55 C62 C80 E40 E45 F35 ◇ REV MR 84g:03050 Zbl 493 # 03011 • ID 34160

GUZICKI, W. [1983] *Definable quantifiers in second order arithmetic and elementary extensions of* ω*-models* (**J** 0202) Diss Math (Warsaw) 208*51pp
◇ C30 C55 C62 C80 E40 E50 F35 ◇ REV MR 85f:03035 Zbl 518 # 03022 • ID 37520

GUZICKI, W. see Vol. III, IV, V for further entries

HAAS, G. [1984] *Konstruktive Einfuehrung in die formale Logik* (**X** 0876) Bibl Inst: Mannheim 268pp
◇ B98 F98 ◇ REV MR 86g:03001 Zbl 562 # 03001 • ID 44527

HABERTHUER, R. [1978] *Choice sequences and reduction processes* (**J** 0009) Arch Math Logik Grundlagenforsch 19*31-49
◇ F05 F35 F50 ◇ REV MR 80e:03071 Zbl 404 # 03046 • ID 29163

HADAMARD, J. [1905] see BAIRE, R.

HADAMARD, J. [1954] *Sur l'impossibilite de demontrer la compatibilite des axiomes de l'arithmetique* (**J** 0423) Pensee NS 58*82
◇ A05 B28 F99 ◇ REV JSL 23.30 • REM Extract from Hadamard: Natures des mathematiques, principes fondamentaux • ID 42157

HADAMARD, J. see Vol. I, V for further entries

HAFNER, I. [1977] *On some subtheory of formal arithmetic* (**J** 3519) Glas Mat, Ser 3 (Zagreb) 12(32)*229-231
◇ F30 ◇ REV MR 58 # 21557 Zbl 376 # 02038 • ID 51680

HAFNER, I. [1979] *On proof length in the equivalential calculus* (**S** 3416) Prepr Ser Dep Math Univ Ljubljana 17*71-90 • REPR [1980] (**J** 3519) Glas Mat, Ser 3 (Zagreb) 15(35)*233-242
◇ B20 F20 ◇ REV MR 82d:03094 Zbl 426 # 03060 Zbl 453 # 03061 • ID 53656

HAFNER, I. see Vol. I, IV, V for further entries

HAGIYA, M. [1983] *A proof description language and its reduction system* (J 0390) Publ Res Inst Math Sci (Kyoto) 19*237-261
◇ B40 F05 ◇ REV MR 85k:03036 Zbl 522 # 03041
• ID 37793

HAHN, S.W. [1969] see BULLOFF, J.J.

HAILPERIN, T. [1969] *A form of Herbrand's theorem* (J 0068) Z Math Logik Grundlagen Math 15*107-120
◇ B10 F05 F07 ◇ REV MR 40 # 23 Zbl 157.335
• ID 05608

HAILPERIN, T. see Vol. I, II, III, V for further entries

HAIMAN, M. [1985] *Linear lattice proof theory: an overview* (P 4178) Universal Algeb & Lattice Th;1984 Charleston 129-141
◇ F05 F07 ◇ REV Zbl 571 # 06009 • ID 47864

HAIMAN, M. [1985] *Proof theory for linear lattices* (J 0345) Adv Math 58*209-242
◇ F05 F07 ◇ ID 48396

HAJEK, P. [1965] *Syntactic models of axiomatic theories* (J 0014) Bull Acad Pol Sci, Ser Math Astron Phys 13*273-278
◇ E30 E70 F25 ◇ REV MR 32 # 2328 Zbl 151.10
• ID 05513

HAJEK, P. [1966] *Generalized interpretability in terms of models. Note to a paper of R.Montague (Czech and Russian summaries)* (J 0086) Cas Pestovani Mat, Ceskoslov Akad Ved 91*352-357
◇ C07 F25 ◇ REV MR 35 # 1456 Zbl 144.245 • REM Montague's paper appeared 1965 in J0028 27*467-476
• ID 05517

HAJEK, P. [1970] *Logische Kategorien* (J 0009) Arch Math Logik Grundlagenforsch 13*168-193
◇ B10 C40 E40 F25 G30 ◇ REV MR 43 # 7386
Zbl 226 # 02043 • ID 05521

HAJEK, P. [1971] *On interpretability in set theories I* (J 0140) Comm Math Univ Carolinae (Prague) 12*73-79
◇ F25 ◇ REV MR 47 # 32 Zbl 231 # 02087 • REM Part II 1972 • ID 05522

HAJEK, P. [1972] *Contributions to the theory of semisets I* (J 0068) Z Math Logik Grundlagen Math 18*241-248
◇ E30 E35 E70 F25 ◇ REV MR 46 # 7029
Zbl 296 # 02037 • REM Part II 1972 by Mlcek,J. & Sochor,A.
• ID 05526

HAJEK, P. & HAJKOVA, M. [1972] *On interpretability in theories containing arithmetic* (J 0027) Fund Math 76*131-137
◇ F25 F30 ◇ REV MR 46 # 7012 Zbl 262 # 02049 • REM Part I. Part II 1981 • ID 05532

HAJEK, P. [1972] *On interpretability in set theories II* (J 0140) Comm Math Univ Carolinae (Prague) 13*445-455
◇ F25 ◇ REV MR 48 # 1922 Zbl 251 # 02064 • REM Part I 1971 • ID 05524

HAJEK, P. [1972] see VOPENKA, P.

HAJEK, P. [1973] *Why semisets ?* (J 0140) Comm Math Univ Carolinae (Prague) 14*397-420
◇ E35 E70 F65 ◇ REV MR 49 # 7139 Zbl 268 # 02005
• ID 05525

HAJEK, P. [1977] *Experimental logics and Π^0_3 theories* (J 0036) J Symb Logic 42*515-522
◇ B60 C62 D55 F30 ◇ REV MR 58 # 16243
Zbl 428 # 03043 • ID 26853

HAJEK, P. [1979] *On partially conservative extensions of arithmetic* (P 2627) Logic Colloq;1978 Mons 225-234
◇ F30 ◇ REV MR 81j:03088 Zbl 453 # 03063 • ID 54193

HAJEK, P. [1981] *Completion closed algebras and models of Peano arithmetic* (J 0140) Comm Math Univ Carolinae (Prague) 22*585-594
◇ C62 F30 ◇ REV MR 83f:03071 Zbl 499 # 03022
• ID 35314

HAJEK, P. [1981] *On interpretability in theories containing arithmetics II* (J 0140) Comm Math Univ Carolinae (Prague) 22*667-688
◇ B28 B45 F25 F30 ◇ REV MR 83j:03094
Zbl 262 # 02049 Zbl 487 # 03032 • REM Part I 1972 by Hajek,P. & Hajkova,M. • ID 35379

HAJEK, P. [1982] see BENDOVA, K.

HAJEK, P. [1984] *On a new notion of partial conservativity* (P 2153) Logic Colloq;1983 Aachen 2*217-232
◇ C62 F30 ◇ REV MR 86d:03054 • ID 43006

HAJEK, P. see Vol. I, II, III, IV, V for further entries

HAJKOVA, M. [1971] *The lattice of bi-numerations of arithmetic. I,II* (J 0140) Comm Math Univ Carolinae (Prague) 12*81-104,281-306
◇ D45 F30 ◇ REV MR 44 # 1565 MR 45 # 64
Zbl 216.289 Zbl 232 # 02025 • ID 05529

HAJKOVA, M. [1972] see HAJEK, P.

HALE, R.L.V. [1974] see BISHOP, P.

HALLALA, P. [1976] see BIELA, A.

HALMOS, P.R. [1977] *Logic from A to G* (J 0497) Math Mag 50*5-11
• TRANSL [1982] (J 1527) Pokroky Mat Fyz Astron (Prague) 27*93-101
◇ A10 F30 ◇ REV MR 55 # 5377 Zbl 354 # 02001
• ID 50141

HALMOS, P.R. see Vol. I, V for further entries

HALPERN, J.D. [1970] see COLLINS, G.E.

HALPERN, J.D. see Vol. III, V for further entries

HALSTEAD, R.H.J. & WARD, S.A. [1980] *A syntactic theory of message passing* (J 0037) ACM J 27*365-383
◇ B40 ◇ REV MR 81b:68013 Zbl 434 # 68065 • ID 83105

HANATANI, Y. [1966] *Calculabilite des fonctionelles recursives primitives de type fini sur les nombres naturels* (J 0260) Ann Jap Ass Phil Sci 3*19-30
◇ D65 F10 F35 F50 ◇ REV MR 33 # 5479 Zbl 143.13
• ID 24779

HANATANI, Y. [1968] *Demonstration de l'ω-non-contradiction de l'arithmetique* (J 0260) Ann Jap Ass Phil Sci 3*105-114
◇ F05 F15 F30 F50 ◇ REV MR 38 # 40 Zbl 179.14
• ID 24821

HANATANI, Y. [1975] *Calculability of the primitive recursive functionals of finite type over the natural numbers* (P 1440) ⊧ ISILC Proof Th Symp (Schuette);1974 Kiel 152-163
◇ D65 F10 F35 F50 ◇ REV MR 53 # 5273
Zbl 317 # 02033 • ID 22904

Author Index

HANAZAWA, M. [1966] *A characterization of axiom schema playing the role of tertium non dactur in intuitionistic logic* (**J** 0081) Proc Japan Acad 42*1007-1010
◇ F50 ◇ REV MR 35 # 4092 Zbl 191.286 JSL 33.607 • ID 05631

HANAZAWA, M. [1979] *An interpretation of Skolem's paradox in the predicate calculus with ε-symbol* (**J** 1472) Sci Rep Saitama Univ, Ser A 9/2*11-13
◇ B10 C07 F05 F25 ◇ REV MR 81h:03110 Zbl 457 # 03004 • ID 54329

HANAZAWA, M. [1979] *On a problem about Skolem's paradox of Takeuti's version* (**J** 2606) Tsukuba J Math 3*53-58
◇ E30 E35 E65 E70 F25 ◇ REV MR 81d:03053 Zbl 448 # 03039 • ID 56639

HANAZAWA, M. see Vol. I, III, V for further entries

HANF, W.P. [1965] *Model-theoretic methods in the study of elementary logic* (**P** 0614) Th Models;1963 Berkeley 132-145
◇ B25 C07 D20 D25 D35 F25 ◇ REV MR 35 # 1457 Zbl 166.258 JSL 34.127 • ID 05635

HANF, W.P. see Vol. I, III, IV, V for further entries

HANSON, N.R. [1961] *The Goedel theorem* (**J** 0047) Notre Dame J Formal Log 2*94-110,228
◇ D20 D35 F30 ◇ REV Zbl 121.254 JSL 27.471 • ID 05642

HANSON, W.H. [1971] *Mechanism and Goedel's theorems* (**J** 0013) Brit J Phil Sci 22*9-16
◇ A05 F30 ◇ REV MR 58 # 27226 Zbl 243 # 02004 • ID 28826

HANSON, W.H. see Vol. II, III for further entries

HAO, K. [1964] *The location in the Kleene hierarchy of some fundamental predicates of constructive analysis (Chinese)* (**J** 0418) Shuxue Xuebao 14*481-493
• TRANSL [1964] (**J** 0419) Chinese Math Acta 5*517-530
◇ F50 ◇ REV MR 31 # 48 • ID 21151

HARDY, G.H. [1904] *A theorem concerning the infinite cardinal numbers* (**J** 0280) Quart J Pure Appl Math 35*87-94
◇ E10 F15 ◇ ID 05655

HARDY, G.H. see Vol. V for further entries

HARMS, S. & KLIX, W.-D. [1977] *Ein Algorithmus zur automatischen Loesung konstruktiver Problemstellungen* (**S** 0410) Math Beitr Univ Halle-Wittenberg 6*55-69
◇ B35 F65 ◇ REV MR 58 # 5101 • ID 73840

HARNIK, V. [1985] *Stability theory and set existence axioms* (**J** 0036) J Symb Logic 50*123-137
◇ C45 F35 ◇ ID 39778

HARNIK, V. see Vol. III, IV, V for further entries

HARRINGTON, L.A. & PARIS, J.B. [1977] *A mathematical incompleteness in Peano arithmetic* (**C** 1523) Handb of Math Logic 1133-1142
◇ C30 C62 E05 F30 ◇ REV MR 58 # 10343 JSL 49.980 • ID 27334

HARRINGTON, L.A. & MORLEY, M.D. & SCEDROV, A. & SIMPSON, S.G. [1985] *Harvey Friedman's research on the foundations of mathematics* (**X** 0809) North Holland: Amsterdam xvi+408pp
◇ B97 C97 D97 E97 F97 ◇ ID 49810

HARRINGTON, L.A. see Vol. III, IV, V for further entries

HARRISON, J. [1963] *Appendix C: Equivalence of the effective operations and the hereditarily continuous functionals* (**C** 4220) Rep Sem Found Anal 2*C1-3
◇ D65 F50 ◇ ID 49833

HARRISON, J. see Vol. IV for further entries

HARROP, R. [1956] *On disjunctions and existential statements in intuitionistic systems of logic* (**J** 0043) Math Ann 132*347-361
◇ F30 F50 ◇ REV MR 18.863 Zbl 71.245 JSL 23.345 • ID 05677

HARROP, R. [1960] *Concerning formulas of the types $A \rightarrow B \vee C$, $A \rightarrow (Ex)B(x)$ in intuitionistic formal systems* (**J** 0036) J Symb Logic 25*27-32
◇ F30 F50 ◇ REV MR 24 # A686 Zbl 98.242 JSL 27.110 • ID 05681

HARROP, R. see Vol. I, III, IV, V for further entries

HART, A.M. & HENDRY, H.E. [1978] *Some observations on a method of McKinsey* (**J** 0047) Notre Dame J Formal Log 19*395-396
◇ B50 F50 ◇ REV MR 58 # 5026 Zbl 368 # 02025 • ID 51664

HART, A.M. see Vol. II for further entries

HARTMANIS, J. [1976] *On effective speed-up and long proofs of trivial theorems in formal theories (French summary)* (**J** 4698) Rev Franc Autom, Inf & Rech Operat, Ser Rouge Inf Th 10/R-1*29-38
◇ B25 B35 D15 D20 F20 ◇ REV MR 54 # 6550 Zbl 399 # 03042 • ID 24160

HARTMANIS, J. [1983] *On Goedel speed-up and succinctness of language representation* (**J** 1426) Theor Comput Sci 26*335-342
◇ D15 F20 ◇ REV MR 85e:68039 Zbl 526 # 68034 • ID 38195

HARTMANIS, J. see Vol. IV for further entries

HASENJAEGER, G. [1952] *Ueber ω-Unvollstaendigkeit in der Peano-Arithmetik* (**J** 0036) J Symb Logic 17*81-97
◇ F30 ◇ REV MR 14.124 Zbl 47.11 JSL 17.271 • ID 05721

HASENJAEGER, G. see Vol. I, II, III, IV, V for further entries

HATCHER, W.S. [1963] *La notion d'equivalence entre systemes formels et une generalisation inductive du systeme dit "New Foundations" de Quine* (**J** 0109) C R Acad Sci, Paris 256*563-566
◇ E70 F25 ◇ REV MR 30 # 3838 Zbl 107.8 JSL 28.172 • ID 05744

HATCHER, W.S. [1964] *La notion d'equivalence entre systemes formels* (**J** 3370) Enseign Math, Ser 2 10*314-315
◇ B22 F25 ◇ ID 32200

HATCHER, W.S. & HODGSON, B.R. [1981] *Complexity bounds on proofs* (**J** 0036) J Symb Logic 46*255-258
◇ D25 F20 ◇ REV MR 82j:03075 Zbl 469 # 03041 • ID 55168

HATCHER, W.S. see Vol. I, II, III, IV, V for further entries

HAUCK, J. [1971] *Ein Kriterium fuer die Annahme des Maximums in der berechenbaren Analysis* (**J** 0068) Z Math Logik Grundlagen Math 17*193-196
◇ F60 ◇ REV MR 44 # 5224 Zbl 225 # 02024 • ID 05751

HAUCK, J. [1971] *Zur Praezisierung des Begriffes berechenbare reelle Funktion* (J 0068) Z Math Logik Grundlagen Math 17*295-300
◇ F60 ◇ REV MR 47 # 8264 Zbl 231 # 02043 • ID 05752

HAUCK, J. [1972] *Berechenbare gleichmaessige Stetigkeit in der konstruktiven Analysis von Klaua (Russian, English and French summaries)* (J 0115) Wiss Z Humboldt-Univ Berlin, Math-Nat Reihe 21*523-526
◇ F60 ◇ REV MR 49 # 2300 Zbl 251 # 02035 • ID 05753

HAUCK, J. [1972] *Zur berechenbaren gleichmaessigen Finitheit rekursiver Funktionale* (J 0115) Wiss Z Humboldt-Univ Berlin, Math-Nat Reihe 21*527-529
◇ D20 F60 ◇ REV MR 49 # 2321 Zbl 251 # 02036 • ID 05754

HAUCK, J. & HERRE, H. & POSEGGA, M. [1972] *Zur Metatheorie formaler Systeme* (C 1533) Quantoren, Modal, Paradox 107-122
◇ B10 C07 F30 ◇ REV Zbl 256 # 02006 • ID 62383

HAUCK, J. [1973] *Berechenbare reelle Funktionen* (J 0068) Z Math Logik Grundlagen Math 19*121-140
◇ F60 ◇ REV MR 48 # 5830 Zbl 298 # 02028 • ID 05756

HAUCK, J. [1975] *Iterationsverfahren in der konstruktiven Analysis* (J 0115) Wiss Z Humboldt-Univ Berlin, Math-Nat Reihe 24*799-800
◇ F60 ◇ REV MR 58 # 5103 Zbl 327 # 02031 • ID 62386

HAUCK, J. [1975] *Turingmaschinen und berechenbare reelle Funktionen* (J 0115) Wiss Z Humboldt-Univ Berlin, Math-Nat Reihe 24*797-799
◇ D10 D20 F60 ◇ REV MR 58 # 5130 Zbl 327 # 02030 • ID 62385

HAUCK, J. [1976] *Berechenbare reelle Funktionenfolgen* (J 0068) Z Math Logik Grundlagen Math 22*265-282
◇ F60 ◇ REV MR 58 # 5104 Zbl 358 # 02039 • ID 18448

HAUCK, J. [1977] *Rekursive Funktionale und der Fixpunktsatz von Banach* (J 0115) Wiss Z Humboldt-Univ Berlin, Math-Nat Reihe 26*635-636
◇ F60 ◇ REV MR 80e:03057 Zbl 423 # 03056 • ID 53568

HAUCK, J. [1977] *Zum Zwischenwertsatz in der konstruktiven Analysis* (J 0115) Wiss Z Humboldt-Univ Berlin, Math-Nat Reihe 26*623-627
◇ F60 ◇ REV MR 80b:03098 Zbl 423 # 03055 • ID 53567

HAUCK, J. [1978] *Konstruktive Darstellungen reeller Zahlen und Folgen* (J 0068) Z Math Logik Grundlagen Math 24*365-374
◇ F60 ◇ REV MR 80b:03099 Zbl 475 # 03034 • ID 55488

HAUCK, J. [1980] *Dezimalbrueche und Berechenbarkeit* (J 0115) Wiss Z Humboldt-Univ Berlin, Math-Nat Reihe 29*409-411
◇ F60 ◇ REV MR 82j:03077 Zbl 468 # 03039 • ID 55104

HAUCK, J. [1980] *Stetigkeitseigenschaften berechenbarer reeller Funktionen* (J 0068) Z Math Logik Grundlagen Math 26*69-76
◇ D20 F60 ◇ REV MR 82b:03107 Zbl 475 # 03035 • ID 55489

HAUCK, J. [1983] *Konstruktive reelle Funktionale und Operatoren* (J 0068) Z Math Logik Grundlagen Math 29*213-218
◇ F60 ◇ REV MR 85i:03181 Zbl 537 # 03041 • ID 43789

HAUCK, J. [1984] *Zur Wellengleichung mit konstruktiven Randbedingungen* (J 0068) Z Math Logik Grundlagen Math 30*561-566
◇ F60 ◇ REV Zbl 537 # 03042 • ID 42280

HAUCK, J. see Vol. III, IV for further entries

HAUSCHILD, K. & RAUTENBERG, W. [1970] *Universelle Interpretierbarkeit in Verbaenden (Russian, English and French summaries)* (J 0115) Wiss Z Humboldt-Univ Berlin, Math-Nat Reihe 19*575-577
◇ D35 F25 G10 ◇ REV MR 48 # 8220 Zbl 275 # 02043 • ID 05774

HAUSCHILD, K. & RAUTENBERG, W. [1971] *Interpretierbarkeit in der Gruppentheorie* (J 0004) Algeb Universalis 1*136-151
◇ C60 D35 F25 ◇ REV MR 46 # 39 Zbl 242 # 02052 • ID 05780

HAUSCHILD, K. & RAUTENBERG, W. [1971] *Interpretierbarkeit und Entscheidbarkeit in der Graphentheorie I* (J 0068) Z Math Logik Grundlagen Math 17*47-55
◇ B25 C65 D35 F25 ◇ REV MR 44 # 6491 Zbl 231 # 02058 • REM Part II 1972 by Hauschild,K. & Herre,H. & Rautenberg,W • ID 27450

HAUSCHILD, K. [1971] *Nichtaxiomatisierbarkeit von Satzmengen durch Ausdruecke spezieller Gestalt* (J 0027) Fund Math 72*245-253
◇ B28 C20 C40 C52 F30 ◇ REV MR 48 # 88 Zbl 215.48 JSL 38.161 • ID 05779

HAUSCHILD, K. & HERRE, H. & RAUTENBERG, W. [1972] *Interpretierbarkeit und Entscheidbarkeit in der Graphentheorie II* (J 0068) Z Math Logik Grundlagen Math 18*457-480
◇ B25 C65 D35 F25 ◇ REV MR 48 # 3737 Zbl 284 # 02023 • REM Part I 1971 • ID 05787

HAUSCHILD, K. [1972] *Universalitaet in der Ringtheorie* (J 0115) Wiss Z Humboldt-Univ Berlin, Math-Nat Reihe 21*505-506
◇ C60 D35 F25 ◇ REV MR 48 # 8221 Zbl 245 # 02041 • ID 05790

HAUSCHILD, K. see Vol. I, II, III, IV, V for further entries

HAYASHI, S. [1977] *A note on provable well-orderings in first order systems with infinitary inference rules* (J 2606) Tsukuba J Math 1*125-135
◇ F15 ◇ REV MR 58 # 16189 Zbl 433 # 03038 • ID 73966

HAYASHI, S. [1977] *On derived rules of intuitionistic second order arithmetic* (J 0407) Comm Math Univ St Pauli (Tokyo) 26*77-103
◇ F05 F35 F50 ◇ REV MR 80h:03086 Zbl 411 # 03055 • ID 52909

HAYASHI, S. [1977] *Some derived rules of intuitionistic second order arithmetic* (J 0081) Proc Japan Acad 53*110-112
◇ F05 F35 F50 ◇ REV MR 55 # 12478 Zbl 381 # 03044 • ID 51912

HAYASHI, S. [1978] *Existence property by means of a normalization method* (J 0407) Comm Math Univ St Pauli (Tokyo) 27*97-100
◇ F05 F50 ◇ REV MR 81i:03095 Zbl 419 # 03036 • ID 53379

HAYASHI, S. [1980] *Derived rules related to a constructive theory of metric spaces in intuitionistic higher order arithmetic without countable choice* (J 0007) Ann Math Logic 19*33-65
◇ F35 F50 G30 ◇ REV MR 82j:03076 Zbl 457 # 03054 • ID 54379

HAYASHI, S. [1981] *On set theory in toposes* (**P** 3201) Logic Symposia;1979/80 Hakone 23-29
◇ E70 F50 G30 ◇ REV MR 84b:18002 Zbl 498 # 03052
• ID 33411

HAYASHI, S. [1982] *A note on the bar induction rule* (**P** 3638) Brouwer Centenary Symp;1981 Noordwijkerhout 149-163
◇ F35 F50 ◇ REV MR 85j:03104 Zbl 514 # 03035
• ID 33412

HAYASHI, S. [1983] *Constructive mathematics and program synthesis (Japanese)* (**P** 4113) Found of Math;1982 Kyoto 33-52
◇ B75 F50 ◇ ID 47675

HAYASHI, S. [1983] *Extracting Lisp programs from constructive proofs: a formal theory of constructive mathematics based on Lisp* (**J** 0390) Publ Res Inst Math Sci (Kyoto) 19*169-191
◇ B35 F35 F50 ◇ REV MR 85f:68066 Zbl 513 # 68022
• ID 37813

HAYASHI, S. [1984] *Constructive mathematics and program synthesis* (**P** 3668) Log & Founds of Math;1983 Kyoto 12-21
◇ B75 F50 ◇ ID 42930

HE, KEGANG [1965] *On the types of some basic predicates of constructive real number theory in the S.C.Kleene classifications (Chinese)* (**P** 4564) Math Logic;1963 Xi-An 119-120
◇ F50 F60 ◇ ID 49354

HEATH, I.J. [1967] *Omitting the replacement schema in recursive arithmetic* (**J** 0047) Notre Dame J Formal Log 8*234-238
◇ F30 ◇ REV MR 38 # 5618 Zbl 212.331 • ID 05833

HEATH, I.J. see Vol. III for further entries

HEIDEMA, J. [1969] *History, principles and methods of intuitionistic mathematics (Africaans) (English summary)* (**J** 1069) Tydskr Natuurwetenskap (Pretoria) 9*10-21
◇ A05 F55 ◇ REV MR 42 # 42 Zbl 184.280 • ID 20936

HEIDEMA, J. see Vol. III for further entries

HEIJENOORT VAN, J. (ED.) [1967] *From Frege to Goedel: a source book in mathematical logic, 1879-1931* (**X** 0858) Harvard Univ Pr: Cambridge x + 660pp
◇ A10 B97 F97 ◇ REV MR 35 # 15 • ID 25903

HEIJENOORT VAN, J. [1967] *Goedel's theorem* (**C** 0601) Encycl of Philos 3*348-357
◇ A05 A10 F30 ◇ REV JSL 35.298 • ID 05871

HEIJENOORT VAN, J. [1979] *Introduction a la semantique des logiques non-classiques* (**X** 3562) Ecole Norm Sup Jeunes Filles: Paris ix + 108pp
◇ B45 C90 F50 ◇ REV MR 81f:03025 • ID 79641

HELLMAN, GEOFFREY [1981] *How to Goedel a Frege-Russell: Goedel's incompleteness theorems and logicism* (**J** 0097) Nous, Quart J Phil 15*451-468
◇ A05 F30 ◇ REV MR 83g:03006 • ID 35987

HELLMAN, GEOFFREY see Vol. II for further entries

HELM, J.P. & MEYER, A.R. & YOUNG, P. [1973] *On orders of translations and enumerations* (**J** 0048) Pac J Math 46*185-195
◇ D15 D20 D25 D45 F40 ◇ REV MR 48 # 1898 Zbl 276 # 02022 • ID 05879

HELM, J.P. see Vol. IV for further entries

HELMAN, GLEN [1977] *Completeness of the normal typed fragment of the λ-system U* (**J** 0122) J Philos Logic 6*33-46
◇ B40 ◇ REV MR 58 # 5079 Zbl 355 # 02017 • ID 50215

HELMAN, GLEN see Vol. I for further entries

HELMER, O. [1937] *Perelman versus Goedel* (**J** 0094) Mind 46*58-60
◇ A05 F30 ◇ REV Zbl 15.338 JSL 2.48 FdM 63.23
• ID 41319

HELMER, O. see Vol. I, II, III for further entries

HENDRY, H.E. [1975] *Another system of natural deduction* (**J** 0047) Notre Dame J Formal Log 16*491-495
◇ B10 F07 ◇ REV MR 53 # 5248 Zbl 254 # 02013
• ID 18186

HENDRY, H.E. [1978] see HART, A.M.

HENDRY, H.E. see Vol. II, III for further entries

HENKIN, L. [1950] *Completeness in the theory of types* (**J** 0036) J Symb Logic 15*81-91
• REPR [1969] (**C** 0569) Phil of Math Oxford Readings 51-63
◇ A05 B15 C85 F35 ◇ REV MR 12.70 Zbl 39.8 JSL 16.72 • ID 05894

HENKIN, L. [1954] *A generalization of the notion of ω-consistency* (**J** 0036) J Symb Logic 19*183-196
◇ B10 C07 F30 ◇ REV MR 16.103 Zbl 56.11 JSL 23.40
• ID 05900

HENKIN, L. [1956] *La structure algebrique des theories mathematiques* (**X** 0834) Gauthier-Villars: Paris 52pp
◇ C07 F50 G05 G10 G15 ◇ REV MR 18.272 Zbl 87.250 JSL 22.215 • ID 05908

HENKIN, L. [1957] *A generalization of the concept of ω-completeness* (**J** 0036) J Symb Logic 22*1-14
◇ B10 C07 F30 ◇ REV MR 20 # 1626 Zbl 81.12 JSL 24.172 • ID 05910

HENKIN, L. [1971] *Mathematical foundations for mathematics* (**J** 0005) Amer Math Mon 78*463-487
◇ A10 C98 E30 E55 F55 G05 G15 ◇ REV MR 44 # 5 Zbl 217.6 JSL 39.333 • ID 05928

HENKIN, L. see Vol. I, II, III, IV, V for further entries

HENNESSY, M. [1976] see ASHCROFT, E.A.

HENNESSY, M. [1980] see ASHCROFT, E.A.

HENNESSY, M. & PLOTKIN, G.D. [1980] *A term model for CCS* (**P** 3210) Math Founds of Comput Sci (9);1980 Rydzyna 261-274
◇ B40 ◇ REV MR 81k:68003 Zbl 479 # 68011 • ID 55697

HENNESSY, M. see Vol. I, III for further entries

HENRICI, P. [1969] see DEJON, B.

HENSEL, G. & PUTNAM, H. [1965] *On the notational independence of various hierarchies of degrees of unsolvability* (**J** 0036) J Symb Logic 30*69-86
◇ D30 D55 F15 ◇ REV MR 37 # 5096 Zbl 137.9 JSL 32.124 • ID 05946

HENSEL, G. see Vol. III, IV for further entries

HENSON, C.W. & RUBEL, L.A. [1984] *Some applications of Nevanlinna theory to mathematical logic:identities of exponential functions* (**J** 0064) Trans Amer Math Soc 282*1-32 • ERR/ADD ibid 294*381
◇ B20 C65 F30 ◇ REV MR 85h:03015 Zbl 533 # 03015
• ID 36539

HENSON, C.W. & KAUFMANN, M. & KEISLER, H.J. [1984] *The strength of nonstandard methods in arithmetic* (**J** 0036) J Symb Logic 49*1039-1058
◇ C62 E30 F30 F35 H05 H15 ◇ REV MR 86h:03115
• ID 39860

HENSON, C.W. see Vol. I, III, IV, V for further entries

HERBRAND, J. [1928] *Sur la theorie de la demonstration* (**J** 0109) C R Acad Sci, Paris 186*1274-1276
• TRANSL [1971] (**C** 0745) Herbrand: Log Writings 29-32
• REPR [1968] (**C** 2486) Herbrand: Ecrits Logiques 21-23
◇ A05 B10 F07 ◇ REV JSL 36.523 JSL 40.94 FdM 54.52
• ID 05972

HERBRAND, J. [1929] *Non-contradiction des axiomes arithmetiques* (**J** 0109) C R Acad Sci, Paris 188*303-304
• TRANSL [1971] (**C** 0745) Herbrand: Log Writings 35-37
◇ F30 ◇ REV JSL 36.523 JSL 40.94 FdM 55.31 • ID 05966

HERBRAND, J. [1929] *Sur le probleme fondamental des mathematiques* (**J** 0109) C R Acad Sci, Paris 189*554-556
• TRANSL [1971] (**C** 0745) Herbrand: Log Writings 41-43
◇ A05 B10 F99 ◇ REV JSL 36.523 JSL 40.94 FdM 55.32
• ID 05968

HERBRAND, J. [1929] *Sur quelques proprietes des propositions vraies et leurs applications* (**J** 0109) C R Acad Sci, Paris 189*1076-1078
• TRANSL [1971] (**C** 0745) Herbrand: Log Writings 38-40
◇ B10 F30 ◇ REV JSL 36.523 JSL 40.94 FdM 55.32
• ID 05967

HERBRAND, J. [1930] *Recherches sur la theorie de la demonstration* (**J** 0459) C R Soc Sci Lett Varsovie Cl 3 33*128pp
• TRANSL [1967] (**C** 0675) From Frege to Goedel 525-581 [1971] (**C** 0745) Herbrand: Log Writings 46-188,272-276
• REPR [1968] (**C** 2486) Herbrand: Ecrits Logiques 35-153 [1931] (**J** 1092) Ann Univ Paris 6*186-189
◇ B10 B25 F05 F07 F25 F30 ◇ REV JSL 36.523 JSL 40.94 FdM 56.824 • ID 20866

HERBRAND, J. [1931] *Sur la non-contradiction de l'arithmetique* (**J** 0127) J Reine Angew Math 166*1-8
• TRANSL [1967] (**C** 0675) From Frege to Goedel 620-628 [1971] (**C** 0745) Herbrand: Log Writings 284-297
◇ F05 F30 ◇ REV Zbl 3.49 JSL 40.95 FdM 57.56
• ID 32733

HERBRAND, J. [1931] *Sur le probleme fondamental de la logique mathematique* (**J** 0459) C R Soc Sci Lett Varsovie Cl 3 24*12-56
• TRANSL [1971] (**C** 0745) Herbrand: Log Writings 215-271
◇ A05 B25 B30 F30 F99 ◇ REV Zbl 3.290 JSL 36.523 FdM 57.1320 • ID 16812

HERBRAND, J. [1968] *Ecrits logiques* (**X** 0840) Pr Univ France: Paris iii + 244pp
• TRANSL [1971] (**X** 0858) Harvard Univ Pr: Cambridge viii + 312pp
◇ A05 A10 B96 F96 ◇ REV MR 37 # 24 MR 51 # 2844 Zbl 194.303 • REM Edited by Heijenoort van,J. • ID 24825

HERBRAND, J. see Vol. I for further entries

HERMANN, G. [1926] *Die Frage der endlich vielen Schritte in der Theorie der Polynomideale* (**J** 0043) Math Ann 95*736-788
◇ C57 C60 D45 F55 ◇ REV FdM 52.127 • ID 05997

HERMES, H. [1937] *Definite Begriffe und berechenbare Zahlen* (**J** 2074) Sem-ber, Muenster 10*110-123
◇ D05 D10 D20 F60 ◇ REV JSL 6.35 FdM 63.823
• ID 32166

HERMES, H. [1955] *Vorlesung ueber Entscheidungsprobleme in Mathematik und Logik* (**X** 0910) Aschendorffsche Verlagsbuchh: Muenster ii + 140pp
◇ B25 B40 B65 D10 D20 D35 D98 ◇ REV MR 17.569 Zbl 67.249 • ID 06008

HERMES, H. [1959] *Zum Inversionsprinzip der operativen Logik* (**P** 0634) Constructivity in Math;1957 Amsterdam 62-68
◇ F50 ◇ REV MR 21 # 4915 Zbl 86.9 JSL 40.501
• ID 06010

HERMES, H. [1961] *Aufzaehlbarkeit, Entscheidbarkeit, Berechenbarkeit: Einfuehrung in die Theorie der rekursiven Funktionen* (**X** 0811) Springer: Heidelberg & New York x + 246pp
• TRANSL [1965] (**X** 0811) Springer: Heidelberg & New York ix + 245pp (English) [1975] (**X** 0905) Boringhieri: Torino 318pp (Italian)
◇ D98 F60 ◇ REV MR 26 # 1252 MR 41 # 8225 MR 49 # 8825 MR 80b:03049 Zbl 383 # 03023 JSL 31.254
• ID 20981

HERMES, H. [1964] *Unentscheidbarkeit der Arithmetik* (**J** 0160) Math-Phys Sem-ber, NS 11*20-34
◇ D35 F30 ◇ REV MR 29 # 3370 Zbl 127.9 JSL 33.469
• ID 06013

HERMES, H. [1969] *On the notion of constructivity* (**P** 1060) Constr Aspects Fund Thm Algeb;1967 Zuerich 115-129
◇ D20 F65 ◇ REV MR 41 # 1529 Zbl 179.316 • ID 20948

HERMES, H. [1976] *Dialog games* (**P** 1619) Coloq Log Simb;1975 Madrid 115-125
◇ A05 B60 F50 ◇ REV MR 56 # 5264 Zbl 366 # 02019
• ID 51102

HERMES, H. see Vol. I, II, III, IV, V for further entries

HERRE, H. [1972] see HAUSCHILD, K.

HERRE, H. [1972] see HAUCK, J.

HERRE, H. see Vol. I, III, IV, V for further entries

HETPER, W. [1934] *Semantische Arithmetik* (**J** 0459) C R Soc Sci Lett Varsovie Cl 3 27*9-26
◇ B28 F30 ◇ REV Zbl 11.1 FdM 61.977 • ID 40820

HETPER, W. [1938] see CHWISTEK, L.B.

HETPER, W. see Vol. I, V for further entries

HEWITT, C. [1974] see BISHOP, P.

HEWITT, C. see Vol. I for further entries

HEYTING, A. [1925] *Intuitionistic axiomatics of projective geometry (Dutch)* (**X** 0812) Wolters-Noordhoff : Groningen iv + 96pp
◇ F55 ◇ REM Trade edition of the thesis. The thesis has viii + 96pp with one loose leaf • ID 49805

HEYTING, A. [1928] *Die Theorie der linearen Gleichungen in einer Zahlenspezies mit nicht-kommutativer Multiplikation* (**J** 0043) Math Ann 98*465-490
◇ F55 ◇ ID 49803

Heyting, A. [1928] *Zur intuitionistischen Axiomatic der projektiven Geometrie* (**J** 0043) Math Ann 98*491-538
◇ F55 ◇ ID 49804

Heyting, A. [1929] *The countability predicates of Professor Brouwer (Dutch)* (**J** 1793) Nieuw Arch Wisk, Ser 2 16*47-58
◇ F55 F99 ◇ REV FdM 55.52 • ID 39243

Heyting, A. [1930] *Die formalen Regeln der intuitionistischen Logik* (**J** 0277) Sitzb Preuss Akad Wiss Phys Math Kl 1930*42-56
◇ F50 ◇ REV FdM 56.823 • ID 20855

Heyting, A. [1930] *Die formalen Regeln der intuitionistischen Mathematik* (**J** 0277) Sitzb Preuss Akad Wiss Phys Math Kl 1930*57-71,158-169
◇ F30 F35 F50 ◇ REV FdM 56.823 • ID 43824

Heyting, A. [1930] *Sur la logique intuitionniste* (**J** 0150) Acad Roy Belg Bull Cl Sci (5) 16*957-963
◇ F50 ◇ REV FdM 56.45 • ID 20844

Heyting, A. [1931] *Die intuitionistische Grundlegung der Mathematik* (**J** 0748) Erkenntnis (Leipzig) 2*106-115
• TRANSL [1964] (**C** 1105) Phil of Math. Sel Readings 42-49 (English)
◇ A05 F50 F55 ◇ REV Zbl 2.321 FdM 57.53 • ID 06046

Heyting, A. [1931] *Die intuitionistische Mathematik* (**J** 0983) Forsch Fortschritte 7*38-39
◇ A05 F98 ◇ REV FdM 57.1323 • ID 43828

Heyting, A. [1932] *A propos d'un article de MM. Barzin et Errera* (**J** 0152) Enseign Math 31*121-122
◇ A05 F99 ◇ REV FdM 59.56 • ID 43829

Heyting, A. [1932] *Anwendung der intuitionistischen Logik auf die Definition der Vollstaendigkeit eines Kalkuels* (**P** 0653) Int Congr Math (II, 4);1932 Zuerich 2*344-345
◇ F50 ◇ REV FdM 58.70 • REM Result incorrect • ID 06047

Heyting, A. [1932] *Reponse a MM. Barzin et Errera* (**J** 0152) Enseign Math 31*271-272,274-275
◇ A05 F99 ◇ REV FdM 59.56 • ID 43830

Heyting, A. [1934] *Mathematische Grundlagenforschung. Intuitionismus. Beweistheorie* (**X** 0811) Springer: Heidelberg & New York iv+73pp
• TRANSL [1955] (**X** 0834) Gauthier-Villars: Paris 91pp (French)
◇ A05 F50 F98 ◇ REV MR 49#8806 Zbl 66.10 Zbl 9.385 JSL 23.33 FdM 60.19 • ID 42160

Heyting, A. [1935] *The debate concerning the foundations of mathematics (Dutch)* (**P** 1486) Handel Natuur- & Geneesk Congr (25);1935 Leiden 94-97
◇ A05 F99 ◇ ID 43847

Heyting, A. [1936] *Intuitionistic mathematics I,II,III (Dutch)* (**J** 1928) Mathematica B 4*72-83,123-128,129-136
◇ F55 ◇ REV Zbl 15.145 FdM 61.48 FdM 61.970 FdM 62.29 • REM Parts IV,V 1937 • ID 43826

Heyting, A. [1937] *Bemerkungen zu dem Aufsatz von Herrn Freudenthal "Zur intuitionistischen Deutung logischer Formeln"* (**J** 0020) Compos Math 4*117-118
◇ A05 F50 ◇ REV Zbl 15.242 JSL 2.48 • ID 06048

Heyting, A. [1937] *Intuitionistic mathematics IV,V (Dutch)* (**J** 1928) Mathematica B 5*62-80,105-112
◇ F55 ◇ REV Zbl 15.145 FdM 65.191 • REM Parts I,II,III 1936. Part VI 1939 • ID 43827

Heyting, A. [1937] *The development of intuitionistic mathematics (Dutch)* (**J** 0290) Euclides 13*129-144
◇ A05 F55 ◇ REV JSL 2.89 FdM 62.1045 • ID 16891

Heyting, A. [1939] *Intuitionistic mathematics VI (Dutch)* (**J** 1928) Mathematica B 7*129-142
◇ F55 ◇ REV Zbl 20.338 JSL 5.73 FdM 65.191 • REM Parts IV,V 1937 • ID 33663

Heyting, A. [1939] *Les fondements des mathematiques du point de vue intuitionniste* (**C** 0656) Phil Mathematique 73-75
◇ A05 F55 ◇ REV JSL 5.78 • ID 06049

Heyting, A. [1941] *Untersuchungen ueber intuitionistische Algebra* (**J** 0258) Nederl Akad Wetensch Verh Tweed Afd Nat 18/2*36pp
◇ F55 ◇ REV MR 7.405 Zbl 27.153 • ID 23367

Heyting, A. [1946] *On weakened quantification* (**J** 0036) J Symb Logic 11*119-121
◇ F50 ◇ REV MR 8.306 JSL 12.59 • ID 06050

Heyting, A. [1950] *Tensions in mathematics (Dutch)* (**J** 0290) Euclides 25*233-236,301-312
◇ A05 F55 ◇ ID 43832

Heyting, A. [1951] *L'axiomatique intuitionniste* (**P** 0642) Congr Int Phil Sci;1949 Paris 2*81-86
◇ F50 F55 ◇ REV MR 13.898 JSL 23.343 • ID 06053

Heyting, A. [1951] *Note on the Riesz-Fischer theorem* (**J** 0028) Indag Math 13*35-40
◇ F55 ◇ REV MR 12.664 Zbl 43.291 • ID 06052

Heyting, A. [1953] *Espace de Hilbert et intuitionnisme* (**P** 0644) Meth Form en Axiom;1950 Paris 59-63
◇ F55 ◇ REV MR 15.91 Zbl 53.7 JSL 23.228 • ID 06054

Heyting, A. [1954] *G.F.C. Griss and his negationless intuitionistic mathematics* (**J** 0154) Synthese 9*91-96
◇ A05 F50 ◇ REV JSL 21.91 • ID 06057

Heyting, A. [1954] *Logique et intuitionnisme* (**P** 0646) Appl Sci de Log Math;1952 Paris 75-83
◇ A05 F50 ◇ REV MR 16.556 Zbl 57.7 JSL 23.33 • ID 06056

Heyting, A. [1954] *Sur la theorie intuitionnististe de la mesure* (**J** 0082) Bull Soc Math Belg 6*70-78
◇ F55 ◇ REV MR 16.910 Zbl 57.44 • ID 06055

Heyting, A. [1955] *On the significance of the mathematical work of G.F.C. Griss (Dutch)* (**J** 0177) Algem Nederl Tijdschr Wijsbeg Psychol 47*8-12
◇ A05 F99 ◇ ID 43834

Heyting, A. [1956] *Intuitionism. An introduction* (**X** 0809) North Holland: Amsterdam viii+133pp
• TRANSL [1965] (**X** 0885) Mir: Moskva 200pp (Russian)
◇ F55 F98 ◇ REV MR 17.698 Zbl 70.8 JSL 21.367 JSL 34.313 • REM 3rd rev. ed. 1971; x+150pp • ID 06059

Heyting, A. [1956] *La conception intuitionniste de la logigue* (**J** 1033) Etud Phil 11*226-233
◇ A05 F50 ◇ ID 49806

Heyting, A. [1957] *Axioms for intuitionistic plane affine geometry. The axiomatic method. With special reference to geometry and physics* (**X** 0809) North Holland: Amsterdam xi+488pp
◇ F55 ◇ REV MR 22#10911 • ID 16948

HEYTING, A. [1958] *Blick von der intuitionistischen Warte* (**J** 0076) Dialectica 12*332-345
• REPR [1959] (**C** 1173) Logica (Bernays) 128-141
◇ A05 F50 ◇ REV MR 20#6360 Zbl 89.245 JSL 27.227
• ID 22141

HEYTING, A. [1958] *Intuitionism in mathematics* (**C** 0742) Phil Mid-Century 101-115
• TRANSL [1967] (**C** 2141) Filos Matematica 249-267
◇ A05 B98 F50 F55 F98 ◇ REV JSL 34.313 JSL 39.609 JSL 40.472 • ID 06062

HEYTING, A. [1958] *Intuitionism and school mathematics (Dutch)* (**J** 0290) Euclides 33*1-12
◇ A05 F55 ◇ REV Zbl 77.2 • ID 49808

HEYTING, A. [1958] *La theorie elementaire de l'integration en mathematiques intuitionnistes* (**P** 0576) Raisonn en Math & Sci Exper;1955 Paris 85-90
◇ F55 ◇ REV MR 21#2588 Zbl 88.248 JSL 25.287
• ID 06061

HEYTING, A. [1959] *Axioms for intuitionistic plane affine geometry* (**P** 0651) Axiomatic Method;1957 Berkeley 160-173
◇ B30 F55 ◇ REV MR 22#10911 Zbl 92.250 • ID 27711

HEYTING, A. (ED.) [1959] *Constructivity in mathematics. Proceedings of the colloquium held at Amsterdam 1957* (**X** 0809) North Holland: Amsterdam viii+297pp
◇ F97 ◇ REV Zbl 86.7 • ID 23441

HEYTING, A. [1959] *Some remarks on intuitionism* (**P** 0634) Constructivity in Math;1957 Amsterdam 69-71
◇ A05 F50 ◇ REV MR 21#2589 Zbl 85.250 JSL 36.673
• ID 06060

HEYTING, A. [1960] *Remarques sur le constructivisme* (**J** 0079) Logique & Anal, NS 3*177-182
◇ A05 F55 ◇ ID 49807

HEYTING, A. [1961] *Axiomatic method and intuitionism* (**C** 0622) Essays Found of Math (Fraenkel) 237-247
◇ B30 F55 ◇ REV MR 29#21 JSL 36.522 • ID 06065

HEYTING, A. [1961] *Infinitistic methods from a finitist point of view* (**P** 0633) Infinitist Meth;1959 Warsaw 185-192
◇ A05 F50 ◇ REV MR 26#2363 Zbl 116.8 JSL 32.515
• ID 06063

HEYTING, A. [1962] *After thirty years* (**P** 0612) Int Congr Log, Meth & Phil of Sci (1,Proc);1960 Stanford 194-197
◇ A05 A10 F99 ◇ REV MR 27#1366 Zbl 166.247 JSL 36.674 • ID 06064

HEYTING, A. [1962] *Methodes et problemes de l'intuitionnisme* (**P** 1606) Colloq Math (Pascal);1962 Clermont-Ferrand 1*101-105
◇ A05 F50 F55 ◇ REV MR 43#6055 JSL 36.674
• ID 22227

HEYTING, A. [1964] *Disputation* (**C** 1105) Phil of Math. Sel Readings 55-65
• TRANSL [1967] (**C** 2141) Filos Matematica 233-248 (Italian) [1972] (**J** 1527) Pokroky Mat Fyz Astron (Prague) 17*4-15 (Czech)
◇ A05 F55 ◇ REV MR 58#10327 • REM This is chapter I of Heyting,A. 1956 'Intuitionism. An Introduction'. The Czech transl. was translated from the Russian • ID 38054

HEYTING, A. [1966] *Remarques sur la theorie intuitionniste des espaces lineaires* (**J** 0154) Synthese 16*47-52
• REPR [1967] (**P** 0683) Log & Founds of Sci (Beth);1964 Paris 4-9
◇ F55 ◇ REV MR 39#1298 Zbl 183.7 • ID 24947

HEYTING, A. [1967] *Informal rigour and intuitionism* (**P** 2268) Int Colloq Philos of Sci;1965 London 1*172-173
◇ A05 F99 ◇ REM This is a discussion remark • ID 43838

HEYTING, A. [1967] *L.E.J. Brouwer (February 27,1881 - December 2,1966) Innovator at the pinnacles and of the ultimate foundations of mathematics (Dutch)* (**J** 4114) De Gids 130*287-294
◇ A05 A10 F99 ◇ ID 43843

HEYTING, A. [1968] *Intuitionism in mathematics* (**C** 0552) Phil Contemp - Chroniques 316-323
◇ A05 F50 F98 ◇ REV JSL 40.472 • ID 14530

HEYTING, A. [1969] *Remark on the paper by Ashvinikumar "The intuitionist contradictority of certain classical set-theoretic results"* (**J** 3077) Nieuw Arch Wisk, Ser 3 17*223
◇ F55 ◇ REV MR 41#1508 Zbl 188.317 • ID 06068

HEYTING, A. [1970] *Recent progress in intuitionistic analysis* (**P** 0603) Intuitionism & Proof Th;1968 Buffalo 95-100
◇ F55 ◇ REV MR 43#3095 Zbl 203.296 JSL 40.500
• ID 21247

HEYTING, A. [1974] *Intuitionistic views on the nature of mathematics* (**J** 0012) Boll Unione Mat Ital, IV Ser 9*122-134
• REPR [1974] (**J** 0154) Synthese 27*79-91
◇ A05 F55 ◇ REV MR 54#2426 MR 58#16195 Zbl 315#02010 Zbl 317#02007 • ID 29798

HEYTING, A. [1981] *Continuum and choice sequence in Brouwer (Dutch)* (**J** 3077) Nieuw Arch Wisk, Ser 3 29*125-139
◇ A10 F55 ◇ REV MR 83k:03002 Zbl 486#03031
• ID 36149

HEYTING, A. see Vol. III, IV for further entries

HICKMAN, J.L. [1974] *Concerning the number of sums obtainable from a countable series of ordinals by permutations that preserve the order-type* (**J** 3172) J London Math Soc, Ser 2 9*239-244
◇ E10 F15 ◇ REV MR 51#165 Zbl 298#04002
• ID 17476

HICKMAN, J.L. [1978] *On R. McBeth's paper: "Fundamental sequences for initial ordinals smaller than a certain Θ_0"* (**J** 0068) Z Math Logik Grundlagen Math 24*441-442
◇ E10 F15 ◇ REV MR 80b:04003 Zbl 397#04003 • REM The article was published ibid. 22(1976)*97-104 • ID 74099

HICKMAN, J.L. see Vol. III, V for further entries

HILBERT, D. [1899] *Grundlagen der Geometrie* (**X** 0823) Teubner: Stuttgart 92pp
• TRANSL [1899] (**X** 1324) Open Court: LaSalle ix+226pp
◇ A05 B30 F99 ◇ REV MR 43#1019 FdM 30.424 • REM 10th ed. 1968;vii+271pp • ID 23454

HILBERT, D. [1900] *Mathematische Probleme* (**J** 1109) Nachr Akad Wiss Goettingen, Math-Phys Kl 1900*253-297
• TRANSL [1900] (**J** 0152) Enseign Math 2*349-354 (French, fragment) [1901] (**J** 0767) Rev Gen Sci Pur Appl 12*168-174 (French, fragment) [1902] (**P** 1484) Int Congr Math (2);1900 Paris 58-114 (French) [1902] (**J** 0015) Bull Amer Math Soc 8*437-479 (English) [1976] (**P** 2957) Math Dev from Hilbert Probl;1974 DeKalb 1*1-35 (English) • REPR [1901] (**J** 3975) Arch Math & Phys 1*44-63,213-237
◇ A05 D35 E30 E50 F25 ◇ REV FdM 31.68 • ID 06078

HILBERT, D. [1931] *Beweis des "tertium non datur"* (**J** 1109) Nachr Akad Wiss Goettingen, Math-Phys Kl 1931*120-125
◇ A05 B28 F50 ◇ REV Zbl 3.49 FdM 57.55 • ID 21186

HILBERT, D. & BERNAYS, P. [1934] *Grundlagen der Mathematik I* (**X** 0811) Springer: Heidelberg & New York xii+471pp
• TRANSL [1979] (**X** 2027) Nauka: Moskva 558pp
◇ A05 B98 F05 F30 F98 ◇ REV MR 81c:03002 Zbl 191.284 Zbl 478 # 03002 JSL 35.321 FdM 60.17 • REM 2nd edition 1968; xv+473pp. Part II 1939 • ID 01098

HILBERT, D. & BERNAYS, P. [1939] *Grundlagen der Mathematik II* (**X** 0811) Springer: Heidelberg & New York xii+498pp
• TRANSL [1982] (**X** 2027) Nauka: Moskva 556pp
◇ A05 B98 F05 F15 F30 F40 F98 ◇ REV MR 42 # 7477 Zbl 20.193 Zbl 211.9 Zbl 518 # 03001 JSL 5.16 FdM 65.21 • REM 2nd edition 1970; xiv+561pp. Part I 1934 • ID 01082

HILBERT, D. see Vol. I, II, III, IV, V for further entries

HINATA, S. [1968] *Calculability of primitive recursive functionals of finite type* (**J** 0350) Sci Rep Tokyo Kyoiku Daigaku Sect A 9*218-235 • ERR/ADD ibid 13*347-365,151
◇ F10 F50 ◇ REV MR 41 # 1530 MR 52 # 5393 Zbl 332 # 02047 • ID 27573

HINATA, S. [1977] *A normalization theorem in formal theories of natural numbers* (**J** 0090) J Math Soc Japan 29*327-340
◇ F05 F30 F50 ◇ REV MR 56 # 99 Zbl 348 # 02034 • ID 62528

HINATA, S. see Vol. IV for further entries

HINDLEY, J.R. [1967] *Axioms for strong reduction in combinatory logic* (**J** 0036) J Symb Logic 32*224-236
◇ B40 ◇ REV MR 35 # 5320 Zbl 153.6 JSL 36.171 • ID 06097

HINDLEY, J.R. [1969] *An abstract form of the Church-Rosser theorem I* (**J** 0036) J Symb Logic 34*545-560
◇ B40 ◇ REV MR 46 # 1578 Zbl 195.21 • REM Part II 1974 • ID 06098

HINDLEY, J.R. [1969] *The principal type-scheme of an object in combinatory logic* (**J** 0064) Trans Amer Math Soc 146*29-60
◇ B40 ◇ REV MR 40 # 7118 Zbl 196.15 • ID 16312

HINDLEY, J.R. & LERCHER, B. [1970] *A short proof of Curry's normal form theorem* (**J** 0053) Proc Amer Math Soc 24*808-810
◇ B40 ◇ REV MR 41 # 67 Zbl 195.21 • ID 06099

HINDLEY, J.R. [1972] see CURRY, H.B.

HINDLEY, J.R. & LERCHER, B. & SELDIN, J.P. [1972] *Introduction to combinatory logic* (**X** 0805) Cambridge Univ Pr: Cambridge, GB 179pp
• TRANSL [1975] (**X** 0905) Boringhieri: Torino 153pp (Italian)
◇ B40 B98 F98 ◇ REV MR 49 # 25 Zbl 269 # 02005 JSL 38.518 • ID 23471

HINDLEY, J.R. [1974] *An abstract Church-Rosser theorem II. Applications* (**J** 0036) J Symb Logic 39*1-21
◇ B40 ◇ REV MR 50 # 61 Zbl 291 # 02015 • REM Part I 1969 • ID 06101

HINDLEY, J.R. [1977] *Combinatory reductions and λ reductions compared* (**J** 0068) Z Math Logik Grundlagen Math 23*169-180
◇ B40 ◇ REV MR 58 # 5080 Zbl 361 # 02035 • ID 26476

HINDLEY, J.R. & MITSCHKE, G. [1977] *Some remarks about the connections between combinatory logic and axiomatic recursion theory* (**J** 0009) Arch Math Logik Grundlagenforsch 18*99-103
◇ B40 D75 ◇ REV MR 57 # 9498 Zbl 365 # 02028 • ID 24332

HINDLEY, J.R. [1977] *The equivalence of complete reductions* (**J** 0064) Trans Amer Math Soc 229*227-248
◇ B40 ◇ REV MR 56 # 2798 Zbl 361 # 02036 • ID 26246

HINDLEY, J.R. [1978] *Reductions of residuals are finite* (**J** 0064) Trans Amer Math Soc 240*345-361
◇ B40 ◇ REV MR 80a:03017 Zbl 387 # 03007 • ID 31141

HINDLEY, J.R. [1978] *Standard and normal reductions* (**J** 0064) Trans Amer Math Soc 241*253-271
◇ B40 ◇ REV MR 80a:03018 Zbl 395 # 03015 • ID 52566

HINDLEY, J.R. [1979] *The discrimination theorem holds for combinatory weak reduction* (**J** 1426) Theor Comput Sci 8*393-394
◇ B40 ◇ REV MR 80g:03012b Zbl 418 # 03011 • ID 53296

HINDLEY, J.R. & LONGO, G. [1980] *Lambda-calculus models and extensionality* (**J** 0068) Z Math Logik Grundlagen Math 26*289-310
◇ B40 ◇ REV MR 81m:03019 Zbl 453 # 03015 • ID 54145

HINDLEY, J.R. & SELDIN, J.P. (EDS.) [1980] *To H. B. Curry: Essays on combinatory logic, lambda-calculus and formalism* (**X** 0801) Academic Pr: New York xxv+606pp
◇ A05 B40 ◇ REV MR 81j:03005 Zbl 469 # 03006 • ID 55134

HINDLEY, J.R. [1982] *The simple semantics for Coppo-Dezani-Salle types* (**P** 3867) Int Symp Progr (5);1982 Turin 212-226
◇ B40 ◇ REV Zbl 514 # 03009 • ID 37403

HINDLEY, J.R. [1983] *Curry's type-rules are complete with respect to the F-semantics too* (**J** 1426) Theor Comput Sci 22*127-133
◇ B40 ◇ REV MR 85e:03030b Zbl 512 # 03010 • ID 36506

HINDLEY, J.R. [1983] *The completeness theorem for typing λ-terms* (**J** 1426) Theor Comput Sci 22*1-17
◇ B40 ◇ REV MR 85e:03030a Zbl 512 # 03009 • ID 36505

HINDLEY, J.R. [1984] *Coppo-Dezani types do not correspond to propositional logic* (**J** 1426) Theor Comput Sci 28*235-236
◇ B40 ◇ REV Zbl 529 # 03003 • ID 37646

HINMAN, P.G. [1984] *Finitely approximable sets* (**P** 2153) Logic Colloq;1983 Aachen 2*233-258
◇ D65 F65 ◇ ID 43007

HINMAN, P.G. see Vol. III, IV, V for further entries

HINTIKKA, K.J.J. [1956] *Identity, variables and impredicative definitions* (**J** 0036) J Symb Logic 21*225-245
◇ A05 B20 F65 ◇ REV MR 18.455 Zbl 71.11 JSL 32.258
• ID 06114

HINTIKKA, K.J.J. & NIINILUOTO, I. [1973] *On the surface semantics of quantificational proof procedures* (**J** 0963) Ajatus (Helsinki) 35*197-215
◇ B35 C07 F07 ◇ REV Zbl 291 # 02005 • ID 32405

HINTIKKA, K.J.J. see Vol. I, II, III, V for further entries

HIRSCHFELD, J. [1980] *A lower bound for Ramsey's theorem* (**J** 0193) Discr Math 32*89-91
◇ F30 ◇ REV MR 81j:05087 Zbl 461 # 05044 • ID 81569

HIRSCHFELD, J. see Vol. I, II, III, IV, V for further entries

HISCHER, HORST & LUCHT, L. [1976] *Zum Verstaendnis des Induktionsaxioms* (**J** 0160) Math-Phys Sem-ber, NS 23*228-236
◇ B28 F30 ◇ REV MR 54 # 9964 • ID 25590

HIZ, H. [1946] *Remarques sur le degre de completude* (**J** 0109) C R Acad Sci, Paris 223*973-974
◇ B22 F50 ◇ REV MR 8.245 Zbl 60.21 JSL 12.57
• ID 06149

HIZ, H. see Vol. I, IV for further entries

HOBSON, E.W. [1906] *On the arithmetic continuum* (**J** 1910) Proc London Math Soc, Ser 2 4*21-28
◇ A05 E47 F35 F55 ◇ REV FdM 37.73 • ID 28776

HOBSON, E.W. see Vol. V for further entries

HODGES, W. [1980] *Interpreting number theory in nilpotent groups* (**J** 0009) Arch Math Logik Grundlagenforsch 20*103-111
◇ C25 C60 C62 F25 F35 ◇ REV MR 82c:03047 Zbl 454 # 03014 • ID 54226

HODGES, W. see Vol. I, II, III, IV, V for further entries

HODGSON, B.R. [1981] see HATCHER, W.S.

HODGSON, B.R. see Vol. III, IV for further entries

HOERING, W. [1962] *Absolut unentscheidbare Saetze der Mathematik* (**C** 0712) Logik & Logikkalkuel 189-194
◇ A05 F30 ◇ REV Zbl 199.29 • ID 06182

HOERING, W. see Vol. V for further entries

HOEVEN VAN DER, G.F. & TROELSTRA, A.S. [1979] *Projections of lawless sequences II* (**P** 2627) Logic Colloq;1978 Mons 265-298
◇ F35 F50 ◇ REV MR 81g:03070 Zbl 443 # 03025 • REM Part I 1970 by Dalen van,D. & Troelstra,A.S. • ID 56431

HOEVEN VAN DER, G.F. [1982] *An application of projections of lawless sequences* (**P** 3638) Brouwer Centenary Symp;1981 Noordwijkerhout 487-503
◇ F35 F50 ◇ REV MR 85g:03087 Zbl 523 # 03046
• ID 37034

HOEVEN VAN DER, G.F. [1982] *Models for the theory of lawless sequences constructed from a single generator* (**J** 0028) Indag Math 44*107-120
◇ F35 F50 ◇ REV MR 84g:03099 Zbl 486 # 03030
• ID 34203

HOEVEN VAN DER, G.F. [1982] *Projections of lawless sequences* (**S** 1605) Math Centr Tracts 152*ii+237pp
◇ F35 F50 ◇ REV MR 84h:03132 Zbl 495 # 03038
• ID 34321

HOEVEN VAN DER, G.F. & MOERDIJK, I. [1983] *On an independence result in the theory of lawless sequences* (**J** 0028) Indag Math 45*185-191
◇ F35 F50 ◇ REV MR 85d:03116 Zbl 523 # 03045
• ID 37033

HOEVEN VAN DER, G.F. & MOERDIJK, I. [1984] *Constructing choice sequences from lawless sequences of neighbourhood functions* (**P** 2153) Logic Colloq;1983 Aachen 1*207-234
◇ C90 F35 F50 ◇ REV MR 86f:03100 • ID 41766

HOEVEN VAN DER, G.F. & MOERDIJK, I. [1984] *On choice sequences determined by spreads* (**J** 0036) J Symb Logic 49*908-916
◇ D30 D55 F35 F50 G30 ◇ ID 42464

HOEVEN VAN DER, G.F. & MOERDIJK, I. [1984] *Sheaf models for choice sequences* (**J** 0073) Ann Pure Appl Logic 27*63-107
◇ C90 F35 F50 G30 ◇ REV Zbl 546 # 03018 • ID 41155

HOLYOKE, T.C. [1969] see BULLOFF, J.J.

HOMAGK, F. [1969] *Ein intuitionistischer Beweis fuer den Graphensatz von D.Koenig* (**J** 0020) Compos Math 21*292-294
◇ F55 ◇ REV MR 42 # 1696 Zbl 205.5 • ID 06197

HOMAGK, F. [1970] *Zur klassischen Deutung der intuitionistischen Speziestheorie und Arithmetik (Russian) (English and French summaries)* (**J** 0115) Wiss Z Humboldt-Univ Berlin, Math-Nat Reihe 19*585-588
◇ F50 ◇ REV MR 47 # 8252 Zbl 276 # 02018 • ID 06198

HOMAGK, F. [1971] *Intuitionistische Kennzeichnung der endlichen Spezies (Polish and Russian summaries)* (**J** 0063) Studia Logica 28*41-64
◇ F55 ◇ REV MR 46 # 31 Zbl 247 # 02031 • ID 14704

HOMAGK, F. [1972] *Zur Axiomatisierung der Nichtidentitaeten des intuitionistischen Aussagenkalkuels (Russian, English and French summaries)* (**J** 0115) Wiss Z Humboldt-Univ Berlin, Math-Nat Reihe 21*465-469
◇ F50 ◇ REV MR 49 # 2272 Zbl 253 # 02023 • ID 06199

HOMAGK, F. [1979] *Zur intuitionistischen Kennzeichnung reeller Funktionen auf substantiellen Intervallen* (**J** 0160) Math-Phys Sem-ber, NS 26*95-113
◇ F55 ◇ REV MR 81f:03069 Zbl 413 # 03035 • ID 53029

HONG, JIAWEI [1984] *The complexity of formal proving (Chinese)* (**J** 3766) Zhongguo Kexue, Xi A 27*1046-1054
◇ F20 ◇ ID 47764

HONG, JIAWEI see Vol. IV for further entries

HONSELL, F. [1984] see COPPO, M.

HONSELL, F. see Vol. III, V for further entries

HOOK, J.L. [1985] *A note on interpretations of many-sorted theories* (**J** 0036) J Symb Logic 50*372-374
◇ B10 C07 F25 ◇ REV Zbl 571 # 03026 • ID 41804

HOOLEY, J. [1960] see GOODSTEIN, R.L.

HORN, A. [1974] see EPSTEIN, G.

HORN, A. [1976] see EPSTEIN, G.

HORN, A. see Vol. I, II, III, V for further entries

HOSOI, T. [1966] *The separation theorem on the classical system* (**J** 0434) J Fac Sci Univ Tokyo, Sect 1 12*223-230 ◇ B10 F50 ◇ REV MR 33 # 5484 Zbl 146.246 JSL 33.128 • ID 06242

HOSOI, T. see Vol. I, II for further entries

HOSONO, C. [1974] *On the cardinality of some lattices* (**J** 0232) Inform Process Lett 3*67-68 ◇ B40 ◇ REV MR 53 # 7880 Zbl 336 # 02022 • ID 62603

HOSONO, C. & SATO, M. [1977] *The retracts in Pω do not form a continuous lattice - a solution to Scott's problem* (**J** 1426) Theor Comput Sci 4*137-142 ◇ B40 B75 G10 ◇ REV MR 58 # 8442 Zbl 366 # 06013 • ID 81594

HOTOMSKI, P. [1982] *An induction law in proofs by contradiction with an application to automatic theorem proving (Russian)* (**J** 0400) Publ Inst Math, NS (Belgrade) 31(45)*51-63 ◇ B35 F30 ◇ REV MR 84k:03049 Zbl 521 # 03007 • ID 37063

HOTOMSKI, P. see Vol. I for further entries

HOWARD, W.A. [1963] *Appendices to section 2, vol. 1* (**C** 4220) Rep Sem Found Anal 2*2.45-2.85 ◇ F10 F35 F50 ◇ ID 49830

HOWARD, W.A. [1963] *The axiom of choice ($\Sigma^1_1 - AC_{01}$), bar induction and bar recursion* (**C** 4220) Rep Sem Found Anal I*2.1-2.44 ◇ F35 F50 ◇ ID 49823

HOWARD, W.A. [1963] *Transfinite induction and transfinite recursion* (**C** 4220) Rep Sem Found Anal 2*6.1-6.55 ◇ F10 F35 F50 ◇ ID 49834

HOWARD, W.A. & KREISEL, G. [1966] *Transfinite induction and bar induction of types zero and one, and the role of continuity in intuitionistic analysis* (**J** 0036) J Symb Logic 31*325-358 ◇ F10 F35 F50 ◇ REV MR 35 # 27 Zbl 156.8 • ID 06267

HOWARD, W.A. [1968] *Functional interpretation of bar induction by bar recursion* (**J** 0020) Compos Math 20*107-124 • REPR [1968] (**C** 0727) Logic Found of Math (Heyting) 107-124 ◇ F10 F35 F50 ◇ REV MR 37 # 6179 Zbl 162.315 JSL 40.499 • ID 06269

HOWARD, W.A. [1970] *Assignment of ordinals to terms for primitive recursive functionals of finite type* (**P** 0603) Intuitionism & Proof Th;1968 Buffalo 443-458 ◇ F10 F15 F50 ◇ REV MR 44 # 6442 Zbl 206.282 JSL 40.501 • ID 21248

HOWARD, W.A. [1972] *A system of abstract constructive ordinals* (**J** 0036) J Symb Logic 37*355-374 ◇ F10 F15 F35 F50 ◇ REV MR 48 # 8209 Zbl 264 # 02026 JSL 50.243 • ID 06270

HOWARD, W.A. [1973] *Appendix: Hereditarily majorizable functionals of finite type* (**C** 1139) Metamath Invest of Intuition Arithm & Anal 454-461 ◇ E25 F10 F35 F50 ◇ REV MR 57 # 9493 • ID 74284

HOWARD, W.A. [1980] *Ordinal analysis of terms of finite type* (**J** 0036) J Symb Logic 45*493-504 ◇ F10 F15 F35 F50 ◇ REV MR 82f:03049 Zbl 444 # 03029 • ID 74282

HOWARD, W.A. [1980] *Ordinal analysis of bar recursion of type zero* (**J** 0020) Compos Math 42*105-119 ◇ F10 F15 F35 F50 ◇ REV MR 82d:03090 Zbl 444 # 03030 • ID 74283

HOWARD, W.A. [1980] *The formulae-as-types notion of construction* (**C** 3050) Essays Combin Log, Lambda Calc & Formalism (Curry) 480-490 ◇ B40 F05 F10 F50 ◇ REV MR 82g:03094 Zbl 469 # 03006 • REM Circulated as preprint since 1969 • ID 74281

HOWARD, W.A. [1981] *Computability of ordinal recursion of type level two* (**P** 3146) Constr Math;1980 Las Cruces 87-104 ◇ D20 F10 F15 F35 F50 ◇ REV MR 83e:03091 Zbl 463 # 03032 • ID 54572

HOWARD, W.A. [1981] *Ordinal analysis of simple cases of bar recursion* (**J** 0036) J Symb Logic 46*17-30 ◇ F10 F15 F35 F50 ◇ REV MR 82g:03095 Zbl 463 # 03031 • ID 54571

HOWARD, W.A. see Vol. IV for further entries

HU, SHIHUA & HUANG, ZULIANG [1963] *Addition and multiplication (Chinese)* (**J** 0420) Shuxue Jinzhan 6*371-378 ◇ D20 F30 ◇ ID 48690

HU, SHIHUA see Vol. I, II, IV for further entries

HUANG, WENGI & NERODE, A. [1985] *Application of pure recursion theory in recursive analysis (Chinese)* (**J** 0418) Shuxue Xuebao 28/5*625-635 ◇ D20 F60 ◇ ID 49393

HUANG, ZULIANG [1963] see HU, SHIHUA

HUBER-DYSON, V. & KREISEL, G. [1961] *Analysis of Beth's semantic construction of intuitionistic logic* (1111) Preprints, Manuscr., Techn. Reports etc. ◇ F50 F55 ◇ REM Technical Report 3, Applied Mathematics and Statistics Laboratories, Standford University, CA • ID 27572

HUBER-DYSON, V. & JONES, JAMES P. & SHEPHERDSON, J.C. [1982] *Some diophantine forms of Goedel's theorem* (**J** 0009) Arch Math Logik Grundlagenforsch 22*51-60 ◇ D35 F30 ◇ REV MR 83k:03073 Zbl 494 # 03043 • ID 35411

HUBER-DYSON, V. see Vol. III, IV for further entries

HUET, G. [1973] *The undecidability of unification on third order logic* (**J** 0194) Inform & Control 22*257-267 ◇ B15 F35 ◇ REV MR 54 # 2442 Zbl 257 # 02038 • ID 06291

HUET, G. [1975] *A unification algorithm for typed $\overline{\lambda}$-calculus* (**J** 1426) Theor Comput Sci 1*27-57 ◇ B40 ◇ REV MR 54 # 6583 Zbl 332 # 02035 Zbl 337 # 68027 • ID 24166

HUET, G. [1975] *Unification in typed lambda calculus* (**P** 1603) λ-Calc & Comput Sci Th;1975 Roma 192-212 ◇ B40 ◇ REV MR 57 # 2886 Zbl 361 # 02037 • ID 50672

HUET, G. [1977] *Confluent reductions: abstract properties and applications to term rewriting systems* (**P** 3572) IEEE Symp Found of Comput Sci (18);1977 Providence 30-45
◇ B40 ◇ REV MR 58 # 32033 Zbl 458 # 68007 • ID 81607

HUET, G. see Vol. I, III, IV for further entries

HULL, R.G. [1969] *Counterexamples in intuitionistic analysis using Kripke's schema* (**J** 0068) Z Math Logik Grundlagen Math 15*241-246
◇ F35 F50 ◇ REV MR 40 # 7090 Zbl 193.290 • ID 06319

HUMPHRIES, J. [1979] *Goedel's proof and the liar paradox* (**J** 0047) Notre Dame J Formal Log 20*535-544
◇ A05 F30 ◇ REV MR 80k:03010 Zbl 315 # 02009 • ID 56091

HUNTINGTON, E.V. [1935] *Effective equality and effective implication in formal logic* (**J** 0054) Proc Nat Acad Sci USA 21*266-271
◇ F50 ◇ REV Zbl 13.98 FdM 61.53 • ID 06339

HUNTINGTON, E.V. see Vol. I, II, III, V for further entries

HYLAND, J.M.E. [1975] *A survey of some useful partial order relations on terms of the lambda calculus* (**P** 1603) λ-Calc & Comput Sci Th;1975 Roma 83-93 • ERR/ADD ibid 94-95
◇ B40 ◇ REV MR 57 # 9496 Zbl 335 # 02013 Zbl 335 # 02014 • ID 62664

HYLAND, J.M.E. [1976] *A syntactic characterization of the equality in some models for the lambda-calculus* (**J** 3172) J London Math Soc, Ser 2 12*361-370
◇ B40 ◇ REV MR 53 # 100 Zbl 329 # 02010 • ID 26097

HYLAND, J.M.E. [1977] *Aspects of constructivity in mathematics* (**P** 1075) Logic Colloq;1976 Oxford 439-454
◇ F50 ◇ REV MR 58 # 5105 Zbl 422 # 03021 • ID 16633

HYLAND, J.M.E. [1979] *Continuity in spatial toposes* (**P** 2901) Appl Sheaves;1977 Durham 442-465
◇ F50 G30 ◇ REV MR 83g:18009 Zbl 429 # 18013 • ID 53888

HYLAND, J.M.E. [1979] see FOURMAN, M.P.

HYLAND, J.M.E. & JOHNSTONE, P.T. & PITTS, A.M. [1980] *Tripos theory* (**J** 0332) Math Proc Cambridge Phil Soc 88*205-232
◇ E70 F50 G30 ◇ REV MR 81i:03102 Zbl 451 # 03027 • ID 54042

HYLAND, J.M.E. [1981] *Function spaces in the category of locales* (**P** 3368) Continuous Lattices;1979 Bremen 264-281
◇ F50 G30 ◇ REV MR 83d:06001 Zbl 483 # 54005 • ID 36809

HYLAND, J.M.E. [1982] *Applications of constructivity* (**P** 3622) Int Congr Log, Meth & Phil of Sci (6,Proc);1979 Hannover 145-152
◇ F35 F50 ◇ REV MR 85b:03105 Zbl 519 # 03043 • ID 37545

HYLAND, J.M.E. [1982] *The effective topos* (**P** 3638) Brouwer Centenary Symp;1981 Noordwijkerhout 165-216
◇ D75 D80 F35 F50 G30 ◇ REV MR 84m:03101 Zbl 522 # 03055 • ID 35795

HYLAND, J.M.E. see Vol. I, III, IV, V for further entries

IBARRA, O.H. & LEININGER, B.S. [1981] *Characterizations of Presburger functions* (**J** 1428) SIAM J Comp 10*22-39
◇ D20 F30 ◇ REV MR 82k:68008 Zbl 471 # 68032 • ID 55261

IBARRA, O.H. [1981] see GURARI, E.M.

IBARRA, O.H. see Vol. III, IV for further entries

IBRAGIMOV, S.G. [1973] *Consequence logic of Paul Lorenzen (Russian)* (**J** 3543) Uch Zap Vopr Prikl Mat Kibern, Univ Baku 1973*3-91
◇ B60 F50 ◇ REV MR 57 # 5692 • ID 74359

IDEL'SON, A.V. [1964] *Calculi of constructive logic with subordinate variables (Russian)* (**S** 0066) Tr Mat Inst Steklov 72*228-343
• TRANSL [1972] (**J** 0225) Amer Math Soc, Transl, Ser 2 99*83-227
◇ F50 ◇ REV MR 33 # 3885 MR 48 # 3 Zbl 175.271 • ID 06371

IDEL'SON, A.V. [1967] *Remarks on the calculi of constructive logic with subordinate variables and the Axiom of complete induction (Russian)* (**S** 0066) Tr Mat Inst Steklov 93*106-112
◇ F50 ◇ REV MR 37 # 3908 Zbl 183.16 • ID 06372

IGARASHI, S. [1974] *Admissibility of fixed-point induction in first-order logic of typed theories* (**P** 1511) Int Symp Th Progr;1972 Novosibirsk 344-383
◇ B40 ◇ REV MR 56 # 5254 Zbl 309 # 68044 • ID 74363

IGARASHI, S. see Vol. IV for further entries

ILSE, D. [1965] *Zur Stetigkeit berechenbarer reeller Funktionen* (**J** 0068) Z Math Logik Grundlagen Math 11*297-342
◇ F60 ◇ REV MR 32 # 41 Zbl 154.6 • ID 06377

IMAI, Y. & ISEKI, K. [1966] *On Griss algebra. I* (**J** 0081) Proc Japan Acad 42*213-216
◇ F50 G10 ◇ REV MR 34 # 1156 Zbl 199.6 • REM Part II 1977 by Iseki,K. • ID 06382

IMAI, Y. see Vol. I for further entries

INDERMARK, K. [1978] see DAMM, W.

INDERMARK, K. see Vol. III, IV for further entries

INOUE, K. & NAKAMURA, A. [1975] *On the expressive power of logical metalanguages I^n and I_+* (**S** 4568) Hiroshima Univ Fac Engin, Mem 6/1*1-8
◇ B15 F35 ◇ REV MR 53 # 2041 • ID 48761

INOUE, K. see Vol. IV for further entries

IONGH DE, J.J. [1949] *Restricted forms of intuitionistic mathematics* (**P** 0682) Int Congr Philos (10);1948 Amsterdam 744-748
◇ A05 F50 ◇ REV MR 10.422 Zbl 34.154 JSL 14.183 • ID 20802

ISAMBERT, E. [1977] see ARTIGUE, M.

ISAMBERT, E. [1978] see ARTIGUE, M.

ISAMBERT, E. see Vol. III, V for further entries

ISEKI, K. [1956] *On the cut operation in Gentzen calculi I* (**J** 0081) Proc Japan Acad 32*719-721
◇ F05 F07 ◇ REV MR 19.626 Zbl 73.7 JSL 36.528 • REM Part II 1957 • ID 24839

ISEKI, K. [1957] *On the cut operation in Gentzen calculi II* (**J** 0081) Proc Japan Acad 33*98-99
◇ F05 F07 ◇ REV MR 19.626 Zbl 81.10 JSL 36.528 • REM Part I 1956 • ID 24840

ISEKI, K. [1966] see IMAI, Y.

ISEKI, K. see Vol. I, II, V for further entries

ISLES, D. [1970] *Regular ordinals and normal forms* (**P** 0603) Intuitionism & Proof Th;1968 Buffalo 339-362
◇ E10 F15 ◇ REV MR 43 # 1842 Zbl 206.283 JSL 38.334 • ID 06457

ISLES, D. [1971] *Natural well-orderings* (**J** 0036) J Symb Logic 36*288-300
◇ F15 ◇ REV MR 45 # 8522 Zbl 278 # 02030 • ID 22258

ISLES, D. [1975] *The accessibility of ε_0* (**J** 0302) Rep Math Logic, Krakow & Katowice 5*37-42
◇ F15 ◇ REV MR 58 # 5091 Zbl 339 # 02028 • ID 21899

ISLES, D. [1981] *Remarks on the notion of standard non-isomorphic natural number series* (**P** 3146) Constr Math;1980 Las Cruces 111-134
◇ A05 F15 F65 ◇ REV MR 83b:03067 Zbl 461 # 03017 • ID 54494

ITO, MAKOTO [1934] *Einige Anwendungen der Theorie des Entscheidungsproblems zur Axiomatik I* (**J** 0261) Tohoku Math J 37*222-235
◇ B30 C75 F25 ◇ REV Zbl 7.385 FdM 61.51 • REM Part II 1935 • ID 32185

ITO, MAKOTO [1935] *Einige Anwendungen der Theorie des Entscheidungsproblems zur Axiomatik. II* (**J** 0261) Tohoku Math J 40*241-251
◇ B30 F25 ◇ REV Zbl 11.97 FdM 61.51 • REM Part I 1934 • ID 28786

ITO, MAKOTO see Vol. I, II, III, IV for further entries

ITO, T. [1974] *A formal approximation theory of semantic data types* (**P** 1511) Int Symp Th Progr;1972 Novosibirsk 384-393
◇ B40 ◇ REV MR 56 # 10108 Zbl 279 # 68025 • ID 81673

ITO, T. [1975] see ANDO, S.

ITO, T. see Vol. IV for further entries

IVANOV, A.A. [1984] *Decidability of extended theories of addition of the natural numbers and the integers (Russian)* (**J** 0092) Sib Mat Zh 25/4*78-81
• TRANSL [1984] (**J** 0475) Sib Math J 25*572-574
◇ B25 C65 F30 ◇ REV MR 86f:03024 • ID 41869

IVANOV, A.A. see Vol. III, IV, V for further entries

IVANOV, L.L. [1978] *Natural combinatory spaces* (**J** 2547) Serdica, Bulgar Math Publ 4*296-310
◇ B40 D75 ◇ REV MR 80m:03081 Zbl 443 # 03020 • ID 56426

IVANOV, L.L. [1981] *P-recursiveness in iterative combinatorial spaces (Russian)* (**J** 2547) Serdica, Bulgar Math Publ 7*281-297
◇ B40 D75 ◇ REV MR 83h:03066 Zbl 499 # 03039 • ID 36073

IVANOV, L.L. [1983] *Iterative operatory spaces and the system of Scott and de Bakker* (**J** 2547) Serdica, Bulgar Math Publ 9*275-288
◇ B40 ◇ REV MR 86c:68054 Zbl 535 # 03023 • ID 36569

IVANOV, L.L. see Vol. IV for further entries

IYANAGA, S. [1984] *Problems in foundations of mathematics. 4. Proof theory (Japanese)* (**J** 4439) Kagaku 1*43-49
◇ A05 F99 ◇ ID 46960

IYANAGA, S. see Vol. III for further entries

IZUMI, Y. [1952] *Ueber den Begriff der ω-Vollstaendigkeit* (**J** 0261) Tohoku Math J 4*314-315
◇ F30 ◇ REV MR 14.1052 Zbl 48.246 • ID 06476

IZUMI, Y. see Vol. I, V for further entries

JACOBS, K. (ED.) [1973] *Selecta mathematica II* (**X** 0811) Springer: Heidelberg & New York xi + 185pp
◇ D97 F30 ◇ ID 23474

JACOBS, K. see Vol. IV for further entries

JACOPINI, G. [1974] *Principio di estensionalita nel calcolo dei combinatori (English summary)* (**J** 0089) Calcolo 11*465-471
◇ B40 ◇ REV MR 53 # 2641 Zbl 325 # 02018 • ID 21527

JACOPINI, G. [1974] *Una condizione per l'identificabilita di due elementi di un generico modello della logica combinatoria* (**J** 3434) Pubbl Ist Appl Calcolo, Ser 3 97*11pp
• TRANSL [1975] (**P** 1603) λ-Calc & Comput Sci Th;1975 Roma 213-219 (English)
◇ B40 ◇ REV MR 57 # 2887 Zbl 357 # 02023 Zbl 432 # 03009 • ID 53967

JACOPINI, G. & VENTURINI ZILLI, M. [1978] *Equating for recurrent terms of lambda-calculus and combinatory logic* (**J** 3436) Quad, Ist Appl Calcolo, Ser 3 85*14pp
◇ B40 ◇ REV Zbl 432 # 03010 • ID 53968

JACOPINI, G. & VENTURINI ZILLI, M. [1985] *Easy terms in the lambda calculus* (**J** 2095) Fund Inform, Ann Soc Math Pol, Ser 4 8*225-233
◇ B40 ◇ REV Zbl 574 # 03005 • ID 47936

JACOPINI, G. see Vol. IV for further entries

JAEGER, G. & SCHUETTE, K. [1979] *Eine syntaktische Abgrenzung der $(\Delta_1^1 - CA)$-Analysis* (**J** 1944) Sitzb, Akad Wiss, Bayern, Math-Nat Kl 1979*15-34
◇ F15 F35 ◇ REV MR 82f:03050 Zbl 449 # 03056 • ID 74452

JAEGER, G. [1980] *Beweistheorie von KPN* (**J** 0009) Arch Math Logik Grundlagenforsch 20*53-63
◇ E30 F05 F15 ◇ REV MR 81h:03117 Zbl 439 # 03043 JSL 48.879 • ID 56034

JAEGER, G. [1982] *Iterating admissibility in proof theory* (**P** 3708) Herbrand Symp Logic Colloq;1981 Marseille 137-146
◇ F35 F50 ◇ REV MR 85k:03040 Zbl 503 # 03028 • ID 36924

JAEGER, G. [1982] *Zur Beweistheorie der Kripke-Platek-Mengenlehre ueber den natuerlichen Zahlen* (**J** 0009) Arch Math Logik Grundlagenforsch 22*121-139
◇ C70 E30 F15 F35 ◇ REV MR 84j:03110 Zbl 503 # 03014 • ID 34700

JAEGER, G. [1983] *A well ordering proof for Feferman's theory T_0* (**J** 0009) Arch Math Logik Grundlagenforsch 23*65-77
◇ F15 F35 ◇ REV MR 85a:03072 Zbl 511 # 03025 • ID 34744

JAEGER, G. [1983] see FEFERMAN, S.

JAEGER, G. & POHLERS, W. [1983] *Eine beweistheoretische Untersuchung von $(\Delta_2^1\text{-}CA) + (BI)$ und verwandter Systeme* (**J** 1944) Sitzb, Akad Wiss, Bayern, Math-Nat Kl 1982*1-28
◇ F15 F35 ◇ REV MR 85f:03060 Zbl 526 # 03035 • ID 38201

JAEGER, G. [1984] *ρ-inaccessible ordinals, collapsing functions and a recursive notation system* (**J** 0009) Arch Math Logik Grundlagenforsch 24*49-62
◇ E10 F15 ◇ REV MR 86g:03094 Zbl 545 # 03031 • ID 39555

JAEGER, G. [1984] *A version of Kripke-Platek set theory which is conservative over Peano arithmetic* (**J** 0068) Z Math Logik Grundlagen Math 30*3-9
◇ E30 F30 F35 ◇ REV MR 85h:03052 Zbl 539 # 03032 • ID 39554

JAEGER, G. [1984] *The strength of admissibility without foundation* (**J** 0036) J Symb Logic 49*867-879
◇ E30 F15 F35 ◇ ID 39557

JAMBON, M. [1975] *Sur la notion de la variabilite locale constructive (English summary)* (**J** 2313) C R Acad Sci, Paris, Ser A-B 281*A423-A426
◇ F55 ◇ REV MR 52 # 2847 Zbl 317 # 26024 • ID 17638

JAMBU-GIRAUDET, M. [1980] *Theorie des modeles de groupes d'automorphismes d'ensembles totalement ordonnes 2-homogenes (English summary)* (**J** 2313) C R Acad Sci, Paris, Ser A-B 290*A1037-A1039
◇ C07 C60 C62 C65 D35 E07 F25 F35 ◇ REV MR 81f:03046 Zbl 453 # 03037 • ID 54167

JAMBU-GIRAUDET, M. [1981] *Interpretations d'arithmetiques dans des groupes et des treillis* (**P** 3404) Model Th & Arithm;1979/80 Paris 143-153
◇ C07 C62 C65 D35 E07 F25 F30 G10 ◇ REV MR 83g:03064 Zbl 476 # 03023 • ID 55535

JAMBU-GIRAUDET, M. see Vol. III, V for further entries

JANICKA-ZUK, I. [1983] *Strong amalgamation property of diagonalizable algebras* (**J** 0387) Bull Sect Logic, Pol Acad Sci 12*105-110
◇ F30 G25 ◇ REV Zbl 542 # 03041 • ID 43706

JANICKA-ZUK, I. see Vol. II for further entries

JANKOWSKI, A.W. & ZAWADOWSKI, M. [1985] *Sheaves over Heyting lattices* (**J** 0063) Studia Logica 44*237-256
◇ C90 F35 F50 G10 G30 ◇ ID 47500

JANKOWSKI, A.W. see Vol. I, II, III, V for further entries

JANSOHN, H.-S. & LANDWEHR, R. & WRIGHTSON, G. [1982] *An interactive proof system for higher order logic* (**C** 3881) Prog in Cybern & Syst Res, Vol 11 343-347
◇ B15 B35 F35 ◇ REV MR 84f:00034 Zbl 484 # 68072 • ID 36620

JASKOWSKI, S. [1936] *Recherches sur le systeme de la logique intuitionniste* (**P** 0632) Congr Int Phil des Sci;1935 Paris 6*58-61
• TRANSL [1967] (**C** 0615) Polish Logic 1920-39 259-263 (English) [1975] (**J** 0063) Studia Logica 34*117-120 (English)
◇ B55 F50 ◇ REV MR 51 # 7832 Zbl 168.247 Zbl 313 # 02014 JSL 2.55 JSL 35.442 FdM 62.1045 • ID 06544

JASKOWSKI, S. see Vol. I, II, IV, V for further entries

JELTSCH, R. [1971] *Der Riemannsche Abbildungssatz in der konstruktiven Analysis. Diss* (**X** 3292) Juris: Zuerich 77pp
◇ F65 ◇ REV MR 57 # 5705 Zbl 268 # 02023 • ID 62757

JENSEN, C.U. [1984] *Theorie des modeles pour des anneaux de fonctions entieres et des corps de fonctions meromorphes* (**S** 3521) Mem Soc Math Fr 16*23-40
◇ C60 C65 D35 F35 ◇ REV Zbl 562 # 12025 • ID 39754

JENSEN, C.U. see Vol. III, IV, V for further entries

JENSEN, D.C. [1973] *On local proof restrictions for strong theories* (**J** 0202) Diss Math (Warsaw) 108*47pp
◇ F07 ◇ REV MR 51 # 120 Zbl 268 # 02019 • ID 15231

JENSEN, D.C. [1976] see EHRENFEUCHT, A.

JENSEN, D.C. see Vol. I for further entries

JENSEN, R.B. [1967] *Unabhaengigkeitsbeweise in Teilsystemen der elementaren Zahlentheorie* (**J** 0043) Math Ann 173*102-126
◇ F30 F35 ◇ REV MR 36 # 2481 Zbl 162.14 • ID 06595

JENSEN, R.B. see Vol. III, IV, V for further entries

JEROSLOW, R. [1971] *Consistency statements in formal theories* (**J** 0027) Fund Math 72*17-40
◇ F30 ◇ REV MR 45 # 8524 Zbl 254 # 02038 • ID 06605

JEROSLOW, R. [1973] *Redundancies in the Hilbert-Bernays derivability conditions for Goedel's second incompleteness theorem* (**J** 0036) J Symb Logic 38*359-367
◇ F30 ◇ REV MR 49 # 2283 Zbl 276 # 02031 JSL 48.875 • ID 06606

JEROSLOW, R. see Vol. II, III, IV for further entries

JERVELL, H.R. [1971] *A normalform in first order arithmetic* (**P** 0604) Scand Logic Symp (2);1970 Oslo 93-108
◇ F05 F30 F50 ◇ REV MR 48 # 10774 Zbl 223 # 02023 JSL 40.503 • ID 06608

JERVELL, H.R. [1972] *Herbrand and Skolem theorems in infinitary languages* (**S** 1626) Oslo Preprint Ser 3
◇ C75 F05 ◇ ID 33262

JERVELL, H.R. [1972] *On Skolem and Herbrand theorems for intuitionistic logic* (**S** 1626) Oslo Preprint Ser 4
◇ F05 F50 ◇ ID 33263

JERVELL, H.R. [1973] *Skolem and Herbrand theorems in first order logic* (**S** 1626) Oslo Preprint Ser 6*77pp
◇ B10 C07 F05 ◇ ID 33265

JERVELL, H.R. [1978] *Constructive universes. I* (**P** 1864) Higher Set Th;1977 Oberwolfach 73-98
◇ F05 F15 F35 F50 ◇ REV MR 80e:03073 Zbl 398 # 03047 • ID 52775

JERVELL, H.R. [1982] *Introducing homogeneous trees* (**P** 3708) Herbrand Symp Logic Colloq;1981 Marseille 147-158
◇ F15 ◇ REV MR 85f:03003 Zbl 503 # 03027 • ID 36923

JERVELL, H.R. [1985] *Gentzen games* (**J** 0068) Z Math Logik Grundlagen Math 31*431-439
◇ F30 ◇ ID 47549

JERVELL, H.R. [1985] *Large finite sets* (**J** 0068) Z Math Logik Grundlagen Math 31*545-549
◇ E20 F15 F30 ◇ ID 47805

JERVELL, H.R. [1985] *Recursion on homogeneous trees* (**J** 0068) Z Math Logik Grundlagen Math 31*295-298
◇ D75 F15 ◇ REV Zbl 567 # 03027 • ID 47555

JERVELL, H.R. see Vol. II, III, V for further entries

JOCKUSCH JR., C.G. & SOARE, R.I. [1972] *Degrees of members of Π_1^0 classes* (**J** 0048) Pac J Math 40*605-616
◇ D25 D30 F30 ◇ REV MR 46 # 8827 Zbl 232 # 02031 • ID 06633

JOCKUSCH JR., C.G. [1975] *Recursiveness of initial segments of Kleene's O* (**J** 0027) Fund Math 87*161-167
◇ D20 F15 ◇ REV MR 58 # 189 Zbl 338 # 02024 • ID 06642

JOCKUSCH JR., C.G. see Vol. I, III, IV, V for further entries

JOHANSSON, I. [1937] *Der Minimalkalkuel, ein reduzierter intuitionistischer Formalismus* (**J** 0020) Compos Math 4*119-136
◇ F50 ◇ REV Zbl 15.241 JSL 2.47 FdM 62.1045 • ID 06643

JOHANSSON, I. [1953] *Sur le concept de "le" (ou de "ce qui") dans le calcul affirmatif et dans les calculs intuitionnistes* (**P** 0644) Meth Form en Axiom;1950 Paris 65-72
◇ B20 F50 ◇ REV MR 15.2 Zbl 53.341 JSL 23.346 • ID 06644

JOHANSSON, I. see Vol. I for further entries

JOHNS, D.L. [1981] see GIBSON, C.G.

JOHNSTONE, P.T. [1977] *Rings, fields, and spectra* (**J** 0032) J Algeb 49*238-260
◇ C90 F50 G30 ◇ REV MR 56 # 8335 Zbl 369 # 13019 • ID 51354

JOHNSTONE, P.T. [1979] *Another condition equivalent to De Morgan's law* (**J** 0394) Commun Algeb 7*1309-1312
◇ C60 C90 F50 G05 G30 ◇ REV MR 81d:03067 Zbl 417 # 18002 • ID 74552

JOHNSTONE, P.T. [1979] *Conditions related to de Morgan's law* (**P** 2901) Appl Sheaves;1977 Durham 479-491
◇ B55 C65 C90 F50 G30 ◇ REV MR 81e:03062 Zbl 445 # 03041 • ID 56505

JOHNSTONE, P.T. [1980] see HYLAND, J.M.E.

JOHNSTONE, P.T. [1981] *Tychonoff's theorem without the axiom of choice* (**J** 0027) Fund Math 113*21-35
◇ E25 E75 F55 G30 ◇ REV MR 84h:03117 Zbl 503 # 54006 • ID 34303

JOHNSTONE, P.T. [1982] *Stone spaces* (**X** 0805) Cambridge Univ Pr: Cambridge, GB xxi + 370pp
◇ F50 G30 ◇ REV MR 85f:54003 Zbl 499 # 54001 • ID 38303

JOHNSTONE, P.T. see Vol. III, V for further entries

JOHNSTONE JR., H.W. [1962] see ANDERSON, J.M.

JOHNSTONE JR., H.W. see Vol. I, II, III for further entries

JONES, JAMES P. [1969] *Independent recursive axiomatizability in arithmetic* (**J** 0053) Proc Amer Math Soc 23*107-113
◇ D25 F30 ◇ REV MR 41 # 1534 Zbl 186.8 • ID 06695

JONES, JAMES P. [1978] *Three universal representations of recursively enumerable sets* (**J** 0036) J Symb Logic 43*335-351
◇ D25 F20 ◇ REV MR 58 # 16226 Zbl 414 # 03025 • ID 29266

JONES, JAMES P. [1982] see HUBER-DYSON, V.

JONES, JAMES P. & SHEPHERDSON, J.C. [1983] *Variants of Robinson's essentially undecidable theory R* (**J** 0009) Arch Math Logik Grundlagenforsch 23*61-64
◇ F30 ◇ REV MR 85d:03088 Zbl 511 # 03015 • ID 37388

JONES, JAMES P. see Vol. IV, V for further entries

JONGH DE, D.H.J. & TROELSTRA, A.S. [1966] *On the connection of partially ordered sets with some pseudo-boolean algebras* (**J** 0028) Indag Math 28*317-329
◇ F50 G10 ◇ REV MR 33 # 5537 Zbl 137.22 • ID 06704

JONGH DE, D.H.J. [1968] *Essay on i-valuations* (**C** 0684) Automation in Lang Transl & Theorem Prov 263-267
◇ F50 ◇ REV MR 39 # 2599 Zbl 213.12 • ID 14710

JONGH DE, D.H.J. [1970] *A characterization of the intuitionistic propositional calculus* (**P** 0603) Intuitionism & Proof Th;1968 Buffalo 211-217
◇ F50 ◇ REV MR 42 # 7482 Zbl 225 # 02020 JSL 40.500 • ID 06706

JONGH DE, D.H.J. & SMORYNSKI, C.A. [1976] *Kripke models and the intuitionistic theory of species* (**J** 0007) Ann Math Logic 9*157-186
◇ C90 F35 F50 ◇ REV MR 53 # 5266 Zbl 299 # 02037 Zbl 317 # 02037 • ID 18214

JONGH DE, D.H.J. [1980] *A class of intuitionistic connectives* (**P** 2058) Kleene Symp;1978 Madison 103-111
◇ B55 C75 C90 F50 ◇ REV MR 81m:03013 Zbl 479 # 03015 • ID 55677

JONGH DE, D.H.J. [1982] *Formulas of one propositional variable in intuitionistic arithmetic* (**P** 3638) Brouwer Centenary Symp;1981 Noordwijkerhout 51-64
◇ F30 F50 ◇ REV MR 85i:03173 Zbl 547 # 03037 • REM A preprint of this paper appeared 1973 • ID 43236

JONGH DE, D.H.J. see Vol. I, II, III, IV, V for further entries

JOSEPH, D. & YOUNG, P. [1981] *A survey of some recent results on computational complexity in weak theories of arithmetic* (**P** 3429) Math Founds of Comput Sci (10);1981 Strbske Pleso 46-60
◇ D15 F30 ◇ REV MR 84j:03111 Zbl 481 # 03025 • ID 34701

JOSEPH, D. [1983] *Polynomial time computations in models of ET* (**J** 0119) J Comp Syst Sci 26*311-338
◇ D15 F30 ◇ REV MR 84k:68034 Zbl 542 # 03019 • ID 39245

JOSEPH, D. see Vol. IV for further entries

JOYAL, A. [1975] *Les theoremes de Chevalley-Tarski et remarque sur l'algebre constructive* (**J** 0306) Cah Topol & Geom Differ 16*256-258
◇ C60 F50 G30 ◇ REV Zbl 354 # 02038 • ID 27668

JOYAL, A. [1981] see BOILEAU, A.

JOYAL, A. see Vol. V for further entries

JOZSA, R. [1984] *Sheaf models and massless fields* (**J** 2736) Int J Theor Phys 23*67-97
◇ B30 C90 F50 G30 ◇ REV MR 85m:81076 • ID 45334

JUKNA, S. [1977] *On cut-elimination in Hoare's system (Russian)* (**J** 2574) Litov Mat Sb (Vil'nyus) 17/3*
◇ B75 F05 F50 ◇ ID 33259

JUKNA, S. [1983] *Arithmetization of calculation (Russian) (English and Lithuanian summaries)* (J 3939) Mat Logika Primen (Akad Nauk Litov SSR) 3*117-125
◇ F99 ◇ REV MR 85i:03128 Zbl 567 # 03015 • ID 45157

JUKNA, S. see Vol. III, IV for further entries

JULIAN, W. [1975] see BERG, G.

JULIAN, W. [1977] see BERG, G.

JULIAN, W. & MINES, R. & RICHMAN, F. [1978] *Algebraic numbers, a constructive development* (J 0048) Pac J Math 74*91-102
◇ F55 ◇ REV MR 81a:03061 Zbl 347 # 12001 • ID 51393

JULIAN, W. [1981] see BRIDGES, D.S.

JULIAN, W. [1982] see BRIDGES, D.S.

JULIAN, W. & MINES, R. & RICHMAN, F. [1983] *Alexander duality* (J 0048) Pac J Math 106*115-127
◇ F55 ◇ REV MR 84j:03115 Zbl 516 # 55005 • ID 34705

JULIAN, W. [1984] *ε-continuity and monotone operators* (J 0048) Pac J Math 115*385-390
◇ F55 ◇ REV MR 86d:46075 Zbl 534 # 03030 • ID 43696

JULIAN, W. & RICHMAN, F. [1984] *A uniformly continuous function on* $[0,1]$ *that is everywhere different from its infimum* (J 0048) Pac J Math 111*333-340
◇ F55 ◇ REV MR 85m:03042 Zbl 527 # 03037 • ID 37582

JULIAN, W. & PHILLIPS, K. [1985] *Constructive bounded sequences and Lipschitz functions* (J 3172) J London Math Soc, Ser 2 31*385-392
◇ F55 ◇ REV Zbl 534 # 03031 • ID 48229

JUSTEN, K. [1981] *A note on regular resolution* (J 0373) Comp Arch Inform & Numerik 26*87-89
◇ B35 F20 ◇ REV MR 82k:03012 Zbl 431 # 03036 • ID 55967

KABAKOV, F.A. [1963] *On the derivability of some realizable formulas of the sentential calculus (Russian) (German summary)* (J 0068) Z Math Logik Grundlagen Math 9*97-104
◇ F50 ◇ REV MR 26 # 4884 Zbl 112.5 • ID 20678

KABAKOV, F.A. [1970] *Intuitionistic deducibility of some realizable formulae of propositional logic (Russian)* (J 0023) Dokl Akad Nauk SSSR 192*269-271
• TRANSL [1970] (J 0062) Sov Math, Dokl 11*612-614
◇ F50 ◇ REV MR 41 # 8209 Zbl 212.312 • ID 06781

KABAKOV, F.A. [1970] *On modelling of pseudo-boolean algebras by realizability (Russian)* (J 0023) Dokl Akad Nauk SSSR 192*16-18
• TRANSL [1970] (J 0062) Sov Math, Dokl 11*562-564
◇ F30 F50 G10 ◇ REV MR 42 # 43 Zbl 278 # 02019 JSL 37.627 • ID 06782

KABZINSKI, J.K. [1973] *The Wajsberg's results connected with separability of the intuitionistic propositional logic* (J 0387) Bull Sect Logic, Pol Acad Sci 2*131-133
◇ F50 ◇ REV MR 55 # 7743 • ID 74589

KABZINSKI, J.K. & WRONSKI, A. [1975] *On equivalential algebras* (P 1805) Int Symp Multi-Val Log (5,Proc);1975 Bloomington 419-428
◇ F50 G25 ◇ REV MR 58 # 10438 • ID 74595

KABZINSKI, J.K. & POREBSKA, M. [1975] *Proof of the separability of the intuitionistic propositional logic by the Wajsberg method* (J 0302) Rep Math Logic, Krakow & Katowice 4*31-37
◇ F50 ◇ REV MR 51 # 12487 Zbl 318 # 02026 • ID 17275

KABZINSKI, J.K. see Vol. I, II for further entries

KALICKI, C. [1980] *Infinitary propositional intuitionistic logic* (J 0047) Notre Dame J Formal Log 21*216-228
◇ B55 C75 F50 ◇ REV MR 81e:03025 Zbl 336 # 02021 • ID 53776

KALISH, D. & MONTAGUE, R. [1957] *Remarks on descriptions and natural deduction I,II* (J 0009) Arch Math Logik Grundlagenforsch 3*50-64,65-73
◇ B10 F07 ◇ REV MR 19.1032 MR 19.724 Zbl 81.245 JSL 23.449 • ID 24907

KALISH, D. see Vol. I, II for further entries

KALMAR, L. [1928] *Eine Bemerkung zur Entscheidungstheorie* (J 0460) Acta Univ Szeged, Sect Mat 4*248-252
◇ B20 B25 F30 ◇ REV FdM 55.32 • ID 06835

KALMAR, L. [1941] *Aims, methods and results of Hilbert proof theory (Hungarian) (German summary)* (J 0461) Mat Fiz Lapok 48*65-119
◇ A05 F99 ◇ REV MR 9.129 Zbl 24.241 JSL 6.110 • ID 06845

KALMAR, L. [1943] *Ein einfaches Beispiel fuer ein unentscheidbares arithmetisches Problem (Hungarian) (German summary)* (J 0461) Mat Fiz Lapok 50*1-23
◇ D20 F30 ◇ REV MR 8.558 JSL 9.24 • ID 19071

KALMAR, L. [1949] *Quelques formes generales du theoreme de Goedel* (J 0109) C R Acad Sci, Paris 229*1047-1049
◇ F30 ◇ REV MR 11.411 Zbl 41.351 JSL 15.230 • ID 06849

KALMAR, L. [1949] *Une forme du theoreme de Goedel sous des hypotheses minimales* (J 0109) C R Acad Sci, Paris 229*963-965
◇ F30 ◇ REV MR 11.411 Zbl 37.2 JSL 15.230 • ID 06848

KALMAR, L. [1950] *A simple construction of undecidable propositions in formal systems (German and English)* (J 0175) Methodos 2*220-226,227-231
◇ F30 ◇ REV MR 12.790 JSL 17.150 • ID 28755

KALMAR, L. [1950] *Another proof of the Goedel-Rosser incompletability theorem* (J 0002) Acta Sci Math (Szeged) 12A*38-43
◇ F30 ◇ REV MR 12.578 Zbl 41.352 JSL 30.386 • ID 06850

KALMAR, L. [1959] *An argument against the plausibility of Church's thesis* (P 0634) Constructivity in Math;1957 Amsterdam 72-80
◇ A05 D20 F99 ◇ REV MR 21 # 5567 Zbl 88.249 JSL 33.471 • ID 06864

KALMAR, L. see Vol. I, II, III, IV, V for further entries

KANDA, A. [1984] *Numeration models of λ-calculus* (P 4007) CAAP'84 Arbres en Algeb & Progr (9);1984 Bordeaux 155-168
◇ B40 D45 ◇ ID 45525

KANDA, A. [1985] *Numeration models of λ-calculus* (J 0068) Z Math Logik Grundlagen Math 31*209-220
◇ B40 B60 C57 D45 ◇ ID 47568

KANDA, A. see Vol. IV, V for further entries

KANGER, S. [1957] *Provability in logic* (**X** 1163) Almqvist & Wiksell: Stockholm 49pp
◇ B10 F05 F07 F98 ◇ REV MR 19.239 Zbl 77.12 JSL 23.37 • ID 22373

KANGER, S. [1963] *A simplified proof method for elementary logic* (**C** 0659) Computer Progr & Formal Syst 87-94
• REPR [1983] (**C** 4659) Autom of Reasoning 1*364-371
◇ B10 F07 ◇ REV MR 27 # 1350 JSL 32.119 • ID 06888

KANGER, S. see Vol. I, II, III for further entries

KANOVEJ, V.G. [1975] *On the independence of some propositions of descriptive set theory and second-order arithmetic (Russian)* (**J** 0023) Dokl Akad Nauk SSSR 223*552-554
• TRANSL [1975] (**J** 0062) Sov Math, Dokl 16*937-940
◇ D55 E15 E35 F35 ◇ REV MR 52 # 10416 Zbl 363 # 02067 • ID 50893

KANOVEJ, V.G. [1976] *Definability with the help of degrees of constructibility (Russian)* (**C** 3271) Issl Teor Mnozh & Neklass Logik 5-95
◇ E25 E35 E45 F35 ◇ REV MR 58 # 5209 Zbl 415 # 03039 • ID 53141

KANOVEJ, V.G. [1978] *The significance of the parameters and of the complexity of the basic formula in the comprehension axiom schema for second order arithmetic (Russian)* (**J** 0023) Dokl Akad Nauk SSSR 243*1384-1386
• TRANSL [1978] (**J** 0062) Sov Math, Dokl 19*1556-1559
◇ F35 ◇ REV MR 80f:03067 Zbl 413 # 03033 • ID 53027

KANOVEJ, V.G. [1979] *On descriptive forms of the countable axiom of choice (Russian)* (**C** 2581) Issl Neklass Log & Teor Mnozh 3-136
◇ D55 E15 E25 E35 F35 ◇ REV MR 83c:03045 Zbl 439 # 03033 • ID 56024

KANOVEJ, V.G. [1979] *The definability of forcing in analysis (Russian) (English summary)* (**J** 0288) Vest Ser Mat Mekh, Univ Moskva 1979/2*3-13,101
• TRANSL [1979] (**J** 0510) Moscow Univ Math Bull 34/2*1-12
◇ D55 E40 F35 ◇ REV MR 80j:03085 Zbl 419 # 03032 • ID 53375

KANOVEJ, V.G. [1981] *The theory of Zermelo without the power set axiom and the theory of Zermelo-Fraenkel without the power set axiom are equiconsistent (Russian)* (**J** 0087) Mat Zametki (Akad Nauk SSSR) 30*407-419,462
• TRANSL [1981] (**J** 1044) Math Notes, Acad Sci USSR 30*695-702
◇ E30 E35 F25 ◇ REV MR 83i:03076 Zbl 504 # 03023 • ID 35529

KANOVEJ, V.G. see Vol. I, III, IV, V for further entries

KANOVICH, M.I. & KUSHNER, B.A. [1969] *Estimating the complexity of certain algorithmic problems of analysis (Russian)* (**S** 0228) Zap Nauch Sem Leningrad Otd Mat Inst Steklov 16*81-90
• TRANSL [1969] (**J** 0521) Semin Math, Inst Steklov 16*40-44
◇ D15 D20 D80 F60 ◇ REV MR 42 # 77 Zbl 267 # 02022 • ID 29887

KANOVICH, M.I. [1973] *Irreducibility of languages of the stepped semantic system (Russian)* (**J** 0023) Dokl Akad Nauk SSSR 212*800-803
• TRANSL [1973] (**J** 0062) Sov Math, Dokl 14*1459-1463
◇ D15 D30 F50 ◇ REV MR 50 # 4275 Zbl 299 # 02050 • ID 06903

KANOVICH, M.I. [1974] *A certain extension of the step-by-step semantic system of A. A. Markov (Russian)* (**C** 1450) Teor Algor & Mat Logika (Markov) 62-70
◇ D15 F50 ◇ REV MR 57 # 16000 Zbl 299 # 02051 • ID 62868

KANOVICH, M.I. [1974] *Complexity of the limit of Specker sequences (Russian)* (**J** 0023) Dokl Akad Nauk SSSR 214*1020-1023
• TRANSL [1974] (**J** 0062) Sov Math, Dokl 15*299-303
◇ D15 D20 F60 ◇ REV MR 49 # 4765 Zbl 299 # 02040 • ID 07592

KANOVICH, M.I. [1975] *A hierachical semantic system with set variables (Russian)* (**J** 0023) Dokl Akad Nauk SSSR 221*1256-1259
• TRANSL [1975] (**J** 0062) Sov Math, Dokl 16*504-509
◇ B20 B65 C40 D05 F50 ◇ REV MR 52 # 5385 Zbl 325 # 02030 • ID 18218

KANOVICH, M.I. [1978] *An estimate of the complexity of arithmetic incompleteness (Russian)* (**J** 0023) Dokl Akad Nauk SSSR 238*1283-1286
• TRANSL [1978] (**J** 0062) Sov Math, Dokl 19*206-210
◇ D15 F20 F30 ◇ REV MR 81i:03091 Zbl 394 # 03041 • ID 52518

KANOVICH, M.I. [1979] *A complexity version of Goedel's incompleteness theorem* (**P** 2935) FCT'79 Fund of Comput Th;1979 Berlin/Wendisch-Rietz 542-543
◇ D15 F20 F30 ◇ REV MR 81e:03035 Zbl 412 # 03014 • ID 52945

KANOVICH, M.I. see Vol. I, III, IV for further entries

KAPITONOVA, YU.V. [1981] see DOMRACHEV, V.N.

KAPITONOVA, YU.V. [1982] see DOMRACHEV, V.N.

KAPITONOVA, YU.V. [1983] see DOMRACHEV, V.N.

KAPITONOVA, YU.V. see Vol. I for further entries

KAPUR, D. & KRISHNAMURTHY, B. [1984] *A natural proof sytem based on rewriting techniques* (**P** 2633) Autom Deduct (7);1984 Napa 53-64
◇ B35 F07 ◇ REV Zbl 546 # 68075 • ID 43561

KAPUR, D. see Vol. IV for further entries

KAROL', A.M. [1978] *Complexity of the set of constructive real numbers (Russian)* (**C** 3211) Mat Ling & Teor Algor 77-82
◇ D15 F60 ◇ REV Zbl 395 # 03028 • ID 52579

KARR, M. [1985] *"Delayability" in proofs of strong normalizability in the typed lambda calculus* (**P** 4627) CAAP'85 Arbres en Algeb & Progr (10);1985 Berlin 1*208-222
◇ B40 ◇ ID 49329

KARR, M. see Vol. III, IV for further entries

KASHAPOVA, F.R. [1984] *Constructive set theory with types, and consistency with Church's thesis (Russian)* (**J** 0288) Vest Ser Mat Mekh, Univ Moskva 1984/4*72-75
• TRANSL [1984] (**J** 0510) Moscow Univ Math Bull 39/4*470-474
◇ D20 E70 F35 F50 ◇ REV MR 85j:03105 • ID 44311

KASHAPOVA, F.R. [1984] *Determination of classes of constructively derivable theorems in a many-sorted intuitionistic set theory equivalent to second-order arithmetic (Russian)* (**J** 0023) Dokl Akad Nauk SSSR 276*782-786
• TRANSL [1984] (**J** 0062) Sov Math, Dokl 29*583-587
◇ F35 F50 ◇ REV MR 86d:03055 • ID 45488

KASHIN, V.I. [1973] see ANISHEV, P.A.

KATERINAK, J. [1979] *The theory of finite structures of G-numbers* (**P** 3035) Ved Konf Vysok Shkol Doprav;1979 Zhilina 97-101
◇ F99 ◇ REV MR 82k:03098 • ID 74702

KAUFMANN, M. [1984] see HENSON, C.W.

KAUFMANN, M. see Vol. III, IV, V for further entries

KAWADA, K. & MUTI, N. [1967] *A generalization of Curry's theorem* (**J** 0081) Proc Japan Acad 43*689-692
◇ F50 ◇ REV MR 38 # 985 Zbl 203.6 • ID 06974

KEARNS, J.T. [1969] *Combinatory logic with discriminators* (**J** 0036) J Symb Logic 34*561-575
◇ B40 ◇ REV MR 42 # 2943 Zbl 195.20 JSL 38.339
• ID 06982

KEARNS, J.T. [1973] *The completeness of combinatory logic with discriminators* (**J** 0047) Notre Dame J Formal Log 14*323-330
◇ B40 ◇ REV MR 47 # 8256 Zbl 258 # 02034 • ID 06985

KEARNS, J.T. [1977] *The logic of calculation* (**J** 0068) Z Math Logik Grundlagen Math 23*45-58
◇ B40 ◇ REV MR 58 # 5081 Zbl 358 # 02026 • ID 26466

KEARNS, J.T. [1978] *Intuitionist logic, a logic of justification* (**J** 0063) Studia Logica 37*243-260
◇ A05 C90 F50 ◇ REV MR 80d:03025 Zbl 408 # 03045
• ID 31151

KEARNS, J.T. [1980] *Fully explicit deductive systems* (**C** 3050) Essays Combin Log, Lambda Calc & Formalism (Curry) 35-54
◇ B40 ◇ REV MR 82g:03020 Zbl 469 # 03006 • ID 74721

KEARNS, J.T. see Vol. I, II, III for further entries

KEISLER, H.J. [1984] see HENSON, C.W.

KEISLER, H.J. see Vol. I, II, III, IV, V for further entries

KELLY, G.M. [1972] *A cut-elimination theorem* (**C** 1211) Coherence in Categories 196-213
◇ F05 G30 ◇ REV MR 49 # 5128 Zbl 243 # 18017
• ID 21272

KELLY, G.M. see Vol. V for further entries

KEMENY, J.G. [1958] *Undecidable problems of elementary number theory* (**J** 0043) Math Ann 135*160-169
◇ F30 H15 ◇ REV MR 20 # 5140 Zbl 81.247 JSL 23.359
• ID 07053

KEMENY, J.G. see Vol. I, II for further entries

KENT, C.F. [1973] *The relation of A to Prov⌜A⌝ in the Lindenbaum sentence algebra* (**J** 0036) J Symb Logic 38*295-298
◇ F30 G05 ◇ REV MR 48 # 10775 Zbl 275 # 02034
• ID 19040

KENT, C.F. [1974] *"Disorder" in lattices of binumerations* (**J** 0140) Comm Math Univ Carolinae (Prague) 15*221-244
◇ F30 G05 ◇ REV MR 50 # 4258 Zbl 291 # 02022
• ID 07061

KENT, C.F. see Vol. I, III, IV for further entries

KETONEN, J. & SOLOVAY, R.M. [1981] *Rapidly growing Ramsey functions* (**J** 0120) Ann of Math, Ser 2 113*267-314
◇ F30 ◇ REV MR 84c:03100 Zbl 494 # 03027 • ID 34034

KETONEN, J. see Vol. I, III, V for further entries

KHACHATRYAN, M.A. [1968] *Constructive series of numbers (Russian) (Armenian summary)* (**S** 0422) Tr Vychisl Tsentra Akad Nauk Armyan SSR & Univ Erevan 5*7-25
◇ F60 ◇ REV MR 41 # 8230 Zbl 235 # 02027 • ID 05465

KHACHATRYAN, M.A. [1969] *An example of a constructive nondifferentiable monotone function (Russian) (Armenian and English summaries)* (**J** 0312) Izv Akad Nauk Armyan SSR, Ser Mat 4*296-299
◇ F60 ◇ REV MR 42 # 7839 Zbl 182.18 • ID 05466

KHACHATRYAN, M.A. [1969] *The differentiation of constructive monotone functions (Russian)* (**P** 2534) Konf Molod Special Vychisl (1);1969 Erevan
◇ F60 ◇ ID 90144

KHACHATRYAN, M.A. [1977] *On certain theorems of analysis in the formal system Kleene-Vesley (Russian)* (**J** 0087) Mat Zametki (Akad Nauk SSSR) 21*109-116
• TRANSL [1977] (**J** 1044) Math Notes, Acad Sci USSR 21*60-64
◇ F50 ◇ REV MR 55 # 12484 Zbl 361 # 02044 • ID 51174

KHACHATRYAN, M.A. [1978] *On the derivation complexity of some formulas in sequential calculi (Russian)* (**J** 0346) Dokl Akad Nauk Armyan SSR 67*198-202
◇ F20 F50 ◇ REV Zbl 399 # 03043 • ID 52839

KHACHATRYAN, M.A. see Vol. I for further entries

KHAKHANYAN, V.KH. [1980] *Comparative strength of variants of Church's thesis at the level of set theory (Russian)* (**J** 0023) Dokl Akad Nauk SSSR 252*1070-1074
• TRANSL [1980] (**J** 0062) Sov Math, Dokl 21*894-898
◇ D20 E70 F50 ◇ REV MR 81g:03069 Zbl 482 # 03026
• ID 73755

KHAKHANYAN, V.KH. [1980] *The consistency of intuitionistic set theory with Church's principle and the uniformization principle (Russian)* (**J** 0288) Vest Ser Mat Mekh, Univ Moskva 1980/5*3-7
• TRANSL [1980] (**J** 0510) Moscow Univ Math Bull 35/5*1-5
◇ D20 E35 E70 F25 F50 ◇ REV MR 81h:03109 MR 82b:03106 Zbl 445 # 03031 • ID 56495

KHAKHANYAN, V.KH. [1980] *The consistency of intuitionistic set theory with formal mathematical analysis (Russian)* (**J** 0023) Dokl Akad Nauk SSSR 253*48-52
• TRANSL [1980] (**J** 0062) Sov Math, Dokl 22*46-50
◇ D20 E35 E70 F25 F35 F50 ◇ REV MR 81h:03109 Zbl 531 # 03036 • ID 73754

KHAKHANYAN, V.KH. [1981] *The consistency of some intuitionistic and constructive principles with a set theory* (**J** 0063) Studia Logica 40*237-248
◇ D20 E35 E70 F50 ◇ REV MR 84e:03061 Zbl 491 # 03019 • ID 36853

KHAKHANYAN, V.KH. [1983] *Set theory and Church's thesis (Russian)* (**C** 3807) Issl Neklass Log & Formal Sist 198-208
◇ D20 E35 E70 F30 F35 F50 ◇ REV MR 85d:03117
• ID 41114

KHINCHIN, A.YA. [1926] *The ideas of intuitionism and the dispute on the object of contemporary mathematics (Russian)* (J 4122) Vest Komm Akad 16*184-192
◇ A05 F99 ◇ ID 43536

KHINCHIN, A.YA. see Vol. I, V for further entries

KHISAMIEV, N.G. [1974] *Strongly constructive models of a decidable theory (Russian)* (J 0403) Izv Akad Nauk Kazak SSR, Ser Fiz-Mat 1974/1*83-84,94
◇ C15 C57 F60 ◇ REV MR 50 # 6824 Zbl 275 # 02051
• ID 06147

KHISAMIEV, N.G. [1978] see DOBRITSA, V.P.

KHISAMIEV, N.G. [1979] *On subgroups of finite index of abelian groups (Russian) (Kazakh summary)* (J 0403) Izv Akad Nauk Kazak SSR, Ser Fiz-Mat 1979/3*43-47,89
◇ C57 C60 D45 F60 ◇ REV MR 81b:20037
Zbl 416 # 20049 • ID 32601

KHISAMIEV, N.G. [1983] *Strongly constructive abelian p-groups (Russian)* (J 0003) Algebra i Logika 22*198-217
• TRANSL [1983] (J 0069) Algeb and Log 22*142-158
◇ C57 C60 D45 F60 ◇ REV MR 85e:03105
Zbl 568 # 20052 • ID 40239

KHISAMIEV, N.G. see Vol. III, IV for further entries

KHLODOVSKIJ, I.N. [1959] *A new proof of the consistency of arithmetic (Russian)* (J 0067) Usp Mat Nauk 14/6*105-140
• TRANSL [1963] (J 0225) Amer Math Soc, Transl, Ser 2 23*191-230
◇ F05 F30 ◇ REV MR 22 # 669 Zbl 129.257 JSL 32.127
• ID 21067

KHOMICH, V.I. [1970] *On the complexity of realization of propositional formulae (Russian)* (J 0023) Dokl Akad Nauk SSSR 195*1050-1051
• TRANSL [1970] (J 0062) Sov Math, Dokl 11*1637-1639
◇ B35 F50 ◇ REV MR 44 # 1576 Zbl 218 # 02028
• ID 06201

KHOMICH, V.I. [1973] *The complexity of recognition of the realizability of logico-arithmetic formulas (Russian)* (J 0068) Z Math Logik Grundlagen Math 19*453-468
◇ F30 F50 ◇ REV MR 49 # 27 Zbl 307 # 02027 • ID 06202

KHOMICH, V.I. [1975] *Weakly and strongly nonrealizable propositional formulae (Russian)* (J 0068) Z Math Logik Grundlagen Math 21*267-288
◇ B20 D15 D20 F50 ◇ REV MR 51 # 10048
Zbl 309 # 02021 • ID 17549

KHOMICH, V.I. see Vol. I, II, IV for further entries

KHOSROVSHAHI, G.B. [1978] *A few words on intuitionism (Persian)* (J 3276) Tehran J Math 18*728-734
◇ F50 ◇ REV Zbl 395 # 03040 • ID 52591

KINO, A. [1958] *A consistency-proof of a formal theory of Ackermann's ordinal numbers* (J 0090) J Math Soc Japan 10*287-303
◇ F15 F25 ◇ REV MR 21 # 3328 Zbl 103.246 JSL 30.392
• ID 07095

KINO, A. [1961] *On ordinal diagrams* (J 0090) J Math Soc Japan 13*346-356
◇ F15 ◇ REV MR 26 # 1268 Zbl 109.7 JSL 37.192
• ID 07096

KINO, A. & TAKEUTI, G. [1962] *On hierarchies of predicates of ordinal numbers* (J 0090) J Math Soc Japan 14*199-232
◇ D55 D60 F15 ◇ REV JSL 33.293 • ID 07097

KINO, A. [1963] *A note on constructive ordinals* (P 1127) Symp Founds of Math;1962 Katada 51-64
• REPR [1964] (J 0260) Ann Jap Ass Phil Sci 2*189-198
◇ F15 ◇ REV MR 31 # 49 Zbl 134.12 JSL 33.294
• ID 21981

KINO, A. & TAKEUTI, G. [1963] *On predicates with constructive infinitely long expressions* (J 0090) J Math Soc Japan 15*176-190
◇ C70 F15 ◇ REV Zbl 118.250 JSL 30.97 • ID 07100

KINO, A. [1968] *Foundations of mathematics: consistency* (C 0552) Phil Contemp - Chroniques 266-274
◇ F35 F98 ◇ ID 14957

KINO, A. [1968] *On provably recursive functions and ordinal recursive functions* (J 0090) J Math Soc Japan 20*456-476
◇ D20 F05 F30 F35 ◇ REV MR 38 # 43 Zbl 195.303
• ID 07103

KINO, A. [1970] *Formalization of the theory of ordinal diagrams of infinite order* (P 0603) Intuitionism & Proof Th;1968 Buffalo 363-376
◇ F15 F35 ◇ REV MR 42 # 5789 Zbl 206.283 JSL 40.501
• ID 21061

KINO, A. & MYHILL, J.R. & VESLEY, R.E. (EDS.) [1970] *Intuitionism and proof theory. Proceedings of the Summer Conference at Buffalo, N.Y., 1968* (X 0809) North Holland: Amsterdam viii+516pp
◇ F97 ◇ REV MR 42 # 5774 Zbl 195.12 • ID 23442

KINO, A. & MYHILL, J.R. [1975] *A hierarchy of languages with infinitely long expressions* (P 0775) Logic Colloq;1973 Bristol 55-71
◇ C40 C75 E10 F15 ◇ REV MR 52 # 2817
Zbl 357 # 02013 • ID 17616

KINO, A. see Vol. III, IV, V for further entries

KIPNIS, M.M. [1968] *The constructive classification of arithmetic predicates and the semantic bases of arithmetic (Russian)* (S 0228) Zap Nauch Sem Leningrad Otd Mat Inst Steklov 8*53-65
• TRANSL [1968] (J 0521) Semin Math, Inst Steklov 8*22-27
◇ D55 F30 F50 ◇ REV MR 43 # 1836 Zbl 262 # 02035
• ID 27486

KIPNIS, M.M. [1971] *On the realizations of predicate formulas (Russian) (English summary)* (S 0228) Zap Nauch Sem Leningrad Otd Mat Inst Steklov 20*40-48,283
• TRANSL [1973] (J 1531) J Sov Math 1*22-27
◇ F50 ◇ REV MR 45 # 1757 MR 48 # 5808
Zbl 222 # 02021 • ID 07110

KIPNIS, M.M. see Vol. I, IV for further entries

KIRBY, L.A.S. & PARIS, J.B. [1977] *Initial segments of models of Peano's axioms* (P 1695) Set Th & Hierarch Th (3);1976 Bierutowice 211-226
◇ C62 F30 ◇ REV MR 58 # 10423 Zbl 364 # 02032
JSL 48.482 • ID 50962

KIRBY, L.A.S. & PARIS, J.B. [1978] *Σ_n-collection schemas in arithmetic* (P 1897) Logic Colloq;1977 Wroclaw 199-209
◇ C62 F30 ◇ REV MR 81e:03056 Zbl 442 # 03042
• ID 56399

Kirby, L.A.S. [1980] *La methode des indicatrices et le theoreme d'incompletude* (**C** 2969) Modeles de l'Arithm; Paris 1977 5-18
◇ C62 F30 ◇ REV MR 84d:03040 Zbl 458 # 03023 JSL 48.483 • ID 54429

Kirby, L.A.S. & Paris, J.B. [1982] *Accessible independence results for Peano arithmetic* (**J** 0161) Bull London Math Soc 14*285-293
◇ C62 F30 ◇ REV MR 83j:03096 Zbl 501 # 03017 • ID 35381

Kirby, L.A.S. [1982] *Flipping properties in arithmetic* (**J** 0036) J Symb Logic 47*416-422
◇ C62 E05 F30 ◇ REV MR 84e:03084 Zbl 488 # 03031 • ID 34416

Kirby, L.A.S. see Vol. III, IV, V for further entries

Kireevskij, N.N. [1958] *Ueber die Allgemeingueltigkeit gewisser Zaehlausdruecke (Russian)* (**J** 0142) Mat Sb, Akad Nauk SSSR, NS 42*669-678
◇ B25 F30 ◇ REV Zbl 14.2 FdM 61.973 • ID 40811

Kireevskij, N.N. see Vol. III, IV for further entries

Kirk, R.E. [1979] *Some classes of Kripke frames characteristic for the intuitionistic logic* (**J** 0068) Z Math Logik Grundlagen Math 25*409-410
◇ B55 C90 F50 ◇ REV MR 81b:03011 Zbl 423 # 03042 • ID 53554

Kirk, R.E. [1980] *A characterization of the classes of finite tree frames which are adequate for the intuitionistic logic* (**J** 0068) Z Math Logik Grundlagen Math 26*497-501
◇ B55 C90 F50 ◇ REV MR 81m:03014 Zbl 446 # 03017 • ID 56546

Kirk, R.E. see Vol. I for further entries

Kirov, K.A. [1984] *An intuitionistic analogue of the modal logic of provability in Peano's arithmetic* (**P** 4392) Mat Logika (Markova);1980 Sofia 65-72
◇ B45 F30 ◇ ID 46974

Kirov, K.A. see Vol. II for further entries

Klaua, D. [1956] *Berechenbare Analysis* (**J** 0068) Z Math Logik Grundlagen Math 2*265-303
◇ F60 ◇ REV MR 19.933 Zbl 74.14 JSL 36.535 • ID 07137

Klaua, D. [1959] *Die Praezisierung des Berechenbarkeitsbegriffs in der Analysis mit Hilfe rationaler Funktionale* (**J** 0068) Z Math Logik Grundlagen Math 5*33-96
◇ F60 ◇ REV MR 21 # 5571 JSL 36.535 • ID 07139

Klaua, D. [1960] *Berechenbare Reihen* (**J** 0068) Z Math Logik Grundlagen Math 6*143-161
◇ F60 ◇ REV MR 23 # A1527 Zbl 96.245 • ID 07141

Klaua, D. [1961] *Konstruktive Analysis* (**X** 0806) Dt Verlag Wiss: Berlin viii+160pp
◇ F60 F98 ◇ REV MR 23 # A62 Zbl 102.251 JSL 38.154 • ID 21010

Klaua, D. [1967] *Bemerkungen zur finiten Mathematik* (**J** 0342) Monatsber Dt Akad Wiss 9*166-169
◇ E30 F65 ◇ REV MR 37 # 3911 Zbl 165.312 • ID 07152

Klaua, D. see Vol. I, II, III, V for further entries

Kleene, S.C. [1934] *Proof by cases in formal logic* (**J** 0120) Ann of Math, Ser 2 35*529-544
◇ B40 ◇ REV Zbl 10.146 FdM 60.847 • ID 07160

Kleene, S.C. [1935] *A theory of positive integers in formal logic I,II* (**J** 0100) Amer J Math 57*153-173,219-244
◇ B28 B40 ◇ REV Zbl 11.2 FdM 61.55 FdM 61.56 • ID 28412

Kleene, S.C. & Rosser, J.B. [1935] *The inconsistency of certain formal logics* (**J** 0120) Ann of Math, Ser 2 36*630-636
◇ B40 ◇ REV Zbl 12.146 FdM 61.56 • ID 07161

Kleene, S.C. [1936] *λ-definability and recursiveness* (**J** 0025) Duke Math J 2*340-353 • ERR/ADD [1937] (**J** 0036) J Symb Logic 2*38
◇ B40 ◇ REV Zbl 14.385 JSL 2.38 FdM 62.45 • ID 07165

Kleene, S.C. [1937] see Church, A.

Kleene, S.C. [1938] *On notation for ordinal numbers* (**J** 0036) J Symb Logic 3*150-155
◇ F15 ◇ REV Zbl 20.338 JSL 4.93 FdM 64.932 • ID 07166

Kleene, S.C. [1944] *On the forms of predicates in the theory of constructive ordinals* (**J** 0100) Amer J Math 66*41-58
◇ D55 D70 F15 ◇ REV MR 5.197 Zbl 61.10 JSL 11.127 • REM ERR in Kleene,S.C.: Introduction to metamathematics, p.527. Part II 1955 • ID 07168

Kleene, S.C. [1945] *On the interpretation of intuitionistic number theory* (**J** 0036) J Symb Logic 10*109-124
◇ B28 F50 ◇ REV MR 7.406 JSL 12.91 • ID 07169

Kleene, S.C. [1949] *On the intuitionistic logic* (**P** 0682) Int Congr Philos (10);1948 Amsterdam 741-743
◇ F50 ◇ REV MR 10.277 Zbl 34.154 JSL 14.63 • ID 20804

Kleene, S.C. [1952] *Finite axiomatizability of theories in the predicate calculus using additional predicate symbols* (**S** 0167) Mem Amer Math Soc 10*27-66
• TRANSL [1967] (**C** 2542) Mat Teor Log Vyvoda 237-283 (Russian)
◇ F05 ◇ REV MR 14.439 Zbl 47.250 JSL 19.62 JSL 35.323 JSL 36.334 • REM 2nd printing, with revisions, 1967. ERR in Kleene,S.C.: Mathematical logic, p.378 • ID 07172

Kleene, S.C. [1952] *Introduction to metamathematics* (**X** 0809) North Holland: Amsterdam x+550pp
• TRANSL [1957] (**X** 1656) Izdat Inostr Lit: Moskva 526pp (Russian) [1984] (**X** 1876) Kexue Chubanshe: Beijing xii+234pp,x+235-688pp (Chinese) [1974] (**X** 1781) Tecnos: Madrid (Spanish)
◇ A05 B98 D98 F30 F50 F98 ◇ REV MR 14.525 Zbl 47.7 JSL 19.215 JSL 25.280 JSL 33.290 JSL 35.350 JSL 38.333 • REM Co-publisher: Wolters-Noordhoff; 8th revised ed. 1980. Chinese transl. in 2 parts • ID 07173

Kleene, S.C. [1952] *Permutability of inferences in Gentzen's calculi LK and LJ* (**S** 0167) Mem Amer Math Soc 10*1-26
• TRANSL [1967] (**C** 2542) Mat Teor Log Vyvoda 208-236
◇ F05 F07 ◇ REV MR 14.439 Zbl 47.250 JSL 19.62 JSL 35.323 • REM 2nd rev. ed. 1967. ERR in Kleene,S.C.: Mathematical logic, p.378 • ID 07171

Kleene, S.C. [1952] *Recursive functions and intuitionistic mathematics* (**P** 0593) Int Congr Math (II, 6);1950 Cambridge MA 1*679-685
◇ F50 ◇ REV MR 13.422 Zbl 49.150 JSL 18.181 • ID 20808

KLEENE, S.C. [1955] *On the forms of the predicates in the theory of constructive ordinals II* (**J** 0100) Amer J Math 77*405-428 • ERR/ADD [1959] (**J** 0064) Trans Amer Math Soc 91*51 ⋄ D55 D70 F15 ⋄ REV MR 17.5 Zbl 67.252 JSL 21.410 • REM Part I 1944 • ID 07178

KLEENE, S.C. [1956] *Decision procedure* (**C** 0647) Encycl Britannica 7*124 ⋄ F05 ⋄ REV JSL 23.22 • ID 07179

KLEENE, S.C. [1957] *Realizability* (**P** 1675) Summer Inst Symb Log;1957 Ithaca 1*100-104 • REPR [1959] (**P** 0634) Constructivity in Math;1957 Amsterdam 285-289 ⋄ D80 F50 ⋄ REV MR 21 # 2590 Zbl 88.249 JSL 27.242 • REM ERR in Kleene,S.C. & Vesley,R.E.: The foundations of intuitionistic mathematics, 192 • ID 22336

KLEENE, S.C. [1960] *Realizability and Shanin's algorithm for the constructive deciphering of mathematical sentences* (**J** 0079) Logique & Anal, NS 3*154-165 ⋄ D80 F50 ⋄ REV JSL 27.243 • REM ERR in Kleene,S.C. & Vesley,R.E.: The foundations of intuitionistic mathematics, p.192-193 • ID 07188

KLEENE, S.C. [1962] *λ-definable functionals of finite types* (**J** 0027) Fund Math 50*281-303 ⋄ B40 D30 D55 D65 ⋄ REV MR 32 # 4003 Zbl 100.250 JSL 29.104 • ID 07191

KLEENE, S.C. [1962] *Disjunction and existence under implication in elementary intuitionistic formalisms* (**J** 0036) J Symb Logic 27*11-18 • ERR/ADD ibid 28*154-156 ⋄ F50 ⋄ REV MR 27 # 1368 MR 32 # 5515 Zbl 112.245 JSL 28.166 JSL 29.146 • ID 07194

KLEENE, S.C. [1965] *Classical extensions of intuitionistic mathematics* (**P** 0623) Int Congr Log, Meth & Phil of Sci (2,Proc);1964 Jerusalem 31-44 • ERR/ADD [1969] (**S** 0167) Mem Amer Math Soc 89*105 ⋄ F35 F50 ⋄ REV MR 35 # 28 Zbl 192.30 JSL 40.508 • ID 07195

KLEENE, S.C. [1965] *Logical calculus and realizability* (**J** 0096) Acta Philos Fenn 18*71-80 ⋄ F50 ⋄ REV MR 33 # 5485 Zbl 133.252 • ID 07196

KLEENE, S.C. & VESLEY, R.E. [1965] *The foundations of intuitionistic mathematics, especially in relation to recursive functions* (**S** 3303) Stud Logic Found Math viii + 206pp • TRANSL [1978] (**X** 2027) Nauka: Moskva 271pp ⋄ F35 F50 F98 ⋄ REV MR 31 # 1190 Zbl 133.246 JSL 31.258 • REM ERR in S0167 89*105, 1965 • ID 23383

KLEENE, S.C. [1967] *Mathematical logic* (**X** 0827) Wiley & Sons: New York xiii + 398pp • TRANSL [1967] (**X** 3636) Tokyo Tosho: Tokyo 200pp + 266pp [1971] (**X** 0850) Colin: Paris [1973] (**X** 0885) Mir: Moskva 480pp ⋄ B98 D20 D25 D55 D98 F30 F98 ⋄ REV MR 36 # 25 Zbl 149.243 JSL 35.438 • ID 45895

KLEENE, S.C. [1968] *Constructive functions in "The foundations of intuitionistic mathematics"* (**P** 0627) Int Congr Log, Meth & Phil of Sci (3,Proc);1967 Amsterdam 137-144 ⋄ F50 ⋄ REV MR 40 # 5424 Zbl 191.287 • ID 07198

KLEENE, S.C. [1968] see DALEN VAN, D.

KLEENE, S.C. [1969] *Formalized recursive functionals and formalized realizability* (**S** 0167) Mem Amer Math Soc 89*106pp ⋄ D65 F35 F50 ⋄ REV MR 39 # 5319 Zbl 184.20 • ID 24941

KLEENE, S.C. [1973] *Realizability: a retrospective survey* (**P** 0713) Cambridge Summer School Math Log;1971 Cambridge GB 95-112 ⋄ F50 ⋄ REV MR 50 # 4263 Zbl 272 # 02036 • ID 07199

KLEENE, S.C. see Vol. I, II, IV for further entries

KLEIN, H. [1937] *Einfuehrung von Ordnung und Rechenoperationen bei natuerlichen Zahlen auf Grund der Peano-Axiome* (**J** 2074) Sem-ber, Muenster 10*52-69 ⋄ F30 ⋄ REV FdM 63.30 • ID 41032

KLEINBERG, E.M. [1970] *Recursion theory and formal deducibility* (**J** 0036) J Symb Logic 35*556-558 ⋄ D25 F07 F40 ⋄ REV MR 43 # 4666 Zbl 221 # 02017 • ID 07211

KLEINBERG, E.M. see Vol. I, III, IV, V for further entries

KLEMKE, D. [1971] *Ein Henkin-Beweis fuer die Vollstaendigkeit eines Kalkuels relativ zur Grzegorczyk-Semantik* (**J** 0009) Arch Math Logik Grundlagenforsch 14*148-161 ⋄ B55 C90 F50 ⋄ REV MR 47 # 8253 Zbl 233 # 02020 • ID 07229

KLIX, W.-D. [1977] see HARMS, S.

KLOP, J.W. [1975] *On solvability by λI-terms* (**P** 1603) λ-Calc & Comput Sci Th;1975 Roma 342-345 ⋄ B40 ⋄ REV MR 57 # 9497 Zbl 333 # 02020 • ID 32376

KLOP, J.W. [1976] see BARENDREGT, H.P.

KLOP, J.W. [1978] see BARENDREGT, H.P.

KLOP, J.W. [1979] see BERGSTRA, J.A.

KLOP, J.W. [1980] *Combinatory reduction systems* (**S** 1605) Math Centr Tracts 127*xiii + 317pp ⋄ B40 ⋄ REV MR 83e:03026 Zbl 466 # 03006 • ID 54957

KLOP, J.W. [1980] see BERGSTRA, J.A.

KLOP, J.W. [1980] *Reduction cycles in combinatory logic* (**C** 3050) Essays Combin Log, Lambda Calc & Formalism (Curry) 193-214 ⋄ B40 ⋄ REV MR 82b:03040 Zbl 469 # 03006 • ID 74883

KLOP, J.W. [1982] see BERGSTRA, J.A.

KLOP, J.W. see Vol. IV for further entries

KMINEK, V. [1980] see DEMUTH, O.

KNESER, H. [1940] *Der Fundamentalsatz der Algebra und der Intuitionismus* (**J** 0044) Math Z 46*287-302 ⋄ F55 ⋄ REV MR 1.322 Zbl 22.301 FdM 66.58 • ID 07250

KNESER, H. see Vol. V for further entries

KNESER, M. [1981] *Ergaenzung zu einer Arbeit von Hellmuth Kneser ueber den Fundamentalsatz der Algebra* (**J** 0044) Math Z 177*285-287 ⋄ F55 ⋄ REV MR 82i:03072 Zbl 439 # 12012 • ID 54115

KOCHUBIEVSKIJ, I.P. & SATAEV, A.G. [1982] *Potential realizability: pragmatic aspect (Russian)* (**J** 4387) Itogi Nauki Tekh, Ser Tekh Kibern 15*305-336 ⋄ F60 ⋄ REV Zbl 556 # 93014 • ID 46167

KOCK, A. & LECOUTURIER, P. & MIKKELSEN, C.J. [1975] *Some topos theoretic concepts of finiteness* (C 0772) Model Th & Topoi 209-283
◇ F50 G30 ◇ REV MR 52 # 2882 Zbl 334 # 18010
• ID 17664

KOCK, A. [1976] *Universal projective geometry via topos theory* (J 0326) J Pure Appl Algebra 9*1-24
◇ E75 F50 G30 ◇ REV MR 55 # 3959 Zbl 375 # 02016
• ID 51593

KOCK, A. & REYES, G.E. [1977] *Doctrines in categorical logic* (C 1523) Handb of Math Logic 283-313
◇ F50 G30 ◇ REV MR 58 # 10395 JSL 49.968 • ID 24203

KOCK, A. & REYES, G.E. [1979] *Connections in formal differential geometry* (S 3462) Var Publ Ser, Aarhus Univ 30*158-195
◇ F50 G30 ◇ REV MR 81h:53030 Zbl 418 # 18008
• ID 53332

KOCK, A. [1981] *Synthetic differential geometry* (S 3306) Lond Math Soc Lect Note Ser 51*viii+311pp
◇ F50 G30 ◇ REV MR 83f:51023 Zbl 466 # 51008
• ID 54988

KOCK, A. see Vol. I, III, V for further entries

KODERA, H. [1980] *Remark on classical logic and intuitionistic logic* (J 2678) Bull Aichi Univ Educ Nat Sci 29*31-37
◇ B20 F05 F50 ◇ REV MR 81m:03015 • ID 74921

KOGALOVSKIJ, S.R. [1968] *Some remarks on higher order logic (Russian)* (J 0023) Dokl Akad Nauk SSSR 178*1007-1009
• TRANSL [1968] (J 0062) Sov Math, Dokl 9*227-229
◇ B15 C40 C85 F35 ◇ REV MR 39 # 43 Zbl 182.321
• ID 07299

KOGALOVSKIJ, S.R. [1970] *Some reduction theorems for higher order logic (Russian)* (J 0023) Dokl Akad Nauk SSSR 190*519-522
• TRANSL [1970] (J 0062) Sov Math, Dokl 11*138-142
◇ B15 F35 ◇ REV MR 41 # 8216 Zbl 212.312 • ID 07300

KOGALOVSKIJ, S.R. see Vol. I, III, IV, V for further entries

KOGAN-BERNSHTEIJN, L.M. [1981] *Simplification of Gentzen's reductions in classical arithmetic (Russian)* (S 0228) Zap Nauch Sem Leningrad Otd Mat Inst Steklov 105*45-52
• TRANSL [1983] (J 1531) J Sov Math 22*1305-1310
◇ F05 F30 ◇ REV MR 82k:03089 Zbl 484 # 03031
• ID 38147

KOHL, H.R. & PARSONS, C. [1960] *Self-reference, truth, and provability* (J 0094) Mind 69*69-73
◇ A05 F25 F30 ◇ ID 28347

KOLATA, G. [1982] *Does Goedel's theorem matter to mathematics?* (J 1154) Science (AAAS) 218*779-780
◇ A05 F99 ◇ REV MR 84m:03089 • ID 35785

KOLETSOS, G. [1982] *A Goedel-functional interpretation of the ω-rule* (J 0465) Bull Greek Math Soc (NS) 23*89-94
◇ F10 F35 ◇ ID 49511

KOLETSOS, G. [1985] *Functional interpretation of the β-rule* (J 0036) J Symb Logic 50*791-805
◇ F10 F30 F35 ◇ ID 47380

KOLETSOS, G. see Vol. II, IV for further entries

KOLMOGOROV, A.N. [1925] *On the principle of excluded middle (Russian)* (J 1404) Mat Sb, Akad Nauk SSSR 32*646-667
• TRANSL [1967] (C 0675) From Frege to Goedel 414-437
◇ F50 ◇ REV JSL 1.191 FdM 51.48 • ID 14927

KOLMOGOROV, A.N. [1932] *Zur Deutung der intuitionistischen Logik* (J 0044) Math Z 35*58-65
◇ A05 F50 ◇ REV Zbl 4.2 JSL 1.191 FdM 58.62
• ID 07316

KOLMOGOROV, A.N. see Vol. I, III, IV, V for further entries

KOLODZIEJ, R. [1973] *On a certain subclass of formulas of recursive arithmetics (Polish) (Russian and English summaries)* (S 4733) Prace Inst Mat, Politech Wroclaw, Ser Stud Mater 5*41-56
◇ F30 ◇ REV MR 50 # 4260 Zbl 264 # 02035 • ID 27507

KOLODZIEJ, R. [1974] *Reducibility of formulae of weak second order arithmetic to pseudo-canonical forms I,II,III* (J 0063) Studia Logica 33*131-152,233-247,247-258
◇ F35 ◇ REV MR 54 # 4935 Zbl 293 # 02025 • ID 24107

KOLODZIEJ, R. see Vol. I for further entries

KONDO, M. [1956] *Sur la nommabilite d'ensembles* (J 0109) C R Acad Sci, Paris 242*1841-1843
◇ D55 E15 F60 F65 ◇ REV MR 17.933 Zbl 70.278 JSL 22.299 • ID 07330

KONDO, M. [1956] *Sur la notion du transfini* (J 0109) C R Acad Sci, Paris 242*2209-2212
◇ E10 E47 F65 ◇ REV MR 18.2 Zbl 70.278 JSL 22.299
• ID 07328

KONDO, M. [1956] *Sur les analyses relatives* (J 0109) C R Acad Sci, Paris 242*2084-2087
◇ D55 E15 F65 ◇ REV MR 17.933 Zbl 70.278 JSL 22.299 • ID 07332

KONDO, M. [1956] *Sur les nombres reels et nommables* (J 0109) C R Acad Sci, Paris 242*1945-1948
◇ B28 D55 E15 F65 ◇ REV MR 17.933 Zbl 70.278 JSL 22.299 • ID 07331

KONDO, M. [1960] *Le fondement constructif du calcul infinitesimal* (J 1770) Osaka Math J 12*61-96
◇ F65 ◇ REV MR 23 # A59 Zbl 115.8 • ID 07336

KONDO, M. [1961] *Sur la nommabilite d'etres mathematiques* (J 0109) C R Acad Sci, Paris 252*3934-3936
◇ E47 F65 ◇ REV MR 23 # A2308 Zbl 108.7 • ID 42825

KONDO, M. [1984] *The completeness theorems for some intuitionistic epistemic logics in terms of interval semantics* (J 0390) Publ Res Inst Math Sci (Kyoto) 20*671-681
◇ B45 F50 ◇ REV MR 85i:03052 • ID 44181

KONDO, M. see Vol. I, II, IV, V for further entries

KOREC, I. & PERETYAT'KIN, M.G. & RAUTENBERG, W. [1974] *Definability in structures of finite valency* (J 0027) Fund Math 81*173-181
◇ C60 C65 F25 ◇ REV MR 49 # 2360 Zbl 276 # 02036
• ID 07371

KOREC, I. & RAUTENBERG, W. [1976] *Model-interpretability into trees and applications* (J 0009) Arch Math Logik Grundlagenforsch 17*97-104
◇ B25 C45 C60 C65 F25 ◇ REV MR 54 # 2443 Zbl 324 # 02041 • ID 07374

KOREC, I. see Vol. IV, V for further entries

KOSHELEVA, O.M. & KREJNOVICH, V.YA. [1981] *What can physics contribute to constructive mathematics? (Russian)* (C 3747) Mat Log & Mat Lingvistika 117-128
◇ A05 F99 ◇ REV MR 84j:03116 • ID 34706

KOSOVSKIJ, N.K. [1967] *Sufficient conditions of incompleteness for the formalization of parts of arithmetic (Russian)* (S 0228) Zap Nauch Sem Leningrad Otd Mat Inst Steklov 4*44-57
• TRANSL [1967] (J 0521) Semin Math, Inst Steklov 4*15-20
◇ F30 ◇ REV MR 39 # 1329 Zbl 254 # 02037 • ID 07385

KOSOVSKIJ, N.K. [1969] *Integrable FR-constructs over a probability space (Russian)* (S 0228) Zap Nauch Sem Leningrad Otd Mat Inst Steklov 16*97-104
• TRANSL [1969] (J 0521) Semin Math, Inst Steklov 16*49-52
◇ F60 ◇ REV MR 41 # 9312 Zbl 272 # 02060 • ID 30392

KOSOVSKIJ, N.K. [1969] *Laws of large numbers in constructive probability theory (Russian)* (S 0228) Zap Nauch Sem Leningrad Otd Mat Inst Steklov 16*105-113
• TRANSL [1969] (J 0521) Semin Math, Inst Steklov 16*53-57
◇ F60 ◇ REV MR 41 # 9313 Zbl 267 # 02024 • ID 29891

KOSOVSKIJ, N.K. [1969] *Necessary and sufficient conditions for a probability space to have Specker properties (Russian)* (S 0228) Zap Nauch Sem Leningrad Otd Mat Inst Steklov 16*91-96
• ERR/ADD ibid 20*292-294
• TRANSL [1969] (J 0521) Semin Math, Inst Steklov 16*45-48
◇ F60 ◇ REV MR 41 # 9311 Zbl 272 # 02059 • ID 30390

KOSOVSKIJ, N.K. [1970] *Some questions in the constructive theory of normed boolean algebras (Russian)* (S 0066) Tr Mat Inst Steklov 113*2-38
• TRANSL [1970] (S 0055) Proc Steklov Inst Math 113*1-41
◇ C57 F60 G05 ◇ REV MR 44 # 5988 Zbl 229 # 02032
• ID 28600

KOSOVSKIJ, N.K. [1973] *Constructive versions of the laws of large numbers (Russian)* (S 0066) Tr Mat Inst Steklov 129*3-23
• TRANSL [1973] (S 0055) Proc Steklov Inst Math 129*1-19
◇ F60 ◇ REV MR 55 # 2523 Zbl 318 # 02037 • ID 63062

KOSOVSKIJ, N.K. [1973] *Some problems in the constructive probability theory* (P 0580) Int Congr Log, Meth & Phil of Sci (4,Sel Pap);1971 Bucharest 83-99
◇ F60 ◇ REV MR 57 # 2890 Zbl 299 # 02041 • ID 63073

KOSOVSKIJ, N.K. [1976] *On constructive distribution functions (Russian) (English summary)* (S 0228) Zap Nauch Sem Leningrad Otd Mat Inst Steklov 60*59-64
• TRANSL [1980] (J 1531) J Sov Math 14*1464-1468
◇ F60 ◇ REV MR 58 # 27380 Zbl 336 # 02030 • ID 63066

KOSOVSKIJ, N.K. [1981] *Elements of mathematical logic and its applications to the theory of subrecursive algorithms (Russian)* (X 0938) Leningrad Univ: Leningrad 192pp
◇ B98 D20 D98 F30 F60 F98 ◇ REV MR 83m:03013 Zbl 479 # 03001 • ID 55663

KOSOVSKIJ, N.K. see Vol. I, IV for further entries

KOSSAK, R. & PARIS, J.B. [1981] *Subsets of models of arithmetic* (P 2614) Open Days in Model Th & Set Th;1981 Jadwisin 159-174
◇ C62 F30 ◇ ID 33723

KOSSAK, R. [1985] *A note on satisfaction classes* (J 0047) Notre Dame J Formal Log 26*1-8
◇ C50 C57 C62 F30 ◇ REV MR 86c:03055 Zbl 562 # 03040 • ID 42582

KOSSAK, R. see Vol. III, IV, V for further entries

KOTAS, J. & PIECZKOWSKI, A. [1966] *On a generalized cylindrical algebra and intuitionistic logic (Polish and Russian summaries)* (J 0063) Studia Logica 18*73-81
◇ F50 G15 ◇ REV MR 34 # 4119 Zbl 304 # 02012
• ID 07397

KOTAS, J. see Vol. I, II, III for further entries

KOVTUN, M.R. [1983] *A system of intuitionistic analysis that is equivalent to classical analysis (Russian) (English summary)* (J 0288) Vest Ser Mat Mekh, Univ Moskva 1983/1*41-44,101
• TRANSL [1983] (J 0510) Moscow Univ Math Bull 38/1*49-52
◇ F35 F50 ◇ REV MR 84m:03091 Zbl 505 # 03028
• ID 35787

KOWALCZYK, W. [1982] *A sufficient condition for the consistency of P = NP with Peano arithmetic* (J 2095) Fund Inform, Ann Soc Math Pol, Ser 4 5*233-245
◇ D15 F25 F30 ◇ REV MR 84b:68043 Zbl 524 # 03026
• ID 37601

KOWALCZYK, W. see Vol. IV for further entries

KOYMANS, C.P.J. [1982] *Models of the lambda-calculus* (J 0194) Inform & Control 52*306-332
◇ B40 ◇ REV MR 85b:03020 Zbl 542 # 03004 • ID 40577

KOYMANS, C.P.J. [1984] *Models of the lambda-calculus* (S 3477) Math Centr Afd Toegepaste Wiskd, Amsterdam iii + 181pp
◇ B40 ◇ REV Zbl 558 # 03005 • ID 44032

KOYMANS, K. [1980] see BARENDREGT, H.P.

KOZEN, D. [1977] *Lower bounds for natural proof systems* (P 3572) IEEE Symp Found of Comput Sci (18);1977 Providence 254-266
◇ D15 F20 ◇ REV MR 58 # 13931 • ID 81928

KOZEN, D. [1981] *Semantics of probabilistic programs* (J 0119) J Comp Syst Sci 22*328-350
◇ B40 B75 ◇ REV MR 82i:68011 Zbl 426 # 68002
• ID 81927

KOZEN, D. see Vol. I, II, III, IV for further entries

KRABBE, E.C.W. [1985] *Noncumulative dialectical models and formal dialectics* (J 0122) J Philos Logic 14*129-168
◇ A05 B46 F99 ◇ REV Zbl 571 # 03003 • ID 42654

KRABBE, E.C.W. see Vol. II, III for further entries

KRAJEWSKI, S. [1974] *Mutually inconsistent satisfaction classes* (J 0014) Bull Acad Pol Sci, Ser Math Astron Phys 22*983-987
◇ C62 F30 ◇ REV MR 52 # 10417 Zbl 298 # 02069
• ID 07447

KRAJEWSKI, S. [1974] *Predicative expansions of axiomatic theories* (J 0068) Z Math Logik Grundlagen Math 20*435-452
◇ C30 C62 E30 E70 F25 ◇ REV MR 51 # 10082 Zbl 298 # 02068 • ID 07445

KRAJEWSKI, S. [1981] *Kurt Goedel and his work (Polish)* (J 0519) Wiad Mat, Ann Soc Math Pol, Ser 2 23*161-187 ◇ A10 C98 D98 E98 F99 ◇ REV MR 84j:01068 Zbl 535 # 01011 • ID 38331

KRAJEWSKI, S. see Vol. II, III, IV, V for further entries

KRAJICEK, J. [1985] *Some theorems on the lattice of local interpretability types* (J 0068) Z Math Logik Grundlagen Math 31*449-460 ◇ F25 ◇ REV Zbl 571 # 03027 • ID 47551

KRAJICEK, J. see Vol. II, V for further entries

KRAMOSIL, I. [1980] *Statistical testing procedure for lengths of formalized proofs* (J 0156) Kybernetika (Prague) 16*209-224 ◇ B35 F20 ◇ REV MR 83b:68097 Zbl 444 # 03007 • ID 56455

KRAMOSIL, I. see Vol. I, II, III, IV for further entries

KRASNER, M. [1957] *Theorie de la definition I* (J 3941) J Math Pures Appl, Ser 9 36*325-357 ◇ A05 E70 F99 ◇ REV MR 19.935 Zbl 81.242 JSL 24.230 • REM Part II 1958 • ID 16974

KRASNER, M. [1958] *Theorie de la definition II* (J 3941) J Math Pures Appl, Ser 9 37*53-101 ◇ A05 E70 F99 ◇ REV MR 21 # 4911 JSL 24.230 • REM Part I 1957 • ID 16975

KRASNER, M. see Vol. III for further entries

KREER, L.I. [1929] *On proofs (Russian)* (J 4376) Ordzhonikidze, Izv Ped Inst, Gor'kij 6*103-113 ◇ A05 F99 ◇ ID 43624

KREIDER, D.L. & ROGERS JR., H. [1961] *Constructive versions of ordinal number classes* (J 0064) Trans Amer Math Soc 100*325-369 ◇ D55 D70 F15 ◇ REV MR 27 # 1381 Zbl 100.12 JSL 31.134 • ID 07467

KREIDER, D.L. see Vol. IV for further entries

KREISEL, G. [1950] *Note on arithmetic models for consistent formulae of the predicate calculus I* (J 0027) Fund Math 37*265-285 ◇ B10 C57 D45 D55 F30 ◇ REV MR 12.790 JSL 18.180 • REM Part II 1953 • ID 07481

KREISEL, G. [1951] *On the interpretation of non-finitist proofs I* (J 0036) J Symb Logic 16*241-267 • ERR/ADD ibid 16*IV ◇ F10 F25 F30 F50 ◇ REV MR 14.122 Zbl 44.3 JSL 18.78 • REM Part II 1952 • ID 28465

KREISEL, G. [1952] *On the concepts of completeness and interpretation of formal systems* (J 0027) Fund Math 39*103-127 ◇ A05 F25 F30 ◇ REV MR 14.937 Zbl 50.6 JSL 24.236 • ID 07484

KREISEL, G. [1952] *On the interpretation of non-finitist proofs II. Interpretation of number theory. Applications* (J 0036) J Symb Logic 17*43-58 ◇ F10 F25 F30 F50 ◇ REV MR 14.440 Zbl 46.7 JSL 18.78 • REM Part I 1951 • ID 07483

KREISEL, G. [1952] *Some concepts concerning formal systems of number theory* (J 0044) Math Z 57*1-12 ◇ F10 F25 F30 ◇ REV MR 14.937 Zbl 48.247 JSL 31.128 • ID 07482

KREISEL, G. [1952] *Some elementary inequalities* (J 0028) Indag Math 14*334-338 ◇ F55 ◇ REV MR 14.458 Zbl 49.170 JSL 18.78 • ID 07479

KREISEL, G. [1953] *Note on arithmetic models for consistent formulae of the predicate calculus II* (P 0645) Int Congr Philos (11);1953 Bruxelles 14*39-49 ◇ B10 C57 D45 D55 F30 ◇ REV MR 15.668 Zbl 53.200 JSL 21.403 • REM Part I 1951 • ID 20815

KREISEL, G. [1953] *On a problem of Henkin's* (J 0028) Indag Math 15*405-406 ◇ F30 ◇ REV MR 15.668 Zbl 53.6 JSL 19.219 • ID 07486

KREISEL, G. [1953] *The diagonal method in formalized arithmetic* (J 0013) Brit J Phil Sci 3*364-373 ◇ F30 ◇ ID 48858

KREISEL, G. [1954] *Discussion sur divers themes* (P 0646) Appl Sci de Log Math;1952 Paris 157-158 ◇ F30 F50 ◇ REM Discussion remarks • ID 27976

KREISEL, G. [1954] *Remark on complete interpretations by models* (J 0009) Arch Math Logik Grundlagenforsch 2*4-9 • REPR [1954] (J 1114) Arch Phil 5*84-89 ◇ B15 F25 F30 ◇ REV MR 17.119 Zbl 55.6 JSL 36.169 • ID 22027

KREISEL, G. [1955] *Models, translations and interpretations* (P 1589) Math Interpr of Formal Systs;1954 Amsterdam 26-50 ◇ F10 F25 ◇ REV MR 17.699 Zbl 68.245 JSL 24.236 • ID 27720

KREISEL, G. & WANG, HAO [1955] *Some applications of formalized consistency proofs* (J 0027) Fund Math 42*101-110 ◇ F25 F30 ◇ REV MR 17.447 Zbl 67.252 JSL 21.404 • REM Part I. Part II 1958 • ID 07488

KREISEL, G. [1956] *Some uses of metamathematics* (J 0013) Brit J Phil Sci 7*161-173 ◇ A05 C98 F99 ◇ ID 90276

KREISEL, G. & LACOMBE, D. & SHOENFIELD, J.R. [1957] *Effective operations and partial recursive functionals* (P 1675) Summer Inst Symb Log;1957 Ithaca 364-365 ◇ D20 F60 ◇ REV JSL 31.261 • ID 07520

KREISEL, G. & PUTNAM, H. [1957] *Eine Unableitbarkeitsbeweismethode fuer den intuitionistischen Aussagenkalkuel* (J 0009) Arch Math Logik Grundlagenforsch 3*74-78 ◇ B55 F50 ◇ REV MR 19.934 Zbl 79.7 JSL 23.229 • ID 07492

KREISEL, G. & LACOMBE, D. [1957] *Ensembles recursivement mesurables et ensembles recursivement ouverts ou fermes* (J 0109) C R Acad Sci, Paris 245*1106-1109 ◇ D80 F60 ◇ REV MR 22 # 3680 Zbl 79.9 JSL 31.113 • ID 07490

KREISEL, G. & LACOMBE, D. & SHOENFIELD, J.R. [1957] *Fonctionnelles recursivement definissables et fonctionnelles recursives* (J 0109) C R Acad Sci, Paris 245*399-402 ◇ D20 F60 ◇ REV MR 19.521 Zbl 78.7 JSL 23.48 • ID 07494

KREISEL, G. [1957] *Goedel's interpretation of Heyting's arithmetic* (P 1675) Summer Inst Symb Log;1957 Ithaca 125-133 ◇ F10 F30 F50 ◇ REV JSL 36.169 • ID 07498

KREISEL, G. [1957] *Relations between classes of constructive functionals* (**P** 1675) Summer Inst Symb Log;1957 Ithaca 292-302
◇ F10 F50 ◇ REV JSL 36.169 • ID 07497

KREISEL, G. [1957] *Sums of squares* (**P** 1675) Summer Inst Symb Log;1957 Ithaca 313-320
◇ C57 C60 D20 F07 F99 ◇ REV JSL 31.128 • ID 29369

KREISEL, G. [1958] *A remark on free choice sequences and the topological completeness proofs* (**J** 0036) J Symb Logic 23*369-388
◇ F50 ◇ REV MR 22#4637 Zbl 91.11 JSL 32.283 • ID 07506

KREISEL, G. [1958] *Constructive mathematics* (1111) Preprints, Manuscr., Techn. Reports etc.
◇ F50 F55 F65 ◇ REM Notes of a Course Given at Stanford • ID 21396

KREISEL, G. [1958] *Elementary completeness properties of intuitionistic logic with a note on negations of prenex formulae* (**J** 0036) J Symb Logic 23*317-330
◇ F50 ◇ REV MR 21#2591 Zbl 86.246 JSL 32.282 • ID 07507

KREISEL, G. [1958] *Hilbert's programme* (**J** 0076) Dialectica 12*346-372
• TRANSL [1967] (**C** 2141) Filos Matematica 185-221 (Italian) • REPR [1959] (**C** 1173) Logica (Bernays) 142-168 [1964] (**C** 1105) Phil of Math. Sel Readings 157-180
◇ A05 F99 ◇ REV MR 21#5550 Zbl 90.10 JSL 27.228 JSL 34.313 • ID 22155

KREISEL, G. [1958] *Mathematical significance of consistency proofs* (**J** 0036) J Symb Logic 23*155-182
◇ C57 C60 F05 F25 F50 ◇ REV MR 22#6710 Zbl 88.15 JSL 31.129 • ID 07514

KREISEL, G. [1959] *Interpretation of analysis by means of constructive functionals of finite types* (**P** 0634) Constructivity in Math;1957 Amsterdam 101-128
◇ D65 F10 F30 F35 F50 ◇ REV MR 21#5568 Zbl 134.10 JSL 36.169 • ID 07499

KREISEL, G. & SHOENFIELD, J.R. & WANG, HAO [1959] *Number theoretic concepts and recursive well-orderings* (**J** 0009) Arch Math Logik Grundlagenforsch 5*42-64
◇ F15 F30 ◇ REV MR 22#6709 Zbl 129.4 JSL 31.511 • ID 07511

KREISEL, G. & LACOMBE, D. & SHOENFIELD, J.R. [1959] *Partial recursive functionals and effective operations* (**P** 0634) Constructivity in Math;1957 Amsterdam 290-297
◇ D20 F60 ◇ REV MR 21#7159 JSL 31.261 • ID 07508

KREISEL, G. [1960] *La predicativite* (**J** 0353) Bull Soc Math Fr 88*371-391
◇ A05 D55 E45 F35 F65 ◇ REV MR 23#A800 Zbl 131.6 JSL 27.79 • ID 07519

KREISEL, G. [1960] *Non-uniqueness results for transfinite progressions* (**J** 0014) Bull Acad Pol Sci, Ser Math Astron Phys 8*287-290
◇ F15 ◇ REV MR 24#A3064 JSL 25.364 • ID 07516

KREISEL, G. [1960] *Ordinal logics and the characterization of informal concepts of proof* (**P** 0660) Int Congr Math (II, 8);1958 Edinburgh 289-299
◇ F07 F15 F30 F65 ◇ REV MR 23#A1511 JSL 27.78 • ID 07500

KREISEL, G. [1960] see GANDY, R.O.

KREISEL, G. [1960] *Wittgenstein's theory and practice of philosophy* (**J** 0013) Brit J Phil Sci 11*238-252
◇ A05 F99 ◇ ID 48859

KREISEL, G. [1961] see HUBER-DYSON, V.

KREISEL, G. [1961] see GANDY, R.O.

KREISEL, G. [1962] *Beweistheorie* (**J** 0148) Math Gaz 46*250-255
◇ F98 ◇ ID 48860

KREISEL, G. [1962] *Foundations of intuitionistic logic* (**P** 0612) Int Congr Log, Meth & Phil of Sci (1,Proc);1960 Stanford 198-210
◇ F50 ◇ REV MR 27#3529 Zbl 133.248 JSL 30.243 • ID 07527

KREISEL, G. [1962] *On weak completeness of intuitionistic predicate logic* (**J** 0036) J Symb Logic 27*139-158
◇ F50 ◇ REV MR 28#5000 Zbl 117.10 JSL 34.119 • ID 07529

KREISEL, G. [1962] *The axiom of choice and the class of hyperarithmetic functions* (**J** 0028) Indag Math 24*307-319
◇ D55 F35 ◇ REV MR 25#3838 Zbl 108.8 JSL 35.333 • ID 07526

KREISEL, G. [1962] see FEFERMAN, S.

KREISEL, G. [1963] *Axiomatic results of second-order arithmetic* (**C** 4220) Rep Sem Found Anal 1*1.1-1.48
◇ C62 F35 ◇ ID 42740

KREISEL, G. [1963] *Generalized inductive definitions* (**C** 4220) Rep Sem Found Anal I*3.1-3.25
◇ F35 F50 ◇ ID 49824

KREISEL, G. [1963] *Introduction (to volume 1)* (**C** 4220) Rep Sem Found Anal 1*0.1-0.22
◇ F35 F50 ◇ ID 49828

KREISEL, G. [1963] *Introduction (to volume 2)* (**C** 4220) Rep Sem Found Anal 2*0.1-0.76
◇ F35 ◇ ID 49829

KREISEL, G. (ED.) [1963] *Reports of the seminar on foundations of analysis* (1111) Preprints, Manuscr., Techn. Reports etc. 171pp
◇ F35 ◇ REM Mimeographed, Stanford University • ID 21397

KREISEL, G. [1963] *Theory of free choice sequences of natural numbers* (**C** 4220) Rep Sem Found Anal I*4.1-4.31
◇ F35 F50 ◇ ID 49825

KREISEL, G. [1965] *Mathematical logic* (**C** 1602) Lect on Modern Math 3*95-195
• TRANSL [1965] (**C** 4652) Lect on Modern Math (Japanese) 128-291
◇ B98 F10 F35 F50 F98 ◇ REV MR 31#2124 Zbl 147.247 JSL 32.419 • ID 27577

KREISEL, G. [1966] see HOWARD, W.A.

KREISEL, G. [1967] *Informal rigour and completeness proofs* (**P** 2268) Int Colloq Philos of Sci;1965 London 1*138-186
• TRANSL [1978] (**C** 4649) Paradiso di Cantor 59-93 • REPR [1969] (**C** 0569) Phil of Math Oxford Readings 78-94
◇ A05 B30 C07 E30 E50 F50 ◇ ID 07533

KREISEL, G. [1968] *A survey of proof theory* (**J** 0036) J Symb Logic 33*321-388
• TRANSL [1981] (**C** 4648) Kreisel: Issl Teor Dokazat 9-113
◇ F98 ◇ REV MR 43 # 7295 Zbl 177.10 JSL 40.498 • REM Part I. Part II 1971 • ID 07541

KREISEL, G. [1968] *Foundations of mathematics: 1900-1950* (**C** 4655) Sci Thought 1900-1960 4-13
◇ A05 A10 F98 ◇ ID 48847

KREISEL, G. [1968] *Functions, ordinals, species* (**P** 0627) Int Congr Log, Meth & Phil of Sci (3,Proc);1967 Amsterdam 145-159
◇ F10 F35 F50 ◇ REV MR 40 # 1262 Zbl 187.265 • ID 07536

KREISEL, G. [1968] *Lawless sequences of natural numbers* (**J** 0020) Compos Math 20*222-248
• REPR [1968] (**C** 0727) Logic Found of Math (Heyting) 222-248
◇ F50 ◇ REV MR 37 # 2577 JSL 40.500 • ID 07537

KREISEL, G. & LEVY, A. [1968] *Reflection principles and their use for establishing the complexity of axiomatic systems* (**J** 0068) Z Math Logik Grundlagen Math 14*97-142
◇ E30 E47 F25 F30 F35 F50 ◇ REV MR 37 # 3914 Zbl 167.13 JSL 36.529 • ID 07538

KREISEL, G. [1969] *Appendix II. Category theory and the foundations of mathematics* (**C** 3599) Midwest Category Sem Rep 3 233-245
◇ A05 F99 G30 ◇ ID 48843

KREISEL, G. [1969] *Axiomatisations of nonstandard analysis that are conservative extensions of formal systems for classical standard analysis* (**P** 0649) Appl Model Th to Algeb, Anal & Probab;1967 Pasadena 93-106
◇ B30 F35 H05 ◇ REV MR 39 # 50 Zbl 188.322 • ID 26648

KREISEL, G. & NEWMAN, M.H.A. [1969] *Luitzen Egbertus Jan Brouwer (1881 - 1966)* (**J** 0480) Roy Soc Bibl Mem Fellows 15*39-68
◇ A05 A10 F99 ◇ REV MR 40 # 5403 • ID 07542

KREISEL, G. [1970] *Church's thesis: a kind of reducibility axiom for constructive mathematics* (**P** 0603) Intuitionism & Proof Th;1968 Buffalo 121-150
◇ F50 ◇ REV MR 43 # 4629 Zbl 199.300 JSL 40.500 • ID 07545

KREISEL, G. & TROELSTRA, A.S. [1970] *Formal systems for some branches of intuitionistic analysis* (**J** 0007) Ann Math Logic 1*229-387 • ERR/ADD ibid 3*437-439
◇ F35 F50 ◇ REV MR 41 # 8210 MR 45 # 1729 Zbl 211.11 Zbl 231 # 02041 • ID 07546

KREISEL, G. [1970] *Hilbert's programme and the search for automatic proof procedures* (**P** 0625) Symp Autom Demonst;1968 Versailles 128-146
◇ A05 B35 F99 ◇ REV MR 44 # 1571 Zbl 206.277 • ID 48003

KREISEL, G. [1970] *Principles of proof and ordinals implicit in given concepts* (**P** 0603) Intuitionism & Proof Th;1968 Buffalo 489-516
◇ A05 F15 ◇ REV MR 44 # 3834 Zbl 208.9 JSL 40.501 • ID 07544

KREISEL, G. [1970] *The formalist-positivist doctrine of mathematical precision in the light of experience* (**J** 0456) L'Age Science 17-46
• TRANSL [1974] (**C** 3214) Mathematiker ueber Math 65-123
◇ A05 F99 ◇ REV Zbl 337 # 02004 • ID 63135

KREISEL, G. [1971] *A survey of proof theory II* (**P** 0604) Scand Logic Symp (2);1970 Oslo 109-170
• TRANSL [1981] (**C** 4648) Kreisel: Issl Teor Dokazat 114-182
◇ F50 F98 ◇ REV MR 54 # 12493 Zbl 227 # 02013 JSL 40.503 • REM Part I 1968 • ID 07549

KREISEL, G. [1971] *The collected works of Gerhard Gentzen* (**J** 0301) J Phil 68*238-265
◇ A10 F99 ◇ ID 48861

KREISEL, G. [1972] *Which number theoretic problems can be solved in recursive progressions on Π^1_1-paths through* $\mathcal{O}$ *?* (**J** 0036) J Symb Logic 37*311-334
◇ A05 D20 D55 F15 F30 F99 ◇ REV MR 51 # 5273 Zbl 255 # 02048 • ID 07553

KREISEL, G. [1973] *Perspectives in the philosophy of pure mathematics* (**P** 0793) Int Congr Log, Meth & Phil of Sci (4,Proc);1971 Bucharest 255-277
◇ A05 F99 ◇ REV MR 56 # 11733 • ID 75082

KREISEL, G. & TAKEUTI, G. [1974] *Formally self-referential propositions for cut free classical analysis and related systems* (**J** 0202) Diss Math (Warsaw) 118*55pp
◇ F05 F35 ◇ REV MR 52 # 5374 Zbl 336 # 02027 • ID 18236

KREISEL, G. [1975] *Observations on a recent generalization of completeness theorems due to Schuette* (**P** 1440) ⊧ ISILC Proof Th Symp (Schuette);1974 Kiel 164-181
◇ A05 C07 C57 F05 F20 F35 F50 ◇ REV MR 54 # 73 Zbl 324 # 02020 • ID 23955

KREISEL, G. & MINTS, G.E. & SIMPSON, S.G. [1975] *The use of abstract language in elementary metamathematics: Some pedagogic examples* (**C** 0758) Logic Colloq Boston 1972-73 38-131
◇ A05 C07 C57 C75 F05 F07 F20 F50 ◇ REV Zbl 318 # 02003 • ID 27698

KREISEL, G. [1976] *What have we learnt from Hilbert's second problem?* (**P** 2957) Math Dev from Hilbert Probl;1974 DeKalb 93-130
◇ A05 F98 ◇ REV MR 55 # 7745 Zbl 366 # 02018 JSL 44.116 • ID 51101

KREISEL, G. [1976] *Wie die Beweistheorie zu ihren Ordinalzahlen kam und kommt* (**J** 0157) Jbuchber Dtsch Math-Ver 78*177-223
• TRANSL [1981] (**C** 4648) Kreisel: Issl Teor Dokazat 183-238
◇ F05 F15 F30 F35 ◇ REV MR 56 # 11765 Zbl 359 # 02024 • ID 50572

KREISEL, G. [1977] *From foundations to science: Justifying and unwinding proofs* (**P** 3269) Set Th Found Math (Kurepa);1977 Beograd 63-72
◇ A05 B35 F07 ◇ REV MR 58 # 16100 Zbl 414 # 03033 • ID 53079

KREISEL, G. [1977] *On the kind of data needed for a theory of proofs* (**P** 1075) Logic Colloq;1976 Oxford 111-128
• TRANSL [1981] (**C** 4648) Kreisel: Issl Teor Dokazat 257-287
◇ A05 A10 F07 ◇ REV MR 58#21397 Zbl 422#03029
• ID 16615

KREISEL, G. [1977] *Review of " L.E.J.Brouwer collected works, Volume I, Philosophy and foundations of mathematics"* (**J** 0015) Bull Amer Math Soc 83*86-93
◇ A05 A10 E30 E70 F99 ◇ ID 48862

KREISEL, G. [1977] *Some uses of proof theory for finding computer programs* (**P** 1729) Colloq Int Log;1975 Clermont-Ferrand 123-134
• TRANSL [1981] (**C** 4648) Kreisel: Issl Teor Dokazat 239-256
◇ B75 F05 F20 ◇ REV MR 81h:03118 Zbl 439#03044
• ID 56035

KREISEL, G. [1978] *The motto of 'Philosophical Investigations' and the philosophy of proofs and rules* (**J** 4128) Grazer Philos Stud 6*13-38
◇ A05 F99 ◇ ID 48846

KREISEL, G. [1978] *Wittgenstein's lectures on the foundations of mathematics, Cambridge 1939* (**J** 0015) Bull Amer Math Soc 84*79-90
◇ A05 F99 ◇ ID 48863

KREISEL, G. [1979] *Comment on Zinoviev's paper* (**J** 0079) Logique & Anal, NS 22*263-264
◇ F30 ◇ REV Zbl 448#03043 • REM The article was published by Zinoviev ibid 22*243-261 • ID 56643

KREISEL, G. [1979] *Formal rules and questions of justifying mathematical practice* (**C** 2537) Spez Wissenschaftsth 2*99-130
◇ A05 F99 ◇ ID 48854

KREISEL, G. [1979] *Some facts from the theory of proofs and some fictions from general proof theory* (**P** 1705) Scand Logic Symp (4);1976 Jyvaeskylae 3-23
◇ A05 F05 F15 ◇ REV MR 80h:03012 Zbl 406#03068
• ID 56151

KREISEL, G. [1980] *Finitismus* (**C** 4082) Handb Wiss Begriffe 233-234
◇ A05 F65 ◇ ID 48848

KREISEL, G. [1980] *Widerspruchsfreiheit* (**C** 4082) Handb Wiss Begriffe 721-722
◇ A05 F99 ◇ ID 48852

KREISEL, G. [1981] *Constructivist approaches to modern logic* (**C** 2617) Modern Log Survey 67-91
◇ A05 F99 ◇ REV MR 82f:03002 Zbl 464#03001
• ID 42765

KREISEL, G. [1981] *Investigations in proof theory (Russian)* (**X** 0885) Mir: Moskva 289pp
◇ F96 ◇ REV MR 83e:03088 Zbl 501#03038 • REM Transl. of several papers from the English • ID 35242

KREISEL, G. [1981] *Kurt Goedel* (**J** 0480) Roy Soc Bibl Mem Fellows 149-224 • ERR/ADD ibid 27*697 & 28*719
◇ A10 E99 F99 ◇ ID 48855

KREISEL, G. [1981] *Monadic operators defined by means of propositional quantification in intuitionistic logic* (**J** 0302) Rep Math Logic, Krakow & Katowice 12*9-15
◇ F50 ◇ REV MR 83a:03011 Zbl 464#03051 • ID 35056

KREISEL, G. [1981] *Zur Bewertung mathematischer Definitionen* (**C** 4092) Philos als Wiss 185-209
◇ A05 F99 ◇ ID 48856

KREISEL, G. [1982] *Brouwer's Cambridge lectures on intuitionism* (**J** 4656) Canad Phil Rev 2*249-251
◇ A05 A10 F55 ◇ ID 48864

KREISEL, G. & MACINTYRE, A. [1982] *Constructive logic versus algebraization I* (**P** 3638) Brouwer Centenary Symp;1981 Noordwijkerhout 217-260
◇ A05 F05 F50 ◇ REV MR 85j:03108 Zbl 522#03046
• ID 37795

KREISEL, G. [1982] *Finiteness theorems in arithmetic: an application of Herbrand's theorem for Σ_2-formulas* (**P** 3708) Herbrand Symp Logic Colloq;1981 Marseille 39-55
◇ F05 F30 ◇ REV MR 86b:03076 Zbl 499#03045
• ID 38127

KREISEL, G. [1983] *Einige Erlaeuterungen zu Wittgensteins Kummer mit Hilbert und Goedel* (**P** 4381) Int Wittgenstein Symp (7);1982 Kirchberg 295-303
◇ A05 F99 ◇ ID 48857

KREISEL, G. [1983] *Wittgenstein on rules and private language* (**J** 4656) Canad Phil Rev 3*287-289
◇ A05 F99 ◇ ID 48865

KREISEL, G. [1984] *Frege's foundations and intuitionistic logic* (**J** 0320) Monist 67*71-91
◇ A05 F50 ◇ ID 48842

KREISEL, G. [1985] *Proof theory and the synthesis of programs: Potential and limitations* (**P** 4601) EUROCAL;1985 Linz 1*136-150
◇ A05 F99 ◇ REV Zbl 574#03044 • ID 48832

KREISEL, G. see Vol. I, II, III, IV, V for further entries

KREITZ, C. & WEIHRAUCH, K. [1984] *A unified approach to constructive and recursive analysis* (**P** 2153) Logic Colloq;1983 Aachen 2*259-278
◇ F60 ◇ ID 43008

KREITZ, C. see Vol. IV for further entries

KREJNOVICH, V.YA. [1974] *Constructivization of the concepts of ε entropy and ε capacity (Russian) (English summary)* (**S** 0228) Zap Nauch Sem Leningrad Otd Mat Inst Steklov 40*38-44,156
• TRANSL [1977] (**J** 1531) J Sov Math 8*271-276
◇ F60 ◇ REV MR 57#78 Zbl 357#02031 • ID 50379

KREJNOVICH, V.YA. [1974] *From what does the law of the excluded middle follow? (Russian) (English summary)* (**S** 0228) Zap Nauch Sem Leningrad Otd Mat Inst Steklov 40*30-37,156
• TRANSL [1977] (**J** 1531) J Sov Math 8*266-271
◇ F35 F50 ◇ REV MR 52#5369 Zbl 361#02045
• ID 18235

KREJNOVICH, V.YA. [1981] see KOSHELEVA, O.M.

KREJNOVICH, V.YA. see Vol. I, III, V for further entries

KRIPKE, S.A. [1962] *"Flexible" predicates of formal number theory* (**J** 0053) Proc Amer Math Soc 13*647-650
◇ C90 F30 ◇ REV MR 25#3827 Zbl 109.9 • ID 07556

KRIPKE, S.A. [1965] *Semantical analysis of intuitionistic logic I* (**P** 0688) Logic Colloq;1963 Oxford 92-130
◇ B55 C90 F50 ◇ REV MR 34#1184 Zbl 137.7 JSL 35.330 • REM Part II never appeared • ID 07557

KRIPKE, S.A. & POUR-EL, M.B. [1967] *Deduction-preserving "recursive isomorphisms" between theories* (J 0015) Bull Amer Math Soc 73*145-148
◇ B30 D25 D35 F30 ◇ REV MR 35 # 6548 Zbl 174.20
• ID 31215

KRIPKE, S.A. & POUR-EL, M.B. [1967] *Deduction-preserving "recursive isomorphisms" between theories* (J 0027) Fund Math 61*141-163
◇ B30 D25 D35 F30 ◇ REV MR 40 # 5447 Zbl 174.20
• ID 31216

KRIPKE, S.A. see Vol. II, III, IV, V for further entries

KRISHNAMURTHY, B. [1984] see KAPUR, D.

KRISHNAMURTHY, B. see Vol. I for further entries

KRIVTSOV, V.N. [1984] *A formal system of negationless arithmetic is conservative with respect to Heyting arithmetic (Russian)* (J 0087) Mat Zametki (Akad Nauk SSSR) 36*583-592
◇ F30 F50 ◇ ID 44479

KRIVTSOV, V.N. [1984] *Deductive possiblitities of intuitionistic analysis without negation (Russian)* (J 0023) Dokl Akad Nauk SSSR 274*786-790
• TRANSL [1984] (J 0062) Sov Math, Dokl 29*94-98
◇ F35 F50 ◇ REV MR 85j:03107 • ID 45343

KRIVTSOV, V.N. [1984] *Imbedding of the intuitionistic theory of types into the negationless intuitionistic theory of types (Russian)* (J 0023) Dokl Akad Nauk SSSR 277*529-533
• TRANSL [1984] (J 0062) Sov Math, Dokl 30*111-114
◇ F35 F50 ◇ REV MR 86b:03077 • ID 45798

KRIVTSOV, V.N. see Vol. I for further entries

KROEGER, F. [1980] *Infinite proof rules for loops* (J 1431) Acta Inf 14*371-389
◇ B75 F07 ◇ REV MR 82d:68011 Zbl 432 # 68007
• ID 69663

KROEGER, F. see Vol. I, II for further entries

KROL', M.D. [1976] *The topological models of intuitionistic analysis. One counter-example (Russian)* (J 0087) Mat Zametki (Akad Nauk SSSR) 19*859-862
• TRANSL [1976] (J 1044) Math Notes, Acad Sci USSR 19*503-504
◇ C90 F35 F50 ◇ REV MR 55 # 2524 Zbl 355 # 02025 JSL 46.660 • ID 50223

KROL', M.D. [1977] *Disjunctive and existential properties of intuitionistic analysis with Kripke's scheme (Russian)* (J 0023) Dokl Akad Nauk SSSR 234*750-753
• TRANSL [1977] (J 0062) Sov Math, Dokl 18*755-758
◇ C90 F35 F50 ◇ REV MR 57 # 9494 Zbl 384 # 03043
• ID 52085

KROL', M.D. [1978] *A topological model for intuitionistic analysis with Kripke's scheme* (J 0068) Z Math Logik Grundlagen Math 24*427-436
◇ C90 F35 F50 ◇ REV MR 80b:03096 Zbl 418 # 03039 JSL 46.660 • ID 53324

KROL', M.D. [1978] *Distinct variants of Kripke's scheme in intuitionistic analysis (Russian)* (J 0023) Dokl Akad Nauk SSSR 239*1048-1051
• TRANSL [1978] (J 0062) Sov Math, Dokl 19*474-477
◇ C90 F35 F50 ◇ REV MR 58 # 5096 Zbl 397 # 03037 JSL 46.660 • ID 52703

KROL', M.D. [1983] *Various forms of the continuity principle (Russian)* (J 0023) Dokl Akad Nauk SSSR 271*33-36
• TRANSL [1983] (J 0062) Sov Math, Dokl 28*27-30
◇ F30 F50 ◇ REV MR 84m:03094 Zbl 549 # 03052
• ID 35789

KRONECKER, L. [1887] *Ueber den Zahlbegriff* (J 0127) J Reine Angew Math 101*337-355
• REPR [1899] (C 4394) Kroneckers Werke 3*1,249-274
◇ B28 F55 F65 ◇ REV FdM 19.63 • ID 37377

KRYL, R. [1974] *The constructive analogue of a certain theorem of Luzin (Russian)* (J 0140) Comm Math Univ Carolinae (Prague) 15*465-480
◇ F60 ◇ REV MR 51 # 3367 Zbl 301 # 02032 • ID 17369

KRYL, R. [1978] see DEMUTH, O.

KRYNICKI, M. [1985] *Interpretations in nonelementary languages* (P 4310) Easter Conf on Model Th (3);1985 Gross Koeris 168-182
◇ C80 C95 F25 ◇ ID 49048

KRYNICKI, M. see Vol. I, III, IV, V for further entries

KUBINSKI, T. [1963] *A proof of consistency of Borkowski's logical system containing Peano's arithmetic (Polish and Russian summaries)* (J 0063) Studia Logica 14*197-225
◇ B15 F25 F30 ◇ REV MR 32 # 3990 Zbl 292 # 02029
• ID 07595

KUBINSKI, T. & MIKOLAJEWICZ, B. [1984] *Finite arithmetics, their categoricity and trees (Polish)* (J 1093) Ruch Filoz 41/1*27-31
◇ F30 ◇ ID 48503

KUBINSKI, T. see Vol. I, II, IV, V for further entries

KUCERA, A. [1970] *Weak convergence in constructive mathematics (Russian)* (J 0140) Comm Math Univ Carolinae (Prague) 11*285-308
◇ F60 ◇ REV MR 42 # 828 Zbl 223 # 02025 • ID 07599

KUCERA, A. [1971] *Sufficient conditions for the normability of linear operators in constructive mathematics (Russian)* (J 0140) Comm Math Univ Carolinae (Prague) 12*377-399
◇ F60 ◇ REV MR 44 # 5800 Zbl 235 # 02028 • ID 27820

KUCERA, A. [1974] *Locally convex topologies of finite-dimensional constructive spaces (Russian)* (C 1450) Teor Algor & Mat Logika (Markov) 71-86,214
◇ F60 ◇ REV MR 53 # 12900 Zbl 309 # 02031 • ID 23152

KUCERA, A. [1977] *The algorithmic nonapproximability of the least upper bounds of constructive pseudosections (Russian)* (J 0140) Comm Math Univ Carolinae (Prague) 18*445-453
◇ F60 ◇ REV MR 57 # 9508 Zbl 367 # 02015 • ID 51175

KUCERA, A. & KUSHNER, B.A. [1978] *A type of recursive isomorphism of certain concepts of constructive analysis (Russian)* (J 0140) Comm Math Univ Carolinae (Prague) 19*97-105
◇ D55 F60 ◇ REV MR 58 # 173 Zbl 398 # 03049
• ID 52777

KUCERA, A. [1978] see DEMUTH, O.

KUCERA, A. [1979] see DEMUTH, O.

KUCERA, A. [1985] *Measure, Π^0_1-classes and complete extensions of PA* (P 3342) Rec Th Week;1984 Oberwolfach 245-259
◇ D30 E15 E75 F30 ◇ ID 45308

KUCERA, A. see Vol. IV for further entries

KUDAJBERGENOV, K.ZH. [1984] *Autostability and extensions of constructivizations (Russian)* (**J** 0092) Sib Mat Zh 25/5*72-78
• TRANSL [1984] (**J** 0475) Sib Math J 25*743-749
◇ C50 C57 F60 ◇ REV MR 86b:03040 • ID 41816

KUDAJBERGENOV, K.ZH. [1984] *Constructivizability of a prime model (Russian)* (**J** 0092) Sib Mat Zh 25/4*93-98
• TRANSL [1984] (**J** 0475) Sib Math J 25*584-588
◇ C50 C57 F60 ◇ REV MR 86d:03030 • ID 41871

KUDAJBERGENOV, K.ZH. see Vol. III, IV for further entries

KURATA, R. [1965] *Recursive progression of intuitionistic number theories* (**J** 0090) J Math Soc Japan 17*140-166
◇ F15 F30 F50 ◇ REV MR 32 # 2327 Zbl 134.246 JSL 38.332 • ID 07650

KURATA, R. [1981] *The reflection principle,transfinite induction, and the Paris-Harrington principle (Japanese)* (**P** 4153) B-Val Anal & Nonstand Anal;1981 Kyoto 1-14
◇ C62 F30 ◇ ID 47747

KURATA, R. [1984] *Paris-Harrington theory and reflection principles* (**P** 3668) Log & Founds of Math;1983 Kyoto 123-131
• REPR [1984] (**J** 3940) Saitama Math J 2*33-45
◇ C62 F30 ◇ REV MR 86g:03095 Zbl 552 # 03038
• ID 42937

KURATA, R. see Vol. III, V for further entries

KUR'EROV, YU.N. [1979] *Normal form of mutual-absorption tactics (Russian)* (**J** 0040) Kibernetika, Akad Nauk Ukr SSR 1979/6*138-141
• TRANSL [1979] (**J** 0021) Cybernetics 15*930-935
◇ B35 F05 ◇ REV MR 81m:68077 Zbl 508 # 68059
• ID 81977

KUR'EROV, YU.N. see Vol. I, IV for further entries

KURKA, P. [1972] *The Heyting doctrines* (**J** 0140) Comm Math Univ Carolinae (Prague) 13*643-657
◇ F50 G30 ◇ REV MR 47 # 3480 Zbl 272 # 02050
• ID 07677

KURKA, P. see Vol. II, III for further entries

KURODA, S. [1948] *On the logic of Aristotle and the logic of Brouwer (Japanese)* (**J** 4439) Kagaku 18*2-10
◇ A05 A10 F07 F50 ◇ REV JSL 21.196 • ID 42620

KURODA, S. [1951] *Intuitionistische Untersuchungen der formalistischen Logik* (**J** 0111) Nagoya Math J 2*35-47
◇ F50 ◇ REV MR 13.4 Zbl 42.6 JSL 21.196 • ID 07680

KURODA, S. [1957] *On the intuitionistic and formalistic theory of real numbers* (**P** 0575) Int Congr Math (II, 7);1954 Amsterdam 1*538-539
◇ F55 F65 ◇ ID 29461

KURODA, S. [1958] *An investigation on the logical structure of mathematics. I. A logical system* (**J** 0107) Abh Math Sem Univ Hamburg 22*242-266
◇ B15 B28 E70 F07 ◇ REV MR 23 # A48 Zbl 81.247
• REM Part II 1959 • ID 07681

KURODA, S. [1958] *An investigation on the logical structure of mathematics. III. Fundamental deductions. IV. Compendium for deductions* (**J** 0111) Nagoya Math J 13*21-52,123-133
◇ B15 B28 E70 F07 ◇ REV MR 23 # A50 MR 23 # A51 Zbl 87.245 • REM Part II 1959. Part V 1958 • ID 07683

KURODA, S. [1959] *An investigation on the logical structure of mathematics. II. Transformation of the proof* (**J** 0107) Abh Math Sem Univ Hamburg 23*201-227
◇ B28 B30 F05 ◇ REV MR 23 # A49 Zbl 87.245 • REM Parts I,III,IV 1958 • ID 25622

KURODA, S. [1959] *An investigation on the logical structure of mathematics. VI. Consistent V-systems T(V) . VII. Set-theoretical contradictions* (**J** 0111) Nagoya Math J 14*95-107,109-127 • ERR/ADD to part XII
◇ B28 B30 E70 F30 ◇ REV MR 23 # A53 MR 23 # A54 Zbl 94.7 • REM Part V 1958. Part VIII 1959 • ID 25455

KURODA, S. [1959] *An investigation on the logical structure of mathematics. VIII. Consistency of the natural-number theory* $T_1(N)$ (**J** 0111) Nagoya Math J 14*129-158
◇ B28 B30 F30 ◇ REV MR 23 # A55a Zbl 94.7 • REM Parts VI,VII,XI,X 1959 • ID 25434

KURODA, S. [1959] *An investigation on the logical struture of mathematics. IX. Deductions in the natural-number theory* $T_1(N)$*. X. Concepts and sets* (**J** 1770) Osaka Math J 11*7-42,213-248
◇ B15 B28 B30 F30 ◇ REV MR 23 # A55b MR 23 # A56 Zbl 87.245 Zbl 92.4 • REM Part VIII 1959
• ID 15552

KURODA, S. see Vol. I, V for further entries

KUSHNER, B.A. [1964] *Riemann integration in constructive analysis (Russian)* (**J** 0023) Dokl Akad Nauk SSSR 156*255-257
• TRANSL [1964] (**J** 0062) Sov Math, Dokl 5*628-630
◇ F60 ◇ REV MR 29 # 22 Zbl 192.59 • ID 19150

KUSHNER, B.A. [1965] *Constructive theory of the Riemann integral (Russian)* (**J** 0023) Dokl Akad Nauk SSSR 165*1238-1240
• TRANSL [1965] (**J** 0062) Sov Math, Dokl 6*1584-1587
◇ F60 ◇ REV MR 33 # 1236 Zbl 192.60 • ID 07693

KUSHNER, B.A. [1965] *On the existence of unbounded analytic constructive functions (Russian)* (**J** 0023) Dokl Akad Nauk SSSR 160*29-31
• TRANSL [1965] (**J** 0062) Sov Math, Dokl 6*26-28
◇ F60 ◇ REV MR 31 # 55 Zbl 192.62 • ID 07694

KUSHNER, B.A. [1966] *Certain properties of quasinumbers and of operators from quasinumbers into quasinumbers (Russian)* (**J** 0023) Dokl Akad Nauk SSSR 171*275-277
• TRANSL [1966] (**J** 0062) Sov Math, Dokl 7*1467-1470
◇ F60 ◇ REV MR 34 # 2450 Zbl 283 # 02031 • ID 07695

KUSHNER, B.A. [1967] *Certain relations among the properties of constructive functions and operators from quasinumbers into quasinumbers (Russian)* (**J** 0023) Dokl Akad Nauk SSSR 177*29-32
• TRANSL [1967] (**J** 0062) Sov Math, Dokl 8*1358-1361
◇ F60 ◇ REV MR 36 # 4993 Zbl 175.274 • ID 14714

KUSHNER, B.A. [1967] *On constructive antiderivatives (Russian)* (**J** 0087) Mat Zametki (Akad Nauk SSSR) 2*157-166
◇ F60 ◇ REV MR 36 # 2499 • ID 07696

KUSHNER, B.A. [1968] *A remark on the domains of definition of constructive functions (Russian)* (**S** 0228) Zap Nauch Sem Leningrad Otd Mat Inst Steklov 8*103-106
• TRANSL [1968] (**J** 0521) Semin Math, Inst Steklov 8*47-48
◇ F60 ◇ REV MR 45 # 61 Zbl 201.13 • ID 07697

KUSHNER, B.A. [1968] *Some examples of quasidense but not dense sets of duplexes (Russian)* (**S** 0228) Zap Nauch Sem Leningrad Otd Mat Inst Steklov 8*95-102
• TRANSL [1968] (**J** 0521) Semin Math, Inst Steklov 8*43-46
◇ F60 ◇ REV MR 46 # 1570 Zbl 198.325 • ID 75207

KUSHNER, B.A. & TSEJTIN, G.S. [1968] *Some properties of F-numbers (Russian)* (**S** 0228) Zap Nauch Sem Leningrad Otd Mat Inst Steklov 8*107-120
• TRANSL [1968] (**J** 0521) Semin Math, Inst Steklov 8*49-55
◇ F60 ◇ REV MR 39 # 61 Zbl 267 # 02023 • ID 28594

KUSHNER, B.A. [1969] see KANOVICH, M.I.

KUSHNER, B.A. [1970] *Some mass problems connected with the integration of constructive functions (Russian)* (**S** 0066) Tr Mat Inst Steklov 113*39-72
• TRANSL [1970] (**S** 0055) Proc Steklov Inst Math 113*42-83
◇ D30 F60 ◇ REV MR 44 # 6917 Zbl 285 # 02033
• ID 07698

KUSHNER, B.A. [1973] *A certain problem of Mostowski (Russian)* (**S** 0554) Issl Teor Algor & Mat Logik (Moskva) 1*257-261
◇ D20 F60 ◇ REV MR 49 # 28 Zbl 285 # 02030 • ID 07702

KUSHNER, B.A. [1973] *Computationally complex real numbers (Russian)* (**J** 0068) Z Math Logik Grundlagen Math 19*447-452
◇ D20 F60 ◇ REV MR 49 # 4764 Zbl 296 # 02019
• ID 07699

KUSHNER, B.A. [1973] *Continuity theorems for some types of computable operators (Russian)* (**J** 0023) Dokl Akad Nauk SSSR 208*1031-1034
• TRANSL [1973] (**J** 0062) Sov Math, Dokl 14*221-225
◇ F60 ◇ REV MR 51 # 7839 Zbl 315 # 02036 • ID 18241

KUSHNER, B.A. [1973] *Coverings of separable sets (Russian)* (**S** 0554) Issl Teor Algor & Mat Logik (Moskva) 1*235-246
◇ F60 ◇ REV MR 49 # 2303 Zbl 285 # 02031 • ID 07700

KUSHNER, B.A. [1973] *Lectures on constructive mathematical analysis (Russian)* (**X** 2027) Nauka: Moskva 447pp
• TRANSL [1984] (**X** 0803) Amer Math Soc: Providence v+346pp
◇ F60 F98 ◇ REV MR 52 # 53 MR 86a:03067
Zbl 285 # 02029 Zbl 547 # 03040 • ID 17218

KUSHNER, B.A. [1973] *Two theorems on Riemann-integrable constructive functions (Russian)* (**S** 0554) Issl Teor Algor & Mat Logik (Moskva) 1*247-256
◇ F60 ◇ REV MR 49 # 490 Zbl 285 # 02032 • ID 07701

KUSHNER, B.A. [1974] *A constructive version of Koenig's theorem; functions that are computable in the sense of Markov, Grzegorczyk and Lacombe (Russian)* (**C** 1450) Teor Algor & Mat Logika (Markov) 87-111,215
◇ F60 ◇ REV MR 53 # 10562 Zbl 356 # 02031 • ID 23071

KUSHNER, B.A. [1974] *On a type of computable real functions (Russian)* (**J** 0023) Dokl Akad Nauk SSSR 215*259-262
• TRANSL [1974] (**J** 0062) Sov Math, Dokl 15*466-470
◇ F60 ◇ REV MR 49 # 2301 Zbl 306 # 02028 • ID 07703

KUSHNER, B.A. [1976] *On the computation of isolated roots of constructive functions (Russian)* (**J** 0068) Z Math Logik Grundlagen Math 22*311-332
◇ F60 ◇ REV MR 58 # 21514 Zbl 356 # 02032 • ID 50304

KUSHNER, B.A. [1976] *On Grzegorczyk's theorem on the computability of an isolated extremum (Russian)* (**S** 0554) Issl Teor Algor & Mat Logik (Moskva) 2*112-121
◇ D20 D80 F60 ◇ REV MR 58 # 16209 • ID 75199

KUSHNER, B.A. [1978] see KUCERA, A.

KUSHNER, B.A. [1978] *On some systems of computable real numbers (Russian)* (**C** 3211) Mat Ling & Teor Algor 105-112
◇ D45 F60 ◇ REV Zbl 411 # 03056 • ID 52910

KUSHNER, B.A. [1979] *Segment coverings and the uniform continuity of constructive functions (Russian)* (**S** 0554) Issl Teor Algor & Mat Logik (Moskva) 3*62-69
◇ F60 ◇ REV MR 81g:03071 Zbl 438 # 03056 • ID 55968

KUSHNER, B.A. [1981] *Behavior of the general term of a Specker series (Russian)* (**C** 3747) Mat Log & Mat Lingvistika 112-116
◇ F60 ◇ REV MR 84i:03111 • ID 34589

KUSHNER, B.A. [1981] *On some topological properties of constructive plane curves (Russian)* (**J** 0023) Dokl Akad Nauk SSSR 260*281-283
• TRANSL [1981] (**J** 0062) Sov Math, Dokl 24*258-261
◇ F60 ◇ REV MR 82k:03096 Zbl 487 # 03035 • ID 75196

KUSHNER, B.A. [1982] *Some extensions of Markov's constructive continuum and their applications to the theory of constructive functions* (**P** 3638) Brouwer Centenary Symp;1981 Noordwijkerhout 261-273
◇ F60 ◇ REV MR 85d:03120 Zbl 526 # 03037 • ID 38178

KUSHNER, B.A. [1983] *A class of Specker sequences (Russian)* (**C** 3798) Mat Log, Mat Ling & Teor Algor 62-65
◇ F60 ◇ REV MR 85e:03152 • ID 40766

KUSHNER, B.A. [1985] *Differentiability and uniform continuity of constructive functions (Russian)* (**J** 0023) Dokl Akad Nauk SSSR 281*1314-1316
• TRANSL [1985] (**J** 0062) Sov Math, Dokl 31*433-435
◇ F60 ◇ ID 45635

KUSHNER, B.A. see Vol. IV for further entries

KUTSCHERA VON, F. [1966] *Zur semantischen Begruendung der klassischen und der intuitionistischen Logik* (**J** 0047) Notre Dame J Formal Log 7*20-47
◇ A05 F50 ◇ REV MR 35 # 4088 Zbl 166.257 • ID 07710

KUTSCHERA VON, F. [1968] *Die Vollstaendigkeit des Operatorensystems* $\{\neg, \wedge, \vee, \rightarrow\}$ *fuer die intuitionistische Aussagenlogik im Rahmen der Gentzensemantik* (**J** 0009) Arch Math Logik Grundlagenforsch 11*3-16
◇ F50 ◇ REV MR 38 # 28 Zbl 165.11 • ID 19148

KUTSCHERA VON, F. [1985] *Der Satz vom ausgeschlossenen Dritten. Untersuchungen ueber die Grundlagen der Logik* (**X** 1174) Gruyter: Berlin
◇ A05 F99 ◇ ID 40459

KUTSCHERA VON, F. see Vol. I, II for further entries

KUZICHEV, A.A. & KUZICHEV, A.S. [1980] *On the embedding of formal arithmetic in combinatorially complete systems (Russian)* (**J** 0023) Dokl Akad Nauk SSSR 250*1310-1315
• TRANSL [1980] (**J** 0062) Sov Math, Dokl 21*311-316
◇ B40 F05 F30 ◇ REV MR 82k:03090 Zbl 452 # 03007
• ID 75215

KUZICHEV, A.A. [1984] *Addition of noncombinatory relations to the calculus of* λ*-conversion (Russian)* (**J** 0288) Vest Ser Mat Mekh, Univ Moskva 1984/6*34-36,111
◇ B40 ◇ ID 45355

KUZICHEV, A.A. [1984] *On the representability of algebras in* λ*-conversion calculus (Russian)* (**X** 2235) VINITI: Moskva 8364-84
◇ B40 ◇ ID 46842

KUZICHEV, A.A. [1984] *The representation of arithmetic in combinatorially complete systems (Russian)* (**J** 0023) Dokl Akad Nauk SSSR 278*537-541
• TRANSL [1984] (**J** 0062) Sov Math, Dokl 30*417-421
◇ B40 ◇ REV MR 86c:03011 • ID 44066

KUZICHEV, A.A. [1985] *On a criterion of λ-representability (Russian)* (**X** 2235) VINITI: Moskva 1548-85
◇ B40 ◇ ID 46697

KUZICHEV, A.S. & NIKOGOSOV, S.L. [1970] *Implicative systems of combinatory logic (Russian) (English summary)* (**J** 0288) Vest Ser Mat Mekh, Univ Moskva 25/3*50-55
• TRANSL [1970] (**J** 0510) Moscow Univ Math Bull 25/3-4*39-43
◇ B40 ◇ REV MR 42 # 5799 Zbl 205.14 • ID 07713

KUZICHEV, A.S. [1971] *F^n-systems of combinatory logic. Generalized arithmetic operators (Russian)* (**J** 0023) Dokl Akad Nauk SSSR 198*759-761 • ERR/ADD [1971] (**J** 0062) Sov Math, Dokl 12*iv
• TRANSL [1971] (**J** 0062) Sov Math, Dokl 12*878-880
◇ B40 ◇ REV MR 47 # 1579 Zbl 234 # 02018 • ID 07717

KUZICHEV, A.S. [1971] *Certain properties of Schoenfinkel-Curry combinators (Russian)* (**J** 0484) Komb Analiz Sb Stat 1*105-119
◇ B40 ◇ REV MR 45 # 8529 Zbl 295 # 02016 • ID 07716

KUZICHEV, A.S. & NIKOGOSOV, S.L. [1971] *Implicative systems of combinatory logic with a universal quantifier (Russian)* (**J** 0288) Vest Ser Mat Mekh, Univ Moskva 26/4*69-74
• TRANSL [1971] (**J** 0510) Moscow Univ Math Bull 26/3-4*121-125
◇ B40 ◇ REV MR 45 # 3210 Zbl 219 # 02016 • ID 27737

KUZICHEV, A.S. [1972] *The functionality operator in implicative systems of combinatory logic (Russian)* (**J** 0288) Vest Ser Mat Mekh, Univ Moskva 27/1*37-43
• TRANSL [1972] (**J** 0510) Moscow Univ Math Bull 27/1-2*26-31
◇ B40 ◇ REV MR 45 # 69 Zbl 229 # 02026 • ID 28958

KUZICHEV, A.S. [1973] *A system of λ-conversion with a deductive formal implication operator (Russian)* (**J** 0023) Dokl Akad Nauk SSSR 212*1290-1292
• TRANSL [1973] (**J** 0062) Sov Math, Dokl 14*1584-1587
◇ B40 ◇ REV MR 49 # 2275 Zbl 295 # 02017 • ID 07722

KUZICHEV, A.S. [1973] *Consistent extensions of pure combinatory logic (Russian) (English summary)* (**J** 0288) Vest Ser Mat Mekh, Univ Moskva 28/3*76-81
• TRANSL [1973] (**J** 0510) Moscow Univ Math Bull 28/3-4*66-72
◇ B40 ◇ REV MR 49 # 2274 Zbl 269 # 02006 • ID 07720

KUZICHEV, A.S. [1973] *Deductive-combinatorial construction of the theory of functionality (Russian)* (**J** 0023) Dokl Akad Nauk SSSR 209*541-543
• TRANSL [1973] (**J** 0062) Sov Math, Dokl 14*440-443
◇ B40 ◇ REV MR 48 # 1890 Zbl 289 # 02017 • ID 07723

KUZICHEV, A.S. [1973] *The subject and the methods of combinatory logic (Russian)* (**S** 2337) Istor Metodol Estest Nauk (Moskva) 14*131-141
◇ B40 ◇ REV MR 58 # 21488 Zbl 272 # 02043 • ID 63228

KUZICHEV, A.S. [1974] *Das Prinzip der kombinatorischen Vollstaendigkeit in der mathematischen Logik (Russian)* (**S** 2337) Istor Metodol Estest Nauk (Moskva) 16*106-127
◇ B40 ◇ REV Zbl 322 # 02025 • ID 63229

KUZICHEV, A.S. [1974] *Deductive operators of combinatory logic (Russian) (English summary)* (**J** 0288) Vest Ser Mat Mekh, Univ Moskva 29/3*13-21
• TRANSL [1974] (**J** 0510) Moscow Univ Math Bull 29/3-4/*8-14
◇ B40 ◇ REV MR 49 # 8820 Zbl 298 # 02016 • ID 07724

KUZICHEV, A.S. [1974] *On the expressive potentialities of deductive systems of λ-conversion and combinatory logic (Russian) (English summary)* (**J** 0288) Vest Ser Mat Mekh, Univ Moskva 29/6*19-26
• TRANSL [1974] (**J** 0510) Moscow Univ Math Bull 29/5-6*58-64
◇ B40 ◇ REV MR 51 # 5267 Zbl 298 # 02017 • ID 17452

KUZICHEV, A.S. [1976] *Combinatorially complete systems with operators $\Xi, F, Q, \Pi, \exists, P, \neg, \&, \vee, \equiv$ (Russian) (English summary)* (**J** 0288) Vest Ser Mat Mekh, Univ Moskva 31/6*53-59
• TRANSL [1976] (**J** 0510) Moscow Univ Math Bull 31/5-6*117-122
◇ B40 F05 ◇ REV MR 57 # 15982a Zbl 358 # 02028
• ID 50482

KUZICHEV, A.S. [1976] *Substitution operation in systems with unrestricted principle of combinatorial completeness (Russian) (English summary)* (**J** 0288) Vest Ser Mat Mekh, Univ Moskva 31/5*56-62
• TRANSL [1976] (**J** 0510) Moscow Univ Math Bull 31/5-6*44-49
◇ B40 ◇ REV MR 54 # 12489 Zbl 346 # 02014 • ID 63232

KUZICHEV, A.S. [1977] *A system of λ-conversion with logical operators and an equality operator (Russian)* (**J** 0023) Dokl Akad Nauk SSSR 236*796-799
• TRANSL [1977] (**J** 0062) Sov Math, Dokl 18*1268-1272
◇ B10 B40 F05 ◇ REV MR 57 # 15983 Zbl 395 # 03009
• ID 52560

KUZICHEV, A.S. [1977] *Cut theorem for R-theories in combinatorially complete systems (Russian) (English summary)* (**J** 0288) Vest Ser Mat Mekh, Univ Moskva 1977/1*62-67
• TRANSL [1977] (**J** 0510) Moscow Univ Math Bull 32/1*53-57
◇ B40 F05 ◇ REV MR 57 # 15982b Zbl 363 # 02030
• ID 50857

KUZICHEV, A.S. [1977] *Formal arithmetic in the $\mathfrak{A}$-system of λ-conversion (Russian)* (**J** 0023) Dokl Akad Nauk SSSR 236*1072-1075
• TRANSL [1977] (**J** 0062) Sov Math, Dokl 18*1327-1330
◇ B40 F05 F30 ◇ REV MR 58 # 5082 Zbl 395 # 03010
• ID 52561

KUZICHEV, A.S. [1978] *A theorem on the consistency of formal arithmetic (Russian)* (**J** 0023) Dokl Akad Nauk SSSR 238*269-272
• TRANSL [1978] (**J** 0062) Sov Math, Dokl 19*30-33
◇ B40 F05 F30 ◇ REV MR 58 # 10335 Zbl 395 # 03011
• ID 52562

KUZICHEV, A.S. [1978] *Formal arithmetic in combinatorially complete systems I,II (Russian) (English summary)* (**J** 0288) Vest Ser Mat Mekh, Univ Moskva 1978/2*13-21,1978/3*60-66
• TRANSL [1978] (**J** 0510) Moscow Univ Math Bull 33/2*9-15,33/3*55-60
◇ B40 F05 F30 ◇ REV MR 80a:03019 Zbl 395 # 03012 Zbl 395 # 03013 • ID 52563

KUZICHEV, A.S. [1978] *On the consistency of formal arithmetic (Russian)* (J 0023) Dokl Akad Nauk SSSR 243*1123-1126
• TRANSL [1978] (J 0062) Sov Math, Dokl 19*1486-1490
◇ B40 F30 ◇ REV MR 80b:03090b Zbl 436 # 03007
• ID 55847

KUZICHEV, A.S. [1978] *The theorem on the midsequent in the 𝔄-system of λ-conversion (Russian)* (J 0023) Dokl Akad Nauk SSSR 243*19-21
• TRANSL [1978] (J 0062) Sov Math, Dokl 19*1305-1308
◇ B40 F05 ◇ REV MR 80b:03090a Zbl 436 # 03006
• ID 55846

KUZICHEV, A.S. [1978] *The Curry paradox in combinatorially complete systems (Russian) (English summary)* (S 2337) Istor Metodol Estest Nauk (Moskva) 20*81-98
◇ B40 ◇ REV MR 81a:03014 • ID 75218

KUZICHEV, A.S. [1979] *Classes of objects having normal forms in the system of λ-conversion with logical operators (Russian)* (J 0023) Dokl Akad Nauk SSSR 249*41-45
• TRANSL [1979] (J 0062) Sov Math, Dokl 20*1202-1207
◇ B40 ◇ REV MR 81h:03031 Zbl 452 # 03006 • ID 54070

KUZICHEV, A.S. [1979] *Formale Arithmetik in einem System der λ-Konversion mit logischen Operatoren (Russisch)* (J 0068) Z Math Logik Grundlagen Math 25*299-314
◇ B40 F05 F30 ◇ REV MR 81a:03058 Zbl 423 # 03011
• ID 53523

KUZICHEV, A.S. [1980] *On consistency proofs for first-order theories (Russian) (English summary)* (S 2337) Istor Metodol Estest Nauk (Moskva) 25*97-103
◇ B40 ◇ REV MR 83m:03065 Zbl 474 # 01010 • ID 35464

KUZICHEV, A.S. [1980] see KUZICHEV, A.A.

KUZICHEV, A.S. [1980] *Permissibility of a cut in combinatorially complete systems with equality (Russian) (English summary)* (J 0288) Vest Ser Mat Mekh, Univ Moskva 1980/5*31-34,84
• TRANSL [1980] (J 0510) Moscow Univ Math Bull 35/5*38-41
◇ B40 F05 ◇ REV MR 81k:03059 Zbl 445 # 03004
• ID 56468

KUZICHEV, A.S. [1980] *Sequential systems of λ-conversion and of combinatory logic* (C 3050) Essays Combin Log, Lambda Calc & Formalism (Curry) 141-155
◇ B40 F30 ◇ REV MR 82b:03041 Zbl 469 # 03006
• ID 75240

KUZICHEV, A.S. [1981] *Arithmetic theories constructed on the basis of λ- conversion (Russian)* (J 0023) Dokl Akad Nauk SSSR 261*792-796
• TRANSL [1981] (J 0062) Sov Math, Dokl 24*584-589
◇ B40 F30 ◇ REV MR 83c:03053 Zbl 495 # 03010
• ID 35147

KUZICHEV, A.S. [1982] *Arithmetically consistent λ-theories (Russian)* (J 0023) Dokl Akad Nauk SSSR 262*795-799
• TRANSL [1982] (J 0062) Sov Math, Dokl 25*122-127
◇ B40 F05 F30 ◇ REV MR 83i:03027 Zbl 523 # 03011
• ID 35499

KUZICHEV, A.S. [1982] *Axiomatic theories in combinatorially complete systems (Russian)* (J 0023) Dokl Akad Nauk SSSR 264*538-542
• TRANSL [1982] (J 0062) Sov Math, Dokl 25*671-675
◇ B40 ◇ REV MR 83k:03020 Zbl 507 # 03003 • ID 36160

KUZICHEV, A.S. [1982] *On the representation of first-order theories in type-free combinatorially complete systems (Russian)* (J 0023) Dokl Akad Nauk SSSR 266*23-27
• TRANSL [1982] (J 0062) Sov Math, Dokl 26*295-300
◇ B40 ◇ REV MR 84c:03032 Zbl 506 # 03003 • ID 35689

KUZICHEV, A.S. [1983] *An arithmetically consistent λ-theory (Russian)* (J 0068) Z Math Logik Grundlagen Math 29*385-416
◇ B40 F30 ◇ REV MR 85d:03026 Zbl 555 # 03005
• ID 40977

KUZICHEV, A.S. [1983] *Arithmetically consistent λ-theories of type-free logic (Russian)* (J 0023) Dokl Akad Nauk SSSR 268*288-292 • ERR/ADD ibid 272*10
• TRANSL [1983] (J 0062) Sov Math, Dokl 27*72-77
◇ B40 F05 F30 ◇ REV MR 85j:03015a MR 85j:03015b Zbl 522 # 03010 • ID 43344

KUZICHEV, A.S. [1983] *Arithmetic completeness of type-free logic (Russian)* (J 0023) Dokl Akad Nauk SSSR 270*1323-1327
• TRANSL [1983] (J 0062) Sov Math, Dokl 27*767-772
◇ F30 ◇ REV MR 85d:03025 Zbl 552 # 03012 • ID 40972

KUZICHEV, A.S. [1983] *Set theory in type-free combinatorially complete systems (Russian) (English summary)* (J 0288) Vest Ser Mat Mekh, Univ Moskva 1983/3*36-42
• TRANSL [1983] (J 0510) Moscow Univ Math Bull 38/3*41-49
◇ B40 E30 E70 ◇ REV MR 85b:03021 Zbl 513 # 03010
• ID 37218

KUZICHEV, A.S. [1983] *The consistency of Quine's system NF (Russian)* (J 0023) Dokl Akad Nauk SSSR 270*537-541
• TRANSL [1983] (J 0062) Sov Math, Dokl 27*633-638
◇ B40 E35 E70 ◇ REV MR 84m:03021 Zbl 552 # 03011
• ID 35727

KUZICHEV, A.S. see Vol. I, V for further entries

KUZICHEVA, Z.A. [1970] *On the history of incompleteness theorems (Russian)* (S 2337) Istor Metodol Estest Nauk (Moskva) 9*182-189
◇ A10 F30 ◇ REV MR 42 # 4370 Zbl 252 # 01010
• ID 21052

KUZNETSOV, A.V. [1957] *Completeness of the axiomatic system of arithmetics with the rules of constructive-infinite induction (Russian)* (J 0067) Usp Mat Nauk 12/4(76)*218-219
◇ F30 ◇ ID 43671

KUZNETSOV, A.V. [1965] *Analogs of the "Sheffer stroke" in constructive logic (Russian)* (J 0023) Dokl Akad Nauk SSSR 160*274-277
• TRANSL [1965] (J 0062) Sov Math, Dokl 6*70-74
◇ F50 ◇ REV MR 34 # 4121 Zbl 147.250 • ID 19142

KUZNETSOV, A.V. [1985] *Proof-intuitionistic propositional calculus (Russian)* (J 0967) Izv Akad Nauk Mold SSR, Ser Fiz-Tekh Mat 1985/1*49-50
◇ F50 ◇ ID 46549

KUZNETSOV, A.V. see Vol. I, II, III, IV for further entries

KUZNETSOV, P.A. [1979] *A method of analysis for consistency in the description of a combinative scheme (Russian)* (S 2850) Tr Ehnerg Inst Moskva 419*91-97,166
◇ B40 ◇ REV MR 81h:03119 • ID 75247

KUZNETSOV, V.E. [1983] *K-systems (Russian)* (J 0023) Dokl Akad Nauk SSSR 273*815-816
• TRANSL [1983] (J 0062) Sov Math, Dokl 28*995-996
◇ F99 ◇ REV MR 85a:03071 Zbl 573 # 03012 • ID 34743

L'ABBE, M. [1953] *Systems of transfinite types involving λ-conversion* (J 0036) J Symb Logic 18*209-224
◇ B15 B40 ◇ REV MR 15.593 Zbl 53.203 JSL 23.361
• ID 07730

L'ABBE, M. see Vol. I for further entries

LACEY, H.M. [1970] *The consequence of Goedel's theorem* (J 1119) Ita-humanidades 6*1-11
◇ A05 F30 ◇ ID 21173

LACEY, H.M. see Vol. II for further entries

LACHLAN, A.H. [1963] *Recursive real numbers* (J 0036) J Symb Logic 28*1-16
◇ F60 ◇ REV MR 30#3013 JSL 30.256 • ID 07737

LACHLAN, A.H. see Vol. I, II, III, IV, V for further entries

LACOMBE, D. [1955] *Extension de la notion de fonction recursive aux fonctions d'une ou plusieurs variables reelles. I* (J 0109) C R Acad Sci, Paris 240*2478-2480
◇ F60 ◇ REV MR 17.225 Zbl 65.2 JSL 24.53 • REM Parts II,III 1955 • ID 47580

LACOMBE, D. [1955] *Extension de la notion de fonction recursive aux fonctions d'une ou plusieurs variables reelles II,III* (J 0109) C R Acad Sci, Paris 241*13-14,151-153
◇ F60 ◇ REV MR 17.225 Zbl 66.261 JSL 24.53 • REM Part I 1955 • ID 11441

LACOMBE, D. [1955] *Remarque sur les operateurs recursifs et sur les fonctions recursives d'une variable reelle* (J 0109) C R Acad Sci, Paris 241*1250-1252
◇ F60 ◇ REV MR 17.701 Zbl 67.2 JSL 24.53 • ID 07766

LACOMBE, D. [1957] see KREISEL, G.

LACOMBE, D. [1957] *Les ensembles recursivement ouverts ou fermes, et leurs applications a l'analyse recursive* (J 0109) C R Acad Sci, Paris 245*1040-1043
◇ D55 F60 ◇ REV MR 21#5572 Zbl 78.7 JSL 24.53
• ID 07768

LACOMBE, D. [1957] *Quelques proprietes d'analyse recursive* (J 0109) C R Acad Sci, Paris 244*838-840,996-997
◇ F60 ◇ REV MR 19.521 Zbl 77.16 JSL 24.53 • ID 16976

LACOMBE, D. [1958] *Sur les possibilites d'extension de la notion de fonction recursive aux fonctions d'une ou plusieurs variables reelles* (P 0576) Raisonn en Math & Sci Exper;1955 Paris 67-71
◇ F60 ◇ REV MR 21#4099 JSL 25.286 • ID 07769

LACOMBE, D. [1959] see KREISEL, G.

LACOMBE, D. & LAUDET, M. & NOLIN, L. & SCHUETZENBERGER, M.-P. (EDS.) [1970] *Symposium on automatic demonstration* (S 3301) Lect Notes Math 125*v+310pp
◇ B35 B97 F97 ◇ REV MR 42#2716 Zbl 196.21
• ID 48620

LACOMBE, D. see Vol. IV for further entries

LADRIERE, J. [1949] *Le role du theoreme de Goedel dans le developpement de la theorie de la demonstration* (J 0252) Rev Philos Louvain 47*469-492
◇ A05 A10 F99 ◇ REV JSL 19.129 • ID 42252

LADRIERE, J. [1951] *Le theoreme fondamental de Gentzen* (J 0252) Rev Philos Louvain 49*357-384
◇ A05 A10 F05 ◇ REV JSL 19.129 • ID 42254

LADRIERE, J. [1956] *La notion de constructivite en metamathematique* (J 0082) Bull Soc Math Belg 8*82-97
◇ A05 F99 ◇ REV MR 19.2 Zbl 74.14 JSL 34.655
• ID 07776

LADRIERE, J. [1957] *Les limitations internes des formalismes. Etude sur la signification du theoreme de Goedel et des theoremes apparentes dans la theorie des fondements des mathematiques* (X 0834) Gauthier-Villars: Paris xv+715pp
◇ A05 B98 F30 F98 ◇ REV MR 20#2 Zbl 78.242 JSL 25.270 • ID 07777

LADRIERE, J. [1961] *Expression de la recursion primitive dans le calcul-λK* (J 0079) Logique & Anal, NS 4*23-54
◇ B40 D20 ◇ REV JSL 30.91 • ID 07778

LADRIERE, J. see Vol. III for further entries

LAEUCHLI, H. [1970] *An abstract notion of realizability for which intuitionistic predicate calculus is complete* (P 0603) Intuitionism & Proof Th;1968 Buffalo 227-234
◇ F50 ◇ REV MR 42#5778 Zbl 216.5 • ID 21060

LAEUCHLI, H. see Vol. I, II, III, V for further entries

LAGASSE, J. [1976] see COURVOISIER, M.

LAKATOS, I. [1963] *Proofs and refutations I,II,III,IV* (J 0013) Brit J Phil Sci 14*1-25,120-139,221-245,296-342
◇ A05 F99 ◇ REV MR 27#4734 MR 27#4735 MR 28#1121 MR 29#2163 • ID 07793

LAKATOS, I. see Vol. I, II for further entries

LAMBEK, J. [1958] *The mathematics of sentence structures* (J 0005) Amer Math Mon 65*154-170
◇ B30 B65 F07 ◇ REV MR 21#4904 Zbl 80.7 JSL 33.627 • ID 07815

LAMBEK, J. [1961] *On the calculus of syntactic types* (P 0701) Struct of Lang & Math Aspects;1960 New York 166-178
◇ F07 ◇ REV JSL 33.627 • ID 07817

LAMBEK, J. [1968] *Deductive systems and categories I: Syntactic calculi and residuated categories* (J 0041) Math Syst Theory 2*287-318
◇ F07 G30 ◇ REV MR 38#4277 Zbl 176.289 • REM Part II 1969 • ID 07819

LAMBEK, J. [1969] *Deductive systems and categories II. Standard constructions and closed categories* (P 0723) Categ Th, Homol Th & Appl;1968 Seattle 1*76-122
◇ F07 G30 ◇ REV MR 39#3967 Zbl 198.337 • REM Part I 1968. Part III 1972 • ID 24975

LAMBEK, J. [1972] *Deductive systems and categories III. Cartesian closed categories, intuitionist propositional calculus, and combinatory logic* (P 0771) Toposes, Algeb Geom & Log;1971 Halifax 57-82
◇ B40 F50 G30 ◇ REV MR 50#1850 Zbl 244#18006
• REM Part II 1969 • ID 27403

LAMBEK, J. [1974] *Functional completeness of cartesian categories* (J 0007) Ann Math Logic 6*259-292
◇ B40 G30 ◇ REV MR 49#5121 Zbl 282#18004
• ID 07820

LAMBEK, J. [1980] *From lambda-calculus to Cartesian closed categories* (C 3050) Essays Combin Log, Lambda Calc & Formalism (Curry) 375-402
◇ B40 F07 G30 ◇ REV MR 81m:03020 Zbl 469#03006
• ID 75326

LAMBEK, J. [1980] *From types to sets* (J 0345) Adv Math 36*113-164
◇ F35 F50 G30 ◇ REV MR 81h:03124 Zbl 436 # 03053 • ID 75328

LAMBEK, J. & SCOTT, P.J. [1980] *Intuitionist type theory and the free topos* (J 0326) J Pure Appl Algebra 19*215-257
◇ F35 F50 G30 ◇ REV MR 82b:03112 Zbl 452 # 03049 • ID 54113

LAMBEK, J. & SCOTT, P.J. [1981] *Independence of premisses and the free topos* (P 3146) Constr Math;1980 Las Cruces 191-207
◇ F35 F50 G30 ◇ REV MR 83e:03098 Zbl 467 # 03054 • ID 55052

LAMBEK, J. & SCOTT, P.J. [1981] *Intuitionist type theory and foundations* (J 0122) J Philos Logic 10*101-115
◇ F35 F50 ◇ REV MR 83j:03111 Zbl 461 # 03012 • ID 54489

LAMBEK, J. & SCOTT, P.J. [1983] *New proofs of some intuitionistic principles* (J 0068) Z Math Logik Grundlagen Math 29*493-504
◇ F35 F50 G30 ◇ REV MR 85e:03149 Zbl 568 # 03030 • ID 40763

LAMBEK, J. & SCOTT, P.J. [1984] *Aspects of higher order categorical logic* (P 2180) Math Appl Categ Th;1983 Denver 145-174
◇ B15 F35 F50 G30 ◇ REV MR 85i:03196 Zbl 549 # 03058 • ID 44302

LAMBEK, J. see Vol. III, IV, V for further entries

LAMBERT, K. & SCHARLE, T.W. [1967] *A translation theorem for two systems of free logic* (J 0079) Logique & Anal, NS 10*328-341
◇ B20 B60 F25 ◇ REV MR 37 # 1229 Zbl 164.309 • ID 07823

LAMBERT, K. see Vol. I, II for further entries

LAMBERT JR., W.M. [1977] *Un tipo de solidez para un sistema de deduccion natural* (J 1680) Cienc Tecnol, Costa Rica 1*35-44
◇ B10 F07 ◇ REV MR 81j:03024 • ID 31153

LAMBERT JR., W.M. see Vol. IV for further entries

LANDWEBER, L.H. [1969] see BUECHI, J.R.

LANDWEBER, L.H. see Vol. IV for further entries

LANDWEHR, R. [1982] see JANSOHN, H.-S.

LAPLAZA, M.L. [1973] see BOLLMAN, D.A.

LAPLAZA, M.L. see Vol. IV, V for further entries

LARGEAULT, J. [1973] *La theorie de la demonstration de Hilbert et les theoremes d'impossibilite de Goedel* (J 0286) Int Logic Rev 7*56-79
◇ A05 A10 F99 ◇ REV Zbl 347 # 02002 • ID 63310

LARGEAULT, J. see Vol. II for further entries

LARGUIER, E.H. [1940] *Brouwerian philosophy of mathematics* (J 0287) Scripta Math 7*69-78
◇ A05 F55 ◇ REV FdM 66.1189 • ID 43816

LASCAR, D. [1980] *Une indicatrice de type "Ramsey" pour l'arithmetique de Peano et la formule de Paris-Harrington* (C 2969) Modeles de l'Arithm; Paris 1977 19-30
◇ C62 F30 ◇ REV MR 83g:03074 Zbl 458 # 03024 JSL 48.483 • ID 54430

LASCAR, D. see Vol. I, III, IV, V for further entries

LATOCHA, P. [1983] *The problem of structural completeness of the intuitionistic propositional logic and its fragments* (J 0302) Rep Math Logic, Krakow & Katowice 16*17-22
◇ F50 ◇ REV MR 86d:03016 Zbl 544 # 03010 • ID 40995

LATREILLE, F. & PAILLET, J.L. [1970] *Rapports entre l'interpretabilite d'une relation par une autre et l'interpretabilite locale* (J 2313) C R Acad Sci, Paris, Ser A-B 270*A629-A632
◇ F25 ◇ REV MR 42 # 2931 Zbl 193.309 • ID 07876

LATREILLE, F. see Vol. V for further entries

LAUDET, M. [1970] see LACOMBE, D.

LAUENER, H. [1978] *Semantique et methode constructiviste* (J 0079) Logique & Anal, NS 21*205-235
◇ A05 F99 ◇ REV MR 80h:03090 Zbl 411 # 03006 • ID 52861

LAVENDHOMME, R. & LUCAS, T. [1983] *A note on intuitionistic models of ZF* (J 0047) Notre Dame J Formal Log 24*54-66
◇ E70 F50 G30 ◇ REV MR 84f:03033 Zbl 487 # 03028 • ID 34453

LAVENDHOMME, R. & LUCAS, T. [1984] *Une interpretation modale de la logique intuitionniste (English summary)* (J 3364) C R Acad Sci, Paris, Ser 1 298*193-196
◇ B45 F50 ◇ REV MR 85b:03016 Zbl 563 # 03008 • ID 40558

LAVENDHOMME, R. see Vol. III, V for further entries

LAVROV, I.A. [1979] *Computability of partial functions and enumerability of sets in Peano's arithmitic (Russian)* (J 0092) Sib Mat Zh 20*1269-1274,1407
• TRANSL [1979] (J 0475) Sib Math J 20*900-904
◇ D20 D25 F30 ◇ REV MR 81d:03044 Zbl 431 # 03030 • ID 53936

LAVROV, I.A. see Vol. I, II, III, IV, V for further entries

LAWSON, J.D. [1980] *Algebraic conditions leading to continuous lattices* (J 0053) Proc Amer Math Soc 78*477-481
◇ B40 ◇ REV MR 81g:06002 Zbl 448 # 06006 • ID 82038

LAWSON, J.D. see Vol. V for further entries

LAWVERE, F.W. (ED.) [1972] *Toposes, algebraic geometry and logic* (X 0811) Springer: Heidelberg & New York vi + 189pp
◇ F50 G30 G97 ◇ REV MR 48 # 8592 Zbl 233 # 00005 Zbl 249 # 18015 • ID 25072

LAWVERE, F.W. & MAURER, C. & WRAITH, G.C. (EDS.) [1975] *Model theory and topoi* (X 0811) Springer: Heidelberg & New York 354pp
◇ C97 F97 G97 ◇ REV MR 51 # 10016 Zbl 299 # 00014 JSL 46.158 • ID 70191

LAWVERE, F.W. see Vol. V for further entries

LEBESGUE, H. [1905] see BAIRE, R.

LEBESGUE, H. see Vol. I, IV, V for further entries

LEBLANC, H. [1960] see BETH, E.W.

LEBLANC, H. [1962] *Etudes sur les regles d'inference dites regles de Gentzen I* (J 0488) Dialogue (Ottawa) 1*56-66
◇ F07 ◇ REV MR 36 # 30 JSL 27.248 • REM Part II 1963 • ID 07920

LEBLANC, H. [1962] see BELNAP JR., N.D.

LEBLANC, H. [1962] *Structural rules of inference* (J 0047) Notre Dame J Formal Log 3*201-205
◇ F07 ◇ REV MR 27 # 3500 Zbl 168.3 JSL 28.256
• ID 07921

LEBLANC, H. [1963] *Etudes sur les regles d'inference dites regles de Gentzen II* (J 0488) Dialogue (Ottawa) 1*355-367
◇ F07 ◇ REV MR 36 # 31 JSL 32.256 • REM Part I 1962
• ID 07922

LEBLANC, H. [1963] see BELNAP JR., N.D.

LEBLANC, H. [1965] *Marginalia on Gentzen's Sequenzen-Kalkule* (C 0749) Contrib Logic & Methodol (Bochenski) 73-83
◇ B10 F07 F50 ◇ REV MR 48 # 3711 Zbl 156.6
• ID 75405

LEBLANC, H. & THOMASON, R.H. [1966] *The demarcation line between intuitionistic logic and classical logic* (J 0068) Z Math Logik Grundlagen Math 12*257-262
◇ A05 F50 ◇ REV MR 35 # 1447 Zbl 143.11 • ID 07924

LEBLANC, H. [1966] *Two separation theorems for natural deduction* (J 0047) Notre Dame J Formal Log 7*159-180
◇ B10 F07 F50 ◇ REV MR 36 # 6263 Zbl 173.3
• ID 07926

LEBLANC, H. [1968] *Subformula theorems for N-sequents* (J 0036) J Symb Logic 33*161-179
◇ F07 F50 ◇ REV MR 42 # 38 Zbl 169.297 • ID 07932

LEBLANC, H. & SNYDER, D.P. [1972] *Duals of Smullyan trees* (J 0047) Notre Dame J Formal Log 13*387-393 • ERR/ADD ibid 15*648
◇ B10 F07 ◇ REV MR 46 # 8816 MR 50 # 1853 Zbl 197.273 • ID 19182

LEBLANC, H. & MEYER, R.K. [1972] *Matters of separation* (J 0047) Notre Dame J Formal Log 13*229-236
◇ B10 F07 F50 ◇ REV MR 45 # 6575 Zbl 234 # 02008
• ID 07940

LEBLANC, H. & MORGAN, C.G. [1983] *Probabilistic semantics for intuitionistic logic* (J 0047) Notre Dame J Formal Log 24*161-180
◇ B48 F50 ◇ REV MR 84c:03027 Zbl 466 # 60003
• ID 54992

LEBLANC, H. & MORGAN, C.G. [1983] *Probability theory, intuitionism, semantics, and the Dutch book argument* (J 0047) Notre Dame J Formal Log 24*289-304
◇ B48 C90 F50 ◇ REV MR 85d:03048 Zbl 531 # 03037
• ID 37685

LEBLANC, H. see Vol. I, II, III for further entries

LECOUTURIER, P. [1975] see KOCK, A.

LEDGARD, H.F. [1972] *Embedding Markov normal algorithms within the λ-calculus* (J 0382) Int J Comput Math 3*131-140
◇ B40 D03 D15 ◇ REV Zbl 261 # 68026 • ID 63344

LEDGARD, H.F. [1977] *Production systems: A notation for defining syntax and translation* (J 2094) IEEE Trans Software Engin SE-3*105-124
◇ B40 ◇ REV Zbl 349 # 68004 • ID 63345

LEE, PENGYEE [1972] *A constructive proof of a Differential inequality* (J 0245) Nanta Math 5*83-84
◇ F55 ◇ REV MR 49 # 5258 Zbl 287 # 26012 • ID 07961

LEE, R.D. [1965] *The substitution schema in recursive arithmetic* (J 0047) Notre Dame J Formal Log 6*193-200
◇ F30 ◇ REV MR 33 # 2539 Zbl 192.53 JSL 32.543
• ID 07962

LEE, R.D. [1966] see GOODSTEIN, R.L.

LEHMAN, R.S. [1960] *On primitive recursive real numbers* (J 0027) Fund Math 49*105-118
◇ F60 ◇ REV MR 23 # A2320 Zbl 199.33 JSL 27.245
• ID 07978

LEHMAN, R.S. see Vol. II for further entries

LEININGER, B.S. [1981] see IBARRA, O.H.

LEININGER, B.S. see Vol. IV for further entries

LEISENRING, A.C. [1969] *Mathematical logic and Hilbert's ε-symbol* (X 0836) Gordon & Breach: New York 142pp
◇ B10 E25 F05 F98 ◇ REV MR 43 # 1807 Zbl 188.315
• ID 07983

LEISENRING, A.C. see Vol. I for further entries

LEIVANT, D. [1971] *A note on translations of C into I* (X 1121) Math Centr: Amsterdam ZW5/71*12pp
◇ F50 ◇ REV Zbl 231 # 02027 • ID 27433

LEIVANT, D. [1975] *Strong normalization for arithmetic (Variations on a theme of Prawitz.)* (P 1440) ⊧ ISILC Proof Th Symp (Schuette);1974 Kiel 500*182-197
◇ F05 F30 F50 ◇ REV MR 56 # 8327 Zbl 322 # 02030
• ID 27277

LEIVANT, D. [1976] *Failure of completeness properties of intuitionistic predicate logic for constructive models* (J 1934) Ann Sci Univ Clermont Math 13*93-107
◇ F50 ◇ REV MR 57 # 15975 Zbl 362 # 02014 • ID 30744

LEIVANT, D. [1979] *Absoluteness of intuitionistic logic* (S 1605) Math Centr Tracts 73*ix + 137pp
◇ F05 F30 F50 ◇ REV MR 80g:03059 Zbl 459 # 03024 JSL 46.873 • ID 54464

LEIVANT, D. [1979] *Assumption classes in natural deduction* (J 0068) Z Math Logik Grundlagen Math 25*1-4
◇ F05 F07 ◇ REV MR 80h:03079 Zbl 408 # 03043
• ID 56281

LEIVANT, D. [1980] *Innocuous substitutions* (J 0036) J Symb Logic 45*363-368
◇ F30 F50 ◇ REV MR 81g:03008 Zbl 434 # 03041
• ID 55745

LEIVANT, D. [1981] *Implicational complexity in intuitionistic arithmetic* (J 0036) J Symb Logic 46*240-248
◇ F30 F50 ◇ REV MR 82h:03066 Zbl 488 # 03032
• ID 75434

LEIVANT, D. [1981] *On the proof theory of the modal logic for arithmetic provability* (J 0036) J Symb Logic 46*531-538
◇ B45 F05 F30 ◇ REV MR 83a:03022 Zbl 464 # 03019
• ID 54609

LEIVANT, D. [1981] *Proof theoretic methodology for propositional dynamic logic* (P 2930) Formal of Progr Concepts;1981 Peniscola 356-373
◇ B75 F99 ◇ REV MR 83a:68030 Zbl 467 # 03015
• ID 55013

LEIVANT, D. [1982] *Unprovability of theorems of complexity theory in weak number theories* (J 1426) Theor Comput Sci 18*259-268
◇ D15 F30 ◇ REV MR 84g:03098 Zbl 482#03025
• ID 34202

LEIVANT, D. [1983] *The optimality of induction as an axiomatization of arithmetic* (J 0036) J Symb Logic 48*182-184
◇ F30 ◇ REV MR 84f:03051 Zbl 515#03018 • ID 34470

LEIVANT, D. [1985] *Syntactic translations and provably recursive functions* (J 0036) J Symb Logic 50*682-688
◇ D20 F30 F50 ◇ ID 39403

LEIVANT, D. see Vol. I for further entries

LERCHER, B. [1967] *Strong reduction and normal form in combinatory logic* (J 0036) J Symb Logic 32*213-223
◇ B40 ◇ REV MR 36#2505 Zbl 153.6 JSL 36.171
• ID 08017

LERCHER, B. [1967] *The decidability of Hindley's axiom for strong reduction* (J 0036) J Symb Logic 32*237-239
◇ B40 ◇ REV MR 35#5321 Zbl 153.7 JSL 36.171
• ID 08016

LERCHER, B. [1970] see HINDLEY, J.R.

LERCHER, B. [1972] see HINDLEY, J.R.

LERCHER, B. [1976] *λ-calculus terms that reduce to themselves* (J 0047) Notre Dame J Formal Log 17*291-292
◇ B40 ◇ REV MR 53#10554 Zbl 313#02017 • ID 18258

LERDA, F. [1959] *Analisi critica dei fondamenti dell'intuizionismo* (J 2038) Rend Sem Mat, Torino 19*121-218
◇ A05 F50 F55 ◇ REV MR 23#A2321 Zbl 95.242
• ID 08018

LEVIN, A.M. [1973] *Some applications of truth definitions (Russian) (English summary)* (J 0288) Vest Ser Mat Mekh, Univ Moskva 28/5*11-17
• TRANSL [1973] (J 0510) Moscow Univ Math Bull 28/5-6*8-13
◇ E35 E70 F35 ◇ REV MR 51#5268 Zbl 272#02052
• ID 17453

LEVIN, A.M. [1975] *A comparison of various forms of the axiom of choice in classical analysis (Russian)* (J 0023) Dokl Akad Nauk SSSR 225*759-7625
• TRANSL [1975] (J 0062) Sov Math, Dokl 16*1532-1535
◇ E25 E35 F35 ◇ REV MR 58#21616 Zbl 338#02040
• ID 63402

LEVIN, A.M. [1975] *The axiom of choice in classical analysis (Russian) (English summary)* (J 0288) Vest Ser Mat Mekh, Univ Moskva 30/4*59-65
• TRANSL [1975] (J 0510) Moscow Univ Math Bull 30/3-4*106-111
◇ B30 E15 E25 E35 E75 F35 ◇ REV MR 54#7254 Zbl 316#02070 • ID 25810

LEVIN, A.M. [1977] *A conservative extension of formal mathematical analysis with a scheme of dependent choice (Russian)* (J 0087) Mat Zametki (Akad Nauk SSSR) 22*61-68
• TRANSL [1977] (J 1044) Math Notes, Acad Sci USSR 22*524-528
◇ E25 F35 ◇ REV MR 58#10346 Zbl 377#02020
• ID 53418

LEVIN, A.M. [1978] *A fragment of classical analysis (Russian) (English summary)* (J 0288) Vest Ser Mat Mekh, Univ Moskva 1978/1*3-9
• TRANSL [1978] (J 0510) Moscow Univ Math Bull 33/1*1-6
◇ E25 E35 F35 ◇ REV MR 58#5092 Zbl 406#03071
• ID 56153

LEVIN, A.M. [1979] *On an interesting axiomatic theory (Russian)* (C 2581) Issl Neklass Log & Teor Mnozh 137-142
◇ B28 F35 ◇ REV MR 81j:03089 Zbl 422#03032
• ID 53493

LEVIN, L.A. [1969] *Some syntactic theorems on the calculus of finite problems of Ju.T.Medvedev (Russian)* (J 0023) Dokl Akad Nauk SSSR 185*32-33
• TRANSL [1969] (J 0062) Sov Math, Dokl 10*288-290
◇ B55 F50 ◇ REV MR 39#1299 • ID 08046

LEVIN, L.A. [1976] *On the principle of conservation of information in intuitionistic mathematics (Russian)* (J 0023) Dokl Akad Nauk SSSR 227*1293-1296
• TRANSL [1976] (J 0062) Sov Math, Dokl 17*601-605
◇ D15 D80 F50 F55 ◇ REV MR 58#21509 Zbl 358#02033 • ID 50487

LEVIN, L.A. see Vol. IV for further entries

LEVITZ, H. [1966] *Ueber die Finslerschen hoeheren arithmetischen Operationen* (J 2022) Comm Math Helvetici 41*273-286
◇ E10 F15 ◇ REV MR 35#4102 Zbl 147.251 • ID 08050

LEVITZ, H. [1969] *A simplification of Takeuti's ordinal diagrams of finite order* (J 0068) Z Math Logik Grundlagen Math 15*141-154
◇ F15 ◇ REV MR 39#3966 Zbl 196.14 • ID 08051

LEVITZ, H. [1969] *On the Finsler and Doner-Tarski arithmetical hierarchies* (J 2022) Comm Math Helvetici 44*89-92
◇ E10 F15 ◇ REV MR 39#5375 Zbl 177.15 • ID 08052

LEVITZ, H. [1970] *On the relationship between Takeuti's ordinal diagrams $O(n)$ and Schuette's system of ordinal notations $\Sigma(n)$* (P 0603) Intuitionism & Proof Th;1968 Buffalo 377-405
◇ F15 ◇ REV MR 42#4377 Zbl 201.7 JSL 40.501
• ID 21053

LEVITZ, H. & SCHUETTE, K. [1971] *A characterisation of Takeuti's ordinal diagrams of finite order* (J 0009) Arch Math Logik Grundlagenforsch 14*75-97
◇ F15 ◇ REV MR 44#2572 Zbl 244#02011 • ID 08053

LEVITZ, H. [1973] *A characterisation of the Veblen-Schuette functions by means of functionals* (J 2022) Comm Math Helvetici 48*382-393
◇ F10 F15 ◇ REV MR 48#3710 Zbl 272#02051
• ID 08055

LEVITZ, H. [1975] *An ordered set of arithmetic functions representing the least ε-number* (J 0068) Z Math Logik Grundlagen Math 21*115-120
◇ C62 F15 F30 ◇ REV MR 51#7845 Zbl 325#04002
• ID 18259

LEVITZ, H. [1978] *An ordinal bound for the set of polynomial functions with exponentiation* (J 0004) Algeb Universalis 8*233-243
◇ D15 D20 E07 E10 F15 F30 ◇ REV MR 81f:04003 Zbl 376#06001 • ID 30732

LEVITZ, H. [1984] see DRIES VAN DEN, L.

LEVITZ, H. see Vol. I, III, V for further entries

LEVY, A. [1958] *Comparison of subtheories* (**J** 0053) Proc Amer Math Soc 9*942-945 • ERR/ADD ibid 10*1000
◇ C07 C40 E30 F30 ◇ REV MR 21 # 1934 JSL 24.226
• ID 24860

LEVY, A. [1958] see BAR-HILLEL, Y.

LEVY, A. [1968] see KREISEL, G.

LEVY, A. [1968] *The effectivity of existential statements in axiomatic set theory* (**J** 0191) Inform Sci 1*119-130
◇ E25 E30 E47 F99 ◇ REV MR 39 # 3988 • ID 08080

LEVY, A. see Vol. III, IV, V for further entries

LEVY, J.-J. [1975] *An algebraic interpretation of the $\lambda\beta k$-calculus and a labelled λ-calculus* (**P** 1603) λ-Calc & Comput Sci Th;1975 Roma 147-165 • ERR/ADD ibid 94-95
◇ B40 ◇ REV MR 58 # 5144 Zbl 335 # 02014
Zbl 335 # 02015 • ID 63425

LEVY, J.-J. [1976] *An algebraic interpretation of the $\lambda\beta K$-calculus and an application of the labelled λ-calculus* (**J** 1426) Theor Comput Sci 2*97-114
◇ B40 ◇ REV MR 53 # 12891 Zbl 335 # 02016 • ID 23145

LEVY, J.-J. [1978] *Le probleme du partage dans l'evaluation des λ-expressions* (**P** 3835) AFCET-SMF Math Appliquees Colloq (1);1978 Palaiseau 1*139-154
◇ B40 ◇ REV Zbl 482 # 03002 • ID 36838

LEVY, J.-J. [1979] see BERRY, GERARD

LEVY, J.-J. [1980] *Optimal reductions in the lambda-calculus* (**C** 3050) Essays Combin Log, Lambda Calc & Formalism (Curry) 159-191
◇ B40 ◇ REV MR 83a:03012 Zbl 469 # 03006 • ID 35057

LEVY, P. [1926] *Sur le principe du tiers exclu et sur les theoremes non susceptibles de demonstration* (**J** 0145) Rev Metaph Morale 33*253-258
◇ A05 F50 ◇ REV FdM 52.46 • ID 08091

LEVY, P. [1927] *Logique classique, logique Brouwerienne et logique mixte* (**J** 0150) Acad Roy Belg Bull Cl Sci (5) 13*256-266
◇ A05 B55 F50 ◇ REV FdM 53.40 • ID 08092

LEVY, P. [1928] *Logique classique, logique Brouwerienne et logique mixte* (**C** 1224) Borel: Lecons sur la Theorie des Fonctions 285-290
◇ A05 F50 ◇ ID 20984

LEVY, P. see Vol. III, V for further entries

LEWIS, H.A. [1973] *Combinators and deep structure, syntactic and semantic functions* (**P** 0580) Int Congr Log, Meth & Phil of Sci (4,Sel Pap);1971 Bucharest 213-222
◇ B40 B65 ◇ REV Zbl 285 # 02012 • ID 29943

LEWIS, H.A. see Vol. II for further entries

LIFSCHITZ, V. [1967] *A normal form for conclusions in the predicate calculus with equality and functional symbols (Russian)* (**S** 0228) Zap Nauch Sem Leningrad Otd Mat Inst Steklov 4*58-64
• TRANSL [1967] (**J** 0521) Semin Math, Inst Steklov 4*21-23
◇ F05 ◇ REV MR 38 # 5583 Zbl 188.11 • ID 08140

LIFSCHITZ, V. [1967] *Constructive groups (Russian)* (**S** 0228) Zap Nauch Sem Leningrad Otd Mat Inst Steklov 4*86-95
• TRANSL [1967] (**J** 0521) Semin Math, Inst Steklov 4*32-35
◇ D45 F60 ◇ REV MR 38 # 4281 Zbl 198.27 • ID 08139

LIFSCHITZ, V. [1967] *Constructive mathematical theories consistent with classical logic (Russian)* (**S** 0066) Tr Mat Inst Steklov 93*113-122
• TRANSL [1967] (**S** 0055) Proc Steklov Inst Math 93*143-155
◇ F50 ◇ REV MR 36 # 6265 Zbl 194.312 • ID 19218

LIFSCHITZ, V. [1967] *The decision problem for some constructive theories of equality (Russian)* (**S** 0228) Zap Nauch Sem Leningrad Otd Mat Inst Steklov 4*78-85
• TRANSL [1967] (**J** 0521) Semin Math, Inst Steklov 4*29-31
◇ B25 D35 F50 ◇ REV MR 39 # 65 Zbl 165.19 • ID 33665

LIFSCHITZ, V. [1968] *A specialization of the form of deduction in the predicate calculus with equality and functional symbols I (Russian)* (**S** 0066) Tr Mat Inst Steklov 98*5-25 • ERR/ADD ibid 121*167
• TRANSL [1968] (**S** 0055) Proc Steklov Inst Math 98*1-23
◇ F07 ◇ REV MR 40 # 5415 Zbl 181.5 • ID 19230

LIFSCHITZ, V. [1968] *Constructive analytic functions of one real variable (Russian)* (**S** 0228) Zap Nauch Sem Leningrad Otd Mat Inst Steklov 8*121-131 • ERR/ADD ibid 16*185-187
◇ F60 ◇ REV MR 43 # 7327 Zbl 191.307 • ID 08141

LIFSCHITZ, V. [1969] *The set of zeros of a constructive power series in the real domain (Russian)* (**S** 0228) Zap Nauch Sem Leningrad Otd Mat Inst Steklov 16*114-125
• TRANSL [1969] (**J** 0521) Semin Math, Inst Steklov 16*58-63
◇ F60 ◇ REV MR 41 # 5193 Zbl 223 # 02026 • ID 08142

LIFSCHITZ, V. [1971] *Investigations of constructive functions by the method of fillings (Russian) (English summary)* (**S** 0228) Zap Nauch Sem Leningrad Otd Mat Inst Steklov 20*67-79,284
• TRANSL [1973] (**J** 1531) J Sov Math 1*41-47
◇ F60 ◇ REV MR 48 # 3714 Zbl 222 # 02030 • ID 08143

LIFSCHITZ, V. [1972] *A locally analytic constructive function that is not analytic (Russian)* (**J** 0023) Dokl Akad Nauk SSSR 202*1265-1267
• TRANSL [1972] (**J** 0062) Sov Math, Dokl 13*281-284
◇ F60 ◇ REV MR 45 # 5320 Zbl 252 # 02038 • ID 08144

LIFSCHITZ, V. [1972] *Metamathematical interpretation of the fan theorem (Russian) (English summary)* (**S** 0228) Zap Nauch Sem Leningrad Otd Mat Inst Steklov 32*45-52,155
• TRANSL [1976] (**J** 1531) J Sov Math 6*383-389
◇ F50 ◇ REV MR 52 # 5386 Zbl 343 # 02021 • ID 18260

LIFSCHITZ, V. [1972] see CHERNOV, V.P.

LIFSCHITZ, V. [1979] CT_0 *is stronger than* CT_0*!* (**J** 0053) Proc Amer Math Soc 73*101-106
◇ F30 F50 ◇ REV MR 80b:03094 Zbl 379 # 02012
• ID 52528

LIFSCHITZ, V. [1979] *An intuitionistic definition of classical natural numbers* (**J** 0053) Proc Amer Math Soc 77*385-388
◇ F30 F50 ◇ REV MR 81c:03052 Zbl 427 # 03051
• ID 53738

LIFSCHITZ, V. [1980] *Semantical completeness theorems in logic and algebra* (**J** 0053) Proc Amer Math Soc 79*89-96
◇ B10 F07 ◇ REV MR 81g:03073 Zbl 453 # 03060
• ID 54190

LIFSCHITZ, V. [1982] *Constructive assertions in an extension of classical mathematics* (**J** 0036) J Symb Logic 47*359-387
◇ F50 ◇ REV MR 84a:03075 Zbl 488 # 03034 • ID 35596

LIFSCHITZ, V. [1985] *Calculable natural numbers* (**C** 3659) Intens Math 173-190
◇ B28 F30 F50 ◇ ID 44843

LIFSCHITZ, V. see Vol. I, IV for further entries

LINDENBAUM, A. [1932] *Sur les constructions non-effectives dans l'arithmetique elementaire* (**J** 0283) Ann Soc Pol Math 10*118-119
◇ F30 ◇ REV FdM 58.1001 • ID 08162

LINDENBAUM, A. see Vol. I, III, V for further entries

LINDNER, R. [1975] *On the theory of inference operators* (**P** 1521) Int Congr Math (II,12);1974 Vancouver 2*471-475
◇ B48 F99 ◇ REV MR 55 # 5417 Zbl 375 # 02029 • ID 51606

LINDNER, R. see Vol. I, II, IV, V for further entries

LINDSTROEM, P. [1979] *Some results on interpretability* (**P** 2615) Scand Logic Symp (5);1979 Aalborg 329-361
◇ F25 F30 ◇ REV MR 82i:03065 Zbl 497 # 03044 • ID 75559

LINDSTROEM, P. [1984] *On certain lattices of degrees of interpretability* (**J** 0047) Notre Dame J Formal Log 25*127-140
◇ F25 F30 ◇ REV Zbl 552 # 03039 • ID 42565

LINDSTROEM, P. [1984] *On faithful interpretability* (**P** 2153) Logic Colloq;1983 Aachen 2*279-288
◇ F25 F30 ◇ ID 43009

LINDSTROEM, P. [1984] *On partially conservative sentences and interpretability* (**J** 0053) Proc Amer Math Soc 91*436-443
◇ F25 F30 ◇ REV MR 85j:03099 • ID 45180

LINDSTROEM, P. [1985] *Provability and interpretability in theories containing arithmetic* (**P** 4646) Atti Incontri Log Mat (2);1983/84 Siena 431-451
◇ F25 F30 ◇ ID 49679

LINDSTROEM, P. see Vol. I, III for further entries

LIPTON, R.J. & MILLER, R.E. & SNYDER, LAWRENCE [1977] *Synchronization and computing capabilities of linear asynchronous structures* (**J** 0119) J Comp Syst Sci 14*49-72
◇ B40 B70 D05 ◇ REV MR 55 # 7632 Zbl 368 # 02038 • ID 51255

LIPTON, R.J. [1978] *Model theoretic aspects of computational complexity* (**P** 3578) IEEE Symp Found of Comput Sci (19);1978 Ann Arbor 193-200
◇ C62 D15 F30 F65 H15 ◇ REV MR 80e:03042 • ID 75578

LIPTON, R.J. [1979] see DEMILLO, R.A.

LIPTON, R.J. [1979] *On the consistency of P = NP and fragments of arithmetic* (**P** 2935) FCT'79 Fund of Comput Th;1979 Berlin/Wendisch-Rietz 269-278
◇ D15 F30 ◇ REV MR 81k:03034 Zbl 419 # 03023 • ID 53366

LIPTON, R.J. see Vol. IV for further entries

LIU, SHICHAO [1973] *Ordering formulas and quasi-unsecuredness* (**J** 0500) Tamkang J Math 4*191-193
◇ F30 ◇ REV MR 51 # 99 Zbl 293 # 02027 • ID 15225

LIU, SHICHAO [1980] *A proof-theoretic approach to nonstandard analysis with emphasis on distinguishing between constructive and nonconstructive results* (**P** 2058) Kleene Symp;1978 Madison 391-414
◇ E70 F99 H05 H20 ◇ REV MR 81m:03076 Zbl 464 # 03060 • ID 75591

LIU, SHICHAO [1984] *A proof-theoretic approach to nonstandard analysis (continued)* (**P** 2153) Logic Colloq;1983 Aachen 1*281-296
◇ E70 F50 H05 ◇ ID 45423

LIU, SHICHAO see Vol. II, IV, V for further entries

LIVCHAK, A.B. [1976] *Definable sets of integers (Russian)* (**J** 0340) Mat Zap (Univ Sverdlovsk) 10/1*54-59,151
◇ C60 C62 F30 ◇ REV MR 58 # 10418 Zbl 435 # 03028 • ID 55788

LIVCHAK, A.B. [1977] *Expressible subsets of the ordered group of integers (Russian)* (**J** 0031) Izv Vyssh Ucheb Zaved, Mat (Kazan) 1977/8*105-107
• TRANSL [1977] (**J** 3449) Sov Math 21/8*81-83
◇ C60 C62 F30 ◇ REV MR 57 # 5742 Zbl 378 # 02025 • ID 33959

LIVCHAK, A.B. see Vol. III, IV for further entries

LODDER, J.S. [1982] see DALEN VAN, D.

LOEB, M.H. [1954] *Solution of a problem by Leon Henkin* (**P** 0575) Int Congr Math (II, 7);1954 Amsterdam 2*405-406
◇ F30 ◇ ID 29470

LOEB, M.H. [1955] *Solution of a problem of Leon Henkin* (**J** 0036) J Symb Logic 20*115-118
◇ F30 ◇ REV MR 17.5 Zbl 67.2 JSL 36.528 • ID 08225

LOEB, M.H. [1956] *Formal systems of constructive mathematics* (**J** 0036) J Symb Logic 21*63-75
◇ F65 ◇ REV MR 17.1037 Zbl 72.1 JSL 35.150 • ID 08226

LOEB, M.H. [1959] *Constructive truth* (**P** 0634) Constructivity in Math;1957 Amsterdam 159-168
◇ A05 F99 ◇ REV MR 21 # 3315 Zbl 116.3 JSL 40.501 • ID 08227

LOEB, M.H. [1968] *Die Vollstaendigkeit der verzweigten Typenlogik mit unendlicher Terminduktion* (**J** 0009) Arch Math Logik Grundlagenforsch 11*68-72
◇ B15 F35 ◇ REV MR 38 # 42 Zbl 182.323 • ID 08229

LOEB, M.H. (ED.) [1968] *Proceedings of the summer school in logic, Leeds, 1967* (**S** 3301) Lect Notes Math 70*iv+331pp
◇ B97 C97 F97 ◇ ID 37293

LOEB, M.H. & WAINER, S.S. [1974] *Hierarchies of number-theoretic functions (Russian)* (**C** 2319) Slozh Vychisl & Algor 33-64
◇ D20 F15 ◇ REV Zbl 288 # 02024 • ID 63480

LOEB, M.H. [1976] *Embedding first order predicate logic in fragments of intuitionistic logic* (**J** 0036) J Symb Logic 41*705-718
◇ B10 D35 F50 ◇ REV MR 56 # 79 Zbl 358 # 02012 • ID 14746

LOEB, M.H. see Vol. I, II, III, IV for further entries

LOEBL, M. [1985] *Hercules and Hydra, a game on rooted finite trees* (**J** 0140) Comm Math Univ Carolinae (Prague) 26*259-267
◇ F30 ◇ REV Zbl 565 # 90090 • ID 48676

LOEWEN, K. [1968] *A standardization theorem for strong reduction* (J 0047) Notre Dame J Formal Log 9*271-283
◇ B40 ◇ REV MR 40 # 1279b Zbl 195.307 • ID 08238

LOEWEN, K. [1968] *Modified strong reduction in combinatory logic* (J 0047) Notre Dame J Formal Log 9*265-270
◇ B40 ◇ REV MR 40 # 1279a Zbl 195.307 • ID 08237

LOEWEN, K. [1968] *The Church-Rosser theorem for strong reduction in combinatory logic* (J 0047) Notre Dame J Formal Log 9*299-302
◇ B40 ◇ REV MR 40 # 1279b Zbl 191.308 • ID 08239

LOEWEN, K. see Vol. III for further entries

LOEWENHEIM, L. [1946] *On making indirect proofs direct* (J 0287) Scripta Math 12*125-139
◇ A05 F99 ◇ REV MR 8.307 Zbl 61.8 • ID 08248

LOEWENHEIM, L. see Vol. I, III, V for further entries

LOLLI, G. [1971] *Formule universali dell'aritmetica di Peano* (J 0319) Matematiche (Sem Mat Catania) 26*12-17
◇ F30 ◇ REV MR 47 # 6467 Zbl 242 # 02060 • ID 08251

LOLLI, G. [1974] *Una dimostrazione di non contraddittorieta per l'aritmetica (English summary)* (J 0012) Boll Unione Mat Ital, IV Ser 9*213-221
◇ F30 ◇ REV MR 50 # 6803 Zbl 312 # 02026 • ID 08254

LOLLI, G. & LONGO, G. & MARCJA, A. (EDS.) [1984] *Logic colloquium '82. Proceedings of the colloquium held in Florence, 23-28 August, 1982* (S 3303) Stud Logic Found Math 112*viii+358pp
◇ B97 C97 D97 F97 ◇ REV MR 85g:03006 Zbl 538 # 00003 • ID 41493

LOLLI, G. see Vol. I, II, III, IV, V for further entries

LONDON, F. [1925] *Ueber die Irreversibilitaet deduktiver Schlussweisen* (J 0157) Jbuchber Dtsch Math-Ver 34*84-86
◇ A05 B10 F07 ◇ REV FdM 51.47 • ID 08256

LONGO, G. [1974] *I problemi di decisione e la loro complessita* (J 3436) Quad, Ist Appl Calcolo, Ser 3 8*87-108
◇ B25 B35 D15 F20 ◇ REV Zbl 427 # 03010 • ID 53697

LONGO, G. [1976] *On the problem of deciding equality in partial combinatory algebras and in a formal system* (J 0063) Studia Logica 35*363-375
◇ B40 D75 ◇ REV MR 56 # 8330 Zbl 353 # 02009 • ID 50092

LONGO, G. [1979] *Ricorsivita nei tipi superiori: un'introduzione alle caratterizzazioni di Ershov ed Hyland* (J 2038) Rend Sem Mat, Torino 37/3*1-29
◇ B40 D65 G30 ◇ REV MR 81i:03073 Zbl 437 # 03024 • ID 55892

LONGO, G. & VENTURINI ZILLI, M. [1980] *A $\lambda\delta$-calculus with an algorithmic δ^** (J 0009) Arch Math Logik Grundlagenforsch 20*41-52
◇ B40 ◇ REV MR 81h:03032 Zbl 432 # 03011 • ID 53969

LONGO, G. [1980] see BARENDREGT, H.P.

LONGO, G. [1980] see HINDLEY, J.R.

LONGO, G. [1983] *Set-theoretical models of λ-calculus: theories, expansions, isomorphisms* (J 0073) Ann Pure Appl Logic 24*153-188
◇ B40 ◇ REV MR 85g:03025 Zbl 513 # 03009 • ID 37217

LONGO, G. [1984] see GIANNINI, P.

LONGO, G. [1984] see COPPO, M.

LONGO, G. & MOGGI, E. [1984] *Goedel numberings, principal morphisms, combinatory algebras* (P 3658) Math Founds of Comput Sci (11);1984 Prague 397-406
◇ B40 F40 G30 ◇ REV Zbl 567 # 03003 • ID 44662

LONGO, G. [1984] see LOLLI, G.

LONGO, G. [1984] see BRUCE, KIM B.

LONGO, G. see Vol. III, IV for further entries

LOOR DE, B. [1924] see BROUWER, L.E.J.

LOPEZ-ESCOBAR, E.G.K. [1967] *A complete, infinitary axiomatization of weak second-order logic* (J 0027) Fund Math 61*93-103
◇ B15 C75 C85 F35 ◇ REV MR 36 # 2482 Zbl 174.13 JSL 35.467 • ID 08262

LOPEZ-ESCOBAR, E.G.K. [1967] *Remarks on an infinitary language with constructive formulas* (J 0036) J Symb Logic 32*305-318
◇ C40 C70 C75 F50 ◇ REV MR 37 # 6168 Zbl 221 # 02005 • ID 08263

LOPEZ-ESCOBAR, E.G.K. [1968] *A decision method for the intuitionistic theory of successor* (J 0028) Indag Math 30*466-467
◇ B25 C10 F30 F50 ◇ REV MR 40 # 4081 Zbl 169.310 • ID 08264

LOPEZ-ESCOBAR, E.G.K. [1972] *Constructions and negationless logic (Polish and Russian summaries)* (J 0063) Studia Logica 30*7-22
◇ B20 E70 F50 ◇ REV MR 48 # 5827 Zbl 287 # 02015 • ID 08267

LOPEZ-ESCOBAR, E.G.K. [1972] *Refutability and elementary number theory* (J 0028) Indag Math 34*362-374
◇ F30 F50 ◇ REV MR 48 # 10772 Zbl 262 # 02027 • ID 08268

LOPEZ-ESCOBAR, E.G.K. [1974] *Elementary interpretations of negationless arithmetic* (J 0027) Fund Math 82*25-38
◇ F30 F50 ◇ REV MR 50 # 12673 Zbl 307 # 02022 • ID 08269

LOPEZ-ESCOBAR, E.G.K. & VELDMAN, W. [1975] *Intuitionistic completeness of a restricted second-order logic* (P 1440) ⊧ ISILC Proof Th Symp (Schuette);1974 Kiel 500*198-232
◇ B15 C90 F50 ◇ REV MR 54 # 67 Zbl 334 # 02015 • ID 24555

LOPEZ-ESCOBAR, E.G.K. [1976] *On an extremely restricted ω-rule* (J 0027) Fund Math 90*159-172
◇ F05 F15 F30 F50 ◇ REV MR 55 # 2517 Zbl 359 # 02026 • ID 18265

LOPEZ-ESCOBAR, E.G.K. [1977] *Infinite rules in finite systems* (P 1076) Latin Amer Symp Math Log (3);1976 Campinas 75-97
◇ F30 F35 ◇ REV MR 57 # 16039 Zbl 386 # 03026 • ID 16599

LOPEZ-ESCOBAR, E.G.K. [1980] *Semantical models for intuitionistic logics* (P 2958) Latin Amer Symp Math Log (4);1978 Santiago 191-207
◇ C90 C95 F50 ◇ REV MR 81j:03090 Zbl 469 # 03022 • ID 55150

LOPEZ-ESCOBAR, E.G.K. [1981] *Equivalence between semantics for intuitionism I* (**J** 0036) J Symb Logic 46*773-780
◇ F50 ◇ REV MR 83k:03075 Zbl 497 # 03047 • ID 33274

LOPEZ-ESCOBAR, E.G.K. [1981] *Integrating intuitionistic and classical theories* (**J** 0027) Fund Math 112*125-140
◇ F50 ◇ REV MR 82k:03092 Zbl 488 # 03033 • ID 75643

LOPEZ-ESCOBAR, E.G.K. [1981] *On the interpolation theorem for the logic of constant domains* (**J** 0036) J Symb Logic 46*87-88
◇ B55 C40 C90 F50 ◇ REV MR 82i:03068 Zbl 469 # 03015 • REM See also 1983 • ID 55143

LOPEZ-ESCOBAR, E.G.K. [1981] *Variations on a system of Gentzen* (**J** 0068) Z Math Logik Grundlagen Math 27*385-389
◇ B55 F50 ◇ REV MR 83e:03037 Zbl 469 # 03014 • ID 55142

LOPEZ-ESCOBAR, E.G.K. [1982] *Further applications of ultra-conservative ω-rules* (**J** 0009) Arch Math Logik Grundlagenforsch 22*89-102
◇ F05 F30 ◇ REV MR 84k:03139 Zbl 535 # 03030 • ID 35031

LOPEZ-ESCOBAR, E.G.K. [1983] *A second paper on the interpolation theorem for the logic of constant domains* (**J** 0036) J Symb Logic 48*595-599
◇ B55 C40 C90 F05 F50 ◇ REV MR 85b:03099 Zbl 547 # 03022 • REM See also 1981 • ID 39786

LOPEZ-ESCOBAR, E.G.K. [1985] *Koenig's lemma, the ω-rule and primitive recursive arithmetic* (**J** 0009) Arch Math Logik Grundlagenforsch 25*67-74
◇ F30 ◇ ID 49121

LOPEZ-ESCOBAR, E.G.K. [1985] *Proof functional connectives* (**P** 2160) Latin Amer Symp Math Log (6);1983 Caracas 208-221
◇ B55 F07 F30 ◇ ID 47264

LOPEZ-ESCOBAR, E.G.K. see Vol. II, III, V for further entries

LORENTS, A.A. [1967] *Certain problems in the theory of finite probabilistic automata (Russian)* (**J** 0474) Avtom Vychis Tekh, Akad Nauk Latv SSR 1967/5*57-80
◇ D05 F50 ◇ REV MR 45 # 6543 Zbl 169.317 • ID 08270

LORENTS, A.A. [1968] *Some problems in the constructive theory of finite probabilistic automata (Russian) (German summary)* (**J** 0068) Z Math Logik Grundlagen Math 14*413-447
◇ D05 F65 ◇ REV MR 40 # 1211 Zbl 169.317 • ID 08275

LORENTS, A.A. [1969] *Elements of the constructive probability theory (Russian) (German summary)* (**J** 0068) Z Math Logik Grundlagen Math 15*437-459
◇ F60 ◇ REV MR 44 # 3805 Zbl 195.19 • ID 16304

LORENTS, A.A. [1972] *Elemente der konstruktiven Theorie stochastischer Automaten (Russian)* (**X** 2230) Zinatne: Riga 235pp
◇ D05 D10 D98 F60 F98 ◇ REV MR 57 # 11977a Zbl 252 # 94036 • ID 63510

LORENTS, A.A. see Vol. IV for further entries

LORENTS, P.P. [1985] *Constructive and enumerable sets (Russian)* (**J** 0067) Usp Mat Nauk 40/3*211-212
• TRANSL [1985] (**J** 1399) Russ Math Surv 40/3*239-240
◇ F60 ◇ ID 46524

LORENTS, P.P. see Vol. IV for further entries

LORENZ, K. [1968] *Dialogspiele als semantische Grundlage von Logikkalkuelen I,II* (**J** 0009) Arch Math Logik Grundlagenforsch 11*32-55,73-100
◇ B60 F50 ◇ REV MR 37 # 1233 MR 39 # 2613 Zbl 179.13 • ID 28288

LORENZ, K. [1973] *Rules versus theorems. A new approach for mediation between intuitionistic and two-valued logic* (**J** 0122) J Philos Logic 2*352-369
◇ A05 B55 F50 ◇ REV MR 55 # 7724 Zbl 262 # 02008 • ID 27473

LORENZ, K. [1977] *On the relation between the partition of a whole into parts and the attribution of properties to an object* (**J** 0063) Studia Logica 36*351-362
◇ A05 F99 ◇ REV MR 80a:03006 Zbl 402 # 03008 • ID 28291

LORENZ, K. & LORENZEN, P. [1978] *Dialogische Logik* (**X** 0890) Wiss Buchges: Darmstadt viii + 238pp
◇ A05 B60 F99 ◇ REV MR 83j:03010 Zbl 435 # 03011 • ID 28287

LORENZ, K. see Vol. II for further entries

LORENZEN, P. [1950] *Einfuehrung in die Logik* (**J** 0008) Arch Math (Basel) 2*60-65
◇ F30 ◇ REV MR 11.411 Zbl 35.1 JSL 15.227 • ID 08279

LORENZEN, P. [1950] *Konstruktive Begruendung der Mathematik* (**J** 0044) Math Z 53*162-202
◇ A05 F65 ◇ REV MR 12.469 Zbl 41.343 JSL 18.260 • ID 08278

LORENZEN, P. [1951] *Algebraische und logistische Untersuchungen ueber freie Verbaende* (**J** 0036) J Symb Logic 16*81-106
◇ A05 F35 G10 ◇ REV MR 13.615 Zbl 45.295 JSL 16.269 • ID 08283

LORENZEN, P. [1951] *Die Widerspruchsfreiheit der klassischen Analysis* (**J** 0044) Math Z 54*1-24
◇ F35 F65 ◇ REV MR 13.310 Zbl 42.10 JSL 18.261 • ID 08281

LORENZEN, P. [1951] *Mass und Integral in der konstruktiven Analysis* (**J** 0044) Math Z 54*275-290
◇ F65 ◇ REV MR 13.615 Zbl 42.247 • ID 08280

LORENZEN, P. [1951] *Ueber das Prinzip "ex falso quodlibet"* (**J** 0175) Methodos 3
◇ A05 F99 ◇ ID 33165

LORENZEN, P. [1952] *Konstruktive Begruendung der klassischen Mathematik* (**P** 0593) Int Congr Math (II, 6);1950 Cambridge MA 1*727-728
◇ F65 ◇ ID 28062

LORENZEN, P. [1953] *Die Allgemeingueltigkeit der logischen Regeln* (**J** 0178) Stud Gen 6*605-609
◇ A05 F50 F65 ◇ REV MR 17.3 Zbl 52.7 JSL 30.104 • ID 08287

LORENZEN, P. [1955] *Einfuehrung in die operative Logik und Mathematik* (**X** 0811) Springer: Heidelberg & New York vii + 298pp
◇ A05 B98 F50 F65 F98 ◇ REV MR 17.223 Zbl 66.248 Zbl 68.8 JSL 22.289 JSL 35.330 • REM 2nd ed. 1969 • ID 08290

LORENZEN, P. [1958] *Formale Logik* (**X** 1174) Gruyter: Berlin 165pp
• TRANSL [1965] (**X** 0835) Reidel: Dordrecht viii+123pp (English)
◇ F50 F98 ◇ REV MR 21 # 645 Zbl 82.14 JSL 32.514 • ID 22149

LORENZEN, P. [1959] *Ueber die Begriffe "Beweis" und "Definition"* (**P** 0634) Constructivity in Math;1957 Amsterdam 169-177
◇ A05 F50 ◇ REV MR 21 # 4103 Zbl 86.9 JSL 40.501 • ID 08299

LORENZEN, P. [1960] *Constructive and axiomatic mathematics* (**J** 0154) Synthese 12*114-119
◇ A05 F65 ◇ REV MR 24 # A1801 Zbl 147.246 • ID 08301

LORENZEN, P. [1961] *Ein dialogisches Konstruktivitaetskriterium* (**P** 0633) Infinitist Meth;1959 Warsaw 193-200
◇ A05 F50 F65 ◇ REV MR 26 # 4880 Zbl 117.13 JSL 32.516 • ID 08300

LORENZEN, P. [1962] *Metamathematik* (**X** 0876) Bibl Inst: Mannheim 173pp
• TRANSL [1967] (**X** 0834) Gauthier-Villars: Paris 162pp (French) [1971] (**X** 1781) Tecnos: Madrid (Spanish)
◇ A05 B98 C60 D20 D35 D98 F98 ◇ REV MR 28 # 3932 Zbl 105.246 JSL 31.106 • ID 08303

LORENZEN, P. [1965] *Die klassische Analysis als eine konstruktive Theorie* (**P** 3696) Logic Colloq;1964 Bristol 18*81-94
◇ B28 F65 ◇ REV MR 34 # 2458 Zbl 158.5 • ID 08304

LORENZEN, P. [1965] *Differential und Integral. Eine konstruktive Einfuehrung in die klassische Analysis (Loose errata)* (**X** 1169) Akad Verlagsges: Wiesbaden v+293pp
• TRANSL [1971] (**X** 1358) Univ Texas Pr: Austin vi+303pp (English)
◇ F65 F98 ◇ REV MR 35 # 308 Zbl 171.271 • ID 22166

LORENZEN, P. [1968] *Constructive mathematics as a philosophical problem* (**J** 0020) Compos Math 20*133-142
• REPR [1968] (**C** 0727) Logic Found of Math (Heyting) 133-142
◇ A05 F65 ◇ REV MR 37 # 25 Zbl 169.293 JSL 40.499 • ID 08305

LORENZEN, P. [1968] *Operative Logik. Eine Uebersicht 1956-1966* (**C** 0552) Phil Contemp - Chroniques 135-140
◇ A05 F50 ◇ ID 14646

LORENZEN, P. [1972] *Dialogkalkuele* (**J** 0009) Arch Math Logik Grundlagenforsch 15*99-102
◇ B60 F50 ◇ REV MR 50 # 12676 Zbl 288 # 02011 • ID 08311

LORENZEN, P. [1972] *Zur konstruktiven Deutung der semantischen Vollstaendigkeit klassischer Quantoren- und Modalkalkuele* (**J** 0009) Arch Math Logik Grundlagenforsch 15*103-117
◇ B45 F50 ◇ REV MR 49 # 4750 Zbl 305 # 02024 • ID 08310

LORENZEN, P. [1978] see LORENZ, K.

LORENZEN, P. see Vol. I, II, IV, V for further entries

LOS, J. & MOSTOWSKI, ANDRZEJ & RASIOWA, H. [1956] *A proof of Herbrand's theorem* (**J** 3941) J Math Pures Appl, Ser 9 35*19-24 • ERR/ADD ibid 40*129-134
◇ B10 C07 F05 F50 ◇ REV MR 17.699 MR 28 # 22 Zbl 199.7 Zbl 73.7 JSL 36.168 • ID 08333

LOS, J. see Vol. I, II, III, IV, V for further entries

LOURENCO, M. (ED.) [1979] *Goedel's theorem and the continuum hypothesis (Portuguese)* (**X** 2719) Fundacao Calouste Gulbenkian: Lisbon xcv+900pp
◇ B97 E50 F30 F97 ◇ REV MR 82b:03003 Zbl 486 # 03005 • REM Transl from several English articles • ID 70030

LOVELAND, D.W. & REDDY, C.R. [1978] *Presburger arithmetic with bounded quantifier alternation* (**P** 1740) ACM Symp Th of Comput (10);1978 San Diego 320-325
◇ B25 C10 D15 F20 F30 ◇ REV MR 80d:68057 • ID 31164

LOVELAND, D.W. see Vol. I, IV for further entries

LUCAS, T. [1983] see LAVENDHOMME, R.

LUCAS, T. [1984] see LAVENDHOMME, R.

LUCAS, T. see Vol. I, III, V for further entries

LUCCHESI, C.L. [1972] *The undecidability of the unification problem for 3rd order languages* (**J** 1388) Rep Univ Waterloo
◇ F35 ◇ ID 28638

LUCCHESI, C.L. see Vol. IV for further entries

LUCHT, L. [1976] see HISCHER, HORST

LUCI, F. [1982] *Una teoria dell'infinito potenziale (English summary)* (**J** 2100) Boll Unione Mat Ital, VI Ser, B 1*305-325
◇ E70 F99 ◇ REV MR 83f:03050 Zbl 507 # 03026 • REM Part I. Part II 1982 • ID 35304

LUCI, F. [1982] *Una teoria dell'infinito potenziale. II (English summary)* (**J** 3768) Boll Unione Mat Ital, VI Ser, D 1*21-39
◇ A05 F99 ◇ REV MR 84h:03133 Zbl 522 # 03050 • REM Part I 1982 • ID 34319

LUCIAN, M. [1972] *Systems of notations and the constructible hierarchy* (**X** 0858) Harvard Univ Pr: Cambridge
◇ D55 E45 F15 ◇ ID 28788

LUCKHAM, D.C. & PUTNAM, H. [1965] *On minimal and almost-minimal systems of notations* (**J** 0064) Trans Amer Math Soc 119*86-100
◇ F15 ◇ REV MR 32 # 2323 Zbl 218 # 02037 • ID 08369

LUCKHAM, D.C. see Vol. I, IV for further entries

LUCKHARDT, H. [1967] *Aussagenlogisch fundierte Theorien* (**J** 0009) Arch Math Logik Grundlagenforsch 10*37-58
◇ B25 F05 ◇ REV MR 35 # 4083 Zbl 165.305 • ID 08378

LUCKHARDT, H. [1969] *Kodifikation und Aussagenlogik* (**J** 0009) Arch Math Logik Grundlagenforsch 12*18-38
◇ B05 B25 F50 ◇ REV MR 41 # 3239 Zbl 182.6 • ID 08380

LUCKHARDT, H. [1970] *Ein Henkin-Vollstaendigkeitsbeweis fuer die intuitionistische Praedikatenlogik bezueglich der Kripke-Semantik* (**J** 0009) Arch Math Logik Grundlagenforsch 13*55-59
◇ C90 F50 ◇ REV MR 44 # 32 Zbl 212.17 • ID 08381

LUCKHARDT, H. [1970] *Kripke-Semantik der derivativen Praedikatenlogik* (**J** 0009) Arch Math Logik Grundlagenforsch 13*134-135
◇ C90 F50 ◇ REV MR 45 # 8499 Zbl 218 # 02019 • ID 08382

LUCKHARDT, H. [1973] *Extensional Goedel functional interpretation. A consistency proof of classical analysis* (**X** 0811) Springer: Heidelberg & New York vi+161pp
◇ F10 F35 F50 F98 ◇ REV MR 49#2281 Zbl 262#02031 • ID 08383

LUCKHARDT, H. [1975] *A short proof of a well-known theorem of intuitionistic analysis* (**J** 0068) Z Math Logik Grundlagen Math 21*185-186
◇ F55 ◇ REV MR 51#10047 Zbl 356#02029 • ID 08384

LUCKHARDT, H. [1975] *The real elements in a consistency proof for simple type theory I* (**P** 1440) ⊧ ISILC Proof Th Symp (Schuette);1974 Kiel 500*233-256
◇ F10 F35 F50 ◇ REV MR 54#12494 Zbl 326#02019 • ID 27278

LUCKHARDT, H. [1975] *Ueber Hilberts reale und ideale Elemente* (**J** 0009) Arch Math Logik Grundlagenforsch 17*61-70
◇ A05 F99 ◇ REV MR 55#5403 Zbl 353#02012 • ID 08385

LUCKHARDT, H. [1976] *Ueber das Markov-Prinzip. I,II* (**J** 0009) Arch Math Logik Grundlagenforsch 18*73-80,147-157
◇ F50 ◇ REV MR 58#10350 Zbl 356#02028 Zbl 381#03045 • ID 23716

LUCKHARDT, H. [1978] *A fundamental effect in computations on real numbers* (**J** 1426) Theor Comput Sci 5*321-324
◇ F50 ◇ REV MR 57#9504 Zbl 374#02018 • ID 51542

LUCKHARDT, H. [1980] see FRIEDRICH, W.

LUCKHARDT, H. [1980] *On constructive functions ranging over propositions* (**J** 0063) Studia Logica 39*371-374
◇ F50 ◇ REV MR 82h:03068 Zbl 463#03036 • ID 54576

LUCKHARDT, H. see Vol. I, II, IV for further entries

LUKAS, J.D. & PUTNAM, H. [1974] *Systems of notations and the ramified analytical hierarchy* (**J** 0036) J Symb Logic 39*243-253
◇ D30 D55 E45 F15 F35 ◇ REV MR 51#12496 Zbl 295#02027 • ID 08387

LUKASIEWICZ, J. [1952] *On the intuitionistic theory of deduction* (**J** 0028) Indag Math 14*202-212
◇ F50 ◇ REV MR 14.4 Zbl 48.4 JSL 19.216 • ID 08399

LUKASIEWICZ, J. [1953] *Comment on K.J.Cohen's remark* (**J** 0028) Indag Math 15*113
◇ B05 F50 ◇ REV MR 14.1053 Zbl 51.6 JSL 19.217 • ID 25951

LUKASIEWICZ, J. see Vol. I, II for further entries

LUZIN, N.N. [1929] *Sur les voies de la theorie des ensembles* (**P** 0741) Int Congr Math (II, 3);1928 Bologna 1*295-299
◇ A05 D55 E15 E50 F55 ◇ REV FdM 55.657 • ID 16811

LUZIN, N.N. see Vol. IV, V for further entries

LYALETSKIJ, A.V. & MALASHONOK, A.I. [1975] *A calculus of k-disjuncts with the rule of a latent clash-resolution (Russian)* (**C** 2962) Mat Voprosy Teor Intell Mashin 3-33
◇ B35 F07 ◇ REV MR 58#21501 Zbl 421#03045 • ID 53444

LYALETSKIJ, A.V. [1975] *A k-disjunct calculus (Russian)* (**C** 2962) Mat Voprosy Teor Intell Mashin 34-48
◇ B35 F07 ◇ REV MR 58#21502 Zbl 421#03046 • ID 53445

LYALETSKIJ, A.V. [1981] *A variant of Herbrand's theorem for formulas in prenex form (Russian)* (**J** 0040) Kibernetika, Akad Nauk Ukr SSR 1981/1*112-116,152
• TRANSL [1981] (**J** 0021) Cybernetics 17*125-129
◇ B35 F05 ◇ REV MR 84a:03072 Zbl 462#68074 • ID 35594

LYALETSKIJ, A.V. see Vol. I for further entries

MAASS, W. [1975] *Church-Rosser-Theorem fuer λ-Kalkuele mit unendlich langen Termen* (**P** 1440) ⊧ ISILC Proof Th Symp (Schuette);1974 Kiel 257-263
◇ B40 ◇ REV MR 53#10555 Zbl 325#02017 • ID 23065

MAASS, W. [1976] *Eine Funktionalinterpretation der praedikativen Analysis* (**J** 0009) Arch Math Logik Grundlagenforsch 18*27-46
◇ D20 F10 F35 F65 ◇ REV MR 58#21503 Zbl 381#03043 • ID 23711

MAASS, W. see Vol. IV, V for further entries

MAC LANE, S. [1975] *Sets, topoi, and internal logic in categories* (**P** 0775) Logic Colloq;1973 Bristol 119-134
◇ A05 E70 F50 G30 ◇ REV MR 52#5419 Zbl 313#18001 • ID 18268

MAC LANE, S. [1982] *Why commutative diagrams coincide with equivalent proofs* (**P** 3817) Algeb Homage:Ring Th & Rel Top (Jacobson);1981 New Haven 387-401
◇ F05 G30 ◇ REV MR 84c:18018 Zbl 504#18004 • ID 38409

MAC LANE, S. see Vol. I, V for further entries

MACHTEY, M. & WINKLMANN, K. & YOUNG, P. [1978] *Simple Goedel numberings, isomorphisms, and programming properties* (**J** 1428) SIAM J Comp 7*39-60
◇ D05 D10 D15 D20 D45 F40 ◇ REV MR 58#178 Zbl 412#03020 • ID 52951

MACHTEY, M. see Vol. IV for further entries

MACINTYRE, A. & SIMMONS, H. [1973] *Goedel's diagonalization technique and related properties of theories* (**S** 0019) Colloq Math (Warsaw) 28*165-180,329
◇ F30 G05 ◇ REV MR 48#10792 Zbl 288#02018 • ID 08471

MACINTYRE, A. [1980] *Ramsey quantifiers in arithmetic* (**P** 2625) Model Th of Algeb & Arithm;1979 Karpacz 186-210
◇ C62 C80 F30 F35 ◇ REV MR 83j:03099 Zbl 464#03031 JSL 50.1078 • ID 54621

MACINTYRE, A. [1982] see KREISEL, G.

MACINTYRE, A. see Vol. I, II, III, IV, V for further entries

MACNAB, D.S. [1977] *Some applications of double-negation sheafification* (**J** 3420) Proc Edinburgh Math Soc, Ser 2 279-285
◇ C30 C90 F50 G30 ◇ REV MR 57#412 Zbl 371#18010 • ID 51394

MACNAB, D.S. see Vol. II for further entries

MADDUX, R. [1983] *A sequent calculus for relation algebras* (**J** 0073) Ann Pure Appl Logic 25*73-101
◇ F07 G15 ◇ REV MR 85h:03067 Zbl 528#03016 • ID 37634

MADDUX, R. see Vol. III, IV, V for further entries

MADISON, E.W. [1970] *A note on computable real fields* (J 0036) J Symb Logic 35*239-241
◇ C57 C60 D45 F60 ◇ REV MR 42 # 7500 Zbl 291 # 02031 • ID 24826

MADISON, E.W. & NELSON, GEORGE C. [1975] *Some examples of constructive and non-constructive extension of the countable atomless boolean algebra* (J 3172) J London Math Soc, Ser 2 11*325-336
◇ C57 F60 G05 ◇ REV MR 52 # 89 Zbl 375 # 06009 • ID 18269

MADISON, E.W. see Vol. III, IV for further entries

MAEHARA, S. [1954] *Eine Darstellung der intuitionistischen Logik in der klassischen* (J 0111) Nagoya Math J 7*45-64
◇ F50 ◇ REV MR 16.325 Zbl 56.247 JSL 22.79 • ID 08512

MAEHARA, S. [1954] *Gentzen's theorem on an extended predicate calculus* (J 0081) Proc Japan Acad 30*923-926
◇ B15 F05 ◇ REV MR 18.271 Zbl 57.247 JSL 27.109 • ID 08513

MAEHARA, S. [1957] *General recursive functions in the number-theoretic formal system* (J 0260) Ann Jap Ass Phil Sci 1*119-130
◇ D20 F30 ◇ REV MR 23 # A61 Zbl 109.9 JSL 27.90 • ID 08518

MAEHARA, S. [1957] *Remark on Skolem's theorem concerning the impossibility of characterization of the natural number sequence* (J 0081) Proc Japan Acad 33*588-590
◇ B28 C62 F30 ◇ REV MR 20 # 6364 Zbl 79.8 JSL 31.659 • ID 08516

MAEHARA, S. [1958] *Another proof of Takeuti's theorems on Skolem's paradox* (J 0434) J Fac Sci Univ Tokyo, Sect 1 7*541-556
◇ C07 F25 F30 ◇ REV MR 20 # 2283 Zbl 102.7 JSL 31.659 • ID 08519

MAEHARA, S. & NISHIMURA, T. & SEKI, S. [1960] *Non-constructive proofs of a metamathematical theorem concerning the consistency of analysis and its extension* (J 0260) Ann Jap Ass Phil Sci 1*269-288
◇ B15 F05 F35 ◇ REV MR 23 # A794 JSL 32.283 • ID 08521

MAEHARA, S. [1960] *On the interpolation theorem of Craig (Japanese)* (J 0091) Sugaku 12*235-237
◇ B10 C40 F05 ◇ REV MR 29 # 3356 Zbl 123.246 • ID 08520

MAEHARA, S. & TAKEUTI, G. [1961] *A formal system of first-order predicate calculus with infinitely long expressions* (J 0090) J Math Soc Japan 13*357-370
◇ C40 C75 F05 ◇ REV MR 29 # 2178 Zbl 108.2 JSL 27.468 • ID 08524

MAEHARA, S. [1962] *Cut-elimination theorem concerning a formal system for ramified theory of types which admits quantification on types* (J 0260) Ann Jap Ass Phil Sci 2*55-64
◇ B15 F05 F30 F35 F65 ◇ REV MR 27 # 3504 Zbl 109.8 JSL 35.325 • ID 08526

MAEHARA, S. [1970] *A general theory of completeness proofs* (J 0260) Ann Jap Ass Phil Sci 3*242-256
◇ C07 C90 F50 ◇ REV MR 44 # 6443 Zbl 226 # 02032 • ID 08455

MAEHARA, S. & TAKEUTI, G. [1971] *Two interpolation theorems for a Π_1^1 predicate calculus* (J 0036) J Symb Logic 36*262-270
◇ B15 C40 F35 ◇ REV MR 46 # 6991 Zbl 278 # 02013 • ID 08528

MAEHARA, S. [1981] *Semi-formal finitist proof of the transfinite induction in an initial segment of Cantor's second number class* (P 3201) Logic Symposia;1979/80 Hakone 67-79
◇ F15 ◇ REV MR 83j:03102 Zbl 499 # 03043 • ID 33394

MAEHARA, S. see Vol. I, II for further entries

MAGARI, R. [1973] *Meaning and truth in the Peano arithmetic (Italian summary)* (J 0149) Atti Accad Naz Lincei Fis Mat Nat, Ser 8 54*902-903
◇ F30 ◇ REV MR 51 # 112 Zbl 294 # 02012 • ID 15209

MAGARI, R. [1974] *Sur certe teorie non enumerabili. Sulle limitazioni dei sistemi formali I* (J 3526) Ann Mat Pura Appl, Ser 4 98*119-152
◇ B53 B60 F30 ◇ REV MR 50 # 6802 Zbl 286 # 02051 • ID 08539

MAGARI, R. [1975] *Metodi algebrici in teoria della dimostrazione* (J 0012) Boll Unione Mat Ital, IV Ser 12*252-261
◇ F30 G05 ◇ REV MR 54 # 7216 Zbl 355 # 02022 • ID 25007

MAGARI, R. [1975] *Representation and duality theory for diagonalizable algebras (The algebraization of theories which express Theor; IV.)* (J 0063) Studia Logica 34*305-313
◇ F30 G25 ◇ REV MR 57 # 107 Zbl 355 # 02021 • REM Part III 1975 by Bernardi,C. Part IV 1975 by Montagna,F. • ID 50219

MAGARI, R. [1975] *Significato e verita nell'aritmetica peaniana (English summary)* (J 3526) Ann Mat Pura Appl, Ser 4 103*343-368
◇ A05 F30 ◇ REV MR 51 # 10075 Zbl 316 # 02037 • ID 17536

MAGARI, R. [1975] *The diagonalizable algebras (The algebraization of the theories which express Theor. II)* (J 0012) Boll Unione Mat Ital, IV Ser 12*117-125
◇ F30 G25 ◇ REV MR 57 # 105 Zbl 352 # 08009 • REM Part I was not published; Part III 1975 by Bernardi,C. • ID 50055

MAGARI, R. [1982] *Primi risultati sulla varieta di Boolos (English summary)* (J 2100) Boll Unione Mat Ital, VI Ser, B 1*359-367
◇ B45 F30 G25 ◇ REV MR 83m:03073 Zbl 487 # 03039 • ID 35470

MAGARI, R. [1983] *Verita non Tarskiana nelle algebre diagonalizzabili (English summary)* (J 2100) Boll Unione Mat Ital, VI Ser, B 2*739-758
◇ F30 G25 ◇ REV MR 86b:03086 Zbl 554 # 03033 • ID 44946

MAGARI, R. [1984] *Algebraic logic and diagonal phenomena* (P 3710) Logic Colloq;1982 Firenze 135-144
◇ F30 G25 ◇ REV MR 86b:03082 • ID 44144

MAGARI, R. see Vol. I, II for further entries

MAHE, L. [1980] see COSTE-ROY, M.-F.

MAHE, L. see Vol. I, V for further entries

MAINZER, K. [1972] *Mathematischer Konstruktivismus im Lichte kantischer Philosophie* (**J** 1099) Phil Math 9*3-26
◇ A05 F99 ◇ REV Zbl 249 # 02003 • ID 30083

MAINZER, K. [1975] *Zur Begruendung von Freges analytischen Begriffen 2.Stufe in konstruktiver Absicht* (**C** 4080) Frege & Mod Grundlagenf 90-110
◇ A05 F99 ◇ REV JSL 44.119 • ID 44566

MAINZER, K. [1977] *Is the intuitionistic bar induction a constructive principle ?* (**J** 0047) Notre Dame J Formal Log 18*583-588
◇ A05 F50 ◇ REV MR 58 # 5097 Zbl 314 # 02044
• ID 24297

MAINZER, K. see Vol. I for further entries

MAKKAI, M. [1965] *Ueber die transfinite Induktion in zahlentheoretischen Formalismen* (**P** 0797) Fonds des Math, Machines Math & Appl;1962 Tihany 39-43
◇ F30 ◇ REV MR 33 # 3910 Zbl 158.249 • ID 08575

MAKKAI, M. & REYES, G.E. [1977] *First order categorical logic. Model-theoretical methods in the theory of topoi and related categories* (**X** 0811) Springer: Heidelberg & New York viii + 301pp
◇ C75 C90 F50 G30 ◇ REV MR 58 # 21600
Zbl 357 # 18002 • ID 50433

MAKKAI, M. see Vol. III, IV, V for further entries

MALASHONOK, A.I. [1975] see LYALETSKIJ, A.V.

MALASHONOK, A.I. see Vol. I for further entries

MALIAUKIENE, L. [1982] *Equivalence of some arithmetical systems (Russian) (English and Lithuanian summaries)* (**J** 3939) Mat Logika Primen (Akad Nauk Litov SSR) 2*9-30
◇ F30 ◇ REV MR 85g:03081 Zbl 516 # 03039 • ID 37268

MALIAUKIENE, L. [1982] *Free variable additive arithmetic with restricted difference (Russian) (English and Lithuanian summaries)* (**J** 3939) Mat Logika Primen (Akad Nauk Litov SSR) 2*31-40
◇ F30 ◇ REV MR 85g:03082 Zbl 516 # 03040 • ID 37269

MALIAUKIENE, L. [1983] *A constructive proof of the replaceability on the induction scheme for quantifier-free formulas in additive arithmetic (Russian) (English and Lithuanian summaries)* (**J** 2574) Litov Mat Sb (Vil'nyus) 23/3*63-77
◇ F30 ◇ REV MR 85e:03141a Zbl 535 # 03032 • ID 38328

MALIAUKIENE, L. [1983] *A constructive proof of the replaceability of the induction axiom in the quantifier-free multiplicative arithmetic (Russian) (English and Lithuanian summaries)* (**J** 2574) Litov Mat Sb (Vil'nyus) 23/3*78-93
◇ F30 ◇ REV MR 85e:03141b Zbl 536 # 03040 • ID 37120

MALIAUKIENE, L. [1983] *On the relationship of certain weak inductions (Russian)(English and Lithuanian summaries)* (**J** 3939) Mat Logika Primen (Akad Nauk Litov SSR) 3*34-37
◇ F30 ◇ REV MR 85h:03062 Zbl 564 # 03046 • ID 43333

MALIAUKIENE, L. [1985] *Elimination of the induction axiom in the multiplicative arithmetic with restricted difference (Russian) (English and Lithuanian summaries)* (**J** 3939) Mat Logika Primen (Akad Nauk Litov SSR) 1985/4*9-15,135
◇ B28 F30 ◇ ID 49123

MALMNAES, P.E. & PRAWITZ, D. [1968] *A survey of some connections between classical, intuitionistic and minimal logic* (**P** 0608) Logic Colloq;1966 Hannover 215-229
◇ B55 F05 F25 F50 ◇ REV MR 38 # 4289 Zbl 188.11
JSL 40.503 • ID 24815

MAL'TSEV, A.I. [1961] *Constructive algebra I (Russian)* (**J** 0067) Usp Mat Nauk 16/3*3-60
• TRANSL [1961] (**J** 1399) Russ Math Surv 16/3*77-129
[1971] (**C** 2621) Mal'tsev: Metamath of Algeb Syst 148-214
◇ C57 C98 F60 F98 ◇ REV MR 27 # 1362 Zbl 129.259
JSL 31.647 • ID 28706

MAL'TSEV, A.I. see Vol. I, II, III, IV, V for further entries

MANCA, V. [1981] *Computational formalism: Abstract combinatory view-point and related first order logical framework* (**J** 2095) Fund Inform, Ann Soc Math Pol, Ser 4 4*3-18
◇ B40 D75 ◇ REV MR 84i:03088 Zbl 468 # 03026
• ID 55091

MANCA, V. see Vol. III for further entries

MANDELKERN, M. [1976] *Connectivity of an interval* (**J** 0053) Proc Amer Math Soc 54*170-172
◇ F55 ◇ REV MR 52 # 10966 Zbl 318 # 26002 • ID 21737

MANDELKERN, M. [1979] *Suprema of located sets* (**J** 3172) J London Math Soc, Ser 2 20*161-164
◇ F55 ◇ REV MR 80k:03066 Zbl 426 # 26002 • ID 75939

MANDELKERN, M. [1980] *Resolutions on the line* (**J** 0048) Pac J Math 88*91-100
◇ F55 ◇ REV MR 81m:54018 Zbl 436 # 26002 • ID 82165

MANDELKERN, M. [1981] *Located sets on the line* (**J** 0048) Pac J Math 95*401-409
◇ F55 ◇ REV MR 83a:03066 Zbl 449 # 26003 • ID 35076

MANDELKERN, M. [1982] *Components of an open set* (**J** 3194) J Austral Math Soc, Ser A 33*249-261
◇ F55 ◇ REV MR 83j:26002 Zbl 502 # 26004 • ID 39873

MANDELKERN, M. [1982] *Continuity of monotone functions* (**J** 0048) Pac J Math 99*413-418
◇ F55 ◇ REV MR 83g:26017 Zbl 537 # 26001 • ID 39159

MANDELKERN, M. [1983] *Constructive continuity* (**S** 2450) Mem Amer Math Soc, NS 42/277*v + 117pp
◇ F55 ◇ REV MR 84j:26007 Zbl 537 # 26002 • ID 39162

MANDELKERN, M. [1985] *Constructive mathematics* (**J** 0497) Math Mag 58/5*272-280
◇ F55 ◇ ID 49128

MANEVITZ, L.M. & STAVI, J. [1980] *Δ_0^2 operators and alternating sentences in arithmetic* (**J** 0036) J Symb Logic 45*144-154
◇ C62 F30 H15 ◇ REV MR 81a:03064 Zbl 458 # 03022
• ID 54428

MANEVITZ, L.M. see Vol. III, V for further entries

MANGIONE, C. [1971] *Su alcune questioni connesse con i principi di riflessione* (**P** 0669) Conv Teor Modelli & Geom;1969/70 Roma 189-201
◇ F30 ◇ REV MR 43 # 4671 Zbl 225 # 02048 • ID 08658

MANGIONE, C. see Vol. I, II, III for further entries

MANIA, B. [1936] *L'infini mathematique et l'evolution de la logique* (**P** 0632) Congr Int Phil des Sci;1935 Paris 6*51-57
◇ A05 F55 F65 ◇ REV JSL 2.83 FdM 62.1045 • ID 41341

MANIN, YU.I. [1977] *A course in mathematical logic* (**X** 0811) Springer: Heidelberg & New York xiii + 286pp
◇ B98 C07 D98 E35 E50 F30 G12 ◇ REV MR 56 # 15345 Zbl 383 # 03002 • REM Translated from Russian • ID 51984

MANIN, YU.I. [1979] *Provable and unprovable (Russian)* (**X** 2643) Sovet Radio: Moskva 168pp
• TRANSL [1981] (**X** 0885) Mir: Moskva 265pp (Spanish)
◇ A05 B98 E35 E50 F30 F98 G12 ◇ REV Zbl 403 # 03002 • ID 54716

MANIN, YU.I. [1981] *Expanding constructive universes* (**P** 3729) Algor in Modern Math & Comput Sci;1979 Urgench 255–260
• TRANSL [1982] (**P** 3803) Algor Sovrem Mat & Prilozh;1979 Urgench II*4–9
◇ A05 F99 ◇ REV MR 83h:03011 Zbl 477 # 68035 • ID 36095

MANIN, YU.I. see Vol. IV, V for further entries

MANN, C.R. [1975] *The connection between equivalence of proofs and Cartesian closed categories* (**J** 3240) Proc London Math Soc, Ser 3 31*289–310
◇ F05 F07 G30 ◇ REV MR 52 # 7869 Zbl 317 # 02036 • ID 18275

MANNOURY, G. [1955] *Finitistic-formalistic development of projective geometry* (**J** 3077) Nieuw Arch Wisk, Ser 3 3*143–147
◇ F65 ◇ REV MR 17.933 Zbl 67.125 • ID 08665

MANUKYAN, S.N. & ZASLAVSKIJ, I.D. [1964] *Jordan theorem in constructive analysis* (**J** 3651) Tr Nauch Sess Ehston (Talinn) 6
◇ F60 ◇ ID 33543

MANUKYAN, S.N. & ZASLAVSKIJ, I.D. [1968] *Partitionings of the plane by constructive curves (Russian)* (**S** 0422) Tr Vychisl Tsentra Akad Nauk Armyan SSR & Univ Erevan 5*26–138
◇ F60 ◇ REV MR 41 # 8211 Zbl 197.6 • ID 14373

MANUKYAN, S.N. [1971] *On interior points of nondegenerate constructive curves (Russian)* (**J** 0023) Dokl Akad Nauk SSSR 196*768–769
• TRANSL [1971] (**J** 0062) Sov Math, Dokl 12*246–248
◇ F60 ◇ REV MR 43 # 1813 Zbl 255 # 02033 • ID 08676

MANUKYAN, S.N. [1973] *Constructive everywhere dense simple arcs (Russian) (Armenian and English summaries)* (**J** 0312) Izv Akad Nauk Armyan SSR, Ser Mat 8*291–305,346
◇ E75 F60 ◇ REV MR 49 # 2304 Zbl 287 # 02020 • ID 08677

MANUKYAN, S.N. [1975] *Some properties of partitioning of the plane by non-degenerate constructive curves (Russian)* (**S** 0422) Tr Vychisl Tsentra Akad Nauk Armyan SSR & Univ Erevan 8*67–95
◇ F60 ◇ REV MR 56 # 5239 Zbl 398 # 03048 • ID 52776

MANUKYAN, S.N. [1975] *The stability of constructive simple arcs with respect to algorithmic transformations (Russian)* (**J** 0023) Dokl Akad Nauk SSSR 220*1267–1270
• TRANSL [1975] (**J** 0062) Sov Math, Dokl 16*238–242
◇ F60 ◇ REV MR 51 # 10049 Zbl 341 # 02022 • ID 17548

MANUKYAN, S.N. [1976] *Some topological peculiarities of constructive simple arcs (Russian)* (**S** 0554) Issl Teor Algor & Mat Logik (Moskva) 2*122–129,158
◇ F60 ◇ REV MR 58 # 16210 • ID 75970

MANUKYAN, S.N. [1977] *On a property of the notion of interior and exterior point with respect to constructive closed curve (Russian)(English and Armenian summaries)* (**J** 0312) Izv Akad Nauk Armyan SSR, Ser Mat 12*303–309
◇ F60 ◇ REV MR 58 # 5107 Zbl 407 # 03047 • ID 56213

MANUKYAN, S.N. [1982] *Algorithms transforming constructive curves into points (Russian) (Armenian summary)* (**S** 0422) Tr Vychisl Tsentra Akad Nauk Armyan SSR & Univ Erevan 10*55–71
◇ F60 ◇ REV MR 85b:03107 • ID 40748

MANUKYAN, S.N. [1984] *The Dirichlet problem in a constructive analysis (Russian) (Armenian summary)* (**J** 0346) Dokl Akad Nauk Armyan SSR 79*58–62
◇ F60 ◇ ID 44897

MARCHINI, C. [1982] *Realizations and witnesses for Kripke models* (**P** 3845) Conf Math Service of Man (2,Proc)(Feriet);1982 Las Palmas 471–476
◇ C90 F50 G30 ◇ REV Zbl 538 # 03046 • ID 41484

MARCHINI, C. see Vol. II, III, V for further entries

MARCJA, A. [1984] see LOLLI, G.

MARCJA, A. see Vol. I, II, III for further entries

MAREK, W. [1974] see APT, K.R.

MAREK, W. & SREBRNY, M. [1975] *No minimal transitive model of* Z^- (**J** 0068) Z Math Logik Grundlagen Math 21*225–228
◇ C62 E30 E70 F35 ◇ REV MR 52 # 2884 Zbl 315 # 02055 • ID 08758

MAREK, W. & SREBRNY, M. [1976] *Urelements and extendability* (**P** 1476) Set Th & Hierarch Th (2) (Mostowski);1975 Bierutowice 203–219
◇ C62 E30 E70 F25 F30 F35 ◇ REV MR 56 # 15415 Zbl 358 # 02067 • ID 24412

MAREK, W. [1978] *ω-models of second order arithmetic and admissible sets* (**J** 0027) Fund Math 98*103–120
◇ C62 E30 E45 F35 ◇ REV MR 57 # 16051 Zbl 385 # 03030 • ID 29210

MAREK, W. [1978] *Some comments on the paper by Artigue, Isambert, Perrin and Zalc: "Some remarks on bicommutability"* (**J** 0027) Fund Math 101*227–228
◇ C62 E30 E45 F25 F35 ◇ REV MR 80j:03019b Zbl 401 # 03025 • REM The paper was published ibid 101*207–226 • ID 76016

MAREK, W. see Vol. I, II, III, IV, V for further entries

MARGARIA, I. & ZACCHI, M. [1983] *Right and left invertibility in $\lambda-\beta$-calculus* (**J** 3441) RAIRO Inform Theor 17*71–88
◇ B40 ◇ REV MR 85d:03027 Zbl 523 # 03010 • ID 37022

MARGARIA, I. [1984] see DEZANI-CIANCAGLINI, M.

MARGENSTERN, M. [1974] *Constructive functionals in spaces of almost periodic functions (Russian) (English summary)* (**S** 0228) Zap Nauch Sem Leningrad Otd Mat Inst Steklov 40*45–62,156–157
• TRANSL [1977] (**J** 1531) J Sov Math 8*276–289
◇ F60 ◇ REV MR 51 # 2884 Zbl 354 # 02029 • ID 17332

MARGENSTERN, M. [1974] *Proprietes topologiques constructives des espaces de fonctions presque periodiques* (**X** 2854) Univ Paris XI UER Math: Paris 119pp
◇ F60 ◇ REV MR 58 # 10353 • ID 76042

MARGENSTERN, M. [1976] *Some constructive topological properties of function spaces* (**J** 1934) Ann Sci Univ Clermont Math 13*109-115
◇ F60 ◇ REV MR 58 # 10354 Zbl 362 # 02024 • ID 50746

MARGENSTERN, M. [1978] *On a variant of constructivisation of the theory of almost periodic functions* (**J** 0068) Z Math Logik Grundlagen Math 24*495-507
◇ F60 ◇ REV MR 80b:03101 Zbl 406 # 03073 • ID 56155

MARGENSTERN, M. [1984] *Sur une extension simple du calcul intuitionniste des predicats du premier ordre applique a l'analyse* (**J** 0068) Z Math Logik Grundlagen Math 30*317-324
◇ F50 ◇ REV Zbl 534 # 03029 • ID 42264

MARGENSTERN, M. see Vol. IV for further entries

MARINI, D. & MIGLIOLI, P.A. & ORNAGHI, M. [1975] *First order logic as a tool to solve and classify problems* (**P** 0784) GI Jahrestag (5);1975 Dortmund 669-679
◇ B10 B25 F50 ◇ REV MR 53 # 5282 Zbl 329 # 68079 • ID 22915

MARKOV, A.A. [1950] *A constructive logic (Russian)* (**J** 0067) Usp Mat Nauk 5/3*187-188
◇ F50 ◇ REV JSL 18.257 • ID 19382

MARKOV, A.A. [1954] *On the continuity of constructive functions (Russian)* (**J** 0067) Usp Mat Nauk 9/3(61)*226-230
◇ F60 ◇ REV MR 16.436 Zbl 56.249 JSL 21.319 • ID 19377

MARKOV, A.A. [1957] *Mathematical logic and numerical analysis (Russian)* (**J** 0476) Vest Akad Nauk SSSR 27*21-25
◇ F60 ◇ REV MR 19.240 Zbl 93.14 JSL 29.209 • ID 19376

MARKOV, A.A. [1958] *On constructive functions (Russian)* (**S** 0066) Tr Mat Inst Steklov 52*315-348
• TRANSL [1963] (**J** 0225) Amer Math Soc, Transl, Ser 2 29*163-195
◇ F60 ◇ REV MR 20 # 5131 Zbl 89.6 JSL 31.258 • ID 19369

MARKOV, A.A. [1959] *On a principle of constructive mathematical logic (Russian)* (**P** 0607) All-Union Math Conf (3);1956 Moskva 2*146-147
◇ F50 ◇ ID 28394

MARKOV, A.A. [1962] *On constructive mathematics (Russian)* (**S** 0066) Tr Mat Inst Steklov 67*8-14
• TRANSL [1971] (**J** 0225) Amer Math Soc, Transl, Ser 2 98*1-9
◇ A05 F60 ◇ REV MR 27 # 3528 Zbl 113.7 • ID 08770

MARKOV, A.A. [1968] *An approach to constructive mathematical logic* (**P** 0627) Int Congr Log, Meth & Phil of Sci (3,Proc);1967 Amsterdam 283-294
◇ F50 ◇ REV MR 41 # 60 Zbl 185.24 JSL 40.85 • ID 08777

MARKOV, A.A. [1970] *On the logic of constructive mathematics (Russian)* (**J** 0288) Vest Ser Mat Mekh, Univ Moskva 25/2*7-29
• TRANSL [1970] (**J** 0510) Moscow Univ Math Bull 25/2*45-60
◇ F50 ◇ REV MR 42 # 5779 Zbl 205.309 JSL 40.85 • ID 19367

MARKOV, A.A. [1971] *Essai de construction d'une logique de la mathematique constructive* (**J** 2076) Rev Int Philos 25*477-507
◇ F50 ◇ REV MR 58 # 27383 JSL 40.85 • ID 76047

MARKOV, A.A. [1974] *On the language* $Я_0$ *(Russian)* (**J** 0023) Dokl Akad Nauk SSSR 214*40-43
• TRANSL [1974] (**J** 0062) Sov Math, Dokl 15*38-42
◇ B25 F50 F65 ◇ REV MR 49 # 2287 Zbl 308 # 02032 • ID 63716

MARKOV, A.A. [1974] *On the language* $Я_1$ *(Russian)* (**J** 0023) Dokl Akad Nauk SSSR 214*279-282
• TRANSL [1974] (**J** 0062) Sov Math, Dokl 15*125-128
◇ D03 D35 F50 ◇ REV MR 49 # 2288 Zbl 308 # 02033 • ID 63719

MARKOV, A.A. [1974] *On the language* $Я_2$ *(Russian)* (**J** 0023) Dokl Akad Nauk SSSR 214*513-516
• TRANSL [1974] (**J** 0062) Sov Math, Dokl 15*184-188
◇ F50 ◇ REV MR 49 # 2289 Zbl 308 # 02034 • ID 63717

MARKOV, A.A. [1974] *On the language* $Я_3$ *(Russian)* (**J** 0023) Dokl Akad Nauk SSSR 214*765-768
• TRANSL [1974] (**J** 0062) Sov Math, Dokl 15*242-246
◇ F50 ◇ REV MR 49 # 2290 Zbl 308 # 02035 • ID 63718

MARKOV, A.A. [1974] *On the languages* $Я_4, Я_5, \ldots$ *(Russian)* (**J** 0023) Dokl Akad Nauk SSSR 214*1031-1034
• TRANSL [1974] (**J** 0062) Sov Math, Dokl 15*313-317
◇ F50 ◇ REV MR 49 # 2291 Zbl 308 # 02036 • ID 63721

MARKOV, A.A. [1974] *On the language* $Я_\omega$ *(Russian)* (**J** 0023) Dokl Akad Nauk SSSR 214*1262-1264
• TRANSL [1974] (**J** 0062) Sov Math, Dokl 15*356-359
◇ F50 ◇ REV MR 49 # 2292 Zbl 308 # 02037 • ID 63720

MARKOV, A.A. [1974] *On the language* $Я_{\omega|}$ *(Russian)* (**J** 0023) Dokl Akad Nauk SSSR 215*57-60
• TRANSL [1974] (**J** 0062) Sov Math, Dokl 15*443-447
◇ F50 ◇ REV MR 49 # 2293 Zbl 342 # 02023 • ID 63715

MARKOV, A.A. [1974] *The completeness of the classical predicate calculus in constructive logic (Russian)* (**J** 0023) Dokl Akad Nauk SSSR 215*266-269
• TRANSL [1974] (**J** 0062) Sov Math, Dokl 15*476-481
◇ C07 F50 ◇ REV MR 49 # 2294 Zbl 306 # 02046 • ID 63722

MARKOV, A.A. [1976] *An attempt to construct a logic of constructive mathematics (Russian)* (**S** 0554) Issl Teor Algor & Mat Logik (Moskva) 2*3-31,157
◇ F50 ◇ REV MR 58 # 21510 • ID 76049

MARKOV, A.A. [1977] *On a semantical language hierarchy in a constructive mathematical logic* (**P** 1704) Int Congr Log, Meth & Phil of Sci (5);1975 London ON 1*299-306
◇ F50 ◇ REV MR 58 # 21511 Zbl 372 # 02019 • ID 51427

MARKOV, A.A. see Vol. I, II, IV, V for further entries

MARKOVIC, Z. [1977] *Reduced products of saturated intuitionistic theories* (**J** 0400) Publ Inst Math, NS (Belgrade) 21(35)*131-133
◇ C20 C90 F50 ◇ REV MR 57 # 16048 Zbl 367 # 02009 • ID 51169

MARKOVIC, Z. [1979] *An intuitionistic omitting types theorem* (**J** 0400) Publ Inst Math, NS (Belgrade) 26(40)*167-169
◇ C07 C90 F50 ◇ REV MR 82g:03103 Zbl 442 # 03027 • ID 56384

MARKOVIC, Z. [1983] *Some preservation results for classical and intuitionistic satisfiability in Kripke models* (**J** 0047) Notre Dame J Formal Log 24*395-398
◇ B20 C25 C90 F50 ◇ REV MR 85a:03016 Zbl 487 # 03015 • ID 37386

MARKOVIC, Z. [1984] *Kripke models for intuitionistic theories with decidable atomic formulas* (**J** 0400) Publ Inst Math, NS (Belgrade) 36(50)*3-7
◇ C90 F50 ◇ REV MR 86h:03105 • ID 45080

MARKOVIC, Z. see Vol. I, III for further entries

MARKWALD, W. [1954] *Zur Theorie der konstruktiven Wohlordnungen* (**J** 0043) Math Ann 127*135-149
◇ C80 D55 F15 ◇ REV MR 15.771 Zbl 56.47 JSL 20.283 • ID 08779

MARKWALD, W. see Vol. I, II, IV for further entries

MARONGIU, G. [1974] *Sequenze di predicati aritmetici di tipo "teor" (English summary)* (**J** 0012) Boll Unione Mat Ital, IV Ser 9*361-375
◇ F30 ◇ REV MR 51 # 5279 Zbl 317 # 02047 • ID 17458

MARONGIU, G. see Vol. III, V for further entries

MARTIN, J.N. [1984] *Epistemic semantics for classical and intuitionistic logic* (**J** 0047) Notre Dame J Formal Log 25*105-116
◇ B50 F50 ◇ REV MR 85e:03058 Zbl 562 # 03005 • ID 40627

MARTIN, J.N. see Vol. I, II for further entries

MARTIN, N.M. [1971] *Systems of logic* (1111) Preprints, Manuscr., Techn. Reports etc. 268pp
◇ F50 ◇ REM Mimeographed Notes • ID 32214

MARTIN, N.M. see Vol. II for further entries

MARTIN-LOEF, P. [1970] *Notes on constructive mathematics* (**X** 1163) Almqvist & Wiksell: Stockholm 109pp
• TRANSL [1975] (**X** 0885) Mir: Moskva 136pp
◇ F50 F60 F98 ◇ REV MR 58 # 5098 Zbl 273 # 02021 • ID 23407

MARTIN-LOEF, P. [1971] *Hauptsatz for the intuitionistic theory of iterated inductive definitions* (**P** 0604) Scand Logic Symp (2);1970 Oslo 179-216
◇ F05 F35 F50 ◇ REV MR 52 # 7870 Zbl 231 # 02040 • ID 15064

MARTIN-LOEF, P. [1971] *Hauptsatz for the theory of species* (**P** 0604) Scand Logic Symp (2);1970 Oslo 217-233
◇ F05 F35 F50 ◇ REV MR 48 # 10776 Zbl 231 # 02039 JSL 40.503 • ID 27439

MARTIN-LOEF, P. [1972] *About models for intuitionistic type theories and the notion of definitional equality* (1111) Preprints, Manuscr., Techn. Reports etc.
◇ C90 F50 ◇ REM Univ. Stockholm, Dep. Math., Report No. 4 • ID 21229

MARTIN-LOEF, P. [1972] *Infinite terms and a system of natural deduction* (**J** 0020) Compos Math 24*93-103
◇ C75 F05 F50 ◇ REV MR 46 # 20 Zbl 237 # 02006 • ID 08833

MARTIN-LOEF, P. [1973] *Hauptsatz for intuitionistic simple type theory* (**P** 0793) Int Congr Log, Meth & Phil of Sci (4,Proc);1971 Bucharest 279-290
◇ B15 F05 F35 F50 ◇ REV MR 56 # 8332 • ID 76088

MARTIN-LOEF, P. [1975] *About models for intuitionistic type theories and the notion of definitional equality* (**P** 0757) Scand Logic Symp (3);1973 Uppsala 81-109
◇ B40 F35 F50 ◇ REV MR 52 # 10376 Zbl 334 # 02017 JSL 43.373 • ID 21678

MARTIN-LOEF, P. [1975] *An intuitionistic theory of types: predicative part* (**P** 0775) Logic Colloq;1973 Bristol 73-118
◇ F05 F35 F50 ◇ REV MR 52 # 7856 Zbl 334 # 02016 JSL 49.311 • ID 18280

MARTIN-LOEF, P. [1982] *Constructive mathematics and computer programming* (**P** 3622) Int Congr Log, Meth & Phil of Sci (6,Proc);1979 Hannover 153-175
• REPR [1984] (**J** 0354) Phil Trans Roy Soc London, Ser A 312*501-518
◇ B75 F35 F50 ◇ REV MR 85d:03112 MR 86c:03051 Zbl 443 # 68039 Zbl 552 # 03040 • ID 41113

MARTIN-LOEF, P. [1984] *Intuitonistic type theory. Notes by Giovanni Sambin of a series of lectures given in Padua, June 1980* (**X** 1732) Bibliopolis: Napoli xii+92pp
◇ F35 F50 F98 ◇ REV Zbl 571 # 03030 • ID 44677

MARTIN-LOEF, P. see Vol. I, IV, V for further entries

MARTINO, E. [1981] see GIARETTA, P.

MARTINO, E. [1982] *Creative subject and bar theorem* (**P** 3638) Brouwer Centenary Symp;1981 Noordwijkerhout 311-318
◇ F50 ◇ REV MR 85g:03088a Zbl 522 # 03047 • ID 37796

MARTINO, E. [1983] *Connection between the "principle of inductive evidence" and the bar theorem* (**J** 0028) Indag Math 45*325-328
◇ F50 ◇ REV MR 85g:03088b Zbl 527 # 03033 • ID 37579

MARTINO, E. [1983] *Semantica intuizionista naturale e semantica di Beth generallizata* (**P** 3829) Atti Incontri Log Mat (1);1982 Siena 373-376
◇ F50 ◇ REV MR 84k:03006 Zbl 522 # 03048 • ID 37797

MARTINO, E. [1985] *On the Brouwerian concept of negative continuity* (**J** 0122) J Philos Logic 14*379-398
◇ A10 F50 F55 ◇ ID 47642

MARTINO, E. [1985] *On Brower's concept of negative continuity (Italian)* (**P** 4646) Atti Incontri Log Mat (2);1983/84 Siena 345-351
◇ A05 A10 F50 F55 ◇ ID 49681

MART'YANOV, V.I. [1977] *Extended universal theories of the integers (Russian)* (**J** 0003) Algebra i Logika 16*588-602,624
• TRANSL [1977] (**J** 0069) Algeb and Log 16*395-405
◇ B25 D35 F30 ◇ REV MR 58 # 27421 Zbl 394 # 03038 • ID 29230

MART'YANOV, V.I. see Vol. I, III, IV for further entries

MASLOV, S.YU. & MINTS, G.E. & OREVKOV, V.P. [1965] *Unsolvability in the constructive predicate calculus of certain classes of formulas containing only monadic predicate variables (Russian)* (**J** 0023) Dokl Akad Nauk SSSR 163*295-297
• TRANSL [1965] (**J** 0062) Sov Math, Dokl 6*918-920
◇ B20 D35 F50 ◇ REV MR 33 # 52 Zbl 173.10 JSL 35.143 • ID 19335

MASLOV, S.YU. [1967] *An invertible sequential version of the constructive predicate calculus (Russian)* (**S** 0228) Zap Nauch Sem Leningrad Otd Mat Inst Steklov 4*96-111
• TRANSL [1967] (**J** 0521) Semin Math, Inst Steklov 4*36-42
◇ F50 ◇ REV MR 39 # 6736 Zbl 161.4 • ID 08845

MASLOV, S.YU. [1972] *Deduction search in calculi of general type (Russian) (English summary)* (**S** 0228) Zap Nauch Sem Leningrad Otd Mat Inst Steklov 32*59-65,156
• TRANSL [1976] (**J** 1531) J Sov Math 6*395-400
◇ B35 D03 F07 ◇ REV MR 49 # 8845 Zbl 344 # 02026
• ID 08850

MASLOV, S.YU. & NORGELA, S.A. [1974] *Cut-type rules for calculi of general type (Russian) (English summary)* (**S** 0228) Zap Nauch Sem Leningrad Otd Mat Inst Steklov 40*63-76,157
• TRANSL [1977] (**J** 1531) J Sov Math 8*289-299
◇ F07 ◇ REV MR 51 # 7835 Zbl 363 # 02039 • ID 17502

MASLOV, S.YU. & NORGELA, S.A. [1977] *Herbrand strategies and the "greater deducibility" relation (Russian) (English summary)* (**S** 0228) Zap Nauch Sem Leningrad Otd Mat Inst Steklov 68*51-61,144
• TRANSL [1981] (**J** 1531) J Sov Math 15*28-33
◇ B25 B35 F05 ◇ REV MR 58 # 21504 Zbl 358 # 02030
• ID 32672

MASLOV, S.YU. [1979] *Deductive systems with thinnings (Russian)* (**P** 2554) All-Union Symp Th Log Infer;1974 Moskva 128-146
◇ F07 ◇ REV MR 84m:03088 • ID 35784

MASLOV, S.YU. see Vol. I, II, III, IV for further entries

MASSERON, M. [1983] *Rungs and trees* (**J** 0036) J Symb Logic 48*847-863
◇ D55 E05 F15 ◇ REV MR 85e:03145 Zbl 567 # 03026
• ID 40760

MATIYASEVICH, YU.V. [1971] *A sufficient condition for the convergence of monotone sequences (Russian) (English summary)* (**S** 0228) Zap Nauch Sem Leningrad Otd Mat Inst Steklov 20*97-103,285
• TRANSL [1973] (**J** 1531) J Sov Math 1*59-63
◇ F60 ◇ REV MR 45 # 62 Zbl 222 # 02029 • ID 08882

MATIYASEVICH, YU.V. & SLISENKO, A.O. (EDS.) [1971] *Studies in constructive mathematics and mathematical logic IV (Russian)* (**S** 0228) Zap Nauch Sem Leningrad Otd Mat Inst Steklov 20*1-294 • ERR/ADD ibid 32*159-162
◇ F97 ◇ REV Zbl 218 # 02002 • REM Part III 1969 by Slisenko,A.O. • ID 26269

MATIYASEVICH, YU.V. & SLISENKO, A.O. (EDS.) [1972] *Studies in constructive mathematics and mathematical logic V (Russian)* (**S** 0228) Zap Nauch Sem Leningrad Otd Mat Inst Steklov 32*1-162
• TRANSL [1976] (**J** 1531) J Sov Math 6*347-474
◇ F97 ◇ REV MR 48 # 50 MR 56 # 8313 • ID 70234

MATIYASEVICH, YU.V. & SLISENKO, A.O. (EDS.) [1974] *Studies in constructive mathematics and mathematical logic VI (Russian)* (**S** 0228) Zap Nauch Sem Leningrad Otd Mat Inst Steklov 40*1-163
◇ F97 ◇ REV MR 49 # 10522 • ID 90018

MATIYASEVICH, YU.V. & SLISENKO, A.O. (EDS.) [1976] *Studies in constructive mathematics and mathematical logic VII (Russian)* (**S** 0228) Zap Nauch Sem Leningrad Otd Mat Inst Steklov 60*1-229
• TRANSL [1980] (**J** 1531) J Sov Math 14*1429-1572
◇ F97 ◇ REV MR 54 # 9966 • ID 70157

MATIYASEVICH, YU.V. & SLISENKO, A.O. (EDS.) [1979] *Studies in constructive mathematics and mathematical logic VIII (Russian)* (**S** 0228) Zap Nauch Sem Leningrad Otd Mat Inst Steklov 88*1-252
• TRANSL [1982] (**J** 1531) J Sov Math 20*2263-2390
◇ B97 F97 ◇ REV MR 80j:03005 • ID 70057

MATIYASEVICH, YU.V. see Vol. I, IV for further entries

MATSON, T. [1974] see BISHOP, P.

MATSUMOTO, K. [1950] *On a lattice relating to the intuitionistic logic* (**J** 0442) J Osaka Inst Sci & Tech, Part 1 2*97-107
◇ F50 G10 ◇ REV MR 15.280 Zbl 41.5 JSL 24.250
• ID 08887

MATSUMOTO, K. [1950] *Sur la structure concernant la logique moderne* (**J** 0442) J Osaka Inst Sci & Tech, Part 1 2*67-78
◇ A05 B45 F50 G05 ◇ REV MR 15.278 Zbl 41.4 JSL 23.443 • ID 08888

MATSUMOTO, K. & SHIRAI, K. [1980] *A note on elimination of function symbols in the intuitionistic predicate calculus* (**J** 1005) Rep Fac Sci, Shizuoka Univ 14*13-20
◇ F50 ◇ REV MR 81f:03018 Zbl 443 # 03026 • ID 56432

MATSUMOTO, K. see Vol. I, II, III, IV for further entries

MATULIS, V.A. [1963] *Variants of the classical predicate calculus with a single deduction tree (Russian)* (**J** 0023) Dokl Akad Nauk SSSR 148*768-770
• TRANSL [1963] (**J** 0062) Sov Math, Dokl 4*176-179
◇ B10 F07 ◇ REV MR 26 # 4894 Zbl 166.249 • ID 08901

MATULIS, V.A. see Vol. I for further entries

MAURER, C. [1975] see LAWVERE, F.W.

MAURER, C. see Vol. V for further entries

MAURI, G. [1984] see BERTONI, A.

MAURI, G. see Vol. III, IV for further entries

MAYBERRY, J. [1985] *Global quantification in Zermelo-Fraenkel set theory* (**J** 0036) J Symb Logic 50*289-301
◇ E30 E70 F10 F50 ◇ ID 42536

MAYBERRY, J. see Vol. V for further entries

MAYOH, B.H. [1965] *Unsolvable problems in the theory of computable numbers* (**P** 0688) Logic Colloq;1963 Oxford 272-279
◇ D20 D35 F60 ◇ REV MR 35 # 56 Zbl 192.59
• ID 08924

MAYOH, B.H. [1968] *Semi-effective numberings and definitions of the computable numbers* (**J** 0009) Arch Math Logik Grundlagenforsch 11*113-125
◇ D25 D30 F60 ◇ REV MR 40 # 4106 Zbl 179.19
• ID 08926

MAYOH, B.H. [1970] *The relation between an object and its name: notation systems and their fixed point theorems* (**P** 0785) Scand Logic Symp (1);1968 Aabo 77-95
◇ A05 B03 D20 D35 D45 F30 F60 ◇ REV MR 52 # 2850 Zbl 318 # 02048 • ID 17641

MAYOH, B.H. [1974] *Extracting information from logical proofs* (**C** 1936) Log Th & Semant Anal (Kanger) 61-72
◇ B35 F07 ◇ REV Zbl 297 # 68073 • ID 63802

MAYOH, B.H. see Vol. I, II, III, IV for further entries

MAZUR, S. [1963] *Computable analysis* (**X** 1034) PWN: Warsaw 110pp
◇ F60 F98 ◇ REV MR 27 # 3517 Zbl 113.243 JSL 36.148 • ID 23408

MAZUR, S. see Vol. I, II, V for further entries

MCALOON, K. [1975] *Applications alternees de theoremes d'incompletude et des theoremes de completude (English summary)* (**J** 2313) C R Acad Sci, Paris, Ser A-B 280*A849-A852
◇ C62 E35 F30 ◇ REV MR 51 # 5295 Zbl 307 # 02036 • ID 17403

MCALOON, K. [1975] *Formules de Rosser pour ZF (English summary)* (**J** 2313) C R Acad Sci, Paris, Ser A-B 281*A669-A672
◇ C62 C70 E30 E35 E45 F25 ◇ REV MR 52 # 5417 Zbl 333 # 02046 • ID 18286

MCALOON, K. [1977] *Consistency statements and number theories* (**P** 1729) Colloq Int Log;1975 Clermont-Ferrand 199-207
◇ C62 F30 ◇ REV MR 80b:03110 Zbl 435 # 03037 • ID 55797

MCALOON, K. [1978] *Completeness theorems, incompleteness theorems and models of arithmetic* (**J** 0064) Trans Amer Math Soc 239*253-277
◇ C25 C62 F30 ◇ REV MR 81e:03066 Zbl 388 # 03028 • ID 52282

MCALOON, K. [1978] *Diagonal methods and strong cuts in models of arithmetic* (**P** 1897) Logic Colloq;1977 Wroclaw 171-181
◇ C57 C62 D80 F30 ◇ REV MR 80k:03068 Zbl 458 # 03021 • ID 54427

MCALOON, K. [1980] *Les rapports entre la methode des indicatrices et la methode de Goedel pour obtenir des resultats d'independance* (**J** 1620) Asterisque 73*31-39
◇ C62 E35 F30 ◇ REV MR 82g:03098a Zbl 462 # 03016 JSL 48.483 • ID 54523

MCALOON, K. (ED.) [1980] *Modeles de l'arithmetique* (**X** 2244) Soc Math France: Paris 155pp
◇ C62 C97 F30 ◇ REV MR 81j:03006 Zbl 424 # 00006 • ID 70045

MCALOON, K. [1980] *Progressions transfinies de theories axiomatiques, formes combinatoires du theoreme d'incompletude et fonctions recursives a croissance rapide* (**J** 1620) Asterisque 73*41-58
◇ C62 D20 F15 F30 ◇ REV MR 82g:03098b Zbl 462 # 03017 JSL 48.483 • ID 54524

MCALOON, K. & RESSAYRE, J.-P. [1981] *Les methodes de Kirby-Paris et la theorie des ensembles* (**P** 3404) Model Th & Arithm;1979/80 Paris 154-184
◇ C62 E05 E30 E55 F30 ◇ REV MR 84c:03067 Zbl 498 # 03041 • ID 34009

MCALOON, K. [1981] see BERLINE, C.

MCALOON, K. [1982] see FRIEDMAN, H.M.

MCALOON, K. [1982] see CEGIELSKI, P.

MCALOON, K. [1983] see ANSHEL, M.

MCALOON, K. [1983] see CLOTE, P.

MCALOON, K. [1985] *Paris-Harrington incompleteness and progressions of theories* (**P** 4046) Rec Th;1982 Ithaca 447-460
◇ F15 F30 ◇ ID 46416

MCALOON, K. see Vol. I, III, IV, V for further entries

MCBETH, R. [1976] *Fundamental sequences for initial ordinals smaller than a certain* Θ_0 (**J** 0068) Z Math Logik Grundlagen Math 22*97-104
◇ E10 F15 ◇ REV MR 56 # 2829 Zbl 331 # 04001 • ID 18494

MCBETH, R. [1979] *A note on Hardy's persistent numbers* (**J** 0068) Z Math Logik Grundlagen Math 25*375-378
◇ E10 F15 ◇ REV MR 81a:04002 Zbl 426 # 03051 • ID 53647

MCBETH, R. [1980] *Fundamental sequences for exponential polynomials* (**J** 0068) Z Math Logik Grundlagen Math 26*115-122
◇ E07 E10 F15 ◇ REV MR 81f:04004 Zbl 436 # 04003 • ID 76194

MCBETH, R. [1981] *A note on exponential polynomials and prime factors* (**J** 0068) Z Math Logik Grundlagen Math 27*213-214
◇ D20 F15 ◇ REV MR 82i:03053 Zbl 471 # 04002 • ID 55251

MCBETH, R. [1984] *A second normal form for functions of the system EP* (**J** 0068) Z Math Logik Grundlagen Math 30*393-400
◇ F15 ◇ REV Zbl 551 # 03024 • ID 42272

MCBETH, R. see Vol. IV, V for further entries

MCCALL, S. [1962] *A simple decision procedure for one-variable implication/negation formulae in intuitionistic logic* (**J** 0047) Notre Dame J Formal Log 3*120-122
◇ B25 F50 ◇ REV MR 26 # 6048 Zbl 118.248 JSL 29.212 • ID 08942

MCCALL, S. [1970] *A non-classical theory of truth, with an application to intuitionism* (**J** 0325) Amer Phil Quart 7*83-88
◇ F50 ◇ ID 31176

MCCALL, S. see Vol. II, III, V for further entries

MCCARTY, C. [1983] *Intuitionism: an introduction to a seminar* (**J** 0122) J Philos Logic 12*105-149
◇ A05 F50 F98 ◇ REV MR 85g:03040 Zbl 527 # 03034 • ID 37580

MCCARTY, C. [1984] *Information systems, continuity and realizability* (**P** 2989) Log of Progr; 1983 Pittsburgh 341-359
◇ B75 F50 ◇ REV Zbl 558 # 03031 • ID 44761

MCCARTY, C. see Vol. III for further entries

MCCULLOUGH, D.P. [1971] *Logical connectives for intuitionistic propositional logic* (**J** 0036) J Symb Logic 36*15-20
◇ F50 ◇ REV MR 44 # 6447 Zbl 216.288 JSL 38.660 • ID 08979

MCKAY, C.G. [1967] *A note on the Jaskowski sequence* (**J** 0068) Z Math Logik Grundlagen Math 13*95-96
◇ B55 F50 ◇ REV MR 37 # 34 Zbl 148.7 JSL 38.520 • REM Result incorrect • ID 08983

MCKAY, C.G. [1967] *Implicationless WFFS in IC* (**J** 0047) Notre Dame J Formal Log 8*227-228
◇ F50 ◇ REV MR 40 # 1263 Zbl 189.281 • ID 08984

McKay, C.G. see Vol. II, III for further entries

McKenzie, R. [1971] *Negative solution of the decision problem for sentences true in every subalgebra of* $\langle N, + \rangle$ (**J** 0036) J Symb Logic 36*607-609
◇ D35 F30 ◇ REV MR 45 # 6625 Zbl 248 # 02052
• ID 08997

McKenzie, R. see Vol. I, III, IV, V for further entries

McKinsey, J.C.C. [1939] *Proof of the independence of the primitive symbols of Heyting's calculus of propositions* (**J** 0036) J Symb Logic 4*155-158
◇ F50 ◇ REV MR 1.131 Zbl 22.193 JSL 5.38
FdM 65.1105 • ID 09016

McKinsey, J.C.C. see Vol. I, II, III, V for further entries

McMinn, T.J. [1974] *A formal number-termed number system based on recursion* (**J** 0308) Rocky Mountain J Math 4*649-672
◇ B28 F30 ◇ REV MR 50 # 7084 Zbl 293 # 02032
• ID 09065

McNaughton, R. [1953] *Some formal relative consistency proofs* (**J** 0036) J Symb Logic 18*136-144
◇ F25 F30 F35 ◇ REV MR 14.1052 Zbl 53.5 JSL 25.353
• ID 09067

McNaughton, R. see Vol. I, II, III, IV, V for further entries

Medvedev, Yu.T. [1962] *Finite problems (Russian)* (**J** 0023) Dokl Akad Nauk SSSR 142*1015-1018
• TRANSL [1962] (**J** 0062) Sov Math, Dokl 3*227-230
◇ D80 F50 ◇ REV MR 24 # A3067 Zbl 286 # 02028
JSL 38.330 • ID 09078

Medvedev, Yu.T. [1963] *Interpretation of logical formulae by means of finite problems and its relation to the realizability theory (Russian)* (**J** 0023) Dokl Akad Nauk SSSR 148*771-774
• TRANSL [1963] (**J** 0062) Sov Math, Dokl 4*180-183
◇ D80 F50 ◇ REV MR 26 # 4904 Zbl 218 # 02048
JSL 38.330 • ID 09079

Medvedev, Yu.T. [1966] *Interpretation of logical formulae by means of finite problems (Russian)* (**J** 0023) Dokl Akad Nauk SSSR 169*20-23
• TRANSL [1966] (**J** 0062) Sov Math, Dokl 7*857-860
◇ D80 F50 ◇ REV MR 38 # 37 Zbl 192.38 JSL 38.330
• ID 09080

Medvedev, Yu.T. [1969] *A method for proving the unsolvability of algorithmic problems (Russian)* (**J** 0023) Dokl Akad Nauk SSSR 185*1232-1235
• TRANSL [1969] (**J** 0062) Sov Math, Dokl 10*495-498
◇ D35 F50 ◇ REV MR 40 # 4112 Zbl 193.316 • ID 09081

Medvedev, Yu.T. [1972] *Locally finitary algorithmic problems (Errata ibid 204*1286) (Russian)* (**J** 0023) Dokl Akad Nauk SSSR 203*285-288
• TRANSL [1972] (**J** 0062) Sov Math, Dokl 13*382-386
◇ D20 F30 F50 ◇ REV MR 47 # 24 Zbl 262 # 02034
• ID 63851

Medvedev, Yu.T. [1973] *An interpretation of intuitionistic number theory* (**P** 0793) Int Congr Log, Meth & Phil of Sci (4,Proc);1971 Bucharest 129-136
◇ F15 F30 F50 ◇ REV MR 56 # 5240 • ID 76260

Medvedev, Yu.T. see Vol. II, IV for further entries

Meigne, M. [1959] *La consistance des theories formelles et le fondement des mathematiques* (**X** 0872) Blanchard: Paris 115pp
◇ A05 F25 F98 ◇ REV MR 22 # 7942 Zbl 94.7 • ID 09085

Melikyan, S.M. [1974] *Constructive transfinite hierarchies of pseudonumbers (Russian)* (**J** 0346) Dokl Akad Nauk Armyan SSR 59*257-260
◇ D55 F60 ◇ REV MR 53 # 7737 Zbl 307 # 02024
• ID 22984

Melikyan, S.M. [1981] *On some properties of classes of pseudonumbers (Russian) (Armenian summary)* (**J** 0312) Izv Akad Nauk Armyan SSR, Ser Mat 73*83-86
◇ F60 ◇ REV MR 83e:03100 Zbl 507 # 03017 • ID 35248

Melikyan, S.M. [1981] *Two theorems on classes of pseudonumbers (Russian) (Armenian summary)* (**J** 0312) Izv Akad Nauk Armyan SSR, Ser Mat 73*273-278
◇ F60 ◇ REV MR 83f:03063 Zbl 507 # 03018 • ID 35310

Melikyan, S.M. [1982] *Classes of pseudonumbers (Russian) (Armenian summary)* (**S** 0422) Tr Vychisl Tsentra Akad Nauk Armyan SSR & Univ Erevan 10*72-89
◇ F60 ◇ REV MR 85i:03182 • ID 44293

Melikyan, S.M. see Vol. IV for further entries

Mellis, W. [1984] *PA-beweisbare ∀∃-Formeln* (**J** 0009) Arch Math Logik Grundlagenforsch 24*23-48
◇ F30 ◇ REV MR 85k:03039 Zbl 547 # 03034 • ID 42604

Meloni, G.C. [1977] see Bozzi, S.

Meloni, G.C. & Radaelli, G. [1979] *Calcolo naturale per categorie con residui* (**J** 0207) Ist Lombardo Accad Sci Rend, A (Milano) 113*45-53
◇ B60 F07 G30 ◇ REV MR 82h:18001 Zbl 449 # 03075
• ID 56742

Meloni, G.C. [1980] see Bozzi, S.

Meloni, G.C. see Vol. I, II, III, V for further entries

Mendelson, E. [1963] *On some recent criticism of Church's thesis* (**J** 0047) Notre Dame J Formal Log 4*201-205
◇ A05 D20 F99 ◇ REV MR 29 # 3369 Zbl 126.20
JSL 33.471 • ID 09116

Mendelson, E. see Vol. I, II, III, V for further entries

Mendler, N.P. [1985] see Constable, R.L.

Menger, K. [1928] *Bemerkungen zu Grundlagenfragen I* (**J** 0157) Jbuchber Dtsch Math-Ver 37*213-226
◇ A05 B30 E30 F99 ◇ REV FdM 54.95 • REM Part II 1928 • ID 39109

Menger, K. [1930] *Der Intuitionismus* (**J** 0168) Blaetter Deutsch Philos 4*311-325
◇ A05 F99 ◇ REV FdM 56.44 • ID 38646

Menger, K. [1931] *Ueber den Konstruktivitaetsbegriff. Zweite Mitteilung* (**J** 0931) Anz Oesterr Akad Wiss, Math-Nat Kl 68*7-9
◇ A05 F99 ◇ REV Zbl 1.51 FdM 57.99 • ID 38647

Menger, K. [1979] *Selected papers in logic and foundations, didactics, economics* (**X** 0835) Reidel: Dordrecht xii+341pp
◇ A05 A10 F96 ◇ REV MR 80f:01026 Zbl 396 # 01025
• ID 52610

Menger, K. see Vol. I, II, IV, V for further entries

MEREDITH, C.A. [1953] *A single axiom of positive logic* (**J** 0093) J Comp Syst 1*169-170
◇ B20 F50 ◇ REV MR 15.1 Zbl 53.197 JSL 19.144 • ID 09134

MEREDITH, C.A. & PRIOR, A.N. [1968] *Equational logic* (**J** 0047) Notre Dame J Formal Log 9*212-226 • ERR/ADD ibid 10*452
◇ B20 F50 G05 ◇ REV MR 40 # 24a Zbl 175.259 • ID 09143

MEREDITH, C.A. see Vol. I, II for further entries

MEREDITH, D. [1974] *Combinatory and propositional logic* (**J** 0047) Notre Dame J Formal Log 15*156-160
◇ B40 ◇ REV MR 49 # 2276 Zbl 272 # 02044 • ID 09149

MEREDITH, D. [1975] *Combinator operations* (**J** 0063) Studia Logica 34*367-385
◇ B40 ◇ REV MR 57 # 15984 Zbl 329 # 02009 • ID 31934

MEREDITH, D. [1978] *Positive logic and λ-constants* (**J** 0063) Studia Logica 37*269-285
◇ B20 B25 B40 ◇ REV MR 80c:03019 Zbl 393 # 03008 • ID 31932

MEREDITH, D. [1979] *Axiomatics for implication* (**J** 0047) Notre Dame J Formal Log 20*89-91
◇ B20 F50 ◇ REV MR 80e:03022 Zbl 314 # 02025 • ID 31933

MEREDITH, D. [1980] *A positive logic proof procedure* (**C** 3050) Essays Combin Log, Lambda Calc & Formalism (Curry) 503-510
◇ B20 B35 B40 F50 ◇ REV MR 82a:03014 Zbl 469 # 03006 • ID 76287

MEREDITH, D. see Vol. I, II, III for further entries

MESCHKOWSKI, H. [1956] *Rekursive reelle Zahlen* (**J** 0044) Math Z 66*189-202
◇ F60 ◇ REV MR 18.552 Zbl 72.5 • ID 09159

MESCHKOWSKI, H. [1956] *Zur rekursiven Funktionentheorie* (**J** 0118) Acta Math 95*9-23
◇ F60 ◇ REV MR 19.238 Zbl 72.5 • ID 09160

MESCHKOWSKI, H. see Vol. V for further entries

METAKIDES, G. & NERODE, A. [1982] *The introduction of non-recursive methods into mathematics* (**P** 3638) Brouwer Centenary Symp;1981 Noordwijkerhout 319-335
◇ C57 D45 D80 F60 ◇ REV MR 85d:03121 Zbl 511 # 03020 • ID 37391

METAKIDES, G. see Vol. I, III, IV, V for further entries

METEAU, G. [1974] *The foundations of mathematics - a new analysis showing Goedel's theorem based on fallacy* (**X** 2721) Meteau: Harlow 12pp
◇ A05 F30 ◇ REV MR 55 # 2475 • ID 76308

METROPOLIS, N. & ROTA, G.-C. & TANNY, S. [1973] *Significance arithmetic: the carrying algorithm* (**J** 0164) J Comb Th, Ser A 14*386-421
◇ B75 F30 ◇ REV MR 48 # 222 Zbl 259 # 00001 • ID 82238

METROPOLIS, N. & ROTA, G.-C. [1974] *Significance arithmetic on the algebra of binary strings* (**C** 1213) Stud in Num Anal (Lanczos) 241-251
◇ D15 F65 G05 ◇ REV MR 50 # 6815 Zbl 311 # 65031 • ID 21270

MEY VAN DER, G. & POEL VAN DER, W.L. & SCHAAP, C.E. [1980] *New arithmetical operators in the theory of combinators. I,II,III* (**J** 0028) Indag Math 42*271-286,287-302,303-325
◇ B40 ◇ REV MR 82a:68163 Zbl 442 # 03015 Zbl 442 # 03016 Zbl 442 # 03017 • ID 56372

MEYER, A.R. [1973] see HELM, J.P.

MEYER, A.R. [1982] *What is a model of the lambda calculus?* (**J** 0194) Inform & Control 52*87-122
◇ B40 ◇ REV MR 84g:03022 Zbl 507 # 03002 • ID 34136

MEYER, A.R. [1984] see BRUCE, KIM B.

MEYER, A.R. & WAND, M. [1985] *Continuation semantics in typed lambda-calculi* (**P** 4571) Log of Progr;1985 Brooklyn 219-224
◇ B40 ◇ REV Zbl 565 # 68028 • ID 48666

MEYER, A.R. [1985] see BREAZU-TANNEN, V.

MEYER, A.R. see Vol. I, II, III, IV for further entries

MEYER, R.K. [1972] see LEBLANC, H.

MEYER, R.K. [1973] *Intuitionism, entailment, negation* (**P** 0783) Truth, Syntax & Modal;1970 Philadelphia 168-198
◇ B46 F50 ◇ REV MR 53 # 95 Zbl 317 # 02016 JSL 42.315 • ID 09200

MEYER, R.K. [1976] *Ackermann, Takeuti, und Schnitt: γ for higher-order relevant logic* (**J** 0387) Bull Sect Logic, Pol Acad Sci 5*138-144
◇ B46 F05 F35 ◇ REV MR 58 # 27341 • ID 27122

MEYER, R.K. [1976] *Metacompleteness* (**J** 0047) Notre Dame J Formal Log 17*501-516
◇ B45 F50 ◇ REV MR 55 # 12479 Zbl 232 # 02015 • ID 21939

MEYER, R.K. [1976] *Relevant arithmetic* (**J** 0387) Bull Sect Logic, Pol Acad Sci 5*133-137
◇ B46 F30 ◇ REV MR 58 # 21477 • ID 27121

MEYER, R.K. [1977] *First degree formulas in Curry's LD* (**J** 0047) Notre Dame J Formal Log 18*181-191
◇ B46 F50 ◇ REV MR 56 # 8328 Zbl 339 # 02021 • ID 21964

MEYER, R.K. [1978] see BUNDER, M.W.

MEYER, R.K. & MORTENSEN, C. [1985] *Relevant quantum arithmetic* (**C** 4181) Math Log & Formal Syst (Costa da) 221-226
◇ B46 F30 ◇ REV Zbl 567 # 03007 • ID 48205

MEYER, R.K. see Vol. I, II, III, IV, V for further entries

MEZGHICHE, M. [1984] *Une nouvelle Cβ-reduction dans la logique combinatoire (English summary)* (**J** 1426) Theor Comput Sci 31*151-163
◇ B40 ◇ REV Zbl 557 # 03009 • ID 45456

MEZHLUMBEKOVA, V.F. [1975] *Cut-elimination in a system of negationless arithmetic (Russian) (English summary)* (**J** 0288) Vest Ser Mat Mekh, Univ Moskva 30/6*10-16 • TRANSL [1975] (**J** 0510) Moscow Univ Math Bull 30/5-6*61-66
◇ B20 F05 F30 F50 ◇ REV MR 53 # 7735 Zbl 322 # 02031 • ID 63913

MEZHLUMBEKOVA, V.F. [1975] *Deductive capabilities of negationless intuitionistic arithmetic (Russian) (English summary)* (**J** 0288) Vest Ser Mat Mekh, Univ Moskva 30/2*3-9
• TRANSL [1975] (**J** 0510) Moscow Univ Math Bull 30/1-2*73-78
◇ F30 F50 ◇ REV MR 52#13336 Zbl 305#02044 • ID 21798

MEZHLUMBEKOVA, V.F. [1975] *On systems of the negationless calculus of predicates (Russian) (English summary)* (**J** 0288) Vest Ser Mat Mekh, Univ Moskva 30/5*3-9
• TRANSL [1975] (**J** 0510) Moscow Univ Math Bull 30/5-6*1-6
◇ B20 F50 ◇ REV MR 52#13337 Zbl 325#02016 • ID 21799

MICALI, S. [1980] see BOEHM, C.

MICALI, S. see Vol. IV for further entries

MICHEL, J.Y. [1973] *APLAMBDA: A lambda-calculus machine for APL-operators* (**P** 3361) APL Congr '73;1973 Copenhagen 325-331
◇ B40 ◇ REV Zbl 271#68035 • ID 63920

MIGLIOLI, P.A. [1975] see MARINI, D.

MIGLIOLI, P.A. & ORNAGHI, M. [1981] *A logically justified model of computation. I,II* (**J** 2095) Fund Inform, Ann Soc Math Pol, Ser 4 4*151-172,277-341
◇ B75 D80 F07 F50 ◇ REV MR 83e:03090 Zbl 473#68017 Zbl 483#68032 • ID 55396

MIGLIOLI, P.A. & MOSCATO, U. & ORNAGHI, M. [1981] *Trees in Kripke models and in an intuitionistic refutation system* (**P** 2923) CAAP'81 Arbres en Algeb & Progr (6);1981 Genova 316-331
◇ C90 F50 ◇ REV MR 83h:03019 Zbl 472#03048 • ID 55311

MIGLIOLI, P.A. & MOSCATO, U. & ORNAGHI, M. & USBERTI, G. [1982] *Constructive validity and classical truth to assign meaning to programs* (**P** 3845) Conf Math Service of Man (2,Proc)(Feriet);1982 Las Palmas 490-500
◇ B75 F65 ◇ REV Zbl 519#03044 • ID 37546

MIGLIOLI, P.A. & ORNAGHI, M. [1982] *Constructive proofs and logical computations* (**J** 3932) Comput Artif Intell (Bratislava) 1*369-388
◇ F65 ◇ REV Zbl 511#68067 • ID 38547

MIGLIOLI, P.A. & MOSCATO, U. & ORNAGHI, M. & USBERTI, G. [1983] *Alcuni calcoli intermedi costruttivi* (**P** 3829) Atti Incontri Log Mat (1);1982 Siena 377-386
◇ B55 F65 ◇ REV MR 84k:03006 Zbl 522#03014 • ID 37781

MIGLIOLI, P.A. [1984] see BERTONI, A.

MIGLIOLI, P.A. see Vol. II, III, IV for further entries

MIHALJINEC, M. [1960] *On the continuity of constructive transformations. I,II (Serbo-Croatian summaries)* (**J** 0371) Glas Mat-Fiz Astron, Ser 2 (Zagreb) 15*21-29,229-235
◇ F60 ◇ REV MR 23#A1528 MR 32#4006 • ID 42824

MIHALJINEC, M. [1965] *Inverse upper bound theorems for constructive real functions (Serbo-Croatian summary)* (**J** 0371) Glas Mat-Fiz Astron, Ser 2 (Zagreb) 20*177-187
◇ F60 ◇ REV MR 35#55 Zbl 192.60 • ID 45438

MIHALJINEC, M. [1972] *Some applications of the theory of categories in partially constructive mathematics (Serbo-Croatian summary)* (**J** 3519) Glas Mat, Ser 3 (Zagreb) 7(27)*3-24
◇ F60 G30 ◇ REV MR 50#9550 Zbl 272#02057 • ID 09238

MIHALJINEC, M. [1978] *Endomorphisms of sheafs of constructive functions* (**J** 3519) Glas Mat, Ser 3 (Zagreb) 13(33)*215-222
◇ F60 G30 ◇ REV MR 80j:03092 Zbl 425#18009 • ID 53588

MIHALJINEC, M. see Vol. I for further entries

MIJOULE, R. [1975] *Une formalisation des fondements par les logiques internes (English summary)* (**J** 2313) C R Acad Sci, Paris, Ser A-B 281*A1071-A1073
◇ F99 G30 ◇ REV MR 53#2665 Zbl 323#18001 • ID 21612

MIJOULE, R. see Vol. I, III, IV, V for further entries

MIKKELSEN, C.J. [1975] see KOCK, A.

MIKKELSEN, C.J. see Vol. I, V for further entries

MIKOLAJEWICZ, B. [1983] *Some classes of models for Peano arithmetic and for some related theories (Polish) (English summary)* (**J** 0481) Acta Univ Wroclaw 605(34)(Logika 10)*31-69
◇ B28 F30 ◇ REV Zbl 549#03022 • ID 43117

MIKOLAJEWICZ, B. [1984] see KUBINSKI, T.

MILLER, DALE A. [1984] *Expansion tree proofs and their conversion to natural deduction proofs* (**P** 2633) Autom Deduct (7);1984 Napa 375-393
◇ B35 F07 ◇ REV Zbl 547#03031 • ID 43218

MILLER, DALE A. see Vol. I for further entries

MILLER, L.W. [1976] *Normal functions and constructive ordinal notations* (**J** 0036) J Symb Logic 41*439-459
◇ E10 F15 ◇ REV MR 53#12894 Zbl 345#02018 • ID 14776

MILLER, R.E. [1977] see LIPTON, R.J.

MILLS, G. [1980] *A tree analysis of unprovable combinatorial statements* (**P** 2625) Model Th of Algeb & Arithm;1979 Karpacz 248-311
◇ F30 ◇ REV MR 84c:03101 Zbl 472#05019 • ID 34035

MILLS, G. [1981] see ERDOES, P.

MILLS, G. & PARIS, J.B. [1984] *Regularity in models of arithmetic* (**J** 0036) J Symb Logic 49*272-280
◇ C62 C80 F30 ◇ REV MR 85h:03035 • ID 42484

MILLS, G. see Vol. III, V for further entries

MILNER, R. [1974] *A calculus for the mathematical theory of computation* (**P** 1511) Int Symp Th Progr;1972 Novosibirsk 332-343
◇ B40 ◇ REV MR 56#1776 Zbl 279#68020 • ID 82262

MILNER, R. [1975] *Processes: a mathematical model of computing agents* (**P** 0775) Logic Colloq;1973 Bristol 157-173
◇ B40 ◇ REV MR 54#4171 Zbl 316#68017 • ID 82263

MILNER, R. [1976] *Models of LCF* (**C** 3517) Found of Comput Sci, Vol 2, Part 2 49-63
◇ B40 D20 ◇ REV MR 58#19304 Zbl 364#02018 • ID 50948

MILNER, R. [1977] *Fully abstract models of typed λ-calculi* (J 1426) Theor Comput Sci 4*1-22
◇ B40 B75 ◇ REV MR 58#16238 Zbl 386#03006
• ID 52182

MILNER, R. [1979] *ICF: A way of doing proofs with a machine* (P 2059) Math Founds of Comput Sci (8);1979 Olomouc 146-159
◇ B35 B40 ◇ REV MR 81g:68124 Zbl 423#68049
• ID 53582

MILNER, R. see Vol. I, II, IV for further entries

MINES, R. [1975] see BERG, G.

MINES, R. [1976] see BERG, G.

MINES, R. [1977] see BERG, G.

MINES, R. [1978] see JULIAN, W.

MINES, R. [1981] see BRIDGES, D.S.

MINES, R. [1982] *Algebraic number theory, a survey* (P 3638) Brouwer Centenary Symp;1981 Noordwijkerhout 337-358
◇ F55 ◇ REV MR 85b:03108 Zbl 529#12001 • ID 40754

MINES, R. [1982] see BRIDGES, D.S.

MINES, R. & RICHMAN, F. [1982] *Separability and factoring polynomials* (J 0308) Rocky Mountain J Math 12*43-54
◇ F55 ◇ REV MR 84e:03075 Zbl 499#12019 • ID 34407

MINES, R. [1983] see JULIAN, W.

MINES, R. & RICHMAN, F. [1984] *Valuation theory: A constructive view* (J 0401) J Number Th 19*40-62
◇ F55 ◇ REV MR 85j:03109 Zbl 549#12015 • ID 43152

MINES, R. [1984] see BRIDGES, D.S.

MINICHIELLO, J.K. [1969] *An extension of negationless logic* (J 0047) Notre Dame J Formal Log 10*298-302
◇ F50 ◇ REV MR 40#5422 Zbl 198.14 JSL 36.689
• ID 09299

MINTS, G.E. [1962] *An analogue of Herbrand's theorem for a constructive predicate calculus (Russian)* (J 0023) Dokl Akad Nauk SSSR 147*783-786
• TRANSL [1962] (J 0062) Sov Math, Dokl 3*1712-1715
◇ F05 F50 ◇ REV MR 26#1242 Zbl 199.34 JSL 36.524
• ID 16221

MINTS, G.E. [1962] *The differentiability predicate and the differentiation operator in constructive mathematical analysis (Russian)* (J 0023) Dokl Akad Nauk SSSR 147*1032-1034
• TRANSL [1962] (J 0062) Sov Math, Dokl 3*1774-1776
◇ F60 ◇ REV MR 26#6055 Zbl 134.10 • ID 09272

MINTS, G.E. & OREVKOV, V.P. [1963] *An extension of the theorems of Glivenko and Kreisel to a certain class of formulae of predicate calculus (Russian)* (J 0023) Dokl Akad Nauk SSSR 152*553-554
• TRANSL [1963] (J 0062) Sov Math, Dokl 4*1365-1367
◇ F50 ◇ REV MR 27#4737 Zbl 133.249 • ID 09274

MINTS, G.E. [1964] *On predicate and operator variants of the formation of theories of constructive mathematics (Russian)* (S 0066) Tr Mat Inst Steklov 72*383-436 • ERR/ADD ibid 93*257-258
• TRANSL [1972] (J 0225) Amer Math Soc, Transl, Ser 2 100*1-68
◇ F50 ◇ REV MR 33#2543 MR 37#35 Zbl 183.15
• ID 19432

MINTS, G.E. [1965] see MASLOV, S.YU.

MINTS, G.E. [1966] *Herbrand's theorem for the predicate calculus with equality and functional symbols (Russian)* (J 0023) Dokl Akad Nauk SSSR 169*273-275
• TRANSL [1966] (J 0062) Sov Math, Dokl 7*911-914
◇ B10 F05 ◇ REV MR 34#7341 Zbl 156.250 JSL 35.325
• ID 09276

MINTS, G.E. [1966] *Skolem's method of elimination of positive quantifiers in sequential calculi (Russian)* (J 0023) Dokl Akad Nauk SSSR 169*24-27
• TRANSL [1966] (J 0062) Sov Math, Dokl 7*861-864
◇ B25 C10 F50 ◇ REV MR 34#5648 Zbl 186.5 JSL 36.526 • ID 19428

MINTS, G.E. [1967] *Analog of Herbrand's theorem for non-prenex formulas of the constructive predicate calculus (Russian)* (S 0228) Zap Nauch Sem Leningrad Otd Mat Inst Steklov 4*123-133
• TRANSL [1967] (J 0521) Semin Math, Inst Steklov 4*47-51
◇ F05 F50 ◇ REV MR 39#2616 Zbl 186.5 JSL 36.525
• ID 28486

MINTS, G.E. [1967] *Choice of terms in quantifier rules of the constructive predicate calculus (Russian)* (S 0228) Zap Nauch Sem Leningrad Otd Mat Inst Steklov 4*112-122
• TRANSL [1967] (J 0521) Semin Math, Inst Steklov 4*43-46
◇ F50 ◇ REV MR 39#2615 Zbl 186.5 JSL 36.525
• ID 09277

MINTS, G.E. [1967] *Herbrand's theorem (Russian)* (C 2542) Mat Teor Log Vyvoda 311-350
◇ B10 F05 F07 ◇ REV MR 36#4958 Zbl 156.250
• ID 37138

MINTS, G.E. & OREVKOV, V.P. [1967] *On imbedding operators (Russian)* (S 0228) Zap Nauch Sem Leningrad Otd Mat Inst Steklov 4*160-167
• TRANSL [1967] (J 0521) Semin Math, Inst Steklov 4*64-66
◇ F25 F30 F50 ◇ REV MR 39#34 Zbl 188.14 • ID 27579

MINTS, G.E. [1967] *Translator's supplements to K.Goedel's paper (Russian)* (C 2542) Mat Teor Log Vyvoda 305-310
◇ F10 ◇ REM The article by Goedel was published in J0076 12(1958)*133-142 • ID 37136

MINTS, G.E. [1967] *Variation in the deduction search tactics in sequential calculi (Russian)* (S 0228) Zap Nauch Sem Leningrad Otd Mat Inst Steklov 4*134-151
• TRANSL [1967] (J 0521) Semin Math, Inst Steklov 4*52-59
◇ B35 F07 F50 ◇ REV MR 41#1505 Zbl 166.249
• ID 19424

MINTS, G.E. [1968] *Admissible and deductive rules (Russian)* (S 0228) Zap Nauch Sem Leningrad Otd Mat Inst Steklov 8*189-191
• TRANSL [1968] (J 0521) Semin Math, Inst Steklov 8*90-91
◇ B10 F07 F50 ◇ REV MR 43#4626 Zbl 207.4
• ID 19422

MINTS, G.E. [1968] *Disjunctive interpretation of the LJ calculus (Russian)* (S 0228) Zap Nauch Sem Leningrad Otd Mat Inst Steklov 8*182-188
• TRANSL [1968] (J 0521) Semin Math, Inst Steklov 8*86-89
◇ F50 ◇ REV MR 43#6056 Zbl 172.291 JSL 36.527
• ID 09282

MINTS, G.E. [1968] *Implicative complexity of axiomatic systems (Russian)* (S 0228) Zap Nauch Sem Leningrad Otd Mat Inst Steklov 8*175-181
• TRANSL [1968] (J 0521) Semin Math, Inst Steklov 8*83-85
◇ F07 F50 ◇ REV MR 43#3088 Zbl 172.291 • ID 09279

MINTS, G.E. [1968] *Independence of the postulates of natural calculi (Russian)* (**S** 0228) Zap Nauch Sem Leningrad Otd Mat Inst Steklov 8*192-195
• TRANSL [1968] (**J** 0521) Semin Math, Inst Steklov 8*92-94
◇ F07 F50 ◇ REV MR 43 # 3091 Zbl 174.9 • ID 09280

MINTS, G.E. [1968] *Solvability of the problem of deducibility in LJ for a class of formulas which do not contain negative occurences of quantors (Russian)* (**S** 0066) Tr Mat Inst Steklov 98*121-130
• TRANSL [1968] (**S** 0055) Proc Steklov Inst Math 98*135-145
◇ B25 F50 ◇ REV MR 41 # 6669 Zbl 179.11 • ID 19418

MINTS, G.E. [1968] *The construction of conservative logical inferences (Russian)* (**S** 0066) Tr Mat Inst Steklov 98*112-120
• TRANSL [1968] (**S** 0055) Proc Steklov Inst Math 98*125-134
◇ F07 F50 ◇ REV MR 40 # 2520 Zbl 179.11 • ID 19420

MINTS, G.E. (ED.) [1970] *Recursive mathematical analysis (Russian)* (**X** 2027) Nauka: Moskva
◇ F60 ◇ ID 48652

MINTS, G.E. [1971] *Exact estimation of the provability of transfinite induction in the initial segments of arithmetic (Russian) (English summary)* (**S** 0228) Zap Nauch Sem Leningrad Otd Mat Inst Steklov 20*134-144,286
• TRANSL [1973] (**J** 1531) J Sov Math 1*85-91
◇ F05 F15 F30 ◇ REV MR 44 # 6487 Zbl 222 # 02023
• ID 09287

MINTS, G.E. [1971] *Quantifier-free and one-quantifier systems (Russian) (English summary)* (**S** 0228) Zap Nauch Sem Leningrad Otd Mat Inst Steklov 20*115-133,285
• TRANSL [1973] (**J** 1531) J Sov Math 1*71-84
◇ B20 D20 F30 ◇ REV MR 44 # 6486 Zbl 222 # 02022
• ID 09286

MINTS, G.E. [1972] *Derivability of addmissible rules (Russian) (English summary)* (**S** 0228) Zap Nauch Sem Leningrad Otd Mat Inst Steklov 32*85-89,156
• TRANSL [1976] (**J** 1531) J Sov Math 6*417-421
◇ B55 F50 ◇ REV MR 49 # 8816 Zbl 358 # 02031
• ID 09288

MINTS, G.E. [1972] *The Skolem method in intuitionistic calculi (Russian)* (**S** 0066) Tr Mat Inst Steklov 121*67-99,165
• TRANSL [1972] (**S** 0055) Proc Steklov Inst Math 121*73-109
◇ F50 ◇ REV MR 49 # 8822 Zbl 282 # 02007 • ID 09290

MINTS, G.E. [1974] *E theorems (Russian) (English summary)* (**S** 0228) Zap Nauch Sem Leningrad Otd Mat Inst Steklov 40*110-118,158-159
• TRANSL [1977] (**J** 1531) J Sov Math 8*323-329
◇ F05 F30 F50 ◇ REV MR 53 # 2646 Zbl 368 # 02034
• ID 21531

MINTS, G.E. [1974] *Functional form. Herbrand's theorem for non-prenex formulas (Russian)* (**C** 4137) Kleene: Mat Logika 447-450
◇ B10 F05 ◇ ID 37140

MINTS, G.E. [1974] *Gentzen's formal systems (Russian)* (**C** 2546) Ehntsikl Kibern 1*222-224
◇ B10 F05 F07 ◇ ID 37142

MINTS, G.E. [1974] *Heyting predicate calculus with ε symbol (Russian) (English summary)* (**S** 0228) Zap Nauch Sem Leningrad Otd Mat Inst Steklov 40*101-109,158
• TRANSL [1977] (**J** 1531) J Sov Math 8*317-323
◇ F05 F50 ◇ REV MR 53 # 2645 Zbl 358 # 02023
• ID 21530

MINTS, G.E. [1974] *Normalization of proofs (Russian)* (**C** 4137) Kleene: Mat Logika 442-447
◇ F05 ◇ ID 37139

MINTS, G.E. [1974] *Proof theory (Russian)* (**C** 2546) Ehntsikl Kibern 1*297-299
◇ A05 F99 ◇ ID 37143

MINTS, G.E. [1975] *Finite investigations of transfinite derivations (Russian) (English summary)* (**S** 0228) Zap Nauch Sem Leningrad Otd Mat Inst Steklov 49*67-122,177-178
• TRANSL [1978] (**J** 1531) J Sov Math 10*548-596
◇ F05 F07 F10 F20 F30 F35 F50 ◇ REV MR 52 # 5375 Zbl 341 # 02020 • ID 18299

MINTS, G.E. [1975] *Proof theory (arithmetic and analysis) (Russian)* (**J** 1501) Itogi Nauki Tekh, Ser Algeb, Topol, Geom 13*5-49,343
• TRANSL [1977] (**J** 1531) J Sov Math 7*501-531
◇ F05 F10 F30 F35 F50 F98 ◇ REV Zbl 413 # 03031
• ID 76418

MINTS, G.E. [1975] see KREISEL, G.

MINTS, G.E. [1975] *Transfinite expansions of arithmetic formulas (Russian) (English summary)* (**S** 0228) Zap Nauch Sem Leningrad Otd Mat Inst Steklov 49*51-66,177
• TRANSL [1978] (**J** 1531) J Sov Math 10*533-547
◇ F30 F50 ◇ REV MR 52 # 2839 Zbl 319 # 02026
• ID 17631

MINTS, G.E. [1976] *The universality of the canonical tree (Russian)* (**J** 0023) Dokl Akad Nauk SSSR 227*808-811
• TRANSL [1976] (**J** 0062) Sov Math, Dokl 17*527-532
◇ F15 F30 ◇ REV MR 56 # 5234 Zbl 391 # 03025
• ID 52352

MINTS, G.E. [1976] *Unification of the two Peano axioms (Russian)* (**X** 2235) VINITI: Moskva 2382-76, 3pp.
◇ F30 ◇ ID 37146

MINTS, G.E. [1976] *What can be done with PRA (Russian) (English summary)* (**S** 0228) Zap Nauch Sem Leningrad Otd Mat Inst Steklov 60*93-102
• TRANSL [1980] (**J** 1531) J Sov Math 14*1487-1492
◇ F05 F30 F35 ◇ REV MR 58 # 27369 Zbl 372 # 02018
• ID 51426

MINTS, G.E. [1977] *Closed categories and the theory of proofs (Russian)* (**S** 0228) Zap Nauch Sem Leningrad Otd Mat Inst Steklov 68*83-114,145
• TRANSL [1981] (**J** 1531) J Sov Math 15*45-62
◇ F05 F07 G30 ◇ REV MR 58 # 22239 Zbl 368 # 02036
• ID 51253

MINTS, G.E. [1977] *Formal arithmetic (Russian)* (**C** 2580) Mat Entsikl 1*319-321
◇ F30 ◇ ID 37147

MINTS, G.E. [1977] *Gentzen's formal system (Russian)* (**C** 2580) Mat Entsikl 1*919-921
◇ F05 F07 ◇ ID 37151

MINTS, G.E. [1977] *Goedel's incompleteness theorem (Russian)* (**C** 2580) Mat Entsikl 1*909-910
◇ F30 ◇ ID 37148

MINTS, G.E. [1977] *Heytings's formal system (Russian)* (**C** 2580) Mat Entsikl 1*911-912
◇ F50 ◇ ID 37150

MINTS, G.E. [1979] *A new reduction sequence for arithmetic (Russian) (English summary)* (**S** 0228) Zap Nauch Sem Leningrad Otd Mat Inst Steklov 88*106-130,240-241
• TRANSL [1982] (**J** 1531) J Sov Math 20*2322-2333
◇ F05 F15 F30 ◇ REV MR 81k:03060 Zbl 439 # 03042
• ID 56033

MINTS, G.E. [1979] *A primitive recursive bound of strong normalization for predicate calculus (Russian) (English summary)* (**S** 0228) Zap Nauch Sem Leningrad Otd Mat Inst Steklov 88*131-136,241-242
• TRANSL [1982] (**J** 1531) J Sov Math 20*2334-2336
◇ F05 ◇ REV MR 81h:03107 Zbl 429 # 03032 • ID 53863

MINTS, G.E. [1979] *Constructive logic (Russian)* (**C** 2580) Mat Entsikl 2*1039-1042
◇ F50 ◇ ID 37153

MINTS, G.E. [1979] *Constructive semantics (Russian)* (**C** 2580) Mat Entsikl 2*1046-1049
◇ F50 ◇ ID 37154

MINTS, G.E. [1979] *Normalisation of natural deduction and effectiveness of classical existence (Russian)* (**P** 2554) All-Union Symp Th Log Infer;1974 Moskva 245-265
◇ F05 F10 F30 F35 ◇ REV MR 84m:03087 • ID 35783

MINTS, G.E. [1980] *Cartesian closed categories and proof theory (Russian)* (**C** 2583) Aktual Probl Log & Metodol Nauki
◇ F05 F07 G30 ◇ ID 37155

MINTS, G.E. [1980] *Category theory and proof theory (Russian)* (**C** 2583) Aktual Probl Log & Metodol Nauki 275-278
◇ F05 F07 G30 ◇ REV Zbl 514 # 03042 • ID 37417

MINTS, G.E. [1982] *A simplified consistency proof for arithmetic (Russian) (English and Estonian summaries)* (**J** 0080) Izv Akad Nauk Ehston SSR, Fiz, Mat 31*376-381
◇ F05 F30 ◇ REV MR 84i:03109 Zbl 523 # 03043
• ID 34587

MINTS, G.E. & TYUGU, E.KH. [1982] *The completeness of structural synthesis rules (Russian)* (**J** 0023) Dokl Akad Nauk SSSR 263*291-295
• TRANSL [1982] (**J** 0062) Sov Math, Dokl 25*343-346
◇ B20 B75 D20 F50 ◇ REV MR 84m:03013
Zbl 517 # 68058 • ID 35718

MINTS, G.E. see Vol. I, II, IV for further entries

MIRASYEDIOGLU, S. [1983] *λ-functional EXCLUSIVE-OR and LOGICAL EQUIVALENCE* (**J** 3980) Karadeniz Univ Math J (Trabzon) 5*64-76
◇ B40 ◇ REV MR 85f:03012 Zbl 525 # 94014 • ID 37498

MIRKOWSKA, G. [1977] *Algorithmic logic and its application in the theory of programs I,II* (**J** 2095) Fund Inform, Ann Soc Math Pol, Ser 4 1*1-17,147-165
◇ B40 B75 ◇ REV MR 58 # 32012 Zbl 358 # 68036
Zbl 384 # 68010 • ID 31914

MIRKOWSKA, G. see Vol. II, IV for further entries

MIROLLI, M. [1981] see BELLISSIMA, F.

MIROLLI, M. see Vol. I, II, III for further entries

MISAWA, T. & YASUDA, Y. [1977] *Suslin logics of Gentzen style* (**J** 3630) Bull Dept of Lib Arts (Numazu) 4*1-12
◇ C75 F07 ◇ ID 33405

MISAWA, T. see Vol. III, V for further entries

MISERCQUE, D. [1980] *Sur le treillis distributif des $\forall_1$-formules fermees de l'arithmetique de Peano (English summary)* (**J** 2313) C R Acad Sci, Paris, Ser A-B 290*A571-A573
◇ B20 F30 G10 ◇ REV MR 81m:03069 Zbl 424 # 03031
• ID 76441

MISERCQUE, D. [1981] *Answer to a problem of D. Guaspari* (**P** 2614) Open Days in Model Th & Set Th;1981 Jadwisin 181-184
◇ F30 ◇ ID 33725

MISERCQUE, D. [1981] *Solutions de deux problemes poses par H. Simmons* (**J** 3133) Bull Soc Math Belg, Ser B 33*65-72
◇ C62 F30 H15 ◇ REV MR 82k:03091 Zbl 482 # 03015
• ID 76440

MISERCQUE, D. [1982] *The nonhomogeneity of the E-tree - answer to a problem raised by D. Jensen and A. Ehrenfeucht* (**J** 0053) Proc Amer Math Soc 84*573-575
◇ C62 F30 ◇ REV MR 83h:03085 Zbl 495 # 03048
• ID 36085

MISERCQUE, D. [1985] *Branches of the E-trees which are not isomorphic* (**J** 0161) Bull London Math Soc 17*513-517
◇ F30 ◇ ID 49512

MISERCQUE, D. see Vol. IV for further entries

MITANI, S. [1983] *Dimension theory constructed on the basis of LJ* (**J** 0352) Math Jap 28*595-599
◇ B35 F55 ◇ REV MR 85h:03019 Zbl 538 # 03010
• ID 41462

MITANI, S. see Vol. I for further entries

MITSCHKE, G. [1972] *λ-definierbare Funktionen auf Peanoalgebren* (**J** 0009) Arch Math Logik Grundlagenforsch 15*31-35
◇ B40 D75 ◇ REV MR 48 # 3720 Zbl 252 # 02041
• ID 09330

MITSCHKE, G. [1972] *Ein algebraischer Beweis fuer das Church-Rosser-Theorem* (**J** 0009) Arch Math Logik Grundlagenforsch 15*146-157
◇ B40 ◇ REV MR 49 # 8843 Zbl 267 # 02017 • ID 09331

MITSCHKE, G. [1977] see HINDLEY, J.R.

MITSCHKE, G. [1979] *The standardization theorem for λ-calculus* (**J** 0068) Z Math Logik Grundlagen Math 25*29-31
◇ B40 ◇ REV MR 80g:03013 Zbl 442 # 03013 • ID 56370

MITSCHKE, G. [1980] *Infinite terms and infinite reductions* (**C** 3050) Essays Combin Log, Lambda Calc & Formalism (Curry) 243-257
◇ B40 C75 ◇ REV MR 82g:03021 Zbl 469 # 03006
• ID 76450

MIYATAKE, T. [1980] *On the length of proofs in formal systems* (**J** 2606) Tsukuba J Math 4*115-125
◇ F20 F30 ◇ REV MR 82h:03065 Zbl 493 # 03034
• ID 76469

MIYATAKE, T. [1981] *On the length of proofs in a formal system of recursive arithmetic* (**P** 3201) Logic Symposia;1979/80 Hakone 81-108
◇ F20 F30 ◇ REV MR 84e:03069 Zbl 499 # 03044
• ID 33393

MIZUTANI, C. [1982] *Definability theorem for the intuitionistic predicate logic with equality* (J 0111) Nagoya Math J 87*1-15
◇ F50 ◇ REV MR 84b:03080 Zbl 497 # 03019 • ID 35662

MIZUTANI, C. see Vol. I, III, V for further entries

MLCEK, J. [1976] *Twin prime problem in an arithmetic without induction* (J 0140) Comm Math Univ Carolinae (Prague) 17*543-555
◇ F30 ◇ REV MR 55 # 5438 Zbl 344 # 02039 • ID 31963

MLCEK, J. [1978] *End-extensions of countable structures and the induction schema* (J 0140) Comm Math Univ Carolinae (Prague) 19*291-308
◇ C15 C62 F30 ◇ REV MR 58 # 10413 Zbl 372 # 02030 • ID 31964

MLCEK, J. see Vol. I, III, V for further entries

MO, SHAOKUI [1957] *Simplified introduction to intuitionistic logic (Chinese)* (J 4364) Dongbei Renmin Daxue Ziran Kexue Xuebao 2*247-265
◇ F50 ◇ REV JSL 25.181 • ID 42546

MO, SHAOKUI [1961] *N-generalizable, intuitionistic, co-denial, pseudo-modal and co-Δ systems (Chinese)* (J 1024) Zhongguo Kexue 10*755-789
◇ F50 ◇ REV MR 27 # 3530 • ID 19398

MO, SHAOKUI & SHEN, BAIYING [1965] *New systems of primitive recursive arithmetic (Chinese)* (P 4564) Math Logic;1963 Xi-An 93-99
◇ F30 ◇ ID 49331

MO, SHAOKUI & SHEN, BAIYING [1965] *Systems of primitive recursive arithmetic (Chinese)* (P 4564) Math Logic;1963 Xi-An 99-109
◇ D20 F30 ◇ ID 48522

MO, SHAOKUI & SHEN, BAIYING [1966] *A new system for primitive recursive arithmetic I,II (Chinese)* (J 0420) Shuxue Jinzhan 9*1-26,103-148
◇ F30 ◇ REV MR 36 # 3650 • ID 19393

MO, SHAOKUI & SHEN, BAIYING [1982] *Inverse functions of number-theoretic functions I,II (Chinese) (English summaries)* (J 3137) Shuxue Niankan 3*103-114,381-394
◇ F30 ◇ REV MR 83g:10007 MR 84d:03053 Zbl 489 # 03014 Zbl 489 # 03015 • REM Part III 1984 by Shen,Baiying • ID 34089

MO, SHAOKUI [1984] see DING, DECHENG

MO, SHAOKUI see Vol. I, II, III, IV, V for further entries

MOERDIJK, I. [1982] *Glueing topoi and higher-order disjunction and existence* (P 3638) Brouwer Centenary Symp;1981 Noordwijkerhout 359-375
◇ F35 F50 G30 ◇ REV MR 84k:03149 Zbl 514 # 03038 • ID 35041

MOERDIJK, I. [1982] *Some topological spaces which are universal for intuitionistic predicate logic* (J 0028) Indag Math 44*227-235
◇ F50 ◇ REV MR 83j:03023 Zbl 487 # 03033 • ID 35336

MOERDIJK, I. [1983] see HOEVEN VAN DER, G.F.

MOERDIJK, I. [1983] *On the Freyd cover of a topos* (J 0047) Notre Dame J Formal Log 24*517-526
◇ F35 F50 G30 ◇ REV MR 85h:03072 Zbl 487 # 03034 • ID 37289

MOERDIJK, I. [1983] see GRAYSON, R.J.

MOERDIJK, I. [1984] see HOEVEN VAN DER, G.F.

MOERDIJK, I. & REYES, G.E. [1984] *De Rham's theorem in a smooth topos* (J 0332) Math Proc Cambridge Phil Soc 96*61-71
◇ F50 G30 ◇ REV MR 85e:18001 • ID 45171

MOERDIJK, I. [1984] *Heine-Borel does not imply the Fan theorem* (J 0036) J Symb Logic 49*514
◇ C90 E35 F35 F50 G30 ◇ ID 42485

MOERDIJK, I. & REYES, G.E. [1984] *Smooth spaces versus continuous spaces in models for synthetic differential geometry* (J 0326) J Pure Appl Algebra 32*143-176
◇ F35 F50 G30 ◇ REV MR 85i:03198 Zbl 535 # 18003 • ID 44304

MOERDIJK, I. see Vol. III, V for further entries

MOGGI, E. [1984] see LONGO, G.

MOGGI, E. see Vol. IV for further entries

MOISIL, G.C. [1938] *Sur le syllogisme hypothetique dans la logique intuitionniste* (J 3941) J Math Pures Appl, Ser 9 17*197-202
◇ F50 ◇ REV Zbl 18.337 FdM 64.30 • ID 41067

MOISIL, G.C. [1975] *Sur une logique intuitionniste* (J 3526) Ann Mat Pura Appl, Ser 4 102*203-212
◇ F50 ◇ REV MR 51 # 2876 Zbl 299 # 02025 • ID 17328

MOISIL, G.C. see Vol. I, II, IV, V for further entries

MOLK, J. [1885] *Sur une notion qui comprend celle de la divisibilite et sur la theorie generale de l'elimination* (J 0118) Acta Math 6*1-166
◇ F55 F65 ◇ ID 49802

MOLLER, S. [1965] *Induction models (Danish)* (J 0311) Nordisk Mat Tidskr 13*127-137
◇ C62 F30 ◇ REV MR 34 # 38 Zbl 131.6 • ID 48096

MOLODSHIJ, V.N. [1938] *Effectivism in mathematics (Russian)* (X 3333) Unknown Publisher: See Remarks 88pp
◇ A05 F99 ◇ REM Moskva • ID 45553

MOLODSHIJ, V.N. [1952] *On interrelations of certain assertions of generality with the induction axiom in Peano's system of axioms (Russian)* (S 1498) Uch Zap Univ, Moskva 155*168-173
◇ C62 F30 ◇ REV MR 17.1040 • ID 09388

MOLODSHIJ, V.N. see Vol. V for further entries

MONRO, G.P. [1974] see BARNES, D.W.

MONRO, G.P. see Vol. III, V for further entries

MONTAGNA, F. [1975] *For every n, the n-freely generated algebra is not functionally free in the equational class of diagonalizable algebras (the algebraization of theories which express Theor. V)* (J 0063) Studia Logica 34*315-319
◇ F30 G05 ◇ REV MR 57 # 108 Zbl 322 # 02032 • REM Part IV 1975 by Magari,R. Part VI 1976 by Bernardi,C. • ID 31910

MONTAGNA, F. [1978] *On the algebraization of a Feferman's predicate (the algebraization of theories, which express Theor. X)* (J 0063) Studia Logica 37*221-236
◇ C05 F30 G05 G25 ◇ REV MR 81h:03123 Zbl 398 # 03054 • REM Part IX 1976 by Sambin,G. Part XI 1978 by Sambin,G. • ID 31911

MONTAGNA, F. [1979] *On the formulas of Peano arithmetic which are provably closed under modus ponens* (**J** 3495) Boll Unione Mat Ital, V Ser, B 16*196-211
◇ F30 ◇ REV MR 80h:03084 Zbl 405 # 03030 • ID 76519

MONTAGNA, F. [1980] *Interpretations of the first-order theory of diagonalizable algebras in Peano arithmetic* (**J** 0063) Studia Logica 39*347-354
◇ F30 G25 ◇ REV MR 82i:03077a Zbl 463 # 03015
• ID 54555

MONTAGNA, F. [1982] *Relatively precomplete numerations and arithmetic* (**J** 0122) J Philos Logic 11*419-430
◇ D45 F30 ◇ REV MR 84f:03052 Zbl 498 # 03046
• ID 34471

MONTAGNA, F. [1983] *Il teorema di Solovay e l'aritmetica di Heyting* (**P** 3829) Atti Incontri Log Mat (1);1982 Siena 99-103
◇ B45 F30 ◇ REV MR 84k:03006 Zbl 521 # 03042
• ID 37078

MONTAGNA, F. [1983] *ZFC-models as Kripke-models* (**J** 0068) Z Math Logik Grundlagen Math 29*163-168
◇ B45 C62 E30 F30 ◇ REV MR 84j:03043
Zbl 519 # 03013 • ID 34635

MONTAGNA, F. [1984] *A completeness result for fixed-point algebras* (**J** 0068) Z Math Logik Grundlagen Math 30*525-532
◇ F30 G10 G25 ◇ REV MR 86f:03106 Zbl 564 # 03044
• ID 42278

MONTAGNA, F. [1984] see BERNARDI, C.

MONTAGNA, F. [1984] *The predicate modal logic of provability* (**J** 0047) Notre Dame J Formal Log 25*179-192
◇ B45 C90 F30 ◇ REV MR 86b:03023 Zbl 549 # 03013
• ID 42567

MONTAGNA, F. see Vol. II, III, IV for further entries

MONTAGUE, R. [1957] *Non-finite axiomatizability* (**P** 1675) Summer Inst Symb Log;1957 Ithaca 256-259
◇ B30 F25 F30 ◇ REV Zbl 148.245 • ID 29356

MONTAGUE, R. [1957] see KALISH, D.

MONTAGUE, R. [1957] *Two theorems on relative interpretability* (**P** 1675) Summer Inst Symb Log;1957 Ithaca 263-264
◇ F25 ◇ ID 29358

MONTAGUE, R. [1961] *Semantical closure and non-finite axiomatizability I* (**P** 0633) Infinitist Meth;1959 Warsaw 45-69
◇ B15 E30 F25 F30 ◇ REV MR 27 # 38 Zbl 116.7
JSL 29.59 • ID 09424

MONTAGUE, R. [1962] *Theories incomparable with respect to relative interpretability* (**J** 0036) J Symb Logic 27*195-211
◇ C07 F25 F30 ◇ REV MR 27 # 5684 Zbl 112.246
JSL 36.688 • ID 09425

MONTAGUE, R. [1963] *Syntactical treatments of modality, with corollaries on reflexion principles and finite axiomatizability* (**J** 0096) Acta Philos Fenn 16*153-167
• REPR [1974] (**C** 4062) Montague: Formal Philos 286-302
◇ B45 B65 C62 F30 ◇ REV MR 29 # 1140 Zbl 117.13
JSL 40.600 JSL 47.210 • ID 14478

MONTAGUE, R. [1965] *Interpretability in terms of models* (**J** 0028) Indag Math 27*467-476
◇ C07 F25 ◇ REV MR 31 # 4724 Zbl 151.11 • ID 09431

MONTAGUE, R. see Vol. I, II, III, IV, V for further entries

MONTEIRO, A. [1955] *Axiomes independants pour les algebres de Brouwer* (**J** 0188) Rev Union Mat Argentina 17*149-160
◇ F50 G10 ◇ REV MR 18.867 Zbl 72.250 JSL 23.444
• ID 09435

MONTEIRO, A. see Vol. I, II, V for further entries

MORALES-LUNA, G. [1985] see ADAMOWICZ, Z.

MORAVEK, J. [1973] *Computational optimality of a dynamic programming method* (**P** 1448) Math Founds of Comput Sci (2);1973 Strbske Pleso 267-270
◇ B40 D15 ◇ REV MR 53 # 2023 • ID 82295

MORGAN, C.G. [1976] *Many-valued propositional intuitionism* (**P** 2011) Int Symp Multi-Val Log (6);1976 Logan 150-156
◇ B50 F50 ◇ REV MR 58 # 16171 • ID 76542

MORGAN, C.G. [1976] *Methods for automated theorem proving in nonclassical logics* (**J** 0187) IEEE Trans Comp C-25*852-862
◇ B35 F50 ◇ REV MR 55 # 9619 Zbl 329 # 68074
• ID 27095

MORGAN, C.G. [1983] see LEBLANC, H.

MORGAN, C.G. see Vol. I, II, III, IV for further entries

MORGENSTERN, C.F. [1982] *On generalized quantifiers in arithmetic* (**J** 0036) J Symb Logic 47*187-190
◇ C62 C80 F30 ◇ REV MR 84g:03051 Zbl 487 # 03019
JSL 50.1078 • ID 34161

MORGENSTERN, C.F. see Vol. III, V for further entries

MORICONI, E. [1981] *Sulla completezza del calcolo dei predicati intuizionista* (**P** 3092) Congr Naz Logica;1979 Montecatini Terme 513-540
◇ F50 ◇ ID 48420

MORICONI, E. [1983] *Hilbert and the "omega rule" (Italian)* (**P** 3829) Atti Incontri Log Mat (1);1982 Siena 387-390
◇ A05 A10 F99 ◇ ID 44857

MORLEY, M.D. [1985] see HARRINGTON, L.A.

MORLEY, M.D. see Vol. III, IV, V for further entries

MOROZOV, A.S. [1983] *Groups of recursive automorphisms of constructive boolean algebras (Russian)* (**J** 0003) Algebra i Logika 22*138-158
• TRANSL [1983] (**J** 0069) Algeb and Log 22*95-112
◇ C07 C57 D45 F60 G05 ◇ REV Zbl 549 # 03031
• ID 43123

MOROZOV, A.S. [1984] *Group* $Aut_r(Q, \leqslant)$ *is not constructivizable (Russian)* (**J** 0087) Mat Zametki (Akad Nauk SSSR) 36*473-478
• TRANSL [1984] (**J** 1044) Math Notes, Acad Sci USSR 36*733-736
◇ C07 C57 F60 ◇ REV MR 86g:20001 Zbl 574 # 03028
• ID 41757

MOROZOV, A.S. see Vol. III, IV for further entries

MORTENSEN, C. [1985] see MEYER, R.K.

MORTENSEN, C. see Vol. I, II, V for further entries

MOSCATO, U. [1981] see MIGLIOLI, P.A.

MOSCATO, U. [1982] see MIGLIOLI, P.A.

MOSCATO, U. [1983] see MIGLIOLI, P.A.

MOSCHOVAKIS, J.R. [1967] *Disjunction and existence in formalized intuitionistic analysis* (**P** 0691) Sets, Models & Recursion Th;1965 Leicester 309-331
◇ F35 F50 ◇ REV MR 36 # 3630 Zbl 207.290 JSL 35.587
• ID 09510

MOSCHOVAKIS, J.R. [1971] *Can there be no nonrecursive functions?* (**J** 0036) J Symb Logic 36*309-315
◇ F35 F50 ◇ REV MR 45 # 3162 Zbl 247 # 02036
• ID 09512

MOSCHOVAKIS, J.R. [1973] *A topological interpretation of second-order intuitionistic arithmetic* (**J** 0020) Compos Math 26*261-275
◇ F35 F50 ◇ REV MR 50 # 9544 Zbl 279 # 02018
• ID 29544

MOSCHOVAKIS, J.R. [1980] *Kleene's realizability and "divides" notions for formalized intuitionistic mathematics* (**P** 2058) Kleene Symp;1978 Madison 167-179
◇ A10 F50 ◇ REV MR 81m:03071 Zbl 517 # 03025
• ID 76577

MOSCHOVAKIS, J.R. [1981] *A disjunctive decomposition theorem for classical theories* (**P** 3146) Constr Math;1980 Las Cruces 250-259
◇ F35 F50 ◇ REV MR 83a:03061 Zbl 472 # 03047
• ID 55310

MOSCHOVAKIS, J.R. see Vol. I for further entries

MOSCHOVAKIS, Y.N. [1964] *Recursive metric spaces* (**J** 0027) Fund Math 55*215-238
◇ C57 C65 D45 F60 ◇ REV MR 32 # 45
Zbl 221 # 02015 JSL 31.651 • ID 09514

MOSCHOVAKIS, Y.N. [1966] *Notation systems and recursive ordered fields* (**J** 0020) Compos Math 17*40-71
◇ C57 C60 D45 F60 ◇ REV MR 31 # 5798 Zbl 143.13
JSL 31.650 • ID 21265

MOSCHOVAKIS, Y.N. see Vol. III, IV, V for further entries

MOSTOWSKI, A.WLODZIMIERZ [1981] *The complexity of automata and subtheories of monadic second order arithmetics* (**P** 3165) FCT'81 Fund of Comput Th;1981 Szeged 453-466
◇ B15 B25 D05 D15 F35 ◇ REV MR 83g:03036
Zbl 469 # 03025 • ID 55152

MOSTOWSKI, A.WLODZIMIERZ see Vol. I, II, III, IV for further entries

MOSTOWSKI, ANDRZEJ [1939] *Bemerkungen zum Begriff der inhaltlichen Widerspruchsfreiheit* (**J** 0036) J Symb Logic 4*113-114
◇ F30 ◇ REV MR 1.34 Zbl 21.385 JSL 5.115 FdM 65.21
• ID 09531

MOSTOWSKI, ANDRZEJ [1946] *On undecidable sentences in formalized mathematical systems (Polish)* (**J** 0479) Kwart Filoz 16*223-277
• TRANSL [1963] (**J** 0225) Amer Math Soc, Transl, Ser 2 29*1-50
◇ F30 F98 ◇ REV MR 9.129 JSL 12.60 • ID 33694

MOSTOWSKI, ANDRZEJ [1947] *On definable sets of positive integers* (**J** 0027) Fund Math 34*81-112
◇ D55 F30 ◇ REV MR 9.129 Zbl 31.194 JSL 13.112
• ID 09537

MOSTOWSKI, ANDRZEJ [1948] *Proofs of non-deducibility in intuitionistic functional calculus* (**J** 0036) J Symb Logic 13*204-207
◇ F50 ◇ REV MR 10.421 Zbl 31.193 JSL 14.137
• ID 09539

MOSTOWSKI, ANDRZEJ [1949] *A classification of logical systems* (**J** 4716) Stud Philos, Leopolis (Poznan) 4*237-274
◇ D55 F30 ◇ REV JSL 17.274 • ID 09547

MOSTOWSKI, ANDRZEJ [1949] *An undecidable arithmetical statement* (**J** 0027) Fund Math 36*143-164
◇ F30 ◇ REV MR 12.2 Zbl 39.8 JSL 19.118 • ID 09546

MOSTOWSKI, ANDRZEJ [1952] *On models of axiomatic systems* (**J** 0027) Fund Math 39*133-158
◇ C62 F25 F30 F35 ◇ REV MR 14.938 Zbl 53.201
JSL 19.220 • ID 09550

MOSTOWSKI, ANDRZEJ [1952] *Sentences undecidable in formalized arithmetic. An exposition of the theory of Kurt Goedel* (**X** 0809) North Holland: Amsterdam viii+117pp
• REPR [1982] (**X** 1270) Greenwood Pr: Westport, CT viii+117pp
◇ F30 F98 ◇ REV MR 14.3 Zbl 47.9 JSL 19.119
• ID 09552

MOSTOWSKI, ANDRZEJ & RASIOWA, H. [1953] *A geometric interpretation of logical formulae (Polish) (English and Russian summaries)* (**J** 0063) Studia Logica 1*254-275,301
◇ B10 F50 G05 G10 G15 ◇ REV MR 19.378 Zbl 59.14
• ID 16376

MOSTOWSKI, ANDRZEJ & ROBINSON, R.M. & TARSKI, A. [1953] *Undecidability and essential undecidability in arithmetic* (**C** 4472) Undecidable Th 39-74
◇ D35 F30 ◇ REV Zbl 53.4 JSL 24.167 • ID 42176

MOSTOWSKI, ANDRZEJ & ROBINSON, R.M. & TARSKI, A. [1953] *Undecidable theories* (**X** 0809) North Holland: Amsterdam xi+98pp
◇ D35 F25 F30 ◇ REV MR 15.384 Zbl 53.4 • REM 2nd ed. 1968; 3rd ed. 1971 • ID 28815

MOSTOWSKI, ANDRZEJ [1956] see LOS, J.

MOSTOWSKI, ANDRZEJ [1957] *On computable sequences* (**J** 0027) Fund Math 44*37-51
◇ D20 F60 ◇ REV MR 19.934 Zbl 79.247 JSL 25.367
• ID 09564

MOSTOWSKI, ANDRZEJ [1958] *Quelques observations sur l'usage des methodes non finitistes dans la meta-mathematique* (**P** 0576) Raisonn en Math & Sci Exper;1955 Paris 19-29 (Disc 29-32)
◇ A05 C30 C62 C80 E30 E35 F99 ◇ REV
MR 21 # 4896 Zbl 105.5 JSL 24.234 • ID 09566

MOSTOWSKI, ANDRZEJ [1958] see GRZEGORCZYK, A.

MOSTOWSKI, ANDRZEJ [1959] *On various degrees of constructivism* (**P** 0634) Constructivity in Math;1957 Amsterdam 178-194
◇ D55 F50 F60 F98 ◇ REV MR 22 # 3688 Zbl 88.250
JSL 35.575 • ID 09568

MOSTOWSKI, ANDRZEJ [1961] *A generalization of the incompleteness theorem* (**J** 0027) Fund Math 49*205-232
◇ F30 ◇ REV MR 24 # A41 Zbl 99.6 • ID 09569

MOSTOWSKI, ANDRZEJ [1961] *Formal systems of analysis based on an infinitistic rule of proof* (**P** 0633) Infinitist Meth;1959 Warsaw 141-166
◇ C62 E45 F35 ◇ REV MR 36 # 3631 Zbl 121.15 JSL 34.128 • ID 24858

MOSTOWSKI, ANDRZEJ [1962] *Representability of sets in formal systems* (**P** 0613) Rec Fct Th;1961 New York V*29-48
◇ D55 E15 F30 F35 ◇ REV MR 26 # 27 Zbl 158.8 • ID 33705

MOSTOWSKI, ANDRZEJ [1965] *Thirty years of foundational studies. Lectures on the development of mathematical logic and the studies of the foundations of mathematics in 1930-1964* (**J** 0096) Acta Philos Fenn 17*1-180
• REPR [1966] (**X** 1096) Blackwell: Oxford 180pp
◇ A10 B98 C98 D98 E98 F98 ◇ REV MR 33 # 18 MR 33 # 5445 Zbl 146.245 JSL 33.111 • ID 09578

MOSTOWSKI, ANDRZEJ see Vol. I, II, III, IV, V for further entries

MOTOHASHI, N. [1972] *A note on Schuette's interpolation theorem* (**J** 0081) Proc Japan Acad 48*56-58
◇ F50 ◇ REV MR 47 # 1577 Zbl 252 # 02017 • ID 09597

MOTOHASHI, N. [1972] *Interpolation theorem and characterization theorem* (**J** 0260) Ann Jap Ass Phil Sci 4*85-150
◇ C40 C75 F05 ◇ REV MR 49 # 10530 Zbl 241 # 02005 • ID 09596

MOTOHASHI, N. [1972] *Object logic and morphism logic* (**J** 0090) J Math Soc Japan 24*684-697
◇ F05 ◇ REV MR 57 # 9502 Zbl 239 # 02018 • ID 27921

MOTOHASHI, N. [1973] *A faithful interpretation of intuitionistic predicate logic in classical predicate logic* (**J** 0407) Comm Math Univ St Pauli (Tokyo) 21*11-23
◇ F50 ◇ REV MR 48 # 3706 Zbl 262 # 02024 • ID 09598

MOTOHASHI, N. [1977] *Partially ordered interpretations* (**J** 0036) J Symb Logic 42*83-93
◇ C90 E40 F25 F50 ◇ REV MR 58 # 201 Zbl 411 # 03030 • ID 23750

MOTOHASHI, N. [1982] *ε-theorems and elimination theorems of uniqueness conditions* (**P** 3634) Patras Logic Symp;1980 Patras 373-387
◇ B10 C40 F05 ◇ REV MR 84h:03027 Zbl 535 # 03009 • ID 33418

MOTOHASHI, N. [1982] *An elimination theorem of uniqueness conditions in the intuitionistic predicate calculus* (**J** 0111) Nagoya Math J 85*223-230
◇ F50 ◇ REV MR 83m:03017 Zbl 526 # 03003 • ID 35425

MOTOHASHI, N. [1984] *Approximation theory of uniqueness conditions by existence conditions* (**J** 0027) Fund Math 120*127-142
◇ C40 C75 C80 F07 F50 ◇ ID 45359

MOTOHASHI, N. see Vol. I, II, III for further entries

MOULOUD, N. [1979] *Quelques hypotheses sur la porte des notions operatoires dans les langues naturelles et techniques* (**J** 0056) Publ Dep Math, Lyon 16/3*1-11
◇ B40 B65 ◇ REV MR 82m:03041 • ID 76632

MUCHNIK, A.A. [1983] *Supplement of the translator to the paper "On alternation. I, II" (Russian)* (**J** 3079) Kiber Sb Perevodov, NS 20*141-158
◇ B45 D05 D10 D15 F50 ◇ REV Zbl 545 # 68041 • REM For the papers see Chandra,A.K. et al. 1976 and 1981 • ID 43519

MUCHNIK, A.A. see Vol. I, II, III, IV for further entries

MUELLER, GERT H. [1961] *Nicht-Standardmodelle der Zahlentheorie* (**J** 0044) Math Z 77*414-438
◇ C62 F30 ◇ REV MR 31 # 3333 Zbl 268 # 02039 JSL 37.405 • ID 09619

MUELLER, GERT H. [1961] *Ueber die unendliche Induktion* (**P** 0633) Infinitist Meth;1959 Warsaw 75-95
◇ F30 ◇ REV MR 26 # 6053 Zbl 124.249 JSL 40.627 • ID 09618

MUELLER, GERT H. [1975] see DILLER, J.

MUELLER, GERT H. see Vol. I, II, III, IV, V for further entries

MUELLER, GUENTER [1970] *Groessen ueber arithmetisierbaren Gesamtheiten* (**J** 0160) Math-Phys Sem-ber, NS 17*233-247
◇ F35 ◇ REV MR 42 # 7506 Zbl 213.290 • ID 09622

MULLIN, A.A. [1967] *On new theorems for elementary number theory* (**J** 0047) Notre Dame J Formal Log 8*353-356
◇ F30 ◇ REV MR 38 # 4397 Zbl 189.42 • ID 09649

MULLIN, A.A. see Vol. I, IV, V for further entries

MULVEY, C.J. [1974] *Intuitionistic algebra and representations of rings* (**S** 0167) Mem Amer Math Soc 148*3-57
◇ C60 C90 F50 F55 G30 ◇ REV MR 53 # 2650 Zbl 274 # 18012 • ID 21534

MULVEY, C.J. see Vol. III, V for further entries

MUNDICI, D. [1983] *Natural limitations of decision procedures for arithmetic with bounded quantifiers* (**J** 0009) Arch Math Logik Grundlagenforsch 23*37-54
◇ B25 D10 D15 F30 ◇ REV MR 84k:03040 Zbl 523 # 03028 • ID 34976

MUNDICI, D. see Vol. I, II, III, IV, V for further entries

MURAVITSKIJ, A.YU. [1981] *Strong equivalence on an intuitionistic Kripke model and assertorically equivolumetric logics (Russian)* (**J** 0003) Algebra i Logika 20*165-182,250
• TRANSL [1981] (**J** 0069) Algeb and Log 20*112-123
◇ B55 C90 F50 ◇ REV MR 83h:03087 Zbl 486 # 03015 • ID 36087

MURAVITSKIJ, A.YU. [1983] *Extensions of the provability logic (Russian)* (**J** 0087) Mat Zametki (Akad Nauk SSSR) 33*915-927
• TRANSL [1983] (**J** 1044) Math Notes, Acad Sci USSR 33*469-475
◇ B45 F30 ◇ REV MR 84j:03044 Zbl 538 # 03023 • ID 34636

MURAVITSKIJ, A.YU. [1985] *Correspondence of proof-intuitionistic logic extensions to proof-logic extensions (Russian)* (**J** 0023) Dokl Akad Nauk SSSR 281*789-793
• TRANSL [1985] (**J** 0062) Sov Math, Dokl 31*345-348
◇ F50 ◇ ID 45122

MURAVITSKIJ, A.YU. see Vol. II, III for further entries

MURAWSKI, R. [1984] *Mathematical incompleteness of arithmetic (Polish)* (**J** 0519) Wiad Mat, Ann Soc Math Pol, Ser 2 26*47-58
◇ D35 F30 ◇ ID 39679

MURAWSKI, R. see Vol. III, IV, V for further entries

MURZIN, F.A. [1979] *One result on finite approximability (Russian)* (**P** 2558) All-Union Conf Math Log (5) (Mal'tsev);1979 Novosibirsk 104
◇ F60 ◇ ID 32634

MURZIN, F.A. see Vol. III for further entries

MUTI, N. [1967] see KAWADA, K.

MUTI, N. [1970] *A characterization of the intuitionistic propositional logic* (**J** 0111) Nagoya Math J 37∗131-136
◇ F50 ◇ REV MR 41 # 1509 Zbl 196.9 • ID 09660

MYCIELSKI, J. [1971] see EHRENFEUCHT, A.

MYCIELSKI, J. [1977] *A lattice of interpretability types of theories* (**J** 0036) J Symb Logic 42∗297-305
◇ F25 G10 ◇ REV MR 58 # 21594 Zbl 371 # 02026
• ID 26462

MYCIELSKI, J. [1981] *Analysis without actual infinity* (**J** 0036) J Symb Logic 46∗625-633
◇ F65 ◇ REV MR 83b:03015 Zbl 466 # 03024 • ID 54975

MYCIELSKI, J. [1981] *Finitistic real analysis* (**S** 2824) Real Anal Exchange 6∗127-130
◇ B28 F65 ◇ REV Zbl 453 # 03064 • ID 54194

MYCIELSKI, J. see Vol. I, III, IV, V for further entries

MYHILL, J.R. [1949] *Note on an idea of Fitch* (**J** 0036) J Symb Logic 14∗175-176
◇ D55 F30 ◇ REV MR 11.151 Zbl 37.297 JSL 14.260
• ID 09698

MYHILL, J.R. [1950] *A system which can define its own truth* (**J** 0027) Fund Math 37∗190-192
◇ B30 D20 F25 F30 ◇ REV MR 13.97 Zbl 41.150
JSL 21.319 • ID 09699

MYHILL, J.R. [1951] *Report on some investigations concerning the consistency of the axiom of reducibility* (**J** 0036) J Symb Logic 16∗35-42
◇ B15 F35 ◇ REV MR 12.791 Zbl 42.8 JSL 16.217
• ID 09702

MYHILL, J.R. [1952] *A finitary metalanguage for extended basic logic* (**J** 0036) J Symb Logic 17∗164-178
◇ B40 ◇ REV MR 14.526 Zbl 49.149 JSL 20.81 • ID 09705

MYHILL, J.R. [1953] *Arithmetic with creative definitions by induction* (**J** 0036) J Symb Logic 18∗115-118
◇ F30 ◇ REV MR 15.924 Zbl 53.200 JSL 22.303
• ID 09707

MYHILL, J.R. [1953] *Criteria of constructibility for real numbers* (**J** 0036) J Symb Logic 18∗7-10
◇ F60 ◇ REV MR 14.938 Zbl 52.251 JSL 20.178
• ID 09709

MYHILL, J.R. [1963] *The invalidity of Markov's schema* (**J** 0068) Z Math Logik Grundlagen Math 9∗359-360
◇ F50 ◇ REV MR 27 # 5686 Zbl 122.11 JSL 39.333
• ID 09727

MYHILL, J.R. [1966] *Notes towards an axiomatization of intuitionistic analysis* (**J** 0079) Logique & Anal, NS 9∗280-297
◇ F35 F50 ◇ REV MR 36 # 35 Zbl 187.263 JSL 33.290
• ID 09730

MYHILL, J.R. [1968] *Formal systems of intuitionistic analysis I* (**P** 0627) Int Congr Log, Meth & Phil of Sci (3,Proc);1967 Amsterdam 161-178
◇ F35 F50 ◇ REV MR 40 # 5425 Zbl 202.6 • REM Part II 1970 • ID 16944

MYHILL, J.R. [1968] *The formalization of intuitionism* (**C** 0552) Phil Contemp - Chroniques 324-341
◇ F50 ◇ REV JSL 40.625 • ID 14689

MYHILL, J.R. [1970] *Formal systems of intuitionistic analysis II. The theory of species* (**P** 0603) Intuitionism & Proof Th;1968 Buffalo 151-162
◇ F35 F50 ◇ REV MR 43 # 6057 Zbl 202.6 JSL 40.500
• REM Part I 1968 • ID 21244

MYHILL, J.R. [1970] see KINO, A.

MYHILL, J.R. [1971] *A recursive function, defined on a compact interval and having a continuous derivative that is not recursive* (**J** 0133) Michigan Math J 18∗97-98
◇ F60 ◇ REV MR 43 # 6093 Zbl 218 # 02029 • ID 09732

MYHILL, J.R. [1971] *Embedding classical type theory in "intuitionistic" type theory* (**P** 0693) Axiomatic Set Th;1967 Los Angeles 1∗267-270 • ERR/ADD ibid 2∗185-188
◇ B15 F35 F50 ◇ REV MR 43 # 7298 Zbl 219 # 02043
• ID 28094

MYHILL, J.R. [1972] *An absolutely independent set of Σ^0_1 sentences* (**J** 0068) Z Math Logik Grundlagen Math 18∗107-109
◇ F30 ◇ REV MR 46 # 1569 Zbl 273 # 02030 • ID 09735

MYHILL, J.R. [1972] see GOODMAN, NICOLAS D.

MYHILL, J.R. [1972] *What is a real number ?* (**J** 0005) Amer Math Mon 79∗748-754
◇ F55 F60 ◇ REV MR 46 # 8819 Zbl 255 # 02030
• ID 09736

MYHILL, J.R. [1973] *A note on indicator functions* (**J** 0053) Proc Amer Math Soc 39∗181-183
◇ F50 ◇ REV MR 55 # 5394 Zbl 262 # 02026 • ID 14718

MYHILL, J.R. [1973] *Embedding classical logic in intuitionistic logic* (**J** 0068) Z Math Logik Grundlagen Math 19∗93-96
◇ F50 ◇ REV MR 47 # 8254 Zbl 255 # 02019 • ID 09738

MYHILL, J.R. [1973] *Some properties of intuitionistic Zermelo-Fraenkel set theory* (**P** 0713) Cambridge Summer School Math Log;1971 Cambridge GB 206-231
◇ E35 E70 F50 ◇ REV MR 57 # 5749 Zbl 272 # 02039
• ID 09739

MYHILL, J.R. [1974] *Embedding classical type theory in "intuitionistic" type theory, a correction* (**P** 0693) Axiomatic Set Th;1967 Los Angeles 2∗185-188
◇ B15 F35 F50 ◇ REV MR 43 # 7298 Zbl 219 # 02043
• REM This is not simply a correction of the 1971 paper, it is a completely new paper • ID 48354

MYHILL, J.R. [1975] see KINO, A.

MYHILL, J.R. [1975] *Constructive set theory* (**J** 0036) J Symb Logic 40∗347-382
◇ E70 F50 ◇ REV MR 52 # 2830 Zbl 314 # 02045
JSL 46.868 • ID 14820

MYHILL, J.R. [1975] *Levels of implication* (**C** 1856) Log Enterprise 179-185
◇ B40 ◇ REV Zbl 362 # 02015 • ID 50737

MYHILL, J.R. [1978] see GOODMAN, NICOLAS D.

MYHILL, J.R. see Vol. I, II, IV, V for further entries

NABEBIN, A.A. [1977] *Expressibility in restricted second-order arithmetic (Russian)* (**J** 0092) Sib Mat Zh 18*830-837,957 • TRANSL [1977] (**J** 0475) Sib Math J 18*588-593 ◇ B15 C40 D05 F35 ◇ REV MR 58 # 16250 Zbl 385 # 03047 • ID 52155

NABEBIN, A.A. see Vol. IV for further entries

NADEL, M.E. [1978] *Infinitary intuitionistic logic from a classical point of view* (**J** 0007) Ann Math Logic 14*159-191 ◇ C75 C90 F50 ◇ REV MR 80f:03027 Zbl 406 # 03055 • ID 29150

NADEL, M.E. [1981] *The completeness of Peano multiplication* (**J** 0029) Israel J Math 39*225-233 ◇ F30 ◇ REV MR 83f:03059 Zbl 472 # 03023 • ID 55289

NADEL, M.E. see Vol. III, IV, V for further entries

NADIU, G.S. [1971] *Sur la logique de Heyting* (**C** 0640) Log, Autom, Inform 41-70 ◇ C90 F50 G10 G30 ◇ REV MR 49 # 21 Zbl 255 # 02018 • ID 09750

NADIU, G.S. see Vol. I, III, IV for further entries

NAGASHIMA, T. [1966] *An extension of the Craig-Schuette interpolation theorem* (**J** 0260) Ann Jap Ass Phil Sci 3*12-18 ◇ C40 F50 ◇ REV MR 33 # 30 Zbl 148.6 JSL 33.291 • ID 09755

NAGASHIMA, T. see Vol. II, III, IV, V for further entries

NAGATA, M. & NAKANISHI, M. & NISHIMURA, T. [1975] *TKP 1.2 - the extension of TKP 1 by adding some facilities* (**J** 0350) Sci Rep Tokyo Kyoiku Daigaku Sect A 19*82-89 ◇ B35 F35 ◇ REV MR 52 # 16171 Zbl 335 # 68063 • ID 21892

NAGATA, M. see Vol. I, II for further entries

NAGATA, S. [1979] see FUNAHASHI, S.

NAGATA, S. see Vol. I, II for further entries

NAGEL, E. & NEWMAN, J.R. [1958] *Goedel's proof* (**X** 0924) New York Univ Pr: New York ix + 118pp ◇ B28 D35 F30 F98 ◇ REV MR 20 # 1625 Zbl 86.246 JSL 24.222 • ID 09765

NAGEL, E. & NEWMAN, J.R. [1961] *Discussion: Putnam's review of Goedel's proof* (**J** 0153) Phil of Sci (East Lansing) 28*209-211 ◇ B28 F30 ◇ REV MR 24 # A5 • ID 09768

NAGEL, E. & NEWMAN, J.R. [1967] *Goedel's proof* (**C** 0721) Contemp Readings in Log Th 51-71 ◇ D35 F30 ◇ REV JSL 21.374 • ID 09763

NAGEL, E. see Vol. I, II for further entries

NAGORNYJ, N.M. [1964] *On realizable and completable logico-arithmetical formulas (Russian)* (**J** 0023) Dokl Akad Nauk SSSR 157*529-531 • TRANSL [1964] (**J** 0062) Sov Math, Dokl 5*981-983 ◇ F30 F50 ◇ REV MR 29 # 3362 Zbl 236 # 02025 • ID 09778

NAGORNYJ, N.M. [1976] *A variant of the definition of realization of a logical arithmetical formula (Russian)* (**S** 0554) Issl Teor Algor & Mat Logik (Moskva) 2*32-45,157 ◇ F30 F50 ◇ REV MR 58 # 16197 • ID 76746

NAGORNYJ, N.M. [1976] *Algorithms and real numbers* (**P** 1401) Math Founds of Comput Sci (5);1976 Gdansk 99-104 ◇ F60 ◇ REV Zbl 358 # 02038 • ID 50492

NAGORNYJ, N.M. see Vol. I, IV for further entries

NAHM, W. [1974] see DILLER, J.

NAKAJIMA, R. [1975] *Infinite normal forms for the λ-calculus* (**P** 1603) λ-Calc & Comput Sci Th;1975 Roma 62-82 ◇ B40 ◇ REV MR 57 # 9495 Zbl 354 # 02023 • ID 27593

NAKAMURA, A. [1975] see INOUE, K.

NAKAMURA, A. & ONO, H. [1977] *Two-dimensional finite automata and their application to the decision problem of monadic first-order arithmetic* $A[P, F(x), G(x)]$ (**P** 2032) Stud on Polyautomata;1976 Zushi & Kyoto 51-71 ◇ D05 D35 F30 F35 ◇ ID 32225

NAKAMURA, A. & ONO, H. [1979] *Undecidability of the first-order arithmetic* $A[P(x), 2x, x+1]$ (**J** 0119) J Comp Syst Sci 18*243-253 ◇ D05 D35 F30 F35 ◇ REV MR 80k:03045 Zbl 414 # 03006 • ID 53052

NAKAMURA, A. see Vol. I, II, III, IV for further entries

NAKANISHI, M. [1975] see NAGATA, M.

NAKANISHI, M. [1981] see AIBA, A.

NAKANISHI, M. see Vol. I, II for further entries

NAMBA, K. [1965] *On a comprehension axiom without negation* (**J** 0260) Ann Jap Ass Phil Sci 2*258-271 ◇ E70 F05 ◇ REV MR 32 # 7397 Zbl 131.246 • ID 09795

NAMBA, K. see Vol. III, V for further entries

NEBESKY, L. [1971] *A combinatory approach to grammar* (**J** 2817) Prague Bull Math Linguist 15*61-65 ◇ B40 ◇ REV Zbl 236 # 68028 • ID 64167

NEBRES, B.F. [1972] *Herbrand uniformity theorems for infinitary languages* (**J** 0090) J Math Soc Japan 24*1-19 ◇ C40 C70 C75 F05 ◇ REV MR 46 # 21 Zbl 226 # 02001 • ID 30771

NEBRES, B.F. see Vol. III for further entries

NEDERPELT, R.P. [1977] *Presentation of natural deduction* (**P** 3269) Set Th Found Math (Kurepa);1977 Beograd 115-126 ◇ B10 F07 ◇ REV MR 58 # 16177 Zbl 415 # 03008 • ID 53110

NEDERPELT, R.P. [1980] *An approach to theorem proving on the basis of a typed lambda-calculus* (**P** 3063) Autom Deduct (5);1980 Les Arcs 182-194 ◇ B35 B40 F60 ◇ REV Zbl 438 # 68053 • ID 55989

NEGRI, M. [1981] *Constructive sequent reduction in Gentzen's first consistency proof for arithmetic* (**C** 3515) Ital Studies in Phil of Sci 153-168 ◇ A10 F05 F30 ◇ REV MR 83f:03057 Zbl 449 # 03053 • ID 56720

NEGRI, M. [1981] *L'aritmetizzazione della computabilita dei funzionali di tipo finito* (**P** 3092) Congr Naz Logica;1979 Montecatini Terme 79-89 ◇ F10 F35 F50 ◇ ID 48423

NEGRI, M. [1984] *An application of recursive saturation* (J 2099) Boll Unione Mat Ital, VI Ser, A 3*449-451
◇ C50 C57 C62 F30 ◇ REV MR 86f:03109 Zbl 562 # 03039 • ID 44715

NEGRI, M. [1985] *On the nonfinite axiomatizability of P (Italian)* (P 4646) Atti Incontri Log Mat (2);1983/84 Siena 621-626
◇ F30 ◇ ID 49684

NELSON, D. [1947] *Recursive functions and intuitionistic number theory* (J 0064) Trans Amer Math Soc 61*307-368,556
◇ F30 F50 ◇ REV MR 10.3 Zbl 58.249 JSL 12.93 • ID 09843

NELSON, D. [1949] *Constructible falsity* (J 0036) J Symb Logic 14*16-26
◇ F50 ◇ REV MR 10.669 Zbl 33.243 JSL 15.228 • ID 09844

NELSON, D. [1959] *Negation and separation of concepts in constructive systems* (P 0634) Constructivity in Math;1957 Amsterdam 208-225
◇ F50 ◇ REV MR 22 # 12044 Zbl 85.250 JSL 40.501 • ID 09845

NELSON, D. [1966] *Non-null implication* (J 0036) J Symb Logic 31*562-572
◇ B20 B28 F50 ◇ REV MR 35 # 26 Zbl 148.244 JSL 33.129 • ID 09846

NELSON, D. [1973] *A complete negationless system (Polish and Russian summaries)* (J 0063) Studia Logica 32*41-49
◇ F50 ◇ REV MR 50 # 4255 Zbl 317 # 02025 • ID 09847

NELSON, D. [1984] see ALMUKDAD, A.

NELSON, GEORGE C. [1971] *A further restricted ω-rule* (S 0019) Colloq Math (Warsaw) 23*1-3
◇ F30 ◇ REV MR 46 # 7010 Zbl 231 # 02032 • ID 09853

NELSON, GEORGE C. [1975] see MADISON, E.W.

NELSON, GEORGE C. see Vol. I, III, IV, V for further entries

NEMECKOVA, L. [1973] see DEMUTH, O.

NEPEJVODA, N.N. [1973] *A new concept of predicative truth and definability (Russian)* (J 0087) Mat Zametki (Akad Nauk SSSR) 13*735-745
• TRANSL [1973] (J 1044) Math Notes, Acad Sci USSR 13*439-445
◇ F35 F65 ◇ REV MR 48 # 1892 Zbl 264 # 02028 • ID 09868

NEPEJVODA, N.N. [1973] *On a generalization of the Kleene-Mostowski hierarchy (Russian)* (J 0023) Dokl Akad Nauk SSSR 212*295-298
• TRANSL [1973] (J 0062) Sov Math, Dokl 14*1365-1368
◇ D55 F15 F35 ◇ REV MR 49 # 4760 Zbl 336 # 02036 • ID 09867

NEPEJVODA, N.N. [1973] *The relationship between predicative significance and the intuition of universality (Russian)* (J 0023) Dokl Akad Nauk SSSR 212*40-43
• TRANSL [1973] (J 0062) Sov Math, Dokl 14*1307-1310
◇ F50 F65 ◇ REV MR 49 # 4759 Zbl 336 # 02029 • ID 09866

NEPEJVODA, N.N. [1975] *The language Δ with intuitionistic connectives (Russian)* (J 0023) Dokl Akad Nauk SSSR 220*41-43
• TRANSL [1975] (J 0062) Sov Math, Dokl 16*39-42
◇ E70 F50 ◇ REV MR 55 # 2518 Zbl 322 # 02034 • ID 64192

NEPEJVODA, N.N. [1977] *Level Beth models and realizability (Russian)* (S 2579) Teor Mnozhestv & Topol (Izhevsk) 1*67-69
◇ F30 F50 ◇ REV Zbl 496 # 03041 • ID 36868

NEPEJVODA, N.N. [1978] *A relation between the natural deduction rules and operators of higher level algorithmic languages (Russian)* (J 0023) Dokl Akad Nauk SSSR 239*526-529
• TRANSL [1978] (J 0062) Sov Math, Dokl 19*360-363
◇ B75 F07 F50 ◇ REV MR 80a:03013 Zbl 392 # 68004 • ID 52413

NEPEJVODA, N.N. [1978] *The construction of correct programs (Russian)* (S 2874) Vopr Kibern, Akad Nauk SSSR 46*88-122
◇ F50 ◇ REV MR 81c:68013 • ID 82346

NEPEJVODA, N.N. [1979] *Application of proof theory to the problem of construction of correct programs (Russian)* (J 0040) Kibernetika, Akad Nauk Ukr SSR 1979/2*43-48
• TRANSL [1979] (J 0021) Cybernetics 15*202-210
◇ B75 F50 ◇ REV MR 81c:68014 Zbl 483 # 68031 • ID 36810

NEPEJVODA, N.N. [1979] *Some remarks about constructive non-standard analysis (Russian)* (S 2579) Teor Mnozhestv & Topol (Izhevsk) 2*74-76
◇ F99 H05 ◇ REV Zbl 496 # 03042 • ID 36869

NEPEJVODA, N.N. [1979] *Stable truth and computability (Russian)* (S 0554) Issl Teor Algor & Mat Logik (Moskva) 3*78-89,133
◇ C90 F35 F50 ◇ REV MR 82f:03052 Zbl 422 # 03036 • ID 53497

NEPEJVODA, N.N. [1982] *Constructive logics (Russian)* (C 3849) Modal & Relevant Log, Vyp 1 91-106
◇ B60 C90 C95 F50 ◇ REV Zbl 557 # 03039 • ID 46228

NEPEJVODA, N.N. [1982] *On some possibilities of applying constructive analysis to classical analysis (Russian)* (J 0023) Dokl Akad Nauk SSSR 266*284-287
• TRANSL [1982] (J 0062) Sov Math, Dokl 26*324-327
◇ A05 F99 ◇ REV MR 84d:03076 Zbl 506 # 03017 • ID 34108

NEPEJVODA, N.N. [1983] *Notes concerning constructive implicative logics (Russian)* (S 0507) Vychisl Sist (Akad Nauk SSSR Novosibirsk) 96*87-108
◇ B55 F50 ◇ REV MR 86e:03058 Zbl 547 # 03036 • ID 43233

NEPEJVODA, N.N. see Vol. II, III, IV for further entries

NERODE, A. & SHORE, R.A. [1980] *Second order logic and first order theories of reducibility orderings* (P 2058) Kleene Symp;1978 Madison 181-200
◇ B10 B15 D30 D35 F35 F40 ◇ REV MR 82g:03078 Zbl 465 # 03024 • ID 54927

NERODE, A. [1982] see METAKIDES, G.

NERODE, A. [1985] see HUANG, WENGI

NERODE, A. see Vol. III, IV, V for further entries

NEUMANN VON, J. [1927] *Zur Hilbertschen Beweistheorie* (J 0044) Math Z 26*1-46
◇ A05 F99 ◇ REV FdM 53.41 • ID 09900

NEUMANN VON, J. [1931] *Bemerkungen zu den Ausfuehrungen von Herrn S. Lesniewski ueber meine Arbeit "Zur Hilbertschen Beweistheorie"* (**J** 0027) Fund Math 17*331-334
◇ A05 F99 ◇ REV Zbl 3.50 FdM 57.55 • REM For original see Neumann von,J. 1927 Math. Z. 26*1-46 • ID 09904

NEUMANN VON, J. see Vol. I, II, IV, V for further entries

NEUMER, W. [1953] *Zur Konstruktion von Ordnungszahlen I,II* (**J** 0044) Math Z 58*391-413,59*434-454
◇ E10 F15 ◇ REV MR 15.512 MR 15.689 Zbl 56.49
• REM Parts III and IV 1954 • ID 09919

NEUMER, W. [1954] *Zur Konstruktion von Ordnungszahlen III,IV* (**J** 0044) Math Z 60*1-16,61*47-69
◇ E10 F15 ◇ REV MR 16.19 MR 16.343 Zbl 56.49 Zbl 57.44 • REM Parts I and II 1953. Part V 1956 • ID 09921

NEUMER, W. [1956] *Zur Konstruktion von Ordnungszahlen V* (**J** 0044) Math Z 64*435-456
◇ E10 F15 ◇ REV MR 18.139 Zbl 71.51 • REM Parts III and IV 1954 • ID 19481

NEUMER, W. [1957] *Algorithmen fuer Ordnungszahlen und Normalfunktionen I* (**J** 0068) Z Math Logik Grundlagen Math 3*108-150
◇ E10 F15 ◇ REV MR 20 # 1631 Zbl 82.44 • REM Part II 1960 • ID 09924

NEUMER, W. [1960] *Algorithmen fuer Ordnungszahlen und Normalfunktionen II* (**J** 0068) Z Math Logik Grundlagen Math 6*1-65
◇ E10 F15 ◇ REV MR 23 # A2314 Zbl 97.40 • REM Part I 1957. Part III 1970 • ID 09926

NEUMER, W. [1970] *Algorithmen fuer Ordnungszahlen und Normalfunktionen III* (**J** 0068) Z Math Logik Grundlagen Math 16*1-112
◇ E10 F15 ◇ REV MR 43 # 7335 Zbl 218 # 04003 • REM Part II 1960 • ID 09927

NEUMER, W. see Vol. V for further entries

NEVES-REAL, L. [1951] *Kurt Goedel and the problems of the foundations of mathematics and the theory of sets (Portuguese)* (**J** 0084) Gaz Mat (Lisboa) 12/48*1-8
◇ A10 E25 E35 E45 E50 F30 ◇ REV MR 13.4 Zbl 42.243 JSL 40.241 • ID 22210

NEWMAN, J.R. [1958] see NAGEL, E.

NEWMAN, J.R. [1961] see NAGEL, E.

NEWMAN, J.R. [1967] see NAGEL, E.

NEWMAN, M.H.A. [1942] *On theories with a combinatorial definition of "equivalence"* (**J** 0120) Ann of Math, Ser 2 43*223-243
◇ B10 B40 ◇ REV MR 4.126 Zbl 60.125 JSL 7.123
• ID 09934

NEWMAN, M.H.A. [1969] see KREISEL, G.

NEWMAN, M.H.A. see Vol. I, V for further entries

NGUEN VAN TIN' [1975] see CHUBARYAN, A.A.

NICOD, J. [1924] *Geometry in the sensible world (French)* (**X** 1071) Alcan:Paris
• LAST ED [1962] (**X** 0840) Pr Univ France: Paris
◇ A05 B80 F99 ◇ REV MR 42 # 33 Zbl 242 # 00023 FdM 50.30 • REM For transl. see Nicod,J. 1930 • ID 25431

NICOD, J. [1930] *Geometry and induction, containing "Geometry in the sensible world", and "The logical problem of induction"* (**X** 0866) Routledge & Kegan Paul: Henley on Thames xx + 245pp
• LAST ED [1970] (**X** 0926) Univ Calif Pr: Berkeley 245pp
◇ A05 B30 F98 ◇ REV MR 42 # 33 Zbl 242 # 00023
• REM Transl. of Nicod,J. 1924 (both) • ID 09941

NICOD, J. see Vol. I, II for further entries

NIEKUS, N.H. & RIEMSDIJK VAN, H. & TROELSTRA, A.S. [1981] *Bibliography of A. Heyting* (**J** 3077) Nieuw Arch Wisk, Ser 3 29*24-35
◇ A10 F99 ◇ REV MR 83g:01091b Zbl 449.01013
• ID 39085

NIELAND, J.J.F. [1967] *Beth's tableau-method* (**P** 0683) Log & Founds of Sci (Beth);1964 Paris 19-38
◇ B05 B25 B50 F07 ◇ REV MR 39 # 1287 Zbl 203.9
• ID 09951

NIELAND, J.J.F. see Vol. II for further entries

NIINILUOTO, I. [1973] see HINTIKKA, K.J.J.

NIINILUOTO, I. see Vol. I, II, V for further entries

NIKOGOSOV, S.L. [1970] see KUZICHEV, A.S.

NIKOGOSOV, S.L. [1971] see KUZICHEV, A.S.

NIKOLENKO, A.B. [1984] *Method of invariant transformations and logical deduction (Russian)* (**J** 0087) Mat Zametki (Akad Nauk SSSR) 36*3-15
• TRANSL [1984] (**J** 1044) Math Notes, Acad Sci USSR 36*493-500
◇ F07 ◇ REV Zbl 554 # 03010 • ID 44185

NINK, C. [1937] *Der Satz vom ausgeschlossenen Dritten. Ein philosophischer Beitrag zur Grundlagenkrise in der Mathematik* (**J** 4539) Scholastik 12*552-558
◇ A05 F99 ◇ REV JSL 3.87 FdM 63.836 • ID 41401

NISHIMURA, I. [1960] *On formulas of one variable in intuitionistic propositional calculus* (**J** 0036) J Symb Logic 25*327-331
◇ F50 ◇ REV MR 26 # 25 Zbl 108.3 JSL 32.396 • ID 09969

NISHIMURA, I. see Vol. II for further entries

NISHIMURA, T. [1959] *On Goedel's theorem (Japanese)* (**J** 0091) Sugaku 11*1-12
• TRANSL [1961] (**J** 0090) J Math Soc Japan 13*1-12
◇ D35 F30 ◇ REV MR 27 # 2416 Zbl 100.11 JSL 29.106 JSL 34.649 • ID 09973

NISHIMURA, T. [1960] see MAEHARA, S.

NISHIMURA, T. [1960] *On a certain system with infinite induction* (**J** 0090) J Math Soc Japan 12*392-400
◇ F30 ◇ REV MR 24 # A3062 Zbl 111.8 JSL 28.293
• ID 09971

NISHIMURA, T. & TANAKA, H. [1965] *On two systems for arithmetic* (**J** 0090) J Math Soc Japan 17*244-280
◇ C75 F30 F35 ◇ REV MR 32 # 37 Zbl 133.251
• ID 09975

NISHIMURA, T. [1974] *Gentzen-style formulation of systems of set-calculus* (**J** 0407) Comm Math Univ St Pauli (Tokyo) 23*29-36
◇ C75 E30 F05 ◇ REV MR 52 # 100 Zbl 361 # 02041
• ID 18312

NISHIMURA, T. [1974] *Gentzen-style formal system representing properties of functions* (**J** 0407) Comm Math Univ St Pauli (Tokyo) 23*37-44
◇ C75 E30 F05 ◇ REV MR 52#101 Zbl 361#02042
• ID 18313

NISHIMURA, T. [1975] *Gentzen-type formal system representing properties of functions* (**S** 2308) Symp Kyoto Univ Res Inst Math Sci (RIMS) 236*76-86
◇ E30 F05 ◇ REV MR 58#5218 Zbl 381#68037
• ID 76854

NISHIMURA, T. [1975] see NAGATA, M.

NISHIMURA, T. [1976] *On cut-elimination in simple type theory (on the work of Moto-o Takahashi, winner of the third Mathematical Society Prize) (Japanese)* (**J** 0091) Sugaku 28*187-191
◇ F05 F35 ◇ REV MR 54#7217 • ID 25008

NISHIMURA, T. see Vol. I, II, III, V for further entries

NJASTAD, O. [1968] *Note on the Peano axioms* (**J** 0121) Kon Norske Vidensk Selsk Forh 41*16-20
◇ F30 ◇ REV MR 38#6 Zbl 184.17 • ID 09979

NOLIN, L. [1970] see LACOMBE, D.

NOLIN, L. [1971] *Logique combinatoire et algorithmes* (**J** 2313) C R Acad Sci, Paris, Ser A-B 272*A1435-A1438,A1485-A1488
◇ B40 ◇ REV MR 44#2612 Zbl 217.8 • ID 22240

NOLIN, L. [1973] *Systemes algorithmiques, systemes fonctionnels* (**P** 0763) Automata, Lang & Progr (1);1972 Rocquencourt 309-318
◇ B40 B75 ◇ REV MR 51#14658 Zbl 276#02027
• ID 29042

NOLIN, L. [1975] *Algorithmes seriels, algorithmes paralleles* (**P** 2974) New Concepts & Tech in Parall Inform Process;1973 Capri 167-186
◇ B40 B75 D20 ◇ REV MR 58#25098 • ID 82368

NOLIN, L. [1975] *Les modeles informatiques des λ-calculs* (**P** 1603) λ-Calc & Comput Sci Th;1975 Roma 166-176
◇ B40 B75 ◇ REV MR 57#1957 Zbl 354#02024
• ID 50164

NOLIN, L. [1975] *Theorie des algorithmes et semantique des langages de programmation* (**P** 1475) Conv Inform Teor & Conv Strutt Corpi Algeb;1973 Roma 283-303
◇ B40 B75 D20 ◇ REV MR 54#1700 Zbl 332#68011
• ID 23994

NOLIN, L. [1981] see BERRE LE, F.

NOLIN, L. see Vol. I, IV for further entries

NORDSTROEM, B. & SMITH, JAN [1984] *Propositions and specifications of programs in Martin-Loef's type theory* (**J** 0130) BIT 24*288-301
◇ B75 F35 F50 ◇ REV MR 86b:03033 Zbl 551#68027
• ID 45466

NORGELA, S.A. [1974] see MASLOV, S.YU.

NORGELA, S.A. [1977] see MASLOV, S.YU.

NORGELA, S.A. [1977] *Minus-normal sequential calculi (Russian) (English summary)* (**J** 0040) Kibernetika, Akad Nauk Ukr SSR 1977/3*33-38
• TRANSL [1977] (**J** 0021) Cybernetics 13*346-351
◇ F20 ◇ REV MR 56#11744 Zbl 359#02006 • ID 32673

NORGELA, S.A. [1978] *Herbrand strategies of deduction-search in predicate calculus I (Russian) (English and Lithuanian summaries)* (**J** 2574) Litov Mat Sb (Vil'nyus) 18/4*95-100,201
• TRANSL [1978] (**J** 3283) Lith Math J 18*513-517
◇ B25 B35 D35 F07 ◇ REV MR 80e:03050
Zbl 394#03017 • REM Part II 1979 • ID 32674

NORGELA, S.A. see Vol. I, III, IV for further entries

NORMANN, D. [1983] *General type-structures of continuous and countable functionals* (**J** 0068) Z Math Logik Grundlagen Math 29*177-192
◇ D55 D65 D75 F35 ◇ REV MR 85d:03095
Zbl 529#03022 • ID 37656

NORMANN, D. [1984] *The infinite - a mathematical necessity* (**J** 3075) Normat 32*63-70
◇ A05 D20 F30 ◇ REV MR 85j:03001 Zbl 536#03013
• ID 37098

NORMANN, D. [1985] see GIRARD, J.-Y.

NORMANN, D. see Vol. I, IV, V for further entries

NORWOOD, F.H. [1982] *Long proofs* (**J** 0005) Amer Math Mon 89*110-112
◇ B35 F20 ◇ REV MR 82m:03002 Zbl 498#03002
• ID 76885

NOVIKOV, P.S. [1949] *On the axiom of complete induction (Russian)* (**J** 0023) Dokl Akad Nauk SSSR 64*457-459
◇ B25 B28 D35 F30 ◇ REV MR 11.304 Zbl 38.151
JSL 14.256 • ID 19476

NOVIKOV, P.S. [1977] *Constructive mathematical logic from the point of view of classical logic (Russian)* (**X** 2027) Nauka: Moskva 328pp
◇ F98 ◇ REV MR 58#21505 • ID 76891

NOVIKOV, P.S. see Vol. I, II, III, IV, V for further entries

NUBER, J.A. [1972] *A constructive ergodic theorem* (**J** 0064) Trans Amer Math Soc 164*115-137 • ERR/ADD ibid 216*393
◇ F55 ◇ REV MR 45#504 MR 52#3479 Zbl 233#28017
• ID 10039

NURTAZIN, A.T. [1978] see DOBRITSA, V.P.

NURTAZIN, A.T. see Vol. III, IV, V for further entries

OBERSCHELP, W. [1984] see BOERGER, E.

OBERSCHELP, W. see Vol. I, III, IV for further entries

OBTULOWICZ, A. [1977] *Functorial semantics of the type free λ-β-η calculus* (**P** 2588) FCT'77 Fund of Comput Th;1977 Poznan 302-307
◇ B40 ◇ REV MR 58#16178 Zbl 367#18001 • ID 76921

OBTULOWICZ, A. & WIWEGER, A. [1978] *Church algebraic theories* (**S** 3125) Ber, Abt Inf, Univ Dortmund 74*73-77
◇ B40 ◇ REV Zbl 458#03020 • ID 54426

OBTULOWICZ, A. & WIWEGER, A. [1981] *Functional interpretation of λ-terms* (**P** 3642) Colloq Math Log in Computer Sci;1978 Salgotarjan 607-610
◇ B40 ◇ REV MR 83g:68007 Zbl 528#03007 • ID 37626

OBTULOWICZ, A. & WIWEGER, A. [1982] *Categorical, functorial and algebraic aspects of the type-free lambda calculus* (**P** 3831) Universal Algeb & Appl;1978 Warsaw 399-422
◇ B40 G30 ◇ REV MR 86f:03026 Zbl 514#03010
• ID 37404

ODIFREDDI, P. [1983] *Forcing and reducibilities* (**J** 0036) J Symb Logic 48*288-310
◇ D30 D55 E40 F30 ◇ REV MR 85m:03034a Zbl 526 # 03025 • REM Part I. Parts II,III 1983 • ID 38171

ODIFREDDI, P. see Vol. III, IV, V for further entries

O'DONNELL, M.J. [1977] *Computing in systems described by equations* (**S** 3302) Lect Notes Comput Sci 58*xiv+111pp
◇ B40 B75 ◇ REV MR 58 # 3632 Zbl 421 # 68038 • ID 53455

O'DONNELL, M.J. [1977] *Subtree replacement systems: a unifying theory for recursive equations, LISP, lucid and combinatory logic* (**P** 1903) ACM Symp Th of Comput (9);1977 Boulder 295-305
◇ B40 B75 D05 D20 ◇ REV MR 57 # 9507 • ID 76932

O'DONNELL, M.J. [1979] *A practical programming theorem which is independent of Peano arithmetic* (**P** 2539) Frege Konferenz (1);1979 Jena 284-298
◇ B40 B75 D20 F05 F30 ◇ REV MR 82i:03066 • ID 76931

O'DONNELL, M.J. [1979] *A programming language theorem which is independent of Peano arithmetic* (**P** 3542) ACM Symp Th of Comput (11);1979 Atlanta 176-188
◇ B75 D15 F30 ◇ REV MR 81d:68015 • ID 82386

O'DONNELL, M.J. see Vol. I for further entries

OGASAWARA, T. [1939] *Relation between intuitionistic logic and lattice* (**J** 0102) Hiroshima Univ J Sci, Ser A Math 9*157-164
◇ F50 G10 ◇ REV MR 1.132 Zbl 22.194 JSL 5.36 FdM 65.30 • ID 10066

OGLESBY, F.C. [1961] *Report: an examination of a decision procedure* (**J** 0015) Bull Amer Math Soc 67*300-304
◇ B25 F07 ◇ REV MR 23 # A795 Zbl 100.10 JSL 28.165 • ID 10067

OGLESBY, F.C. see Vol. I, III for further entries

OHAMA, S. [1979] *On the results of Paris and Harrington (Japanese)* (**P** 4108) Found of Math;1979 Kyoto 124-135
◇ F30 ◇ ID 47663

OHAMA, S. see Vol. I, II for further entries

OHASHI, K. [1957] *On undecidable theorems (Japanese)* (**J** 0091) Sugaku 9*96-97
◇ F30 ◇ REV MR 22 # 10910 JSL 34.131 • ID 10070

OHASHI, K. see Vol. IV for further entries

OHNISHI, M. [1953] *On intuitionistic functional calculus* (**J** 1770) Osaka Math J 5*203-209
◇ F50 ◇ REV MR 15.494 Zbl 53.342 JSL 22.299 • ID 10084

OHNISHI, M. see Vol. II, III for further entries

OHYA, T. [1970] *On recursive restriction of proofs in a system with constructive infinitely long expressions* (**J** 0407) Comm Math Univ St Pauli (Tokyo) 18*43-52
◇ F05 F35 ◇ REV MR 41 # 3254 Zbl 203.299 • ID 10088

OHYA, T. see Vol. II for further entries

OKEE, J. [1972] *Untersuchungen ueber den einstelligen intuitionistischen Praedikatenkalkuel der ersten Stufe* (**J** 0068) Z Math Logik Grundlagen Math 18*37-48
◇ F50 ◇ REV MR 57 # 74 Zbl 251 # 02030 • ID 10089

OKEE, J. [1975] *A semantical proof of the undecidability of the monadic intuitionistic predicate calculus of first order* (**J** 0047) Notre Dame J Formal Log 16*552-554
◇ D35 F50 ◇ REV MR 52 # 5370 Zbl 258 # 02032 • ID 18315

OKEE, J. [1976] *A species algebraic interpretation of the intuitionistic propositional calculus* (**J** 0047) Notre Dame J Formal Log 17*222-232
◇ F50 ◇ REV MR 54 # 7209 Zbl 258 # 02032 • ID 18317

OKEE, J. [1976] *Completeness of the algebra of species* (**J** 0047) Notre Dame J Formal Log 17*392-394
◇ F50 G15 ◇ REV MR 56 # 8329 Zbl 271 # 02022 • ID 18316

OKEE, J. [1976] *On the independence of the fundamental operations of the algebra of species* (**J** 0047) Notre Dame J Formal Log 17*526-530
◇ F50 ◇ REV MR 58 # 5199 Zbl 271 # 02023 • ID 21941

OKITA, K. [1982] *A valuation for the primitive logic* (1111) Preprints, Manuscr., Techn. Reports etc. 22*172-179
◇ B40 ◇ REV Zbl 541 # 03007 • REM Meijyo Univ., Fac. Sci. Technol. • ID 41370

ONO, H. [1977] see NAKAMURA, A.

ONO, H. [1979] see NAKAMURA, A.

ONO, H. [1985] *Semantical analysis of predicate logics without the contraction rule* (**J** 0063) Studia Logica 44*187-196
◇ B10 C90 F50 ◇ ID 47510

ONO, H. see Vol. I, II, III, IV, V for further entries

ONO, K. [1938] *Logische Untersuchungen ueber die Grundlagen der Mathematik* (**J** 0434) J Fac Sci Univ Tokyo, Sect 1 3*329-389
◇ F05 F30 ◇ REV Zbl 19.242 JSL 4.89 FdM 64.27 • ID 10114

ONO, K. [1962] *On a practical way of describing formal deductions* (**J** 0111) Nagoya Math J 21*115-121
◇ F07 ◇ REV MR 26 # 3580 Zbl 111.7 JSL 34.307 • ID 10117

ONO, K. [1966] *Taboo versus axiom* (**J** 0111) Nagoya Math J 28*113-117
◇ F50 ◇ REV MR 34 # 7346 Zbl 163.6 • ID 90070

ONO, K. [1969] *On a method of describing formal deductions convenient for theoretical purposes* (**J** 0111) Nagoya Math J 36*83-97
◇ F07 ◇ REV MR 40 # 7085 Zbl 188.316 • ID 10137

ONO, K. see Vol. I, II, V for further entries

OPPEN, D.C. [1973] *Elementary bounds for Presburger arithmetic* (**P** 1482) ACM Symp Th of Comput (5);1973 Austin 34-37
◇ B25 C10 D15 F30 ◇ REV MR 54 # 4187 Zbl 306 # 02044 • ID 24077

OPPEN, D.C. [1978] *A $2^{2^{2^{pn}}}$ upper bound on the complexity of Presburger arithmetic* (**J** 0119) J Comp Syst Sci 16*323-332
◇ B25 C10 D15 F30 ◇ REV MR 57 # 18224 Zbl 381 # 03021 • ID 51889

OPPEN, D.C. see Vol. I, III, IV for further entries

OREVKOV, V.P. [1963] *A constructive mapping of the square onto itself displacing every constructive point (Russian)* (J 0023) Dokl Akad Nauk SSSR 152*55-58
• TRANSL [1963] (J 0062) Sov Math, Dokl 4*1253-1256
◇ F60 ◇ REV MR 29 # 19 • ID 10146

OREVKOV, V.P. [1963] see MINTS, G.E.

OREVKOV, V.P. [1963] *Constructive mappings of polyhedra (Russian)* (J 0023) Dokl Akad Nauk SSSR 152*278-281
• TRANSL [1963] (J 0062) Sov Math, Dokl 4*1299-1302
◇ F60 ◇ REV MR 29 # 20 • ID 10147

OREVKOV, V.P. [1964] *Certain questions of the theory of polynomials with constructive real coefficients (Russian)* (S 0066) Tr Mat Inst Steklov 72*462-487 • ERR/ADD ibid 93*261
• TRANSL [1972] (J 0225) Amer Math Soc, Transl, Ser 2 100*101-134
◇ F60 ◇ REV MR 34 # 2459 Zbl 243 # 02031 • ID 19541

OREVKOV, V.P. [1964] *On constructive mappings of a disk into itself (Russian)* (S 0066) Tr Mat Inst Steklov 72*437-461
• ERR/ADD ibid 93*260-261
• TRANSL [1972] (J 0225) Amer Math Soc, Transl, Ser 2 100*69-100
◇ F60 ◇ REV MR 34 # 5670 Zbl 243 # 02030 • ID 14722

OREVKOV, V.P. [1965] *Certain reduction classes and solvable classes of sequents for the constructive predicate calculus (Russian)* (J 0023) Dokl Akad Nauk SSSR 163*30-32
• TRANSL [1965] (J 0062) Sov Math, Dokl 6*888-891
◇ F50 ◇ REV MR 33 # 51 Zbl 173.10 • ID 19540

OREVKOV, V.P. [1965] *Unsolvability of the class of formulas of the type ¬ ¬ ∀∃ in the constructive predicate calculus (Russian)* (J 0023) Dokl Akad Nauk SSSR 163*581-583
• TRANSL [1965] (J 0062) Sov Math, Dokl 6*977-980
◇ F50 ◇ REV MR 33 # 50 Zbl 173.10 • ID 19222

OREVKOV, V.P. [1965] see MASLOV, S.YU.

OREVKOV, V.P. [1967] *Constructive mappings of finite polyhedra (Russian)* (S 0066) Tr Mat Inst Steklov 93*142-163
• TRANSL [1967] (S 0055) Proc Steklov Inst Math 93*183-209
◇ F60 ◇ REV MR 37 # 2579 Zbl 195.19 • ID 10150

OREVKOV, V.P. [1967] *On certain types of continuity of constructive operators (Russian)* (S 0066) Tr Mat Inst Steklov 93*164-186
• TRANSL [1967] (S 0055) Proc Steklov Inst Math 93*211-239
◇ F60 ◇ REV MR 37 # 1242 Zbl 194.312 • ID 10151

OREVKOV, V.P. [1967] see MINTS, G.E.

OREVKOV, V.P. & SHANIN, N.A. (EDS.) [1967] *Problems in the constructive trend in mathematics IV (Russian)* (S 0066) Tr Mat Inst Steklov 93*269pp
• TRANSL [1967] (S 0055) Proc Steklov Inst Math 93*viii + 329pp
◇ F60 F97 ◇ REV Zbl 192.311 • ID 23447

OREVKOV, V.P. [1968] *Glivenko classes of sequents (Russian)* (S 0228) Zap Nauch Sem Leningrad Otd Mat Inst Steklov 8*196-201
• TRANSL [1968] (J 0521) Semin Math, Inst Steklov 8*95-97
◇ B10 F50 ◇ REV MR 42 # 5775 Zbl 198.318 • ID 10152

OREVKOV, V.P. [1968] *Glivenko's sequence classes (Russian)* (S 0066) Tr Mat Inst Steklov 98*131-154 • ERR/ADD ibid 121*167
• TRANSL [1968] (S 0055) Proc Steklov Inst Math 98*147-173
◇ B10 B20 F50 ◇ REV MR 39 # 2606 Zbl 172.290
• ID 19539

OREVKOV, V.P. (ED.) [1968] *The calculi of symbolic logic I (Russian)* (S 0066) Tr Mat Inst Steklov 98*5-202 • ERR/ADD ibid 121*167
• TRANSL [1968] (S 0055) Proc Steklov Inst Math 98*iv + 229pp
◇ B97 F97 ◇ REV MR 43 # 4620 MR 49 # 18 • ID 37197

OREVKOV, V.P. [1969] *On nonlengthening applications of equality rules (Russian)* (S 0228) Zap Nauch Sem Leningrad Otd Mat Inst Steklov 16*152-156 • ERR/ADD ibid 20*292-294
• TRANSL [1969] (J 0521) Semin Math, Inst Steklov 16*77-79
◇ B25 B35 F05 F07 ◇ REV MR 41 # 6665 Zbl 197.273
• ID 10154

OREVKOV, V.P. [1969] *Some properties of homeomorphisms of constructive metric spaces (Russian)* (S 0228) Zap Nauch Sem Leningrad Otd Mat Inst Steklov 16*157-164 • ERR/ADD ibid 20*292-294
• TRANSL [1969] (J 0521) Semin Math, Inst Steklov 16*80-83
◇ F60 ◇ REV MR 42 # 5780 Zbl 282 # 02009 • ID 10155

OREVKOV, V.P. & SHANIN, N.A. (EDS.) [1970] *Problems in the constructive trend in mathematics V (Russian)* (S 0066) Tr Mat Inst Steklov 113*259pp
• TRANSL [1970] (S 0055) Proc Steklov Inst Math 113*iii + 287pp
◇ F60 F97 ◇ REV MR 49 # 4761 • ID 23449

OREVKOV, V.P. [1971] *Equivalence of two definitions of continuity (Russian) (English summary)* (S 0228) Zap Nauch Sem Leningrad Otd Mat Inst Steklov 20*145-159,286
• TRANSL [1973] (J 1531) J Sov Math 1*92-99
◇ F60 ◇ REV MR 45 # 3644 MR 48 # 5808 Zbl 223 # 02027 • ID 10158

OREVKOV, V.P. [1971] *On the continuity of constructive functionals (Russian) (English summary)* (S 0228) Zap Nauch Sem Leningrad Otd Mat Inst Steklov 20*160-169,287
◇ F60 ◇ REV MR 45 # 63 Zbl 223 # 02028 • ID 10159

OREVKOV, V.P. [1972] *A specialization of the form of deductions in Gentzen calculi and its applications (Russian)* (S 0228) Zap Nauch Sem Leningrad Otd Mat Inst Steklov 32*98-104,156-157
• TRANSL [1976] (J 1531) J Sov Math 6*429-433
◇ F05 F07 F50 ◇ REV MR 49 # 8824 Zbl 345 # 02019
• ID 10160

OREVKOV, V.P. [1972] *Undecidable classes of formulas for the constructive predicate calculus I (Russian)* (S 0066) Tr Mat Inst Steklov 121*100-108,165
• TRANSL [1972] (S 0055) Proc Steklov Inst Math 121*111-119
◇ D35 F50 ◇ REV MR 49 # 7132 Zbl 286 # 02031
• ID 10161

OREVKOV, V.P. & SHANIN, N.A. (EDS.) [1973] *Problems in the constructive trend in mathematics VI (Russian)* (**S** 0066) Tr Mat Inst Steklov 129*269pp
• TRANSL [1975] (**X** 0803) Amer Math Soc: Providence iii+272pp
◇ F60 F97 ◇ REV MR 54 # 7190 MR 55 # 5379 Zbl 486 # 03018 • ID 70161

OREVKOV, V.P. [1974] *A new proof of the uniqueness theorem for constructive differentiable functions of a complex variable (Russian) (English summary)* (**S** 0228) Zap Nauch Sem Leningrad Otd Mat Inst Steklov 40*119-126
• TRANSL [1977] (**J** 1531) J Sov Math 8*329-334
◇ F60 ◇ REV MR 52 # 2845 Zbl 358 # 02040 • ID 17636

OREVKOV, V.P. [1976] *Solvable classes of pseudoprenex formulas (Russian) (English summary)* (**S** 0228) Zap Nauch Sem Leningrad Otd Mat Inst Steklov 60*109-170,225 • ERR/ADD ibid 88*248-249
• TRANSL [1980] (**J** 1531) J Sov Math 14*1497-1538
◇ B25 F50 ◇ REV MR 58 # 27423 Zbl 342 # 02034 • ID 64291

OREVKOV, V.P. [1979] *Lower bounds for lengthening of proofs after cut-elimination (Russian) (English summary)* (**S** 0228) Zap Nauch Sem Leningrad Otd Mat Inst Steklov 88*137-162,242-243
• TRANSL [1982] (**J** 1531) J Sov Math 22*237-2350
◇ F05 F07 F20 ◇ REV MR 81i:03087 Zbl 429 # 03033 • ID 53864

OREVKOV, V.P. [1979] *Three ways of recognizing unessential formulas in sequences (Russian) (English summary)* (**S** 0228) Zap Nauch Sem Leningrad Otd Mat Inst Steklov 88*163-175,243-244
• TRANSL [1982] (**J** 1531) J Sov Math 20*2351-2357
◇ F50 ◇ REV MR 81i:03096 Zbl 429 # 03034 • ID 53865

OREVKOV, V.P. [1984] *Upper bounds for lengthening of proofs after cut elimination (Russian) (English summary)* (**S** 0228) Zap Nauch Sem Leningrad Otd Mat Inst Steklov 137*87-98
◇ F05 F07 F20 ◇ REV MR 86d:03053 Zbl 562 # 03027 • ID 44140

OREVKOV, V.P. see Vol. I, II, III, IV for further entries

OREY, S. [1956] *On ω-consistency and related properties* (**J** 0036) J Symb Logic 21*246-252
◇ C07 F30 ◇ REV MR 18.632 Zbl 71.8 JSL 23.40 • ID 10163

OREY, S. [1961] *Relative interpretations* (**J** 0068) Z Math Logik Grundlagen Math 7*146-153
◇ C62 F25 ◇ REV MR 26 # 3608 Zbl 121.255 JSL 40.627 • ID 10166

OREY, S. [1962] see FEFERMAN, S.

OREY, S. see Vol. I, III, V for further entries

ORGASS, R.J. [1969] see FITCH, F.B.

ORGASS, R.J. [1975] *Extended basic logic and ordinal numbers* (**C** 1856) Log Enterprise 87-108
◇ B40 F15 ◇ REV Zbl 356 # 02023 • ID 50295

ORLOWSKA, E. [1977] *On a problem posed by Horn* (**J** 0014) Bull Acad Pol Sci, Ser Math Astron Phys 25*1067-1068
◇ F50 ◇ REV MR 58 # 5036 Zbl 369 # 02036 • ID 27126

ORLOWSKA, E. see Vol. I, II for further entries

ORNAGHI, M. [1975] see MARINI, D.

ORNAGHI, M. [1981] see MIGLIOLI, P.A.

ORNAGHI, M. [1982] see MIGLIOLI, P.A.

ORNAGHI, M. [1983] see MIGLIOLI, P.A.

ORNAGHI, M. [1984] see BERTONI, A.

OSHIBA, T. [1961] *On elimination of function-types of GLC* (**J** 0407) Comm Math Univ St Pauli (Tokyo) 10*75-109
◇ F35 ◇ REV MR 27 # 1356 Zbl 112.6 • ID 10176

OSHIBA, T. [1979] see FUNAHASHI, S.

OSHIBA, T. see Vol. I for further entries

OSIUS, G. [1973] *The internal and external aspect of logic and set theory in elementary topoi* (**J** 0306) Cah Topol & Geom Differ 14*47-49
◇ C90 E70 E75 F50 G30 ◇ REV MR 52 # 7896 Zbl 362 # 18001 • ID 27669

OSIUS, G. [1975] *A note on Kripke-Joyal semantics for the internal language of topoi* (**C** 0772) Model Th & Topoi 349-354
◇ C90 F50 G30 ◇ REV MR 52 # 7898 Zbl 348 # 18003 JSL 46.158 • ID 18321

OSIUS, G. [1975] *Logical and set theoretical tools in elementary topoi* (**C** 0772) Model Th & Topoi 297-346
◇ C90 E70 F50 G30 ◇ REV MR 52 # 7897 Zbl 348 # 18002 JSL 46.158 • ID 18320

OSIUS, G. see Vol. V for further entries

OSSWALD, H. [1970] *Modelltheoretische Untersuchungen in der Kripke-Semantik* (**J** 0009) Arch Math Logik Grundlagenforsch 13*3-21
◇ C20 C90 F50 ◇ REV MR 43 # 7299 Zbl 216.12 • ID 10184

OSSWALD, H. [1972] *Homomorphie-invariante Formeln in der intuitionistischen Logik* (**J** 0009) Arch Math Logik Grundlagenforsch 15*86-96
◇ C40 C90 F50 ◇ REV MR 48 # 3707 Zbl 251 # 02031 • ID 10186

OSSWALD, H. [1972] *Unterstruktur-invariante Formeln in der intuitionistischen Logik* (**J** 0009) Arch Math Logik Grundlagenforsch 15*118-128
◇ F50 ◇ REV MR 51 # 68 Zbl 263 # 02014 • ID 10187

OSSWALD, H. [1972] *Vollstaendigkeit und Schnittelimination in der intuitionistischen Typenlogik (English summary)* (**J** 0504) Manuscr Math 6*17-31
◇ F05 F35 F50 ◇ REV MR 45 # 4955 Zbl 239 # 02015 • ID 10185

OSSWALD, H. [1973] *Ein syntaktischer Beweis fuer die Zuverlaessigkeit der Schnittregel im Kalkuel von Schuette fuer die intuitionistische Typenlogik* (**J** 0504) Manuscr Math 8*243-249
◇ F05 F35 F50 ◇ REV MR 53 # 103 Zbl 255 # 02024 • ID 10188

OSSWALD, H. [1975] *Ueber Skolemerweiterungen in der intuitionistischen Logik mit Gleichheit* (**P** 1440) ⊧ ISILC Proof Th Symp (Schuette);1974 Kiel 264-266
◇ F50 ◇ REV MR 53 # 12884 Zbl 325 # 02021 • ID 27279

OSSWALD, H. see Vol. I, III, V for further entries

OSTROUKHOV, D.A. [1973] *Linearization of constructive sequences of normal algorithms (Russian)* (**S** 0554) Issl Teor Algor & Mat Logik (Moskva) 1*97-133
◇ D03 F60 ◇ REV MR 49 # 2299 Zbl 284 # 02011
• ID 10192

OSTROUKHOV, D.A. see Vol. I, IV for further entries

OWINGS JR., J.C. [1973] *Diagonalization and the recursion theorem* (**J** 0047) Notre Dame J Formal Log 14*95-99
◇ B40 D20 F30 ◇ REV MR 48 # 84 Zbl 247 # 02038
• ID 10206

OWINGS JR., J.C. see Vol. IV, V for further entries

PABION, J.F. [1974] *L'axiomatisation de la syntaxe et le second theoreme de Goedel* (**J** 0056) Publ Dep Math, Lyon 11/4*27-87
◇ B30 F30 ◇ REV MR 52 # 2862 Zbl 331 # 02017
• ID 17648

PABION, J.F. [1980] *TT_3I est equivalent a l'arithmetique du second ordre (English summary)* (**J** 2313) C R Acad Sci, Paris, Ser A-B 290*A1117-A1118
◇ B15 F25 F35 ◇ REV MR 82f:03051 Zbl 437 # 03004
• ID 55872

PABION, J.F. & RICHARD, D. [1981] *Synonymy and re-interpretation for some sublanguages of Peano arithmetic* (**P** 2614) Open Days in Model Th & Set Th;1981 Jadwisin 231-236
◇ B28 C62 F30 ◇ ID 33728

PABION, J.F. see Vol. I, II, III, V for further entries

PAEPPINGHAUS, P. [1980] *A version of the Σ_1-reflection principle for CFA provable in PRA* (**J** 0009) Arch Math Logik Grundlagenforsch 20*27-40
◇ F30 F35 ◇ REV MR 82b:03105 Zbl 429 # 03040
JSL 50.244 • ID 53871

PAEPPINGHAUS, P. [1981] see GIRARD, J.-Y.

PAEPPINGHAUS, P. [1981] *Was ist konstruktive Mathematik?* (**C** 2981) Phil der Math Erkenntnis Muenster 1978/79 27-48
◇ A05 F99 ◇ REV MR 82g:03104 Zbl 464 # 03004
• ID 77109

PAEPPINGHAUS, P. [1983] *Completeness properties of classical theories of finite type and the normal form theorem* (**J** 0202) Diss Math (Warsaw) 207,62pp
◇ B15 F05 F35 ◇ REV MR 85b:03100 Zbl 537 # 03038
• ID 40742

PAEPPINGHAUS, P. [1985] *A typed λ-calculus and Girard's model of ptykes* (**P** 4180) Int Congr Log, Meth & Phil of Sci (7,Pap);1983 Salzburg 245-279
◇ B40 F10 F15 F35 ◇ ID 47945

PAEPPINGHAUS, P. see Vol. II, IV for further entries

PAGLI, P. [1982] *Alcune estensioni del lemma di diagonalizzazione (English summary)* (**J** 3768) Boll Unione Mat Ital, VI Ser, D 1*1-12
◇ F30 G25 ◇ REV MR 84e:03070 Zbl 503 # 03029
• ID 34403

PAGLI, P. [1982] see BERNARDI, C.

PAGLI, P. see Vol. I for further entries

PAILLET, J.L. [1970] see LATREILLE, F.

PAILLET, J.L. see Vol. I, III for further entries

PAKHOMOV, S.V. [1976] *Approximability of operators in constructive metric spaces (Russian) (English summary)* (**S** 0228) Zap Nauch Sem Leningrad Otd Mat Inst Steklov 60*171-182,225-226
• TRANSL [1980] (**J** 1531) J Sov Math 14*1539-1546
◇ F60 ◇ REV MR 58 # 21515a Zbl 338 # 02015 • ID 64343

PAKHOMOV, S.V. [1976] *Continuity of operators in separable constructive metric spaces (Russian) (English summary)* (**S** 0228) Zap Nauch Sem Leningrad Otd Mat Inst Steklov 60*194-196,226
• TRANSL [1980] (**J** 1531) J Sov Math 14*1554-1556
◇ F60 ◇ REV MR 58 # 21515c Zbl 338 # 02017 • ID 64345

PAKHOMOV, S.V. [1976] *Hierarchies of operators in constructive metric spaces (Russian) (English summary)* (**S** 0228) Zap Nauch Sem Leningrad Otd Mat Inst Steklov 60*183-193,226
• TRANSL [1980] (**J** 1531) J Sov Math 14*1547-1554
◇ D80 F60 ◇ REV MR 58 # 21515 Zbl 338 # 02016
• ID 64344

PAKHOMOV, S.V. see Vol. IV for further entries

PALUCH, S. [1973] *The lattices of numerations of theories containing Peano's arithmetic* (**J** 0140) Comm Math Univ Carolinae (Prague) 14*339-359
◇ F30 G10 ◇ REV MR 50 # 12703 Zbl 268 # 02033
• ID 10264

PANOV, M.I. [1984] *Methodological problems of intuitionistic mathematics (Russian)* (**X** 2027) Nauka: Moskva 224pp
◇ A05 F98 ◇ ID 45123

PAOLA DI, R.A. [1966] *On sets represented by the same formula in distinct consistent axiomatizable Rosser theories* (**J** 0048) Pac J Math 18*455-456
◇ D25 F30 ◇ REV MR 37 # 6181 Zbl 203.11 • ID 03047

PAOLA DI, R.A. [1966] *Pseudo-complements and ordinal logics based on consistency statements* (**J** 0036) J Symb Logic 31*359-364
◇ D25 F15 F30 ◇ REV MR 38 # 2022 Zbl 212.329
JSL 37.406 • ID 03046

PAOLA DI, R.A. [1966] *Some properties of pseudo-complements of recursively enumerable sets* (**J** 0064) Trans Amer Math Soc 121*296-308
◇ D25 D30 F30 ◇ REV MR 33 # 3920 Zbl 203.11
JSL 37.406 • ID 03045

PAOLA DI, R.A. [1967] *Some theorems on extensions of arithmetic* (**J** 0036) J Symb Logic 32*180-189
◇ D25 D35 F15 F30 ◇ REV MR 36 # 2493 • ID 03048

PAOLA DI, R.A. [1975] *A theorem on shortening the length of proof in formal systems of arithmetic* (**J** 0036) J Symb Logic 40*398-400
◇ F20 F30 ◇ REV MR 52 # 2838 Zbl 316 # 02046
• ID 14815

PAOLA DI, R.A. [1984] *A uniformly, extremely nonextensional formula of arithmetic with many undecidable fixed points in many theories* (**J** 0053) Proc Amer Math Soc 92*291-297
• REPR [1985] (**P** 4646) Atti Incontri Log Mat (2);1983/84 Siena 171-183
◇ F30 G25 ◇ REV MR 85h:03069 Zbl 564 # 03042
Zbl 574 # 03046 • ID 43347

PAOLA DI, R.A. see Vol. I, III, IV for further entries

PARIKH, R. [1963] *Some generalisations of the notion of well-ordering* (**C** 4220) Rep Sem Found Anal 2*7.1-7.13
◇ F35 F50 ◇ ID 49835

PARIKH, R. [1966] *Some generalisations of the notion of well ordering* (**J** 0068) Z Math Logik Grundlagen Math 12*333-340
◇ D20 D45 E07 F35 ◇ REV MR 34 # 4131 Zbl 202.310 • ID 10279

PARIKH, R. [1967] *On nonuniqueness in transfinite progressions* (**J** 0441) J Indian Math Soc, NS 31*23-32
◇ F15 ◇ REV MR 37 # 1248 Zbl 189.288 • ID 10280

PARIKH, R. [1971] *Existence and feasibility in arithmetic* (**J** 0036) J Symb Logic 36*494-508
◇ A05 D15 F30 F65 H15 ◇ REV MR 46 # 3287 Zbl 243 # 02037 • ID 10282

PARIKH, R. [1973] *A note on paths through O* (**J** 0053) Proc Amer Math Soc 39*178-180
◇ F15 ◇ REV MR 47 # 30 Zbl 354 # 02033 • ID 10287

PARIKH, R. [1973] *Some results on the length of proofs* (**J** 0064) Trans Amer Math Soc 177*29-36
◇ B25 F20 F30 ◇ REV MR 55 # 5404 Zbl 269 # 02011 • ID 10286

PARIKH, R. see Vol. I, II, III, IV, V for further entries

PARIS, J.B. [1977] see HARRINGTON, L.A.

PARIS, J.B. [1977] see KIRBY, L.A.S.

PARIS, J.B. [1978] see KIRBY, L.A.S.

PARIS, J.B. [1978] *Note on an induction axiom* (**J** 0036) J Symb Logic 43*113-117
◇ B28 C62 F30 ◇ REV MR 81e:03057 Zbl 399 # 03009 • ID 29247

PARIS, J.B. [1978] *Some independence results for Peano arithmetic* (**J** 0036) J Symb Logic 43*725-731
◇ C62 F30 ◇ REV MR 82d:03095 Zbl 408 # 03048 JSL 48.482 • ID 31967

PARIS, J.B. [1980] *A hierarchy of cuts in models of arithmetic* (**P** 2625) Model Th of Algeb & Arithm;1979 Karpacz 312-337
◇ C15 C62 F30 ◇ REV MR 84e:03085 Zbl 448 # 03054 • ID 56654

PARIS, J.B. & WILKIE, A.J. [1981] *Δ_0 sets and induction* (**P** 2614) Open Days in Model Th & Set Th;1981 Jadwisin 237-248
◇ C62 F30 ◇ ID 33729

PARIS, J.B. [1981] *Some conservation results for fragments of arithmetic* (**P** 3404) Model Th & Arithm;1979/80 Paris 251-262
◇ C62 F30 ◇ REV MR 83f:03060 Zbl 475 # 03041 • ID 55495

PARIS, J.B. [1981] see KOSSAK, R.

PARIS, J.B. [1982] see KIRBY, L.A.S.

PARIS, J.B. [1983] see DIMITRACOPOULOS, C.

PARIS, J.B. [1984] *On the structure of models with restricted E_1-induction (Czech) (Russian and English summaries)* (**J** 0086) Cas Pestovani Mat, Ceskoslov Akad Ved 109*372-379
◇ C62 F30 ◇ REV MR 86e:03036 Zbl 567 # 03028 • ID 46698

PARIS, J.B. [1984] see MILLS, G.

PARIS, J.B. & WILKIE, A.J. [1984] *Some results on bounded induction* (**P** 1545) Easter Conf on Model Th (2);1984 Wittenberg 223-228
◇ F30 ◇ REV MR 85k:03003 Zbl 553 # 03036 • ID 43348

PARIS, J.B. & WILKIE, A.J. [1985] *Counting problems in bounded arithmetic* (**P** 2160) Latin Amer Symp Math Log (6);1983 Caracas 317-340
◇ B28 C62 F30 ◇ REV Zbl 572 # 03034 • ID 41796

PARIS, J.B. see Vol. I, II, III, IV, V for further entries

PARKS, R.Z. & RESCHER, N. [1971] *Restricted inference* (**J** 0079) Logique & Anal, NS 14*675-683
◇ B10 F07 ◇ REV MR 46 # 17 Zbl 232 # 02022 • ID 11123

PARKS, R.Z. see Vol. II for further entries

PARLAMENTO, F. [1978] *Sui sistemi metaformali e le progressioni ricorsive di teorie* (**J** 3285) Boll Unione Mat Ital, V Ser, A 15*243-247
◇ F15 F30 ◇ REV MR 58 # 10347 Zbl 405 # 03029 • ID 77138

PARLAMENTO, F. [1978] *Sulla completezza della ω-regola (English summary)* (**J** 2038) Rend Sem Mat, Torino 36*235-249
◇ F20 F30 ◇ REV MR 80h:03085 Zbl 415 # 03036 • ID 53138

PARLAMENTO, F. [1981] *Binumerability in a sequence of theories* (**J** 0144) Rend Sem Mat Univ Padova 65*9-12
◇ F30 ◇ REV MR 83h:03086 Zbl 484 # 03033 • ID 36086

PARLAMENTO, F. [1984] *PRA provability of Schmerl's fine structure theorem* (**J** 2038) Rend Sem Mat, Torino 42/2*87-106
◇ F30 ◇ ID 49560

PARLAMENTO, F. [1985] *Rules and principles of restricted induction relative to PRA (Italian)* (**P** 4646) Atti Incontri Log Mat (2);1983/84 Siena 513-517
◇ F30 ◇ ID 49685

PARRY, W.T. [1965] *Comments on a variant form of natural deduction* (**J** 0036) J Symb Logic 30*119-122
◇ B10 F07 ◇ REV Zbl 173.4 JSL 31.286 • ID 10301

PARRY, W.T. see Vol. I, II for further entries

PARSONS, C. [1960] see KOHL, H.R.

PARSONS, C. [1962] *The ω-consistency of ramified analysis* (**J** 0009) Arch Math Logik Grundlagenforsch 6*30-34
◇ F05 F10 F35 F65 ◇ REV MR 25 # 3834 Zbl 106.4 JSL 30.94 JSL 31.98 • ID 21096

PARSONS, C. [1970] *On a number-theoretic choice schema and its relation to induction* (**P** 0603) Intuitionism & Proof Th;1968 Buffalo 459-473
◇ F10 F30 ◇ REV MR 43 # 6050 Zbl 202.12 JSL 39.342 • ID 10310

PARSONS, C. [1972] *On n-quantifier induction* (**J** 0036) J Symb Logic 37*466-482
◇ F10 F30 ◇ REV MR 48 # 3712 Zbl 264 # 02027 JSL 39.342 • ID 10311

PARSONS, C. see Vol. I, II, IV, V for further entries

PARTIS, M.T. [1963] *Commutative partially ordered recursive arithmetics* (**J** 0132) Math Scand 13*199-216
◇ D20 F60 ◇ REV MR 29 # 4675 JSL 34.117 • ID 10312

PARTIS, M.T. [1967] *Limited universal and existential quantifiers in commutative partially ordered recursive arithmetics* (**J** 0047) Notre Dame J Formal Log 8*17-23
◇ D20 F60 ◇ REV MR 39 # 63 Zbl 211.312 • ID 10313

PAUN, G. [1983] see CALUDE, C.

PAUN, G. see Vol. I, II, IV for further entries

PEACOCKE, C. [1981] *Hacking on logic: two comments* (**J** 0301) J Phil 78*168-175
◇ A05 F50 ◇ REV JSL 47.689 • ID 47405

PEACOCKE, C. see Vol. II for further entries

PEANO, G. [1889] *The principles of arithmetic, presented by a new method (Latin)* (**X** 4721) Bocca & Clausen: Torino xvi+20pp
• TRANSL [1967] (**C** 0675) From Frege to Goedel 85-97 (English)
◇ B03 B28 E75 F30 ◇ REV FdM 21.51 • ID 19497

PEANO, G. see Vol. I, V for further entries

PEARCE, D.A. [1984] see BENTHEM VAN, J.F.A.K.

PEARCE, J. [1984] *A constructive consistency proof of a fragment of set theory* (**J** 0073) Ann Pure Appl Logic 27*25-62
◇ E30 E35 F05 F15 F35 ◇ ID 45489

PENON, J. [1981] *Infinitesimaux et intuitionnisme* (**J** 0306) Cah Topol & Geom Differ 22*67-72
◇ E75 F50 G30 ◇ REV MR 82m:51021 Zbl 463 # 18005
• ID 82458

PENZIN, YU.G. [1973] *Decidability of certain theories of integers (Russian)* (**J** 0092) Sib Mat Zh 14*1139-1143,1160
• TRANSL [1973] (**J** 0475) Sib Math J 14*796-799
◇ B25 F30 ◇ REV MR 48 # 8222 Zbl 272 # 02070
• ID 10358

PENZIN, YU.G. [1973] *Decidability of the theory of integers with addition, order and multiplication by an arbitrary number (Russian)* (**J** 0087) Mat Zametki (Akad Nauk SSSR) 13*667-675
• TRANSL [1973] (**J** 1044) Math Notes, Acad Sci USSR 13*401-405
◇ B25 F30 ◇ REV MR 48 # 1907 Zbl 267 # 02035
• ID 10357

PENZIN, YU.G. [1973] *Decidability of a theory of the integers with addition, order and predicates that distinguish a chain of subgroups (Russian)* (**C** 1443) Algebra, Vyp 2 (Irkutsk) 138-153
◇ B25 F30 ◇ REV MR 53 # 7755 • ID 23000

PENZIN, YU.G. [1976] *Undecidability of a theory of the integers with addition and predicate mutually disjoint (Russian)* (**C** 2555) Algeb Sistemy (Irkutsk) 149-153
◇ D35 F30 ◇ ID 32592

PENZIN, YU.G. [1979] *Twins problem in formal arithmetic (Russian)* (**J** 0087) Mat Zametki (Akad Nauk SSSR) 26*505-511,653
• TRANSL [1979] (**J** 1044) Math Notes, Acad Sci USSR 26*743-746
◇ F30 H15 ◇ REV MR 80m:03115 Zbl 422 # 03034
• ID 32595

PENZIN, YU.G. see Vol. I, III, IV for further entries

PERETTI, P. [1979] see BOEHM, C.

PERETYAT'KIN, M.G. [1974] see KOREC, I.

PERETYAT'KIN, M.G. see Vol. III, IV for further entries

PERLATOV, G.N. [1968] *The law of the excluded middle, and intuitionistic propositional logic (Russian)* (**J** 0789) Uch Zap Mat Ped Inst, Tula 1968*54-70
◇ F50 ◇ REV MR 44 # 5207 • ID 22239

PERLIS, D. [1985] *Languages with self-reference. I. Foundations* (**J** 0503) Artif Intell 25*301-322
◇ F99 ◇ ID 45533

PERLIS, D. see Vol. III, V for further entries

PERRIN, M.J. & ZALC, A. [1974] *Sous-systemes bicommutables d'analyse et de theorie des ensembles* (**J** 2313) C R Acad Sci, Paris, Ser A-B 279*A805-A808
◇ E25 E30 F25 F35 ◇ REV MR 52 # 104
Zbl 301 # 02055 • ID 18331

PERRIN, M.J. [1978] see ARTIGUE, M.

PERRIN, M.J. see Vol. I, III, V for further entries

PERZANOWSKI, J. [1970] *On the origin of Hilbert's programme (Polish) (English summary)* (**S** 0458) Zesz Nauk, Prace Log, Uniw Krakow 5*39-41
◇ A10 F99 ◇ REV MR 44 # 72 • ID 10385

PERZANOWSKI, J. see Vol. I, II, V for further entries

PETER, R. [1940] *Contribution to recursive number theory* (**J** 0460) Acta Univ Szeged, Sect Mat 9*233-238
◇ D20 F30 ◇ REV MR 1.132 Zbl 22.194 JSL 5.70
FdM 66.32 • ID 10396

PETER, R. [1950] *Zum Begriff der rekursiven reellen Zahl* (**J** 0002) Acta Sci Math (Szeged) 12*239-245
◇ F60 ◇ REV MR 12.385 Zbl 41.150 JSL 16.216
• ID 10399

PETER, R. [1959] *Rekursivitaet und Konstruktivitaet* (**P** 0634) Constructivity in Math;1957 Amsterdam 226-233
◇ A05 D20 F60 ◇ REV MR 21 # 4916 Zbl 85.250
JSL 33.471 • ID 10410

PETER, R. also published under the name POLITZER, R.

PETER, R. see Vol. I, III, IV, V for further entries

PETKOV, P.P. [1971] *On the question of the possibility of introducing disjunction on the lower levels of A.A. Markov's hierarchy of mathematical logic (Russian)* (**J** 0137) C R Acad Bulgar Sci 24*851-853
◇ B60 F50 ◇ REV MR 45 # 1726 Zbl 242 # 02036
• ID 10430

PETKOV, P.P. [1973] *The possibility of introducing disjunction at the level ω in A.A. Markov's graduated semantic system* (**J** 0137) C R Acad Bulgar Sci 26*995-998
◇ F50 ◇ REV MR 49 # 2295 Zbl 353 # 02013 • ID 10431

PETKOV, P.P. [1984] *The uniqueness of syntactical analysis for some calculi that are similar to Post's calculus. A generalization of the cut elimination theorem for classical propositional calculus (Russian)* (**P** 4117) Conf Math Log (Markov);1980 Sofia 73-88
◇ B50 F05 ◇ ID 46576

PETRI, N.V. [1974] *Effektive Nichtaufzaehlbarkeit von Pseudozahlen (Russian)* (**C** 1450) Teor Algor & Mat Logika (Markov) 143-147
◇ F60 ◇ REV MR 55 # 7749 Zbl 319 # 02028 • ID 29679

PETRI, N.V. see Vol. I, IV, V for further entries

PETROV, V.P. & SKORDEV, D.G. [1979] *Combinatory structures* (**J** 2547) Serdica, Bulgar Math Publ 5*128–148
◇ B40 D75 ◇ REV MR 81j:03079 Zbl 438 # 03047
• ID 55959

PETTOROSSI, A. [1975] see BATINI, C.

PETTOROSSI, A. [1977] *Combinators as tree transducers* (**P** 3199) CAAP'77 Arbres en Algeb & Progr (2);1977 Lille 213–223
◇ B40 ◇ REV Zbl 364 # 94033 • ID 50996

PETTOROSSI, A. [1979] *A hierarchy of infinite computations in type free subtree replacement systems* (**P** 2952) CAAP'79 Arbres en Algeb & Progr (4);1979 Lille 181–195
◇ B40 ◇ REV MR 81b:68008 • ID 82478

PETTOROSSI, A. [1979] *On the definition of hierarchies of infinite sequential computations* (**P** 2935) FCT'79 Fund of Comput Th;1979 Berlin/Wendisch-Rietz 335–341
◇ B40 D20 ◇ REV MR 81j:68021 Zbl 419 # 68077
• ID 53394

PETTOROSSI, A. [1980] *Synthesis of subtree rewriting systems behaviour by solving equations* (**P** 3057) CAAP'80 Arbres en Algeb & Progr (5);1980 Lille 63–74
◇ B40 ◇ REV Zbl 452 # 68057 • ID 54124

PETTOROSSI, A. [1981] *A property which guarantees termination in weak combinatory logic and subtree replacement systems* (**J** 0047) Notre Dame J Formal Log 22*344–356
◇ B40 ◇ REV MR 83e:03027 Zbl 416 # 03015 • ID 54446

PETTOROSSI, A. [1981] *Comparing and putting together recursive path ordering, simplification orderings and non-ascending property for termination proofs of term rewriting systems* (**P** 2903) Automata, Lang & Progr (8);1981 Akko 432–447
◇ B40 ◇ REV MR 83g:03016 Zbl 475 # 68011 • ID 55507

PETTOROSSI, A. [1982] *Tree rewriting systems, combinatory weak reduction systems and type free languages* (**J** 2095) Fund Inform, Ann Soc Math Pol, Ser 4 5*279–299
◇ B40 ◇ REV MR 85c:68042 Zbl 507 # 68047 • ID 37723

PETTOROSSI, A. see Vol. IV for further entries

PFEIFFER, H. [1969] *Ein Bezeichnungssystem fuer Ordinalzahlen* (**J** 0009) Arch Math Logik Grundlagenforsch 12*12–17
◇ F15 ◇ REV MR 40 # 5419 Zbl 193.311 JSL 39.186 • REM For another article with the same title see 1970 • ID 10448

PFEIFFER, H. [1970] *Ein Bezeichnungssystem fuer Ordinalzahlen* (**J** 0009) Arch Math Logik Grundlagenforsch 13*74–90
◇ F15 ◇ REV MR 44 # 60 Zbl 209.22 JSL 39.342 • REM For another article with the same title see 1969 • ID 10449

PFEIFFER, H. [1972] *Vergleich zweier Bezeichnungssysteme fuer Ordinalzahlen* (**J** 0009) Arch Math Logik Grundlagenforsch 15*41–56
◇ F15 ◇ REV MR 48 # 5829 Zbl 278 # 02029 JSL 39.342
• ID 10450

PFEIFFER, H. [1973] *Bezeichnungssysteme fuer Ordinalzahlen* (**X** 1051) Univ Utrecht Math Inst: Utrecht ii + 106pp
◇ E10 F15 ◇ REV MR 54 # 7212 • ID 25003

PFEIFFER, H. [1974] *Ueber zwei Bezeichnungssysteme fuer Ordinalzahlen* (**J** 0009) Arch Math Logik Grundlagenforsch 16*23–36
◇ F15 ◇ REV MR 50 # 1886 Zbl 336 # 02025 • ID 10451

PFEIFFER, H. [1975] *Eine Variante des Bezeichnungssystems $W(X)$ fuer Ordinalzahlen* (**P** 1440) ⊧ ISILC Proof Th Symp (Schuette);1974 Kiel 500*267–270
◇ F15 ◇ REV MR 54 # 2431 Zbl 341 # 02019 • ID 24021

PFEIFFER, H. see Vol. I, II, IV for further entries

PFENDER, M. & REITER, R. & SARTORIUS, M. [1982] *Constructive arithmetics* (**P** 3885) Categor Th.Appl to Algeb Log & Topol;1981 Gummersbach 228–236
◇ F30 G30 ◇ REV MR 84d:18003 Zbl 523 # 03049
• ID 37036

PFENDER, M. see Vol. IV for further entries

PHAN DINH DIEU [1965] *Constructive locally convex linear topological spaces (Russian)* (**J** 0023) Dokl Akad Nauk SSSR 162*766–769
• TRANSL [1965] (**J** 0062) Sov Math, Dokl 6*764–768
◇ F60 ◇ REV MR 33 # 6337 Zbl 284 # 02015 • ID 30520

PHAN DINH DIEU [1965] *The metrizability, normability and multinormability of constructive locally convex spaces (Russian)* (**J** 0023) Dokl Akad Nauk SSSR 162*1011–1014
• TRANSL [1965] (**J** 0062) Sov Math, Dokl 6*814–819
◇ F60 ◇ REV MR 33 # 6338 Zbl 284 # 02016 • ID 30522

PHAN DINH DIEU [1966] *On spaces adjoint to constructive locally convex spaces (Russian)* (**J** 0023) Dokl Akad Nauk SSSR 166*45–48
• TRANSL [1966] (**J** 0062) Sov Math, Dokl 7*43–47
◇ F60 ◇ REV MR 33 # 1702 Zbl 196.16 • ID 16317

PHAN DINH DIEU [1967] *A language of constructive mathematics involving systems of sets (Russian)* (**S** 0066) Tr Mat Inst Steklov 93*123–141
• TRANSL [1967] (**S** 0055) Proc Steklov Inst Math 93*157–182
◇ F50 ◇ REV MR 36 # 4994 • ID 14733

PHAN DINH DIEU [1967] *Certain properties of constructive generalized functions (Russian)* (**J** 0023) Dokl Akad Nauk SSSR 174*298–301
• TRANSL [1967] (**J** 0062) Sov Math, Dokl 8*643–647
◇ F60 ◇ REV MR 36 # 44 Zbl 196.16 • ID 03656

PHAN DINH DIEU [1967] *Constructive generalized functions (Russian)* (**J** 0023) Dokl Akad Nauk SSSR 174*37–40
• TRANSL [1967] (**J** 0062) Sov Math, Dokl 8*598–602
◇ F60 ◇ REV MR 36 # 43 Zbl 196.15 • ID 03654

PHAN DINH DIEU [1967] *On closed and open sets in constructive topological spaces (Russian)* (**S** 0066) Tr Mat Inst Steklov 93*250–256
• TRANSL [1967] (**S** 0055) Proc Steklov Inst Math 93*319–326
◇ F60 ◇ REV MR 37 # 2580 Zbl 191.307 • ID 03655

PHAN DINH DIEU [1968] *On spaces of constructive infinitely differentiable functions and on functionals in them (Russian)* (**J** 0023) Dokl Akad Nauk SSSR 180*799–802
• TRANSL [1968] (**J** 0062) Sov Math, Dokl 9*695–698
◇ F60 ◇ REV MR 38 # 529 Zbl 286 # 02039 • ID 14732

PHAN DINH DIEU [1970] *Some questions in constructive functional analysis (Russian)* (**S** 0066) Tr Mat Inst Steklov 114*222pp
• TRANSL [1970] (**S** 0055) Proc Steklov Inst Math 114*1–228
◇ F60 ◇ REV MR 46 # 2383 Zbl 275 # 02038 • ID 03657

PHAN DINH DIEU see Vol. IV for further entries

PHILLIPS, K. [1985] see JULIAN, W.

PIECZKOWSKI, A. [1966] see KOTAS, J.

PIECZKOWSKI, A. see Vol. I, II, IV for further entries

PIETRZYKOWSKI, T. [1973] *A complete mechanization of second-order type theory* (**J** 0037) ACM J 20*333-364
◇ B15 B35 F35 ◇ REV MR 49 # 8443 Zbl 253 # 68021
• ID 10491

PIETRZYKOWSKI, T. see Vol. I for further entries

PIL'CHAK, B.YU. [1950] *On the decision problem for the calculus of problems (Russian)* (**J** 0023) Dokl Akad Nauk SSSR 75*773-776
◇ B25 F50 ◇ REV MR 12.661 Zbl 39.7 JSL 16.226
• ID 16889

PIL'CHAK, B.YU. [1952] *On the calculus of problems (Russian)* (**J** 0265) Ukr Mat Zh, Akad Nauk Ukr SSR 4*174-194
◇ F50 ◇ REV MR 17.932 Zbl 49.150 JSL 21.372
• ID 33692

PIL'CHAK, B.YU. [1952] *On the role of the law of the excluded middle in mathematics (Russian)* (**J** 4148) Uch Zap Ped Inst, Novozybkov 1*115-131
◇ F50 ◇ ID 45575

PINTER, C. [1978] *Properties preserved under definitional equivalence and interpretations* (**J** 0068) Z Math Logik Grundlagen Math 24*481-488
◇ C25 C35 C50 F25 ◇ REV MR 80b:03039
Zbl 408 # 03028 • ID 56267

PINTER, C. see Vol. I, II, III, V for further entries

PINUS, A.G. [1975] *Effective linear orders (Russian)* (**J** 0092) Sib Mat Zh 16*1246-1254,1371
• TRANSL [1975] (**J** 0475) Sib Math J 16*956-962
◇ C57 D45 E07 F15 ◇ REV MR 53 # 137
Zbl 333 # 02035 • ID 16662

PINUS, A.G. [1981] *Constructivization of Boolean algebras (Russian)* (**J** 0092) Sib Mat Zh 22/4*169-175,231
• TRANSL [1981] (**J** 0475) Sib Math J 22*616-620
◇ F60 G05 ◇ REV MR 83b:03052 Zbl 499 # 03034
• ID 35104

PINUS, A.G. [1982] *Complete embeddings of categories of algebraic systems and definability of a model for its semigroup of endomorphisms (Russian)* (**J** 0031) Izv Vyssh Ucheb Zaved, Mat (Kazan) 1982/1*80-83
• TRANSL [1982] (**J** 3449) Sov Math 26/1*97-101
◇ C07 F25 G30 ◇ REV MR 84a:03036 Zbl 514 # 03023
• ID 35574

PINUS, A.G. see Vol. I, III, IV, V for further entries

PITTS, A.M. [1980] see HYLAND, J.M.E.

PITTS, A.M. see Vol. III, V for further entries

PLA I CARRERA, J. [1981] see BATLE, N.

PLA I CARRERA, J. [1982] *On the R-representability of primitive recursive functions (Catalan)* (**P** 3870) Congr Catala de Log Mat (1);1982 Barcelona 101-105
◇ D20 F30 ◇ REV MR 84i:03003 Zbl 521 # 03025
• ID 37071

PLA I CARRERA, J. see Vol. I, II, III, V for further entries

PLAISTED, D.A. [1980] *The application of multivariate polynomials to inference rules and partial tests for unsatisfiability* (**J** 1428) SIAM J Comp 9*698-705
◇ B35 D15 F20 ◇ REV MR 82g:68045 Zbl 448 # 68022
• ID 56667

PLAISTED, D.A. see Vol. I, II, IV for further entries

PLESNEVICH, G.S. [1984] *The semantics of schemes of structure-free notions (Russian)* (**J** 4404) Vopr Mat Logiki & Pril 1984*14-35
◇ B40 B75 ◇ ID 46526

PLISKO, V.E. [1973] *On realizable predicate formulae (Russian)* (**J** 0023) Dokl Akad Nauk SSSR 212*553-556
• TRANSL [1973] (**J** 0062) Sov Math, Dokl 14*1420-1424
◇ F50 ◇ REV MR 48 # 10779 Zbl 295 # 02009 • ID 10537

PLISKO, V.E. [1974] *A certain formal system that is connected with realizability (Russian)* (**C** 1450) Teor Algor & Mat Logika (Markov) 148-158,215
◇ F50 ◇ REV MR 53 # 10561 Zbl 298 # 02024 • ID 23070

PLISKO, V.E. [1974] *On interpretations of predicate formulae that are connected with constructive logic (Russian)* (**P** 1590) All-Union Conf Math Log (3);1974 Novosibirsk 170-172
◇ F50 ◇ ID 33042

PLISKO, V.E. [1974] *Recursive realizability and constructive predicate logic (Russian)* (**J** 0023) Dokl Akad Nauk SSSR 214*520-523
• TRANSL [1974] (**J** 0062) Sov Math, Dokl 15*193-197
◇ F50 ◇ REV MR 54 # 2433 Zbl 299 # 02027 • ID 24023

PLISKO, V.E. [1976] *Some variants of the notion of realizability for predicate formulas (Russian)* (**J** 0023) Dokl Akad Nauk SSSR 226*61-64
• TRANSL [1976] (**J** 0062) Sov Math, Dokl 17*59-63
◇ F50 ◇ REV MR 58 # 172 Zbl 358 # 02035 • ID 50489

PLISKO, V.E. [1977] *The nonarithmeticity of the class of realizable predicate formulas (Russian)* (**J** 0216) Izv Akad Nauk SSSR, Ser Mat 41*483-502
• TRANSL [1977] (**J** 0448) Math of USSR, Izv 11*453-471
◇ D35 D55 F50 ◇ REV MR 57 # 16031 Zbl 373 # 02032
• ID 51967

PLISKO, V.E. [1978] *Some variants of the notion of realizability for predicate formulas (Russian)* (**J** 0216) Izv Akad Nauk SSSR, Ser Mat 42*637-653
• TRANSL [1978] (**J** 0448) Math of USSR, Izv 12*588-604
◇ B10 F50 ◇ REV MR 80b:03035 Zbl 384 # 03042
• ID 52084

PLISKO, V.E. [1983] *Absolute realizability of predicate formulas (Russian)* (**J** 0216) Izv Akad Nauk SSSR, Ser Mat 47*315-334
• TRANSL [1983] (**J** 0448) Math of USSR, Izv 22*291-308
◇ F50 ◇ REV MR 85f:03063 Zbl 554 # 03029 • ID 40806

PLIUSKEVICIENE, A. [1969] *Elimination of cut type rules in axiomatic theories with equality (Russian)* (**S** 0228) Zap Nauch Sem Leningrad Otd Mat Inst Steklov 16*175-184
• ERR/ADD ibid 20*292-294
• TRANSL [1969] (**J** 0521) Semin Math, Inst Steklov 16*90-94
◇ F05 ◇ REV MR 42 # 39 Zbl 231 # 02034. • ID 10538

PLIUSKEVICIENE, A. [1971] *Elimination of cut-type rules from the Robinson and Presburger axiomatic systems (Russian)* (**S** 0228) Zap Nauch Sem Leningrad Otd Mat Inst Steklov 20*186-199,287
• TRANSL [1973] (**J** 1531) J Sov Math 1*117-125
◇ F05 F30 ◇ REV MR 45 # 4962 Zbl 231 # 02036
• ID 10539

PLIUSKEVICIENE, A. [1972] *A sequential variant of R. M. Robinson's arithmetic system not containing cut rules (Russian)* (S 0066) Tr Mat Inst Steklov 121*109-135,166
• TRANSL [1972] (S 0055) Proc Steklov Inst Math 121*121-150
◇ F05 F30 ◇ REV MR 49 # 4746 Zbl 286 # 02036
• ID 64557

PLIUSKEVICIENE, A. see Vol. I, II for further entries

PLIUSKEVICIUS, R. [1965] *A version of the constructive calculus of predicates without structural deduction rules (Russian)* (J 0023) Dokl Akad Nauk SSSR 161*292-295
• TRANSL [1965] (J 0062) Sov Math, Dokl 6*416-419
◇ F50 ◇ REV MR 30 # 4673 Zbl 143.251 • ID 24801

PLIUSKEVICIUS, R. [1967] *A sequential version of the calculus of constructive logic for normal formulas without structural rules of interference (Russian)* (S 0228) Zap Nauch Sem Leningrad Otd Mat Inst Steklov 4*174-188
◇ F50 ◇ REV MR 38 # 5585 • ID 10541

PLIUSKEVICIUS, R. [1968] *A sequential variant of constructive logic calculi for normal formulas not containing structural rules (Russian)* (S 0066) Tr Mat Inst Steklov 98*155-202
• TRANSL [1968] (S 0055) Proc Steklov Inst Math 98*175-221
◇ F50 ◇ REV MR 38 # 4282 Zbl 175.273 • ID 10542

PLIUSKEVICIUS, R. [1981] *On the Gentzen type proof theory for program analysis* (P 3642) Colloq Math Log in Computer Sci;1978 Salgotarjan 659-674
◇ B75 F07 F35 F50 ◇ REV MR 83g:68007
Zbl 497 # 03048 • ID 38112

PLIUSKEVICIUS, R. [1983] *Some disjunction properties for propositional classical and constructive dynamic logic (Russian) (English and Lithuanian summaries)* (J 3939) Mat Logika Primen (Akad Nauk Litov SSR) 3*61-100
◇ B75 F07 ◇ REV MR 85g:03046 Zbl 572 # 03009
• ID 43884

PLIUSKEVICIUS, R. see Vol. I, II for further entries

PLOTKIN, G.D. [1974] *The λ-calculus is ω-incomplete* (J 0036) J Symb Logic 39*313-317
◇ B40 ◇ REV MR 50 # 6799 Zbl 299 # 02029 • ID 10565

PLOTKIN, G.D. [1975] *Call-by-name, call-by-value and the λ-calculus* (J 1426) Theor Comput Sci 1*125-159
◇ B40 ◇ REV MR 55 # 2514 Zbl 325 # 68006 • ID 64559

PLOTKIN, G.D. [1978] *$\mathbb{T}^{\omega}$ as a universal domain* (J 0119) J Comp Syst Sci 17*209-236
◇ B40 B75 ◇ REV MR 80d:68105 Zbl 419 # 03007
• ID 53350

PLOTKIN, G.D. [1980] *λ-definability in the full type hierarchy* (C 3050) Essays Combin Log, Lambda Calc & Formalism (Curry) 363-373
◇ B15 B40 ◇ REV MR 82a:03016 Zbl 469 # 03006
• ID 77373

PLOTKIN, G.D. [1980] see HENNESSY, M.

PLOTKIN, G.D. see Vol. I, IV for further entries

PLOTKIN, J.M. & ROSENTHAL, J.W. [1982] *The expected complexity of analytic tableaux analysis in propositional calculus* (J 0047) Notre Dame J Formal Log 23*409-426
◇ B05 B35 D15 F20 ◇ REV MR 83k:03044
Zbl 464 # 03011 • ID 35398

PLOTKIN, J.M. see Vol. II, III, V for further entries

PODNIEKS, K.M. [1972] *Constructive decomposition of stochastic matrices (Russian)* (J 0474) Avtom Vychis Tekh, Akad Nauk Latv SSR 1972/3*18-20
• TRANSL [1972] (J 2666) Autom Control Comput Sci 1972/3*17-19
◇ F60 ◇ REV MR 47 # 8577 Zbl 238 # 02030 • ID 27902

PODNIEKS, K.M. [1975] *The double-incompleteness theorem (Russian) (English summary)* (S 2587) Teor Algor & Progr (Riga) 2*191-200,209
◇ D35 F30 ◇ REV MR 57 # 16036 Zbl 347 # 02021
• ID 64585

PODNIEKS, K.M. [1981] *Around Goedel's theorem (Russian)* (X 0895) Latv Valsts (Gos) Univ : Riga 106pp
◇ F30 ◇ REV Zbl 506 # 03018 • ID 36965

PODNIEKS, K.M. see Vol. IV for further entries

POEL VAN DER, W.L. [1975] *Combinators and lambda forms (Dutch)* (J 0358) Versl Gewone Vergad Afd Natuurkd 84*172-179
◇ B40 ◇ REV MR 53 # 5268 Zbl 337 # 02019 • ID 22899

POEL VAN DER, W.L. [1980] see MEY VAN DER, G.

POEL VAN DER, W.L. see Vol. I for further entries

POGORZELSKI, H.A. [1962] *Recursive arithmetic of Skolem. I,II* (J 0132) Math Scand 11*33-36,156-160
◇ F30 ◇ REV MR 27 # 3514 MR 27 # 3515 Zbl 109.241
JSL 29.101 • ID 19593

POGORZELSKI, H.A. [1969] *Goldbach sentences in abstract arithmetics $\mathscr{A}^{k}(A)$ I* (J 0127) J Reine Angew Math 237*65-96
◇ D20 F30 ◇ REV MR 40 # 4107 Zbl 179.348 • REM Part II 1970 • ID 19592

POGORZELSKI, H.A. [1970] *Goldbach sentences in abstract arithmetics $\mathscr{A}^{k}(A)$ II* (J 0127) J Reine Angew Math 243*32-54
◇ F30 ◇ REV MR 41 # 8235 Zbl 202.12 • REM Part I 1969
• ID 19591

POGORZELSKI, H.A. [1974] *On the Goldbach conjecture and the consistency of general recursive arithmetic* (J 0127) J Reine Angew Math 268/269*1-16
◇ F30 ◇ REV MR 50 # 69 Zbl 319 # 02027 • ID 10596

POGORZELSKI, H.A. [1976] *Dirichlet theorems and prime number hypotheses of a conditional Goldbach theorem* (J 0127) J Reine Angew Math 286-287*33-45
◇ F30 ◇ REV MR 58 # 21554 Zbl 333 # 02025 • ID 64598

POGORZELSKI, H.A. [1977] *Semisemiological structure of the prime numbers and conditional Goldbach theorems* (J 0127) J Reine Angew Math 290*77-92
◇ F30 ◇ REV MR 55 # 7961 Zbl 344 # 02023 • ID 64596

POGORZELSKI, H.A. see Vol. I, IV for further entries

POGORZELSKI, W.A. & SLUPECKI, J. [1960] *Basic properties of deductive systems based on nonclassical logics I,II (Polish) (Russian and English summaries)* (J 0063) Studia Logica 9*163-176,10*77-95
◇ B22 F50 ◇ REV MR 26 # 6039 Zbl 129.256 • ID 21001

POGORZELSKI, W.A. see Vol. I, II, III for further entries

POHLERS, W. [1973] *Ein starker Normalisationssatz fuer die intuitionistische Typentheorie* (**J** 0504) Manuscr Math 8*371-387
◇ F05 F35 F50 ◇ REV MR 53 # 12896 Zbl 256 # 02013 • ID 14735

POHLERS, W. [1975] *An upper bound for the provability of transfinite induction in systems with N-times iterated inductive definitions* (**P** 1440) ⊧ ISILC Proof Th Symp (Schuette);1974 Kiel 271-289
◇ F15 F35 F50 ◇ REV MR 54 # 4929 Zbl 351 # 02026 JSL 48.878 • ID 24104

POHLERS, W. [1978] *Ordinals connected with formal theories for transfinitely iterated inductive definitions* (**J** 0036) J Symb Logic 43*161-182
◇ F05 F15 F35 ◇ REV MR 81k:03061 Zbl 394 # 03049 JSL 48.878 • ID 29251

POHLERS, W. [1978] see BUCHHOLZ, W.

POHLERS, W. [1981] *Cut-elimination for impredicative infinitary systems I. Ordinal-analysis for ID_1* (**J** 0009) Arch Math Logik Grundlagenforsch 21*113-129
◇ C75 F05 F15 F35 ◇ REV MR 84a:03070a Zbl 484 # 03030 JSL 48.879 • REM Part II 1982 • ID 33310

POHLERS, W. [1981] see BUCHHOLZ, W.

POHLERS, W. [1981] *Proof-theoretical analysis of ID_ν by the method of local predicativity* (**C** 3588) Iter Induct Def & Subsyst of Anal: Rec Proof-Th Stud 261-357
◇ F05 F15 F35 F50 ◇ ID 45808

POHLERS, W. [1982] *Admissibility in proof theory; a survey* (**P** 3622) Int Congr Log, Meth & Phil of Sci (6,Proc);1979 Hannover 123-139
◇ C70 F15 F35 F98 ◇ REV MR 84h:03129 Zbl 506 # 03015 • ID 34316

POHLERS, W. [1982] *Cut elimination for impredicative infinitary systems II. Ordinal analysis for iterated inductive definitions* (**J** 0009) Arch Math Logik Grundlagenforsch 22*69-87
◇ C75 F05 F15 F35 ◇ REV MR 84a:03070b Zbl 497 # 03043 JSL 48.879 • REM Part I 1981 • ID 35591

POHLERS, W. [1983] see JAEGER, G.

POIGNE, A. [1984] *Higher order data structures - Cartesian closure versus λ-calculus* (**P** 3565) Symp of Th Aspects of Comput Sci (1);1984 Paris 174-185
◇ B40 B75 G30 ◇ REV MR 86b:68038 Zbl 575 # 68018 • ID 47093

POIGNE, A. see Vol. IV for further entries

POLIVKA, I. [1979] see DEMUTH, O.

POLLACK, P.L. [1971] see CAVINESS, B.F.

POPOV, S.V. [1974] *An equivalence relation and a complete system of schemes of equivalent transformations of deductions in propositional calculus (Russian)* (**J** 0052) Probl Kibern 29*103-150,245
◇ B05 F07 ◇ REV MR 51 # 79 Zbl 309 # 02029 • ID 15197

POPOV, S.V. [1975] *The complexity of deductions in certain propositional calculi (Russian)* (**J** 0052) Probl Kibern 30*57-146
◇ B05 F20 ◇ REV MR 52 # 5377 Zbl 414 # 03035 • ID 18344

POPOV, S.V. [1976] *On the complexity of derivations in classical propositional calculus (Russian)* (**J** 0023) Dokl Akad Nauk SSSR 228*1041-1044
• TRANSL [1976] (**J** 0062) Sov Math, Dokl 17*876-880
◇ B05 F20 ◇ REV MR 54 # 9972 Zbl 357 # 02009 • ID 25598

POPOV, S.V. [1977] *Some complexity characteristics of derivations in propositional calculus (Russian) (English summary)* (**S** 2651) Prepr Inst Prikl Mat, Akad Nauk SSSR 116*63pp
◇ B05 B70 F20 ◇ REV MR 58 # 10284 • ID 77440

POPOV, S.V. [1977] *The complexity of derivations in intuitionistic propositional calculus (Russian)* (**S** 2651) Prepr Inst Prikl Mat, Akad Nauk SSSR 120*55pp
◇ F20 F50 ◇ REV MR 58 # 169 • ID 77441

POPOV, S.V. & ZAKHAR'YASHCHEV, M.V. [1980] *On the power of countermodels in intuitionistic calculus (Russian)* (**S** 2651) Prepr Inst Prikl Mat, Akad Nauk SSSR 80/45*28pp
◇ C90 F50 ◇ REV MR 82m:03022 • ID 80198

POPOV, S.V. [1984] *Diagrams of deductions in sequential calculi (Russian)* (**J** 0052) Probl Kibern 41*49-100
◇ F05 F07 ◇ REV MR 85j:03096 • ID 45849

POPOV, S.V. see Vol. I, II, IV for further entries

POPPER, K.R. [1947] *Functional logic without axioms or primitive rules of inference* (**J** 0028) Indag Math 9*561-571
◇ A05 B20 F50 ◇ REV MR 9.321 JSL 13.173 • ID 10648

POPPER, K.R. see Vol. I, II for further entries

POREBSKA, M. & WRONSKI, A. [1974] *A characterization of fragments of the intuitionistic propositional logic* (**J** 0387) Bull Sect Logic, Pol Acad Sci 3/1*19-20
◇ F50 ◇ REV MR 55 # 2510 • ID 77451

POREBSKA, M. & WRONSKI, A. [1975] *A characterization of fragments of the intuitionistic propositional logic* (**J** 0302) Rep Math Logic, Krakow & Katowice 4*39-42
◇ B55 F50 ◇ REV MR 51 # 12488 Zbl 318 # 02027 • ID 17277

POREBSKA, M. [1975] see KABZINSKI, J.K.

POREBSKA, M. see Vol. II, III for further entries

PORTE, J. [1958] *Une propriete du calcul propositionnel intuitionniste* (**J** 0028) Indag Math 20*362-365
◇ F50 ◇ REV MR 20 # 5138 Zbl 84.7 JSL 24.68 • ID 10664

PORTE, J. see Vol. I, II, IV for further entries

POSEGGA, M. [1972] see HAUCK, J.

POSY, C.J. [1974] *Brouwer's constructivism* (**J** 0154) Synthese 27*125-159
◇ A05 F99 ◇ REV MR 57 # 15976 Zbl 315 # 02011 • ID 29799

POSY, C.J. [1976] *Varieties of indeterminacy in the theory of general choice sequences* (**J** 0122) J Philos Logic 5*91-132
◇ F35 F50 ◇ REV MR 55 # 10247 Zbl 349 # 02027 • ID 64637

POSY, C.J. [1980] *On Brouwer's definition of unextendable order* (**J** 2028) Hist & Phil Log 139-149
◇ A10 F55 ◇ REV MR 83i:03095 Zbl 518 # 03024 • ID 35542

POSY, C.J. & SWART DE, H.C.M. [1981] *Validity and quantification in intuitionism* (J 0122) J Philos Logic 10*117-126
◇ A05 F50 ◇ REV MR 82a:03059 Zbl 462 # 03001
• ID 54508

POSY, C.J. see Vol. II for further entries

POTTINGER, G. [1976] *A new way of normalizing intuitionist propositional logic* (J 0063) Studia Logica 35*387-408
◇ B40 F05 F50 ◇ REV MR 58 # 10329 Zbl 355 # 02019
• ID 31427

POTTINGER, G. [1977] *Normalization as a homomorphic image of cut-elimination* (J 0007) Ann Math Logic 12*323-357
◇ F05 F07 F50 ◇ REV MR 57 # 15995 Zbl 378 # 02017
• ID 27949

POTTINGER, G. [1978] *Proofs of the normalization and Church-Rosser theorems for the typed λ-calculus* (J 0047) Notre Dame J Formal Log 19*445-451
◇ B40 ◇ REV MR 58 # 5083 Zbl 351 # 02019 • ID 31430

POTTINGER, G. [1980] *A type assignment for the strongly normalizable lambda-terms* (C 3050) Essays Combin Log, Lambda Calc & Formalism (Curry) 561-577
◇ B40 F07 ◇ REV MR 82j:03014 Zbl 469 # 03006
• ID 77480

POTTINGER, G. [1981] *The Church-Rosser theorem for the typed λ-calculus with surjective pairing* (J 0047) Notre Dame J Formal Log 22*264-268
◇ B40 ◇ REV MR 82j:03015 Zbl 423 # 03009 • ID 53521

POTTINGER, G. see Vol. II for further entries

POTTS, T.C. [1979] *"The grossest confusion possible?" - Frege and the lambda-calculus* (J 2076) Rev Int Philos 33*761-785
◇ A05 B40 ◇ REV MR 82d:03004 • ID 77486

POTTS, T.C. see Vol. I for further entries

POUR-EL, M.B. [1967] see KRIPKE, S.A.

POUR-EL, M.B. [1973] *Abstract computability versus analog-generability. A survey* (P 0713) Cambridge Summer School Math Log;1971 Cambridge GB 345-360
◇ D75 D80 F65 ◇ REV MR 49 # 17 Zbl 272 # 02068
• ID 10709

POUR-EL, M.B. [1973] *Analog computers, digital computers, mathematical logic, differential equations - interrelations* (P 1753) Int Congr AICA Hybrid Comput (7); 122-124
◇ D20 D75 D80 F65 ◇ ID 31218

POUR-EL, M.B. [1974] *Abstract computability and its relation to the general purpose analog computer (some connections between logic, differential equations and analog computers)* (J 0064) Trans Amer Math Soc 199*1-28
◇ D20 D75 D80 F65 ◇ REV MR 50 # 78
Zbl 296 # 02022 • ID 10706

POUR-EL, M.B. & RICHARDS, I. [1978] *Differentiability properties of computable functions - a summary* (J 0380) Acta Cybern (Szeged) 4*123-125
◇ D20 D80 F60 ◇ REV MR 80d:03059 Zbl 397 # 03038
• ID 52704

POUR-EL, M.B. & RICHARDS, I. [1979] *A computable ordinary differential equation which possesses no computable solution* (J 0007) Ann Math Logic 17*61-90
◇ D80 F60 ◇ REV MR 81k:03064 Zbl 424 # 68028
JSL 47.900 • ID 77493

POUR-EL, M.B. & RICHARDS, I. [1984] *L^p-computability in recursive analysis* (J 0053) Proc Amer Math Soc 92*93-97
◇ D80 F60 ◇ REV MR 86f:03102 Zbl 558 # 03030
• ID 45841

POUR-EL, M.B. see Vol. I, III, IV for further entries

POWELL, W.C. [1975] *Extending Goedel's negative interpretation to ZF* (J 0036) J Symb Logic 40*221-229
◇ E35 E70 F50 ◇ REV MR 52 # 103 Zbl 356 # 02049
• ID 10718

POWELL, W.C. see Vol. III, V for further entries

POZSGAY, L.J. [1968] *Goedel's second theorem for elementary arithmetic* (J 0068) Z Math Logik Grundlagen Math 14*67-80
◇ F30 ◇ REV MR 36 # 4991 Zbl 184.20 • ID 10719

POZSGAY, L.J. [1971] *Liberal intuitionism as a basis for set theory* (P 0693) Axiomatic Set Th;1967 Los Angeles 1*321-330
◇ A05 E30 F65 ◇ REV MR 44 # 5219 Zbl 223 # 02003
JSL 40.506 • ID 10720

POZSGAY, L.J. [1972] *Semi-intuitionistic set theory* (J 0047) Notre Dame J Formal Log 13*546-550
◇ A05 E30 E70 F65 ◇ REV MR 47 # 6489
Zbl 242 # 02070 • ID 10721

PRANK, R.K. [1979] *Expressibility in the elementary theory of recursive sets with realizability logic (Russian) (English summary)* (S 3468) Tr Mat & Mekh (Tartu) 25(500)*119-129
◇ B60 D20 F50 ◇ REV MR 81a:03043 Zbl 421 # 03038
• ID 53437

PRANK, R.K. [1980] *Semantics of realizability for a language with variables for recursively enumerable sets (Russian)* (J 2852) Tr Vychisl Tsentra, Univ Tartu 43*112-131
◇ B60 D25 F35 F50 ◇ REV MR 81h:03111 • ID 77510

PRANK, R.K. [1981] *Expressibility in the elementary theory of recursively enumerable sets with realizability logic (Russian)* (J 0003) Algebra i Logika 20*427-439,484-485
• TRANSL [1981] (J 0069) Algeb and Log 20*282-291
◇ B60 D25 F50 ◇ REV MR 83i:03068 • ID 35552

PRANK, R.K. see Vol. IV for further entries

PRAWITZ, D. [1963] *Concerning constructive logic and the concept of implication (Swedish)* (C 4234) Sju Filos Studier 9-32
◇ A05 F07 F50 ◇ REV JSL 33.605 • ID 43209

PRAWITZ, D. [1965] *Natural deduction. A proof-theoretical study* (X 1163) Almqvist & Wiksell: Stockholm 113pp
◇ B15 B98 F05 F07 F50 F98 ◇ REV MR 33 # 1227
Zbl 173.2 JSL 32.255 • ID 22164

PRAWITZ, D. [1967] *Completeness and Hauptsatz for second order logic* (J 0105) Theoria (Lund) 33*246-258
◇ F05 F35 ◇ REV MR 39 # 2607 Zbl 175.267 JSL 39.607
• ID 10729

PRAWITZ, D. [1968] see MALMNAES, P.E.

PRAWITZ, D. [1968] *Hauptsatz for higher order logic* (J 0036) J Symb Logic 33*452-457
◇ B15 F05 F35 ◇ REV MR 39 # 44 Zbl 164.310
JSL 39.607 • ID 10733

PRAWITZ, D. [1970] *Constructive semantics* (P 0785) Scand Logic Symp (1);1968 Aabo 96-114
◇ F50 ◇ REV MR 51 # 5266 Zbl 318 # 02032 • ID 15202

PRAWITZ, D. [1970] *On the proof theory of mathematical analysis* (**C** 0735) Logic & Value (Dahlquist) 169-180
◇ F35 ◇ REV MR 51 # 80 • ID 15201

PRAWITZ, D. [1970] *Some results for intuitionistic logic with second order quantification rules* (**P** 0603) Intuitionism & Proof Th;1968 Buffalo 259-269
◇ F05 F35 F50 ◇ REV MR 43 # 4630 Zbl 212.13 JSL 40.500 • ID 10735

PRAWITZ, D. [1971] *Ideas and results in proof theory* (**P** 0604) Scand Logic Symp (2);1970 Oslo 235-307
◇ F05 F07 F10 F30 F35 F50 F98 ◇ REV MR 52 # 7871 Zbl 226 # 02031 JSL 40.232 • ID 10737

PRAWITZ, D. [1972] *The philosophical position of proof theory* (**C** 0558) Contemp Phil Scand 123-134
◇ A05 F98 ◇ ID 14900

PRAWITZ, D. [1973] *Towards a foundation of a general proof theory* (**P** 0793) Int Congr Log, Meth & Phil of Sci (4,Proc);1971 Bucharest 225-250
◇ A05 F05 F07 ◇ REV MR 57 # 5699 • ID 15074

PRAWITZ, D. [1974] *On the idea of a general proof theory* (**J** 0012) Boll Unione Mat Ital, IV Ser 9/2Suppl*108-121 • REPR [1974] (**J** 0154) Synthese 27*63-77
◇ A05 F05 F07 ◇ REV MR 54 # 7218 MR 57 # 15996 Zbl 319 # 02033 Zbl 336 # 02026 • ID 25009

PRAWITZ, D. [1975] *Comments on Gentzen-type procedures and the classical notion of truth* (**P** 1440) ⊧ ISILC Proof Th Symp (Schuette);1974 Kiel 500*290-319
◇ B10 F05 F07 F30 ◇ REV MR 55 # 12482 Zbl 342 # 02022 • ID 27280

PRAWITZ, D. [1977] *Meaning and proofs: On the conflict between classical and intuitionistic logic* (**J** 0105) Theoria (Lund) 43*2-40
◇ A05 F99 ◇ REV MR 56 # 15354 Zbl 361 # 02008 JSL 47.689 • ID 50643

PRAWITZ, D. [1979] *Proofs and the meaning and completeness of the logical constants* (**P** 1705) Scand Logic Symp (4);1976 Jyvaeskylae 25-40
◇ A05 F07 F50 ◇ REV MR 81g:03075 Zbl 406 # 03069 JSL 47.689 • ID 77515

PRAWITZ, D. [1981] *Validity and normalizability of proofs in 1st and 2nd order classical and intuitionistic logic* (**P** 3092) Congr Naz Logica;1979 Montecatini Terme 11-36
◇ F05 F07 F35 F50 ◇ ID 48432

PRAWITZ, D. [1982] *Beweise und die Bedeutung und Vollstaendigkeit der logischen Konstanten* (**J** 2688) Conceptus (Wien) 16*31-44
◇ A05 B20 F07 ◇ REV MR 84c:03021 • ID 34933

PRAWITZ, D. [1985] *Normalizations of proofs in set theory* (**P** 4646) Atti Incontri Log Mat (2);1983/84 Siena 357-371
◇ E30 F05 ◇ ID 49658

PRAWITZ, D. [1985] *Remarks on some approaches to the concept of logical consequence* (**J** 0154) Synthese 62*153-171
◇ A05 B22 F07 ◇ ID 45636

PRAWITZ, D. see Vol. I for further entries

PRAZMOWSKI, K. & SZCZERBA, L.W. [1976] *Interpretability and categoricity (Russian summary)* (**J** 0014) Bull Acad Pol Sci, Ser Math Astron Phys 24*309-312
◇ C35 F25 ◇ REV MR 56 # 8356 Zbl 342 # 02038 • ID 18347

PRAZMOWSKI, K. [1978] *Conciseness and hierarchy of theories (Russian summary)* (**J** 0014) Bull Acad Pol Sci, Ser Math Astron Phys 26*577-583
◇ C07 F25 ◇ REV MR 80j:03021 Zbl 408 # 03033 • ID 29177

PRESBURGER, M. [1930] *Ueber die Vollstaendigkeit eines gewissen Systems der Arithmetik ganzer Zahlen, in welchem die Addition als einzige Operation hervortritt* (**P** 0796) Congr Math Pays Slaves (1);1929 Warsaw 92-101,395
◇ B25 B28 C10 C35 F30 ◇ REV FdM 56.825 • ID 10749

PREVIALE, F. [1967] *Teorie equipollenti e K-equipollenti* (**J** 0220) Atti Accad Sci Torino, Fis Mat Nat 102*769-794
◇ F25 ◇ REV MR 41 # 3259 Zbl 175.268 • ID 10752

PREVIALE, F. [1975] *Tavole semantiche per sistemi astratti di logica estensionale* (**J** 0144) Rend Sem Mat Univ Padova 54*31-57
◇ B50 F50 ◇ REV MR 55 # 2522 Zbl 357 # 02020 • ID 31421

PREVIALE, F. see Vol. I, III for further entries

PRICE, ROBERT [1961] *The stroke function in natural deduction* (**J** 0068) Z Math Logik Grundlagen Math 7*117-123
◇ B05 F07 ◇ REV MR 28 # 2049 Zbl 107.6 JSL 38.149 • ID 10756

PRICE, ROBERT see Vol. I for further entries

PRIDA, J.F. [1978] *Una nueva prueba del teorema de incompletitud de la aritmetica* (**J** 4320) Rev Centro Calc Univ Complutense (Madrid) 32*1-3
◇ F30 ◇ ID 46650

PRIDA, J.F. see Vol. I, IV for further entries

PRIEST, G. [1983] *The logical paradoxes and the law of excluded middle* (**J** 2811) Phil Quart (St Andrews) 33*160-165
◇ A05 F99 ◇ REV MR 86c:03005a • ID 45133

PRIEST, G. see Vol. I, II, III, V for further entries

PRINI, G. [1981] see AIELLO, L.

PRIOR, A.N. [1968] see MEREDITH, C.A.

PRIOR, A.N. see Vol. I, II, V for further entries

PUDLAK, P. [1983] *A definition of exponentiation by a bounded arithmetical formula* (**J** 0140) Comm Math Univ Carolinae (Prague) 24*667-671
◇ B28 C62 F30 ◇ REV MR 85e:03142 Zbl 533 # 03032 • ID 36550

PUDLAK, P. [1983] *Some prime elements in the lattice of interpretability types* (**J** 0064) Trans Amer Math Soc 280*255-275
◇ E30 F25 F30 G10 ◇ REV MR 85k:03038 Zbl 561 # 03014 • ID 45314

PUDLAK, P. [1985] *Cuts, consistency statements and interpretations* (**J** 0036) J Symb Logic 50*423-441
◇ C62 E30 F25 ◇ REV Zbl 569 # 03024 • ID 42556

PUDLAK, P. see Vol. I, III, IV, V for further entries

PUTNAM, H. [1957] see KREISEL, G.

PUTNAM, H. [1961] *Uniqueness ordinals in higher constructive number classes* (**C** 0622) Essays Found of Math (Fraenkel) 190-206
◇ D55 F15 ◇ REV MR 29 # 4686 Zbl 143.254 JSL 31.135 • ID 10835

PUTNAM, H. [1964] *On families of sets represented in theories* (J 0009) Arch Math Logik Grundlagenforsch 6*66-70
◇ C40 D25 F30 ◇ REV MR 31 # 3329 Zbl 126.21
• ID 10838

PUTNAM, H. [1964] *On hierarchies and systems of notations* (J 0053) Proc Amer Math Soc 15*44-50
◇ D55 D70 F15 ◇ REV MR 28 # 1126 Zbl 237 # 02011 JSL 31.136 • ID 10839

PUTNAM, H. [1965] see LUCKHAM, D.C.

PUTNAM, H. [1965] see HENSEL, G.

PUTNAM, H. [1974] see LUKAS, J.D.

PUTNAM, H. see Vol. I, II, III, IV, V for further entries

QUINE, W.V.O. [1936] *A reinterpretation of Schoenfinkel's logical operators* (J 0015) Bull Amer Math Soc 42*87-89
◇ B40 ◇ REV Zbl 14.3 JSL 1 FdM 62.37 • ID 10856

QUINE, W.V.O. [1947] see GOODMAN, NELSON

QUINE, W.V.O. [1949] *On decidability and completeness* (J 0154) Synthese 7*441-446
◇ B25 D35 F30 ◇ REV MR 12.70 JSL 16.76 • ID 10881

QUINE, W.V.O. [1953] *On ω-inconsistency and a so-called axiom of infinity* (J 0036) J Symb Logic 18*119-124
◇ E70 F30 ◇ REV MR 15.1 Zbl 53.5 JSL 19.128
• ID 10889

QUINE, W.V.O. [1954] *Interpretations of sets of conditions* (J 0036) J Symb Logic 19*97-102
◇ F25 ◇ REV MR 15.924 Zbl 58.246 JSL 24.242
• ID 10893

QUINE, W.V.O. [1954] *Reduction to a dyadic predicate* (J 0036) J Symb Logic 19*180-182
◇ F25 ◇ REV MR 16.208 Zbl 56.11 JSL 22.297 • ID 10894

QUINE, W.V.O. [1956] *On formulas with valid cases* (J 0036) J Symb Logic 21*148
◇ F25 ◇ REV MR 17.1171 Zbl 72.250 • ID 10901

QUINE, W.V.O. [1972] *Algebraic logic and predicate functors* (C 1698) Logic & Art (Goodman) 214-238
• REPR [1976] (C 1587) Quine: Ways of Paradox & Essays 283-307
◇ A05 B40 G15 ◇ REV MR 58 # 5201 • ID 25050

QUINE, W.V.O. [1981] *Predicate functors revisited* (J 0036) J Symb Logic 46*649-652
◇ B10 B40 ◇ REV MR 83e:03028 Zbl 472 # 03007
• ID 55273

QUINE, W.V.O. see Vol. I, II, III, IV, V for further entries

RABIN, M.O. [1961] *Non-standard models and independence of the induction axiom* (C 0622) Essays Found of Math (Fraenkel) 287-299
◇ B28 C62 F30 H15 ◇ REV MR 28 # 4999 Zbl 143.10 JSL 38.159 • ID 10934

RABIN, M.O. [1965] *A simple method for undecidability proofs and some applications* (P 0623) Int Congr Log, Meth & Phil of Sci (2,Proc);1964 Jerusalem 58-68
◇ D35 F25 ◇ REV MR 36 # 4976 Zbl 192.55 JSL 36.150
• ID 16270

RABIN, M.O. [1966] see ELGOT, C.C.

RABIN, M.O. [1974] see FISCHER, MICHAEL J.

RABIN, M.O. see Vol. I, III, IV, V for further entries

RABINOWICZ, W. [1985] *Intuitionistic truth* (J 0122) J Philos Logic 14*191-228
◇ A05 F50 ◇ REV Zbl 574 # 03047 • ID 42656

RABINOWICZ, W. see Vol. II for further entries

RACKOFF, C.W. [1975] see FERRANTE, J.

RACKOFF, C.W. see Vol. III, IV for further entries

RADAELLI, G. [1979] see MELONI, G.C.

RADU, E. [1973] *On Goedel's argument for the completeness of intuitionistic logic (Romanian) (French summary)* (J 0197) Stud Cercet Mat Acad Romana 25*557-566
◇ F50 ◇ REV MR 51 # 126 Zbl 262 # 02025 • ID 17442

RADU, E. [1975] *Le "sujet creatif" dans les mathematiques intuitionnistes* (J 0060) Rev Roumaine Math Pures Appl 20*803-818
◇ A05 F50 ◇ REV MR 52 # 2831 Zbl 313 # 02015
• ID 17625

RADU, E. see Vol. I, II for further entries

RAGGIO, A.R. [1964] *Direct consistency proof of Gentzen's system of natural deduction* (J 0047) Notre Dame J Formal Log 5*27-30
◇ B10 F07 ◇ REV MR 31 # 29 Zbl 148.6 • ID 10969

RAGGIO, A.R. [1965] *Gentzen's Hauptsatz for the systems NI and NK* (J 0079) Logique & Anal, NS 8*91-100
◇ F05 F50 ◇ REV MR 33 # 31 Zbl 178.311 • ID 10970

RAGGIO, A.R. [1974] *A simple proof of Herbrand's theorem* (J 0047) Notre Dame J Formal Log 15*487-488
◇ B10 F05 F07 ◇ REV MR 51 # 5270 Zbl 272 # 02017
• ID 10972

RAGGIO, A.R. [1977] *Semi-formal Beth tableaux* (P 1076) Latin Amer Symp Math Log (3);1976 Campinas 239-241
◇ B10 F07 ◇ REV MR 57 # 15949 Zbl 385 # 03050
• ID 16607

RAGGIO, A.R. see Vol. II for further entries

RASIOWA, H. [1951] *Algebraic treatment of the functional calculi of Heyting and Lewis* (J 0027) Fund Math 38*99-126
◇ B45 F50 G10 G25 ◇ REV MR 15.385 Zbl 44.249 JSL 18.72 • ID 11001

RASIOWA, H. [1953] see MOSTOWSKI, ANDRZEJ

RASIOWA, H. & SIKORSKI, R. [1953] *On satisfiability and decidability in non-classical functional calculi* (J 0014) Bull Acad Pol Sci, Ser Math Astron Phys 1*229-231
◇ B10 B25 B45 C90 F50 ◇ REV MR 15.668 Zbl 51.245
• ID 11004

RASIOWA, H. [1954] *Constructive theories* (J 0014) Bull Acad Pol Sci, Ser Math Astron Phys 2*121-124
◇ F50 G05 G10 G25 ◇ REV MR 16.556 Zbl 59.16 JSL 33.285 • ID 11009

RASIOWA, H. [1955] *Algebraic models of axiomatic theories* (J 0027) Fund Math 41*291-310
◇ B30 C90 F50 G05 G10 G25 ◇ REV MR 19.111 Zbl 65.4 JSL 33.285 • ID 11010

RASIOWA, H. & SIKORSKI, R. [1955] *On existential theorems in non-classical functional calculi* (J 0027) Fund Math 41*21-28
◇ B45 C90 F50 G05 ◇ REV MR 16.987 Zbl 56.11 JSL 20.80 • ID 11011

RASIOWA, H. [1956] see LOS, J.

RASIOWA, H. [1958] $\mathcal{N}$-lattices and constructive logic with strong negation (J 0027) Fund Math 46*61-80
◇ F50 G10 ◇ REV MR 20#05137 Zbl 87.9 JSL 34.118
• ID 11022

RASIOWA, H. [1958] see BIALYNICKI-BIRULA, A.

RASIOWA, H. [1959] *Algebraische Charakterisierung der intuitionistischen Logik mit starker Negation* (P 0634) Constructivity in Math;1957 Amsterdam 234-240
◇ F50 G10 ◇ REV MR 21#5569 Zbl 88.11 JSL 34.118
• ID 11023

RASIOWA, H. & SIKORSKI, R. [1959] *Formalisierte intuitionistische elementare Theorien* (P 0634) Constructivity in Math;1957 Amsterdam 241-249
◇ F50 ◇ REV MR 21#4105 Zbl 86.7 JSL 40.501
• ID 11018

RASIOWA, H. & SIKORSKI, R. [1960] *On the Gentzen theorem* (J 0027) Fund Math 48*57-69
◇ B10 C07 F05 ◇ REV MR 22#1510 Zbl 99.6 • ID 11024

RASIOWA, H. & SIKORSKI, R. [1963] *The mathematics of metamathematics* (S 0257) Monograf Mat 41*522pp
• TRANSL [1972] (X 2027) Nauka: Moskva 591pp (Russian)
◇ B98 C98 F50 F98 G05 G10 G98 ◇ REV MR 29#1149 MR 50#4232 Zbl 122.243 JSL 32.274
• REM 3rd ed 1979 • ID 11028

RASIOWA, H. [1974] *An algebraic approach to non-classical logics* (X 0809) North Holland: Amsterdam xv+403pp
◇ B98 F50 G05 G10 G20 G25 G98 ◇ REV MR 56#5285 Zbl 299#02069 JSL 42.432 • ID 31451

RASIOWA, H. see Vol. I, II, III, IV, V for further entries

RATAJCZYK, Z. [1982] *Satisfaction classes and combinatorial sentences independent from PA* (J 0068) Z Math Logik Grundlagen Math 28*149-165
◇ C62 E05 F30 ◇ REV MR 83j:03097 Zbl 522#03043
• ID 35382

RATAJCZYK, Z. see Vol. III, V for further entries

RATSA, M.F. [1971] *A criterion for functional completeness in the intuitionistic propositional logic (Russian)* (J 0023) Dokl Akad Nauk SSSR 201*794-797 • ERR/ADD ibid 206(4,5,6)*7)
• TRANSL [1971] (J 0062) Sov Math, Dokl 12*1732-1737
◇ F50 ◇ REV MR 54#68 Zbl 244#02008 • ID 77625

RATSA, M.F. [1982] *Functional completeness in intuitionistic propositional logic (Russian)* (J 0052) Probl Kibern 39*107-150
◇ F50 ◇ REV MR 84h:03028 Zbl 543#03014 • ID 34241

RATSA, M.F. see Vol. I, II for further entries

RAULEFS, P. [1975] *The D-calculus: a system to describe the semantics of programs involving complex data types* (P 3527) GI Jahrestag (4);1974 Berlin 140-152
◇ B40 B75 ◇ REV MR 57#2898 Zbl 327#68027
• ID 64766

RAULEFS, P. see Vol. I for further entries

RAUSZER, C. [1974] *A formalization of the propositional calculus of H-B logic* (J 0063) Studia Logica 33*23-34
◇ B55 F50 G10 ◇ REV MR 51#69 Zbl 289#02015
• ID 15188

RAUSZER, C. [1974] *Semi-boolean algebras and their applications to intuitionistic logic with dual operations* (J 0027) Fund Math 83*219-249
◇ F50 G10 ◇ REV MR 49#4763 Zbl 298#02064
• ID 64771

RAUSZER, C. [1976] *On the strong semantical completeness of any extension of the intuitionistic predicate calculus* (J 0014) Bull Acad Pol Sci, Ser Math Astron Phys 24*81-87
◇ B55 F50 ◇ REV MR 53#2640 Zbl 343#02015
• ID 18356

RAUSZER, C. [1977] *An algebraic approach to the Heyting-Brouwer predicate calculus* (J 0027) Fund Math 96*127-135
◇ F50 G10 ◇ REV MR 58#21596 Zbl 365#02047
• ID 26533

RAUSZER, C. [1977] *Applications of Kripke models to Heyting-Brouwer logic* (J 0063) Studia Logica 36*61-71
◇ B55 C90 F50 ◇ REV MR 57#15977 Zbl 361#02033
• ID 50668

RAUSZER, C. [1977] *Model theory for an extension of intuitionistic logic* (J 0063) Studia Logica 36*73-87
◇ B55 C90 F50 ◇ REV MR 57#15978 Zbl 361#02034
• ID 50669

RAUSZER, C. [1977] *The Craig interpolation theorem for an extension of intuitionistic logic* (J 0014) Bull Acad Pol Sci, Ser Math Astron Phys 25*337-341
◇ F50 ◇ REV MR 58#21487 Zbl 361#02032 • ID 26559

RAUSZER, C. [1980] *An algebraic and Kripke-style approach to a certain extension of intuitionistic logic* (J 0202) Diss Math (Warsaw) 167*62pp
◇ B55 C20 C90 F50 G25 ◇ REV MR 82k:03100 Zbl 442#03024 • ID 56381

RAUSZER, C. see Vol. II, III for further entries

RAUTENBERG, W. [1970] see HAUSCHILD, K.

RAUTENBERG, W. [1971] see HAUSCHILD, K.

RAUTENBERG, W. [1972] see HAUSCHILD, K.

RAUTENBERG, W. [1974] see KOREC, I.

RAUTENBERG, W. [1976] see KOREC, I.

RAUTENBERG, W. see Vol. I, II, III, IV for further entries

RAUZY, G. [1985] *Mots infinis en arithmetique* (P 4622) Autom on Infinite Words;1984 Le Mont-Dore 165-171
◇ C05 F30 ◇ ID 49459

RECKHOW, R.A. [1974] see COOK, S.A.

RECKHOW, R.A. [1979] see COOK, S.A.

RECKHOW, R.A. see Vol. IV for further entries

REDDY, C.R. [1978] see LOVELAND, D.W.

REDDY, C.R. see Vol. I for further entries

REICHBACH, J. [1967] *On generalizations of the satisfiability definition and Gentzen-Jaskowski's sequent proof rules* (X 3333) Unknown Publisher: See Remarks 21pp
◇ B10 F07 ◇ REV Zbl 189.8 • REM Tel-Aviv • ID 19564

REICHBACH, J. [1967] *Some methods of formal proofs. II: Generalization of the satisfiability definition* (J 0537) Yokohama Math J 15*35-43
◇ F07 ◇ REV MR 37#6174 Zbl 157.14 • ID 11071

REICHBACH, J. [1973] *Generalized models for intuitionistic and classical predicate calculi with ultraproducts* (J 1550) Creation Math 6*9-44
◇ C20 C25 C80 C90 F50 ◇ REV Zbl 299 # 02068
• ID 31998

REICHBACH, J. [1975] *Generalized models for classical and intuitionistic predicate calculi* (J 0537) Yokohama Math J 23*5-30
◇ B10 B25 C90 F50 ◇ REV MR 55 # 2556
Zbl 343 # 02040 • ID 32003

REICHBACH, J. see Vol. I, II, III, IV for further entries

REINHARDT, W.N. [1985] *Absolute versions of incompleteness theorems* (J 0097) Nous, Quart J Phil 19*317-346
◇ F30 ◇ ID 49124

REINHARDT, W.N. see Vol. II, III, V for further entries

REITER, R. [1982] see PFENDER, M.

REITER, R. see Vol. I for further entries

RENARDEL DE LAVALETTE, G.R. [1981] *The interpolation theorem in fragments of logics* (J 0028) Indag Math 43*71-86
◇ B20 B55 C40 F50 ◇ REV MR 82d:03019
Zbl 471 # 03012 • ID 55207

RENARDEL DE LAVALETTE, G.R. [1984] *Descriptions in mathematical logic* (J 0063) Studia Logica 43*281-294
◇ B10 F50 ◇ ID 46669

RENARDEL DE LAVALETTE, G.R. [1985] *A type-free system for constructive metamathematics* (P 4646) Atti Incontri Log Mat (2);1983/84 Siena 373-376
◇ F50 ◇ ID 49687

RESCHER, N. [1966] *On modal renderings of intuitionistic propositional logic* (J 0047) Notre Dame J Formal Log 7*277-280
• REPR [1969] (C 0555) Topic Philos Logic 18-23
◇ B45 F50 ◇ REV MR 35 # 2729 Zbl 163.4 • ID 11118

RESCHER, N. [1971] see PARKS, R.Z.

RESCHER, N. see Vol. I, II, III, V for further entries

RESL, M. & SOCHOR, A. [1981] *Provability in the alternative set theory* (J 0140) Comm Math Univ Carolinae (Prague) 22*655-660
◇ E70 F07 ◇ REV MR 84h:03123 Zbl 493 # 03027
• ID 34310

RESL, M. see Vol. III, V for further entries

RESNIK, M.D. [1974] *On the philosophical significance of consistency proofs* (J 0122) J Philos Logic 3*133-147
◇ A05 F25 F30 ◇ REV MR 54 # 12473 Zbl 278 # 02008
• ID 29052

RESNIK, M.D. see Vol. I, II, V for further entries

RESSAYRE, J.-P. [1981] see MCALOON, K.

RESSAYRE, J.-P. [1981] see BERLINE, C.

RESSAYRE, J.-P. [1982] *Bounding generalized recursive functions of ordinals by effective functors: a complement to the Girard theorem* (P 3708) Herbrand Symp Logic Colloq;1981 Marseille 251-279
◇ D60 F15 ◇ REV MR 86f:03071 Zbl 513 # 03020
• ID 37224

RESSAYRE, J.-P. [1985] see GIRARD, J.-Y.

RESSAYRE, J.-P. see Vol. III, IV, V for further entries

REVESZ, G. [1981] *Axioms for the theory of lambda-conversion* (1111) Preprints, Manuscr., Techn. Reports etc. 83-81*18pp
◇ B40 ◇ REV Zbl 486 # 03012 • REM University of Kentucky, Department of Computer Science. Lexington
• ID 38077

REVESZ, G. [1984] *An extension of lambda-calculus for functional programming* (J 2551) J Log Progr 1/3*241-251
◇ B40 B75 ◇ REV MR 86g:68111 • ID 44698

REVESZ, G. [1985] *Axioms for the theory of lambda-conversion* (J 1428) SIAM J Comp 14*373-382
◇ B40 ◇ ID 45063

REVESZ, G. see Vol. IV for further entries

REYES, G.E. [1974] *From sheaves to logic* (C 0768) Stud Algeb Logic 143-204
◇ C90 F50 G30 ◇ REV MR 50 # 13182 Zbl 344 # 02042
JSL 43.146 • ID 11131

REYES, G.E. [1975] *Faisceaux et concepts* (J 0306) Cah Topol & Geom Differ 16*307
◇ C90 F50 G30 ◇ REV Zbl 342 # 02039 • ID 64812

REYES, G.E. [1977] see KOCK, A.

REYES, G.E. [1977] see MAKKAI, M.

REYES, G.E. [1977] *Sheaves and concepts: a model-theoretic interpretation of Grothendieck topoi* (J 0306) Cah Topol & Geom Differ 18*105-137
◇ C75 C90 F50 G30 ◇ REV MR 58 # 5184
Zbl 396 # 18002 • ID 52660

REYES, G.E. [1978] *Theorie des modeles et faisceaux* (J 0345) Adv Math 30*156-170
◇ C90 F50 G30 ◇ REV MR 81f:03050 Zbl 409 # 03040
• ID 56342

REYES, G.E. [1979] see KOCK, A.

REYES, G.E. [1979] *Cramer's rule in the Zariski topos* (P 2901) Appl Sheaves;1977 Durham 586-594
◇ F65 G30 ◇ REV MR 81e:03063 Zbl 422 # 14002
• ID 77765

REYES, G.E. [1981] *Logic and category theory* (C 2617) Modern Log Survey 235-252
◇ F35 F50 G30 ◇ REV MR 82f:03002 Zbl 464 # 03001
• ID 42770

REYES, G.E. [1984] see MOERDIJK, I.

REYES, G.E. see Vol. III, V for further entries

REYMOND, A. [1928] *L'axiomatique logique et le principe du tiers exclu* (J 0951) Bull Soc Fr Phil 27*1-18
◇ A05 F50 ◇ REV JSL 1.164 • ID 11133

REYMOND, A. [1936] *La negation et le principe du tiers exclu* (P 0632) Congr Int Phil des Sci;1935 Paris 6*62-68
◇ A05 F50 ◇ REV FdM 62.1045 • ID 40872

REYMOND, A. see Vol. V for further entries

REYNOLDS, J.C. [1974] *Towards a theory of type structure* (P 3013) Progr Symp;1974 Paris 408-425
◇ B40 ◇ REV MR 56 # 17187 Zbl 309 # 68016 • ID 82602

REYNOLDS, J.C. see Vol. I, IV for further entries

REYNVAAN, H.H. [1976] see BREMER, H.

REZNIKOFF, I. [1963] *Chaines de formules* (J 0109) C R Acad Sci, Paris 256*5021-5023
◇ B05 C40 C90 F50 ◇ REV MR 27 # 27 Zbl 143.7
• ID 11135

REZNIKOFF, I. [1966] *Sur les ensembles denombrables de formules en logique intuitionniste* (J 2313) C R Acad Sci, Paris, Ser A-B 262*A415-A418
◇ C40 C90 F50 ◇ REV MR 33 # 2544 Zbl 143.7
• ID 11137

REZNIKOFF, I. [1967] *On independent recursive axiomatisation in intuitionistic logic (Russian)* (J 0003) Algebra i Logika 6/6*21-27
◇ F50 ◇ REV MR 37 # 2581 Zbl 203.7 • ID 11138

REZNIKOFF, I. [1968] *Axiomatisation independante des ensembles denombrables de formules en logique intuitionniste* (J 0020) Compos Math 20*170-187
• REPR [1968] (C 0727) Logic Found of Math (Heyting) 170-187
◇ F50 ◇ REV MR 37 # 1231 Zbl 157.14 JSL 40.500
• ID 11139

REZNIKOFF, I. [1971] *Logique mathematique. II Theorie de la demonstration et intuitionnisme* (C 1495) Encycl Universalis 10*57-64
◇ F50 F98 ◇ REV JSL 38.341 • REM Parts I,III 1971 by Sabbagh,G. • ID 28611

REZNIKOFF, I. see Vol. I, III for further entries

REZUS, A. [1982] *A bibliography of lambda-calculi, combinatory logics and related topics* (X 1121) Math Centr: Amsterdam i+86pp
◇ A10 B40 ◇ REV MR 83c:03010 Zbl 481 # 03010
• ID 35123

REZUS, A. [1983] see BARENDREGT, H.P.

REZUS, A. see Vol. I for further entries

RICCI, G. [1979] *P-algebras and combinatory notation* (J 0549) Riv Mat Univ Parma, Ser 4 5*577-589
◇ B40 C05 G30 ◇ REV MR 83d:18004 • ID 39604

RICE, H.G. [1954] *Recursive real numbers* (J 0053) Proc Amer Math Soc 5*784-791
◇ F60 ◇ REV MR 16.104 Zbl 58.6 JSL 20.177 • ID 11151

RICE, H.G. see Vol. IV for further entries

RICHARD, D. [1981] see PABION, J.F.

RICHARD, D. [1982] *La theorie sans egalite du successeur et de la coprimarite des entiers naturels est indecidable. Le predicat de primarite est definissable dans le langage de cette theorie (English summary)* (J 3364) C R Acad Sci, Paris, Ser 1 294*143-146
◇ D35 F30 ◇ REV MR 84c:03084 Zbl 486 # 03027
• ID 34019

RICHARD, D. [1984] *Les relations arithmetiques sur les entiers primaires sont definissables au premier ordre par successeur et coprimarite (English summary)* (J 3364) C R Acad Sci, Paris, Ser 1 299*795-798
◇ B28 F30 ◇ REV MR 85m:03040 FdM 61.1028
• ID 39634

RICHARD, D. [1984] *The arithmetics as theories of two orders (English and French summaries)* (P 2167) Orders: Descr & Roles;1982 L'Arbresle 287-311
◇ B28 C62 D35 F30 ◇ REV MR 85k:06001 MR 86h:03102 Zbl 555 # 03026 • ID 41779

RICHARD, D. [1985] *All arithmetical sets of powers of primes are first-order definable in terms of the successor function and the coprimeness predicate* (J 0193) Discr Math 53*221-247
◇ B28 F30 ◇ REV MR 86h:03103 Zbl 562 # 03006
• ID 45095

RICHARD, D. [1985] *Answer to a problem raised by J.Robinson: the arithmetic of positive or negative integers is definable from successor and divisibility* (J 0036) J Symb Logic 50*927-935
◇ F30 ◇ ID 49671

RICHARD, D. [1985] *Definissabilite de l'arithmetique par successeur, coprimarite et puissance* (J 3364) C R Acad Sci, Paris, Ser 1 300*415-418
◇ B28 F30 ◇ ID 46289

RICHARD, D. see Vol. I, III, V for further entries

RICHARD, J. [1976] see COURVOISIER, M.

RICHARD, J. see Vol. V for further entries

RICHARDS, I. [1978] see POUR-EL, M.B.

RICHARDS, I. [1979] see POUR-EL, M.B.

RICHARDS, I. [1984] see POUR-EL, M.B.

RICHARDS, I. see Vol. IV for further entries

RICHARDSON, D.B. [1971] *The simple exponential constant problem* (J 0068) Z Math Logik Grundlagen Math 17*133-136
◇ D80 F30 ◇ REV MR 44 # 3872 Zbl 231 # 02053
• ID 11164

RICHARDSON, D.B. [1974] *Sets of theorems with short proofs* (J 0036) J Symb Logic 39*235-242
◇ F20 F30 ◇ REV MR 52 # 2840 Zbl 291 # 02021
• ID 11165

RICHARDSON, D.B. see Vol. III, IV for further entries

RICHMAN, F. [1973] *The constructive theory of countable abelian p-groups* (J 0048) Pac J Math 45*621-637
◇ F55 ◇ REV MR 49 # 8828 Zbl 244 # 02012 • ID 11167

RICHMAN, F. [1974] *Constructive aspects of noetherian rings* (J 0053) Proc Amer Math Soc 44*436-441
◇ F55 ◇ REV MR 54 # 4937 Zbl 265 # 13011 • ID 24110

RICHMAN, F. [1975] *The constructive theory of KT-modules* (J 0048) Pac J Math 61*263-274
◇ F55 ◇ REV MR 53 # 2920 Zbl 305 # 02046 • ID 21651

RICHMAN, F. [1975] see BERG, G.

RICHMAN, F. [1976] *A constructive modification of Vietoris homology* (J 0027) Fund Math 91*231-240
◇ F55 ◇ REV MR 54 # 13898 Zbl 357 # 55006 • ID 26495

RICHMAN, F. [1976] see BERG, G.

RICHMAN, F. [1977] *Computing heights in Tor* (J 1447) Houston J Math 3*267-270
◇ F55 ◇ REV MR 58 # 5977 Zbl 357 # 02032 • ID 50380

RICHMAN, F. [1977] see BERG, G.

RICHMAN, F. [1978] see JULIAN, W.

RICHMAN, F. [1981] see BRIDGES, D.S.

RICHMAN, F. (ED.) [1981] *Constructive mathematics. Proceedings of the New Mexico State University Conference* (S 3301) Lect Notes Math 873*vii+347pp
◇ F65 F97 ◇ REV MR 82m:03003 Zbl 453 # 00009
• ID 54129

RICHMAN, F. [1981] *Seidenberg's condition P* (**P** 3146) Constr Math;1980 Las Cruces 1-11
◇ F55 ◇ REV MR 84k:12017 Zbl 461 # 03016 • ID 54493

RICHMAN, F. [1982] *Finite dimensional algebras over discrete fields* (**P** 3638) Brouwer Centenary Symp;1981 Noordwijkerhout 397-411
◇ F55 ◇ REV MR 84k:16025 Zbl 521 # 16016 • ID 37465

RICHMAN, F. [1982] *Meaning and information in constructive mathematics* (**J** 0005) Amer Math Mon 89*385-388
◇ A05 F55 ◇ REV MR 83f:03006 Zbl 524 # 03048 • ID 35290

RICHMAN, F. [1982] see BRIDGES, D.S.

RICHMAN, F. [1982] see MINES, R.

RICHMAN, F. [1983] see JULIAN, W.

RICHMAN, F. [1983] *Church's thesis without tears* (**J** 0036) J Symb Logic 48*797-803
◇ D20 D75 F55 F60 ◇ REV MR 84j:03084 Zbl 527 # 03036 • ID 34676

RICHMAN, F. [1984] see JULIAN, W.

RICHMAN, F. [1984] see MINES, R.

RICHTER, M.M. [1978] *Logikkalkuele* (**X** 0823) Teubner: Stuttgart 232pp
◇ B35 B98 F05 G10 ◇ REV MR 80g:03002 Zbl 381 # 03002 • ID 51870

RICHTER, M.M. [1984] see BOERGER, E.

RICHTER, M.M. [1984] *Some reordering properties for inequality proof trees* (**P** 2342) Symp Rek Kombin;1983 Muenster 183-197
◇ F07 ◇ REV Zbl 574 # 68076 • ID 45374

RICHTER, M.M. see Vol. I, II, III, IV, V for further entries

RICHTER, W.H. [1965] *Extensions of the constructive ordinals* (**J** 0036) J Symb Logic 30*193-211
◇ D70 F15 ◇ REV MR 36 # 2500 Zbl 134.12 JSL 36.341 • ID 11170

RICHTER, W.H. [1967] *Constructive transfinite number classes* (**J** 0015) Bull Amer Math Soc 73*261-265
◇ D65 D70 F15 ◇ REV MR 34 # 7372 Zbl 155.341 JSL 36.341 • ID 11171

RICHTER, W.H. [1968] *Constructively accessible ordinal numbers* (**J** 0036) J Symb Logic 33*43-55
◇ D65 D70 F15 ◇ REV MR 38 # 4305 Zbl 197.5 JSL 36.341 • ID 11172

RICHTER, W.H. [1971] *Recursively Mahlo ordinals and inductive definitions* (**P** 0638) Logic Colloq;1969 Manchester 273-288
◇ D60 D65 D70 E55 F15 ◇ REV MR 43 # 7331 Zbl 252 # 02024 • ID 11173

RICHTER, W.H. see Vol. IV, V for further entries

RIDDER, J. [1950] *Formalistische Betrachtungen ueber intuitionistische und verwandte logische Systeme I, II, III, IV* (**J** 0028) Indag Math 12*75-84,98-107,231-243,445-459
◇ F50 ◇ REV MR 11.636 Zbl 40.147 JSL 24.175 • REM Parts V,VI,VII 1951 • ID 28429

RIDDER, J. [1951] *Formalistische Betrachtungen ueber intuitionistische und verwandte logische Systeme V, VI, VII* (**J** 0028) Indag Math 13*94-105,169-177,226-236
◇ F50 ◇ REV MR 12.71 MR 13.310 Zbl 40.147 Zbl 42.247 Zbl 45.151 JSL 24.175 • REM Parts I,II,III,IV 1950 • ID 28430

RIDDER, J. see Vol. I, II for further entries

RIEGER, L. [1949] *On the lattice theory of brouwerian propositional logic* (**J** 0165) Acta Univ Carolinae Math Phys (Prague) 189*40pp
◇ F50 G10 ◇ REV MR 12.663 JSL 17.146 • ID 11184

RIEGER, L. [1961] *On a critique of Church's thesis concerning general recursive functions in arithmetic (Czech)* (**J** 0086) Cas Pestovani Mat, Ceskoslov Akad Ved 86*480-481
◇ A05 D20 F99 ◇ REV JSL 35.489 • ID 22008

RIEGER, L. [1961] *Sur le probleme des nombres naturels* (**P** 0633) Infinitist Meth;1959 Warsaw 225-233
◇ A05 A10 F30 ◇ REV MR 26 # 6052 Zbl 122.11 • ID 11190

RIEGER, L. see Vol. I, II, III, IV, V for further entries

RIEMSDIJK VAN, H. [1981] see NIEKUS, N.H.

RIMSCHA VON, M. [1980] *Mengentheoretische Modelle des λK-Kalkuels* (**J** 0009) Arch Math Logik Grundlagenforsch 20*65-73
◇ B40 C65 E65 E70 ◇ REV MR 81g:03012 Zbl 428 # 03045 • ID 53804

RIMSCHA VON, M. see Vol. III, V for further entries

RITCHIE, R.W. & YOUNG, P. [1968] *Strong representability of partial functions in arithmetic theories* (**J** 0191) Inform Sci 1*189-204
◇ D20 F30 ◇ REV MR 41 # 3276 • ID 11208

RITCHIE, R.W. see Vol. I, IV for further entries

RITTER, W.E. [1966] *Notation systems and an effective fixed point property* (**J** 0053) Proc Amer Math Soc 17*390-395
◇ D20 D30 F15 ◇ REV MR 34 # 1181 Zbl 216.290 JSL 40.626 • ID 11210

RITTER, W.E. [1967] *Representability of partial recursive functions in formal theories* (**J** 0053) Proc Amer Math Soc 18*647-651
◇ D20 F30 ◇ REV MR 37 # 6177 Zbl 207.305 • ID 11211

RIVETTI BARBO, F. [1964] *Il teorema e il corollario di Goedel* (**X** 1364) Vita e Pensiero: Milano 66pp
◇ A05 F30 F98 ◇ ID 23481

ROBERTSON, E.L. [1974] *Structure of complexity in the weak monadic second-order theories of the natural numbers* (**P** 1464) ACM Symp Th of Comput (6);1974 Seattle 161-171
◇ D15 D55 F35 ◇ REV MR 54 # 11850 Zbl 361 # 68071 • ID 25884

ROBERTSON, E.L. see Vol. IV for further entries

ROBINET, B. [1973] *Un modele semantique pour un langage simple de programmation* (**P** 1630) GI Fachtag Automatenth & Form Sprach (1);1973 Bonn 301-310
◇ B40 B75 ◇ REV MR 56 # 1780 Zbl 277 # 68012 • ID 82621

ROBINET, B. [1974] *Permutations et logique combinatoire* (**P** 0776) Permutations;1972 Paris 257-267
◇ B40 B70 ◇ REV MR 51 # 105 Zbl 291 # 02016
• ID 24792

ROBINET, B. [1976] *About the logical foundations of data types* (**P** 3225) New Direct in Algor Lang;1976 St Pierre 63-79
◇ B40 B75 ◇ REV Zbl 383 # 68024 • ID 52032

ROBINET, B. [1977] *Un modele fonctionnel des structures de controle* (**J** 3441) RAIRO Inform Theor 11*213-236
◇ B40 B75 ◇ REV MR 58 # 19315 Zbl 389 # 68015
• ID 52323

ROBINET, B. [1978] *Logique combinatoire et programmation: Une autre approche de la theorie de la programmation* (**J** 2038) Rend Sem Mat, Torino 36*35-37
◇ B40 B75 ◇ REV MR 80b:03017 Zbl 394 # 03020
• ID 52497

ROBINET, B. [1980] *Les langages de Backus sont des systemes de manipulation d'arbres* (**P** 3057) CAAP'80 Arbres en Algeb & Progr (5);1980 Lille 83-94
◇ B40 ◇ REV MR 82g:68018 Zbl 444 # 68025 • ID 82619

ROBINET, B. [1982] *Sur des sequences iteratives de combinateurs (English summary)* (**J** 3364) C R Acad Sci, Paris, Ser 1 295*29-30
◇ B40 ◇ REV MR 83h:03020 Zbl 496 # 03005 • ID 36040

ROBINSON, JOHN ALAN [1969] *Mechanizing higher-order logic* (**J** 0508) Machine Intelligence 4*151-170
◇ B35 B40 F35 ◇ REV MR 46 # 6992 Zbl 228 # 68025
• ID 11291

ROBINSON, JOHN ALAN see Vol. I, II for further entries

ROBINSON, JULIA [1949] *Definability and decision problems in arithmetic* (**J** 0036) J Symb Logic 14*98-114
◇ D35 F30 ◇ REV MR 11.151 Zbl 34.8 JSL 15.68
• ID 11292

ROBINSON, JULIA [1973] *Axioms for number theoretic functions (Russian)* (**C** 0733) Izbr Vopr Algeb & Log (Mal'tsev) 253-263
◇ C62 D20 D75 F30 ◇ REV MR 48 # 8224
Zbl 279 # 02035 • ID 29547

ROBINSON, JULIA see Vol. III, IV, V for further entries

ROBINSON, R.M. [1951] *Arithmetical definitions in the ring of integers* (**J** 0053) Proc Amer Math Soc 2*279-284
◇ C65 F30 ◇ REV MR 12.791 Zbl 54.7 JSL 17.269
• ID 11312

ROBINSON, R.M. [1952] *An essentially undecidable axiom system* (**P** 0593) Int Congr Math (II, 6);1950 Cambridge MA 1*729-730
◇ D35 F30 ◇ ID 11309

ROBINSON, R.M. [1953] see MOSTOWSKI, ANDRZEJ

ROBINSON, R.M. [1956] *Arithmetical representation of recursively enumerable sets* (**J** 0036) J Symb Logic 21*162-187
◇ D25 F30 ◇ REV MR 18.272 Zbl 73.252 JSL 24.170
• ID 11314

ROBINSON, R.M. [1957] *Restricted set-theoretical definitions in arithmetic* (**P** 1675) Summer Inst Symb Log;1957 Ithaca 139-140
◇ B28 F30 F35 ◇ REV MR 20 # 3 Zbl 112.7 JSL 31.659
• REM Summary. See 1958 • ID 11315

ROBINSON, R.M. [1958] *Restricted set-theoretical definitions in arithmetic* (**J** 0053) Proc Amer Math Soc 9*238-242
◇ B28 F30 F35 ◇ REV MR 20 # 3 Zbl 112.7 JSL 31.659
• REM For a summary see 1957 • ID 48434

ROBINSON, R.M. see Vol. III, IV, V for further entries

ROBINSON, T.T. [1965] *Interpretations of Kleene's metamathematical predicate* $\Gamma | A$ *in intuitionistic arithmetic* (**J** 0036) J Symb Logic 30*140-154
◇ F30 F50 ◇ REV MR 33 # 7254 Zbl 133.249 • ID 11327

ROBINSON, T.T. see Vol. I for further entries

ROBITASHVILI, N.G. [1971] *A combination of the inverse method and the method of resolution (Russian) (Georgian and English summaries)* (**J** 0233) Soobshch Akad Nauk Gruz SSR 64*269-272
◇ B35 F07 ◇ REV MR 45 # 4945 Zbl 251 # 68048
• ID 11331

ROBITASHVILI, N.G. see Vol. I, II, III for further entries

ROCHE LA, P. [1981] *Effective Galois theory* (**J** 0036) J Symb Logic 46*385-392
◇ C57 C60 D45 F60 ◇ REV MR 82j:03053
Zbl 464 # 03039 • ID 75255

ROEDDING, W. [1968] *Eine Art von Gleichgewicht zahlentheoretischer und mengentheoretischer Axiomensysteme* (**J** 0009) Arch Math Logik Grundlagenforsch 11*17-31
◇ B30 E30 F30 ◇ REV MR 37 # 2602 Zbl 169.12
• ID 11344

ROEDDING, W. see Vol. IV for further entries

ROESSLER, K. [1934] *Beweis der Widerspruchsfreiheit des Funktionenkalkuels der mathematischen Logik (French summary)* (**J** 0945) Riv Filos 18*8pp
◇ B10 F25 ◇ REV Zbl 11.3 FdM 60.850 • ID 19547

ROEVER DE, W.P. [1974] *Recursion and parameter mechanisms: An axiomatic approach* (**P** 1869) Automata, Lang & Progr (2);1974 Saarbruecken 14*34-65
◇ B40 B75 D75 ◇ REV MR 55 # 1788 Zbl 302 # 68019
• ID 64875

ROGAVA, M.G. [1972] *Sequential variants of applied predicate calculi without structural deductive rules (Russian)* (**S** 0066) Tr Mat Inst Steklov 121*136-164,167
• TRANSL [1972] (**S** 0055) Proc Steklov Inst Math 121*151-181
◇ B10 F07 ◇ REV MR 54 # 12495 Zbl 286 # 02037
• ID 64879

ROGAVA, M.G. [1975] *Cut elimination in SCI* (**J** 0387) Bull Sect Logic, Pol Acad Sci 4*119-124
◇ B60 F05 ◇ REV MR 52 # 7872 • ID 18361

ROGAVA, M.G. [1979] *A new decision procedure for SCI (Russian)* (**P** 2539) Frege Konferenz (1);1979 Jena 361-364
◇ B25 B60 F05 ◇ REV MR 82h:03012 • ID 77850

ROGAVA, M.G. see Vol. I, II for further entries

ROGERS, L. [1980] *Basic subgroups from a constructive viewpoint* (**J** 0394) Commun Algeb 8*1903-1925
◇ F55 ◇ REV MR 81j:20073 Zbl 439 # 03048 • ID 56039

ROGERS, L. see Vol. I for further entries

ROGERS JR., H. [1958] *Goedel numberings of partial recursive functions* (**J** 0036) J Symb Logic 23*331-341
◇ D20 D45 F40 ◇ REV MR 21 # 2585 Zbl 88.16
JSL 29.146 • ID 11357

ROGERS JR., H. [1961] see KREIDER, D.L.

ROGERS JR., H. [1967] *Theory of recursive functions and effective computability* (**X** 0822) McGraw-Hill: New York xx+482pp
• TRANSL [1972] (**X** 0885) Mir: Moskva 624pp
◇ D25 D30 D55 D98 F15 ◇ REV MR 37 # 61 MR 50 # 4262 Zbl 183.14 JSL 24.70 JSL 36.141 • ID 24823

ROGERS JR., H. see Vol. I, III, IV for further entries

ROMANSKI, J. [1977] *On connections between algebraic and Kripke semantics* (**J** 1008) Demonstr Math (Warsaw) 10*123-127
◇ C90 F50 G10 G25 ◇ REV MR 56 # 5231 Zbl 369 # 02032 • ID 51332

RONCHI DELLA ROCCA, S. [1978] see COPPO, M.

RONCHI DELLA ROCCA, S. [1979] see BOEHM, C.

RONCHI DELLA ROCCA, S. [1979] see DEZANI-CIANCAGLINI, M.

RONCHI DELLA ROCCA, S. [1981] *Discriminability of infinite sets of terms in the D_∞-models of the λ-calculus* (**P** 2923) CAAP'81 Arbres en Algeb & Progr (6);1981 Genova 350-364
◇ B40 ◇ REV MR 83c:03011 Zbl 469 # 03008 • ID 55136

RONCHI DELLA ROCCA, S. [1982] *Characterization theorems for a filter lambda model* (**J** 0194) Inform & Control 54*201-216
◇ B40 ◇ REV MR 85k:03009 Zbl 513 # 03008 • ID 37216

RONCHI DELLA ROCCA, S. see Vol. I for further entries

ROOTSELAAR VAN, B. [1952] *Un probleme de M. Dijkman* (**J** 0028) Indag Math 14*405-407
◇ F55 ◇ REV MR 14.441 Zbl 47.285 • ID 11390

ROOTSELAAR VAN, B. [1955] *Generating schemes for full mappings* (**J** 0028) Indag Math 17*646-649
◇ F55 ◇ REV MR 17.772 Zbl 68.368 • ID 11391

ROOTSELAAR VAN, B. [1955] *On the mapping of spreads* (**J** 0028) Indag Math 17*557-563
◇ F55 ◇ REV MR 17.772 Zbl 68.368 • ID 11392

ROOTSELAAR VAN, B. [1956] *A remark on Brouwer measurable functions* (**J** 0028) Indag Math 18*579-580
◇ F55 ◇ REV MR 18.632 Zbl 75.6 • ID 11393

ROOTSELAAR VAN, B. [1957] *Intuitionismus und Arithmetik (Dutch)* (**J** 0290) Euclides 32*193-198
◇ A05 F55 ◇ REV Zbl 113.5 • ID 48049

ROOTSELAAR VAN, B. [1960] *On intuitionistic difference relations* (**J** 0028) Indag Math 22*316-322 • ERR/ADD ibid 25*132-133
◇ F55 ◇ REV MR 23 # A1531 Zbl 95.242 JSL 34.520 • ID 19544

ROOTSELAAR VAN, B. [1964] *On intuitionistic notions of divergence and convergence* (**J** 3077) Nieuw Arch Wisk, Ser 3 12*19-24
◇ F55 ◇ REV MR 32 # 50 Zbl 156.250 • ID 11396

ROOTSELAAR VAN, B. [1965] *Intuitives ueber den Intuitionismus* (**J** 0160) Math-Phys Sem-ber, NS 12*157-163
◇ A05 F55 ◇ REV Zbl 185.7 • ID 11398

ROOTSELAAR VAN, B. [1966] *Intuition und Konstruktion* (**J** 0178) Stud Gen 19*175-181
◇ A05 F98 ◇ REV Zbl 192.20 JSL 34.656 • ID 16281

ROOTSELAAR VAN, B. [1970] *On subjective mathematical assertions* (**P** 0603) Intuitionism & Proof Th;1968 Buffalo 187-196
◇ F50 ◇ REV MR 43 # 1814 Zbl 205.4 JSL 40.500 • ID 21245

ROOTSELAAR VAN, B. see Vol. I, II, IV for further entries

ROSE, A. [1955] *A Goedel theorem for an infinite-valued erweiterter Aussagenkalkuel* (**J** 0068) Z Math Logik Grundlagen Math 1*89-90
◇ B50 F30 ◇ REV MR 18.866 Zbl 67.250 • ID 24903

ROSE, A. see Vol. I, II, IV, V for further entries

ROSE, G.F. [1953] *Propositional calculus and realizibility* (**J** 0064) Trans Amer Math Soc 75*1-19
◇ F50 ◇ REV MR 15.1 Zbl 53.199 JSL 19.126 • ID 11502

ROSE, G.F. see Vol. IV for further entries

ROSE, H.E. [1960] *Independence of induction schemas in recursive arithmetic* (**J** 0036) J Symb Logic 25*95
◇ F30 ◇ ID 11506

ROSE, H.E. [1961] *On the consistency and undecidability of recursive arithmetic* (**J** 0068) Z Math Logik Grundlagen Math 7*124-135
◇ D35 F30 ◇ REV MR 25 # 3833 Zbl 109.10 • ID 11507

ROSE, H.E. [1962] *Ternary recursive arithmetic* (**J** 0132) Math Scand 10*201-216
◇ F30 ◇ REV MR 26 # 23 Zbl 118.15 • ID 11508

ROSE, H.E. [1965] *A note on reducible induction schemata* (**J** 0068) Z Math Logik Grundlagen Math 11*121-126
◇ F30 ◇ REV MR 31 # 53 Zbl 148.246 • ID 11509

ROSE, H.E. [1967] see CLEAVE, J.P.

ROSE, H.E. [1967] *Some metamathematical results in recursive arithmetic* (**J** 0068) Z Math Logik Grundlagen Math 13*381-384
◇ F30 ◇ REV MR 36 # 47 Zbl 189.10 • ID 11510

ROSE, H.E. [1972] *$\mathscr{E}^{\alpha}$-arithmetic and transfinite induction* (**J** 0036) J Symb Logic 37*19-30
◇ D20 F30 ◇ REV MR 54 # 12500 Zbl 262 # 02043 • ID 11511

ROSE, H.E. see Vol. I, II, IV for further entries

ROSEN, B.K. [1973] *Tree-manipulating systems and Church-Rosser-theorems* (**J** 0037) ACM J 20*160-187
◇ B40 ◇ REV MR 48 # 10182 Zbl 267 # 68013 • ID 11512

ROSEN, B.K. see Vol. IV, V for further entries

ROSENBERG, R. [1982] *Recursively enumerable images of arithmetic sets* (**J** 0068) Z Math Logik Grundlagen Math 28*189-201
◇ D25 D55 F30 ◇ REV MR 84e:03074 Zbl 535 # 03019 • ID 34406

ROSENBLOOM, P.C. [1945] *An elementary constructive proof of the fundamental theorem of algebra* (**J** 0005) Amer Math Mon 52*562-570
◇ F55 ◇ REV Zbl 60.47 • ID 38715

ROSENBLOOM, P.C. [1955] *Konstruktive Aequivalente fuer Saetze aus der klassischen Analysis* (**P** 0664) Congr Int Union Phil of Sci (2);1954 Zuerich 2*135-137
◇ F55 ◇ REV MR 17.934 JSL 24.238 • ID 42190

ROSENBLOOM, P.C. see Vol. I, II, IV for further entries

ROSENTHAL, J.W. [1982] see PLOTKIN, J.M.

ROSENTHAL, J.W. see Vol. III, IV, V for further entries

ROSENZWEIG, D. [1985] *Inductive definitions in ML_0* (**P** 4661) Algeb & Log;1984 Zagreb 135-154 ◇ F35 ◇ ID 49126

ROSINGER, E.E. [1982] *The algebraic uniqueness of the addition of natural numbers* (**J** 0529) Aequationes Math 25*269-273 ◇ F30 ◇ REV MR 85i:03034 Zbl 527 # 08001 • ID 38554

ROSOLINI, G. [1983] *Un modello per la teoria intuizionista degli insiemi* (**P** 3829) Atti Incontri Log Mat (1);1982 Siena 227-230 ◇ E70 F50 G30 ◇ REV MR 84k:03006 Zbl 521 # 03038 • ID 37075

ROSOLINI, G. see Vol. V for further entries

ROSSER, J.B. [1935] *A mathematical logic without variables. I* (**J** 0120) Ann of Math, Ser 2 36*127-150 ◇ B40 ◇ REV Zbl 11.2 FdM 61.56 • REM Part II 1935 • ID 40784

ROSSER, J.B. [1935] see KLEENE, S.C.

ROSSER, J.B. [1936] *Extensions of some theorems of Goedel and Church* (**J** 0036) J Symb Logic 1*87-91 ◇ D25 D35 F30 ◇ REV JSL 2.52 FdM 62.1058 • ID 11542

ROSSER, J.B. [1936] see CHURCH, A.

ROSSER, J.B. [1937] *Goedel theorems for non-constructive logics* (**J** 0036) J Symb Logic 2*129-137 • ERR/ADD ibid 2*IV ◇ D70 F30 ◇ REV Zbl 17.242 JSL 3.50 FdM 63.824 • ID 11544

ROSSER, J.B. [1939] *An informal exposition of proofs of Goedel's theorems and Church's theorem* (**J** 0036) J Symb Logic 4*53-60 ◇ A05 B20 D35 F30 ◇ REV Zbl 22.292 JSL 4.165 FdM 65.27 • ID 11547

ROSSER, J.B. [1942] *New sets of postulates for combinatory logics* (**J** 0036) J Symb Logic 7*18-27 ◇ B40 ◇ REV MR 3.289 JSL 7.122 • ID 11550

ROSSER, J.B. & WANG, HAO [1950] *Non-standard models for formal logics* (**J** 0036) J Symb Logic 15*113-129 ◇ B15 C62 E30 E70 F25 H15 H20 ◇ REV MR 12.384 Zbl 37.295 JSL 16.145 • ID 11558

ROSSER, J.B. [1955] *Deux esquisses de logique* (**X** 0834) Gauthier-Villars: Paris 69pp ◇ B10 B40 D20 E30 ◇ REV MR 16.661 Zbl 64.7 • ID 23483

ROSSER, J.B. [1955] *Logique combinatoire et λ-conversion* (**C** 4157) Rosser:Deux Esquisses Log 3-31 ◇ B40 D20 ◇ REV JSL 22.293 • ID 42147

ROSSER, J.B. [1958] see GAL, I.L.

ROSSER, J.B. [1984] *Highlights of the history of the lambda-calculus* (**J** 3789) Ann Hist of Comp 6*337-349 ◇ A10 B40 ◇ REV MR 86f:01030 • ID 43957

ROSSER, J.B. see Vol. I, II, III, IV, V for further entries

ROTA, G.-C. [1973] see METROPOLIS, N.

ROTA, G.-C. [1974] see METROPOLIS, N.

ROTA, G.-C. see Vol. I, III for further entries

ROUSSEAU, C. [1978] *Topos theory and complex analysis* (**J** 0326) J Pure Appl Algebra 10*299-313 ◇ F50 G30 H05 ◇ REV MR 57 # 413 Zbl 378 # 02028 • ID 51810

ROUSSEAU, C. [1979] *Topos theory and complex analysis* (**P** 2901) Appl Sheaves;1977 Durham 623-659 ◇ F50 F55 G30 H05 ◇ REV MR 81d:32014 Zbl 433 # 32003 • ID 82662

ROUSSEAU, C. see Vol. V for further entries

ROUSSEAU, G. [1966] *A decidable class of number theoretic equations* (**J** 0039) J London Math Soc 41*737-741 ◇ B20 B25 F30 ◇ REV MR 34 # 1182 Zbl 166.5 • ID 11593

ROUSSEAU, G. [1968] *Note on a problem of Porte* (**J** 0171) Proc Cambridge Phil Soc Math Phys 64*1 ◇ F50 ◇ REV MR 36 # 3632 Zbl 165.12 • ID 11599

ROUSSEAU, G. [1968] *Note on the decidability of a certain class of number theoretic equations* (**J** 0039) J London Math Soc 43*385-386 ◇ B20 B25 F30 ◇ REV MR 37 # 1244 Zbl 162.317 • ID 11598

ROUSSEAU, G. [1968] *Sheffer functions in intuitionistic logic* (**J** 0068) Z Math Logik Grundlagen Math 14*279-282 ◇ F50 ◇ REV MR 37 # 6161 Zbl 175.262 • ID 11597

ROUSSEAU, G. [1970] *The separation theorem for fragments of the intuitionistic propositional calculus* (**J** 0068) Z Math Logik Grundlagen Math 16*469-474 ◇ B55 F50 ◇ REV MR 43 # 1815 Zbl 211.10 • ID 11601

ROUSSEAU, G. see Vol. II, V for further entries

ROUTLEDGE, N.A. [1953] *Ordinal recursion* (**J** 0171) Proc Cambridge Phil Soc Math Phys 49*175-182 ◇ D20 F15 ◇ REV MR 14.714 Zbl 52.13 JSL 24.69 • ID 11604

ROUTLEDGE, N.A. see Vol. I, IV for further entries

ROUTLEY, R. [1974] *Semantical analyses of propositional systems of Fitch and Nelson* (**J** 0063) Studia Logica 33*283-298 ◇ B25 B46 C90 F50 ◇ REV MR 51 # 70 Zbl 356 # 02022 • ID 15192

ROUTLEY, R. [1978] *An inadequacy in Kripke-semantics for intuitionistic quantificational logic* (**J** 0387) Bull Sect Logic, Pol Acad Sci 7*61-67 ◇ B55 C90 F50 ◇ REV MR 58 # 16172 Zbl 414 # 03037 • ID 53083

ROUTLEY, R. see Vol. I, II, III, IV, V for further entries

ROYDEN, H.L. [1985] *Aspects of constructive analysis* (**P** 4271) E.Bishop-Reflection on Him & Research;1983 San Diego 57-64 ◇ A05 F98 ◇ ID 45637

ROYSE, JAMES R. [1969] *Mathematical induction in ramified type theory* (**J** 0068) Z Math Logik Grundlagen Math 15*7-10 ◇ B15 F35 F65 ◇ REV MR 38 # 5613 Zbl 184.7 • ID 11634

RUBALD, C.M. [1971] see CAVINESS, B.F.

RUBEL, L.A. [1984] see HENSON, C.W.

RUBEL, L.A. see Vol. I, III, IV, V for further entries

RUGGIU, G. [1973] *Les types et les appels de procedures* (**P** 0763) Automata, Lang & Progr (1);1972 Rocquencourt 319-330
◇ B40 ◇ REV MR 53 # 7105 Zbl 279 # 02014 • ID 22964

RUGGIU, G. & VASSEUR, J. [1973] *Structure et proprietes d'une famille de langages de programmation, EXEL* (**J** 2313) C R Acad Sci, Paris, Ser A-B 277*A345-A347
◇ B40 ◇ REV MR 49 # 11872 Zbl 277 # 68037 • ID 65898

RUITENBURG, W. [1981] *Field extensions* (**P** 3146) Constr Math;1980 Las Cruces 12-15
◇ F55 ◇ REV MR 83e:03099 Zbl 461 # 03013 • ID 54490

RUITENBURG, W. [1982] *Primality and invertibility of polynomials* (**P** 3638) Brouwer Centenary Symp;1981 Noordwijkerhout 413-434
◇ F55 ◇ REV MR 85e:03150 Zbl 523 # 03047 • ID 37035

RUITENBURG, W. [1984] *On the period of sequences $(A^n(p))$ in intuitionistic propositional calculus* (**J** 0036) J Symb Logic 49*892-899
◇ F50 ◇ REV MR 86b:03016 • ID 42497

RUTLEDGE, J.D. [1964] see ELGOT, C.C.

RUTLEDGE, J.D. see Vol. II, IV for further entries

RYAN, W.J. [1976] *The equivalence of equational and sentential general recursive arithmetics* (**J** 3172) J London Math Soc, Ser 2 14*463-475
◇ F30 ◇ REV MR 55 # 7756 Zbl 351 # 02031 • ID 78027

RYAN, W.J. [1978] *Goedel's second incompleteness theorem for general recursive arithmetic* (**J** 0068) Z Math Logik Grundlagen Math 24*457-459
◇ D20 F30 ◇ REV MR 80b:03092 Zbl 427 # 03031 • ID 53718

RYAN, W.J. [1979] *Proof of the quadratic reciprocity law in primitive recursive arithmetic* (**J** 0132) Math Scand 45*177-197
◇ F30 ◇ REV MR 82b:03104 Zbl 433 # 03024 • ID 78026

RYAN, W.J. see Vol. I for further entries

RYLL-NARDZEWSKI, C. [1958] see GRZEGORCZYK, A.

RYLL-NARDZEWSKI, C. see Vol. I, III, IV, V for further entries

SAARNIO, U. [1962] *Von den Rechenoperationen hoeherer Ordnung bei der Darstellung der transfiniten Ordnungszahlen* (**J** 0043) Math Ann 146*217-225
◇ E10 F15 ◇ REV MR 25 # 19 Zbl 119.15 • ID 11717

SAARNIO, U. [1967] *Ueber die Positionsdarstellung der Ordnungszahlen der Cantorschen zweiten Zahlenklasse* (**J** 0044) Math Z 100*396-413
◇ E10 F15 ◇ REV MR 37 # 6187 Zbl 157.28 • ID 11718

SAARNIO, U. [1968] *Die kritischen Zahlen hoeherer Ordnung innerhalb der zweiten Cantorschen Zahlenklasse* (**J** 0043) Math Ann 178*173-183
◇ E10 F15 ◇ REV MR 38 # 1004 Zbl 165.314 • ID 11720

SAARNIO, U. [1968] *Ein Zeichensystem fuer die Darstellung der transfiniten Ordnungszahlen der zweiten Zahlenklasse* (**J** 0068) Z Math Logik Grundlagen Math 14*321-328
◇ E10 F15 ◇ REV MR 39 # 72 Zbl 165.314 • ID 11721

SAARNIO, U. see Vol. V for further entries

SAHYOUN, A.H. [1981] see GORGY, F.W.

SAHYOUN, A.H. [1983] see GORGY, F.W.

SAITTA, L. [1979] see DEZANI-CIANCAGLINI, M.

SAKAI, H. [1967] *Ueber den Relativisationssatz* (**J** 0260) Ann Jap Ass Phil Sci 3*72-76
◇ F05 ◇ REV MR 35 # 2731 Zbl 158.7 • ID 11771

SAKAI, H. [1974] *On necessary but not-sufficient conditions* (**J** 0009) Arch Math Logik Grundlagenforsch 16*143-146
◇ B10 F30 F35 G05 ◇ REV MR 51 # 2881 Zbl 291 # 02024 • ID 11772

SAKAI, H. [1974] *On numerations of a formal system* (**J** 0260) Ann Jap Ass Phil Sci 4*227-230
◇ D45 F30 ◇ REV MR 56 # 100 Zbl 291 # 02023 • ID 65025

SALES VALLES, F.A. [1981] see BATLE, N.

SALES VALLES, F.A. see Vol. V for further entries

SALLE, P. [1978] *Une extension de la theorie des types en λ-calcul* (**P** 1872) Automata, Lang & Progr (5);1978 Udine 398-410
◇ B15 B40 ◇ REV MR 80k:03019 Zbl 384 # 03008 • ID 52050

SALLE, P. [1979] see COPPO, M.

SALLE, P. [1980] *Une generalisation de la theorie des types en λ-calcul I,II* (**J** 3441) RAIRO Inform Theor 14*143-167,301-314
◇ B15 B40 ◇ REV MR 82f:03012 Zbl 461 # 03002 • ID 56539

SAMBIN, G. [1974] *Un'estensione del teorema di Loeb (English summary)* (**J** 0144) Rend Sem Mat Univ Padova 52*193-199
◇ F30 ◇ REV MR 52 # 5378 Zbl 324 # 02019 • ID 18373

SAMBIN, G. [1976] *An effective fixed-point theorem in intuitionistic diagonalizable algebras (The algebraisation of the theories which express Theor. IX)* (**J** 0063) Studia Logica 35*345-361
◇ F30 F50 G10 G25 ◇ REV MR 57 # 112 Zbl 357 # 02028 • REM Part VIII 1976 by Bernardi,C. Part X 1978 by Montagna,F. • ID 50376

SAMBIN, G. [1978] *Fixed points through the finite model property (The algebraization of the theories which express Theor. XI)* (**J** 0063) Studia Logica 37*287-289
◇ B45 F30 G25 ◇ REV MR 80b:03107 Zbl 398 # 03055 • REM Part X 1978 by Montagna,F. • ID 52783

SAMBIN, G. & VALENTINI, S. [1980] *A modal sequent calculus for a fragment of arithmetic* (**J** 0063) Studia Logica 39*245-256
◇ B45 F05 F30 ◇ REV MR 81m:03025 Zbl 457 # 03016 • ID 54341

SAMBIN, G. & VALENTINI, S. [1982] *The modal logic of provability. The sequential approach* (**J** 0122) J Philos Logic 11*311-342
◇ B45 F05 F30 ◇ REV MR 84b:03033 Zbl 523 # 03014 • ID 35627

SAMBIN, G. see Vol. II for further entries

SAMOJLENKO, L.G. [1981] see DOMRACHEV, V.N.

SAMOJLENKO, L.G. [1982] see DOMRACHEV, V.N.

SAMOJLENKO, L.G. [1983] see DOMRACHEV, V.N.

SAMOJLENKO, L.G. see Vol. IV for further entries

SAMOKHVALOV, K.F. [1979] *Hilbert's program and Goedel's theorems (Russian)* (**C** 2967) Metodol Probl Mat 65-76
◇ A05 F99 ◇ REV MR 83j:03013 • ID 35328

SAMOKHVALOV, K.F. [1982] *Refinement of the usual interpretation of Goedel's theorems on incompleteness and concepts of recursive enumerability (Russian)* (**C** 3743) Probl Log & Metodol Nauk 42-57
◇ F30 ◇ REV MR 83j:03100 Zbl 516 # 03022 • ID 34868

SAMOKHVALOV, K.F. see Vol. I, II for further entries

SANCHIS, L.E. [1964] *Types in combinatory logic* (**J** 0047) Notre Dame J Formal Log 5*161-180
◇ B40 ◇ REV MR 34 # 5675 Zbl 158.247 JSL 32.420 • ID 11823

SANCHIS, L.E. [1965] *A predicative extension of elementary logic I* (**J** 0188) Rev Union Mat Argentina 22*123-138
◇ B15 F05 F65 ◇ REV MR 32 # 7398 Zbl 242 # 02009 • ID 11824

SANCHIS, L.E. [1965] *A property of combinations in normal form* (**J** 0188) Rev Union Mat Argentina 22*168-172
◇ B40 ◇ REV MR 34 # 52 • ID 11829

SANCHIS, L.E. [1967] *Functionals defined by recursion* (**J** 0047) Notre Dame J Formal Log 8*161-174
◇ D20 F10 ◇ REV MR 39 # 1294 Zbl 183.14 • ID 11825

SANCHIS, L.E. [1971] *A generalization of the Gentzen Hauptsatz* (**J** 0047) Notre Dame J Formal Log 12*499-504
◇ B10 F05 ◇ REV MR 46 # 30 Zbl 188.11 • ID 11827

SANCHIS, L.E. [1971] *Cut elimination, consistency and completeness in classical logic* (**J** 0079) Logique & Anal, NS 14*715-723
◇ B10 F05 ◇ REV MR 47 # 4771 Zbl 242 # 02032 • ID 11826

SANCHIS, L.E. [1973] *Formally defined operations in Kripke models* (**J** 0047) Notre Dame J Formal Log 14*467-480
◇ C90 F50 ◇ REV MR 48 # 10770 Zbl 226 # 02024 • ID 11828

SANCHIS, L.E. [1977] *Data types as lattices: Retractions, closures and projections* (**J** 3441) RAIRO Inform Theor 11*329-344
◇ B40 B75 D25 G10 ◇ REV MR 58 # 13876 Zbl 394 # 03019 • ID 52496

SANCHIS, L.E. [1979] *Reducibilities in two models for combinatory logic* (**J** 0036) J Symb Logic 44*221-234
◇ B40 D30 ◇ REV MR 80k:03044 Zbl 417 # 03016 • ID 53253

SANCHIS, L.E. [1980] *Reflexive domains* (**C** 3050) Essays Combin Log, Lambda Calc & Formalism (Curry) 339-361
◇ B40 ◇ REV MR 83c:03012 Zbl 469 # 03006 • ID 35124

SANCHIS, L.E. see Vol. I, IV for further entries

SARANTOPOULOS, S. [1935] *Un theoreme se rattachant a la methode par recurrence (induction complete)* (**P** 1660) Congr Interbalkan Math;1934 Athens 261-263
◇ F99 ◇ REV Zbl 15.145 FdM 61.975 • ID 40815

SARKARIA, K.S. [1985] *Formal theories are acyclic* (**J** 0068) Z Math Logik Grundlagen Math 31*363-368
◇ F07 ◇ ID 47564

SARTORIUS, M. [1982] see PFENDER, M.

SATAEV, A.G. [1982] see KOCHUBIEVSKIJ, I.P.

SATINDER, P.K. [1971] *Completeness and cut-elimination in constructive ω-rule arithmetics* (**P** 2335) Canad Math Congr (25);1971 Thunder Bay 563
◇ F05 F30 ◇ REV Zbl 319 # 02025 • ID 29678

SATO, M. [1977] see HOSONO, C.

SATO, M. see Vol. I, II, III for further entries

SAZONOV, V.YU. [1975] *Sequentially and parallelly computable functionals* (**P** 1603) λ-Calc & Comput Sci Th;1975 Roma 312-318
◇ B40 B75 ◇ REV MR 57 # 15985 Zbl 332 # 02036 • ID 65085

SAZONOV, V.YU. [1976] *Degrees of parallelism in computations* (**P** 1401) Math Founds of Comput Sci (5);1976 Gdansk 45*517-523
◇ B40 B75 D30 D65 ◇ REV Zbl 341 # 02035 • ID 29755

SAZONOV, V.YU. [1976] *Expressibility of functions in D.Scott's LFC language (Russian)* (**J** 0003) Algebra i Logika 15*308-330,366
• TRANSL [1976] (**J** 0069) Algeb and Log 15*192-206
◇ B40 D20 ◇ REV MR 55 # 12480 Zbl 415 # 03011 • ID 26049

SAZONOV, V.YU. [1976] *Functionals computable in series and in parallel (Russian)* (**J** 0092) Sib Mat Zh 17*648-672,717
• TRANSL [1976] (**J** 0475) Sib Math J 17*498-516
◇ B40 D20 ◇ REV MR 54 # 7211 Zbl 342 # 02016 • ID 25002

SAZONOV, V.YU. [1981] *On existence of complete predicate calculus in metamathematics without exponentiation* (**P** 3429) Math Founds of Comput Sci (10);1981 Strbske Pleso 483-490
◇ F20 F30 ◇ REV MR 83j:03093 Zbl 473 # 03049 • ID 55377

SAZONOV, V.YU. see Vol. IV for further entries

SCANLON, T.M. [1973] *The consistency of number theory via Herbrand's theorem* (**J** 0036) J Symb Logic 38*29-58
◇ B10 F05 F30 ◇ REV MR 58 # 16191 Zbl 255 # 02027 • ID 11874

SCANLON, T.M. [1974] see GOLDFARB, W.D.

SCARPELLINI, B. [1969] *Some applications of Gentzen's second consistency proof* (**J** 0043) Math Ann 181*325-354
◇ F05 F30 F35 F50 ◇ REV MR 40 # 4099 Zbl 184.12 • ID 11881

SCARPELLINI, B. [1970] *A model for intuitionistic analysis* (**J** 2022) Comm Math Helvetici 45*440-471
◇ F35 F50 ◇ REV MR 43 # 7300 Zbl 213.12 • ID 11884

SCARPELLINI, B. [1970] *On cut elimination in intuitionistic systems of analysis* (**P** 0603) Intuitionism & Proof Th;1968 Buffalo 271-285
◇ F05 F35 F50 ◇ REV MR 43 # 7296 Zbl 206.274 • ID 11883

SCARPELLINI, B. [1971] *A model for barrecursion of higher types* (**J** 0020) Compos Math 23*123
◇ F35 F50 ◇ REV MR 44 # 6448 Zbl 212.19 • ID 22241

SCARPELLINI, B. [1971] *Proof theory and intuitionistic systems* (**X** 0811) Springer: Heidelberg & New York vii + 290pp
◇ F05 F15 F30 F35 F50 F98 ◇ REV MR 48 # 10771 Zbl 253 # 02026 JSL 39.607 • ID 11885

SCARPELLINI, B. [1972] *Formally constructive model for barrecursion of higher types* (**J** 0068) Z Math Logik Grundlagen Math 18*321-383
◇ F35 F50 ◇ REV MR 49 # 2280 Zbl 255 # 02025 • ID 11886

SCARPELLINI, B. [1972] *Induction and transfinite induction in intuitionistic systems* (J 0007) Ann Math Logic 4*173-227
◇ F35 F50 ◇ REV MR 45 # 8500 Zbl 243 # 02021
• ID 22256

SCARPELLINI, B. [1973] *On barinduction of higher types for decidable predicates* (J 0007) Ann Math Logic 5*77-164
• ERR/ADD ibid 6*421
◇ F35 F50 ◇ REV MR 48 # 73 MR 49 # 2284
Zbl 255 # 02026 • ID 19640

SCARPELLINI, B. [1974] *Disjunctive properties of intuitionistic systems* (J 0009) Arch Math Logik Grundlagenforsch 16*147-158
◇ F50 ◇ REV MR 53 # 2647 Zbl 299 # 02035 • ID 11887

SCARPELLINI, B. [1975] *Bemerkungen zu Regel und Schema* (P 1440) ⊧ ISILC Proof Th Symp (Schuette);1974 Kiel 320-340
◇ F07 F30 F50 ◇ REV MR 54 # 7219 Zbl 334 # 02019
• ID 25010

SCARPELLINI, B. [1977] *A new realizability notion for intuitionistic analysis* (J 0068) Z Math Logik Grundlagen Math 23*137-167
◇ F35 F50 ◇ REV MR 58 # 5099 Zbl 366 # 02020
• ID 26475

SCARPELLINI, B. [1984] *Complexity of subcases of Presburger arithmetic* (J 0064) Trans Amer Math Soc 284*203-218
◇ B25 D15 F20 F30 ◇ REV MR 86c:03010
Zbl 548 # 03018 • ID 43196

SCARPELLINI, B. [1985] *Lower bound results on lengths of second-order formulas* (J 0073) Ann Pure Appl Logic 29*29-58
◇ B15 C13 D10 F20 F35 ◇ ID 47473

SCARPELLINI, B. see Vol. I, II, III, IV, V for further entries

SCEDROV, A. [1981] *Consistency and independence results in intuitionistic set theory* (P 3146) Constr Math;1980 Las Cruces 54-86
◇ E35 E70 F50 G30 ◇ REV MR 83a:03062
Zbl 463 # 03033 • ID 54573

SCEDROV, A. & SCOTT, P.J. [1982] *A note on the Friedman slash and Freyd covers* (P 3638) Brouwer Centenary Symp;1981 Noordwijkerhout 443-452
◇ F35 F50 G30 ◇ REV MR 85d:03118 Zbl 529 # 03039
• ID 37668

SCEDROV, A. [1982] *Independence of the fan theorem in the presence of continuity principles* (P 3638) Brouwer Centenary Symp;1981 Noordwijkerhout 435-442
◇ F35 F50 G30 ◇ REV MR 85b:03106 Zbl 529 # 03038
• ID 37667

SCEDROV, A. [1982] see FOURMAN, M.P.

SCEDROV, A. & VESLEY, R.E. [1983] *On a weakening of Markov's principle* (J 0009) Arch Math Logik Grundlagenforsch 23*153-160
◇ F35 F50 ◇ REV MR 85d:03113 Zbl 533 # 03033
• ID 36551

SCEDROV, A. [1983] see FRIEDMAN, H.M.

SCEDROV, A. [1984] see BEESON, M.J.

SCEDROV, A. [1984] *Differential equations in constructive analysis and in the recursive realizability topos* (J 0326) J Pure Appl Algebra 33*69-80
◇ F60 G30 ◇ REV MR 85h:03063 Zbl 539 # 03041
• ID 41241

SCEDROV, A. [1984] *Forcing and classifying topoi* (S 0167) Mem Amer Math Soc 295*93pp
◇ C90 E35 E40 F50 G30 ◇ REV MR 86d:03057
Zbl 536 # 03048 JSL 50.852 • ID 37128

SCEDROV, A. [1984] see FRIEDMAN, H.M.

SCEDROV, A. [1984] *On some nonclassical extensions of second-order intuitionistic propositional calculus* (J 0073) Ann Pure Appl Logic 27*155-164
◇ B55 F50 ◇ REV MR 86f:03022 Zbl 569 # 03026
• ID 44085

SCEDROV, A. [1985] see FRIEDMAN, H.M.

SCEDROV, A. [1985] *Extending Goedel's modal interpretation to type theory and set theory* (C 3659) Intens Math 81-119
◇ B15 B45 E70 F35 F50 ◇ REV Zbl 569 # 03025
• ID 44771

SCEDROV, A. [1985] see HARRINGTON, L.A.

SCEDROV, A. see Vol. III, V for further entries

SCHAAP, C.E. [1980] see MEY VAN DER, G.

SCHACH, A. [1958] *Two forms of mathematical induction* (J 0497) Math Mag 32*83-85
◇ F30 ◇ REV MR 22 # 15 • ID 12856

SCHARLE, T.W. [1967] see LAMBERT, K.

SCHARLE, T.W. see Vol. I, II for further entries

SCHINZEL, B. [1984] see BOERGER, E.

SCHINZEL, B. see Vol. IV for further entries

SCHLOMIUK, D.I. [1977] *Logique des topos* (X 0893) Pr Univ Montreal: Montreal 132pp
◇ E70 F50 G30 ◇ REV MR 56 # 11790 Zbl 363 # 18001
• ID 21543

SCHLOMIUK, D.I. see Vol. V for further entries

SCHMERL, U.R. [1979] *A fine structure generated by reflection formulas over primitive recursive arithmetic* (P 2627) Logic Colloq;1978 Mons 335-350
◇ F30 ◇ REV MR 81m:03070 Zbl 429 # 03039 • ID 53870

SCHMERL, U.R. [1982] *A proof theoretical fine structure in systems of ramified analysis* (J 0009) Arch Math Logik Grundlagenforsch 22*167-186
◇ F15 F35 F65 ◇ REV MR 85g:03083 Zbl 506 # 03016
• ID 36964

SCHMERL, U.R. [1982] *Iterated reflection principles and the ω-rule* (J 0036) J Symb Logic 47*721-733
◇ F30 ◇ REV MR 85e:03143 Zbl 501 # 03039 • ID 36953

SCHMERL, U.R. [1982] *Number theory and the Bachmann/Howard ordinal* (P 3708) Herbrand Symp Logic Colloq;1981 Marseille 287-298
◇ F15 F30 ◇ REV MR 85j:03100 Zbl 496 # 03040
• ID 36867

SCHMERL, U.R. [1984] *Diophantine equations in a fragment of number theory* (P 2153) Logic Colloq;1983 Aachen 2*389-398
◇ C62 F30 ◇ REV MR 86g:03096 • ID 43011

SCHMERL, U.R. see Vol. IV for further entries

SCHMIDT, DIANA [1975] *Bounds for the closure ordinals of replete monotonic increasing functions* (**J** 0036) J Symb Logic 40*305-316
◇ E10 F15 ◇ REV MR 52 # 7865 Zbl 325 # 02019
• ID 14823

SCHMIDT, DIANA [1976] *Built-up systems of fundamental sequences and hierarchies of number theoretic functions* (**J** 0009) Arch Math Logik Grundlagenforsch 18*47-53
• ERR/ADD ibid 18*145-146
◇ D20 E10 F15 ◇ REV MR 57 # 16025 Zbl 358 # 02061
• ID 24318

SCHMIDT, DIANA [1978] *Associative ordinal functions, well partial orderings and a problem of Skolem* (**J** 0068) Z Math Logik Grundlagen Math 24*297-302
◇ E10 F15 ◇ REV MR 80b:04004 Zbl 403 # 03045
• ID 54758

SCHMIDT, DIANA [1979] *A partition theorem for ordinals* (**J** 0164) J Comb Th, Ser A 27*382-384
◇ E05 E10 F15 ◇ REV MR 81d:04004 Zbl 426 # 04003
• ID 78208

SCHMIDT, DIANA see Vol. IV, V for further entries

SCHMIDT, H.A. [1950] *Wie duerfen wir mit dem Unendlichen umgehen? (die Grundlage des mathematischen Intuitionismus)* (**J** 0160) Math-Phys Sem-ber, NS 1*200-212
◇ A05 F55 ◇ REV MR 12.3 Zbl 37.6 JSL 17.62 • ID 12885

SCHMIDT, H.A. [1958] *Un procede maniable de decision pour la logique propositionnelle intuitionniste* (**P** 0576) Raisonn en Math & Sci Exper;1955 Paris 57-65
◇ B25 F50 ◇ REV MR 21 # 2587 Zbl 88.248 JSL 25.286
• ID 12891

SCHMIDT, H.A. [1960] *Mathematische Gesetze der Logik. I. Vorlesungen ueber Aussagenlogik* (**X** 0811) Springer: Heidelberg & New York xxiv+555pp
◇ B05 B98 F50 F98 ◇ REV MR 29 # 1135
Zbl 253 # 08002 Zbl 88.246 • ID 12892

SCHMIDT, H.A. see Vol. I, II for further entries

SCHMIDT, J. [1958] *Einige Prinzipien der doppelten Induktion* (**J** 0160) Math-Phys Sem-ber, NS 6*137-147
◇ F30 ◇ REV MR 20 # 3772 Zbl 87.11 • ID 12910

SCHMIDT, J. see Vol. I, II, III, V for further entries

SCHNEIDER, B. [1984] *Klassische und intuitionistische Mathematik bei L.E.J. Brouwer- dargestellt am Cantorschen Haupttheorem* (**J** 1670) Mitt Math Ges DDR 1984/1*83-90
◇ A10 F55 ◇ REV Zbl 556 # 03044 • ID 46155

SCHNORR, C.-P. [1969] *Eine Bemerkung zum Begriff der zufaelligen Folge (English summary)* (**J** 0982) Z Wahrscheinltheor & Verw Geb 14*27-35
◇ D20 D80 F65 ◇ REV MR 41 # 9315 Zbl 188.28
• ID 12929

SCHNORR, C.-P. see Vol. I, III, IV for further entries

SCHOENFELD, W. [1982] *Upper bounds for proof-search in a sequent calculus for relational equations* (**J** 0068) Z Math Logik Grundlagen Math 28*239-246
◇ B35 F20 ◇ REV MR 84d:03072 Zbl 506 # 03001
• ID 34105

SCHOENFELD, W. [1983] *Proof search for unprovable formulas* (**P** 3087) Germ Worksh on Artif Intell (7);1983 Dassel 207-215
◇ B35 F07 F20 ◇ REV MR 85g:68007 Zbl 548 # 68088
• ID 43253

SCHOENFELD, W. see Vol. I, III, IV, V for further entries

SCHOENFINKEL, M. [1924] *Ueber die Bausteine der mathematischen Logik* (**J** 0043) Math Ann 92*305-316
• TRANSL [1967] (**C** 0675) From Frege to Goedel 355-366
◇ A05 B10 B40 ◇ REV FdM 50.23 • ID 12955

SCHOENFINKEL, M. see Vol. I, III, IV for further entries

SCHREIBER, J. [1975] see BIBEL, W.

SCHREIBER, P. [1971] *Konstruktive Teiltheorien* (**J** 0068) Z Math Logik Grundlagen Math 17*197-204
◇ F65 ◇ REV MR 51 # 10051 Zbl 223 # 02024 • ID 12982

SCHREIBER, P. [1978] *Zur optimalen Loesung von Konstruktionsaufgaben* (**P** 3060) Tagung Geom & Anw;1977 Gaussig 114-126
◇ F99 ◇ REV MR 82b:03108 • ID 78242

SCHREIBER, P. [1984] *Constructivity in arbitrary mathematical theories* (**P** 3621) Frege Konferenz (2);1984 Schwerin 94-100
◇ A05 F99 ◇ REV MR 85m:03006 Zbl 562 # 03031
• ID 42921

SCHREIBER, P. see Vol. I, II, III, IV for further entries

SCHROEDER-HEISTER, P. [1982] *Logische Konstanten und Regeln. Zur Deutung von Aussagenoperatoren* (**J** 2688) Conceptus (Wien) 16*45-59
◇ F07 F50 ◇ REV MR 84a:03071 • ID 35592

SCHROEDER-HEISTER, P. [1983] *The completeness of intuitionistic logic with respect to a validity concept based on an inversion principle* (**J** 0122) J Philos Logic 12*359-377
◇ F07 F50 ◇ REV MR 85d:03018 Zbl 523 # 03005
• ID 37018

SCHROEDER-HEISTER, P. [1984] *A natural extension of natural deduction* (**J** 0036) J Symb Logic 49*1284-1300
◇ B22 F07 F50 ◇ REV Zbl 574 # 03045 • ID 42500

SCHROEDER-HEISTER, P. [1984] *Generalized rules for quantifiers and the completeness of the intuitionistic operators* &, ∨, ⊃, ⊥, ∀, ∃ (**P** 2153) Logic Colloq;1983 Aachen 2*399-426
◇ F07 F50 ◇ REV Zbl 562 # 03028 • ID 43012

SCHROEDER-HEISTER, P. [1985] *Proof-theoretic validity and the completeness of intuitionistic logic* (**P** 4180) Int Congr Log, Meth & Phil of Sci (7,Pap);1983 Salzburg 43-87
◇ F50 ◇ ID 48113

SCHROETER, K. [1948] *Die Arithmetik der natuerlichen Zahlen im Rahmen der Theorie der Verbaende* (**J** 0043) Math Ann 120*197-201
◇ F30 G10 ◇ REV MR 9.404 Zbl 30.193 • ID 12986

SCHROETER, K. [1955] *Theorie des logischen Schliessens I* (**J** 0068) Z Math Logik Grundlagen Math 1*37-86
◇ B10 F07 F50 ◇ REV MR 17.814 Zbl 64.7 • REM Part II 1958 • ID 12989

SCHROETER, K. [1956] *Ueber den Zusammenhang der in den Implikationsaxiomen vollstaendigen Axiomensysteme des zweiwertigen mit denen des intuitionistischen Aussagenkalkuel* (**J** 0068) Z Math Logik Grundlagen Math 2*173-176
◇ B20 F50 ◇ REV MR 18.866 Zbl 75.6 JSL 35.583
• ID 12993

SCHROETER, K. [1957] *Eine Umformung des heytingschen Axiomensystems fuer den intuitionistischen Aussagenkalkuel* (**J** 0068) Z Math Logik Grundlagen Math 3*18-29
◇ F50 ◇ REV MR 19.1151 Zbl 81.10 JSL 35.583
• ID 12994

SCHROETER, K. [1958] *Theorie des logischen Schliessens II* (**J** 0068) Z Math Logik Grundlagen Math 4*10-65
◇ B10 C07 F05 F07 ◇ REV MR 20#6979 Zbl 80.8 JSL 32.418 • REM Part I 1955 • ID 24841

SCHROETER, K. [1973] *Interpretation der intuitionistischen Mathematik mit Hilfe des Beweisbarkeitsbegriffs* (**J** 0963) Ajatus (Helsinki) 35*222-245
◇ A05 F50 ◇ REV Zbl 302#02007 • ID 65133

SCHROETER, K. see Vol. I, III for further entries

SCHUETTE, K. [1950] *Beweistheoretische Erfassung der unendlichen Induktion in der Zahlentheorie* (**J** 0043) Math Ann 122*369-389
◇ F05 F30 ◇ REV MR 13.615 Zbl 42.8 JSL 16.221
• ID 13000

SCHUETTE, K. [1950] *Schlussweisen-Kalkuele der Praedikatenlogik* (**J** 0043) Math Ann 122*47-65
◇ B10 B25 F05 F07 F50 ◇ REV MR 12.233 Zbl 36.148 JSL 16.155 • ID 12999

SCHUETTE, K. [1951] *Die Eliminierbarkeit des bestimmten Artikels in Kodifikaten der Analysis* (**J** 0043) Math Ann 123*166-186
◇ F35 ◇ REV MR 13.4 Zbl 42.9 JSL 21.402 • ID 13001

SCHUETTE, K. [1952] *Beweistheoretische Untersuchungen der verzweigten Analysis* (**J** 0043) Math Ann 124*123-147
◇ F05 F35 F65 ◇ REV MR 13.615 Zbl 46.6 JSL 18.76
• ID 13002

SCHUETTE, K. [1953] *Zur Widerspruchsfreiheit einer typenfreien Logik* (**J** 0043) Math Ann 125*394-400
◇ B15 B40 F05 F65 ◇ REV MR 15.386 Zbl 50.244 JSL 20.67 • ID 13003

SCHUETTE, K. [1954] *Ein widerspruchsloses System der Analysis auf typenfreier Grundlage* (**J** 0044) Math Z 61*160-179
◇ E70 F35 F65 ◇ REV MR 16.662 Zbl 56.246 JSL 20.277
• ID 13004

SCHUETTE, K. [1954] *Kennzeichnung von Ordnungszahlen durch rekursiv erklaerte Funktionen* (**J** 0043) Math Ann 127*15-32
◇ D20 F15 ◇ REV MR 15.689 Zbl 55.49 JSL 19.217
• ID 13005

SCHUETTE, K. [1956] *Ein System des verknuepfenden Schliessens* (**J** 1114) Arch Phil 5*375-387
• REPR [1956] (**J** 0009) Arch Math Logik Grundlagenforsch 2*55-67
◇ B10 F07 F30 ◇ REV MR 19.3 Zbl 71.8 JSL 22.297
• ID 19693

SCHUETTE, K. [1958] *Aussagenlogische Grundeigenschaften formaler Systeme* (**J** 0076) Dialectica 12*422-442
◇ B05 F05 F07 ◇ REV MR 21#3324 Zbl 90.9 JSL 27.230 • ID 13008

SCHUETTE, K. [1960] *Beweistheorie* (**X** 0811) Springer: Heidelberg & New York x+355pp
• TRANSL [1977] (**X** 0811) Springer: Heidelberg & New York xii+299pp
◇ B98 F05 F15 F30 F35 F98 ◇ REV MR 22#9438 MR 58#21497 Zbl 102.247 Zbl 367#02012 JSL 25.243 JSL 47.218 • REM 2nd rev. ed. Comment 1960 by Frey,G.
• ID 13010

SCHUETTE, K. [1960] *Syntactical and semantical properties of simple type theory* (**J** 0036) J Symb Logic 25*305-326
◇ B15 F05 F35 ◇ REV MR 26#2354 Zbl 109.5 JSL 32.418 • ID 13011

SCHUETTE, K. [1962] *Der Interpolationssatz der intuitionistischen Praedikatenlogik* (**J** 0043) Math Ann 148*192-200
◇ C40 F50 ◇ REV MR 26#24 Zbl 108.3 JSL 29.145
• ID 13012

SCHUETTE, K. [1962] *Lecture notes in mathematical logic. Vol.I* (**X** 1328) Penn State Math: University Park 105pp
◇ B98 F15 ◇ REM Vol.II 1963 • ID 22382

SCHUETTE, K. [1962] *Logische Abgrenzung des Transfiniten* (**C** 0712) Logik & Logikkalkuel 105-114
◇ A05 E10 F15 ◇ REV Zbl 121.14 JSL 32.284 • ID 13013

SCHUETTE, K. [1963] *Lecture notes in mathematical logic. Vol.II* (**X** 1328) Penn State Math: University Park 109pp
◇ B98 F15 ◇ REM Vol.I 1962 • ID 22383

SCHUETTE, K. [1965] *Eine Grenze fuer die Beweisbarkeit der transfiniten Induktion in der verzweigten Typenlogik* (**J** 0009) Arch Math Logik Grundlagenforsch 7*45-60
◇ B15 F15 F35 F65 ◇ REV MR 33#7240 Zbl 158.7 JSL 32.284 • ID 13015

SCHUETTE, K. [1965] *Predicative well-orderings* (**P** 0688) Logic Colloq;1963 Oxford 280-303
◇ F15 F35 F65 ◇ REV MR 33#5466 Zbl 134.12 JSL 32.284 • ID 13014

SCHUETTE, K. [1965] *Probleme und Methoden der Beweistheorie* (**J** 0178) Stud Gen 18*562-567
◇ A05 F98 ◇ REV JSL 33.604 • ID 13016

SCHUETTE, K. [1966] see CROSSLEY, J.N.

SCHUETTE, K. [1968] *Ein konstruktives System von Ordinalzahlen I* (**J** 0009) Arch Math Logik Grundlagenforsch 11*126-137
◇ F15 ◇ REV MR 40#2534 Zbl 193.311 JSL 39.186 • REM Part II 1969 • ID 24961

SCHUETTE, K. [1968] *Neuere Ergebnisse der Beweistheorie* (**P** 0657) Int Congr Math (II,10);1966 Moskva 130-138
◇ F05 F10 F15 F30 F35 F98 ◇ REV MR 38#2002 Zbl 208.12 • ID 13017

SCHUETTE, K. [1968] *Vollstaendige Systeme modaler und intuitionistischer Logik* (**X** 0811) Springer: Heidelberg & New York vii+87pp
• TRANSL [1974] (**C** 4138) Feys: Modal Logika
◇ B45 C90 F50 F98 ◇ REV MR 37#2587 Zbl 157.16 JSL 36.522 • ID 24822

SCHUETTE, K. [1968] *Zur Semantik der intuitionistischen Aussagenlogik* (**P** 0608) Logic Colloq;1966 Hannover 231-235
◇ F50 ◇ REV MR 38#5589 Zbl 209.9 JSL 40.503
• ID 13019

SCHUETTE, K. [1969] *Ein konstruktives System von Ordinalzahlen II* (**J** 0009) Arch Math Logik Grundlagenforsch 12*3-11
◇ F15 ◇ REV MR 40 # 5418 Zbl 193.311 JSL 39.186 • REM Part I 1968 • ID 24962

SCHUETTE, K. [1971] see LEVITZ, H.

SCHUETTE, K. [1971] see DILLER, J.

SCHUETTE, K. [1976] see BUCHHOLZ, W.

SCHUETTE, K. [1976] *Einfuehrung der Normalfunktionen* Θ_α *ohne Auswahlaxiom und ohne Regularitaetsbedingung* (**J** 0009) Arch Math Logik Grundlagenforsch 17*171-178
◇ E10 E25 F15 ◇ REV MR 54 # 7213 Zbl 338 # 02013 • ID 13021

SCHUETTE, K. [1976] *Primitiv-rekursive Ordinalzahlfunktionen* (**J** 1944) Sitzb, Akad Wiss, Bayern, Math-Nat Kl 1975*143-153
◇ D60 F15 ◇ REV MR 54 # 12503 Zbl 358 # 02055 • ID 31510

SCHUETTE, K. [1978] *Die Entwicklung der Beweistheorie* (**J** 0157) Jbuchber Dtsch Math-Ver 81*3-12
◇ A10 F98 ◇ REV MR 81i:03003 Zbl 389 # 03001 • ID 31511

SCHUETTE, K. [1979] see JAEGER, G.

SCHUETTE, K. [1980] *Beweistheoretische Abgrenzung von Teilsystemen der Analysis* (**J** 0045) Monatsh Math 90*1-12
◇ F15 F35 ◇ REV MR 82g:03096 Zbl 434 # 03040 • ID 55744

SCHUETTE, K. [1981] *Proof theory* (**C** 2617) Modern Log Survey 37-43
◇ F98 ◇ REV MR 82f:03002 Zbl 464 # 03001 • ID 39602

SCHUETTE, K. [1981] see BUCHHOLZ, W.

SCHUETTE, K. [1983] see BUCHHOLZ, W.

SCHUETTE, K. & SIMPSON, S.G. [1985] *Ein in der reinen Zahlentheorie unbeweisbarer Satz ueber endliche Folgen von natuerlichen Zahlen* (**J** 0009) Arch Math Logik Grundlagenforsch 25*75-89
◇ F30 ◇ ID 49129

SCHUETTE, K. see Vol. I, III, IV, V for further entries

SCHUETZENBERGER, M.-P. [1970] see LACOMBE, D.

SCHUETZENBERGER, M.-P. see Vol. IV for further entries

SCHULTE-MOENTING, J. [1981] *Cut elimination and word problems for varieties of lattices* (**J** 0004) Algeb Universalis 12*290-321
◇ F05 G10 ◇ REV MR 82j:08005 Zbl 528.03029 • ID 82282

SCHULTE-MOENTING, J. see Vol. I, III for further entries

SCHULTZ, KONRAD [1966] *Zur Darstellung der intuitionistischen Mathematik im Rahmen eines Beweisbarkeitskalkuels* (**J** 0342) Monatsber Dt Akad Wiss 8*399-402
◇ F50 ◇ REV MR 35 # 6525 Zbl 156.9 • ID 13031

SCHULTZ, KONRAD [1967] *Speziestheorie und intuitionistische Topologie* (**J** 0068) Z Math Logik Grundlagen Math 13*377-378
◇ F55 ◇ REV MR 36 # 36 Zbl 164.308 • ID 13033

SCHULTZ, KONRAD [1967] *Spreadtheorie und intuitionistische Topologie* (**J** 0068) Z Math Logik Grundlagen Math 13*251-264
◇ F55 ◇ REV MR 35 # 6526 Zbl 164.308 • ID 13032

SCHULTZ, KONRAD [1980] *A topological model for Troelstra's system CS of intuitionistic analysis* (**J** 0068) Z Math Logik Grundlagen Math 26*349-354
◇ C90 E70 F35 F50 ◇ REV MR 83d:03066 Zbl 457 # 03055 • ID 54380

SCHULTZ, KONRAD see Vol. II, III, V for further entries

SCHWABHAEUSER, W. [1979] *Non-finitizability of a weak second-order theory* (**J** 0027) Fund Math 103*83-102
◇ B15 C62 C65 C85 F35 ◇ REV MR 81h:03079 Zbl 422 # 03010 • ID 53471

SCHWABHAEUSER, W. see Vol. I, III, IV, V for further entries

SCHWICHTENBERG, H. [1972] *Beweistheoretische Charakterisierung einer Erweiterung der Grzegorczyk-Hierarchie* (**J** 0009) Arch Math Logik Grundlagenforsch 15*129-145
◇ D20 F15 F30 ◇ REV MR 48 # 3717 Zbl 257 # 02021 • ID 13077

SCHWICHTENBERG, H. [1975] *Elimination of higher type levels in definitions of primitive recursive functionals by means of transfinite recursion* (**P** 0775) Logic Colloq;1973 Bristol 279-303
◇ C75 F10 F15 F35 F50 ◇ REV MR 52 # 7866 Zbl 318 # 02047 • ID 18407

SCHWICHTENBERG, H. & WAINER, S.S. [1975] *Infinite terms and recursion in higher types* (**P** 1440) ⊧ ISILC Proof Th Symp (Schuette);1974 Kiel 341-364
◇ C75 D65 F10 ◇ REV MR 54 # 7231 Zbl 341 # 02033 • ID 25022

SCHWICHTENBERG, H. [1976] *Definierbare Funktionen im* λ*-Kalkuel mit Typen* (**J** 0009) Arch Math Logik Grundlagenforsch 17*113-114
◇ B40 ◇ REV MR 54 # 4928 Zbl 329 # 02011 • ID 13078

SCHWICHTENBERG, H. [1977] *Proof theory: Some applications of cut-elimination* (**C** 1523) Handb of Math Logic 867-895
◇ F05 F10 F15 F30 F35 F98 ◇ REV MR 58 # 10343 JSL 49.980 • ID 27328

SCHWICHTENBERG, H. [1979] *Logic and the axiom of choice* (**P** 2627) Logic Colloq;1978 Mons 351-356
◇ E25 F05 F50 ◇ REV MR 81h:03112 Zbl 435 # 03007 • ID 55768

SCHWICHTENBERG, H. [1979] *On bar recursion of types 0 and 1* (**J** 0036) J Symb Logic 44*325-329
◇ F10 F35 F50 ◇ REV MR 80h:03081 Zbl 433 # 03037 • ID 78307

SCHWICHTENBERG, H. [1982] *Complexity of normalization in the pure typed lambda-calculus* (**P** 3638) Brouwer Centenary Symp;1981 Noordwijkerhout 453-457
◇ B40 ◇ REV MR 85f:03013 Zbl 537 # 03028 • ID 40088

SCHWICHTENBERG, H. [1983] *On Martin-Loef's theory of types* (**P** 3829) Atti Incontri Log Mat (1);1982 Siena 299-325
◇ B15 F35 F50 ◇ REV Zbl 537 # 03039 • ID 40090

SCHWICHTENBERG, H. [1985] *A normal form for natural deductions in a type theory with realizing terms* (**P** 4180) Int Congr Log, Meth & Phil of Sci (7,Pap);1983 Salzburg
◇ F05 F35 F50 ◇ ID 40093

SCHWICHTENBERG, H. see Vol. I, III, IV for further entries

SCIENZA, G. [1979] *Elementary mathematics from an information-theoretic viewpoint: the real numbers* (**J** 2825) Rend Sem Fac Sci Univ Cagliari 49*475-504
◇ B28 F60 ◇ REV MR 82f:03055 • ID 78312

SCIENZA, G. see Vol. V for further entries

SCIORE, E. & TANG, A. [1978] *Admissible coherent c.p.o.'s* (**P** 1872) Automata, Lang & Progr (5);1978 Udine 440-456
◇ B40 B75 E07 ◇ REV MR 80d:68106 Zbl 384 # 03009
• ID 52051

SCIORE, E. see Vol. IV for further entries

SCOTT, D.S. [1957] *Completeness proofs for the intuitionistic sentential calculus* (**P** 1675) Summer Inst Symb Log;1957 Ithaca 231-241
◇ F50 ◇ REV Zbl 201.324 JSL 25.351 • ID 11912

SCOTT, D.S. [1958] see GAL, I.L.

SCOTT, D.S. [1962] *Algebras of sets binumerable in complete extensions of arithmetic* (**P** 0613) Rec Fct Th;1961 New York 117-121
◇ C40 C62 F30 ◇ REV MR 25 # 4993 Zbl 199.26
• ID 11914

SCOTT, D.S. [1967] *A proof of the independence of the continuum hypothesis* (**J** 0041) Math Syst Theory 1*89-111
◇ E35 E50 F35 ◇ REV MR 36 # 1321 Zbl 149.253
JSL 33.293 • ID 11919

SCOTT, D.S. [1968] *Extending the topological interpretation to intuitionistic analysis I* (**J** 0020) Compos Math 20*194-210
• REPR [1968] (**C** 0727) Logic Found of Math (Heyting) 194-210
◇ F35 F50 ◇ REV MR 37 # 3912 Zbl 197.2 • REM Part II 1970 • ID 16831

SCOTT, D.S. [1970] *Constructive validity* (**P** 0625) Symp Autom Demonst;1968 Versailles 237-275
◇ F50 ◇ REV MR 43 # 4631 Zbl 206.284 • ID 11921

SCOTT, D.S. [1970] *Extending the topological interpretation to intuitionistic analysis, II* (**P** 0603) Intuitionism & Proof Th;1968 Buffalo 235-255
◇ F35 F50 ◇ REV MR 43 # 25 Zbl 213.12 • REM Part I 1968 • ID 14739

SCOTT, D.S. [1972] *Continuous lattices* (**P** 0771) Toposes, Algeb Geom & Log;1971 Halifax 97-136
◇ B40 G10 ◇ REV MR 53 # 7879 Zbl 239 # 54006
• ID 23027

SCOTT, D.S. [1972] *Lattice theory, data types and semantics* (**P** 3078) Form Semant Progr Lang, Courant Comput Sci (2);1970 New York 65-106
◇ B40 B75 G10 ◇ REV MR 56 # 7304 Zbl 279 # 68042
• ID 65168

SCOTT, D.S. [1973] *Models for various type-free calculi* (**P** 0793) Int Congr Log, Meth & Phil of Sci (4,Proc);1971 Bucharest 157-187
◇ B40 C90 ◇ REV MR 57 # 15987 • ID 78313

SCOTT, D.S. [1975] *Combinators and classes* (**P** 1603) λ-Calc & Comput Sci Th;1975 Roma 1-26
◇ B40 B50 E70 ◇ REV MR 58 # 21489 Zbl 342 # 02018
• ID 27594

SCOTT, D.S. [1975] *Data types as lattices* (**P** 1442) ⊧ ISILC Logic Conf;1974 Kiel 579-651
• REPR [1976] (**J** 1428) SIAM J Comp 5*522-587
◇ B40 B70 G10 ◇ REV MR 53 # 9700 MR 55 # 10262
Zbl 322 # 02024 • ID 23048

SCOTT, D.S. [1975] *Lambda calculus and recursion theory* (**P** 0757) Scand Logic Symp (3);1973 Uppsala 154-193
◇ B40 B75 D25 D75 ◇ REV MR 52 # 5372
Zbl 322 # 02023 JSL 43.373 • ID 18375

SCOTT, D.S. [1975] *Some philosophical issues concerning theories of combinators* (**P** 1603) λ-Calc & Comput Sci Th;1975 Roma 346-366
◇ A05 B40 ◇ REV MR 57 # 15986 Zbl 342 # 02019
• ID 27595

SCOTT, D.S. [1979] *Identity and existence in intuitionistic logic* (**P** 2901) Appl Sheaves;1977 Durham 660-696
◇ F35 F50 ◇ REV MR 81a:03060 Zbl 418 # 03016
JSL 50.548 • ID 53301

SCOTT, D.S. [1979] see FOURMAN, M.P.

SCOTT, D.S. [1980] *Lambda calculus: some models, some philosophy* (**P** 2058) Kleene Symp;1978 Madison 223-265
◇ A05 B40 C90 ◇ REV MR 82d:03024 Zbl 515 # 03004
• ID 78314

SCOTT, D.S. [1980] *Relating theories of the λ-calculus* (**C** 3050) Essays Combin Log, Lambda Calc & Formalism (Curry) 403-450
◇ B40 C90 F50 G30 ◇ REV MR 82d:03025
Zbl 469 # 03006 • ID 78323

SCOTT, D.S. [1982] *Domains for denotational semantics* (**P** 3836) Automata, Lang & Progr (9);1982 Aarhus 577-613
◇ B22 B40 ◇ REV MR 83m:68029 Zbl 495 # 68025
• ID 38520

SCOTT, D.S. [1982] *Lectures on a mathematical theory of computation* (**P** 3906) Th Found of Progr Methodol;1981 Marktoberdorf 145-292
◇ B40 B98 D75 ◇ REV MR 85g:68043 Zbl 516 # 68064
• ID 38593

SCOTT, D.S. see Vol. I, II, III, IV, V for further entries

SCOTT, P.J. [1978] *The "Dialectica" interpretation and categories* (**J** 0068) Z Math Logik Grundlagen Math 24*553-575
◇ F10 F35 F50 G30 ◇ REV MR 80m:03091
Zbl 409 # 03034 • ID 56336

SCOTT, P.J. [1980] see LAMBEK, J.

SCOTT, P.J. [1981] see LAMBEK, J.

SCOTT, P.J. [1982] see SCEDROV, A.

SCOTT, P.J. [1983] see LAMBEK, J.

SCOTT, P.J. [1984] see LAMBEK, J.

SCOTT, P.J. see Vol. V for further entries

SCOWCROFT, P. [1984] *The real algebraic structure of Scott's model of intuitionistic analysis* (**J** 0073) Ann Pure Appl Logic 27*275-308
◇ B25 C90 F35 F50 ◇ REV MR 86f:03101
Zbl 566 # 03034 • ID 44277

SECKENDORFF VON, V. [1937] *Beweis des Induktionsschlusses der natuerlichen Zahlen aus der Dedekindschen Definition endlicher Mengen* (**J** 0366) Sitzber Berlin Math Ges 36*16-24
◇ B28 E10 F30 ◇ REV Zbl 17.156 FdM 63.30 • ID 41030

SECKENDORFF VON, V. see Vol. I, V for further entries

SEELY, R.A.G. [1979] *Weak adjointness in proof theory* (**P** 2901) Appl Sheaves;1977 Durham 697-701
◇ F07 G30 ◇ REV MR 81b:03063 Zbl 438 # 18002 • ID 55974

SEELY, R.A.G. [1982] *Locally Cartesian closed categories and type theory 1* (**J** 2128) C R Math Acad Sci, Soc Roy Canada 4*271-275
◇ B15 F35 F50 G30 ◇ REV MR 83m:03076 Zbl 509 # 03034 • ID 34890

SEELY, R.A.G. [1983] *Hyperdoctrines, natural deduction and the Beck condition* (**J** 0068) Z Math Logik Grundlagen Math 29*505-542
◇ F07 F50 G30 ◇ REV MR 86c:03054 Zbl 565 # 03032 • ID 42213

SEELY, R.A.G. [1984] *Locally Cartesian closed categories and type theory* (**J** 0332) Math Proc Cambridge Phil Soc 95*33-48
◇ F35 F50 ◇ REV MR 86b:18008 Zbl 539 # 03048 • ID 48798

SEGERBERG, K. [1968] *Propositional logics related to Heyting and Johansson* (**J** 0105) Theoria (Lund) 34*26-61
◇ B55 F50 ◇ REV MR 39 # 2617 • ID 11934

SEGERBERG, K. [1974] *Proof of a conjecture of McKay* (**J** 0027) Fund Math 81*267-270
◇ B55 F50 ◇ REV MR 50 # 1849 Zbl 287 # 02013 • ID 11941

SEGERBERG, K. see Vol. I, II, III for further entries

SEIDENBERG, A. [1970] *Construction of the integral closure of a finite integral domain* (**J** 0059) Rend Sem Mat Fis Milano 40*100-120
◇ C57 C60 F55 ◇ REV MR 45 # 3396 Zbl 218 # 14023 • REM Part II 1975 • ID 41695

SEIDENBERG, A. [1971] *On the length of a Hilbert ascending chain* (**J** 0053) Proc Amer Math Soc 29*443-450
◇ C60 F55 ◇ REV MR 43 # 6193 Zbl 216.328 • ID 41694

SEIDENBERG, A. [1972] *Constructive proof of Hilbert's theorem on ascending chains* (**J** 0064) Trans Amer Math Soc 174*305-312
◇ F55 ◇ REV MR 47 # 3379 Zbl 256 # 13012 • ID 11946

SEIDENBERG, A. [1973] *On the impossibility of some constructions in polynomial rings* (**X** 2121) Accad Naz Linc: Roma
◇ C60 F55 ◇ ID 41700

SEIDENBERG, A. [1974] *Constructions in algebra* (**J** 0064) Trans Amer Math Soc 197*273-313
◇ F55 ◇ REV MR 50 # 2141 Zbl 356 # 13007 • ID 11947

SEIDENBERG, A. [1974] *What is noetherian ? (Italian summary)* (**J** 0059) Rend Sem Mat Fis Milano 44*55-61
◇ F55 ◇ REV MR 54 # 4938 Zbl 345 # 13010 • ID 24717

SEIDENBERG, A. [1975] *Construction of the integral closure of a finite integral domain. II* (**J** 0053) Proc Amer Math Soc 52*368-372
◇ C57 C60 F55 ◇ REV MR 54 # 12741 Zbl 333 # 13004 • REM Part I 1970 • ID 41698

SEIDENBERG, A. [1978] *Constructions in a polynomial ring over the ring of integers* (**J** 0100) Amer J Math 100*685-706
◇ D45 F55 ◇ REV MR 81g:13016 Zbl 416 # 13013 • ID 53225

SEIDENBERG, A. [1984] *On the Lasker-Noether decomposition theorem* (**J** 0100) Amer J Math 106*611-638
◇ F55 ◇ REV MR 86b:03078 Zbl 567 # 13006 • ID 45181

SEIDENBERG, A. see Vol. III for further entries

SEKI, S. [1954] *A metatheorem on SLK* (**J** 0407) Comm Math Univ St Pauli (Tokyo) 3*31-36
◇ F35 ◇ REV MR 16.436 Zbl 53.344 • ID 11955

SEKI, S. [1960] see MAEHARA, S.

SEKI, S. see Vol. I, IV, V for further entries

SELDIN, J.P. [1968] *Note on definitional reductions* (**J** 0047) Notre Dame J Formal Log 9*4-6 • ERR/ADD ibid 10*412; Corr ibid 21*728
◇ B40 ◇ REV MR 38 # 5629 MR 81k:03030 Zbl 179.316 • ID 19634

SELDIN, J.P. [1972] see CURRY, H.B.

SELDIN, J.P. [1972] see HINDLEY, J.R.

SELDIN, J.P. [1973] *Equality in* $\mathfrak{F}_{21}$ (**J** 0036) J Symb Logic 38*571-575
◇ B40 F05 ◇ REV MR 49 # 8821 Zbl 285 # 02022 • ID 11959

SELDIN, J.P. [1975] *Equality in* $\mathfrak{F}_{22}$ (**P** 0775) Logic Colloq;1973 Bristol 433-444
◇ B40 F05 ◇ REV MR 52 # 2835 Zbl 311 # 02039 • ID 17629

SELDIN, J.P. [1976] *Recent advances in Curry's program* (**J** 2038) Rend Sem Mat, Torino 35*77-88
◇ B40 F05 ◇ REV MR 80a:03002b Zbl 382 # 03014 • ID 32241

SELDIN, J.P. [1977] *A sequent calculus for type assignment* (**J** 0036) J Symb Logic 42*11-28
◇ B40 ◇ REV MR 58 # 10336 Zbl 369 # 02014 • ID 23740

SELDIN, J.P. [1977] *The Q-consistency of* F_{22} (**J** 0047) Notre Dame J Formal Log 18*117-127
◇ B40 F05 ◇ REV MR 57 # 76 Zbl 283 # 02027 • ID 21954

SELDIN, J.P. [1978] *A sequent calculus formulation of type assignment with equality rules for the* $\lambda\beta$*-calculus* (**J** 0036) J Symb Logic 43*643-649
◇ B40 F05 ◇ REV MR 80e:03012 Zbl 395 # 03014 • ID 32006

SELDIN, J.P. [1978] see BUNDER, M.W.

SELDIN, J.P. [1979] *Progress report on generalized functionality* (**J** 0007) Ann Math Logic 17*29-59
◇ B40 ◇ REV MR 82a:03017 Zbl 442 # 03014 • ID 56371

SELDIN, J.P. [1980] *Curry's program* (**C** 3050) Essays Combin Log, Lambda Calc & Formalism (Curry) 3-33
◇ A05 B40 ◇ REV MR 82j:03013 Zbl 469 # 03006 • ID 78366

SELDIN, J.P. [1980] see HINDLEY, J.R.

SELDIN, J.P. [1981] *A constructive approach to classical mathematics* (**P** 3146) Constr Math;1980 Las Cruces 105-110
◇ F99 ◇ REV MR 83e:03105 Zbl 487 # 03036 • ID 35252

SELDIN, J.P. see Vol. I for further entries

SEMENENKO, M.I. [1970] *Properties of some subsystems of classical and intuitionistic propositional calculi (Russian)* (**J** 0068) Z Math Logik Grundlagen Math 16*201-238 • TRANSL [1979] (**J** 0225) Amer Math Soc, Transl, Ser 2 113*1-35
◇ B55 F50 ◇ REV MR 44 # 29 Zbl 212.12 JSL 39.351 • ID 11972

SEMENOV, A.L. [1977] *Presburgerness of predicates regular in two number systems (Russian)* (**J** 0092) Sib Mat Zh 18*403-418,479
• TRANSL [1977] (**J** 0475) Sib Math J 18*289-300
◇ B10 D20 F30 ◇ REV MR 56 # 8349 Zbl 369 # 02023 • ID 51323

SEMENOV, A.L. [1979] *On certain extensions of the arithmetic of addition of natural numbers (Russian)* (**J** 0216) Izv Akad Nauk SSSR, Ser Mat 43*1175-1195,1199
• TRANSL [1980] (**J** 0448) Math of USSR, Izv 15*401-418
◇ B25 F30 ◇ REV MR 81e:03009 Zbl 417 # 03005 • ID 53242

SEMENOV, A.L. [1982] *On the definability of arithmetic in its fragments (Russian)* (**J** 0023) Dokl Akad Nauk SSSR 263*44-47
• TRANSL [1982] (**J** 0062) Sov Math, Dokl 25*300-303
◇ F30 ◇ REV MR 84a:03074 Zbl 515 # 03014 • ID 35595

SEMENOV, A.L. see Vol. III, IV for further entries

SENDOV, B. & VACHOV, D. (EDS.) [1980] *Constructive function theory '77. Proceedings* (**X** 2237) Publ Bulg Acad Sci: Sofia 560pp
◇ F97 ◇ REV Zbl 446 # 00007 • ID 56526

SENDOV, B. see Vol. IV for further entries

SEPER, K. [1977] *Constructive mathematics - an essay, I, II* (**J** 1942) Matematika (Belgrade) 3*69-75,4*19-24
◇ A05 F55 F65 ◇ ID 31507

SEPER, K. [1977] *Contribution to the discussion of Markov principle* (**J** 1943) Recueil Travaux Inst Math (Beograd) 2(10)*120-125
◇ A05 F65 ◇ ID 31508

SEPER, K. see Vol. II, IV for further entries

SEREBRYANNIKOV, O.F. [1979] *Normal forms of logical proofs (Russian)* (**P** 2554) All-Union Symp Th Log Infer;1974 Moskva 267-273
◇ F05 ◇ REV MR 84k:03137 • ID 35029

SEREBRYANNIKOV, O.F. see Vol. I, II for further entries

SETHI, R. [1974] *Testing for the Church-Rosser property* (**J** 0037) ACM J 21*671-679 • ERR/ADD ibid 22*424
◇ B40 E07 ◇ REV MR 51 # 10053 Zbl 308 # 68077 • ID 78402

SETHI, R. see Vol. IV for further entries

SETTE, A.-M. & SZCZERBA, L.W. [1978] *Algebraic characterization of interpretability* (**J** 1573) Notas Commun Mat (Recife) 94*10pp
◇ F25 G30 ◇ REV Zbl 419 # 03019 • ID 53362

SETTE, A.-M. [1980] *Functorial approach to interpretability* (**P** 2958) Latin Amer Symp Math Log (4);1978 Santiago 365-376
◇ C90 F25 G30 ◇ REV MR 83k:03043 Zbl 419 # 03020 Zbl 464 # 03029 • ID 53363

SETTE, A.-M. & SZCZERBA, L.W. [1985] *Characterizations of elementary interpretations in category theory* (**C** 4181) Math Log & Formal Syst (Costa da) 243-292
◇ C07 F25 G30 ◇ REV Zbl 574 # 03013 • ID 48209

SETTE, A.-M. see Vol. I, II, III for further entries

SHABUNIN, L.V. [1971] *On the consistency of certain systems of combinatory logic (Russian) (English summary)* (**J** 0288) Vest Ser Mat Mekh, Univ Moskva 26/6*71-75
• TRANSL [1971] (**J** 0510) Moscow Univ Math Bull 26/5-6*103-107
◇ B40 ◇ REV MR 46 # 3273 Zbl 252 # 02019 • ID 11731

SHABUNIN, L.V. [1972] *A variant of combinatory logic with discriminators (Russian)* (**J** 0288) Vest Ser Mat Mekh, Univ Moskva 27/4*53-58
• TRANSL [1972] (**J** 0510) Moscow Univ Math Bull 27/3-4*115-120
◇ B40 ◇ REV MR 47 # 31 Zbl 238 # 02028 Zbl 259 # 02021 • ID 11732

SHABUNIN, L.V. [1973] *Simple combinatory calculi (Russian) (English summary)* (**J** 0288) Vest Ser Mat Mekh, Univ Moskva 28/6*30-35
• TRANSL [1973] (**J** 0510) Moscow Univ Math Bull 28/5-6*78-83
◇ B40 ◇ REV MR 49 # 2277 Zbl 278 # 02020 • ID 11734

SHABUNIN, L.V. [1973] *The undecidability of certain formal systems of combinatory logic (Russian)* (**J** 0288) Vest Ser Mat Mekh, Univ Moskva 28/5*36-40
• TRANSL [1973] (**J** 0510) Moscow Univ Math Bull 28/5-6*29-33
◇ B40 D35 ◇ REV MR 49 # 8846 Zbl 281 # 02034 • ID 11733

SHABUNIN, L.V. [1974] *Some algorithmic problems of calculi of combinatory logic (Russian) (English summary)* (**J** 0288) Vest Ser Mat Mekh, Univ Moskva 29/6*36-41
• TRANSL [1974] (**J** 0510) Moscow Univ Math Bull 29/5-6*72-77
◇ B40 D35 ◇ REV MR 52 # 2834 Zbl 298 # 02018 • ID 17628

SHABUNIN, L.V. [1975] *Combinatory calculi I,II (Russian)* (**J** 0288) Vest Ser Mat Mekh, Univ Moskva 30/1*12-17,30/2*10-14
• TRANSL [1975] (**J** 0510) Moscow Univ Math Bull 30/1-2*9-13,79-83
◇ B40 D35 ◇ REV MR 52 # 7863 MR 52 # 7864 Zbl 314 # 02040 Zbl 314 # 02041 • ID 18369

SHABUNIN, L.V. [1975] *On an interpretation of α-words that is connected with simple combinatory calculi I (Russian) (English summary)* (**J** 0288) Vest Ser Mat Mekh, Univ Moskva 30/6*3-9
• TRANSL [1975] (**J** 0510) Moscow Univ Math Bull 30/6*55-60
◇ B40 ◇ REV MR 54 # 7210a Zbl 327 # 02023 • REM Part II 1976 • ID 25000

SHABUNIN, L.V. [1976] *An interpretation of α-words connected with simple combinatoral calculi. II (Russian) (English summary)* (**J** 0288) Vest Ser Mat Mekh, Univ Moskva 31/1*12-27
• TRANSL [1976] (**J** 0510) Moscow Univ Math Bull 31/1-2*9-14
◇ B40 ◇ REV MR 54 # 7210b Zbl 327 # 02024 • REM Part I 1975 • ID 25001

SHABUNIN, L.V. [1983] *On the interpretaion of combinators with weak reduction* (J 0036) J Symb Logic 48*558-563
◇ B40 ◇ REV MR 85g:03026 Zbl 536 # 03003 • ID 37091

SHANIN, N.A. [1953] *On some operations on logico-arithmetical formulae (Russian)* (J 0023) Dokl Akad Nauk SSSR 93*779-782
◇ F30 F50 ◇ REV MR 15.593 Zbl 52.248 JSL 19.297
• ID 19650

SHANIN, N.A. [1954] *On imbeddings of the classical logico-arithmetical calculus into the constructive logico-arithmetical calculus (Russian)* (J 0023) Dokl Akad Nauk SSSR 94*193-196
◇ F30 F50 ◇ REV MR 15.771 Zbl 59.16 JSL 19.297
• ID 19649

SHANIN, N.A. [1955] *On some logical problems of arithmetic (Russian)* (S 0066) Tr Mat Inst Steklov 43*112pp
◇ D20 F30 F50 ◇ REV MR 19.4 Zbl 68.12 JSL 22.79
• ID 19648

SHANIN, N.A. [1956] *Some problems of mathematical analysis in the light of constructive logic (Russian) (German summary)* (J 0068) Z Math Logik Grundlagen Math 2*27-36
◇ F60 ◇ REV MR 18.2 Zbl 71.246 JSL 24.54 • ID 21082

SHANIN, N.A. [1958] *On the constructive interpretation of mathematical judgements (Russian)* (S 0066) Tr Mat Inst Steklov 52*226-311
• TRANSL [1963] (J 0225) Amer Math Soc, Transl, Ser 2 23*109-189
◇ A05 F50 ◇ REV MR 21 # 2 Zbl 92.254 JSL 31.258
• ID 11833

SHANIN, N.A. [1958] *Ueber einen Algorithmus zur konstruktiven Dechiffrierung mathematischer Urteile (Russian) (German summary)* (J 0068) Z Math Logik Grundlagen Math 4*293-303
◇ A05 F50 F60 ◇ REV MR 21 # 5566 Zbl 92.11
• ID 11834

SHANIN, N.A. [1959] *On constructive mathematical analysis (Russian)* (P 0607) All-Union Math Conf (3);1956 Moskva 2*69-71
◇ F60 ◇ ID 28391

SHANIN, N.A. [1959] *On the constructive concept of mathematical judgements (Russian)* (P 0607) All-Union Math Conf (3);1956 Moskva 1*189-190
◇ F50 ◇ ID 28245

SHANIN, N.A. [1959] *Ueber konstruktive lineare Funktionale in einem konstruktiven Hilbertschen Raum (Russian) (German summary)* (J 0068) Z Math Logik Grundlagen Math 5*1-8
◇ F60 ◇ REV MR 21 # 5573 Zbl 92.11 JSL 29.107
• ID 24895

SHANIN, N.A. [1962] *Constructive real numbers and constructive function spaces (Russian)* (S 0066) Tr Mat Inst Steklov 67*15-294
• TRANSL [1968] (X 0803) Amer Math Soc: Providence iv + 325pp
◇ F60 ◇ REV MR 28 # 30 MR 37 # 5082 Zbl 113.8 JSL 34.290 JSL 35.487 • ID 19647

SHANIN, N.A. [1964] *Concerning the constructive interpretation of auxiliary formulas I (Russian)* (S 0066) Tr Mat Inst Steklov 72*348-379
• TRANSL [1972] (J 0225) Amer Math Soc, Transl, Ser 2 99*233-275
◇ F50 ◇ REV MR 36 # 1324 Zbl 249 # 02013 • ID 28883

SHANIN, N.A. [1967] see OREVKOV, V.P.

SHANIN, N.A. & TSEJTIN, G.S. & ZASLAVSKIJ, I.D. [1968] *Peculiarities in constructive mathematical analysis (Russian)* (P 0657) Int Congr Math (II,10);1966 Moskva 253-261
◇ F60 ◇ REV MR 38 # 1000 Zbl 195.305 • ID 01906

SHANIN, N.A. [1970] see OREVKOV, V.P.

SHANIN, N.A. [1973] *On a hierarchy of methods of interpreting propositions in constructive mathematics (Russian)* (S 0066) Tr Mat Inst Steklov 129*203-266,267
• TRANSL [1973] (S 0055) Proc Steklov Inst Math 129*209-271
◇ F50 ◇ REV MR 58 # 27384 Zbl 312 # 02028 • ID 29099

SHANIN, N.A. [1973] see OREVKOV, V.P.

SHANIN, N.A. [1974] *The hierarchy of constructive Brouwer functionals (Russian) (English summary)* (S 0228) Zap Nauch Sem Leningrad Otd Mat Inst Steklov 40*142-147,160
• TRANSL [1977] (J 1531) J Sov Math 8*345-349
◇ F50 ◇ REV MR 54 # 74 Zbl 361 # 02046 Zbl 404 # 03052
• ID 23956

SHANIN, N.A. [1976] *On the quantifier of limiting realizability (Russian)* (S 0228) Zap Nauch Sem Leningrad Otd Mat Inst Steklov 60*209-220,227
• TRANSL [1980] (J 1531) J Sov Math 14*1565-1572
◇ C57 C80 F50 ◇ REV MR 58 # 27418 Zbl 344 # 02025
• ID 65058

SHANIN, N.A. [1977] *On the quantifier of limiting realizability* (P 3269) Set Th Found Math (Kurepa);1977 Beograd 127
◇ B55 C57 C80 F50 ◇ REV MR 58 # 21516 Zbl 361 # 02046 • ID 53812

SHANIN, N.A. [1981] *Role of a notion algorithm in the arithmetic language semantics* (P 3729) Algor in Modern Math & Comput Sci;1979 Urgench 361-389
• TRANSL [1982] (P 3803) Algor Sovrem Mat & Prilozh;1979 Urgench II*147-180
◇ B75 F30 F50 ◇ REV MR 84j:03113 Zbl 477 # 68035
• ID 34703

SHANIN, N.A. see Vol. I, IV, V for further entries

SHAPIRO, N.Z. [1969] *Real numbers and functions in the Kleene hierarchy and limits of recursive, rational functions* (J 0036) J Symb Logic 34*207-214
◇ D55 F60 ◇ REV MR 45 # 67 Zbl 185.23 • ID 12020

SHAPIRO, N.Z. [1969] *Recursively countable subsets of recursive metric spaces (Russian summary)* (J 0014) Bull Acad Pol Sci, Ser Math Astron Phys 17*603-608
◇ F60 ◇ REV MR 41 # 1535 Zbl 193.314 • ID 12021

SHAPIRO, N.Z. see Vol. IV for further entries

SHAPIRO, S. [1977] *On Church's thesis* (P 1704) Int Congr Log, Meth & Phil of Sci (5);1975 London ON 4*23-24
◇ A05 D20 F99 ◇ ID 32450

SHAPIRO, S. [1981] *Understanding Church's thesis* (J 0122) J Philos Logic 10*353-365
◇ A05 D20 F99 ◇ REV MR 82j:03006 • ID 78442

SHAPIRO, S. [1983] *Remarks on the development of computability* (J 2028) Hist & Phil Log 4*203-220
◇ A10 D98 F99 ◇ REV MR 85c:01037 Zbl 529 # 03015
• ID 37650

SHAPIRO, S. [1985] *Epistemic and intuitionistic arithmetic* (**C** 3659) Intens Math 11-46
◇ B28 F30 F50 ◇ REV Zbl 559 # 03036 • ID 44774

SHAPIRO, S. [1985] *Introduction - intensional mathematics and constructive mathematics* (**C** 3659) Intens Math 1-10
◇ A05 F50 ◇ REV Zbl 559 # 03035 • ID 44898

SHAPIRO, S. see Vol. I, II, III, IV for further entries

SHELAH, S. [1976] *Interpreting set theory in the endomorphism semi-group of a free algebra or in a category* (**J** 1934) Ann Sci Univ Clermont Math 13*1-29
◇ C55 E05 F25 G30 ◇ REV MR 58 # 21622 Zbl 372 # 02026 • ID 51434

SHELAH, S. [1983] see GUREVICH, Y.

SHELAH, S. [1984] *On logical sentences in PA* (**P** 3710) Logic Colloq;1982 Firenze 145-160
◇ C62 C80 E05 F25 F30 ◇ REV MR 86g:03097 Zbl 564 # 03045 • ID 38758

SHELAH, S. [1984] see GUREVICH, Y.

SHELAH, S. see Vol. I, II, III, IV, V for further entries

SHEN', A.KH. [1984] *Algorithmic variants of the notion of entropy (Russian)* (**J** 0023) Dokl Akad Nauk SSSR 276*563-566 • TRANSL [1984] (**J** 0062) Sov Math, Dokl 29*569-573
◇ F99 ◇ REV MR 86f:94023 • ID 45481

SHEN', A.KH. see Vol. IV for further entries

SHEN, BAIYING [1965] see MO, SHAOKUI

SHEN, BAIYING [1966] see MO, SHAOKUI

SHEN, BAIYING [1982] see MO, SHAOKUI

SHEN, BAIYING [1983] *Studies on higher rules (Chinese) (English summary)* (**J** 3187) Shuxue Niankan, Xi A 4*743-754
◇ F30 F99 ◇ REV MR 85i:03174 Zbl 548 # 03020 • REM English summary in J4719 4*535 • ID 43200

SHEN, BAIYING [1984] *Primitive recursive arithmetic in the second class $A^0(D)$ (Chinese)* (**J** 0418) Shuxue Xuebao 27*345-363
◇ D20 F30 ◇ REV Zbl 564 # 03036 • ID 46417

SHEN, BAIYING [1984] *The primitive recursive arithmetic $B^1 1$ of the first class (Chinese)* (**J** 0418) Shuxue Xuebao 27*203-207
◇ F30 ◇ REV MR 85i:03175 Zbl 543 # 03033 • ID 40929

SHEN, BAIYING [1985] *Primitive recursive arithmetic in the first class A^o. I (Chinese)* (**J** 0418) Shuxue Xuebao 28*294-307
◇ D20 F30 ◇ ID 46288

SHEN, BAIYING see Vol. I, II, IV, V for further entries

SHEN, YOUDING [1955] *Two semantical paradoxes* (**J** 0036) J Symb Logic 20*119-120
◇ A05 F99 ◇ REV Zbl 66.10 • ID 18077

SHEN, YOUDING see Vol. I, II, V for further entries

SHEPARD, P.T. [1973] *A finite arithmetic* (**J** 0036) J Symb Logic 38*232-248
◇ F30 F65 ◇ REV MR 48 # 8210 Zbl 281 # 02055 • ID 12065

SHEPHERDSON, J.C. [1961] *Representability of recursively enumerable sets in formal theories* (**J** 0009) Arch Math Logik Grundlagenforsch 5*119-127
◇ D25 F30 ◇ REV MR 23 # A3674 Zbl 113.243 JSL 34.117 • ID 12071

SHEPHERDSON, J.C. [1964] *A non-standard model for a free variable fragment of number theory* (**J** 0014) Bull Acad Pol Sci, Ser Math Astron Phys 12*79-86
◇ C62 F30 ◇ REV MR 28 # 5002 Zbl 132.247 JSL 30.389 • ID 12075

SHEPHERDSON, J.C. [1965] *Non-standard models for fragments of number theory* (**P** 0614) Th Models;1963 Berkeley 342-358
◇ C62 F30 ◇ REV MR 33 # 5483 Zbl 154.262 • ID 12072

SHEPHERDSON, J.C. [1967] *The rule of induction in the free variable arithmetic based on + and ** (**J** 0179) Ann Fac Sci Clermont 35*25-31
◇ C62 F30 ◇ REV MR 43 # 1837 • ID 12077

SHEPHERDSON, J.C. [1969] *Weak and strong induction* (**J** 0005) Amer Math Mon 76*989-1004
◇ C62 F30 ◇ REV MR 40 # 2518 Zbl 188.24 • ID 12078

SHEPHERDSON, J.C. [1976] see BEZBORUAH, A.

SHEPHERDSON, J.C. [1976] *On the definition of computable function of a real variable* (**J** 0068) Z Math Logik Grundlagen Math 22*391-402
◇ D20 F60 ◇ REV MR 56 # 102 Zbl 359 # 02029 • ID 23677

SHEPHERDSON, J.C. [1982] see HUBER-DYSON, V.

SHEPHERDSON, J.C. [1983] see JONES, JAMES P.

SHEPHERDSON, J.C. see Vol. I, II, III, IV, V for further entries

SHESTOPAL, G.A. [1968] see FOL'K, N.F.

SHESTOPAL, G.A. see Vol. I for further entries

SHIMADA, K. [1974] *A theorem prover for intuitionistic propositional logic* (**J** 0381) J Tsuda College (Tokyo) 6*39-44
◇ B35 F50 ◇ REV MR 58 # 10331 • ID 78527

SHIMODA, M. [1981] *Categorical aspects of Heyting-valued models for intuitionistic set theory* (**J** 0407) Comm Math Univ St Pauli (Tokyo) 30*17-35
◇ C90 E40 E70 F50 G30 ◇ REV MR 83a:03070 Zbl 466 # 03020 • ID 54971

SHIMODA, M. see Vol. III, V for further entries

SHIRAI, K. [1971] *Intuitionistic predicate calculus with ε-symbol* (**J** 0260) Ann Jap Ass Phil Sci 4*49-67
◇ F50 ◇ REV MR 45 # 3164 Zbl 287 # 02014 • ID 12085

SHIRAI, K. [1972] *On the intuitionistic predicate calculus with the ε-symbol* (**J** 0091) Sugaku 24*269-279
◇ F50 ◇ REV MR 58 # 27353 • ID 78532

SHIRAI, K. [1975] *Intuitionistic version of the Los-Tarski-Robinson theorem* (**J** 0260) Ann Jap Ass Phil Sci 4*323-332
◇ C40 F05 F50 ◇ REV MR 58 # 10332 Zbl 317 # 02024 • ID 29640

SHIRAI, K. [1977] *A relation between transfinite induction and mathematical induction in elementary number theory* (**J** 2606) Tsukuba J Math 1*91-124
◇ F15 F30 ◇ REV MR 58 # 16188 Zbl 415 # 03048
• ID 53150

SHIRAI, K. [1980] see MATSUMOTO, K.

SHIRAI, K. see Vol. II for further entries

SHMAIN, I.KH. [1970] *Imbeddings of classical systems in intuitionistic and minimal ones (Russian)* (**C** 1530) Issl Log Sist (Yanovskaya) 51-67
◇ F30 F50 ◇ REV MR 49 # 22 Zbl 223 # 02015 • ID 12501

SHMAIN, I.KH. [1974] *Extended calculus of recursive functions I (Russian)* (**C** 2577) Issl Formaliz Yazyk & Neklass Log 50-81
◇ D20 D75 F50 ◇ REV MR 57 # 16028 • ID 78543

SHOENFIELD, J.R. [1954] *A relative consistency proof* (**J** 0036) J Symb Logic 19*21-28
◇ B30 E30 E35 E70 F25 ◇ REV MR 15.668 Zbl 55.4 JSL 22.367 • ID 12087

SHOENFIELD, J.R. [1957] see KREISEL, G.

SHOENFIELD, J.R. [1958] *Open sentences and the induction axiom* (**J** 0036) J Symb Logic 23*7-12
◇ F30 ◇ REV MR 20 # 5726 Zbl 85.245 JSL 27.90
• ID 12091

SHOENFIELD, J.R. [1959] see KREISEL, G.

SHOENFIELD, J.R. [1959] *On a restricted ω-rule (Russian summary)* (**J** 0014) Bull Acad Pol Sci, Ser Math Astron Phys 7*405-407
◇ F30 ◇ REV MR 22 # 6707 Zbl 95.8 JSL 34.130
• ID 12093

SHOENFIELD, J.R. [1967] *Mathematical logic* (**X** 0832) Addison-Wesley: Reading viii+344pp
• TRANSL [1975] (**X** 2027) Nauka: Moskva 527pp
◇ B98 C98 D98 E98 F98 ◇ REV MR 37 # 1224 MR 53 # 87 Zbl 155.11 JSL 40.234 • ID 22384

SHOENFIELD, J.R. see Vol. III, IV, V for further entries

SHOESMITH, D.J. & SMILEY, T.J. [1978] *Multiple-conclusion logic* (**X** 0805) Cambridge Univ Pr: Cambridge, GB xiii+396pp
◇ B22 F07 F98 ◇ REV MR 80k:03001 Zbl 381 # 03001 JSL 46.161 • ID 51869

SHOESMITH, D.J. see Vol. II for further entries

SHORE, R.A. [1980] see NERODE, A.

SHORE, R.A. [1981] *The degrees of unsolvability: global results* (**P** 2628) Log Year;1979/80 Storrs 283-301
◇ C40 D30 D35 F35 ◇ REV MR 82k:03065 Zbl 474 # 03022 • ID 55426

SHORE, R.A. [1981] *The theory of the degrees below O'* (**J** 3172) J London Math Soc, Ser 2 24*1-14
◇ D25 D30 D35 F30 ◇ REV MR 83m:03051 Zbl 469 # 03027 JSL 50.550 • ID 55154

SHORE, R.A. see Vol. III, IV, V for further entries

SHOSTAK, R.E. [1977] *On the SUP-INF method for proving Presburger formulas* (**J** 0037) ACM J 24*529-543
◇ B25 B35 F30 ◇ REV MR 58 # 13992 Zbl 423 # 68052
• ID 69916

SHOSTAK, R.E. [1979] *A practical decision procedure for arithmetic with function symbols* (**J** 0037) ACM J 26*351-360
◇ B25 B35 F30 ◇ REV MR 80f:68020 Zbl 496 # 03003
• ID 82799

SHOSTAK, R.E. see Vol. I, III for further entries

SHUKLA, S.L. [1971] see ASHVINIKUMAR

SHUKLA, S.L. [1972] *Intuitionist treatment of some spaces of sequences* (**J** 0028) Indag Math 34*139-151
◇ F55 ◇ REV MR 47 # 9239 Zbl 233 # 46018 • ID 12128

SHUKLA, S.L. [1972] *On intuitionist analogues of classically inseparable spaces* (**J** 0028) Indag Math 34*221-226
◇ F55 ◇ REV MR 46 # 4161 Zbl 248 # 02034 • ID 12126

SHUKLA, S.L. [1972] *On some linear spaces which coincide classically but are different intuitionistically* (**J** 0028) Indag Math 34*275-277
◇ F55 ◇ REV MR 46 # 4162 Zbl 248 # 02035 • ID 12127

SHUKLA, S.L. [1973] *On a new method in intuitionist linear analysis* (**J** 0020) Compos Math 26*181-202
◇ F55 ◇ REV MR 48 # 10778 Zbl 275 # 02037 • ID 12129

SHUMIKHIN, V.P. [1978] *About the inconsistency of the system $\mathscr{F}_{21}^{*A}$ of combinatory logic (Russian)* (**J** 0288) Vest Ser Mat Mekh, Univ Moskva 1978/3*3-8
• TRANSL [1978] (**J** 0510) Moscow Univ Math Bull 33/3*1-6
◇ B40 ◇ REV MR 80d:03013 Zbl 403 # 03010 • ID 54724

SHURYGIN, V.A. [1966] *Nontrivial constructive mappings of certain sets (Russian)* (**J** 0023) Dokl Akad Nauk SSSR 168*40-42 • ERR/ADD ibid 171*510
• TRANSL [1966] (**J** 0062) Sov Math, Dokl 7*604-607
◇ D20 D45 F60 ◇ REV MR 34 # 57 Zbl 219 # 02019
• REM Transl-err ibid 7*below table of contents • ID 12786

SHURYGIN, V.A. [1967] *Constructive sets with equality and their mappings (Russian)* (**J** 0023) Dokl Akad Nauk SSSR 173*54-57
• TRANSL [1967] (**J** 0062) Sov Math, Dokl 8*348-351
◇ D03 F60 ◇ REV MR 35 # 1478 Zbl 234 # 02020
• ID 12761

SHURYGIN, V.A. [1968] *Complete constructive sets with equality, and some of their properties (Russian)* (**S** 0228) Zap Nauch Sem Leningrad Otd Mat Inst Steklov 8*272-280
• TRANSL [1968] (**J** 0521) Semin Math, Inst Steklov 8*133-135
◇ D03 F60 ◇ REV MR 40 # 5448 Zbl 233 # 02012
• ID 12762

SHURYGIN, V.A. [1970] *Constructive sets with equality and their mappings (Russian)* (**S** 0066) Tr Mat Inst Steklov 113*173-259
• TRANSL [1970] (**S** 0055) Proc Steklov Inst Math 113*195-287
◇ D03 F60 ◇ REV MR 45 # 58 Zbl 247 # 02032 • ID 12763

SHURYGIN, V.A. [1973] *Invertibility and regularity of constructive operators (Russian)* (**S** 0554) Issl Teor Algor & Mat Logik (Moskva) 1*262-286
◇ F60 ◇ REV MR 49 # 2305 Zbl 287 # 02021 • ID 12764

SHURYGIN, V.A. [1974] *Einige Eigenschaften der Kompliziertheit konstruktiver reeller Zahlen (Russian)* (**C** 1450) Teor Algor & Mat Logika (Markov) 177-194
◇ D15 F60 ◇ REV MR 58 # 16211 Zbl 311 # 02045
• ID 29609

SHURYGIN, V.A. [1976] *Pseudoregular and quasiregular constructive operators (Russian)* (**S** 0554) Issl Teor Algor & Mat Logik (Moskva) 2*130-139,160
◇ F60 ◇ REV MR 58 # 16212 • ID 78593

SHURYGIN, V.A. [1976] *Traceable operators in constructive analysis (Russian)* (**J** 0023) Dokl Akad Nauk SSSR 230*795-798
• TRANSL [1976] (**J** 0062) Sov Math, Dokl 17*1410-1413
◇ F60 ◇ REV MR 55 # 86 Zbl 399 # 03052 • ID 52848

SHURYGIN, V.A. [1977] *On estimates of the complexity of algorithmic problems in constructive analysis (Russian)* (**J** 0023) Dokl Akad Nauk SSSR 233*1064-1067
• TRANSL [1977] (**J** 0062) Sov Math, Dokl 18*577-581
◇ D15 D20 F60 ◇ REV MR 57 # 16003 Zbl 378 # 02018
• ID 51800

SHURYGIN, V.A. [1979] *On the structure of constructive sets with equality (Russian)* (**S** 0554) Issl Teor Algor & Mat Logik (Moskva) 3*115-130
◇ F60 ◇ REV MR 81h:03116 Zbl 431 # 03039 • ID 53945

SHURYGIN, V.A. [1982] *Construction of eigenvectors of completely continuous selfadjoint linear operators in constructive analysis for an eigenvalue of given multiplicity (Russian)* (**J** 0031) Izv Vyssh Ucheb Zaved, Mat (Kazan) 1982/7*86-88
• TRANSL [1982] (**J** 0062) Sov Math, Dokl 26*116-119
◇ F60 ◇ REV MR 84j:47033 Zbl 545 # 03039 • ID 39170

SHURYGIN, V.A. [1982] *On eigenvectors of completely continuous linear operators in constructive analysis (Russian)* (**J** 0031) Izv Vyssh Ucheb Zaved, Mat (Kazan) 1982/6*84-86
• TRANSL [1982] (**J** 3449) Sov Math 26/6*112-115
◇ F60 ◇ REV MR 85k:47039 Zbl 545 # 03038 • ID 43514

SHURYGIN, V.A. [1982] *On the eigenvalue problem for linear operators in constructive analysis (Russian)* (**J** 0031) Izv Vyssh Ucheb Zaved, Mat (Kazan) 1982/5*83-85
• TRANSL [1982] (**J** 3449) Sov Math 26/5*113-116
◇ F60 ◇ REV MR 85k:47038 Zbl 514 # 03040 • ID 37416

SHURYGIN, V.A. see Vol. IV for further entries

SHVARTS, G.F. [1979] *Some extensions of intuitionistic type theory (Russian) (English summary)* (**J** 0288) Vest Ser Mat Mekh, Univ Moskva 1979/3*31-34
• TRANSL [1979] (**J** 0510) Moscow Univ Math Bull 34/3*32-36
◇ B15 F35 F50 ◇ REV MR 80i:03068 Zbl 415 # 03049
• ID 56037

SHVARTS, G.F. [1980] *The existence property with parameters for some extensions of the intuitionistic theory of types (Russian)* (**J** 0023) Dokl Akad Nauk SSSR 251*562-565
• TRANSL [1980] (**J** 0062) Sov Math, Dokl 21*493-496
◇ F35 F50 ◇ REV MR 82a:03058 Zbl 531 # 03035
• ID 78599

SHVARTS, G.F. [1983] *Properties of effectiveness of logical connectives in the intuitionistic theory of types (Russian)* (**C** 3807) Issl Neklass Log & Formal Sist 173-198
◇ F35 F50 ◇ REV MR 85i:03032 • ID 44068

SIDRISTYJ, B. [1973] see ANISHEV, P.A.

SIEFKES, D. [1970] *Decidable theories I. Buechi's monadic second order successor arithmetic* (**S** 3301) Lect Notes Math 120*xii + 130pp
◇ B15 B25 C85 D05 F35 F98 ◇ REV MR 44 # 6488
Zbl 399 # 03011 • ID 12145

SIEFKES, D. [1970] *Decidable extensions of monadic second order successor arithmetic* (**P** 0577) Automatenth & Formale Sprachen;1969 Oberwolfach 441-472
◇ B15 B25 C85 D05 F35 ◇ REV MR 56 # 8354
Zbl 213.19 • ID 27984

SIEFKES, D. [1971] *Undecidable extensions of monadic second order successor arithmetic* (**J** 0068) Z Math Logik Grundlagen Math 17*385-394
◇ B15 B25 C85 D35 F35 ◇ REV MR 45 # 1763
Zbl 193.312 • ID 12146

SIEFKES, D. [1975] *The recursive sets in certain monadic second order fragments of arithmetic* (**J** 0009) Arch Math Logik Grundlagenforsch 17*71-80
◇ D20 F35 ◇ REV MR 52 # 13362 Zbl 325 # 02033
• ID 12147

SIEFKES, D. see Vol. I, III, IV, V for further entries

SIEG, W. [1981] see FEFERMAN, S.

SIEG, W. [1981] *Inductive definitions, constructive ordinals, and normal derivations* (**C** 3588) Iter Induct Def & Subsyst of Anal: Rec Proof-Th Stud 143-187
◇ F05 F15 F35 F50 ◇ REV MR 84d:03074 • ID 39785

SIEG, W. [1981] see BUCHHOLZ, W.

SIEG, W. [1984] *Foundations of analysis and proof theory* (**J** 0154) Synthese 60*159-200
◇ A10 F30 F35 F98 ◇ REV MR 86e:03055
Zbl 547 # 01013 • ID 39787

SIEG, W. [1985] *Fragments of arithmetic* (**J** 0073) Ann Pure Appl Logic 28*33-71
◇ F30 F35 ◇ REV Zbl 558 # 03029 • ID 39795

SIEG, W. [1985] *Reductions of theories for classical analysis* (**P** 4180) Int Congr Log, Meth & Phil of Sci (7,Pap);1983 Salzburg 199-232
◇ F35 ◇ ID 39794

SIERPINSKI, W. [1932] *Sur les ensembles de la meme puissance qui ne sont pas effectivement de la meme puissance* (**J** 0027) Fund Math 18*189-192
◇ E10 F65 ◇ REV Zbl 4.294 FdM 58.81 • ID 39936

SIERPINSKI, W. see Vol. II, IV, V for further entries

SIKIC, Z. [1979] *Towards the distinction of pure and applied mathematics in light of the problem of mathematical existence (Croatian)* (**C** 1932) Probl of Exist in Math
◇ A05 F99 ◇ ID 31455

SIKIC, Z. [1980] *A note on sum-eliminator* (**J** 3519) Glas Mat, Ser 3 (Zagreb) 15(35)*249-250
◇ F15 ◇ REV MR 82d:03091 Zbl 458 # 03014 • ID 54420

SIKIC, Z. [1984] *Multiple forms of Gentzen's rules and some intermediate forms* (**J** 0068) Z Math Logik Grundlagen Math 30*335-338
◇ B55 F07 ◇ REV MR 86a:03024 • ID 42266

SIKORSKI, R. [1953] see RASIOWA, H.

SIKORSKI, R. [1955] see RASIOWA, H.

SIKORSKI, R. [1956] *A theorem on non-classical functional calculi* (J 0014) Bull Acad Pol Sci, Ser Math Astron Phys 4*649-650
◇ B45 F50 G10 ◇ REV MR 19.830 Zbl 75.4 JSL 32.537 • ID 12298

SIKORSKI, R. [1958] *On Herbrand's theorem* (S 0019) Colloq Math (Warsaw) 6*55-58
◇ B10 F05 G05 ◇ REV MR 20 # 6352 Zbl 88.10 JSL 35.587 • ID 12303

SIKORSKI, R. [1958] *Some applications of interior mappings* (J 0027) Fund Math 45*200-212
◇ B45 E75 F50 G05 G10 G25 ◇ REV MR 24 # A2371 Zbl 83.386 JSL 32.537 • ID 12300

SIKORSKI, R. [1959] *Der Heyting'sche Praedikatenkalkuel und metrische Raeume* (P 0634) Constructivity in Math;1957 Amsterdam 250-253
◇ F50 ◇ REV MR 21 # 4106 Zbl 86.7 JSL 40.501 • ID 12299

SIKORSKI, R. [1959] see RASIOWA, H.

SIKORSKI, R. [1960] see RASIOWA, H.

SIKORSKI, R. [1962] *Applications of topology to foundations of mathematics* (P 1766) Gen Topol & Rel to Mod Anal & Algeb (1);1961 Prague 322-330
◇ B30 E75 F50 ◇ REV MR 26 # 3583 • ID 42875

SIKORSKI, R. [1963] see RASIOWA, H.

SIKORSKI, R. see Vol. I, II, III, IV, V for further entries

SILVERSTEIN, A. [1978] *A generalization of combinatorial operators* (J 0047) Notre Dame J Formal Log 19*639-645
◇ B40 D50 ◇ REV MR 84d:03058 Zbl 363 # 02049 • ID 52141

SIMAUTI, T. [1956] *Proof of a special case of the fundamental conjecture of Takeuti's GLC* (J 0090) J Math Soc Japan 8*135-144
◇ B15 F05 F35 ◇ REV MR 18.1 Zbl 75.234 JSL 24.64 • ID 12080

SIMAUTI, T. [1964] *A note on the construction of ordinal numbers* (J 0407) Comm Math Univ St Pauli (Tokyo) 12*37-39
◇ E10 F15 ◇ REV MR 29 # 1152 Zbl 199.21 JSL 32.396 • ID 12324

SIMAUTI, T. see Vol. I for further entries

SIMMONS, H. [1973] see MACINTYRE, A.

SIMMONS, H. [1975] *Topological aspects of suitable theories* (J 3420) Proc Edinburgh Math Soc, Ser 2 19*383-391
◇ B45 F30 G05 ◇ REV MR 52 # 7894 Zbl 327 # 02028 • ID 18388

SIMMONS, H. [1980] *The lattice of universal sentences modulo Peano arithmetic* (J 1650) Rapp, Sem Math Pure, Univ Cathol Louvain 28pp
◇ F30 G10 ◇ REV Zbl 471 # 03023 • ID 55218

SIMMONS, H. see Vol. III, IV for further entries

SIMONE, C. [1974] *Un modello di λ-calcolo fortemente equivalente agli schemi ricorsivi (English summary)* (J 0089) Calcolo 11*111-125
◇ B40 ◇ REV MR 54 # 69 Zbl 295 # 68012 • ID 24557

SIMPSON, S.G. [1975] see KREISEL, G.

SIMPSON, S.G. [1977] *Degrees of unsolvability: A survey of results* (C 1523) Handb of Math Logic 631-652
◇ D25 D30 D35 F35 ◇ REV MR 58 # 5109 JSL 49.975 • ID 27322

SIMPSON, S.G. [1977] *First order theory of the degrees of recursive unsolvability* (J 0120) Ann of Math, Ser 2 105*121-139
◇ D30 D35 F35 ◇ REV MR 55 # 5423 Zbl 349 # 02035 • ID 30798

SIMPSON, S.G. [1982] *Σ^1_1 and Π^1_1 transfinite induction* (P 3623) Logic Colloq;1980 Prague 239-253
◇ F35 ◇ REV MR 84b:03078a Zbl 497 # 03045 • ID 35659

SIMPSON, S.G. [1982] see FRIEDMAN, H.M.

SIMPSON, S.G. [1982] *Set theoretic aspects of ATR_0* (P 3623) Logic Colloq;1980 Prague 255-271
◇ C62 E30 F35 ◇ REV MR 84b:03078b Zbl 497 # 03046 • ID 35660

SIMPSON, S.G. [1983] see FRIEDMAN, H.M.

SIMPSON, S.G. [1984] *Which set existence axioms are needed to prove the Cauchy/Peano theorem for ordinary differential equations?* (J 0036) J Symb Logic 49*783-802
◇ E30 F35 ◇ REV MR 86a:03066 • ID 42501

SIMPSON, S.G. [1985] see SCHUETTE, K.

SIMPSON, S.G. [1985] *Friedman's research on subsystems of second order arithmetic* (C 4699) Friedman's Res on Found of Math 137-159
◇ C62 C98 F35 F98 ◇ ID 49879

SIMPSON, S.G. [1985] see HARRINGTON, L.A.

SIMPSON, S.G. [1985] *Nichtbeweisbarkeit von gewissen kombinatorischen Eigenschaften endlicher Baeume (English summary)* (J 0009) Arch Math Logik Grundlagenforsch 25*45-65
◇ F30 ◇ ID 49130

SIMPSON, S.G. [1985] *Reverse mathematics* (P 4046) Rec Th;1982 Ithaca 461-471
◇ E30 F35 ◇ ID 45809

SIMPSON, S.G. see Vol. I, III, IV, V for further entries

SINGH, S. [1967] *The natural number arithmetic in Goedel's axiomatic set theory* (J 0545) Proc Indian Acad Sci, Sect A 37*221-228
◇ B28 F30 ◇ REV MR 41 # 6684 Zbl 241 # 02026 • ID 12349

SINGH, S. see Vol. V for further entries

SKARBEK, W. & ZEMBRZUSKI, K. [1972] *Computable real functions and their relation to the analog computer* (J 1929) Prace Centr Oblicz Pol Akad Nauk 83*40pp
◇ D20 F60 ◇ REV Zbl 263 # 02018 • ID 29864

SKARBEK, W. see Vol. IV for further entries

SKOLEM, T.A. [1923] *Begruendung der elementaren Arithmetik durch die rekurrierende Denkweise ohne Anwendung scheinbarer Veraenderlichen mit unendlichem Ausdehnungsbereich* (J 1145) Vidensk Selsk Kristiana Skrifter Ser 1 6*1-38
• TRANSL [1967] (C 0675) From Frege to Goedel 303-333 (English) [1970] (C 1098) Skolem: Select Works in Logic 153-188
◇ A05 B28 D20 F30 ◇ REV Zbl 228 # 02001 • ID 38689

SKOLEM, T.A. [1926] *The most important recent discussions on the foundations of mathematics (Norwegian)* (**J** 4510) Norsk Mat Tidsskr 8*1-13
◇ A05 F99 ◇ REV FdM 52.51 • ID 41681

SKOLEM, T.A. [1931] *Ueber einige Satzfunktionen in der Arithmetik* (**J** 0974) Norsk Vid-Akad Oslo Mat-Natur Kl Skr 7*28pp
• REPR [1970] (**C** 1098) Skolem: Select Works in Logic 281-306
◇ B25 B28 C10 F30 ◇ REV Zbl 2.3 FdM 57.1320 • ID 12388

SKOLEM, T.A. [1933] *Ueber die Unmoeglichkeit einer Charakterisierung der Zahlenreihe mittels eines endlichen Axiomensystems* (**J** 0975) Norsk Mat Forenings Skr 10*73-82
• REPR [1970] (**C** 1098) Skolem: Select Works in Logic 345-354
◇ B28 C20 C62 F30 H15 ◇ REV Zbl 7.193 FdM 59.53 • ID 12392

SKOLEM, T.A. [1934] *Ueber die Nichtcharakterisierbarkeit der Zahlenreihe mittels endlich oder abzaehlbar unendlich vieler Aussagen mit ausschliesslich Zahlenvariablen* (**J** 0027) Fund Math 23*150-161
• REPR [1970] (**C** 1098) Skolem: Select Works in Logic 355-366
◇ B28 C20 C62 F30 H15 ◇ REV Zbl 10.49 FdM 60.25 • ID 12393

SKOLEM, T.A. [1936] *Ueber die Zurueckfuehrbarkeit einiger durch Rekursionen definierter Relationen auf "arithmetische"* (**J** 0460) Acta Univ Szeged, Sect Mat 8*73-88
• REPR [1970] (**C** 1098) Skolem: Select Works in Logic 425-440
◇ B28 F30 ◇ REV Zbl 16.194 JSL 2.85 FdM 63.25 • ID 12396

SKOLEM, T.A. [1939] *Eine Bemerkung ueber die Induktionsschemata in der rekursiven Zahlentheorie* (**J** 0124) Monatsh Math-Phys 48*268-276
• REPR [1970] (**C** 1098) Skolem: Select Works in Logic 441-449
◇ F30 ◇ REV MR 1.132 Zbl 21.385 JSL 5.34 FdM 65.31 • ID 12397

SKOLEM, T.A. [1940] *Einfacher Beweis der Unmoeglichkeit eines allgemeinen Loesungsverfahren fuer arithmetische Probleme* (**J** 0121) Kon Norske Vidensk Selsk Forh 13*1-4
• REPR [1970] (**C** 1098) Skolem: Select Works in Logic 451-454
◇ D35 F30 ◇ REV MR 2.210 Zbl 61.10 JSL 9.21 FdM 66.32 • ID 12398

SKOLEM, T.A. [1944] *A note on recursive arithmetic* (**J** 0121) Kon Norske Vidensk Selsk Forh 17*107-109
◇ D20 F30 ◇ REV MR 8.4 JSL 11.26 • ID 12402

SKOLEM, T.A. [1944] *Some remarks on recursive arithmetic* (**J** 0121) Kon Norske Vidensk Selsk Forh 17*103-106
• REPR [1970] (**C** 1098) Skolem: Select Works in Logic 487-493
◇ D20 F30 ◇ REV MR 8.4 JSL 11.26 • ID 12401

SKOLEM, T.A. [1946] *Recursive arithmetic (Norwegian)* (**J** 4510) Norsk Mat Tidsskr 28*1-12
◇ F30 ◇ REV MR 8.125 JSL 13.169 • ID 28458

SKOLEM, T.A. [1947] *The development of recursive arithmetic* (**P** 1403) Skand Mat Kongr (10);1946 Copenhagen 1-46
• REPR [1970] (**C** 1098) Skolem: Select Works in Logic 499-514
◇ F30 ◇ REV MR 8.558 Zbl 30.101 JSL 13.54 • ID 17000

SKOLEM, T.A. [1950] *A remark on the induction scheme* (**J** 0121) Kon Norske Vidensk Selsk Forh 22*167-170
• REPR [1970] (**C** 1098) Skolem: Select Works in Logic 515-518
◇ F30 ◇ REV MR 11.635 Zbl 36.7 JSL 16.220 • ID 12404

SKOLEM, T.A. [1952] *Consideraciones sobre los fundamentos de la matematica I* (**J** 0236) Rev Mat Hisp-Amer, Ser 4 12*169-200
◇ B98 F30 F50 ◇ REV MR 15.92 Zbl 52.8 JSL 21.373 • REM Part II 1953 • ID 21071

SKOLEM, T.A. [1953] *Consideraciones sobre los fundamentos de la matematica II* (**J** 0236) Rev Mat Hisp-Amer, Ser 4 13*149-174
◇ B98 F30 F50 ◇ REV MR 15.92 Zbl 52.8 JSL 21.373 • REM Part I 1952 • ID 21072

SKOLEM, T.A. [1953] *Some considerations concerning recursive arithmetic* (**J** 0082) Bull Soc Math Belg 6*35-46
• REPR [1970] (**C** 1098) Skolem: Select Works in Logic 563-574
◇ F30 ◇ REV MR 16.553 Zbl 56.248 JSL 23.228 • ID 12409

SKOLEM, T.A. [1954] *Results in investigations in the foundations (Norwegian)* (**P** 0788) Skand Mat Kongr (12);1953 Lund 273-289
◇ A05 B98 D03 F55 ◇ REV MR 16.553 Zbl 56.245 • ID 21133

SKOLEM, T.A. [1955] *Peano's axioms and models of arithmetic* (**P** 1589) Math Interpr of Formal Systs;1954 Amsterdam 1-14
• REPR [1970] (**C** 1098) Skolem: Select Works in Logic 587-600
◇ B28 C20 C62 F30 H15 ◇ REV MR 17.699 Zbl 68.246 JSL 22.306 • ID 27718

SKOLEM, T.A. [1956] *A version of the proof of equivalence between complete induction and the uniqueness of primitive recursion* (**J** 0121) Kon Norske Vidensk Selsk Forh 29*10-15
• REPR [1970] (**C** 1098) Skolem: Select Works in Logic 606-606
◇ B28 F30 ◇ REV MR 18.2 Zbl 72.6 JSL 24.69 • ID 12416

SKOLEM, T.A. [1956] *An ordered set of arithmetic functions representing the least ε-number* (**J** 0121) Kon Norske Vidensk Selsk Forh 29*54-59
◇ F15 F30 ◇ REV MR 18.785 Zbl 77.13 • ID 12417

SKOLEM, T.A. [1958] *Remarks on the connection between intuitionistic logic and a certain class of lattices* (**J** 0132) Math Scand 6*231-236
• REPR [1970] (**C** 1098) Skolem: Select Works in Logic 639-644
◇ F50 G10 ◇ REV MR 21 # 5574 Zbl 92.5 • ID 12422

SKOLEM, T.A. see Vol. I, II, III, IV, V for further entries

SKORDEV, D.G. [1963] *Rekursiv vollstaendige arithmetische Operationen (Russian summary)* (**J** 0137) C R Acad Bulgar Sci 16*469
◇ B40 ◇ REV MR 29 # 4676 Zbl 156.9 • ID 12438

SKORDEV, D.G. [1975] *Some topological examples of iterative combinatory spaces (Russian)* (**J** 0137) C R Acad Bulgar Sci 28*1575-1578
◇ B40 D75 ◇ REV MR 55 # 7754 Zbl 341 # 02032 • ID 29752

SKORDEV, D.G. [1976] *Certain combinatory spaces that are connected with the complexity of data processing (Russian)* (**J** 0137) C R Acad Bulgar Sci 29*7-10
◇ B40 B75 D15 D75 ◇ REV MR 53 # 12905 Zbl 358 # 02056 • ID 23157

SKORDEV, D.G. [1976] *Recursion theory on iterative combinatory spaces* (**J** 0014) Bull Acad Pol Sci, Ser Math Astron Phys 24*23-31
◇ B40 D75 ◇ REV MR 54 # 4950 Zbl 328 # 02026 • ID 18392

SKORDEV, D.G. [1976] *Some models of combinatory logic (Russian)* (**J** 0087) Mat Zametki (Akad Nauk SSSR) 19*149-154
• TRANSL [1976] (**J** 1044) Math Notes, Acad Sci USSR 19*88-90
◇ B40 ◇ REV MR 53 # 12893 Zbl 411 # 03012 • ID 23147

SKORDEV, D.G. [1976] *The concept of search computability from the point of view of the theory of combinatory spaces (Russian)* (**J** 2547) Serdica, Bulgar Math Publ 2*343-349
◇ B40 D75 ◇ REV MR 57 # 2894 Zbl 437 # 03025 • ID 55893

SKORDEV, D.G. [1976] *The partial order on the set* $\mathfrak{C}$ *in combinatory spaces (Russian)* (**J** 0137) C R Acad Bulgar Sci 29*151-154
◇ B40 ◇ REV MR 53 # 12892 Zbl 358 # 02057 • ID 23146

SKORDEV, D.G. [1977] *Simplification of some definitions in the theory of combinatory spaces* (**J** 0137) C R Acad Bulgar Sci 30*947-950
◇ B40 D75 ◇ REV MR 58 # 5084 Zbl 364 # 02025 • ID 50955

SKORDEV, D.G. [1978] *A normal form theorem for recursive operators in iterative combinatory spaces* (**J** 0068) Z Math Logik Grundlagen Math 24*115-124
◇ B40 D75 ◇ REV MR 80b:03062 Zbl 405 # 03024 • ID 54899

SKORDEV, D.G. [1979] see PETROV, V.P.

SKORDEV, D.G. [1979] *The first recursion theorem for iterative combinatory spaces* (**J** 0068) Z Math Logik Grundlagen Math 25*69-77
◇ B40 D75 ◇ REV MR 80g:03049 Zbl 412 # 03032 • ID 52963

SKORDEV, D.G. [1980] *Combinatory spaces and recursiveness in them (Russian) (English summary)* (**P** 4117) Conf Math Log (Markov);1980 Sofia 456pp
◇ B40 D75 ◇ REV Zbl 537 # 03029 • ID 41127

SKORDEV, D.G. [1980] *Semi-combinatory spaces (Russian)* (**J** 0137) C R Acad Bulgar Sci 33*739-742
◇ B40 D75 ◇ REV MR 82d:03076b Zbl 452 # 03037 • ID 54101

SKORDEV, D.G. [1982] *An algebraic treatment of flow diagrams and its application to generalized recursion theory* (**P** 3831) Universal Algeb & Appl;1978 Warsaw 277-287
◇ B40 B75 D75 ◇ REV MR 85f:68007 Zbl 518 # 03017 • ID 37517

SKORDEV, D.G. [1984] *First reduction theorem for iterative semicombinatory spaces (Russian)* (**P** 4117) Conf Math Log (Markov);1980 Sofia 89-111
◇ B40 ◇ ID 41129

SKORDEV, D.G. see Vol. II, III, IV, V for further entries

SKVORTSOV, D.P. [1976] *The occurrence of an implication in finitely valid, intuitively unprovable formulas of propositional logic (Russian)* (**J** 0087) Mat Zametki (Akad Nauk SSSR) 20*383-390
• TRANSL [1976] (**J** 1044) Math Notes, Acad Sci USSR 20*771-775
◇ F50 ◇ REV MR 55 # 2511 Zbl 349 # 02026 • ID 65355

SKVORTSOV, D.P. [1979] *Logic of infinite problems and Kripke models on atomic semilattices of sets (Russian)* (**J** 0023) Dokl Akad Nauk SSSR 245*798-801
• TRANSL [1979] (**J** 0062) Sov Math, Dokl 20*360-363
◇ B55 C90 F50 G20 ◇ REV MR 80i:03040 Zbl 438 # 03028 • ID 55940

SKVORTSOV, D.P. [1979] *Two generalizations of the concept of a finite problem (Russian)* (**C** 2581) Issl Neklass Log & Teor Mnozh 201-240
◇ F50 ◇ REV MR 81j:03060 • ID 78691

SKVORTSOV, D.P. [1980] *On the connection of finitary general validity of certain propositional formulas with derivability in the Kreisel-Putnam logic (Russian)* (**J** 0288) Vest Ser Mat Mekh, Univ Moskva 1980/3*29-32
• TRANSL [1980] (**J** 0510) Moscow Univ Math Bull 35/3*30-33
◇ B55 F50 ◇ REV MR 81i:03088 Zbl 438 # 03029 • ID 55941

SKVORTSOV, D.P. [1983] *The intuitionistic propositional calculus with an additional logical connective (Russian)* (**C** 3807) Issl Neklass Log & Formal Sist 154-173
◇ F50 ◇ REV MR 85e:03025 • ID 40599

SKVORTSOV, D.P. see Vol. I, II, III, V for further entries

SLAGLE, J.R. [1972] *An approach for finding C-linear complete inference systems* (**J** 0037) ACM J 19*496-516
◇ F07 ◇ REV MR 47 # 3145 Zbl 246 # 68018 • ID 12453

SLAGLE, J.R. see Vol. I, III for further entries

SLESSENGER, P.H. [1984] *On subsets of the Skolem class of exponential polynomials* (**P** 2153) Logic Colloq;1983 Aachen 2*427-436
◇ F15 F30 ◇ ID 43013

SLESSENGER, P.H. [1985] *A height restricted generation of a set of arithmetic functions of order-type* ε_0 (**J** 0068) Z Math Logik Grundlagen Math 31*117-122
◇ F15 F30 ◇ ID 42292

SLEZAK, P. [1982] *Goedel's theorem and the mind* (**J** 0013) Brit J Phil Sci 33*41-52
◇ A05 F99 ◇ REV MR 83g:03013 • ID 35993

SLEZAK, P. [1984] *Minds, machines and self-reference (French and German summaries)* (**J** 0076) Dialectica 38*17-34
◇ A05 D99 F30 ◇ REV MR 85b:03011 • ID 40543

SLISENKO, A.O. [1963] *Certain properties of arithmetic operations on duplexes (Russian)* (**J** 0023) Dokl Akad Nauk SSSR 152*292-295
• TRANSL [1963] (**J** 0062) Sov Math, Dokl 4*1313-1315
◇ F60 ◇ REV MR 28 # 31 Zbl 235 # 02029 • ID 12462

SLISENKO, A.O. [1964] *Example of a non-discontinuous but not continuous constructive operator in a metric space (Russian)* (S 0066) Tr Mat Inst Steklov 72*524-532
• TRANSL [1972] (J 0225) Amer Math Soc, Transl, Ser 2 100*184-194
◇ F60 ◇ REV MR 33 # 3928 Zbl 235 # 02031 • ID 65364

SLISENKO, A.O. [1964] *On certain algorithmic problems connected with arithmetic operations on duplexes (Russian)* (S 0066) Tr Mat Inst Steklov 72*488-523
• TRANSL [1972] (J 0225) Amer Math Soc, Transl, Ser 2 100*135-184
◇ F60 ◇ REV MR 34 # 2460 Zbl 235 # 02030 • ID 19659

SLISENKO, A.O. [1964] *On constructive non-separable spaces (Russian)* (S 0066) Tr Mat Inst Steklov 72*533-536
• TRANSL [1972] (J 0225) Amer Math Soc, Transl, Ser 2 100*195-199
◇ F60 ◇ REV MR 33 # 3929 Zbl 235 # 02032 • ID 19663

SLISENKO, A.O. [1967] *Arithmetical operations on certain sets of duplexes (Russian)* (S 0066) Tr Mat Inst Steklov 93*187-207
• TRANSL [1967] (S 0055) Proc Steklov Inst Math 93*241-267
◇ F60 ◇ REV MR 37 # 3920 Zbl 283 # 02032 • ID 12464

SLISENKO, A.O. [1967] *On constructing of maximal continuity regulators of constructive functions (Russian)* (S 0066) Tr Mat Inst Steklov 93*208-249
• TRANSL [1967] (S 0055) Proc Steklov Inst Math 93*269-317
◇ F60 ◇ REV MR 39 # 1871 Zbl 205.309 • ID 90023

SLISENKO, A.O. [1967] *On maximal continuity regulators of constructive functions (Russian)* (S 0228) Zap Nauch Sem Leningrad Otd Mat Inst Steklov 4*201-208
• TRANSL [1967] (J 0521) Semin Math, Inst Steklov 4*82-84
◇ F60 ◇ REV MR 39 # 4000 Zbl 213.20 • ID 12463

SLISENKO, A.O. (ED.) [1967] *Studies in constructive mathematics and mathematical logic I (Russian)* (S 0228) Zap Nauch Sem Leningrad Otd Mat Inst Steklov 4*1-218 • ERR/ADD ibid 8*insidebackcover
• TRANSL [1967] (J 0521) Semin Math, Inst Steklov 4*1-88
◇ F97 ◇ REV MR 39 # 2595 MR 48 # 5805 • ID 90014

SLISENKO, A.O. (ED.) [1968] *Studies in constructive mathematics and mathematical logic II (Russian)* (S 0228) Zap Nauch Sem Leningrad Otd Mat Inst Steklov 8*1-280 • ERR/ADD ibid 8*insidebackcover & 16*185-187 & 20*292-294
• TRANSL [1968] (J 0521) Semin Math, Inst Steklov 8*1-136
◇ F97 ◇ REV MR 46 # 8817 MR 48 # 5805 Zbl 167.1 • ID 90015

SLISENKO, A.O. (ED.) [1969] *Studies in constructive mathematics and mathematical logic III (Russian)* (S 0228) Zap Nauch Sem Leningrad Otd Mat Inst Steklov 16*1-188 • ERR/ADD ibid 20*292-294 & 32*159-162
• TRANSL [1969] (J 0521) Semin Math, Inst Steklov 16*1-97
◇ F97 ◇ REV MR 46 # 8818 Zbl 191.306 • ID 27418

SLISENKO, A.O. [1970] *Some questions of the approximation of the maximal regulators of continuity (Russian)* (S 0066) Tr Mat Inst Steklov 113*73-78
• TRANSL [1970] (S 0055) Proc Steklov Inst Math 113*84-90
◇ F60 ◇ REV MR 46 # 1969 Zbl 302 # 02011 • ID 22275

SLISENKO, A.O. [1971] *A property of enumerable sets containing "complexly deducible" formulas (Russian) (English summary)* (S 0228) Zap Nauch Sem Leningrad Otd Mat Inst Steklov 20*200-207,288
• TRANSL [1973] (J 1531) J Sov Math 1*126-131
◇ D25 F20 ◇ REV MR 44 # 6501 Zbl 222 # 02045 • ID 12465

SLISENKO, A.O. [1971] see MATIYASEVICH, YU.V.

SLISENKO, A.O. [1972] see MATIYASEVICH, YU.V.

SLISENKO, A.O. [1974] see MATIYASEVICH, YU.V.

SLISENKO, A.O. [1976] see MATIYASEVICH, YU.V.

SLISENKO, A.O. [1979] see MATIYASEVICH, YU.V.

SLISENKO, A.O. see Vol. I, IV for further entries

SLOBODSKOJ, A.M. [1973] see FRIDMAN, EH.I.

SLOBODSKOJ, A.M. see Vol. III, IV for further entries

SLONNEGER, K. [1976] *A complete infinitary logic* (J 0036) J Symb Logic 41*730-746
◇ C75 F05 ◇ REV MR 55 # 7739 Zbl 357 # 02015 • ID 14575

SLUPECKI, J. [1960] see POGORZELSKI, W.A.

SLUPECKI, J. see Vol. I, II, III, V for further entries

SMART, J.J.C. [1961] *Goedel's theorem, Church's theorem and mechanism* (J 0154) Synthese 13*105-110
◇ A05 B10 D35 F30 ◇ REV Zbl 104.242 • ID 31260

SMART, J.J.C. see Vol. I for further entries

SMETANICH, YA.S. [1959] *Ueber eine Eigenschaft der Formeln des konstruktiven Aussagenkalkuels, die von einer Variablen abhaengen (Russisch)* (J 4170) Nauch Dokl Vys Shk, Fiz Mat (Moskva) 1959/2*74-77
◇ F50 ◇ REV Zbl 115.5 • ID 48060

SMETANICH, YA.S. see Vol. I for further entries

SMILEY, T.J. [1978] see SHOESMITH, D.J.

SMILEY, T.J. see Vol. I, II, IV for further entries

SMIRNOV, V.A. [1970] *A syllogistics without the law of the excluded middle and its inclusion into the predicate calculus (Russian)* (C 1530) Issl Log Sist (Yanovskaya) 68-77
◇ F50 ◇ REV MR 48 # 1877 Zbl 223 # 02005 • ID 12519

SMIRNOV, V.A. [1971] *Elimination des termes ε dans la logique intuitionniste* (J 2076) Rev Int Philos 25*512-519
◇ F50 ◇ REV MR 58 # 27371 • ID 78739

SMIRNOV, V.A. [1972] *Formal derivation and logical calculi (Russian)* (X 2027) Nauka: Moskva 272pp
◇ B98 C07 C40 C98 F05 F07 F98 ◇ REV Zbl 269 # 02004 • ID 30359

SMIRNOV, V.A. [1973] *An absolute first order predicate calulus* (J 0387) Bull Sect Logic, Pol Acad Sci 2*38-45
◇ B10 F05 ◇ REV MR 52 # 13339 • ID 21800

SMIRNOV, V.A. [1979] *Theory of quantification and 𝔈-calculi* (P 1705) Scand Logic Symp (4);1976 Jyvaeskylae 41-47
◇ B10 F50 ◇ REV MR 81c:03010 Zbl 431 # 03015 • ID 53921

SMIRNOV, V.A. see Vol. I, II, III, V for further entries

SMIRNOVA, E.D. [1974] *Consistency and eliminability in proof theory (Russian)* (**C** 2578) Filos & Logika 84-101
◇ A05 B10 F99 ◇ REV MR 58 # 27265 • ID 78744

SMIRNOVA, E.D. see Vol. I, II for further entries

SMITH, BRIAN [1974] see BISHOP, P.

SMITH, BRIAN see Vol. I for further entries

SMITH, JAN [1983] *The identification of propositions and types in Martin-Loef's type theory: a programming example* (**P** 3864) FCT'83 Found of Comput Th;1983 Borgholm 445-456
◇ B75 F35 F50 ◇ REV MR 86d:03014 Zbl 522 # 68034
• ID 37058

SMITH, JAN [1984] see NORDSTROEM, B.

SMITH, JAN see Vol. I for further entries

SMITH, RICK L. [1981] *Effective valuation theory* (**P** 2902) Aspects Effective Algeb;1979 Clayton 232-245
◇ C57 C60 D45 F55 ◇ REV MR 83b:03053
Zbl 479 # 03023 • ID 55685

SMITH, RICK L. [1983] see FRIEDMAN, H.M.

SMITH, RICK L. see Vol. III, IV for further entries

SMORYNSKI, C.A. [1973] *Applications of Kripke models* (**C** 1139) Metamath Invest of Intuition Arithm & Anal 324-391
◇ F30 F50 ◇ REV MR 56 # 2795 • ID 22097

SMORYNSKI, C.A. [1973] *Elementary intuitionistic theories* (**J** 0036) J Symb Logic 38*102-134
◇ B25 C90 D35 F50 ◇ REV MR 48 # 5842
Zbl 261 # 02033 • ID 12535

SMORYNSKI, C.A. [1973] *Investigations of intuitionistic formal systems by means of Kripke models* (0000) Diss., Habil. etc
◇ B55 C90 F50 ◇ REM Diss., University of Chicago
• ID 27058

SMORYNSKI, C.A. [1976] see JONGH DE, D.H.J.

SMORYNSKI, C.A. [1977] *ω-consistency and reflection* (**P** 1729) Colloq Int Log;1975 Clermont-Ferrand 167-181
◇ F30 ◇ REV MR 80g:03057 • ID 78762

SMORYNSKI, C.A. [1977] *On axiomatizing fragments* (**J** 0036) J Symb Logic 42*530-544
◇ C90 F50 ◇ REV MR 57 # 15998 Zbl 381 # 03046
• ID 26856

SMORYNSKI, C.A. [1977] *The incompleteness theorems* (**C** 1523) Handb of Math Logic 821-865
◇ C62 F25 F30 F98 ◇ REV MR 58 # 10343 JSL 49.980
• ID 27327

SMORYNSKI, C.A. [1978] *Avoiding self-referential statements* (**J** 0053) Proc Amer Math Soc 70*181-184
◇ D25 F30 ◇ REV MR 57 # 16015 Zbl 392 # 03037
• ID 28384

SMORYNSKI, C.A. [1978] *Beth's theorem and self-referential sentences* (**P** 1897) Logic Colloq;1977 Wroclaw 253-261
◇ B45 C40 F30 ◇ REV MR 80c:03023 Zbl 453 # 03018
• ID 54148

SMORYNSKI, C.A. [1978] *The axiomatization problem for fragments* (**J** 0007) Ann Math Logic 14*193-221
◇ B55 F50 ◇ REV MR 80e:03040 Zbl 395 # 03039
• ID 29151

SMORYNSKI, C.A. [1979] *Calculating self-referential statements. I: Explicit calculations* (**J** 0063) Studia Logica 38*17-36
◇ B45 F30 G25 ◇ REV MR 80m:03030 Zbl 412 # 03037
• ID 52968

SMORYNSKI, C.A. [1979] *Some rapidly growing functions* (**J** 2789) Math Intell 2*149-154
◇ D20 D80 F30 ◇ REV MR 82c:03064 Zbl 453 # 03049
• ID 54179

SMORYNSKI, C.A. [1980] *Calculating self-referential statements* (**J** 0027) Fund Math 109*189-210
◇ B45 F30 G25 ◇ REV MR 82g:03099 Zbl 449 # 03067
• ID 56734

SMORYNSKI, C.A. [1981] *Calculating self-referential statements: Guaspari sentences of the first kind* (**J** 0036) J Symb Logic 46*329-344
◇ F30 G25 ◇ REV MR 83a:03059 Zbl 475 # 03032
• ID 55486

SMORYNSKI, C.A. [1981] *Fifty years of self-reference in arithmetic* (**J** 0047) Notre Dame J Formal Log 22*357-374
◇ A10 F30 F40 ◇ REV MR 83i:03094 Zbl 481 # 03033
• ID 33513

SMORYNSKI, C.A. [1982] *Commutativity and self-reference* (**J** 0047) Notre Dame J Formal Log 23*443-452
◇ B45 F30 ◇ REV MR 83k:03027 Zbl 528 # 03009
• ID 36167

SMORYNSKI, C.A. [1982] *Fixed point algebras* (**J** 0589) Bull Amer Math Soc (NS) 6*317-356 • ERR/ADD ibid 8*407-408
◇ E30 F30 G25 ◇ REV MR 83f:03066 MR 84b:03081
Zbl 544 # 03032 • ID 35670

SMORYNSKI, C.A. [1982] *Nonstandard models and constructivity* (**P** 3638) Brouwer Centenary Symp;1981 Noordwijkerhout 459-464
◇ C62 F30 F50 ◇ REV MR 85d:03114 Zbl 518 # 03026
• ID 37523

SMORYNSKI, C.A. [1982] *The finite inseparability of the first-order theory of diagonalisable algebras* (**J** 0063) Studia Logica 41*347-349
◇ B45 C13 D35 F30 ◇ REV Zbl 542 # 03024 • ID 43687

SMORYNSKI, C.A. [1982] *The varieties of arboreal experience* (**J** 2789) Math Intell 4*182-189
◇ D20 F30 ◇ REV MR 84i:03110 Zbl 521 # 03041
• ID 34588

SMORYNSKI, C.A. [1984] *Lectures on nonstandard models of arithmetic* (**P** 3710) Logic Colloq;1982 Firenze 1-70
◇ A10 C62 C98 F30 F98 H15 ◇ REV MR 86c:03035
Zbl 554 # 03036 • ID 41837

SMORYNSKI, C.A. [1984] *Modal logic and self-reference* (**C** 4085) Handb Philos Log 2*441-496
◇ B45 F30 ◇ ID 44891

SMORYNSKI, C.A. [1985] *Nonstandard models and related developments* (**C** 4699) Friedman's Res on Found of Math 179-229
◇ C62 C98 F30 F98 ◇ ID 49880

SMORYNSKI, C.A. [1985] *Self-reference and modal logic* (**X** 0811) Springer: Heidelberg & New York vii+333pp
◇ B45 F30 F98 ◇ ID 44663

SMORYNSKI, C.A. see Vol. III, IV, V for further entries

SMULLYAN, R.M. [1957] *Languages in which self reference is possible* (J 0036) J Symb Logic 22*55-67
• REPR [1969] (C 0569) Phil of Math Oxford Readings 64-77
◇ A05 B10 B28 F30 F40 ◇ REV MR 19.830 Zbl 218 # 02026 JSL 24.228 • ID 12538

SMULLYAN, R.M. [1963] *A unifying principle in quantification theory* (J 0054) Proc Nat Acad Sci USA 49*828-832
◇ B10 F05 F07 ◇ REV MR 27 # 2410 Zbl 118.249 • ID 12547

SMULLYAN, R.M. [1963] *First order logic* (X 0811) Springer: Heidelberg & New York xii + 158pp
◇ B98 F07 F98 ◇ REV MR 39 # 5311 Zbl 172.289 JSL 40.237 • REM 3rd edition 1971 • ID 28562

SMULLYAN, R.M. [1965] *A unifying principle in quantification theory* (P 0614) Th Models;1963 Berkeley 433-434
◇ B10 F05 F07 ◇ REV Zbl 199.7 • ID 16212

SMULLYAN, R.M. [1965] *Analytic natural deduction* (J 0036) J Symb Logic 30*123-139
◇ B10 F05 F07 ◇ REV MR 34 # 34 Zbl 149.4 • ID 12551

SMULLYAN, R.M. [1968] *Analytic cut* (J 0036) J Symb Logic 33*560-564
◇ B10 F05 F07 ◇ REV MR 39 # 3969 Zbl 181.5 • ID 12555

SMULLYAN, R.M. [1970] *Abstract quantification theory* (P 0603) Intuitionism & Proof Th;1968 Buffalo 79-91
◇ B10 C07 F07 ◇ REV MR 42 # 2928 Zbl 206.272 • ID 12557

SMULLYAN, R.M. [1973] *A generalization of intuitionistic and modal logics* (P 0783) Truth, Syntax & Modal;1970 Philadelphia 274-293
◇ B45 F50 ◇ REV MR 53 # 7733 Zbl 268 # 02017 JSL 42.316 • ID 14740

SMULLYAN, R.M. [1979] *Trees and ball games* (J 1377) Ann New York Acad Sci 321*86-90
◇ E05 F05 ◇ REV MR 80i:04008 Zbl 432 # 90100 • ID 78773

SMULLYAN, R.M. [1983] *Dame oder Tiger? Logische Denkspiele und eine mathematische Novelle ueber Goedels grosse Entdeckung* (X 1265) Fischer: Stuttgart 239pp
◇ B98 F99 ◇ REV Zbl 545 # 00007 • REM Transl. from English • ID 41200

SMULLYAN, R.M. [1984] *Fixed points and self-reference* (J 3386) Int J Math & Math Sci 7*283-289
◇ F30 ◇ REV MR 85h:03002 • ID 43268

SMULLYAN, R.M. [1985] *Modality and self-reference* (C 3659) Intens Math 191-211
◇ B45 F30 ◇ ID 44678

SMULLYAN, R.M. [1985] *Some principles related to Loeb's theorem* (C 3659) Intens Math 213-230
◇ B45 F30 ◇ ID 44783

SMULLYAN, R.M. see Vol. I, II, IV, V for further entries

SMYTH, M.B. [1974] *Involution as a basis for propositional calculi* (J 0047) Notre Dame J Formal Log 15*569-588
◇ B05 F50 ◇ REV MR 51 # 10021 Zbl 225 # 02018 • ID 12559

SMYTH, M.B. see Vol. I, IV for further entries

SNYDER, D.P. [1972] see LEBLANC, H.

SNYDER, D.P. see Vol. II for further entries

SNYDER, LAWRENCE [1977] see LIPTON, R.J.

SOARE, R.I. [1969] *Recursion theory and Dedekind cuts* (J 0064) Trans Amer Math Soc 140*271-294
◇ D25 D30 F60 ◇ REV MR 39 # 3997 Zbl 181.305 JSL 36.148 • ID 12565

SOARE, R.I. [1972] see JOCKUSCH JR., C.G.

SOARE, R.I. see Vol. I, III, IV for further entries

SOBOLEV, S.K. [1977] *On finite-dimensional superintuitionistic logics (Russian)* (J 0216) Izv Akad Nauk SSSR, Ser Mat 41*963-986,1199
• TRANSL [1977] (J 0448) Math of USSR, Izv 11*909-935
◇ B55 D05 F50 G10 G25 ◇ REV MR 58 # 10333 Zbl 368 # 02062 • ID 51279

SOBOLEV, S.K. [1977] *The intuitionistic propositional calculus with quantifiers (Russian)* (J 0087) Mat Zametki (Akad Nauk SSSR) 22*69-76
• TRANSL [1977] (J 1044) Math Notes, Acad Sci USSR 22*528-532
◇ B55 C80 C90 D35 F50 ◇ REV MR 56 # 15371 Zbl 365 # 02013 • ID 51014

SOBOLEV, S.K. see Vol. II for further entries

SOCHOR, A. [1981] see RESL, M.

SOCHOR, A. see Vol. I, III, V for further entries

SOLITRO, U. & VALENTINI, S. [1983] *The modal logic of consistency assertions of Peano arithmetic* (J 0068) Z Math Logik Grundlagen Math 29*25-32
◇ B45 F30 ◇ REV MR 84c:03043 Zbl 521 # 03010 • ID 35696

SOLOMON, M.K. [1978] *Some results on measure independent Goedel speed-ups* (J 0036) J Symb Logic 43*667-672
◇ D15 F20 ◇ REV MR 80b:03091 Zbl 398 # 03027 • ID 52755

SOLOVAY, R.M. [1976] *Provability interpretations of modal logic* (J 0029) Israel J Math 25*287-304
◇ B45 E30 F30 ◇ REV MR 56 # 15369 Zbl 352 # 02019 JSL 46.661 • ID 26088

SOLOVAY, R.M. [1979] see GUASPARI, D.

SOLOVAY, R.M. [1981] see KETONEN, J.

SOLOVAY, R.M. [1985] *Explicit Henkin sentences* (J 0036) J Symb Logic 50*91-93
◇ F30 F40 ◇ ID 42525

SOLOVAY, R.M. see Vol. III, IV, V for further entries

SOLOV'EV, S.V. [1979] see BABAEV, A.A.

SOLOV'EV, S.V. [1979] *Preservation of the equivalence of proofs under reduction of the formula depth (Russian) (English summary)* (S 0228) Zap Nauch Sem Leningrad Otd Mat Inst Steklov 88*197-208,245-246
• TRANSL [1982] (J 1531) J Sov Math 20*2370-2376
◇ F07 F20 ◇ REV MR 81g:03066 Zbl 429 # 03036 • ID 78848

SOLOV'EV, S.V. [1979] *The increase in length of an L-derivation transformed into a natural deduction (Russian) (English summary)* (**S** 0228) Zap Nauch Sem Leningrad Otd Mat Inst Steklov 88*192-196,245
• TRANSL [1982] (**J** 1531) J Sov Math 20*2367-2369
◇ F07 F20 F50 ◇ REV MR 81g:03065 Zbl 429 # 03038
• ID 78849

SOLOV'EV, S.V. [1981] *The category of finite sets and Cartesian closed categories (Russian) (English summary)* (**S** 0228) Zap Nauch Sem Leningrad Otd Mat Inst Steklov 105*174-194,200
◇ F07 F25 G30 ◇ REV MR 83d:18011 Zbl 474 # 18003
• ID 40077

SONOBE, O. [1975] *Bi-relational frameworks for minimal and intuitionistic logics* (**P** 3299) Progr Kiso Riron, Algor Okeru Shomei Ron;1973/74 Kyoto 174-189
◇ B45 C90 F50 ◇ REV MR 58 # 5075 Zbl 375 # 02009
• ID 51586

SONOBE, O. see Vol. II for further entries

SORBI, A. [1982] *Σ_0^n-equivalence relations* (**J** 0063) Studia Logica 41*351-358
◇ D55 F30 ◇ REV MR 85j:03071 Zbl 539 # 03023 • REM
The title of this article contains a misprint: Σ_0^n should be Σ_n^0
• ID 41231

SORBI, A. [1982] *Numerazioni positive, r.e. classi e formule (English summary)* (**J** 3768) Boll Unione Mat Ital, VI Ser, D 1*79-88
• REPR [1982] (**J** 2100) Boll Unione Mat Ital, VI Ser, B 1/1*403-411
◇ D20 F30 ◇ REV MR 83e:03071 MR 84c:03085 Zbl 502 # 03023 Zbl 508 # 03022 • ID 46829

SORBI, A. [1983] see BERNARDI, C.

SORBI, A. see Vol. III, IV for further entries

SORNIKOV, YA.A. [1977] *On quantifier-free ε_0-recursive arithmetic (Russian) (English summary)* (**J** 0288) Vest Ser Mat Mekh, Univ Moskva 1977/6*19-25
• TRANSL [1977] (**J** 0510) Moscow Univ Math Bull 32/6*15-20
◇ F15 F30 ◇ REV Zbl 394 # 03050 • ID 53380

SOTO-ANDRADE, J. & VARELA GARCIA, F.J. [1984] *Self-reference and fixed points: a discussion and an extension of Lawvere's theorem* (**J** 3376) Acta Appl Math 2/1*1-19
◇ F99 G30 ◇ REV MR 85h:03073 Zbl 538 # 03052
• ID 41485

SOUDIEUX, C. [1960] *De l'infini arithmetique* (**X** 0978) Schulthess: Zuerich iii+115pp
◇ B28 F30 F98 ◇ REV MR 24 # A1823 Zbl 91.50
• ID 12654

SOYKA, D. [1980] *Metamathematische Methoden in der konstruktiven Mathematik (Russian) (English and French summaries)* (**J** 0115) Wiss Z Humboldt-Univ Berlin, Math-Nat Reihe 29*433-440
◇ F50 ◇ REV MR 83f:03061 Zbl 481 # 03037 • ID 35309

SPANIER, E.H. [1966] see GINSBURG, S.

SPANIER, E.H. see Vol. IV for further entries

SPECKER, E. [1949] *Nicht konstruktiv beweisbare Saetze der Analysis* (**J** 0036) J Symb Logic 14*145-158
◇ F60 ◇ REV MR 11.151 Zbl 33.341 JSL 15.67 • ID 12658

SPECKER, E. [1957] *Eine Verschaerfung des Unvollstaendigkeitssatzes der Zahlentheorie (Russian summary)* (**J** 0014) Bull Acad Pol Sci, Ser Math Astron Phys 5*1041-1045
◇ F30 ◇ REV MR 19.934 Zbl 78.246 • ID 12665

SPECKER, E. [1959] *Der Satz vom Maximum in der rekursiven Analysis* (**P** 0634) Constructivity in Math;1957 Amsterdam 254-265
◇ F60 ◇ REV MR 21 # 6328 Zbl 88.17 JSL 40.501
• ID 12666

SPECKER, E. [1969] *The fundamental theorem of algebra in recursive analysis* (**P** 1060) Constr Aspects Fund Thm Algeb;1967 Zuerich 321-329
◇ F60 ◇ REV MR 41 # 1536 Zbl 221 # 02014 • ID 20949

SPECKER, E. [1971] *Ramsey's theorem does not hold in recursive set theory* (**P** 0638) Logic Colloq;1969 Manchester 439-442
◇ D25 D80 E05 F60 ◇ REV MR 43 # 4667
Zbl 285 # 02038 • ID 29960

SPECKER, E. see Vol. I, II, III, IV, V for further entries

SPECTOR, C. [1955] *Recursive well-orderings* (**J** 0036) J Symb Logic 20*151-163
◇ D30 D45 D55 F15 ◇ REV MR 17.570 Zbl 67.3
JSL 21.412 • ID 12672

SPECTOR, C. [1957] *Recursive ordinals and predicative set theory* (**P** 1675) Summer Inst Symb Log;1957 Ithaca 377-382
◇ D20 D55 E70 F15 F65 ◇ REV JSL 31.138 • ID 12676

SPECTOR, C. [1962] see FEFERMAN, S.

SPECTOR, C. [1962] *Provably recursive functionals of analysis: a consistency proof of analysis by an extension of principles formulated in current intuitionistic mathematics* (**P** 0613) Rec Fct Th;1961 New York 1-27
◇ F10 F35 F50 ◇ REV MR 27 # 4745 Zbl 143.255
JSL 32.128 • ID 12678

SPECTOR, C. see Vol. IV for further entries

SPENCER, J.H. [1983] *Large numbers and unprovable theorems* (**J** 0005) Amer Math Mon 90*669-675
◇ D20 F30 ◇ REV MR 85a:03073 Zbl 539 # 03042
• ID 34745

SPENCER, J.H. [1983] *Short theorems with long proofs* (**J** 0005) Amer Math Mon 90*365-366
◇ F20 ◇ REV MR 85b:03001 • ID 40518

SPENCER, J.H. see Vol. V for further entries

SPRING, D. [1981] *The Riemann integral in constructive mathematics* (**J** 0053) Proc Amer Math Soc 83*839-840
◇ F50 ◇ REV MR 82k:03097 Zbl 484 # 03034 • ID 78900

SREBRNY, M. [1975] see MAREK, W.

SREBRNY, M. [1976] see MAREK, W.

SREBRNY, M. see Vol. II, III, IV, V for further entries

STAHL, G. [1958] *Ideas generales de la logica combinatoria* (**J** 1718) Rev Mat Santiago 5*20-27
◇ B40 ◇ REV JSL 35.321 • ID 32469

STAHL, G. [1981] *The diagonal method in set theory and metamathematics (Spanish)* (**J** 0162) Teorema (Valencia) 11*27-35
◇ A05 E10 F99 ◇ REV MR 83g:03001 • ID 35983

STAHL, G. see Vol. I, II, IV, V for further entries

STANFORD, P.H. [1976] *A formalisation of the integers in a multi-successor arithmetic* (**J** 0068) Z Math Logik Grundlagen Math 22*119-121
◇ F30 ◇ REV MR 56#11774 Zbl 339#02041 • ID 65486

STANFORD, P.H. see Vol. I for further entries

STAPLES, J. [1971] *On constructive fields* (**J** 3240) Proc London Math Soc, Ser 3 23*753-768
◇ F55 ◇ REV MR 45#6587 Zbl 238#02029 • ID 13115

STAPLES, J. [1973] *Axioms for constructive fields* (**J** 0016) Bull Austral Math Soc 8*221-232
◇ F55 ◇ REV MR 48#74 Zbl 253#02028 • ID 13117

STAPLES, J. [1973] *Combinator realizability of constructive finite type analysis* (**P** 0713) Cambridge Summer School Math Log;1971 Cambridge GB 253-273
◇ B40 F35 F50 ◇ REV MR 50#63 Zbl 272#02037 • ID 13116

STAPLES, J. [1974] *Combinator realizability of constructive Morse theory* (**J** 0036) J Symb Logic 39*226-234
◇ B40 F35 F50 ◇ REV MR 51#74 Zbl 306#02023 • ID 13118

STAPLES, J. [1975] *Church-Rosser theorems for replacement systems* (**P** 0765) Algeb & Log;1974 Clayton 291-307
◇ B40 ◇ REV MR 51#10040 Zbl 306#02022 • ID 17554

STAPLES, J. [1977] *A class of replacement systems with simple optimality theory* (**J** 0016) Bull Austral Math Soc 17*335-350
◇ B40 ◇ REV MR 58#32041 Zbl 383#68062 • ID 30957

STAPLES, J. [1977] *Combinatory logics and lambda calculi as model programming languages* (**J** 0387) Bull Sect Logic, Pol Acad Sci 6*6141-6142
◇ B40 B75 ◇ ID 30961

STAPLES, J. [1977] *Optimal reductions in replacement systems* (**J** 0016) Bull Austral Math Soc 16*341-349
◇ B40 ◇ REV MR 57#11159 Zbl 357#68051 • ID 30956

STAPLES, J. [1978] *Truth in constructive metamathematics* (**J** 0047) Notre Dame J Formal Log 19*489-494
◇ F50 ◇ REV MR 57#15999 Zbl 332#02032 • ID 30955

STAPLES, J. [1979] *A graph-like lambda-calculus for which leftmost-outermost reduction is optimal* (**P** 3379) Graph-Gram & Appl to Comput Sci & Biol;1978 Bad Honnef 440-455
◇ B40 ◇ REV MR 81i:03017 Zbl 406#03017 • ID 56100

STAPLES, J. [1979] *A lambda calculus with naive substitution* (**J** 3194) J Austral Math Soc, Ser A 28*269-282
◇ B40 ◇ REV MR 81b:03015 Zbl 396#03016 • ID 53055

STAPLES, J. [1980] *Computation on graph-like expressions* (**J** 1426) Theor Comput Sci 10*171-185
◇ B40 ◇ REV MR 82c:68019a Zbl 423#68007 • ID 30958

STAPLES, J. [1980] *Optimal evaluations of graph-like expressions* (**J** 1426) Theor Comput Sci 10*297-316
◇ B40 ◇ REV MR 82c:68019b Zbl 426#68006 • ID 30959

STAPLES, J. [1980] *Speeding up subtree replacement systems* (**J** 1426) Theor Comput Sci 11*39-47
◇ B40 ◇ REV MR 82e:68019c Zbl 438#68032 • ID 30960

STAPLES, J. [1981] *Efficient combinatory reduction* (**J** 0068) Z Math Logik Grundlagen Math 27*391-402
◇ B40 ◇ REV MR 83a:03013 Zbl 467#03011 • ID 55009

STAPLES, J. [1983] *Two-level expression representation for faster evaluation* (**P** 3908) Graph-Gram & Appl to Comput Sci (2);1982 Neunkirchen 392-404
◇ B35 B40 D03 ◇ REV Zbl 522#68036 • ID 37059

STATMAN, R. [1977] *Complexity and derivations from quantifier-free Horn formulae, mechanical introduction of explicit definitions, and refinement of completeness theorems* (**P** 1075) Logic Colloq;1976 Oxford 7*505-518
◇ B35 F20 ◇ REV MR 58#170 Zbl 441#03021 • ID 16639

STATMAN, R. [1977] *Herbrand's theorem and Gentzen's notion of a direct proof* (**C** 1523) Handb of Math Logic 897-912
◇ D15 F05 F07 F20 ◇ REV MR 58#10343 JSL 49.980 • ID 27329

STATMAN, R. [1977] *The typed λ-calculus is not elementary recursive* (**P** 3572) IEEE Symp Found of Comput Sci (18);1977 Providence 90-94
◇ B40 D15 D20 ◇ REV MR 58#27356 • ID 78954

STATMAN, R. [1978] *Bounds for proof-search and speed-up in the predicate calculus* (**J** 0007) Ann Math Logic 15*225-287
◇ B35 F05 F20 ◇ REV MR 81c:03049 Zbl 411#03047 • ID 52901

STATMAN, R. [1979] see DALEN VAN, D.

STATMAN, R. [1979] *Intuitionistic propositional logic is polynomial-space complete* (**J** 1426) Theor Comput Sci 9*67-72
◇ B40 D15 F20 F50 ◇ REV MR 80m:68042 Zbl 411#03049 • ID 52903

STATMAN, R. [1979] *Lower bounds on Herbrand's theorem* (**J** 0053) Proc Amer Math Soc 75*104-107
◇ F05 F07 F20 ◇ REV MR 80h:03080 Zbl 411#03048 • ID 52902

STATMAN, R. [1979] *The typed λ-calculus is not elementary recursive* (**J** 1426) Theor Comput Sci 9*73-81
◇ B40 D15 D20 ◇ REV MR 80m:68043 Zbl 411#03050 • ID 52904

STATMAN, R. [1980] *On the existence of closed terms in the typed λ-calculus. I* (**C** 3050) Essays Combin Log, Lambda Calc & Formalism (Curry) 511-534
◇ B40 ◇ REV MR 82m:03014a Zbl 469#03006 • REM Part II 1981 • ID 78952

STATMAN, R. [1981] *On the existence of closed terms in the typed λ calculus. II: Transformations of unification problems* (**J** 1426) Theor Comput Sci 15*329-338
◇ B40 ◇ REV MR 82m:03014b Zbl 486#03011 • REM Part I 1980 • ID 78948

STATMAN, R. [1981] *Speed-up by theories with infinite models* (**J** 0053) Proc Amer Math Soc 81*465-469
◇ F20 ◇ REV MR 82c:03075 Zbl 469#03040 • ID 55167

STATMAN, R. [1982] *Completeness, invariance and λ-definability* (**J** 0036) J Symb Logic 47*17-26
◇ B40 D15 D20 ◇ REV MR 84e:03020 Zbl 487#03006 • ID 34359

STATMAN, R. [1983] *λ-definable functionals and βη conversion* (**J** 0009) Arch Math Logik Grundlagenforsch 23*21-26
◇ B40 ◇ REV MR 84m:03022 Zbl 537#03009 • ID 35801

STATMAN, R. [1985] *Logical relations and the typed λ-calculus* (**J** 0194) Inform & Control 65*85-97
◇ B40 ◇ ID 49357

STATMAN, R. see Vol. I, IV for further entries

STAVI, J. [1980] see MANEVITZ, L.M.

STAVI, J. see Vol. I, II, III, IV, V for further entries

STEEL, J.R. [1978] *Forcing with tagged trees* (**J** 0007) Ann Math Logic 15*55-74
◇ C62 D30 D55 E40 F35 ◇ REV MR 81c:03044 Zbl 404 # 03020 • ID 29156

STEEL, J.R. see Vol. III, IV, V for further entries

STEEN, S.W.P. [1972] *Mathematical logic, with special reference to the natural numbers* (**X** 0805) Cambridge Univ Pr: Cambridge, GB xvi+638pp
◇ B98 F30 F98 ◇ REV MR 57 # 9451 Zbl 275 # 02002 • ID 22386

STEGMUELLER, W. [1959] *Unvollstaendigkeit und Unentscheidbarkeit. Die metamathematischen Resultate von Goedel, Church, Kleene, Rosser und ihre erkenntnistheoretische Bedeutung* (**X** 0902) Springer: Wien iii+114pp
◇ A05 D35 F30 F98 ◇ REV MR 22 # 2538 Zbl 86.244 • REM 2nd ed. 1970; iii+114pp • ID 13145

STEGMUELLER, W. [1964] *Remarks on the completeness of logical systems relative to the validity concepts of P.Lorenzen and K.Lorenz* (**J** 0047) Notre Dame J Formal Log 5*81-112
◇ A05 F99 ◇ REV MR 31 # 60 Zbl 166.257 • ID 13148

STEGMUELLER, W. [1965] *Die Aequivalenz des klassischen und intuitionistischen Ableitungsbegriffs im Gentzen-Quine-Kalkuel und in Kleenes Kalkuel H* (**J** 0009) Arch Math Logik Grundlagenforsch 8*3-27
◇ F50 ◇ REV MR 31 # 4715 Zbl 178.306 • ID 13149

STEGMUELLER, W. see Vol. I, II, III, V for further entries

STEIGER, R. [1974] see BISHOP, P.

STEIN, M. [1979] *Interpretationen der Heyting-Arithmetik endlicher Typen* (**J** 0009) Arch Math Logik Grundlagenforsch 19*175-189
◇ F10 F35 F50 ◇ REV MR 81i:03092 Zbl 411 # 03051 • ID 52905

STEIN, M. [1980] *Interpretations of Heyting's arithmetic - An analysis by means of a language with set symbols* (**J** 0007) Ann Math Logic 19*1-31
◇ F10 F30 F50 ◇ REV MR 82i:03067 Zbl 453 # 03062 • ID 54192

STEIN, M. [1981] *A general theorem on existence theorems* (**J** 0068) Z Math Logik Grundlagen Math 27*435-452
◇ B40 F10 F35 F50 ◇ REV MR 83e:03089 Zbl 473 # 03050 • ID 55378

STEIN, M. see Vol. I for further entries

STEINER, H.-G. [1966] *Aequivalente Fassungen des Vollstaendigkeitsaxioms fuer die Theorie der reellen Zahlen* (**J** 0160) Math-Phys Sem-ber, NS 13*180-201
◇ F35 ◇ REV MR 34 # 2461 Zbl 147.42 • ID 13158

STEINER, H.-G. see Vol. I, IV for further entries

STENIUS, E. [1952] *Das Interpretationsproblem der formalisierten Zahlentheorie und ihre formale Widerspruchsfreiheit* (**J** 0991) Acta Acad Aabo, Ser B 18/3*102pp
◇ F30 ◇ REV MR 14.345 Zbl 47.14 JSL 18.262 • ID 13165

STENIUS, E. see Vol. I, II, III for further entries

STENLUND, S. [1970] *Combinators as effectively calculable functions* (**C** 0735) Logic & Value (Dahlquist) 67-74
◇ B40 D20 ◇ REV MR 58 # 21490 • ID 78994

STENLUND, S. [1971] *Introduction to combinatory logic* (**X** 0882) Univ Filos Foeren: Uppsala 104pp
◇ B40 ◇ REV MR 58 # 21491 Zbl 315 # 02032 • REM Filosofiska Studier No.11 • ID 29808

STENLUND, S. [1972] *Combinators, λ-terms and proof theory* (**X** 0835) Reidel: Dordrecht 184pp
◇ B40 F05 F10 F50 F98 ◇ REV MR 58 # 21492 Zbl 248 # 02032 • ID 23494

STENLUND, S. [1975] *Description in intuitionistic logic* (**P** 0757) Scand Logic Symp (3);1973 Uppsala 197-212
◇ F50 ◇ REV MR 52 # 7857 Zbl 322 # 02022 JSL 43.373 • ID 15180

STENLUND, S. see Vol. II for further entries

STEPANOV, V.A. [1985] *Propositional logic of reflexive sentences. II (Russian)* (**X** 2235) VINITI: Moskva 4276-85
◇ B05 F30 ◇ REM Part I 1984 • ID 46515

STEPANOV, V.A. see Vol. I for further entries

STEPANOV, V.I. [1981] *Self-referential propositions for standard and Rosser provability predicates (Russian)* (**J** 0067) Usp Mat Nauk 36/1*227-228
◇ F30 ◇ REV MR 83j:03101 Zbl 495 # 03037 Zbl 509 # 03032 • ID 35384

STEPANOV, V.I. [1982] *Second-order arithmetic and the consistency of first-order theories (Russian)* (**J** 0067) Usp Mat Nauk 37/4*179-180
◇ F30 F35 ◇ REV MR 83j:03098 Zbl 509 # 03032 • ID 35383

STEPANOV, V.I. [1983] *On model theory for intuitionistic logic (Russian)* (**J** 0142) Mat Sb, Akad Nauk SSSR, NS 120*227-239,288
• TRANSL [1984] (**J** 0349) Math of USSR, Sbor 48*223-235
◇ C90 F50 ◇ REV MR 85b:03058 Zbl 568 # 03018 • ID 39758

STEPANOV, V.I. [1985] *The incompleteness theorems and related results for nonconstructive theories* (**P** 4180) Int Congr Log, Meth & Phil of Sci (7,Pap);1983 Salzburg 233-243
◇ F30 ◇ ID 48126

STIGT VAN, W.P. [1979] *L.E.J.Brouwer: intuitionism and topology* (**P** 1947) Bicent Congr Wisk Genootschap;1978 Amsterdam 2*359-374
◇ A10 F99 ◇ REV MR 81d:01023 Zbl 413 # 01008 • ID 83055

STIGT VAN, W.P. [1979] *The rejected parts of Brouwer's dissertation on the foundations of mathematics* (**J** 1648) Hist Math 6*385-404
◇ A10 F99 ◇ REV MR 80k:01048 Zbl 426 # 01006 • ID 83056

STIGT VAN, W.P. [1982] *L.E.J.Brouwer, the signific interlude* (**P** 3638) Brouwer Centenary Symp;1981 Noordwijkerhout 505-512
◇ A05 A10 F99 ◇ REV MR 85d:01026 • ID 38942

STOLZENBERG, G. [1968] *Constructive normalization of an algebraic variety* (**J** 0015) Bull Amer Math Soc 74*595-599
◇ F55 ◇ REV MR 37 # 201 • ID 41692

Stolzenberg, G. [1970] *Review: Foundations of constructive analysis by Errett Bishop* (**J** 0015) Bull Amer Math Soc 76*301-323
◇ A05 F55 ◇ rev MR 41 # 2337 • id 13198

Stone, M.H. [1937] *Topological representations of distributive lattices and Brouwerian logics* (**J** 0086) Cas Pestovani Mat, Ceskoslov Akad Ved 67*1-25
◇ F50 G10 ◇ rev Zbl 18.3 JSL 3.90 FdM 63.830
• id 13212

Stone, M.H. see Vol. I for further entries

Stout, L.N. [1975] *Topological space objects in a topos. II: 𝔈-completeness and 𝔈-cocompleteness* (**J** 0504) Manuscr Math 17*1-14
◇ E75 F50 G30 ◇ rev MR 54 # 2751 Zbl 319 # 18008
• id 24059

Stout, L.N. [1976] *Topological properties of the real numbers object in a topos* (**J** 0332) Math Proc Cambridge Phil Soc 17*295-326
◇ E70 E75 F50 G30 ◇ rev MR 54 # 10015
Zbl 354 # 18010 • id 25859

Stout, L.N. [1978] *Quels sont les espaces topologiques dans les topos?* (**J** 2660) Ann Sci Math Quebec 2*123-141
◇ F50 G30 ◇ rev MR 80d:18007 Zbl 394 # 18005
• id 52544

Stout, L.N. see Vol. II, V for further entries

Stoy, J.E. [1981] *Denotational semantics: the Scott-Strachey approach to programming language theory* (**X** 0865) MIT Pr: Cambridge, MA xxx + 414pp
◇ B40 B75 ◇ rev MR 82m:68003 Zbl 503 # 68059
• id 82900

Stralberg, A.H. [1965] *On the reduction of classical logic. An extension of some theorems of Glivenko and Goedel* (**J** 0121) Kon Norske Vidensk Selsk Forh 38*94-99
◇ B55 F30 F50 ◇ rev MR 33 # 2536 Zbl 148.244
• id 13218

Strauch, O. [1972] *Minimal covering of a closed interval (Slovak) (English summary)* (**J** 0128) Acta Math Univ Comenianae (Bratislava) 28*1-15
◇ E75 F60 ◇ rev MR 47 # 3603 Zbl 251 # 54012
• id 13221

Strauch, O. see Vol. V for further entries

Strauss, P. [1985] *Number-theoretic set theories* (**J** 0047) Notre Dame J Formal Log 26*81-95
◇ E70 F35 ◇ rev MR 86f:03096 Zbl 558 # 03025
• id 42588

Strauss, P. see Vol. I for further entries

Streinu, I. [1983] *Grammar directed Goedel numberings* (**J** 0382) Int J Comput Math 14*223-237
◇ F40 ◇ rev MR 85b:68029 Zbl 517 # 03023 • id 37288

Suetuna, Z. [1951] *Ueber die Grundlagen der Mathematik* (**J** 0090) J Math Soc Japan 3*59-68
◇ A05 F65 ◇ rev MR 13.310 Zbl 43.245 JSL 21.94
• rem Part I. Part II 1951 • id 12711

Sundholm, G. [1981] *Hacking's logic* (**J** 0301) J Phil 78*160-168
◇ F05 F55 ◇ rev JSL 47.689 • id 47404

Sundholm, G. [1983] *Constructions, proofs and the meaning of logical constants* (**J** 0122) J Philos Logic 12*151-172
◇ A05 F50 ◇ rev MR 85h:03027 Zbl 539 # 03038
• id 41240

Sundholm, G. see Vol. II, III for further entries

Sureshkumar [1977] *A note on Ashvinikumar's paper:"The intuitionist contradictority of certain classical set-theoretic results"* (**J** 3077) Nieuw Arch Wisk, Ser 3 25*40-47
◇ E70 E75 F55 ◇ rev MR 56 # 15374 Zbl 358 # 02036
• rem The article was published ibid 17(1969)*218-222
• id 79102

Surma, S.J. [1971] *Jaskowski's matrix criterion for the intuitionistic propositional calculus (Polish summary)* (**S** 0458) Zesz Nauk, Prace Log, Uniw Krakow 6*21-54
◇ F50 ◇ rev MR 45 # 3165 • id 12757

Surma, S.J. [1973] *Jaskowski's matrix criterion for the intuitionistic propositional calculus* (**C** 2084) Stud Hist of Math Log 87-121
◇ B55 F50 ◇ id 42072

Surma, S.J. & Wronski, A. & Zachorowski, S. [1974] *On Jaskowski-type semantics for the intuitionistic propositional logic* (**J** 0387) Bull Sect Logic, Pol Acad Sci 3/2*6-10
◇ B55 F50 ◇ rev MR 53 # 98 Zbl 341 # 02016 • id 16561

Surma, S.J. & Wronski, A. & Zachorowski, S. [1975] *On Jaskowski-type semantics for the intuitionistic propositional logic* (**J** 0063) Studia Logica 34*145-148
◇ B55 F50 ◇ rev MR 53 # 12885 Zbl 341 # 02016
• id 23141

Surma, S.J. see Vol. I, II, III, V for further entries

Suszko, R. [1974] *A note on intuitionistic calculus (ISC)* (**J** 0387) Bull Sect Logic, Pol Acad Sci 3/1*20-21
◇ F50 G05 ◇ rev MR 53 # 5269 • id 22900

Suszko, R. see Vol. I, II, III, V for further entries

Suvorov, P.Yu. [1979] *Representation of proofs by colored graphs and the Hadwiger conjecture (Russian)* (**S** 0228) Zap Nauch Sem Leningrad Otd Mat Inst Steklov 88*209-217,246
• transl [1982] (**J** 1531) J Sov Math 20*2376-2381
◇ B10 F07 ◇ rev MR 81g:05060 Zbl 494 # 05022
• id 36649

Suvorov, P.Yu. see Vol. I for further entries

Svejdar, V. [1978] *Degrees of interpretability* (**J** 0140) Comm Math Univ Carolinae (Prague) 19*789-813
◇ F25 F30 G10 ◇ rev MR 80e:03072 Zbl 407 # 03020
• id 56186

Svejdar, V. [1981] *A sentence that is difficult to interpret* (**J** 0140) Comm Math Univ Carolinae (Prague) 22*661-666
◇ E30 F25 ◇ rev MR 83f:03058 Zbl 484 # 03032
• id 35308

Svejdar, V. [1983] *Modal analysis of generalized Rosser sentences* (**J** 0036) J Symb Logic 48*986-999
◇ B45 F25 F30 ◇ rev MR 85k:03041 Zbl 543 # 03010
• id 40906

Svenonius, L. [1979] *Two kinds of extensions of primitive recursive arithmetic* (**P** 1705) Scand Logic Symp (4);1976 Jyvaeskylae 49-94
◇ D20 F30 F35 ◇ rev MR 81e:03058 Zbl 399 # 03047
• id 52843

SVENONIUS, L. see Vol. I, III for further entries

SWARD, G.L. [1971] *Transfinite sequences of axiom systems for set theory* (**P** 0693) Axiomatic Set Th;1967 Los Angeles 1*429-438
◇ E30 E65 F15 ◇ REV MR 44 # 6479 Zbl 225 # 02035 • ID 12803

SWARD, G.L. see Vol. I for further entries

SWART DE, H.C.M. [1976] *Another intuitionistic completeness proof* (**J** 0036) J Symb Logic 41*644-662
◇ F50 ◇ REV MR 58 # 5176 Zbl 342 # 02013 • ID 14587

SWART DE, H.C.M. [1976] *Elements of intuitionistic analysis I: Rolle's theorem and complete totally bounded, metric spaces. II: The Stone-Weierstrass theorem and Ascoli's theorem* (**J** 0068) Z Math Logik Grundlagen Math 22*289-298,501-508
◇ F55 ◇ REV MR 57 # 12189 Zbl 362 # 02021 Zbl 362 # 02022 • ID 18446

SWART DE, H.C.M. [1977] *An intuitionistically plausible interpretation of intuitionistic logic* (**J** 0036) J Symb Logic 42*564-578
◇ C90 F50 ◇ REV MR 58 # 27354 Zbl 383 # 03040 • ID 26858

SWART DE, H.C.M. [1978] *First steps in intuitionistic model theory* (**J** 0036) J Symb Logic 43*3-12
◇ C90 F50 ◇ REV MR 80c:03059 Zbl 395 # 03037 • ID 29237

SWART DE, H.C.M. [1981] see GIELEN, W.

SWART DE, H.C.M. [1981] see POSY, C.J.

SWART DE, H.C.M. see Vol. II, III, V for further entries

SZABO, M.E. [1969] *Introduction* (**C** 1409) Gentzen: Collected Papers 1-28
◇ A10 F05 F30 F98 ◇ REV MR 41 # 6660 Zbl 209.300 JSL 37.752 • ID 45629

SZABO, M.E. [1974] *A categorical equivalence of proofs* (**J** 0047) Notre Dame J Formal Log 15*177-191 • ERR/ADD ibid 17*78
◇ F05 F07 F50 G05 G30 ◇ REV MR 50 # 6805 MR 52 # 5379 Zbl 225 # 02023 • ID 29700

SZABO, M.E. [1975] *Polycategories* (**J** 0394) Commun Algeb 3*663-689
◇ F07 G30 ◇ REV MR 51 # 10046 Zbl 353 # 18008 • ID 17550

SZABO, M.E. [1976] *Quantifier complete categories* (**J** 0326) J Pure Appl Algebra 7*97-114
◇ F05 F50 G10 G30 ◇ REV MR 53 # 2648 Zbl 359 # 02028 • ID 21532

SZABO, M.E. [1977] *The logic of closed categories* (**J** 0047) Notre Dame J Formal Log 18*441-457
◇ F50 G30 ◇ REV MR 58 # 5093 Zbl 258 # 02038 • ID 23639

SZABO, M.E. [1978] *Algebra of proofs* (**S** 3303) Stud Logic Found Math 88*xii+297pp
◇ F05 F07 F50 F98 G30 ◇ REV MR 80b:03097 Zbl 532 # 03030 JSL 47.904 • ID 31249

SZABO, M.E. [1983] *A sequent calculus for Kroeger logic* (**P** 3830) Logics of Progr & Appl;1980 Poznan 295-303
◇ B75 F05 F07 ◇ REV MR 84i:68025 Zbl 541 # 68011 • ID 40144

SZABO, M.E. see Vol. I, II, IV for further entries

SZCZERBA, L.W. [1976] see PRAZMOWSKI, K.

SZCZERBA, L.W. [1977] *Interpretability of elementary theories* (**P** 1704) Int Congr Log, Meth & Phil of Sci (5);1975 London ON 1*129-145
◇ B30 F25 ◇ REV MR 57 # 16040 Zbl 375 # 02005 • ID 51583

SZCZERBA, L.W. [1978] see SETTE, A.-M.

SZCZERBA, L.W. [1979] *Interpretability and axiomatizability* (**J** 3293) Bull Acad Pol Sci, Ser Math 27*425-429
◇ B30 F25 ◇ REV MR 81f:03017 Zbl 445 # 03013 • ID 56477

SZCZERBA, L.W. [1979] see EPSTEIN, R.L.

SZCZERBA, L.W. [1980] *Interpretations with parameters* (**J** 0068) Z Math Logik Grundlagen Math 26*35-39
◇ B25 C07 C35 C45 F25 ◇ REV MR 81k:03062 Zbl 444 # 03017 • ID 79172

SZCZERBA, L.W. [1983] *Interpretations* (**J** 0387) Bull Sect Logic, Pol Acad Sci 12*208-213
◇ B30 C07 F25 ◇ ID 39510

SZCZERBA, L.W. [1984] *Some remarks on interpretations with parameters* (**P** 3621) Frege Konferenz (2);1984 Schwerin 194-199
◇ F25 ◇ ID 45380

SZCZERBA, L.W. [1985] see SETTE, A.-M.

SZCZERBA, L.W. see Vol. I, III, IV for further entries

SZMIELEW, W. & TARSKI, A. [1952] *Mutual interpretability of some essentially undecidable theories* (**P** 0593) Int Congr Math (II, 6);1950 Cambridge MA 734
◇ D35 F25 ◇ ID 12847

SZMIELEW, W. see Vol. I, III, IV, V for further entries

TABAKOV, M. [1984] *Goedel's theorem in retrospect* (**J** 0387) Bull Sect Logic, Pol Acad Sci 13*132-136
◇ A05 F99 ◇ REV MR 86g:03098 Zbl 561 # 03029 • ID 44647

TABAKOV, M. see Vol. I for further entries

TAIT, W.W. [1960] see GANDY, R.O.

TAIT, W.W. [1961] see GANDY, R.O.

TAIT, W.W. [1963] *A second order theory of functionals of higher type* (**C** 4220) Rep Sem Found Anal II*5.1-5.20
◇ F10 F35 F50 ◇ ID 49826

TAIT, W.W. [1963] *Appendix A: Intensional functionals* (**C** 4220) Rep Sem Found Anal 2*A1-3
◇ F10 F35 F50 ◇ ID 49831

TAIT, W.W. [1963] *Appendix B: An interpretation of functionals by convertible terms* (**C** 4220) Rep Sem Found Anal 2*B1-10
◇ F05 F35 F50 ◇ ID 49832

TAIT, W.W. [1965] *Functionals defined by transfinite recursion* (**J** 0036) J Symb Logic 30*155-174
◇ E30 F10 F15 F35 ◇ REV MR 34 # 1161 Zbl 133.252 JSL 31.509 • ID 13282

TAIT, W.W. [1965] *Infinitely long terms of transfinite type* (**P** 0688) Logic Colloq;1963 Oxford 176-185
◇ C75 F10 F15 F50 ◇ REV MR 33 # 3925 Zbl 154.5 JSL 40.623 • ID 13281

TAIT, W.W. [1965] *The substitution method* (J 0036) J Symb Logic 30*175-192
◇ F10 F15 F30 F35 ◇ REV MR 34 # 1162 Zbl 133.253 JSL 38.660 • ID 13280

TAIT, W.W. [1966] *A nonconstructive proof of Gentzen's Hauptsatz for second order predicate logic* (J 0015) Bull Amer Math Soc 72*980-983
◇ B15 F05 F35 ◇ REV MR 34 # 5655 Zbl 199.8 JSL 33.289 • ID 13283

TAIT, W.W. [1967] *Intensional interpretation of functionals of finite type I* (J 0036) J Symb Logic 32*198-212
◇ F10 F30 F35 F50 ◇ REV MR 36 # 2483 Zbl 174.12 JSL 40.624 • ID 13284

TAIT, W.W. [1968] *Constructive reasoning* (P 0627) Int Congr Log, Meth & Phil of Sci (3,Proc);1967 Amsterdam 185-199
◇ B40 F10 F15 F30 F35 F50 F65 ◇ REV MR 41 # 1533 Zbl 191.308 • ID 13285

TAIT, W.W. [1968] *Normal derivability in classical logic* (P 0637) Syntax & Semant Infinitary Lang;1967 Los Angeles 204-236
◇ C75 F05 F15 F30 F35 F60 ◇ REV Zbl 206.5 • ID 13286

TAIT, W.W. [1970] *Applications of the cut elimination theorem to some subsystems of classical analysis* (P 0603) Intuitionism & Proof Th;1968 Buffalo 475-488
◇ F05 F35 ◇ REV MR 44 # 6462 Zbl 231 # 02033 JSL 40.501 • ID 21249

TAIT, W.W. [1971] *Normal form theorem for bar recursive functions of finite type* (P 0604) Scand Logic Symp (2);1970 Oslo 353-367
◇ F05 F10 F35 F50 ◇ REV MR 55 # 7747 Zbl 281 # 02036 • ID 29572

TAIT, W.W. [1975] *A realizability interpretation of the theory of species* (C 0758) Logic Colloq Boston 1972-73 240-251
◇ D65 F05 F35 F50 ◇ REV MR 52 # 5380 Zbl 328 # 02014 • ID 18413

TAIT, W.W. [1981] *Finitism* (J 0301) J Phil 78*524-546
◇ A05 F65 ◇ ID 41689

TAIT, W.W. [1983] *Against intuitionism: constructive mathematics is part of classical mathematics* (J 0122) J Philos Logic 12*173-195
◇ A05 F99 ◇ REV MR 85c:03002 Zbl 574 # 03002 • ID 40474

TAIT, W.W. see Vol. I, III, IV for further entries

TAKAHASHI, A. [1972] *Mathematical concepts I,II,III,IV,V (Spanish)* (J 0377) Bol Mat (Bogota) 6/1*23-42,6/2*36-51,6/4*33-47,6/5*33-45,6/6*59-71
◇ A10 E98 F98 F99 ◇ REV MR 51 # 5231 • ID 45445

TAKAHASHI, A. see Vol. V for further entries

TAKAHASHI, MOTO-O [1967] *A proof of cut-elimination in simple type theory* (J 0090) J Math Soc Japan 19*399-410
◇ F05 F35 ◇ REV MR 36 # 2491 Zbl 206.275 JSL 39.607 • ID 13291

TAKAHASHI, MOTO-O [1968] *On simple type theory (Japanese)* (J 0091) Sugaku 20*129-141
◇ B15 F05 F35 ◇ REV MR 40 # 2535 • ID 13295

TAKAHASHI, MOTO-O [1968] *Simple type theory of Gentzen style with the inference of extensionality* (J 0081) Proc Japan Acad 44*43-45
◇ F05 F35 ◇ REV MR 40 # 5442 Zbl 223 # 02022 • ID 13294

TAKAHASHI, MOTO-O [1970] *A system of simple type theory of Gentzen style with inference on extensionality, and the cut-elimination in it* (J 0407) Comm Math Univ St Pauli (Tokyo) 18*129-147
◇ F05 F35 ◇ REV MR 43 # 44 Zbl 223 # 02021 • ID 13297

TAKAHASHI, MOTO-O [1970] *A theorem on the second order arithmetic with the ω-rule* (J 0090) J Math Soc Japan 22*15-24
◇ F35 ◇ REV MR 40 # 7086 • ID 13298

TAKAHASHI, MOTO-O [1971] *Cut-elimination theorem and Brouwerian-valued models for intuitionistic type theory* (J 0407) Comm Math Univ St Pauli (Tokyo) 19*55-72
◇ F05 F35 F50 ◇ REV MR 48 # 1889 Zbl 233 # 02011 • ID 22300

TAKAHASHI, MOTO-O [1977] *A foundation of finite mathematics* (J 0390) Publ Res Inst Math Sci (Kyoto) 12*577-708
◇ F30 ◇ REV MR 56 # 101 Zbl 357 # 02033 • ID 50381

TAKAHASHI, MOTO-O [1977] *An abstract form of Goedel's theorem on consistency and Loeb's* (J 0350) Sci Rep Tokyo Kyoiku Daigaku Sect A 13*233-237
◇ F30 ◇ REV MR 57 # 5700 Zbl 371 # 02007 • ID 51368

TAKAHASHI, MOTO-O see Vol. II, III, IV, V for further entries

TAKAMATSU, T. [1974] *Elements of linear algebra (Japanese)* (X 1726) Bunkashobo-Hakubu-Sha: Tokyo 246pp
◇ F50 ◇ ID 32492

TAKAMATSU, T. see Vol. I, II for further entries

TAKANO, M. [1977] *On Goedel's primitive recursive functionals (Japanese)* (J 0091) Sugaku 29*289-298
◇ F10 ◇ REV MR 82j:03074 • ID 79203

TAKANO, M. [1985] *Completeness of a cut-free calculus with equality and function constants* (J 0009) Arch Math Logik Grundlagenforsch 25*37-41
◇ F05 ◇ ID 49131

TAKANO, M. see Vol. II, V for further entries

TAKEUTI, G. [1952] *A metamathematical theorem on the theory of ordinal numbers* (J 0090) J Math Soc Japan 4*146-165
◇ E10 F15 ◇ REV MR 14.714 Zbl 48.36 JSL 24.62 • ID 13306

TAKEUTI, G. [1953] *On a generalized logic calculus* (J 2307) Japan J Math 23*39-96 • ERR/ADD ibid 24*149-156
◇ B15 F35 ◇ REV MR 17.701 Zbl 53.202 JSL 22.351 • ID 37768

TAKEUTI, G. [1955] *On the fundamental conjecture of GLC I,II* (J 0090) J Math Soc Japan 7*249-275,394-408
◇ B15 F05 F35 ◇ REV MR 18.1 Zbl 86.7 JSL 24.62 • REM Parts III,IV 1956 • ID 28446

TAKEUTI, G. [1956] *A metamathematical theorem on functions* (J 0090) J Math Soc Japan 8*65-78
◇ B15 F05 F10 F35 ◇ REV Zbl 86.7 JSL 24.65 • ID 13308

TAKEUTI, G. [1956] *Construction of ramified real numbers* (**J** 0260) Ann Jap Ass Phil Sci 1*41-61
◇ B15 F05 F35 F65 ◇ REV MR 18.271 Zbl 75.233 JSL 22.352 • ID 13307

TAKEUTI, G. [1956] *On the fundamental conjecture of GLC III,IV* (**J** 0090) J Math Soc Japan 8*54-64,145-155
◇ B15 F05 F35 ◇ REV MR 18.1 Zbl 75.233 JSL 24.62 • REM Parts I,II 1955. Part V 1958 • ID 28447

TAKEUTI, G. [1957] *On the theory of ordinal numbers* (**J** 0090) J Math Soc Japan 9*93-113 • ERR/ADD ibid 12*127
◇ B28 E10 E30 E35 E45 F05 ◇ REV MR 19.237 Zbl 84.8 JSL 24.67 • REM Part I. Part II 1958 • ID 19761

TAKEUTI, G. [1957] *On Skolem's theorem* (**J** 0090) J Math Soc Japan 9*71-76
◇ C07 F25 ◇ REV MR 19.4 Zbl 86.8 JSL 24.66 • ID 13310

TAKEUTI, G. [1957] *Ordinal diagrams* (**J** 0090) J Math Soc Japan 9*386-394
◇ F07 F15 ◇ REV MR 20 # 6346 Zbl 79.244 JSL 24.64 • REM Part I. Part II 1960 • ID 13309

TAKEUTI, G. [1957] *Remark on my paper: On Skolem's theorem* (**J** 0090) J Math Soc Japan 9*192-194
◇ C07 F25 F30 ◇ REV MR 19.829 Zbl 105.6 JSL 24.66 • REM The paper was published ibid 9*71-76 • ID 13311

TAKEUTI, G. [1958] *On the formal theory of the ordinal diagrams* (**J** 0260) Ann Jap Ass Phil Sci 1*151-170
◇ F15 F35 ◇ REV MR 23 # A3081 Zbl 75.234 JSL 24.65 • ID 13314

TAKEUTI, G. [1958] *On the fundamental conjecture of GLC V* (**J** 0090) J Math Soc Japan 10*121-134
◇ B15 F05 F35 ◇ REV MR 20 # 6348 Zbl 75.233 JSL 24.62 • REM Parts III,IV 1956. Part VI 1961 • ID 28448

TAKEUTI, G. [1958] *Remark on the fundamental conjecture of GLC* (**J** 0090) J Math Soc Japan 10*44-45
◇ B15 F05 F35 ◇ REV MR 20 # 6347 Zbl 75.234 • ID 13312

TAKEUTI, G. [1960] *An example on the fundamental conjecture of GLC* (**J** 0090) J Math Soc Japan 12*238-242
◇ B15 F05 F35 ◇ REV MR 24 # A3056 Zbl 75.234 JSL 28.173 • ID 13316

TAKEUTI, G. [1960] *Ordinal diagrams II* (**J** 0090) J Math Soc Japan 12*385-391
◇ F15 ◇ REV MR 26 # 1267 Zbl 97.246 JSL 29.146 • REM Part I 1957 • ID 13317

TAKEUTI, G. [1961] see MAEHARA, S.

TAKEUTI, G. [1961] *On the fundamental conjecture of GLC VI* (**J** 0081) Proc Japan Acad 37*440-443
◇ B15 F05 F35 ◇ REV MR 25 # 1995 Zbl 106.237 JSL 29.147 • REM Part V 1958 • ID 13321

TAKEUTI, G. [1961] *On the inductive definition with quantifiers of second order* (**J** 0090) J Math Soc Japan 13*333-341
◇ B15 F15 F35 ◇ REV MR 26 # 1269 Zbl 109.6 JSL 29.147 • ID 13320

TAKEUTI, G. [1961] *Remarks on the truth definition* (**J** 0090) J Math Soc Japan 13*207-209
◇ F25 F30 ◇ REV MR 24 # A679 Zbl 126.11 JSL 27.110 • ID 13323

TAKEUTI, G. [1962] see KINO, A.

TAKEUTI, G. [1963] *A remark on Gentzen's paper "Beweisbarkeit und Unbeweisbarkeit von Anfangsfaellen der transfiniten Induktion in der reinen Zahlentheorie" I-II* (**J** 0081) Proc Japan Acad 39*263-269
◇ F15 F35 ◇ REV MR 27 # 3542 MR 27 # 3542 Zbl 126.11 JSL 29.58 • REM The article was published in J0043 119(1943)*140-161 • ID 13327

TAKEUTI, G. [1963] see KINO, A.

TAKEUTI, G. [1967] *Consistency proofs of subsystems of classical analysis* (**J** 0120) Ann of Math, Ser 2 86*299-348
◇ F05 F15 F35 ◇ REV MR 36 # 2492 Zbl 159.9 • ID 13331

TAKEUTI, G. & YASUGI, M. [1968] *Reflection principles of subsystems of analysis* (**P** 0608) Logic Colloq;1966 Hannover 255-273
◇ F15 F35 ◇ REV MR 39 # 3970 Zbl 186.10 JSL 40.504 • ID 22204

TAKEUTI, G. [1968] *The Π^1_1-comprehension schema and ω-rules* (**P** 0692) Summer School in Logic;1967 Leeds 303-331
◇ F05 F15 F35 ◇ REV MR 40 # 7087 Zbl 195.301 • ID 13333

TAKEUTI, G. [1971] see MAEHARA, S.

TAKEUTI, G. & YASUGI, M. [1973] *The ordinals of the systems of second order arithmetic with the provably Δ^1_2-comprehension axiom and with the Δ^1_2-comprehension axiom respectively* (**J** 2307) Japan J Math 41*1-67
◇ F05 F15 F35 ◇ REV MR 58 # 10344 Zbl 287 # 02016 JSL 48.877 • ID 30529

TAKEUTI, G. [1974] see KREISEL, G.

TAKEUTI, G. [1975] *Consistency proofs and ordinals* (**P** 1440) ⊧ ISILC Proof Th Symp (Schuette);1974 Kiel 500*365-369
◇ A05 F15 ◇ REV MR 53 # 5272 Zbl 325 # 02020 • ID 22903

TAKEUTI, G. [1975] *Proof theory* (**X** 0809) North Holland: Amsterdam vi + 372pp
• TRANSL [1978] (**X** 0885) Mir: Moskva 412pp
◇ F05 F07 F30 F35 F98 ◇ REV MR 58 # 27366ab Zbl 354 # 02027 Zbl 355 # 02023 • ID 27569

TAKEUTI, G. & YASUGI, M. [1976] *Fundamental sequences of ordinal diagrams* (**J** 0407) Comm Math Univ St Pauli (Tokyo) 25*1-80
◇ F15 ◇ REV MR 54 # 12492 Zbl 357 # 02026 • ID 50374

TAKEUTI, G. [1978] *Two applications of logic to mathematics* (**X** 3552) Iwanami Shoten: Tokyo viii + 137pp
• LAST ED [1978] (**X** 0857) Princeton Univ Pr: Princeton viii + 137pp
◇ B98 C90 E40 E75 F05 F30 F35 G12 ◇ REV MR 58 # 21591 Zbl 393 # 03027 • ID 52447

TAKEUTI, G. & TITANI, S. [1979] *Heyting valued set theory (Japanese)* (**P** 4108) Found of Math;1979 Kyoto 1-24
◇ E40 E70 F50 G30 ◇ ID 47655

TAKEUTI, G. & YASUGI, M. [1981] *An accessibility proof of ordinal diagrams* (**J** 0090) J Math Soc Japan 33*1-21
◇ F15 ◇ REV MR 82d:03092 Zbl 472 # 03045 • ID 55309

TAKEUTI, G. [1981] *Logic and set theory* (C 2617) Modern Log Survey 167-171
◇ B51 B98 E40 E70 F50 ◇ REV MR 82f:03002 Zbl 464 # 03001 • ID 39565

TAKEUTI, G. & TITANI, S. [1984] *Intuitionistic fuzzy logic and intuitionistic fuzzy set theory* (J 0036) J Symb Logic 49*851-866
◇ E70 F99 ◇ REV MR 86f:03043 Zbl 575 # 03015 • ID 39577

TAKEUTI, G. [1985] *Proof theory and set theory* (J 0154) Synthese 62*255-263
◇ E55 F15 F98 ◇ ID 39568

TAKEUTI, G. see Vol. I, II, III, IV, V for further entries

TALYSHLY, S.M. [1973] *The problem of inverse logical operations (Russian)* (C 1662) Teor Log Vyvoda 249-252
◇ F07 ◇ REV Zbl 263 # 02009 • ID 29860

TANAKA, H. [1965] see NISHIMURA, T.

TANAKA, H. [1974] *Some analytical rules of inference in the second-order arithmetic* (J 0407) Comm Math Univ St Pauli (Tokyo) 23/1*71-81
◇ C62 D55 F35 ◇ REV MR 53 # 10559 Zbl 357 # 02012 • ID 23068

TANAKA, H. see Vol. I, II, IV, V for further entries

TANG, A. [1978] see SCIORE, E.

TANG, A. [1979] *Chain properties in Pω* (J 1426) Theor Comput Sci 9*153-172
◇ B40 B75 E20 ◇ REV MR 80j:68009 Zbl 471 # 68009 • ID 69945

TANG, A. see Vol. IV, V for further entries

TANG, TONGGAO [1981] *A note on Herbrand's theorem (Chinese) (English summary)* (J 2754) Huazhong Gongxueyuan Xuebao 9/5*1-4
◇ B10 F05 ◇ REV MR 84d:03017 • ID 34057

TANG, TONGGAO see Vol. II, IV for further entries

TANNY, S. [1973] see METROPOLIS, N.

TARSKI, A. [1931] *Sur les ensembles definissables de nombres reels I* (J 0027) Fund Math 17*210-239
• TRANSL [1956] (C 1159) Tarski: Logic, Semantics, Metamathematics 110-142 (English)
◇ B25 B28 C10 C40 C60 C65 D55 E15 E47 F35 ◇ REV Zbl 75.4 JSL 34.99 FdM 57.60 • ID 16924

TARSKI, A. [1932] *Der Wahrheitsbegriff in den Sprachen der deduktiven Disziplinen* (J 0931) Anz Oesterr Akad Wiss, Math-Nat Kl 69*23-25
◇ A05 B30 C07 C40 F30 ◇ REV Zbl 4.1 FdM 58.997 • REM Summary. The full article appeared in 1933 • ID 13406

TARSKI, A. [1933] *Einige Betrachtungen ueber die Begriffe der ω-Widerspruchsfreiheit und der ω-Vollstaendigkeit* (J 0124) Monatsh Math-Phys 40*97-112
• TRANSL [1956] (C 1159) Tarski: Logic, Semantics, Metamathematics 279-295 (English)
◇ B28 C62 F30 ◇ REV MR 17.1171 Zbl 7.97 JSL 34.99 FdM 59.53 • ID 13407

TARSKI, A. [1933] *On the notion of truth in reference to formalized deductive sciences (Polish)* (J 0459) C R Soc Sci Lett Varsovie Cl 3 34*vii+116pp
• TRANSL [1935] (J 4716) Stud Philos, Leopolis (Poznan) 1*261-405 (German) [1956] (C 1159) Tarski: Logic, Semantics, Metamathematics 152-278 (English) [1961] (C 0769) L'Antinom Ment nel Pensiero Contemp 391-677 (Italian)
◇ A05 B15 B30 C07 C40 F30 ◇ REV Zbl 13.289 Zbl 4.1 Zbl 75.7 JSL 34.99 FdM 58.997 FdM 62.1051 • REM Polish summary in J1093, 1930/31, 210-211; German summary in J0931, 1932 • ID 28816

TARSKI, A. [1938] *Der Aussagenkalkuel und die Topologie* (J 0027) Fund Math 31*103-134
• TRANSL [1956] (C 1159) Tarski: Logic, Semantics, Metamathematics 421-454
◇ B05 F50 G05 G10 ◇ REV Zbl 20.337 JSL 34.100 JSL 4.26 FdM 64.928 • ID 13420

TARSKI, A. [1939] *On undecidable statements in enlarged systems of logic and the concept of truth* (J 0036) J Symb Logic 4*105-112
◇ B10 F30 ◇ REV MR 1.34 Zbl 21.385 JSL 5.115 FdM 65.20 • ID 13428

TARSKI, A. [1952] see SZMIELEW, W.

TARSKI, A. [1953] *A general method in proofs of undecidability* (C 4472) Undecidable Th 3-35
◇ D35 F25 ◇ REV JSL 24.167 • ID 42175

TARSKI, A. [1953] see MOSTOWSKI, ANDRZEJ

TARSKI, A. also published under the name TAJTELBAUM, A.

TARSKI, A. see Vol. I, II, III, IV, V for further entries

TAUTS, A. [1972] *A connection between generalized Beth models and topological pseudo-Boolean algebras (Russian) (Estonian and German summaries)* (S 3468) Tr Mat & Mekh (Tartu) 12(305)*3-8
◇ C90 F50 G10 ◇ REV MR 51 # 7862 Zbl 321 # 02013 • ID 65673

TAUTS, A. [1972] *A semantic interpretation of formulas in generalized Beth models and in pseudo-boolean algebras (Russian) (Estonian and German summaries)* (S 3468) Tr Mat & Mekh (Tartu) 12(305)*9-20
◇ C90 F50 G10 ◇ REV MR 51 # 7863 Zbl 321 # 02014 • ID 17695

TAUTS, A. [1974] *A formal deduction of tautological formulas in pseudo-boolean algebras (Russian) (Estonian and German summaries)* (S 3468) Tr Mat & Mekh (Tartu) 13(336)*3-30
◇ B60 C90 F50 G05 G10 ◇ REV MR 51 # 7864 Zbl 356 # 02010 • ID 17694

TAUTS, A. [1974] *A game for the construction of a propositional semantics in generalized Beth models (Russian) (Estonian and German summaries)* (S 3468) Tr Mat & Mekh (Tartu) 14(342)*13-28
◇ C75 C90 E60 F50 ◇ REV MR 51 # 7865 Zbl 362 # 02042 • ID 17693

TAUTS, A. [1975] *A semantic model for infinite formulas (Russian) (Estonian and German summaries)* (S 3468) Tr Mat & Mekh (Tartu) 15(355)*7-19
◇ C75 C90 F50 G10 ◇ REV MR 52 # 7846 Zbl 354 # 02018 • ID 18414

TAUTS, A. [1975] *Search for deduction by means of a semantic model (Russian) (Estonian and German summaries)* (S 3468) Tr Mat & Mekh (Tartu) 17(374)*3-28
◇ B55 C75 C90 F50 ◇ REV MR 58 # 5076 Zbl 393 # 03039 • ID 52459

TAUTS, A. [1975] *Strong tautology and construction of a countermodel for nonderivable propositions (Russian) (Estonian and German summaries)* (S 3468) Tr Mat & Mekh (Tartu) 17(374)*29-42
◇ C75 C90 F50 ◇ REV MR 57 # 16049 Zbl 393 # 03040 • ID 52460

TAUTS, A. [1981] *The connection between semantic models and pseudo-boolean algebras (Russian) (Estonian and German summaries)* (S 0393) Uch Zap Univ, Tartu 556*3-10
◇ C75 C90 F50 G10 ◇ REV MR 83b:03070 Zbl 487 # 03011 • ID 38114

TAUTS, A. see Vol. I, II, III, IV, V for further entries

TEENSMA, E. [1955] *The intuitionistic interpretation of analysis* (P 0664) Congr Int Union Phil of Sci (2);1954 Zuerich 2*144-150
◇ F50 ◇ REV MR 17.934 • ID 13509

TEIXEIRA, M.T. [1961] *Der Tarskische Konsequenz-Operator und mit dem intuitionistischen Aussagenkalkuel verknuepfte algebraische Strukturen (Portugiesisch)* (J 0108) Bol Soc Mat Sao Paulo 13*67-74
◇ F50 G10 ◇ REV Zbl 104.6 • ID 47997

TEIXEIRA, M.T. see Vol. IV for further entries

TENNANT, N. [1980] *A proof-theoretic approach to entailment* (J 0122) J Philos Logic 9*185-209
◇ B46 F05 ◇ REV MR 81j:03035 Zbl 432 # 03032 • ID 53989

TENNANT, N. see Vol. I, II, III for further entries

TERLOUW, J. [1985] *Reduction of higher type levels by means of an ordinal analysis of finite terms* (J 0073) Ann Pure Appl Logic 28*73-102
◇ F10 F15 ◇ REV Zbl 554 # 03024 • ID 42600

TERLOUW, J. see Vol. IV for further entries

THARP, L.H. [1971] *A quasi-intuitionistic set theory* (J 0036) J Symb Logic 36*456-460
◇ E35 E70 F50 ◇ REV MR 45 # 6617 Zbl 247 # 02055 • ID 13526

THARP, L.H. see Vol. I, III, IV, V for further entries

THIEL, C. [1970] *Was will die operative Mathematik ?* (J 1515) Archimede 22*51-53
◇ A05 F65 ◇ ID 30583

THIEL, C. [1977] *Que significa "constructivismo" ?* (J 0162) Teorema (Valencia) 7*5-21
◇ A05 F65 ◇ REV MR 55 # 10224 • ID 30591

THIEL, C. [1978] *Duality lost? Transforming Gentzen derivations into winning strategies for dialogue games* (J 0162) Teorema (Valencia) 8*57-66
◇ B10 F07 ◇ REV MR 58 # 133 • ID 30592

THIEL, C. see Vol. I, V for further entries

THOMAS, WILLIAM J. [1973] *Doubts about some standard arguments for Church's Thesis* (P 0580) Int Congr Log, Meth & Phil of Sci (4,Sel Pap);1971 Bucharest 55-62
◇ A05 D10 D20 F99 ◇ REV MR 57 # 2869 Zbl 286 # 02009 • ID 65709

THOMAS, WILLIAM J. see Vol. I, IV for further entries

THOMAS, WOLFGANG [1975] *A note on undecidable extensions of monadic second order successor arithmetic* (J 0009) Arch Math Logik Grundlagenforsch 17*43-44
◇ D35 F35 ◇ REV MR 56 # 110 Zbl 325 # 02032 • ID 13578

THOMAS, WOLFGANG [1984] see BOERGER, E.

THOMAS, WOLFGANG see Vol. I, II, III, IV, V for further entries

THOMASON, R.H. [1963] see BELNAP JR., N.D.

THOMASON, R.H. [1966] see LEBLANC, H.

THOMASON, R.H. [1968] *On the strong semantical completeness of the intuitionistic predicate calculus* (J 0036) J Symb Logic 33*1-7
◇ F50 ◇ REV MR 39 # 1301 Zbl 155.14 • ID 13580

THOMASON, R.H. see Vol. I, II, III for further entries

TIRNOVEANU, M. [1964] *Ueber ein logisches semikonstruktives globales System (Rumaenisch)* (J 0440) Bul Inst Politeh Bucuresti, Ser Mec 26/5*17-22
◇ F99 ◇ REV MR 32 # 1123 • ID 16291

TIRNOVEANU, M. [1970] *Sur quelques notions de proto-logistique semi-constructive* (J 0147) An Univ Bucuresti, Acta Logica 13*133-145
◇ B60 F99 ◇ REV Zbl 272 # 02049 • ID 30384

TIRNOVEANU, M. [1970] *Ueber einen Begriff der semikonstruktiven Protologik. I (Rumaenisch)* (S 1613) Probl Logic (Bucharest) 2*181-202
◇ F99 ◇ REV Zbl 218 # 02020 • ID 26274

TIRNOVEANU, M. see Vol. I, II for further entries

TITANI, S. [1965] *An algebraic formulation of cut-elimination theorem* (J 0090) J Math Soc Japan 17*72-83
◇ F05 G05 G25 ◇ REV MR 33 # 32 Zbl 148.7 JSL 35.145 • ID 13614

TITANI, S. [1973] *A proof of the cut-elimination theorem in simple type theory* (J 0036) J Symb Logic 38*215-226
◇ B15 F05 F35 ◇ REV MR 48 # 10777 Zbl 275 # 02017 • ID 13615

TITANI, S. [1979] see TAKEUTI, G.

TITANI, S. [1984] see TAKEUTI, G.

TITANI, S. see Vol. III, V for further entries

TITGEMEYER, R. [1961] *Ueber einen Widerspruch in Cogans Darstellung der Mengenlehre* (J 0068) Z Math Logik Grundlagen Math 7*161-163
◇ B40 E70 ◇ REV MR 25 # 3844 Zbl 101.14 JSL 35.147 • ID 13616

TOKARZ, M. & WOJCICKI, R. [1971] *The problem of reconstructability of propositional calculi (Polish and Russian summaries)* (J 0063) Studia Logica 28*119-129
◇ B22 F50 ◇ REV MR 45 # 8497 Zbl 257 # 02012 • ID 13627

TOKARZ, M. [1976] *Definitions by context in propositional calculi* (J 0387) Bull Sect Logic, Pol Acad Sci 5*54-56
◇ F50 ◇ REV MR 55 # 7733 • ID 31575

TOKARZ, M. see Vol. I, II for further entries

TOLEDO, S. [1975] *Tableau systems for first order number theory and certain higher order theories* (S 3301) Lect Notes Math 447*iii+339pp
◇ F05 F07 F30 F35 F98 ◇ REV MR 52 # 10378 Zbl 309 # 02028 JSL 45.636 • ID 21679

TOLSTOVA, YU.N. [1971] *A weakening of intuitionistic logic (Russian) (English summary)* (S 0228) Zap Nauch Sem Leningrad Otd Mat Inst Steklov 20*208-219,288
• TRANSL [1973] (J 1531) J Sov Math 1*132-138
◇ B60 F50 ◇ REV MR 45 # 4951 MR 48 # 5808 Zbl 231 # 02028 • ID 13633

TOLSTOVA, YU.N. see Vol. II for further entries

TOMAS, F. [1975] *A test for consistency and its application to recursive arithmetic (Spanish)* (J 1507) An Inst Mat, Univ Nac Aut Mexico 15*133-190
◇ B28 F30 ◇ REV MR 54 # 7238 Zbl 345 # 02031 • ID 25028

TOMAS, F. [1977] *A free formalism for the general recursive arithmetic (Spanish)* (J 1507) An Inst Mat, Univ Nac Aut Mexico 17*165-204
◇ F30 ◇ REV MR 58 # 10384 Zbl 399 # 03046 • ID 52842

TOMAS, F. [1982] *A recursive, finitarily consistent formalization of a fragment of nonconstructive arithmetic, and its relation to recursive analysis (Spanish)* (J 1507) An Inst Mat, Univ Nac Aut Mexico 20*205-258
◇ F30 ◇ REV MR 84e:03071 Zbl 529 # 03034 • ID 34404

TOMAS, F. [1982] *About a recursive formalism for arithmetic and its relation to a recursive construction of analysis (Catalan)* (P 3870) Congr Catala de Log Mat (1);1982 Barcelona 117-120
◇ F30 ◇ REV MR 84i:03003 Zbl 512 # 03029 • ID 36515

TOMAS, F. [1984] *Formally recursive arithmetic and analysis (Catalan)* (J 2562) Publ Sec Mat Univ Autonoma Barcelona 28*19-78
◇ F30 F35 ◇ REV MR 85j:03101 • ID 44279

TOMAS, F. see Vol. I for further entries

TOMPKINS, R.R. [1968] *On Kleene's recursive realizability as an interpretation for intuitionistic elementary number theory* (J 0047) Notre Dame J Formal Log 9*289-293
◇ F30 F50 ◇ REV MR 39 # 3975 Zbl 179.11 JSL 35.475 • ID 13639

TOMS, E. [1941] *The law of excluded middle* (J 0153) Phil of Sci (East Lansing) 8*33-38
◇ A05 F99 ◇ ID 38698

TORTORA, R. [1978] see CRISCUOLO, G.

TORTORA, R. see Vol. I, II for further entries

TOSI, P. [1980] *Normal derivability and first-order arithmetic* (J 0047) Notre Dame J Formal Log 21*449-466
◇ F05 F30 F50 ◇ REV MR 81g:03067 Zbl 332 # 02039 • ID 53811

TOSI, P. [1981] *Forme normali* (P 3092) Congr Naz Logica;1979 Montecatini Terme 37-45
◇ F05 F30 ◇ ID 48450

TOSI, P. [1981] *Intuitive proofs and first-order derivations: Some notes on the metamathematics of first-order number theory* (C 3515) Ital Studies in Phil of Sci 137-152
◇ F05 F30 ◇ REV MR 82j:03073 Zbl 455 # 03024 • ID 54285

TRAGESSER, R.S. & ZUCKER, J.I. [1978] *The adequacy problem for inferential logic* (J 0122) J Philos Logic 7*501-516
◇ A05 B55 F50 ◇ REV MR 80c:03031 Zbl 408 # 03020 JSL 47.689 • ID 30597

TRAKHTENBROT, B.A. [1959] *Descriptive classifications in recursive arithmetics (Russian)* (P 0607) All-Union Math Conf (3);1956 Moskva 1*185
◇ D20 F30 ◇ ID 45721

TRAKHTENBROT, B.A. [1961] *Certain constructions in the logic of one-place predicates (Russian)* (J 0023) Dokl Akad Nauk SSSR 138*320-321
• TRANSL [1961] (J 0062) Sov Math, Dokl 2*623-625
◇ B20 F35 ◇ REV MR 25 # 1996 JSL 29.100 • ID 19735

TRAKHTENBROT, B.A. [1961] *Finite automata and the logic of one-place predicates (Russian)* (J 0023) Dokl Akad Nauk SSSR 140*326-329
• TRANSL [1961] (J 0470) Sov Phys, Dokl 6*753-755
◇ B20 D05 F35 ◇ REV MR 26 # 1246 Zbl 115.7 JSL 29.100 • ID 19734

TRAKHTENBROT, B.A. [1962] *Finite automata and the logic of one-place predicates (Russian)* (J 0092) Sib Mat Zh 3*103-131
◇ B20 D05 F35 ◇ REV MR 26 # 4908 Zbl 115.7 JSL 29.98 • ID 19733

TRAKHTENBROT, B.A. see Vol. I, II, III, IV, V for further entries

TROELSTRA, A.S. [1966] see JONGH DE, D.H.J.

TROELSTRA, A.S. [1967] *Finite and infinite in intuitionistic mathematics* (J 0020) Compos Math 18*94-116
◇ E10 F55 ◇ REV MR 37 # 2582 Zbl 163.5 • ID 13678

TROELSTRA, A.S. [1967] *Intuitionistic continuity* (J 3077) Nieuw Arch Wisk, Ser 3 15*2-6
◇ F55 ◇ REV MR 36 # 3633 Zbl 153.7 • ID 13677

TROELSTRA, A.S. [1967] *Intuitionistic connectedness* (J 0028) Indag Math 29*96-105
◇ F55 ◇ REV MR 35 # 4868 Zbl 165.305 • ID 14741

TROELSTRA, A.S. [1968] see DALEN VAN, D.

TROELSTRA, A.S. [1968] *New sets of postulates for intuitionistic topology* (J 0020) Compos Math 20*211-221
• REPR [1968] (C 0727) Logic Found of Math (Heyting) 211-221
◇ F55 ◇ REV MR 38 # 4283 Zbl 164.308 JSL 40.500 • ID 13680

TROELSTRA, A.S. [1968] *One-point compactifications of intuitionistic locally compact spaces* (J 0027) Fund Math 62*75-93
◇ F55 ◇ REV MR 38 # 4284 Zbl 153.523 • ID 13679

TROELSTRA, A.S. [1968] *The scientific work of A.Heyting* (J 0020) Compos Math 20*3-12
• REPR [1968] (C 0727) Logic Found of Math (Heyting) 3-12
◇ A10 F99 ◇ REV MR 37 # 26 Zbl 155.9 JSL 40.499 • ID 13681

TROELSTRA, A.S. [1968] *The theory of choice sequences* (P 0627) Int Congr Log, Meth & Phil of Sci (3,Proc);1967 Amsterdam 201-223
◇ F35 F50 ◇ REV MR 40 # 5427 Zbl 185.10 JSL 38.332 • ID 24834

TROELSTRA, A.S. [1968] *The use of "Brouwer's principle" in intuitionistic topology* (**P** 0608) Logic Colloq;1966 Hannover 289–298
◇ F55 ◇ REV MR 40 # 5426 Zbl 181.6 JSL 40.504 • ID 16943

TROELSTRA, A.S. [1969] *Informal theory of choice sequences (Polish and Russian summaries)* (**J** 0063) Studia Logica 25*31–54
◇ F35 F50 F55 ◇ REV MR 43 # 4632 Zbl 263 # 02017 • ID 13682

TROELSTRA, A.S. [1969] *Notes on the intuitionistic theory of sequences I* (**J** 0028) Indag Math 31*430–440
◇ F35 F50 ◇ REV MR 41 # 3243 Zbl 180.316 • REM Parts II,III 1970 • ID 13683

TROELSTRA, A.S. [1969] *Principles of intuitionism* (**X** 0811) Springer: Heidelberg & New York ii + 111pp
◇ F35 F50 F55 F98 ◇ REV MR 39 # 5320 Zbl 181.5 JSL 40.447 • ID 21145

TROELSTRA, A.S. [1970] see KREISEL, G.

TROELSTRA, A.S. [1970] *Notes on the intuitionistic theory of sequences II,III* (**J** 0028) Indag Math 32*99–109,245–252
◇ F35 F50 ◇ REV MR 42 # 2929 MR 43 # 4633 Zbl 207.5 Zbl 211.10 • REM Part I 1969 • ID 13684

TROELSTRA, A.S. [1970] see DALEN VAN, D.

TROELSTRA, A.S. [1971] *Computability of terms and notions of realizability for intuitionistic analysis* (**J** 1583) Rep Dept Math, Univ Amsterdam 70–02
◇ F35 F50 ◇ ID 28612

TROELSTRA, A.S. [1971] *Intuitionisme (Dutch)* (**J** 1728) Wijsg Persp Weten Maatsch 11*235–267
◇ A05 F99 ◇ ID 32498

TROELSTRA, A.S. [1971] *Notions of realizability for intuitionistic arithmetic and intuitionistic arithmetic in all finite types* (**P** 0604) Scand Logic Symp (2);1970 Oslo 369–405
◇ F30 F35 F50 ◇ REV MR 49 # 23 Zbl 227 # 02015 JSL 40.625 • ID 22203

TROELSTRA, A.S. [1973] *Intuitionistic formal systems* (**C** 1139) Metamath Invest of Intuition Arithm & Anal 1–96
◇ F30 F35 F50 ◇ REV MR 56 # 2792 • ID 79458

TROELSTRA, A.S. (ED.) [1973] *Metamathematical investigation of intuitionistic arithmetic and analysis* (**S** 3301) Lect Notes Math 344*xvii + 485pp • ERR/ADD [1974] (**J** 1583) Rep Dept Math, Univ Amsterdam 74–16*ii + 26pp
◇ F05 F10 F30 F35 F50 F55 F98 ◇ REV MR 48 # 3699 MR 56 # 8334 Zbl 275 # 02025 • REM With contributions by A.S.Troelstra, C.A.Smorynski, J.I.Zucker and W.A.Howard • ID 29695

TROELSTRA, A.S. [1973] *Models and computability* (**C** 1139) Metamath Invest of Intuition Arithm & Anal 97–174
◇ F05 F10 F30 F35 F50 ◇ REV MR 56 # 2793 • ID 79457

TROELSTRA, A.S. [1973] *Normalization theorems for systems of natural deduction* (**C** 1139) Metamath Invest of Intuition Arithm & Anal 275–323
◇ F05 F30 F35 F50 ◇ REV MR 56 # 8333 • ID 79455

TROELSTRA, A.S. [1973] *Notes on intuitionistic second order arithmetic* (**P** 0713) Cambridge Summer School Math Log;1971 Cambridge GB 171–205
◇ F35 F50 ◇ REV MR 51 # 81 Zbl 275 # 02036 • ID 13686

TROELSTRA, A.S. [1973] *Realizability and functional interpretations* (**C** 1139) Metamath Invest of Intuition Arithm & Anal 175–274
◇ F10 F30 F35 F50 ◇ REV MR 56 # 2794 • ID 79456

TROELSTRA, A.S. [1974] *Non-extensional equality* (**J** 0027) Fund Math 82*307-322
◇ F50 ◇ REV MR 51 # 7837 Zbl 299 # 02036 • ID 13688

TROELSTRA, A.S. [1974] *Note on the fan theorem* (**J** 0036) J Symb Logic 39*584–596
◇ F30 F35 F50 ◇ REV MR 52 # 5371 Zbl 306 # 02026 • ID 13687

TROELSTRA, A.S. [1975] *Markov's principle and Markov's rule for theories of choice sequences* (**P** 1440) ⊧ ISILC Proof Th Symp (Schuette);1974 Kiel 370–383
◇ F35 F50 ◇ REV MR 54 # 9982 Zbl 325 # 02015 • ID 27281

TROELSTRA, A.S. [1977] *A note on non-extensional operations in connection with continuity and recursiveness* (**J** 0028) Indag Math 39*455–462
◇ F35 F50 ◇ REV MR 57 # 12186 Zbl 368 # 02026 • ID 32502

TROELSTRA, A.S. [1977] *Aspects of constructive mathematics* (**C** 1523) Handb of Math Logic 973–1052
◇ F30 F35 F50 F55 F60 F98 ◇ REV MR 58 # 10343 JSL 49.980 • ID 27331

TROELSTRA, A.S. [1977] *Axioms for intuitionistic mathematics incompatible with classical logic* (**P** 1704) Int Congr Log, Meth & Phil of Sci (5);1975 London ON 1*59–84
◇ A05 F35 F50 ◇ REV MR 57 # 12185 Zbl 375 # 02010 • ID 32500

TROELSTRA, A.S. [1977] *Choice sequences, a chapter of intuitionistic mathematics* (**X** 0815) Clarendon Pr: Oxford ix + 170pp
◇ A05 A10 F35 F50 F55 F98 ◇ REV MR 57 # 15979 Zbl 355 # 02026 • ID 27586

TROELSTRA, A.S. [1977] *Completeness and validity for intuitionistic predicate logic* (**P** 1729) Colloq Int Log;1975 Clermont-Ferrand 39–58
◇ C90 F50 ◇ REV MR 80k:03017 Zbl 439 # 03009 • ID 32501

TROELSTRA, A.S. [1977] *Some models for intuitionistic finite type arithmetic with fan functional* (**J** 0036) J Symb Logic 42*194–202
◇ F35 F50 ◇ REV MR 58 # 10334 Zbl 385 # 03048 • ID 26443

TROELSTRA, A.S. [1977] *Special instances of generalized continuity which are conservative over intuitionistic arithmetic* (**J** 0028) Indag Math 39*55–65
◇ F35 F50 ◇ REV MR 55 # 5395 Zbl 361 # 02047 • ID 26062

TROELSTRA, A.S. [1978] *A. Heyting on the formalization of intuitionistic mathematics* (**P** 1730) Two Decades Math Netherlands 1920–40;1978 Amsterdam 153–175
◇ A10 F50 ◇ ID 32504

TROELSTRA, A.S. [1978] *Appendix to D. van Dalen's "An interpretation of intuitionistic analysis"* (**J** 0007) Ann Math Logic 13*38–43
◇ F35 F50 ◇ REV MR 80i:03067 • REM For van Dalen's paper see ibid 13*1–43 • ID 32503

TROELSTRA, A.S. [1978] *Some remarks on the complexity of Henkin-Kripke models* (J 0028) Indag Math 40*296-302
◇ C90 D55 F50 ◇ REV MR 58 # 21575 Zbl 389 # 03024
• ID 27952

TROELSTRA, A.S. [1979] *On Ashvinikumar's principle of microscopic completeness* (J 0028) Indag Math 41*77-81
◇ F35 F50 ◇ REV MR 80i:03066b Zbl 405 # 03032 • REM Ashvinikumar's paper appeared ibid 41*95-103 • ID 79448

TROELSTRA, A.S. [1979] see HOEVEN VAN DER, G.F.

TROELSTRA, A.S. [1980] *Extended bar induction of type zero* (P 2058) Kleene Symp;1978 Madison 277-316
◇ F35 F50 ◇ REV MR 82f:03054 Zbl 444 # 03031
• ID 79445

TROELSTRA, A.S. [1980] *Intuitionistic extensions of the reals* (J 3077) Nieuw Arch Wisk, Ser 3 28*63-113
◇ F35 F50 F55 ◇ REV MR 82a:03060 Zbl 432 # 03035
• REM Part I. Part II 1982 • ID 53992

TROELSTRA, A.S. [1981] *Arend Heyting and his contribution to intuitionism* (J 3077) Nieuw Arch Wisk, Ser 3 29*1-23
◇ A10 F55 ◇ REV MR 83g:01091a Zbl 452 # 01010
• ID 54064

TROELSTRA, A.S. [1981] see NIEKUS, N.H.

TROELSTRA, A.S. [1981] *Lawless sequences and their uses* (P 3092) Congr Naz Logica;1979 Montecatini Terme 165-173
◇ F35 F50 ◇ ID 43821

TROELSTRA, A.S. [1981] *On a second order propositional operator in intuitionistic logic* (J 0063) Studia Logica 40*113-139
◇ B55 F35 F50 ◇ REV MR 84a:03015 Zbl 473 # 03022
• ID 55352

TROELSTRA, A.S. [1981] *The interplay between logic and mathematics: Intuitionism* (C 2617) Modern Log Survey 197-221
◇ A05 F50 F98 ◇ REV MR 82f:03002 Zbl 464 # 03001
• ID 43820

TROELSTRA, A.S. [1982] *Intuitionistic extensions of the reals II* (P 3623) Logic Colloq;1980 Prague 279-310
◇ F35 F50 F55 ◇ REV MR 84k:03140 Zbl 494 # 03039
• REM Part I 1980 • ID 35032

TROELSTRA, A.S. [1982] *On the origin and development of Brouwer's concept of choice sequence* (P 3638) Brouwer Centenary Symp;1981 Noordwijkerhout 465-486
◇ A10 F55 ◇ REV MR 85d:03002 Zbl 522 # 03049
• ID 37798

TROELSTRA, A.S. & DALEN VAN, D. (EDS.) [1982] *The L.E.J. Brouwer centenary symposium* (S 3303) Stud Logic Found Math 110*ix+523pp
◇ A05 A10 B97 F50 F55 F97 ◇ REV MR 84i:03010 Zbl 505 # 00008 • ID 34494

TROELSTRA, A.S. [1983] *Analysing choice sequences* (J 0122) J Philos Logic 12*197-260
◇ A05 F35 F50 ◇ REV MR 85d:03119 Zbl 518 # 03023
• ID 37521

TROELSTRA, A.S. [1983] *Definability of finite sum types in Martin-Loef's type theories* (J 0028) Indag Math 45*475-481
◇ F35 F50 ◇ REV MR 85i:03177 Zbl 532 # 03027
• ID 38284

TROELSTRA, A.S. [1983] *Logic in the writings of Brouwer and Heyting* (P 3091) Conv Int Storica Logica;1982 San Gimignano 193-210
◇ A10 F99 ◇ ID 43822

TROELSTRA, A.S. [1984] see DILLER, J.

TROELSTRA, A.S. [1985] *Choice sequences and informal rigour* (J 0154) Synthese 62*217-227
◇ A05 F35 F50 ◇ ID 43823

TROELSTRA, A.S. see Vol. II for further entries

TSEJTIN, G.S. [1955] *On Cauchy's theorem in constructive analysis (Russian)* (J 0067) Usp Mat Nauk 10/4(66)*207-209
◇ F60 ◇ REV Zbl 65.5 JSL 21.408 • ID 01899

TSEJTIN, G.S. [1959] *Algorithmic operators in constructive complete separable metric spaces (Russian)* (J 0023) Dokl Akad Nauk SSSR 128*49-52
◇ F60 ◇ REV MR 22 # 6708 Zbl 124.4 • ID 01903

TSEJTIN, G.S. & ZASLAVSKIJ, I.D. [1959] *On relations between fundamental properties of constructive functions (Russian)* (P 0607) All-Union Math Conf (3);1956 Moskva 1*180-181
◇ F60 ◇ ID 28234

TSEJTIN, G.S. [1959] *The theorem concerning nested segments, Cauchy's theorem, and Rolle's theorem in constructive analysis (Russian)* (P 0607) All-Union Math Conf (3);1956 Moskva 1*186-187
◇ F60 ◇ ID 28241

TSEJTIN, G.S. [1959] *Uniform recursiveness of algorithmic operators on general recursive functions and a canonical representation for constructive functions of a real argument (Russian)* (P 0607) All-Union Math Conf (3);1956 Moskva 1*188-189
◇ D20 F60 ◇ ID 28243

TSEJTIN, G.S. [1962] *Algorithmic operators in constructive metric spaces (Russian)* (S 0066) Tr Mat Inst Steklov 67*295-361
• TRANSL [1967] (J 0225) Amer Math Soc, Transl, Ser 2 64*1-80
◇ F60 ◇ REV MR 27 # 2406 Zbl 113.10 • ID 01905

TSEJTIN, G.S. [1962] *Mean value theorems in constructive analysis (Russian)* (S 0066) Tr Mat Inst Steklov 67*362-384
• TRANSL [1971] (J 0225) Amer Math Soc, Transl, Ser 2 98*11-40
◇ F60 ◇ REV MR 27 # 2407 Zbl 113.11 • ID 01904

TSEJTIN, G.S. & ZASLAVSKIJ, I.D. [1962] *On singular coverings and properties of constructive functions connected with them (Russian)* (S 0066) Tr Mat Inst Steklov 67*458-502
• TRANSL [1971] (J 0225) Amer Math Soc, Transl, Ser 2 98*41-89
◇ F60 ◇ REV Zbl 113.13 • ID 19516

TSEJTIN, G.S. & ZASLAVSKIJ, I.D. [1964] *Concerning a generalized principle of constructive selection (Russian)* (S 0066) Tr Mat Inst Steklov 72*344-347
• TRANSL [1972] (J 0225) Amer Math Soc, Transl, Ser 2 99*228-232
◇ F60 ◇ REV Zbl 199.33 • ID 28885

TSEJTIN, G.S. [1964] *Three theorems on constructive functions (Russian)* (S 0066) Tr Mat Inst Steklov 72*537-543
• TRANSL [1972] (J 0225) Amer Math Soc, Transl, Ser 2 100*201-209
◇ F60 ◇ REV MR 34 # 5669 Zbl 236 # 02030 • ID 14496

TSEJTIN, G.S. [1968] see SHANIN, N.A.

TSEJTIN, G.S. [1968] see KUSHNER, B.A.

TSEJTIN, G.S. [1968] *The complexity of a deduction in the propositional calculus (Russian)* (S 0228) Zap Nauch Sem Leningrad Otd Mat Inst Steklov 8*234-259 • ERR/ADD ibid 16*185-187
• TRANSL [1968] (J 0521) Semin Math, Inst Steklov 8*115-125 [1983] (C 4659) Autom of Reasoning 2*466-486
◇ F20 ◇ REV MR 43 # 6098 Zbl 197.1 • ID 02193

TSEJTIN, G.S. [1968] *The disjunctive rank of the formulas of constructive arithmetic (Russian)* (S 0228) Zap Nauch Sem Leningrad Otd Mat Inst Steklov 8*260-271
◇ F30 F50 ◇ REV MR 43 # 3118 Zbl 191.307 • ID 02194

TSEJTIN, G.S. & ZASLAVSKIJ, I.D. [1970] *A criterion of the rectifiability of constructive plane curves (Russian) (Armenian and English summaries)* (J 0312) Izv Akad Nauk Armyan SSR, Ser Mat 5*434-440
◇ F60 ◇ REV MR 44 # 2577 Zbl 222 # 02032 • ID 14745

TSEJTIN, G.S. [1970] *On upper bounds of recursively enumerable sets of constructive real numbers (Russian)* (S 0066) Tr Mat Inst Steklov 113*102-172
• TRANSL [1970] (S 0055) Proc Steklov Inst Math 113*119-194
◇ D25 F60 ◇ REV MR 48 # 77 Zbl 229 # 02031 • ID 01909

TSEJTIN, G.S. [1971] *A pseudofundamental sequence that is not equivalent to any monotone sequence (Russian) (English summary)* (S 0228) Zap Nauch Sem Leningrad Otd Mat Inst Steklov 20*263-271,290
• TRANSL [1973] (J 1531) J Sov Math 1*169-175
◇ F60 ◇ REV MR 44 # 5223 Zbl 222 # 02046 • ID 26317

TSEJTIN, G.S. & ZASLAVSKIJ, I.D. [1971] *Another constructive version of the Cauchy theorems (Russian) (English summary)* (S 0228) Zap Nauch Sem Leningrad Otd Mat Inst Steklov 20*36-39
• TRANSL [1973] (J 1531) J Sov Math 1*19-21
◇ F60 ◇ REV MR 45 # 6984 Zbl 222 # 02031 • ID 26306

TSEJTIN, G.S. [1972] see CHUBARYAN, A.A.

TSEJTIN, G.S. see Vol. I, IV for further entries

TSINMAN, L.L. [1967] *On the complete induction axiom (Russian)* (J 0023) Dokl Akad Nauk SSSR 173*273-274
• TRANSL [1967] (J 0062) Sov Math, Dokl 8*381-383
◇ B28 F30 ◇ REV MR 35 # 1468 Zbl 159.10 • ID 02176

TSINMAN, L.L. [1968] *The role of the principle of induction in a formal arithmetical system (Russian)* (J 0142) Mat Sb, Akad Nauk SSSR, NS 77(119)*71-104
• TRANSL [1968] (J 0349) Math of USSR, Sbor 6*65-95
◇ B28 F30 ◇ REV MR 39 # 5312 Zbl 165.18 • ID 02177

TSINMAN, L.L. [1969] *A theorem on normal proofs in the theory of regular formulae (Russian)* (J 0031) Izv Vyssh Ucheb Zaved, Mat (Kazan) 1969/10(89)*90-96
◇ F30 ◇ REV MR 41 # 3270 Zbl 197.280 • ID 02179

TSINMAN, L.L. [1969] *Certain algorithms in a formal arithmetic system (Russian)* (J 0023) Dokl Akad Nauk SSSR 189*489-490
• TRANSL [1969] (J 0062) Sov Math, Dokl 10*1452-1454
◇ B25 F30 ◇ REV MR 41 # 64 Zbl 208.16 • ID 02178

TSINMAN, L.L. see Vol. IV for further entries

TSITKIN, A.I. [1977] *On admissible rules of intuitionistic propositional logic (Russian)* (J 0142) Mat Sb, Akad Nauk SSSR, NS 102(144)*314-323,328
• TRANSL [1977] (J 0349) Math of USSR, Sbor 31*279-288
◇ B55 F50 ◇ REV MR 57 # 5689 Zbl 355 # 02016
• ID 50214

TSITKIN, A.I. [1979] *On the question of an error in a famous work due to Wajsberg (Russian)* (C 2581) Issl Neklass Log & Teor Mnozh 240-256
◇ A10 B20 F50 ◇ REV MR 82d:03018 Zbl 422 # 03035
• ID 53496

TSITKIN, A.I. see Vol. II for further entries

TSUBOI, A. [1982] *On reflection principles* (J 2606) Tsukuba J Math 6*201-213
◇ F30 ◇ REV MR 85b:03110 Zbl 539 # 03036 • ID 37187

TSUBOI, A. see Vol. III, IV for further entries

TUCKER, J.V. [1963] *Constructivity and grammar* (J 0113) Proc Aristotelian Soc 63*45-66
◇ A05 B65 F99 ◇ REV JSL 28.107 • ID 43028

TUCKER, J.V. [1983] see BERGSTRA, J.A.

TUCKER, J.V. see Vol. I, III, IV for further entries

TURING, A.M. [1936] *On computable numbers, with an application to the "Entscheidungsproblem"* (J 1910) Proc London Math Soc, Ser 2 42*230-265 • ERR/ADD ibid 43*544-546
◇ D05 D10 D20 D35 F60 ◇ REV Zbl 16.97 JSL 2.42 FdM 62.1059 • ID 13723

TURING, A.M. [1937] *Computability and λ-definability* (J 0036) J Symb Logic 2*153-163
◇ A05 B40 D05 D10 D20 F99 ◇ REV Zbl 18.193 JSL 3.89 FdM 63.824 • ID 13725

TURING, A.M. [1937] *The 𝔭-function in λ-K-conversion* (J 0036) J Symb Logic 2*164
◇ B40 ◇ REV Zbl 18.194 FdM 63.825 • ID 13724

TURING, A.M. [1939] *Systems of logic based on ordinals* (J 1910) Proc London Math Soc, Ser 2 45*161-228
◇ B40 D25 D30 F15 F30 ◇ REV Zbl 21.97 JSL 4.128 FdM 65.1102 • ID 13726

TURING, A.M. see Vol. I, III, IV for further entries

TURNER, D.A. [1979] *Another algorithm for bracket abstraction* (J 0036) J Symb Logic 44*267-271
◇ B40 ◇ REV MR 80f:03021 Zbl 408 # 03013 • ID 56252

TVERSKOJ, A.A. [1980] *A sequence of combinatorial judgements which are independent of Peano arithmetic (Russian) (English summary)* (J 0288) Vest Ser Mat Mekh, Univ Moskva 1980/5*7-13,83
• TRANSL [1980] (J 0510) Moscow Univ Math Bull 35/5*6-13
◇ C62 E05 F30 ◇ REV MR 83e:03096 Zbl 453 # 03045
• ID 35246

TVERSKOJ, A.A. see Vol. III, IV, V for further entries

TWER VON DER, T. [1981] *Some remarks on the mathematical incompleteness of Peano's arithmetic found by Paris and Harrington* (P 3268) Set Th & Model Th;1979 Bonn 157-174
◇ C62 E05 F30 ◇ REV MR 83a:03060 Zbl 522 # 03056
• ID 35072

TWER VON DER, T. see Vol. III for further entries

TYUGU, E.KH. [1982] see MINTS, G.E.

UESU, T. [1967] *Zermelo's set theory and G^*LC* (**J** 0407) Comm Math Univ St Pauli (Tokyo) 16*69-88 • ERR/ADD ibid 19*47-49
◇ E30 E35 F35 ◇ REV MR 38 # 41 MR 45 # 35 Zbl 268 # 02020 • ID 16356

UESU, T. [1969] *Gentzen's Hauptsatz for simple type theory with the infinite induction* (**J** 0350) Sci Rep Tokyo Kyoiku Daigaku Sect A 10*124-126
◇ F05 F35 ◇ REV MR 41 # 34 Zbl 216.11 • ID 13762

UESU, T. [1970] *On the recursively restricted rules* (**J** 0407) Comm Math Univ St Pauli (Tokyo) 18*31-42
◇ C75 F30 ◇ REV MR 42 # 71 Zbl 193.302 • ID 13763

UESU, T. [1971] *Simple type theory with constructive infinitely long expressions* (**J** 0407) Comm Math Univ St Pauli (Tokyo) 19*131-163
◇ B15 C75 F05 F35 ◇ REV MR 45 # 8509 Zbl 255 # 02023 • ID 13764

UESU, T. [1971] *Two formal systems of simple type theory with type variables* (**J** 0407) Comm Math Univ St Pauli (Tokyo) 19*13-46
◇ F05 F35 ◇ REV MR 46 # 8810 Zbl 238 # 02057 • ID 13765

UESU, T. [1981] *Intuitionistic theories and toposes* (**P** 3201) Logic Symposia;1979/80 Hakone 323-358
◇ E70 F50 G30 ◇ REV MR 84f:03057 Zbl 478 # 03028 • ID 34476

UESU, T. [1981] *Intuitionistic theories and topoi (Japanese)* (**P** 4153) B-Val Anal & Nonstand Anal;1981 Kyoto 93-115
◇ E70 F50 G30 ◇ ID 47768

UESU, T. [1983] *On a problem of Smorynski (Japanese)* (**P** 4113) Found of Math;1982 Kyoto 192-203
◇ F30 ◇ ID 47684

UESU, T. [1984] *A method of axiomatizing fragments of intuitionistic theories* (**P** 3668) Log & Founds of Math;1983 Kyoto 132-141
◇ F50 ◇ ID 42938

UESU, T. [1984] *An axiomatization of the apartness fragment of the theory DLO^+ of dense linear order* (**P** 2153) Logic Colloq;1983 Aachen 2*453-475
◇ F05 F50 ◇ REV Zbl 574 # 03049 • ID 43015

UESU, T. see Vol. I, III, IV, V for further entries

ULRICH, D. [1976] *On a property of matrices for subsystems of IC^+* (**J** 0068) Z Math Logik Grundlagen Math 22*193-194
◇ B20 F50 ◇ REV MR 58 # 16173 Zbl 347 # 02018 • ID 18459

ULRICH, D. see Vol. I, II, III for further entries

UMEZAWA, T. [1974] *Disjunction property in higher order number theory with intuitionistic rules of inference I* (**J** 1005) Rep Fac Sci, Shizuoka Univ 9*17-27
◇ F05 F35 F50 ◇ REV MR 56 # 5232a Zbl 361 # 02048 • REM Part II 1975 • ID 31275

UMEZAWA, T. [1975] *Disjunction property in higher order number theory with intuitionistic rules of inference II* (**J** 1005) Rep Fac Sci, Shizuoka Univ 10*17-35
◇ F05 F35 F50 ◇ REV MR 56 # 5232b Zbl 369 # 02013 • REM Part I 1974 • ID 31276

UMEZAWA, T. [1979] *A method for cut elimination in intuitionistic predicate logic and classical predicate logic* (**J** 1005) Rep Fac Sci, Shizuoka Univ 13*1-13
◇ B10 F05 F50 ◇ REV MR 80k:03063 Zbl 415 # 03047 • ID 53149

UMEZAWA, T. [1980] *Cut elimination in intuitionistic and some intermediate predicate logics* (**J** 1005) Rep Fac Sci, Shizuoka Univ 14*1-12
◇ B55 F05 F50 ◇ REV MR 81h:03108 Zbl 442 # 03038 • ID 56395

UMEZAWA, T. [1982] *Definability and strict definability of logical symbols in intuitionistic predicate logics of first and higher order* (**J** 1005) Rep Fac Sci, Shizuoka Univ 16*1-16
◇ F50 ◇ REV MR 83m:03018 Zbl 497 # 03006 • ID 35426

UMEZAWA, T. see Vol. I, II, III, IV, V for further entries

URQUHART, A.I.F. [1974] *Implicational formulas in intuitionistic logic* (**J** 0036) J Symb Logic 39*661-664
◇ F50 G05 G25 ◇ REV MR 50 # 12674 Zbl 317 # 02022 • ID 13797

URQUHART, A.I.F. [1974] *Proofs, snakes and ladders* (**J** 0488) Dialogue (Ottawa) 13*723-731
◇ B22 F07 ◇ REV MR 58 # 162 • ID 79550

URQUHART, A.I.F. see Vol. I, II, III, IV for further entries

URSINI, A. [1977] *A sequence of theories for arithmetic whose union is complete* (**J** 0144) Rend Sem Mat Univ Padova 57*75-92
◇ C62 F30 ◇ REV MR 80d:03058 Zbl 411 # 03053 • ID 52907

URSINI, A. [1978] *On the set of "meaningful" sentences of arithmetic* (**J** 0063) Studia Logica 37*237-241
◇ F30 ◇ REV MR 80j:03084 Zbl 404 # 03043 • ID 54830

URSINI, A. see Vol. I, II, III, IV for further entries

USBERTI, G. [1982] see MIGLIOLI, P.A.

USBERTI, G. [1983] see MIGLIOLI, P.A.

USBERTI, G. see Vol. II for further entries

USPENSKIJ, V.A. [1953] *The Goedel theorem and the theory of algorithms (Russian)* (**J** 0067) Usp Mat Nauk 8/4(56)*176-178
◇ D20 D35 F30 ◇ REV Zbl 51.245 • ID 45765

USPENSKIJ, V.A. [1953] *Theorem of Goedel and theory of algorithms (Russian)* (**J** 0023) Dokl Akad Nauk SSSR 91*737-740
◇ D20 D25 F30 ◇ REV MR 17.4 Zbl 52.251 JSL 19.218 • ID 16353

USPENSKIJ, V.A. [1957] *On the theorem of uniform continuity (Russian)* (**J** 0067) Usp Mat Nauk 12/1(73)*99-142
◇ F60 ◇ REV MR 19.2 Zbl 79.8 JSL 29.62 • ID 16347

USPENSKIJ, V.A. [1959] *Computable operations, computable operators and constructively continuous functions (Russian)* (**P** 0607) All-Union Math Conf (3);1956 Moskva 1*185
◇ D20 F60 ◇ ID 28239

USPENSKIJ, V.A. [1960] *On the relation between various systems of constructive real numbers (Russian)* (**J** 0031) Izv Vyssh Ucheb Zaved, Mat (Kazan) 1960/2(15)*199-208
◇ F60 ◇ REV MR 25 # 3822 Zbl 113.13 • ID 13803

USPENSKIJ, V.A. [1974] *An elementary exposition of Goedel's incompleteness theorem (Russian)* (J 0067) Usp Mat Nauk 29/1*3-47
• TRANSL [1974] (J 1399) Russ Math Surv 29/1*63-106
◇ F30 ◇ REV MR 53 # 2612 Zbl 291 # 02001 • ID 21505

USPENSKIJ, V.A. [1982] *Goedel's incompleteness theorem (Russian)* (X 2027) Nauka: Moskva 112pp
◇ F30 F98 ◇ REV MR 83k:03001 Zbl 559 # 03001
• ID 45922

USPENSKIJ, V.A. see Vol. I, IV, V for further entries

VACHOV, D. [1980] see SENDOV, B.

VAEAENAENEN, J. [1979] *A new incompleteness in arithmetic (Finnish) (English summary)* (J 1108) Arkhimedes (Helsinki) 31*30-37
◇ A10 C62 D25 E05 F30 ◇ REV MR 80c:03058 Zbl 398 # 03021 • ID 52749

VAEAENAENEN, J. see Vol. I, III, V for further entries

VAIDYANATHASWAMY, R. [1943] *On disjunction in intuitionistic logic* (J 0545) Proc Indian Acad Sci, Sect A 17*41-45
◇ F50 ◇ REV MR 5.86 JSL 9.48 • ID 13807

VAIDYANATHASWAMY, R. see Vol. I, II for further entries

VAJNBERG, YU.R. [1976] *On the domains of definition of computable mappings of metric spaces (Russian)* (J 0023) Dokl Akad Nauk SSSR 227*1037-1040
• TRANSL [1976] (J 0062) Sov Math, Dokl 17*548-552
◇ F60 ◇ REV MR 53 # 12903 Zbl 353 # 02017 • ID 23155

VAJNBERG, YU.R. see Vol. IV for further entries

VAKARELOV, D. [1974] *Representation theorems for semi-boolean algebras and semantics for Heyting-Brouwer predicate logic* (J 0014) Bull Acad Pol Sci, Ser Math Astron Phys 22*1087-1095
◇ F50 G05 G10 ◇ REV MR 51 # 71 Zbl 301 # 02066
• ID 13812

VAKARELOV, D. [1977] *Notes on $\mathcal{N}$-lattices and constructive logic with strong negation* (J 0063) Studia Logica 36*109-125
◇ B55 F50 G10 ◇ REV MR 57 # 12216 Zbl 385 # 03055
• ID 31280

VAKARELOV, D. see Vol. I, II, III for further entries

VALENTINI, S. [1980] see SAMBIN, G.

VALENTINI, S. [1982] see SAMBIN, G.

VALENTINI, S. [1983] *A "canonical" model for GL (Italian summary)* (J 2099) Boll Unione Mat Ital, VI Ser, A 2*361-368
◇ B45 F30 ◇ REV MR 85a:03027 Zbl 525 # 03006
• ID 34778

VALENTINI, S. [1983] see SOLITRO, U.

VALENTINI, S. [1983] *The modal logic of provability: cut-elimination* (J 0122) J Philos Logic 12*471-476
◇ B45 F05 F30 ◇ REV MR 86a:03019 Zbl 535 # 03031
• ID 38327

VALENTINI, S. see Vol. II for further entries

VALPOLA, V. [1947] *The position of negation in a language which expresses knowledge (Finnish)* (J 0963) Ajatus (Helsinki) 14*325-381
◇ A05 B28 F30 F50 ◇ REV JSL 13.125 • ID 42606

VALPOLA, V. [1955] *Ein System der negationslosen Logik mit ausschliesslich realisierbaren Praedikaten* (J 0096) Acta Philos Fenn 9*247pp
◇ B20 F50 ◇ REV MR 17.699 Zbl 66.8 JSL 22.355
• ID 16346

VALPOLA, V. [1955] *Eine Eigenschaft gewoehnlicher negationsloser Kalkuele der Propositionen- und Praedikatenlogik* (J 0132) Math Scand 3*107-114
◇ B20 F50 ◇ REV MR 17.447 Zbl 65.1 JSL 22.380
• ID 13817

VANDERVEKEN, D.R. [1975] *An extension of Lesniewski-Curry's formal theory of syntactical categories adequate for the categorially open functors* (J 0387) Bull Sect Logic, Pol Acad Sci 4*78-79
◇ B40 B60 ◇ REV MR 53 # 7722 • ID 22976

VANDERVEKEN, D.R. see Vol. I, II, IV for further entries

VANDIVER, H.S. [1934] *On the foundations of a constructive theory of discrete commutative algebra* (J 0054) Proc Nat Acad Sci USA 20*579-584
◇ F55 ◇ REV Zbl 10.195 FdM 60.898 • ID 38700

VANDIVER, H.S. [1935] *On the foundations of a constructive theory of discrete commutative algebra II* (J 0054) Proc Nat Acad Sci USA 21*162-165
◇ F55 ◇ REV FdM 61.1027 • REM Part I 1934 • ID 48453

VANDIVER, H.S. [1936] *Constructive derivation of the decomposition-field of a polynomial* (J 0120) Ann of Math, Ser 2 37*1-6
◇ F55 ◇ REV FdM 62.1150 • ID 48454

VANDIVER, H.S. [1936] *On the ordering of real algebraic numbers by constructive methods* (J 0120) Ann of Math, Ser 2 37*7-16
◇ F55 ◇ REV FdM 62.1151 • ID 48455

VARELA GARCIA, F.J. [1984] see SOTO-ANDRADE, J.

VARELA GARCIA, F.J. see Vol. II for further entries

VARPAKHOVSKIJ, F.L. [1965] *The nonrealizability of a disjunction of nonrealizable formulas of propositional logic (Russian)* (J 0023) Dokl Akad Nauk SSSR 161*1257-1258
• TRANSL [1965] (J 0062) Sov Math, Dokl 6*568-570
◇ B55 F50 ◇ REV MR 31 # 2132 Zbl 236 # 02026 JSL 32.396 • ID 27855

VARPAKHOVSKIJ, F.L. [1971] *A class of realizable propositional formulae (Russian) (English summary)* (S 0228) Zap Nauch Sem Leningrad Otd Mat Inst Steklov 20*8-23,282
• TRANSL [1973] (J 1531) J Sov Math 1*1-11
◇ F50 ◇ REV MR 44 # 6449 Zbl 222 # 02020 Zbl 259 # 02020 • ID 13829

VARPAKHOVSKIJ, F.L. [1971] *The nonrealizability of the disjunction of nonrealizable formulas of propositional logic (Russian)* (S 0208) Uch Zap, Ped Inst, Moskva 277*214-217
◇ F50 ◇ REV MR 46 # 5117 • ID 13828

VASSEUR, J. [1973] see RUGGIU, G.

VAUZEILLES, J. [1982] *Functors and ordinal notations. III: dilators and gardens* (P 3708) Herbrand Symp Logic Colloq;1981 Marseille 333-364
◇ E10 F15 G30 ◇ REV Zbl 568 # 03029 • REM Part II 1984 by Vauzeilles,J. & Girard,J.-Y. Part IV 1985. • ID 43062

VAUZEILLES, J. [1984] see GIRARD, J.-Y.

VAUZEILLES, J. [1985] *Functors and ordinal notations. IV: the Howard ordinal and the functor Λ* (J 0036) J Symb Logic 50*331-338
◇ F15 G30 ◇ REM Part III 1982 • ID 42537

VEBLEN, O. [1908] *Continuous increasing functions of finite and transfinite ordinals* (J 0064) Trans Amer Math Soc 9*280-292
◇ E05 E10 F15 ◇ REV FdM 39.102 • ID 13917

VEBLEN, O. see Vol. I, V for further entries

VELDMAN, W. [1975] see LOPEZ-ESCOBAR, E.G.K.

VELDMAN, W. [1976] *An intuitionistic completeness theorem for intuitionistic predicate logic* (J 0036) J Symb Logic 41*159-166
◇ F50 ◇ REV MR 53 # 12886 Zbl 355 # 02018 • ID 14794

VELDMAN, W. [1981] see GIELEN, W.

VELDMAN, W. [1982] *On the constructive contrapositions of two axioms of countable choice* (P 3638) Brouwer Centenary Symp;1981 Noordwijkerhout 513-523
◇ E25 E70 F50 F55 ◇ REV MR 85g:03089 Zbl 532 # 03026 • ID 38283

VENNE, M. [1966] *Langues et theories d'ordres superieurs* (J 2313) C R Acad Sci, Paris, Ser A-B 262*A1142-A1143
◇ F35 ◇ REV MR 33 # 7241 Zbl 173.5 • ID 13926

VENNE, M. see Vol. I, III for further entries

VENNERI, B.M. [1979] *Arbres de deduction naturelle et programmation en LCF: possibilites de "normalisation"* (P 2952) CAAP'79 Arbres en Algeb & Progr (4);1979 Lille 261-274
◇ F05 ◇ REV MR 81c:68074 • ID 83064

VENNERI, B.M. [1980] see COPPO, M.

VENNERI, B.M. [1981] see COPPO, M.

VENNERI, B.M. see Vol. I, III for further entries

VENTURINI ZILLI, M. [1974] *On different kinds of indefinite* (J 0089) Calcolo 11*67-77
◇ B40 D75 ◇ REV MR 54 # 70 Zbl 294 # 02016 • ID 23952

VENTURINI ZILLI, M. [1978] see JACOPINI, G.

VENTURINI ZILLI, M. [1978] *Head recurrent terms in combinatory logic: A generalization of the notion of head normal form* (P 1872) Automata, Lang & Progr (5);1978 Udine 477-493
◇ B40 ◇ REV MR 80g:03014 Zbl 393 # 03010 • ID 52430

VENTURINI ZILLI, M. [1980] see LONGO, G.

VENTURINI ZILLI, M. [1983] *Cofinality in reduction graphs* (P 3889) CAAP'83 Arbres en Algeb & Progr (8);1983 L'Aquila 405-416
◇ B40 ◇ REV MR 85j:68071 Zbl 536 # 03004 • ID 37092

VENTURINI ZILLI, M. [1984] *Reduction graphs in the lambda calculus* (J 1426) Theor Comput Sci 29*251-275
◇ B40 ◇ REV MR 85k:68063 Zbl 538 # 03011 • ID 41463

VENTURINI ZILLI, M. [1985] see JACOPINI, G.

VENTURINI ZILLI, M. see Vol. I, IV for further entries

VERDU I SOLANS, V. [1981] see BATLE, N.

VERDU I SOLANS, V. see Vol. I, II, III, V for further entries

VERKHOZINA, M.I. [1973] *The undecidability of the separation problem for positive fragments of logical calculi (Russian)* (C 1443) Algebra, Vyp 2 (Irkutsk) 69-83
◇ D35 F50 ◇ REV MR 54 # 4959 • ID 24127

VERSHININ, V.A. [1975] *On the question of superposability of constructive sets with equality (Russian)* (J 0023) Dokl Akad Nauk SSSR 223*781-784
• TRANSL [1975] (J 0062) Sov Math, Dokl 16*976-980
◇ D25 F60 ◇ REV MR 53 # 7743 Zbl 338 # 02022 • ID 22991

VERSHININ, V.A. [1977] *On certain operations over constructive sets with equality (Russian)* (J 0023) Dokl Akad Nauk SSSR 234*269-272
• TRANSL [1977] (J 0062) Sov Math, Dokl 18*620-624
◇ F60 ◇ REV MR 57 # 5703 Zbl 399 # 03050 • ID 52846

VERSHININ, V.A. [1982] *Some limit theorems in the constructive probability theory (Russian)* (S 2874) Vopr Kibern, Akad Nauk SSSR 75*117-142
◇ F60 ◇ REV Zbl 516 # 03041 • ID 37270

VESLEY, R.E. [1963] *On strengthening intuitionistic logic* (J 0047) Notre Dame J Formal Log 4*80
◇ F50 ◇ REV MR 27 # 29 Zbl 118.13 JSL 34.307 JSL 35.307 • ID 13935

VESLEY, R.E. [1965] see KLEENE, S.C.

VESLEY, R.E. [1970] *A palatable substitute for Kripke's schema* (P 0603) Intuitionism & Proof Th;1968 Buffalo 197-207
◇ F35 F50 ◇ REV MR 44 # 5208 Zbl 199.299 JSL 39.334 • ID 13936

VESLEY, R.E. [1970] see KINO, A.

VESLEY, R.E. [1972] *Choice sequences and Markov's principle* (J 0020) Compos Math 24*33-53
◇ F35 F50 ◇ REV MR 47 # 1578 Zbl 237 # 02005 • ID 13937

VESLEY, R.E. [1980] *Intuitionistic analysis: the search for axiomatization and understanding* (P 2058) Kleene Symp;1978 Madison 317-331
◇ A10 F35 F50 ◇ REV MR 81m:03072 Zbl 473 # 03005 • ID 55336

VESLEY, R.E. [1983] see SCEDROV, A.

VESLEY, R.E. see Vol. I for further entries

VESSEL', KH.A. [1973] *Ueber eine logische explizite Beschreibung der Entwicklungsterme (Russian)* (C 1662) Teor Log Vyvoda 259-270
◇ F05 ◇ REV Zbl 263 # 02007 • ID 29859

VESSEL', KH.A. see Vol. II for further entries

VETULANI, Z. [1984] *Ramified analysis and the minimal β-models of higher order arithmetics* (J 0027) Fund Math 121*1-15
◇ C62 D55 E45 F35 F65 ◇ REV Zbl 576 # 03021 • ID 44142

VETULANI, Z. see Vol. III, V for further entries

VIER, L.C. [1972] *Church's thesis in northern Dutch constructivism* (S 3303) Stud Logic Found Math x+321pp
◇ D20 F50 ◇ ID 49913

VISSER, A. [1980] *Numerations, λ-calculus & arithmetic* (**C** 3050) Essays Combin Log, Lambda Calc & Formalism (Curry) 259–284
◇ B40 D45 F30 ◇ REV MR 83c:03041 Zbl 469 # 03006 • ID 35141

VISSER, A. [1981] *An incompleteness result for paths through or within O* (**J** 0028) Indag Math 43*237–243
◇ F15 ◇ REV MR 84g:03094 Zbl 459 # 03025 • ID 54465

VISSER, A. [1982] *On the completeness principle: a study of provability in Heyting's arithmetic and extensions* (**J** 0007) Ann Math Logic 22*263–295
◇ F30 F50 ◇ REV MR 84h:03130 Zbl 505 # 03026 • ID 34317

VISSER, A. [1984] *The provability logic of recursively enumerable theories extending Peano arithmetic and arbitrary theories extending Peano arithmetic* (**J** 0122) J Philos Logic 13*97–113
◇ B45 C62 F30 ◇ REV MR 85j:03102 • ID 42643

VISSER, A. see Vol. II for further entries

VOGEL, HELMUT [1975] see DILLER, J.

VOGEL, HELMUT [1976] *Ein starker Normalisationssatz fuer die bar-rekursiven Funktionale* (**J** 0009) Arch Math Logik Grundlagenforsch 18*81–84
◇ F10 F35 ◇ REV MR 58 # 171 Zbl 352 # 02035 • ID 23717

VOGEL, HELMUT [1977] *Ausgezeichnete Folgen fuer praedikative Ordinalzahlen und praedikativ-rekursive Funktionen* (**J** 0068) Z Math Logik Grundlagen Math 23*435–438
◇ D20 F15 ◇ REV MR 58 # 16232 Zbl 453 # 03043 • ID 54173

VOGEL, HELMUT [1978] *Eine beweistheoretische Anwendung partieller stetiger Funktionale* (**J** 0009) Arch Math Logik Grundlagenforsch 19*65–78
◇ F10 F35 F50 ◇ REV MR 80h:03087 Zbl 401 # 03021 • ID 29165

VOGEL, HELMUT [1979] *Ueber die mit dem Bar-Rekursor vom Typ 0 definierbaren Ordinalzahlen* (**J** 0009) Arch Math Logik Grundlagenforsch 19*165–173
◇ F10 F15 ◇ REV MR 80m:03092 Zbl 414 # 03034 • ID 53080

VOGEL, HELMUT [1979] *Ueber ein mit der Bar-Induktion verwandtes Schema* (**J** 0068) Z Math Logik Grundlagen Math 25*465–473
◇ F10 F35 F50 ◇ REV MR 80m:03095 Zbl 426 # 03059 • ID 53655

VOGEL, HELMUT [1981] *Der finite Standpunkt in Hilberts Programm* (**C** 2981) Phil der Math Erkenntnis Muenster 1978/79 11–25
◇ A05 F65 F99 ◇ REV MR 82h:03064 Zbl 464030004 • ID 79736

VOGEL, HELMUT [1982] *Eine Variante des Ordinalzahlbezeichungssystems Σ* (**J** 0009) Arch Math Logik Grundlagenforsch 22*27–41
◇ F15 ◇ REV MR 85b:03102 Zbl 496 # 03039 • ID 36866

VOGEL, HELMUT see Vol. IV for further entries

VOLKEN, H. [1976] see BARENDREGT, H.P.

VOLKEN, H. [1978] see BARENDREGT, H.P.

VOPENKA, P. [1966] *A new proof of the Goedel's result on non-provability of consistency (Russian summary)* (**J** 0014) Bull Acad Pol Sci, Ser Math Astron Phys 14*111–116
◇ E30 E70 F30 ◇ REV MR 34 # 7361 Zbl 156.250 • ID 46939

VOPENKA, P. & HAJEK, P. [1972] *The theory of semisets* (**X** 0809) North Holland: Amsterdam 332pp
◇ E35 E40 E70 E98 F25 ◇ REV MR 56 # 2824 Zbl 332 # 02064 JSL 49.1422 • ID 23308

VOPENKA, P. see Vol. III, IV, V for further entries

VOROB'EV, N.N. [1952] *A constructive propositional calculus with strong negation (Russian)* (**J** 0023) Dokl Akad Nauk SSSR 85*465–468
◇ F50 ◇ REV MR 14.233 Zbl 47.251 JSL 18.257 • ID 16334

VOROB'EV, N.N. [1952] *The problem of deducibility in the constructive propositional calculus with strong negation (Russian)* (**J** 0023) Dokl Akad Nauk SSSR 85*689–692
◇ F50 ◇ REV MR 14.233 Zbl 47.252 JSL 18.258 • ID 19800

VOROB'EV, N.N. [1958] *A new algorithm of deducibility in the constructive propositional calculus (Russian)* (**S** 0066) Tr Mat Inst Steklov 52*193–225
◇ F50 ◇ REV MR 20 # 6361 Zbl 87.253 JSL 29.109 • ID 16332

VOROB'EV, N.N. [1964] *A constructive calculus of statements with strong negation (Russian)* (**S** 0066) Tr Mat Inst Steklov 72*195–227
• TRANSL [1972] (**J** 0225) Amer Math Soc, Transl, Ser 2 99*40–82
◇ B55 F50 ◇ REV MR 33 # 44 Zbl 202.295 • ID 16331

VREDENDUIN, P.G.J. [1949] *The constructive method* (**P** 0682) Int Congr Philos (10);1948 Amsterdam 749–751
◇ A05 F99 ◇ REV MR 10.585 Zbl 33.6 JSL 14.247 • ID 13890

VREDENDUIN, P.G.J. [1954] *The logic of negationless mathematics* (**J** 0020) Compos Math 11*204–270
◇ F50 ◇ REV MR 15.846 Zbl 53.202 JSL 21.92 • ID 13891

VREDENDUIN, P.G.J. see Vol. II for further entries

VRIJER DE, R. [1975] *Big trees in a λ-calculus with λ-expressions as types* (**P** 1603) λ-Calc & Comput Sci Th;1975 Roma 252–271
◇ B40 ◇ REV MR 58 # 5085 Zbl 356 # 02024 • ID 50296

VRIJER DE, R. [1985] *A direct proof of the finite developments theorem* (**J** 0036) J Symb Logic 50*339–343
◇ B40 ◇ ID 42538

VUCKOVIC, V. [1959] *Partially ordered recursive arithmetics* (**J** 0132) Math Scand 7*305–320
◇ D20 F30 ◇ REV MR 22 # 7939 Zbl 93.12 JSL 28.251 • ID 13892

VUCKOVIC, V. [1960] *Rekursive Modelle einiger nichtklassischer Aussagenkalkuele (Serbo-Croatian)* (**J** 1179) Filoz Jugos Chas (Belgrade) 4*69–84
◇ B22 F30 ◇ REV JSL 28.291 • ID 22133

VUCKOVIC, V. [1962] *Einfuehrung von $\Sigma_f(x)$ und $\Pi_f(x)$ in der rekursiven Gitterpunktarithmetik* (**J** 1156) Izv Bulgar Akad Nauk Mat Inst 6*15–25
◇ D20 F30 ◇ REV MR 26 # 1250 JSL 28.251 • ID 22057

VUCKOVIC, V. [1967] *A recursive model for the extended system A of B. Sobocinski* (J 0047) Notre Dame J Formal Log 8*154-158
◇ B45 F30 ◇ REV MR 38 # 2015 Zbl 262 # 02022
• ID 27479

VUCKOVIC, V. [1967] *Mathematics of incompleteness and undecidability* (J 0068) Z Math Logik Grundlagen Math 13*123-150
◇ D35 F30 ◇ REV MR 35 # 5303 Zbl 183.10 JSL 37.195
• ID 13899

VUCKOVIC, V. [1967] *Recursive models for three-valued propositional calculi with classical implication* (J 0047) Notre Dame J Formal Log 8*148-153
◇ B50 C57 C90 F30 ◇ REV Zbl 262 # 02021 • ID 27478

VUCKOVIC, V. [1971] *Combinatorial operators and their quasi-inverses* (J 0047) Notre Dame J Formal Log 12*305-308
◇ B40 ◇ REV MR 48 # 1891 Zbl 197.281 • ID 13905

VUCKOVIC, V. see Vol. III, IV, V for further entries

WADSWORTH, C.P. [1976] *The relation between computational and denotational properties for Scott's D_∞-models of the lambda-calculus* (J 1428) SIAM J Comp 5*488-521
◇ B40 ◇ REV MR 58 # 21493 Zbl 346 # 02013 • ID 27597

WADSWORTH, C.P. [1978] *Approximate reduction and λ calculus models* (J 1428) SIAM J Comp 7*337-356
◇ B40 ◇ REV MR 81f:03020 Zbl 407 # 03021 • ID 56187

WADSWORTH, C.P. [1980] *Some unusual lambda-calculus numeral systems* (C 3050) Essays Combin Log, Lambda Calc & Formalism (Curry) 215-230
◇ B40 ◇ REV MR 82b:03043 Zbl 469 # 03006 • ID 79783

WAERDEN VAN DER, B.L. [1930] *Eine Bemerkung ueber die Unzerlegbarkeit von Polynomen* (J 0043) Math Ann 102*738-739
◇ B25 C60 D45 F55 ◇ REV FdM 56.825 • ID 38703

WAERDEN VAN DER, B.L. see Vol. V for further entries

WAGNER, E.G. [1969] *Uniformly reflexive structures: on the nature of Goedelization and relative computability* (J 0064) Trans Amer Math Soc 144*1-41
◇ B40 D45 D75 F40 ◇ REV MR 40 # 2543
Zbl 265 # 02029 • ID 13965

WAGNER, E.G. see Vol. II, IV, V for further entries

WAINER, S.S. [1974] see LOEB, M.H.

WAINER, S.S. [1975] see SCHWICHTENBERG, H.

WAINER, S.S. [1983] see CICHON, E.A.

WAINER, S.S. [1984] see DENNIS-JONES, E.C.

WAINER, S.S. [1985] *Subrecursive ordinals* (P 3342) Rec Th Week;1984 Oberwolfach 405-418
◇ D20 F15 ◇ ID 45318

WAINER, S.S. [1985] *The "slow-growing" Π^1_2 approach to hierarchies* (P 4046) Rec Th;1982 Ithaca 487-502
◇ D20 F15 ◇ ID 46381

WAINER, S.S. see Vol. III, IV, V for further entries

WAJSBERG, M. [1933] *Beitrag zur Metamathematik* (J 0043) Math Ann 109*200-229
• TRANSL [1977] (C 4055) Wajsberg: Logical Works 62-88 [1935] (J 4710) Pol Tow Mat, Wiad Mat 39*43-84
◇ A05 B10 B30 B96 C35 F99 ◇ REV Zbl 8.97
FdM 59.53 FdM 61.972 • ID 40808

WAJSBERG, M. [1938] *Untersuchungen ueber den Aussagenkalkuel von A.Heyting* (J 4710) Pol Tow Mat, Wiad Mat 46*45-101 • ERR/ADD ibid 47*139
• TRANSL [1977] (C 4055) Wajsberg: Logical Works 132-171
◇ B25 F50 ◇ REV Zbl 19.385 JSL 3.169 JSL 48.873
FdM 65.1104 • ID 13982

WAJSBERG, M. see Vol. I, II, III, V for further entries

WAND, M. [1978] see FRIEDMAN, D.P.

WAND, M. [1985] see MEYER, A.R.

WAND, M. see Vol. II, IV for further entries

WANDSCHNEIDER, D. [1975] *Zur Eliminierung des Goedelschen Unvollstaendigkeitsproblems im Zusammenhang mit dem Antinomien-Problem* (J 0989) Z Allg Wissth 6*65-81
◇ A05 E30 F30 ◇ REV MR 57 # 91 • ID 79815

WANDSCHNEIDER, D. [1976] *Reflexive Unbeweisbarkeitsaussagen. Anmerkungen zur Grundsatzdiskussion um Goedels Unvollstaendigkeitsproblem* (J 0989) Z Allg Wissth 7*133-140
◇ A05 E30 F30 ◇ REV MR 58 # 16295c • ID 79813

WANG, HAO [1949] *A theory of constructive types* (J 0175) Methodos 1*374-384
◇ B15 F35 F65 ◇ REV MR 12.384 Zbl 36.165 JSL 19.288
• ID 14006

WANG, HAO [1950] *A formal system of logic* (J 0036) J Symb Logic 15*25-32
◇ E70 F25 ◇ REV MR 11.636 Zbl 39.246 JSL 15.228
• ID 14013

WANG, HAO [1950] see ROSSER, J.B.

WANG, HAO [1950] *Remarks on the comparison of axiom systems* (J 0054) Proc Nat Acad Sci USA 36*448-453
◇ B15 F25 F30 ◇ REV MR 12.578 Zbl 37.296 JSL 16.142
• ID 14011

WANG, HAO [1951] *Arithmetic models for formal systems* (J 0175) Methodos 3*217-232
◇ F25 F30 ◇ REV MR 14.439 Zbl 45.296 JSL 20.76
• ID 14015

WANG, HAO [1951] *Arithmetic translation of axiom systems* (J 0064) Trans Amer Math Soc 71*283-293
◇ F25 F30 ◇ REV MR 13.198 Zbl 43.9 JSL 21.402
• ID 14016

WANG, HAO [1952] *The irreducibility of impredicative principles* (J 0043) Math Ann 125*56-66
◇ E30 E70 F65 ◇ REV MR 14.439 Zbl 49.165
JSL 18.265 • ID 14020

WANG, HAO [1952] *Truth definitions and consistency proofs* (J 0064) Trans Amer Math Soc 73*243-275
• REPR [1963] (C 1009) Wang: Survey Math Logic 443-477
◇ F25 F30 F65 ◇ REV MR 14.123 Zbl 47.13 JSL 22.365
JSL 30.100 • ID 14019

WANG, HAO [1953] *Between number theory and set theory* (J 0043) Math Ann 126*385-409
• REPR [1963] (C 1009) Wang: Survey Math Logic 478-506
◇ B28 E30 F30 ◇ REV MR 15.670 Zbl 51.246 JSL 22.82 JSL 30.100 • ID 14023

WANG, HAO [1953] *Certain predicates defined by induction schemata* (J 0036) J Symb Logic 18*49-59
◇ D70 F30 F35 ◇ REV MR 14.936 Zbl 51.5 JSL 30.99 • ID 14021

WANG, HAO [1953] *Quelques notions d'axiomatique* (J 0252) Rev Philos Louvain 51*409-443
◇ B25 B30 C07 C35 E30 F30 ◇ REV JSL 20.289 • ID 14024

WANG, HAO [1954] *The formalization of mathematics* (J 0036) J Symb Logic 19*241-266
• REPR [1963] (C 1009) Wang: Survey Math Logic 559-584
◇ A05 B30 E70 F35 F65 ◇ REV MR 16.661 Zbl 56.245 JSL 22.290 JSL 30.100 • ID 14025

WANG, HAO [1955] see KREISEL, G.

WANG, HAO [1955] *Undecidable sentences generated by semantic paradoxes* (J 0036) J Symb Logic 20*31-43
• REPR [1963] (C 1009) Wang: Survey Math Logic 68-81
◇ A05 B28 F30 ◇ REV MR 16.988 Zbl 64.245 JSL 30.100 • ID 14026

WANG, HAO [1957] *Remarks on constructive ordinals and set theory* (P 1675) Summer Inst Symb Log;1957 Ithaca 383-390
◇ D55 E45 F15 ◇ ID 29384

WANG, HAO [1957] *Symbolic representations of calculating machines* (P 1675) Summer Inst Symb Log;1957 Ithaca 181-188
◇ B70 D05 D10 F30 ◇ REV Zbl 145.407 JSL 27.103 • ID 14033

WANG, HAO [1957] *The axiomatization of arithmetic* (J 0036) J Symb Logic 22*145-158
◇ A05 B28 F30 ◇ REV MR 20 # 6963 Zbl 78.5 JSL 27.77 • ID 33305

WANG, HAO [1958] *Alternative proof of a theorem of Kleene* (J 0036) J Symb Logic 23*250
◇ F15 ◇ REV MR 21 # 2584 JSL 27.81 • ID 14031

WANG, HAO [1959] see KREISEL, G.

WANG, HAO [1959] *Ordinal numbers and predicative set theory* (J 0068) Z Math Logik Grundlagen Math 5*216-239
• REPR [1963] (C 1009) Wang: Survey Math Logic 624-651
◇ A05 E70 F15 F65 ◇ REV MR 24 # A2534 Zbl 90.9 JSL 30.250 • ID 14032

WANG, HAO [1963] *Ackermann's consistency proof* (C 1009) Wang: Survey Math Logic 362-375
◇ F25 F30 ◇ REV JSL 29.106 • ID 14048

WANG, HAO [1963] *Partial systems of number theory* (C 1009) Wang: Survey Math Logic 376-382
◇ C62 F30 ◇ REV JSL 29.147 • ID 14049

WANG, HAO [1963] *Relative strength and reducibility* (C 1009) Wang: Survey Math Logic 432-442
◇ B10 E30 E70 F25 F30 ◇ REV JSL 29.208 • ID 14050

WANG, HAO [1963] *Some formal details on predicative set theories* (C 1009) Wang: Survey Math Logic 585-623
◇ B15 E70 F65 ◇ REV JSL 30.250 • ID 14044

WANG, HAO [1963] *The arithmetization of metamathematics* (C 1009) Wang: Survey Math Logic 334-361
◇ F25 F30 ◇ REV JSL 29.106 • ID 14046

WANG, HAO [1967] *Introductory note* (C 0675) From Frege to Goedel 414-416
◇ A10 F50 ◇ ID 14926

WANG, HAO [1981] *Popular lectures on mathematical logic (Chinese)* (X 1876) Kexue Chubanshe: Beijing vii+257pp
• TRANSL [1981] (X 0864) Van Nostrand: New York ix+273pp
◇ A05 B98 C98 E98 F30 F98 ◇ REV MR 82e:03001 MR 84g:03002 JSL 47.908 • ID 34116

WANG, HAO see Vol. I, II, III, IV, V for further entries

WARD, S.A. [1980] see HALSTEAD, R.H.J.

WARREN, D.S. [1980] see FRIEDMAN, JOYCE

WARREN, D.S. see Vol. IV for further entries

WASILEWSKA, A. [1973] *The diagrams of formulas of the intuitionistic propositional calculus (Polish and Russian summaries)* (J 0063) Studia Logica 32*109-115
◇ F50 ◇ REV MR 50 # 4256 Zbl 322 # 02021 • ID 14075

WASILEWSKA, A. [1976] *A sequence formalization for SCI* (J 0063) Studia Logica 35*213-217
◇ B25 F07 ◇ REV MR 55 # 10238 Zbl 346 # 02006 • ID 65985

WASILEWSKA, A. [1976] *On the decidability theorems* (J 1929) Prace Centr Oblicz Pol Akad Nauk 246*16pp
◇ B25 F07 ◇ REV Zbl 338 # 68066 • ID 65986

WASILEWSKA, A. [1980] *On the Gentzen-type formalizations* (J 0068) Z Math Logik Grundlagen Math 26*439-444
◇ B25 D05 F07 ◇ REV MR 82g:03102 Zbl 471 # 03021 • ID 55216

WASILEWSKA, A. see Vol. I, II, III, IV for further entries

WAUW-DE KINDER VAN DE, G. [1975] *Arithmetique de premier ordre dans les topos (English summary)* (J 2313) C R Acad Sci, Paris, Ser A-B 280*A1579-A1582
◇ F30 F50 G30 H15 ◇ REV MR 52 # 8217 Zbl 307 # 18001 • ID 18421

WAVRE, R. [1926] *Sur le principe du tiers exclu* (J 0145) Rev Metaph Morale 33*425-430
◇ A05 F99 ◇ REV FdM 52.47 • ID 38710

WEAVER, G.E. [1969] see CORCORAN, J.

WEAVER, G.E. see Vol. I, II, III, V for further entries

WEBB, J.C. [1980] *Mechanism, mentalism, and metamathematics. An essay on finitism* (X 0835) Reidel: Dordrecht xiii+277pp
◇ A05 D20 F99 ◇ REV MR 83j:03015 Zbl 454 # 03001 JSL 51.472 • ID 54214

WEBB, J.C. [1983] *Goedel's theorems and Church's thesis: A prologue to mechanism* (C 3834) Lang, Logic and Method 309-353
◇ A05 D20 F99 ◇ REV Zbl 537 # 03002 • ID 43711

WEESE, M. [1976] *The universality of boolean algebras with the Haertig quantifier* (P 1476) Set Th & Hierarch Th (2) (Mostowski);1975 Bierutowice 291-296
◇ C55 C80 D35 F25 G05 ◇ REV MR 54 # 12478 Zbl 331 # 02027 • ID 23814

WEESE, M. see Vol. I, II, III, IV, V for further entries

WEIHRAUCH, K. [1984] see KREITZ, C.

WEIHRAUCH, K. see Vol. IV for further entries

WEINSTEIN, S. [1979] *Some applications of Kripke models to formal systems of intuitionistic analysis* (**J** 0007) Ann Math Logic 16*1-32
◇ C90 F35 F50 ◇ REV MR 80h:03088 Zbl 431 # 03037 • ID 53943

WEINSTEIN, S. [1983] *The intended interpretation of intuitionistic logic* (**J** 0122) J Philos Logic 12*261-270
◇ A05 F50 ◇ REV MR 85g:03090 Zbl 539 # 03037 • ID 41239

WELCH, P. [1975] *Continuous semantics and inside-out reductions* (**P** 1603) λ-Calc & Comput Sci Th;1975 Roma 122-146
◇ B40 ◇ REV MR 58 # 27357 Zbl 399 # 03012 • ID 52809

WELCH, P. see Vol. IV, V for further entries

WELLS, C. [1985] see BARR, M.

WENDEL, N. [1978] *The inconsistency of Bernini's very strong intuitionistic theory* (**J** 0063) Studia Logica 37*341-347
◇ B55 F35 F50 ◇ REV MR 80m:03096 Zbl 416 # 03051 • ID 53217

WENDEL, N. [1983] *Kontextuelle Definition neuer Wahlfolgen* (**X** 4636) Goethe Univ, Fachber Math: Frankfurt 132pp
◇ F35 F50 ◇ REV Zbl 572 # 03035 • ID 49235

WERMUS, H. [1961] *Eine konstruktiv-figuerliche Begruendung eines Abschnittes der zweiten Zahlklasse* (**J** 2022) Comm Math Helvetici 35*263-318
◇ F15 ◇ REV MR 23 # A2309 Zbl 127.244 • ID 14139

WERNER, G. [1979] see COMYN, G.

WESSEL, H. [1984] *Kritische Bemerkungen zur inuitionistischen Logikkonzeption* (**P** 3621) Frege Konferenz (2);1984 Schwerin 284-288
◇ A05 F99 ◇ REV MR 85m:03006 • ID 45365

WESSEL, H. see Vol. I, II for further entries

WESSELS, L. [1977] *Cut elimination in a Gentzen-style ε-calculus without identity* (**J** 0068) Z Math Logik Grundlagen Math 23*527-538
◇ F05 ◇ REV MR 58 # 27372 Zbl 381 # 03042 • ID 51910

WESSELS, L. see Vol. II for further entries

WETTE, E. [1959] *Von operativen Modellen der axiomatischen Mengenlehre* (**P** 0634) Constructivity in Math;1957 Amsterdam 266-277
◇ C62 E35 F99 ◇ REV MR 24 # A1832 Zbl 86.10 JSL 40.501 • ID 14162

WETTE, E. [1960] *Intuitionistisch-rekursiver Konsistenzbeweis fuer die axiomatische Mengenlehre* (**P** 1953) Int Congr Log, Meth & Phil of Sci (1;Abstr);1960 Stanford 31-33
◇ E30 E35 F50 ◇ ID 31556

WETTE, E. [1969] *Definition eines (relativ vollstaendigen) formalen Systems konstruktiver Arithmetik* (**C** 0705) Found of Math (Goedel) 130-195
◇ F10 F15 F30 F35 F65 ◇ REV MR 42 # 2941 Zbl 181.307 JSL 37.203 • ID 14164

WETTE, E. [1970] *Vom Unendlichen zum Endlichen* (**J** 0076) Dialectica 24*303-323
◇ A05 B28 E30 F30 ◇ REV Zbl 263 # 02008 • ID 14165

WETTE, E. [1974] *Contradiction within pure number theory because of a system-internal "consistency"-deduction* (**J** 0286) Int Logic Rev 9*51-62
◇ B10 F30 ◇ REV Zbl 335 # 02033 • ID 14955

WETTE, E. [1974] *The refutation of number theory. I* (**J** 0286) Int Logic Rev 10*8pp
◇ D10 F30 ◇ REM Without use of any natural language. Printed in two colours • ID 31553

WETTE, E. [1980] *Inconsistenza dell'induzione matematica* (**J** 4058) Nominazione 1*153-158
◇ F30 ◇ ID 25950

WETTE, E. see Vol. II, IV, V for further entries

WEYL, H. [1910] *Ueber die Definition der mathematischen Grundbegriffe* (**J** 3978) Math-Nat Blaetter 7*93-95,109-113
◇ A05 B30 F99 ◇ REV FdM 41.89 • ID 42999

WEYL, H. [1918] *Das Kontinuum* (**X** 2636) Veit: Leipzig iv+83pp
◇ A05 B28 E30 F65 ◇ REV MR 22 # 10886 JSL 25.282 FdM 46.56 • REM Also reprinted in: Weyl,H. 1966 "Das Kontinuum und andere Monographien" Chelsea: New York • ID 19791

WEYL, H. [1919] *Der "Circulus vitiosus" in der heutigen Begruendung der Analysis* (**J** 0157) Jbuchber Dtsch Math-Ver 28*85-92
◇ A05 B28 F65 ◇ REV FdM 47.895 • ID 14166

WEYL, H. [1921] *Ueber die neue Grundlagenkrise der Mathematik* (**J** 0044) Math Z 10*39-70
• REPR [1956] (**C** 4557) Weyl: Selecta 211-247
◇ A05 E70 F55 ◇ REV JSL 22.81 FdM 48.220 FdM 48.47 • ID 14167

WEYL, H. [1924] *Randbemerkungen zu Hauptproblemen der Mathematik* (**J** 0044) Math Z 20*131-150
◇ B30 F55 ◇ REV FdM 50.37 • ID 38712

WEYL, H. [1925] *Die heutige Erkenntnislage in der Mathematik* (**J** 4171) Symposion 1*1-32
◇ A05 F99 ◇ REV FdM 51.43 • ID 41682

WEYL, H. [1928] *Diskussionsbemerkungen zu dem zweiten Hilbertschen Vortrag ueber die Grundlagen der Mathematik* (**J** 0107) Abh Math Sem Univ Hamburg 6*86-88
• TRANSL [1967] (**C** 0675) From Frege to Goedel 482-484
◇ A05 F50 ◇ REV FdM 54.56 • ID 14168

WEYL, H. [1929] *Consistency in mathematics* (**J** 1012) Rice Inst Pamphlet 16*245-265
◇ A05 F99 ◇ REV FdM 55.625 • ID 14169

WEYL, H. [1931] *Die Stufen des Unendlichen* (1111) Preprints, Manuscr., Techn. Reports etc. 19pp
◇ A05 E10 F99 ◇ REV FdM 57.1315 • REM Vortrag am 27.Okt.1930 bei der Gaestetagung der Math. Gesellsch. der Univ. Jena in Abbeanum • ID 22091

WEYL, H. [1944] *David Hilbert and his mathematical work* (**J** 0015) Bull Amer Math Soc 50*612-654
◇ A10 F98 ◇ REV MR 6.142 Zbl 60.15 JSL 9.98 • ID 41510

WEYL, H. [1985] *Axiomatic versus constructive procedures in mathematics* (**J** 2789) Math Intell 7/4*10-17,38
◇ A05 B30 F99 ◇ REV Zbl 569 # 01011 • ID 49136

WEYL, H. see Vol. I, II, V for further entries

WHITE, R.B. [1979] *The consistency of the axiom of comprehension in the infinite-valued predicate logic of Lukasiewicz* (J 0122) J Philos Logic 8*509-534
◇ B50 E35 E70 F05 ◇ REV MR 81b:03059
Zbl 418 # 03037 • ID 53322

WHITE, R.B. see Vol. II for further entries

WIEDMER, E. [1974] *Ein neuer negationsloser Beweis eines Satzes von G.F.C.Griss* (J 0028) Indag Math 36*89-93 • ERR/ADD ibid 77(A)*412)
◇ B20 F55 ◇ REV MR 49 # 26a MR 49 # 26b
Zbl 276 # 02017 • ID 14190

WIEDMER, E. see Vol. IV for further entries

WIELE VAN DE, J. [1982] *Recursive dilators and generalized recursions* (P 3708) Herbrand Symp Logic Colloq;1981 Marseille 325-332
◇ D60 D65 F15 F35 ◇ REV MR 85i:03184
Zbl 499 # 03035 • ID 38124

WILKIE, A.J. [1978] *Some results and problems on weak systems of arithmetic* (P 1897) Logic Colloq;1977 Wroclaw 285-296
◇ C62 F30 H15 ◇ REV MR 81c:03050 Zbl 449 # 03076
• ID 56743

WILKIE, A.J. [1980] *Applications of complexity theory to Σ_0-definability problems in arithmetic* (P 2625) Model Th of Algeb & Arithm;1979 Karpacz 363-369
◇ C62 D15 F30 ◇ REV MR 82b:03085 Zbl 483 # 03024
• ID 79968

WILKIE, A.J. [1981] see PARIS, J.B.

WILKIE, A.J. [1982] *On core structures for Peano arithmetic* (P 3623) Logic Colloq;1980 Prague 311-314
◇ C50 C62 F30 H15 ◇ REV MR 84j:03142
Zbl 517 # 03030 • ID 34735

WILKIE, A.J. [1984] see PARIS, J.B.

WILKIE, A.J. [1985] see PARIS, J.B.

WILKIE, A.J. see Vol. III, IV, V for further entries

WILMERS, G.M. [1982] see CEGIELSKI, P.

WILMERS, G.M. [1985] *Bounded existential induction* (J 0036) J Symb Logic 50*72-90
◇ C62 F30 ◇ REV MR 86h:03068 • ID 40251

WILMERS, G.M. see Vol. III, V for further entries

WINKLMANN, K. [1978] see MACHTEY, M.

WINKLMANN, K. see Vol. II, IV for further entries

WIWEGER, A. [1978] see OBTULOWICZ, A.

WIWEGER, A. [1981] see OBTULOWICZ, A.

WIWEGER, A. [1982] see OBTULOWICZ, A.

WIWEGER, A. [1984] *Pre-adjunctions and lambda-algebraic theories* (S 0019) Colloq Math (Warsaw) 48*153-165
◇ B40 G30 ◇ REV MR 86d:18001 Zbl 564 # 18003
• ID 44205

WIWEGER, A. see Vol. V for further entries

WOEHL, K. [1979] *Zur Komplexitaet der Presburger Arithmetik und des Aequivalenz-Problems einfacher Programme* (P 3488) Theor Comput Sci (4);1979 Aachen 310-318
◇ B25 B75 D15 F20 F30 ◇ REV MR 81k:03015
Zbl 419 # 03024 • ID 53367

WOJCICKI, R. [1970] *On reconstructability of the classical propositional logic in intuitionistic logic (Russian summary)* (J 0014) Bull Acad Pol Sci, Ser Math Astron Phys 18*421-422
◇ B55 F50 ◇ REV MR 43 # 1816 Zbl 218 # 02009
• ID 14254

WOJCICKI, R. [1971] see TOKARZ, M.

WOJCICKI, R. see Vol. I, II, III, V for further entries

WOJTYLAK, P. [1981] *Mutual interpretability of sentential logics. I,II* (J 0302) Rep Math Logic, Krakow & Katowice 11*69-89,12*51-66
◇ B22 F25 ◇ REV MR 82k:03033 Zbl 468 # 03006
Zbl 468 # 03007 • ID 55071

WOJTYLAK, P. [1983] *Collapse of a class of infinite disjunctions in intuitionistic propositional logic* (J 0302) Rep Math Logic, Krakow & Katowice 16*37-49
◇ F50 ◇ REV MR 86b:03017 • ID 44013

WOJTYLAK, P. [1984] *A proof of Herbrand's theorem* (J 0302) Rep Math Logic, Krakow & Katowice 17*13-17
◇ C10 F05 ◇ REV MR 85m:03007 Zbl 573 # 03002
• ID 44096

WOJTYLAK, P. [1984] *A recursive theory for the $\{\neg, \wedge, \vee, \rightarrow, o\}$ fragment of intuitionistic logic* (J 0302) Rep Math Logic, Krakow & Katowice 18*3-35
◇ F50 ◇ REV MR 86h:03045 • ID 45541

WOJTYLAK, P. see Vol. I, II for further entries

WOLTER, H. [1973] *Eine Erweiterung der elementaren Praedikatenlogik: Anwendungen in der Arithmetik und anderen mathematischen Theorien* (J 0068) Z Math Logik Grundlagen Math 19*181-190
◇ B28 C10 C80 F30 ◇ REV MR 51 # 7815
Zbl 302 # 02001 • ID 14280

WOLTER, H. [1975] *Entscheidbarkeit der Arithmetik mit Addition und Ordnung in Logiken mit verallgemeinerten Quantoren* (J 0068) Z Math Logik Grundlagen Math 21*321-330
◇ B25 C10 C55 C80 F30 ◇ REV MR 51 # 7816
Zbl 337 # 02032 • ID 66137

WOLTER, H. see Vol. I, III, IV, V for further entries

WRAITH, G.C. [1975] see LAWVERE, F.W.

WRAITH, G.C. [1976] *Logic from topology: a survey of topoi* (J 2679) Bull Inst Math Appl (Southend oS) 12*115-119
◇ C98 F98 G30 ◇ REV MR 58 # 27176 • ID 80097

WRAITH, G.C. [1980] *Intuitionistic algebra: some recent developments in topos theory* (P 1959) Int Congr Math (II,13);1978 Helsinki 1*331-337
◇ F50 G30 ◇ REV MR 81f:18021 Zbl 429 # 03049
• ID 53879

WRAITH, G.C. see Vol. III, V for further entries

WRIGHT, C. [1982] *Strict finitism* (J 0154) Synthese 51*202-282
◇ A05 F65 ◇ REV MR 84b:03018 Zbl 497 # 03002
• ID 35617

WRIGHT, C. see Vol. I for further entries

WRIGHTSON, G. [1982] see JANSOHN, H.-S.

WRIGHTSON, G. see Vol. I, II for further entries

WRONSKI, A. [1971] *Axiomatization of the implicational Goedel's matrices by Kalmar's method (Polish summary)* (S 0458) Zesz Nauk, Prace Log, Uniw Krakow 6*89-98
◇ B55 F50 ◇ REV MR 45 # 6590 • ID 14337

WRONSKI, A. [1974] see POREBSKA, M.

WRONSKI, A. [1974] *On cardinalities of matrices strongly adequate for the intuitionistic propositional logic* (J 0302) Rep Math Logic, Krakow & Katowice 3*67-72
◇ B55 F50 ◇ REV MR 52 # 7858 Zbl 342 # 02011
• ID 18433

WRONSKI, A. [1974] see SURMA, S.J.

WRONSKI, A. & ZYGMUNT, J. [1974] *Remarks on the free pseudo-boolean algebra with one-element free-generating set* (J 0302) Rep Math Logic, Krakow & Katowice 2*77-81
◇ F50 G10 ◇ REV MR 51 # 12520 Zbl 312 # 02002
• ID 17238

WRONSKI, A. [1974] *The degree of completeness of some fragments of the intuitionistic propositional logic* (J 0302) Rep Math Logic, Krakow & Katowice 2*55-61
◇ B20 F50 ◇ REV MR 52 # 5401 Zbl 312 # 02023
• ID 18432

WRONSKI, A. [1975] see POREBSKA, M.

WRONSKI, A. [1975] see KABZINSKI, J.K.

WRONSKI, A. [1975] see SURMA, S.J.

WRONSKI, A. see Vol. I, II, III for further entries

WU, WENJUN [1981] *A constructive theory of algebraic topology I. Notions of measure and calculability (Chinese summary)* (J 4573) Xilong Kexue yu Shuxue Xuebao 1/1*53-68
◇ F60 ◇ ID 49676

WU, WENJUN [1985] *Renaisance of constructive mathematics (Chinese)* (J 0420) Shuxue Jinzhan 14/4*334-339
◇ A05 F55 ◇ ID 49382

WU, WENJUN see Vol. I for further entries

XENAKIS, J. [1975] *Natural deduction "puzzle"* (J 0079) Logique & Anal, NS 18*59-60
◇ A05 F07 ◇ REV MR 56 # 11738 Zbl 316 # 02012
• ID 66166

XU, LIZHI & YUAN, XIANGWAN & ZHENG, YUXIN & ZHU, WUJIA [1981] *On Goedel's incompleteness theorem (Chinese) (English summary)* (J 3742) Shuxue Yanjiu yu Pinglun 1/1*151-162
◇ A05 F30 ◇ REV MR 83m:03001 • ID 34884

XU, LIZHI see Vol. I, II for further entries

YAKUBOVICH, A.M. [1981] *On the consistency of the theory of types with the axiom of choice relative to type theory (Russian)* (J 0023) Dokl Akad Nauk SSSR 261*825-828
• TRANSL [1981] (J 0062) Sov Math, Dokl 24*621-624
◇ B15 E25 E35 F35 ◇ REV MR 83e:03092
Zbl 495 # 03008 • ID 46491

YAKUBOVICH, A.M. see Vol. I, V for further entries

YANKOV, V.A. [1963] *Realizable formulas of propositional logic (Russian)* (J 0023) Dokl Akad Nauk SSSR 151*1035-1037
• TRANSL [1963] (J 0062) Sov Math, Dokl 4*1146-1148
◇ B55 F50 ◇ REV MR 27 # 4746 Zbl 143.251 JSL 35.138
• ID 19006

YANKOV, V.A. [1963] *The relationship between deducibility in the intuitionistic propositional calculus and finite implicational structures (Russian)* (J 0023) Dokl Akad Nauk SSSR 151*1293-1294
• TRANSL [1963] (J 0062) Sov Math, Dokl 4*1203-1204
◇ B55 F50 G25 ◇ REV MR 27 # 5685 Zbl 143.252
JSL 35.138 • ID 16539

YANKOV, V.A. [1967] *Finite validity of formulas of a special form (Russian)* (J 0023) Dokl Akad Nauk SSSR 174*302-304
• TRANSL [1967] (J 0062) Sov Math, Dokl 8*648-650
◇ B55 C13 F50 ◇ REV MR 36 # 2468 Zbl 209.304
JSL 38.331 • ID 06513

YANKOV, V.A. [1968] *On the extension of the intuitionist propositional calculus to the classical calculus, and the minimal calculus to the intuitionist calculus* (J 0216) Izv Akad Nauk SSSR, Ser Mat 32*208-211
• TRANSL [1968] (J 0448) Math of USSR, Izv 2*205-208
◇ B55 F50 ◇ REV MR 37 # 31 Zbl 191.286 JSL 38.331
• ID 06514

YANKOV, V.A. [1969] *Conjunctively indecomposable formulas in propositional calculi (Russian)* (J 0216) Izv Akad Nauk SSSR, Ser Mat 33*18-38
• TRANSL [1969] (J 0448) Math of USSR, Izv 3*17-35
◇ B55 F50 G10 ◇ REV MR 39 # 5318 Zbl 181.4
JSL 37.186 • ID 06518

YANKOV, V.A. see Vol. II, IV, V for further entries

YANOV, YU.I. [1977] *Computations in a class of programs (Russian)* (J 0052) Probl Kibern 32*237-245,247
◇ B75 D20 F30 ◇ REV MR 58 # 10382 Zbl 415 # 03009
• ID 74466

YANOV, YU.I. see Vol. I, II, III, IV for further entries

YASHIN, A.D. [1984] *Completeness of the intuitionistic predicate calculus with the concept of "bar" (Russian)* (J 0288) Vest Ser Mat Mekh, Univ Moskva 1984/4*67-69
• TRANSL [1984] (J 0510) Moscow Univ Math Bull 39/4*62-65
◇ C90 F05 F50 ◇ REV MR 85i:03078 Zbl 573 # 03006
• ID 44180

YASHIN, A.D. [1984] *Intuitionistic logical connectives on linear structures (Russian)* (J 0087) Mat Zametki (Akad Nauk SSSR) 35*663-675
• TRANSL [1984] (J 1044) Math Notes, Acad Sci USSR 35*343-355
◇ F50 ◇ REV MR 85g:03022 • ID 43495

YASHIN, A.D. [1984] *Nishimura's formulas as one-place logical connectives in the elementary theory of Kripke models (Russian)* (J 0288) Vest Ser Mat Mekh, Univ Moskva 1984/5*12-15
◇ C90 F50 ◇ REV MR 86a:03010 • ID 44104

YASHIN, A.D. see Vol. II for further entries

YASUDA, Y. [1977] see MISAWA, T.

YASUDA, Y. see Vol. IV, V for further entries

YASUGI, M. [1963] *Intuitionistic analysis and Goedel's interpretation* (J 0090) J Math Soc Japan 15*101-112
◇ F10 F35 F50 ◇ REV MR 27 # 2418 Zbl 117.257
JSL 37.404 • ID 18023

YASUGI, M. [1968] see TAKEUTI, G.

YASUGI, M. [1970] *Cut elimination theorem for second order arithmetic with the Π^1_1 comprehension axiom and the ω-rule* (**J** 0090) J Math Soc Japan 22*308-324
◇ F05 F35 ◇ REV MR 42 # 1653 Zbl 191.294 • ID 18025

YASUGI, M. [1973] see TAKEUTI, G.

YASUGI, M. [1974] *On ordinal diagrams (Japanese)* (**J** 0091) Sugaku 26*121-136 • ERR/ADD ibid 28*383-384
◇ F15 ◇ REV MR 54 # 7214 • ID 25004

YASUGI, M. [1976] see TAKEUTI, G.

YASUGI, M. [1978] *A formalization of $Od(\Omega)$* (**J** 0407) Comm Math Univ St Pauli (Tokyo) 27*133-154
◇ F15 F35 ◇ REV MR 81f:03068 Zbl 473 # 03048 • ID 80159

YASUGI, M. [1978] *Self-iterating schemes of ordinal diagrams* (**J** 0407) Comm Math Univ St Pauli (Tokyo) 27*11-24
◇ F15 ◇ REV MR 80h:03082b Zbl 394 # 03048 • ID 52525

YASUGI, M. [1978] *Some properties of ordinal diagrams* (**J** 0111) Nagoya Math J 70*143-155
◇ F15 ◇ REV MR 80h:03082a Zbl 351 # 02025 • ID 51798

YASUGI, M. [1980] *A progression of consistency proofs* (**J** 0407) Comm Math Univ St Pauli (Tokyo) 28*31-49
◇ F05 F15 F35 ◇ REV MR 82d:03093 Zbl 433 # 03036 • ID 80158

YASUGI, M. [1980] *Gentzen reduction revisited* (**J** 0390) Publ Res Inst Math Sci (Kyoto) 16*1-33
◇ F05 F15 F30 F35 ◇ REV MR 83e:03093 Zbl 436 # 03049 • ID 35245

YASUGI, M. [1981] see TAKEUTI, G.

YASUGI, M. [1981] *Definability problems in metric spaces; a summary* (**P** 4153) B-Val Anal & Nonstand Anal;1981 Kyoto 66-82
◇ E15 F99 ◇ ID 47763

YASUGI, M. [1981] *The Hahn-Banach theorem and a restricted inductive definition* (**P** 3201) Logic Symposia;1979/80 Hakone 359-394
◇ F35 F65 ◇ REV MR 84m:03095 Zbl 475 # 03036 • ID 55490

YASUGI, M. [1982] *Construction principle and transfinite induction up to ε_0* (**J** 3194) J Austral Math Soc, Ser A 32*24-47
◇ F10 F15 ◇ REV MR 84a:03073 Zbl 482 # 03024 • ID 35593

YASUGI, M. [1983] *Definability problems in elementary topology* (**J** 3194) J Austral Math Soc, Ser A 34*399-420
◇ F65 ◇ REV MR 85b:03109 Zbl 524 # 03049 • ID 37613

YASUGI, M. [1983] *Definability in the abstract theory of integration* (**J** 0407) Comm Math Univ St Pauli (Tokyo) 32*137-161
◇ E30 F65 ◇ REV MR 85k:03043 Zbl 524 # 03050 • ID 37614

YASUGI, M. [1983] *Definability in L^p-spaces* (**P** 4113) Found of Math;1982 Kyoto 209-229
◇ F35 ◇ REV Zbl 547 # 03042 • ID 47687

YASUGI, M. [1984] *A definable interpretation of metric spaces* (**J** 0407) Comm Math Univ St Pauli (Tokyo) 33*127-142
◇ F65 ◇ ID 44016

YASUGI, M. [1984] *Definability in L^p-spaces* (**J** 0407) Comm Math Univ St Pauli (Tokyo) 33*1-18
◇ F65 ◇ REV Zbl 547 # 03042 • ID 43243

YASUGI, M. [1984] *Projection and elevation of ordinal diagrams* (**P** 3668) Log & Founds of Math;1983 Kyoto 1-9
◇ F15 ◇ ID 42928

YASUGI, M. [1985] *Groundedness property and accessibility of ordinal diagrams* (**J** 0090) J Math Soc Japan 37*1-16
◇ F15 ◇ REV MR 86f:03095 Zbl 565 # 03027 • ID 44973

YASUGI, M. [1985] *Hyperprinciple and the functional structure of ordinal diagrams. I* (**J** 0407) Comm Math Univ St Pauli (Tokyo) 34/2*227-263
◇ F15 ◇ ID 49514

YASUGI, M. see Vol. II, III, V for further entries

YASUHARA, M. [1966] *The cut elimination theorem in the unary second order language* (**J** 0053) Proc Amer Math Soc 17*1069-1073
◇ F05 ◇ REV MR 34 # 4117 Zbl 168.7 • ID 18030

YASUHARA, M. [1974] *Completeness of cut-free type theories* (**J** 0009) Arch Math Logik Grundlagenforsch 16*137-141
◇ F05 ◇ REV MR 51 # 2900 Zbl 291 # 02020 • ID 18036

YASUHARA, M. [1982] *Cut elimination in ε-calculi* (**J** 0068) Z Math Logik Grundlagen Math 28*311-316
◇ F05 ◇ REV MR 84b:03076 Zbl 548 # 03033 • ID 35658

YASUHARA, M. see Vol. III, V for further entries

YEVICK, M.L. [1975] *Holographic or Fourier logic* (**J** 3227) Pattern Recognition 7*197-213
◇ F50 ◇ REV MR 52 # 12442 Zbl 331 # 02024 • ID 66178

YOUNG, P. [1968] see RITCHIE, R.W.

YOUNG, P. [1973] see HELM, J.P.

YOUNG, P. [1978] see MACHTEY, M.

YOUNG, P. [1981] see JOSEPH, D.

YOUNG, P. [1983] see COLLINS, W.J.

YOUNG, P. [1985] *Goedel theorems, exponential difficulty and undecidability of arithmetic theories: an exposition* (**P** 4046) Rec Th;1982 Ithaca 503-522
◇ D15 D35 F20 F30 ◇ REV Zbl 577 # 03023 • ID 46382

YOUNG, P. see Vol. IV for further entries

YUAN, XIANGWAN [1981] see XU, LIZHI

YUAN, XIANGWAN see Vol. I, II, V for further entries

YUKAMI, T. [1977] *A theorem on the formalized arithmetic with function symbols ' and +* (**J** 2606) Tsukuba J Math 1*195-211
◇ F20 F30 ◇ REV MR 58 # 10348 Zbl 392 # 03036 • ID 52403

YUKAMI, T. [1978] *A note on a formalized arithmetic with function symbols ' and +* (**J** 2606) Tsukuba J Math 2*69-73
◇ F20 F30 ◇ REV MR 80g:03058 Zbl 411 # 03052 • ID 52906

YUKAMI, T. [1983] *A theorem on lengths of proof of Presburger formulas* (**J** 2606) Tsukuba J Math 7*169-184
◇ B35 F20 ◇ REV MR 85b:03103 Zbl 524 # 03047 • ID 37612

YUKAMI, T. [1984] *Some results on speed-up* (**J** 0260) Ann Jap Ass Phil Sci 6*195-205
◇ D15 F20 F30 ◇ REV MR 86c:03050 Zbl 545 # 03032 • ID 43511

YUKAMI, T. see Vol. I for further entries

ZACCHI, M. [1983] see MARGARIA, I.

ZACHOROWSKI, S. [1974] see SURMA, S.J.

ZACHOROWSKI, S. [1975] *A proof of a conjecture of Roman Suszko* (**J** 0063) Studia Logica 34*253-256
◇ B45 B50 F50 ◇ REV MR 54 # 65 Zbl 325 # 02013 • ID 23947

ZACHOROWSKI, S. [1975] see SURMA, S.J.

ZACHOROWSKI, S. see Vol. II, III, V for further entries

ZAHN, P. [1967] *Eine Einfuehrung der reellen Zahlen in der operativen Mathematik ohne die Unterscheidung von Sprachschichten* (**S** 0405) Mitt Math Sem Giessen 72*ii + 75pp
◇ B28 F35 F65 ◇ REV MR 36 # 2469 Zbl 201.332 • ID 24851

ZAHN, P. [1969] *Eine Verallgemeinerung des Hauptsatzes von Gentzen* (**J** 0009) Arch Math Logik Grundlagenforsch 12*134-150
◇ F05 ◇ REV MR 42 # 40 Zbl 198.14 • ID 18090

ZAHN, P. [1978] *Ein konstruktiver Weg zur Masstheorie und Funktionalanalysis* (**X** 0890) Wiss Buchges: Darmstadt 350pp
◇ F65 F98 ◇ REV MR 80c:03060 Zbl 393 # 03043 JSL 47.703 • ID 52463

ZAHN, P. [1985] *Gilt das Lemma von Koenig "konstruktiv"?* (**J** 0009) Arch Math Logik Grundlagenforsch 25*33-35
◇ F65 ◇ ID 49132

ZAHN, P. see Vol. I, II for further entries

ZAIONC, M. [1985] *Nondeterministic programs definable in typed λ-calculus* (**J** 2095) Fund Inform, Ann Soc Math Pol, Ser 4 8*63-72
◇ B40 ◇ ID 47252

ZAIONC, M. [1985] *The set of unifers in typed λ-calculus as regular expression* (**P** 4244) Rewriting Techn & Appl (1);1985 Dijon 430-440
◇ B40 ◇ ID 49792

ZAKHAR'YASHCHEV, M.V. [1980] see POPOV, S.V.

ZAKHAR'YASHCHEV, M.V. see Vol. I, II, III for further entries

ZALC, A. [1974] see PERRIN, M.J.

ZALC, A. [1978] see ARTIGUE, M.

ZALC, A. see Vol. I, III, V for further entries

ZAMOV, N.K. [1974] *Beschraenkung der Kompliziertheit der Terme in Ableitungen und aufloesbare Fragmente der Praedikatenkalkuele (Russisch)* (**S** 2738) Issl Prikl Mat (Univ Kazan') 2*56-62
◇ B35 F20 ◇ REV MR 57 # 9503 Zbl 357 # 02011 • ID 50359

ZAMOV, N.K. see Vol. I, IV for further entries

ZARACH, A. [1976] *Generic extension of admissible sets* (**P** 1476) Set Th & Hierarch Th (2) (Mostowski);1975 Bierutowice 321-333
◇ C62 E25 E35 F35 ◇ REV MR 55 # 2574 Zbl 357 # 02060 • ID 23816

ZARACH, A. see Vol. II, III, IV, V for further entries

ZARISKI, O. [1925] *Gli sviluppi piu recenti della teoria degli insiemi e il principio di Zermelo* (**J** 0313) Period Mat (Bologna) 5*57-80
◇ A05 B10 E25 F99 ◇ REV FdM 51.164 • ID 41579

ZARNECKA-BIALY, E. [1972] *The deduction theorems for propositional calculi when implication and falsum is present (Polish summary)* (**S** 0458) Zesz Nauk, Prace Log, Uniw Krakow 7*93-100
◇ B20 F50 ◇ REV MR 47 # 4757 • ID 16532

ZARNECKA-BIALY, E. [1973] *Modal functors and their definability in propositional calculus systems (Polish) (English summary)* (**J** 0302) Rep Math Logic, Krakow & Katowice 1*68-98
◇ B45 F50 ◇ REV MR 49 # 4757 Zbl 292 # 02020 • ID 14364

ZARNECKA-BIALY, E. [1973] *Negation in Ch.S.Peirce's propositional calculus* (**J** 0302) Rep Math Logic, Krakow & Katowice 1*99-101
◇ A10 B20 F50 ◇ REV MR 50 # 6782 Zbl 291 # 02002 • ID 14363

ZARNECKA-BIALY, E. see Vol. I, II for further entries

ZASHEV, J.A. [1979] *On a stepwise semantic system with deductive understanding of the second order universal quantifier (Russian)* (**J** 0137) C R Acad Bulgar Sci 32*851-854
◇ F35 ◇ REV MR 81i:03097 Zbl 423 # 03054 • ID 53566

ZASHEV, J.A. [1984] *∇-combinatory algebras and abstract Plotkin's models* (**J** 0137) C R Acad Bulgar Sci 37*711-714
◇ B40 ◇ ID 44074

ZASHEV, J.A. [1984] *Basic recursion theory in partially ordered models of some fragments of the combinatory logic* (**J** 0137) C R Acad Bulgar Sci 37*561-564
◇ B40 D20 ◇ REV MR 85j:03077 Zbl 544 # 03020 • ID 41008

ZASLAVSKIJ, I.D. [1955] *Disproof of some theorems of classical analysis in constructive analysis (Russian)* (**J** 0067) Usp Mat Nauk 10/4*209-210
◇ F60 ◇ REV Zbl 65.5 JSL 21.409 • ID 19787

ZASLAVSKIJ, I.D. [1959] *On constructive Dedekind cuts (Russian)* (**P** 0607) All-Union Math Conf (3);1956 Moskva 1*182-183
◇ F60 ◇ ID 28236

ZASLAVSKIJ, I.D. [1959] see TSEJTIN, G.S.

ZASLAVSKIJ, I.D. [1959] *Some peculiarities of constructive functions of a real variable in comparison with classical ones (Russian)* (**P** 0607) All-Union Math Conf (3);1956 Moskva 1*181-182
◇ F60 ◇ ID 28235

ZASLAVSKIJ, I.D. [1962] see TSEJTIN, G.S.

ZASLAVSKIJ, I.D. [1962] *Some properties of constructive real numbers and constructive functions (Russian)* (S 0066) Tr Mat Inst Steklov 67*385-457
• TRANSL [1966] (J 0225) Amer Math Soc, Transl, Ser 2 57*1-84
◇ F60 ◇ REV MR 27 # 3519 Zbl 113.12 JSL 40.454
• ID 14369

ZASLAVSKIJ, I.D. [1963] *Ueber gewisse Unterschiede zwischen den Basis- und den untergeordneten Variablen in logisch-mathematischen Sprachen (Russisch)* (S 0422) Tr Vychisl Tsentra Akad Nauk Armyan SSR & Univ Erevan 1*13-29
◇ F50 ◇ REV Zbl 192.57 • ID 16268

ZASLAVSKIJ, I.D. [1964] see TSEJTIN, G.S.

ZASLAVSKIJ, I.D. [1964] *Differentation and integration of constructive functions (Russian)* (J 0023) Dokl Akad Nauk SSSR 156*25-27
• TRANSL [1964] (J 0062) Sov Math, Dokl 5*599-601
◇ F60 ◇ REV MR 31 # 2147 Zbl 192.61 • ID 14370

ZASLAVSKIJ, I.D. [1964] see MANUKYAN, S.N.

ZASLAVSKIJ, I.D. [1967] *Rectifiability of constructive plane curves (Russian) (English and Armenian summaries)* (J 0312) Izv Akad Nauk Armyan SSR, Ser Mat 2*69-82
◇ F60 ◇ REV MR 36 # 48 Zbl 164.318 • ID 14371

ZASLAVSKIJ, I.D. [1968] see MANUKYAN, S.N.

ZASLAVSKIJ, I.D. [1968] see SHANIN, N.A.

ZASLAVSKIJ, I.D. [1969] *On Shannon pseudofunctions (Russian)* (S 0228) Zap Nauch Sem Leningrad Otd Mat Inst Steklov 16*65-76
• TRANSL [1969] (J 0521) Semin Math, Inst Steklov 16*32-37
◇ D05 D15 F60 ◇ REV MR 42 # 78 Zbl 199.309
• ID 27441

ZASLAVSKIJ, I.D. [1969] *The axiomatic determination of constructive objects and operations (Russian) (Armenian and English summaries)* (J 0312) Izv Akad Nauk Armyan SSR, Ser Mat 4*153-181
◇ D20 F50 ◇ REV MR 42 # 2942 Zbl 176.278 • ID 14375

ZASLAVSKIJ, I.D. [1970] see TSEJTIN, G.S.

ZASLAVSKIJ, I.D. [1971] see TSEJTIN, G.S.

ZASLAVSKIJ, I.D. [1973] *On predicate and arithmetic calculi of symmetric constructive logic (Russian)* (J 0023) Dokl Akad Nauk SSSR 210*517-520
• TRANSL [1973] (J 0062) Sov Math, Dokl 14*751-754
◇ F50 ◇ REV MR 50 # 1851 Zbl 291 # 02014 • ID 66221

ZASLAVSKIJ, I.D. [1975] *On constructive truth of judgements and some untraditional systems of constructive logic (Russian)* (S 0422) Tr Vychisl Tsentra Akad Nauk Armyan SSR & Univ Erevan 8*99-153
◇ F50 ◇ REV MR 57 # 2891 Zbl 412 # 03038 • ID 52969

ZASLAVSKIJ, I.D. [1978] *Symmetric constructive logic (Russian)* (X 2225) Akad Nauk Armyan SSR : Erevan 281pp
◇ B55 F05 F30 F50 ◇ REV MR 81a:03059 • ID 80231

ZASLAVSKIJ, I.D. see Vol. II, IV for further entries

ZAWADOWSKI, M. [1985] see JANKOWSKI, A.W.

ZAWADOWSKI, M. see Vol. III, V for further entries

ZAWIRSKI, Z. [1946] *Origin and development of intuitionistic logic (Polish)* (J 0479) Kwart Filoz 16*165-222
◇ A05 A10 F50 ◇ REV MR 10.421 • ID 14368

ZAWIRSKI, Z. see Vol. II for further entries

ZBIERSKI, P. [1971] *Models for higher order arithmetics (Russian summary)* (J 0014) Bull Acad Pol Sci, Ser Math Astron Phys 19*557-562
◇ C62 E30 F35 ◇ REV MR 46 # 7022 Zbl 236 # 02043
• ID 14379

ZBIERSKI, P. [1978] *Axiomatizability of second order arithmetic with ω-rule* (J 0027) Fund Math 100*51-57
◇ B15 C62 E45 E70 F35 ◇ REV MR 58 # 5161 Zbl 387 # 03021 • ID 29200

ZBIERSKI, P. [1980] *Indicators and incompleteness of Peano arithmetic* (J 2085) Acta Cient Venez 31*487-495
◇ C62 F30 ◇ REV MR 84g:03095 Zbl 526 # 03019
• ID 34199

ZBIERSKI, P. [1981] *Nonstandard interpretations of higher order theories* (J 0027) Fund Math 112*175-186
◇ C62 F35 H15 ◇ REV MR 84d:03073 Zbl 481 # 03046
• ID 34106

ZBIERSKI, P. see Vol. III, V for further entries

ZEMAN, J.J. [1968] *Some calculi with strong negation primitive* (J 0036) J Symb Logic 33*97-100
◇ B45 F50 ◇ REV MR 38 # 3140 Zbl 167.8 • ID 14395

ZEMAN, J.J. see Vol. II, III for further entries

ZEMBRZUSKI, K. [1972] see SKARBEK, W.

ZEMBRZUSKI, K. see Vol. IV for further entries

ZEMKE, F. [1977] *P.R.-regulated systems of notation and the subrecursive hierarchy equivalence property* (J 0064) Trans Amer Math Soc 234*89-118
◇ D20 F15 ◇ REV MR 58 # 16222 Zbl 366 # 02032
• ID 51115

ZHANG, JINWEN [1981] *Higher order and transfinite incompleteness (Chinese)* (J 3793) Jisuanjii Xuebao 4*476-478
◇ F35 ◇ ID 48552

ZHANG, JINWEN see Vol. II, III, IV, V for further entries

ZHAROV, V.G. [1974] *On an analog of a theorem of Specker (Russian)* (J 0023) Dokl Akad Nauk SSSR 215*526-528
• TRANSL [1974] (J 0062) Sov Math, Dokl 15*538-541
◇ D20 D80 F60 ◇ REV MR 50 # 6810 Zbl 301 # 02031
• ID 66215

ZHAROV, V.G. [1979] *On the complexity of the terms of the constructive sequences of Turing machines (Russian)* (S 0554) Issl Teor Algor & Mat Logik (Moskva) 3*39-52
◇ D10 D45 F60 ◇ REV MR 81g:03045 Zbl 454 # 03016
• ID 54228

ZHAROV, V.G. see Vol. IV for further entries

ZHEGALKIN, I.I. [1934] *On the decision problem in Brouwer's propositional logic (Russian)* (P 4095) All-Union Math Conf (2);1934 Moskva 437
◇ F50 ◇ ID 45788

ZHEGALKIN, I.I. see Vol. I, III, IV, V for further entries

ZHENG, YUXIN [1981] see XU, LIZHI

ZHENG, YUXIN see Vol. I, II for further entries

ZHU, SHUILIN [1984] *An essay on Goedel's incompleteness theorem (Chinese)* (**J** 4430) Ziran Zazhi 7*595-602
◇ A05 F30 ◇ ID 46887

ZHU, WUJIA [1981] see XU, LIZHI

ZHU, WUJIA see Vol. I, II, V for further entries

ZINOV'EV, A.A. [1974] *Certain systems of formal arithmetic (Russian)* (**J** 0302) Rep Math Logic, Krakow & Katowice 3*73-90
◇ B45 F30 ◇ REV MR 51 # 7812 Zbl 324 # 02033
• ID 17312

ZINOV'EV, A.A. [1979] *Complete (rigorous) induction & Fermat's great theorem* (**J** 0079) Logique & Anal, NS 22*243-261
◇ F30 ◇ REV MR 81i:03098 Zbl 448 # 03042 • ID 56642

ZINOV'EV, A.A. see Vol. I, II, III, V for further entries

ZLATUSKA, J. [1985] *Normal forms in the typed λ-calculus with tuple type* (**J** 0156) Kybernetika (Prague) 21*366-381
◇ B40 ◇ ID 49370

ZUCKER, J.I. [1973] *Iterated inductive definitions, trees, and ordinals* (**C** 1139) Metamath Invest of Intuition Arithm & Anal 392-453
◇ F10 F15 F35 ◇ REV MR 56 # 2796 • ID 80298

ZUCKER, J.I. [1974] *The correspondence between cut-elimination and normalization I,II* (**J** 0007) Ann Math Logic 7*1-112,113-156
◇ F05 F30 F50 ◇ REV MR 51 # 5271 MR 51 # 5272 Zbl 298 # 02023 JSL 48.1204 • ID 14432

ZUCKER, J.I. [1978] see TRAGESSER, R.S.

ZUCKER, J.I. see Vol. I, IV for further entries

ZYGMUNT, J. [1974] see WRONSKI, A.

ZYGMUNT, J. see Vol. I, II, III for further entries

Source Index

Journals

J 0002 Acta Sci Math (Szeged) • H
Acta Scientiarum Mathematicarum [1947ff] ISSN 0001-6969
• CONT OF (J 0460) Acta Univ Szeged, Sect Mat

J 0003 Algebra i Logika • SU
Algebra i Logika (Algebra and Logic) [1962ff] ISSN 0373-9252
• TRANSL IN (J 0069) Algeb and Log

J 0004 Algeb Universalis • CDN
Algebra Universalis [1970ff] ISSN 0002-5240

J 0005 Amer Math Mon • USA
American Mathematical Monthly [1894ff] ISSN 0002-9890

J 0007 Ann Math Logic • NL
Annals of Mathematical Logic [1970-1982] ISSN 0003-4843
• CONT AS (J 0073) Ann Pure Appl Logic

J 0008 Arch Math (Basel) • CH
*Archiv der Mathematik * Archives of Mathematics * Archives Mathematiques* [1948ff] ISSN 0003-889X

J 0009 Arch Math Logik Grundlagenforsch • D
Archiv fuer Mathematische Logik und Grundlagenforschung [1950ff] ISSN 0003-9268

J 0010 Autom & Remote Control • USA
Automation and Remote Control [1958ff] ISSN 0005-1179
• TRANSL OF (J 0011) Avtom Telemekh

J 0011 Avtom Telemekh • SU
Avtomatika i Telemekhanika (Automation and Telemechanics) [1934ff] ISSN 0005-2310
• TRANSL IN (J 0010) Autom & Remote Control

J 0012 Boll Unione Mat Ital, IV Ser • I
Bolletino della Unione Matematica Italiana. Serie IV [1968-1975] ISSN 0041-7084
• CONT OF (J 4408) Boll Unione Mat Ital, III Ser • CONT AS (J 3285) Boll Unione Mat Ital, V Ser, A & (J 3495) Boll Unione Mat Ital, V Ser, B

J 0013 Brit J Phil Sci • GB
British Journal for the Philosophy of Science [1950ff] ISSN 0007-0882

J 0014 Bull Acad Pol Sci, Ser Math Astron Phys • PL
Bulletin de l'Academie Polonaise des Sciences. Serie des Sciences Mathematiques, Astronomiques et Physiques [1953-1978] ISSN 0001-4117
• CONT AS (J 3293) Bull Acad Pol Sci, Ser Math

J 0015 Bull Amer Math Soc • USA
Bulletin of the American Mathematical Society [1894-1978] ISSN 0002-9904
• CONT AS (J 0589) Bull Amer Math Soc (NS)

J 0016 Bull Austral Math Soc • AUS
Bulletin of the Australian Mathematical Society [1969ff] ISSN 0004-9727

J 0017 Canad J Math • CDN
*Canadian Journal of Mathematics * Journal Canadien de Mathematiques* [1949ff] ISSN 0008-414X

J 0020 Compos Math • NL
Compositio Mathematica [1933ff] ISSN 0010-437X

J 0021 Cybernetics • USA
Cybernetics [1965ff] ISSN 0011-4235
• TRANSL OF (J 0040) Kibernetika, Akad Nauk Ukr SSR

J 0023 Dokl Akad Nauk SSSR • SU
Doklady Akademii Nauk SSSR (Reports of the Academy of Sciences of the USSR) [1933ff] ISSN 0002-3264
• TRANSL IN (J 0062) Sov Math, Dokl & (J 0470) Sov Phys, Dokl

J 0025 Duke Math J • USA
Duke Mathematical Journal [1935ff] ISSN 0012-7094

J 0026 Elem Math • CH
*Elemente der Mathematik * Revue de Mathematiques Elementaires * Rivista di Matematica Elementare* [1946ff] ISSN 0013-6018

J 0027 Fund Math • PL
Fundamenta Mathematicae [1920ff] ISSN 0016-2736

J 0028 Indag Math • NL
*Indagationes Mathematicae * Nederlandse Akademie van Wetenschappen. Proceedings* [1939ff] ISSN 0019-3577, ISSN 0023-3358
• REM Until 1950 part of Koninklijke Nederlandsche Akademie van Wetenschappen, Proceedings of the Section of Sciences; vol n+41 with separate pagination. Since 1951 same as Koninklijke Nederlandse Akademie van Wetenschappen, Proceedings of the Section of Sciences, Series A; vol n+41. Before 1951 page numbers in Proceedings and Indagationes different. Since 1951 the same page numbers as Proceedings Series A.

J 0029 Israel J Math • IL
Israel Journal of Mathematics [1963ff] ISSN 0021-2172
• CONT OF (J 0493) Bull Res Counc Israel Sect F

J 0031 Izv Vyssh Ucheb Zaved, Mat (Kazan) • SU
Izvestiya Vysshikh Uchebnykh Zavedenij. Matematika (Proceedings of the University. Mathematics) [1957ff] ISSN 0021-3446
• TRANSL IN (J 3449) Sov Math

J 0032 J Algeb • USA
Journal of Algebra [1964ff] ISSN 0021-8693

J 0033 J Comb Th, Ser B • USA
Journal of Combinatorial Theory. Series B [1971ff] ISSN 0095-8956
• CONT OF (J 1669) J Comb Th

J 0036 J Symb Logic • USA
The Journal of Symbolic Logic [1936ff] ISSN 0022-4812

J 0037 ACM J • USA
Journal of the ACM (= Association for Computing Machinery) [1954ff] ISSN 0004-5411

J 0038 J Austral Math Soc • AUS
Journal of the Australian Mathematical Society [1959-1975] ISSN 0004-9735
• CONT AS (**J** 3194) J Austral Math Soc, Ser A

J 0039 J London Math Soc • GB
The Journal of the London Mathematical Society [1926-1968]
• CONT AS (**J** 3172) J London Math Soc, Ser 2

J 0040 Kibernetika, Akad Nauk Ukr SSR • SU
Kibernetika. Akademiya Nauk Ukrainskoj SSR (Cybernetics. Academy of Sciences of the Ukrainian SSR) [1965ff] ISSN 0023-1274
• TRANSL IN (**J** 0021) Cybernetics

J 0041 Math Syst Theory • D
Mathematical Systems Theory. An International Journal [1967ff] ISSN 0025-5661

J 0043 Math Ann • D
Mathematische Annalen [1868ff] ISSN 0025-5831

J 0044 Math Z • D
Mathematische Zeitschrift [1918ff] ISSN 0025-5874

J 0045 Monatsh Math • A
Monatshefte fuer Mathematik [1943ff] ISSN 0026-9255
• CONT OF (**J** 0124) Monatsh Math-Phys

J 0046 Nieuw Arch Wisk • NL
Nieuw Archief voor Wiskunde [1875-1893]
• CONT AS (**J** 1793) Nieuw Arch Wisk, Ser 2

J 0047 Notre Dame J Formal Log • USA
Notre Dame Journal of Formal Logic [1960ff] ISSN 0029-4527

J 0048 Pac J Math • USA
Pacific Journal of Mathematics [1951ff] ISSN 0030-8730

J 0050 Port Math • P
Portugaliae Mathematica [1937ff] ISSN 0032-5155

J 0051 Commentat Math, Ann Soc Math Pol, Ser 1 • PL
*Annales Societatis Mathematicae.Polonae. Series I. Commentationes Mathematicae * Roczniki Polskiego Towarzystwa Matematycznego. Seria I. Prace Matematyczne.* [1955ff] ISSN 0079-368X
• CONT OF (**J** 0611) Pol Tow Mat, Prace Mat-Fiz

J 0052 Probl Kibern • SU
Problemy Kibernetiki. Glavnaya Redaktsiya Fiziko-Matematicheskoj Literatury [1958ff] ISSN 0555-277X
• TRANSL IN (**J** 0471) Syst Th Res & (**J** 0449) Probl Kybern & (**J** 1195) Probl Cybernet

J 0053 Proc Amer Math Soc • USA
Proceedings of the American Mathematical Society [1950ff] ISSN 0002-9939

J 0054 Proc Nat Acad Sci USA • USA
Proceedings of the National Academy of Sciences of the United States of America [1915ff] ISSN 0027-8424

J 0056 Publ Dep Math, Lyon • F
Publications du Departement de Mathematiques. Faculte des Sciences de Lyon. [1964-1981] ISSN 0076-1656
• CONT AS (**J** 2107) Publ Dep Math, Lyon, NS

J 0059 Rend Sem Mat Fis Milano • I
Rendiconti del Seminario Matematico e Fisico di Milano. Sotto gli Auspici dell'Universita e del Politecnico [1927ff]

J 0060 Rev Roumaine Math Pures Appl • RO
Revue Roumaine de Mathematiques Pures et Appliquees. Academia Republicii Socialiste Romania [1956ff] ISSN 0035-3965

J 0061 Simon Stevin • B
Simon Stevin: A Quarterly Journal of Pure and Applied Mathematics [1921ff] ISSN 0037-5454
• CONT OF (**J** 1928) Mathematica B & (**J** 0291) Wis-natuur Tijdsch & (**J** 1379) Christiaan Huygens Internat Math Tijdschr
• REM The journal has incorporated J1928, J1379, J0291

J 0062 Sov Math, Dokl • USA
Soviet Mathematics. Doklady. [1960ff] ISSN 0038-5573
• TRANSL OF (**J** 0023) Dokl Akad Nauk SSSR

J 0063 Studia Logica • PL
Studia Logica [1953ff] ISSN 0039-3215

J 0064 Trans Amer Math Soc • USA
Transactions of the American Mathematical Society [1900ff] ISSN 0002-9947

J 0067 Usp Mat Nauk • SU
Uspekhi Matematicheskikh Nauk (Advances in Mathematical Sciences) [1936ff] ISSN 0042-1316
• TRANSL IN (**J** 1399) Russ Math Surv

J 0068 Z Math Logik Grundlagen Math • DDR
Zeitschrift fuer Mathematische Logik und Grundlagen der Mathematik [1955ff] ISSN 0044-3050

J 0069 Algeb and Log • USA
Algebra and Logic [1968ff] ISSN 0002-5232
• TRANSL OF (**J** 0003) Algebra i Logika

J 0070 Bull Soc Sci Math Roumanie, NS • RO
Bulletin Mathematique de la Societe des Sciences Mathematiques de la Republique Socialiste de Roumanie. Nouvelle Serie. [1957ff] ISSN 0007-4691
• CONT OF (**J** 0494) Bull Math Soc Sci Roumanie

J 0072 IRE Trans Electr Comp • USA
Transactions on Electronic Computers. IRE (= Institute of Radio Engineers) [1952-1962]
• CONT AS (**J** 4305) IEEE Trans Electr Comp

J 0073 Ann Pure Appl Logic • NL
Annals of Pure and Applied Logic [1983ff] ISSN 0168-0072
• CONT OF (**J** 0007) Ann Math Logic

J 0076 Dialectica • CH
Dialectica. International Review of Philosophy of Knowledge [1947ff] ISSN 0012-2017

J 0077 Proc London Math Soc • GB
Proceedings of the London Mathematical Society [1865-1903]
• CONT AS (**J** 1910) Proc London Math Soc, Ser 2

J 0079 Logique & Anal, NS • B
Logique et Analyse. Nouvelle Serie. Publication Trimestrielle du Centre National Belge de Recherche de Logique [1958ff] ISSN 0024-5836

J 0080 Izv Akad Nauk Ehston SSR, Fiz, Mat • SU
*Izvestiya Akademiya Nauk Ehstonskoj SSR. Fizika. Matematika * Eesti NSV Teaduste Akadeemia Toimetised. Fueuesika-Matemaatika (Proceedings of the Academy of Sciences of the Estonian SSR. Physics. Mathematics)* [1956ff] ISSN 0002-3140

J 0081 Proc Japan Acad • J
Proceedings of the Japan Academy [1925-1977] ISSN 0021-4280
• CONT AS (**J** 3239) Proc Japan Acad, Ser A

J 0082 Bull Soc Math Belg • B
Bulletin de la Societe Mathematique de Belgique [1948-1976] ISSN 0037-9476
• CONT AS (**J** 3133) Bull Soc Math Belg, Ser B & (**J** 3824) Bull Soc Math Belg, Ser A

J 0084 Gaz Mat (Lisboa) • P
Gazeta de Matematica. Jornal dos Concorrentes ao Exame de Aptidao e dos Estudantes de Matematica das Escolas Superiores [1940ff]

J 0086 Cas Pestovani Mat, Ceskoslov Akad Ved • CS
Casopis pro Pestovani Matematiky. Ceskoslovenska Akademie Ved (Journal for the Cultivation of Mathematics. Czechoslovak Academy of Sciences) [1872ff]

J 0087 Mat Zametki (Akad Nauk SSSR) • SU
Matematicheskie Zametki (Mathematical Notes) [1967ff] ISSN 0025-567X
• TRANSL IN (**J** 1044) Math Notes, Acad Sci USSR

J 0088 Ann Univ Ferrara, NS, Sez 7 • I
Annali dell'Universita di Ferrara. Nuova Serie. Sezione 7. Scienze Matematiche [1966ff]

J 0089 Calcolo • I
Calcolo [1964ff] ISSN 0008-0624

J 0090 J Math Soc Japan • J
Journal of the Mathematical Society of Japan [1885ff] ISSN 0025-5645

J 0091 Sugaku • J
Sugaku (Mathematics) [1947ff] ISSN 0039-470X

J 0092 Sib Mat Zh • SU
Sibirskij Matematicheskij Zhurnal. Akademiya Nauk SSSR. Sibirskoe Otdelenie (Siberian Mathematical Journal. Academy of Sciences of the USSR. Siberian Section) [1960ff] ISSN 0037-4474
• TRANSL IN (**J** 0475) Sib Math J

J 0093 J Comp Syst • USA
The Journal of Computing Systems [1952-1954]

J 0094 Mind • GB
Mind. A Quarterly Review of Philosophy [1876ff] ISSN 0026-4423

J 0095 Philos Stud • NL
Philosophical Studies. An International Journal for Philosophy in the Analytic Tradition [1950ff] ISSN 0031-8116

J 0096 Acta Philos Fenn • SF
Acta Philosophica Fennica [1948ff] ISSN 0355-1792

J 0097 Nous, Quart J Phil • USA
Nous. A Quarterly Journal of Philosophy [1967ff] ISSN 0029-4624

J 0100 Amer J Math • USA
American Journal of Mathematics [1878ff] ISSN 0002-9327

J 0102 Hiroshima Univ J Sci, Ser A Math • J
Hiroshima University. Journal of Science. Series A. Section 1. Mathematics [1930-1970] ISSN 0018-2079
• CONT AS (**J** 0431) Hiroshima Math J

J 0105 Theoria (Lund) • S
Theoria. A Swedish Journal of Philosophy [1934ff] ISSN 0040-5825

J 0107 Abh Math Sem Univ Hamburg • D
Abhandlungen aus dem Mathematischen Seminar der Universitaet Hamburg [1922ff] ISSN 0025-5858

J 0108 Bol Soc Mat Sao Paulo • BR
Sociedade de Matematica de Sao Paulo. Boletim. [1946ff]

J 0109 C R Acad Sci, Paris • F
Academie des Sciences de Paris. Comptes Rendus Hebdomadaires des Seances [1835-1965]
• CONT AS (**J** 2313) C R Acad Sci, Paris, Ser A-B

J 0111 Nagoya Math J • J
Nagoya Sugaku Zashi (Nagoya Mathematical Journal) [1950ff] ISSN 0027-7630

J 0113 Proc Aristotelian Soc • GB
Aristotelian Society. Proceedings [1900ff] ISSN 0066-7374
• REL PUBL (**J** 2119) Suppl Aristotelian Soc

J 0115 Wiss Z Humboldt-Univ Berlin, Math-Nat Reihe • DDR
Wissenschaftliche Zeitschrift der Humboldt-Universitaet Berlin. Mathematisch-Naturwissenschaftliche Reihe [1951ff] ISSN 0043-6852

J 0118 Acta Math • S
Acta Mathematica [1882ff] ISSN 0001-5962

J 0119 J Comp Syst Sci • USA
Journal of Computer and System Sciences [1967ff] ISSN 0022-0000

J 0120 Ann of Math, Ser 2 • USA
Annals of Mathematics. 2nd Series [1899ff] ISSN 0003-486X

J 0121 Kon Norske Vidensk Selsk Forh • N
Kongelige Norske Videnskabers Selskabs. Forhandlinger. (Proceedings of the Royal Scandinavian Society of Sciences) [1926ff] ISSN 0368-6302

J 0122 J Philos Logic • NL
Journal of Philosophical Logic [1972ff] ISSN 0022-3611

J 0123 Vierteljahrschr Naturforsch Ges Zuerich • CH
Naturforschende Gesellschaft in Zuerich. Vierteljahresschrift [1856ff] ISSN 0042-5672

J 0124 Monatsh Math-Phys • A
Monatshefte fuer Mathematik und Physik [1900-1942]
• CONT AS (**J** 0045) Monatsh Math

J 0127 J Reine Angew Math • D
Journal fuer die Reine und Angewandte Mathematik [1826ff] ISSN 0075-4102

J 0128 Acta Math Univ Comenianae (Bratislava) • CS
Universitas Comeniana. Acta Facultatis Rerum Naturalium. Mathematica [1956ff]

J 0129 Elektr Informationsverarbeitung & Kybern • DDR
Elektronische Informationsverarbeitung und Kybernetik [1965ff] ISSN 0013-5712

J 0130 BIT • DK
BIT. Nordisk Tidskrift for Informationsbehandling (BIT. Scandinavian Journal for Informatics) [1961ff] ISSN 0006-3835

J 0132 Math Scand • DK
Mathematica Scandinavica [1953ff] ISSN 0025-5521

J 0133 Michigan Math J • USA
The Michigan Mathematical Journal [1952ff] ISSN 0026-2285

J 0135 Izv Akad Nauk Azerb SSR, Ser Fiz-Tekh Mat • SU
Izvestiya Akademii Nauk Azerbajdzhanskoj SSR. Seriya Fiziko-Tekhnicheskikh i Matematicheskikh Nauk (Proceedings of the Academy of Sciences of the Azerbaijan SSR. Series: Physical-Technical and Mathematical Sciences) [1958ff] ISSN 0002-3108

J 0137 C R Acad Bulgar Sci • BG
Doklady Bolgarskoi Akademii Nauk (Comptes Rendus de l'Academie Bulgare des Sciences) [1948ff] ISSN 0001-3978

J 0140 Comm Math Univ Carolinae (Prague) • CS
Commentationes Mathematicae Universitatis Carolinae [1960ff] ISSN 0010-2628

J 0142 Mat Sb, Akad Nauk SSSR, NS • SU
Matematicheskij Sbornik. Novaya Seriya. Akademiya Nauk SSSR i Moskovskoe Matematicheskoe Obshchestvo (Mathematical Collected Articles. New Series. Academy of Sciences of the USSR and Moskovian Mathematical Society) [1936ff] ISSN 0025-5157
• CONT OF (**J** 1404) Mat Sb, Akad Nauk SSSR • TRANSL IN (**J** 0349) Math of USSR, Sbor

J 0144 Rend Sem Mat Univ Padova • I
Rendiconti del Seminario Matematico dell'Universita di Padova [1930ff] ISSN 0041-8994

J 0145 Rev Metaph Morale • F
Revue de Metaphysique et de Morale [1893ff] ISSN 0035-1571

J 0147 An Univ Bucuresti, Acta Logica • RO
Universitatea Bucuresti. Analele. Acta Logica (Universitaet Bukarest. Annalen. Acta Logica) [1958ff] ISSN 0524-823X
• REM From 1962-1967 Analele Universitatii Bucuresti. Seria Acta Logica. From 1958-1961 Analele Universitatii C.I. Parhon. Seria Acta Logica

J 0148 Math Gaz • GB
The Mathematical Gazette [1894ff] ISSN 0025-5572

J 0149 Atti Accad Naz Lincei Fis Mat Nat, Ser 8 • I
Atti della Accademia Nazionale dei Lincei. Rendiconti. Classe di Scienze Fisiche, Matematiche e Naturali. Serie VIII [1946ff] ISSN 0001-4435

J 0150 Acad Roy Belg Bull Cl Sci (5) • B
*Academie Royale des Sciences, des Lettres et des Beaux Arts de Belgique. Bulletin de la Classe des Sciences. Cinquieme Serie * Koninklijke Academie voor Wetenschappen. Mededeelingen van de Afdeeling Wetenschappen. 5. Serie* [1915ff] ISSN 0001-4141

J 0151 Arch Soc Belg Philos • B
Archives de la Societe Belge de Philosophie. [1928-1937]
• CONT AS (**J** 2076) Rev Int Philos

J 0152 Enseign Math • CH
L'Enseignement Mathematique: Revue Internationale [1899-1954]
• CONT AS (**J** 3370) Enseign Math, Ser 2

J 0153 Phil of Sci (East Lansing) • USA
Philosophy of Science [1934ff] ISSN 0031-8248

J 0154 Synthese • NL
Synthese. An International Journal for Epistemology, Methodology and Philosophy of Science [1936ff] ISSN 0039-7857

J 0156 Kybernetika (Prague) • CS
Kybernetika (Cybernetics) [1965ff] ISSN 0023-5954
• REL PUBL (**J** 3524) Kybernetika Suppl (Prague)

J 0157 Jbuchber Dtsch Math-Ver • D
Jahresbericht der Deutschen Mathematiker-Vereinigung [1890ff] ISSN 0012-0456

J 0158 Erkenntnis (Dordrecht) • NL
Erkenntnis. The Journal of Unified Science: An International Journal of Analytic Philosophy [1975ff] ISSN 0165-0106
• CONT OF (**J** 3597) J Unif Sci

J 0160 Math-Phys Sem-ber, NS • D
Mathematisch-Physikalische Semesterberichte: Zur Pflege des Zusammenhangs von Schule und Universitaet. Neue Folge [1950-1979] ISSN 0340-4897
• CONT AS (**J** 2790) Math Sem-ber

J 0161 Bull London Math Soc • GB
The Bulletin of the London Mathematical Society [1926ff] ISSN 0024-6093

J 0162 Teorema (Valencia) • E
Teorema [1971ff]

J 0164 J Comb Th, Ser A • USA
Journal of Combinatorial Theory. Series A [1971ff] ISSN 0097-3165
• CONT OF (**J** 1669) J Comb Th

J 0165 Acta Univ Carolinae Math Phys (Prague) • CS
Acta Universitatis Carolinae. Mathematica et Physica [1959ff] ISSN 0001-7140

J 0168 Blaetter Deutsch Philos • D
Blaetter fuer Deutsche Philosophie [1927-1944]
• CONT OF (**J** 1095) Beitr Phil Deutsch Idealismus

J 0169 Theoria (Madrid) • E
Theoria [1953ff] ISSN 0040-5817

J 0171 Proc Cambridge Phil Soc Math Phys • GB
Proceedings of the Cambridge Philosophical Society. Mathematical and Physical Sciences [1843-1974] ISSN 0008-1981
• CONT AS (**J** 0332) Math Proc Cambridge Phil Soc

J 0175 Methodos • I
Methodos [1949ff]
• REM Does not seem to appear anymore

J 0177 Algem Nederl Tijdschr Wijsbeg Psychol • NL
Algemeen Nederlands Tijdschrift voor Wijsbegeerte en Psychologie [1933-1969]
• CONT OF (**J** 0292) Tijdsch Wijsbegeerte • CONT AS (**J** 1829) Algem Nederl Tijdschr Wijsbeg
• REM Incorporating: Annalen der Critische Philosophie

J 0178 Stud Gen • D
Studium Generale: Zeitschrift fuer die Einheit der Wissenschaften. Zusammenhang ihrer Begriffsbildungen und Forschungsmethoden. [1947-1971] ISSN 0039-4149

J 0179 Ann Fac Sci Clermont • F
Universite de Clermont. Faculte des Sciences. Annales [1952-1972]
• CONT AS (J 1934) Ann Sci Univ Clermont Math

J 0180 Rev Phil France & Etranger • F
Revue Philosophique de la France et de l'Etranger [1876ff] ISSN 0035-3833

J 0182 Kon Nederl Akad Wetensch Afd Let Med N S • NL
Koninklijke Nederlandse Akademie van Wetenschappen. Afdeeling Letterkunde: Mededelingen. N.S. [1921ff]
• REL PUBL (J 0028) Indag Math Mededelingen

J 0187 IEEE Trans Comp • USA
Transactions on Computers. IEEE (= Institute of Electrical and Electronics Engineers) [1968ff] ISSN 0018-9340
• CONT OF (J 4305) IEEE Trans Electr Comp

J 0188 Rev Union Mat Argentina • RA
Revista de la Union Matematica Argentina [1936ff] ISSN 0041-6932

J 0189 Ark Mat • S
Arkiv foer Matematik (Archiv fuer Mathematik) [1949ff] ISSN 0004-2080
• CONT OF (J 3970) Ark Mat, Astron & Fys

J 0191 Inform Sci • USA
Information Sciences. An International Journal [1957ff] ISSN 0020-0255

J 0193 Discr Math • NL
Discrete Mathematics [1971ff] ISSN 0012-365X

J 0194 Inform & Control • USA
Information and Control [1958ff] ISSN 0019-9958

J 0197 Stud Cercet Mat Acad Romana • RO
Studii si Cercetari Matematice. Academia Republicii Socialiste Romania. (Mathematische Studien und Untersuchungen. Akademie der Sozialistischen Republik Rumaenien) [1950ff] ISSN 0567-6401
• CONT OF (J 0524) Disq Math Phys

J 0202 Diss Math (Warsaw) • PL
*Dissertationes Mathematicae. Polska Akademia Nauk, Instytut Matematyczny * Rozprawy Matematyczne* [1952ff] ISSN 0012-3862

J 0205 Rev Franc Autom, Inf & Rech Operat • F
Revue Francaise d'Automatique, Informatique et Recherche Operationelle (RAIRO). Series: Bleue, Jaune, Rouge, Verte [1972-1976] ISSN 0399-0559
• CONT OF (J 3954) Rev Franc Inf & Rech Operat • CONT AS (J 4698) Rev Franc Autom, Inf & Rech Operat, Ser Rouge Inf Th & (J 2831) RAIRO Autom & (J 2832) RAIRO Inform
• REM In 1975 the Serie Rouge split into: Serie Rouge Analyse Numerique & J4698

J 0207 Ist Lombardo Accad Sci Rend, A (Milano) • I
Istituto Lombardo. Accademia di Science e Lettere. Rendiconti. A. Scienze Matematiche, Fisiche, Chimichze e Geologiche [1937ff] ISSN 0021-2504
• CONT OF (J 3986) Ist Lombardo Rend, Ser 2 (Milano)

J 0211 Rev Gen Sci Pur Appl & Bull Soc Philomatique • F
Revue Generale des Sciences Pures et Appliquees et Bulletin de la Societe Philomatique. [1953-1959]
• CONT OF (J 0767) Rev Gen Sci Pur Appl

J 0215 Proc Irish Acad, Sect A • IRL
Proceedings of the Royal Irish Academy. Section A. Mathematical and Physical Sciences [1899ff] ISSN 0557-4056

J 0216 Izv Akad Nauk SSSR, Ser Mat • SU
Izvestiya Akademii Nauk SSSR. Seriya Matematicheskaya (Proceedings of the Academy of Sciences of the USSR. Mathematical Series) [1937ff] ISSN 0373-2436
• CONT OF (J 4717) Izv Akad Nauk SSSR • TRANSL IN (J 0448) Math of USSR, Izv

J 0220 Atti Accad Sci Torino, Fis Mat Nat • I
*Atti della Accademia delle Scienze di Torino. Classi di Scienze Fisiche, Matematiche e Naturali. Parte 1. * Acta Academiae Scientiarum Taurinensis* [1940ff] ISSN 0373-3033
• CONT OF (J 1742) Atti Accad Sci Torino, Fis Mat Nat

J 0225 Amer Math Soc, Transl, Ser 2 • USA
American Mathematical Society. Translations. Series 2 [1955ff] ISSN 0065-9290

J 0229 Ann Mat Pura Appl, Ser 3 • I
Annali di Matematica Pura ed Applicata [1898-1923]
• CONT AS (J 3526) Ann Mat Pura Appl, Ser 4

J 0232 Inform Process Lett • NL
Information Processing Letters. Devoted to the Rapid Publication of Short Contributions to Information Processing [1971ff] ISSN 0020-0190

J 0233 Soobshch Akad Nauk Gruz SSR • SU
*Soobshcheniya Akademii Nauk Gruzinskoj SSR * Sakaharth SSR Mecnierebatha Akademia Moambe (Communications of the Academy of Sciences of the Georgian SSR)* [1940ff] ISSN 0002-3167

J 0234 Rev Acad Cienc Exact Fis Nat Madrid • E
Revista de la Real Academia de Ciencias Exactas, Fisicas y Naturales de Madrid [1904ff] ISSN 0034-0596

J 0236 Rev Mat Hisp-Amer, Ser 4 • E
Revista Matematica Hispano-Americana. 4a Serie. Real Sociedad Matematica Espanola. [1941ff] ISSN 0373-0999
• CONT OF (J 3993) Rev Mat Hisp-Amer, Ser 2

J 0241 Bull Math Soc Nanyang Univ • SGP
Shu Hsuen Nien K'An (Nan-Yang Ta Hsuen Shu Shueh Hui) (Nanyang University. Mathematical Society. Bulletin) [1959-1965]
• CONT AS (J 0245) Nanta Math

J 0245 Nanta Math • SGP
Nanta Mathematica [1966ff] ISSN 0077-2739
• CONT OF (J 0241) Bull Math Soc Nanyang Univ

J 0248 Math Student • IND
The Mathematics Student [1933ff] ISSN 0025-5742

J 0249 J Math Pures Appl • F
Journal de Mathematiques Pures et Appliquees [1836-1921]
• CONT AS (J 3941) J Math Pures Appl, Ser 9

J 0252 Rev Philos Louvain • B
Revue Philosophique de Louvain [1946ff] ISSN 0035-3841
• CONT OF (J 1720) Rev Neoscolast Philos, Ser 2

J 0254 Gen Topology Appl • NL
General Topology and its Applications. A Journal Devoted to Set Theoretic, Axiomatic and Geometric Topology [1971-1979] ISSN 0016-660X
• CONT AS (**J** 2635) Topology Appl

J 0258 Nederl Akad Wetensch Verh Tweed Afd Nat • NL
Verhandelingen der Koninklijke Nederlandse Akademie van Wetenschappen Tweede Reeks. Afdeling Natuurkunde [1892ff] ISSN 0065-552X

J 0259 Nyt Tidsskr Mat • DK
Nyt Tidsskrift for Matematik (New Journal for Mathematics) [1890-1918]
• CONT AS (**J** 4510) Norsk Mat Tidsskr

J 0260 Ann Jap Ass Phil Sci • J
Annals of the Japan Association for Philosophy of Science [1956ff]

J 0261 Tohoku Math J • J
Tohoku Mathematical Journal (Tohoku Sugaku Zashi) [1911ff] ISSN 0040-8735

J 0265 Ukr Mat Zh, Akad Nauk Ukr SSR • SU
Ukrainskij Matematicheskij Zhurnal. Akademiya Nauk Ukrainskoj SSR. Institut Matematiki (Ukrainian Mathematical Journal. Academy of Sciences of the Ukrainian SSR. Institute of Mathematics) [1949ff] ISSN 0041-6053
• TRANSL IN (**J** 3281) Ukr Math J

J 0277 Sitzb Preuss Akad Wiss Phys Math Kl • DDR
Die Preussische Akademie der Wissenschaften. Sitzungsberichte. Physikalisch-Mathematische Klasse [1922-1949]

J 0278 Nederl Akad Wetensch Proc • NL
Koninklijke Nederlandse Akademie van Wetenschappen. Proceedings of the Section of Sciences. * *(J 0028)* Indag Math [1898-1950] ISSN 0023-3358
• REM Indagationes Mathematicae was part of the journal from 1939 until 1950. Since 1951, Series A of the journal has been the same as Indagationes Mathematicae.

J 0279 S African J Sci • ZA
Suid Afrikaanse Tydskrif vir Wetenskap (South African Journal of Science) [1903ff] ISSN 0038-2353

J 0280 Quart J Pure Appl Math • GB
Quarterly Journal of Pure and Applied Mathematics [1857-1927]

J 0283 Ann Soc Pol Math • PL
Societe Polonaise de Mathematique. Annales. * *Rocznik i Polskiego Towarzystwa Matematycznego* [1922-1952]
• CONT AS (**J** 1405) Ann Pol Math

J 0286 Int Logic Rev • I
International Logic Review. * *Rassegna Internazionale di Logica* [1970ff] ISSN 0048-6779

J 0287 Scripta Math • USA
Scripta Mathematica. [1932ff] ISSN 0036-9713

J 0288 Vest Ser Mat Mekh, Univ Moskva • SU
Vestnik Moskovskogo Universiteta. Seriya I. Matematika, Mekhanika (Publications of the Moscow University. Series I. Mathematics. Mechanics) [1946ff] ISSN 0201-7385, ISSN 0579-9368
• TRANSL IN (**J** 0510) Moscow Univ Math Bull & Moscow University. Mechanics Bulletin

J 0290 Euclides • NL
Euclides: Maandblad voor de Didactiek van de Wiskunde [1925ff]

J 0291 Wis-natuur Tijdsch • B
Wis- en Natuurkundig Tijdschrift: Orgaan van her Vlaamsch Natuur-, Wis- en Geneeskund Congres [1921-1944]
• CONT AS (**J** 0061) Simon Stevin

J 0292 Tijdsch Wijsbegeerte • NL
Tijdschrift Voor Wijsbegeerte [1907-1932]
• CONT AS (**J** 0177) Algem Nederl Tijdschr Wijsbeg Psychol

J 0294 Anuar Soc Paranaense Mat • BR
Sociedade Paranaense de Matematica. Anuario. [1954-1957]
• CONT AS (**J** 1831) Anuar Soc Paranaense Mat, Ser 2

J 0297 Rend Mat Appl, Ser 5 • I
Rendiconti di Matematica e delle sue Applicazioni. Seria 5. Universita di Roma. Istituto Nazionale di Alta Matematica (INDAM) [1942-1967]
• CONT AS (**J** 3741) Rend Mat Appl, Ser 7

J 0300 Rozpr Akad Krakow Histor-Filoz Ser 2 • PL
Akademia Umiejetnosci. Krakow. Rozprawy i Sprawozdania z Posiedzen. Wydzialu Historyczno-Filozoficzny. Seria 2 (Academy of Sciences. Cracow. Papers of the Academy of Sciences. Department of History and Philosophy. 2. Series) [1891-1952]

J 0301 J Phil • USA
The Journal of Philosophy [1904ff] ISSN 0022-362X

J 0302 Rep Math Logic, Krakow & Katowice • PL
Reports on Mathematical Logic. The Jagiellonian University of Cracow. The Silesian University of Katowice [1973ff] ISSN 0083-4432
• CONT OF (**S** 0458) Zesz Nauk, Prace Log, Uniw Krakow

J 0305 Invent Math • D
Inventiones Mathematicae [1966ff] ISSN 0020-9910

J 0306 Cah Topol & Geom Differ • F
Cahiers de Topologie et Geometrie Differentielle [1959ff] ISSN 0008-0004

J 0307 Rev Colomb Mat • CO
Revista Colombiana de Matematicas [1967ff] ISSN 0034-7426
• CONT OF (**J** 0348) Rev Mat Elementales

J 0308 Rocky Mountain J Math • USA
The Rocky Mountain Journal of Mathematics [1971ff] ISSN 0035-7596

J 0311 Nordisk Mat Tidskr • N
Nordisk Matematisk Tidskrift (Scandinavian Mathematical Journal) [1953-1978] ISSN 0029-1412
• CONT OF (**J** 4510) Norsk Mat Tidsskr • CONT AS (**J** 3075) Normat

J 0312 Izv Akad Nauk Armyan SSR, Ser Mat • SU
Izvestiya Akademii Nauk Armyanskoj SSR. Seriya Matematika (Proceedings of the Academy of Sciences of the Armenian SSR. Series: Mathematics) [1965ff] ISSN 0002-3043
• TRANSL IN (**J** 3265) Sov J Contemp Math Anal, Armen Acad Sci

J 0313 Period Mat (Bologna) • I
Periodico di Matematiche

J 0315 Ann Sc Norm Sup Pisa Fis Mat, Ser 3 • I
Annali della Scuola Normale Superiore di Pisa. Classe di Science. Fisiche e Matematiche. Seria III [1947-1973] ISSN 0036-9918
• CONT OF (J 1568) Ann Sc Norm Sup Pisa, Fis Mat, Ser 2
• CONT AS (J 4702) Ann Sc Norm Sup Pisa Fis Mat, Ser 4

J 0319 Matematiche (Sem Mat Catania) • I
Le Matematiche [1946ff]

J 0320 Monist • USA
The Monist: International Quarterly of General Philosophical Inquiry [1890ff] ISSN 0026-9662

J 0323 Uch Zap Ped Inst, Ryazan • SU
Ryazanskij Gosudarstvennyj Pedagogicheskij Institut. Uchenye Zapiski. (State Institute of Education in Ryazan. Scientific Notes)

J 0324 Izv Akad Nauk Latv SSR • SU
*Izvestiya Akademii Nauk Latviskoj SSR * Latvijas PSR Zinatnu. Akademijas Vestis (Proceedings of the Academy of Sciences of the Latvian SSR)* [1947ff] ISSN 0023-8929

J 0325 Amer Phil Quart • GB
American Philosophical Quarterly [1964ff] ISSN 0003-0481

J 0326 J Pure Appl Algebra • NL
Journal of Pure and Applied Algebra [1971ff] ISSN 0022-4049

J 0332 Math Proc Cambridge Phil Soc • GB
Mathematical Proceedings of the Cambridge Philosophical Society [1975ff] ISSN 0305-0041
• CONT OF (J 0171) Proc Cambridge Phil Soc Math Phys

J 0335 Atti Accad Ligure Sci Lett (Genova) • I
Atti della Accademia Ligure di Scienze e Lettere [1890ff] ISSN 0392-2219

J 0340 Mat Zap (Univ Sverdlovsk) • SU
Matematicheskie Zapiski (Mathematical Notes) ISSN 0076-5368

J 0342 Monatsber Dt Akad Wiss • DDR
Die Deutsche Akademie der Wissenschaften zu Berlin. Monatsberichte: Mitteilungen aus Mathematik, Naturwissenschaft, Medizin und Technik [1959-1971] ISSN 0011-9814

J 0345 Adv Math • USA
Advances in Mathematics [1964ff] ISSN 0001-8708
• REL PUBL (S 3105) Adv Math, Suppl Stud

J 0346 Dokl Akad Nauk Armyan SSR • SU
Doklady Akademii Nauk Armyanskoj SSR (Reports of the Academy of Sciences of the Armenian SSR) [1944ff] ISSN 0321-1339

J 0348 Rev Mat Elementales • CO
Revista de Matematicas Elementales [1952-1966]
• CONT AS (J 0307) Rev Colomb Mat

J 0349 Math of USSR, Sbor • USA
Mathematics of the USSR, Sbornik [1967ff] ISSN 0025-5734
• TRANSL OF (J 0142) Mat Sb, Akad Nauk SSSR, NS

J 0350 Sci Rep Tokyo Kyoiku Daigaku Sect A • J
Tokyo Kyoiku Daigaku (Tokyo University of Education. Science Reports. Section A.)

J 0351 Osaka J Math • J
Osaka Journal of Mathematics [1964ff] ISSN 0030-6126
• CONT OF (J 1770) Osaka Math J

J 0352 Math Jap • J
Mathematica Japonica [1948ff] ISSN 0025-5513

J 0353 Bull Soc Math Fr • F
Bulletin de la Societe Mathematique de France [1873ff] ISSN 0037-9484

J 0354 Phil Trans Roy Soc London, Ser A • GB
Philosophical Transactions of the Royal Society of London. Series A. Mathematical and Physical Sciences. ISSN 0080-4614

J 0358 Versl Gewone Vergad Afd Natuurkd • NL
Koninklijke Nederlandse Akademie van Wetenschappen. Verslagee van de Gewone Vergaderingen der Afdling Natuurkunde (Royal Dutch Academy of Sciences. Reports of the Meetings of the Physics Section) [1892ff] ISSN 0023-3382

J 0366 Sitzber Berlin Math Ges • DDR
Die Berliner Mathematische Gesellschaft. Sitzungsberichte

J 0371 Glas Mat-Fiz Astron, Ser 2 (Zagreb) • YU
Glasnik Matematichko-Fizichki i Astronomichki. Ser II (Publications of Mathematics, Physics, Astronomy. Ser II) [1946-1965]
• CONT AS (J 3519) Glas Mat, Ser 3 (Zagreb)

J 0373 Comp Arch Inform & Numerik • A
*Computing: Archiv fuer Informatik und Numerik * Computing: Archives for Informatics and Numerical Computation* [1966ff] ISSN 0010-485X

J 0377 Bol Mat (Bogota) • CO
Boletin de Matematicas [1967ff]

J 0380 Acta Cybern (Szeged) • H
Acta Cybernetica. Forum Centrale Publicationum Cyberneticarum Hungaricum [1969ff] ISSN 0324-721X

J 0381 J Tsuda College (Tokyo) • J
Journal of Tsuda College [1969ff]

J 0382 Int J Comput Math • GB
International Journal of Computer Mathematics. Section A: Programming Languages; Theory and Methods. Section B: Computational Methods [1964ff] ISSN 0020-7160

J 0385 Atti Sem Mat Fis Univ Modena • I
Atti del Seminario Matematico e Fisico dell'Universita di Modena [1947ff] ISSN 0041-8986

J 0386 Proc Edinburgh Math Soc • GB
Proceedings of the Edinburgh Mathematical Society [1883-1926]
• CONT AS (J 3420) Proc Edinburgh Math Soc, Ser 2

J 0387 Bull Sect Logic, Pol Acad Sci • PL
Bulletin of the Section of Logic. Polish Academy of Sciences. Institute of Philosophy and Sociology. [1972ff]
• REM Papers Published in the Bulletin are Generally: 1. Abstracts or preprints of papers submitted to other journals e.g. Studia Logica. 2. Abstracts of papers read at seminars or local conferences

J 0390 Publ Res Inst Math Sci (Kyoto) • J
Publications of the Research Institute for Mathematical Sciences [1965ff] ISSN 0034-5318
• REM Vols 1-4 Issued as: Kyoto Univ. Research Institute for Mathematical Sciences. Publications. Series A.

J 0391 Riv Mat Univ Parma • I
Rivista di Matematica della Universita di Parma [1891-1915]
• CONT AS (J 1526) Riv Mat Univ Parma, Ser 2

J 0392 Math Sci Hum • F
Mathematiques et Sciences Humaines [1962ff] ISSN 0025-5815

J 0394 Commun Algeb • USA
Communications in Algebra [1974ff] ISSN 0092-7872

J 0396 Mat Lapok • H
Matematikai Lapok (Mathematical Papers) [1949ff] ISSN 0025-519X
• CONT OF (**J** 0461) Mat Fiz Lapok

J 0397 Proc Math Phys Soc Egypt • ET
Proceedings of the Mathematical and Physical Society of Egypt [1937ff] ISSN 0076-5317

J 0400 Publ Inst Math, NS (Belgrade) • YU
Institut Mathematique. Publications de l'Institut Mathematique. Nouvelle Serie [1961ff] ISSN 0522-828X
• CONT OF (**J** 4706) Publ Inst Math (Belgrade)

J 0401 J Number Th • USA
Journal of Number Theory [1968ff] ISSN 0022-314X

J 0403 Izv Akad Nauk Kazak SSR, Ser Fiz-Mat • SU
Izvestiya Akademii Nauk Kazakhskoj SSR. Seriya Fiziko-Matematicheskaya (Proceedings of the Academy of Sciences of the Kazakh SSR. Series: Physics & Mathematics) [1963ff] ISSN 0002-3191

J 0407 Comm Math Univ St Pauli (Tokyo) • J
Commentarii Mathematici Universitatis Sancti Pauli [1952ff] ISSN 0010-258X

J 0418 Shuxue Xuebao • TJ
Shuxue Xuebao (Acta Mathematica Sinica) [1951ff]
• TRANSL IN (**J** 0419) Chinese Math Acta
• REM In 1951 published as: Journal of the Chinese Mathematical Society (N.S.)

J 0419 Chinese Math Acta • USA
Chinese Mathematics. Acta [1962-1967]
• TRANSL OF (**J** 0418) Shuxue Xuebao

J 0420 Shuxue Jinzhan • TJ
Shuxue Jinzhan (Advances in Mathematics) [1955ff] ISSN 0559-9326

J 0423 Pensee NS • F
La Pensee. Revue de Bon Rationalisme Moderne. Nouvelle Serie [1944ff]

J 0426 Dt Math • DDR
Deutsche Mathematik [1936-1942]

J 0431 Hiroshima Math J • J
Hiroshima Mathematical Journal [1971ff] ISSN 0018-2079
• CONT OF (**J** 0102) Hiroshima Univ J Sci, Ser A Math

J 0434 J Fac Sci Univ Tokyo, Sect 1 • J
Journal of the Faculty of Science. University of Tokyo. Section 1 Mathematics, Astronomy, Physics, Chemistry [1925-1970] ISSN 0040-8980
• CONT AS (**J** 2332) J Fac Sci, Univ Tokyo, Sect 1 A

J 0435 Int J Comput & Inf Sci • USA
International Journal of Computer and Information Sciences [1972ff] ISSN 0091-7036

J 0440 Bul Inst Politeh Bucuresti, Ser Mec • RO
Buletinul Institutului Politehnic "Gheorghe Gheorghiu-Dej" Bucuresti. Seria Mecanica (Bulletin des Polytechnischen Instituts "Gheorghe Gheorghiu-Dej" Bukarest. Serie Mechanik)

J 0441 J Indian Math Soc, NS • IND
The Journal of the Indian Mathematical Society. New Series [1934ff] ISSN 0019-5839

J 0442 J Osaka Inst Sci & Tech, Part 1 • J
Osaka Institute of Science and Technology (The Kinki University). Journal. Part 1. Mathematics and Physics [1949ff]

J 0448 Math of USSR, Izv • USA
Mathematics of the USSR, Izvestiya [1967ff] ISSN 0025-5726
• TRANSL OF (**J** 0216) Izv Akad Nauk SSSR, Ser Mat

J 0449 Probl Kybern • DDR
Probleme der Kybernetik [1958-1965]
• CONT AS (**J** 0471) Syst Th Res • TRANSL OF (**J** 0052) Probl Kibern

J 0455 Phil Naturalis • D
Philosophia Naturalis: Archiv fuer Naturphilosphie und die Philosophischen Grenzgebiete der Exakten Wissenschaften und Wissenschaftsgeschichte [1950ff] ISSN 0031-8027

J 0456 L'Age Science • F
L'Age de la Science [1968ff] ISSN 0002-0737

J 0459 C R Soc Sci Lett Varsovie Cl 3 • PL
*Societe des Sciences et des Lettres de Varsovie. Comptes Rendus des Seances. Classe III: Sciences Mathematiques et Physiques * Towarzystwo Naukowe Warszawskie. Sprawozdania z Posiedze. Wydzialu III: Nauk Matematyczno-Fizycznych (Warschauer Sitzungsberichte)* [1908-1950]

J 0460 Acta Univ Szeged, Sect Mat • H
Acta Litterarum ac Scientiarum Regiae Universitatis Hungaricae Francisco-Josephinae, Sectio Scientiarum Mathematicarum [1922-1946]
• CONT AS (**J** 0002) Acta Sci Math (Szeged)

J 0461 Mat Fiz Lapok • H
Matematikai es Fizikai Lapok (Mathematical and Physical Papers) [1892-1948]
• CONT AS (**J** 0396) Mat Lapok

J 0465 Bull Greek Math Soc (NS) • GR
Bulletin of the Greek Mathematical Society.New Series (Hellenike Mathematike Hetaireia. Deltion. Nea Seira.) [1960ff] ISSN 0072-7466
• CONT OF (**J** 1699) Bull Soc Math Grece

J 0470 Sov Phys, Dokl • USA
Soviet Physics. Doklady. [1956ff] ISSN 0038-5689
• TRANSL OF (**J** 0023) Dokl Akad Nauk SSSR

J 0471 Syst Th Res • USA
Systems Theory Research [1966ff] ISSN 0082-1255
• CONT OF (**J** 1195) Probl Cybernet & (**J** 0449) Probl Kybern
• TRANSL OF (**J** 0052) Probl Kibern

J 0474 Avtom Vychis Tekh, Akad Nauk Latv SSR • SU
Avtomatika i Vychislitel'naya Tekhnika. Akademiya Nauk Latvijskoj SSR (Automation and Computer Science. Academy of Sciences of the Latvian SSR) [1967ff] ISSN 0572-4538
• TRANSL IN (**J** 2666) Autom Control Comput Sci

J 0475 Sib Math J • USA
Siberian Mathematical Journal [1966ff] ISSN 0037-4466
• TRANSL OF (**J** 0092) Sib Mat Zh

J 0476 Vest Akad Nauk SSSR • SU
Vestnik Akademii Nauk SSSR (Publication of the Academy of Sciences of the USSR) [1931ff] ISSN 0002-3442

J 0479 Kwart Filoz • PL
Kwartalnik Filozoficzny (Philosophical Quaterly)

J 0480 Roy Soc Bibl Mem Fellows • GB
The Royal Society. Bibliographical Memoirs of Fellows. [1955ff] ISSN 0080-4606

J 0481 Acta Univ Wroclaw • PL
Acta Universitatis Wratislaviensis

J 0482 Publ Math Univ Belgrade • YU
Universite de Belgrade. Publications Mathematiques [1932ff]

J 0483 Uch Zap, Univ Riga • SU
*Uchenye Zapiski Latvijskogo Gosudarstvennogo Universiteta imeni Petra Stutski. * Zinatniskie Raksti. Latvijas Valsts Universitate. (Scientific Notes of the Latvian State University.)*
• REM This journal consists of two series: Teoriya Algoritmov i Programm & Teoreticheskie Voprosy Avtomaticheskikh Sistem Upravleniya

J 0484 Komb Analiz Sb Stat • SU
Kombinatornyj Analiz. Sbornik Statej. (Combinatorial Analysis. Collected Articles) [1971]

J 0488 Dialogue (Ottawa) • CDN
*Dialogue: Canadian Philosophical Review * Dialogue: Revue Canadienne de Philosophie.* [1962ff] ISSN 0012-2173

J 0493 Bull Res Counc Israel Sect F • IL
Research Council of Israel. Bulletin. Section F. Mathematics and Physics [1952-1962]
• CONT AS (J 0029) Israel J Math

J 0494 Bull Math Soc Sci Roumanie • RO
Societatea Romana de Stiinte, Sectia Mathematica. Bulletin Mathematiques de la Societe Roumaine des Sciences [1908-1956]
• CONT AS (J 0070) Bull Soc Sci Math Roumanie, NS

J 0497 Math Mag • USA
Mathematics Magazine [1947ff] ISSN 0025-570X
• CONT OF (J 1737) Nat Math Magazine (Louisiana)

J 0500 Tamkang J Math • RC
Tamkang Journal of Mathematics [1970ff] ISSN 0049-2930

J 0503 Artif Intell • NL
Artificial Intelligence [1970ff] ISSN 0004-3702

J 0504 Manuscr Math • D
Manuscripta Mathematica [1969ff] ISSN 0025-2611

J 0508 Machine Intelligence • GB
Machine Intelligence [1967ff] ISSN 0541-6418

J 0510 Moscow Univ Math Bull • USA
Moscow University Mathematics Bulletin [1969ff] ISSN 0027-1322
• TRANSL OF (J 0288) Vest Ser Mat Mekh, Univ Moskva

J 0517 Mathematica (Cluj) • RO
Mathematica. Revue d'Analyse Numerique et de Theorie de l'Approximation [1929ff] ISSN 0025-5505

J 0519 Wiad Mat, Ann Soc Math Pol, Ser 2 • PL
*Annales Societatis Mathematicae Polonae. Seria 2. Wiadomosci Matematyczne * Roczniki Polskiego Towarzystwa Matematycznego. Seria 2. Wiadomosci Matematyczne* [1955ff] ISSN 0079-3698
• CONT OF (J 4710) Pol Tow Mat, Wiad Mat

J 0521 Semin Math, Inst Steklov • USA
Seminars in Mathematics. V.A.Steklov Mathematical Institute Leningrad [1967ff]
• TRANSL OF (S 0228) Zap Nauch Sem Leningrad Otd Mat Inst Steklov

J 0523 Bull Nagoya Inst Tech • J
Bulletin of Nagoya Institute of Technology [1949ff]

J 0524 Disq Math Phys • RO
Disquisitiones Mathematicae et Physicae [1940-1949]
• CONT AS (J 0197) Stud Cercet Mat Acad Romana

J 0529 Aequationes Math • CH
Aequationes Mathematicae [1968ff] ISSN 0001-9054

J 0537 Yokohama Math J • J
The Yokohama Mathematical Journal [1953ff] ISSN 0044-0523

J 0545 Proc Indian Acad Sci, Sect A • IND
Indian Academy of Sciences. Proceedings. Section A: Physical Sciences [1931ff] ISSN 0019-428X

J 0549 Riv Mat Univ Parma, Ser 4 • I
Rivista di Matematica della Universita di Parma. Serie 4 [1975ff] ISSN 0035-6298
• CONT OF (J 3254) Riv Mat Univ Parma, Ser 3

J 0589 Bull Amer Math Soc (NS) • USA
Bulletin of the American Mathematical Society. New Series [1979ff] ISSN 0273-0979
• CONT OF (J 0015) Bull Amer Math Soc

J 0597 Afr Tydsk Wetensk Kuns • ZA
Afrikaans Tydskrif vir Wetenskap en Kuns

J 0611 Pol Tow Mat, Prace Mat-Fiz • PL
Polskie Towarzystwo Matematyczene. Prace Matematyczno-Fizyczne (Mathematische und Physikalische Abhandlungen) [1887-1954]
• CONT AS (J 0051) Commentat Math, Ann Soc Math Pol, Ser 1 & (J 2095) Fund Inform, Ann Soc Math Pol, Ser 4

J 0748 Erkenntnis (Leipzig) • DDR
Erkenntnis [1930-1939]
• CONT OF (J 1380) Ann Philos & Philos Kritik • CONT AS (J 3597) J Unif Sci

J 0767 Rev Gen Sci Pur Appl • F
Revue Generale des Sciences Pures et Appliquees [1890-1952]
• CONT AS (J 0211) Rev Gen Sci Pur Appl & Bull Soc Philomatique

J 0789 Uch Zap Mat Ped Inst, Tula • SU
Tul'skij Gosudarstvennyj Pedagogicheskij Institut im. L.N.Tolstogo. Uchenye Zapiski Matematicheskikh Kafedr (Tolstoi State Institute of Education in Tula. Scientific Notes of the Mathematical Faculties)

J 0931 Anz Oesterr Akad Wiss, Math-Nat Kl • A
Anzeiger der Oesterreichischen Akademie der Wissenschaften. Mathematisch-Naturwissenschaftliche Klasse [1864ff] ISSN 0065-535X

J 0945 Riv Filos • I
Rivista di Filosofia [1909ff] ISSN 0035-6239

J 0951 Bull Soc Fr Phil • F
Societe Francaise de Philosophie. Bulletin. [1901ff] ISSN 0037-9352

J 0956 Forsch Logik Grundl exakt Wiss • D
Forschungen zur Logik und zur Grundlegung der exakten Wissenschaften [1937-1943]

J 0963 Ajatus (Helsinki) • SF
Ajatus. Suomen Filosofisen Yhdislyksen Vuosikirja (Yearbook of the Philosopical Society of Finland) [1926ff]

J 0967 Izv Akad Nauk Mold SSR, Ser Fiz-Tekh Mat • SU
*Izvestiya Akademii Nauk Moldavskoj SSR. Seriya Fiziko-Tekhnicheskikh i Matematicheskikh Nauk * Buletinul Akademiei de Shtiince a RSS Moldovensht (Proceedings of the Academy of Sciences of the Moldavian SSR. Series: Physical-Technical and Mathematical Sciences)* [1951ff] ISSN 0321-169X

J 0974 Norsk Vid-Akad Oslo Mat-Natur Kl Skr • N
Norske Videnskaps - Akademi i Oslo. Matematisk-Naturvidenskapelig Klasse. Skrifter. (Monographs of the Scandinavian Academy of Sciences. Mathematical and Natural Sciences Class) [1929ff] ISSN 0029-2338
• CONT OF (J 1145) Vidensk Selsk Kristiana Skrifter Ser 1

J 0975 Norsk Mat Forenings Skr • N
Norsk Matematisk Forenings Skrifter

J 0982 Z Wahrscheinltheor & Verw Geb • D
Zeitschrift fuer Wahrscheinlichkeitstheorie und Verwandte Gebiete [1962ff] ISSN 0044-3719

J 0983 Forsch Fortschritte • DDR
Forschungen und Fortschritte. Nachrichtenblatt der Deutschen Wissenschaft und Technik [1924-1967]

J 0989 Z Allg Wissth • D
Zeitschrift fuer Allgemeine Wissenschaftstheorie (Journal for General Philosophy of Science) [1969ff] ISSN 0044-2216

J 0991 Acta Acad Aabo, Ser B • SF
Acta Academiae Aaboensis, Ser. B. Mathematica et Physica. Matematik, Naturvetenskaper, Teknik [1936ff] ISSN 0001-5105

J 1005 Rep Fac Sci, Shizuoka Univ • J
Reports of the Faculty of Science. Shizuoka University. [1965ff] ISSN 0583-0923

J 1008 Demonstr Math (Warsaw) • PL
Demonstratio Mathematica [1969ff] ISSN 0420-1213

J 1012 Rice Inst Pamphlet • USA
The Rice Institute Pamphlet. Monographs in Mathematics.

J 1021 Itogi Nauki Ser Mat • SU
Itogi Nauki. Seriya Matematiki (Progress in Science. Mathematical Series) [1962-1971] ISSN 0579-1731
• CONT AS (J 1488) Itogi Nauki Tekh, Ser Probl Geom & (J 1501) Itogi Nauki Tekh, Ser Algeb, Topol, Geom & (J 1452) Itogi Nauki Tekh, Ser Sovrem Probl Mat & (J 3188) Itogi Nauki Tekh, Ser Teor Veroyat Mat Stat Teor Kibern & (J 4387) Itogi Nauki Tekh, Ser Tekh Kibern • TRANSL IN (J 1531) J Sov Math
• REM J1531 contains only selected translations

J 1024 Zhongguo Kexue • TJ
Zhongguo Kexue (Scientia Sinica) [1950-1981]
• CONT AS (J 3766) Zhongguo Kexue, Xi A

J 1033 Etud Phil • F
Les Etudes Philosophiques [1926ff] ISSN 0014-2166

J 1044 Math Notes, Acad Sci USSR • USA
Mathematical Notes of the Academy of Sciences of the USSR [1967ff] ISSN 0001-4346
• TRANSL OF (J 0087) Mat Zametki (Akad Nauk SSSR)

J 1048 Kiber Sb Perevodov • SU
Kiberneticheskij Sbornik: Sbornik Perevodov. (Collected Articles on Cybernetics: Collected Translations) [1960-1964]
• CONT AS (J 3079) Kiber Sb Perevodov, NS

J 1069 Tydskr Natuurwetenskap (Pretoria) • ZA
Tydskrif vir Naturwetenskappe (Journal for Natural Sciences) [1961ff] ISSN 0041-4786

J 1092 Ann Univ Paris • F
Universite de Paris. Annales ISSN 0041-9176

J 1093 Ruch Filoz • PL
Ruch Filozoficzny. Polskie Towarzystwo Filozoficzne (Philosophical Movement. Polish Philosophical Society) ISSN 0035-9599

J 1095 Beitr Phil Deutsch Idealismus • D
Beitraege zur Philosophie des Deutschen Idealismus [1918-1926]
• CONT AS (J 0168) Blaetter Deutsch Philos

J 1099 Phil Math • USA
Philosophia Mathematica. An International Journal for Philosophy of Modern Mathematics [1964ff] ISSN 0031-8019

J 1108 Arkhimedes (Helsinki) • SF
Arkhimedes [1949ff] ISSN 0004-1920

J 1109 Nachr Akad Wiss Goettingen, Math-Phys Kl • D
Nachrichten der Akademie der Wissenschaften in Goettingen. II. Mathematisch-Physikalische Klasse [1893ff] ISSN 0065-5295

J 1114 Arch Phil • D
Archiv fuer Philosophie [1911-1930, 1947-1964]

J 1119 Ita-humanidades • BR
Ita-Humanidades

J 1124 Ergebn Math Kolloquium • A
Ergebnisse eines Mathematischen Kolloquiums [1929-1936]

J 1145 Vidensk Selsk Kristiana Skrifter Ser 1 • N
Videnskaps Selskapet i Kristiana. Skrifter Utgit. 1 Matematisk-Naturvidenskapelig Klasse [?-1928]
• CONT AS (J 0974) Norsk Vid-Akad Oslo Mat-Natur Kl Skr

J 1150 Proc Roy Soc London, Ser A • GB
Proceedings of the Royal Society of London. Series A. Mathematical and Physical Sciences. [1905ff] ISSN 0080-4630

J 1154 Science (AAAS) • USA
Science. AAAS (= American Association for Advancement of Science) [1880ff] ISSN 0036-8075

J 1156 Izv Bulgar Akad Nauk Mat Inst • BG
B"lgarska Akademiya na Naukite. Otdelenie Za Fiziko-Matematicheski i Tekhnicheski Nauki. Izvestiya na Matematicheski Institut (Bulgarian Academy of Sciences. Department of Physics, Mathematics & Engineering. Reports of the Mathematical Institute) [1957-1974]
• CONT AS (J 2547) Serdica, Bulgar Math Publ

J 1157 Rend Accad Lincei Roma, Ser 5 • I
Atti della Reale Accademia dei Lincei. Rendiconti. Classe di Scienze Fisiche, Matematiche e Naturali. Serie 5 [1892-1924]
• CONT AS (J 3990) Rend Accad Lincei Roma, Ser 6

J 1179 Filoz Jugos Chas (Belgrade) • YU
Filozofija: Jugoslovenski Chasopis Za Filozofiju (Philosophy: Yugoslavian Journal of Philosophy) [1957ff] ISSN 0015-1866

J 1195 Probl Cybernet • GB
Problems of Cybernetics [1958-1965]
• CONT AS (**J** 0471) Syst Th Res • TRANSL OF (**J** 0052) Probl Kibern

J 1377 Ann New York Acad Sci • USA
New York Academy of Sciences. Annals [1877ff] ISSN 0077-8923

J 1379 Christiaan Huygens Internat Math Tijdschr • NL
Christiaan Huygens. Internationaal Mathematisch Tijdschrift [1868-1942]
• CONT AS (**J** 0061) Simon Stevin

J 1380 Ann Philos & Philos Kritik • DDR
Annalen der Philosophie und Philosophischen Kritik [1925-1929]
• CONT AS (**J** 0748) Erkenntnis (Leipzig)

J 1388 Rep Univ Waterloo • CDN
University of Waterloo. Reports

J 1393 Ann Acad Nac Cienc Exact Fis Nat Buenos Aires • RA
Academia Nacional de Ciencias Exactas, Fisicas y Naturales de Buenos Aires. Annales

J 1399 Russ Math Surv • GB
Russian Mathematical Surveys [1946ff] ISSN 0036-0279
• TRANSL OF (**J** 0067) Usp Mat Nauk

J 1404 Mat Sb, Akad Nauk SSSR • SU
Matematicheskij Sbornik. Akademiya Nauk SSSR i Moskovskoe Matematicheskoe Obshchestvo (Mathematical Collected Articles. New Series. Academy of Sciences of the USSR and the Moscovian Mathematical Society) [1866-1935]
• CONT AS (**J** 0142) Mat Sb, Akad Nauk SSSR, NS

J 1405 Ann Pol Math • PL
Annales Polonici Mathematici. Polska Akademia Nauk, Instytut Matematyczny [1953ff] ISSN 0066-2216
• CONT OF (**J** 0283) Ann Soc Pol Math

J 1426 Theor Comput Sci • NL
Theoretical Computer Science [1975ff] ISSN 0304-3975

J 1428 SIAM J Comp • USA
Journal on Computing. SIAM (= Society for Industrial and Applied Mathematics) [1972ff] ISSN 0097-5397

J 1431 Acta Inf • D
Acta Informatica [1971ff] ISSN 0001-5903

J 1447 Houston J Math • USA
Houston Journal of Mathematics [1975ff] ISSN 0362-1588

J 1452 Itogi Nauki Tekh, Ser Sovrem Probl Mat • SU
Itogi Nauki i Tekhniki: Seriya Sovremennye Problemy Matematiki (Progress in Science and Technology: Series on Current Problems in Mathematics) [1972ff]
• CONT OF (**J** 1021) Itogi Nauki Ser Mat • TRANSL IN (**J** 1531) J Sov Math

J 1472 Sci Rep Saitama Univ, Ser A • J
The Science Reports of the Saitama University. Series A. Mathematics [1952-1982] ISSN 0558-2431
• CONT AS (**J** 3940) Saitama Math J Physics, and Chemistry

J 1488 Itogi Nauki Tekh, Ser Probl Geom • SU
Itogi Nauki i Tekhnike. Seriya Problemy Geometrii (Progress in Science and Technology. Series Problems in Geometry) [1972ff] ISSN 0202-7461
• CONT OF (**J** 1021) Itogi Nauki Ser Mat • TRANSL IN (**J** 1531) J Sov Math

J 1501 Itogi Nauki Tekh, Ser Algeb, Topol, Geom • SU
Itogi Nauki i Tekhniki. Seriya Algebra, Topologiya, Geometriya. (Progress in Science and Technology. Series Algebra, Topology, Geometry) [1972ff] ISSN 0202-7445
• CONT OF (**J** 1021) Itogi Nauki Ser Mat • TRANSL IN (**J** 1531) J Sov Math & (**C** 4688) Prog in Math, Vol 12
• REM C4688 is Volume 1968

J 1507 An Inst Mat, Univ Nac Aut Mexico • MEX
Anales del Instituto de Matematicas. Universidad Nacional Autonoma de Mexico [1961ff] ISSN 0076-7441

J 1515 Archimede • I
Archimede. Rivista per gli Insegnanti e i Cultori di Matematiche Pure e Applicate [1949ff] ISSN 0003-8369

J 1526 Riv Mat Univ Parma, Ser 2 • I
Rivista di Matematica della Universita di Parma. Serie 2 [1960-1971] ISSN 0035-6298
• CONT OF (**J** 0391) Riv Mat Univ Parma • CONT AS (**J** 3254) Riv Mat Univ Parma, Ser 3

J 1527 Pokroky Mat Fyz Astron (Prague) • CS
Pokroky Matematiky, Fyziky a Astronomie (Progress in Mathematics, Physics and Astronomy) [1956ff] ISSN 0032-2423

J 1531 J Sov Math • USA
Journal of Soviet Mathematics [1973ff] ISSN 0090-4104
• TRANSL OF (**J** 1021) Itogi Nauki Ser Mat & (**S** 0228) Zap Nauch Sem Leningrad Otd Mat Inst Steklov & Problemy Matimaticheskogo Analiza & (**J** 1452) Itogi Nauki Tekh, Ser Sovrem Probl Mat & (**J** 1501) Itogi Nauki Tekh, Ser Algeb, Topol, Geom & (**J** 3188) Itogi Nauki Tekh, Ser Teor Veroyat Mat Stat Teor Kibern & (**J** 1488) Itogi Nauki Tekh, Ser Probl Geom
• REM This contains selected translations from each of the Russian Journals listed

J 1550 Creation Math • IL
Creation in Mathematics [1970ff]

J 1568 Ann Sc Norm Sup Pisa, Fis Mat, Ser 2 • I
Annali della R. Scuola Normale Superiore di Pisa. Serie 2 Scienze Fisiche e Matematiche [1932-1946]
• CONT OF (**J** 1908) Ann Sc Norm Sup Pisa, Fis Mat • CONT AS (**J** 0315) Ann Sc Norm Sup Pisa Fis Mat, Ser 3

J 1573 Notas Commun Mat (Recife) • BR
Notas e Comunicacoes de Matematica [1965ff] ISSN 0085-5413

J 1576 Publ Sem Mat Garcia Galdeano (Zaragoza) • E
Publicaciones del Seminario Matematico Garcia de Galdeano. Consejo Superior de Investigaciones Cientificas. Faculdad de Ciencias de Zaragoza. [1959ff] ISSN 0085-6029

J 1583 Rep Dept Math, Univ Amsterdam • NL
Report. Department of Mathematics, University of Amsterdam
• REM Preprint Series

J 1604 J Struct Learning • USA
Journal of Structural Learning ISSN 0022-4774

J 1620 Asterisque • F
Asterisque [1973ff] ISSN 0303-1179

J 1622 Arch Int Hist Sci (NS) • D
Archives Internationales d'Histoire des Sciences. Nouvelle Serie [1972ff] ISSN 0003-9810
• CONT OF (**J** 2341) Arch Int Hist Sci

J 1648 Hist Math • USA
Historia Mathematica. International Journal of History of Mathematics [1974ff] ISSN 0315-0860

J 1650 Rapp, Sem Math Pure, Univ Cathol Louvain • B
Rapport. Seminaire de Mathematique Pure

J 1669 J Comb Th • USA
Journal of Combinatorial Theory [1966-1970] ISSN 0021-9800
• CONT AS (**J** 0164) J Comb Th, Ser A & (**J** 0033) J Comb Th, Ser B

J 1670 Mitt Math Ges DDR • DDR
Mitteilungen der Mathematischen Gesellschaft der DDR

J 1680 Cienc Tecnol, Costa Rica • CR
Ciencia y Tecnologia, Revista de la Universidad de Costa Rica [1976ff] ISSN 0378-052X

J 1699 Bull Soc Math Grece • GR
(Bulletin de la Societe Mathematique Grece) [1921-1959]
• CONT AS (**J** 0465) Bull Greek Math Soc (NS)

J 1718 Rev Mat Santiago • RCH
Revista de Matematicas, Santiago

J 1720 Rev Neoscolast Philos, Ser 2 • B
Revue Neo-Scolastique de Philosophie. Serie 2 [1910-1945]
• CONT AS (**J** 0252) Rev Philos Louvain

J 1728 Wijsg Persp Weten Maatsch • NL
Wijsgerig Perspectief op Wetenschap en Maatschappij (Philosophical Perspective on Society and Science.) [1960ff] ISSN 0043-5414

J 1737 Nat Math Magazine (Louisiana) • USA
National Mathematics Magazine [1926-1946]
• CONT AS (**J** 0497) Math Mag

J 1742 Atti Accad Sci Torino, Fis Mat Nat • I
Atti della Reale Accademia delle Scienze di Torino. Classe I: Scienze Fisiche, Matematiche e Naturali [1865-1939]
• CONT AS (**J** 0220) Atti Accad Sci Torino, Fis Mat Nat

J 1770 Osaka Math J • J
Osaka Mathematical Journal [1949-1963]
• CONT AS (**J** 0351) Osaka J Math

J 1784 Wisk Tijdsch • NL
Wiskundig Tijdschrift [1904-1921]

J 1793 Nieuw Arch Wisk, Ser 2 • NL
Nieuw Archief voor Wiskunde. Reeks 2 [1894-1952]
• CONT OF (**J** 0046) Nieuw Arch Wisk • CONT AS (**J** 3077) Nieuw Arch Wisk, Ser 3

J 1829 Algem Nederl Tijdschr Wijsbeg • NL
Algemeen Nederlands Tijdschrift Voor Wijsbegeerte [1970ff] ISSN 0002-5275
• CONT OF (**J** 0177) Algem Nederl Tijdschr Wijsbeg Psychol

J 1831 Anuar Soc Paranaense Mat, Ser 2 • BR
Sociedade Paranaense de Matematica. Anuario. Serie 2 [1958ff]
• CONT OF (**J** 0294) Anuar Soc Paranaense Mat

J 1854 Bull Sec Mat Soc Catalana Cienc • E
Bulleti de la Seccio de Matematiques de la Societat de Ciencies Fisiques, Quimiques i Matematiques (Bericht der Abteilung Mathematik der Wissenschaftsgesellschaft fuer Physik, Chemie und Mathematik) [1981ff]

J 1908 Ann Sc Norm Sup Pisa, Fis Mat • I
Annali della Reale Scuola Normale Superiore Universitaria di Pisa. Scienze Fisiche e Matematiche [1913-1931]
• CONT AS (**J** 1568) Ann Sc Norm Sup Pisa, Fis Mat, Ser 2

J 1910 Proc London Math Soc, Ser 2 • GB
Proceedings of the London Mathematical Society. Serie 2 [1904-1951]
• CONT OF (**J** 0077) Proc London Math Soc • CONT AS (**J** 3240) Proc London Math Soc, Ser 3

J 1928 Mathematica B • NL
Mathematica : Tijdschrift voor Allen die de Hoogere Wiskunde Beoefenen. Afdeling B [1934-1948]
• CONT AS (**J** 0061) Simon Stevin
• REM Vols 1 - 3/1 appeared under the title 'Mathematica', which from 3/2 onwards was split into 'Mathematica A' and 'Mathematica B'

J 1929 Prace Centr Oblicz Pol Akad Nauk • PL
Polska Akademija Nauk. Centrum Obliczeniowe. Prace. (Polish Academy of Sciences. Computation Centre. Reports) ISSN 0079-3175

J 1934 Ann Sci Univ Clermont Math • F
Annales Scientifiques de l'Universite de Clermont-Ferrand II, Section Mathematiques (Clermont Ferrand) [1973ff] ISSN 0069-472X
• CONT OF (**J** 0179) Ann Fac Sci Clermont

J 1942 Matematika (Belgrade) • YU
Matematika (Beograd= Belgrade) (Mathematics)

J 1943 Recueil Travaux Inst Math (Beograd) • YU
Recueil des Travaux de l'Institut Mathematique (Beograd)

J 1944 Sitzb, Akad Wiss, Bayern, Math-Nat Kl • D
Bayerische Akademie der Wissenschaften. Sitzungsberichte [1931ff] ISSN 0340-7586

J 2022 Comm Math Helvetici • CH
Commentarii Mathematici Helvetici [1929ff] ISSN 0010-2571

J 2028 Hist & Phil Log • GB
History and Philosophy of Logic [1980ff] ISSN 0144-5340

J 2038 Rend Sem Mat, Torino • I
Rendiconti del Seminario Matematico (gia "Conferenze di Fisica e di Matematica"). Universita e Politecnico di Torino

J 2053 Math Medley • SGP
Mathematical Medley [1975ff]

J 2074 Sem-ber, Muenster • D
Semesterbericht Muenster [1931-1939]

J 2076 Rev Int Philos • B
Revue Internationale de Philosophie [1938ff] ISSN 0048-8143
• CONT OF (**J** 0151) Arch Soc Belg Philos

J 2085 Acta Cient Venez • YV
Acta Cientifica Venezolana [1950ff] ISSN 0001-5504

J 2094 IEEE Trans Software Engin • USA
Transactions on Software Engineering. IEEE (= Institute of Electrical and Electronics Engineers) [1975ff] ISSN 0098-5589

J 2095 Fund Inform, Ann Soc Math Pol, Ser 4 • PL
*Fundamenta Informaticae. Annales Societatis Mathematicae Polonae. Series 4 * Roczniki Polskiego Towarzystwa Matematycznego. Seria 4* [1977ff] ISSN 0324-8429
• CONT OF (**J** 0611) Pol Tow Mat, Prace Mat-Fiz

J 2099 Boll Unione Mat Ital, VI Ser, A • I
Bolletino della Unione Matematica Italiana. Serie VI. A [1982ff] ISSN 0041-7084
• CONT OF (**J** 3285) Boll Unione Mat Ital, V Ser, A

J 2100 Boll Unione Mat Ital, VI Ser, B • I
Bolletino della Unione Matematica Italiana. Serie VI. B [1982ff] ISSN 0041-7084
• CONT OF (**J** 3495) Boll Unione Mat Ital, V Ser, B

J 2107 Publ Dep Math, Lyon, NS • F
Publications du Departement de Mathematiques. Nouvelle Serie. Faculte des Sciences de Lyon. [1982ff] ISSN 0076-1656
• CONT OF (**J** 0056) Publ Dep Math, Lyon

J 2119 Suppl Aristotelian Soc • GB
The Aristotelian Society Supplementary [1918ff]
• REL PUBL (**J** 0113) Proc Aristotelian Soc

J 2128 C R Math Acad Sci, Soc Roy Canada • CDN
*Comptes Rendus Mathematiques de l'Academie des Sciences. La Societe Royale du Canada * Mathematical Reports of the Academy of Sciences* [1979ff] ISSN 0706-1994

J 2307 Japan J Math • J
Japanese Journal of Mathematics [1924-1974]
• CONT AS (**J** 2347) Japan J Math, NS

J 2313 C R Acad Sci, Paris, Ser A-B • F
Academie des Sciences de Paris. Comptes Rendus Hebdomadaires des Seances. Serie A: Sciences Mathematiques, Serie B: Sciences Physiques [1966-1980] ISSN 0001-4036
• CONT OF (**J** 0109) C R Acad Sci, Paris • CONT AS (**J** 3364) C R Acad Sci, Paris, Ser 1 & (**J** 2314) C R Acad Sci, Paris, Ser 2

J 2314 C R Acad Sci, Paris, Ser 2 • F
Comptes Rendus des Seances de l'Academie des Sciences. Serie II. Mecanique-Physique, Chimie, Sciences de la Terre, Sciences de l'Univers [1981ff]
• CONT OF (**J** 2313) C R Acad Sci, Paris, Ser A-B

J 2332 J Fac Sci, Univ Tokyo, Sect 1 A • J
Journal of the Faculty of Science. University of Tokyo. Section 1 A. Mathematics [1971ff] ISSN 0040-8980
• CONT OF (**J** 0434) J Fac Sci Univ Tokyo, Sect 1

J 2341 Arch Int Hist Sci • F
Archives Internationales d'Histoire des Sciences. Academie Internationale d'Histoire des Sciences [1919-1971] ISSN 0003-9810
• CONT AS (**J** 1622) Arch Int Hist Sci (NS)

J 2347 Japan J Math, NS • J
Japanese Journal of Mathematics. New Series [1975ff] ISSN 0075-3432
• CONT OF (**J** 2307) Japan J Math

J 2547 Serdica, Bulgar Math Publ • BG
Serdica. Bulgaricae Mathematicae Publicationes [1975ff] ISSN 0204-4110
• CONT OF (**J** 1156) Izv Bulgar Akad Nauk Mat Inst

J 2551 J Log Progr • USA
Journal of Logic Programming [1984ff] ISSN 0743-1066

J 2562 Publ Sec Mat Univ Autonoma Barcelona • E
Publications de la Seccio de Matematiques. Universitat Autonoma de Barcelona (Veroeffentlichunge der Abteilung Mathematik)

J 2574 Litov Mat Sb (Vil'nyus) • SU
*Litovskij Matematicheskij Sbornik * Lietuvos Matematikos Rinkinys (Lithuanian Mathematical Collected Articles)* [1961ff] ISSN 0132-2818
• TRANSL IN (**J** 3283) Lith Math J

J 2606 Tsukuba J Math • J
Tsukuba Journal of Mathematics [1977ff] ISSN 0387-4982

J 2635 Topology Appl • NL
Topology and its Applications. A Journal Devoted to General, Geometric, Set-Theoretic and Algebraic Topology [1980ff] ISSN 0166-8641
• CONT OF (**J** 0254) Gen Topology Appl

J 2660 Ann Sci Math Quebec • CDN
Les Annales des Sciences Mathematiques du Quebec. [1977ff] ISSN 0707-9109

J 2661 Ann Probab • USA
The Annals of Probability. An Official Journal of the Institute of Mathematical Statistics. [1973ff] ISSN 0091-1798

J 2666 Autom Control Comput Sci • USA
Automatic Control and Computer Sciences [1969ff] ISSN 0146-4116
• TRANSL OF (**J** 0474) Avtom Vychis Tekh, Akad Nauk Latv SSR

J 2678 Bull Aichi Univ Educ Nat Sci • J
Bulletin of Aichi University of Education. Natural Science

J 2679 Bull Inst Math Appl (Southend oS) • GB
Bulletin of the Institute of Mathematics and its Applications [1965ff]

J 2684 J Huazhong Inst Tech (Engl Ed) • TJ
Journal of Huazhong Institute of Technology. English Edition [1979-1981]
• CONT AS (**J** 3218) J Huazhong Univ Sci Tech (Engl Ed)
• TRANSL OF (**J** 2754) Huazhong Gongxueyuan Xuebao

J 2688 Conceptus (Wien) • A
Conceptus. Zeitschrift fuer Philosophie [1967ff] ISSN 0010-5155

J 2736 Int J Theor Phys • USA
International Journal of Theoretical Physics [1968ff] ISSN 0020-7748

J 2747 J Approx Th • USA
Journal of Approximation Theory [1968ff] ISSN 0021-9045

J 2751 J Fac Lib Art Yamaguchi Univ • J
Journal of the Faculty of Liberal Arts. Yamaguchi University. Natural Science [1967ff]

J 2752 J Fct Anal • USA
Journal of Functional Analysis [1967ff] ISSN 0022-1236

J 2754 Huazhong Gongxueyuan Xuebao • TJ
*Huazhong Gongxueyuan Xuebao * Zhongguo Kexue Shuxue Zhuanji (Journal Huazhong (Central China) University of Science and Technology)*
• TRANSL IN (**J** 2684) J Huazhong Inst Tech (Engl Ed) & (**J** 3218) J Huazhong Univ Sci Tech (Engl Ed)

J 2757 J Math Econ • NL
Journal of Mathematical Economics [1974ff] ISSN 0304-4068

J 2770 Keio Math Sem Rep (Yokohama) • J
Keio Mathematical Seminar Reports [1976ff]

J 2778 Lett Nuovo Cimento Ser 2 • I
Lettere al Nuovo Cimento. Rivista Internazionale della Societa Italiana di Fisica. Serie 2 [1969ff] ISSN 0375-930X

J 2789 Math Intell • D
The Mathematical Intelligencer [1978ff] ISSN 0343-6993

J 2790 Math Sem-ber • D
Mathematische Semesterberichte [1980ff] ISSN 0720-728X
• CONT OF (**J** 0160) Math-Phys Sem-ber, NS

J 2811 Phil Quart (St Andrews) • GB
The Philosophical Quarterly [1950ff] ISSN 0031-8094

J 2817 Prague Bull Math Linguist • CS
The Prague Bulletin of Mathematical Linguistics [1964ff] ISSN 0032-6585

J 2825 Rend Sem Fac Sci Univ Cagliari • I
Rendiconti del Seminario della Facolta di Scienze della Universita di Cagliari

J 2827 Ricerche Mat • I
Ricerche di Matematica [1952ff] ISSN 0035-5038

J 2831 RAIRO Autom • F
RAIRO Automatique. RAIRO (= Revue Francaise d'Automatique, d'Informatique et de Recherche Operationnelle). Series Automatique [1977ff] ISSN 0399-0524
• CONT OF (**J** 0205) Rev Franc Autom, Inf & Rech Operat Ser Jaune

J 2832 RAIRO Inform • F
RAIRO Informatique. Revue Francaise d'Automatique, d'Informatique et de Recherche Operationnelle. Series Informatique [1977ff] ISSN 0399-0532
• CONT OF (**J** 0205) Rev Franc Autom, Inf & Rech Operat Ser Bleue

J 2852 Tr Vychisl Tsentra, Univ Tartu • SU
Trudy Vychislitel'nogo Tsentra. Tartuskij Gosudarstvennyj Universitet (Publications of the Computational Centre of the State University of Tartu) [1893ff]

J 2887 Zbor Radova, NS • YU
Zbornik Radova. Nova Serija. (Collected Papers. New Series)

J 2892 Molodoj Nauch Rabotnik, Erevan • SU
Molodoj Nauchnyj Rabotnik. Erevanskij Gosudarstvennyj Universitet (The Young Scientist. University of Erevan)

J 3075 Normat • N
Normat. Nordisk Matematisk Tidskrift (Normat. Scandinavian Mathematical Journal) [1979ff]
• CONT OF (**J** 0311) Nordisk Mat Tidskr

J 3077 Nieuw Arch Wisk, Ser 3 • NL
Nieuw Archief voor Wiskunde. Derde Serie [1953-1982] ISSN 0028-9825
• CONT OF (**J** 1793) Nieuw Arch Wisk, Ser 2 • CONT AS (**J** 3929) Nieuw Arch Wisk, Ser 4

J 3079 Kiber Sb Perevodov, NS • SU
Kiberneticheskij Sbornik. Novaya Seriya. Sbornik Perevodov (Collected Articles on Cybernetics: New Series. Collected Translations) [1965ff] ISSN 0453-8382
• CONT OF (**J** 1048) Kiber Sb Perevodov

J 3099 Theoretic Papers • N
Theoretic Papers [1984ff]

J 3116 Austral Comp J • AUS
The Australian Computer Journal [1967ff] ISSN 0004-8917

J 3120 ACM Trans Program Lang & Syst • USA
ACM (= Association for Computing Machinery) Transactions on Programming Languages and Systems [1979ff] ISSN 0164-0925

J 3133 Bull Soc Math Belg, Ser B • B
Bulletin de la Societe Mathematique de Belgique. Serie B [1977ff] ISSN 0037-9476
• CONT OF (**J** 0082) Bull Soc Math Belg

J 3137 Shuxue Niankan • TJ
Shuxue Niankan (Chinese Annals of Mathematics) [1980-1982]
• CONT AS (**J** 3187) Shuxue Niankan, Xi A & (**J** 4719) Chinese Ann Math Ser B

J 3144 Comput Lang, Int J • GB
Computer Languages. An International Journal [1975ff] ISSN 0096-0551

J 3172 J London Math Soc, Ser 2 • GB
Journal of the London Mathematical Society. 2nd Series [1969ff] ISSN 0024-6107
• CONT OF (**J** 0039) J London Math Soc

J 3187 Shuxue Niankan, Xi A • TJ
Shuxue Niankan. Xi A (Chinese Annals of Mathematics. Series A) [1983ff]
• CONT OF (**J** 3137) Shuxue Niankan

J 3188 Itogi Nauki Tekh, Ser Teor Veroyat Mat Stat Teor Kibern • SU
Itogi Nauki i Tekhniki. Seriya Teoriya Veroyatnostej, Matematicheskaya Statistika, Teoreticheskaya Kibernetika (Progress in Science and Technology. Series Probability Theory, Mathematical Statistics, Theoretical Cybernetics) [1972ff] ISSN 0202-7488
• CONT OF (**J** 1021) Itogi Nauki Ser Mat • TRANSL IN (**J** 1531) J Sov Math

J 3194 J Austral Math Soc, Ser A • AUS
Journal of the Australian Mathematical Society. Series A [1976ff] ISSN 0263-6115
• CONT OF (**J** 0038) J Austral Math Soc

J 3197 Rass Mat Log Mat Appl Didat • I
Rassegna di Matematica: Logica Matematica, Matematica Applicata, Didattica della Matematica

J 3218 J Huazhong Univ Sci Tech (Engl Ed) • TJ
Journal of Huazhong University of Science and Technology [1982ff]
• CONT OF (**J** 2684) J Huazhong Inst Tech (Engl Ed) • TRANSL OF (**J** 2754) Huazhong Gongxueyuan Xuebao

J 3227 Pattern Recognition • GB
Pattern Recognition. The Journal of the Pattern Recognition Society [1968ff] ISSN 0031-3203

J 3239 Proc Japan Acad, Ser A • J
Proceedings of the Japan Academy. Series A. Mathematical Sciences [1978ff] ISSN 0386-2194

J 3240 Proc London Math Soc, Ser 3 • GB
Proceedings of the London Mathematical Society. 3rd Series [1951ff] ISSN 0024-6115
• CONT OF (**J** 1910) Proc London Math Soc, Ser 2

J 3254 Riv Mat Univ Parma, Ser 3 • I
Rivista di Matematica della Universita di Parma. Serie 3 [1972-1974] ISSN 0035-6298
• CONT OF (J 1526) Riv Mat Univ Parma, Ser 2 • CONT AS (J 0549) Riv Mat Univ Parma, Ser 4

J 3265 Sov J Contemp Math Anal, Armen Acad Sci • USA
Soviet Journal of Contemporary Mathematical Analysis. Armenian Academy of Sciences [1979ff] ISSN 0735-2719
• TRANSL OF (J 0312) Izv Akad Nauk Armyan SSR, Ser Mat

J 3273 Stud Hist Philos Sci • GB
Studies in History and Philosophy of Science [1970ff] ISSN 0039-3681

J 3276 Tehran J Math • IR
Tehran Journal of Mathematics

J 3281 Ukr Math J • USA
Ukrainian Mathematical Journal [1967ff] ISSN 0041-5995
• TRANSL OF (J 0265) Ukr Mat Zh, Akad Nauk Ukr SSR

J 3283 Lith Math J • USA
Lithuanian Mathematical Journal. [1975ff] ISSN 0363-1672
• TRANSL OF (J 2574) Litov Mat Sb (Vil'nyus)
• REM Only selected translations of J2574

J 3285 Boll Unione Mat Ital, V Ser, A • I
Bollettino della Unione Matematica Italiana. Serie V. A [1976-1981] ISSN 0041-7084
• CONT OF (J 0012) Boll Unione Mat Ital, IV Ser • CONT AS (J 2099) Boll Unione Mat Ital, VI Ser, A

J 3293 Bull Acad Pol Sci, Ser Math • PL
Bulletin de l'Academie Polonaise des Sciences. Serie des Sciences Mathematiques [1979-1982] ISSN 0001-4117
• CONT OF (J 0014) Bull Acad Pol Sci, Ser Math Astron Phys
• CONT AS (J 3417) Bull Pol Acad Sci, Math

J 3364 C R Acad Sci, Paris, Ser 1 • F
Comptes Rendus des Seances de l'Academie des Sciences. Serie I: Science Mathematique [1981ff] ISSN 0151-0509
• CONT OF (J 2313) C R Acad Sci, Paris, Ser A-B

J 3370 Enseign Math, Ser 2 • CH
L'Enseignement Mathematique. Revue Internationale. Serie 2 [1955ff] ISSN 0013-8584
• CONT OF (J 0152) Enseign Math

J 3376 Acta Appl Math • NL
Acta Applicandae Mathematicae. An International Journal on Applying Mathematics and Mathematical Applications [1983ff]

J 3386 Int J Math & Math Sci • USA
International Journal of Mathematics and Mathematical Sciences [1978ff] ISSN 0161-1712

J 3417 Bull Pol Acad Sci, Math • PL
Bulletin of the Polish Academy of Sciences. Mathematics [1983ff] ISSN 0001-4117
• CONT OF (J 3293) Bull Acad Pol Sci, Ser Math

J 3420 Proc Edinburgh Math Soc, Ser 2 • GB
Proceedings of the Edinburgh Mathematical Society. Series 2. [1927ff] ISSN 0013-0915
• CONT OF (J 0386) Proc Edinburgh Math Soc

J 3434 Pubbl Ist Appl Calcolo, Ser 3 • I
Pubblicazioni. Serie III. Istituto per le Applicazioni del Calcolo "Mauro Picone" (IAC) Consiglio Nazionale delle Ricerche

J 3436 Quad, Ist Appl Calcolo, Ser 3 • I
Quaderni Serie III. Istituto Applicazione Calcolo

J 3441 RAIRO Inform Theor • F
Revue Francaise d'Automatique, d'Informatique et de Recherche Operationnelle (RAIRO), Informatique Theorique [1977ff] ISSN 0399-0540
• CONT OF (J 4698) Rev Franc Autom, Inf & Rech Operat, Ser Rouge Inf Th

J 3449 Sov Math • USA
Soviet Mathematics [1974ff] ISSN 0197-7156
• TRANSL OF (J 0031) Izv Vyssh Ucheb Zaved, Mat (Kazan)

J 3480 Informatik & Philos • A
Informatik und Philosophie

J 3495 Boll Unione Mat Ital, V Ser, B • I
Bolletino della Unione Matematica Italiana. Serie V. B [1976-1981] ISSN 0041-7084
• CONT OF (J 0012) Boll Unione Mat Ital, IV Ser • CONT AS (J 2100) Boll Unione Mat Ital, VI Ser, B

J 3519 Glas Mat, Ser 3 (Zagreb) • YU
Glasnik Matematichki. Serija 3 (Publications of Mathematics. Series 3) [1966ff] ISSN 0017-095X
• CONT OF (J 0371) Glas Mat-Fiz Astron, Ser 2 (Zagreb)

J 3524 Kybernetika Suppl (Prague) • CS
Kybernetika. Supplement. (Prague) (Cybernetics. Supplement. (Prague))
• REL PUBL (J 0156) Kybernetika (Prague)

J 3526 Ann Mat Pura Appl, Ser 4 • I
Annali di Matematica Pura ed Applicata. Serie Quarta. Sotto gli Auspici del Consiglio Nazionale delle Ricerche ISSN 0003-4622
• CONT OF (J 0229) Ann Mat Pura Appl, Ser 3

J 3543 Uch Zap Vopr Prikl Mat Kibern, Univ Baku • SU
Azerbajdzhanskij Gosudarstvennyj Universitet. Uchenye Zapiski. Voprosy Prikladnoj Matematiki i Kibernetiki (State University of Azerbaidjan. Scientific Notes. Problems of Applied Mathematics and Cybernetics)

J 3597 J Unif Sci • USA
The Journal of Unified Science [1940-1974]
• CONT OF (J 0748) Erkenntnis (Leipzig) • CONT AS (J 0158) Erkenntnis (Dordrecht)

J 3630 Bull Dept of Lib Arts (Numazu) • J
Bulletin of the Department of Liberal Arts [1974ff]

J 3651 Tr Nauch Sess Ehston (Talinn) • SU
Trudy Nauchnykh Sessiya Ehstonskogo Gosudarstvennogo Universiteta (Publications of Scientific Sessions. Estonian State University) [1959ff]

J 3741 Rend Mat Appl, Ser 7 • I
Rendiconti di Matematica e delle sue Applicazioni. Seria 7 [1981ff] ISSN 0034-4427
• CONT OF (J 0297) Rend Mat Appl, Ser 5

J 3742 Shuxue Yanjiu yu Pinglun • TJ
Shuxue Yanjiu yu Pinglun (Journal of Mathematical Research & Exposition) [1981ff]

J 3746 Note Math (Lecce) • I
Note di Matematica. Pubblicazione Semestrale [1981ff]

J 3750 Jilin Daxue, Ziran Kexue Xuebao • TJ
*Jilin Daxue. Ziran Kexue Xuebao (Acta Scientiarum Naturalium Universitatis Jilinensis * Jilin University. Natural Sciences Journal)* [1975ff]
• CONT OF (J 4364) Dongbei Renmin Daxue Ziran Kexue Xuebao

J 3766 Zhongguo Kexue, Xi A • TJ
*Zhongguo Kexue. Xi A. Mathematical, Physical, Astronomical & Technical Sciences * Scientia Sinica. Series A* [1982ff]
• CONT OF (**J** 1024) Zhongguo Kexue

J 3768 Boll Unione Mat Ital, VI Ser, D • I
Bolletino della Unione Matematica Italiana. Serie VI. D. Algebra e Geometria [1982ff] ISSN 0041-7084

J 3783 J Non-Classical Log (Univ Campinas) • BR
The Journal of Non-Classical Logic [1982ff]

J 3789 Ann Hist of Comp • USA
Annals of the History of Computing [1979ff] ISSN 0164-1239

J 3793 Jisuanjii Xuebao • TJ
Jisuanji Xuebao (Chinese Journal of Computers) [1978ff]

J 3824 Bull Soc Math Belg, Ser A • B
Bulletin de la Societe Mathematique de Belgique. Serie A [1977ff]
• CONT OF (**J** 0082) Bull Soc Math Belg

J 3915 Pap Michigan Acad Sci • USA
Papers of the Michigan Academy of Science, Arts and Letters ISSN 0026-2005

J 3929 Nieuw Arch Wisk, Ser 4 • NL
Nieuw Archief voor Wiskunde. Vierde Serie. Uitgegeven door het Wiskundid Genootschap te Amsterdam [1983ff] ISSN 0028-9825
• CONT OF (**J** 3077) Nieuw Arch Wisk, Ser 3

J 3932 Comput Artif Intell (Bratislava) • CS
Pocitace a Umela Inteligencia (Computers and Artificial Intelligence) [1982ff]

J 3939 Mat Logika Primen (Akad Nauk Litov SSR) • SU
Matematicheskaya Logika i Ee Primeneniya (Mathematical Logic and its Applications) [1981ff]

J 3940 Saitama Math J • J
Saitama Mathematical Journal [1983ff]
• CONT OF (**J** 1472) Sci Rep Saitama Univ, Ser A

J 3941 J Math Pures Appl, Ser 9 • F
Journal de Mathematiques Pures et Appliquees. Neuvieme Serie [1922ff] ISSN 0021-7824
• CONT OF (**J** 0249) J Math Pures Appl

J 3954 Rev Franc Inf & Rech Operat • F
Association Francaise pour la Cybernetique Economique et Technique. Revue Francaise d'Informatique et de Recherche Operationnelle [1967-1971]
• CONT AS (**J** 0205) Rev Franc Autom, Inf & Rech Operat

J 3962 Handel Nederl Natuur- & Geneesk Congr • NL
Handelingen van het Vlaamsche Natuur- en Geneeskundig Congress [1887ff]

J 3970 Ark Mat, Astron & Fys • S
Arkiv foer Matematik, Astronomi och Fysik (Archiv fuer Mathematik, Astronomie und Physik) [1904-1948]
• CONT AS (**J** 0189) Ark Mat

J 3975 Arch Math & Phys • D
Archiv der Mathematik und Physik mit Besonderer Ruecksicht auf die Beduerfnisse der Lehrer an Hoeheren Unterrichtsanstalten. 3.Reihe [1841ff]

J 3978 Math-Nat Blaetter • D
Mathematisch-Naturwissenschaftliche Blaetter. Organ des Verbandes Mathematischer und Naturwissenschaftlicher Vereine an Deutschen Hochchulen

J 3980 Karadeniz Univ Math J (Trabzon) • TR
Karadeniz University Mathematical Journal

J 3986 Ist Lombardo Rend, Ser 2 (Milano) • I
Reale Istituto Lombardo di Scienze e Lettre. Rendiconti. Series 2 [1868-1936]
• CONT AS (**J** 0207) Ist Lombardo Accad Sci Rend, A (Milano)

J 3990 Rend Accad Lincei Roma, Ser 6 • I
Atti della Reale Accademia Nazionale dei Lincei. Rendiconti. Classe di Scienze Fisiche, Matematiche e Naturali. Ser. 6 [1925ff] ISSN 0001-4435
• CONT OF (**J** 1157) Rend Accad Lincei Roma, Ser 5

J 3993 Rev Mat Hisp-Amer, Ser 2 • E
Revista Matematica Hispano-Americana. Serie 2 ISSN 0373-0999
• CONT AS (**J** 0236) Rev Mat Hisp-Amer, Ser 4

J 3996 Boll Unione Mat Ital • I
Bollettino della Unione Matematica Italiana [1922-1945]
• CONT AS (**J** 4408) Boll Unione Mat Ital, III Ser

J 4058 Nominazione • I
Nominazione [1980ff]

J 4110 Zap Inst Nar Prosv, Dnepropetrovsk • SU
Dnepropetrovsk, Zapiski Instituta Narodnogo Prosveshcheniya (Notes of the Institute of Education. Dnepropetrovsk)

J 4114 De Gids • NL
De Gids [1837ff] ISSN 0016-9730

J 4122 Vest Komm Akad • SU
Vestnik Kommunisticheskoj Akademii (Publications of the Communist Academy)

J 4128 Grazer Philos Stud • A
Grazer Philosophische Studien. Internationale Zeitschrift fuer Analytische Philosophie [1975ff]

J 4148 Uch Zap Ped Inst, Novozybkov • SU
Novozybkov Uchenye Zapiski Ped. Instituta (Scientific Notes of the Institute of Education in Novozybkov)

J 4170 Nauch Dokl Vys Shk, Fiz Mat (Moskva) • SU
Nauchnye Doklady Vysshej Shkoly, Fiziko-Matematicheskie Nauki (Scientific Reports of the University. Physic-Mathematics)

J 4171 Symposion • D
Symposion. Philosophische Zeitschrift fuer Forschung und Aussprache [1925ff]

J 4274 Proc British Acad • GB
Proceedings of the British Academy

J 4305 IEEE Trans Electr Comp • USA
Transactions of Electronic Computers. IEEE (= Institute of Electrical and Electronics Engineers) [1963-1967]
• CONT OF (**J** 0072) IRE Trans Electr Comp • CONT AS (**J** 0187) IEEE Trans Comp

J 4320 Rev Centro Calc Univ Complutense (Madrid) • E
Revista del Centro de Calculo de la Universidad Complutense de Madrid

J 4364 Dongbei Renmin Daxue Ziran Kexue Xuebao • TJ
*Dongbei Renmin Daxue Ziran Kexue Xuebao * Acta Scientiarum Naturalium (Natural Science Journal of Northeast Peoples's University)* [?-1974]
• CONT AS (**J** 3750) Jilin Daxue, Ziran Kexue Xuebao

J 4376 Ordzhonikidze, Izv Ped Inst, Gor'kij • SU
Ordzhonikidze, Izvestiya Gor'kov Ped. Instituta (Ordzhonikidze. Proceedings of the Institute of Education, Gor'kij)

J 4387 Itogi Nauki Tekh, Ser Tekh Kibern • SU
Itogi Nauki i Tekhniki. Seriya Tekhnicheskaya Kibernetika. Gosudarstvennyj Komitet SSSR po Nauke i Tekhnike. Akademiya Nauk SSSR. Vsesoyuznyj Institut Nauchnoj i Tekhnicheskoj Informatsii. (Progress in Science and Technology. Series: Engineering Cybernetics) [1972ff]
• CONT OF (**J** 1021) Itogi Nauki Ser Mat

J 4404 Vopr Mat Logiki & Pril • SU
Voprosy Matematicheskoj Logiki i Ee Prilozheniya (Problems of Mathematical Logic and its Applications)

J 4408 Boll Unione Mat Ital, III Ser • I
Bolletino della Unione Matematica Italiana, Ser III [1946-1967]
• CONT OF (**J** 3996) Boll Unione Mat Ital • CONT AS (**J** 0012) Boll Unione Mat Ital, IV Ser

J 4430 Ziran Zazhi • TJ
Ziran Zazhi (Nature Journal) [1978ff]

J 4439 Kagaku • J
Kagaku (Chemical Sciences)

J 4510 Norsk Mat Tidsskr • N
Norsk Matematisk Tidsskrift (Scandinavian Mathematical Journal) [1919-1952]
• CONT OF (**J** 0259) Nyt Tidsskr Mat • CONT AS (**J** 0311) Nordisk Mat Tidskr

J 4539 Scholastik • D
Scholastik

J 4573 Xilong Kexue yu Shuxue Xuebao • TJ
Xilong Kexue yu Shuxue Xuebao (Journal of Systems Sciences and Mathematical Sciences)

J 4656 Canad Phil Rev • CDN
Canadian Philosophical Review [1981ff]

J 4698 Rev Franc Autom, Inf & Rech Operat, Ser Rouge Inf Th • F
Revue Francaise d'Automatique, d'Informatique et de Recherche Operationnelle (RAIRO). Serie Rouge Informatique Theorique [1975-1976] ISSN 0399-0540
• CONT OF (**J** 0205) Rev Franc Autom, Inf & Rech Operat Ser Rouge • CONT AS (**J** 3441) RAIRO Inform Theor

J 4702 Ann Sc Norm Sup Pisa Fis Mat, Ser 4 • I
Annali della Scuola Normale Superiore di Pisa. Classe di Science. Fisiche e Matematiche. Seria IV [1974ff]
• CONT OF (**J** 0315) Ann Sc Norm Sup Pisa Fis Mat, Ser 3

J 4706 Publ Inst Math (Belgrade) • YU
Academie Serbe des Sciences. Publications de l'Institut Mathematique [1948-1960]
• CONT AS (**J** 0400) Publ Inst Math, NS (Belgrade)

J 4710 Pol Tow Mat, Wiad Mat • PL
Polskie Towarzystwo Matematyczne. Wiadomosci Matematiczne [1899-1954]
• CONT AS (**J** 0519) Wiad Mat, Ann Soc Math Pol, Ser 2

J 4716 Stud Philos, Leopolis (Poznan) • PL
Studia Philosophica. Commentarii Societatis Polonorum. Leopolis [1935ff]

J 4717 Izv Akad Nauk SSSR • SU
Izvestiya Akademii Nauk SSSR (Bulletin de l'Academie des Sciences Mathematiques et Naturelles. Leningrad) [?-1936]
• CONT AS (**J** 0216) Izv Akad Nauk SSSR, Ser Mat

J 4719 Chinese Ann Math Ser B • TJ
Shuxue Niankan. Xi B (Chinese Annals of Mathematics. Series B) [1983ff]
• CONT OF (**J** 3137) Shuxue Niankan

Series

S 0019 Colloq Math (Warsaw) • PL
Colloquium Mathematicum [1947ff] • PUBL Academie Polonaise des Sciences, Institut Mathematique: Warsaw
• ISSN 0010-1354

S 0055 Proc Steklov Inst Math • USA
Proceedings of the Steklov Institute of Mathematics [1967ff]
• PUBL (**X** 0803) Amer Math Soc: Providence
• TRANSL OF (**S** 0066) Tr Mat Inst Steklov
• ISSN 0081-5438

S 0066 Tr Mat Inst Steklov • SU
Trudy Ordena Lenina Matematicheskogo Instituta imeni V.A.Steklova. Akademiya Nauk SSSR (Proceedings of the Mathematical Steklov-Institute of the Academy of Sciences SSSR) [1938ff] • PUBL (**X** 0899) Akad Nauk SSSR : Moskva
• CONT OF (**S** 1644) Tr Fiz-Mat Inst Steklov • TRANSL IN (**S** 0055) Proc Steklov Inst Math

S 0167 Mem Amer Math Soc • USA
Memoirs of the American Mathematical Society [1950-1974]
• PUBL (**X** 0803) Amer Math Soc: Providence
• CONT AS (**S** 2450) Mem Amer Math Soc, NS
• ISSN 0065-9266

S 0200 Acad Roy Belg Cl Sci Mem, Ser 2 • B
*Academie Royale des Sciences, des Lettres et des Beaux Arts de Belgique. Memoires de la Classe des Sciences. Collection Octavo. Deuxieme Serie * Koninklijke Vlaamse Academie van Belgie. Klasse der Wetenschappen. Verhandelingen. Verzameling in Octavo. Tweede Reeks* [1904ff] • PUBL (**X** 2210) Acad Roy Belg: Bruxelles
• ISSN 0065-0595

S 0208 Uch Zap, Ped Inst, Moskva • SU
Uchenye Zapiski Moskovskogo Gosudarstvennogo Pedagogicheskogo Instituta imeni V.I.Lenina (Scientific Notes of the Moscow State Institute of Education) [1950ff] • PUBL (**X** 2802) Moskov Ped Inst: Moskva

S 0228 Zap Nauch Sem Leningrad Otd Mat Inst Steklov • SU
Zapiski Nauchnykh Seminarov Leningradskogo Otdeleniya Ordena Lenina Matematicheskogo Instituta imeni V.A.Steklova Akademii Nauk SSSR (LOMI) (Reports of the Scientific Seminars of the Leningrad Steklov Institute of Mathematics) PUBL (**X** 2641) Nauka: Leningrad
• TRANSL IN (**J** 1531) J Sov Math & (**J** 0521) Semin Math, Inst Steklov
• REM Transl. in J0521 up to vol. 19

S 0257 Monograf Mat • PL
Monografie Matematyczne (Mathematical Monography) [1932ff] • PUBL (**X** 1034) PWN: Warsaw • ALT PUBL (**X** 1758) Ars Polona: Warsaw
• ISSN 0077-0507

S 0393 Uch Zap Univ, Tartu • SU
*Uchenye Zapiski Tartuskogo Gosudarstvennogo Universiteta. * Tartu Riikliku Uelikooli Toimetised. * Acta et Commentationes Universitatis Tartuensis (Scientific Notes of the Tartu State University.)* [1961ff] • PUBL (**X** 2463) Tartusk Gos Univ: Tartu

S 0405 Mitt Math Sem Giessen • D
Mitteilungen aus dem Mathematischen Seminar Giessen PUBL (**X** 2464) Math Sem Selbstverl: Giessen

S 0410 Math Beitr Univ Halle-Wittenberg • DDR
Die Martin Luther-Universitaet Halle-Wittenberg. Mathematische Beitraege [1968ff] • PUBL (**X** 2446) Univ Halle-Wittenberg: Halle Wissenschaftliche Beitraege. ISSN 0440-1298
• ISSN 0441-6228

S 0422 Tr Vychisl Tsentra Akad Nauk Armyan SSR & Univ Erevan • SU
Trudy Vychislitel'nogo Tsentra Akademij Nauk Armyanskoj SSR i Erevanskogo Gosudarstvennogo Universiteta. Matematicheskie Voprosy Kibernetiki i Vychislitel'noj Tekhniki (Publications of the Computing Centre of the Academy of Sciences of the Armyan SSR and of the Erevan State University. Mathematical Problems of Cybernetics and Computational Techniques) [1963ff] • PUBL (**X** 2225) Akad Nauk Armyan SSR : Erevan

S 0458 Zesz Nauk, Prace Log, Uniw Krakow • PL
Zeszyty Naukowe Uniwersytetu Jagiellonskiego Prace z Logiki (Scientific Papers. Jagielleonian University. Reports on Logic) [1965-1972] • PUBL (**X** 1034) PWN: Warsaw
• CONT AS (**J** 0302) Rep Math Logic, Krakow & Katowice

S 0507 Vychisl Sist (Akad Nauk SSSR Novosibirsk) • SU
Vychislitel'nye Sistemy. Sbornik Trudov. Akademiya Nauk SSSR. Sibirskoe Otdelenie (Computer Systems. Collected Articles) [1962ff] • PUBL (**X** 2652) Akad Nauk Sibirsk Otd Inst Mat: Novosibirsk
• ISSN 0568-661X

S 0554 Issl Teor Algor & Mat Logik (Moskva) • SU
Issledovaniya po Teorii Algorifmov i Matematicheskoj Logike (Studies in the Theory of Algorithms and Mathematical Logic) [1973,1976,1979] • ED: MARKOV, A.A. & PETRI, N.V. (VOL 1). MARKOV, A.A. & KUSHNER, B.A. (VOL 2). MARKOV, A.A. & KHOMICH, V.I. (VOL 3) • PUBL (**X** 2265) Akad Nauk Vychis Tsentr: Moskva 287pp,160pp,133pp
• ISSN 0302-9085

S 1415 Sel Math • D
Selecta Mathematica [1969ff] • PUBL (**X** 0811) Springer: Heidelberg & New York
• ISSN 0586-4313

S 1498 Uch Zap Univ, Moskva • SU
Uchenye Zapiski Moskovskogo Gosudarstvennogo Universiteta (Scientific Notes of the Moscow State University) PUBL (**X** 0898) Moskov Gos Univ: Moskva

S 1562 Prace Inst Mat Fis, Politech Wroclaw, Ser Stud Mater • PL
Prace Naukowe. Instytut Matematyki i Fizyki Teoretycznej. Politechnika Wroclawska. Seria Studia i Materialy. (Scientific Publications. Institute of Mathematics and Theoretical Physics. Technical University of Wroclaw) [?-1969] • PUBL Politechnika Wroclawska: Wroclaw
• CONT AS (S 4733) Prace Inst Mat, Politech Wroclaw, Ser Stud Mater

S 1567 Semin Bourbaki • F
Seminaire Bourbaki [1948ff] • SER (S 3301) Lect Notes Math, (J 1620) Asterisque • PUBL (X 0811) Springer: Heidelberg & New York, (X 2244) Soc Math France: Paris, (X 1623) Univ Paris VI Inst Poincare: Paris Secretariat Mathematique: Paris

S 1582 Matematika - Period Sb Perevodov Inostran Statej • SU
Matematika. Periodicheskij Sbornik Perevodov Inostrannykh Statej (Mathematics. Series of Collections of Translations of Foreign Articles) [1957ff] • PUBL (X 0885) Mir: Moskva
• REM Does not appear any more

S 1605 Math Centr Tracts • NL
Mathematical Centre Tracts [1963-1983] • PUBL (X 1121) Math Centr: Amsterdam

S 1613 Probl Logic (Bucharest) • RO
Probleme de Logica [1956ff] • PUBL (X 0871) Acad Rep Soc Romania: Bucharest
• ISSN 0556-1655

S 1626 Oslo Preprint Ser • N
Oslo Preprint Series [1970ff] • PUBL (X 2786) Univ Oslo Mat Inst: Oslo

S 1642 Schr Inf Angew Math, Ber (Aachen) • D
Schriften zur Informatik und Angewandten Mathematik. Bericht PUBL (X 3215) TH Aachen Math Nat Fak: Aachen

S 1644 Tr Fiz-Mat Inst Steklov • SU
Trudy Fiziko-Matematicheskogo Instituta imeni V.A.Steklova Akademiya Nauk SSSR (Travaux de l'Institut Physico-Mathematique V.A.Stekloff de l'Academie des Sciences SSSR) [1930-1937] • PUBL (X 0899) Akad Nauk SSSR : Moskva
• CONT AS (S 0066) Tr Mat Inst Steklov

S 1802 Colloq Int CNRS • F
Colloques Internationaux du Centre National de la Recherche Scientifique (CNRS) PUBL (X 0999) CNRS Inst B Pascal: Paris

S 1926 Banach Cent Publ • PL
Banach Center Publications. Polish Academy of Science. Institut of Mathematics [1976ff] • PUBL (X 1034) PWN: Warsaw
• ISSN 0137-6934

S 1956 Tagungsbericht, Oberwolfach • D
Tagungsbericht. Mathematisches Forschungsinstitut Oberwolfach PUBL (X 0876) Bibl Inst: Mannheim

S 2073 Actualites Sci Indust • F
Actualites Scientifiques et Industrielles PUBL (X 0859) Hermann: Paris

S 2308 Symp Kyoto Univ Res Inst Math Sci (RIMS) • J
Surikaisekikenkyusho Kokyuroku (Kyoto University. Research Institute for Mathematical Sciences (RIMS). Proceedings of Symposia) PUBL (X 2441) Kyoto Univ Res Inst Math Sci: Kyoto

S 2337 Istor Metodol Estest Nauk (Moskva) • SU
Istoriya i Metodologiya Estestvennykh Nauk. (History and Methodology of Natural Science) [1960ff] • PUBL (X 0898) Moskov Gos Univ: Moskva
• ISSN 0579-0204

S 2450 Mem Amer Math Soc, NS • USA
Memoirs of the American Mathematical Society. New Series [1975ff] • PUBL (X 0803) Amer Math Soc: Providence
• CONT OF (S 0167) Mem Amer Math Soc
• ISSN 0065-9266

S 2579 Teor Mnozhestv & Topol (Izhevsk) • SU
Teoriya Mnozhestv i Topologiya (Set Theory and Topology) [1977,1979,1982] • ED: GRYZLOV, A.A. • PUBL (X 4562) Udmurtskij Gos Univ: Izhevsk 114pp,116pp,116pp
• REM Title of Vol.2: Sovremennaya Topologiya i Teoriya Mnozhestv, Vyp.2

S 2582 Semiotika & Inf, Akad Nauk SSSR • SU
Semiotika i Informatika. Gosudarstvennyj Komitet SSSR po Nauke i Tekhnike. Akademiya Nauk SSSR. Vsesoyuznyj Institut Nauchnoj i Tekhnicheskoj Informatsii (Semiotics and Information Science) ED: MIKHAJLOV, A.I. • PUBL (X 2235) VINITI: Moskva

S 2587 Teor Algor & Progr (Riga) • SU
Teoriya Algoritmov i Programm (Theory of Algorithms and Programs) [1974,1975,1977] • ED: BARZDIN', YA.M. (VOL.2 & 3) & IKAUNIEKS, E. & FREIVALDS, R. (VOL.2) • SER (J 0483) Uch Zap, Univ Riga 210 (Vol.1), 233 (Vol.2) • PUBL (X 0895) Latv Valsts (Gos) Univ : Riga

S 2651 Prepr Inst Prikl Mat, Akad Nauk SSSR • SU
Preprint. Akademiya Nauk SSSR. Institut Prikladnoj Matematiki. (Preprint. Academy of Sciences of the USSR. Institute of Applied Mathematics) PUBL Akad Nauk SSSR, Inst Prikl Mat: Moskva

S 2738 Issl Prikl Mat (Univ Kazan') • SU
Issledovaniya po Prikladnoj Matematike (Studies in Applied Mathematics) [1973ff] • PUBL (X 3605) Kazan Gos Univ: Kazan'

S 2824 Real Anal Exchange • USA
Real Analysis Exchange [1976ff] • PUBL Western Illinois Univ: Macomb
• ISSN 0147-1937

S 2850 Tr Ehnerg Inst Moskva • SU
Trudy Moskovskogo Ordena Lenina Ehnergeticheskogo Instituta Tematicheskij Sbornik (Proceedings of the Moscow Institute of Energetics) PUBL Moskovsk Ehnergetichesk Instituta: Moskva

S 2874 Vopr Kibern, Akad Nauk SSSR • SU
Akademiya Nauk SSSR Nauchnyj Sovet po Kompleksnoj Probleme. "Kibernetika". Voprosy Kibernetiki (Problems of Cybernetics. Academy of Sciences. Scientific Council for Complexity Problems. Cybernetics) PUBL Akad Nauk SSSR Nauch Sovet Komplek Probl Kibern: Moskva

S 3105 Adv Math, Suppl Stud • USA
Advances in Mathematics. Supplementary Studies. [1965ff]
• PUBL (X 0801) Academic Pr: New York
• REL PUBL (J 0345) Adv Math

S 3125 Ber, Abt Inf, Univ Dortmund • D
Bericht. Abteilung Informatik. Universitaet Dortmund, Dortmund PUBL Univ Dortmund, Abt Inf: Dortmund

S 3301 Lect Notes Math • D
Lecture Notes in Mathematics [1964ff] • PUBL (**X** 0811) Springer: Heidelberg & New York
• ISSN 0075-8434

S 3302 Lect Notes Comput Sci • D
Lecture Notes in Computer Science [1973ff] • PUBL (**X** 0811) Springer: Heidelberg & New York
• ISSN 0302-9743

S 3303 Stud Logic Found Math • NL
Studies in Logic and the Foundations of Mathematics [1954ff] • PUBL (**X** 0809) North Holland: Amsterdam
• ISSN 0049-237X

S 3304 Proc Symp Pure Math • USA
Proceedings of Symposia in Pure Mathematics PUBL (**X** 0803) Amer Math Soc: Providence

S 3305 Symposia Matematica • I
Symposia Matematica [1969ff] • PUBL (**X** 3604) INDAM: Roma
• ISSN 0082-0725

S 3306 Lond Math Soc Lect Note Ser • GB
London Mathematical Society Lecture Note Series [1971ff] • PUBL (**X** 0805) Cambridge Univ Pr: Cambridge, GB
• ISSN 0076-0552

S 3307 Synth Libr • NL
Synthese Library. Studies in Epistemology, Logic, Methodology, and Philosophy of Science [1959ff] • PUBL (**X** 0835) Reidel: Dordrecht
• ISSN 0082-1128

S 3308 Univ Western Ontario Ser in Philos of Sci • NL
University of Western Ontario Series in Philosophy of Science [1972ff] • PUBL (**X** 0835) Reidel: Dordrecht

S 3309 AFIPS Conference Proc • USA
AFIPS Conference Proceedings PUBL (**X** 1354) Spartan Books : Sutton

S 3310 Lect Notes Pure Appl Math • USA
Lecture Notes in Pure and Applied Mathematics [1971ff] • PUBL (**X** 1684) Dekker: New York
• ISSN 0075-8469

S 3311 Boston St Philos Sci • NL
Boston Studies in the Philosophy of Science [1963ff] • PUBL (**X** 0835) Reidel: Dordrecht
• ISSN 0068-0346

S 3312 Coll Math Soc Janos Bolyai • H
Colloquia Mathematica Societatis Janos Bolyai PUBL (**X** 0809) North Holland: Amsterdam

S 3313 Contemp Math • USA
Contemporary Mathematics [1980ff] • PUBL (**X** 0803) Amer Math Soc: Providence
• ISSN 0271-4132

S 3358 Ann Discrete Math • NL
Annals of Discrete Mathematics [1977ff] • PUBL (**X** 0809) North Holland: Amsterdam

S 3382 Sem-ber, Humboldt-Univ Berlin, Sekt Math • DDR
Seminarberichte. Humboldt-Universitaet zu Berlin, Sektion Mathematik PUBL (**X** 2219) Humboldt-Univ Berlin: Berlin

S 3416 Prepr Ser Dep Math Univ Ljubljana • YU
Preprint Series. University of Ljubljana, Department of Mathematics PUBL Univ Ljubljana, Dept Math: Ljubljana

S 3462 Var Publ Ser, Aarhus Univ • DK
Various Publications Series [1962ff] • PUBL (**X** 1599) Aarhus Univ Mat Inst: Aarhus
• ISSN 0065-0188

S 3468 Tr Mat & Mekh (Tartu) • SU
*Matemaatika - ja Mekhaanika-Alaseid Toeid. * Trudy po Matematike i Mekhanike. (Works about Mathematics and Mechanics.)* SER (**S** 0393) Uch Zap Univ, Tartu • PUBL (**X** 2463) Tartusk Gos Univ: Tartu

S 3477 Math Centr Afd Toegepaste Wiskd, Amsterdam • NL
Stichting Mathematisch Centrum. Afdeling Toegepaste Wiskunde PUBL (**X** 1121) Math Centr: Amsterdam
• ISSN 0517-6425

S 3521 Mem Soc Math Fr • F
Memoire de la Mathematique de France. Supplement au Bulletin [1927ff] • PUBL (**X** 0834) Gauthier-Villars: Paris
• REM Since 1981: Nouvelle Serie

S 4568 Hiroshima Univ Fac Engin, Mem • J
Memoirs of the Faculty of Engineering Hiroshima University

S 4733 Prace Inst Mat, Politech Wroclaw, Ser Stud Mater • PL
Politechnika Wroclawska Instytut Matematyki. Prace Naukowe. Studia i Materialy (Scientific Publications. Institute of Mathematics. Technical University of Wroclaw) [1970ff] • PUBL Politechnika Wroclawska: Wroclaw
• CONT OF (**S** 1562) Prace Inst Mat Fis, Politech Wroclaw, Ser Stud Mater

Proceedings

P 0098 Skand Mat Kongr (11);1949 Trondheim • N
[1952] *den 11te Skandinaviske Matematikerkongress (Comptes-Rendus du 11eme Congres des Mathematiciens Scandinaves)* ED: LYCHE, R.T. • PUBL Johan Grund Tanums: Oslo xxxi+300pp
• DAT&PL 1949 Aug; Trondheim, N

P 0454 Math Founds of Comput Sci (4);1975 Marianske Lazne • CS
[1975] *Mathematical Foundations of Computer Science. Proceedings of the 4th Symposium* ED: BECVAR, J. • SER (S 3302) Lect Notes Comput Sci 32 • PUBL (X 0811) Springer: Heidelberg & New York x+476pp
• DAT&PL 1975 Sep; Marianske Lazne, CS • ISBN 3-540-07389-2, LC-No 75-22406

P 0560 Int Congr Philos (12);1958 Venezia • I
[1958-1961] *Atti del 12 Congresso Internazionale di Filosofia (Actes du 12eme Congres International de Philosophie * Proceedings of the 12th International Congress of Philosophy)* PUBL (X 1346) Sansoni: Firenze 12 Vols
• DAT&PL 1958 Sep; Venezia, I

P 0575 Int Congr Math (II, 7);1954 Amsterdam • NL
[1954-1957] *Proceedings of the International Congress of Mathematicians 1954* ED: GERRETSEN, J.C.H. & GROOT DE, J. • PUBL (X 0809) North Holland: Amsterdam 3 Vols: 582pp,440pp,560pp • ALT PUBL (X 1317) Noordhoff: Groningen 1954-1957 & (X 3602) Kraus: Vaduz 1967
• DAT&PL 1954 Sep; Amsterdam, NL • LC-No 52-1808

P 0576 Raisonn en Math & Sci Exper;1955 Paris • F
[1958] *La Raisonnement en Mathematiques et en Sciences Experimentales.* SER (S 1802) Colloq Int CNRS 70 • PUBL (X 0999) CNRS Inst B Pascal: Paris 140pp
• DAT&PL 1955 Sep; Paris, F • LC-No 63-24106

P 0577 Automatenth & Formale Sprachen;1969 Oberwolfach • D
[1970] *Automatentheorie und Formale Sprachen.* ED: DOERR, J. & HOTZ, G. • SER (S 1956) Tagungsbericht, Oberwolfach 3 • PUBL (X 0876) Bibl Inst: Mannheim 505pp
• DAT&PL 1969 Oct; Oberwolfach, D • LC-No 76-857074

P 0580 Int Congr Log, Meth & Phil of Sci (4,Sel Pap);1971 Bucharest • RO
[1973] *Logic, Language and Probability.* ED: BOGDAN, R.J. & NIINILUOTO, I. • SER (S 3307) Synth Libr • PUBL (X 0835) Reidel: Dordrecht x+323pp
• DAT&PL 1971 Aug; Bucharest, RO • ISBN 90-277-0312-4, LC-No 72-95892 • REL PUBL (P 0793) Int Congr Log, Meth & Phil of Sci (4,Proc);1971 Bucharest
• REM A Selection of Papers Contributed to Sections IV, VI and XI of P0793

P 0593 Int Congr Math (II, 6);1950 Cambridge MA • USA
[1952] *Proceedings of the International Congress of Mathematicians* ED: GRAVES, L.M & HILLE, E. & SMITH, P.A. & ZARISKI, O. • PUBL (X 0803) Amer Math Soc: Providence 2 Vols: 769pp,461pp • ALT PUBL (X 3602) Kraus: Vaduz 1967 2 Vols
• DAT&PL 1950 Aug; Cambridge, MA, USA • LC-No 52-1808

P 0603 Intuitionism & Proof Th;1968 Buffalo • USA
[1970] *Intuitionism and Proof Theory* ED: KINO, A. & MYHILL, J. & VESLEY, R.E. • PUBL (X 0809) North Holland: Amsterdam viii+516pp
• DAT&PL 1968 Aug,Buffalo, NY, USA • ISBN 0-7204-2257-4, LC-No 77-97196

P 0604 Scand Logic Symp (2);1970 Oslo • N
[1971] *Proceedings of the 2nd Scandinavian Logic Symposium* ED: FENSTAD, J.E. • SER (S 3303) Stud Logic Found Math 63 • PUBL (X 0809) North Holland: Amsterdam ii+405pp
• DAT&PL 1970 Jun; Oslo, N • ISBN 0-7204-2259-0, LC-No 71-153401

P 0607 All-Union Math Conf (3);1956 Moskva • SU
[1959] *Trudy 3'go Vsesoyuznogo Matematicheskogo S'ezda (Proceedings of the 3rd All Union Mathematical Conference)* ED: NIKOL'SKIJ, S.M. & ABRAMOV, A.A. & BOLTYANSKIJ, V.G. • PUBL (X 0899) Akad Nauk SSSR : Moskva 4 Vols
• DAT&PL 1956 Jun; Moskva, SU

P 0608 Logic Colloq;1966 Hannover • D
[1968] *Contributions to Mathematical Logic. Proceedings of the Logic Colloquium* ED: SCHMIDT, H.A. & SCHUETTE, K. & THIELE, H.-J. • SER (S 3303) Stud Logic Found Math • PUBL (X 0809) North Holland: Amsterdam ix+298pp
• DAT&PL 1966 Aug; Hannover, D • LC-No 68-24434

P 0610 Tarski Symp;1971 Berkeley • USA
[1974] *Proceedings of the Tarski Symposium. An International Symposium held to Honor Alfred Tarski on the Occasion of His 70th Birthday* ED: HENKIN, L. & ADDISON, J. & CHANG, C.C. & CRAIG, W. & SCOTT, D.S. & VAUGHT, R. • SER (S 3304) Proc Symp Pure Math 25 • PUBL (X 0803) Amer Math Soc: Providence ix+498pp
• DAT&PL 1971 Jun; Berkeley, CA, USA • ISBN 0-8218-1425-7, LC-No 74-8666
• REM Corrected Reprint 1979; xx+498pp

P 0612 Int Congr Log, Meth & Phil of Sci (1,Proc);1960 Stanford • USA
[1962] *Proceedings of the 1st International Congress for Logic, Methodology and Philosophy of Science* ED: NAGEL, E. & SUPPES, P. & TARSKI, A. • PUBL (X 1355) Stanford Univ Pr: Stanford ix+661pp
• DAT&PL 1960 Aug; Stanford, CA, USA • LC-No 62-9620
• TRANSL IN [1965] (P 2251) Mat Log & Primen;1960 Stanford

P 0613 Rec Fct Th;1961 New York • USA
[1962] *Recursive Function Theory* ED: DEKKER, J.C.E. • SER (S 3304) Proc Symp Pure Math 5 • PUBL (X 0803) Amer Math Soc: Providence vii+247pp
• DAT&PL 1961 Apr; New York, NY, USA • ISBN 0-8218-1405-2, LC-No 50-1183
• REM 2nd ed. 1970

P 0614 Th Models;1963 Berkeley • USA
[1965] *The Theory of Models.* ED: ADDISON, J.W. & HENKIN, L. & TARSKI, A. • SER (S 3303) Stud Logic Found Math • PUBL (X 0809) North Holland: Amsterdam xv+494pp
• DAT&PL 1963 Jun; Berkeley, CA, USA • LC-No 66-7051

P 0623 Int Congr Log, Meth & Phil of Sci (2,Proc);1964 Jerusalem • IL
[1965] *Proceedings of the 2nd International Congress for Logic, Methodology and Philosophy of Science* ED: BAR-HILLEL, Y. • SER (S 3303) Stud Logic Found Math • PUBL (X 0809) North Holland: Amsterdam viii+440pp
• DAT&PL 1964 Aug; Jerusalem, IL • ISBN 0-7204-2235-3, LC-No 66-7008
• REM 2nd ed. 1972

P 0625 Symp Autom Demonst;1968 Versailles • F
[1970] *Symposium on Automatic Demonstration.* ED: LAUDET, M. & LACOMBE, D. & NOLIN, L. & SCHUETZENBERGER, M. • SER (S 3301) Lect Notes Math 125 • PUBL (X 0811) Springer: Heidelberg & New York v+310pp
• DAT&PL 1968 Dec; Versailles, F • ISBN 3-540-04914-2, LC-No 79-117526

P 0627 Int Congr Log, Meth & Phil of Sci (3,Proc);1967 Amsterdam • NL
[1968] *Proceedings of the 3rd International Congress for Logic, Methodology and Philosophy of Science* ED: ROOTSELAAR VAN, B. & STAAL, J.F. • SER (S 3303) Stud Logic Found Math • PUBL (X 0809) North Holland: Amsterdam xii+553pp
• DAT&PL 1967 Aug; Amsterdam, NL • ISBN 0-444-85423-1, LC-No 68-29768

P 0632 Congr Int Phil des Sci;1935 Paris • F
[1936] *Actes du Congres International de Philosophie Scientifique* SER (S 2073) Actualites Sci Indust 388-395 • PUBL (X 0859) Hermann: Paris 8 Vol
• DAT&PL 1935 Sep; Paris, F
• REM Vol. II: Unite de la Science. Vol. III: Language et Pseudo-Problemes. Vol. IV: Induction et Probabilite. Vol. VI: Philosophie des Mathematiques. Vol. VII: Logique. Vol. VIII: Histoire de la Logique et de la Philosophie Scientifique.

P 0633 Infinitist Meth;1959 Warsaw • PL
[1961] *Infinitistic Methods. Proceedings of the Symposium on Foundations of Mathematics* PUBL (X 1034) PWN: Warsaw 362pp • ALT PUBL (X 0869) Pergamon Pr: Oxford 1961
• DAT&PL 1959 Sep; Warsaw, PL • LC-No 61-11351

P 0634 Constructivity in Math;1957 Amsterdam • NL
[1959] *Constructivity in Mathematics* ED: HEYTING, A. • SER (S 3303) Stud Logic Found Math • PUBL (X 0809) North Holland: Amsterdam viii+298pp
• DAT&PL 1957 Aug; Amsterdam, NL • LC-No 63-522

P 0637 Syntax & Semant Infinitary Lang;1967 Los Angeles • USA
[1968] *The Syntax and Semantics of Infinitary Languages* ED: BARWISE, K.J. • SER (S 3301) Lect Notes Math 72 • PUBL (X 0811) Springer: Heidelberg & New York iv+268pp
• DAT&PL 1967 Dec; Los Angeles, CA, USA • ISBN 3-540-04242-3, LC-No 68-57175

P 0638 Logic Colloq;1969 Manchester • GB
[1971] *Logic Colloquium '69* ED: GANDY, R.O. & YATES, C.M.E. • SER (S 3303) Stud Logic Found Math 61 • PUBL (X 0809) North Holland: Amsterdam xiv+457pp
• DAT&PL 1969 Aug; Manchester, GB • ISBN 0-7204-2261-2, LC-No 71-146188

P 0639 Congr Nat des Sci (2);1935 Bruxelles • B
[1935] *2me Congres National des Sciences, Comptes Rendus*
• DAT&PL 1935 Jun; Bruxelles, B • LC-No 37-39347

P 0642 Congr Int Phil Sci;1949 Paris • F
[1951-1952] *Congres International de Philosophie des Sciences* SER (S 2073) Actualites Sci Indust 1126,1134,1137,1146,1153,1155,1156,1166,1167, Philosophie 15-23 • PUBL (X 0859) Hermann: Paris 9 Vols
• DAT&PL 1949 -?-; Paris, F

P 0644 Meth Form en Axiom;1950 Paris • F
[1953] *Les Methodes Formelles en Axiomatiques* ED: DESTOUCHES, J.-L. & DESTOUCHES-FEVRIER, P. & JOFFRE, D. • SER (S 1802) Colloq Int CNRS 36 • PUBL (X 0999) CNRS Inst B Pascal: Paris 197pp
• DAT&PL 1950 Dec; Paris, F

P 0645 Int Congr Philos (11);1953 Bruxelles • B
[1953] *Actes du 11eme Congres International de Philosophie.* * *Proceedings of the 11th International Congress of Philosophy* PUBL (X 1313) Nauwelaerts: Louvain • ALT PUBL (X 0809) North Holland: Amsterdam 1953;14 Vols
• DAT&PL 1953 Aug; Bruxelles, B

P 0646 Appl Sci de Log Math;1952 Paris • F
[1954] *Applications Scientifiques de la Logique Mathematique. Actes du 2eme Colloque International de Logique Mathematique* ED: DESTOUCHES, J.-L. & DESTOUCHES-FEVRIER, P. • SER Collection de Logique Mathematique, Serie A 5 • PUBL (X 0834) Gauthier-Villars: Paris 176pp • ALT PUBL (X 1313) Nauwelaerts: Louvain 1954
• DAT&PL 1952 Aug; Paris, F

P 0649 Appl Model Th to Algeb, Anal & Probab;1967 Pasadena • USA
[1969] *Applications of Model Theory to Algebra, Analysis and Probability* ED: LUXEMBURG, W.A.J. • PUBL (X 0818) Holt Rinehart & Winston: New York vii+307pp
• DAT&PL 1967 May; Pasadena, CA, USA • LC-No 69-11203

P 0651 Axiomatic Method;1957 Berkeley • USA
[1959] *The Axiomatic Method. With Special Reference to Geometry and Physics* ED: HENKIN, L. & SUPPES, P. & TARSKI, A. • SER (S 3303) Stud Logic Found Math • PUBL (X 0809) North Holland: Amsterdam xi+488pp
• DAT&PL 1957 Dec; Berkeley, CA, USA • LC-No 58-63025

P 0652 Entretiens Zuerich Fond & Method Sci Math;1938 Zuerich • CH
[1941] *Les Entretiens de Zuerich sur les Fondements et la Methode des Sciences Mathematiques: Exposes et Discussions* ED: GONSETH, F. • PUBL (X 2220) Leemen: Zuerich 209pp
• DAT&PL 1938 Dec; Zuerich, CH • LC-No 42-650

P 0653 Int Congr Math (II, 4);1932 Zuerich • CH
[1932] *Verhandlungen des Internationalen Mathematiker-Kongresses Zuerich* ED: SAX, W. • PUBL (X 1268) Orell Fuessli: Zuerich 2 Vols: 335pp,365pp
• DAT&PL 1932 Sep; Zuerich, CH • LC-No 52-1808

P 0657 Int Congr Math (II,10);1966 Moskva • SU
[1968] *Trudy Mezhdunarodnogo Kongressa Matematikov * Proceedings of the International Congress of Mathematicians * Travaux du Congres International des Mathematiciens * Berichte des Internationalen Mathematikerkongresses* ED: PETROVSKIJ, YU.G. • PUBL (**X** 0885) Mir: Moskva 726pp
• DAT&PL 1966 Aug; Moskva, SU • LC-No 70-437326

P 0660 Int Congr Math (II, 8);1958 Edinburgh • GB
[1960] *Proceedings of the International Congress of Mathematicians* ED: TODD, J.A. • PUBL (**X** 0805) Cambridge Univ Pr: Cambridge, GB lxiv+573pp
• DAT&PL 1958 Aug; Edinburgh, GB • LC-No 52-1808

P 0664 Congr Int Union Phil of Sci (2);1954 Zuerich • CH
[1955] *Actes du 2eme Congres International de L'Union Internationale de la Philosophie des Sciences * Proceedings of the 2nd International Congress of the International Union for the Philosophy of Science* PUBL (**X** 0272) Griffon: Neuchatel 5 Vols
• DAT&PL 1954 -?-; Zuerich, CH • LC-No 64-52910

P 0669 Conv Teor Modelli & Geom;1969/70 Roma • I
[1971] *Convegni Teoria dei Modelli & Geometria* SER (**S** 3305) Symposia Matematica 5 • PUBL (**X** 3604) INDAM: Roma 475pp • ALT PUBL (**X** 0801) Academic Pr: New York
• DAT&PL 1969 Nov; Rome, I, 1970 Apr; Rome, I

P 0682 Int Congr Philos (10);1948 Amsterdam • NL
[1949] *Library of the 10th International Congress of Philosophy* ED: BETH, E.W. & POS, H.J. & HOLLAK, H.J.A.
• PUBL (**X** 0809) North Holland: Amsterdam Vol 1, L.J.Veen: Amsterdam Vol 2
• DAT&PL 1948 Aug; Amsterdam, NL • LC-No 50-35721

P 0683 Log & Founds of Sci (Beth);1964 Paris • F
[1967] *Logic and Foundations of Science. E.W.Beth Memorial Colloquium.* ED: DESTOUCHES, J.-L. • PUBL (**X** 0835) Reidel: Dordrecht viii+140pp
• DAT&PL 1964 May; Paris, F • LC-No 68-107520

P 0688 Logic Colloq;1963 Oxford • GB
[1965] *Formal Systems and Recursive Functions. Proceedings of the 8th Logic Colloquium* ED: CROSSLEY, J.N. & DUMMETT, M.A.E. • SER (**S** 3303) Stud Logic Found Math • PUBL (**X** 0809) North Holland: Amsterdam 320pp
• DAT&PL 1963 Jul; Oxford, GB • LC-No 66-2289

P 0689 Int Congr Math (4);1908 Roma • I
[1909] *Atti del 4 Congresso Internazionale dei Matematici* ED: CASTELNUOVO, G. • PUBL (**X** 2121) Accad Naz Linc: Roma 3 Vols: iv+216pp,319pp,588pp • ALT PUBL (**X** 3602) Kraus: Vaduz 1967
• DAT&PL 1908 Apr; Roma, I

P 0691 Sets, Models & Recursion Th;1965 Leicester • GB
[1967] *Sets, Models and Recursion Theory. Proceedings of the Summer School in Mathematical Logic and 10th Logic Colloquium* ED: CROSSLEY, J.N. • SER (**S** 3303) Stud Logic Found Math • PUBL (**X** 0809) North Holland: Amsterdam v+331pp • ALT PUBL (**X** 0838) Amer Elsevier: New York 1967
• DAT&PL 1965 Aug; Leicester, GB • ISBN 0-7204-2242-6, ISBN 0-444-10696-0, LC-No 67-21973
• REM 2nd ed. 1974

P 0692 Summer School in Logic;1967 Leeds • GB
[1968] *Proceedings of the Summer School in Logic* ED: LOEB, M.H. • SER (**S** 3301) Lect Notes Math 70 • PUBL (**X** 0811) Springer: Heidelberg & New York iv+331pp
• DAT&PL 1967 Aug; Leeds, GB • ISBN 3-540-04240-7, LC-No 68-56951

P 0693 Axiomatic Set Th;1967 Los Angeles • USA
[1971-1974] *Axiomatic Set Theory* ED: SCOTT, D. (V 1) & JECH, T.J. (V 2) • SER (**S** 3304) Proc Symp Pure Math 13
• PUBL (**X** 0803) Amer Math Soc: Providence 2 Vols: vi+474pp,viii+222pp
• DAT&PL 1967 Jul; Los Angeles, CA, USA • ISBN 0-8218-0245-3 (V1), ISBN 0-8218-0246-1 (V2), LC-No 78-125172

P 0701 Struct of Lang & Math Aspects;1960 New York • USA
[1961] *Structure of Language and Its Mathematical Aspects* ED: JAKOBSON, R. • SER Proc Symp Appl Math 12 • PUBL (**X** 0803) Amer Math Soc: Providence vi+279pp
• DAT&PL 1960 Apr; New York, NY, USA • LC-No 50-1183

P 0713 Cambridge Summer School Math Log;1971 Cambridge GB • GB
[1973] *Cambridge Summer School in Mathematical Logic* ED: ROGERS, H. & MATHIAS, A.R.D. • SER (**S** 3301) Lect Notes Math 337 • PUBL (**X** 0811) Springer: Heidelberg & New York ix+660pp
• DAT&PL 1971 Aug; Cambridge, GB • ISBN 3-540-05569-X, LC-No 73-12410

P 0723 Categ Th, Homol Th & Appl;1968 Seattle • USA
[1969] *Category Theory, Homology Theory and Their Applications* ED: HILTON, P.J. • SER (**S** 3301) Lect Notes Math 86,92,99 • PUBL (**X** 0811) Springer: Heidelberg & New York 3 Vols: vi+216pp,v+308pp,iv+498pp
• DAT&PL 1968 Jun; Seattle, WA, USA • ISBN 3-540-04605-4 (V1), ISBN 3-540-04611-9 (V2), ISBN 3-540-04618-6 (V3), LC-No 75-75931

P 0741 Int Congr Math (II, 3);1928 Bologna • I
[1929-1932] *Atti del Congresso Internazionale dei Matematici* PUBL (**X** 1375) Zanichelli: Bologna 6 Vols: 338pp,365pp,472pp,429pp,494pp,554pp
• DAT&PL 1928 Sep; Bologna, I • LC-No 52-1808

P 0743 Int Congr Math (II,11,Proc);1970 Nice • F
[1971] *Actes du Congres International de Mathematiciens 1970* ED: BERGER, M. & DIEUDONNE, J. & LERAY, J. & LIONS, J.-L. & MALLIAVIN, M.P. & SERRE, J.-P. • PUBL (**X** 0834) Gauthier-Villars: Paris 3 Vols: xxxiii+532pp,959pp,iii+371pp
• DAT&PL 1970 Sep; Nice, F • REL PUBL (**P** 1158) Int Congr Math (II,11,Comm Ind);1970 Nice
• REM Vol 1: Documents.Medailles Fields.Conferences Generales.Logique.Algebre. Vol 2: Geometrie et Topologie. Analyse. Vol 3: Mathematiques Appliquees.Historie et Enseignement.

P 0746 Automata Th;1964 Ravello • I
[1966] *Automata Theory. International School of Physics* ED: CAIANIELLO, E.R. • PUBL (**X** 0801) Academic Pr: New York xiv+342pp
• DAT&PL 1964 Jun; Ravello, I • LC-No 65-22775
• REM 2nd ed. 1968

Source Index

P 0747 Form Lang, Descript Langs for Comput Progr;1964 Wien • A
[1966] *Formal Language Description. Languages for Computer Programming. Proceedings of IFIP Working Conference* ED: STEEL JR., T.B. • PUBL (**X** 0809) North Holland: Amsterdam 330pp
• DAT&PL 1964 Sep; Wien, A • LC-No 66-8170

P 0756 Congr Int Phil (9);1937 Paris • F
[1937] *Travaux du IXe Congres International de Philosophie: Congres Descartes* ED: BAYER, R. • SER (**S** 2073) Actualites Sci Indust 530-541 • PUBL (**X** 0859) Hermann: Paris 12 Vols
• DAT&PL 1937 -?-; Paris, F

P 0757 Scand Logic Symp (3);1973 Uppsala • S
[1975] *Proceedings of the 3rd Scandinavian Logic Symposium* ED: KANGER, S. • SER (**S** 3303) Stud Logic Found Math 82 • PUBL (**X** 0809) North Holland: Amsterdam vii+214pp • ALT PUBL (**X** 0838) Amer Elsevier: New York
• DAT&PL 1973 Apr; Uppsala, S • ISBN 0-444-10679-0, LC-No 74-80113

P 0761 Compl of Computation;1973 New York • USA
[1974] *Complexity of Computation. Proceedings of a Symposium in Applied Mathematics of the American Mathematical Society (AMS) and the Society for Industrial and Applied Mathematics (SIAM)* ED: KARP, R.M. • SER SIAM-AMS Proceedings: 7 • PUBL (**X** 0803) Amer Math Soc: Providence viii+166pp
• DAT&PL 1973 Apr; New York, NY, USA • ISBN 0-8218-1327-7, LC-No 74-22062

P 0763 Automata, Lang & Progr (1);1972 Rocquencourt • F
[1973] *Automata, Languages and Programming. Proceedings of a Symposium Organised by IRIA (Institut de Recherche d'Informatique et d'Automatique)* ED: NIVAT, M. • PUBL (**X** 0809) North Holland: Amsterdam 638pp • ALT PUBL (**X** 0838) Amer Elsevier: New York
• DAT&PL 1972 Jul; Versailles-Rocquencourt, F • ISBN 0-444-10426-7, LC-No 72-93498
• REM Also Abbreviated as ICALP 72

P 0765 Algeb & Log;1974 Clayton • AUS
[1975] *Algebra and Logic. Papers from the 1974 Summer Research Institute of the Australian Mathematical Society* ED: CROSSLEY, J.N. • SER (**S** 3301) Lect Notes Math 450 • PUBL (**X** 0811) Springer: Heidelberg & New York viii+307pp
• DAT&PL 1974 Jan; Clayton, Vic, AUS • ISBN 3-540-07152-0, LC-No 75-9903

P 0771 Toposes, Algeb Geom & Log;1971 Halifax • CDN
[1972] *Toposes, Algebraic Geometry and Logic. Partial Report on a Conference on Connections Between Category Theory and Algebraic Geometry and Intuitionistic Logic* ED: LAWVERE, F.W. • SER (**S** 3301) Lect Notes Math 274 • PUBL (**X** 0811) Springer: Heidelberg & New York 189pp
• DAT&PL 1971 Jan; Halifax, NS, CDN • ISBN 3-540-05920-2, LC-No 72-86101

P 0775 Logic Colloq;1973 Bristol • GB
[1975] *Logic Colloquium '73* ED: ROSE, H.E. & SHEPHERDSON, J.C. • SER (**S** 3303) Stud Logic Found Math 80 • PUBL (**X** 0809) North Holland: Amsterdam viii+513pp • ALT PUBL (**X** 0838) Amer Elsevier: New York
• DAT&PL 1973 Jul; Bristol, GB • ISBN 0-444-10642-1, LC-No 74-79302

P 0776 Permutations;1972 Paris • F
[1974] *Permutations. Actes du Colloque sur les Permutations* SER Mathematiques et Sciences de L'Homme 20 • PUBL (**X** 0834) Gauthier-Villars: Paris xiv+289pp • ALT PUBL (**X** 0873) Mouton: Paris
• DAT&PL 1972 Jul; Paris, F • ISBN 2-04-009681-7, ISBN 2-7193-0897-8, LC-No 74-189927

P 0783 Truth, Syntax & Modal;1970 Philadelphia • USA
[1973] *Truth, Syntax and Modality: Proceedings of the Temple University Conference on Alternative Semantics* ED: LEBLANC, H. • SER (**S** 3303) Stud Logic Found Math 68 • PUBL (**X** 0809) North Holland: Amsterdam vii+317pp
• DAT&PL 1970 Dec; Philadelphia, PA, USA • ISBN 0-7204-2269-8, LC-No 72-79730

P 0784 GI Jahrestag (5);1975 Dortmund • D
[1975] *GI - 5. Jahrestagung (Gesellschaft fuer Informatik)* ED: MUEHLBACHER, J. • SER (**S** 3302) Lect Notes Comput Sci 34 • PUBL (**X** 0811) Springer: Heidelberg & New York x+755pp
• DAT&PL 1975 Oct; Dortmund, D • ISBN 3-540-07410-4

P 0785 Scand Logic Symp (1);1968 Aabo • S
[1970] *Proceedings of the 1st Scandinavian Logic Symposium* SER Filosofiska Studier 8 • PUBL (**X** 0882) Univ Filos Foeren: Uppsala 171pp
• DAT&PL 1968 Sep; Aabo, S • LC-No 72-186670

P 0788 Skand Mat Kongr (12);1953 Lund • S
[1954] *12te Skandinaviska Matematikerkongressen (Comptes-Rendus du 12eme Congres des Mathematiciens Scandinaves)* PUBL Hakan Oh Issons Boktryckeri: Lund xvi+337pp
• DAT&PL 1953 Aug; Lund, S • LC-No 55-58514

P 0793 Int Congr Log, Meth & Phil of Sci (4,Proc);1971 Bucharest • RO
[1973] *Proceedings of the 4th International Congress for Logic, Methodology and Philosophy of Science* ED: SUPPES, P. & HENKIN, L. & MOISIL, G.C. & JOJA, A. • SER (**S** 3303) Stud Logic Found Math 74 • PUBL (**X** 0809) North Holland: Amsterdam x+981pp • ALT PUBL (**X** 0838) Amer Elsevier: New York & (**X** 1034) PWN: Warsaw
• DAT&PL 1971 Aug; Bucharest, RO • ISBN 0-444-10491-7, LC-No 72-88505 • REL PUBL (**P** 0580) Int Congr Log, Meth & Phil of Sci (4,Sel Pap);1971 Bucharest

P 0794 Congr Unione Mat Ital (7);1963 Genova • I
[1965] *Atti del 7 Congresso dell'Unione Matematica Italiana* PUBL (**X** 0860) Cremonese: Firenze 415pp
• DAT&PL 1963 Sep; Genova, I

P 0796 Congr Math Pays Slaves (1);1929 Warsaw • PL
[1930] *Sprawozdanie z 1 Kongresu Matematykow Krajow Slowianskich (Comptes Rendus du 1. Congres des Mathematiciens des Pays Slaves)* ED: LEJA, F. • PUBL (**X** 1764) Ksiaznica Atlas: Warsaw iii+395pp
• DAT&PL 1929 Sep; Warsaw, PL • LC-No 32-17553

P 0797 Fonds des Math, Machines Math & Appl;1962 Tihany • H
[1965] *Colloque sur les Fondements des Mathematiques, les Machines Mathematiques, et leurs Applications* ED: KALMAR, L. • SER Collection de Logique Mathematique, Serie A 19 • PUBL (**X** 0928) Akad Kiado: Budapest 320pp • ALT PUBL (**X** 0834) Gauthier-Villars: Paris & (**X** 1313) Nauwelaerts: Louvain
• DAT&PL 1962 Sep; Tihany, H

P 1060 Constr Aspects Fund Thm Algeb;1967 Zuerich • CH
[1969] *Constructive Aspects of the Fundamental Theorem of Algebra. Proceedings of a Symposium Conducted at the IBM Research Laboratory.* ED: DEJON, B. & HENRICI, P. • PUBL (**X** 0827) Wiley & Sons: New York vii+337pp
• DAT&PL 1967 Jun; Zuerich, CH • ISBN 0-471-20300-9, LC-No 69-19380

P 1075 Logic Colloq;1976 Oxford • GB
[1977] *Logic Colloquium 76* ED: GANDY, R.O. & HYLAND, J.M.E. • SER (**S** 3303) Stud Logic Found Math 87 • PUBL (**X** 0809) North Holland: Amsterdam x+612pp • ALT PUBL (**X** 0838) Amer Elsevier: New York
• DAT&PL 1976 Jul; Oxford, GB • ISBN 0-7204-0691-9, LC-No 77-8943

P 1076 Latin Amer Symp Math Log (3);1976 Campinas • BR
[1977] *Non-Classical Logic, Model Theory and Computability. 3rd Latin American Symposium on Mathematical Logic* ED: ARRUDA, A.I. & COSTA DA, N.C.A & CHUAQUI, R. • SER (**S** 3303) Stud Logic Found Math 89 • PUBL (**X** 0809) North Holland: Amsterdam xviii+307pp • ALT PUBL (**X** 0838) Amer Elsevier: New York
• DAT&PL 1976 Jul; Campinas, BR • ISBN 0-7204-0752-4, LC-No 77-7366

P 1127 Symp Founds of Math;1962 Katada • J
[1963] *Proceedings of the Symposium on the Foundations of Mathematics* ED: TAKEUTI, G. • PUBL Tokyo University of Education: Tokyo vi+144pp
• DAT&PL 1962 Oct; Katada, J

P 1158 Int Congr Math (II,11,Comm Ind);1970 Nice • F
[1970] *Congres International des Mathematiciens 1970. Les 265 Communications Individuels* PUBL (**X** 0834) Gauthier-Villars: Paris vii+290pp
• DAT&PL 1970 Sep; Nice, F • LC-No 72-374601 • REL PUBL (**P** 0743) Int Congr Math (II,11,Proc);1970 Nice

P 1197 AFIPS Spring Jt Computer Conf (21);1962 San Francisco • USA
[1962] *AFIPS Conference Proceedings: 1962: Spring Joint Computer Conference* SER (**S** 3309) AFIPS Conference Proc 21 • PUBL (**X** 1354) Spartan Books : Sutton xv+392pp • ALT PUBL (**X** 0874) Nat Pr Books : Palo Alto
• DAT&PL 1962 May; San Francisco, CA, USA • LC-No 55-44701

P 1401 Math Founds of Comput Sci (5);1976 Gdansk • PL
[1976] *Mathematical Foundations of Computer Science. Proceedings of the 5th Symposium* ED: MAZURKIEWICZ, A. • SER (**S** 3302) Lect Notes Comput Sci 45 • PUBL (**X** 0811) Springer: Heidelberg & New York xi+606pp
• DAT&PL 1976 Sep; Gdansk, PL • ISBN 3-540-07854-1, LC-No 76-25494

P 1403 Skand Mat Kongr (10);1946 Copenhagen • DK
[1947] *den 10. Skandinavske Matematiker Kongress (Comptes-Rendus du 10eme Congres des Mathematiciens Scandinaves)* PUBL Jul. Gjellerups Forlag: Copenhagen xxv+383pp
• DAT&PL 1946 Aug; Copenhagen, DK

P 1440 ⊧ ISILC Proof Th Symp (Schuette);1974 Kiel • D
[1975] ⊧ *ISILC Proof Theory Symposium. Dedicated to Kurt Schuette on the Occasion of His 65th Birthday. Proceedings of the International Summer Institute and Logic Colloquium* ED: DILLER, J. & MUELLER, GERT H. • SER (**S** 3301) Lect Notes Math 500 • PUBL (**X** 0811) Springer: Heidelberg & New York viii+383pp
• DAT&PL 1974 Jul; Kiel, D • ISBN 3-540-07533-X, LC-No 75-40482 • REL PUBL (**P** 1442) ⊧ ISILC Logic Conf;1974 Kiel
• REM This Volume Contains Only the Proof Theory Part of the Conference.

P 1442 ⊧ ISILC Logic Conf;1974 Kiel • D
[1975] ⊧ *ISILC Logic Conference. Proceedings of the International Summer Institute and Logic Colloquium* ED: MUELLER, GERT H. & OBERSCHELP, A. & POTTHOFF, K. • SER (**S** 3301) Lect Notes Math 499 • PUBL (**X** 0811) Springer: Heidelberg & New York iv+651pp
• DAT&PL 1974 Jul; Kiel, D • ISBN 3-540-07534-8, LC-No 75-40431 • REL PUBL (**P** 1440) ⊧ ISILC Proof Th Symp (Schuette);1974 Kiel

P 1448 Math Founds of Comput Sci (2);1973 Strbske Pleso • CS
[1973] *Mathematical Foundations of Computer Science. Proceedings of the 2nd Symposium* ED: HAVEL, I.M. • PUBL (**X** 1773) Vydat Slov Akad: Bratislava 338pp
• DAT&PL 1973 Sep; Strbske Pleso, CS

P 1449 Automata Th & Formal Lang;1975 Kaiserslautern • D
[1975] *Automata Theory and Formal Languages. 2nd GI-Conference (Gesellschaft fuer Informatik)* ED: BRAKHAGE, H. • SER (**S** 3302) Lect Notes Comput Sci 33 • PUBL (**X** 0811) Springer: Heidelberg & New York viii+292pp
• DAT&PL 1975 May; Kaiserslautern, D • ISBN 3-540-07407-4, LC-No 75-28494

P 1455 Inform Processing (5);1971 Ljubljana • YU
[1972] *Information Processing '71. Proceedings of IFIP Congress* ED: FREIMAN, C.V. & GRIFFITH, J.E. & ROSENFELD, J.L. • PUBL (**X** 0809) North Holland: Amsterdam 2 Vols: xviii+1621pp • ALT PUBL (**X** 2403) AFIPS Pr: Montvale
• DAT&PL 1971 Aug; Ljubljana, YU • ISBN 0-7204-2063-6, LC-No 76-184997
• REM Vol 1: Foundations and Systems. Vol 2: Applications

P 1464 ACM Symp Th of Comput (6);1974 Seattle • USA
[1974] *Proceedings of 6th Annual ACM Symposium on Theory of Computing (Association for Computing Machinery)* PUBL (**X** 2205) ACM: New York iv+347pp
• DAT&PL 1974 Apr; Seattle, WA, USA • LC-No 82-642181

P 1475 Conv Inform Teor & Conv Strutt Corpi Algeb;1973 Roma • I
[1975] *Convegno di Informatica Teoretica. Convegno di Strutture in Corpi Algebrici* SER (**S** 3305) Symposia Matematica 15 • PUBL (**X** 3604) INDAM: Roma 604pp
• ALT PUBL (**X** 0801) Academic Pr: New York
• DAT&PL 1973 Feb; Rome, I, 1973 Apr; Rome, I

P 1476 Set Th & Hierarch Th (2) (Mostowski);1975 Bierutowice • PL
[1976] *Set Theory and Hierarchy Theory. A Memorial Tribute to Andrzej Mostowski. Proceedings of the 2nd Conference on Set Theory and Hierarchy Theory* ED: MAREK, W. & SREBRNY, M. & ZARACH, A. • SER (**S** 3301) Lect Notes Math 537 • PUBL (**X** 0811) Springer: Heidelberg & New York xiii+345pp
• DAT&PL 1975 Sep; Bierutowice, PL • ISBN 3-540-07856-8, LC-No 76-26536

P 1482 ACM Symp Th of Comput (5);1973 Austin • USA
[1973] *5th Annual ACM Symposium on Theory of Computing (Association for Computing Machinery)* PUBL (**X** 2205) ACM: New York iv+277pp
• DAT&PL 1973 May; Austin, TX, USA • LC-No 82-642181

P 1483 Austral Conf Combin Math (1);1972 Newcastle • AUS
[1972] *Proceedings of the 1st Australian Conference on Combinatorial Mathematics.* ED: WALLIS, J. & WALLIS, W.D. • PUBL Tunra: Newcastle x+239pp
• DAT&PL 1972 Jun; Newcastle, NSW, AUS • LC-No 74-160419

P 1484 Int Congr Math (2);1900 Paris • F
[1902] *Comptes Rendus du 2eme Congres International des Mathematiciens. Proces Verbaux et Communications* ED: DUPORCQ, E. • PUBL (**X** 0834) Gauthier-Villars: Paris 455pp
• DAT&PL 1900 Aug; Paris, F

P 1486 Handel Natuur- & Geneesk Congr (25);1935 Leiden • NL
[1935] *Handelingen van het XXVe Natuur- en Geneeskundig Congres* PUBL Ruygrok: Haarlem
• DAT&PL 1935 Apr; Leiden, NL

P 1511 Int Symp Th Progr;1972 Novosibirsk • SU
[1974] *International Symposium on Theoretical Programming* ED: ERSHOV, A.P. & NEPOMNYASHCHIJ, V.A. • SER (**S** 3302) Lect Notes Comput Sci 5 • PUBL (**X** 0811) Springer: Heidelberg & New York vi+407pp
• DAT&PL 1972 Aug; Novosibirsk, SU • ISBN 3-540-06720-5, LC-No 74-176124

P 1521 Int Congr Math (II,12);1974 Vancouver • CDN
[1975] *Proceedings of the International Congress of Mathematicians* ED: JAMES, R.D. • PUBL Canadian Mathematical Congress: Montreal 2 Vols: xlix+552pp,viii+600pp
• DAT&PL 1974 Aug; Vancouver, BC, CDN • ISBN 0-919558-04-6, LC-No 74-34533

P 1545 Easter Conf on Model Th (2);1984 Wittenberg • DDR
[1984] *Proceedings of the 2nd Easter Conference on Model Theory* SER (**S** 3382) Sem-ber, Humboldt-Univ Berlin, Sekt Math 60 • PUBL (**X** 2219) Humboldt-Univ Berlin: Berlin ii+243pp
• DAT&PL 1984 Apr; Wittenberg, DDR

P 1589 Math Interpr of Formal Systs;1954 Amsterdam • NL
[1955] *Mathematical Interpretations of Formal Systems* SER (**S** 3303) Stud Logic Found Math 10 • PUBL (**X** 0809) North Holland: Amsterdam viii+113pp
• DAT&PL 1954 Sep; Amsterdam, NL • LC-No 56-3127
• REM 2nd ed. 1971

P 1590 All-Union Conf Math Log (3);1974 Novosibirsk • SU
[1974] *3 Vsesoyuznaya Konferentsiya po Matematicheskoj Logike. Tezitsy Doklady i Soobshcheniya (3rd All-Union Conference on Mathematical Logic)*
• DAT&PL 1974 -?-; Novosibirsk, SU • LC-No 75-558515

P 1601 Easter Conf on Model Th (1);1983 Diedrichshagen • DDR
[1983] *Proceedings of the 1st Easter Conference on Model Theory* ED: DAHN, B.I. • SER (**S** 3382) Sem-ber, Humboldt-Univ Berlin, Sekt Math 49 • PUBL (**X** 2219) Humboldt-Univ Berlin: Berlin 154pp
• DAT&PL 1983 Apr; Diedrichshagen, DDR

P 1603 λ-Calc & Comput Sci Th;1975 Roma • I
[1975] *λ Calculus and Computer Science Theory* ED: BOEHM, C. • SER (**S** 3302) Lect Notes Comput Sci 37 • PUBL (**X** 0811) Springer: Heidelberg & New York xii+370pp
• DAT&PL 1975 Mar; Roma, I • ISBN 3-540-07416-3, LC-No 75-33375

P 1606 Colloq Math (Pascal);1962 Clermont-Ferrand • F
[1962] *Actes du Colloque de Mathematiques Reuni a Clermont a l'Occasion du Tricentenaire de la Mort de Blaise Pascal* SER (**J** 0179) Ann Fac Sci Clermont 7,8 • PUBL Univ Clermont, Fac. Sci.: Clermont 2 Vols: 123pp,189pp
• DAT&PL 1962 Jun; Clermont-Ferrand, F
• REM Vol 1: Introduction et Logique Mathematique. Vol 2: Calcul des Probabilites, Analyse Numerique et Calcul Automatique, Geometrie et Physique Mathematique

P 1608 Int Congr Math (II, 5);1936 Oslo • N
[1937] *Comptes Rendus du Congres International des Mathematiciens* PUBL A.M. Broeggers Boktrykkeri a/S: Oslo 2 Vols: 316pp,vv+289pp • ALT PUBL (**X** 3602) Kraus: Vaduz 1967 2 Vols
• DAT&PL 1936 Jul; Oslo, N • LC-No 52-1808

P 1618 ACM Symp Th of Comput (7);1975 Albuquerque • USA
[1975] *Proceedings of the 7th Annual ACM Symposium on the Theory of Computing (Association for Computing Machinery)* PUBL (**X** 2205) ACM: New York 265pp
• DAT&PL 1975 May; Albuquerque, NM, USA • LC-No 82-642181

P 1619 Coloq Log Simb;1975 Madrid • E
[1976] *Coloquio Sobre Logica Simbolica* PUBL Centro Calculo Univ. Complutense: Madrid 176pp
• DAT&PL 1975 Feb; Madrid, E • LC-No 77-555677

P 1628 Generalized Recursion Th (2);1977 Oslo • N
[1978] *Generalized Recursion Theory II* ED: FENSTAD, J.E. & GANDY, R.O. & SACKS, G.E. • SER (**S** 3303) Stud Logic Found Math 94 • PUBL (**X** 0809) North Holland: Amsterdam vii+466pp • ALT PUBL (**X** 0838) Amer Elsevier: New York
• DAT&PL 1977 Jun; Oslo, N • ISBN 0-444-85163-1, LC-No 78-5366

P 1629 Symp Math Log;1974 Oulo;1975 Helsinki • SF
[1977] *Proceedings of the Symposia on Mathematical Logic* ED: MIETTINEN, S. & VAEAENAENEN, J. • PUBL University of Helsinki, Department of Philosophy: Helsinki iv+103pp
• DAT&PL 1974 -?-; Oulo, SF, 1975 -?-; Helsinki, SF

P 1630 GI Fachtag Automatenth & Form Sprach (1);1973 Bonn • D
[1973] *1. GI-Fachtagung ueber Automatentheorie und Formale Sprachen. (Gesellschaft fuer Informatik)* ED: BOEHLING, K.H. & INDERMARK, K. • SER (**S** 3302) Lect Notes Comput Sci 2 • PUBL (**X** 0811) Springer: Heidelberg & New York vii+322pp
• DAT&PL 1973 Jul; Bonn, D • ISBN 3-540-06527-X, LC-No 74-649668

P 1632 Automata, Lang & Progr (4);1977 Turku SF • SF [1977] *Automata, Languages and Programming. 4th Colloquium* ED: SALOMAA, A. & STEINBY, M. • SER (S 3302) Lect Notes Comput Sci 52 • PUBL (X 0811) Springer: Heidelberg & New York x+569pp • DAT&PL 1977 Jul; Turku, SF • ISBN 3-540-08342-1, LC-No 78-337949 • REM Also Abbreviated as ICALP 77

P 1635 Math Founds of Comput Sci (6);1977 Tatranska Lomnica • CS [1977] *Mathematical Foundations of Computer Science. Proceedings of the 6th Symposium* ED: GRUSKA, J. • SER (S 3302) Lect Notes Comput Sci 53 • PUBL (X 0811) Springer: Heidelberg & New York xi+595pp • DAT&PL 1977 Sep; Tatranska Lomnica, CS • ISBN 3-540-08353-7, LC-No 77-10135

P 1645 Oesterr Math Kongr (4);1956 Wien • A [1956] *4ter Oesterreichischer Mathematikerkongress* PUBL Oesterr. Mathematische Gesellschaft: Wien • DAT&PL 1956 Sep; Wien, A

P 1660 Congr Interbalkan Math;1934 Athens • GR [1935] *Actes du Congres Interbalkanique de Mathematique* PUBL Imprimerie Nationale: Athens • DAT&PL 1934 Sep; Athens, GR

P 1675 Summer Inst Symb Log;1957 Ithaca • USA [1957] *Summaries of Talks Presented at the Summer Institute for Symbolic Logic* PUBL Institute for Defense Analyses, Communications Research Division: Princeton; xvi+427pp • DAT&PL 1957 Jul; Ithaca, NY, USA • LC-No 65-4418 • REM 2nd ed. 1960

P 1694 Inform Processing (7);1977 Toronto • CDN [1977] *Information Processing '77. Proceedings of IFIP Congress* ED: GILCHRIST, B. • SER IFIP Congress Series 7 • PUBL (X 0809) North Holland: Amsterdam xix+1004pp • DAT&PL 1977 Aug; Toronto, ON, CDN • ISBN 0-7204-0755-9, LC-No 77-80624

P 1695 Set Th & Hierarch Th (3);1976 Bierutowice • PL [1977] *Set Theory and Hierarchy Theory V. Proceedings of the 3rd Conference on Set Theory and Hierarchy Theory* ED: LACHLAN, A. & SREBRNY, M. & ZARACH, A. • SER (S 3301) Lect Notes Math 619 • PUBL (X 0811) Springer: Heidelberg & New York viii+358pp • DAT&PL 1976 Sep; Bierutowice, PL • ISBN 3-540-08521-1, LC-No 78-309663

P 1704 Int Congr Log, Meth & Phil of Sci (5);1975 London ON • CDN [1977] *Proceedings of 5th International Congress of Logic, Methodology and Philosophy of Science* ED: BUTTS, R.E. & HINTIKKA, J. • SER (S 3308) Univ Western Ontario Ser in Philos of Sci 9-12 • PUBL (X 0835) Reidel: Dordrecht 4 Vols: x+406pp, x+427pp, x+321pp, x+336pp • DAT&PL 1975 Aug; London, ON, CDN • ISBN 90-277-0708-1 (V1), ISBN 90-277-0710-3 (V2), ISBN 90-277-0829-0 (V3), ISBN 90-277-0831-2 (V4), ISBN 90-277-0706-5 (Set of the 4 Vols), LC-No 77-22429 (V1), LC-No 77-22431 (V2), LC-No 77-22432 (V3), LC-No 77-22433 (V4) • REM Vol 1: Logic, Foundations of Mathematics, and Computability Theory. Vol 2: Foundational problems in the Special Sciences. Vol 3: Basic Problems in Methodology and Linguistics. Vol 4: Historical and Philosophical Dimensions of Logic, Methodology and Philosophy of Science.

P 1705 Scand Logic Symp (4);1976 Jyvaeskylae • SF [1979] *Essays on Mathematical and Philosophical Logic. Proceedings of the 4th Scandinavian Logic Symposium and of the 1st Soviet-Finnish Logic Conference* ED: HINTIKKA, J. & NIINILUOTO, I. & SAARINEN, E. • SER (S 3307) Synth Libr 122 • PUBL (X 0835) Reidel: Dordrecht viii+462pp • DAT&PL 1976 Jun; Jyvaeskylae, SF • ISBN 90-277-0879-7, LC-No 78-14736

P 1729 Colloq Int Log;1975 Clermont-Ferrand • F [1977] *Colloque International de Logique* SER (S 1802) Colloq Int CNRS 249 • PUBL (X 0999) CNRS Inst B Pascal: Paris 224pp • DAT&PL 1975 Jul; Clermont-Ferrand, F • ISBN 2-222-02019-0, LC-No 78-367483

P 1730 Two Decades Math Netherlands 1920-40;1978 Amsterdam • NL [1978] *Two Decades of Mathematics in the Netherlands 1920-1940. A Retrospection on the Occasion of the Bicentennial of the Wiskundig Genootschap. Part 1,2.* ED: BERTIN, E.M.J. & BOS, H.J.M. & GROOTENDORST, A.W. • PUBL (X 1121) Math Centr: Amsterdam 2 Vols • DAT&PL 1978 Mar; Amsterdam, NL

P 1740 ACM Symp Th of Comput (10);1978 San Diego • USA [1978] *Conference Record of the 10th Annual ACM Symposium on Theory of Computing (Association for Computing Machinery)* PUBL (X 2205) ACM: New York 346pp • DAT&PL 1978 May; San Diego, CA, USA • LC-No 79-101797

P 1748 Lang & Pensee Math;1976 Luxembourg • L [1976] *Langage et Pensee Mathematiques. Actes du Colloque International* PUBL (X 3203) Cent Univ Luxembourg: Luxembourg 454pp • DAT&PL 1976 Jun; Luxembourg, L • LC-No 79-380450

P 1753 Int Congr AICA Hybrid Comput (7); • USA [1973] *Proceedings of the 7th International Congress of AICA on Hybrid Computation, Part 1*

P 1766 Gen Topol & Rel to Mod Anal & Algeb (1);1961 Prague • CS [1962] *General Topology and Its Relations to Modern Analysis and Algebra I. Proceedings of the 1st Prague Topological Symposium* ED: NOVAK, J. • PUBL (X 1034) PWN: Warsaw 363pp • ALT PUBL (X 0801) Academic Pr: New York • DAT&PL 1961 Sep; Prague, CS • LC-No 63-751

P 1789 Congr Soc Espan Hist de Cien (1);1978 Madrid • E [1980] *1 Congreso de la Sociedad Espanola de Historia de las Ciencias. El Cientifico Espanol ante su Historia. La Ciencia on Espana entre 1750-1850.* ED: GARMA, S. • PUBL Diput Prov Madrid, Serv Extensio Cult y Divulgacion: Madrid 589pp • DAT&PL 1978 Dec; Madrid, E • ISBN 84-500-4142-2, LC-No 81-162317

P 1792 Struct Learning;1968 Philadelphia • USA [1973-1976] *Structural Learning* ED: SCANDURA, J.M. • SER Structural Learning Series • PUBL (X 0836) Gordon & Breach: New York 2 Vols • DAT&PL 1968 Apr; Philadelphia, PA, USA • ISBN 0-677-15110-1 (V2), LC-No 73-76710

P 1800 Brazil Conf Math Log (1);1977 Campinas • BR [1978] *Proceedings of 1st Brazilian Conference on Mathematical Logic* ED: ARRUDA, A.I. & CHAQUI, R. & COSTA DA, N.C.A. • SER (S 3310) Lect Notes Pure Appl Math 39 • PUBL (X 1684) Dekker: New York xii+303pp • DAT&PL 1977 Jul; Campinas, BR • LC-No 78-14488

P 1805 Int Symp Multi-Val Log (5,Proc);1975 Bloomington • USA
[1975] *Proceedings of the 1975 International Symposium on Multiple-Valued Logic* PUBL (**X** 2179) IEEE: New York iv+475pp
• DAT&PL 1975 May; Bloomington, IN, USA • LC-No 76-370321 • REL PUBL (**P** 1894) Int Symp Multi-Val Log (5,Inv Pap);1975 Bloomington

P 1841 Fct Recurs & Appl;1967 Tihany • H
[1969] *Les Fonctions Recursives et leurs Applications* PUBL (**X** 0999) CNRS Inst B Pascal: Paris • ALT PUBL (**X** 3725) Bolyai Janos Mat Tars: Budapest
• DAT&PL 1967 Sep; Tihany, H

P 1864 Higher Set Th;1977 Oberwolfach • D
[1978] *Higher Set Theory* ED: MUELLER, GERT H. & SCOTT, D.S. • SER (**S** 3301) Lect Notes Math 669 • PUBL (**X** 0811) Springer: Heidelberg & New York xii+476pp
• DAT&PL 1977 Apr; Oberwolfach, D • ISBN 3-540-08926-8, LC-No 79-312135

P 1869 Automata, Lang & Progr (2);1974 Saarbruecken • D
[1974] *Automata, Languages and Programming: 2nd Colloquium* ED: LOECKX, J. • SER (**S** 3302) Lect Notes Comput Sci 14 • PUBL (**X** 0811) Springer: Heidelberg & New York viii+611pp
• DAT&PL 1974 Jul; Saarbruecken, D • ISBN 3-540-06841-4, LC-No 74-180345
• REM Also Abbreviated as ICALP 74

P 1870 Automata, Lang & Progr (3);1976 Edinburgh • GB
[1976] *Automata, Languages and Programming. 3rd Colloquium* ED: MICHAELSON, S. & MILNER, R. • PUBL (**X** 1261) Edinburgh Univ Pr: Edinburgh vi+559pp
• DAT&PL 1976 Jul; Edinburgh, GB • ISBN 0-85224-308-1, LC-No 77-359145
• REM Also Abbreviated as ICALP 76

P 1872 Automata, Lang & Progr (5);1978 Udine • I
[1978] *Automata, Languages and Programming. 5th Colloquium* ED: AUSIELLO, G. & BOEHM, C. • SER (**S** 3302) Lect Notes Comput Sci 62 • PUBL (**X** 0811) Springer: Heidelberg & New York viii+508pp
• DAT&PL 1978 Jul; Udine, I • ISBN 3-540-08860-1, LC-No 79-303999
• REM Also Abbreviated as ICALP 78

P 1873 Automata, Lang & Progr (6);1979 Graz • A
[1979] *Automata, Languages and Programming. 6th Colloquium* ED: MAURER, H.A. • SER (**S** 3302) Lect Notes Comput Sci 71 • PUBL (**X** 0811) Springer: Heidelberg & New York ix+682pp
• DAT&PL 1979 Jul; Graz, A • ISBN 3-540-09510-1, LC-No 79-15859
• REM Also Abbreviated as ICALP 79

P 1894 Int Symp Multi-Val Log (5,Inv Pap);1975 Bloomington • USA
[1977] *Modern Uses of Multiple-Valued Logic. Invited Papers from the 5th International Symposium on Multiple-Valued Logic* ED: DUNN, J.M. & EPSTEIN, G. • SER Epistime 2 • PUBL (**X** 0835) Reidel: Dordrecht x+338pp
• DAT&PL 1975 May; Bloomington, IN, USA • ISBN 90-277-0747-2, LC-No 77-23098 • REL PUBL (**P** 1805) Int Symp Multi-Val Log (5,Proc);1975 Bloomington
• REM With a Bibliography of Many-Valued Logic

P 1897 Logic Colloq;1977 Wroclaw • PL
[1978] *Logic Colloquium 77* ED: MACINTYRE, A. & PACHOLSKI, L. & PARIS, J. • SER (**S** 3303) Stud Logic Found Math 96 • PUBL (**X** 0809) North Holland: Amsterdam x+311pp
• DAT&PL 1977 Aug; Wroclaw, PL • ISBN 0-444-85178-X, LC-No 78-13396

P 1903 ACM Symp Th of Comput (9);1977 Boulder • USA
[1977] *Conference Records of the 9th Annual ACM Symposium on the Theory of Computing (Association for Computing Machinery)* PUBL (**X** 2205) ACM: New York v+314pp
• DAT&PL 1977 May; Boulder, CO, USA • LC-No 82-642181

P 1947 Bicent Congr Wisk Genootschap;1978 Amsterdam • NL
[1979] *Proceedings of the Bicentennial Congress of the Wiskundig Genootschap, Part I,II* ED: BAAYEN, P.C. & DULST VAN, D. & OOSTERHOFF, J. • SER (**S** 1605) Math Centr Tracts 100,101 • PUBL (**X** 1121) Math Centr: Amsterdam xiv+431pp
• DAT&PL 1978 Mar; Amsterdam, NL • ISBN 90-6196-168-8 (V1), ISBN 90-6196-169-6 (V2), LC-No 79-317324

P 1953 Int Congr Log, Meth & Phil of Sci (1;Abstr);1960 Stanford • USA
[1960] *1st International Congress for Logic, Methodology and Philosophy of Science. Abstracts*
• DAT&PL 1960 Aug; Stanford, CA, USA

P 1959 Int Congr Math (II,13);1978 Helsinki • SF
[1980] *Proceedings of the International Congress of Mathematicians* ED: LEHTO, O. • PUBL Academia Scientiarum Fennica: Helsinki 2 Vols: 1022pp
• DAT&PL 1978 Aug; Helsinki, SF • ISBN 951-41-0352-1

P 2011 Int Symp Multi-Val Log (6);1976 Logan • USA
[1976] *Proceedings of the 6th International Symposium on Multiple-Valued Logic* PUBL (**X** 2205) ACM: New York vii+273pp • ALT PUBL (**X** 2179) IEEE: New York
• DAT&PL 1976 May; Logan, UT, USA • LC-No 79-641110
• REM IEEE Publication No. 76CH1111-4C

P 2032 Stud on Polyautomata;1976 Zushi & Kyoto • J
[1977] *Studies on Polyautomata* ED: NISHIO, H.
• DAT&PL 1976 Aug; Zushi, J, 1977 Feb; Kyoto, J

P 2058 Kleene Symp;1978 Madison • USA
[1980] *The Kleene Symposium* ED: BARWISE, K.J. & KEISLER, H.J. & KUNEN, K. • SER (**S** 3303) Stud Logic Found Math 101 • PUBL (**X** 0809) North Holland: Amsterdam xx+425pp
• DAT&PL 1978 Jun; Madison, WI, USA • ISBN 0-444-85345-6, LC-No 79-20792

P 2059 Math Founds of Comput Sci (8);1979 Olomouc • CS
[1979] *Mathematical Foundations of Computer Science. Proceedings of the 8th Symposium* ED: BECVAR, J. • SER (**S** 3302) Lect Notes Comput Sci 74 • PUBL (**X** 0811) Springer: Heidelberg & New York ix+580pp
• DAT&PL 1979 Sep; Olomouc, CS • ISBN 3-540-09526-8, LC-No 79-17801

P 2064 All-Union Conf Math Log (4);1976 Kishinev • SU
[1976] *4 Vsesoyuznaya Konferentsiya po Matematicheskoj Logike. Tezitsy Doklady i Soobshcheniya (4th All-Union Conference on Mathematical Logic)* ED: KUZNETSOV, A.V.
• PUBL (**X** 2741) Shtiintsa: Kishinev 170pp
• DAT&PL 1976 -?-; Kishinev, SU • LC-No 78-410667

P 2080 Conf Math Log;1970 London • GB
[1972] *Conference on Mathematical Logic - London '70* ED: HODGES, W. • SER (**S** 3301) Lect Notes Math 255 • PUBL (**X** 0811) Springer: Heidelberg & New York viii+351pp
• DAT&PL 1970 Aug; London, GB • ISBN 3-540-05744-7, LC-No 70-189457

P 2153 Logic Colloq;1983 Aachen • D
[1984] *Proceedings of the Logic Colloquium, Part 1, 2* ED: MUELLER, GERT H. & RICHTER, M.M. (V1). BOERGER, E. & OBERSCHELP, W. & RICHTER, M.M. & SCHINZEL, B. & THOMAS, W. (V2) • SER (**S** 3301) Lect Notes Math 1103,1104
• PUBL (**X** 0811) Springer: Heidelberg & New York 2 Vols: viii+484pp,viii+475pp
• DAT&PL 1983 Jul; Aachen, D • ISBN 3-540-13900-1 (V1), ISBN 3-540-13901-X (V2), LC-No 84-26704
• REM Vol 1: Models and Sets. Vol 2: Computation and Proof Theory.

P 2160 Latin Amer Symp Math Log (6);1983 Caracas • YV
[1985] *Methods in Mathematical Logic. Proceedings of the 6th Latin American Symposium on Mathematical Logic* ED: PRISCO DI, C.A. • SER (**S** 3301) Lect Notes Math 1130
• PUBL (**X** 0811) Springer: Heidelberg & New York vii+407pp
• DAT&PL 1983 Aug; Caracas, YV • ISBN 3-540-15236-9, LC-No 85-14779

P 2167 Orders: Descr & Roles;1982 L'Arbresle • F
[1984] *Orders: Description and Roles in Set Theory, Lattices, Ordered Groups, Topology, Theory of Models and Relations, Combinatorics, Effectiveness, Social Sciences. Proceedings of a Conference on Ordered Sets and Their Applications* ED: POUZET, M. & RICHARD, D. • SER (**S** 3358) Ann Discrete Math 23, North Holland Mathematics Studies 99 • PUBL (**X** 0809) North Holland: Amsterdam xxviii+548pp
• DAT&PL 1982 Jul; L'Arbresle, F • ISBN 0-444-87601-4, LC-No 84-13749

P 2180 Math Appl Categ Th;1983 Denver • USA
[1984] *Mathematical Applications of Category Theory. Proceedings of the Special Session of the 89th Annual Meeting of the American Mathematical Society (AMS)* ED: GRAY, J.W.
• SER (**S** 3313) Contemp Math 30 • PUBL (**X** 0803) Amer Math Soc: Providence vii+307pp
• DAT&PL 1983 Jan; Denver, CO, USA • ISBN 0-8218-5032-6, LC-No 84-9371

P 2251 Mat Log & Primen;1960 Stanford • USA
[1965] *Matematicheskaya Logika i Ee Primeneniya: Sbornik Statei (Mathematical Logic and Its Applications. Logic, Methodology and Philosophy of Science)* ED: MAL'TSEV, A.I. & NAGEL, E. & SUPPES, P. & TARSKI, A. • PUBL (**X** 0885) Mir: Moskva 341pp
• DAT&PL 1960 Aug; Stanford, CA, USA
• TRANSL OF [1962] (**P** 0612) Int Congr Log, Meth & Phil of Sci (1,Proc);1960 Stanford
• REM Only Parts of P0612 are translated.

P 2268 Int Colloq Philos of Sci;1965 London • GB
[1967-1970] *Proceedings of the International Colloquium in the Philosophy of Science. 4 Volumes* ED: LAKATOS, I. (V 1 - V 4) & MUSGRAVE, A. (V 3, V 4) • SER (**S** 3303) Stud Logic Found Math V 1 - V 3 • PUBL (**X** 0809) North Holland: Amsterdam V 1: xv+241pp, V 2: viii+420pp, V 3: ix+448pp, (**X** 0805) Cambridge Univ Pr: Cambridge, GB V 4: viii+282pp
• DAT&PL 1965, Jul; London, GB • ISBN 0-521-07826-1 (V4), LC-No 67-20007 (V1), LC-No 67-28648 (V2), LC-No 67-28649 (V3), LC-No 78-105496 (V4)
• REM Vol 1: Problems in the Philosophy of Mathematics. Vol 2: The Problem of Inductive Logic. Vol 3: Problems in the Philosophy of Science. Vol 4: Criticism and the Growth of Knowledge.

P 2335 Canad Math Congr (25);1971 Thunder Bay • CDN
[1971] *Proceedings of the 25th Summer Meeting of the Canadian Mathematical Congress* ED: EAMES, W.R. & STANTON, R.G. & THOMAS, R.S.D. • PUBL Lakehead University: Thunder Bay viii+648pp
• DAT&PL 1971 Jun; Thunder Bay, ON, CDN • LC-No 48-791

P 2342 Symp Rek Kombin;1983 Muenster • D
[1984] *Logic and Machines: Decision Problems and Complexity. Proceedings of the Symposium Rekursive Kombinatorik* ED: BOERGER, E. & HASENJAEGER, G. & ROEDDING, D. • SER (**S** 3302) Lect Notes Comput Sci 171
• PUBL (**X** 0811) Springer: Heidelberg & New York vi+456pp
• DAT&PL 1983 May; Muenster, D • ISBN 3-540-13331-3, LC-No 84-55

P 2534 Konf Molod Special Vychisl (1);1969 Erevan • SU
[1969] *Trudy 1 Konferentsiya Molodych Spetsialistov Vychisleniya (Proceedings of the 1st Conference of Young Computer Specialists)* PUBL (**X** 2225) Akad Nauk Armyan SSR : Erevan
• DAT&PL 1969 Apr; Erevan, SU

P 2539 Frege Konferenz (1);1979 Jena • DDR
[1979] *"Begriffsschrift". Jenaer Frege-Konferenz* ED: BOLCK, F. • PUBL (**X** 2211) Schiller Univ: Jena iii+548pp
• DAT&PL 1979 May; Jena, DDR

P 2554 All-Union Symp Th Log Infer;1974 Moskva • SU
[1979] *Logicheskij Vyvod. (Logical Inference. Proceedings of the All-Union Symposium on the Theory of Logical Inference.)* ED: SMIRNOV, V.A. • PUBL (**X** 2027) Nauka: Moskva 312pp
• DAT&PL 1974 Mar; Moskva, SU • LC-No 79-380844

P 2558 All-Union Conf Math Log (5) (Mal'tsev);1979 Novosibirsk • SU
[1979] *5 Vsesoyuznaya Konferentsiya po Matematicheskoj Logike. Tezitsy Doklady i Soobshcheniya (5th All-Union Conference on Mathematical Logic. Dedicated to the 70th Anniversary of the Academician A.I. Mal'tsev)*
• DAT&PL 1979 -?-; Novosibirsk, SU

P 2570 Intens Log & Log Analiz Estestv Yazyk;1979 Moskva • SU
[1979] *2. Sovetsko-Finskij Kollokvium "Intensional'nye Logiki i Logicheskij Analiz Estestvennykh Yazykov" (Intensional Logics and Logic Analysis of Natural Languages)* PUBL (**X** 3353) Akad Nauk SSSR Inst Istor Filol & Filos: Novosibirsk
• DAT&PL 1979 -?-; Moskva, SU

P 2588 FCT'77 Fund of Comput Th;1977 Poznan • PL
[1977] *Fundamentals of Computation Theory. Proceedings of the International FCT '77 - Conference* ED: KARPINSKI, M. • SER (S 3302) Lect Notes Comput Sci 56 • PUBL (X 0811) Springer: Heidelberg & New York 542pp
• DAT&PL 1977 Sep; Poznan-Kornik, PL • ISBN 3-540-08442-8, LC-No 77-14022

P 2599 CAAP'78 Arbres en Algeb & Progr (3);1978 Lille • F
[1978] *Les Arbres en Algebre et en Programmation. 3eme Colloquium* PUBL Universite de Lille I: Lille ii+253pp
• DAT&PL 1978 Feb; Lille, F

P 2614 Open Days in Model Th & Set Th;1981 Jadwisin • PL
[1981] *Proceedings of the International Conference "Open Days in Model Theory and Set Theory"* ED: GUZICKI, W. & MAREK, W. & PELC, A. & RAUSZER, C. • PUBL Selbstverlag 333pp
• DAT&PL 1981 Sep; Jadwisin, PL
• REM Available from John Derrick, School of Mathematics, University of Leeds, GB, or from Cecylia Rauszer, Institute of Mathematics, University of Warsaw, PL

P 2615 Scand Logic Symp (5);1979 Aalborg • DK
[1979] *Proceedings of the 5th Scandinavian Logic Symposium* ED: JENSEN, F.V. & MAYOH, B.H. & MOLLER, K.K. • PUBL (X 2646) Aalborg Univ Pr: Aalborg vii+361pp
• DAT&PL 1979 Jan; Aalborg, DK • ISBN 87-7307-037-8, LC-No 80-464603

P 2625 Model Th of Algeb & Arithm;1979 Karpacz • PL
[1980] *Model Theory of Algebra and Arithmetic. Proceedings of the Conference on Applications of Logic to Algebra and Arithmetic* ED: PACHOLSKI, L. & WIERZEJEWSKI, J. & WILKIE, A.J. • SER (S 3301) Lect Notes Math 834 • PUBL (X 0811) Springer: Heidelberg & New York vi+410pp
• DAT&PL 1979 Sep; Karpacz, PL • ISBN 3-540-10269-8, LC-No 82-131180

P 2627 Logic Colloq;1978 Mons • B
[1979] *Logic Colloquium '78* ED: BOFFA, M. & DALEN VAN, D. & MCALOON, K. • SER (S 3303) Stud Logic Found Math 97
• PUBL (X 0809) North Holland: Amsterdam x+434pp
• DAT&PL 1978 Aug; Mons, B • LC-No 79-21152

P 2628 Log Year;1979/80 Storrs • USA
[1981] *Logic Year 1979-80. The University of Connecticut* ED: LERMAN, M. & SCHMERL, J.H. & SOARE, R.I. • SER (S 3301) Lect Notes Math 859 • PUBL (X 0811) Springer: Heidelberg & New York viii+326pp
• DAT&PL 1979 Nov; Storrs, CT, USA • ISBN 3-540-10708-8, LC-No 81-5628

P 2633 Autom Deduct (7);1984 Napa • USA
[1984] *7th International Conference on Automated Deduction* ED: SHOSTAK, R.E. • SER (S 3302) Lect Notes Comput Sci 170 • PUBL (X 0811) Springer: Heidelberg & New York vi+508pp
• DAT&PL 1984 May; Napa, CA, USA • ISBN 3-540-96022-8, LC-No 84-5441

P 2901 Appl Sheaves;1977 Durham • GB
[1979] *Applications of Sheaves. Proceedings of the Research Symposium on Applications of Sheaf Theory to Logic, Algebra and Analysis* ED: FOURMAN, M.P. & MULVEY, C.J. & SCOTT, D.S. • SER (S 3301) Lect Notes Math 753 • PUBL (X 0811) Springer: Heidelberg & New York xiv+779pp
• DAT&PL 1977 Jul; Durham, GB • ISBN 3-540-09564-0, LC-No 79-23219

P 2902 Aspects Effective Algeb;1979 Clayton • AUS
[1981] *Aspects of Effective Algebra. Proceedings of a Conference at Monash University, Australia* ED: CROSSLEY, J.N. • PUBL (X 2863) Upside Down A Book: Yarra Glen x+290pp
• DAT&PL 1979 Aug; Clayton, Vic, AUS • ISBN 0-949865-01-X

P 2903 Automata, Lang & Progr (8);1981 Akko • IL
[1981] *Automata, Languages and Programming. 8th Colloquium* ED: EVEN, S. & KARIV, O. • SER (S 3302) Lect Notes Comput Sci 115 • PUBL (X 0811) Springer: Heidelberg & New York viii+552pp
• DAT&PL 1981 Jul; Akko, IL • ISBN 3-540-10843-2, LC-No 81-9053
• REM Also Abbreviated as ICALP 81

P 2904 Automata, Lang & Progr (7);1980 Noordwijkerhout • NL
[1980] *Automata, Languages and Programming. 7th Colloquium* ED: BAKKER DE, J.W. & LEEUWEN VAN, J. • SER (S 3302) Lect Notes Comput Sci 85 • PUBL (X 0811) Springer: Heidelberg & New York viii+671pp
• DAT&PL 1980 Jul; Noordwijkerhout, NL • ISBN 3-540-10003-2, LC-No 81-156919
• REM Also Abbreviated as ICALP 80

P 2923 CAAP'81 Arbres en Algeb & Progr (6);1981 Genova • I
[1981] *CAAP '81. Les Arbres en Algebre et en Programmation. 6eme Colloque* ED: ASTESIANO, E. & BOEHM, C. • SER (S 3302) Lect Notes Comput Sci 112 • PUBL (X 0811) Springer: Heidelberg & New York vi+364pp
• DAT&PL 1981 Mar; Genova, I • ISBN 3-540-10828-9, LC-No 81-8959

P 2929 Form Descr of Progr Concepts (1);1977 St.Andrews • CDN
[1978] *Formal Description of Programming Concepts. Proceedings of the IFIP Working Conference* ED: NEUHOLD, E.J. • PUBL (X 0809) North Holland: Amsterdam xviii+648pp
• DAT&PL 1977 Aug; St. Andrews, NB, CDN • ISBN 0-444-85107-0, LC-No 81-455275

P 2930 Formal of Progr Concepts;1981 Peniscola • E
[1981] *Formalization of Programming Concepts.* ED: DIAZ, J. & RAMOS, I. • SER (S 3302) Lect Notes Comput Sci 107
• PUBL (X 0811) Springer: Heidelberg & New York vii+478pp
• DAT&PL 1981 Apr; Peniscola, E • ISBN 3-540-10699-5, LC-No 81-5715

P 2935 FCT'79 Fund of Comput Th;1979 Berlin/Wendisch-Rietz • DDR
[1979] *Fundamentals of Computation Theory - FCT '79. Proceedings of the Conference on Algebraic, Arithmetic, and Categorical Methods in Computation Theory* ED: BUDACH, L.
• SER Mathematical Research - Mathematische Forschung 2 • PUBL (X 0911) Akademie Verlag: Berlin 576pp
• DAT&PL 1979 Sep; Berlin/Wendisch-Rietz, DDR • LC-No 82-460828

P 2952 CAAP'79 Arbres en Algeb & Progr (4);1979 Lille • F
[1979] *Les Arbres en Algebre et en Programmation. 4eme Colloquium* PUBL Universite de Lille I: Lille iv+327pp
• DAT&PL 1979 Feb; Lille, F

P 2957 Math Dev from Hilbert Probl;1974 DeKalb • USA
[1976] *Mathematical Developments Arising from Hilbert Problems* ED: BROWDER, F.E. • SER (S 3304) Proc Symp Pure Math 28 • PUBL (X 0803) Amer Math Soc: Providence xii+628pp
• DAT&PL 1974 May; DeKalb, IL, USA • LC-No 76-20437

P 2958 Latin Amer Symp Math Log (4);1978 Santiago • RCH
[1980] *Mathematical Logic in Latin America. Proceedings of the 4th Latin American Symposium on Mathematical Logic* ED: ARRUDA, A.I. & CHUAQUI, R. & COSTA DA, N.C.A. • SER (S 3303) Stud Logic Found Math 99 • PUBL (X 0809) North Holland: Amsterdam xii+392pp
• DAT&PL 1978 Dec; Santiago, RCH • ISBN 0-444-85402-9, LC-No 79-20797

P 2974 New Concepts & Tech in Parall Inform Process;1973 Capri • I
[1975] *New Concepts and Technologies in Parallel Information Processing. Proceedings of the NATO Advanced Study Institute* ED: CAIANIELLO, E.R. • SER NATO Advanced Study Institute, Series E 9 • PUBL (X 1317) Noordhoff: Groningen ix+401pp
• DAT&PL 1973 Jun; Capri, I • ISBN 90-286-0553-3, LC-No 75-319021

P 2989 Log of Progr; 1983 Pittsburgh • USA
[1984] *Logics of Programs* ED: CLARKE, E. & KOZEN, D. • SER (S 3302) Lect Notes Comput Sci 164 • PUBL (X 0811) Springer: Heidelberg & New York vi+527pp
• DAT&PL 1983 Jan; Pittsburgh, PA, USA • ISBN 3-540-12896-4, LC-No 84-3123

P 3013 Progr Symp;1974 Paris • F
[1974] *Programming Symposium. Proceedings. Colloque sur la Programmation* ED: ROBINET, B. • SER (S 3302) Lect Notes Comput Sci 19 • PUBL (X 0811) Springer: Heidelberg & New York v+425pp
• DAT&PL 1974 Apr; Paris, F • ISBN 3-540-06859-7, LC-No 75-19256

P 3035 Ved Konf Vysok Shkol Doprav;1979 Zhilina • CS
[1979] *6 Konferentsiya Vysokej Shkoly Dopravnej (6th Scientific Conference of the College of Transport Engineering)* ED: GRESHAK, P. • PUBL College of Transport Engineering:Zhilina 201pp
• DAT&PL 1979 -?-; Zhilina, CS

P 3057 CAAP'80 Arbres en Algeb & Progr (5);1980 Lille • F
[1980] *Les Arbres en Algebre et en Programmation. 5eme Colloquium* PUBL Universite de Lille I: Lille vi+268pp
• DAT&PL 1980 Feb; Lille, F

P 3060 Tagung Geom & Anw;1977 Gaussig • DDR
[1978] *Vortraege von der Tagung Geometrie und Anwendungen* SER Weiterbildungszentrum fuer Mathematische Kybernetik und Rechentechnik; 28 • PUBL Tech. Univ. Dresden:Dresden 142pp
• DAT&PL 1977 Oct; Gaussig, DDR

P 3062 IEEE Symp Switch & Automata Th (14);1973 Iowa City • USA
[1973] *14th Annual IEEE Symposium on Switching and Automata Theory* PUBL (X 2179) IEEE: New York v+213pp
• DAT&PL 1973 Oct; Iowa City, IA, USA • LC-No 80-646635

P 3063 Autom Deduct (5);1980 Les Arcs • F
[1980] *5th Conference on Automated Deduction* ED: BIBEL, W. & KOWALSKI, R. • SER (S 3302) Lect Notes Comput Sci 87 • PUBL (X 0811) Springer: Heidelberg & New York vii+385pp
• DAT&PL 1980 Jul; Les Arcs, F • ISBN 3-540-10009-1, LC-No 80-18708

P 3078 Form Semant Progr Lang, Courant Comput Sci (2);1970 New York • USA
[1972] *Formal Semantics of Programming Languages. Courant Computer Science Symposium 2* ED: RUSTIN, R. • SER Prentice-Hall Series in Automatic Computation • PUBL (X 0819) Prentice Hall: Englewood Cliffs xiii+248pp
• DAT&PL 1970 Sep; New York, NY, USA • LC-No 75-39372

P 3083 FCT'83 Found of Comput Th (Sel Pap);1983 Borgholm • S
[1985] *Topics in the Theory of Computation. Selected Papers of the International Conference on Foundations of Computation Theory, FCT '83* ED: KARPINSKI, M. & LEEUWEN VAN, J. • SER (S 3358) Ann Discrete Math 24, North-Holland Math Stud 102 • PUBL (X 0809) North Holland: Amsterdam ix+187pp
• DAT&PL 1983 Aug; Borgholm, S • LC-No 84-21089

P 3087 Germ Worksh on Artif Intell (7);1983 Dassel • D
[1983] *GWAI-83. Proceedings of the 7th German Workshop on Artificial Intelligence* ED: NEUMANN, B. • SER Informatik-Fachberichte 76 • PUBL (X 0811) Springer: Heidelberg & New York vi+240pp
• DAT&PL 1983 Sep; Dassel/Solling, D • ISBN 3-540-12871-9, LC-No 84-130170

P 3090 Semant of Data Types;1984 Sophia-Antipolis • F
[1984] *Semantics of Data Types* ED: KAHN, G. & MACQUEEN, D.B. & PLOTKIN, G. • SER (S 3302) Lect Notes Comput Sci 173 • PUBL (X 0811) Springer: Heidelberg & New York vi+391pp
• DAT&PL 1984 Jun; Sophia-Antipolis, F • ISBN 3-540-13346-1, LC-No 84-10575

P 3091 Conv Int Storica Logica;1982 San Gimignano • I
[1983] *Atti del Convegno Internazionale di Storica della Logica* ED: ABRUSCI, V.M. & CASARI, E. & MUGNAI, M. • PUBL Cooperativa Libraria, Universitaria Ed: Bologna x+401pp
• DAT&PL 1982 Dec; San Gimignano, I • LC-No 84-181758

P 3092 Congr Naz Logica;1979 Montecatini Terme • I
[1981] *Atti del Congresso Nazionale di Logica* ED: BERNINI, S. • PUBL (X 1732) Bibliopolis: Napoli 735pp
• DAT&PL 1979 Oct; Montecatini Terme, I • LC-No 81-198713

P 3146 Constr Math;1980 Las Cruces • USA
[1981] *Constructive Mathematics. Proceedings of the New Mexico State University Conference* ED: RICHMAN, F. • SER (S 3301) Lect Notes Math 873 • PUBL (X 0811) Springer: Heidelberg & New York vii+347pp
• DAT&PL 1980 Aug; Las Cruces, NM, USA • ISBN 3-540-10850-5, LC-No 81-9345

P 3165 FCT'81 Fund of Comput Th;1981 Szeged • H
[1981] *Fundamentals of Computation Theory. Proceedings of the 1981 International FCT-Conference* ED: GECSEG, F. • SER (S 3302) Lect Notes Comput Sci 117 • PUBL (X 0811) Springer: Heidelberg & New York xi+471pp
• DAT&PL 1981 Aug; Szeged, H • ISBN 3-540-10854-8, LC-No 81-13533

P 3199 CAAP'77 Arbres en Algeb & Progr (2);1977 Lille • F
[1977] *Les Arbres en Algebre et en Programmation. 2eme Colloquium* PUBL Universite de Lille I: Lille 273pp
• DAT&PL 1977 Feb; Lille, F

P 3201 Logic Symposia;1979/80 Hakone • J
[1981] *Logic Symposia Hakone 1979,1980* ED: MUELLER, GERT H. & TAKEUTI, G. & TUGUE, T. • SER (S 3301) Lect Notes Math 891 • PUBL (X 0811) Springer: Heidelberg & New York xi+394pp
• DAT&PL 1979 Mar; Hakone, J, 1980 Feb; Hakone, J • ISBN 3-540-11161-1, LC-No 81-18424

P 3210 Math Founds of Comput Sci (9);1980 Rydzyna • PL
[1980] *Mathematical Foundations of Computer Science. Proceedings of the 9th Symposium* ED: DEMBINSKI, P. • SER (S 3302) Lect Notes Comput Sci 88 • PUBL (X 0811) Springer: Heidelberg & New York viii+723pp
• DAT&PL 1980 Sep; Rydzyna, PL • ISBN 3-540-10027-X, LC-No 80-20087

P 3225 New Direct in Algor Lang;1976 St Pierre • F
[1976] *New Directions in Algorithmic Languages 1976. Prepared for IFIP Working Group 2.1 on ALGOL* ED: SCHUMAN, S.A. • PUBL (X 2732) INRIA: Le Chesnay Cedex 145pp
• DAT&PL 1976 -?-; St. Pierre de Chartreuse, F • LC-No 80-503668

P 3247 Transform de Progr;1978 Paris • F
[1978] *Transformations de Programmes. Actes du 3eme Colloque International sur la Programmation* ED: ROBINET, B. • PUBL (X 1422) Bordas: Paris xxii+426pp
• DAT&PL 1978 Mar; Paris, F • LC-No 79-371673

P 3268 Set Th & Model Th;1979 Bonn • D
[1981] *Set Theory and Model Theory* ED: JENSEN, R.B. & PRESTEL, A. • SER (S 3301) Lect Notes Math 872 • PUBL (X 0811) Springer: Heidelberg & New York v+174pp
• DAT&PL 1979 Jun; Bonn, D • ISBN 3-540-10849-1, LC-No 81-9234

P 3269 Set Th Found Math (Kurepa);1977 Beograd • YU
[1977] *Set Theory. Foundations of Mathematics* SER (J 2887) Zbor Radova, NS 2(10) • PUBL (X 3727) Beograd Mat Inst: Belgrade 152pp
• DAT&PL 1977 Aug; Beograd, YU • LC-No 79-373865

P 3299 Progr Kiso Riron, Algor Okeru Shomei Ron;1973/74 Kyoto • J
[1975] *"Program no Kiso Riron" Kenkyu Shugo Oyobi Tanki Kyodo Kenkyu "Algorithm ni Okeru Shomei Ron". Hokoku Shu (The 3rd Symposium on Basic Theory of Programs and the Workshop for Proof Theory about Algorithms. Proceedings)* ED: IGARASHI, S. • SER (S 2308) Symp Kyoto Univ Res Inst Math Sci (RIMS) 236 • PUBL (X 2441) Kyoto Univ Res Inst Math Sci: Kyoto 215pp
• DAT&PL 1973 May; Kyoto, J, 1974 Nov; Kyoto, J

P 3342 Rec Th Week;1984 Oberwolfach • D
[1985] *Recursion Theory Week* ED: EBBINGHAUS, H.-D. & MUELLER, GERT H. & SACKS, G.E. • SER (S 3301) Lect Notes Math 1141 • PUBL (X 0811) Springer: Heidelberg & New York viii+418pp
• DAT&PL 1984 Apr; Oberwolfach, D • ISBN 3-540-15673-9, LC-No 85-20808

P 3361 APL Congr '73;1973 Copenhagen • DK
[1973] *APL Congress 73* ED: GJERLOV, P. & HELMS, H.J. & NIELSEN, J. • PUBL (X 0809) North Holland: Amsterdam xv+506pp • ALT PUBL (X 0838) Amer Elsevier: New York
• DAT&PL 1973 Aug; Copenhagen, DK • LC-No 73-84650

P 3368 Continuous Lattices;1979 Bremen • D
[1981] *Continuous Lattices. Proceedings of the Conference on Topological and Categorical Aspects of Continuous Lattices (Workshop IV)* ED: BANASCHEWSKI, B. & HOFFMANN, R.-E. • SER (S 3301) Lect Notes Math 871 • PUBL (X 0811) Springer: Heidelberg & New York x+413pp
• DAT&PL 1979 Nov; Bremen, D • ISBN 3-540-10848-3, LC-No 81-9291

P 3379 Graph-Gram & Appl to Comput Sci & Biol;1978 Bad Honnef • D
[1979] *Graph-Grammars and Their Application to Computer Science and Biology* ED: CLAUS, V. & EHRIG, H. & ROZENBERG, G. • SER (S 3302) Lect Notes Comput Sci 73 • PUBL (X 0811) Springer: Heidelberg & New York vii+477pp
• DAT&PL 1978 Oct; Bad Honnef, D • ISBN 3-540-09525-X, LC-No 79-9560

P 3404 Model Th & Arithm;1979/80 Paris • F
[1981] *Model Theory and Arithmetic. Comptes Rendus d'une Action Thematique Programmee du CNRS sur la Theorie des Modeles et l'Arithmetique* ED: BERLINE, C. & MCALOON, K. & RESSAYRE, J.-P. • SER (S 3301) Lect Notes Math 890 • PUBL (X 0811) Springer: Heidelberg & New York vi+306pp
• DAT&PL 1979 -?-; Paris, F • ISBN 3-540-11159-X, LC-No 82-137454

P 3429 Math Founds of Comput Sci (10);1981 Strbske Pleso • CS
[1981] *Mathematical Foundations of Computer Science 1981. Proceedings of the 10th Symposium* ED: GRUSKA, J. & CHYTIL, M. • SER (S 3302) Lect Notes Comput Sci 118 • PUBL (X 0811) Springer: Heidelberg & New York xi+589pp
• DAT&PL 1981 Aug; Strbske Pleso, CS • ISBN 3-540-10856-4, LC-No 81-9302

P 3479 Math Stud of Inform Process;1978 Kyoto • J
[1979] *Mathematical Studies of Information Processing* ED: BLUM, E.K. & PAUL, M. & TAKASU, S. • SER (S 3302) Lect Notes Comput Sci 75 • PUBL (X 0811) Springer: Heidelberg & New York viii+629pp
• DAT&PL 1978 Aug; Kyoto, J • ISBN 3-540-09541-1, LC-No 80-494262

P 3482 Logic & Algor (Specker);1980 Zuerich • CH
[1982] *Logic and Algorithmic. An International Symposium Held in Honour of Ernst Specker* ED: ENGELER, E. & LAEUCHLI, H. & STRASSEN, V. • SER Monogr l'Enseign Math 30 • PUBL (X 3718) Enseign Math, Univ Geneve: Geneve 392pp
• DAT&PL 1980 Feb; Zuerich, CH • LC-No 82-184564

P 3488 Theor Comput Sci (4);1979 Aachen • D
[1979] *Theoretical Computer Science. 4th GI Conference (Gesellschaft fuer Informatik)* ED: WEIHRAUCH, K. • SER (S 3302) Lect Notes Comput Sci 67 • PUBL (X 0811) Springer: Heidelberg & New York vii+324pp
• DAT&PL 1979 Mar; Aachen D • ISBN 3-540-09118-1, LC-No 79-9707

P 3525 Int Comput Symp;1975 Antibes • F
[1975] *International Computing Symposium 1975* ED: GELENBE, E. & POTIER, D. • PUBL (**X** 0809) North Holland: Amsterdam vii+266pp • ALT PUBL (**X** 0838) Amer Elsevier: New York
• DAT&PL 1975 Jun; Antibes, F • ISBN 0-7204-2839-4, LC-No 75-38988

P 3527 GI Jahrestag (4);1974 Berlin • D
[1975] *GI - 4. Jahrestagung (Gesellschaft fuer Informatik)* ED: SIEFKES, D. • SER (S 3302) Lect Notes Comput Sci 26 • PUBL (**X** 0811) Springer: Heidelberg & New York ix+748pp
• DAT&PL 1974 Oct; Berlin, D • ISBN 3-540-07141-5

P 3542 ACM Symp Th of Comput (11);1979 Atlanta • USA
[1979] *Conference Record of the 11th Annual ACM Symposium on Theory of Computing (Association for Computing Machinery)* PUBL (**X** 2205) ACM: New York vii+368pp
• DAT&PL 1979 Apr; Atlanta, GA, USA • ISBN 0-89791-003-6, LC-No 82-642181

P 3565 Symp of Th Aspects of Comput Sci (1);1984 Paris • F
[1984] *STACS 84. Symposium of Theoretical Aspects of Computer Science by AFCET (Association Francaise pour la Cybernetique Economique et Technique) and GI (Gesellschaft fuer Informatik)* ED: FONTET, M. & MEHLHORN, K. • SER (S 3302) Lect Notes Comput Sci 166 • PUBL (**X** 0811) Springer: Heidelberg & New York vi+338pp
• DAT&PL 1984 Apr; Paris, F • ISBN 3-540-12920-0, LC-No 84-5299

P 3572 IEEE Symp Found of Comput Sci (18);1977 Providence • USA
[1977] *18th Annual IEEE Symposium on Foundations of Computer Science* PUBL (**X** 2179) IEEE: New York v+269pp
• DAT&PL 1977 Oct; Providence, RI, USA • LC-No 80-646634

P 3578 IEEE Symp Found of Comput Sci (19);1978 Ann Arbor • USA
[1978] *19th Annual IEEE Symposium on Foundations of Computer Science* PUBL (**X** 2179) IEEE: New York v+290pp
• DAT&PL 1978 Oct; Ann Arbor, MI, USA • LC-No 80-646634

P 3621 Frege Konferenz (2);1984 Schwerin • DDR
[1984] *Frege Konferenz* ED: WECHSUNG, G. • SER Mathematical Research - Mathematische Forschungen 20 • PUBL (**X** 0911) Akademie Verlag: Berlin 408pp
• DAT&PL 1984 Sep; Schwerin, DDR • LC-No 84-248695

P 3622 Int Congr Log, Meth & Phil of Sci (6,Proc);1979 Hannover • D
[1982] *Proceedings of the 6th International Congress for Logic, Methodology and Philosophy of Science* ED: COHEN, L.J. & LOS, J. & PFEIFFER, H. & PODEWSKI, K.-P. • SER (S 3303) Stud Logic Found Math 104 • PUBL (**X** 0809) North Holland: Amsterdam xiv+842pp
• DAT&PL 1979 Aug; Hannover, D • LC-No 80-12713

P 3623 Logic Colloq;1980 Prague • CS
[1982] *Logic Colloquium '80* ED: DALEN VAN, D. & LASCAR, D. & SMILEY, T.J. • SER (S 3303) Stud Logic Found Math 108 • PUBL (**X** 0809) North Holland: Amsterdam x+342pp
• ISBN 0-444-86465-2, LC-No 82-12611

P 3634 Patras Logic Symp;1980 Patras • GR
[1982] *Patras Logic Symposium* ED: METAKIDES, G. • SER (S 3303) Stud Logic Found Math 109 • PUBL (**X** 0809) North Holland: Amsterdam ix+391pp
• DAT&PL 1980 Aug; Patras, GR • ISBN 0-444-86476-8, LC-No 82-14107

P 3638 Brouwer Centenary Symp;1981 Noordwijkerhout • NL
[1982] *The L.E.J. Brouwer Centenary Symposium* ED: TROELSTRA, A.S. & DALEN VAN, D. • SER (S 3303) Stud Logic Found Math 110 • PUBL (**X** 0809) North Holland: Amsterdam x+524pp
• DAT&PL 1981 Jun; Noordwijkerhout, NL • ISBN 0-444-86494-6, LC-No 82-239358

P 3642 Colloq Math Log in Computer Sci;1978 Salgotarjan • H
[1981] *Proceedings of the Colloquium on Mathematical Logic in Computer Science* ED: DOEMALKI, B. & GERGELY, T. • SER (S 3312) Coll Math Soc Janos Bolyai 26 • PUBL (**X** 0809) North Holland: Amsterdam 758pp • ALT PUBL (**X** 3725) Bolyai Janos Mat Tars: Budapest
• DAT&PL 1978 Sep; Salgotarjan, H • ISBN 0-444-85440-1, LC-No 82-101557

P 3658 Math Founds of Comput Sci (11);1984 Prague • CS
[1984] *Mathematical Foundations of Computer Science 1984. Proceedings of the 11th Symposium* ED: CHYTIL, M.P. & KOUBEK, V. • SER (S 3302) Lect Notes Comput Sci 176 • PUBL (**X** 0811) Springer: Heidelberg & New York 581pp
• DAT&PL 1984 Sep; Prague, CS • ISBN 3-540-13372-0, LC-No 83-13980

P 3668 Log & Founds of Math;1983 Kyoto • J
[1984] *Logic and the Foundations of Mathematics* ED: TUGUE, T. • SER (S 2308) Symp Kyoto Univ Res Inst Math Sci (RIMS) 516 • PUBL (**X** 2441) Kyoto Univ Res Inst Math Sci: Kyoto ii+195pp
• DAT&PL 1983 Oct; Kyoto, J

P 3696 Logic Colloq;1964 Bristol • GB
[1965] *Logic Colloquium 1964* SER (**J** 0096) Acta Philos Fenn 18 • PUBL (**X** 0809) North Holland: Amsterdam
• DAT&PL 1964 Aug; Bristol, GB

P 3708 Herbrand Symp Logic Colloq;1981 Marseille • F
[1982] *Proceedings of the Herbrand Symposium Logic. Colloquium '81* ED: STERN, J. • SER (S 3303) Stud Logic Found Math 107 • PUBL (**X** 0809) North Holland: Amsterdam xi+384pp
• DAT&PL 1981 Jul; Marseille, F • ISBN 0-444-86417-2, LC-No 82-6433

P 3710 Logic Colloq;1982 Firenze • I
[1984] *Proceedings of the Logic Colloquium '82* ED: LOLLI, G. & LONGO, G. & MARCJA, A. • SER (S 3303) Stud Logic Found Math 112 • PUBL (**X** 0809) North Holland: Amsterdam viii+358pp
• DAT&PL 1982 Aug; Firenze, I • ISBN 0-444-86876-3, LC-No 84-1630

P 3729 Algor in Modern Math & Comput Sci;1979 Urgench • SU
[1981] *Algorithms in Modern Mathematics and Computer Science* ED: ERSHOV, A.P. & KNUTH, D.E. • SER (S 3302) Lect Notes Comput Sci 122 • PUBL (**X** 0811) Springer: Heidelberg & New York xi+487pp
• DAT&PL 1979 Sep; Urgench, SU • ISBN 3-540-11157-3, LC-NO 81-18418
• TRANSL IN [1982] (**P** 3803) Algor Sovrem Mat & Prilozh;1979 Urgench

P 3753 Penser Math;1981 Paris • F
[1982] *Penser les Mathematiques. Lectures from a Seminar on Philosophy and Mathematics* ED: GUENARD, F. & LELIEVRE, G. • SER Collection Points: Serie Sciences 29 • PUBL (**X** 1349) Seuil: Paris 277pp
• DAT&PL 1981; Paris, F • ISBN 2-02-006061-2, LC-No 82-144038

P 3803 Algor Sovrem Mat & Prilozh;1979 Urgench • SU
[1982] *Algoritmy v Sovremennoj Matematike i ee Prilozheniyakh. Chast' I, II (Algorithms in Modern Mathematics and Computer Science Part I, II)* ED: ERSHOV, A.P. & KNUTH, D.E. • PUBL (**X** 2652) Akad Nauk Sibirsk Otd Inst Mat: Novosibirsk 2 Vols: 364pp,316pp
• DAT&PL 1979 Sep; Urgench, SU • LC-No 84-128439
• TRANSL OF [1981] (**P** 3729) Algor in Modern Math & Comput Sci;1979 Urgench
• REM With Additions to the Original Collection

P 3817 Algeb Homage:Ring Th & Rel Top (Jacobson);1981 New Haven • USA
[1982] *Algebraists' Homage: Papers in Ring Theory and Related Topics. Proceedings of the Conference on Algebra in Honor of Nathan Jacobson* ED: AMITSUR, S.A. & SALTMAN, D.J. & SELIGMAN, G.B. • SER (**S** 3313) Contemp Math 13
• PUBL (**X** 0803) Amer Math Soc: Providence viii+409pp
• DAT&PL 1981 Jun; New Haven, CT, USA • ISBN 0-8218-5013-X, LC-No 82-18934

P 3829 Atti Incontri Log Mat (1);1982 Siena • I
[1983] *Atti Degli Incontri di Logica Matematica* ED: BERNARDI, C. • PUBL (**X** 3812) Univ Siena, Dip Mat: Siena 398pp
• DAT&PL 1982 Jan; Siena, I, 1982 Apr; Siena, I, 1982 Jun; Siena, I

P 3830 Logics of Progr & Appl;1980 Poznan • PL
[1983] *Logics of Programs and Their Applications* ED: SALWICKI, A. • SER (**S** 3302) Lect Notes Comput Sci 148
• PUBL (**X** 0811) Springer: Heidelberg & New York vi+324pp
• DAT&PL 1980 Aug; Poznan, PL • ISBN 3-540-11981-7, LC-No 83-158934

P 3831 Universal Algeb & Appl;1978 Warsaw • PL
[1982] *Universal Algebra and Applications* ED: TRACZYK, T.
• SER (**S** 1926) Banach Cent Publ 9 • PUBL (**X** 1034) PWN: Warsaw 454pp
• DAT&PL 1978 Feb-Sep; Warsaw, PL • ISBN 83-01-02145-4, LC-No 83-167533

P 3835 AFCET-SMF Math Appliquees Colloq (1);1978 Palaiseau • F
[1978] *1er Colloque AFCET-SMF de Mathematiques Appliquees. Tome I, II, III. Association Francaise pour la Cybernetique Economique et Technique (AFCET) et Societe Mathematique de France (SMF)* PUBL Ecole Polytechnique: Palaiseau 3 Vols: 375pp, 323pp, 85pp
• DAT&PL 1978 -?-; Palaiseau, F

P 3836 Automata, Lang & Progr (9);1982 Aarhus • DK
[1982] *Automata, Languages and Programming. 9th Colloquium* ED: NIELSEN, M. & SCHMIDT, E.M. • SER (**S** 3302) Lect Notes Comput Sci 140 • PUBL (**X** 0811) Springer: Heidelberg & New York vii+614pp
• DAT&PL 1982 Jul; Aarhus, DK • ISBN 3-540-11576-5, LC-No 83-10430
• REM Also Abbreviated as ICALP '82

P 3840 Autom Deduct (6);1982 New York • USA
[1982] *6th Conference on Automated Deduction* ED: LOVELAND, D.W. • SER (**S** 3302) Lect Notes Comput Sci 138
• PUBL (**X** 0811) Springer: Heidelberg & New York vii+389pp
• DAT&PL 1982 Jun; New York, NY, USA • ISBN 3-540-11558-7, LC-No 82-5948

P 3845 Conf Math Service of Man (2,Proc)(Feriet);1982 Las Palmas • E
[1982] *2nd World Conference on Mathematics at the Service of Man. Dedicated to the Remembrance of Joseph Kampe de Feriet (1893-1982)* ED: BALLESTER, A. & CARDUS, D. & TRILLAS, E. • PUBL Universidad Politecnica: Las Palmas xxx+696pp
• DAT&PL 1982 Jun; Las Palmas, E

P 3855 Algeb Conf (3);1982 Beograd • YU
[1983] *Proceedings of the 3rd Algebraic Conference* ED: PRESIC, S.B. & MILIC, . & MIJAJLOVIC, Z. & PERIC, V. & CUPONA, G. • PUBL (**X** 4030) Univ Novom Sadu, Inst Mat: Novi Sad vi+157pp
• DAT&PL 1982 Oct; Beograd, YU • LC-No 84-233209

P 3862 Theor Comput Sci (6);1983 Dortmund • D
[1982] *Theoretical Computer Science. 6th GI Conference. (Gesellschaft fuer Informatik)* ED: CREMERS, A.B. & KRIEGEL, H.P. • SER (**S** 3302) Lect Notes Comput Sci 145 • PUBL (**X** 0811) Springer: Heidelberg & New York x+365pp
• DAT&PL 1983 Jan; Dortmund, D • ISBN 3-540-11973-6, LC-No 82-19673

P 3864 FCT'83 Found of Comput Th;1983 Borgholm • S
[1983] *Foundation of Computation Theory. 1983 FCT-Conference* ED: KARPINSKI, M. • SER (**S** 3302) Lect Notes Comput Sci 158 • PUBL (**X** 0811) Springer: Heidelberg & New York xi+517pp
• DAT&PL 1983 Aug; Borgholm, S • ISBN 3-540-12689-9, LC-No 83-16736

P 3867 Int Symp Progr (5);1982 Turin • I
[1982] *International Symposium on Programming, ISI'82* ED: DEZANI-CIANCAGLINI, M. & MONTANARI, U. • SER (**S** 3302) Lect Notes Comput Sci 137 • PUBL (**X** 0811) Springer: Heidelberg & New York vi+406pp
• DAT&PL 1982 Apr; Turin, I • ISBN 3-540-11494-7, LC-No 82-161894

P 3870 Congr Catala de Log Mat (1);1982 Barcelona • E
[1982] *1er Congres Catala de Logica Matematica. Actes* PUBL Univ Politecnica & Univ Barcelona: Barcelona 130pp
• DAT&PL 1982 -?-; Barcelona, E

P 3885 Categor Th.Appl to Algeb Log & Topol;1981 Gummersbach • D
[1982] *Category Theory. Applications to Algebra, Logic and Topology* ED: KAMPS, K.H. & PUMPLUEN, D. & THOLEN, W.
• SER (**S** 3301) Lect Notes Math 962 • PUBL (**X** 0811) Springer: Heidelberg & New York xv+322pp
• DAT&PL 1981 Jul; Gummersbach, D • ISBN 3-540-11961-2, LC-No 82-19602

P 3889 CAAP'83 Arbres en Algeb & Progr (8);1983 L'Aquila • I
[1983] *CAAP'83. Les Arbres en Algebre et en Programmation. 8eme Colloque* ED: AUSIELLO, G. & PROTASI, M. • SER (**S** 3302) Lect Notes Comput Sci 159 • PUBL (**X** 0811) Springer: Heidelberg & New York vi+416pp
• DAT&PL 1983 Mar; L'Aquila, I • ISBN 3-540-12727-5, LC-No 83-16909

P 3906 Th Found of Progr Methodol;1981 Marktoberdorf • D
[1982] *Theoretical Foundations of Programming Methodology* ED: BROY, M. & SCHMIDT, G. • SER NATO Adv Study Inst Ser C 91 • PUBL (**X** 0835) Reidel: Dordrecht xiii+658pp
• DAT&PL 1981 -?-; Marktoberdorf, D • LC-No 82-12347

P 3908 Graph-Gram & Appl to Comput Sci (2);1982 Neunkirchen • D
[1983] *Graph-Grammars and Their Application to Computer Science.* ED: EHRIG, H. & NAGL, M. & ROZENBERG, G. • SER (**S** 3302) Lect Notes Comput Sci 153 • PUBL (**X** 0811) Springer: Heidelberg & New York vii+452pp
• DAT&PL 1982 Oct; Neunkirchen, D • ISBN 3-540-12310-5, LC-No 83-6677

P 4007 CAAP'84 Arbres en Algeb & Progr (9);1984 Bordeaux • F
[1984] *Les Arbres en Algebre et en Programmation. 9eme Colloquium* PUBL (**X** 0805) Cambridge Univ Pr: Cambridge, GB vi+326pp
• DAT&PL 1984 Mar; Bordeaux, F

P 4046 Rec Th;1982 Ithaca • USA
[1985] *Recursion Theory. Proceedings of the AMS-ASL Summer Institute* ED: NERODE, A. & SHORE, R.A. • SER (**S** 3304) Proc Symp Pure Math 42 • PUBL (**X** 0803) Amer Math Soc: Providence vii+528pp
• DAT&PL 1982 Jun; Ithaca, NY, USA • ISBN 0-8218-1447-8, LC-No 84-18525

P 4072 Kuenstl Intell;1982 Teisendorf • D
[1982] *Kuenstliche Intelligenz* ED: BIBEL, W. & SIEKMANN, J.H. • SER Informatik-Fachberichte 59 • PUBL (**X** 0811) Springer: Heidelberg & New York xiii+383pp
• DAT&PL 1982 Mar: Teisendorf, D • ISBN 3-540-11974-4, LC-No 84-207604

P 4095 All-Union Math Conf (2);1934 Moskva • SU
[1934] *Trudy 2-go Vsesoyuznogo Matematicheskogo S'ezda (Proceedings of the 2nd All-Union Conference)*
• DAT&PL 1934 -?-; Moskva, SU

P 4108 Found of Math;1979 Kyoto • J
[1979] *Sugaku Kisoron (Foundations of Mathematics)* ED: TUGUE, T. • SER (**S** 2308) Symp Kyoto Univ Res Inst Math Sci (RIMS) 362 • PUBL (**X** 2441) Kyoto Univ Res Inst Math Sci: Kyoto ii+185pp
• DAT&PL 1979 May; Kyoto, J

P 4113 Found of Math;1982 Kyoto • J
[1983] *Foundations of Mathematics* ED: SHINODA, J. • SER (**S** 2308) Symp Kyoto Univ Res Inst Math Sci (RIMS) 480 • PUBL (**X** 2441) Kyoto Univ Res Inst Math Sci: Kyoto iii+236pp
• DAT&PL 1982 Oct; Kyoto, J

P 4117 Conf Math Log (Markov);1980 Sofia • BG
[1984] *Matematicheskaya Logika: Trudy Konferentsij po Matematicheskoj Logike, Posvyashchennoj Pamiati A.A.Markova 1903-1979 (Proceedings of a Conference on Mathematical Logic)* ED: SKORDEV, D. • PUBL Bulgar Akad Nauk 172pp
• DAT&PL 1980 Sep; Sofia, BG • LC-No 84-215513

P 4153 B-Val Anal & Nonstand Anal;1981 Kyoto • J
[1981] *Boolean-Algebra-Valued Analysis and Nonstandard Analysis* ED: NANBA, K. • SER (**S** 2308) Symp Kyoto Univ Res Inst Math Sci (RIMS) 441 • PUBL (**X** 2441) Kyoto Univ Res Inst Math Sci: Kyoto ii+158pp
• DAT&PL 1981 May; Kyoto, J

P 4178 Universal Algeb & Lattice Th;1984 Charleston • USA
[1985] *Universal Algebra and Lattice Theory* ED: COMER, S.D. • SER (**S** 3301) Lect Notes Math 1149 • PUBL (**X** 0811) Springer: Heidelberg & New York vi+282pp
• DAT&PL 1984 Jul; Charleston, SC, USA • ISBN 3-540-15691-7, LC-No 85-17329

P 4180 Int Congr Log, Meth & Phil of Sci (7,Pap);1983 Salzburg • A
[1985] *Foundations of Logic and Linguistics. Problems and Their Solutions. Papers from the 7th International Congress of Logic, Methodology and Philosophy of Science* ED: DORN, G. & WEINGARTNER, P. • PUBL (**X** 1332) Plenum Publ: New York xi+715pp
• DAT&PL 1983 Jul; Salzburg, A • ISBN 0-306-41916-5, LC-No 84-26518

P 4244 Rewriting Techn & Appl (1);1985 Dijon • F
[1985] *Rewriting Techniques and Applications. Papers from the 1st Conference* ED: JOUANNAUD, J.-P. • SER (**S** 3302) Lect Notes Comput Sci 202 • PUBL (**X** 0811) Springer: Heidelberg & New York vi+441pp
• DAT&PL 1985 May; Dijon, F • ISBN 3-540-15976-2, LC-No 85-22164

P 4271 E.Bishop-Reflection on Him & Research;1983 San Diego • USA
[1985] *Errett Bishop: Reflections on Him and His Research* ED: ROSENBLATT, M. • SER (**S** 3313) Contemp Math 39 • PUBL (**X** 0803) Amer Math Soc: Providence xvii+91pp
• DAT&PL 1983 Sep; San Diego, CA, USA • ISBN 0-8218-5040-7, LC-No 85-752

P 4310 Easter Conf on Model Th (3);1985 Gross Koeris • DDR
[1985] *Proceedings of the 3rd Easter Conference on Model Theory* SER Seminarbericht 70 • PUBL (**X** 2219) Humboldt-Univ Berlin: Berlin iv+226pp
• DAT&PL 1985 Apr; Gross Koeris, DDR

P 4313 Int Congr Math (II,14);1983 Warsaw • PL
[1984] *Proceedings of the International Congress of Mathematicians, Vol 1,2* ED: CIESIELSKI, Z. & OLECH, C. • PUBL (**X** 1034) PWN: Warsaw lxii+1730pp • ALT PUBL (**X** 0809) North Holland: Amsterdam
• DAT&PL 1983 Aug; Warsaw, PL • ISBN 83-01-05523-5, LC-No 84-13788

P 4381 Int Wittgenstein Symp (7);1982 Kirchberg • A
[1983] *Akten des 7ten Internationalen Wittgenstein Symposium* ED: CZERMAK, H. & WEINGARTNER, P. • PUBL (**X** 2728) Hoelder-Pichler-Tempsky: Wien 573pp
• DAT&PL 1982 Aug; Kirchberg, A • LC-No 83-227181, ISBN 3-209-00499-4

P 4392 Mat Logika (Markov);1980 Sofia • BG
identical with P 4117

P 4564 Math Logic;1963 Xi-An • TJ
[1965] *1963 Nian Cueanguo Shuli Luoji Zhuanye Xueshu Huiyi Lunwen Xuanji (Mathematical Logic. Proceedings of the National Symposium)* PUBL Defence Industry Press: Beijing
• DAT&PL 1963 Oct; Xi-An, TJ

P 4571 Log of Progr;1985 Brooklyn • USA
Logic of Programs ED: PARIKH, R. • SER (**S** 3302) Lect Notes Comput Sci 193 • PUBL (**X** 0811) Springer: Heidelberg & New York vi+424pp
• DAT&PL 1985 Jun; Brooklyn, NY, USA • ISBN 3-540-15648-8

P 4601 EUROCAL;1985 Linz • A
[1985] *Proceedings of European Conference on Computer Algebra (EUROCAL)* ED: BUCHBERGER, B. (V1) & CAVINESS, B.F. (V2) • SER (S 3302) Lect Notes Comput Sci 203,204 • PUBL (**X** 0811) Springer: Heidelberg & New York vi + 233pp, xvi + 650pp
• DAT&PL 1985 Apr; Linz, A • ISBN 3-540-15983-5 (V1), ISBN 3-540-15984-3 (V2)
• REM Vol 1: Invited Lectures. Vol 2: Research Contributions

P 4622 Autom on Infinite Words;1984 Le Mont-Dore • F
[1985] *Automata on Infinite Words* ED: NIVAT, M. & PERRIN, D. • SER (S 3302) Lect Notes Comput Sci 192 • PUBL (**X** 0811) Springer: Heidelberg & New York v + 216pp
• DAT&PL 1984 May; Le Mont-Dore, F • ISBN 3-540-15641-0

P 4627 CAAP'85 Arbres en Algeb & Progr (10);1985 Berlin • D
[1985] *Mathematical Foundations of Software Development, Vol 1. Colloquium on Trees in Algebra and Programming (CAAP '85)* ED: EHRIG, H. & FLOYD, C. & NIVAT, M. & THATCHER, J. • SER (S 3302) Lect Notes Comput Sci 185 • PUBL (**X** 0811) Springer: Heidelberg & New York xiii + 418pp
• DAT&PL 1985 Mar; Berlin, D • ISBN 3-540-15198-2

P 4628 Automata, Lang & Progr (12);1985 Nafplion • GR
[1985] *Automata, Languages and Programming.* ED: BRAUER, W. • SER (S 3302) Lect Notes Comput Sci 194 • PUBL (**X** 0811) Springer: Heidelberg & New York viii + 520pp
• DAT&PL 1985 Jul; Nafplion, GR • ISBN 3-540-15650-X

P 4646 Atti Incontri Log Mat (2);1983/84 Siena • I
[1985] *Atti degli Incontri di Logica Matematica* ED: BERNARDI, C. & PAGLI, P. • PUBL (**X** 3812) Univ Siena, Dip Mat: Siena 648pp
• DAT&PL 1983 Jan; Siena, I, 1983 Apr; Siena, I, 1984 Jan; Siena, I, 1984 Apr; Siena, I

P 4661 Algeb & Log;1984 Zagreb • YU
[1985] *Proceedings of the Conference 'Algebra and Logic'* ED: STOJAKOVIC, Z. • PUBL (**X** 4030) Univ Novom Sadu, Inst Mat: Novi Sad vi + 193pp
• DAT&PL 1984 Jun; Zagreb, YU

P 4670 Comput Th (5);1984 Zaborow • SU
[1985] *Computation Theory. Proceedings of the 5th Symposium* ED: SKOWRON, A. • SER (S 3302) Lect Notes Comput Sci 208 • PUBL (**X** 0811) Springer: Heidelberg & New York vii + 397pp
• DAT&PL 1984 Dec; Zaborow, SU • ISBN 3-540-16066-3

Collection volumes

C 0552 Phil Contemp - Chroniques • I
[1968] *La Philosophie Contemporaine: Chroniques (Contemporary Philosophy: A Survey)* ED: KLIBANSKY, R. • PUBL (**X** 1319) La Nuova Italia: Firenze 4 Vols
• LC-No 68-55649

C 0555 Topic Philos Logic • NL
[1969] *Topics in Philosophical Logic* ED: RESCHER, N. • SER (**S** 3307) Synth Libr • PUBL (**X** 0835) Reidel: Dordrecht xiv+350pp • ALT PUBL (**X** 2696) Humanities Pr: Atlantic Highlands
• LC-No 74-141565

C 0558 Contemp Phil Scand • USA
[1972] *Contemporary Philosophy in Scandinavia* ED: OLSON, R.E. & PAUL, A.M. • PUBL (**X** 1291) Johns Hopkins Univ Pr: Baltimore x+508pp
• ISBN 0-8018-1315-8, LC-No 70-148242

C 0569 Phil of Math Oxford Readings • GB
[1969] *The Philosophy of Mathematics.* ED: HINTIKKA, K.J.J. • SER Oxford Readings in Philosophy • PUBL (**X** 0894) Oxford Univ Pr: Oxford v+186pp
• ISBN 0-19-875011-0, LC-No 71-441791

C 0601 Encycl of Philos • USA
[1967] *Encyclopedia of Philosophy* ED: EDWARDS, P. • PUBL (**X** 0843) Macmillan : New York & London 8 Vols
• LC-No 67-10059

C 0615 Polish Logic 1920-39 • GB
[1967] *Polish Logic 1920-39.* ED: MCCALL, S. • PUBL (**X** 0815) Clarendon Pr: Oxford viii+406pp
• LC-No 67-106639
• REM Translations of Several Articles

C 0621 Kontrolliertes Denken (Britzelmayr) • D
[1951] *Kontrolliertes Denken. Untersuchungen zum Logikkalkuel und zur Logik der Einzelwissenschaften. Festschrift fuer Wilhelm Britzelmayr* ED: MENNE, A. & WILHELMY, A. & ANGSTL, H. • PUBL (**X** 0826) Alber: Freiburg 122pp
• LC-No 56-30120

C 0622 Essays Found of Math (Fraenkel) • IL
[1961] *Essays on the Foundations of Mathematics: Dedicated to A.A.Fraenkel on His 70th Anniversary* ED: BAR-HILLEL, Y. & POZNANSKI, E.I.J. & RABIN, M.O. & ROBINSON, A. • PUBL (**X** 1299) Magnes Pr: Jerusalem x+351pp • ALT PUBL (**X** 0809) North Holland: Amsterdam & (**X** 0833) Acad Pr: Jerusalem & (**X** 0894) Oxford Univ Pr: Oxford
• LC-No 63-753

C 0640 Log, Autom, Inform • RO
[1971] *Logique, Automatique, Informatique* ED: MOISIL, G.C. • PUBL (**X** 0871) Acad Rep Soc Romania: Bucharest 456pp
• LC-No 77-880254

C 0647 Encycl Britannica • GB
[1950ff] *Encyclopaedia Britannica: A New Survey of Universal Knowledge* ED: YUST, W. • PUBL Encyclopaedia Britannica: Chicago-London-Toronto 24 Vols
• LC-No 50-4840

C 0656 Phil Mathematique • F
[1939] *Philosophie Mathematique.* ED: GONSETH, F. • SER (**S** 2073) Actualites Sci Indust 837 • PUBL (**X** 0859) Hermann: Paris 100pp

C 0659 Computer Progr & Formal Syst • NL
[1963] *Computer Programming and Formal Systems* ED: BRAFFORT, P. & HIRSCHBERG, D. • SER (**S** 3303) Stud Logic Found Math • PUBL (**X** 0809) North Holland: Amsterdam vi+161pp • ALT PUBL (**X** 2696) Humanities Pr: Atlantic Highlands
• LC-No 63-3816

C 0675 From Frege to Goedel • USA
[1967] *From Frege to Goedel: A Source Book in Mathematical Logic 1879-1931* ED: HEIJENOORT VAN, J. • PUBL (**X** 0858) Harvard Univ Pr: Cambridge x+660pp • ALT PUBL (**X** 0894) Oxford Univ Pr: Oxford
• LC-No 67-10905
• REM 2nd ed. 1971; xi+660pp. Some Articles have been Reprinted 1970; iv+117pp

C 0684 Automation in Lang Transl & Theorem Prov • B
[1968] *Automation in Language Translation and Theorem Proving: Some Applications of Mathematical Logic* ED: BRAFFORT, P. & SCHEEPEN VAN, F. • PUBL (**X** 1714) Commiss Europ Comm : Bruxelles xv+295pp
• LC-No 77-467790

C 0705 Found of Math (Goedel) • D
[1969] *Foundations of Mathematics. Symposium Papers Commemorating the 60th Birthday of Kurt Goedel.* ED: BULLOFF, J.J. & HOLYOKE, T.C. & HAHN, S.W. • PUBL (**X** 0811) Springer: Heidelberg & New York xii+195pp
• ISBN 3-540-04490-6, LC-No 68-28757

C 0712 Logik & Logikkalkuel • D
[1962] *Logik und Logikkalkuel* ED: KAESBAUER, M. & KUTSCHERA VON, F. • PUBL (**X** 0826) Alber: Freiburg 249pp

C 0718 The Undecidable • USA
[1965] *The Undecidable. Basic Papers on Undecidable Propositions, Unsolvable Problems and Computable Functions* ED: DAVIS, M. • PUBL (**X** 0887) Raven Pr: New York 440pp
• LC-No 65-3996

C 0721 Contemp Readings in Log Th • USA
[1967] *Contemporary Readings in Logical Theory* ED: COPI, I.M. & GOULD, J.A. • PUBL (**X** 0843) Macmillan : New York & London 342pp
• LC-No 67-15535

C 0727 Logic Found of Math (Heyting) • NL
[1968] *Logic and Foundations of Mathematics. Papers Dedicated to A.Heyting on the Occasion of His 70th Birthday* ED: DALEN VAN, D. & DYKMAN, J.G. & KLEENE, S.C. & TROELSTRA, A.S. • PUBL (**X** 0812) Wolters-Noordhoff : Groningen 249pp
• LC-No 70-408259
• REM Also published as J0020 vol. 20

C 0733 Izbr Vopr Algeb & Log (Mal'tsev) • SU
[1973] *Izbrannye Voprosy Algebry i Logiki: Sbornik Posvyashch Pamyati A.I.Mal'tsev (Selected Questions of Algebra and Logic: Volume Dedicated to the Memory of A.I.Mal'tsev)* ED: ERSHOV, YU.L. & KARGAPOLOV, M.I. & MERZLYAKOV, YU.I. & SMIRNOV, D.M. & SHIRSHOV, A.L. • PUBL (**X** 2642) Nauka: Novosibirsk 339pp
• LC-No 73-360316

C 0735 Logic & Value (Dahlquist) • S
[1970] *Logic and Value. Essays Dedicated to Thorild Dahlquist on His 50th Birthday* ED: PAULI, T. • SER Filosofiska Studier 9 • PUBL (**X** 0882) Univ Filos Foeren: Uppsala vi+247pp
• LC-No 72-186683

C 0742 Phil Mid-Century • I
[1958-1959] *Philosophy in the Mid-Century. A Survey * La Philosophie au Milieu du Vingtieme Siecle. Chroniques* ED: KLIBANSKY, R. • PUBL (**X** 1319) La Nuova Italia: Firenze 4 Vols

C 0745 Herbrand: Log Writings • NL
[1971] *Logical Writings by J.Herbrand* ED: GOLDFARB, W.D. • PUBL (**X** 0835) Reidel: Dordrecht vii+312pp • ALT PUBL (**X** 0858) Harvard Univ Pr: Cambridge vii+312pp
• ISBN 90-277-0176-8, LC-No 74-146963
• TRANSL OF [1968] (**C** 2486) Herbrand: Ecrits Logiques

C 0749 Contrib Logic & Methodol (Bochenski) • NL
[1965] *Contributions to Logic and Methodology in Honour of J.M.Bochenski* ED: TYMIENIECKA, A.T. & PARSONS, C. • PUBL (**X** 0809) North Holland: Amsterdam xviii+326pp
• LC-No 66-1566

C 0758 Logic Colloq Boston 1972-73 • D
[1975] *Logic Colloquium. Symposium on Logic Held at Boston, MA, USA, 1972-1973* ED: PARIKH, R.J. • SER (**S** 3301) Lect Notes Math 453 • PUBL (**X** 0811) Springer: Heidelberg & New York iv+251pp
• ISBN 3-540-07155-5, LC-No 75-11528

C 0768 Stud Algeb Logic • USA
[1974] *Studies in Algebraic Logic* ED: DAIGNEAULT, A. • SER Studies in Mathematics 9 • PUBL (**X** 1298) Math Ass Amer: Washington vii+207pp
• ISBN 0-88385-109-1, LC-No 74-84580

C 0769 L'Antinom Ment nel Pensiero Contemp • I
[1961] *L'Antinomia del Mentitiore nel Pensiero Contemporaneo, Da Peirce a Tarski* ED: RIVETTI BARBO, F. • PUBL (**X** 1364) Vita e Pensiero: Milano xxxvii+740pp
• LC-No 65-69081

C 0772 Model Th & Topoi • D
[1975] *Model Theory and Topoi* ED: LAWVERE, F.W. & MAURER, C. & WRAITH, G.C. • SER (**S** 3301) Lect Notes Math 445 • PUBL (**X** 0811) Springer: Heidelberg & New York 354pp
• ISBN 3-540-07164-4, LC-No 75-20007

C 0782 Model Th & Algeb (A. Robinson) • D
[1975] *Model Theory and Algebra. A Memorial Tribute to Abraham Robinson.* ED: SARACINO, D.H. & WEISPFENNING, V.B. • SER (**S** 3301) Lect Notes Math 498 • PUBL (**X** 0811) Springer: Heidelberg & New York 436pp
• ISBN 3-540-07538-0, LC-No 75-40483

C 1009 Wang: Survey Math Logic • TJ
[1963] *Wang,H.: Survey of Mathematical Logic* PUBL (**X** 1876) Kexue Chubanshe: Beijing • ALT PUBL (**X** 0809) North Holland: Amsterdam 1963

C 1098 Skolem: Select Works in Logic • N
[1970] *Skolem,T.A.: Selected Works in Logic* ED: FENSTAD, J.E. • SER Scandinavian University Books • PUBL (**X** 1554) Universitesforlaget: Oslo 732pp
• LC-No 74-485971

C 1102 Critic Approach to Sci & Philos (Popper) • USA
[1964] *The Critical Approach to Science and Philosophy. In Honour of Karl R.Popper* ED: BUNGE, MARIO • PUBL (**X** 0824) Free Press: New York xv+480pp
• LC-No 64-11220

C 1105 Phil of Math. Sel Readings • USA
[1964] *Philosophy of Mathematics. Selected Readings.* ED: BENACERRAF, P. & PUTNAM, H. • SER Prentice Hall Philosophy Series • PUBL (**X** 0819) Prentice Hall: Englewood Cliffs vii+563pp • ALT PUBL (**X** 1096) Blackwell: Oxford
• LC-No 64-13252

C 1139 Metamath Invest of Intuition Arithm & Anal • D
[1973] *Metamathematical Investigation of Intuitionistic Arithmetic and Analysis* ED: TROELSTRA, A.S. • SER (**S** 3301) Lect Notes Math 344 • PUBL (**X** 0811) Springer: Heidelberg & New York xvii+486pp
• ISBN 3-540-06491-5, LC-No 73-14239
• REM Corrigenda and Addenda 1974 in J1583 74-16*26pp

C 1159 Tarski: Logic, Semantics, Metamathematics • GB
[1956] *Logic, Semantics, Metamathematics. Papers from 1923 to 1938 by Alfred Tarski* PUBL (**X** 0815) Clarendon Pr: Oxford 471pp
• LC-No 56-4171
• TRANSL IN [1964] (**C** 1186) Tarski: Logique, Semantique, Metamath
• REM Translations of Several Articles

C 1173 Logica (Bernays) • CH
[1959] *Logica: Studia Paul Bernays Dedicata* PUBL (**X** 0272) Griffon: Neuchatel

C 1186 Tarski: Logique, Semantique, Metamath • F
[1964] *Tarski,A.: Logique, Semantique, Metamathematique. 1923-1944. Tome 1* PUBL (**X** 0850) Colin: Paris
• TRANSL OF [1956] (**C** 1159) Tarski: Logic, Semantics, Metamathematics
• REM Revised and extended French ed. 1972

C 1211 Coherence in Categories • D
[1972] *Coherence in Categories* ED: MAC LANE, S. • SER (**S** 3301) Lect Notes Math 281 • PUBL (**X** 0811) Springer: Heidelberg & New York vii+235pp
• ISBN 3-540-05963-6, LC-No 72-87920

C 1213 Stud in Num Anal (Lanczos) • USA
[1974] *Studies in Numerical Analysis. Papers in Honour of Cornelius Lanczos* ED: SCAIFE, B.K.P. • PUBL (**X** 0801) Academic Pr: New York xxii+333pp
• ISBN 0-12-621150-7, LC-No 72-12280

C 1224 Borel: Lecons sur la Theorie des Fonctions • F
[1898] *Lecons sur la Theorie des Fonctions, par E.Borel. Elements et Principes de la Theorie des Ensembles. Applications a la Theorie des Fonctions* PUBL (**X** 0834) Gauthier-Villars: Paris viii+136pp
• LC-No 11-31157, LC-No 50-4143
• REM Later editions 1928, 1950

C 1409 Gentzen: Collected Papers • NL
[1969] *The Collected Papers of Gentzen* ED: SZABO, M.E.
• SER (**S** 3303) Stud Logic Found Math • PUBL (**X** 0809) North Holland: Amsterdam xii+338pp
• ISBN 0-7204-2254-X, LC-No 71-97201

C 1443 Algebra, Vyp 2 (Irkutsk) • SU
[1973] *Algebra. Vyp 2* ED: FRIDMAN, E.I. • PUBL (**X** 1006) Irkutsk Gos Univ: Irkutsk 164pp
• LC-No 74-645580 • REL PUBL (**C** 3549) Algebra, Vyp 1 (Irkutsk)

C 1450 Teor Algor & Mat Logika (Markov) • SU
[1974] *Teoriya Algorifmov i Matematicheskaya Logika: Sbornik Statej (Theory of Algorithms and Mathematical Logic. Collection of Articles Dedicated to Andrej Andrejevich Markov)* ED: KUSHNER, B.A. & NAGORNYJ, N.M. • PUBL (**X** 2265) Akad Nauk Vychis Tsentr: Moskva 216pp
• LC-No 75-554732

C 1468 Sets & Classes (Bernays) • NL
[1976] *Sets and Classes. On the Work of Paul Bernays* ED: MUELLER, GERT H. • SER (**S** 3303) Stud Logic Found Math 84 • PUBL (**X** 0809) North Holland: Amsterdam xxiii+358pp • ALT PUBL (**X** 0838) Amer Elsevier: New York
• ISBN 0-7204-2284-1, LC-No 76-16791

C 1495 Encycl Universalis • F
[1968-1976] *Encyclopaedia Universalis* PUBL (**X** 2524) Encyclopaedia Universalis: Paris 20 Vols
• ISBN 2-85229-281-5, LC-No 75-516014

C 1523 Handb of Math Logic • NL
[1977 & 2nd ed. 1978] *Handbook of Mathematical Logic* ED: BARWISE, J. • SER (**S** 3303) Stud Logic Found Math 90
• PUBL (**X** 0809) North Holland: Amsterdam xi+1165pp
• ALT PUBL (**X** 0838) Amer Elsevier: New York
• ISBN 0-7204-2285-X, LC-No 76-26032
• TRANSL IN [1982] (**C** 1920) Spravochnaya Kniga po Mat Logike, Chast 1-4

C 1530 Issl Log Sist (Yanovskaya) • SU
[1970] *Issledovaniya Logicheskikh Sistem (Investigations of Logical Systems)* ED: TAVANETS, P.V. • PUBL (**X** 2027) Nauka: Moskva 336pp
• LC-No 70-554105

C 1533 Quantoren, Modal, Paradox • DDR
[1972] *Quantoren, Modalitaeten, Paradoxien. Beitraege zur Logik* ED: WESSEL, H. • PUBL (**X** 0806) Dt Verlag Wiss: Berlin 524pp
• LC-No 72-3662278

C 1534 Mash Turing & Rek Funk • SU
[1972] *Mashiny Tyuringa i Rekursivnye Funktsii (Turing Machines and Recursive Functions)* PUBL (**X** 0885) Mir: Moskva 264pp
• TRANSL IN Turing Machines and Recursive Functions

C 1587 Quine: Ways of Paradox & Essays • USA
[1966] *Quine, W.V.C.: The Ways of Paradox and Other Essays* PUBL (**X** 0981) Random House: New York x+258pp • ALT PUBL (**X** 0858) Harvard Univ Pr: Cambridge x+335pp
• ISBN 0-674-94835-1 (0858), LC-No 66-11146, LC-No 76-4200,
• REM 2nd revised and enlarged ed. 1976

C 1602 Lect on Modern Math • USA
[1963-1965] *Lectures on Modern Mathematics* ED: SAATY, T.L. • PUBL (**X** 0827) Wiley & Sons: New York 3 Vols
• LC-No 63-20369
• TRANSL IN [1965] (**C** 4652) Lect on Modern Math (Japanese)

C 1627 Etud Phil Sci (Gonseth) • CH
[1950] *Etudes de Philosophie des Sciences, en Hommage a F. Gonseth a l'Occasion de son 60eme Anniversaire* SER Bibliotheque Scientifique,Serie Dialectica 20 • PUBL (**X** 0272) Griffon: Neuchatel 176pp

C 1662 Teor Log Vyvoda • SU
[1973] *Teoriya Logicheskogo Vyvoda (Theory of Logical Deductions)* ED: TAVANETS, P.V. • PUBL (**X** 2027) Nauka: Moskva 272pp
• LC-No 73-353812

C 1698 Logic & Art (Goodman) • USA
[1972] *Logic and Art. Essays in Honor of Nelson Goodman* ED: RUDNER, R. & SCHEFFLER, J. • PUBL (**X** 1238) Bobbs-Merril: Indianapolis ix+330pp
• LC-No 76-140799

C 1856 Log Enterprise • USA
[1975] *The Logical Enterprise* ED: MARCUS, R.B. & ANDERSON, A.R. & MARTIN, R.M. • PUBL (**X** 0875) Yale Univ Pr: New Haven x+261pp
• ISBN 0-300-01790-1, LC-No 74-20084

C 1920 Spravochnaya Kniga po Mat Logike, Chast 1-4 • SU
[1982] *Spravochnaya Kniga po Matematicheskoj Logike. Chast. 1-4 (Handbook of Mathematical Logic. Part 1-4)* ED: ERSHOV, YU.L. & PALYUTIN, E.A. & TAJMANOV, A.D. • PUBL (**X** 2027) Nauka: Moskva 4 Vols: 392pp, 375pp, 360pp, 391pp
• TRANSL OF [1977] (**C** 1523) Handb of Math Logic
• REM The Translation Contains a New Supplement by Palyutin,E.A. Part 1: Model Theory. Part 2: Set Theory. Part 3: Recursion Theory. Part 4: Proof Theory and Constructive Mathematics.

C 1932 Probl of Exist in Math • YU
[1979] *The Problem of Existence in Mathematics* SER Biblioteka Matematicki Vidicki • PUBL (**X** 3727) Beograd Mat Inst: Belgrade

C 1936 Log Th & Semant Anal (Kanger) • NL
[1974] *Logical Theory and Semantic Analysis. Essays Dedicated to Stig Kanger on His 50th Birthday* ED: STENLUND, S. & HENSCHEN-DAHLQUIST, A.-M. & LINDAHL, L. & NORDENFELT, L. & ODELSTAD, J. • SER (**S** 3307) Synth Libr 63 • PUBL (**X** 0835) Reidel: Dordrecht v+217pp
• ISBN 90-277-0438-4, LC-No 73-94456

C 2084 Stud Hist of Math Log • PL
[1973] *Studies in the History of Mathematical Logic* ED: SURMA, S.J. • PUBL (**X** 2212) Ossolineum: Wroclaw 288pp
• LC-No 73-170578

C 2141 Filos Matematica • I
[1967] *La Filosofia della Matematica* ED: CELLUCCI, C. • PUBL Laterza: Bari 320pp • LC-No 68-120024

C 2295 Stud Founds & Combin • USA
[1978] *Studies in Foundations and Combinatorics* ED: ROTA, G.-C. • SER (S 3105) Adv Math, Suppl Stud 1 • PUBL (X 0801) Academic Pr: New York x+265pp • ISBN 0-12-599101-0, LC-No 78-12921

C 2319 Slozh Vychisl & Algor • SU
[1974] *Slozhnost' Vychislennij i Algorifmov. Sbornik Peredov (Die Kompliziertheit von Berechnungen und Algorithmen. Sammlung von Uebersetzungen)* ED: KOZMIDIADI, V.A. & MASLOV, A.N. & PETRI, N.V. • PUBL (X 0885) Mir: Moskva 392pp • LC-No 75-568382

C 2486 Herbrand: Ecrits Logiques • F
[1968] *Herbrand,J.: Ecrits Logiques* ED: HEIJENOORT VAN, J. • PUBL (X 0840) Pr Univ France: Paris • LC-No 68-104359 • TRANSL IN [1971] (C 0745) Herbrand: Log Writings

C 2537 Spez Wissenschaftsth • D
[1979] *Spezielle Wissenschaftstheorie* ED: LORENZ, K. • PUBL (X 1174) Gruyter: Berlin 2 Vols • REM Vol 1: Konstruktionen versus Positionen.

C 2542 Mat Teor Log Vyvoda • SU
[1967] *Matematicheskaya Teoriya Logicheskogo Vyvoda (Mathematical Theory of Logical Deduction)* ED: IDEL'SON, A.V. & MINTS, G.E. • PUBL (X 2027) Nauka: Moskva 351pp • LC-No 68-41857

C 2546 Ehntsikl Kibern • SU
[1974] *Ehntsiklopediya Kibernetiki* ED: GLUSHKOV, V.M. • PUBL Glavnaya Redaktsiya Ukrainskoj Sov. Ehntsiklopedii: Kiev • LC-No 74-359262 • TRANSL IN Lexikon der Kybernetik

C 2555 Algeb Sistemy (Irkutsk) • SU
[1976] *Algebraicheskie Sistemy (Algebraic Systems)* ED: KOKORIN, A.I. • PUBL (X 1006) Irkutsk Gos Univ: Irkutsk 170pp • LC-No 79-410179

C 2577 Issl Formaliz Yazyk & Neklass Log • SU
[1974] *Issledovaniya po Formalizovannym Yazykam i Neklassicheskim Logikam (Investigations on Formalized Languages and Non-Classical Logics)* ED: BOCHVAR, D.A. • PUBL (X 2027) Nauka: Moskva 275pp • LC-No 75-554105

C 2578 Filos & Logika • SU
[1974] *Filosofiya i Logika. Filosofiya v Sovremennom Mire (Philosophy and Logic. Philosophy in the Modern World)* ED: TAVANETS, P.V. & SMIRNOV, V.A. • SER Filosofiya v Sovremennom Mire • PUBL (X 2027) Nauka: Moskva 479pp • LC-No 74-358872

C 2580 Mat Entsikl • SU
[1977ff] *Matematicheskaya Ehntsiklopediya (Mathematical Encyclopedia)* ED: VINOGRADOV, I.M. (VOL 1) • PUBL Izdatel'stvo Sovetskaya Ehntsiklopediya: Moskva • LC-No 77-515331

C 2581 Issl Neklass Log & Teor Mnozh • SU
[1979] *Issledovaniya po Neklassicheskim Logikam i Teorii Mnozhestv (Investigations on Non-Classical Logics and Set Theory)* ED: MIKHAJLOV, A.I. ET AL. • PUBL (X 2027) Nauka: Moskva 374pp • LC-No 80-475529

C 2583 Aktual Probl Log & Metodol Nauki • SU
[1980] *Aktual'nye Problemy Logiki i Metodologii Nauki (Current Problems of Logic and the Methodology of Science. Collection of Scientific Works)* ED: POPOVICH, M.V. • PUBL (X 2199) Naukova Dumka: Kiev 336pp • LC-No 80-506762

C 2617 Modern Log Survey • NL
[1981] *Modern Logic - A Survey. Historical, Philosophical and Mathematical Aspects of Modern Logic and Its Applications.* ED: AGAZZI, E. • SER (S 3307) Synth Libr 149 • PUBL (X 0835) Reidel: Dordrecht viii+475pp • ISBN 90-277-1137-2, LC-No 80-22027

C 2621 Mal'tsev: Metamath of Algeb Syst • NL
[1971] *The Metamathematics of Algebraic Systems. Mal'tsev,A.I: Collected Papers, 1936-1967* ED: WELLS III, B.F. • SER (S 3303) Stud Logic Found Math 66 • PUBL (X 0809) North Holland: Amsterdam xviii+494pp • LC-No 73-157020

C 2962 Mat Voprosy Teor Intell Mashin • SU
[1975] *Matematicheskie Voprosy Teorii Intellektual'nykh Mashin. Sbornik Trudov (Mathematical Questions of the Theory of Machine Intelligence. Collection of Papers)* ED: KAPITONOVA, YU.V. & ASEL'DEROV, Z.M. • PUBL (X 2522) Akad Nauk Inst Kibernet: Kiev 84pp • LC-No 77-507084

C 2967 Metodol Probl Mat • SU
[1979] *Metodologicheskie Problemy Matematiki (Methodological Problems of Mathematics)* ED: BORISOV, YU.F. • PUBL (X 2642) Nauka: Novosibirsk 303pp

C 2969 Modeles de l'Arithm; Paris 1977 • F
[1980] *Modeles de l'Arithmetique* ED: MCALOON, K. • SER (J 1620) Asterisque 73 • PUBL (X 2244) Soc Math France: Paris 155pp

C 2981 Phil der Math Erkenntnis Muenster 1978/79 • D
[1981] *Zur Philosophie der Mathematischen Erkenntnis* ED: BOERGER, E. & BARNOCCHI, D. & KAULBACH, F. • PUBL (X 2773) Koenigshausen & Neumann: Wuerzburg 159pp • ISBN 3-88479-022-6, LC-No 81-135346

C 3050 Essays Combin Log, Lambda Calc & Formalism (Curry) • USA
[1980] *To H.B. Curry: Essays on Combinatory Logic, Lambda Calculus and Formalism* ED: SELDIN, J.P. & HINDLEY, J.R. • PUBL (X 0801) Academic Pr: New York xxv+606pp • ISBN 0-12-349050-2, LC-No 80-40139

C 3211 Mat Ling & Teor Algor • SU
[1978] *Matematicheskaya Lingvistika i Teoriya Algoritmov. Mezhvuzovskij Tematicheskij Sbornik (Mathematical Linguistics and Theory of Algorithms. Interuniversity Thematic Collection)* ED: GLADKIJ, A.V. • PUBL (X 1434) Kalinin Gos Univ: Kalinin 155pp • LC-No 80-458139

C 3214 Mathematiker ueber Math • D
[1974] *Mathematiker ueber die Mathematik* ED: OTTE, M. • PUBL (X 0811) Springer: Heidelberg & New York 481pp • ISBN 3-540-06898-8, LC-No 74-13965

C 3271 Issl Teor Mnozh & Neklass Logik • SU
[1976] *Issledovaniya po Teorij Mnozhestv i Neklassicheskim Logikam. Sbornik Trudov (Studies in Set Theory and Nonclassical Logics. Collection of Papers)* ED: BOCHVAR, D.A. & GRISHIN, V.N. • PUBL (**X** 2027) Nauka: Moskva 328pp • LC-No 77-501571

C 3351 Teorem Goedel & Hipotese Cont • P
[1979] *O Teorema de Goedel e a Hipotese do Continuo* PUBL (**X** 2719) Fundacao Calouste Gulbenkian: Lisbon

C 3515 Ital Studies in Phil of Sci • NL
[1981] *Italian Studies in the Philosophy of Science* ED: CHIARA SCABIA DALLA, M.L. • SER (**S** 3311) Boston St Philos Sci 47 • PUBL (**X** 0835) Reidel: Dordrecht xi+525pp • ISBN 90-277-0735-9, LC-No 80-16665

C 3517 Found of Comput Sci, Vol 2, Part 2 • NL
[1976] *Foundations of Computer Science Vol 2, Part 2.* ED: APT, K.R. & BAKKER DE, J.W. • SER (**S** 1605) Math Centr Tracts 82 • PUBL (**X** 1121) Math Centr: Amsterdam 149pp • ISBN 90-6196-141-6, LC-No 77-368695

C 3549 Algebra, Vyp 1 (Irkutsk) • SU
[1972] *Algebra. Vyp. 1* ED: KOKORIN, A.I. & PENZIN, YU.G. • PUBL (**X** 1006) Irkutsk Gos Univ: Irkutsk 135pp • LC-No 74-645580 • REL PUBL (**C** 1443) Algebra, Vyp 2 (Irkutsk)

C 3588 Iter Induct Def & Subsyst of Anal: Rec Proof-Th Stud • D
[1981] *Iterated Inductive Definitions and Subsystems of Analysis: Recent Proof-Theoretical Studies* ED: BUCHHOLZ, W. & FEFERMAN, S. & POHLERS, W. & SIEG, W. • SER (**S** 3301) Lect Notes Math 897 • PUBL (**X** 0811) Springer: Heidelberg & New York vi+383pp • ISBN 3-540-11170-0, LC-No 81-21364

C 3599 Midwest Category Sem Rep 3 • D
[1969] *Reports of the Midwest Category Seminar. III* ED: MAC LANE, S. • SER (**S** 3301) Lect Notes Math 106 • PUBL (**X** 0811) Springer: Heidelberg & New York iii+247pp • ISBN 3-540-04625-9, LC-No 70-96694

C 3626 Rekursiv Mat Analiz • SU
[1970] *Rekursivnyj Matematicheskij Analiz (Recursive Mathematical Analysis)* ED: MINTS, G.E. • SER Prilozhenije 3 • PUBL (**X** 2027) Nauka: Moskva

C 3659 Intens Math • NL
[1985] *Intensional Mathematics* ED: SHAPIRO, S. • SER (**S** 3303) Stud Logic Found Math 113 • PUBL (**X** 0809) North Holland: Amsterdam • ISBN 0-444-87632-4, LC-No 84-18856

C 3724 Probl Gil'berta • SU
[1969] *Problemy Gil'berta (Hilbert's Problems)* PUBL (**X** 2027) Nauka: Moskva 240pp • LC-No 75-444471

C 3743 Probl Log & Metodol Nauk • SU
[1982] *Problemy Logiki i Metodologii Nauki (Problems of Logic and the Methodology of Science)* ED: TSELISHCHEV, V.V. • PUBL (**X** 2027) Nauka: Moskva 336pp • LC-No 82-197677

C 3747 Mat Log & Mat Lingvistika • SU
[1981] *Matematicheskaya Logika i Matematicheskaya Lingvistika (Mathematical Logic and Mathematical Linguistics)* ED: GLADKIJ, M. • PUBL (**X** 1434) Kalinin Gos Univ: Kalinin 172pp

C 3798 Mat Log, Mat Ling & Teor Algor • SU
[1983] *Matematicheskaya Logika, Matematicheskaya Lingvistike i Teoriya Algoritmov (Mathematical Logic, Mathematical Linguistics and Theory of Algorithms)* ED: GLADKIJ, A.V. • PUBL (**X** 1434) Kalinin Gos Univ: Kalinin 116pp • LC-No 84-157035

C 3807 Issl Neklass Log & Formal Sist • SU
[1983] *Issledovaniya po Neklassicheskim Logikam i Formal'nym Sistemam (Studies in Nonclassical Logics and Formal Systems)* ED: MIKHAJLOV, A.I. • PUBL (**X** 2027) Nauka: Moskva 360pp • LC-No 83-181942

C 3810 Log-Semat Issl (Tbilisi) • SU
[1981] *Logiko-Semanticheskie Issledovaniya (Studies in Logic and Semantics)* ED: MIKELADZE, Z.N. • PUBL (**X** 3790) Metsniereba: Tbilisi 144pp • LC-No 82-159969

C 3832 Avtom Issl Mat • SU
[1982] *Avtomatizatsiya Issledovanij v Matematike. Sbornik Nauchnykh Trudov (Automatization of Investigations in Mathematics. Collection of Scientific Works)* ED: KAPITONOVA, YU.V. • PUBL (**X** 2522) Akad Nauk Inst Kibernet: Kiev 98pp • LC-No 83-165025

C 3834 Lang, Logic and Method • NL
[1983] *Language, Logic, and Method* ED: COHEN, R.S. & WARTOFSKY, M.W. • SER (**S** 3311) Boston St Philos Sci 31 • PUBL (**X** 0835) Reidel: Dordrecht viii+464pp • LC-No 82-7558

C 3849 Modal & Relevant Log, Vyp 1 • SU
[1982] *Modal'nye i Relevantnye Logiki. Trudy Nauchno-Issledovatel'skogo Seminara po Logike Institute Filosofii AN SSSR Vyp.1 (Modal and Relevant Logics. Vol.1)* ED: SMIRNOV, V.A. & KARPENKO, A.S. • PUBL (**X** 0899) Akad Nauk SSSR : Moskva 107pp

C 3881 Prog in Cybern & Syst Res, Vol 11 • USA
[1982] *Progress in Cybernetics and Systems Research Vol. 11. Data Base Design, International Information Systems, Semiotic Systems, Artificial Intelligence, Cybernetics and Philosophy, Special Aspects* ED: TRAPPL, R. & FINDLER, N.V. & HORN, W. • PUBL (**X** 2437) Hemisphere Publ: Washington xv+601pp • ALT PUBL (**X** 0822) McGraw-Hill: New York • ISBN 0-89116-240-2, LC-No 75-6641

C 3953 Mat Log & Teor Algor • SU
[1982] *Matematicheskaya Logika i Teoriya Algoritmov (Mathematical Logic and the Theory of Algorithms)* ED: SOBOLEV, S.L. • SER Trudy Inst Mat 2 • PUBL (**X** 2642) Nauka: Novosibirsk 176pp

C 4055 Wajsberg: Logical Works • PL
[1977] *Wajsberg,M.: Logical Works* ED: SURMA, S.J. • PUBL (**X** 2882) Pol Akad Nauk: Wroclaw 216pp • LC-No 84-246718

C 4062 Montague: Formal Philos • USA
[1974] *Formal Philosophy. Selected Papers of Richard Montague* ED: THOMASON, R.H. • PUBL (**X** 0875) Yale Univ Pr: New Haven 369pp • ISBN 0-300-01527-5, LC-No 73-77159

Source Index

C 4080 Frege & Mod Grundlagenf • D
[1975] *Frege und die Moderne Grundlagenforschung* ED: THIEL, C. • PUBL (**X** 1275) Hain: Koenigstein viii + 168pp • LC-No 76-450988

C 4082 Handb Wiss Begriffe • D
[1980] *Handbuch Wissenschaftstheoretischer Begriffe, 3 Baende* ED: SPECK, J. • PUBL (**X** 0903) Vandenhoeck & Ruprecht: Goettingen 3 Vols: 780pp
• ISBN 3-525-03313-3 (V 1), ISBN 3-525-03314-1 (V 2), ISBN 3-525-03316-8 (V 3)

C 4085 Handb Philos Log • NL
[1983 & 1984] *Handbook of Philosophical Logic* ED: GABBAY, D. & GUENTHNER, F. • SER (**S** 3307) Synth Libr 164,165
• PUBL (**X** 0835) Reidel: Dordrecht 2 Vols: xi + 493pp,xi + 776pp
• ISBN 90-277-1542-4 (V1), ISBN 90-277-1604-8 (V2), LC-No 83-4277

C 4092 Philos als Wiss • D
[1981] *Philosophie als Wissenschaft. Essays in Scientific Philosophie* ED: MORSCHER, E. & NEUMAIER, O. & ZECHA, G.
• PUBL Comes Verlag: Bad Reichenhall

C 4137 Kleene: Mat Logika • SU
[1974] *Kleene,S.C.: Matematicheskaya Logika (Mathematical Logic)* PUBL (**X** 0885) Mir: Moskva 480pp
• TRANSL OF Kleene: Mathematical Logic
• REM Contains 2 additional articles of Mints,G.E.

C 4138 Feys: Modal Logika • SU
[1974] *Feys,R.: Modal'naya Logika (Modal Logic)* ED: MINTS, G.E. • PUBL (**X** 2027) Nauka: Moskva 520pp
• TRANSL OF Modal Logic
• REM Contains translations of Kripke,S.A. 1959 ID 07554, 1962 ID 07555, 1963 ID 07558, 1965 ID 07559 and of Schuette,K. 1968 ID 24822

C 4157 Rosser:Deux Esquisses Log • F
[1955] *Rosser,J.B.: Deux Esquisses de Logique.* SER Collection de Logique Mathematique, Ser A 7 • PUBL (**X** 0834) Gauthier-Villars: Paris 67pp • ALT PUBL (**X** 1313) Nauwelaerts: Louvain
• LC-No 55-3012

C 4181 Math Log & Formal Syst (Costa da) • USA
[1985] *Mathematical Logic and Formal Systems. A Collection of Papers in Honor of Newton C.A. da Costa* ED: ALCANTARA DE, L.P. • SER (**S** 3310) Lect Notes Pure Appl Math 94 • PUBL (**X** 1684) Dekker: New York xv + 297pp
• ISBN 0-8247-7330-6, LC-No 84-25984

C 4219 Feestbundel an H.J.Pos • NL
[1948] *Feestbundel aangeboden door Vrienden en Leerlingen aan Prof. Dr. H.J. Pos* PUBL (**X** 0809) North Holland: Amsterdam

C 4220 Rep Sem Found Anal • USA
[1963] *Reports of the Seminar on Foundation of Analysis* ED: KREISEL, G. • PUBL Stanford Univ: Stanford
• REM Mimeographed

C 4234 Sju Filos Studier • S
[1963] *Sju Filosofiska Studier Tillaegnade Anders Wedberg* PUBL Stockholm Univ: Stockholm

C 4297 Curry & Feys: Combin Logic • NL
[1958] *Curry,H.B. & Feys,R.: Combinatory Logic* SER (**S** 3303) Stud Logic Found Math • PUBL (**X** 0809) North Holland: Amsterdam xvi + 415pp

C 4349 Spec Design & Model Tools for PROEKT-ES Syst (Kiev) • SU
[1981] *(Special Design and Modeling Tools for PROEKT-ES Systems)* PUBL (**X** 2522) Akad Nauk Inst Kibernet: Kiev

C 4394 Kroneckers Werke • DDR
[1895-1931] *Kroneckers Werke, Vol I, II, III.1, III.2, IV, V* ED: HENSEL, K. • PUBL (**X** 1079) Teubner: Leipzig vi + 474pp, viii + 541pp, vii + 474pp, i + 216pp, x + 509pp, x + 528pp

C 4428 Way of Express of Sci • NL
[1933] *De Uitdrukkingswijze der Wetenschap (The Way of Expression of Science)* PUBL (**X** 1317) Noordhoff: Groningen

C 4429 Brouwer: Coll Works • NL
[1975] *Brouwer,L.E.J.: Collected Works. Vol 1* ED: HEYTING, A. • PUBL (**X** 0809) North Holland: Amsterdam xv + 628pp

C 4464 Husserl Stud • NL
[1984] *Husserl Studies* PUBL (**X** 1316) Nijhoff: Leiden & s'Gravenhage

C 4472 Undecidable Th • NL
[1953] *Undecidable Theories* SER (**S** 3303) Stud Logic Found Math • PUBL (**X** 0809) North Holland: Amsterdam

C 4557 Weyl: Selecta • CH
[1956] *Selecta Herman Weyl* PUBL (**X** 0804) Birkhaeuser: Basel

C 4648 Kreisel: Issl Teor Dokazat • SU
[1981] *Issledovaniya po Teorii Dokazatel'stv (Investigations in Proof Theory)* ED: MASLOV, S.YU. • PUBL (**X** 0885) Mir: Moskva 289pp
• REM Russian transl. of several papers of Kreisel,G.

C 4649 Paradiso di Cantor • I
[1978] *Il Paradiso di Cantor* ED: CELLUCCI, C. • PUBL (**X** 1732) Bibliopolis: Napoli

C 4652 Lect on Modern Math (Japanese) • J
[1965] *(Lectures on Modern Mathematics)* ED: SAATY, T.L.
• PUBL (**X** 3552) Iwanami Shoten: Tokyo
• TRANSL OF [1963] (**C** 1602) Lect on Modern Math

C 4655 Sci Thought 1900-1960 • GB
[1968] *Scientific Thought 1900-1960* ED: HARRE, R. • PUBL (**X** 0894) Oxford Univ Pr: Oxford

C 4659 Autom of Reasoning • D
[1983] *Automation of Reasoning. Vol 1, 2* ED: SIEKMANN, J. & WRIGHTSON, G. • PUBL (**X** 0811) Springer: Heidelberg & New York xii + 525pp,xii + 637pp
• ISBN 3-540-12043-2 (V1), ISBN 3-540-12044-0 (V2)
• REM Vol 1: Classical Papers on Computational Logic 1957-1966. Vol 2: Classical Papers on Computational Logic 1967-1970

C 4688 Prog in Math, Vol 12 • USA
[1972] *Progress in Mathematics. Vol 12: Algebra and Geometry* ED: GAMKRELIDZE, R.V. • PUBL (**X** 1332) Plenum Publ: New York ix + 254pp
• TRANSL OF [1972] (**J** 1501) Itogi Nauki Tekh, Ser Algeb, Topol, Geom 1968

C 4699 Friedman's Res on Found of Math • NL
[1985] *Harvey Friedman's Research on the Foundations of Mathematics* ED: HARRINGTON, L.A. & MORLEY, M. & SCEDROV, A. & SIMPSON, S.G. • PUBL (**X** 0809) North Holland: Amsterdam xvi+408pp
• IBSN 0-444-87834-3

Publishers

X 0272 *Editions du Griffon* (Neuchatel, CH) ISBN 2-88006

X 0740 *Shanghai Kexue Jishu Chubanshe (Scientific and Technical Press)* (Shanghai, TJ)

X 0801 *Academic Press* (New York, NY, USA & London, GB) ISBN 0-12

X 0803 *American Mathematical Society* (Providence, RI, USA) ISBN 0-8218

X 0804 *Birkhaeuser Verlag* (Basel, CH & Stuttgart, D & Cambridge, MA, USA) ISBN 3-7643

X 0805 *The Cambridge University Press.* (Cambridge, GB & New York, NY, USA & Melbourne, Vic, AUS) ISBN 0-521

X 0806 *VEB Deutscher Verlag der Wissenschaften* (Berlin, DDR) ISBN 3-326

X 0807 *Duke University Press* (Durham, NC, USA) ISBN 0-8223

X 0808 *W. Kohlhammer* (Stuttgart, D & Koeln, D & Berlin, D & Mainz, D) ISBN 3-17

X 0809 *North-Holland Publishing Company.* (Amsterdam, NL & Oxford, GB) ISBN 0-7204 • REL PUBL (**X** 0838) Amer Elsevier: New York

X 0811 *Springer-Verlag* (Heidelberg, D & Berlin, D & New York, NY, USA & Tokyo, J) ISBN 3-540, ISBN 0-387 • REL PUBL (**X** 1231) Barth: Leipzig & (**X** 0902) Springer: Wien

X 0812 *Wolters-Noordhoff* (Groningen, NL) ISBN 90-01 • REL PUBL (**X** 1317) Noordhoff: Groningen

X 0815 *The Clarendon Press* (Oxford, GB) ISBN 0-19 • REL PUBL (**X** 0894) Oxford Univ Pr: Oxford • REM This Imprint is Used for Academic Books Published by X0894.

X 0816 *Methuen & Company* (London, GB & New York, NY, USA & Agincourt, M, CDN & North Ryde, AUS) ISBN 0-416

X 0818 *Holt Rinehart & Winston* (New York, NY, USA & Toronto, ON, CDN & Artarmon, NSW, AUS & & Eastbourne,GB) ISBN 0-03 • REM In Australia & United Kingdom: Holt-Saunders: Eastbourne, GB & Artarmon, NSW, AUS

X 0819 *Prentice Hall* (Englewood Cliffs, NJ, USA & Brookvale, NSW, AUS & Scarborough, ON, CDN) ISBN 0-13 • REL PUBL (**X** 2040) Winthrop: Cambridge

X 0820 *Interscience Publishers* (New York, NY, USA & Chichester, GB) ISBN 0-470 • REL PUBL (**X** 0827) Wiley & Sons: New York

X 0821 *Wadsworth Publishing Co.* (Belmont, CA, USA & Crows Nest, NSW, AUS & Artamon, NSW, AUS) ISBN 0-534

X 0822 *McGraw-Hill Book Company* (New York, NY, USA & Roseville, NSW, AUS & Isando, ZA & Maidenhead, GB & Singapore, SGP & Scarborough, CDN & Sao Paulo, BR) ISBN 0-07 • REM Member Firms: 1) CRM/McGraw-Hill, Del Mar, CA, USA. 2) CTB/McGraw-Hill, Monterey, CA, USA. 3) Edutronics/McGraw-Hill, Los Angeles, CA, USA. 4) Instruto/McGraw-Hill, Paoli, PA, USA. 5) McGraw-Hill Continuing Education Center, Washington, DC, USA. 6) McGraw-Hill International Book Company, Singapore, SGP. 7) Sheperd's/McGraw-Hill, Colorado Springs, CO, USA. 8) McGraw-Hill do Brasil, Sao Paulo, BR.

X 0823 *B.G.Teubner* (Stuttgart, D) ISBN 3-519 • REM See also X1079

X 0824 *The Free Press* (New York, NY, USA) ISBN 0-02 • REL PUBL (**X** 0843) Macmillan : New York & London

X 0826 *Verlag Karl Alber* (Freiburg, D & Muenchen, D) ISBN 3-495 • REL PUBL (**X** 1279) Herder: Freiburg

X 0827 *J. Wiley & Sons* (New York, NY, USA & Chichester, GB & Rexdale, ON, CDN & Auckland, NZ) ISBN 0-471 • REL PUBL (**X** 0942) Norton: New York & (**X** 0820) Intersci Publ: New York & (**X** 0880) Ronald Press: New York & (**X** 2737) Israel Progr Sci Transl: Jerusalem

X 0832 *Addison-Wesley Publishing Co.* (Reading, MA, USA & London, GB & Don Mills, ON, CDN & North Ryde, NSW, AUS) ISBN 0-201 • REL PUBL (**X** 0867) Benjamin: Reading

X 0833 *Jerusalem Academic Press* (Jerusalem, IL)

X 0834 *Gauthier-Villars Editeur* (Paris, F) ISBN 2-04

X 0835 *D.Reidel Publishing Company* (Dordrecht, NL & Hingham, MA, USA) ISBN 90-277

X 0836 *Gordon & Breach, Science Publishers* (New York, NY, USA & London, GB & Paris, F) ISBN 0-677

X 0838 *American Elsevier Publishing Co.* (New York, NY, USA & Amsterdam, NL & London, GB) ISBN 0-444, ISBN 0-525, ISBN 0-87690 • REL PUBL (**X** 0809) North Holland: Amsterdam

X 0840 *Editions Presses Universitaires de France* (Paris, F) ISBN 2-13 • REL PUBL (**X** 2435) Pr Univ France Period: Paris

X 0842 *University of Wisconsin Press* (Madison, WI, USA & London, GB) ISBN 0-299 • REL PUBL (**X** 3828) Amer Univ Publ Group: London

X 0843 *Macmillan Publishing Company* (New York, NY, USA & Melbourne, Vic, AUS & London, GB & Toronto, Ont, CDN) ISBN 0-02 • REL PUBL (**X** 2375) Macmillan Journal: London & (**X** 0824) Free Press: New York

X 0844 *Giangiacomo Feltrinelli Editore* (Milano, I) ISBN 88-07

X 0845 *University of Notre Dame Press* (Notre Dame, IN, USA & London, GB) ISBN 0-268

X 0850 *Armand Colin, Editeur* (Paris, F) ISBN 2-200

X 0856 *Dunod, Editeur* (Paris, F) ISBN 2-04

X 0857 *Princeton University Press* (Princeton, NJ, USA & Guildford, GB) ISBN 0-691

X 0858 *Harvard University Press* (Cambridge, MA, USA & London, GB) ISBN 0-674

X 0859 *Hermann, Editeurs des Sciences et des Arts* (Paris, F) ISBN 2-7056

X 0860 *Cremonese Edizioni* (Firenze, I) ISBN 88-7083

X 0864 *Van Nostrand, Reinhold* (New York, NY, USA & Scarborough, ON, CDN & Mitcham, Vic, AUS & & Wokingham, GB & Florence, KY, USA) ISBN 0-442

X 0865 *The MIT Press* (Cambridge, MA, USA & London, GB) ISBN 0-262

X 0866 *Routledge & Kegan Paul* (Henley on Thames, GB & Boston, MA, USA) ISBN 0-7100, ISBN 0-7102

X 0867 *W.A. Benjamin* (Reading, MA, USA) ISBN 0-8053 • REL PUBL (**X** 0832) Addison-Wesley: Reading

X 0869 *Pergamon Press* (Oxford, GB & Elmsford, NY, USA & Rushcutters Bay, NSW, AUS & & Willowdale, ON, CDN & Paris, F) ISBN 0-08 • REL PUBL (**X** 0900) Vieweg: Wiesbaden

X 0871 *Academiei Republicii Socialiste Romania Editura (RSR)* (Bucharest, RO)

X 0872 *Albert Blanchard* (Paris, F)

X 0873 *Mouton et Cie.* (Paris, F) ISBN 2-7193

X 0874 *National Press Books* (Palo Alto, CA, USA) ISBN 0-87484

X 0875 *Yale University Press* (New Haven, CT, USA & London, GB) ISBN 0-300

X 0876 *Bibliographisches Institut* (Mannheim, D & Wien, A & Zuerich, CH) ISBN 3-411

X 0879 *Institut Superieur de Philosophie* (Louvain, B)

X 0880 *Ronald Press & Co.* (New York, NY, USA) ISBN 0-8260 • REL PUBL (**X** 0827) Wiley & Sons: New York

X 0882 *Uppsala Universitet, Filosofiska Foereningen och Filosofiska Institutionen* (Uppsala, S)

X 0884 *Paideia Press* (Memphis, TN, USA) ISBN 0-912490

X 0885 *Izdatel'stvo Mir* (Moskva, SU)

X 0886 *Leicester University Press* (Leicester, GB) ISBN 0-7185

X 0887 *Raven Press* (New York, NY, USA) ISBN 0-89004, ISBN 0-911216

X 0890 *Wissenschaftliche Buchgesellschaft* (Darmstadt, D) ISBN 3-534

X 0892 *Georg Olms Verlag* (Hildesheim, D & New York, NY, USA & Zuerich, CH) ISBN 3-487

X 0893 *Les Presses de l'Universite de Montreal* (Montreal, PQ, CDN) ISBN 2-7606

X 0894 *Oxford University Press* (Oxford, GB & London, GB & Melbourne, Vic, AUS & Don Mills, ON, CDN & Nairobi, EAK & Auckland, NZ & Petaling Jaya, MAL & New York, NY, USA & Karachi, PAK & Harare, ZW) ISBN 0-19 • REL PUBL (**X** 0815) Clarendon Pr: Oxford

X 0895 *Latvijas Valsts Universitate* (Riga, SU)

X 0898 *Izdatel'stvo Moskovskogo Gosudarstvennogo Universiteta* (Moskva, SU)

X 0899 *Izdatel'stvo Akademii Nauk SSSR* (Moskva, SU)

X 0900 *Vieweg, Friedrich & Sohn Verlagsgesellschaft* (Wiesbaden, D) ISBN 3-528 • REL PUBL (**X** 0869) Pergamon Pr: Oxford

X 0902 *Springer-Verlag* (Wien, A) ISBN 3-211 • REL PUBL (**X** 0811) Springer: Heidelberg & New York

X 0903 *Vandenhoeck & Ruprecht* (Goettingen, D) ISBN 3-525 • REM Member Firms: 1) E. Klotz Verlag, Goettingen, D. 2) Verlag der Medizinischen Psychologie Goettingen, Dr D.& Dr A. Ruprecht, Goettingen, D

X 0905 *Boringhieri Editore* (Torino, I) ISBN 88-339

X 0908 *Universitaet Bonn, Mathematisches Institut* (Bonn, D)

X 0909 *Cedam* (Padova, I)

X 0910 *Aschendorffsche Verlagsbuchhandlung* (Muenster, D) ISBN 3-402

X 0911 *Akademie Verlag* (Berlin, DDR)

X 0913 *Novosibirskij Gosudarstvennyj Universitet* (Novosibirsk, SU)

X 0918 *Ernst Klett Verlag* (Stuttgart, D) ISBN 3-12 • REL PUBL (**X** 1255) Deuticke: Wien

X 0923 *Universidade Federal de Pernambuco, Instituto de Matematica* (Recife, BR)

X 0924 *New York University Press* (New York, NY, USA) ISBN 0-8147

X 0926 *University of California Press* (Berkeley, CA, USA & London, GB) ISBN 0-520 • REL PUBL (**X** 1291) Johns Hopkins Univ Pr: Baltimore

X 0928 *Akademiai Kiado, Publishing House of the Hungarian Academy of Sciences.* (Budapest, H) ISBN 963-05

X 0938 *Izdatel'stvo Leningradskogo Universiteta* (Leningrad, SU)

X 0942 *W.W.Norton & Co.* (New York, NY, USA & London, GB) ISBN 0-393 • REL PUBL (**X** 0827) Wiley & Sons: New York

X 0978 *Schulthess Polygraphischer Verlag* (Zuerich, CH) ISBN 3-7255

X 0981 *Random House* (New York, NY, USA & Mississauga, ON, CDN) ISBN 0-394

X 0999 *Centre National de la Recherche Scientifique (CNRS), Institut Blaise Pascal* (Paris, F)

X 1006 *Irkutskij Gosudarstvennyj Universitet* (Irkutsk, SU)

X 1015 *University of Toronto Press* (Toronto, ON, CDN & Buffalo, NY, USA) ISBN 0-8020

X 1034 *Panstwowe Wydawnictwo Naukowe (PWN)* (Warsaw, PL) ISBN 83-01

X 1051 *Rijksuniversiteit Utrecht, Mathematisch Instituut.* (Utrecht, NL)

X 1071 *F. Alcan* (Paris, F)

X 1079 *B.G.Teubner Verlagsgesellschaft* (Leipzig, DDR) ISBN 3-519 • REM See also X0823

X 1080 *Scottish Academic Press* (Edinburgh, GB) ISBN 0-7073

X 1096 *Basil Blackwell* (Oxford, GB) ISBN 0-631, ISBN 0-632, ISBN 0-86286, ISBN 0-86793 • REM Also: Blackwell Scientific Publications: Oxford, GB & Carlton, Vic, AUS

X 1121 *Mathematisch Centrum* (Amsterdam, NL) ISBN 90-6196

X 1163 *Almqvist & Wiksell Foerlag* (Stockholm, S & Bromma, S & Goeteborg, S & Malmoe, S) ISBN 91-20

X 1169 *Akademische Verlagsgesellschaft* (Wiesbaden, D & Leipzig, DDR) ISBN 3-400

X 1170 *Sonne Verlag* (New Dehli, IND)

X 1172 *Les Editions de l'Office Central de Librairie S.A.R.L. (O.C.D.L.)* (Paris, F) ISBN 2-7043

X 1174 *Walter de Gruyter* (Berlin, D) ISBN 3-11 • REM Member Firms: 1) de Gruyter: Hawthorne, NY, USA. 2) Mouton Publishers: Berlin, D

X 1205 *Junker und Duennhaupt Verlag* (Berlin, D)

X 1226 *Academia* (Prague, CS) • REM Publishing House of the Czechoslovak Academy of Science

X 1231 *Johann Ambrosius Barth* (Leipzig, DDR) ISBN 3-335 • REL PUBL (**X** 0811) Springer: Heidelberg & New York

X 1234 *G.Bell & Sons* (London, GB) ISBN 0-7135

X 1238 *The Bobbs-Merril Company* (Indianapolis, IN, USA) ISBN 0-672 • REL PUBL (**X** 1320) Odyssey Pr: Indianapolis

X 1240 *G. Braun* (Karlsruhe, D) ISBN 3-7650 • REM Formerly: G. Braun'sche Hofbuchdruckerei und Verlag: Karlsruhe, D

X 1241 *British Academy* (London, GB) ISBN 0-85672, ISBN 0-902732

X 1255 *Franz Deuticke, Verlagsgesellschaft* (Wien, A) ISBN 3-7005 • REL PUBL (**X** 0918) Klett: Stuttgart

X 1261 *Edinburgh University Press* (Edinburgh, GB) ISBN 0-85224

X 1262 *Giuliu Einaudi, Editore* (Torino, I) ISBN 88-06

X 1265 *Gustav Fischer Verlag* (Stuttgart, D) ISBN 3-437

X 1268 *Orell Fuessli Verlag* (Zuerich, CH) ISBN 3-280

X 1270 *Greenwood Press* (Westport, CT, USA & San Francisco, CA, USA & London, GB) ISBN 0-313, ISBN 0-8371, ISBN 0-89930

X 1275 *Anton Hain K.G. Meisenheim Verlag* (Koenigstein, D & Meisenheim, D) ISBN 3-445 • REL PUBL (**X** 1588) Athenaeum/Hain/Hanstein: Koenigstein

X 1277 *Hayden Book Company* (Rochelle Park, NY, USA) ISBN 0-8104 • REL PUBL (**X** 1354) Spartan Books : Sutton

X 1279 *Herder & Co.* (Freiburg, D & Roma, I) ISBN 3-451

X 1282 *S.Hirzel Verlag* (Stuttgart, D) ISBN 3-7776

X 1285 *University of Illinois Press* (Urbana, IL, USA & London, GB) ISBN 0-252 • REL PUBL (**X** 3828) Amer Univ Publ Group: London

X 1286 *Indiana University Press* (Bloomington, IN, USA & London, GB) ISBN 0-253 • REL PUBL (**X** 3828) Amer Univ Publ Group: London

X 1291 *Johns Hopkins University Press* (Baltimore, MD, USA) ISBN 0-8018

X 1298 *Mathematical Association of America* (Washington, DC, USA) ISBN 0-88385

X 1299 *Magnes Press* (Jerusalem, IL) ISBN 965-223

X 1313 *Nauwelaerts, Beatrice* (Louvain, B) ISBN 2-900014

X 1316 *Martinus Nijhoff Social Sciences Devision* (Leiden, NL & s'Gravenhage, NL) ISBN 90-207, ISBN 90-247 • REL PUBL (**X** 4624) Kluwer-Nijhoff: Boston

X 1317 *P.Noordhoff International Publishing* (Groningen, NL & Leiden, NL) ISBN 90-01 • REL PUBL (**X** 0812) Wolters-Noordhoff : Groningen & (**X** 1352) Sijthoff: Leiden • REM Now: Sijthoff & Noordhoff International Publishers: Leiden, NL

X 1319 *La Nuova Italia Editrice* (Firenze, I) ISBN 88-221

X 1320 *Odyssey Press* (Indianapolis, IN, USA) ISBN 0-8399 • REL PUBL (**X** 1238) Bobbs-Merril: Indianapolis

X 1323 *Oliver & Boyd* (Edinburgh, GB) ISBN 0-05

X 1324 *Open Court Publishing Co.* (LaSalle, IL, USA) ISBN 0-87548, ISBN 0-89688

X 1328 *Pennsylvania State University, Department of Mathematics.* (University Park, PA, USA)

X 1330 *Pitman Publishing* (Belmont, CA, USA & London, GB & Toronto, ON, CDN & & Carlton, Vic, AUS & Johannesburg, ZA & Wellington, NZ & & Auckland, NZ) ISBN 0-273, ISBN 0-8224, ISBN 0-915092, ISBN 0-272

X 1332 *Plenum Publishing Corporation* (New York, NY, USA & London, GB) ISBN 0-306 • REM Also Called Plenum Press

X 1342 *Royal Irish Academy* (Dublin, IRL) ISBN 0-901714

X 1346 *G.C.Sansoni, Editrice* (Firenze, I) ISBN 88-383

X 1349 *Editions du Seuil* (Paris, F) ISBN 2-02

X 1352 *A.W.Sijthoff International Publishing Co.* (Leiden, NL) ISBN 90-218, ISBN 90-286 • REL PUBL (**X** 1317) Noordhoff: Groningen

X 1354 *Spartan Books* (Sutton, GB & Rochelle Park, NJ, USA) ISBN 0-905532 • REL PUBL (**X** 1277) Hayden: Rochelle Park

X 1355 *Stanford University Press* (Stanford, CA, USA) ISBN 0-8047

X 1358 *University of Texas Press* (Austin, TX, USA & London, GB) ISBN 0-292 • REL PUBL (**X** 3828) Amer Univ Publ Group: London

X 1364 *Vita e Pensiero, Publicazioni della Universita Cattolica* (Milano, I) ISBN 88-343

X 1367 *University of Washington Press* (Seattle, WA, USA & London, GB) ISBN 0-295 • REL PUBL (**X** 3828) Amer Univ Publ Group: London

X 1375 *Nicola Zanichelli Editore* (Bologna, I) ISBN 88-08

X 1422 *Editions Bordas* (Paris, F) ISBN 2-04

X 1434 *Kalininskij Gosudarstvennyj Universitet* (Kalinin, SU)

X 1493 *C.W.K.Gleerup Bokfoerlag a.B.* (Lund, S) ISBN 91-40

X 1512 *Rheinische Friedrich-Wilhelms-Universitaet Bonn, Institut fuer Informatik.* (Bonn, D)

X 1554 *Universitetsforlaget* (Oslo, N & Bergen, N & Tromsoe, N & New York, NY, USA) ISBN 82-00 • REM Also: UB-Forlaget: Oslo, N

X 1588 *Verlagsgruppe Athenaeum/ Hain/ Hanstein* (Koenigstein/Ts, D) • REL PUBL (**X** 1275) Hain: Koenigstein & (**X** 2665) Athenaeum: Frankfurt

X 1599 *Aarhus Universitet, Matematisk Institut* (Aarhus, DK) ISBN 87-87436

X 1623 *Universite de Paris VI, Institut Henri Poincare* (Paris, F) • REM The University Has Now Been Divided Into Smaller Units. The Mathematics Department Is Now: Universite de Paris VII-Pierre et Marie Curie, Secretariat Mathematique

X 1656 *Izdatel'stvo Inostr. Lit.* (Moskva, SU)

X 1684 *Marcel Dekker* (New York, NY, USA & Basel, CH) ISBN 0-8247 • REL PUBL (**X** 2442) Dekker Journal: New York

X 1714 *Commission of the European Communities.* (Bruxelles, B) ISBN 92-825, ISBN 92-826, ISBN 92-827, ISBN 92-828

X 1726 *Bunkashobo-Hakubu-Sha* (Tokyo, J) ISBN 4-8301

X 1732 *Bibliopolis* (Napoli, I) ISBN 88-7088

X 1758 *Ars Polona* (Warsaw, PL)

X 1761 *Hekdosis Hellinikis Mathematikes Hetaireias-Greek Mathematical Society* (Athens, GR)

X 1763 *Royal Society* (London, GB) ISBN 0-85403

X 1764 *Ksiaznica Atlas T.N.S.W.* (Warsaw, PL & Lwow, PL)

X 1773 *VEDA - Vydavatelstvo Slovenskej Akademie Vied* (Bratislava, CS)

X 1776 *Les editions Sociales* (Paris, F) ISBN 2-209

X 1781 *Tecnos* (Madrid, E) ISBN 84-309

X 1821 *Real Academia de Ciencias Exactas, Fisicas y Naturales* (Madrid, E)

X 1876 *Kexue Chubanshe (Science Press)* (Beijing, TJ)

X 1941 *Nanyang University Publications* (Singapore, MAL)

X 1994 *Koninklijke van Gorcum (Royal van Gorcum)* (Assen, NL & Amsterdam, NL) ISBN 90-232

X 2027 *Izdatel'stvo Nauka* (Moskva, SU & Alma-Ata, SU & Leningrad, SU & Novosibirsk, SU)

X 2040 *Winthrop* (Cambridge, MA, USA) ISBN 0-87626 • REL PUBL (**X** 0819) Prentice Hall: Englewood Cliffs

X 2091 *Consejo Superior de Investigaciones Cientificas.* (Madrid, E) ISBN 84-00

X 2121 *Accademia Nazionale dei Lincei* (Roma, I) ISBN 88-218

X 2152 *Tokai University Press* (Tokyo, J) ISBN 4-486

X 2179 *IEEE (Institute of Electrical and Electronics Engineers* (New York, NY, USA & Long Beach, CA, USA & Piscataway, CA, USA) ISBN 0-87942 • REM Section: IEEE Computer Society: Long Beach, CA, USA. Section: IEEE United States Activities Commitee: Piscataway, CA, USA

X 2193 *Indian Mathematical Society* (Bombay, IND)

X 2195 *Peeters S.P.R.L.* (Louvain, B) ISBN 2-8017

X 2197 *Tohoku University, Mathematical Institute* (Sendai, J)

X 2199 *Izdatel'stvo Naukova Dumka* (Kiev, SU)

X 2203 *Philosophy of Science Association* (East Lansing, MI, USA) ISBN 0-917586

X 2204 *American Institute of Physics* (New York, NY, USA) ISBN 0-88318

X 2205 *(ACM) Association for Computing Machinery* (New York, NY, USA) ISBN 0-89791

X 2206 *Accademia delle Science* (Torino, I)

X 2208 *Taylor & Francis* (London, GB) ISBN 0-85066

X 2209 *Izdatel'stvo Akademii Nauk Gruzinskoj SSR* (Tbilisi, SU)

X 2210 *Academie Royale de Belgique* (Bruxelles, B) ISBN 2-8031

X 2211 *Friedrich Schiller Universitaet* (Jena, DDR)

X 2212 *Ossolineum, Publishing House of the Polish Academy of Sciences* (Wroclaw, PL & Warsaw, PL) • REL PUBL (**X** 2885) Zakl Narod Wyd Pol Ak: Wroclaw & (**X** 2882) Pol Akad Nauk: Wroclaw

X 2214 *Dialectica* (Bienne/Biel, CH)

X 2215 *Indiana University, Department of Philosophy* (Bloomington, IN, USA)

X 2219 *Humboldt-Universitaet zu Berlin* (Berlin, DDR)

X 2220 *Buchdruckerei und Verlag Leemen* (Zuerich, CH)

X 2224 *Matematisk Institut* (Bergen, N)

X 2225 *Izdatel'stvo Akademii Nauk Armyanskoj SSR* (Erevan, SU)

X 2227 *Scuola Normale Superiore di Pisa* (Pisa, I) ISBN 88-7642

X 2228 *Hegler Institute* (La Salle, IL, USA)

X 2230 *Isdevnieciba Zinatne* (Riga, SU)

X 2235 *Vsesoyuznyj Institut Nauchnoj i Tekhnicheskoj Informatsii (VINITI), Gosudarstvennyj Komitet SSSR po Nauke: Tekhnike (GKNT SSSR), Akademiya Nauk (AN) SSSR* (Moskva, SU)

X 2237 *Izdatelstvo na Bulgarskata Akademia na Naukite (Publishing House of the Bulgarian Academy of Sciences.)* (Sofia, BG)

X 2238 *Sociedad Colombiana de Matematicas* (Bogota, CO) ISBN 958-95081

X 2241 *New York Academy of Sciences* (New York, NY, USA) ISBN 0-89072, ISBN 0-89766

X 2242 *Osaka University, Department of Mathematics* (Osaka, J)

X 2243 *University of Osaka Prefecture, Department of Mathematics* (Osaka, J)

X 2244 *Societe Mathematique de France* (Paris, F) ISBN 2-85629

X 2248 *Drushtvo Matematichara i Fizichara Sr Hrvatske.* (Zagreb, YU)

X 2254 *Universidad Nacional Autonoma de Mexico, Instituto de Matematicas* (Mexico City, MEX)

X 2256 *Casa Editrice Felice le Monnier* (Firenze, I)

X 2257 *John Benjamins B.V.* (Amsterdam, NL & Philadelphia, PA, USA) ISBN 90-272

X 2258 *Periodika* (Tallinn, SU)

Source Index

X 2265 *Vychislitel'nyj Tsentr Akademii Nauk SSSR* (Moskva, SU)

X 2299 *Universidade de Campinas "EDUC" (State University of Campinas)* (Sao Paulo, BR)

X 2364 *Meulenhoff-Bruna B.V.* (Amsterdam, NL) ISBN 90-280

X 2375 *MacMillan Journals* (London, GB) ISBN 0-333 • REL PUBL (**X** 0843) Macmillan : New York & London

X 2378 *Society for Industrial and Applied Mathematics (SIAM)* (Philadelphia, PA, USA) ISBN 0-89871

X 2379 *University of Houston, Department of Mathematics* (Houston, TX, USA)

X 2382 *Institute of Mathematics and its Applications* (Southend-on-Sea, GB) ISBN 0-905091, ISBN 0-9501159

X 2390 *Institut Mittag-Leffler* (Djursholm, S)

X 2403 *AFIPS Press* (Montvale, NJ, USA) ISBN 0-88283

X 2421 *University of Tokyo, Faculty of Science* (Tokyo, J)

X 2422 *Canadian Philosophical Association* (Montreal, PQ, CDN) ISBN 0-9690153

X 2426 *Universa, P.V.B.A.* (Wetteren, B) ISBN 90-6281

X 2427 *Asociacion Venezolana para el Avance de la Ciencia* (Caracas, YV)

X 2434 *University of Tokyo College of General Education* (Tokyo, J)

X 2435 *Presses Universitaires de France, Service Periodiques* (Paris, F) ISBN 2-13 • REL PUBL (**X** 0840) Pr Univ France: Paris

X 2436 *Indian Academy of Sciences, Publications Department* (Bangalore, IND)

X 2437 *Hemisphere Publishing* (Washington, DC, USA & New York, NY, USA & London, GB) ISBN 0-89116

X 2440 *Universita Degli Studi di Modena, Seminario Matematico e Fisico* (Modena, I)

X 2441 *Kyoto University, Research Institute for Mathematical Sciences* (Kyoto, J)

X 2442 *Marcel Dekker Journals.* (New York, NY, USA) ISBN 0-8247 • REL PUBL (**X** 1684) Dekker: New York

X 2443 *Izdatel'stvo Nauka, Otdelenie v Kazakhstane SSR* (Alma-Ata, SU)

X 2446 *Martin-Luther-Universitaet Halle-Wittenberg* (Halle, DDR)

X 2451 *Uniwersitet Jagiellonskie, Instytut Matematiyczny* (Krakow, PL)

X 2455 *Yokohama City University, Department of Mathematics* (Yokohama, J)

X 2456 *Tamkang College of Arts and Sciences, Research Institute of Mathematics* (Taipei, RC)

X 2457 *Allerton Press* (New York, NY, USA) ISBN 0-89864

X 2462 *Cairo University Press* (Cairo, ETH)

X 2463 *Tartuskij Gosudarstvennyj Universitet* (Tartu, SU)

X 2464 *Mathematisches Seminar, Selbstverlag* (Giessen, D)

X 2465 *Kinokuniya Company* (Tokyo, J) ISBN 4-314

X 2467 *Societatea de Stiinte Matematice din Republica Socialista Romania (RSR)* (Bucharest, RO)

X 2469 *Data, A/S af 2. april 1971* (Copenhagen, DK) ISBN 87-980512

X 2470 *South African Association for the Advancement of Science - Suid Afrikaanse Genootschap vir die Bevordering van die Wetenschap.* (Pretoria, ZA)

X 2472 *Centro Superiore di Logica e Scienze Comporate, Editrice Franco Spisani* (Bologna, I)

X 2473 *Belfort Graduate School of Science; Yeshiva University* (New York, NY, USA)

X 2476 *Journal of Philosophy Inc.* (New York, NY, USA) ISBN 0-931206

X 2479 *Rocky Mountain Mathematics Consortium* (Tempe, AZ, USA)

X 2483 *Hiroshima University, Department of Mathematics* (Hiroshima, J)

X 2492 *Japan Association for the Philosophy of Science* (Tokyo, J)

X 2495 *Suid Afrikaanse Akademie vir Wetenskap en Kuns* (Pretoria, ZA) ISBN 0-949976

X 2496 *Suomen Fyysikkoseura* (Helsinki, SF)

X 2499 *American Association for the Advancement of Science* (Washington, DC, USA) ISBN 0-87168

X 2510 *Elm (Izdatel'stvo Akademiya Nauk Azerbajdzhanskoj SSR)* (Baku, SU)

X 2511 *Universita Karlova, Matematicky Ustav* (Prague, CS)

X 2512 *Institut Matematiki Akademii Nauk SSSR* (Moskva, SU)

X 2513 *Mathematical Association* (Leicester, GB) ISBN 0-906588

X 2514 *Institut de Mathematiques* (Geneve, CH)

X 2515 *Universita Karlova, Fakulta Matematiky a Fyziky* (Prague, CS)

X 2516 *Union Matematica Argentina* (Buenos Aires, RA)

X 2518 *Aberdeen University Press* (Aberdeen, GB) ISBN 0-900015, ISBN 0-08

X 2520 *National Academy of Sciences, Transportation Research Board* (Washington, DC, USA) ISBN 0-309

X 2522 *Akademiya Nauk USSR, Nauchnyj Sovet po Kibernetike, Institut Kibernetiki.* (Kiev, SU)

X 2524 *Encyclopaedia Universalis* (Paris, F) ISBN 2-85229

X 2526 *Gazeta de Matematica* (Lisboa, P)

X 2529 *Consultants Bureau* (New York, NY, USA) • REL PUBL (**X** 1332) Plenum Publ: New York

X 2530 *Societe Mathematique de Belgique* (Bruxelles, B)

X 2531 *Japan Academy* (Tokyo, J)

X 2532 *Fratelli Fusi, Tipografia Successon* (Pavia, I)

X 2636 *Verlag von Veit & Co.* (Leipzig, DDR)

X 2641 *Izdatel'stvo Nauka Leningradskoe Otdelenie* (Leningrad, SU)

X 2642 *Izdatel'stvo Nauka Sibirskoe Otdelenie* (Novosibirsk, SU)

X 2643 *Izdatel'stvo Sovetskoe Radio* (Moskva, SU)

X 2646 *Aalborg University Press = Aalborg Universitetsforlag* (Aalborg, DK) ISBN 87-7307

X 2652 *Institut Matematiki Sibirskogo Otdeleniya Akademii Nauk SSSR (SOAN SSSR)* (Novosibirsk, SU)

X 2665 *Athenaeum Verlag* (Frankfurt, D) ISBN 3-7610 • REL PUBL (**X** 1588) Athenaeum/Hain/Hanstein: Koenigstein

X 2692 *University of Queensland Press* (St. Lucia, Qld, AUS) ISBN 0-7022

X 2696 *Humanities Press* (Atlantic Highlands, NY, USA) ISBN 0-391

X 2698 *Departamento de Logica y Filosofia de la Ciencia, Universidad de Valencia* (Valencia, E)

X 2717 *Franz Steiner Verlag* (Wiesbaden, D & Stuttgart, D) ISBN 3-515

X 2719 *Fundacao Calouste Gulbenkian* (Lisbon, P) ISBN 972-15

X 2721 *G. Meteau* (Harlow, GB)

X 2728 *Hoelder-Pichler-Tempsky* (Wien, A) ISBN 3-209

X 2732 *Institut National de Recherche en Informatique et en Automatique (INRIA)* (Le Chesnay Cedex, F) ISBN 2-7261 • REM Also Called: Institut de Recherche en Informatique et en Automatique (IRIA)

X 2737 *Israel Program for Scientific Translations* (Jerusalem, IL) ISBN 0-7065 • REL PUBL (**X** 0827) Wiley & Sons: New York

X 2741 *Shtiintsa* (Kishinev, SU)

X 2773 *Dr.J. Koenigshausen & Dr.Th. Neumann* (Wuerzburg, D) ISBN 3-88479

X 2786 *Universitet i Oslo, Matematisk Institut* (Oslo, N) ISBN 82-553

X 2802 *Gosudarstvennyj Pedagogicheskij Institut* (Moskva, SU)

X 2812 *The Philosophical Society of Finland* (Helsinki, SF) ISBN 951-95053

X 2854 *U.E.R. (UER) Mathematique, Universite Paris XI* (Paris-Orsay, F)

X 2863 *Upside Down A Book Company* (Yarra Glen, Vic, AUS) ISBN 0-949865

X 2882 *Wydawnictwo Polskiej Akademii Nauk (Publisher of the Polish Academy of Science)* (Wroclaw, PL) • REL PUBL (**X** 2885) Zakl Narod Wyd Pol Ak: Wroclaw & (**X** 2212) Ossolineum: Wroclaw

X 2885 *Zaklad Narodowy imienia Ossolinskich, Wydawnictwo Polskiej Akademii Nauk* (Wroclaw, PL) ISBN 83-04 • REL PUBL (**X** 2882) Pol Akad Nauk: Wroclaw & (**X** 2212) Ossolineum: Wroclaw

X 3152 *Technische Hogeschool Eindhoven* (Eindhoven, NL)

X 3203 *Centre Universitaire de Luxembourg* (Luxembourg, L)

X 3215 *Mathematisch-Naturwissenschaftliche Fakultaet der Rheinisch-Westfaelischen Technischen Hochschule Aachen.* (Aachen, D)

X 3249 *The Weizmann Institute of Science* (Rehovot, IL) ISBN 965-281

X 3292 *Juris Druck & Verlag* (Zuerich, CH) ISBN 3-260

X 3333 *Unknown Publisher*

X 3353 *Akademiya Nauk SSSR. Institut Istor Filol Filos* (Novosibirsk, SU)

X 3407 *"Vysshaya Shkola"* (Moskva, SU)

X 3552 *Iwanami Shoten Publishers* (Tokyo, J) ISBN 4-00

X 3559 *Erevanskij Gosudarstvennyj Universitet* (Erevan, SU)

X 3562 *Ecole Normale Superieure de Jeunes Filles* (Paris, F) ISBN 2-85929

X 3602 *Kraus Reprint* (Vaduz, FL & Nendeln, FL) ISBN 3-262 • REM Parent Firm: Kraus Thomson Organisation: Vaduz, FL

X 3604 *Istituto Nazionale di Alta Matematica (INDAM)* (Roma, I)

X 3605 *Izdatel'stvo Kazanskogo Gosudarstvennogo Universiteta* (Kazan', SU)

X 3636 *Tokyo Tosho* (Tokyo, J) ISBN 4-489

X 3718 *L'Enseignement Mathematique, Universite de Geneve* (Geneve, CH)

X 3725 *Bolyai Janos Matematikai Tarsulat (Janos Bolyai Mathematical Society)* (Budapest, H)

X 3727 *Beograd. Matematicki Institut* (Belgrade, YU)

X 3790 *Metsniereba* (Tbilisi, SU)

X 3805 *The Blindern Theoretic Research Team* (Oslo, N)

X 3812 *Universita di Siena, Dipartimento di Matematica, Scuola di Specializzazione in Logica Matematica* (Siena, I)

X 3828 *American University Publishers Group* (London, GB) • REL PUBL (**X** 0842) Univ Wisconsin Pr: Madison & (**X** 1285) Univ Ill Pr: Urbana & (**X** 1286) Indiana Univ Pr : Bloomington & (**X** 1358) Univ Texas Pr: Austin & (**X** 1367) Univ Washington: Seattle

X 4030 *Univerzitet u Novom Sadu, Institut za Matematiku* (Novi Sad, YU)

X 4562 *Udmurtskij Gosudarstvennij Universiteta* (Izhevsk, SU)

X 4624 *Kluwer - Nijhoff Publishing* (Boston, MA, USA) ISBN 0-89838 • REL PUBL (**X** 1316) Nijhoff: Leiden & s'Gravenhage

X 4636 *Johann Wolfgang Goethe Universitaet zu Frankfurt am Main. Fachbereich Mathematik* (Frankfurt, D)

X 4721 *Libreria Fratelli Bocca & C. Clausen* (Torino, I & Roma, I)

Miscellaneous Indexes

External classifications

This index complements the Subject Index at the beginning of this volume; it lists the items which, in addition to classifications in the present volume, have classifications *external to this volume*. These items are ordered by external classification code and within each code by author (the first alphabetically in the case of multi-author items), year and identification number (thus an item with, for example, two external classifications occurs twice in this listing). This index provides another way to search the bibliography. With it, the user can easily identify those items in this volume classified also in some area external to this volume.

B03

Calude, C. [1978] 52071
Mayoh, B.H. [1970] 17641
Peano, G. [1889] 19497

B05

Beth, E.W. [1960] 01171
Brouwer, L.E.J. [1955] 02051
Call, R.L. [1984] 40582
Chubaryan, A.A. [1972] 17157
Chubaryan, A.A. [1975] 76827
Church, A. [1928] 02122
Cook, S.A. [1974] 25005
Cook, S.A. [1975] 28181
Cook, S.A. [1979] 56282
Destouches-Fevrier, P. [1945] 02962
Gleason, G.G. [1974] 23612
Goodstein, R.L. [1958] 05182
Greniewski, H. [1951] 05333
Luckhardt, H. [1969] 08380
Lukasiewicz, J. [1953] 25951
Nieland, J.J.F. [1967] 09951
Plotkin, J.M. [1982] 35398
Popov, S.V. [1974] 15197
Popov, S.V. [1975] 18344
Popov, S.V. [1976] 25598
Popov, S.V. [1977] 77440
Price, Robert [1961] 10756
Reznikoff, I. [1963] 11135
Schmidt, H.A. [1960] 12892
Schuette, K. [1958] 13008
Smyth, M.B. [1974] 12559
Stepanov, V.A. [1985] 46515
Tarski, A. [1938] 13420

B10

Ackermann, W. [1925] 00104
Ackermann, W. [1952] 00124
Ackermann, W. [1953] 00125
Bellot, P. [1985] 42554
Beneyto, R. [1971] 01031
Bennett, D.W. [1973] 01038
Bennett, D.W. [1977] 21957
Benthem van, J.F.A.K. [1974] 03889
Beth, E.W. [1955] 01163
Beth, E.W. [1962] 04250
Bing, K. [1968] 01210
Boolos, G. [1984] 40442
Bowen, K.A. [1972] 01505
Bunder, M.W. [1981] 55277
Chubaryan, A.A. [1977] 52021
Chubaryan, A.A. [1982] 40752
Church, A. [1932] 02123
Church, A. [1933] 02124
Craig, W. [1957] 02508
Craig, W. [1957] 02509
Craig, W. [1957] 02510
Craig, W. [1960] 02513
Curry, H.B. [1950] 04222
Curry, H.B. [1952] 02620
Dragalin, A.G. [1972] 03610
Dreben, B. [1971] 17043
Dreben, B. [1971] 17045
Funahashi, S. [1979] 77023
Gentzen, G. [1935] 24221
Giovannetti, E. [1983] 35624
Glubrecht, J.-M. [1982] 35619
Goedel, K. [1934] 35971
Gregg, J.R. [1970] 05317
Gruenberg, T. [1983] 43736
Gupta, H.N. [1968] 14682
Hailperin, T. [1969] 05608
Hajek, P. [1970] 05521
Hanazawa, M. [1979] 54329
Hauck, J. [1972] 62383
Hendry, H.E. [1975] 18186
Henkin, L. [1954] 05900
Henkin, L. [1957] 05910
Herbrand, J. [1928] 05972
Herbrand, J. [1929] 05967
Herbrand, J. [1929] 05968
Herbrand, J. [1930] 20866
Hook, J.L. [1985] 41804
Hosoi, T. [1966] 06242
Jervell, H.R. [1973] 33265
Kalish, D. [1957] 24907
Kanger, S. [1957] 22373
Kanger, S. [1963] 06888
Kreisel, G. [1950] 07481
Kreisel, G. [1953] 20815
Kuzichev, A.S. [1977] 52560
Lambert Jr., W.M. [1977] 31153
Leblanc, H. [1965] 75405
Leblanc, H. [1966] 07926
Leblanc, H. [1972] 07940
Leblanc, H. [1972] 19182
Leisenring, A.C. [1969] 07983
Lifschitz, V. [1980] 54190
Loeb, M.H. [1976] 14746
London, F. [1925] 08256
Los, J. [1956] 08333
Maehara, S. [1960] 08520
Marini, D. [1975] 22915
Matulis, V.A. [1963] 08901
Mints, G.E. [1966] 09276
Mints, G.E. [1967] 37138
Mints, G.E. [1968] 19422
Mints, G.E. [1974] 37140
Mints, G.E. [1974] 37142
Mostowski, Andrzej [1953] 16376
Motohashi, N. [1982] 33418
Nederpelt, R.P. [1977] 53110
Nerode, A. [1980] 54927
Newman, M.H.A. [1942] 09934
Ono, H. [1985] 47510
Orevkov, V.P. [1968] 10152
Orevkov, V.P. [1968] 19539
Parks, R.Z. [1971] 11123
Parry, W.T. [1965] 10301
Plisko, V.E. [1978] 52084
Prawitz, D. [1975] 27280
Quine, W.V.O. [1981] 55273
Raggio, A.R. [1964] 10969
Raggio, A.R. [1974] 10972
Raggio, A.R. [1977] 16607
Rasiowa, H. [1953] 11004
Rasiowa, H. [1960] 11024
Reichbach, J. [1967] 19564
Reichbach, J. [1975] 32003
Renardel de Lavalette, G.R. [1984] 46669
Roessler, K. [1934] 19547
Rogava, M.G. [1972] 64879
Rosser, J.B. [1955] 23483
Sakai, H. [1974] 11772
Sanchis, L.E. [1971] 11826
Sanchis, L.E. [1971] 11827
Scanlon, T.M. [1973] 11874
Schoenfinkel, M. [1924] 12955
Schroeter, K. [1955] 12989
Schroeter, K. [1958] 24841
Schuette, K. [1950] 12999
Schuette, K. [1956] 19693
Semenov, A.L. [1977] 51323
Sikorski, R. [1958] 12303
Smart, J.J.C. [1961] 31260
Smirnov, V.A. [1973] 21800
Smirnov, V.A. [1979] 53921

Smirnova, E.D. [1974] 78744
Smullyan, R.M. [1957] 12538
Smullyan, R.M. [1963] 12547
Smullyan, R.M. [1965] 12551
Smullyan, R.M. [1965] 16212
Smullyan, R.M. [1968] 12555
Smullyan, R.M. [1970] 12557
Suvorov, P.Yu. [1979] 36649
Tang, Tonggao [1981] 34057
Tarski, A. [1939] 13428
Thiel, C. [1978] 30592
Umezawa, T. [1979] 53149
Wajsberg, M. [1933] 40808
Wang, Hao [1963] 14050
Wette, E. [1974] 14955
Zariski, O. [1925] 41579

B15

Ackermann, W. [1950] 00122
Ackermann, W. [1952] 00124
Ackermann, W. [1953] 00125
Ackermann, W. [1965] 00135
Amer, M.A. [1976] 17946
Amer, M.A. [1977] 30619
Anderson, A.R. [1975] 50675
Andrews, P.B. [1965] 00402
Andrews, P.B. [1971] 00365
Andrews, P.B. [1972] 00366
Andrews, P.B. [1972] 00367
Andrews, P.B. [1974] 00368
Andrews, P.B. [1974] 03832
Anikeev, A.S. [1972] 00375
Baxter, L.D. [1978] 52222
Belyakin, N.V. [1983] 41479
Beth, E.W. [1937] 41045
Bibel, W. [1969] 01198
Boffa, M. [1983] 37638
Boffa, M. [1984] 42423
Borkowski, L. [1958] 01479
Bowen, K.A. [1973] 01507
Bowen, K.A. [1974] 03943
Buchholz, W. [1975] 01681
Buechi, J.R. [1960] 01690
Buechi, J.R. [1969] 01700
Bunder, M.W. [1980] 53699
Bunder, M.W. [1983] 31631
Chauvin, A. [1949] 01996
Chwistek, L.B. [1912] 04231
Chwistek, L.B. [1924] 02670
Cocchiarella, N.B. [1974] 04010
Cocchiarella, N.B. [1985] 47532
Constable, R.L. [1985] 49190
Crabbe, M. [1978] 29204
Dragalin, A.G. [1980] 36552
Elgot, C.C. [1966] 03291
Feferman, S. [1977] 27330
Fitch, F.B. [1938] 04321
Friedman, H.M. [1973] 04650
Friedman, H.M. [1978] 31762
Gastev, Yu.A. [1970] 17919
Gentzen, G. [1936] 04851
Girard, J.-Y. [1984] 48565
Goguadze, D.F. [1970] 05093
Goldfarb, W.D. [1981] 54331
Grayson, R.J. [1984] 41840
Gurevich, Y. [1983] 33765
Henkin, L. [1950] 05894
Huet, G. [1973] 06291
Inoue, K. [1975] 48761
Jansohn, H.-S. [1982] 36620
Kogalovskij, S.R. [1968] 07299
Kogalovskij, S.R. [1970] 07300
Kreisel, G. [1954] 22027
Kubinski, T. [1963] 07595
Kuroda, S. [1958] 07681
Kuroda, S. [1958] 07683
Kuroda, S. [1959] 15552
L'Abbe, M. [1953] 07730
Lambek, J. [1984] 44302
Loeb, M.H. [1968] 08229
Lopez-Escobar, E.G.K. [1967] 08262
Lopez-Escobar, E.G.K. [1975] 24555
Maehara, S. [1954] 08513
Maehara, S. [1960] 08521
Maehara, S. [1962] 08526
Maehara, S. [1971] 08528
Martin-Loef, P. [1973] 76088
Montague, R. [1961] 09424
Mostowski, A.Wlodzimierz [1981] 55152
Myhill, J.R. [1951] 09702
Myhill, J.R. [1971] 28094
Myhill, J.R. [1974] 48354
Nabebin, A.A. [1977] 52155
Nerode, A. [1980] 54927
Pabion, J.F. [1980] 55872
Paeppinghaus, P. [1983] 40742
Pietrzykowski, T. [1973] 10491
Plotkin, G.D. [1980] 77373
Prawitz, D. [1965] 22164
Prawitz, D. [1968] 10733
Rosser, J.B. [1950] 11558
Royse, James R. [1969] 11634
Salle, P. [1978] 52050
Salle, P. [1980] 56539
Sanchis, L.E. [1965] 11824
Scarpellini, B. [1985] 47473
Scedrov, A. [1985] 44771
Schuette, K. [1953] 13003
Schuette, K. [1960] 13011
Schuette, K. [1965] 13015
Schwabhaeuser, W. [1979] 53471
Schwichtenberg, H. [1983] 40090
Seely, R.A.G. [1982] 34890
Shvarts, G.F. [1979] 56037
Siefkes, D. [1970] 12145
Siefkes, D. [1970] 27984
Siefkes, D. [1971] 12146
Simauti, T. [1956] 12080
Tait, W.W. [1966] 13283
Takahashi, Moto-o [1968] 13295
Takeuti, G. [1953] 37768
Takeuti, G. [1955] 28446
Takeuti, G. [1956] 13307
Takeuti, G. [1956] 13308
Takeuti, G. [1956] 28447
Takeuti, G. [1958] 13312
Takeuti, G. [1958] 28448
Takeuti, G. [1960] 13316
Takeuti, G. [1961] 13320
Takeuti, G. [1961] 13321
Tarski, A. [1933] 28816
Titani, S. [1973] 13615
Uesu, T. [1971] 13764
Wang, Hao [1949] 14006
Wang, Hao [1950] 14011
Wang, Hao [1963] 14044
Yakubovich, A.M. [1981] 46491
Zbierski, P. [1978] 29200

B20

Barthelemy, J.-P. [1974] 04140
Bernays, P. [1965] 16287
Craig, W. [1967] 02516
Curry, H.B. [1952] 28059
Czermak, J. [1977] 24249
Dywan, Z. [1984] 45756
Fitch, F.B. [1944] 04326
Fitch, F.B. [1944] 04328
Fitch, F.B. [1948] 04329
Fitch, F.B. [1949] 04334
Fraenkel, A.A. [1951] 47908
Goodstein, R.L. [1963] 05189
Goodstein, R.L. [1966] 05190
Guillaume, M. [1958] 42813
Gurevich, Y. [1965] 05452
Gurevich, Y. [1982] 33757
Hafner, I. [1979] 53656
Henson, C.W. [1984] 36539
Hintikka, K.J.J. [1956] 06114
Johansson, I. [1953] 06644
Kalmar, L. [1928] 06835
Kanovich, M.I. [1975] 18218
Khomich, V.I. [1975] 17549
Kodera, H. [1980] 74921
Lambert, K. [1967] 07823
Lopez-Escobar, E.G.K. [1972] 08267
Markovic, Z. [1983] 37386
Maslov, S.Yu. [1965] 19335
Meredith, C.A. [1953] 09134
Meredith, C.A. [1968] 09143
Meredith, D. [1978] 31932
Meredith, D. [1979] 31933
Meredith, D. [1980] 76287
Mezhlumbekova, V.F. [1975] 21799
Mezhlumbekova, V.F. [1975] 63913
Mints, G.E. [1971] 09286
Mints, G.E. [1982] 35718
Misercque, D. [1980] 76441
Nelson, D. [1966] 09846
Orevkov, V.P. [1968] 19539
Popper, K.R. [1947] 10648
Prawitz, D. [1982] 34933
Renardel de Lavalette, G.R. [1981] 55207
Rosser, J.B. [1939] 11547
Rousseau, G. [1966] 11593
Rousseau, G. [1968] 11598
Schroeter, K. [1956] 12993
Trakhtenbrot, B.A. [1961] 19734
Trakhtenbrot, B.A. [1961] 19735
Trakhtenbrot, B.A. [1962] 19733
Tsitkin, A.I. [1979] 53496
Ulrich, D. [1976] 18459
Valpola, V. [1955] 13817
Valpola, V. [1955] 16346
Wiedmer, E. [1974] 14190
Wronski, A. [1974] 18432
Zarnecka-Bialy, E. [1972] 16532
Zarnecka-Bialy, E. [1973] 14363

B22

Belnap Jr., N.D. [1963] 01010
Bloom, S.L. [1977] 28146
Dzik, W. [1975] 21897
Furmanowski, T. [1982] 42298
Gabbay, D.M. [1978] 32163
Grzegorczyk, A. [1972] 05418
Hatcher, W.S. [1964] 32200
Hiz, H. [1946] 06149
Pogorzelski, W.A. [1960] 21001
Prawitz, D. [1985] 45636
Schroeder-Heister, P. [1984] 42500
Scott, D.S. [1982] 38520
Shoesmith, D.J. [1978] 51869
Tokarz, M. [1971] 13627
Urquhart, A.I.F. [1974] 79550
Vuckovic, V. [1960] 22133
Wojtylak, P. [1981] 55071

B25

Andrews, P.B. [1974] 03832
Bel'tyukov, A.P. [1976] 29789
Berman, L. [1977] 70972
Berman, L. [1980] 55471
Bernardi, C. [1975] 31012
Boolos, G. [1976] 14578
Boolos, G. [1977] 26442
Boolos, G. [1979] 56312
Bozzi, S. [1980] 54221
Buechi, J.R. [1960] 01690
Buechi, J.R. [1962] 01693
Buechi, J.R. [1969] 01700
Cegielski, P. [1980] 55770
Cegielski, P. [1981] 55626
Cegielski, P. [1984] 46773
Cobham, A. [1957] 29385
Cooper, D.C. [1972] 61080
Danko, W. [1983] 39522
Dries van den, L. [1980] 54246
Elgot, C.C. [1961] 03288
Elgot, C.C. [1966] 03291
Ershov, Yu.L. [1974] 25764
Ferrante, J. [1975] 21702
Fischer, Michael J. [1974] 17344
Fol'k, N.F. [1968] 17126
Friedrich, U. [1972] 04668
Gabbay, D.M. [1970] 04721
Gabbay, D.M. [1972] 04729
Gabbay, D.M. [1973] 17911
Gabbay, D.M. [1973] 61880
Gabbay, D.M. [1975] 04738
Gabbay, D.M. [1975] 17286
Gentzen, G. [1935] 24221
Goad, C.A. [1978] 73424
Goldblatt, R.I. [1981] 55422
Golota, Ya.Ya. [1969] 05124
Golota, Ya.Ya. [1974] 15236
Goncharov, S.S. [1976] 52189
Goodstein, R.L. [1955] 05177
Goodstein, R.L. [1963] 05189
Goodstein, R.L. [1966] 05190
Goodstein, R.L. [1969] 32559
Grzegorczyk, A. [1957] 24864
Gurari, E.M. [1981] 54218
Gurevich, Y. [1965] 05452
Hanf, W.P. [1965] 05635
Hartmanis, J. [1976] 24160
Hauschild, K. [1971] 27450
Hauschild, K. [1972] 05787
Herbrand, J. [1930] 20866
Herbrand, J. [1931] 16812
Hermes, H. [1955] 06008
Ivanov, A.A. [1984] 41869
Kalmar, L. [1928] 06835
Kireevskij, N.N. [1958] 40811
Korec, I. [1976] 07374
Lifschitz, V. [1967] 33665
Longo, G. [1974] 53697
Lopez-Escobar, E.G.K. [1968] 08264
Loveland, D.W. [1978] 31164
Luckhardt, H. [1967] 08378
Luckhardt, H. [1969] 08380
Marini, D. [1975] 22915
Markov, A.A. [1974] 63716
Mart'yanov, V.I. [1977] 29230
Maslov, S.Yu. [1977] 32672
McCall, S. [1962] 08942
Meredith, D. [1978] 31932
Mints, G.E. [1966] 19428
Mints, G.E. [1968] 19418
Mostowski, A.Wlodzimierz [1981] 55152
Mundici, D. [1983] 34976
Nieland, J.J.F. [1967] 09951
Norgela, S.A. [1978] 32674
Novikov, P.S. [1949] 19476
Oglesby, F.C. [1961] 10067
Oppen, D.C. [1973] 24077
Oppen, D.C. [1978] 51889
Orevkov, V.P. [1969] 10154
Orevkov, V.P. [1976] 64291
Parikh, R. [1973] 10286
Penzin, Yu.G. [1973] 10357
Penzin, Yu.G. [1973] 10358
Penzin, Yu.G. [1973] 23000
Pil'chak, B.Yu. [1950] 16889
Presburger, M. [1930] 10749
Quine, W.V.O. [1949] 10881
Rasiowa, H. [1953] 11004
Reichbach, J. [1975] 32003
Rogava, M.G. [1979] 77850
Rousseau, G. [1966] 11593
Rousseau, G. [1968] 11598
Routley, R. [1974] 15192
Scarpellini, B. [1984] 43196
Schmidt, H.A. [1958] 12891
Schuette, K. [1950] 12999
Scowcroft, P. [1984] 44277
Semenov, A.L. [1979] 53242
Shostak, R.E. [1977] 69916
Shostak, R.E. [1979] 82799
Siefkes, D. [1970] 12145
Siefkes, D. [1970] 27984
Siefkes, D. [1971] 12146
Skolem, T.A. [1931] 12388
Smorynski, C.A. [1973] 12535
Szczerba, L.W. [1980] 79172
Tarski, A. [1931] 16924
Tsinman, L.L. [1969] 02178
Waerden van der, B.L. [1930] 38703
Wajsberg, M. [1938] 13982
Wang, Hao [1953] 14024
Wasilewska, A. [1976] 65985
Wasilewska, A. [1976] 65986
Wasilewska, A. [1980] 55216
Woehl, K. [1979] 53367
Wolter, H. [1975] 66137

B28

Ackermann, W. [1928] 00106
Apelt, H. [1966] 00403
Beck, Jon M. [1979] 55969
Bel'tyukov, A.P. [1976] 29789
Bernays, P. [1970] 01284
Beth, E.W. [1962] 01176
Bollman, D.A. [1973] 01422
Boolos, G. [1979] 56312
Boolos, G. [1980] 71208
Bosch, J.E. [1985] 48160
Both, N. [1967] 01491
Cegielski, P. [1984] 46773
Christian, C.C. [1972] 71771
Christian, C.C. [1976] 60952
Clay, R.E. [1968] 02255
Clote, P. [1983] 40361
Collins, G.E. [1970] 02332
Curry, H.B. [1941] 02612
Dapueto, C. [1982] 45889
Dejon, B. [1969] 23438
Deutsch, M. [1975] 04086
Dienes, P. [1938] 03008
Ding, Decheng [1984] 44972
Fitch, F.B. [1949] 04333
Fogelis, E. [1950] 17125
Friedman, H.M. [1980] 55240
Gel'fond, M.G. [1972] 04841
Gentzen, G. [1936] 22402
Gentzen, G. [1954] 04853
Goedel, K. [1931] 15052
Goodstein, R.L. [1945] 05166
Goodstein, R.L. [1955] 05177
Goodstein, R.L. [1957] 05180
Gruzintsev, G.A. [1927] 40886
Grzegorczyk, A. [1971] 23119
Hadamard, J. [1954] 42157
Hajek, P. [1981] 35379
Hauschild, K. [1971] 05779
Hetper, W. [1934] 40820
Hilbert, D. [1931] 21186
Hischer, Horst [1976] 25590
Kleene, S.C. [1935] 28412
Kleene, S.C. [1945] 07169
Kondo, M. [1956] 07331
Kronecker, L. [1887] 37377
Kuroda, S. [1958] 07681
Kuroda, S. [1958] 07683
Kuroda, S. [1959] 15552
Kuroda, S. [1959] 25434
Kuroda, S. [1959] 25455
Kuroda, S. [1959] 25622
Levin, A.M. [1979] 53493
Lifschitz, V. [1985] 44843
Lorenzen, P. [1965] 08304
Maehara, S. [1957] 08516
Maliaukiene, L. [1985] 49123
McMinn, T.J. [1974] 09065
Mikolajewicz, B. [1983] 43117
Mycielski, J. [1981] 54194
Nagel, E. [1958] 09765
Nagel, E. [1961] 09768
Nelson, D. [1966] 09846

Novikov, P.S. [1949] 19476
Pabion, J.F. [1981] 33728
Paris, J.B. [1978] 29247
Paris, J.B. [1985] 41796
Peano, G. [1889] 19497
Presburger, M. [1930] 10749
Pudlak, P. [1983] 36550
Rabin, M.O. [1961] 10934
Richard, D. [1984] 39634
Richard, D. [1984] 41779
Richard, D. [1985] 45095
Richard, D. [1985] 46289
Robinson, R.M. [1957] 11315
Robinson, R.M. [1958] 48434
Scienza, G. [1979] 78312
Seckendorff von, V. [1937] 41030
Shapiro, S. [1985] 44774
Singh, S. [1967] 12349
Skolem, T.A. [1923] 38689
Skolem, T.A. [1931] 12388
Skolem, T.A. [1933] 12392
Skolem, T.A. [1934] 12393
Skolem, T.A. [1936] 12396
Skolem, T.A. [1955] 27718
Skolem, T.A. [1956] 12416
Smullyan, R.M. [1957] 12538
Soudieux, C. [1960] 12654
Takeuti, G. [1957] 19761
Tarski, A. [1931] 16924
Tarski, A. [1933] 13407
Tomas, F. [1975] 25028
Tsinman, L.L. [1967] 02176
Tsinman, L.L. [1968] 02177
Valpola, V. [1947] 42606
Wang, Hao [1953] 14023
Wang, Hao [1955] 14026
Wang, Hao [1957] 33305
Wette, E. [1970] 14165
Weyl, H. [1918] 19791
Weyl, H. [1919] 14166
Wolter, H. [1973] 14280
Zahn, P. [1967] 24851

B30

Barzin, M. [1935] 00914
Bernays, P. [1932] 01289
Bernays, P. [1950] 37320
Bernays, P. [1953] 47892
Carnap, R. [1935] 03962
Chwistek, L.B. [1929] 02160
Chwistek, L.B. [1938] 02162
Fitch, F.B. [1950] 17113
Fitch, F.B. [1951] 17114
Gentzen, G. [1974] 04855
Germano, G. [1970] 04906
Goedel, K. [1931] 15052
Goetlind, E. [1952] 21134
Herbrand, J. [1931] 16812
Heyting, A. [1959] 27711
Heyting, A. [1961] 06065
Hilbert, D. [1899] 23454
Ito, Makoto [1934] 32185
Ito, Makoto [1935] 28786
Josza, R. [1984] 45334
Kreisel, G. [1967] 07533
Kreisel, G. [1969] 26648
Kripke, S.A. [1967] 31215
Kripke, S.A. [1967] 31216
Kuroda, S. [1959] 15552
Kuroda, S. [1959] 25434
Kuroda, S. [1959] 25455
Kuroda, S. [1959] 25622
Lambek, J. [1958] 07815
Levin, A.M. [1975] 25810
Menger, K. [1928] 39109
Montague, R. [1957] 29356
Myhill, J.R. [1950] 09699
Nicod, J. [1930] 09941
Pabion, J.F. [1974] 17648
Rasiowa, H. [1955] 11010
Roedding, W. [1968] 11344
Shoenfield, J.R. [1954] 12087
Sikorski, R. [1962] 42875
Szczerba, L.W. [1977] 51583
Szczerba, L.W. [1979] 56477
Szczerba, L.W. [1983] 39510
Tarski, A. [1932] 13406
Tarski, A. [1933] 28816
Wajsberg, M. [1933] 40808
Wang, Hao [1953] 14024
Wang, Hao [1954] 14025
Weyl, H. [1910] 42999
Weyl, H. [1924] 38712
Weyl, H. [1985] 49136

B35

Ben-Ari, M. [1980] 55966
Bibel, W. [1975] 60499
Bibel, W. [1979] 53410
Bibel, W. [1982] 40212
Bibel, W. [1982] 40242
Bishop, P. [1974] 62501
Boehm, C. [1972] 60576
Bruijn de, N.G. [1978] 29191
Bruijn de, N.G. [1980] 72186
Cannonito, F.B. [1962] 21294
Cooper, D.C. [1972] 61080
Dantsin, E.Ya. [1981] 55532
Dardzhaniya, G.K. [1979] 53026
Gilbert, M.A. [1976] 62011
Goad, C.A. [1980] 55988
Harms, S. [1977] 73840
Hartmanis, J. [1976] 24160
Hayashi, S. [1983] 37813
Hintikka, K.J.J. [1973] 32405
Hotomski, P. [1982] 37063
Jansohn, H.-S. [1982] 36620
Justen, K. [1981] 55967
Kapur, D. [1984] 43561
Khomich, V.I. [1970] 06201
Kramosil, I. [1980] 56455
Kreisel, G. [1970] 48003
Kreisel, G. [1977] 53079
Kur'erov, Yu.N. [1979] 81977
Lacombe, D. [1970] 48620
Longo, G. [1974] 53697
Lyaletskij, A.V. [1975] 53444
Lyaletskij, A.V. [1975] 53445
Lyaletskij, A.V. [1981] 35594
Maslov, S.Yu. [1972] 08850
Maslov, S.Yu. [1977] 32672
Mayoh, B.H. [1974] 63802
Meredith, D. [1980] 76287
Miller, Dale A. [1984] 43218
Milner, R. [1979] 53582
Mints, G.E. [1967] 19424
Mitani, S. [1983] 41462
Morgan, C.G. [1976] 27095
Nagata, M. [1975] 21892
Nederpelt, R.P. [1980] 55989
Norgela, S.A. [1978] 32674
Norwood, F.H. [1982] 76885
Orevkov, V.P. [1969] 10154
Pietrzykowski, T. [1973] 10491
Plaisted, D.A. [1980] 56667
Plotkin, J.M. [1982] 35398
Richter, M.M. [1978] 51870
Robinson, John Alan [1969] 11291
Robitashvili, N.G. [1971] 11331
Schoenfeld, W. [1982] 34105
Schoenfeld, W. [1983] 43253
Shimada, K. [1974] 78527
Shostak, R.E. [1977] 69916
Shostak, R.E. [1979] 82799
Staples, J. [1983] 37059
Statman, R. [1977] 16639
Statman, R. [1978] 52901
Yukami, T. [1983] 37612
Zamov, N.K. [1974] 50359

B45

Akimov, A.P. [1979] 70347
Artemov, S.N. [1980] 54546
Artemov, S.N. [1982] 38128
Artemov, S.N. [1984] 45832
Artemov, S.N. [1985] 48972
Artemov, S.N. [1985] 49322
Avron, A. [1984] 42415
Bendova, K. [1982] 34402
Bernardi, C. [1975] 31011
Bernardi, C. [1976] 60478
Bernardi, C. [1984] 42581
Boolos, G. [1976] 14578
Boolos, G. [1977] 26442
Boolos, G. [1979] 56312
Boolos, G. [1979] 56313
Boolos, G. [1980] 55341
Boolos, G. [1982] 35340
Boolos, G. [1984] 43982
Boolos, G. [1985] 47529
Borga, M. [1983] 42318
Bunder, M.W. [1980] 53770
Bunder, M.W. [1980] 54073
Cherniavsky, J.C. [1973] 71692
Corcoran, J. [1969] 02388
Dosen, K. [1981] 35691
Dosen, K. [1984] 44540
Fischer Servi, G. [1977] 50945
Fischer Servi, G. [1981] 54078
Fitch, F.B. [1948] 04332
Fitch, F.B. [1966] 17115
Fitting, M. [1973] 04355
Fitting, M. [1983] 34628
Flagg, R.C. [1985] 49605
Gabbay, D.M. [1973] 17911
Gabbay, D.M. [1975] 04738
Gabbay, D.M. [1977] 50853
Gargov, G.K. [1984] 46969
Goble, L.F. [1974] 05060
Goedel, K. [1931] 22120
Goldblatt, R.I. [1978] 56314

Goldblatt, R.I. [1981] 55422
Goodman, Nicolas D. [1984] 42452
Gorgy, F.W. [1979] 53991
Grzegorczyk, A. [1967] 62216
Guaspari, D. [1979] 53658
Hajek, P. [1981] 35379
Heijenoort van, J. [1979] 79641
Kirov, K.A. [1984] 46974
Kondo, M. [1984] 44181
Lavendhomme, R. [1984] 40558
Leivant, D. [1981] 54609
Lorenzen, P. [1972] 08310
Magari, R. [1982] 35470
Matsumoto, K. [1950] 08888
Meyer, R.K. [1976] 21939
Montagna, F. [1983] 34635
Montagna, F. [1983] 37078
Montagna, F. [1984] 42567
Montague, R. [1963] 14478
Muchnik, A.A. [1983] 43519
Muravitskij, A.Yu. [1983] 34636
Rasiowa, H. [1951] 11001
Rasiowa, H. [1953] 11004
Rasiowa, H. [1955] 11011
Rescher, N. [1966] 11118
Sambin, G. [1978] 52783
Sambin, G. [1980] 54341
Sambin, G. [1982] 35627
Scedrov, A. [1985] 44771
Schuette, K. [1968] 24822
Sikorski, R. [1956] 12298
Sikorski, R. [1958] 12300
Simmons, H. [1975] 18388
Smorynski, C.A. [1978] 54148
Smorynski, C.A. [1979] 52968
Smorynski, C.A. [1980] 56734
Smorynski, C.A. [1982] 36167
Smorynski, C.A. [1982] 43687
Smorynski, C.A. [1984] 44891
Smorynski, C.A. [1985] 44663
Smullyan, R.M. [1973] 14740
Smullyan, R.M. [1985] 44678
Smullyan, R.M. [1985] 44783
Solitro, U. [1983] 35696
Solovay, R.M. [1976] 26088
Sonobe, O. [1975] 51586
Svejdar, V. [1983] 40906
Valentini, S. [1983] 34778
Valentini, S. [1983] 38327
Visser, A. [1984] 42643
Vuckovic, V. [1967] 27479
Zachorowski, S. [1975] 23947
Zarnecka-Bialy, E. [1973] 14364
Zeman, J.J. [1968] 14395
Zinov'ev, A.A. [1974] 17312

B46

Dosen, K. [1981] 55339
Giambrone, S. [1985] 47499
Krabbe, E.C.W. [1985] 42654
Meyer, R.K. [1973] 09200
Meyer, R.K. [1976] 27121
Meyer, R.K. [1976] 27122
Meyer, R.K. [1977] 21964
Meyer, R.K. [1985] 48205
Routley, R. [1974] 15192
Tennant, N. [1980] 53989

B48

Leblanc, H. [1983] 37685
Leblanc, H. [1983] 54992
Lindner, R. [1975] 51606

B50

Bertolini, F. [1971] 01565
Bunder, M.W. [1979] 56104
Cantini, A. [1980] 54349
Epstein, G. [1974] 22895
Epstein, G. [1976] 18456
Frink, O. [1938] 04671
Gabbay, D.M. [1976] 50854
Girard, J.-Y. [1976] 50375
Goodstein, R.L. [1958] 05182
Hart, A.M. [1978] 51664
Martin, J.N. [1984] 40627
Morgan, C.G. [1976] 76542
Nieland, J.J.F. [1967] 09951
Petkov, P.P. [1984] 46576
Previale, F. [1975] 31421
Rose, A. [1955] 24903
Scott, D.S. [1975] 27594
Vuckovic, V. [1967] 27478
White, R.B. [1979] 53322
Zachorowski, S. [1975] 23947

B51

Takeuti, G. [1981] 39565

B52

Almukdad, A. [1984] 42414

B53

Grishin, V.N. [1981] 54617
Magari, R. [1974] 08539

B55

Arruda, A.I. [1978] 31617
Belding, W.R. [1971] 00957
Bessonov, A.V. [1977] 53417
Boricic, B.R. [1981] 35211
Bowen, K.A. [1971] 01504
Bozic, M. [1983] 46894
Bozzi, S. [1980] 54221
Burgess, J.P. [1981] 56494
Dardzhaniya, G.K. [1977] 53196
Dardzhaniya, G.K. [1979] 53268
Destouches-Fevrier, P. [1948] 02965
Dosen, K. [1981] 55339
Dosen, K. [1984] 44539
Dosen, K. [1985] 47541
Drugush, Ya.M. [1978] 52635
Fidel, M.M. [1978] 52314
Fischer Servi, G. [1976] 55901
Fischer Servi, G. [1978] 52127
Fischer Servi, G. [1981] 54078
Fitting, M. [1983] 34628
Gabbay, D.M. [1970] 04721
Gabbay, D.M. [1971] 04726
Gabbay, D.M. [1972] 04730
Gabbay, D.M. [1973] 17911
Gabbay, D.M. [1976] 14748
Gabbay, D.M. [1977] 26456
Gabbay, D.M. [1977] 50853
Gabbay, D.M. [1981] 54131
Gabbay, D.M. [1982] 38464
Gargov, G.K. [1975] 21677
Georgacarakos, G.N. [1982] 43512
Gimon, V.V. [1979] 53494
Goernemann, S. [1971] 39863
Golota, Ya.Ya. [1969] 05123
Jaskowski, S. [1936] 06544
Johnstone, P.T. [1979] 56505
Jongh de, D.H.J. [1980] 55677
Kalicki, C. [1980] 53776
Kirk, R.E. [1979] 53554
Kirk, R.E. [1980] 56546
Klemke, D. [1971] 07229
Kreisel, G. [1957] 07492
Kripke, S.A. [1965] 07557
Levin, L.A. [1969] 08046
Levy, P. [1927] 08092
Lopez-Escobar, E.G.K. [1981] 55142
Lopez-Escobar, E.G.K. [1981] 55143
Lopez-Escobar, E.G.K. [1983] 39786
Lopez-Escobar, E.G.K. [1985] 47264
Lorenz, K. [1973] 27473
Malmnaes, P.E. [1968] 24815
McKay, C.G. [1967] 08983
Miglioli, P.A. [1983] 37781
Mints, G.E. [1972] 09288
Muravitskij, A.Yu. [1981] 36087
Nepejvoda, N.N. [1983] 43233
Porebska, M. [1975] 17277
Rauszer, C. [1974] 15188
Rauszer, C. [1976] 18356
Rauszer, C. [1977] 50668
Rauszer, C. [1977] 50669
Rauszer, C. [1980] 56381
Renardel de Lavalette, G.R. [1981] 55207
Rousseau, G. [1970] 11601
Routley, R. [1978] 53083
Scedrov, A. [1984] 44085
Segerberg, K. [1968] 11934
Segerberg, K. [1974] 11941
Semenenko, M.I. [1970] 11972
Shanin, N.A. [1977] 53812
Sikic, Z. [1984] 42266
Skvortsov, D.P. [1979] 55940
Skvortsov, D.P. [1980] 55941
Smorynski, C.A. [1973] 27058
Smorynski, C.A. [1978] 29151
Sobolev, S.K. [1977] 51014
Sobolev, S.K. [1977] 51279
Stralberg, A.H. [1965] 13218
Surma, S.J. [1973] 42072
Surma, S.J. [1974] 16561
Surma, S.J. [1975] 23141
Tauts, A. [1975] 52459
Tragesser, R.S. [1978] 30597
Troelstra, A.S. [1981] 55352
Tsitkin, A.I. [1977] 50214
Umezawa, T. [1980] 56395
Vakarelov, D. [1977] 31280
Varpakhovskij, F.L. [1965] 27855
Vorob'ev, N.N. [1964] 16331
Wendel, N. [1978] 53217
Wojcicki, R. [1970] 14254
Wronski, A. [1971] 14337

Wronski, A. [1974] 18433
Yankov, V.A. [1963] 16539
Yankov, V.A. [1963] 19006
Yankov, V.A. [1967] 06513
Yankov, V.A. [1968] 06514
Yankov, V.A. [1969] 06518
Zaslavskij, I.D. [1978] 80231

B60

Asanidze, G.Z. [1981] 34653
Berkling, K.J. [1982] 43431
Ehrenfeucht, A. [1974] 03261
Goad, C.A. [1978] 73424
Grishin, V.N. [1974] 73648
Grishin, V.N. [1981] 54617
Hajek, P. [1977] 26853
Hermes, H. [1976] 51102
Ibragimov, S.G. [1973] 74359
Kanda, A. [1985] 47568
Lambert, K. [1967] 07823
Lorenz, K. [1968] 28288
Lorenz, K. [1978] 28287
Lorenzen, P. [1972] 08311
Magari, R. [1974] 08539
Meloni, G.C. [1979] 56742
Nepejvoda, N.N. [1982] 46228
Petkov, P.P. [1971] 10430
Prank, R.K. [1979] 53437
Prank, R.K. [1980] 77510
Prank, R.K. [1981] 35552
Rogava, M.G. [1975] 18361
Rogava, M.G. [1979] 77850
Tauts, A. [1974] 17694
Tirnoveanu, M. [1970] 30384
Tolstova, Yu.N. [1971] 13633
Vanderveken, D.R. [1975] 22976

B75

Ausiello, G. [1974] 80587
Bakker de, J.W. [1971] 28688
Berkling, K.J. [1982] 43431
Cherniavsky, J.C. [1976] 31025
Constable, R.L. [1972] 61071
Damm, W. [1978] 51522
Danko, W. [1983] 39522
Golota, Ya.Ya. [1969] 05123
Golota, Ya.Ya. [1969] 05124
Leivant, D. [1981] 55013
Miglioli, P.A. [1981] 55396
Mints, G.E. [1982] 35718
Mirkowska, G. [1977] 31914
Nolin, L. [1975] 23994
Nolin, L. [1975] 82368
O'Donnell, M.J. [1977] 76932
O'Donnell, M.J. [1979] 76931
O'Donnell, M.J. [1979] 82386
Pliuskevicius, R. [1983] 43884
Poigne, A. [1984] 47093
Roever de, W.P. [1974] 64875
Sanchis, L.E. [1977] 52496
Sazonov, V.Yu. [1976] 29755
Sciore, E. [1978] 52051
Scott, D.S. [1975] 18375
Skordev, D.G. [1976] 23157
Skordev, D.G. [1982] 37517
Tang, A. [1979] 69945
Woehl, K. [1979] 53367
Yanov, Yu.I. [1977] 74466

B80

Evangelist, M. [1982] 54618
Nicod, J. [1924] 25431

B96

Gentzen, G. [1969] 29483
Herbrand, J. [1968] 24825
Wajsberg, M. [1933] 40808

B97

Boerger, E. [1984] 40062
Gonseth, F. [1941] 48625
Harrington, L.A. [1985] 49810
Heijenoort van, J. [1967] 25903
Lacombe, D. [1970] 48620
Loeb, M.H. [1968] 37293
Lolli, G. [1984] 41493
Lourenco, M. [1979] 70030
Matiyasevich, Yu.V. [1979] 70057
Orevkov, V.P. [1968] 37197
Dalen van, D. [1982] 34494

B98

Anderson, J.M. [1962] 00341
Bar-Hillel, Y. [1958] 00805
Barendregt, H.P. [1981] 55008
Barwise, J. [1977] 70117
Beth, E.W. [1956] 01166
Beth, E.W. [1962] 01176
Blanche, R. [1957] 01538
Boolos, G. [1979] 56312
Church, A. [1936] 23466
Courvoisier, M. [1976] 51687
Crossley, J.N. [1975] 31689
Curry, H.B. [1957] 29324
Curry, H.B. [1958] 02627
Curry, H.B. [1967] 02637
Dopp, J. [1962] 03085
Enderton, H.B. [1972] 03355
Fisher, A. [1982] 36967
Fitch, F.B. [1952] 04335
Fitch, F.B. [1974] 24019
Fitting, M. [1969] 04349
Friedman, H.M. [1975] 04296
Goldblatt, R.I. [1979] 55754
Goodstein, R.L. [1951] 25669
Haas, G. [1984] 44527
Heyting, A. [1958] 06062
Bernays, P. [1934] 01098
Bernays, P. [1939] 01082
Hindley, J.R. [1972] 23471
Kleene, S.C. [1952] 07173
Kleene, S.C. [1967] 45895
Kosovskij, N.K. [1981] 55663
Kreisel, G. [1965] 27577
Ladriere, J. [1957] 07777
Lorenzen, P. [1955] 08290
Lorenzen, P. [1962] 08303
Manin, Yu.I. [1977] 51984
Manin, Yu.I. [1979] 54716
Mostowski, Andrzej [1965] 09578
Prawitz, D. [1965] 22164
Rasiowa, H. [1963] 11028
Rasiowa, H. [1974] 31451
Richter, M.M. [1978] 51870
Schmidt, H.A. [1960] 12892
Schuette, K. [1960] 13010
Schuette, K. [1962] 22382
Schuette, K. [1963] 22383
Scott, D.S. [1982] 38593
Shoenfield, J.R. [1967] 22384
Skolem, T.A. [1952] 21071
Skolem, T.A. [1953] 21072
Skolem, T.A. [1954] 21133
Smirnov, V.A. [1972] 30359
Smullyan, R.M. [1963] 28562
Smullyan, R.M. [1983] 41200
Steen, S.W.P. [1972] 22386
Takeuti, G. [1978] 52447
Takeuti, G. [1981] 39565
Wang, Hao [1981] 34116

C05

Blass, A.R. [1983] 40316

C07

Bennett, D.W. [1973] 01038
Bennett, D.W. [1977] 21957
Benthem van, J.F.A.K. [1974] 03889
Bowen, K.A. [1972] 01505
Craig, W. [1957] 02508
Craig, W. [1957] 02510
Dalla Chiara Scabia, M.L. [1968] 04285
Glubrecht, J.-M. [1982] 35619
Hajek, P. [1966] 05517
Hanazawa, M. [1979] 54329
Hanf, W.P. [1965] 05635
Hauck, J. [1972] 62383
Henkin, L. [1954] 05900
Henkin, L. [1956] 05908
Henkin, L. [1957] 05910
Hintikka, K.J.J. [1973] 32405
Hook, J.L. [1985] 41804
Jambu-Giraudet, M. [1980] 54167
Jambu-Giraudet, M. [1981] 55535
Jervell, H.R. [1973] 33265
Kreisel, G. [1967] 07533
Kreisel, G. [1975] 23955
Kreisel, G. [1975] 27698
Levy, A. [1958] 24860
Los, J. [1956] 08333
Maehara, S. [1958] 08519
Maehara, S. [1970] 08455
Manin, Yu.I. [1977] 51984
Markov, A.A. [1974] 63722
Markovic, Z. [1979] 56384
Montague, R. [1962] 09425
Montague, R. [1965] 09431
Morozov, A.S. [1983] 43123
Morozov, A.S. [1984] 41757
Orey, S. [1956] 10163
Pinus, A.G. [1982] 35574
Prazmowski, K. [1978] 29177
Rasiowa, H. [1960] 11024
Schroeter, K. [1958] 24841
Sette, A.-M. [1985] 48209
Smirnov, V.A. [1972] 30359
Smullyan, R.M. [1970] 12557

Szczerba, L.W. [1980] 79172
Szczerba, L.W. [1983] 39510
Takeuti, G. [1957] 13310
Takeuti, G. [1957] 13311
Tarski, A. [1932] 13406
Tarski, A. [1933] 28816
Wang, Hao [1953] 14024

C10

Cegielski, P. [1980] 55770
Cegielski, P. [1981] 55626
Ferrante, J. [1975] 21702
Fol'k, N.F. [1968] 17126
Fridman, Eh.I. [1973] 23081
Grzegorczyk, A. [1957] 24864
Lopez-Escobar, E.G.K. [1968] 08264
Loveland, D.W. [1978] 31164
Mints, G.E. [1966] 19428
Oppen, D.C. [1973] 24077
Oppen, D.C. [1978] 51889
Presburger, M. [1930] 10749
Skolem, T.A. [1931] 12388
Tarski, A. [1931] 16924
Wojtylak, P. [1984] 44096
Wolter, H. [1973] 14280
Wolter, H. [1975] 66137

C13

Boolos, G. [1984] 40442
Scarpellini, B. [1985] 47473
Smorynski, C.A. [1982] 43687
Yankov, V.A. [1967] 06513

C15

Friedman, H.M. [1973] 04654
Khisamiev, N.G. [1974] 06147
Mlcek, J. [1978] 31964
Paris, J.B. [1980] 56654

C20

Benthem van, J.F.A.K. [1984] 39892
Gabbay, D.M. [1972] 04730
Hauschild, K. [1971] 05779
Markovic, Z. [1977] 51169
Osswald, H. [1970] 10184
Rauszer, C. [1980] 56381
Reichbach, J. [1973] 31998
Skolem, T.A. [1933] 12392
Skolem, T.A. [1934] 12393
Skolem, T.A. [1955] 27718

C25

Friedrichsdorf, U. [1976] 24288
Hodges, W. [1980] 54226
Markovic, Z. [1983] 37386
McAloon, K. [1978] 52282
Pinter, C. [1978] 56267
Reichbach, J. [1973] 31998

C30

Feferman, S. [1968] 03703
Feferman, S. [1972] 03706
Guzicki, W. [1981] 34160
Guzicki, W. [1983] 37520
Harrington, L.A. [1977] 27334
Krajewski, S. [1974] 07445
Macnab, D.S. [1977] 51394
Mostowski, Andrzej [1958] 09566

C35

Apelt, H. [1966] 00403
Danko, W. [1983] 39522
Pinter, C. [1978] 56267
Prazmowski, K. [1976] 18347
Presburger, M. [1930] 10749
Szczerba, L.W. [1980] 79172
Wajsberg, M. [1933] 40808
Wang, Hao [1953] 14024

C40

Africk, H. [1972] 00199
Apt, K.R. [1978] 29131
Barwise, J. [1973] 04141
Bouvere de, K.L. [1959] 01501
Buechi, J.R. [1969] 01700
Cantini, A. [1980] 54349
Carstengerdes, W. [1971] 01873
Corcoran, J. [1980] 35994
Craig, W. [1957] 02508
Craig, W. [1957] 02510
Csirmaz, L. [1979] 72011
Dimitracopoulos, C. [1983] 35043
Feferman, S. [1968] 03700
Feferman, S. [1968] 03703
Feferman, S. [1972] 03706
Felscher, W. [1976] 04239
Gabbay, D.M. [1971] 04726
Gabbay, D.M. [1977] 26456
Georgieva, N.V. [1973] 20996
Germano, G. [1971] 04907
Germano, G. [1973] 04910
Hajek, P. [1970] 05521
Hauschild, K. [1971] 05779
Kanovich, M.I. [1975] 18218
Kino, A. [1975] 17616
Kogalovskij, S.R. [1968] 07299
Levy, A. [1958] 24860
Lopez-Escobar, E.G.K. [1967] 08263
Lopez-Escobar, E.G.K. [1981] 55143
Lopez-Escobar, E.G.K. [1983] 39786
Maehara, S. [1960] 08520
Maehara, S. [1961] 08524
Maehara, S. [1971] 08528
Motohashi, N. [1972] 09596
Motohashi, N. [1982] 33418
Motohashi, N. [1984] 45359
Nabebin, A.A. [1977] 52155
Nagashima, T. [1966] 09755
Nebres, B.F. [1972] 30771
Osswald, H. [1972] 10186
Putnam, H. [1964] 10838
Renardel de Lavalette, G.R. [1981] 55207
Reznikoff, I. [1963] 11135
Reznikoff, I. [1966] 11137
Schuette, K. [1962] 13012
Scott, D.S. [1962] 11914
Shirai, K. [1975] 29640
Shore, R.A. [1981] 55426
Smirnov, V.A. [1972] 30359
Smorynski, C.A. [1978] 54148
Tarski, A. [1931] 16924
Tarski, A. [1932] 13406
Tarski, A. [1933] 28816

C45

Harnik, V. [1985] 39778
Korec, I. [1976] 07374
Szczerba, L.W. [1980] 79172

C50

Cegielski, P. [1982] 34553
Kossak, R. [1985] 42582
Kudajbergenov, K.Zh. [1984] 41816
Kudajbergenov, K.Zh. [1984] 41871
Negri, M. [1984] 44715
Pinter, C. [1978] 56267
Wilkie, A.J. [1982] 34735

C52

Carstens, H.G. [1977] 51961
Friedrichsdorf, U. [1976] 24288
Hauschild, K. [1971] 05779

C55

Friedman, H.M. [1974] 04656
Guzicki, W. [1974] 62249
Guzicki, W. [1981] 34160
Guzicki, W. [1983] 37520
Shelah, S. [1976] 51434
Weese, M. [1976] 23814
Wolter, H. [1975] 66137

C57

Adamowicz, Z. [1985] 39780
Cegielski, P. [1982] 34553
DeMillo, R.A. [1979] 72280
Dobritsa, V.P. [1978] 32600
Ershov, Yu.L. [1973] 30533
Ershov, Yu.L. [1974] 25764
Feferman, S. [1975] 22905
Girstmair, K. [1979] 56289
Goncharov, S.S. [1976] 52189
Goncharov, S.S. [1982] 40426
Hermann, G. [1926] 05997
Kanda, A. [1985] 47568
Khisamiev, N.G. [1974] 06147
Khisamiev, N.G. [1979] 32601
Khisamiev, N.G. [1983] 40239
Kosovskij, N.K. [1970] 28600
Kossak, R. [1985] 42582
Kreisel, G. [1950] 07481
Kreisel, G. [1953] 20815
Kreisel, G. [1957] 29369
Kreisel, G. [1958] 07514
Kreisel, G. [1975] 23955
Kreisel, G. [1975] 27698
Kudajbergenov, K.Zh. [1984] 41816
Kudajbergenov, K.Zh. [1984] 41871
Madison, E.W. [1970] 24826
Madison, E.W. [1975] 18269
Mal'tsev, A.I. [1961] 28706
McAloon, K. [1978] 54427
Metakides, G. [1982] 37391
Morozov, A.S. [1983] 43123

Morozov, A.S. [1984] 41757
Moschovakis, Y.N. [1964] 09514
Moschovakis, Y.N. [1966] 21265
Negri, M. [1984] 44715
Pinus, A.G. [1975] 16662
Roche la, P. [1981] 75255
Seidenberg, A. [1970] 41695
Seidenberg, A. [1975] 41698
Shanin, N.A. [1976] 65058
Shanin, N.A. [1977] 53812
Smith, Rick L. [1981] 55685
Vuckovic, V. [1967] 27478

C60

Adamowicz, Z. [1985] 49053
Bauval, A. [1985] 48368
Bozzi, S. [1977] 52446
Delon, F. [1978] 27313
Delzell, C.N. [1982] 36898
Delzell, C.N. [1982] 38556
Delzell, C.N. [1984] 43271
Dobritsa, V.P. [1978] 32600
Feferman, S. [1975] 22905
Ferrante, J. [1975] 21702
Gabbay, D.M. [1973] 61880
Girstmair, K. [1979] 56289
Gurevich, Y. [1965] 05452
Hauschild, K. [1971] 05780
Hauschild, K. [1972] 05790
Hermann, G. [1926] 05997
Hodges, W. [1980] 54226
Jambu-Giraudet, M. [1980] 54167
Jensen, C.U. [1984] 39754
Johnstone, P.T. [1979] 74552
Joyal, A. [1975] 27668
Khisamiev, N.G. [1979] 32601
Khisamiev, N.G. [1983] 40239
Korec, I. [1974] 07371
Korec, I. [1976] 07374
Kreisel, G. [1957] 29369
Kreisel, G. [1958] 07514
Livchak, A.B. [1976] 55788
Livchak, A.B. [1977] 33959
Lorenzen, P. [1962] 08303
Madison, E.W. [1970] 24826
Moschovakis, Y.N. [1966] 21265
Mulvey, C.J. [1974] 21534
Roche la, P. [1981] 75255
Seidenberg, A. [1970] 41695
Seidenberg, A. [1971] 41694
Seidenberg, A. [1973] 41700
Seidenberg, A. [1975] 41698
Smith, Rick L. [1981] 55685
Tarski, A. [1931] 16924
Waerden van der, B.L. [1930] 38703

C62

Aczel, P. [1981] 34552
Adamowicz, Z. [1984] 45406
Adamowicz, Z. [1985] 49053
Apt, K.R. [1973] 00431
Apt, K.R. [1974] 00432
Apt, K.R. [1976] 14789
Apt, K.R. [1978] 29131
Artigue, M. [1978] 70505
Barnes, D.W. [1974] 03863
Belyakin, N.V. [1970] 29065
Belyakin, N.V. [1983] 41479
Berline, C. [1981] 54902
Bezboruah, A. [1976] 14785
Bily, J. [1974] 01205
Boffa, M. [1983] 37638
Cantini, A. [1983] 41846
Cegielski, P. [1982] 34553
Clote, P. [1980] 56493
Clote, P. [1981] 55425
Clote, P. [1983] 40361
Clote, P. [1985] 41801
Clote, P. [1985] 45298
Csirmaz, L. [1979] 72011
Dimitracopoulos, C. [1983] 35043
Dimitracopoulos, C. [1985] 47569
Dries van den, L. [1980] 54246
Ehrenfeucht, A. [1976] 25797
Enderton, H.B. [1967] 03347
Enderton, H.B. [1971] 03353
Feferman, S. [1965] 03696
Fourman, M.P. [1980] 56570
Friedman, H.M. [1970] 14546
Friedman, H.M. [1971] 17088
Friedman, H.M. [1973] 04654
Friedrichsdorf, U. [1976] 24288
Gandy, R.O. [1960] 15061
Gandy, R.O. [1961] 22330
Girard, J.-Y. [1984] 48565
Goodstein, R.L. [1965] 05187
Gordeev, L.N. [1982] 37615
Grzegorczyk, A. [1958] 05398
Grzegorczyk, A. [1971] 23119
Guaspari, D. [1979] 53266
Gurevich, Y. [1984] 38749
Guzicki, W. [1974] 62249
Guzicki, W. [1981] 34160
Guzicki, W. [1983] 37520
Hajek, P. [1977] 26853
Hajek, P. [1981] 35314
Hajek, P. [1984] 43006
Harrington, L.A. [1977] 27334
Henson, C.W. [1984] 39860
Hodges, W. [1980] 54226
Jambu-Giraudet, M. [1980] 54167
Jambu-Giraudet, M. [1981] 55535
Kirby, L.A.S. [1977] 50962
Kirby, L.A.S. [1978] 56399
Kirby, L.A.S. [1980] 54429
Kirby, L.A.S. [1982] 34416
Kirby, L.A.S. [1982] 35381
Kossak, R. [1981] 33723
Kossak, R. [1985] 42582
Krajewski, S. [1974] 07445
Krajewski, S. [1974] 07447
Kreisel, G. [1963] 42740
Kurata, R. [1981] 47747
Kurata, R. [1984] 42937
Lascar, D. [1980] 54430
Levitz, H. [1975] 18259
Lipton, R.J. [1978] 75578
Livchak, A.B. [1976] 55788
Livchak, A.B. [1977] 33959
Macintyre, A. [1980] 54621
Maehara, S. [1957] 08516
Manevitz, L.M. [1980] 54428
Marek, W. [1975] 08758
Marek, W. [1976] 24412
Marek, W. [1978] 29210
Marek, W. [1978] 76016
McAloon, K. [1975] 17403
McAloon, K. [1975] 18286
McAloon, K. [1977] 55797
McAloon, K. [1978] 52282
McAloon, K. [1978] 54427
McAloon, K. [1980] 54523
McAloon, K. [1980] 54524
McAloon, K. [1980] 70045
McAloon, K. [1981] 34009
Mills, G. [1984] 42484
Misercque, D. [1981] 76440
Misercque, D. [1982] 36085
Mlcek, J. [1978] 31964
Moller, S. [1965] 48096
Molodshij, V.N. [1952] 09388
Montagna, F. [1983] 34635
Montague, R. [1963] 14478
Morgenstern, C.F. [1982] 34161
Mostowski, Andrzej [1952] 09550
Mostowski, Andrzej [1958] 09566
Mostowski, Andrzej [1961] 24858
Mueller, Gert H. [1961] 09619
Negri, M. [1984] 44715
Orey, S. [1961] 10166
Pabion, J.F. [1981] 33728
Paris, J.B. [1978] 29247
Paris, J.B. [1978] 31967
Paris, J.B. [1980] 56654
Paris, J.B. [1981] 33729
Paris, J.B. [1981] 55495
Paris, J.B. [1984] 46698
Paris, J.B. [1985] 41796
Pudlak, P. [1983] 36550
Pudlak, P. [1985] 42556
Rabin, M.O. [1961] 10934
Ratajczyk, Z. [1982] 35382
Richard, D. [1984] 41779
Robinson, Julia [1973] 29547
Rosser, J.B. [1950] 11558
Schmerl, U.R. [1984] 43011
Schwabhaeuser, W. [1979] 53471
Scott, D.S. [1962] 11914
Shelah, S. [1984] 38758
Shepherdson, J.C. [1964] 12075
Shepherdson, J.C. [1965] 12072
Shepherdson, J.C. [1967] 12077
Shepherdson, J.C. [1969] 12078
Simpson, S.G. [1982] 35660
Simpson, S.G. [1985] 49879
Skolem, T.A. [1933] 12392
Skolem, T.A. [1934] 12393
Skolem, T.A. [1955] 27718
Smorynski, C.A. [1977] 27327
Smorynski, C.A. [1982] 37523
Smorynski, C.A. [1984] 41837
Smorynski, C.A. [1985] 49880
Steel, J.R. [1978] 29156
Tanaka, H. [1974] 23068
Tarski, A. [1933] 13407
Tverskoj, A.A. [1980] 35246
Twer von der, T. [1981] 35072
Ursini, A. [1977] 52907
Vaeaenaenen, J. [1979] 52749
Vetulani, Z. [1984] 44142

Visser, A. [1984] 42643
Wang, Hao [1963] 14049
Wette, E. [1959] 14162
Wilkie, A.J. [1978] 56743
Wilkie, A.J. [1980] 79968
Wilkie, A.J. [1982] 34735
Wilmers, G.M. [1985] 40251
Zarach, A. [1976] 23816
Zbierski, P. [1971] 14379
Zbierski, P. [1978] 29200
Zbierski, P. [1980] 34199
Zbierski, P. [1981] 34106

C65

Dalen van, D. [1984] 39878
Delzell, C.N. [1982] 36898
Delzell, C.N. [1982] 38556
Ershov, Yu.L. [1973] 30533
Fourman, M.P. [1979] 53715
Gurevich, Y. [1980] 53951
Gurevich, Y. [1983] 33765
Gurevich, Y. [1984] 38749
Hauschild, K. [1971] 27450
Hauschild, K. [1972] 05787
Henson, C.W. [1984] 36539
Ivanov, A.A. [1984] 41869
Jambu-Giraudet, M. [1980] 54167
Jambu-Giraudet, M. [1981] 55535
Jensen, C.U. [1984] 39754
Johnstone, P.T. [1979] 56505
Korec, I. [1974] 07371
Korec, I. [1976] 07374
Moschovakis, Y.N. [1964] 09514
Rimscha von, M. [1980] 53804
Robinson, R.M. [1951] 11312
Schwabhaeuser, W. [1979] 53471
Tarski, A. [1931] 16924

C70

Feferman, S. [1968] 03700
Feferman, S. [1968] 03703
Jaeger, G. [1982] 34700
Kino, A. [1963] 07100
Lopez-Escobar, E.G.K. [1967] 08263
McAloon, K. [1975] 18286
Nebres, B.F. [1972] 30771
Pohlers, W. [1982] 34316

C75

Arzarello, F. [1975] 51166
Arzarello, F. [1976] 50556
Blass, A.R. [1983] 40316
Bowen, K.A. [1974] 03942
Carstengerdes, W. [1971] 01873
Dahn, B.I. [1981] 35717
Engeler, E. [1961] 03362
Engeler, E. [1963] 03364
Feferman, S. [1968] 03700
Feferman, S. [1968] 03703
Feferman, S. [1972] 03706
Felscher, W. [1976] 04239
Ferbus, M.-C. [1984] 42605
Friedman, H.M. [1974] 04656
Geiser, J.R. [1974] 04831
Girard, J.-Y. [1984] 48565
Goad, C.A. [1978] 73424
Ito, Makoto [1934] 32185
Jervell, H.R. [1972] 33262
Jongh de, D.H.J. [1980] 55677
Kalicki, C. [1980] 53776
Kino, A. [1975] 17616
Kreisel, G. [1975] 27698
Lopez-Escobar, E.G.K. [1967] 08262
Lopez-Escobar, E.G.K. [1967] 08263
Maehara, S. [1961] 08524
Makkai, M. [1977] 50433
Martin-Loef, P. [1972] 08833
Misawa, T. [1977] 33405
Mitschke, G. [1980] 76450
Motohashi, N. [1972] 09596
Motohashi, N. [1984] 45359
Nadel, M.E. [1978] 29150
Nebres, B.F. [1972] 30771
Nishimura, T. [1965] 09975
Nishimura, T. [1974] 18312
Nishimura, T. [1974] 18313
Pohlers, W. [1981] 33310
Pohlers, W. [1982] 35591
Reyes, G.E. [1977] 52660
Schwichtenberg, H. [1975] 18407
Schwichtenberg, H. [1975] 25022
Slonneger, K. [1976] 14575
Tait, W.W. [1965] 13281
Tait, W.W. [1968] 13286
Tauts, A. [1974] 17693
Tauts, A. [1975] 18414
Tauts, A. [1975] 52459
Tauts, A. [1975] 52460
Tauts, A. [1981] 38114
Uesu, T. [1970] 13763
Uesu, T. [1971] 13764

C80

Apelt, H. [1966] 00403
Cegielski, P. [1980] 55770
Cegielski, P. [1981] 55626
Guzicki, W. [1974] 62249
Guzicki, W. [1981] 34160
Guzicki, W. [1983] 37520
Krynicki, M. [1985] 49048
Macintyre, A. [1980] 54621
Markwald, W. [1954] 08779
Mills, G. [1984] 42484
Morgenstern, C.F. [1982] 34161
Mostowski, Andrzej [1958] 09566
Motohashi, N. [1984] 45359
Reichbach, J. [1973] 31998
Shanin, N.A. [1976] 65058
Shanin, N.A. [1977] 53812
Shelah, S. [1984] 38758
Sobolev, S.K. [1977] 51014
Weese, M. [1976] 23814
Wolter, H. [1973] 14280
Wolter, H. [1975] 66137

C85

Andrews, P.B. [1972] 00366
Andrews, P.B. [1972] 00367
Bauval, A. [1985] 48368
Buechi, J.R. [1960] 01690
Buechi, J.R. [1962] 01693
Buechi, J.R. [1969] 01700
Elgot, C.C. [1966] 03291
Friedrich, U. [1972] 04668
Girard, J.-Y. [1976] 50375
Gurevich, Y. [1980] 53951
Gurevich, Y. [1983] 33765
Gurevich, Y. [1984] 38749
Henkin, L. [1950] 05894
Kogalovskij, S.R. [1968] 07299
Lopez-Escobar, E.G.K. [1967] 08262
Schwabhaeuser, W. [1979] 53471
Siefkes, D. [1970] 12145
Siefkes, D. [1970] 27984
Siefkes, D. [1971] 12146

C90

Amer, M.A. [1976] 17946
Amer, M.A. [1977] 30619
Artemov, S.N. [1980] 54546
Bozzi, S. [1977] 52446
Bozzi, S. [1980] 54221
Bozzi, S. [1982] 40429
Dahn, B.I. [1981] 35717
Dalen van, D. [1978] 27959
Dalen van, D. [1984] 39878
Dragalin, A.G. [1973] 29696
Fitting, M. [1969] 04349
Fitting, M. [1970] 14547
Fourman, M.P. [1979] 53155
Fourman, M.P. [1979] 53715
Fourman, M.P. [1980] 56570
Gabbay, D.M. [1971] 04726
Gabbay, D.M. [1972] 04729
Gabbay, D.M. [1972] 04730
Gabbay, D.M. [1973] 61880
Gabbay, D.M. [1975] 04738
Gabbay, D.M. [1977] 26456
Gabbay, D.M. [1977] 26463
Gabbay, D.M. [1982] 38464
Girard, J.-Y. [1976] 50375
Girard, J.-Y. [1977] 53265
Goernemann, S. [1971] 39863
Goldblatt, R.I. [1981] 55422
Grayson, R.J. [1981] 54421
Grayson, R.J. [1984] 41840
Heijenoort van, J. [1979] 79641
Hoeven van der, G.F. [1984] 41155
Hoeven van der, G.F. [1984] 41766
Jankowski, A.W. [1985] 47500
Johnstone, P.T. [1977] 51354
Johnstone, P.T. [1979] 56505
Johnstone, P.T. [1979] 74552
Jongh de, D.H.J. [1976] 18214
Jongh de, D.H.J. [1980] 55677
Josza, R. [1984] 45334
Kearns, J.T. [1978] 31151
Kirk, R.E. [1979] 53554
Kirk, R.E. [1980] 56546
Klemke, D. [1971] 07229
Kripke, S.A. [1962] 07556
Kripke, S.A. [1965] 07557
Krol', M.D. [1976] 50223
Krol', M.D. [1977] 52085
Krol', M.D. [1978] 52703
Krol', M.D. [1978] 53324
Leblanc, H. [1983] 37685
Lopez-Escobar, E.G.K. [1975] 24555
Lopez-Escobar, E.G.K. [1980] 55150

Lopez-Escobar, E.G.K. [1981] 55143
Lopez-Escobar, E.G.K. [1983] 39786
Luckhardt, H. [1970] 08381
Luckhardt, H. [1970] 08382
Macnab, D.S. [1977] 51394
Maehara, S. [1970] 08455
Makkai, M. [1977] 50433
Marchini, C. [1982] 41484
Markovic, Z. [1977] 51169
Markovic, Z. [1979] 56384
Markovic, Z. [1983] 37386
Markovic, Z. [1984] 45080
Martin-Loef, P. [1972] 21229
Miglioli, P.A. [1981] 55311
Moerdijk, I. [1984] 42485
Montagna, F. [1984] 42567
Motohashi, N. [1977] 23750
Mulvey, C.J. [1974] 21534
Muravitskij, A.Yu. [1981] 36087
Nadel, M.E. [1978] 29150
Nadiu, G.S. [1971] 09750
Nepejvoda, N.N. [1979] 53497
Nepejvoda, N.N. [1982] 46228
Ono, H. [1985] 47510
Osius, G. [1973] 27669
Osius, G. [1975] 18320
Osius, G. [1975] 18321
Osswald, H. [1970] 10184
Osswald, H. [1972] 10186
Popov, S.V. [1980] 80198
Rasiowa, H. [1953] 11004
Rasiowa, H. [1955] 11010
Rasiowa, H. [1955] 11011
Rauszer, C. [1977] 50668
Rauszer, C. [1977] 50669
Rauszer, C. [1980] 56381
Reichbach, J. [1973] 31998
Reichbach, J. [1975] 32003
Reyes, G.E. [1974] 11131
Reyes, G.E. [1975] 64812
Reyes, G.E. [1977] 52660
Reyes, G.E. [1978] 56342
Reznikoff, I. [1963] 11135
Reznikoff, I. [1966] 11137
Romanski, J. [1977] 51332
Routley, R. [1974] 15192
Routley, R. [1978] 53083
Sanchis, L.E. [1973] 11828
Scedrov, A. [1984] 37128
Schuette, K. [1968] 24822
Schultz, Konrad [1980] 54380
Scott, D.S. [1973] 78313
Scott, D.S. [1980] 78314
Scott, D.S. [1980] 78323
Scowcroft, P. [1984] 44277
Sette, A.-M. [1980] 53363
Shimoda, M. [1981] 54971
Skvortsov, D.P. [1979] 55940
Smorynski, C.A. [1973] 12535
Smorynski, C.A. [1973] 27058
Smorynski, C.A. [1977] 26856
Sobolev, S.K. [1977] 51014
Sonobe, O. [1975] 51586
Stepanov, V.I. [1983] 39758
Swart de, H.C.M. [1977] 26858
Swart de, H.C.M. [1978] 29237
Takeuti, G. [1978] 52447
Tauts, A. [1972] 17695
Tauts, A. [1972] 65673
Tauts, A. [1974] 17693
Tauts, A. [1974] 17694
Tauts, A. [1975] 18414
Tauts, A. [1975] 52459
Tauts, A. [1975] 52460
Tauts, A. [1981] 38114
Troelstra, A.S. [1977] 32501
Troelstra, A.S. [1978] 27952
Vuckovic, V. [1967] 27478
Weinstein, S. [1979] 53943
Yashin, A.D. [1984] 44104
Yashin, A.D. [1984] 44180

C95

Krynicki, M. [1985] 49048
Lopez-Escobar, E.G.K. [1980] 55150
Nepejvoda, N.N. [1982] 46228

C97

Berline, C. [1981] 54902
Dejon, B. [1969] 23438
Diller, J. [1975] 21771
Harrington, L.A. [1985] 49810
Lawvere, F.W. [1975] 70191
Loeb, M.H. [1968] 37293
Lolli, G. [1984] 41493
McAloon, K. [1980] 70045

C98

Barr, M. [1985] 40111
Barwise, J. [1977] 70117
Clote, P. [1981] 55425
Dalla Chiara Scabia, M.L. [1968] 04285
Enderton, H.B. [1972] 03355
Ershov, Yu.L. [1974] 25764
Fitting, M. [1969] 04349
Fourman, M.P. [1979] 53155
Friedman, H.M. [1975] 04296
Gabbay, D.M. [1973] 17911
Goldblatt, R.I. [1979] 55754
Henkin, L. [1971] 05928
Krajewski, S. [1981] 38331
Kreisel, G. [1956] 90276
Mal'tsev, A.I. [1961] 28706
Mostowski, Andrzej [1965] 09578
Rasiowa, H. [1963] 11028
Shoenfield, J.R. [1967] 22384
Simpson, S.G. [1985] 49879
Smirnov, V.A. [1972] 30359
Smorynski, C.A. [1984] 41837
Smorynski, C.A. [1985] 49880
Wang, Hao [1981] 34116
Wraith, G.C. [1976] 80097

D03

Curry, H.B. [1975] 50738
Ebbinghaus, H.-D. [1970] 28196
Elgot, C.C. [1961] 03288
Gabbay, D.M. [1976] 50854
Ledgard, H.F. [1972] 63344
Markov, A.A. [1974] 63719
Maslov, S.Yu. [1972] 08850
Ostroukhov, D.A. [1973] 10192
Shurygin, V.A. [1967] 12761
Shurygin, V.A. [1968] 12762
Shurygin, V.A. [1970] 12763
Skolem, T.A. [1954] 21133
Staples, J. [1983] 37059

D05

Ando, S. [1975] 52049
Anikeev, A.S. [1972] 00375
Arbib, M.A. [1964] 00454
Baralt-Torrijos, J. [1975] 17965
Buechi, J.R. [1960] 01690
Buechi, J.R. [1962] 01693
Buechi, J.R. [1969] 01700
Dalen van, D. [1968] 02761
Elgot, C.C. [1961] 03288
Elgot, C.C. [1964] 03289
Elgot, C.C. [1966] 03291
Fal'kovich, M.A. [1980] 69607
Friedrich, U. [1972] 04668
Gandy, R.O. [1980] 54925
Ginsburg, S. [1966] 04980
Gurari, E.M. [1981] 54218
Hermes, H. [1937] 32166
Kanovich, M.I. [1975] 18218
Lipton, R.J. [1977] 51255
Lorents, A.A. [1967] 08270
Lorents, A.A. [1968] 08275
Lorents, A.A. [1972] 63510
Machtey, M. [1978] 52951
Mostowski, A.Wlodzimierz [1981] 55152
Muchnik, A.A. [1983] 43519
Nabebin, A.A. [1977] 52155
Nakamura, A. [1977] 32225
Nakamura, A. [1979] 53052
O'Donnell, M.J. [1977] 76932
Siefkes, D. [1970] 12145
Siefkes, D. [1970] 27984
Sobolev, S.K. [1977] 51279
Trakhtenbrot, B.A. [1961] 19734
Trakhtenbrot, B.A. [1962] 19733
Turing, A.M. [1936] 13723
Turing, A.M. [1937] 13725
Wang, Hao [1957] 14033
Wasilewska, A. [1980] 55216
Zaslavskij, I.D. [1969] 27441

D10

Arbib, M.A. [1964] 00454
Belyakin, N.V. [1970] 29065
Berman, L. [1980] 55471
Cannonito, F.B. [1962] 21294
Caporaso, S. [1979] 56152
Charlesworth, A. [1981] 35267
Dekhtyar', M.I. [1979] 52952
Ebbinghaus, H.-D. [1970] 28196
Elgot, C.C. [1964] 03289
Fitch, F.B. [1969] 14672
Friedman, H.M. [1971] 04648
Gandy, R.O. [1980] 54925
Hauck, J. [1975] 62385
Hermes, H. [1937] 32166
Hermes, H. [1955] 06008
Lorents, A.A. [1972] 63510
Machtey, M. [1978] 52951
Muchnik, A.A. [1983] 43519

Mundici, D. [1983] 34976
Scarpellini, B. [1985] 47473
Thomas, William J. [1973] 65709
Turing, A.M. [1936] 13723
Turing, A.M. [1937] 13725
Wang, Hao [1957] 14033
Wette, E. [1974] 31553
Zharov, V.G. [1979] 54228

D15

Arbib, M.A. [1966] 27533
Ausiello, G. [1974] 80587
Baker, T.P. [1979] 80606
Ben-Ari, M. [1980] 55966
Berman, L. [1977] 70972
Berman, L. [1980] 55471
Bibel, W. [1979] 53410
Cellucci, C. [1985] 45633
Chaitin, G.J. [1974] 60893
Cherniavsky, J.C. [1973] 71692
Cherniavsky, J.C. [1976] 31025
Constable, R.L. [1972] 61071
Cook, S.A. [1974] 25005
Cook, S.A. [1979] 56282
Dantsin, E.Ya. [1981] 55532
Dekhtyar', M.I. [1979] 52952
Dezani-Ciancaglini, M. [1979] 81091
DeMillo, R.A. [1979] 72280
Dimitracopoulos, C. [1985] 47569
Emde Boas van, P. [1975] 21889
Evangelist, M. [1982] 54618
Ferrante, J. [1975] 21702
Fischer, Michael J. [1974] 17344
Gordon, D. [1979] 53557
Gurari, E.M. [1981] 54218
Hartmanis, J. [1976] 24160
Hartmanis, J. [1983] 38195
Helm, J.P. [1973] 05879
Joseph, D. [1981] 34701
Joseph, D. [1983] 39245
Kanovich, M.I. [1969] 29887
Kanovich, M.I. [1973] 06903
Kanovich, M.I. [1974] 07592
Kanovich, M.I. [1974] 62868
Kanovich, M.I. [1978] 52518
Kanovich, M.I. [1979] 52945
Karol', A.M. [1978] 52579
Khomich, V.I. [1975] 17549
Kowalczyk, W. [1982] 37601
Kozen, D. [1977] 81928
Ledgard, H.F. [1972] 63344
Leivant, D. [1982] 34202
Levin, L.A. [1976] 50487
Levitz, H. [1978] 30732
Lipton, R.J. [1978] 75578
Lipton, R.J. [1979] 53366
Longo, G. [1974] 53697
Loveland, D.W. [1978] 31164
Machtey, M. [1978] 52951
Metropolis, N. [1974] 21270
Moravek, J. [1973] 82295
Mostowski, A.Wlodzimierz [1981] 55152
Muchnik, A.A. [1983] 43519
Mundici, D. [1983] 34976
O'Donnell, M.J. [1979] 82386
Oppen, D.C. [1973] 24077
Oppen, D.C. [1978] 51889
Parikh, R. [1971] 10282
Plaisted, D.A. [1980] 56667
Plotkin, J.M. [1982] 35398
Robertson, E.L. [1974] 25884
Scarpellini, B. [1984] 43196
Shurygin, V.A. [1974] 29609
Shurygin, V.A. [1977] 51800
Skordev, D.G. [1976] 23157
Solomon, M.K. [1978] 52755
Statman, R. [1977] 27329
Statman, R. [1977] 78954
Statman, R. [1979] 52903
Statman, R. [1979] 52904
Statman, R. [1982] 34359
Wilkie, A.J. [1980] 79968
Woehl, K. [1979] 53367
Young, P. [1985] 46382
Yukami, T. [1984] 43511
Zaslavskij, I.D. [1969] 27441

D20

Ackermann, W. [1928] 00106
Anshel, M. [1983] 36994
Ausiello, G. [1974] 80587
Bakker de, J.W. [1971] 28688
Beeson, M.J. [1975] 17547
Beeson, M.J. [1976] 14764
Beeson, M.J. [1984] 42422
Berg, J. [1975] 70955
Bowie, G.L. [1973] 27103
Bundy, A. [1973] 01765
Byerly, R.E. [1982] 35525
Calude, C. [1983] 37287
Cannonito, F.B. [1971] 71525
Cantini, A. [1985] 49071
Cherniavsky, J.C. [1976] 31025
Chernov, V.P. [1972] 17992
Church, A. [1932] 02123
Church, A. [1933] 02124
Church, A. [1935] 02126
Church, A. [1935] 27542
Church, A. [1936] 02130
Church, A. [1936] 23466
Church, A. [1941] 02182
Church, A. [1957] 02402
Cichon, E.A. [1983] 40697
Cleave, J.P. [1967] 02682
Cleave, J.P. [1969] 02266
Clote, P. [1985] 45298
Cohen, D.E. [1984] 43120
Collins, W.J. [1978] 52157
Collins, W.J. [1983] 43537
Coste-Roy, M.-F. [1980] 53743
Crossley, J.N. [1966] 02555
Curry, H.B. [1941] 02612
Damm, W. [1978] 51522
Davis, Martin D. [1965] 04186
Davis, Martin D. [1965] 25888
Demuth, O. [1976] 29783
Dennis-Jones, E.C. [1984] 45385
Ebbinghaus, H.-D. [1970] 28196
Emde Boas van, P. [1975] 21889
Enderton, H.B. [1968] 03348
Engeler, E. [1981] 36907
Fahmy, M.H. [1982] 34459
Fischer, P.C. [1965] 04301
Fisher, A. [1982] 36967
Fitch, F.B. [1968] 17117
Frejdzon, R.I. [1972] 04594
Friedman, H.M. [1971] 04648
Friedman, H.M. [1978] 31762
Gandy, R.O. [1980] 54925
Goedel, K. [1931] 15052
Goedel, K. [1965] 21353
Goodman, Nicolas D. [1978] 29277
Goodstein, R.L. [1954] 05175
Gordon, D. [1979] 53557
Grzegorczyk, A. [1955] 05392
Grzegorczyk, A. [1957] 05396
Hanf, W.P. [1965] 05635
Hanson, N.R. [1961] 05642
Hartmanis, J. [1976] 24160
Hauck, J. [1972] 05754
Hauck, J. [1975] 62385
Hauck, J. [1980] 55489
Helm, J.P. [1973] 05879
Hermes, H. [1937] 32166
Hermes, H. [1955] 06008
Hermes, H. [1969] 20948
Howard, W.A. [1981] 54572
Hu, Shihua [1963] 48690
Huang, Wengi [1985] 49393
Ibarra, O.H. [1981] 55261
Jockusch Jr., C.G. [1975] 06642
Kalmar, L. [1943] 19071
Kalmar, L. [1959] 06864
Kanovich, M.I. [1969] 29887
Kanovich, M.I. [1974] 07592
Kashapova, F.R. [1984] 44311
Khakhanyan, V.Kh. [1980] 56495
Khakhanyan, V.Kh. [1980] 73754
Khakhanyan, V.Kh. [1980] 73755
Khakhanyan, V.Kh. [1981] 36853
Khakhanyan, V.Kh. [1983] 41114
Khomich, V.I. [1975] 17549
Kino, A. [1968] 07103
Kleene, S.C. [1967] 45895
Kosovskij, N.K. [1981] 55663
Kreisel, G. [1957] 07494
Kreisel, G. [1957] 07520
Kreisel, G. [1957] 29369
Kreisel, G. [1959] 07508
Kreisel, G. [1972] 07553
Kushner, B.A. [1973] 07699
Kushner, B.A. [1973] 07702
Kushner, B.A. [1976] 75199
Ladriere, J. [1961] 07778
Lavrov, I.A. [1979] 53936
Leivant, D. [1985] 39403
Levitz, H. [1978] 30732
Loeb, M.H. [1974] 63480
Lorenzen, P. [1962] 08303
Maass, W. [1976] 23711
Machtey, M. [1978] 52951
Maehara, S. [1957] 08518
Mayoh, B.H. [1965] 08924
Mayoh, B.H. [1970] 17641
McAloon, K. [1980] 54524
McBeth, R. [1981] 55251
Medvedev, Yu.T. [1972] 63851
Mendelson, E. [1963] 09116
Milner, R. [1976] 50948
Mints, G.E. [1971] 09286
Mints, G.E. [1982] 35718

Mo, Shaokui [1965] 48522
Mostowski, Andrzej [1957] 09564
Myhill, J.R. [1950] 09699
Nolin, L. [1975] 23994
Nolin, L. [1975] 82368
Normann, D. [1984] 37098
O'Donnell, M.J. [1977] 76932
O'Donnell, M.J. [1979] 76931
Owings Jr., J.C. [1973] 10206
Parikh, R. [1966] 10279
Partis, M.T. [1963] 10312
Partis, M.T. [1967] 10313
Peter, R. [1940] 10396
Peter, R. [1959] 10410
Pettorossi, A. [1979] 53394
Pla i Carrera, J. [1982] 37071
Pogorzelski, H.A. [1969] 19592
Pour-El, M.B. [1973] 31218
Pour-El, M.B. [1974] 10706
Pour-El, M.B. [1978] 52704
Prank, R.K. [1979] 53437
Richman, F. [1983] 34676
Rieger, L. [1961] 22008
Ritchie, R.W. [1968] 11208
Ritter, W.E. [1966] 11210
Ritter, W.E. [1967] 11211
Robinson, Julia [1973] 29547
Rogers Jr., H. [1958] 11357
Rose, H.E. [1972] 11511
Rosser, J.B. [1955] 23483
Rosser, J.B. [1955] 42147
Routledge, N.A. [1953] 11604
Ryan, W.J. [1978] 53718
Sanchis, L.E. [1967] 11825
Sazonov, V.Yu. [1976] 25002
Sazonov, V.Yu. [1976] 26049
Schmidt, Diana [1976] 24318
Schnorr, C.-P. [1969] 12929
Schuette, K. [1954] 13005
Schwichtenberg, H. [1972] 13077
Semenov, A.L. [1977] 51323
Shanin, N.A. [1955] 19648
Shapiro, S. [1977] 32450
Shapiro, S. [1981] 78442
Shen, Baiying [1984] 46417
Shen, Baiying [1985] 46288
Shepherdson, J.C. [1976] 23677
Shmain, I.Kh. [1974] 78543
Shurygin, V.A. [1966] 12786
Shurygin, V.A. [1977] 51800
Siefkes, D. [1975] 12147
Skarbek, W. [1972] 29864
Skolem, T.A. [1923] 38689
Skolem, T.A. [1944] 12401
Skolem, T.A. [1944] 12402
Smorynski, C.A. [1979] 54179
Smorynski, C.A. [1982] 34588
Sorbi, A. [1982] 46829
Spector, C. [1957] 12676
Spencer, J.H. [1983] 34745
Statman, R. [1977] 78954
Statman, R. [1979] 52904
Statman, R. [1982] 34359
Stenlund, S. [1970] 78994
Svenonius, L. [1979] 52843
Thomas, William J. [1973] 65709
Trakhtenbrot, B.A. [1959] 45721
Tsejtin, G.S. [1959] 28243
Turing, A.M. [1936] 13723
Turing, A.M. [1937] 13725
Uspenskij, V.A. [1953] 16353
Uspenskij, V.A. [1953] 45765
Uspenskij, V.A. [1959] 28239
Vier, L.C. [1972] 49913
Vogel, Helmut [1977] 54173
Vuckovic, V. [1959] 13892
Vuckovic, V. [1962] 22057
Wainer, S.S. [1985] 45318
Wainer, S.S. [1985] 46381
Webb, J.C. [1980] 54214
Webb, J.C. [1983] 43711
Yanov, Yu.I. [1977] 74466
Zashev, J.A. [1984] 41008
Zaslavskij, I.D. [1969] 14375
Zemke, F. [1977] 51115
Zharov, V.G. [1974] 66215

D25

Bernardi, C. [1981] 55085
Cannonito, F.B. [1962] 21294
Carstens, H.G. [1977] 51961
Caviness, B.F. [1971] 02214
Cohen, D.E. [1984] 43120
Craig, W. [1960] 02513
Davis, Martin D. [1965] 25888
Dawes, A.M. [1975] 04287
Deutsch, M. [1975] 04085
Ebbinghaus, H.-D. [1970] 28196
Ehrenfeucht, A. [1959] 03237
Ellentuck, E. [1971] 03306
Hanf, W.P. [1965] 05635
Hatcher, W.S. [1981] 55168
Helm, J.P. [1973] 05879
Jockusch Jr., C.G. [1972] 06633
Jones, James P. [1969] 06695
Jones, James P. [1978] 29266
Kleene, S.C. [1967] 45895
Kleinberg, E.M. [1970] 07211
Kripke, S.A. [1967] 31215
Kripke, S.A. [1967] 31216
Lavrov, I.A. [1979] 53936
Mayoh, B.H. [1968] 08926
Paola di, R.A. [1966] 03045
Paola di, R.A. [1966] 03046
Paola di, R.A. [1966] 03047
Paola di, R.A. [1967] 03048
Prank, R.K. [1980] 77510
Prank, R.K. [1981] 35552
Putnam, H. [1964] 10838
Robinson, R.M. [1956] 11314
Rogers Jr., H. [1967] 24823
Rosenberg, R. [1982] 34406
Rosser, J.B. [1936] 11542
Sanchis, L.E. [1977] 52496
Scott, D.S. [1975] 18375
Shepherdson, J.C. [1961] 12071
Shore, R.A. [1981] 55154
Simpson, S.G. [1977] 27322
Slisenko, A.O. [1971] 12465
Smorynski, C.A. [1978] 28384
Soare, R.I. [1969] 12565
Specker, E. [1971] 29960
Tsejtin, G.S. [1970] 01909
Turing, A.M. [1939] 13726
Uspenskij, V.A. [1953] 16353
Vaeaenaenen, J. [1979] 52749
Vershinin, V.A. [1975] 22991

D30

Barendregt, H.P. [1978] 31152
Clote, P. [1985] 45298
Dekhtyar', M.I. [1979] 52952
Demuth, O. [1978] 31808
Ershov, Yu.L. [1970] 03540
Farrington, Paddy [1984] 42273
Fischer, P.C. [1965] 04301
Friedman, H.M. [1976] 14767
Friedman, H.M. [1976] 21610
Goodman, Nicolas D. [1978] 29277
Hensel, G. [1965] 05946
Hoeven van der, G.F. [1984] 42464
Jockusch Jr., C.G. [1972] 06633
Kanovich, M.I. [1973] 06903
Kleene, S.C. [1962] 07191
Kucera, A. [1985] 45308
Kushner, B.A. [1970] 07698
Lukas, J.D. [1974] 08387
Mayoh, B.H. [1968] 08926
Nerode, A. [1980] 54927
Odifreddi, P. [1983] 38171
Paola di, R.A. [1966] 03045
Ritter, W.E. [1966] 11210
Rogers Jr., H. [1967] 24823
Sanchis, L.E. [1979] 53253
Sazonov, V.Yu. [1976] 29755
Shore, R.A. [1981] 55154
Shore, R.A. [1981] 55426
Simpson, S.G. [1977] 27322
Simpson, S.G. [1977] 30798
Soare, R.I. [1969] 12565
Spector, C. [1955] 12672
Steel, J.R. [1978] 29156
Turing, A.M. [1939] 13726

D35

Andrews, P.B. [1974] 03832
Artemov, S.N. [1985] 48972
Barendregt, H.P. [1976] 21504
Bernardi, C. [1981] 55085
Carstens, H.G. [1977] 51961
Christian, C.C. [1981] 55456
Church, A. [1935] 27542
Cohen, D.E. [1984] 43120
Collins, G.E. [1970] 02332
Davis, Martin D. [1965] 25888
Delon, F. [1978] 27313
Ehrenfeucht, A. [1971] 29998
Elgot, C.C. [1961] 03288
Elgot, C.C. [1966] 03291
Farrington, Paddy [1984] 42273
Finsler, P. [1926] 03793
Gabbay, D.M. [1972] 04731
Gabbay, D.M. [1973] 61880
Gabbay, D.M. [1977] 27197
Germano, G. [1971] 04907
Goedel, K. [1931] 15052
Goedel, K. [1965] 21353
Goldfarb, W.D. [1981] 54331
Goncharov, S.S. [1976] 52189
Grandy, R.E. [1966] 05303

Grzegorczyk, A. [1956] 05395
Grzegorczyk, A. [1957] 24864
Gurevich, Y. [1965] 05452
Gurevich, Y. [1982] 33757
Gurevich, Y. [1983] 33765
Hanf, W.P. [1965] 05635
Hanson, N.R. [1961] 05642
Hauschild, K. [1970] 05774
Hauschild, K. [1971] 05780
Hauschild, K. [1971] 27450
Hauschild, K. [1972] 05787
Hauschild, K. [1972] 05790
Hermes, H. [1955] 06008
Hermes, H. [1964] 06013
Hilbert, D. [1900] 06078
Huber-Dyson, V. [1982] 35411
Jambu-Giraudet, M. [1980] 54167
Jambu-Giraudet, M. [1981] 55535
Jensen, C.U. [1984] 39754
Kripke, S.A. [1967] 31215
Kripke, S.A. [1967] 31216
Lifschitz, V. [1967] 33665
Loeb, M.H. [1976] 14746
Lorenzen, P. [1962] 08303
Markov, A.A. [1974] 63719
Mart'yanov, V.I. [1977] 29230
Maslov, S.Yu. [1965] 19335
Mayoh, B.H. [1965] 08924
Mayoh, B.H. [1970] 17641
McKenzie, R. [1971] 08997
Medvedev, Yu.T. [1969] 09081
Mostowski, Andrzej [1953] 28815
Mostowski, Andrzej [1953] 42176
Murawski, R. [1984] 39679
Nagel, E. [1958] 09765
Nagel, E. [1967] 09763
Nakamura, A. [1977] 32225
Nakamura, A. [1979] 53052
Nerode, A. [1980] 54927
Nishimura, T. [1959] 09973
Norgela, S.A. [1978] 32674
Novikov, P.S. [1949] 19476
Okee, J. [1975] 18315
Orevkov, V.P. [1972] 10161
Paola di, R.A. [1967] 03048
Penzin, Yu.G. [1976] 32592
Plisko, V.E. [1977] 51967
Podnieks, K.M. [1975] 64585
Quine, W.V.O. [1949] 10881
Rabin, M.O. [1965] 16270
Richard, D. [1982] 34019
Richard, D. [1984] 41779
Robinson, Julia [1949] 11292
Robinson, R.M. [1952] 11309
Rose, H.E. [1961] 11507
Rosser, J.B. [1936] 11542
Rosser, J.B. [1939] 11547
Shabunin, L.V. [1973] 11733
Shabunin, L.V. [1974] 17628
Shabunin, L.V. [1975] 18369
Shore, R.A. [1981] 55154
Shore, R.A. [1981] 55426
Siefkes, D. [1971] 12146
Simpson, S.G. [1977] 27322
Simpson, S.G. [1977] 30798
Skolem, T.A. [1940] 12398
Smart, J.J.C. [1961] 31260
Smorynski, C.A. [1973] 12535
Smorynski, C.A. [1982] 43687
Sobolev, S.K. [1977] 51014
Stegmueller, W. [1959] 13145
Szmielew, W. [1952] 12847
Tarski, A. [1953] 42175
Thomas, Wolfgang [1975] 13578
Turing, A.M. [1936] 13723
Uspenskij, V.A. [1953] 45765
Verkhozina, M.I. [1973] 24127
Vuckovic, V. [1967] 13899
Weese, M. [1976] 23814
Young, P. [1985] 46382

D40

Anshel, M. [1983] 36994

D45

Aczel, P. [1966] 00137
Aczel, P. [1975] 21303
Belyakin, N.V. [1970] 29065
Calude, C. [1978] 52071
Cannonito, F.B. [1971] 71525
Chen, Kehsun [1978] 27962
Chernov, V.P. [1972] 17990
Chernov, V.P. [1972] 17991
Crossley, J.N. [1966] 02555
Crossley, J.N. [1966] 02556
Ershov, Yu.L. [1974] 25764
Ershov, Yu.L. [1977] 16634
Friedman, H.M. [1985] 47284
Goncharov, S.S. [1982] 40426
Hajkova, M. [1971] 05529
Helm, J.P. [1973] 05879
Hermann, G. [1926] 05997
Kanda, A. [1984] 45525
Kanda, A. [1985] 47568
Khisamiev, N.G. [1979] 32601
Khisamiev, N.G. [1983] 40239
Kreisel, G. [1950] 07481
Kreisel, G. [1953] 20815
Kushner, B.A. [1978] 52910
Lifschitz, V. [1967] 08139
Machtey, M. [1978] 52951
Madison, E.W. [1970] 24826
Mayoh, B.H. [1970] 17641
Metakides, G. [1982] 37391
Montagna, F. [1982] 34471
Morozov, A.S. [1983] 43123
Moschovakis, Y.N. [1964] 09514
Moschovakis, Y.N. [1966] 21265
Parikh, R. [1966] 10279
Pinus, A.G. [1975] 16662
Roche la, P. [1981] 75255
Rogers Jr., H. [1958] 11357
Sakai, H. [1974] 65025
Seidenberg, A. [1978] 53225
Shurygin, V.A. [1966] 12786
Smith, Rick L. [1981] 55685
Spector, C. [1955] 12672
Visser, A. [1980] 35141
Waerden van der, B.L. [1930] 38703
Wagner, E.G. [1969] 13965
Zharov, V.G. [1979] 54228

D50

Aczel, P. [1975] 21303
Silverstein, A. [1978] 52141

D55

Amstislavskij, V.I. [1966] 00303
Apt, K.R. [1974] 00432
Apt, K.R. [1976] 14789
Apt, K.R. [1978] 29131
Artemov, S.N. [1985] 48972
Baire, R. [1905] 05473
Belyakin, N.V. [1983] 41479
Chauvin, A. [1962] 02061
Clote, P. [1985] 45298
Demuth, O. [1975] 17635
Demuth, O. [1977] 72287
Demuth, O. [1978] 31808
Demuth, O. [1983] 44288
Dombrovskij-Kabanchenko, M.N. [1979] 56430
Ershov, Yu.L. [1970] 03540
Feferman, S. [1960] 03686
Feferman, S. [1962] 03688
Feferman, S. [1962] 03693
Feferman, S. [1965] 03696
Feferman, S. [1975] 23151
Feferman, S. [1977] 27330
Friedman, H.M. [1970] 03773
Friedman, H.M. [1970] 14546
Friedman, H.M. [1976] 14767
Friedman, H.M. [1976] 21610
Gandy, R.O. [1960] 15061
Gandy, R.O. [1961] 22330
Gass, F.S. [1971] 04808
Gass, F.S. [1972] 04809
Girard, J.-Y. [1981] 35541
Goodman, Nicolas D. [1978] 29277
Gordeev, L.N. [1977] 50580
Grzegorczyk, A. [1958] 05398
Hajek, P. [1977] 26853
Hensel, G. [1965] 05946
Hoeven van der, G.F. [1984] 42464
Kanovej, V.G. [1975] 50893
Kanovej, V.G. [1979] 53375
Kanovej, V.G. [1979] 56024
Kino, A. [1962] 07097
Kipnis, M.M. [1968] 27486
Kleene, S.C. [1944] 07168
Kleene, S.C. [1955] 07178
Kleene, S.C. [1962] 07191
Kleene, S.C. [1967] 45895
Kondo, M. [1956] 07330
Kondo, M. [1956] 07331
Kondo, M. [1956] 07332
Kreider, D.L. [1961] 07467
Kreisel, G. [1950] 07481
Kreisel, G. [1953] 20815
Kreisel, G. [1960] 07519
Kreisel, G. [1962] 07526
Kreisel, G. [1972] 07553
Kucera, A. [1978] 52777
Lacombe, D. [1957] 07768
Lucian, M. [1972] 28788
Lukas, J.D. [1974] 08387
Luzin, N.N. [1929] 16811
Markwald, W. [1954] 08779

Masseron, M. [1983] 40760
Melikyan, S.M. [1974] 22984
Mostowski, Andrzej [1947] 09537
Mostowski, Andrzej [1949] 09547
Mostowski, Andrzej [1959] 09568
Mostowski, Andrzej [1962] 33705
Myhill, J.R. [1949] 09698
Nepejvoda, N.N. [1973] 09867
Normann, D. [1983] 37656
Odifreddi, P. [1983] 38171
Plisko, V.E. [1977] 51967
Putnam, H. [1961] 10835
Putnam, H. [1964] 10839
Robertson, E.L. [1974] 25884
Rogers Jr., H. [1967] 24823
Rosenberg, R. [1982] 34406
Shapiro, N.Z. [1969] 12020
Sorbi, A. [1982] 41231
Spector, C. [1955] 12672
Spector, C. [1957] 12676
Steel, J.R. [1978] 29156
Tanaka, H. [1974] 23068
Tarski, A. [1931] 16924
Troelstra, A.S. [1978] 27952
Vetulani, Z. [1984] 44142
Wang, Hao [1957] 29384

D60

Girard, J.-Y. [1984] 42404
Girard, J.-Y. [1985] 46368
Kino, A. [1962] 07097
Ressayre, J.-P. [1982] 37224
Richter, W.H. [1971] 11173
Schuette, K. [1976] 31510
Wiele van de, J. [1982] 38124

D65

Aczel, P. [1970] 00141
Belyakin, N.V. [1970] 29065
Belyakin, N.V. [1974] 17342
Chernov, V.P. [1972] 17990
Chernov, V.P. [1972] 17991
Curry, H.B. [1964] 02635
Ershov, Yu.L. [1972] 61604
Ershov, Yu.L. [1977] 16634
Feferman, S. [1977] 27330
Girard, J.-Y. [1977] 56065
Girard, J.-Y. [1985] 42597
Grzegorczyk, A. [1964] 05411
Hanatani, Y. [1966] 24779
Hanatani, Y. [1975] 22904
Harrison, J. [1963] 49833
Hinman, P.G. [1984] 43007
Kleene, S.C. [1962] 07191
Kleene, S.C. [1969] 24941
Kreisel, G. [1959] 07499
Longo, G. [1979] 55892
Normann, D. [1983] 37656
Richter, W.H. [1967] 11171
Richter, W.H. [1968] 11172
Richter, W.H. [1971] 11173
Sazonov, V.Yu. [1976] 29755
Schwichtenberg, H. [1975] 25022
Tait, W.W. [1975] 18413
Wiele van de, J. [1982] 38124

D70

Aczel, P. [1970] 00141
Apt, K.R. [1978] 29131
Belyakin, N.V. [1969] 00982
Belyakin, N.V. [1970] 29065
Bracho, F. [1980] 69241
Grzegorczyk, A. [1958] 05398
Kleene, S.C. [1944] 07168
Kleene, S.C. [1955] 07178
Kreider, D.L. [1961] 07467
Putnam, H. [1964] 10839
Richter, W.H. [1965] 11170
Richter, W.H. [1967] 11171
Richter, W.H. [1968] 11172
Richter, W.H. [1971] 11173
Rosser, J.B. [1937] 11544
Wang, Hao [1953] 14021

D75

Barendregt, H.P. [1975] 27590
Feferman, S. [1978] 72806
Friedman, H.M. [1971] 04648
Hindley, J.R. [1977] 24332
Hyland, J.M.E. [1982] 35795
Ivanov, L.L. [1978] 56426
Ivanov, L.L. [1981] 36073
Jervell, H.R. [1985] 47555
Longo, G. [1976] 50092
Manca, V. [1981] 55091
Mitschke, G. [1972] 09330
Normann, D. [1983] 37656
Petrov, V.P. [1979] 55959
Pour-El, M.B. [1973] 10709
Pour-El, M.B. [1973] 31218
Pour-El, M.B. [1974] 10706
Richman, F. [1983] 34676
Robinson, Julia [1973] 29547
Roever de, W.P. [1974] 64875
Scott, D.S. [1975] 18375
Scott, D.S. [1982] 38593
Shmain, I.Kh. [1974] 78543
Skordev, D.G. [1975] 29752
Skordev, D.G. [1976] 18392
Skordev, D.G. [1976] 23157
Skordev, D.G. [1976] 55893
Skordev, D.G. [1977] 50955
Skordev, D.G. [1978] 54899
Skordev, D.G. [1979] 52963
Skordev, D.G. [1980] 41127
Skordev, D.G. [1980] 54101
Skordev, D.G. [1982] 37517
Venturini Zilli, M. [1974] 23952
Wagner, E.G. [1969] 13965

D80

Anshel, M. [1983] 36994
Arbib, M.A. [1964] 00454
Baxter, L.D. [1978] 52222
Bishop, E.A. [1985] 48575
Burger, E. [1964] 01771
Chaitin, G.J. [1974] 60893
Chernyakhovskij, N.P. [1976] 71693
Christian, C.C. [1981] 55456
Cleave, J.P. [1969] 02266
Hyland, J.M.E. [1982] 35795
Kanovich, M.I. [1969] 29887
Kleene, S.C. [1957] 22336
Kleene, S.C. [1960] 07188
Kreisel, G. [1957] 07490
Kushner, B.A. [1976] 75199
Levin, L.A. [1976] 50487
McAloon, K. [1978] 54427
Medvedev, Yu.T. [1962] 09078
Medvedev, Yu.T. [1963] 09079
Medvedev, Yu.T. [1966] 09080
Metakides, G. [1982] 37391
Miglioli, P.A. [1981] 55396
Pakhomov, S.V. [1976] 64344
Pour-El, M.B. [1973] 10709
Pour-El, M.B. [1973] 31218
Pour-El, M.B. [1974] 10706
Pour-El, M.B. [1978] 52704
Pour-El, M.B. [1979] 77493
Pour-El, M.B. [1984] 45841
Richardson, D.B. [1971] 11164
Schnorr, C.-P. [1969] 12929
Smorynski, C.A. [1979] 54179
Specker, E. [1971] 29960
Zharov, V.G. [1974] 66215

D97

Boerger, E. [1984] 40062
Davis, Martin D. [1965] 25888
Harrington, L.A. [1985] 49810
Jacobs, K. [1973] 23474
Lolli, G. [1984] 41493

D98

Barwise, J. [1977] 70117
Friedman, H.M. [1975] 04296
Grzegorczyk, A. [1957] 24864
Hermes, H. [1955] 06008
Hermes, H. [1961] 20981
Kleene, S.C. [1952] 07173
Kleene, S.C. [1967] 45895
Kosovskij, N.K. [1981] 55663
Krajewski, S. [1981] 38331
Lorents, A.A. [1972] 63510
Lorenzen, P. [1962] 08303
Manin, Yu.I. [1977] 51984
Mostowski, Andrzej [1965] 09578
Rogers Jr., H. [1967] 24823
Shapiro, S. [1983] 37650
Shoenfield, J.R. [1967] 22384

D99

Bulloff, J.J. [1969] 14484
Slezak, P. [1984] 40543

E05

Aczel, P. [1981] 34552
Clote, P. [1980] 56493
Clote, P. [1981] 55425
Clote, P. [1985] 41801
Harrington, L.A. [1977] 27334
Kirby, L.A.S. [1982] 34416
Masseron, M. [1983] 40760
McAloon, K. [1981] 34009
Ratajczyk, Z. [1982] 35382
Schmidt, Diana [1979] 78208
Shelah, S. [1976] 51434

Shelah, S. [1984] 38758
Smullyan, R.M. [1979] 78773
Specker, E. [1971] 29960
Tverskoj, A.A. [1980] 35246
Twer von der, T. [1981] 35072
Vaeaenaenen, J. [1979] 52749
Veblen, O. [1908] 13917

E07

Chandrasekharan, K. [1942] 02407
Dries van den, L. [1984] 46154
Ehrenfeucht, A. [1973] 23826
Gurevich, Y. [1983] 33765
Jambu-Giraudet, M. [1980] 54167
Jambu-Giraudet, M. [1981] 55535
Levitz, H. [1978] 30732
McBeth, R. [1980] 76194
Parikh, R. [1966] 10279
Pinus, A.G. [1975] 16662
Sciore, E. [1978] 52051
Sethi, R. [1974] 78402

E10

Bachmann, H. [1950] 00608
Bachmann, H. [1952] 00609
Bachmann, H. [1954] 00610
Bachmann, H. [1955] 00611
Bridge, J. [1975] 04262
Brouwer, L.E.J. [1909] 02690
Buchholz, W. [1975] 23954
Dalen van, D. [1968] 02759
Feferman, S. [1972] 03706
Felscher, W. [1979] 72846
Finsler, P. [1951] 03799
Girard, J.-Y. [1981] 35541
Girard, J.-Y. [1984] 42447
Girard, J.-Y. [1985] 45634
Goodstein, R.L. [1947] 05167
Grayson, R.J. [1982] 34472
Hardy, G.H. [1904] 05655
Hickman, J.L. [1974] 17476
Hickman, J.L. [1978] 74099
Isles, D. [1970] 06457
Jaeger, G. [1984] 39555
Kino, A. [1975] 17616
Kondo, M. [1956] 07328
Levitz, H. [1966] 08050
Levitz, H. [1969] 08052
Levitz, H. [1978] 30732
McBeth, R. [1976] 18494
McBeth, R. [1979] 53647
McBeth, R. [1980] 76194
Miller, L.W. [1976] 14776
Neumer, W. [1953] 09919
Neumer, W. [1954] 09921
Neumer, W. [1956] 19481
Neumer, W. [1957] 09924
Neumer, W. [1960] 09926
Neumer, W. [1970] 09927
Pfeiffer, H. [1973] 25003
Saarnio, U. [1962] 11717
Saarnio, U. [1967] 11718
Saarnio, U. [1968] 11720
Saarnio, U. [1968] 11721
Schmidt, Diana [1975] 14823
Schmidt, Diana [1976] 24318
Schmidt, Diana [1978] 54758
Schmidt, Diana [1979] 78208
Schuette, K. [1962] 13013
Schuette, K. [1976] 13021
Seckendorff von, V. [1937] 41030
Sierpinski, W. [1932] 39936
Simauti, T. [1964] 12324
Stahl, G. [1981] 35983
Takeuti, G. [1952] 13306
Takeuti, G. [1957] 19761
Troelstra, A.S. [1967] 13678
Vauzeilles, J. [1982] 43062
Veblen, O. [1908] 13917
Weyl, H. [1931] 22091

E15

Amstislavskij, V.I. [1966] 00303
Baire, R. [1905] 05473
Burgess, J.P. [1980] 53988
Demuth, O. [1977] 72287
Friedman, H.M. [1970] 03773
Gielen, W. [1981] 54491
Kanovej, V.G. [1975] 50893
Kanovej, V.G. [1979] 56024
Kondo, M. [1956] 07330
Kondo, M. [1956] 07331
Kondo, M. [1956] 07332
Kucera, A. [1985] 45308
Levin, A.M. [1975] 25810
Luzin, N.N. [1929] 16811
Mostowski, Andrzej [1962] 33705
Tarski, A. [1931] 16924
Yasugi, M. [1981] 47763

E20

Dantzig van, D. [1942] 02772
Jervell, H.R. [1985] 47805
Tang, A. [1979] 69945

E25

Aczel, P. [1982] 37664
Andrews, P.B. [1972] 00367
Ashvinikumar [1971] 03841
Bachmann, H. [1954] 00610
Bachmann, H. [1955] 00611
Baire, R. [1905] 05473
Barendregt, H.P. [1973] 29015
Beeson, M.J. [1980] 54110
Bily, J. [1974] 01205
Blass, A.R. [1983] 40316
Borel, E. [1948] 01466
Diaconescu, R. [1975] 17526
Feferman, S. [1965] 03696
Feferman, S. [1974] 21714
Fitting, M. [1969] 04349
Fourman, M.P. [1980] 56570
Fourman, M.P. [1982] 33227
Friedman, H.M. [1970] 14546
Goodman, Nicolas D. [1978] 52233
Guzicki, W. [1974] 62249
Guzicki, W. [1976] 26507
Howard, W.A. [1973] 74284
Johnstone, P.T. [1981] 34303
Kanovej, V.G. [1976] 53141
Kanovej, V.G. [1979] 56024
Leisenring, A.C. [1969] 07983
Levin, A.M. [1975] 25810
Levin, A.M. [1975] 63402
Levin, A.M. [1977] 53418
Levin, A.M. [1978] 56153
Levy, A. [1968] 08080
Neves-Real, L. [1951] 22210
Perrin, M.J. [1974] 18331
Schuette, K. [1976] 13021
Schwichtenberg, H. [1979] 55768
Veldman, W. [1982] 38283
Yakubovich, A.M. [1981] 46491
Zarach, A. [1976] 23816
Zariski, O. [1925] 41579

E30

Artigue, M. [1977] 27310
Artigue, M. [1978] 70505
Bar-Hillel, Y. [1958] 00805
Bernays, P. [1935] 01079
Bernstein, F. [1919] 01136
Brouwer, L.E.J. [1907] 16756
Bunder, M.W. [1974] 01764
Cantini, A. [1983] 41846
Cellucci, C. [1974] 29670
Chauvin, A. [1979] 31368
Christian, C.C. [1972] 71771
Christian, C.C. [1973] 71773
Collins, G.E. [1970] 02332
Esenin-Vol'pin, A.S. [1981] 35110
Feferman, S. [1966] 03699
Feferman, S. [1974] 21714
Feferman, S. [1978] 72806
Flannagan, T.B. [1976] 23736
Friedman, H.M. [1971] 17088
Friedman, H.M. [1973] 04650
Friedman, H.M. [1974] 04656
Friedman, H.M. [1975] 61836
Friedman, H.M. [1978] 31765
Friedman, H.M. [1983] 40758
Friedman, H.M. [1983] 44230
Grishin, V.N. [1981] 54617
Hajek, P. [1965] 05513
Hajek, P. [1972] 05526
Hanazawa, M. [1979] 56639
Henkin, L. [1971] 05928
Henson, C.W. [1984] 39860
Hilbert, D. [1900] 06078
Jaeger, G. [1980] 56034
Jaeger, G. [1982] 34700
Jaeger, G. [1984] 39554
Jaeger, G. [1984] 39557
Kanovej, V.G. [1981] 35529
Klaua, D. [1967] 07152
Krajewski, S. [1974] 07445
Kreisel, G. [1967] 07533
Kreisel, G. [1968] 07538
Kreisel, G. [1977] 48862
Kuzichev, A.S. [1983] 37218
Levy, A. [1958] 24860
Levy, A. [1968] 08080
Marek, W. [1975] 08758
Marek, W. [1976] 24412
Marek, W. [1978] 29210
Marek, W. [1978] 76016
Mayberry, J. [1985] 42536
McAloon, K. [1975] 18286
McAloon, K. [1981] 34009

Menger, K. [1928] 39109
Montagna, F. [1983] 34635
Montague, R. [1961] 09424
Mostowski, Andrzej [1958] 09566
Nishimura, T. [1974] 18312
Nishimura, T. [1974] 18313
Nishimura, T. [1975] 76854
Pearce, J. [1984] 45489
Perrin, M.J. [1974] 18331
Pozsgay, L.J. [1971] 10720
Pozsgay, L.J. [1972] 10721
Prawitz, D. [1985] 49658
Pudlak, P. [1983] 45314
Pudlak, P. [1985] 42556
Roedding, W. [1968] 11344
Rosser, J.B. [1950] 11558
Rosser, J.B. [1955] 23483
Shoenfield, J.R. [1954] 12087
Simpson, S.G. [1982] 35660
Simpson, S.G. [1984] 42501
Simpson, S.G. [1985] 45809
Smorynski, C.A. [1982] 35670
Solovay, R.M. [1976] 26088
Svejdar, V. [1981] 35308
Sward, G.L. [1971] 12803
Tait, W.W. [1965] 13282
Takeuti, G. [1957] 19761
Uesu, T. [1967] 16356
Vopenka, P. [1966] 46939
Wandschneider, D. [1975] 79815
Wandschneider, D. [1976] 79813
Wang, Hao [1952] 14020
Wang, Hao [1953] 14023
Wang, Hao [1953] 14024
Wang, Hao [1963] 14050
Wette, E. [1960] 31556
Wette, E. [1970] 14165
Weyl, H. [1918] 19791
Yasugi, M. [1983] 37614
Zbierski, P. [1971] 14379

E35

Ackermann, W. [1953] 00125
Andrews, P.B. [1972] 00366
Andrews, P.B. [1972] 00367
Batle, N. [1981] 40048
Beeson, M.J. [1979] 53872
Benda, M. [1980] 53861
Blass, A.R. [1983] 40316
Cartier, P. [1979] 56156
Christian, C.C. [1973] 71773
Clay, R.E. [1968] 02255
Crabbe, M. [1978] 29204
Dalla Chiara Scabia, M.L. [1968] 04285
Feferman, S. [1965] 03696
Fenstad, J.E. [1967] 17152
Fitting, M. [1969] 04349
Fitting, M. [1970] 14547
Flagg, R.C. [1985] 49605
Fourman, M.P. [1980] 56570
Fourman, M.P. [1982] 33227
Friedman, H.M. [1970] 03773
Friedman, H.M. [1971] 17088
Friedman, H.M. [1973] 04650
Goodman, Nicolas D. [1985] 42539
Grayson, R.J. [1979] 53376
Grayson, R.J. [1981] 54421
Guzicki, W. [1974] 62249
Guzicki, W. [1976] 26507
Hajek, P. [1972] 05526
Hajek, P. [1973] 05525
Hanazawa, M. [1979] 56639
Kanovej, V.G. [1975] 50893
Kanovej, V.G. [1976] 53141
Kanovej, V.G. [1979] 56024
Kanovej, V.G. [1981] 35529
Khakhanyan, V.Kh. [1980] 56495
Khakhanyan, V.Kh. [1980] 73754
Khakhanyan, V.Kh. [1981] 36853
Khakhanyan, V.Kh. [1983] 41114
Kuzichev, A.S. [1983] 35727
Levin, A.M. [1973] 17453
Levin, A.M. [1975] 25810
Levin, A.M. [1975] 63402
Levin, A.M. [1978] 56153
Manin, Yu.I. [1977] 51984
Manin, Yu.I. [1979] 54716
McAloon, K. [1975] 17403
McAloon, K. [1975] 18286
McAloon, K. [1980] 54523
Moerdijk, I. [1984] 42485
Mostowski, Andrzej [1958] 09566
Myhill, J.R. [1973] 09739
Neves-Real, L. [1951] 22210
Pearce, J. [1984] 45489
Powell, W.C. [1975] 10718
Scedrov, A. [1981] 54573
Scedrov, A. [1984] 37128
Scott, D.S. [1967] 11919
Shoenfield, J.R. [1954] 12087
Takeuti, G. [1957] 19761
Tharp, L.H. [1971] 13526
Uesu, T. [1967] 16356
Hajek, P. [1972] 23308
Wette, E. [1959] 14162
Wette, E. [1960] 31556
White, R.B. [1979] 53322
Yakubovich, A.M. [1981] 46491
Zarach, A. [1976] 23816

E40

Adamowicz, Z. [1984] 45406
Amer, M.A. [1976] 17946
Amer, M.A. [1977] 30619
Bily, J. [1974] 01205
Feferman, S. [1965] 03696
Fitting, M. [1970] 14547
Friedman, H.M. [1983] 40758
Grayson, R.J. [1983] 35033
Gurevich, Y. [1984] 38749
Guzicki, W. [1974] 62249
Guzicki, W. [1981] 34160
Guzicki, W. [1983] 37520
Hajek, P. [1970] 05521
Kanovej, V.G. [1979] 53375
Motohashi, N. [1977] 23750
Odifreddi, P. [1983] 38171
Scedrov, A. [1984] 37128
Shimoda, M. [1981] 54971
Steel, J.R. [1978] 29156
Takeuti, G. [1978] 52447
Takeuti, G. [1979] 47655
Takeuti, G. [1981] 39565
Hajek, P. [1972] 23308

E45

Apt, K.R. [1974] 00432
Apt, K.R. [1976] 14789
Cantini, A. [1983] 41846
Chauvin, A. [1962] 02061
Christian, C.C. [1973] 71773
Dalla Chiara Scabia, M.L. [1968] 04285
Enderton, H.B. [1971] 03353
Farrington, Paddy [1984] 42273
Feferman, S. [1965] 03696
Fitting, M. [1969] 04349
Friedman, H.M. [1970] 03773
Friedman, H.M. [1971] 17088
Friedman, H.M. [1973] 04654
Guzicki, W. [1981] 34160
Kanovej, V.G. [1976] 53141
Kreisel, G. [1960] 07519
Lucian, M. [1972] 28788
Lukas, J.D. [1974] 08387
Marek, W. [1978] 29210
Marek, W. [1978] 76016
McAloon, K. [1975] 18286
Mostowski, Andrzej [1961] 24858
Neves-Real, L. [1951] 22210
Takeuti, G. [1957] 19761
Vetulani, Z. [1984] 44142
Wang, Hao [1957] 29384
Zbierski, P. [1978] 29200

E47

Borel, E. [1919] 41669
Girard, J.-Y. [1984] 42404
Hobson, E.W. [1906] 28776
Kondo, M. [1956] 07328
Kondo, M. [1961] 42825
Kreisel, G. [1968] 07538
Levy, A. [1968] 08080
Tarski, A. [1931] 16924

E50

Batle, N. [1981] 40048
Cartier, P. [1979] 56156
Fenstad, J.E. [1967] 17152
Fitting, M. [1969] 04349
Gielen, W. [1981] 54491
Gurevich, Y. [1983] 33765
Guzicki, W. [1983] 37520
Hilbert, D. [1900] 06078
Kreisel, G. [1967] 07533
Lourenco, M. [1979] 70030
Luzin, N.N. [1929] 16811
Manin, Yu.I. [1977] 51984
Manin, Yu.I. [1979] 54716
Neves-Real, L. [1951] 22210
Scott, D.S. [1967] 11919

E55

Farrington, Paddy [1984] 42273
Friedman, H.M. [1984] 41019
Girard, J.-Y. [1984] 42404
Henkin, L. [1971] 05928
McAloon, K. [1981] 34009
Richter, W.H. [1971] 11173
Takeuti, G. [1985] 39568

E60

Euwe, M. [1929] 39332
Friedman, H.M. [1970] 03773
Tauts, A. [1974] 17693

E65

Hanazawa, M. [1979] 56639
Rimscha von, M. [1980] 53804
Sward, G.L. [1971] 12803

E70

Ackermann, W. [1950] 00122
Ackermann, W. [1952] 00124
Ackermann, W. [1953] 00125
Ackermann, W. [1965] 00135
Aczel, P. [1978] 70298
Aczel, P. [1982] 37664
Bar-Hillel, Y. [1958] 00805
Beeson, M.J. [1979] 53872
Beeson, M.J. [1980] 54110
Beeson, M.J. [1982] 37415
Beeson, M.J. [1984] 42422
Behmann, H. [1960] 14635
Bernays, P. [1935] 01079
Bily, J. [1974] 01205
Boffa, M. [1983] 37638
Brouwer, L.E.J. [1908] 01616
Brouwer, L.E.J. [1914] 01617
Brouwer, L.E.J. [1918] 41665
Brouwer, L.E.J. [1919] 01619
Brouwer, L.E.J. [1919] 41406
Brouwer, L.E.J. [1928] 02697
Brouwer, L.E.J. [1942] 01627
Bunder, M.W. [1974] 01764
Bunder, M.W. [1981] 55277
Bunder, M.W. [1983] 31631
Cantini, A. [1983] 41846
Chandrasekharan, K. [1941] 02408
Chandrasekharan, K. [1942] 02407
Chandrasekharan, K. [1945] 02409
Chauvin, A. [1979] 31368
Christian, C.C. [1977] 50790
Clay, R.E. [1968] 02255
Cogan, E.J. [1955] 02295
Crabbe, M. [1978] 29204
Cuda, K. [1983] 37690
Esenin-Vol'pin, A.S. [1961] 03557
Feferman, S. [1974] 21714
Feferman, S. [1978] 72806
Fitch, F.B. [1949] 04333
Fitch, F.B. [1950] 17113
Fitch, F.B. [1951] 17114
Flagg, R.C. [1985] 49605
Flannagan, T.B. [1976] 23736
Fourman, M.P. [1979] 53155
Fourman, M.P. [1980] 56570
Fourman, M.P. [1984] 43977
Friedman, H.M. [1974] 04656
Friedman, H.M. [1977] 31761
Friedman, H.M. [1978] 31762
Friedman, H.M. [1983] 40803
Friedman, H.M. [1984] 41019
Friedman, H.M. [1985] 48256
Goodman, Nicolas D. [1985] 42539
Gordeev, L.N. [1977] 50745
Gordeev, L.N. [1982] 37615
Grayson, R.J. [1979] 53376
Grayson, R.J. [1981] 54421
Grayson, R.J. [1982] 34472
Grayson, R.J. [1983] 35033
Greenleaf, N. [1981] 35148
Grishin, V.N. [1974] 73648
Grishin, V.N. [1981] 54617
Hajek, P. [1965] 05513
Hajek, P. [1972] 05526
Hajek, P. [1973] 05525
Hanazawa, M. [1979] 56639
Hatcher, W.S. [1963] 05744
Hayashi, S. [1981] 33411
Hyland, J.M.E. [1980] 54042
Kashapova, F.R. [1984] 44311
Khakhanyan, V.Kh. [1980] 56495
Khakhanyan, V.Kh. [1980] 73754
Khakhanyan, V.Kh. [1980] 73755
Khakhanyan, V.Kh. [1981] 36853
Khakhanyan, V.Kh. [1983] 41114
Krajewski, S. [1974] 07445
Krasner, M. [1957] 16974
Krasner, M. [1958] 16975
Kreisel, G. [1977] 48862
Kuroda, S. [1958] 07681
Kuroda, S. [1958] 07683
Kuroda, S. [1959] 25455
Kuzichev, A.S. [1983] 35727
Kuzichev, A.S. [1983] 37218
Lavendhomme, R. [1983] 34453
Levin, A.M. [1973] 17453
Liu, Shichao [1980] 75591
Liu, Shichao [1984] 45423
Lopez-Escobar, E.G.K. [1972] 08267
Luci, F. [1982] 35304
Mac Lane, S. [1975] 18268
Marek, W. [1975] 08758
Marek, W. [1976] 24412
Mayberry, J. [1985] 42536
Myhill, J.R. [1973] 09739
Myhill, J.R. [1975] 14820
Namba, K. [1965] 09795
Nepejvoda, N.N. [1975] 64192
Osius, G. [1973] 27669
Osius, G. [1975] 18320
Powell, W.C. [1975] 10718
Pozsgay, L.J. [1972] 10721
Quine, W.V.O. [1953] 10889
Resl, M. [1981] 34310
Rimscha von, M. [1980] 53804
Rosolini, G. [1983] 37075
Rosser, J.B. [1950] 11558
Scedrov, A. [1981] 54573
Scedrov, A. [1985] 44771
Schlomiuk, D.I. [1977] 21543
Schuette, K. [1954] 13004
Schultz, Konrad [1980] 54380
Scott, D.S. [1975] 27594
Shimoda, M. [1981] 54971
Shoenfield, J.R. [1954] 12087
Spector, C. [1957] 12676
Stout, L.N. [1976] 25859
Strauss, P. [1985] 42588
Sureshkumar [1977] 79102
Takeuti, G. [1979] 47655
Takeuti, G. [1981] 39565
Takeuti, G. [1984] 39577
Tharp, L.H. [1971] 13526
Titgemeyer, R. [1961] 13616
Uesu, T. [1981] 34476
Uesu, T. [1981] 47768
Veldman, W. [1982] 38283
Vopenka, P. [1966] 46939
Hajek, P. [1972] 23308
Wang, Hao [1950] 14013
Wang, Hao [1952] 14020
Wang, Hao [1954] 14025
Wang, Hao [1959] 14032
Wang, Hao [1963] 14044
Wang, Hao [1963] 14050
Weyl, H. [1921] 14167
White, R.B. [1979] 53322
Zbierski, P. [1978] 29200

E75

Bockwinkel, H.B.A. [1911] 37966
Bollman, D.A. [1973] 01422
Chernov, V.P. [1974] 17333
Dries van den, L. [1984] 46154
Grayson, R.J. [1981] 54421
Grayson, R.J. [1982] 34036
Johnstone, P.T. [1981] 34303
Kock, A. [1976] 51593
Kucera, A. [1985] 45308
Levin, A.M. [1975] 25810
Manukyan, S.N. [1973] 08677
Osius, G. [1973] 27669
Peano, G. [1889] 19497
Penon, J. [1981] 82458
Sikorski, R. [1958] 12300
Sikorski, R. [1962] 42875
Stout, L.N. [1975] 24059
Stout, L.N. [1976] 25859
Strauch, O. [1972] 13221
Sureshkumar [1977] 79102
Takeuti, G. [1978] 52447

E97

Gonseth, F. [1941] 48625
Harrington, L.A. [1985] 49810

E98

Bachmann, H. [1955] 00611
Bar-Hillel, Y. [1958] 00805
Barr, M. [1985] 40111
Barwise, J. [1977] 70117
Brouwer, L.E.J. [1914] 01617
Friedman, H.M. [1975] 04296
Goldblatt, R.I. [1979] 55754
Krajewski, S. [1981] 38331
Mostowski, Andrzej [1965] 09578
Shoenfield, J.R. [1967] 22384
Takahashi, A. [1972] 45445
Hajek, P. [1972] 23308
Wang, Hao [1981] 34116

E99

Bulloff, J.J. [1969] 14484
Kreisel, G. [1981] 48855

F96

Gentzen, G. [1969] 29483
Herbrand, J. [1968] 24825

G30

Abrusci, V.M. [1984] 44760
Arzarello, F. [1983] 43237
Barr, M. [1985] 40111
Blass, A.R. [1983] 40316
Boileau, A. [1981] 71191
Bunder, M.W. [1984] 41220
Calude, C. [1978] 52071
Cartier, P. [1979] 56156
Chernov, V.P. [1974] 17333
Coste-Roy, M.-F. [1980] 53743
Diaconescu, R. [1975] 17526
Espanol, L. [1983] 39927
Feferman, S. [1972] 03706
Ferbus, M.-C. [1984] 42605
Fourman, M.P. [1977] 27332
Fourman, M.P. [1979] 53155
Fourman, M.P. [1980] 56570
Fourman, M.P. [1982] 33227
Fourman, M.P. [1982] 33228
Fourman, M.P. [1982] 33229
Fourman, M.P. [1984] 43977
Girard, J.-Y. [1981] 35541
Girard, J.-Y. [1984] 42447
Goldblatt, R.I. [1979] 55754
Goldblatt, R.I. [1981] 55422
Grayson, R.J. [1979] 53376
Grayson, R.J. [1981] 54421
Grayson, R.J. [1982] 34036
Grayson, R.J. [1984] 41840
Hajek, P. [1970] 05521
Hayashi, S. [1980] 54379
Hayashi, S. [1981] 33411
Hoeven van der, G.F. [1984] 41155
Hoeven van der, G.F. [1984] 42464
Hyland, J.M.E. [1979] 53888
Hyland, J.M.E. [1980] 54042
Hyland, J.M.E. [1981] 36809
Hyland, J.M.E. [1982] 35795
Jankowski, A.W. [1985] 47500
Johnstone, P.T. [1977] 51354
Johnstone, P.T. [1979] 56505
Johnstone, P.T. [1979] 74552
Johnstone, P.T. [1981] 34303
Johnstone, P.T. [1982] 38303
Josza, R. [1984] 45334
Joyal, A. [1975] 27668
Kock, A. [1975] 17664
Kock, A. [1976] 51593
Kock, A. [1977] 24203
Kock, A. [1979] 53332
Kock, A. [1981] 54988
Lambek, J. [1968] 07819
Lambek, J. [1974] 07820
Lambek, J. [1980] 54113
Lambek, J. [1980] 75326
Lambek, J. [1980] 75328
Lambek, J. [1981] 55052
Lambek, J. [1983] 40763
Lambek, J. [1984] 44302
Lavendhomme, R. [1983] 34453
Lawvere, F.W. [1972] 25072
Longo, G. [1979] 55892
Longo, G. [1984] 44662
Mac Lane, S. [1975] 18268
Mac Lane, S. [1982] 38409
Macnab, D.S. [1977] 51394
Makkai, M. [1977] 50433
Marchini, C. [1982] 41484
Mihaljinec, M. [1972] 09238
Mints, G.E. [1980] 37155
Mints, G.E. [1980] 37417
Moerdijk, I. [1982] 35041
Moerdijk, I. [1983] 37289
Moerdijk, I. [1984] 42485
Moerdijk, I. [1984] 44304
Moerdijk, I. [1984] 45171
Mulvey, C.J. [1974] 21534
Osius, G. [1973] 27669
Osius, G. [1975] 18320
Osius, G. [1975] 18321
Penon, J. [1981] 82458
Pfender, M. [1982] 37036
Poigne, A. [1984] 47093
Reyes, G.E. [1974] 11131
Reyes, G.E. [1975] 64812
Reyes, G.E. [1977] 52660
Reyes, G.E. [1978] 56342
Reyes, G.E. [1979] 77765
Reyes, G.E. [1981] 42770
Rosolini, G. [1983] 37075
Rousseau, C. [1978] 51810
Rousseau, C. [1979] 82662
Scedrov, A. [1981] 54573
Scedrov, A. [1982] 37667
Scedrov, A. [1982] 37668
Scedrov, A. [1984] 37128
Scedrov, A. [1984] 41241
Schlomiuk, D.I. [1977] 21543
Scott, D.S. [1980] 78323
Seely, R.A.G. [1982] 34890
Seely, R.A.G. [1983] 42213
Shelah, S. [1976] 51434
Shimoda, M. [1981] 54971
Solov'ev, S.V. [1981] 40077
Soto-Andrade, J. [1984] 41485
Stout, L.N. [1975] 24059
Stout, L.N. [1976] 25859
Stout, L.N. [1978] 52544
Takeuti, G. [1979] 47655
Uesu, T. [1981] 34476
Uesu, T. [1981] 47768
Vauzeilles, J. [1982] 43062
Vauzeilles, J. [1985] 42537
Wauw-de Kinder van de, G. [1975] 18421
Wraith, G.C. [1976] 80097
Wraith, G.C. [1980] 53879

H05

Demuth, O. [1967] 02915
Demuth, O. [1968] 16949
Demuth, O. [1968] 28592
Henson, C.W. [1984] 39860
Kreisel, G. [1969] 26648
Liu, Shichao [1980] 75591
Liu, Shichao [1984] 45423
Nepejvoda, N.N. [1979] 36869
Rousseau, C. [1978] 51810
Rousseau, C. [1979] 82662

H10

Geiser, J.R. [1974] 04831

H15

Adamowicz, Z. [1985] 39780
Carstens, H.G. [1977] 51961
Cegielski, P. [1982] 34553
Cuda, K. [1983] 37690
Dawson Jr., J.W. [1979] 53657
DeMillo, R.A. [1979] 72280
Dimitracopoulos, C. [1983] 35043
Dries van den, L. [1980] 54246
Ehrenfeucht, A. [1976] 25797
Geiser, J.R. [1974] 04831
Henson, C.W. [1984] 39860
Kemeny, J.G. [1958] 07053
Lipton, R.J. [1978] 75578
Manevitz, L.M. [1980] 54428
Misercque, D. [1981] 76440
Parikh, R. [1971] 10282
Penzin, Yu.G. [1979] 32595
Rabin, M.O. [1961] 10934
Rosser, J.B. [1950] 11558
Skolem, T.A. [1933] 12392
Skolem, T.A. [1934] 12393
Skolem, T.A. [1955] 27718
Smorynski, C.A. [1984] 41837
Wauw-de Kinder van de, G. [1975] 18421
Wilkie, A.J. [1978] 56743
Wilkie, A.J. [1982] 34735
Zbierski, P. [1981] 34106

Alphabetization and alternative spellings of author names

The purpose of this index is to help the user find an author in whom he is interested. We begin by outlining both the general principles of alphabetization followed in the Author Index and the systems of transliteration used. The second half of this index addresses the problems which arise with author names for which there may be many variants in the literature. How do you find the primary form of a name used in the Bibliography? The ideal would be to have a table linking all the 'imaginable' versions of an author name to the unique primary form used here, but the obstacles to realizing this are obvious: one 'imaginable' form may correspond to two different authors and, worse, 'imaginable' itself depends on the linguistic background of the user. We have instead suggested some guidelines for identifying the primary form of a name from one of its variants. Finally, there is a list of alternative forms of names for those cases in which the difference between the alternative and the primary forms is particularly striking. For an author whose name has changed, each publication is listed under the name form used on that publication. Pointers to the other name form are given in the Author Index.

The Roman alphabet is as usual alphabetized in the following form:

A B C D E F G H I J K L M N O P Q R S T U V W X Y Z

Within this general framework, the ordering for hyphenated and double names is illustrated by the following example:

Ab,G. ; Ab-Aa,G. ; Ab Aa,G. ; Aba,G.

Apostrophes in a name are disregarded: Mal'tsev, for example, is treated as Maltsev for alphabetical purposes.

Titular prefixes such as von, du, de la, etc. come immediately after the surname (family name), and before the given name (or initials); so, e. g., J. von Neumann appears as Neumann von, J. Similarly J. Smith, Jr., and C.F. Miller III are given as Smith Jr, J. and Miller III, C.F., respectively.

In general, initials are used for given names. The full given name(s) are used only where necessary or helpful to distinguish between authors with the same surnames and initials.

As has been mentioned in the Preface, diacritical marks have, for practical reasons, mostly been disregarded. The following lists those diacritical marks of Scandinavian and German languages that have been transliterated:

æ to ae,
ø to oe,
å to aa,
ä to ae,
ö to oe,
ü to ue.

By the way one cannot infer that every ae, oe, or ue in German comes from ä, ö, or ü; e.g. Gloede is the correct spelling, not Glöde!

Note that the hacek in languages written in the Roman alphabet (e. g. Serbo-Croatian) has not been transliterated (so, for example, Šešelja appears here as Seselja).

The transliteration used for Cyrillic is explained in another index. (For a Russian author who has emigrated to the West the primary name is usually the form used by the author in Western publications. This form does not always agree with the transliteration of the Cyrillic name.) For Chinese names, the Pinyin system of transliteration has been used as far as possible, and commas have been added to separate the surname and given names (which are not abbreviated to initials) to accord with Western style. However, for Korean names no commas are used.

Over the last hundred years there have been in general use several different systems for transliterating Cyrillic into the Roman alphabet. This has given rise to many variants for author names originally written in Cyrillic. We list here our transliteration of those Cyrillic letters for which there have been several variants and give the most common alternative transliterations. If you are searching for an author name you suspect may be of Slavic origin this list will help you to find the most likely form used here: simply replace each (block of) letter(s) on the right occurring in your version by the appropriate letters given on the left.

Our transliteration	Possible alternatives
ya	ja a
yu	ju
eh	e
e	je ye
ts	c
ch	c tsh tch tsch
sh	s sch
zh	z

Our transliteration	Possible alternatives
z	s
j	i y
kh	h
v	w ff
”	y
ks	x
u	ou

The following is a selective listing of alternative forms of author names. It contains only those alternative forms from which the primary may not be guessed by using the guidelines above.

for	see
Abellanas Cebollero, P.	Abellanas, P.F.
Adams, M.M.	McCord Adams, M.
Albuquerque, J.	Ribeiro de Albuquerque, J.
Angulin, D.	Angluin, D.
Artalejo, R.M.	Rodriguez Artalejo, M.
Asjwiniekoemaar	Ashvinikumar
Avraham, U.	Abraham, U.
Barzdin', Ya.M.	Barzdins, J.
Benlahcen, D.	Benhalcen, D.
Bhaskara Rao, K.P.S.	Rao, K.P.S.Bhaskara
Bhaskara Rao, M.	Rao, M.Bhaskara
Bloch, A.S.	Blokh, A.Sh.
Blochina, G.N.	Blokhina, G.N.
Carroll, L.	Dodgson, C.L.
Chakan, B.	Csakany, B.
Char-Tung, R.	Lee, R.C.-T.
Chen, T.T.	Tang, Caozhen
Choodnovsky, D.V.	Chudnovsky, D.V.
Chu, W.J.	Zhu, Wujia
Cohen, E.L.	Longini Cohen, E.
Colburn, C.J.	Colbourn, C.J.
Colburn, M.J.	Colbourn, M.J.
Coppola, L.G.	Gonzalez Coppola, L.
Costa, A.A.	Almeida Costa, A.
Cresswell, M.M.	Meyerhoff Cresswell, Mary
Dao, D.H.	Dang Huu Dao
Decew, J.W.	Wagner Decew, J.
Dieu, P.D.	Phan Dinh Dieu
Duncan Luce, R.	Luce, R.D.
Dyson, V.H.	Huber-Dyson, V.
Fan Din' Zieu'	Phan Dinh Dieu
Foellesdal, D.	Follesdal, D.
Frejvald, R.V.	Freivalds, R.
Gegalkine, I.	Zhegalkin, I.I.
Gibbelato Valabrega, E.	Valabrega, E.G.
Greendlinger, M.	Grindlinger, M.
Hoo, T.-H.	Hu, Shihua
Hsu, L.C.	Xu, Lizhi
Hsueh, Yuang Cheh	Xueh, Yuangche
Jutting, L.S.B.	Benthem Jutting van, L.S.
Kao, H.	Gao, Hengshan
Kapinska, E.	Capinska, E.
Keldych, L.	Keldysh, L.V.
Khunyadvari, L.	Hunyadvari, L.
Kister, J.E.	Bridge, J.
Klein, F.	Klein-Barmen, F.
Kroonenberg, A.V.	Verbeek-Kroonenberg, A.
Kurepa, G.	Kurepa, D.
Kurkova-Pohlova, V.	Pohlova, V.
Kwei, M.S.	Mo, Shaokui
Lifshits, V.	Lifschitz, V.
Lo, Li Bo	Luo, Libo
Loewenthal, F.	Lowenthal, F.
Macdonald, S.O.	Oates MacDonald, S.
Malyaukene, L.K.	Maliaukiene, L.
Markus, S.	Marcus, S.
Moenting, J.S.	Schulte-Moenting, J.
Moh, S.-K.	Mo, Shaokui
Moura, J.E.A.	Almeida Moura de, J.E.
Nardzewski, C.R.	Ryll-Nardzewski, C.
Nash, W.C.S.J.A.	Nash-Williams, C.St.J.A.
Oates-Williams, S.	Oates MacDonald, S.
Plattner, A.	Pieczkowski, A.
Plyushkevichus, R.A.	Pliuskevicius, R.
Plyushkevichene, A.Yu.	Pliuskeviciene, A.
Poprougenko, G.	Popruzenko, J.
Puzio-Pol, E.	Pol, E.
R.-Salinas, B.	Rodriguez-Salinas, B.
Reymond, A.	Virieux-Reymond, A.
Riccioli, B.V.	Veit Riccioli, B.
Rucker, R.	Bitter-Rucker von, R.
Russi, G.Z.	Zubieta Russi, G.
Salinas, B.	Rodriguez-Salinas, B.
Schmir-Hay, L.	Hay, L.
Shain, B.M.	Schein, B.M.
Shaw, M.K.	Mo, Shaokui
Shih-Hua, H.	Hu, Shihua
Shlyakhovaya, N.I.	Slyakhova, N.I.
Solans, V.	Verdu i Solans, V.
Strazdin', I.Eh.	Strazdins, I.E.
Themaat, W.A.v.	Verloren van Themaat, W.A.
Toa van, T.	Tran van Toan
Toth, P.	Ecsedi-Toth, P.
Tsao-Chen, T.	Tang, Caozhen
Tseng, Y.X.	Zheng, Yuxin
Tsirulis, Ya.P.	Cirulis, J.
Tulcea, C.	Ionescu Tulcea, C.
Turksen, I.B.	Tuerksen, I.B.
Tzeng, O.C.	Tseng, O.C.
Vinter, H.	Winter, H.
Williams, C.St.J.A.N.	Nash-Williams, C.St.J.A.
Wou, Shou Zhi	Wu, Shouzhi
Wu, K.J.	Johnson Wu, K.
Yukna, S.P.	Jukna, S.
Yuting, S.	Shen, Y.-T.
Zhay, B.	Zhang, Bosheng
Zhen, Z.	Zhao, Zhen
Zilli, M.V.	Venturini Zilli, M.
Zou, Juan	Zhou, Juan

International vehicle codes

The following abbreviations are used as *codes for the country* in which a conference took place or in which a publishing company is located. (These abbreviations are those used internationally for vehicles.)

Code	Country
A	Austria
ADN	People's Dem. Rep. Yemen (South Yemen)
AFG	Afghanistan
AL	Albania
AND	Andorra
AUS	Australia
B	Belgium
BD	Bangladesh
BDS	Barbados
BG	Bulgaria
BH	Belize
BOL	Bolivia
BR	Brazil
BRN	Bahrain
BRU	Brunei
BS	Bahamas
BU	Burundi
BUR	Burma
C	Cuba
CDN	Canada
CH	Switzerland
CI	Ivory Coast
CL	Sri Lanka
CO	Columbia
CR	Costa Rica
CS	Czechoslovakia
CY	Cyprus
D	Fed. Rep. Germany (West Germany)
DDR	German Dem. Rep. (East Germany)
DK	Denmark
DOM	Dominican Republic
DZ	Algeria
E	Spain
EAK	Kenya
EAT	Tanzania
EAU	Uganda
EC	Ecuador
ES	El Salvador
ET	Egypt
ETH	Ethiopia
F	France
FJL	Fiji Islands
FL	Liechtenstein
FR	Faeroes
GB	Great Britain and Northern Ireland
GBA	Alderney
GBG	Guernsey
GBJ	Jersey
GBM	Isle of Man

Code	Country
GBZ	Gibraltar
GCA	Guatemala
GH	Ghana
GR	Greece
GUY	Guyana
H	Hungary
HK	Hong Kong
HV	Upper Volta
I	Italy
IL	Israel
IND	India
IR	Iran
IRL	Ireland (Eire)
IRQ	Iraq
IS	Iceland
J	Japan
JA	Jamaica
JOR	Jordan
K	Cambodia
KWT	Kuwait
L	Luxembourg
LAO	Laos
LAR	Libya
LB	Liberia
LS	Lesotho
M	Malta
MA	Morocco
MAL	Malaysia
MC	Monaco
MEX	Mexico
MS	Mauritius
MW	Malawi
N	Norway
NA	Netherlands Antilles
NIC	Nicaragua
NL	Netherlands
NZ	New Zealand
P	Portugal
PA	Panama
PAK	Pakistan
PE	Peru
PL	Poland
PNG	Papua-New Guinea
PRI	Puerto Rico
PRK	People's Rep. Korea (North Korea)
PY	Paraguay
Q	Qatar
RA	Argentina
RB	Botswana
RC	Taiwan
RCA	Central African Republic
RCB	Congo

Code	Country
RCH	Chile
RFC	Cameroon
RH	Haiti
RI	Indonesia
RIM	Mauritania
RL	Lebanon
RM	Madagascar
RMM	Mali
RN	Niger
RO	Romania
ROK	South Korea
ROU	Uruguay
RP	Philippines
RPB	Benin
RSM	San Marino
RWA	Ruanda
S	Sweden
SA	Saudi Arabia
SCV	Vatican
SD	Swaziland
SF	Finland
SGP	Singapore
SME	Surinam
SN	Senegal
SP	Somalia
STL	Windward Islands St. Lucia
SU	Soviet Union
SY	Seychelles
SYR	Syria
TG	Togo
THA	Thailand
TJ	People's Rep. China
TN	Tunisia
TR	Turkey
TT	Trinidad and Tobago
USA	United States of America
VN	Vietnam
WAG	Gambia
WAL	Sierra Leone
WAN	Nigeria
WD	Dominica
WG	Grenada
WS	Samoa
WV	Windward Islands St. Vincent
Y	Arabic Rep. Yemen (North Yemen)
YU	Yugoslavia
YV	Venezuela
Z	Zambia
ZA	South Africa
ZRE	Zaire
ZW	Zimbabwe

Transliteration scheme for Cyrillic

Author names and titles originally in *Cyrillic* have been transliterated into the Roman alphabet using the following scheme. (It is the same as the scheme curently used by Zbl and differs only slightly from that used by MR.)

Cyrillic		Roman
а	А	a
б	Б	b
в	В	v
г	Г	g
д	Д	d
е(ё)	Е(Ё)	e
ж	Ж	zh
з	З	z
и	И	i
й	Й	j
к	К	k

Cyrillic		Roman
л	Л	l
м	М	m
н	Н	n
о	О	o
п	П	p
р	Р	r
с	С	s
т	Т	t
у	У	u
ф	Ф	f
х	Х	kh

Cyrillic		Roman
ц	Ц	ts
ч	Ч	ch
ш	Ш	sh
щ	Щ	shch
ъ	Ъ	”
ы	Ы	y
ь	Ь	’
э	Э	eh
ю	Ю	yu
я	Я	ya